# DISCRETE COMMUNICATION SYSTEMS

# Discrete Communication Systems

Stevan Berber

*The University of Auckland*

# OXFORD
UNIVERSITY PRESS

Great Clarendon Street, Oxford, OX2 6DP,
United Kingdom

Oxford University Press is a department of the University of Oxford.
It furthers the University's objective of excellence in research, scholarship,
and education by publishing worldwide. Oxford is a registered trade mark of
Oxford University Press in the UK and in certain other countries

First Edition published in 2021

Impression: 1

Published in the United States of America by Oxford University Press
198 Madison Avenue, New York, NY 10016, United States of America

British Library Cataloguing in Publication Data

Data available

Library of Congress Control Number: 2021931155

ISBN 978–0–19–886079–2

DOI: 10.1093/oso/9780198860792.001.0001

Printed in Great Britain by
Bell & Bain Ltd., Glasgow

*This book is dedicated to all authors who committed to write an original book in their profession and passed through a demanding and exhausting process full of uncertainties and challenges.*

# Preface

A comunication system is a set of devices that work together allowing the transmission of messages at the distance. In that sense the system can be also called a telecommunication system that allows generation of a message at one point by means of a transmitter and its sufficiently accurate reproduction at another distant point by means of a receiver. A communication network can be considered as a set of communication systems that are inteconnected and work in such a way to allow the transmission of messages between any two or more points belonging to the network. The messages can be of various forms and types containing the information content defined in the sence of information theory. Following these definitions we can say that the existing telecommunication network is the biggest machine ever invented. It is composed of an enormously high number of devices interconnected by transmission media, which work synchronously allowing transmision of messages around the globe and in the space.

We can say that the early developed world terrestrial networks have been two-dimensional covering the continents and interconnecting them by undersea cables that are deployed in the last two centuries. In the twentieth century, due to the deployment of communication satellites, this world network became three-dimensional and global, characterized by the transmission systems of extremely high capacity. In addition, tremendous efforts were undertaken by the mankind to develop wireless mobile communication systems as an integral part of the global stationary high capacity telecommunication network. Finally, the development of the space technology required new communication devices that are able to achieve deep space communications. Due to the existing theoretical and technologial advancement in telecommunicaiton systems' theory and practice, it has been relativelly easy to develop the systems for this kind of communications.

Since the early attempts to develop an electric telecommunication system in the nineteenth century, there have been two, I dare to say, dreams of us, telecomumunication engineers and scientists. The first dream was to communicate while we are on the move. That dream was achieved by the development of wireless mobile communication systems as integral parts of the global stationary telecommunication network. The second dream has been and still is, to develop wireless communication systems that have the transmission capacity equivalent to the capacity of the stationary network. This dream has not been achieved yet. However, we cannot say that the second dream will not be achieved. The development of the wideband radio systems can contribute to that. In addition, we can believe in the possible development of new means of signal transmission that will contribute to the achievement of this dream.

This book aims to present theoretical base of discrete communication systems and does not have any intention to address any aspect of communication networks design. Therefore, the contents of the book will solely address problems of the design of a communication system that includes a transmitter, a transmission channel, and a receiver. From the network point of view, the book will focus on so called a physical layer design. The signals processed in the communication system will be presented in two domains, the continuous time domain and the discrete time domain represented by the continuous variable $t$ and the discrete variable $n$, respectively. To distinguish the two systems in this book, the system operating in the continuous time domain is named *the digital system*, while the system operating in the discrete time domain is named *the dicrete system*.

The first objective of writing this book is to present the fundamental theory of discrete communication systems that every student studying communication systems needs to acquire and relate it to the existing theory of digital communication systems. The second obective of writing this book is to relate the acquired theory and mathematical models to the practical simulation, emulation, and design of communication system devices in digital DSP and FPGA technology.

I have been teaching courses in Communication Systems, Digital Communications, Information Theory, and Digital Signal Processing for over 20 years. In addition, my research has mainly been in the field of communication systems and networks, where I worked on the mathematical modelling, simulation, and implementation of various transmitter and receiver structures in digital technology (primarily DSP and FPGA) and their testing for different channels. I noticed that the theory of continuous time communication systems is important to understand operation of the systems in the presence of noise and fading; however, it is not sufficient to work on the design and implementation of the communication system blocks in modern DSP technology.

For that reason, I began to work on the theory of discrete communication systems, primarily in my research, where the building blocks inside the transceiver and the simulated/emulated channel perform discrete time signal processing. In parallel, I was upgrading my DSP course reaching the point of lecturing the discrete time stochastic processes, which are, among the other things, the essential theoretical background for the discrete communication systems analysis and design.

I have also noticed that the theoretical models and structures of transceiver building blocks cannot be directly used as such when we come to their digital design. However, the principles of operation would be preserved in practice. Due to this fact, the interest in the book will be far reaching across not only students, but also researchers, design engineers, and practitioners in the industry.

The writing of this book is additionally motivated by modern trends in the design of communication systems. The design is based on digital technology, primarily on FPGA and DSP platforms, which is in extensive use replacing analog hardware that has been traditionally used, to implement necessary signal processing functions inside the baseband and intermediate frequency (IF) transceiver blocks. These trends were heavily supported by the advances in the theory of discrete time signal processing, primarily by the mathematical theory of discrete time stochastic processes. These trends will continue in the future supported by the everlasting increase in the processing ability of digital technology allowing the development of sophisticated communication algorithms we could not dream of in the past.

For these reasons, it is necessary to teach students how to use the discrete time signal processing theory and how to apply it in the design of modern communication devices. Even more important is to make available the theory of discrete time communication systems to the researchers, practicing engineers, and designers of communication devices. The design of these devices is impossible without deep understanding of the theoretical principals and concepts related to their operation in the discrete time domain. Practically all modern communication devices, like wireless and cable modems, TV modems, consumer entertainment systems, satellite modems, and similar, are based on the use of digital processing technology and the principles of the discrete time signal processing theory. In addition to the mentioned main purposes of the book, it is important to apostrophize the importance of the discrete time signal processing theory for the researchers, designers, and manufacturers in the field of instrumentation because the main functions of modern instruments are implemented using principles of the discrete time signal processing.

This book is the first of its kind. Nearly all the books written in the field of communication systems present signals and their processing in continuous time domain to explain the operation of a communication system. In this book, the signals are presented in discrete time domain for the main building blocks of a communication system.

The book is dedicated to the undergraduate and graduate students doing courses in communication systems and also to the practicing engineers working on the design of transceivers in discrete communication systems. A one-semester senior level course, for students who have had prior exposure to classical analog communication systems covering passband and baseband signal transmission, can use material in Chapters 1–6, and Chapter 9 supported with related complementary Chapters 11–15, and Chapter 19 and Projects 1, 2, or 3.

In a first-year postgraduate level course, the first six chapters provide the students with a good review of the digital and discrete communication systems theory and the main lecturing will cover Chapters 2–5 and Chapters 7–10, which present the discrete communication systems and their design, and related Projects 1–5. The background theory for this course is contained in complementary Chapters 13–18.

For practicing engineers, who are experienced in theory of digital communication systems, the material covered in Sections 2.2 and 2.3, Chapters 3–5, and then Chapters 7–10 supported by complementary Chapters 13–18 is a good base for understanding the vital concepts in discrete communication systems. All projects are relevant for them, in particular Projects 4 and 5.

To master the theory, each chapter contains a set of problems for students' exercises. The solutions to problems are inside a separate book belonging to the Supplementary Material. Majority of the problem solutions can be confirmed using available software packages.

The book contains two parts. The first part of the book contains ten chapters and presents essential theory of the discrete and digital communication systems, and operation of their building blocks, at the first place the operation of modulators and demodulators/detectors. Due to the importance of the theory of discrete and continuous time signal processing, for both the deterministic and random signals, nine chapters containing this theory are incorporated into the second part of the book.

The distinguishing features of the book can be summarized as follows:

1. The book presents essential theory and practice of the discrete communication systems design, based on the theory of discrete time stochastic processes, and their relation to the existing theory of digital communication systems.

2. Based on the presented orthogonality principles, a generic structure of a communication system, based on correlation demodulation and optimum detection, is developed and presented in the form of mathematical operators with precisely defined inputs and outputs and related functions.

3. Due to the random nature of the signals processed, starting with the randomly generated messages at the transmitter source, the theory of stochastic signal processing is extensively and consequently applied to present the signals at the inputs and outputs of the system blocks. This approach gives an opportunity to the reader to understand complete signal processing procedures inside the transmitter, channel, and receiver.

4. Based on the developed generic system structure, the traditionally defined phase shift keying (PSK), frequency shift keying (FSK), quadrature amplitude modulation (QAM), orthogonal frequency division multiplexing (OFDM), and code division multiple access (CDMA) systems are deduced as special cases of the generic system. The signals are presented in the time and frequency domain, which requires precise derivatives of their amplitude spectral density functions, correlation functions, and related energy and power spectral densities.

5. Having in mind the controversial nature of the continuous time white Gaussian noise process having the infinite power, a separate chapter is dedicated to the noise discretization by introducing notions of the noise entropy and the truncated Gaussian density function to avoid limitations in applying the Nyquist criterion.

6. The book is self-sufficient because it uses a unified notation whenever is possible, both in the main ten chapters explaining communication systems theory and nine supplementary chapters dealing with the continuous and discrete time signal processing for both the deterministic and stochastic signals.

7. For the sake of explanation and clarity, the theory of digital communication systems is presented at certain extent and related to the main theory of discrete communication systems. In this way, the reader can see complete theory of modern communication systems.

8. The unified notation and unified terminology allow clear distinction of the deterministic signals from stochastic, power signals from energy signals, discrete time signal and processes from continuous time signals and processes. Consequently, this approach allows an easy way of understanding the related differences in defining the correlation functions, power and energy spectral densities, amplitude, and power spectra of the mentioned signals and processes.

9. The text of the book is accompanied by solutions of about 300 problems and five design projects with the defined projects' topics and tasks.

The book chapters are closely interconnected. Their relationship is presented here in the form of a diagram. The main chapters, presenting the theory of communication systems, are in the middle of the diagram. The chapters containing the theory of signal sampling and reconstruction, and necessary theory in discrete time signal processing, are on the left hand side. The chapters containing the theory of the continuous time signal processing are on the right hand side. The chapters on the left and right contain the theory of both the deterministic and stochastic signals, which are extensively used in explaining the theory of communication systems in the middle chapters. Chapters 16–18, at the bottom of the diagram, contain the essential theory of digital filters and multirate signal processing that is relevant for nearly all chapters of the book and in particular for Chapters 7 and 10. The chapters on the diagram are interconnected by the input arrow and output diamond lines, which show what the necessary background theory for a chapter is and where the theory of a chapter can be used, respectively. We will apostrophize the importance of some of these connections for each chapter.

Chapter 1 introduces the subject of the book, defines the main terms in communication systems that will be used in the book chapters, and presents the main objectives of the book writing from the communication systems theory point of view. The chapter presents various classifications of signals and systems and theoretical concepts related to the signal conversions in time domain that will be used in the subsequent chapters. The signals are classified using various criteria including periodicity and symmetry, continuity and discreteness, power and energy properties, randomness, and physical realizability of signals. The analog-to-digital (AD) and digital-to-analog (DA) conversions, and their places and importance in the processing of signals in relation to the application in discrete and digital communication systems, are briefly explained. The final contents return back to the definition of the signals related to the continuity and discreteness in time and in their values, due to the importance of distinguishing them in the theoretical analysis and design of digital and discrete communication systems. The terms defined in this chapter are used in Chapters 4 and 6–10. The relation of the chapter to other chapters is presented in the diagram.

Chapter 2 is dedicated to the principle of discrete time signals orthogonalization because the orthogonal signals are widely used in telecommunication theory and practice, like the carriers of baseband signals, subcarriers in OFDM systems, and the spreading sequences in spread-spectrum and CDMA systems. The orthonormal discrete basis functions are defined, and the procedure of the vector representation of signals is demonstrated. The Gram-Schmidt orthogonalization procedure and construction of the space diagram are presented in detail. Based on orthonormal signals, the signal synthesizers and analysers are theoretically founded, which can be used to form the discrete time transmitters and receivers. Understanding of this chapter is a prerequisite to understand Chapters 4–10 because the orthonormal signals defined in this chapter will be used throughout the book. In addition, the contents of Chapters 5, 6, 7, 12, and 15 are related to the use of the orthogonal signals, which contribute to understanding of this chapter from the practical point of view.

Chapter 3 contains the theory of discrete time stochastic processes, including their mathematical presentation in the time and frequency domain. The typical processes, relevant for the discrete communication systems design, including Gaussian process, white noise, binary, and harmonic processes, are presented. A comprehensive analysis of stationary and ergodic processes and the linear time invariant (LTI) systems with stochastic inputs is presented in this chapter. The processes are analysed in terms of their autocorrelation functions (ACFs) and power spectral densities that are related by the Wiener-Khintchine theorem. The chapter is placed at the beginning of the book because its contents are considered as a prerequisite for the chapters that follow, in particular, for the chapter related to the theory of discrete communication systems. A unique notation used in this chapter is used in the rest of the book, which makes the book to be self-sufficient. For book readers, it is highly advisable to read this chapter first and acquire its notation. The theory presented in this chapter is an unavoidable base for Chapters 2–10 to understand relationships between the continuous and the discrete time signal processing needed in any analysis of communication systems.

Chapter 4 addressed the issues related to the theory of noise in communication systems. The problem of discretization of the white Gaussian noise process is raised due to the noise strict definition implying that it has theoretically infinite power. If we start with this definition, it would be impossible to generate discrete noise process because the sampling theorem requires that the sampled signal must be physically realizable, i.e. the sampled noise needs to have a finite power. To overcome this problem, the noise entropy is defined as an additional measure of the noise properties, and the truncated Gaussian probability density function is used to define the distribution of noise amplitudes. Addition of the entropy and truncated density function to the noise autocorrelation and power spectral density (PSD) functions allowed mathematical modelling of the discrete noise source and the design of both baseband and bandpass noise generators and regenerators. The developed noise theory and designed noise generators are essential for the theoretical explanation of the operation of digital and discrete communications systems, their design, simulation, emulation, and testing. This chapter is in close relations with Chapters 3, 19, 13, 16, and 17. The presented approach for the noise characterization is an attempt to overcome controversial interpretations of the white Gaussian noise that can be found in the existing books in communication theory and signal processing.

The Project 3 in the Supplementary Material presents the design of noise generators and their application in communication system.

Chapter 5 is a vital part of this book presenting a generic communication system operating in discrete time domain, which is based on implementation of the orthogonal modulators, correlation demodulators, and optimum detectors. Based on definition of the signal synthesizers and analysers in Chapter 2, this generic discrete system is developed to be used to deduce the practical systems as its special cases. The signal synthesizer is transferred into a discrete transmitter, and the signal analyser is used as a correlation receiver followed by an optimum detector. The system structure is presented in terms of mathematical operators and supported by exact mathematical expressions based on the theory of stochastic processes presented in Chapters 3 and 19. The likelihood function is derived, and the maximum likelihood rule is applied to specify the decision process and construct the optimum detector. An example multilevel system is deduced as a special case of the generic system, and the bit error probability expression is derived. For the sake of continuity and completeness in presenting communication systems theory, a generic digital communication system is developed and related to its discrete counterpart. Notwithstanding that the generic model is developed for the assumed correlation receiver the same model can be easily replaced with the matched filter receiver. Due to the simplicity of this replacement, the matched filter

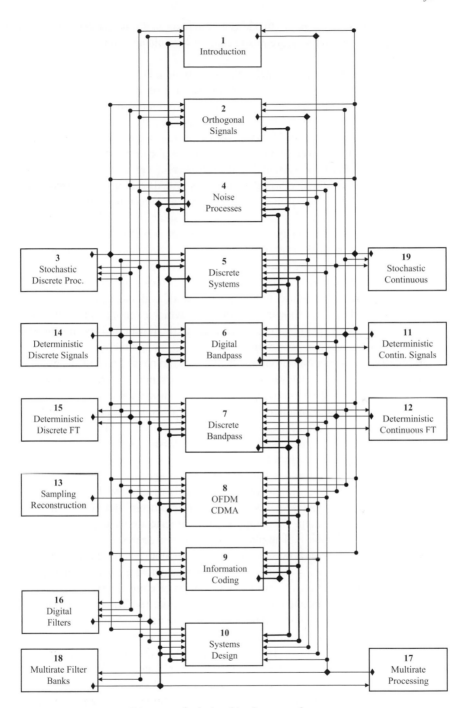

*Diagram of relationships between chapters.*

receivers are not analysed in the book. In addition, the results of this chapter are needed at the first place in Chapters 6–8, and its prerequisite chapters are mostly supplementary chapters as shown in the diagram.

Chapter 6 presents mathematical models of the baseband and bandpass digital communication systems based on the binary and quaternary phase shift keying (BPSK and QPSK), FSK, and QAM methods. It is dedicated to the readers who are not familiar at all with the digital communications systems. The systems are deduced as special cases of the generic system structure presented in Section 5.8.2. The systems are uniquely presented using mathematical operators as their blocks, followed by detailed derivatives for signals in time and frequency domain at the vital points of the systems' transmitters, receivers, and the noise generators, using the concepts of both the stochastic (continuous and discrete) signal processing from Chapters 3, 4, 5, and 19 and the deterministic (continuous and discrete) signal processing from Chapters 2 and 11–15. The vital characteristics of the system and its blocks are expressed in terms of the amplitude spectral density, ACFs, power and energy spectral densities, and the bit error probability. The theoretical results obtained are heavily used in the subsequent Chapters 7, 8, 9, and 10. Projects 1 and 2 are mainly related to this chapter.

We mostly used the notion of ACF and PSD to express inputs and outputs of the signal processing blocks in the transceiver. For the sake of understanding, the mathematical derivatives are presented in detail showing several steps, which cannot be found in other books in communication theory. A specific attention is made to the power and energy calculation of the signals in both frequency and time domain.

From modulation methods point of view, Chapter 7 is the central part of the book because it presents the operation of transceiver blocks that process pure discrete time signals. The inputs for this chapter are all previous chapters and succeeding Chapters 11–15 and Chapters 17 and 19. This chapter presents mathematical models of the discrete baseband and bandpass communication systems based on BPSK and QPSK, FSK, and QAM modulation methods. The operation of IF systems is presented where all processing is performed in the discrete time domain. The systems are deduced as special cases from the generic system structure presented in Section 5.6.4, Figure 5.6.4, that is based on application of orthonormal basis functions presented in Chapter 2, which confirms the basic idea of this book that the existing communication systems are special cases of the generic system. The idea will be further demonstrated on the development of complex communications systems, like OFDM and CDMA, as presented in Chapter 8. The block schematics of analysed systems are presented using mathematical operators, followed by detailed mathematical expressions for signals in discrete time domain at the vital points of the system's structure using the concepts of both the discrete stochastic and discrete deterministic signal processing. The key characteristics of discrete communication systems and their blocks are expressed in terms of amplitude spectral density, ACFs, power and energy spectral densities, and the bit error probability.

Chapter 8 presents modern multiuser and multicarrier communication systems based on the CDMA and OFDM technology. The analog, digital, and discrete OFDM systems are separately presented and interrelated from the theoretical and practical point of view. A precise mathematical model of the discrete baseband and IF blocks is presented, including the procedures of signal mapping and the discrete Fourier transform (DFT) application, and then related to the model of analog radio-frequency block to make the whole OFDM system.

The basic theory of binary and non-binary CDMA system operation is presented. To support deeper theoretical understanding of the CDMA system operation and design, Project 4 in the Supplementary Material demonstrates the procedure of mathematical modelling, simulation, and design of this system in FPGA technology and presents required development tools. The fundamental concepts that are needed to support understanding of this chapter are presented in Chapters 1–7 in communication theory and Chapters 11–17 in the theory of processing deterministic signals.

Chapter 9 presents the fundamentals of information theory, which are required for understanding of the information measure, entropy, and limits in signal transmission including the definition and derivation of the communication channel capacity. The proof of coding theorem is separately presented. The chapter contains a part that defines the entropy of continuous and discrete Gaussian and uniform stochastic processes. The results of this unique analysis are essential to understand the notion of continuous and discrete white Gaussian noise processes presented in Chapter 4. The block and convolutional codes, including hard decision Viterbi algorithm, are presented. The theory of iterative and turbo channel coding is presented in Project 5 in the Supplementary Material, where several topics are defined and the related solutions are offered. The background theory that supports the understanding of this chapter can be found mainly in Chapters 1–7 and Chapters 11–15.

Chapter 10 presents practical aspects of discrete communication systems design in digital technology, primarily in DSP and FPGA. The systems are presented at the level of block schematics apostrophizing the main issues in their design and discussing expected advantages and disadvantages of the systems' design in digital technology. Designs of systems based on the QPSK and QAM modulation are separately presented. The operation of each designed system is explained using the related theoretical structure of the system, which allowed a clear understanding of the relationships between the system's theoretical model and its practical design. The structures of the first, second, and third generation of the discrete transceiver designs are presented. This chapter relays on the theoretical principles presented in Chapters 5–9. Due to the size of this book we could not go into details of various discrete system implementations based on the multirate signal processing and application of digital filters and bank of filters that are presented in Chapters 16, 17, and 18.

In addition to the main ten chapters, the book contains nine complementary chapters. These chapters are added for two reasons. Firstly, they contain the basic theory of continuous time and discrete time signal processing that is essential for the understanding of mathematical models and operations of the digital and discrete transceivers, where we are using the theory of deterministic and stochastic signal processing. Secondly, the notation of these nine chapters is preserved in the main chapters, which will help students and experts to understand in depth the communication systems theory. Therefore, even though a reader can be very familiar with the content of these nine chapters, it is advisable to read them before starting to work on the main ten chapters. I am apostrophizing this for one additional simple reason. Namely, we unified the notation in all chapters that simplifies understanding of their content.

Due to importance of the concept of independent variable modification and definition of LTI system based on its impulse response, Chapter 11 presents basic deterministic signals and systems in the continuous time domain. These signals, expressed in the form of functions and functionals, like Dirac's delta function, are used in the whole book for the deterministic and stochastic signal analysis. The definition of the ACF and explanation of the convolution procedure in LTI systems are presented in detail due to their importance in communication systems analysis and synthesis. A linear modification of the independent continuous variable is presented for specific cases, like time shift, time reversal, time, and amplitude scaling. Content of this chapter is important input to the comprehensive analysis of LTI systems dealing with the deterministic signals in Chapter 12, stochastic signals in Chapters 3 and 19, and for the analysis of discrete time deterministic signals in Chapters 14 and 15.

Chapter 12 presents a detailed analysis of continuous time signals and systems in frequency domain, including the theory of Fourier series and Fourier transforms, and presenting key examples relevant for the analysis and synthesis of signals processed in a digital transceiver blocks of a communication system. The amplitude, magnitude, phase, and power spectra are defined and calculated for typical signal examples. In particular, the Fourier transform of periodic signals is presented due to its importance in communication systems theory and practice. Using a unique notation that distinguishes the energy and power signals, the correlation, power, or energy spectral density functions are interrelated by proving the Winer-Khintchine theorem. A comprehensive analysis of the LTI system, using concepts of the impulse response, system correlation function, and PSD, both for power and energy signals, is presented. The content of this chapter is of a vital importance for understanding mathematical presentations of transceiver blocks in communications systems analysed in Chapters 1–9 and also for acquiring the signal processing theory in Chapters 13–15 and 19.

Chapter 13 presents the theory of transferring a continuous time signal into discrete time form by sampling, and then converting the obtained samples to digital signal suitable for processing in a processing machine using the procedure of a sample quantizing and coding. Then, the procedure of converting a digitally processed signal into discrete signal samples and the reconstruction of the initial continuous time signal using a low-pass reconstruction filter is presented. The presented theory makes a mathematical base of both the AD and DA conversions that are extensively used in processing signals in discrete communication systems. It was shown that the Nyquist criterion must be fulfilled to eliminate the signal aliasing in frequency domain. The mathematical model of transferring continuous time signal into its discrete time form and vice versa is presented and demonstrated on a sinusoidal signal. These concepts presented in this chapter are used in main Chapters 4–8 and also in complementary Chapters 3 and 19.

Due to the importance of the concept of independent discrete variable modification and the definition of discrete LTI systems, Chapter 14 presents basic deterministic discrete time signals and systems. These discrete signals, expressed in the form of functions, including the Kronecker delta function and discrete rectangular pulse, are used in the whole book for the deterministic discrete signal analysis. The definition of the ACF and the explanation of the convolution procedure in LTI systems for the discrete time signals are presented in detail due to their importance in the analysis and synthesis of discrete communication systems. As such, this chapter is closely related to both Chapter 15 and Chapter 3 dealing with discrete time deterministic signal processing in the frequency domain and discrete time stochastic processes, respectively. The content of this chapter is important input to the comprehensive analysis of LTI systems dealing with the stochastic signals.

Chapter 15 presents a detailed analysis of deterministic discrete time signals and systems in frequency domain, including the theory of discrete Fourier series, discrete time Fourier transform (DTFT), and DFT, and key examples relevant for the analysis and synthesis of signals processed in the discrete transceiver blocks of a communication system. The amplitude, magnitude, phase, and power spectra are defined and calculated for typical signals relevant for communication systems. Using a unique notation, which distinguishes the energy and power signals, the correlation, power, or energy spectral density functions are interrelated by proving the Winer-Khintchine theorem. A comprehensive analysis of the LTI systems, using the concepts of impulse responses, correlation, and convolution, both for power and energy signals, is presented. The presented theory is vital for understanding mathematical modelling of transceiver blocks in communication systems presented in Chapters 4–10. The chapter finishes with the $z$-transform containing the theory required to understand the contents of Chapters 16, 17, and 18 that are dealing with digital filters and multirate signal processing.

Chapter 16 presents the theoretical base of the digital filters including the issues related to their designs. Basic characteristics and structures of the finite impulse response (FIR) and the infinite impulse response (IIF) filters are presented and discussed. The methods of filter design and related algorithms, which are based on bilinear transformation method, windowed Fourier series, and algorithms based on iterative optimization, are presented. The chapter serves as an introduction to Chapters 17 and 18 and is relevant for understanding Chapters 1, 2, 4, and 7–10. It is important to acquire the knowledge from Chapters 13 to 15 to read this chapter.

Chapter 17 presents the multirate signal processing staring with the explanation of the upsampling and downsampling procedures on a discrete signal in time domain. The operations of a downsampler (decimator) and an upsampler (interpolator) are analysed in the frequency domain emphasizing the problem of possible aliasing. Complex systems that include both the upsampling and downsampling are analysed, and the problem of the complexity reduction is mentioned. The operation of systems that combine an interpolator and an interpolation filter, and a decimator and a decimation low-pass (LP) filter, is presented in time and frequency domain. In particular, the problem of reducing complexity of a multirate system is addressed, and a poly-phase decomposition for both the FIR and IIF filters is offered as an efficient solution. The content of this chapter is relevant for understanding Chapters 4–8 and 10. The prerequisites for this chapter are Chapters 13–15 and 16.

Chapter 18, which is based on theory in Chapters 16 and 17, and chapters related to the discrete time signal processing, presents the theoretical description and principals of operation of the analysis and synthesis filter banks, which is essential material for understanding modern design of the transceivers that are based on the discrete time signal processing. The structure of a quadrature mirror filter (QMF) bank is presented, and the operation of the analysis and synthesis component filters is explained. The condition for a perfect reconstruction of two-channel filter banks is derived. Based on a two-channel QMF filter bank, the procedure of making multichannel QMF bank is presented. The theory presented in this chapter is important in the design of modulators and demodulators in discrete communication systems.

Chapter 19 contains the theory of continuous time stochastic processes, including their mathematical presentation in the time and frequency domain. The typical processes relevant for the communication systems theory, including Gaussian process, white noise, binary, and harmonic processes, are analysed in the time and frequency domain. A comprehensive analysis of stationary and ergodic processes and the LTI systems with stochastic inputs is presented. The processes are analysed in terms of their ACFs and power spectral densities. The notation used in this chapter complies with the notation used in other chapters of the book. For the book readers who are not familiar with the continuous time stochastic processes, it is highly advisable to read this chapter and acquire its notation, due to its importance for understanding the content of Chapters 1–9. The prerequisites for this chapter are Chapters 11 and 12.

The book contains the Supplementary Material that is composed of two parts: the Solutions of the Problems and Research Projects with offered solutions. To master the theory, key chapters contain a set of problems for students' exercises. The solutions to the problems are inside a separate book belonging to the Supplementary Material for the readers. In addition to the solved problems, the book contains several real-world case studies in the form of projects related to the advanced modelling and designs of modern communication systems based on digital and discrete time signal processing and application of modern technologies like DSP and FPGA. The purpose of the projects is two-fold: firstly, the reader will reinforce understanding of the theory learned and, secondly, understand the importance of theoretical knowledge for the practical design, where exact mathematical forms need to be used to implement functions of a communication system's blocks in digital processing technology like DSP or FPGA, i.e. the exact mathematical expressions are usually directly transferred into precise lines of code. The projects contain two parts: the first part defines Topics and Tasks to be supplied to the project executors, students, for example, and the second part contains the offered solutions of the project. The implementation of the systems defined in the projects can be performed using simulation tools like MATLAB or software development tools used for designs in DSP and FPGA technologies. A project is not a laboratory exercise but a self-contained piece of research work related to a particular book chapter, and as such can be a part of one semester project inside the course in discrete and digital communication systems.

In summary, the book contains the theory of discrete and digital communication systems and practical advices related to their design. The book includes the essential background theory in deterministic and stochastic signal processing, with an emphasis on the discrete time signal processing, with the adopted unified notations and the definition of terms to be directly used in the main chapters of the book related to the communication systems theory.

# Acknowlegements

This book has two origins. Firstly, the book originates from courses in Communication Systems, Digital Communications, and Digital Signal Processing that my colleagues Kevin Sowerby, Bernard Guillemin, Mark Andrews, and myself were developing and teaching at the University of Auckland. I thank them all for their enormous help because their ideas greatly influenced my thinking about how these courses should be taught. Secondly, the content of the book has been inspired by my research in digital and discrete communication systems conducted at the Department of Electrical, Computer, and Software Engineering at the University of Auckland. I am grateful to my Department, Faculty, and the University that made an environment where the book writing was possible.

I am thankful to the staff at Oxford University Press, in particular Harriet Konishi, Francesca McMahon and Sumintra Gaur, for their support and perfect cooperation during the preparation of the book manuscript. It was my greatest pleasure to work with them.

Special thanks for support and encouragement go to my beloved family; my wife Zorka, son Pavle, and daughters Ana and Marina with their families.

# Contents

List of Symbols, Functions, Operators, and Abbreviations     xxxv

1    **Introduction to Communication Systems**     1
    1.1    Communication Systems and Networks     1
    1.2    Classification of Signals and Systems     4
       1.2.1    Classification of Signals with Respect to Time and Value     5
       1.2.2    Periodic and Symmetric Signals     8
       1.2.3    Deterministic and Stochastic Signals     9
       1.2.4    Classification of Signals with Respect to Power and Energy     9
       1.2.5    Classification of Signals with Respect to Realizability     10
       1.2.6    Classification of Systems     11
    1.3    Conversions of Analogue and Digital Signals     13
       1.3.1    Analogue-to-Digital Conversion     14
       1.3.2    Digital-to-Analogue Conversion     15
       1.3.3    Application of Signals in Digital and Discrete Communication Systems     17

2    **Orthogonal Signals and the Orthogonalization Procedure**     18
    2.1    Introduction     18
    2.2    Geometric Representation of Signals     19
       2.2.1    Orthonormal Basis Functions     19
       2.2.2    Vector Representation of Signals     24
    2.3    The Gram–Schmidt Orthogonalization Procedure     26
    2.4    Continuous-Time Orthogonal Signals     35
       2.4.1    Continuous-Time Versus Discrete-Time Basis Signals     35
       2.4.2    Orthonormal Signals     36
       2.4.3    The Gram–Schmidt Orthogonalization Procedure     37
    2.5    Orthogonal Signals in Code Division Multiple Access Communication Systems     38
       Problems     42

3    **Discrete-Time Stochastic Processes**     56
    3.1    Definition and Analysis of Discrete-Time Stochastic Processes     56
       3.1.1    Introduction     56
       3.1.2    Definition of a Stochastic Process     57
       3.1.3    Mathematical Analysis of Stochastic Processes     59
    3.2    Statistical Properties of Stochastic Processes     61
       3.2.1    First-Order Statistics     61
       3.2.2    Second-Order Statistics     63
       3.2.3    Higher-Order Statistics     72
       3.2.4    Types of Discrete-Time Stochastic Processes     72

3.3 The Stationarity of Discrete-Time Stochastic Processes 75
    3.3.1 The Stationarity of One Discrete-Time Stochastic Process 75
    3.3.2 Properties of the Autocorrelation Function 81
    3.3.3 The Stationarity of Two Discrete-Time Stochastic Processes 83
3.4 Ergodic Processes 84
    3.4.1 Ensemble Averages and Time Averages 84
    3.4.2 Ergodic Processes 85
    3.4.3 Estimate of the Mean across the Ensemble of Realizations of $X(n)$ 85
    3.4.4 Estimate of the Mean across a realization of $X(n)$ 86
    3.4.5 Estimate of the Mean of an Ergodic Process $X(n)$ 87
    3.4.6 Summary of Ergodic Stochastic Processes 88
3.5 The Frequency-Domain Representation of Discrete-Time Stochastic Processes 89
    3.5.1 Continuous-Time Stochastic Processes in the Frequency Domain 90
    3.5.2 Discrete-Time Stochastic Processes in the Frequency Domain 91
    3.5.3 Cross-Spectrum Functions 92
3.6 Typical Stochastic Processes 94
    3.6.1 Noise Processes 94
    3.6.2 General Gaussian Noise Processes 96
    3.6.3 Harmonic Processes 98
    3.6.4 Stochastic Binary Processes 101
3.7 Linear Systems with Stationary Random Inputs 103
    3.7.1 An LTI System with Stationary Random Inputs in the Time Domain 104
    3.7.2 Frequency-Domain Analysis of an LTI System 108
3.8 Summary 112
    Problems 112

**4 Noise Processes in Discrete Communication Systems** 121

4.1 Gaussian Noise Processes in the Continuous-Time Domain 121
    4.1.1 Continuous White Gaussian Noise Processes 121
    4.1.2 The Entropy of White Gaussian Noise Processes 126
    4.1.3 Truncated Gaussian Noise Processes 128
    4.1.4 Concluding Notes on Gaussian Noise Processes 131
4.2 Gaussian Noise Processes in the Discrete-Time Domain 133
    4.2.1 White Gaussian Noise Processes with Discrete-Time and
    Continuous-Valued Samples 133
    4.2.2 Discrete-Time White Gaussian Noise Processes with
    Discrete-Valued Samples 139
    4.2.3 White Gaussian Noise Processes with Quantized Samples
    in a Strictly Limited Interval 140
    4.2.4 Band-Limited Continuous- and Discrete-Time Signals and Noise 141
4.3 Operation of a Baseband Noise Generator 143
    4.3.1 Band-Limited Continuous-Time Noise Generators 143
    4.3.2 Band-Limited Discrete-Time Noise Generators 144

| | | |
|---|---|---|
| 4.3.3 | Spectral Analysis of Continuous-Time Baseband Noise | 147 |
| 4.3.4 | Spectral Analysis of Discrete-Time Baseband Noise | 148 |
| 4.4 | Operation of a Bandpass Noise Generator | 151 |
| 4.4.1 | Ideal Bandpass Continuous-Time Gaussian Noise | 151 |
| 4.4.2 | Ideal Bandpass Discrete Gaussian Noise | 154 |
| 4.4.3 | Modulators and Demodulators of Ideal Bandpass Discrete Gaussian Noise | 156 |
| 4.5 | Practical Design of a Band-Limited Discrete-Time Noise Modulator | 162 |
| 4.6 | Design of an Ordinary Band-Limited Discrete-Time Noise Modulator | 166 |
| | Problems | 168 |

**5 Operation of a Discrete Communication System** — 172

| | | |
|---|---|---|
| 5.1 | Structure of a Discrete System | 172 |
| 5.2 | Operation of a Discrete Message Source | 172 |
| 5.3 | Operation of a Discrete Modulator | 176 |
| 5.4 | Additive White Gaussian Noise Channels in a Discrete-Time Domain | 178 |
| 5.5 | Correlation Demodulators | 180 |
| 5.5.1 | Operation of a Correlator | 180 |
| 5.5.2 | Statistical Characterization of Correlator Output | 185 |
| 5.5.3 | Signal Constellation | 189 |
| 5.6 | Optimum Detectors | 190 |
| 5.6.1 | The Maximum Likelihood Estimator of a Transmitted Signal | 190 |
| 5.6.2 | Application of the Maximum Likelihood Rule | 194 |
| 5.6.3 | Design of an Optimum Detector | 194 |
| 5.6.4 | Generic Structure of a Discrete Communication System | 198 |
| 5.7 | Multilevel Systems with a Binary Source | 198 |
| 5.7.1 | Transmitter Operation | 201 |
| 5.7.2 | Radio Frequency Blocks and Additive White Gaussian Noise Waveform Channels | 202 |
| 5.7.3 | Operation of a Bandpass Noise Generator | 203 |
| 5.7.4 | Intermediate-Frequency Optimum Receivers | 204 |
| 5.7.5 | Intermediate-Frequency Optimum Detectors | 207 |
| 5.8 | Operation of a Digital Communication System | 207 |
| 5.8.1 | Digital versus Discrete Communication Systems | 207 |
| 5.8.2 | Generic Structure of a Digital Communication System | 208 |
| | Appendix: Operation of a Correlator in the Presence of Discrete White Gaussian Noise | 212 |

**6 Digital Bandpass Modulation Methods** — 215

| | | |
|---|---|---|
| 6.1 | Introduction | 215 |
| 6.2 | Coherent Binary Phase-Shift Keying Systems | 220 |
| 6.2.1 | Operation of a Binary Phase-Shift Keying System | 220 |
| 6.2.2 | Transmitter Operation | 221 |

6.2.2.1  Modulating Signal Presentation                                        221
6.2.2.2  Modulated Signals in Time and Frequency Domains                       224
6.2.3    Receiver Operation                                                    231
6.2.3.1  Correlation Demodulator Operation                                     231
6.2.3.2  Operation of the Optimum Detector, and Structure
         of the Receiver                                                       233
6.2.3.3  Bit Error Probability Calculation                                     238
6.3  Quadriphase-Shift Keying                                                  245
6.3.1    Operation of a Quadrature Phase-Shift Keying System                   245
6.3.2    Transmitter Operation                                                 246
6.3.2.1  Modulating Signals in Time and Frequency Domains                      246
6.3.2.2  Modulated Signals in the Time Domain                                  250
6.3.2.3  Modulated Signals in the Frequency Domain                            253
6.3.2.4  The Power Spectral Density of Signals in a Quadriphase-Shift
         Keying System                                                         253
6.3.3    Receiver Operation                                                    256
6.3.3.1  Operation of the Correlation Demodulator and the
         Optimum Detector                                                      256
6.3.3.2  Bit Error Probability Calculation                                     257
6.3.3.3  Signal Analysis and Transceiver Structure in a Quadrature
         Phase-Shift Keying System                                             259
6.4  Coherent Binary Frequency-Shift Keying with a Continuous Phase            260
6.4.1    Operation of a Binary Frequency-Shift Keying System                   260
6.4.2    Transmitter Operation                                                 262
6.4.2.1  Modulating Signals in Time and Frequency Domains                      262
6.4.2.2  Modulated Signals in the Time Domain and the
         Signal-Space Diagram                                                  263
6.4.2.3  Modulating and Modulated Signals in Time and Frequency
         Domains                                                               264
6.4.2.4  Modulated Signals in the Frequency Domain                            270
6.4.3    Receiver Operation                                                    272
6.4.3.1  Operation of a Correlation Demodulator                                272
6.4.3.2  Operation of an Optimum Detector                                      274
6.4.3.3  Calculation of the Bit Error Probability                             276
6.4.3.4  Design of a Transceiver for a Binary Frequency-Shift
         Keying Signal                                                         278
6.5  $M$-ary Quadrature Amplitude Modulation                                   280
6.5.1    System Operation                                                      280
6.5.2    Transmitter Operation                                                 282
6.5.3    Receiver Operation                                                    285
Appendix A: Densities of the Correlation Variables $X_1$ and $X_2$ in a Quadrature
         Phase-Shift Keying System                                             289

Appendix B: Derivatives of Density Functions for a Binary Frequency-Shift
      Keying System     290
Appendix C: Precise Derivation of the Bit Error Probability for a Binary
      Frequency-Shift Keying System     292
Appendix D: Power Spectral Density of a Quadrature Component in a
      Frequency-Shift Keying Signal     294
Problems     296

**7   Discrete Bandpass Modulation Methods**     **305**

7.1   Introduction     305
7.2   Coherent Binary Phase-Shift Keying Systems     308
    7.2.1   Operation of a Binary Phase-Shift Keying System     308
    7.2.2   Transmitter Operation     309
        7.2.2.1  Presentation of a Modulating Signal     309
        7.2.2.2  Modulated Signals in Time and Frequency Domains     312
        7.2.2.3  The Power Spectral Density of Binary Phase-Shift Keying
                 Modulated Signals     316
    7.2.3   Receiver Operation     319
        7.2.3.1  Operation of a Correlation Demodulator     319
        7.2.3.2  Operation of an Optimum Detector, and Structure of a Receiver     321
        7.2.3.3  Calculation of the Bit Error Probability     326
7.3   Quadriphase-Shift Keying     328
    7.3.1   System Operation     328
    7.3.2   Transmitter Operation     330
        7.3.2.1  Modulating Signals in Time and Frequency Domains     330
        7.3.2.2  Modulated Signals in the Time Domain     333
        7.3.2.3  Modulated Signals in the Frequency Domain     336
    7.3.3   Receiver Operation     338
        7.3.3.1  Operation of the Correlation Demodulator and the
                 Optimum Detector     338
        7.3.3.2  Calculation of the Bit Error Probability     340
        7.3.3.3  Signal Analysis and Structure of the Transceiver in a
                 Quadriphase-Shift Keying System     341
7.4   Coherent Binary Frequency-Shift Keying with Continuous Phase     343
    7.4.1   Operation of a Binary Frequency-Shift Keying System     343
    7.4.2   Transmitter Operation     345
        7.4.2.1  Modulating Signals in Time and Frequency Domains     345
        7.4.2.2  Modulated Signal Analysis in the Time Domain and a
                 Signal-Space Diagram     346
        7.4.2.3  Modulated Signal Analysis in Time and Frequency Domains     349
        7.4.2.4  Modulated Signals in the Frequency Domain     356
    7.4.3   Receiver Operation     358
        7.4.3.1  Operation of the Correlation Demodulator     358

|  |  |  |
|---|---|---|
| | 7.4.3.2 Operation of the Optimum Detector | 359 |
| | 7.4.3.3 Calculation of the Bit Error Probability | 361 |
| | 7.4.3.4 Transceiver Design for a Binary Frequency-Shift Keying Signal | 364 |
| 7.5 | $M$-ary Discrete Quadrature Amplitude Modulation | 365 |
| | 7.5.1 Operation of a Discrete $M$-ary Quadrature Amplitude Modulation System | 365 |
| | 7.5.2 Transmitter Operation | 367 |
| | 7.5.3 Operation of a Correlation Demodulator | 371 |
| | 7.5.4 Operation of an Optimum Detector | 374 |
| | Appendix A: Power Spectral Density of a Quadriphase-Shift Keying Modulating Signal | 375 |
| | Appendix B: Probability Density Functions for a Quadriphase-Shift Keying System | 376 |
| | Appendix C: Density Functions for $X_1$ and $X_2$ in a Frequency-Shift Keying System | 377 |
| | Appendix D: Statistics of the Decision Variable $X = X_1 - X_2$ | 380 |
| | Problems | 382 |

**8  Orthogonal Frequency Division Multiplexing and Code Division Multiple Access Systems**                                                  **386**

| | | |
|---|---|---|
| 8.1 | Introduction | 386 |
| 8.2 | Digital Orthogonal Frequency Division Multiplexing Systems | 388 |
| | 8.2.1 Introduction | 388 |
| | 8.2.2 Transmitter Operation | 390 |
| | 8.2.3 Receiver Operation | 396 |
| | 8.2.4 Operation of a Receiver in the Presence of Noise | 399 |
| 8.3 | Discrete Orthogonal Frequency Division Multiple Access Systems | 401 |
| | 8.3.1 Principles of Discrete Signal Processing in an Orthogonal Frequency Division Multiple Access System | 401 |
| | 8.3.2 A Discrete Baseband Orthogonal Frequency Division Multiple Access System Based on Binary Phase-Shift Keying | 404 |
| | 8.3.3 Structure and Operation of a Discrete Orthogonal Frequency Division Multiple Access System | 406 |
| | 8.3.4 Operation of the Receiver in an Orthogonal Frequency Division Multiple Access System | 412 |
| | 8.3.5 Operation of the Receiver in the Presence of Noise | 415 |
| 8.4 | Discrete Code Division Multiple Access Systems | 418 |
| | 8.4.1 Principles of Operation of a Discrete Code Division Multiple Access System | 418 |
| | 8.4.2 Derivation of the Probability of Error | 422 |
| | Problems | 423 |

**9  Information Theory and Channel Coding**                                  **427**

| | | |
|---|---|---|
| 9.1 | Characterization of a Discrete Source | 428 |

9.2   Characterization of a Discrete Channel                                                432
    9.2.1   A Discrete Memoryless Channel                                         432
    9.2.2   Discrete Binary Channels with and without Memory                      434
        9.2.2.1   Discrete Binary Channels                                   434
        9.2.2.2   Discrete Binary Memoryless Channels                        436
        9.2.2.3   Discrete Binary Channels with Memory                       438
    9.2.3   Capacity of a Discrete Memoryless Channel                             440
        9.2.3.1   Capacity of a Discrete Channel                             440
        9.2.3.2   Example of the Capacity of a Binary Memoryless Channel      444
9.3   Characterization of Continuous Channels                                             448
    9.3.1   Differential Entropy                                                  448
    9.3.2   Channel Information for Random Vectors                                450
    9.3.3   Definition of the Capacity of a Continuous Channel                    451
    9.3.4   Proof of the Channel Capacity Theorem                                 452
9.4   Capacity Limits and the Coding Theorem                                              459
    9.4.1   Capacity Limits                                                       459
    9.4.2   The Coding Theorem and Coding Channel Capacity                        461
9.5   Information and Entropy of Uniform Density Functions                                465
    9.5.1   Continuous Uniform Density Functions                                  466
    9.5.2   Discrete Uniform Density Functions                                    468
9.6   Information and Entropy of Gaussian Density Functions                               471
    9.6.1   Continuous Gaussian Density Functions                                 471
    9.6.2   Discrete Gaussian Density Functions                                   473
9.7   Block Error Control Codes                                                           477
    9.7.1   Theoretical Basis and Definitions of Block Code Terms                 477
    9.7.2   Coding Procedure Using a Generator Matrix                             479
    9.7.3   Error Detection Using a Parity Check Matrix                           481
    9.7.4   Standard Array Decoding                                               484
    9.7.5   Syndrome Table Decoding                                               485
9.8   Convolutional Codes                                                                 488
    9.8.1   Linear Convolutional Codes                                            488
    9.8.2   Operation of a Coding Communication System                            498
    9.8.3   Operation of a Decoder                                                501
    9.8.4   Decoding Algorithms                                                   504
9.9   Introduction to Iterative Decoding and Turbo Decoding                               512
    9.9.1   Coding Models for Communication Systems                               512
    9.9.2   The Hard-Output Viterbi Algorithm                                     514
    9.9.3   Iterative Algorithms and Turbo Coding                                 518
    Appendix A: Derivation of Mutual Information                                 519
    Appendix B: Entropy of a Truncated Discrete Gaussian Density Function        521
    Problems                                                                     522

**10   Designing Discrete and Digital Communication Systems**                     542

   10.1   Introduction                                                           542

   10.2   Designing Quadriphase-Shift Keying Transceivers                        544
          10.2.1   Quadriphase-Shift Keying Systems with Digital-to-Analogue
                   and Analogue-to-Digital Conversion of Modulating Signals      544
                   10.2.1.1   Quadriphase-Shift Keying Transmitters with
                              Baseband Discrete-Time Signal Processing           544
                   10.2.1.2   Designing a Quadriphase-Shift Keying Receiver      545
                   10.2.1.3   Practical Design of a Quadriphase-Shift Keying Receiver   547
          10.2.2   Quadriphase-Shift Keying Systems with Digital-to-Analogue and
                   Analogue-to-Digital Conversion of Intermediate-Frequency Signals   548
                   10.2.2.1   Designing a Digital Quadriphase-Shift Keying
                              Transmitter at Intermediate Frequency              548
                   10.2.2.2   Design of a Digital Quadriphase-Shift Keying
                              Receiver at Intermediate Frequency                 550
                   10.2.2.3   Practical Design of a Discrete Quadriphase-Shift Keying
                              Receiver at Intermediate Frequency                 551

   10.3   Designing Quadrature Amplitude Modulation Transceivers                 552
          10.3.1   Quadrature Amplitude Modulation Systems with
                   Digital-to-Analogue and Analogue-to-Digital Conversion
                   of Modulating Signals                                         552
                   10.3.1.1   Quadrature Amplitude Modulation Transmitters
                              with Baseband Discrete-Time Signal Processing      553
                   10.3.1.2   Designing a Quadrature Amplitude
                              Modulation Receiver                                554
                   10.3.1.3   Practical Design of a Quadrature Amplitude
                              Modulation Receiver                                555
          10.3.2   Quadrature Amplitude Modulation Systems with
                   Digital-to-Analogue and Analogue-to-Digital Conversion
                   of Intermediate-Frequency Signals                            556
                   10.3.2.1   Digital Design of a Quadrature Amplitude Modulation
                              Transmitter at Intermediate Frequency              556
                   10.3.2.2   Digital Design of a Quadrature Amplitude Modulation
                              Receiver at Intermediate Frequency                 557
                   10.3.2.3   Practical Design of a Discrete Quadrature Amplitude
                              Modulation Receiver at Intermediate Frequency      558

   10.4   Overview of Discrete Transceiver Design                                559
          10.4.1   Introduction                                                 559
          10.4.2   Designing Quadrature Amplitude Modulation Systems            560

**11   Deterministic Continuous-Time Signals and Systems**                       562

   11.1   Basic Continuous-Time Signals                                         562
   11.2   Modification of the Independent Variable                              571

11.3  Combined Modifications of the Independent Variable                575
11.4  Cross-Correlation and Autocorrelation                            576
    11.4.1  The Cross-Correlation Function                         576
    11.4.2  The Autocorrelation Function                           576
11.5  System Classification                                           579
11.6  Continuous-Time Linear-Time-Invariant Systems                   581
    11.6.1  Modelling of Systems in the Time Domain                581
    11.6.2  Representation of an Input Signal                      581
    11.6.3  Basic Representation of a Linear-Time-Invariant System 583
    11.6.4  Representation of an Output Signal                     584
    11.6.5  Properties of Convolution                             585
11.7  Properties of Linear-Time-Invariant Systems                     586
    Problems                                                      588

**12  Transforms of Deterministic Continuous-Time Signals**           599

12.1  Introduction                                                    599
12.2  The Fourier Series                                              600
    12.2.1  The Fourier Series in Trigonometric Form              600
    12.2.2  An Example of Periodic Signal Analysis                 603
    12.2.3  The Fourier Series in Complex Exponential Form        609
    12.2.4  Amplitude Spectra, Magnitude Spectra, and Phase
        Spectra of Periodic Signals                            612
    12.2.5  The Power and Energy of Signals, and Parseval's Theorem 615
    12.2.6  Existence of Fourier Series                          619
    12.2.7  Orthogonality Characteristics of the Fourier Series  620
    12.2.8  Table of the Fourier Series                          621
12.3  Fourier Transform of Continuous Signals                         622
    12.3.1  Derivative and Application of Fourier Transform Pairs  622
    12.3.2  Convergence Conditions                                627
    12.3.3  The Rayleigh Theorem and the Energy of Signals        627
    12.3.4  Properties of the Fourier Transform                   628
    12.3.5  Important Problems and Solutions                      630
    12.3.6  Tables of the Fourier Transform                      640
12.4  Fourier Transform of Periodic Signals                           640
12.5  Correlation Functions, Power Spectral Densities, and
    Linear-Time-Invariant Systems                                646
    12.5.1  Correlation of Real-Valued Energy Signals             646
    12.5.2  Correlation of Real-Valued Power Signals              648
    12.5.3  Comprehensive Analysis of Linear-Time-Invariant Systems 650
        12.5.3.1 System Presentation                            650
        12.5.3.2 Correlation and Energy Spectral Density of Complex
            Energy Signals                                      655

12.5.3.3 Correlation and Power Spectral Density of Complex
Power Signals 657
12.5.3.4 Analysis of a Linear-Time-Invariant System with
Deterministic Energy Signals 662
12.5.4 Tables of Correlation Functions and Related Spectral Densities 665
Problems 666

**13 Sampling and Reconstruction of Continuous-Time Signals** 674

13.1 Introduction 674
13.2 Sampling of Continuous-Time Signals 675
13.3 Reconstruction of Analogue Signals 679
13.4 Operation of a Lowpass Reconstruction Filter 682
13.5 Generation of Discrete-Time Signals 685

**14 Deterministic Discrete-Time Signals and Systems** 690

14.1 Discrete-Time Signals 690
14.1.1 Elementary Discrete-Time Signals 690
14.1.2 Modification of Independent Variables 697
14.1.3 Cross-Correlation and Autocorrelation Functions 700
14.2 Discrete-Time Systems 702
14.2.1 Systems Classification 702
14.2.2 Discrete-Time Linear-Time-Invariant Systems 705
14.3 Properties of Linear-Time-Invariant Systems 707
14.4 Analysis of Linear-Time-Invariant Systems in Time and
Frequency Domains 710
Problems 710

**15 Deterministic Discrete-Time Signal Transforms** 714

15.1 Introduction 714
15.2 The Discrete-Time Fourier Series 714
15.2.1 Continuous-Time Fourier Series and Transforms 714
15.2.2 The Discrete-Time Fourier Series 715
15.2.3 Fourier Series Examples Important for Communication Systems 718
15.3 The Discrete-Time Fourier Transform 725
15.3.1 Derivation of the Discrete-Time Fourier Transform Pair 725
15.3.2 The Problem of Convergence 727
15.3.3 Properties of the Discrete-Time Fourier Transform 736
15.3.4 Tables for the Discrete-Time Fourier Transform 739
15.4 Discrete Fourier Transforms 740
15.4.1 Fundamentals of Frequency-Domain Sampling 740
15.4.2 Discrete Fourier Transforms 741
15.4.3 Inverse Discrete Fourier Transforms 744
15.4.4 Three Typical Cases of Discrete Fourier Transforms 747

15.5 Algorithms for Discrete Fourier Transforms 751
  15.5.1 Goertzel's Algorithm 751
  15.5.2 Discrete Fourier Transforms as Linear Transformations 756
  15.5.3 The Radix-2 Fast Fourier Transform Algorithm 758
15.6 Correlation and Spectral Densities of Discrete-Time Signals 763
  15.6.1 Cross-Correlation and Correlation of Real-Valued Energy Signals 763
  15.6.2 Cross-Correlation and Correlation of Real-Valued Power Signals 766
  15.6.3 Parseval's Theorem and the Wiener–Khintchine Theorem 768
  15.6.4 Comprehensive Analysis of Discrete Linear-Time-Invariant Systems 769
    15.6.4.1 System Presentation 769
    15.6.4.2 Correlation and Power Spectral Density of Complex Energy Signals 773
    15.6.4.3 Correlation of Complex Power Signals 776
    15.6.4.4 Analysis of a Linear-Time-Invariant System with Energy Signals 777
  15.6.5 Tables of Correlation Functions and Related Spectral Density Functions 783
15.7 The $z$-Transform 783
  15.7.1 Introduction 783
  15.7.2 Derivation of Expressions for the $z$-Transform 784
  15.7.3 Properties of the $z$-Transform 787
  15.7.4 The Inverse $z$-Transform 790
  Problems 790

**16 Theory of the Design, and Operation of Digital Filters** 797

16.1 The Basic Concept of Filtering 797
16.2 Ideal and Real Transfer Functions 799
16.3 Representation of Digital Filters 801
16.4 Basic Finite Impulse Response Filters 803
16.5 Structures of Finite Impulse Response Filters 805
16.6 Basic Infinite Impulse Response Filters 808
16.7 Structures of Infinite Impulse Response Filters 811
  16.7.1 Introduction 811
  16.7.2 Conventional Description of Block Diagrams 811
  16.7.3 Direct Forms of Infinite Impulse Response Filters 813
16.8 Algorithms for the Design of Digital Filters 815
  16.8.1 Ideal and Real Frequency Responses 815
  16.8.2 Basic Methods for the Design of Digital Filters 817
  16.8.3 Algorithms Based on Iterative Optimization 822

**17 Multi-Rate Discrete-Time Signal Processing** 824

17.1 Multi-Rate Signals in Time and Frequency Domains 825
  17.1.1 Time-Domain Analysis 825

|  |  | 17.1.2 | Frequency-Domain Analysis | 829 |
|  |  | 17.1.3 | Complex Multi-Rate Systems | 334 |
|  |  | 17.1.4 | Complexity Reduction | 836 |
|  | 17.2 | Multi-Rate Systems |  | 836 |
|  |  | 17.2.1 | Basic System Structures | 836 |
|  |  | 17.2.2 | System Analysis in Time and Frequency Domains | 838 |
|  | 17.3 | Reduction of Computational Complexity |  | 843 |
|  |  | 17.3.1 | Multistage Decimators and Interpolators | 843 |
|  |  | 17.3.2 | Polyphase Decomposition of a Decimation Filter | 843 |
|  |  | 17.3.3 | Polyphase Decomposition of a Finite Impulse Response Transfer Function | 844 |
|  |  | 17.3.4 | Polyphase Decomposition of an Infinite Impulse Response Transfer Function | 847 |
|  |  | Problems |  | 848 |

**18   Multi-Rate Filter Banks**   853

|  | 18.1 | Digital Filter Banks |  | 853 |
|  | 18.2 | Two-Channel Quadrature Mirror Filter Banks |  | 858 |
|  |  | 18.2.1 | Basic Theory | 858 |
|  |  | 18.2.2 | Elimination of Aliasing in Two-Channel Quadrature Mirror Filter Banks | 861 |
|  | 18.3 | Perfect Reconstruction of Two-Channel Filter Banks |  | 863 |
|  | 18.4 | Multichannel Quadrature Mirror Filter Banks |  | 865 |
|  | 18.5 | Multilevel Filter Banks and Adaptive Filter Banks |  | 867 |
|  |  | 18.5.1 | Banks with Equal or Unequal Passband Widths | 867 |
|  |  | 18.5.2 | Adaptive Filter Banks | 871 |
|  |  | Problems |  | 872 |

**19   Continuous-Time Stochastic Processes**   874

|  | 19.1 | Continuous-Time Stochastic Processes |  | 874 |
|  |  | 19.1.1 | Probability, Random Variables, and Stochastic Processes | 874 |
|  |  | 19.1.2 | Statistical Analysis of Stochastic Processes | 876 |
|  | 19.2 | Statistical Properties of Stochastic Processes |  | 879 |
|  |  | 19.2.1 | First- and Second-Order Properties of Stochastic Processes | 879 |
|  |  | 19.2.2 | Types of Stochastic Processes | 886 |
|  |  | 19.2.3 | Entropy of Stochastic Processes and White Noise | 888 |
|  | 19.3 | Stationary and Ergodic Stochastic Processes |  | 893 |
|  |  | 19.3.1 | Stationary Stochastic Processes in Time and Frequency Domains | 893 |
|  |  |  | 19.3.1.1   Time Domain Analysis | 893 |
|  |  |  | 19.3.1.2   Frequency Domain Analysis | 895 |
|  |  | 19.3.2 | Ergodic Stochastic Processes | 898 |
|  |  | 19.3.3 | Characterization of White Noise Processes | 900 |
|  |  | 19.3.4 | Gaussian Correlated Processes | 903 |

19.4  Linear-Time Invariant Systems with Stationary Stochastic Inputs          905
    19.4.1  Analysis of Linear-Time Invariant Systems in Time and
        Frequency Domains          905
    19.4.2  Definition of a System Correlation Function for Stochastic Input          911
    19.4.3  Application of the Theory of Linear-Time-Invariant Systems
        to the Analysis of the Operation of a Lowpass Filter          914
    19.4.4  Analysis of the Operation of a Bandpass Filter          918
    Problems          922

**Bibliography**          925
**Index**          929

# List of Symbols, Functions, Operators, and Abbreviations

## Symbols

| | |
|---|---|
| $A$, A | the alphabet of the discrete source of information, amplitude of a signal |
| $A_b$ | amplitude of a binary signal |
| $a_k$ | coefficient, Fourier series coefficient |
| $A_k$ | amplitude value of a QAM signal |
| $a_m$ | source symbols |
| $A(f)$ | amplitude spectrum, amplitude spectral density |
| $A_X, A_x$ | input alphabet of a channel |
| $A_Y, A_y$ | output alphabet of a channel |
| $B$ | baseband bandwidth, the bandwidth of a modulating signal |
| $\mathbf{B}$ | basis in a vector space |
| $b_k$ | coefficient, Fourier series coefficient |
| $B_k$ | amplitude value of a QAM signal |
| $BP$ | band-pass |
| $B_W$ | required bandwidth, required channel bandwidth |
| $C$ | channel capacity, a constant |
| $\mathbf{c}$ | code word |
| $c_i$ | bit in a code word |
| $c_k$ | complex Fourier series coefficient for the discrete or continuous-time signals |
| $C_{xx}$ | autocovariance function of the discrete or continuous-time ergodic process |
| $\mathbf{C}_{XX}$ | matrix of the autocovariance function of stochastic processes $X(n)$ |
| $C_{XX}(t_1, t_2)$ | autocovariance function of the continuous-time stochastic process $X(t)$ |
| $C_{XX}(m,n)$ | autocovariance function of the discrete-time stochastic process $X(n)$ |
| $C_{xy}$ | crosscovariance function of the discrete or continuous-time ergodic process |
| $C_{XY}(m,n)$ | crosscovariance function of stochastic processes $X(n)$ and $Y(n)$ |
| $D$ | distance function, Hamming distance |
| $dB$ | decibel |
| $d_{ik}$ | Euclidean distance |
| $e$ | natural number 2.7183 |
| $\mathbf{e}$ | error sequence |
| $E$ | energy, linear expectation operator |
| $E_b$ | energy of a bit |

| | |
|---|---|
| $E_b/N_0$ | bit energy to noise power spectral density ratio |
| $E_i$ | energy of a symbol |
| $E_m(f)$ | energy spectral density of a continuous-time signal $m(t)$ |
| $E_m(\Omega)$ | energy spectral density of a discrete-time signal $m(n)$ |
| $E_{sc}$ | energy of the carrier signal |
| $E_x(f)$ | energy spectral density of the continuous-time energy signal $x(t)$ |
| $E_{xx}(f)$ | energy spectral density of the continuous-time power signal $x(t)$ |
| $E_{xy}(f)$ | cross-energy spectral density of the continuous-time energy and power signals |
| $E\{X\}$ | expectation of random variable $X$ |
| $f$ | cyclic frequency |
| $F$ | normalised cyclic frequency for the disrete time signals |
| $f_c$ | carrier frequency |
| $f_i$ | frequency of a subcarrier |
| $f_j$ | frequency of a modulated signal |
| $f_{Nyq}$ | Nyquist sampling frequency |
| $f_{RF}$ | radio frequency |
| $f_s$ | sampling frequency |
| $F_x, F_y$ | sampling frequencies in multi-rate signal processing |
| $f_X(x)$ | probability density function of the random variable $X$ |
| $F_X(x)$ | probability distribution function of the random variable $X$ |
| $f_X(x_1; t_1)$ | the probability density function of the random variable $X(t_1)$ at any time instant $t_1$ of continuous-time process $X(t)$ |
| $F_X(x_1; t_1)$ | the probability distribution function of random variable $X(t_1)$ |
| $F_X(x_1, x_2; t_1, t_2)$ | the second order joint probability distribution function of continuous-time process $X(t)$ |
| $f_X(x_1, x_2; t_1, t_2)$ | the second order joint probability density function of the process $X(t)$ |
| $f_X(x_1, x_2, ..., x_k; t_1, t_2, ..., t_k)$ | $k^{\text{th}}$ order joint probability density function |
| $f_X(x; n) = f_X(x(n)) = f_X(x)$ | probability density function of random variable $X(n)$ of discrete-time process $X(n)$ |
| $F_X(x; n) = F_X(x(n)) = F_X(x)$ | probability distribution function of the random variable $X(n)$ at any time instant $n$ of discrete-time process $X(n)$ |
| $f_X(x_1, x_2, ..., x_k; n_1, n_2, ..., n_k)$ | joint probability density function of random variables defining discrete-time process $X(n)$ at time instants $n_1, n_2, ..., n_k$ |
| $F_X(x_1, x_2, ..., x_k; n_1, n_2, ..., n_k)$ | joint probability distribution function of random variables defined at discrete-time instants $n_1, n_2, ..., n_k$. |

| | |
|---|---|
| $f_{\mathbf{X}}(\mathbf{x}\|m_i)$ | conditional probability density function of (observation) vector $\mathbf{x}$ given the message symbol $m_i$ |
| $f_{X_j}(x_j\|m_i)$ | conditional probability density of $x_j$ given the message symbol $m_i$ |
| $f_{W_m W_n}(w_m, w_n)$ | the joint probability density function of discrete-time random variables $W_m$ and $W_n$ |
| | |
| $\mathbf{G}$ | generator matrix |
| $g_i(n)$ | $i^{\text{th}}$ intermediate discrete-time function |
| $g_i(t)$ | $i^{\text{th}}$ intermediate continuous-time function |
| | |
| $H$ | entropy |
| $\mathbf{H}$ | parity-check matrix |
| $H(f)$ | FT of the impulse response (transfer function) of continuous-time LTI system |
| $h(n)$ | impulse response of discrete-time LTI system |
| $h(t)$ | impulse response of continuous-time LTI system |
| $h(X)$ | differential entropy of continuous random variable $X$ |
| $H(X)$ | entropy of the channel input, entropy of a stochastic process $X(t)$ or $X(n)$ |
| $h(\mathbf{X})$ | differential entropy of continuous random vector $\mathbf{X}$ |
| $h(X/Y)$ | conditional differential entropy of $X$ given $Y$ |
| $H(X\|Y)$ | channel conditional entropy |
| $H(\omega) = H(j\omega)$ | FT of the impulse response (transfer function) of continuous-time LTI system |
| $H(\Omega) = H(e^{j\Omega})$ | FT of the impulse response of discrete-time LTI system |
| $H_k(z)$ | frequency response of a filter |
| $H_{LP}(\omega)$ | LP filter transfer characteristics for continuous-time signals |
| $H_{LP}(\Omega)$ | LP filter transfer characteristics for discrete-time signals |
| $H(z)$ | filter transfer function |
| | |
| $I$ | measure of information, information (content), information bits per symbol |
| $\mathbf{I}$ | identity matrix |
| $i$ | an integer index |
| $I(m_k)$ | amount (content) of information of symbol $m_k$ |
| $I(n)$ | in-phase discrete-time signal component |
| $I(t)$ | in-phase continuous-time signal component |
| $I(X;Y)$ | mutual information |
| $I(\mathbf{X};\mathbf{Y})$ | mutual information for two random vectors $\mathbf{X}$ and $\mathbf{Y}$ |
| | |
| $j$ | imaginary number $\sqrt{-1}$, an integer index |
| $J$ | space dimension, number of alphabet elements in a message symbol or a message vector, number of coefficients in a signal vector, number of modulating signals in a symbol interval $N$, number of orthogonal carriers |
| | |
| $k$ | an integer index, time instants at the correlator output |
| | |
| $l$ | correlation lag, an integer index |

| | |
|---|---|
| $L$ | linear operator, constraint length of a convolutional code, up-sampling factor |
| $l(m_i)$ | log-likelihood function |
| $L(m_i)$ | likelihood function |
| | |
| $\mathbf{m}$ | message matrix containing all possible message symbol vectors |
| $M$ | down-sampling factor |
| $\boldsymbol{m}$ | message symbol in information theory |
| $\boldsymbol{m}'$ | estimate of a message symbol in information theory |
| $\mathbf{m_i}$ | message symbol vector that contains message symbols $m_i$, column vector in $\mathbf{m}$, message point in $J$ dimensional space, modulating signal of $J$ coefficients |
| $m_i$ | message bit |
| $M_i$ | message points |
| $m_k$ | estimated received message bit |
| $m_{ij}$ | message symbol in a vector $\mathbf{m_i}$ |
| $M(f)$ | amplitude spectral density of deterministic $m(t)$ |
| $m(n)$ | modulating discrete-time signal |
| $m(t)$ | modulating continuous-time signal |
| $M(t)$ | modulating stochastic process ( for random signal $m(t)$), modulating rectangular pulse ( for deterministic $m(t)$) |
| $M(\Omega)$ | amplitude spectral density of discrete-time message signal |
| $m_i(n)$ | modulating discrete-time signal in a symbol interval |
| $m_i(t)$ | modulating continuous-time signal in a symbol interval |
| $m_k(n)$ | received discrete-time message bit |
| $m_k(t)$ | received continuous-time message bit |
| $m_{ij}(n)$ | message symbol in the discrete-time |
| $m_{ij}(t)$ | message symbol in the continuous-time |
| | |
| $n$ | discrete-time variable, corresponds to variable $t$ in the continuous-time domain |
| $\mathbf{n}$ | noise sequence |
| $N$ | discrete-time duration of energy signals, the fundamental period in DT Fourier series, band-limited noise process |
| $\boldsymbol{N}$ | vector of noise random variables in a channel, band-limited noise process |
| $N_0/2$ | two-sided PSD of the white Gaussian noise process |
| $N_b$ | discrete-time duration of a bit |
| $N(n)$ | discrete-time bandpass correlated Gaussian noise process |
| $n(n)$ | one realization of $N(n)$ |
| $N(t)$ | continuous-time bandpass correlated Gaussian noise process |
| $n(t)$ | one realization of $N(t)$ |
| $N_B(t)$ | continuous-time baseband noise process and related random variable defined at time instant $t$ |
| $n_B(t)$ | one realization of continuous-time baseband noise process $N_B(t)$ |
| $N_B(n)$ | discrete-time baseband noise process and related discrete-time baseband noise variable defined at time instant $n$ |

| | |
|---|---|
| $n_B(n)$ | realization of discrete-time baseband noise process $N_B(n)$ |
| $n_{GI}, n_{GQ}$ | inphase and quadrature noise components |
| $N_{Nq}(n)$ | discrete-time baseband noise process defined by Nyquist sampling |
| $n_{Nq}(n)$ | one realisation of the discrete-time baseband noise process $N_{Nq}(n)$ |
| $p$ | bit error probability |
| $P$ | probability, power |
| $P_{av}$ | average power |
| $P_e$ | average probability of symbol error |
| $p_k$ | message symbol probability, probability of generating symbol $m_k$ at the source |
| $P_s$ | power of a modulated signal |
| $P_{sc}$ | power of carrier signal |
| $P_T$ | word error probability |
| $P_X$ | power of WSS discrete-time process $X(n)$ or continuous-time process $X(t)$ |
| $P(t)$ | instanteneous power of a continuous-time signal |
| $P(m_i)$ | a priori probability |
| $p(x_k)$ | marginal probabilities of channel inputs |
| $p(y_k)$ | marginal probabilities of channel outputs |
| $p(\zeta_1)$ | probabily of outcome (elementary event) $\zeta_1$ |
| $P(m_i|\mathbf{x})$ | a posteriori conditional probability |
| $p(x_j|y_k)$ | channel input-output conditional probability |
| $p(x_j,y_k)$ | channel input-output joint probability |
| $\mathbf{P_{XY}}, \mathbf{P_{YX}}$ | channel matrix probabilities or transition probabilities |
| $p(y_k|x_j)$ | channel output-input conditional (transitional) probability |
| $Q(n)$ | quadrature discrete-time signal component |
| $Q(t)$ | quadrature continuous-time signal component |
| $r$ | radix of the fast Fourier transform algorithm |
| $R$ | digital signal rate, information bits per symbol, sampling rate conversion |
| $R_c$ | encoded bit rate |
| $R_s$ | symbol rate |
| $R_{xx}$ | autocorrelation function of discrete-time or continuous-time ergodic process |
| $\mathbf{R}_{XX}$ | matrix of an autocorrelation function |
| $\mathbf{R}_{XX}^{-1}$ | inverse matrix of an autocorrelation function |
| $R_{xy}$ | crosscorrelation function of discrete-time or continuous-time ergodic process |
| $R(n)$ | received discrete-time stochastic noisy process |
| $r(n)$ | received discrete-time random noisy signal, one realization of $R(n)$ |
| $R(t)$ | continuous-time WSS received process |
| $r(t)$ | received noisy continuous-time signal, a realization of a continuous-time WSS process $R(t)$ |
| $R_h(l), R_{HH}(l)$ | system correlation function of discrete-time LTI system |
| $R_h(\tau)$ | system correlation function of a continuous-time LTI system |
| $R_x(l)$ | autocorrelation function of discrete-time energy signal |
| $R_x(\tau)$ | autocorrelation function of continuous-time energy signal |

| | |
|---|---|
| $R_{BB}(\tau)$ | ACF of the baseband process |
| $R_{NB}(\tau)$ | ACF of the baseband continuous-time noise process $N_B(t)$ |
| $R_{NN}(\tau)$ | ACF of the bandpass continuous-time noise process $N(t)$ |
| $R_{NN}(l)$ | ACF of the bandpass discrete-time noise process $N(n)$ |
| $R_{sc}(\tau)$ | ACF of continuous-time carrier signal |
| $R_{sc}(l)$ | ACF of discrete-time carrier signal |
| $R_{sI}(\tau)$ | ACF of continuous-time in-phase carrier |
| $R_{sQ}(\tau)$ | ACF of continuous-time quadrature carrier |
| $R_{SS}(l)$ | ACF of a discrete-time modulated signal |
| $R_{SS}(\tau)$ | ACF of a continuous-time modulated signal |
| $R_{xx}(l)$ | autocorrelation function of discrete-time power signal |
| $R_{xx}(\tau)$ | autocorrelation function of a power signal $x(t)$, or an ergodic process |
| $R_{XX}(l)$ | autocorrelation function of WSS process $X(n)$ |
| $R_{XY}(l)$ | crosscorrelation function of WSS processes $X(n)$ and $Y(n)$ |
| $R_{XX}(m,n)$ | autocorrelation function of a discrete-time stochastic process $X(n)$ |
| $R_{XX}(t_1,t_2)$ | autocorrelacion function of a continuous-time stochastic process $X(t)$ |
| $R_{XX}(\tau)$ | autocorrelation function of continuous-time WSS stochastic process $X(t)$ |
| $R_{xy}(l),\ R_{yx}(l)$ | cross-correlation functions of discrete-time energy and power signal |
| $R_{xy}(l),\ R_{yx}(l)$ | cross-correlation function of discrete-time energy and power signal |
| $R_{XY}(\tau),\ R_{YX}(\tau)$ | crosscorrelation of continuous-time WSS stochastic processes $X(t)$ and $Y(t)$ |
| $R_{xy}(\tau),\ R_{yx}(\tau)$ | cross-correlation function of continuous-time energy and power signal |
| $R_{XY}(m,n)$ | crosscorrelation function of stochastic processes $X(n)$ and $Y(n)$ |
| $R_{WW}(\tau)$ | ACF of a continuous-time noise process $W(t)$ |
| $R_{WW}(l)$ | autocorrelation function of a discrete-time noise process $W(n)$ |
| $R_{W_1 W_1}(\tau)$ | autocorrelation function of a continuous-time unit noise process $W_1(t)$ |
| | |
| **s** | sets of $I$ modulating signals $s_i$ |
| $S$ | sample space, set of outcomes (elementary events) |
| **S** | vector subspace |
| $\mathbf{s}_i$ | modulating signal vector defined by $J$ signal coefficients, $i^{\text{th}}$ column vector in **s**, signal points in $J$ dimensional space |
| $s_i$ | complex signal in OFDM systems |
| $s_{ij}$ | signal coefficients, signal part |
| $s_n$ | spread signal in CDMA systems |
| $S/N$ | signal-to-noise ratio |
| $s_B(n)$ | complex modulated OFDM signal in discrete time |
| $s_B(t)$ | modulated OFDM signal in continuous-time |
| $s_c(n)$ | discete-time deterministic or random carrier signal |
| $S_c(n)$ | discrete-time carrier stochastic process |

| | |
|---|---|
| $s_c(t)$ | carrier as an continuous-time orthonormal function |
| $S_i(n)$ | modulated discrete-time stochastic process |
| $s_i(n)$ | modulated discrete-time signal |
| $S_i(t)$ | modulated contrinuous-time stochastic process |
| $s_i(t)$ | modulated continuous-time signal, one realization of $S_i(t)$ |
| $S_h(f)$ | energy spectral density of continuous-time LTI system response |
| $S_h(\Omega)$ | energy spectral density of discrete-time LTI system response |
| $s_{cI}(t)$ | inphase carrier |
| $s_{cQ}(t)$ | quadrature carrier |
| $\{s_{ij}(t)\}$ | sets of possible modulating signals |
| $s_{ij}(t), m_{ij}(t)$ | modulating signal for $j$th symbol |
| $S_x(f) = E_x(f) = \|X(f)\|^2$ | energy spectral density of continuous-time energy signal |
| $S_x(\Omega) = E_x(\Omega) = \|X(\Omega)\|^2$ | energy spectral density of discrete-time energy signal |
| $S_{HH}(f)$ | power spectral density of the impulse response of a stochastic LTI system |
| $S_{MM}(\omega)$ | power spectral density of a modulated continuous-time signal |
| $S_{MM}(\Omega)$ | power spectral density of a modulated discrete-time signal |
| $S_{NB}(f)$ | power spectral density of a baseband continuous-time noise process $N_B(t)$ |
| $S_{NN}(f)$ | PSD function of the bandpass noise process $N(t)$ |
| $s_{RF}(t)$ | radio-frequency signal |
| $S_{RF}(f)$ | power spectral density of a radio-frequency signal |
| $S_{sc}(f)$ | PSD of a carrier signal |
| $S_{sI}(f)$ | PSD of an inphase carrier |
| $S_{sQ}(f)$ | PSD of a quadrature carrier |
| $S_{SS}(f)$ | PSD of a continuous-time modulated signal |
| $S_{SS}(\Omega)$ | PSD of a discrete-time modulated signal |
| $S_{xx}(f)$ | power spectral density of continuous-time power signal |
| $S_{XX}(\omega)$ | power spectral density of continuous-time WSS stochastic process $X(t)$ |
| $S_{xx}(\Omega)$ | power spectral density of discrete-time power signal |
| $S_{xx}(\Omega) = S_{xx}(e^{j\Omega})$ | PSD of discrete-time process $X(n)$ |
| $S_{XX}(z)$ | $z$-transform of correlation function $R_{XX}(l)$ |
| $S_{xy}(f)$ | crossenergy spectral density of continuous-time power and energy signal |
| $S_{xy}(\Omega)$ | crossenergy spectral density of discrete-time power and energy signal |
| $S_{XY}(\omega)$ | crosspower spectral density of continuous-time WSSS processes $X(t)$ and $Y(t)$ |

$S_{XY}(\Omega)$    cross-spectral function of WSS discrete processes $X(n)$ and $Y(n)$
$S_{XY}(z)$    complex cross-power spectral densities
$S_{WW}(f)$    PSD of a continuous-time noise process $W(t)$
$S_{WW}(\Omega)$    power spectral density of discrete-time stochastic process $W(n)$
$S_{W_1 W_1}(f)$    power spectral density of continuous-time unit noise process $W_1(t)$

$t$    continuous-time variable in the continuous-time domain
$T$    symbol duration
$T_b$    bit duration
$T_c$    encoded bit duration
$T_{Nq}$    Nyquist sampling interval
$T_s$    sampling interval
$T_x, T_y$    sampling intervals in multi-rate signal processing

$\mathbf{V}$    vector space

$\mathbf{w}$    demodulated noise vector
$\mathbf{W}$    matrix of the demodulated noise vector variables
$\mathbf{w_i}$    noise sample vector (a part of the noise $\mathbf{w}$ at the output of the correlator interfering with the signal)
$W_j$    noise random variable at $j$th correlator output, the sum of Gaussians
$w_j$    noise part, a realization of the noise random variable $W_j$ at $j$th correlator
$W(n)$    discrete-time noise process, discrete-time white Gaussian noise process, and white Gaussian random variable defined at time instant $n$
$w(n)$    one realization of discrete-time white Gaussian noise process $W(n)$
$W(t)$    continuous-time noise process, white Gaussian noise process, and white Gaussian random variable defined at time instant $t$
$w(t)$    one realization of white Gaussian noise process $W(t)$
$W_d(n)$    discrete-time discrete-valued (quantized) noise stochastic process
$w_d(n)$    one realization of discrete-time and discrete-valued (quantized) noise stochastic process $W_d(n)$
$W_j(n)$    $j$th discrete-time Walsh function, Walsh sequence
$W_j(t)$    $j$th continuous-time Walsh function

$x$    correlator output, one value in $x(kT)$, decision variable
$\boldsymbol{x}$    transmitted signal, input of a channel in information theory
$\mathbf{x}$    demodulation vector or demodulated sample vector, vector of random values
$\mathbf{X}$    array of random variables $X_j$ (a vector of $J$ demodulation variables)
$X$    random variable, correlator output random variable, channel input random variable, input random variable of a channel in information theory
$X_i = X(t_i)$    random variable of process $X(t)$ defined at time instant $t = t_i$
$x_i = x(t_i)$    random value of the random variable $X_i(t)$ at time instant $t = t_i$
$X_j$    demodulation variable for $j$th demodulator
$x_j$    demodulated sample at $j$th correlator output, a realization of $X_j$
$x_n$    a realization of random variable $X(n)$, spreading sequence
$x(t)$    continuous-time deterministic or random signal
$x(n)$    discrete-time continuous valued deterministic or random signal

| $X(n)$ | discrete-time stochastic process, discrete-time random variable defined at time $n$ |
|---|---|
| $X(f)$ | continuous-time Fourier transform, amplitude spectral density |
| $X(t)$ | continuous-time stochastic process, random $X(t) = X$ defined at time instant $t$ |
| $x(t)$ | continuous-time deterministic or random signal |
| $x(t) = x(t, \zeta_k)$ | a realisation of the continuous-time stochastic process $X(t)$ for the outcome $\zeta_k$ |
| $X(\omega) = X(j\omega)$ | continuous-time Fourier transform, amplitude spectral density |
| $|X(\omega)|$ | magnitude spectral density |
| $X(\Omega) = X(e^{j\Omega})$ | DTFT of discrete-time signal $x(n)$, amplitude spectral density |
| $X(z)$ | $z$-transform |
| $X(kT)$ | discrete-time stochastic process at the output of each correlator |
| $x(kT)$ | realization of $X(kT)$ as a series of random values |
| $X_j(kN)$ | $k$th demodulation variable for $j$th demodulator |
| $X(n_i) = X(i) = X_i$ | random variable defined at the time instant $n = n_i = i$ |
| $x(n_i) = x(i) = x_i$ | $i$th realisation of the random variable $X(n)$ at time instant $n_i = i$ |
| $|X(e^{j\Omega})|$ | magnitude spectrum of DTFT, magnitude spectral density |
| $X(2\pi k/N) = X(k)$ | discrete Fourier transform |
| $x_D(n)$ | discrete-time discrete valued signal |
| $x_u(n)$ | up-sampled signal |
| $x_d(n)$ | down-sampled signal |
| $x(n, \zeta_k) = x(n)$ | one realisation of $X(n)$ for the outcome $\zeta_k$ |
| $y$ | received signal, receiver decision value, the random output of a channel in information theory |
| $\mathbf{y}$ | received binary signal, vector of random values $y$ |
| $Y$ | receiver decision random variable, channel output random variable in information theory |
| $\mathbf{Y}$ | vector of random variables $Y$ |
| $Z_i$ | decision variable value representing the estimate of symbol $m_i$ |
| $Z_k$ | decision variable values pointing the MLE of the symbol $m_i = m_k$ |

# Greek and Cyrillic Symbols

| $\alpha$ | fading coefficient |
|---|---|
| $2\beta$ | processing gain |
| $\delta(n) = \delta_{ij}$ | Kronecker delta function, impulse discrete-time signal |
| $\delta(t)$ | Dirac delta function, impulse signal |
| $\delta_s$ | stopband peak ripple values |
| $\delta_p$ | passband peak ripple values |
| $\Delta(f)$ | Fourier transform of the Dirac delta signal |
| $\Delta(\Omega)$ | Fourier transform of the Kronecker delta signal |

| | |
|---|---|
| $\mathscr{E}$ | random experiment |
| $\zeta_i$ | $i$th outcome of a random experiment |
| $\eta$ | spectral efficiency |
| $\eta_X$ | mean value of a random variable $X$, or of a stochastic process $X(t)$ |
| $\eta_X(n)$ | mean value of a discrete-time stochastic process $X(n)$ |
| $\boldsymbol{\eta}_X$ | vector of mean values, array of mean values |
| $\eta_X(t)$ | mean value function of a continuous-time stochastic process $X(t)$ |
| $\eta_X(t_1)$ | mean value of a continuous-time stochastic process $X(t)$ defined at $t = t_1$ |
| $\eta_x$ | mean value of ergodic process |
| $\theta_{ik}$ | the angle between two signal vectors $\mathbf{s}_i$ and $\mathbf{s}_k$ |
| $\lambda$ | dummy variable of integration |
| $\mu$ | efficiency of a communication system |
| $\xi$ | noise in CDMA systems |
| $\pi$ | 3.14159 |
| $\rho$ | correlation coefficient |
| $\sigma$ | standard deviation |
| $\sigma_t^2$ | variance of truncated density function |
| $\sigma_X^2(n)$ | variance of discrete-time stochastic process $X(n)$ |
| $\sigma_X^2(t)$ | variance function of continuous-time stochastic process $X(t)$ |
| $\sigma_X^2$ | variance of random variable $X$ |
| $\sigma_x^2$ | variance of discrete or continuous time ergodic process |
| $\tau$ | independent variable of ACF |
| $\phi_k$ | phase angle, phase spectrum of a periodic discrete-time signal |
| $\phi(f), \phi(\omega)$ | phase spectrum, phase spectral density |
| $\phi_j(n)$ | orthonormal discrete-time basis function, a carrier signal, subcarriers |
| $\varphi_j(t)$ | orthonormal basis function, a carrier signal, subcarriers |
| $\omega = 2\pi f$ | radian frequency |
| $\omega_c = 2\pi f_c$ | radian carrier frequency |
| $\omega_0$ | fundamental frequency |
| $\Omega = 2\pi F$ | normalised radian frequency for discrete-time signals |
| $\Omega_k = 2\pi k/N$ | normalised angular frequency in rad/sample |
| $\Omega_p$ | passband frequency for discrete-time filter |
| $\Omega_s$ | stopband edge frequency for discrete-time filter |
| $\hbar_k$ | coefficients of FT of a WSS periodic process |

## Defined Functions

| | |
|---|---|
| $\mathrm{erfc}(\cdot)$ | error function complementary |
| $\mathit{Erfc}(\cdot)$ | function of erfc functions |
| $\ln(\cdot)$ | natural logarithm |
| $\log(\cdot)$ | base 10 logarithm |
| $\log_2(\cdot)$ | base 2 logarithm |
| $\mathrm{sinc}(y)$ | $(\sin\pi y)/\pi y$, or $(\sin y)/y$ |
| $u(\cdot)$ | unit step function |

# Operators

| | |
|---|---|
| Im{·} | imaginary part |
| Re{·} | real part |
| ⟨·⟩ | time averaging |
| (·)*(·) | convolution |
| (·)⊗(·) | periodic (or circular) convolution |
| (·)* | conjugate |
| \|(·)\| | absolute value, first norm |
| FT{·} | Fourier transform, discrete-time Fourier transform |

# Abbreviations

| | |
|---|---|
| ACF | autocorrelation function |
| AD | analog-to-digital |
| ADC | analog-to-digital converter |
| APF | allpass filter |
| ARQ | automatic repeat request |
| ASK | amplitude-shift keying |
| AWGN | additive white Gaussian noise |
| BB | baseband |
| BEP | bit error probability |
| BER | bit error rate |
| BFSK | binary frequency-shift keying (modulation) |
| BIBO | bounded-input bounded-output |
| BP | band-pass |
| BPF | bandpass filter |
| BPSK | binary phase shift-keying (modulation) |
| BSMC | binary symmetric memoryless channel |
| CDMA | code division multiple access |
| CLT | central limit theorem |
| CP-BFSK | continuous-phase binary frequency shift keying |
| ct | continuous time |
| DA | digital-to-analog |
| DAC | digital-to-analog converter |
| DC | direct current |
| DDS | direct digital synthesizer |
| DEMUX | de-multiplexer |
| DFT | discrete Fourier transform |
| DMC | discrete memoryless channel |
| DSP | digital signal processing, digital signal processors |
| DTFS | discete-time Fourier series |
| DTFT | discrete-time Fourier transform |
| ESD | energy spectral density |

| | |
|---|---|
| FEC | forward error correction |
| FFT | fast Fourier transform |
| FIR | finite impulse response |
| FPGA | field programmable gate arrays |
| FS | Fourier series |
| FSK | frequency shift keying |
| FT | Fourier transform |
| GL(2) | Galois field |
| GN | Gaussian noise |
| GSOP | Gram-Schmidt orthogonalization procedure |
| HOVA | hard output Viterbi algorithm |
| HPF | high-pass filter |
| IDFT | inverse discrete Fourier transform |
| IF | intermediate frequency |
| IFF | intermediate frequency filter |
| IFT | inverse Fourier transform |
| iid | identical and independent random variables |
| IIR | infinite impulse response |
| IPF | interpolation filter |
| LF | likelihood function |
| LLF | log-likelihood function |
| LO | local oscillator |
| LPF | lowpass filter |
| LTI | linear time invariant |
| LUT | look-up-table |
| MAP | maximum a posteriori probability |
| MC | multicarrier systems |
| ML | maximum likelihood |
| MPSK | multilevel phase-shift keying |
| MQAM | M-ary quadrature amplitude modulation |
| MS | means square |
| MUX | multiplexer |
| NRZ | non-return-to-zero |
| OFDM | orthogonal frequency division multiplexing |
| PLL | phase-locked loop |
| PS | pulse-shaping |
| PSD | power spectral density |
| PSK | phase-shift keying |

| | |
|---|---|
| QAM | quadrature-amplitude modulation |
| QMF | quadrature-mirror filter |
| QPSK | quadri-phase shift keying, quadrature phase-shift keying |
| | |
| RF | radiofrequency |
| ROC | region of convergence |
| RTL | Register Transfer Languages |
| | |
| SOVA | soft output Viterbi algorithm |
| S/P | serial-to-parallel (conversion) |
| SSS | strict sense stationary |
| | |
| VCO | voltage-controlled oscillator |
| | |
| WGN | white Gaussian noise |
| WSS | wide sense stationary |

# 1

# Introduction to Communication Systems

## 1.1 Communication Systems and Networks

A communication system is a set of devices that work together to allow the transmission of messages at a distance. A communication network can be considered to be a set of communication systems that are interconnected and work in such a way so as to allow the transmission of messages between any two or more points belonging to the network.

This book does not intend to address all aspects of communication systems and network design. Rather, it will solely address problems relating to the design of a communication system that includes a transceiver, a communication channel, and a receiver. From the network point of view, the book will focus on the so-called physical layer design. A simplified block schematic of the system, which will be the subject of our analysis and synthesis, is presented in Figure 1.1.1. In this block schematic, signals are presented in two domains: the continuous-time domain and the discrete-time domain, which are represented by the continuous variable $t$ and the discrete variable $n$, respectively. The chapter will discuss system operation in both domains but will focus on the discrete-time domain. To distinguish the two types of systems, in this book, the system operating in the continuous-time domain will be called a *digital system*, while the system operating in the discrete-time domain will be called a *discrete system*. Signals will be also presented in the frequency domain using the Fourier transforms. The spectra of continuous-time signals will be consistently expressed as functions of the normalized radian frequency $\Omega = 2\pi F$, and the spectra of discrete-time signals as functions of the radian frequency $\omega = 2\pi f$. The structure of both systems with their related signals is presented in Figure 1.1.1. The operation of both systems will be briefly explained below.

**Digital communication systems**. In a digital communication system, the transmitter is composed of a message source, a source encoder, an encryption block, a channel encoder, and a modulator with amplifier. The receiver includes an amplifier, a demodulator with a detector/decision device, a channel decoder, a decryption block, a source decoder, and the message sink. We will assume that the source generates message bits $m_i$ which are encoded using a source encoder, protected for secure transmission in the encryption block, and then encoded in the channel encoder for reliable transmission through the channel. The output of the channel encoder is made up of encoded bits. In the binary case, these encoded bits are represented by two distinct binary values: 0 and 1. For the sake of clarity, in the channel

*Discrete Communication Systems*. Stevan Berber, Oxford University Press. © Stevan Berber 2021.
DOI: 10.1093/oso/9780198860792.003.0001

TRANSMITTER

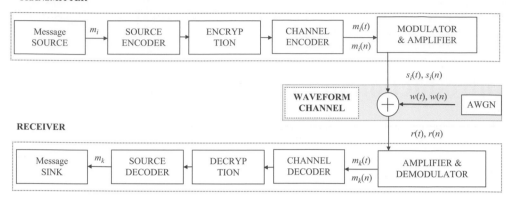

**Fig. 1.1.1** *The structure of digital and discrete communication systems.*

coding theory, we will also use the symbol $c_i$ to represent encoded bits. These encoded bits form a modulating signal $m_i(t)$ that modulates a carrier to obtain the modulated signal $s_i(t)$ as the output of the transmitter.

The modulated signal $s_i(t)$ is transmitted through the communication channel characterized by the additive white Gaussian noise $w(t)$, which is called the waveform channel. We can define the channel at various levels of abstraction which depend on the intended theoretical and practical analysis of the system. This abstraction will be necessary for the investigation of various modulation methods and channel encoder–decoder structures. From the theoretical point of view, we can define a *discrete-input continuous-output baseband channel* as a channel that incorporates a modulator, a waveform channel, and a demodulator. The output of this channel is a continuous-valued discrete-time signal that is usually called a *correlation sample value*. If the output of the demodulator is used to determine a discrete binary value of the encoded bit, the channel will also include a detector/decision block. The channel that includes this block as well as a discrete-input continuous-output channel is called a *discrete-input discrete-output channel*. When these discrete-output values are expressed in a digital form, the channel is called a *digital channel*. In the case of a binary signal transmission, the encoder output values and the discrete decision-device output values are represented by members of the additive group (0, 1) and sometimes by members of the multiplicative group (–1, 1). As this channel has both binary inputs and binary outputs, is called a *binary digital channel*. This channel will be analysed in detail in Chapter 9, which is dedicated to information theory.

The received noisy signal $r(t)$ is amplified and demodulated, resulting in an estimate $m_k(t)$ of the modulating signal $m_i(t)$, and then sent to the channel decoder. After decoding, the signal is decrypted, it is subsequently decoded in the source decoder and then an estimate $m_k$ of the source message $m_i$ is made available to the recipient in the message sink.

**Discrete communication systems.** If the transmitter operates in the discrete-time domain, the modulating signal will be expressed in the discrete-time form as $m_i(n)$ and it will modulate the discrete-time carrier to obtain the modulated signal $s_i(n)$. This modulation is usually performed at an intermediate frequency. Therefore, all baseband and intermediate-frequency processing will be in the discrete-time domain. The discrete-time signal $s_i(n)$ will be converted into the analogue form by means of an analogue-to-digital converter to produce a continuous-time-modulated signal $s(t)$ that will be transmitted through a channel characterized by the additive white Gaussian noise $w(t)$, as shown in Figure 1.1.1. In principle, a mixer is used after a digital-to-analogue converter to up-convert the modulated signal to the carrier radiofrequency. (The analogue-to-digital and digital-to-analogue converters are not shown in Figure 1.1.1.)

In the same figure, the noise is presented in the discrete-time domain as $w(n)$. This noise will be used in the mathematical modelling, simulation, and emulation of the discrete systems, as will be discussed in later chapters. In that case, the transmitter and receiver are modelled and implemented in the discrete-time domain, and the noise will be generated from the bandpass discrete noise generator.

If the intermediate-frequency signal is up-converted at the transmitter side and affected by the continuous-time additive white Gaussian noise $w(t)$, the received noisy signal $r(t)$ is firstly down-converted to the intermediate frequency and then converted to the discrete-time domain by an analogue-to-digital converter. (This processing is not shown in Figure 1.1.1.) The received discrete-time signal $r(n)$ is further processed in the discrete demodulator, resulting in an estimate $m_k(n)$ of the modulating signal $m_i(n)$, and then sent to the channel decoder. After decoding, the signal is decrypted, it is subsequently decoded in the source decoder and then an estimate $m_k$ of the source message $m_i$ is made available to the recipient in the message sink.

Note that Figure 1.1.1 does not show either the up-conversion or the down-conversion of the intermediate-frequency modulated signal. These operations will be presented in detail in later chapters.

We assume that the carrier frequency is much larger than the highest message signal frequency. Therefore, the modulated signal contains the message signal in a narrow band around the carrier frequency, and the system is called the passband communication system.

When the message signal is transmitted in its frequency band, the system is called a baseband communication system. In this system, we do not need to modulate the baseband message signal. Instead, we use a baseband filter to limit the bandwidth of the modulating signal, and an amplifier to adapt the signal for transmission through the baseband channel.

In a communication system, we process signals. These signals are defined in various ways depending on which criterion of classification of the signals we define. Therefore, in this book, we will define the signals precisely and use the same definitions whenever possible. In this context, we will be able to distinguish, for example, continuous-time from discrete-time signals and their variants, power signals from energy signals, deterministic signals from stochastic and random signals, and so on.

This book is the first of its kind. Nearly all the books written in the communication systems field present signals and their processing in continuous-time domains to explain the operation of a communication system. The main objective of this book is to present signals in a discrete-time domain for all of the building blocks of a communication system where this kind of processing takes place.

## 1.2 Classification of Signals and Systems

A *signal* is defined as a function whose values convey information about a phenomenon it represents. Depending on the criterion applied, we may have various ways to classify signals. Signals can be digital or analogue, as traditionally defined; continuous or discrete, with respect to their independent variables or values; power signals or energy signals, with respect to the definitions and values of their power and energy; deterministic or stochastic (random), with respect to their randomness behaviour; physically realizable or non-realizable, with respect to their existence; periodic or aperiodic; real or complex; and so on. In this book, we will classify signals in such a way so that we can apply and use the same definitions in most of the chapters; in some cases, we will be able to use the same definitions throughout the book. One method for classifying signals, based on various criteria, is shown in Figure 1.2.1. A function representing a signal can be of one or more independent variables, that is, the signal can

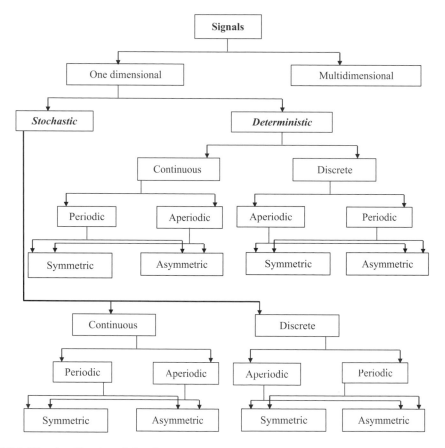

**Fig. 1.2.1** *The classification of signals.*

generally be *one dimensional* or *multidimensional*. A signal is multidimensional if its value is a function of multiple independent variables.

Most of the signals discussed in this book are one dimensional. They can be divided into two broad classes, deterministic or stochastic, depending on the probability of generating their values. Deterministic signals can be processed in the time domain and the corresponding frequency domain, in both their discrete and their continuous forms. Therefore, deterministic signals are separated into two groups: continuous signals and discrete signals. The continuity and discreteness in their definitions can be related to their presentations in various domains, but it is primarily the time and frequency domains that are of particular interest in this book.

Both discrete and continuous signals can be periodic or aperiodic. The concept of periodicity includes the controversial premise that the signal exists in an infinite interval of time, which is not the case for physically realizable signals. However, both the theoretical analysis of these signals and the establishment of their relation to physical reality are of high importance. Symmetry is an essential characteristic of natural objects, including living creatures. Signals can be also symmetric and asymmetric. In the classification presented, periodic and aperiodic signals are subdivided into asymmetric and symmetric signals. In the definition of symmetry, it is important to define the criterion by which the symmetry is determined, that is to say, with respect to what a particular signal is symmetric or asymmetric. For example, a signal can be symmetric with respect to the origin; alternatively, it can be symmetric with respect to the ordinate. In subsequent chapters, we will define signals precisely, using their mathematical presentations.

## 1.2.1   Classification of Signals with Respect to Time and Value

The focus of this book is on the operation of communication systems that use discrete-time signal processing. Therefore, this section will present the signals that will most frequently be used in this book and will be subject of the analysis and synthesis of the building blocks of both digital and discrete communication systems.

A *continuous-time* signal $x(t)$ is defined as a function that takes on any value in the continuous interval of its amplitudes $(x_a, x_b)$ for all values of the *continuous* independent variable $t$ taken in the interval $(a, b)$, where $a$ may be $-\infty$, and $b$ may be $\infty$. In Figure 1.2.2, for any $t$ in the interval $(a, b) = (-\infty, \infty)$, there exist values of the signal $x(t)$ in the infinite interval $(x_a, x_b) = (0, \infty)$.

A *continuous-valued signal* $x(t)$ can take on any possible value in a finite or an infinite interval of its amplitudes $(x_a, x_b)$. For example, for any value of the variable $t$, the signal in Figure 1.2.2 can take on values in the continuous interval $(x_a, x_b) = (0, \infty)$.

A *discrete-time* signal $x(n)$ is defined as a function that takes any value of its amplitudes (values) for *discrete* values of time $t = n$ of the discrete independent variable $n$. For example, for every $n$, we can have a continuum of values $x(n)$ in the interval $(x_a, x_b) = (0, \infty)$, as can be seen from Figure 1.2.3.

A *discrete-valued signal* $x(t)$ takes on a countable number of possible values over a finite or infinite interval of its amplitudes $(x_a, x_b)$.

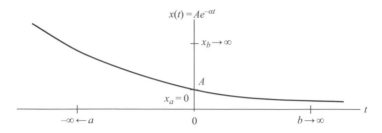

**Fig. 1.2.2** *A continuous-time continuous-valued signal.*

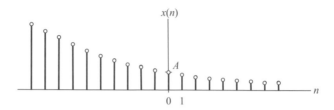

**Fig. 1.2.3** *A discrete-time signal on the continuum of its values.*

In summary, if we combine these four signals, we can get the following possible signal types: a continuous-time continuous-valued signal, a continuous-time discrete-valued signal, a discrete-time continuous-valued signal, and a discrete-time discrete-valued signal. Having a precise definition for each of these signals is extremely important, because they will be extensively used in the theoretical analysis, mathematical modelling, and design of communication systems. The signals are presented in Figure 1.2.4. They can be deterministic or stochastic.

A *continuous-time continuous-valued signal* $x(t)$ (shown in Figure 1.2.4 in black) is a function of time that takes any value on a continuum of amplitudes and for any value of the independent variable $t$. These signals are mostly present in nature and are usually generated at the output of the message source. In analogue communication systems, such signals were processed inside transceivers. However, with the development of discrete technology and digital-processing machines, this signal is adapted to these technologies by transforming it into a discrete form.

An important note on the definition of analogue signals is as follows. A continuous-time continuous-valued signal is traditionally termed an *analogue signal*. However, an analogue signal fulfils the conditions of a continuous function while, in this book, a continuous-time continuous-valued signal is not strictly defined as a continuous function. That is the reason why, in this book, the term 'analogue signal' will be, in principle, defined as a continuous-time continuous-valued signal. Precise definitions of analogue and digital signals will be given in Section 1.3.1.

A *continuous-time discrete-valued signal* $x(t_i)$ (shown in Figure 1.2.4 as a staircase graph) is a function of continuous time that takes its discrete values from a set of countable number

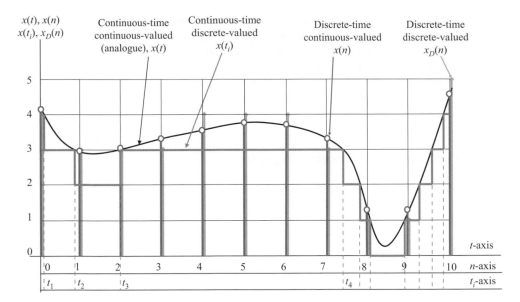

**Fig. 1.2.4** *Definition of continuity and discreteness for signals.*

of values and for any value of the independent variable $t_i$. This signal can be obtained from $x(t)$ by discretizing the ordinate and assigning, say, the lower discrete amplitude whenever the signal 'drops' into the discrete interval. If the number of possible signal values is 2, the generated signal is a so-called binary signal. If the number of possible values is greater than 2, the generated signal is a multilevel signal.

A *discrete-time continuous-valued signal* $x(n)$ (shown in Figure 1.2.4 with circles on top) is a function of discrete time that takes any value on a continuum of amplitudes and for any discrete value of the independent variable $n$. These signals are mostly generated by sampling continuous-time signals and are generated at the output of a sampler, as was done in Figure 1.2.4. In discrete communication systems, sampling of the processed signals can be performed at various levels: sometimes at the baseband, sometimes in the bandpass signals, and sometimes both. However, it is important to note that the analogue source signal needs to be reconstructed at the receiver side in both the time domain and the frequency domain. Therefore, processing of the signals in a discrete domain, where sometimes the notion of real time is lost, needs to be controllable, and the relationship between the generated discrete-time signal and its counterpart continuous-time signal has to be maintained all the time.

A *discrete-time discrete-valued signal* $x(n)$ (shown in Figure 1.2.4 with diamonds on top) is a function of discrete time that takes any value from a countable set of discrete amplitudes and for any discrete value of the independent variable $n$. These signals are mostly obtained by the sampling procedure (time discretization) and quantization (amplitude discretization that can be performed by rounding or truncation) of continuous-time signals and are generated at the output of a quantizer. In a discrete communication system, this type of signal plays the most important role. They are suitable for processing in digital technology, because their discrete values can be numerically expressed as decimal numbers, and then transferred into binary form for processing in digital technology. The assignment of binary processing bits to discrete signal values is called signal coding, which represents the final step in one of the most important processing blocks in modern communication devices: an analogue-to-digital converter. The majority of these signals in communication systems are stochastic, and the related theory of communications is heavily based on classical mathematics, the theory of random variables, stochastic processes and statistics.

## 1.2.2 Periodic and Symmetric Signals

When signals are classified using periodicity as the criterion, they can be periodic or aperiodic. A continuous-time signal is periodic if there is some positive number $T$ such that

$$x(t) = x(t+nT), \qquad (1.2.1)$$

where $T$ is the fundamental period of $x(t)$, and $n$ is an integer. This signal repeats an exact pattern, which is defined in the time interval $T$, for an infinite time from $-\infty$ to $t = 0$, and will continue to repeat that exact pattern for an infinite time interval from $t = 0$ to $+\infty$. The time interval $T$ is called the fundamental period, and its reciprocal value is the fundamental frequency $f$. The frequency $f$ is called the *cyclic frequency* because it expresses the number of cycles (periods) per second. The *radian frequency* is $\omega = 2\pi f$ and it shows the number of radians per second.

A continuous-time signal is aperiodic if it is not periodic.

In the case when signals are classified using their symmetry as the criterion, they can be even or odd. A continuous-time signal $x(t)$ is *even* if

$$x(-t) = x(t), \qquad (1.2.2)$$

that is, it is symmetric with respect to the $y$-axis. A continuous-time signal $x(t)$ is *odd* if

$$x(-t) = -x(t), \qquad (1.2.3)$$

that is, it is symmetric with respect to the origin. Given a signal $x(t)$, an even signal can be expressed as

$$x_e(t) = \frac{1}{2}[x(t) + x(-t)] \qquad (1.2.4)$$

and an odd signal can be expressed as

$$x_o(t) = \frac{1}{2}[x(t) - x(-t)]. \qquad (1.2.5)$$

Here, $x_e(t)$ and $x_o(t)$ are called the even and odd parts of $x(t)$. Then, any real-valued signal $x(t)$ can be written as the sum of its even and odd parts, that is,

$$x(t) = x_e(t) + x_o(t). \tag{1.2.6}$$

The signal-processing methods and the procedures for the analysis of an analogue or a digital system basically depend on the characteristics of specific signals. One example of a physical signal is the voltage that is measured across a resistor.

### 1.2.3 Deterministic and Stochastic Signals

A signal is *deterministic* if all its values are precisely defined by specifying the functional dependence on its independent variable. This signal can be described by a functional relationship in a closed form, as an arbitrary graph, or in a tabular form.

In this book, a realization of a stochastic process is termed a *random* or *stochastic signal*. This random signal will be taken from the ensemble of random signals that define the stochastic process. A signal is *random* (*stochastic*) if all its values cannot be precisely defined by specifying the functional dependence on the random variable. It is normally obtained as a realization of a stochastic process. In principle, these signals should be analysed using mathematical and statistical techniques instead of formulas in closed form. In spite of their random nature, these signals can be classified in the same way as deterministic signals, as can be seen from Figure 1.2.1. They can be continuous-time discrete-valued signals, or they can be discrete-time continuous-valued signals. Due to the importance of these signals and the methods of their processing for the design of communication systems, part of this book is dedicated to their analysis.

A *continuous-time random (stochastic) signal* is a realization of a continuous-time *stochastic process* that takes the random value at each time instant $t$ that cannot be predicted in advance or found using any expression, graphical presentation or table. An example of such a signal is the noise generated in a communication system, or a speech signal generated at the output of a microphone.

A *discrete-time random (stochastic) signal is* a single realization of a discrete-time stochastic process which has a random value at each time instant $n$ that cannot be predicted in advance or found using any expression, graphical presentation or table. An example of such a signal is the discrete noise affecting the signal processed in a discrete communication system or a speech signal generated and sampled at the output of a microphone.

All of the signals presented in Section 1.2.1 can be deterministic or stochastic. Due to the importance of these signals in the modelling and design of communication systems, two chapters of this book will be dedicated to their analysis.

### 1.2.4 Classification of Signals with Respect to Power and Energy

Sometimes we need to analyse signals in the entire range of an independent variable, from minus to plus infinity. In communication systems, the independent variable is usually time, which can be continuous, denoted by $t$, or discrete, denoted by $n$. The classification of these signals is made using the criterion of their power and energy values, and the signals are divided into two broad classes: power signals and energy signals.

A real or complex signal $x(t)$ is a *power signal* if and only if its power, averaged over an infinite interval, is finite and non-zero, that is, for power expressed as

$$P = \lim_{a \to \infty} \frac{1}{2a} \int_{-a}^{a} |x(t)|^2 dt, \tag{1.2.7}$$

the condition $0 < P < \infty$ is fulfilled, so that the energy is $E = \infty$.

A real or complex signal $x(t)$ is an *energy signal* if and only if its total energy, calculated in the infinite interval, is finite and non-zero, that is, for energy expressed as

$$E = \lim_{a \to \infty} \int_{-a}^{a} |x(t)|^2 dt, \tag{1.2.8}$$

the condition $0 < E < \infty$ is fulfilled, so that the defined power is $P = 0$. In summary:

1. $x(t)$ is a power signal if and only if its power fulfils the condition $0 < P < \infty$, which implies that its energy is $E = \infty$.

2. $x(t)$ is an energy signal if and only if its energy fulfils the condition $0 < E < \infty$, so that its power is $P = 0$.

3. The power and energy are calculated as average values, where the averaging is done over the hypothetical infinite interval.

4. A signal cannot be an energy signal and a power signal at the same time.

5. There are signals that are neither energy signals nor power signals. In addition, it is possible to construct a signal that has both infinite energy and infinite power, as is the case with the white Gaussian noise process.

Physically realizable signals are energy signals. However, we often model them as signals of infinite duration, that is, as power signals. For example, a periodic sinusoidal signal at the output of a generator can be modelled as a power signal, assuming that it can last forever. However, the generator's lifetime is finite, and the generated signal is not periodic in the strict mathematical sense. Therefore, this signal is an energy signal having finite energy and zero power averaged over an infinite interval of time. In our analysis, the power and energy are calculated assuming that the voltage $V$ is applied across a 1 Ohm resistor. In that case, the power is expressed as the squared voltage $V^2$ or watts (W), and the energy is expressed in watt-seconds (W sec) or joules (J).

## 1.2.5   Classification of Signals with Respect to Realizability

In reality, we deal with signals that are physically realizable. This property of a signal is of particular importance when we are performing its analogue-to-digital transform. We say that a signal is physically realizable if, by definition, it fulfils the following conditions:

1. The signal is defined in a limited time interval in a time domain and has a significant number of non-zero values in the interval.

2. The signal can be continuous time or discrete time in the interval of its values.

3. The number of peak values is finite in the interval of possible signal values.

4. The signal we are dealing with will have a real value.

5. The signal is defined in a finite composite bandwidth in a frequency domain.

The definition of realisability is applicable for real signals. However, if the signal is complex, with both real and imaginary parts, these parts need to be analysed separately. There are some signals that can be theoretically defined and successfully manipulated (processed) even though the conditions of this definition are not fulfilled. These signals are required in theoretical analyses, and their widespread use cannot be avoided. In their applications, which occur very often, the analysis of such signals needs to include tools and explanations to relate them to their corresponding physically realizable signals. In this book, this issue is addressed in a large number of places. In particular, proper use of these signals is important when the signals are defined in an infinite interval of time or frequency, or when the property of a signal in infinity has to be investigated in order to understand the behaviour of the signal in a finite interval.

A separate problem is how to relate signal properties to the physical world. Sometimes it is necessary to work with signals that do not comply with the above definition of physical realisability, particularly when we are dealing with stochastic processes. One example of a signal that does not comply with the above definition is the impulse delta function, which is widely used in signal analysis and synthesis. This signal is generated in practice as a pulse of very short duration and very large amplitude. It is used to find the impulse response of a linear time-invariant (LTI) system.

The notion of realizable signals is important for the theoretical explanation of the way a discrete-time communication system operates. A discrete-time domain signal is usually obtained by an analogue-to-digital conversion and needs to fulfil the conditions of the Nyquist sampling theorem, which requires the signal to be physically realizable. However, sometimes we have to deal with signals which are not physically realizable, for example if we need to present white Gaussian noise in a discrete-time domain and generate it using the analogue-to-digital conversion base with the Nyquist theorem. That problem will need to be solved theoretically, but the solution will need to lead to the practical design of a discrete-time noise generator. A similar problem is when a statistical description of discrete noise obtained from continuous-time unrealizable noise (with infinite power) needs to be presented in exact mathematical form. The latter problem has been addressed and, in Chapter 4, a solution is offered that includes the entropy of the noise source.

## 1.2.6 Classification of Systems

A *system* is defined as a device (a processor, manipulator, apparatus or algorithm) that performs an operation or some operations on a signal or signals. A system is specified by a set of processing rules for the defined input signal(s). It can have one or more inputs and

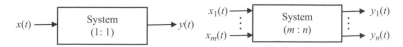

**Fig. 1.2.5** *The two main types of systems.*

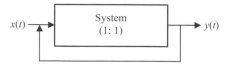

**Fig. 1.2.6** *A system with feedback.*

outputs, as shown in Figure 1.2.5. In other words, a system processes its inputs to obtain the outputs according to a set of rules. Thus, the system can be designed as a physical device or a computer program. An example of a physical system is an audio amplifier. The input of the amplifier is an audio signal, and the output is the amplified version of the input signal. Both the input and the output signals are continuous-time continuous-valued signals. Within communication systems, a signal-processing block inside a transceiver, for example, will sometimes be considered to constitute a system, in the sense of this definition. In this book, particular attention will be paid to LTI systems that process deterministic and/or stochastic signals. The set of processing rules can be specified mathematically and then implemented in hardware and/or software. The method for transferring a set of rules from a mathematical representation to a form suitable for implementation in the appropriate technology is called an algorithm. An algorithm should be simple, hardware/software efficient, easy to implement and fast. One of the main goals in signal processing is the development of such algorithms.

The characteristics of a system are expressed by their *attributes*. The aim is to define these attributes mathematically and then relate them to physical phenomena in order to understand the concepts involved. A special class of systems of great importance are systems with feedback. The output signal (or some intermediate signal) in the form of 'feedback' is an additional type of input for the system, as shown in Figure 1.2.6.

The systems are classified according to their general properties. If a system has a particular property, then this property must hold for every possible input signal for this system. One possible division of systems is shown in Figure 1.2.7.

Systems are divided into two broad classes: systems with distributed parameters (like a coaxial cable in telecommunication systems) and systems with lumped parameters (like an RLC network). The systems can then be divided in a manner similar to that used for the classification of signals. Precise definitions of systems relevant for the mathematical modelling and design of discrete communication systems will be given in this book. For our analysis of communications systems, we will particularly be interested in deterministic and stochastic LTI system operations in discrete- and continuous-time domains. Detailed descriptions of the main systems relevant for this book will be presented in Chapters 11 and 14.

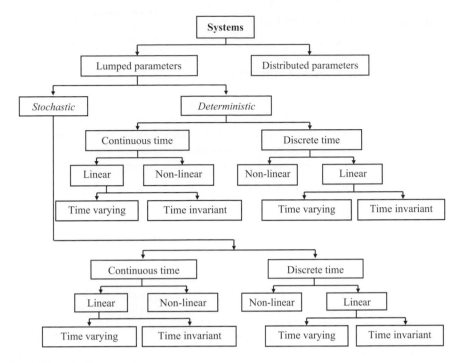

**Fig. 1.2.7** *Classification of systems.*

The classification of systems follows the classification of signals presented in Figure 1.2.1. A particular system can be deterministic or stochastic. Both types of systems can process signals of continuous or discrete time and be linear or non-linear, and time varying or time invariant. Using the system classification diagram in Figure 1.2.7, we can define various types of systems.

## 1.3 Conversions of Analogue and Digital Signals

Analogue-to digital and digital-to-analogue conversions are widely used procedures in communication systems and signal processing. They make it possible to get a digital signal suitable for digital signal processing (DSP) and then return the digital signal into its analogue form. In this section, these conversions will be briefly explained in order to understand the principal ideas of their operation. In addition, more precise definitions of analogue and digital signals will be presented in the context of the definitions of the four basic signals presented in Section 1.2.1.

We say that a digital signal can be obtained from an analogue signal via analogue-to-digital conversion, which involves performing the following processes on the signal:

1. sampling (discretization in time)

2. quantization (values discretization)
3. coding of the quantized values.

### 1.3.1   Analogue-to-Digital Conversion

A simplified system that presents analogue-to-digital and digital-to-analogue conversions is shown in Figure 1.3.1. The DSP block is used as a general term, meaning that it can incorporate a very complex system like a communication system. Analogue-to-digital and digital-to-analogue conversions will be demonstrated and explained via one simple example of a continuous-time continuous-valued signal $x(t)$, which is shown in Figure 1.3.2.

**Sampling.** Discretization in time, or sampling, is the process of taking samples from a continuous-time signal $x(t)$ at discrete instants of time $T_s$ and then producing a discrete-time continuous-valued signal $x(nT_s) = x(n)$, which is expressed as a function of the discrete variable $n$. The resulting signal is a series of samples represented by their amplitudes as teen lines with small circles on top, as shown in Figure 1.3.2.

**Quantization.** Value discretization, or quantization is the process of approximating the values of continuous-valued samples $x(n)$ obtained in the sampling procedure. This approximation procedure is usually performed by rounding or truncating the sampled values. In principle, the quantized samples are obtained by rounding the sampled values to the closest discrete

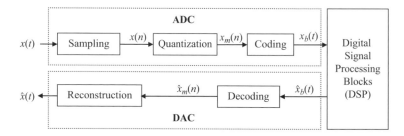

**Fig. 1.3.1** *Analogue-to-digital converter (ADC) and digital-to-analogue converter (DAC) conversions.*

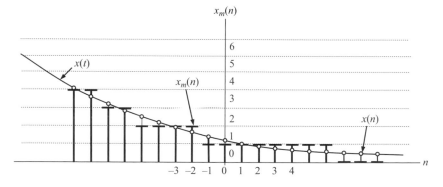

**Fig. 1.3.2** *An example of the sampling and quantization of a signal x(t).*

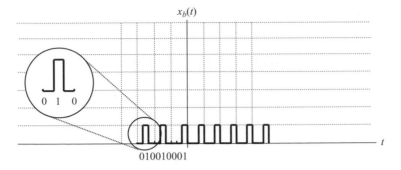

**Fig. 1.3.3** *A coded signal in its binary form.*

amplitude value. Quantized samples are represented by their amplitudes with small horizontal lines on top, as shown in Figure 1.3.2. These samples form the signal $x_m(n)$, which is a discrete-time (discrete-time points from $-\infty$ to $+\infty$, denoted by $n$) discrete-valued ($m = 8$ discrete levels) signal.

**Coding.** The quantized values of a signal are not suitable for processing via existing digital technology. Therefore, we need to transfer them into streams of bits or bytes for further processing in digital devices, which can be digital signal processors or field-programmable gate-array (FPGA) devices, for example. If the number of different discrete values (also called the dynamic range of a signal) is finite, then they can be encoded into a digital signal (usually binary) via the process of *coding*, the third operation performed on the signal in the process of analogue-to-digital conversion. Through coding, the decimal values of the quantized samples can be transformed into binary numbers and represented by a binary signal. That binary signal is defined by two logic levels, ordinary defined as 0 and 1.

For the eight levels of quantized samples assumed possible, as shown in Figure 1.3.2, the number of binary digits needed to represent each quantized sample is 3. For example, the discrete value $x_8(-3) = +2$, taken in the time interval that corresponds to the discrete-time instant $n = -3$, is represented in its binary form as 010 (the enlarged triplet in Figure 1.3.3). Thus, in this way, a discrete-time discrete-valued signal $x_m(n)$ can be represented by a stream of binary digits in the form of the continuous-time discrete-valued binary signal $x_b(t)$, as shown in Figure 1.3.3. This binary signal is a function of a new, continuous, time-independent variable $t$, which is suitable for further processing in digital technology.

## 1.3.2 Digital-to-Analogue Conversion

An analogue signal can be reconstructed from a processed digital binary signal, such as that shown in Figure 1.3.4, via the procedure of digital-to-analogue conversion, as shown in Figure 1.3.5, which includes the following two steps:

1. decoding of a binary digital signal, resulting in discrete samples, and
2. low-pass filtering that reconstructs an analogue signal.

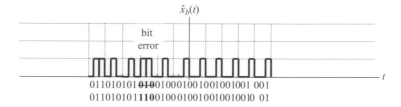

Fig. 1.3.4 *Input of the decoder in its binary form.*

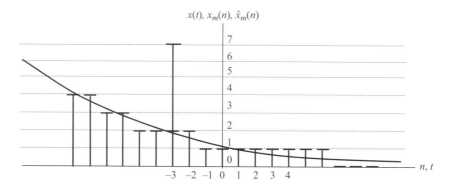

**Fig. 1.3.5** *Discrete samples of decoded signal in decimal form.*

Suppose an error occurred inside one triplet, as shown in Figure 1.3.4 in bold, that is, instead of 010, the input at the decoder is 110, meaning that one error occurred inside the DSP block. Therefore, at the output of the decoder, there will be the quantized, discrete signal $\hat{x}_m(n)$, which is identical to the output of an analogue-to-digital converter except that the amplitude is defined for $n = -3$, which has the value 7 instead of 2, as shown in Figure 1.3.5.

The samples are then applied to the input of a low-pass reconstruction filter that generates a sinc function for each sample, as shown in Figure 1.3.6. Due to a single binary error, the reconstructed signal $\hat{x}(t)$, obtained as a sum of sinc functions and indicated in the figure by dashed lines, does not follow the original continuous-time signal $x(t)$ presented by the thick black continuous line in Figure 1.3.6. Details related to sampling and the related mathematical description are presented in Chapter 13.

From the point of view of this analysis, and bearing in mind the definitions given in Section 1.2.1 of the four main types of signals, let us now revisit the definitions of analogue and digital signals, using examples from the analysis given above.

A signal is *analogue* if it is a continuous-time continuous-valued signal represented by a continuous function of time $x(t)$. A signal is *digital* if it is a continuous-time discrete-valued signal that is presented in digital form, like $x_b(t)$ in Figure 1.3.3. A digital signal is usually a binary signal with two possible values. If a digital signal has more than two levels, it is called a multilevel digital signal. If a binary signal is generated at the output of a shift register or a clock generator, it is a continuous-time discrete-valued signal. However, if the signal

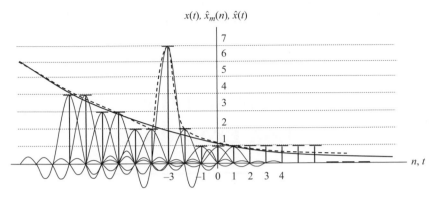

**Fig. 1.3.6** *Discrete samples of a decoded signal with the related sinc functions.*

is generated inside a simulator, it will be represented by a stream of ones and zeros, which represent a discrete-time discrete-valued signal. Therefore, the signals in our analyses need to be strictly defined in the context of the main four types of signals precisely defined in Section 1.2.1.

## 1.3.3   Application of Signals in Digital and Discrete Communication Systems

A digital binary signal, as shown in Figure 1.3.4, is used as a modulating signal in digital communication systems. This binary signal, presented in the discrete-time domain by a number of samples inside a bit interval, is used as a modulating signal in discrete communication systems. The discrete-time bit, represented by one sample per bit, is used in the analysis of channel coding methods. These digital and discrete forms of the binary signal will be extensively used in Chapters 2–10.

# 2

# Orthogonal Signals and the Orthogonalization Procedure

## 2.1 Introduction

Analysis of linear discrete-time invariant systems showed that any discrete-time signal $x(n)$ can be decomposed into a sum of elementary (basis) functions represented by the shifted impulse $\delta$-functions, which are expressed as the convolution

$$x(n) = \sum_{k=-\infty}^{\infty} x(k)\delta\,(n-k)\,. \tag{2.1.1}$$

Thus, no matter how complicated the function $x(n)$ may be, the sum of products $x(k)\delta\,(n-k)$ over all $k$ simply equals the value of $x(n)$ at the point $n = k$. Analysis of discrete-time signals in frequency domains showed that a discrete-time signal could be decomposed into the sum of the complex exponentials:

$$x(n) = \sum_{k=0}^{N-1} c_k e^{+jk2\pi n/N}\,. \tag{2.1.2}$$

In other words, the discrete-time Fourier series obtained is able to express any periodic signal as a linear combination of exponential signals. In the case of an aperiodic signal, the discrete-time Fourier transform should be used and the signal $x(n)$ can be interpreted as a continuous sum of complex exponentials $e^{j\Omega n}$, each with the infinitesimally small amplitude $X(e^{j\Omega})d\Omega$, which is expressed as

$$x(n) = \frac{1}{2\pi} \int_{-\pi}^{\pi} X\left(e^{j\Omega}\right) e^{j\Omega n} d\Omega\,. \tag{2.1.3}$$

The above-mentioned transforms make it possible for a signal in the discrete-time domain to be represented as a linear combination of the basis functions, where the basis functions are sinusoids or exponentials. These linear combinations contain a countable infinite or infinite uncountable number of terms in the sums. Thus, in order to represent and process a real signal

*Discrete Communication Systems.* Stevan Berber, Oxford University Press. © Stevan Berber 2021.
DOI: 10.1093/oso/9780198860792.003.0002

in a finite interval of time, we must approximate the signal with a finite number of components from these sums, which in turn introduces an error of approximation that must be controlled.

However, the question of theoretical and practical interests is whether it is possible to find a finite set of basis functions (the base) that is sufficient to represent any discrete-time function by a given finite set of arbitrary functions. The answer is yes. Namely, the Gram–Schmidt orthogonalization procedure allows us to construct a set of discrete-time basis functions (the base of orthonormal waveforms) $\{\phi_j(n)\}$ for the given set of arbitrary signals $\{s_i(n)\}$. Then, using the base, any signal in the set $\{s_i(n)\}$ can be expressed as a linear combination of *orthonormal basis functions* in the base $\{\phi_j(n)\}$. How to construct the set of basis functions $\{\phi_j(n)\}$, how to use it to analyse and synthesize the signals $\{s_i(n)\}$, and how to apply this theory in practice is the subject of this chapter.

The problem of the orthogonalization of signals, and their geometric representation, has been mentioned in nearly every book related to digital communications. However, the existing books have not mentioned discrete-time signals and their geometric representation and orthogonalization. To proceed towards an explanation of transceiver operation based on discrete-time signal-processing blocks, this chapter is dedicated to the geometric representation and orthogonalization of discrete-time-domain signals. The analysis of signals in the continuous-time domain can be found in numerous books on digital communications.

The main result of this chapter, relevant for later chapters, is the development of a synthesizer and an analyser of discrete- and continuous-time signals, based on the concept of orthonormal functions. They will be the building blocks for the development of the general model of a communication system in Chapter 5. This model will be used to develop modern transceivers operating in discrete- and continuous-time domains and using various modulation methods. It will be shown that the systems developed here are special cases of the general model developed in Chapter 5. The receivers developed here will operate on the concept of correlation rather than on the concept of matched filters. However, the models developed here can be easily adapted to the matched filter application.

## 2.2 Geometric Representation of Signals

### 2.2.1 Orthonormal Basis Functions

Let us define a set of $I$ real-valued discrete-time energy signals $\{s_i(n), i = 1, \ldots, I\} = \{s_1(n), s_2(n), \ldots, s_I(n)\}$, each of duration $N$. Any signal in the set can be expressed as a linear combination of $J$ *orthonormal discrete-time basis functions* $\phi_j(n)$, that is,

$$s_i(n) = \sum_{j=1}^{J} s_{ij}\phi_j(n), \text{ for } 0 \leq j \leq J, \text{and } i = 1, 2, \ldots, I. \tag{2.2.1}$$

The coefficients of the expansion are defined by

$$s_{ij} = \sum_{n=0}^{N-1} s_i(n)\phi_j(n), \qquad \text{for } j = 1, 2, \ldots, J, \text{and } J \leq I. \tag{2.2.2}$$

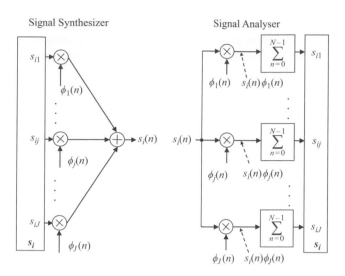

**Fig. 2.2.1** *Synthesizer and analyser of the discrete-time signal $s_i(n)$.*

By definition, the basis functions are orthonormal if

$$\sum_{n=0}^{N-1} \phi_i(n)\phi_j(n) = \delta_{ij} = \begin{cases} 1 & \text{if } i=j \\ 0 & \text{if } i \neq j \end{cases}, \qquad (2.2.3)$$

where $\delta_{ij}$ is the Kronecker delta function. Two characteristics of the basis functions can be observed from this equation. Firstly, each basis function is *normalized* to have unit energy and, secondly, the basis functions are *orthogonal* with respect to each other over the interval $0 \leq n < N$. Therefore, we say that these functions are *orthonormal*.

Based on the above equations, we can construct a *synthesizer* to generate the signal $s_i(n)$, and an *analyser* to generate the set of coefficients, which are shown in Figure 2.2.1. The inputs of the synthesizer are the coefficients of the signal, which are multiplied by appropriate basis functions, and the resultant products are added together. The output of the adder is the synthesized signal $s_i(n)$.

The *analyser* consists of a parallel connection of *J product accumulators or correlators* with a common input. In each correlator, the input signal $s_i(n)$ is multiplied by the corresponding basis function, and the products obtained are added to each other. The result of the addition is a coefficient that corresponds to the basis function applied to the input of the multiplier.

Any signal in the set $\{s_i(n), i = 1, \ldots, I\} = \{s_1(n), s_2(n), \ldots, s_I(n)\}$ is completely determined by a set of its coefficients. These coefficients can be viewed as a *J*-dimensional vector called a *signal vector*, which can be denoted by $\mathbf{s}_i$ and expressed as the transposed vector

$$\mathbf{s}_i = [s_{i1} \quad s_{i2} \quad \cdots \quad s_{iJ}]^T, \qquad (2.2.4)$$

where the coefficients are the coordinates of the signal vector $s_i$. The whole set of discrete-time signals $\{s_i(n)\}$ can be represented by a set of *signal vectors* $\{s_i | i = 1, 2, \ldots, I\}$. This set of signal vectors can then be visualized as defining a set of $I$ points in a $J$-dimensional Euclidean space called the *signal space diagram* of the signal $s_i$ or *constellation diagram*, with $J$ mutually perpendicular axes labelled by $\phi_1, \phi_2, \ldots, \phi_J$.

## Example

a. Formulate the signal space diagram (signal constellation) for a discrete-time polar non-return-to-zero signal, which is obtained from the continuous-time signal presented in Figure 2.2.2. Find the discrete-time signal, assuming that the highest frequency of the signal $s(t)$ corresponds to the first zero crossing of its amplitude spectral density. The bit duration is $T_b$ in continuous time, and $N_b$ in the discrete-time domain.

b. Design a synthesizer and an analyser for this signal.

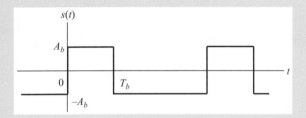

**Fig. 2.2.2** *Continuous-time polar non-return-to-zero signal.*

## Solution.

a. The first zero crossing in the amplitude spectra density occurs at the frequency $f_g = 1/T_b$. Therefore, the Nyquist sampling interval is $T_s = 1/2f_g = T_b/2$, and the corresponding discrete-time domain signal is obtained as presented in Figure 2.2.3.

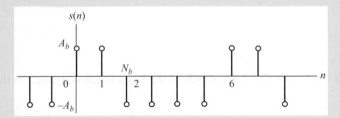

**Fig. 2.2.3** *Discrete-time polar non-return-to-zero signal.*

The binary symbols $m_i = (1, 0)$ are represented by two signals $s_i$, $i = (1, 2)$, in the discrete-time domain and defined by

$$s_1(n) = A_b = \sqrt{\frac{E_b}{N_b}}, \qquad s_2(n) = -A_b = -\sqrt{\frac{E_b}{N_b}},$$

for $0 \leq n \leq N_b - 1$, where the energy of the signal is

$$E_b = \sum_{n=0}^{N_b-1} s_i^2(n) = \sum_{n=0}^{N_b-1} A_b^2 = N_b A_b^2.$$

These symbols in the time domain are shown in Figure 2.2.4.

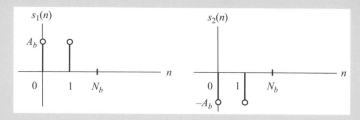

**Fig. 2.2.4**  *Binary symbols.*

Let us define the basis function to be

$$\phi_1(n) = \sqrt{1/N_b},$$

which has a graph as presented in Figure 2.2.5.

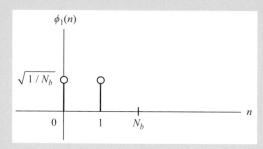

**Fig. 2.2.5**  *The first basis function.*

Therefore, the symbols 1 and 0 can be expressed in terms of the basis function as

$$s_1(n) = s_{11}\phi_1(n) = \sqrt{E_b}\phi_1(n),$$

and

$$s_2(n) = s_{21}\phi_1(n) = -\sqrt{E_b}\phi_1(n),$$

as shown in Figure 2.2.4. All of the above-mentioned symbols and the basis function are defined in the interval $0 \leq n \leq N_b$. The signal space diagram is defined by two points having the two values $\left(+\sqrt{E_b}, -\sqrt{E_b}\right)$, as shown in Figure 2.2.6.

**Fig. 2.2.6** *Signal space diagram.*

b. Designs of the synthesizer and the analyser are shown in Figure 2.2.7.

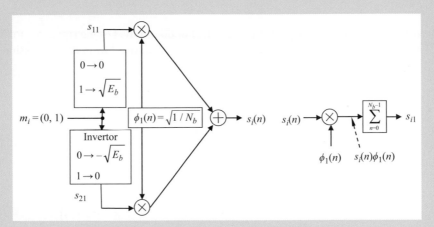

**Fig. 2.2.7** *Synthesizer and analyser of the signal.*

The message source generates a binary signal defined by the values $(0, 1)$. When the binary 0 is generated at the output of the source, it is transferred into the lower branch as an inverted value of the signal coefficient $-\sqrt{E_b}$, which multiplies the basis function $\phi_1(n)$. At the same time, in the upper branch, the binary 0 is transferred to the multiplier. At the output of the multiplier is 0, which is added to the signal from the lower branch. Therefore, at the output of the synthesizer is the signal $s_2(n) = s_{21}\phi_1(n) = -\sqrt{E_b}\phi_1(n)$. If the source generates binary 1, the output of the synthesizer will be the first signal $s_1(n) = s_{11}\phi_1(n) = \sqrt{E_b}\phi_1(n)$.

## 2.2.2 Vector Representation of Signals

It has been shown that a signal $s_i(n)$ in the set $\{s_i(n)\} = \{s_1(n), s_2(n), \ldots, s_I(n)\}$ is completely determined by its coefficients. Therefore, there is a one-to-one correspondence between the signal and its coefficients, and the signal $s_i(n)$ can be represented as the *signal vector* $\mathbf{s}_i$ of its coefficients in a vector-transposed form as

$$\mathbf{s}_i^T = [s_{i1} \ \ s_{i2} \ \ \cdots \ \ s_{iJ}]. \tag{2.2.5}$$

The vector representation of the set of $I$ signals $\{s_i(n)\}$ in the vector form $\{\mathbf{s}_i\} = \{\mathbf{s}_1, \mathbf{s}_2, \ldots, \mathbf{s}_I\}$ is called the signal space diagram or the signal constellation.

The length of a signal vector, and the angles and Euclidean distances between two signal vectors can be defined. The *length* (the absolute value or norm) squared of a signal vector $\mathbf{s}_i$ is defined as the product of the signal transpose vector $\mathbf{s}_i^T$ with itself, according to the relation

$$\|\mathbf{s}_i\|^2 = \mathbf{s}_i^T \mathbf{s}_i == s_{i1}^2 + s_{i2}^2 + \cdots + s_{iJ}^2 = \sum_{j=1}^{J} s_{ij}^2 = \sum_{n=0}^{N-1} s_i^2(n). \tag{2.2.6}$$

This product can be understood as the inner product or dot product of the signal vector $\mathbf{s}_i$ with itself. The *angle* between two signal vectors $\mathbf{s}_i$ and $\mathbf{s}_k$ can be found from the expression

$$\cos\theta_{ik} = \frac{\mathbf{s}_i^T \mathbf{s}_k}{\|\mathbf{s}_i\| \|\mathbf{s}_k\|}. \tag{2.2.7}$$

We say then that two vectors are orthogonal or perpendicular to each other if their inner product is zero, resulting in a 90° angle. The *Euclidean distance* $d_{ik}$ between two vectors $\mathbf{s}_i$ and $\mathbf{s}_k$ is defined as

$$d_{ik} = \|\mathbf{s}_i - \mathbf{s}_k\| = \sqrt{\|\mathbf{s}_i - \mathbf{s}_k\|^2} = \sqrt{\sum_{j=1}^{J} \left(s_{ij} - s_{kj}\right)^2} = \sqrt{\sum_{n=0}^{N-1} \left[s_i(n) - s_k(n)\right]^2}. \tag{2.2.8}$$

**Example**

Suppose a set of two signals is defined to be $\{s_i(n)\} = \{s_1(n), s_2(n)\}$, or in a vector form as

$$\mathbf{s}_1 = \begin{bmatrix} s_{i1} \\ s_{i2} \end{bmatrix} = \begin{bmatrix} s_{11} \\ s_{12} \end{bmatrix} = \begin{bmatrix} 2 \\ 1 \end{bmatrix},$$

and

$$\mathbf{s}_2 = \begin{bmatrix} s_{i1} \\ s_{i2} \end{bmatrix} = \begin{bmatrix} s_{21} \\ s_{22} \end{bmatrix} = \begin{bmatrix} 1 \\ 2 \end{bmatrix}.$$

Find the lengths of the signal vectors, and the angles and the Euclidean distances between the two signal vectors.

**Solution.** The squared length of signal vector $s_1$ is

$$\|\mathbf{s}_i\|^2 = \mathbf{s}_i^T \mathbf{s}_i = \mathbf{s}_1^T \mathbf{s}_1 = [s_{11} \ \ s_{12}] \cdot \begin{bmatrix} s_{11} \\ s_{12} \end{bmatrix} = \begin{bmatrix} 2 & 1 \end{bmatrix} \cdot \begin{bmatrix} 2 \\ 1 \end{bmatrix} = 4 + 1 = 5,$$

and the squared length of signal vector $s_2$ is

$$\|\mathbf{s}_2\|^2 = \mathbf{s}_2^T \mathbf{s}_2 = [s_{21} \ \ s_{22}] \cdot \begin{bmatrix} s_{21} \\ s_{22} \end{bmatrix} = \begin{bmatrix} 1 & 2 \end{bmatrix} \cdot \begin{bmatrix} 1 \\ 2 \end{bmatrix} = 1 + 4 = 5.$$

The angle between signal vectors $s_i$ and $s_k$ can be found from the expression

$$\cos\theta_{ik} = \frac{\mathbf{s}_i^T \mathbf{s}_k}{\|\mathbf{s}_i\| \, \|\mathbf{s}_k\|} \Rightarrow \cos\theta_{12} = \frac{\mathbf{s}_1^T \mathbf{s}_2}{\|\mathbf{s}_1\| \, \|\mathbf{s}_2\|} = \frac{[s_{11} \ \ s_{12}] \cdot \begin{bmatrix} s_{21} \\ s_{22} \end{bmatrix}}{\sqrt{5}\sqrt{5}} = \frac{\begin{bmatrix} 2 & 1 \end{bmatrix} \begin{bmatrix} 1 \\ 2 \end{bmatrix}}{\sqrt{5}\sqrt{5}} = \frac{4}{\sqrt{5}\sqrt{5}} = 0.8$$

to be

$$\theta_{12} = 36.87^{\circ}.$$

The Euclidean distance $d_{12}$ can be obtained as follows:

$$d_{ik}^2 = \|\mathbf{s}_i - \mathbf{s}_k\|^2 = [\mathbf{s}_i - \mathbf{s}_k]^T \cdot [\mathbf{s}_i - \mathbf{s}_k] = \sum_{j=1}^{N} (s_{ij} - s_{kj})^2 \Rightarrow$$

$$d_{12}^2 = \sum_{j=1}^{2} (s_{1j} - s_{2j})^2 = (s_{11} - s_{21})^2 + (s_{12} - s_{22})^2 = (2-1)^2 + (1-2)^2 = 2,$$

that is, $d_{12} = \sqrt{2}$. The vector representation of these signals is given in Figure 2.2.8.

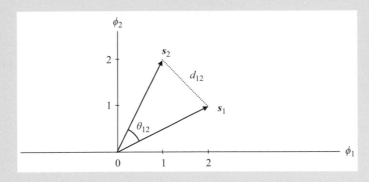

**Fig. 2.2.8** *Vector representation of the signals.*

## 2.3   The Gram–Schmidt Orthogonalization Procedure

The number of basis functions is expected to be at most equal to the number of signals, that is, we expect $I \leq J$. In practice, the number of basis functions needs to be smaller than the number of signals they can generate. In modulation theory and practice, for example, the basis functions are related to the number of carriers. Therefore, the smaller the number of carriers is, the simpler the signal-processing procedures inside the transceiver are.

Any signal in a set of $I$ real-valued energy signals $\{s_i(n), i = 1, ..., I\} = \{s_1(n), s_2(n), ..., s_I(n)\}$, each of duration $N$, can be expressed as a linear combination of $J$ *orthonormal basis functions* $\phi_j(n)$. The problem of particular theoretical and practical interest is how to construct this set of basis functions for a given set of signals. It is the Gram–Schmidt orthogonalization procedure that allows us to construct a set of basis functions (or orthonormal signals) having a set of arbitrary signals $\{s_i(n)\}$ available. Due to the arbitrary wave shape of these signals, the signals in the set are also called the symbols. The procedure is based on the Gram–Schmidt orthogonalization procedure and will be explained in the following paragraphs.

Suppose that there exists a set of finite energy discrete-time signal waveforms or symbols $\{s_i(n), i = 1, 2, ..., I\}$. The first basis function corresponds to the first energy signal normalized by the square root of its energy $E_1$, that is,

$$\phi_1(n) = \frac{s_1(n)}{\sqrt{E_1}}, \tag{2.3.1}$$

because the basis function should have unit energy (normalization). Then, the signal can be expressed as a function of the basis function as

$$s_1(n) = s_{11}\phi_1(n) = \sqrt{E_1}\phi_1(n) \quad \rightarrow \quad s_{11} = \sqrt{E_1}, \tag{2.3.2}$$

where the coefficient $s_{11}$ is the coordinate along the $\phi_1$ axis of the signal $s_1(n)$ expressed as the vector $\mathbf{s_1}$. For the second basis function, the projection of $s_2(n)$ onto $\phi_1(n)$ must be found first, that is,

$$s_{21} = \sum_{n=0}^{N-1} s_2(n)\phi_1(n), \tag{2.3.3}$$

and then an *intermediate function* $g_2(n)$, which is *orthogonal* to $\phi_1(n)$, is defined as

$$g_2(n) = s_2(n) - s_{21}\phi_1(n). \tag{2.3.4}$$

Now, the second basis function is defined as a normalized $g_2(n)$, that is,

$$\phi_2(n) = \frac{g_2(n)}{\sqrt{E_{g2}}} = \frac{s_2(n) - s_{21}\phi_1(n)}{\sqrt{E_2 - s_{21}^2}}, \tag{2.3.5}$$

where the energy of the intermediate function is

$$E_{g_2} = \sum_{n=0}^{N-1} g_2^2(n) = \sum_{n=0}^{N-1} [s_2(n) - s_{21}\phi_1(n)]^2 = E_2 - s_{21}^2. \qquad (2.3.6)$$

Then, since we know the basis functions, the signal can be expressed as a function of the two basis functions (two-dimensional signal) in the form

$$s_2(n) = s_{21}\phi_1(n) + s_{22}\phi_2(n), \qquad (2.3.7)$$

where the coefficients are the coordinates of the signal on both axes, that is,

$$s_{21} = \sum_{n=0}^{N-1} s_2(n)\phi_1(n), \quad \text{and} \quad s_{22} = \sum_{n=0}^{N-1} s_2(n)\phi_2(n). \qquad (2.3.8)$$

The basis functions $\phi_1(n)$ and $\phi_2(n)$ form an orthonormal pair because

$$\sum_{n=0}^{N-1} \phi_2^2(n) = 1, \qquad (2.3.9)$$

and

$$\sum_{n=0}^{N-1} \phi_1(n)\phi_2(n) = 0. \qquad (2.3.10)$$

In general, following the same procedure, it is possible to define the *i*th *intermediate function* as

$$g_i(n) = s_i(n) - \sum_{j=1}^{i-1} s_{ij}\phi_j(n), \qquad (2.3.11)$$

where the coefficients, as defined before, are

$$s_{ij} = \sum_{n=0}^{N-1} s_i(n)\phi_j(n), \qquad (2.3.12)$$

$j = 1, 2, \ldots, i - 1$, which gives the orthonormal set of basis functions

$$\phi_i(n) = \frac{g_i(n)}{\sqrt{E_{g_i}}}, \qquad (2.3.13)$$

for $i = 1, 2, \ldots, I$. It is important to notice that the number of orthonormal functions is not necessarily equal to the number of signals. Therefore, it is expected that some of the functions calculated from eqn (2.3.13) will be zero and the number of basis functions will be $J$ instead of $I$, where $J < I$. Following the previous steps, we can find the base, or the set of basis functions $\{\phi_j(n), j = 1, 2, \ldots, J\}$. That set is not unique. It depends on the order of the signals $\{s_i(n)\}$ that are used to find the basis.

Note that the dimension of $J$ can be:

1. $J = I$, where the signals in the set $\{s_i(n), i = 1, 2, \ldots, I\}$ form a *linearly independent set*, or

2. $J < I$, where the signals in the set $\{s_i(n), i = 1, 2, \ldots, I\}$ are not *linearly independent* and the intermediate function $g_i(n) = 0$ for $i > J$.

The Gram–Schmidt orthogonalization procedure possesses the following properties:

1. The waveforms of the basis functions are not specified. Thus, the Gram–Schmidt orthogonalization procedure is not restricted to present signals in terms of sinusoidal or exponential functions as is the case, for example, in the Fourier series expansion of a periodic signal.

2. The signal $s_i(n)$ is expanded in terms of a finite number of basis functions. However, this is not an approximation but rather an exact expression where $J$, *and only J*, terms are significant.

**Example**

Formulate the signal space diagram (signal constellation) for the polar non-return-to-zero discrete-time line code. Suppose this signal in the discrete-time domain is as shown in Figure 2.3.1.

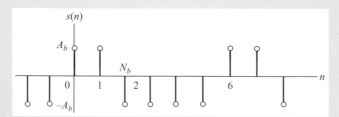

**Fig. 2.3.1** *Discrete-time polar non-return-to-zero line-code signal.*

In this case, a message source generates two possible messages (symbols), as presented in Figure 2.3.2. Each symbol has the period $N_b$. Each symbol can be generated from the orthonormal basis function $\phi_1(n)$.

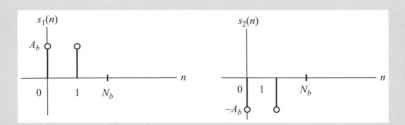

**Fig. 2.3.2** *Binary symbols in a discrete-time domain.*

a. Use the symbol $s_1(n)$ and the Gram–Schmidt orthogonalization procedure to determine the basis function $\phi_1(n)$.

b. Plot the symbols $s_1(n)$ and $s_2(n)$ on a signal space diagram and indicate the energy of each symbol.

**Solution.**

a. The symbols are defined as

$$s_1(n) = A_b = \sqrt{\frac{E_b}{N_b}}, \quad 0 \le n \le N_b - 1,$$
$$s_2(n) = -A_b = -\sqrt{\frac{E_b}{N_b}}, \quad 0 \le n \le N_b - 1.$$

Thus, the first basis function corresponds to the first energy signal $s_1(n)$ normalized by its energy $E_b$, that is,

$$\phi_1(n) = \frac{s_1(n)}{\sqrt{E_b}} = \frac{s_1(n)}{\sqrt{\sum_{n=0}^{N_b-1} s_1^2(n)}} = \frac{A_b}{\sqrt{A_b^2 N_b}} = \sqrt{\frac{1}{N_b}},$$

where the energy is $E_b = A_b^2 N_b$, because the basis function should have unit energy. The first basis function is shown in Figure 2.3.3. This function obviously fulfils the conditions of orthogonality and normality.

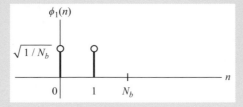

**Fig. 2.3.3** *The first basis function in the time domain.*

Thus, the first signal and its coefficient can be expressed as

$$s_1(n) = \sqrt{E_b}\phi_1(n) = s_{11}\phi_1(n) \quad \rightarrow \quad s_{11} = \sqrt{E_b}.$$

The second signal and its coefficient are

$$s_2(n) = -\sqrt{E_b}\phi_1(n) = s_{21}\phi_1(n) \quad \rightarrow \quad s_{21} = -\sqrt{E_b},$$

and the signals are as has already been presented in Figure 2.3.2.

b. The signal space diagram is defined by the two points $\left(+\sqrt{E_b}, -\sqrt{E_b}\right)$, as shown in Figure 2.3.4.

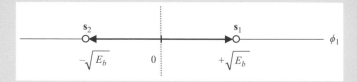

**Fig. 2.3.4** *Signal space diagram.*

The energies of the signals are $E_1 = \left(\sqrt{E_b}\right)^2 = E_b \ J$ and $E_2 = \left(-\sqrt{E_b}\right)^2 = E_b \ J$.

## Example

A communication system can transmit three possible messages (symbols), as presented in Figure 2.3.5. Each symbol has the period $N_b = 4$. Each symbol can be generated from the orthonormal basis functions $\phi_1(n)$ and $\phi_2(n)$.

a. Use the symbol $s_1(n)$ and the Gram–Schmidt orthogonalization procedure to determine the basis function $\phi_1(n)$.

b. Use the symbol $s_2(n)$ and the Gram–Schmidt orthogonalization procedure to determine the basis function $\phi_2(n)$.

c. Show that $\phi_1(n)$ and $\phi_2(n)$ are orthonormal.

d. Express the signals $s_1(n)$, $s_2(n)$, and $s_3(n)$ in a vector form, plot them on a signal space diagram, and indicate the energy of each symbol.

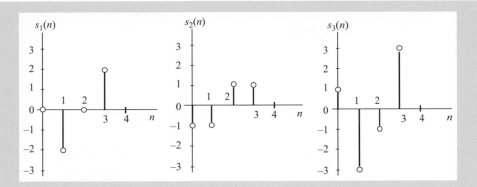

**Fig. 2.3.5** *Signals in time domain.*

**Solution.**

a. The first basis function corresponds to the first energy signal normalized by its energy $E_1$, that is,

$$\phi_1(n) = \frac{s_1(n)}{\sqrt{E_1}} = \frac{s_1(n)}{2\sqrt{2}},$$

because the basis function should have unit energy. The energy of the first signal is calculated as

$$E_1 = \sum_{n=0}^{N_b-1} [s_1(n)]^2 = 2^2 + 2^2 = 8.$$

The first basis function is shown in Figure 2.3.6. Thus, the first signal can be expressed as a function of the first basis function as follows:

$$s_1(n) = s_{11}\phi_1(n) = 2\sqrt{2}\phi_1(n).$$

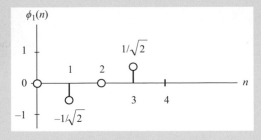

**Fig. 2.3.6** *The first basis function in the time domain.*

b. For the second basis function, the projection of $s_2(n)$ onto $\phi_1(n)$ must be found first. The coefficient $s_{21}$ can be calculated as

$$s_{21} = \sum_{n=0}^{N_b-1} s_2(n)\phi_1(n) = \sqrt{2},$$

using the product of functions inside the sum that is presented in Figure 2.3.7.

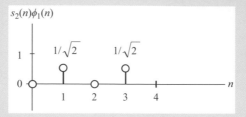

**Fig. 2.3.7** *The product of functions in the discrete-time domain.*

The first *intermediate function* $g_2(n)$, which is *orthogonal* to $\phi_1(n)$, is defined as

$$g_2(n) = s_2(n) - s_{21}\phi_1(n) = s_2(n) - \sqrt{2}\phi_1(n),$$

and shown in Figure 2.3.8.

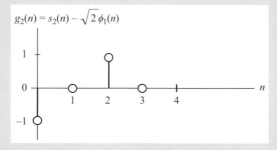

**Fig. 2.3.8** *An intermediate function in the time domain.*

The second basis function is defined as the normalized $g_2(n)$, that is,

$$\phi_2(n) = \frac{g_2(n)}{\sqrt{E_{g_2}}} = \frac{g_2(n)}{\sqrt{2}},$$

where the energy of the intermediate function is calculated as

$$E_{g_2} = \sum_{n=0}^{N_b-1} g_2^2(n) = 2.$$

The second basis function is shown in Figure 2.3.9.

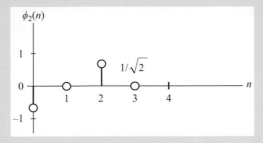

**Fig. 2.3.9** *The second basis function in the discrete-time domain.*

c. The basis functions $\phi_1(n)$ and $\phi_2(n)$ form an orthonormal pair because

$$E_{\phi_1} = \sum_{n=0}^{N_b-1} \phi_1^2(n) = 1, E_{\phi_2} = \sum_{n=0}^{N_b-1} \phi_2^2(n) = 1 \text{ and } E_{\phi_1\phi_2} = \sum_{n=0}^{N_b-1} \phi_1(n)\,\phi_2(n) = 0.$$

d. From a., we have the coefficients $s_{11} = 2\sqrt{2}$ and $s_{12} = 0$ and, from b., we have $s_{21} = \sqrt{2}$ and then can calculate the second coefficient $s_{22}$ by using the graph in Figure 2.3.10 as follows:

$$s_{22} = \sum_{n=0}^{N_b-1} s_2(n)\phi_2(n) = \sqrt{2}.$$

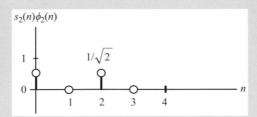

**Fig. 2.3.10** *The product of functions in the discrete-time domain.*

The second *intermediate function* is zero, that is,

$$g_3(n) = s_3(n) - \sum_{j=1}^{3-1} s_{3j}\phi_j(n) = 0.$$

The third basis function is zero, and the third signal is a linear combination of the first two basis functions with the coefficients $s_{31}$ and $s_{32}$, calculated using the graphs in Figure 2.3.11 as

$$s_{31} = \sum_{n=0}^{3} s_3(n)\phi_1(n) = 3\sqrt{2}, \text{ and } s_{32} = \sum_{n=0}^{3} s_3(n)\phi_2(n) = -\sqrt{2}.$$

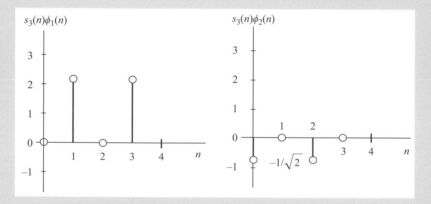

**Fig. 2.3.11** *Addends in the time domain.*

In summary, the analysed signals can be represented as the following vectors in the two-dimensional space defined by their coordinates:

$$s_1 = \begin{bmatrix} s_{11} \\ s_{12} \end{bmatrix} = \begin{bmatrix} 2\sqrt{2} \\ 0 \end{bmatrix}, s_2 = \begin{bmatrix} s_{21} \\ s_{22} \end{bmatrix} = \begin{bmatrix} \sqrt{2} \\ \sqrt{2} \end{bmatrix}, s_3 = \begin{bmatrix} s_{31} \\ s_{32} \end{bmatrix} = \begin{bmatrix} 3\sqrt{2} \\ -\sqrt{2} \end{bmatrix},$$

and drawn in the coordinate system specified by the basis vectors as shown in Figure 2.3.12.

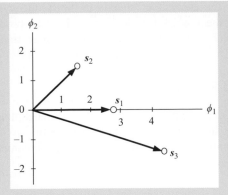

**Fig. 2.3.12** *Vector representation of signals.*

Now, the energies of signals are

$$E_1 = (s_{11})^2 = 8\,J, E_2 = (s_{21})^2 + (s_{22})^2 = 4\,J, \text{and } E_3 = (s_{31})^2 + (s_{32})^2 = 20\,J.$$

## 2.4 Continuous-Time Orthogonal Signals

### 2.4.1 Continuous-Time Versus Discrete-Time Basis Signals

The main objective of this book is to present the theory and design of discrete communication systems, or more precisely, the systems where most of the signal processing is in the discrete-time domain. It is assumed that the readers are familiar with the elementary theory of digital communication systems which is based on continuous-time signal processing. In spite of that, part of this book contains a review of the theory of digital communication systems, to help the reader connect the theory of discrete communications to the classical theory of digital communications. To this end, for the sake of completeness and to gain an understanding of orthogonal signals and the procedure of their orthogonalization, this section will give a brief review of the orthogonalization procedure for continuous-time signals and its relation to discrete communication systems. The reader of the book is advised to read this section before proceeding with the theory of discrete communication systems.

The formal differences between the digital and discrete communication systems can be, in principle, summarized as follows:

1. The continuous independent time variable $t$ in digital systems is replaced by the discrete-time variable $n$ in discrete systems.

2. The integrals in the continuous-time domain are replaced by sums in the discrete-time domain.

3. In digital systems, the analysis of signals in the frequency domain is based on the Fourier series and the Fourier transform. For discrete systems, we need to use the discrete-time Fourier series, the discrete-time Fourier transform, and the discrete Fourier transform.

## 2.4.2    Orthonormal Signals

The definition of orthonormal functions is very similar to the definition for the functions in the discrete-time domain: any signal in a set of real-valued continuous-time energy signals $s_1(t)$, $s_2(t)$, ..., $s_J(t)$, each of duration $T$, can be expressed as a linear combination of $J$ *orthonormal basis functions* $\phi_j(t)$, that is,

$$s_i(t) = \sum_{j=1}^{J} s_{ij}\phi_j(t), \tag{2.4.1}$$

for $0 \leq t \leq T$, and $i = 1, 2, ..., I$, where the coefficients of expansion are defined by

$$s_{ij} = \int_0^T s_i(t)\phi_j(t)dt, \tag{2.4.2}$$

for $j = 1, 2, ..., J$, and $J \leq I$, and the basis functions fulfil the condition of orthonormality expressed as

$$\int_0^T \phi_i(t)\phi_j(t)dt = \delta_{ij} = \begin{cases} 1 & \text{if } i=j \\ 0 & \text{if } i \neq j \end{cases}, \tag{2.4.3}$$

where $\delta_{ij}$ is the Kronecker delta function. The basis functions are *normalized* to have unit energy and are *orthogonal* with respect to each other over the continuous-time interval $0 \leq t \leq T$.

The structure of the synthesizer for generating the signal $s_i(t)$, and that of the analyser for generating the set of related coefficients, are the same as shown in Figure 2.2.1, with the change of the discrete variable $n$ to the continuous variable $t$, and the sums to integrals in the signal analyser. Here, as for the discrete case, any signal in the orthonormal set $\{s_i(t), i = 1, ..., I\} = \{s_1(t), s_2(t), ..., s_I(t)\}$ is completely defined by a set of its coefficients that can be presented as a $J$-dimensional *signal vector*, denoted by $\mathbf{s}_i$, and expressed as

$$\mathbf{s}_i = [s_{i1} \ s_{i2} \ \cdots \ s_{iJ}]^T, \tag{2.4.4}$$

where the coefficients are the coordinates of the signal vector $\mathbf{s}_i$. All signals in the set $\{s_i(t)\}$ are represented by a set of *signal vectors* $\{\mathbf{s}_i | i = 1, 2, ..., I\}$. The vectors are presented by a set of $I$ points, which is called the *signal space*, in the $J$-dimensional Euclidian space with $J$ mutually perpendicular axes labelled as $\phi_1, \phi_2, ..., \phi_J$.

### 2.4.3   The Gram–Schmidt Orthogonalization Procedure

The problem is how to find the set of orthonormal basis functions for a given set of energy signals. The theoretical foundation and practical procedure are defined by the Gram–Schmidt orthogonalization procedure, which is defined as for discrete-time signals as follows.

For the set of finite energy signals $\{s_i(t), i = 1, 2, \ldots, I\}$, the first basis function is the first energy signal normalized by its energy $E_1$, that is,

$$\phi_1(t) = \frac{s_1(t)}{\sqrt{E_1}}, \tag{2.4.5}$$

because the basis function should have unit energy. For the second basis function, the projection of $s_2(t)$ onto $\phi_1(t)$ has to be found first, which is defined as the coordinate of the second signal on the $\phi_1$ axis

$$s_{21} = \int_0^T s_2(t)\phi_1(t)dt. \tag{2.4.6}$$

A new *intermediate function* $g_2(t)$, which is *orthogonal* to $\phi_1(t)$, is defined as

$$g_2(t) = s_2(t) - s_{21}\phi_1(t), \tag{2.4.7}$$

and the second basis function is defined as the normalized $g_2(t)$, that is,

$$\phi_2(t) = \frac{g_2(t)}{\sqrt{E_{g2}}} = \frac{g_2(t)}{\sqrt{\int_0^T g_2^2(t)dt}} = \frac{s_2(t) - s_{21}\phi_1(t)}{\sqrt{E_2 - s_{21}^2}}, \tag{2.4.8}$$

where the energy of the intermediate function is

$$E_{g2} = \int_0^T g_2^2(t)dt = \int_0^T [s_2(t) - s_{21}\phi_1(t)]^2 dt = E_2 - 2s_{21}s_{21} + s_{21}^2 = E_2 - s_{21}^2. \tag{2.4.9}$$

The basis functions $\phi_1(t)$ and $\phi_2(t)$ form an orthonormal pair. In general, it is possible to define the *i*th *intermediate function* as

$$g_i(t) = s_i(t) - \sum_{j=1}^{i-1} s_{ij}\phi_j(t), \tag{2.4.10}$$

where the coefficients are

$$s_{ij} = \int_0^T s_i(t)\phi_j(t)dt, \tag{2.4.11}$$

for $j = 1, 2, \ldots, i - 1$. The orthonormal set of basis functions is expressed as

$$\phi_i(t) = \frac{g_i(t)}{\sqrt{\int_0^T g_i^2(t)dt}}, \qquad (2.4.12)$$

for $i = 1, 2, \ldots, I$. The properties of the functions are the same as for the discrete case. It is important to note that the expansion of the signal $s_i(t)$ in terms of a finite number of terms $J$ is not an approximation but rather an exact expression where $J$, *and only J*, terms are significant. Based on the Gram–Schmidt orthogonalization procedure, the signals in the set $\{s_i(t)\} = \{s_1(t), s_2(t), \ldots, s_I(t)\}$ can be expressed as *signal vectors* $s_i$ with defined coefficients as the vectors' coordinates and presented in matrix form in eqn (2.4.4).

The set of signals $\{s_i(t)\}$ can be represented by a set of *signal vectors* $\{s_i | i = 1, 2, \ldots, I\}$. The set of signal vectors can then be visualized as defining a set of $I$ points in a $J$-dimensional Euclidean space, which is called the *signal space*, with $J$ mutually perpendicular axes labelled as $\phi_1, \phi_2, \ldots, \phi_J$. The length of a signal vector, the angles, and the Euclidean distances between two signal vectors can be defined as for discrete-time signals.

## 2.5   Orthogonal Signals in Code Division Multiple Access Communication Systems

Sinusoidal orthogonal signals are extensively used as carriers in communication systems. Their orthonormality is used to efficiently demodulate the signals in single-user, multi-user, and multi-carrier communication systems. In addition to sinusoidal signals, due to the development of spread-spectrum systems, orthogonal binary sequences have been used in communication systems, like pseudorandom and Walsh signals and sequences. In particular, these sequences are used in direct-sequence spread-spectrum systems and in code division multiple access (CDMA) systems. In order to increase the security in signal transmission, some non-binary orthogonal sequences have been widely investigated in the last 20 years for possible application in spread-spectrum and CDMA systems. In this section, a brief presentation of typical binary and non-binary orthogonal signals and their application in the CDMA system will be presented.

**Binary orthogonal Walsh functions and sequences.** Orthogonal binary signals, named Walsh functions, are used in CDMA systems. They are used in the IS-95 system, which was the first commercial mobile system based on the application of CDMA technology. The Walsh functions $W_j(t)$ of the order $J$ are defined as a set of binary continuous-time signals with duration $T$ and the following properties:

1. $W_j(t)$ takes on the values from a binary set $(-1, 1)$.
2. $W_j(0) = 1$ for any $j = 0, 1, 2, \ldots, J - 1$.
3. $W_j(t)$ has $j$ zero crossings.

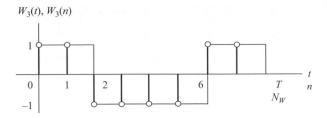

**Fig. 2.5.1** *The fourth Walsh function, continuous and discrete.*

4. The functions fulfil the orthogonality condition, that is,

$$\int_0^T W_j(t)W_k(t)dt = \begin{Bmatrix} T & j=k \\ 0 & j \neq k \end{Bmatrix}. \qquad (2.5.1)$$

5. The functions are odd or even with respect to their midpoint.

An example of such a function, $W_3(t)$, is presented in Figure 2.5.1 as a continuous-time discrete-valued binary signal. This function can also be presented as a discrete-time function. If we perform sampling at time instants $nT/8$, for $n = 0, 1, 2, \ldots, T$, a discrete-time signal can be obtained, as presented by its samples in Figure 2.5.1. We can perform up-sampling and a linear interpolation and generate more samples in each binary interval.

The discrete-time signal obtained is represented by a set of binary values, that is, $W_3(n) = [1\ 1\ -1\ -1\ -1\ -1\ 1\ 1]$ and corresponds to the discrete-time Walsh function that fulfils the condition of orthogonality now expressed in discrete form as

$$\sum_{n=0}^{N-1} W_j(n)W_k(n) = \begin{Bmatrix} N & j=k \\ 0 & j \neq k \end{Bmatrix} = N\delta_{jk}, \qquad (2.5.2)$$

where $N$ is the duration of the discrete signal, and $\delta_{jk}$ is the Kronecker delta function. The normalized autocorrelation function $W_{12}(\tau)$ and the cross-correlation function $W_{12-8}(\tau)$, for discrete-time signals $W_{12}(n)$ and $W_8(n)$, respectively, are presented in Figure 2.5.2. For the discrete delay $\tau = 0$, the autocorrelation function is defined by the Kronecker's delta function and the cross-correlation function has zero value, as we expect due to their orthogonality.

The Walsh function can be expressed in terms of values taken from the additive group $(0, 1)$ or the multiplicative group $(1, -1)$ where all ones are converted to zeros, and minus ones are converted to ones. Then, the Walsh function $W_3(n)$, now called the Walsh sequence, can be expressed as $w_3 = [0\ 0\ 1\ 1\ 1\ 1\ 0\ 0]$ or $w_3 = [1\ 1\ -1\ -1\ -1\ -1\ 1\ 1]$. The Walsh sequences, expressed in terms of the multiplicative group $(1, -1)$, form a group under the multiplication operation. Furthermore, the Walsh sequences expressed in terms of the additive group $(0, 1)$ form a group under modulo-2 addition. Therefore, the modulo-2 sum of two sequences is a Walsh sequence, while the product of two functions $W_j(t)$ and $W_k(t)$, or $W_j(n)$ and $W_k(n)$,

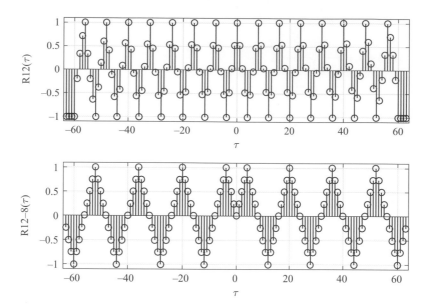

**Fig. 2.5.2** *The autocorrelation and cross-corelation functions.*

is a Walsh function. The Walsh functions and sequences can be generated using Rademacher functions, Hadamard matrices, or basis vectors.

**Non-binary orthogonal signals and sequences.** The number of orthogonal Walsh functions or Walsh sequences is finite. In addition, their structure is deterministic and easy to be generated. If we want to hide the content of a transmitted message, it will be required to generate nondeterministic signals and implement some elements of randomness in their structure. For this reason, the orthogonal non-binary signals are investigated and applied for the design of spread-spectrum and CDMA communication systems. These sequences can be chaotic and random sequences. We will demonstrate the generation of orthogonal chaotic sequences and their application in CDMA communication systems.

They are, in principle, generated using recursive expression, and their wave shapes depend heavily on the initial values. We will demonstrate the operation of generators or mappers of the chaotic signals that are based on the Chebyshev degree-$M$ maps. For a given initial value $x_0$, the generator calculates the chaotic values according to the recursive expression

$$x_{n+1} = \cos\left(M\cos^{-1}x_n\right), \tag{2.5.3}$$

where $-1 \leq x_n \leq 1$ is the $n$th chaotic sample inside the generated chaotic sequence. Therefore, the sequence has the values at the discrete-time instants defined by the natural numbers $n = 1$, 2, 3, …. For degree $M = 2$, the Chebyshev map is called the logistic map, and the recursive expression is

$$x_{n+1} = 2 \cdot (x_n)^2 - 1. \tag{2.5.4}$$

Likewise, for degree $M = 3$, the Chebyshev map is called the cubic map, which is defined by the recursive expression as

$$x_{n+1} = 4 \cdot (x_n)^2 - 3x_n. \tag{2.5.5}$$

A small variation in the initial value results in generating a new chaotic sequence that is orthogonal to the previously generated one. Therefore, a pair of chaotic signals, generated from different generators or from a single generator with different initial conditions, will be characterized with a cross-correlation function that has values close to zero. At the same time, their autocorrelation functions can be approximated by a delta function. This function is presented in Figure 2.5.3 for an example of a chaotic discrete-time signal that is generated using a logistic map for the initial condition $x_0 = 0.344$. Due to the impulse nature of the autocorrelation function, the power spectral density of the chaotic sequence will have a constant value in a wide interval of frequencies like the power spectral density of noise process.

Chaotic signals can be classified as semi-deterministic signals. The probability density function of a chaotic sequence generated by the Chebyshev map is expressed as

$$p(x_n) = \frac{1}{\pi \sqrt{1 - x_n^2}}, \tag{2.5.6}$$

and defined in the interval of the values $-1 \leq x_n \leq 1$. The mean value of the chaotic process is zero, and the variance is 0.5. These sequences will be a subject of the theoretical analysis of a CDMA system and will be applied in the design of a chaos-based CDMA system.

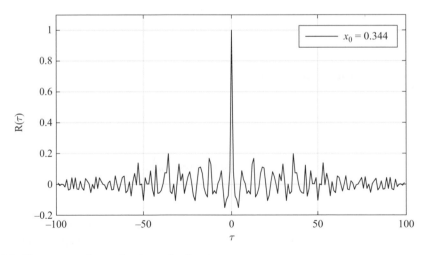

**Fig. 2.5.3** *The autocorrelation function of a chaotic sequence.*

PROBLEMS

1. A message system can transmit four possible messages (symbols), as presented in Figure 1. Each symbol has the period $N = 3$. Each symbol can be generated from the orthonormal basis functions $\phi_1(n)$, $\phi_2(n)$, and $\phi_3(n)$.

   a. Use the symbol $s_1(n)$ and carry out the Gram–Schmidt orthogonalization procedure to determine the basis function $\phi_1(n)$.
   b. Use the symbol $s_2(n)$, and the Gram–Schmidt orthogonalization procedure to determine the basis function $\phi_2(n)$.
   c. Use the symbol $s_3(n)$, and the Gram–Schmidt orthogonalization procedure to determine the basis function $\phi_3(n)$.
   d. Show that the fourth basis function $\phi_4(n)$ is zero.
   e. Find the coordinates of symbols $s_1(n)$, $s_2(n)$, $s_3(n)$, and $s_4(n)$ and plot these symbols on a signal space diagram. Calculate the energy of each symbol.

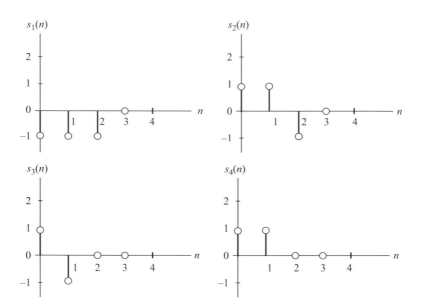

**Fig. 1** *Symbols in the time domain.*

2. A message system can transmit four possible symbols (messages), as presented in Figure 2 below. Each symbol has the period $N = 3$. Each symbol can be generated from a set of the orthonormal basis functions $\phi_1(n)$, $\phi_2(n)$, and $\phi_3(n)$.

a. Use the symbol $s_1(n)$ and the Gram–Schmidt orthogonalization procedure to determine the basis function $\phi_1(n)$.

b. Use the symbol $s_2(n)$, and the Gram–Schmidt orthogonalization procedure to determine the basis function $\phi_2(n)$.

c. Use the symbol $s_3(n)$, and the Gram–Schmidt orthogonalization procedure to determine the basis function $\phi_3(n)$.

d. Show that $\phi_1(n)$, $\phi_2(n)$, and $\phi_3(n)$ are orthonormal.

e. Plot the symbols $s_1(n)$, $s_2(n)$, $s_3(n)$, and $s_4(n)$ on a signal space diagram and indicate the energy of each symbol.

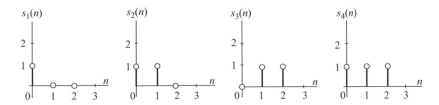

**Fig. 2** *Symbols in the time domain.*

3. Formulate the signal space diagram (signal constellation) for the binary signal that modulates the carrier and produces the signal presented in Figure 3.1 (binary 1 is a cosine function, and binary 0 is the negative cosine function). For defined $\phi = 0$ and $\Omega_0 = 2\pi f_0 = 2\pi/N = 2\pi/16$, each symbol is represented by $N = 16$ samples.

In this case, a message system transmits two possible messages (symbols), as presented in Figure 3.2. Each symbol can be generated from the basis function $\phi_1(n)$.

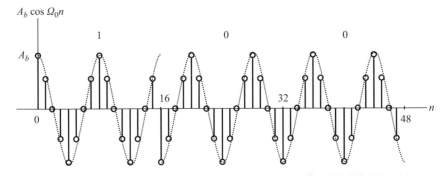

**Fig. 3.1** *Signal in the time domain.*

a. Use the symbol $s_1(n)$ and the Gram–Schmidt orthogonalization procedure to determine the basis function $\phi_1(n)$.

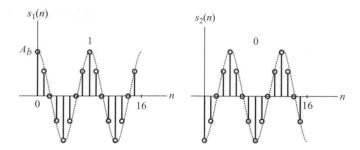

**Fig. 3.2** *Symbols in the time domain.*

   b.  Plot the symbols $s_1(n)$ and $s_2(n)$ on a signal space diagram and indicate the energy of
each symbol.

4.  A set of basis functions $\phi_1(n)$, $\phi_2(n)$, and $\phi_3(n)$ is shown in Figure 4.1. They are defined
in the interval $N = 3$.

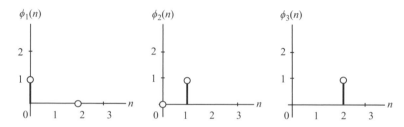

**Fig. 4.1** *Basis functions in the time domain.*

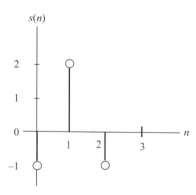

**Fig. 4.2** *The signal in the discrete-time domain.*

a. Show that $\phi_1(n)$, $\phi_2(n)$, and $\phi_3(n)$ are orthogonal. Are they orthonormal?

b. Express these functions as three-dimensional vectors. Plot the signals $\phi_1(n)$, $\phi_2(n)$, and $\phi_3(n)$ on a signal space diagram and indicate the energy of each symbol.

c. A signal $s(n)$ is shown in Figure 4.2. Express this signal as a linear combination of the basis functions $\phi_1(n)$, $\phi_2(n)$, and $\phi_3(n)$. Sketch this signal as a vector in a vector space and find its energy using this presentation.

5. A set of symbols is shown in Figure 5. For $\phi = 0$ and $\Omega_c = 2\pi f_c = 2\pi/N = 2\pi/8$, each symbol is represented by $N = 8$ samples and defined as

$$s_1(n) = \begin{cases} A_b \sin(\Omega_c n) & 0 \le n < N/2 \\ 0 & otherwise \end{cases}$$

$$s_2(n) = \begin{cases} A_b \sin(\Omega_c n) & 0 \le n < N \\ 0 & otherwise \end{cases}$$

$$s_3(n) = \begin{cases} A_b \sin(\Omega_c n) & N/2 \le n < N \\ 0 & otherwise \end{cases}.$$

Prove that the basis functions are expressed as

$$\phi_1(n) = \begin{cases} \frac{2}{\sqrt{N}} \sin(\Omega_c n) & 0 \le n < N/2 \\ 0 & otherwise \end{cases}$$

$$\phi_2(n) = \begin{cases} \frac{-2}{\sqrt{N}} \sin(\Omega_c n) & N/2 \le n < N \\ 0 & otherwise \end{cases}.$$

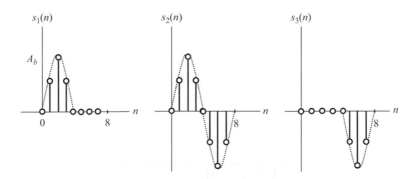

$s_1(n)$      $s_2(n)$      $s_3(n)$

**Fig. 5** *Symbols in the time domain.*

6. A message system can transmit four possible symbols, presented in Figure 6. Each symbol has the period $T = 3$. Each symbol can be generated from the orthonormal basis functions $\phi_1(t)$, $\phi_2(t)$, and $\phi_3(t)$.

a. Using symbols $s_1(t)$, $s_2(t)$, and $s_3(t)$, carry out the Gram–Schmidt orthogonalization procedure to determine the related basis functions. Show that the fourth basis function $\phi_4(t)$ is zero.

b. Find the coordinates of symbols $s_1(t)$, $s_2(t)$, $s_3(t)$, and $s_4(t)$ and plot these symbols on a signal space diagram. Calculate the energy of each symbol.

c. Compare the results obtained for these continuous-time signals with the results for the discrete-time signals defined in Problem 1.

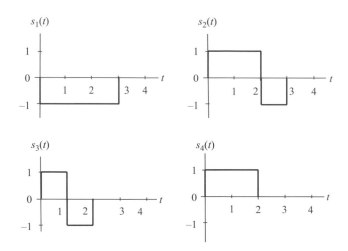

**Fig. 6** *Symbols in the time domain.*

7. A message system can transmit four possible symbols, as presented in Figure 7. Each symbol has the period $T = 3$. Each symbol can be generated from a set of the orthonormal basis functions $\phi_1(t)$, $\phi_2(t)$, and $\phi_3(t)$.

a. Use the symbol $s_1(t)$ and the Gram–Schmidt orthogonalization procedure to determine the basis function $\phi_1(t)$.

b. Use the symbol $s_2(t)$, and the Gram–Schmidt orthogonalization procedure to determine the basis function $\phi_2(t)$.

c. Use the symbol $s_3(t)$, and the Gram–Schmidt orthogonalization procedure to determine the basis function $\phi_3(t)$.

d. Show that $\phi_1(t)$, $\phi_2(t)$, and $\phi_3(t)$ arc orthonormal.

e. Plot the symbols $s_1(t)$, $s_2(t)$, $s_3(t)$, and $s_4(t)$ on a signal space diagram and indicate the energy of each symbol.

f. Present graphs for all the signal and basis functions in the time domain. Note the signals' orthogonality as in the graphs presented in this chapter.

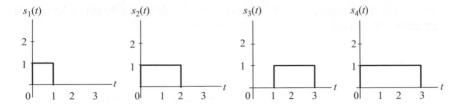

**Fig. 7** *Symbols in the time domain.*

8. Formulate the signal space diagram (signal constellation) for the binary signal modulated as shown in Figure 8.1 (where 1 is the cosine function, and 0 is the negative cosine function).

   In this case, a message system transmits two possible symbols (messages), as presented in Figure 8.2. Each symbol can be generated using a single basis function, $\phi_1(t)$.

   a. Use the symbol 1 (i.e. $s_1(t)$) and the Gram–Schmidt orthogonalization procedure to determine the basis function $\phi_1(t)$.

   b. Present the symbols $s_1(t)$ and $s_2(t)$ as vectors in a signal space diagram and calculate the energy of each symbol.

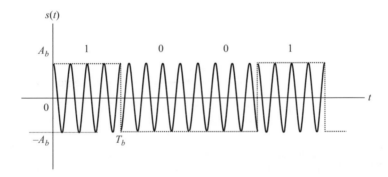

**Fig. 8.1** *Signal in the time domain.*

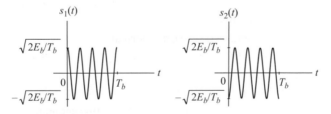

**Fig. 8.2** *Symbols in the time domain.*

9. A set of basis functions $\phi_1(t)$, $\phi_2(t)$, and $\phi_3(t)$ is shown in Figure 9.1. They are defined in the interval $T = 3$.

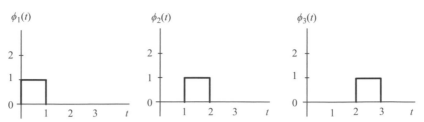

**Fig. 9.1** *Basis functions in the time domain.*

a. Show that $\phi_1(t)$, $\phi_2(t)$, and $\phi_3(t)$ are orthogonal. Are they orthonormal?

b. Express these functions as three-dimensional vectors. Present signals $\phi_1(t)$, $\phi_2(t)$, and $\phi_3(t)$ in the signal space and calculate the energy of each symbol.

c. A signal $s(t)$ is shown in Figure 9.2. Express this signal as a linear combination of the functions $\phi_1(t)$, $\phi_2(t)$, and $\phi_3(t)$. Sketch this signal as a vector in a vector space and find its energy.

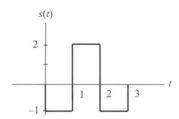

**Fig. 9.2** *Signal in the time domain.*

10. A message system can transmit four possible signals, which are presented in Figure 10. Each signal has the period $T = 3$. Each signal can be generated from a set of the orthonormal basis functions $\phi_1(t)$, $\phi_2(t)$, and $\phi_3(t)$.

a. Use signals $s_1(t)$, $s_2(t)$, and $s_3(t)$ and perform the Gram–Schmidt orthogonalization procedure to determine the basis functions $\phi_1(t)$, $\phi_2(t)$, and $\phi_3(t)$.

b. Plot the signals $s_1(t)$, $s_2(t)$, $s_3(t)$, and $s_4(t)$ on a signal space diagram and calculate the energy of each signal.

c. Present graphs for all the signal and basis functions in the time domain. Note the signals' orthogonality, as in the graphs presented in this chapter.

d. Sample the signals, generate their discrete-time forms, and repeat a., b., and c..

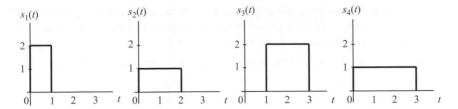

**Fig. 10** *Signals in the time domain.*

11. Formulate the signal space diagram (signal constellation) for the polar non-return-to-zero line code. This signal is shown in Figure 11.

In this case, a message system transmits two possible messages (symbols). Each symbol has the period $T_b$ and the amplitude $A_b$ and $-A_b$. Each symbol can be generated from the orthonormal basis function $\phi_1(t)$.

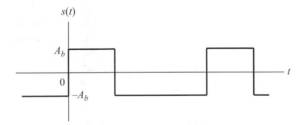

**Fig. 11** *Polar non-return-to-zero signal.*

a. Use the symbol $s_1(t)$ and the Gram–Schmidt orthogonalization procedure to determine the basis function $\phi_1(t)$.

b. Plot the symbols $s_1(t)$ and $s_2(t)$ on a signal space diagram and indicate the energy of each symbol.

12. A source generates three possible symbols, which are presented in Figure 12. Each symbol has the period $T = 4$ seconds.

a. Use the signal $s_1(t)$ and the Gram–Schmidt orthogonalization procedure to determine the basis function $\phi_1(t)$. Express the signal $s_1(t)$ as a function of the basis function and calculate its energy. (Strong suggestion: Use the calculated numbers in square root form. Do not calculate them in decimal form.)

b. Use the signal $s_2(t)$, and the Gram–Schmidt orthogonalization procedure to determine the basis function $\phi_2(t)$. Express the signal $s_2(t)$ as a function of the basis functions.

c.  Show that $\phi_1(t)$ and $\phi_2(t)$ are orthonormal. Find the coefficients $s_{31}$ and $s_{32}$. Plot the signals $s_1(t)$ and $s_2(t)$ on a signal space diagram and calculate the energy of each signal from their vector representations.

d.  Sample the signals, make their discrete-time forms, and repeat a., b., and c..

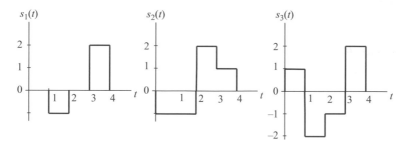

**Fig. 12** *Three signals in the time domain.*

13. A message system can transmit three possible messages represented by signals presented in Figure 13. Each signal has the period $T = 4$, and can be generated from the orthonormal basis functions, $\phi_1(t)$ and $\phi_2(t)$.

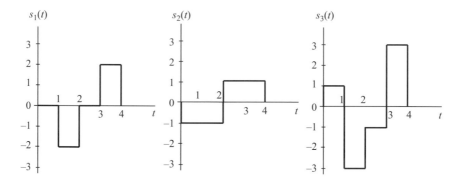

**Fig. 13** *Signals in the time domain.*

a.  Use signals $s_1(t)$ and $s_2(t)$ and the Gram–Schmidt orthogonalization procedure to determine the basis function $\phi_1(t)$ and $\phi_2(t)$. Show that $\phi_1(t)$ and $\phi_2(t)$ are orthonormal.

b.  Present all signals in *vector form*, plot them in a signal space diagram, and calculate the energy of each signal.

c. Present graphs for all the signal and basis functions in the time domain. Note the signals' orthogonality, as in the graphs presented in this chapter.

d. Suppose the signals are sampled in the time domain, and discrete-time signals are obtained. Suppose that the bandwidth of the signals does not exceed two decades of their spectrum. Present the signals in the time domain. Perform the whole procedure on the discrete-time signals as you did in a. to c. for the continuous-time signals to find the basis functions, the vector representations of all the discrete signals, and the signals' energies.

14. A set of basis functions $\phi_1(t)$, $\phi_2(t)$, and $\phi_3(t)$ is shown in Figure 14.1. They are defined in the interval $T = 4$.

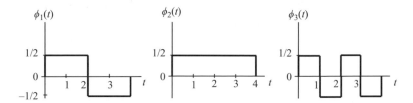

**Fig. 14.1** *Basis functions in the time domain.*

a. Show that $\phi_1(t)$, $\phi_2(t)$, and $\phi_3(t)$ are orthogonal. Are they orthonormal? Prove the answer.

b. Express these functions as three-dimensional vectors. Plot the signals $\phi_1(t)$, $\phi_2(t)$, and $\phi_3(t)$ on a signal space diagram and indicate the energy of each symbol.

c. A signal $s_1(t)$ is shown in Figure 14.2. Express this signal as a linear combination of the basis functions $\phi_1(t)$, $\phi_2(t)$, and $\phi_3(t)$.

d. Sketch signal $s_1(t)$ as a vector in a vector space.

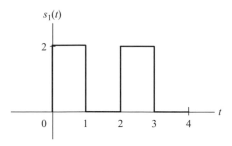

**Fig. 14.2** *Signal in the time domain.*

15. A source generates three possible symbols, as presented in Figure 15. Each symbol has the period $T = 4$ seconds.

a. Use the symbol $s_1(t)$ and the Gram–Schmidt orthogonalization procedure to determine the basis function $\phi_1(t)$. Express the signal $s_1(t)$ as a function of the basis function and calculate its energy.

b. Use the symbol $s_2(t)$ and the Gram–Schmidt orthogonalization procedure to determine the basis function $\phi_2(t)$. Express the signal $s_2(t)$ as a function of the basis functions.

c. Show that $\phi_1(t)$ and $\phi_2(t)$ are orthonormal.

d. Use the symbol $s_3(t)$ and the Gram–Schmidt orthogonalization procedure to confirm that the basis function $\phi_3(t) = 0$. Express signals $s_1(t)$, $s_2(t)$, and $s_3(t)$ as functions of the basis functions.

e. Plot the symbols $s_1(t)$, $s_2(t)$, and $s_3(t)$ on a signal space diagram and calculate the energy of each symbol from their vector representations.

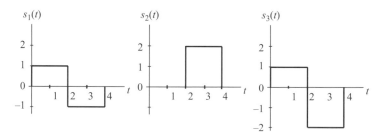

**Fig. 15** *Symbols in the time domain.*

16. Analyse an orthogonal frequency division multiplexing (OFDM) system as specified in this problem.

a. Two functions are given: $\phi_i(t)$ and $\phi_j(t)$. Suppose they are orthonormal. Explain the term 'orthonormal'. Express this orthonormality in a mathematical form.

b. In OFDM systems, a set of orthogonal carriers is used to transmit a wideband digital message signal. Suppose the two carriers are the functions

$$\phi_i(t) = \sqrt{\frac{2}{T}}\cos{(i\omega_c t)} \text{ and } \phi_j(t) = \sqrt{\frac{2}{T}}\cos{(j\omega_c t)}.$$

c. Prove that these two carriers are orthogonal. What conditions should $i$ and $j$ fulfil in that case? Suppose the functions represent the voltages on a unit resistor. Prove that the first function is normalized (with respect to energy). Suppose the basic carrier frequency is $f_c = 1$ MHz. Define a set of at least three carrier frequencies that satisfy the condition of the orthogonality of these two functions.

d. Suppose the carrier frequencies $\phi_1(t)$ and $\phi_2(t)$ are $f_{c1} = 1$ MHz and $f_{c2} = 2$ MHz. One symbol interval is equal to $T = 1/f_{c1}$. Plot the two carriers in the $T$ interval. Observe and explain their orthogonality, in the case that orthogonality exists. You can make this explanation without any calculations.

17. Suppose a message source generates three possible symbols, as presented in Figure 17. Each symbol has the period $T = 4$ seconds.

a. Use the symbol $s_1(t)$ and the Gram–Schmidt orthogonalization procedure to determine the basis function $\phi_1(t)$. Plot the graph of this basis function. Express the signal $s_1(t)$ as a function of the basis function and calculate its energy.

b. Use symbol $s_2(t)$ and the Gram–Schmidt orthogonalization procedure to determine the basis function $\phi_2(t)$. Firstly, calculate coefficient $s_{21}$ and then calculate the intermediate function. Express the signal $s_2(t)$ as a function of the basis functions.

c. Use the symbol $s_3(t)$ and the Gram–Schmidt orthogonalization procedure to calculate the third basis function $\phi_3(t)$. Express the signal $s_3(t)$ as a function of the basis functions.

d. Define the orthonormality of two signals. Prove that the signals $\phi_1(t)$ and $\phi_2(t)$ are orthonormal.

e. Present the signals $s_1(t)$, $s_2(t)$, and $s_3(t)$ in vector form. Present the obtained vectors in a signal space diagram. Calculate the energy of all symbols from their vector representations.

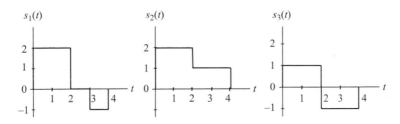

**Fig. 17** *Symbols in the time domain.*

18. An OFDM transmitter generates four subcarriers having the frequencies

$$f_i = \frac{i}{T_b}, \quad \text{for } i = 1, 2, 3, \text{and } 4,$$

where $T_b$ are the time durations of the subcarriers. All the subcarrier signals generated at the output of the transmitter have the same amplitude and can be expressed in the form

$$s_i(t) = \begin{cases} A_b \cos(2\pi f_i t) & 0 \leq t \leq T_b \\ 0 & \text{elsewhere} \end{cases}, \text{for } i = 1, 2, 3, \text{and } 4.$$

a. Find the energy of signals $s_1(t)$ and $s_2(t)$. Prove that these signals are orthogonal. Are the energies of all subcarriers equal to each other?

b. There are four orthonormal basis functions that can be used to generate the OFDM signal. Calculate the basis function for the first signal $s_1(t)$.

c. Suppose the first subcarrier has the energy $E_b = 1$ $\mu$J, and its frequency is $f_1 = 1$ MHz. Calculate the bit duration $T_b$ and then the frequencies of subcarriers $f_2, f_3$, and $f_4$. Calculate the amplitude of the first subcarrier. Sketch the graphs of all four subcarriers inside the interval of $T_b$.

d. Suppose that the subcarriers are phase modulated using binary phase-shift keying and that the two-sided power spectral density of noise in the channel is $N_0/2 = 2.5 \times 10^{-7}$ W/Hz. Calculate the probability of error in this system. Express the signal-to-noise ratio in decibels. What will be the probability of error for other subcarriers?

19. A set of three signals $\phi_1(t)$, $\phi_2(t)$, and $\phi_3(t)$ is shown in Figure 19.1. They are defined in the interval $T = 3$.

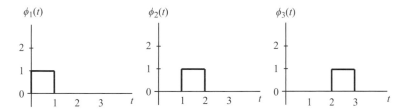

**Fig. 19.1** *Basis functions.*

a. Calculate the energies of these signals. Show that $\phi_1(t)$, $\phi_2(t)$, and $\phi_3(t)$ are orthonormal.

b. These three signals can be represented as unit vectors in the three-dimensional space. Express them in matrix form as three-dimensional vectors. Plot the signals $\phi_1(t)$, $\phi_2(t)$, and $\phi_3(t)$ on a signal space diagram and calculate the energy of each signal using this vector presentation.

c. A signal $s(t)$ is shown in Figure 19.2. Express this signal as a linear combination of the basis functions $\phi_1(t)$, $\phi_2(t)$, and $\phi_3(t)$. Sketch this signal as a vector in a vector space and find its energy using this presentation. Confirm the calculated energy value of the signal by recalculating the energy value, using the presentation of the signal in the time domain in Figure 19.2.

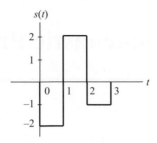

**Fig. 19.2** *Signal wave shape.*

# 3

# Discrete-Time Stochastic Processes

## 3.1 Definition and Analysis of Discrete-Time Stochastic Processes

### 3.1.1 Introduction

The early theory developed in digital signal processing was related to the class of signals that are discrete and deterministic in nature, which are called *discrete-time deterministic signals*. The discrete-time values of these signals can be found using a mathematical expression or their tabular form. An example of such a signal is a discrete-time sinusoidal signal. All the values of this signal can be calculated using a simple expression in a closed form. A *signal* is defined as a function whose values convey information about a phenomenon it represents. The function representing a signal can be of one or more independent variables, that is, the signal can be *one dimensional* or *multidimensional*. We will deal mostly with the signals that are solely a function of time, that is, those with one-dimensional signals.

A *discrete-time stochastic process* is represented by an ensemble of discrete-time stochastic or random signals, which have the random values at all time instants $n$ that cannot be predicted in advance or found using any expression, graphical presentation or table. An example of such a signal is the discrete-time noise generated in a communication system or a speech signal generated and sampled at the output of a microphone.

Most practical signals are, in some sense, random. A discrete-time stochastic process can be defined as an indexed sequence of random variables. We will characterize this process in terms of the multivariate density and distribution functions. However, this description of a stochastic process, based on the density and distribution functions, requires a lot of information about the process that is difficult to obtain in practice. Fortunately, many properties of a stochastic process can be described in terms of averages associated with its first-order and second-order densities. For this reason, we will analyse the stochastic processes in terms of ensemble averages like the mean, the variance and the correlation and covariance functions.

Because discrete random signals are assumed to be infinite in length, they do not have finite energy, and the summations of the Fourier transform and the z-transform do not converge. Therefore, we say that these transforms do not exist and these kinds of signals need to be expressed using the theory of probability, random variables and statistical analysis. For this reason, a class of stochastic processes is defined, the stationary and ergodic processes, for

*Discrete Communication Systems*. Stevan Berber, Oxford University Press. © Stevan Berber 2021.
DOI: 10.1093/oso/9780198860792.003.0003

which it is possible to define the power spectral density. The analysis of random signals based on the autocorrelation function and the power spectral density allows a relatively simple characterization of the input–output relationship of LTI systems. This relationship can be expressed in terms of input and output correlation functions. In particular, we analyse a class of stochastic processes that can be generated by filtering noise using a LTI filter.

The theory of discrete-time stochastic processes is essential for the theoretical development of discrete-time communication systems, in addition to the optimal and adaptive signal processing. For this reason, the theory presented in this chapter is relevant for the theory of discrete and digital communication systems, in particular for modulation, detection, information and coding theory.

## 3.1.2    Definition of a Stochastic Process

To obtain a formal definition of a stochastic process, let us consider a random experiment $\mathcal{E}$ with a finite or an infinite number of unpredictable outcomes $\{\zeta_i\}$ from a sample space $S = \{\zeta_1, \zeta_2, \ldots, \zeta_k, \ldots, \zeta_K\}$, where assigned probabilities to all outcomes are $p(\zeta_1), p(\zeta_2), \ldots, p(\zeta_K)$ and $K$ can take an infinite value. Also, according to a certain rule, a deterministic or a random discrete-time signal $x(n, \zeta_k)$ is assigned to each outcome $\zeta_k$, for $-\infty < n < \infty$. Thus, a discrete-time stochastic process is defined by:

- the sample space $S$, a set of outcomes or elementary events of a random experiment,
- a set of associated probabilities $P = \{p(\zeta_k)\}$ that are assumed to comply with the axioms of the probability theory and the theory of random variables, and
- an ensemble of associated discrete-time signals (sequences) $\{x(n, \zeta_k)\}$, which are realizations of the stochastic process $X(n)$ defined for $k = 1, 2, \ldots, K$, where $K$ is finite or infinite and $-\infty < n < \infty$.

According to the notation in this chapter, a discrete-time stochastic process $X(n)$ generally represents an infinite set or ensemble of discrete-time random signals that are defined for all outcomes $\zeta_k$, $k = 1, 2, \ldots, K$, and all $n$ values. Each discrete-time stochastic signal in this ensemble, denoted by $x(n, \zeta_k)$, is called a realization of the stochastic process $X(n)$.

Any value at the time instant $n = n_i$ of any discrete-time stochastic signal $x(n, \zeta_i)$, denoted by $x(n_i)$, is a realization of the random variable $X(n_i)$ that is defined at time instant $n = n_i$ for a particular outcome $\zeta_i$. In this sense, a stochastic process $X(n)$ can be treated as a series (sequence) of random variables defined for each value $n$.

In the following analysis, we will use $X(n)$ to denote the stochastic process that is defined for all values $n$ and the random variable $X(n)$ defined at the time instant $n$. A graphical description of a stochastic process, showing the process $X(n)$, the random variable $X(n_i)$ defined for $n = n_i$, samples (realizations) of the random variable $X(n_i)$ and three realizations of the stochastic process, is given in Figure 3.1.1.

Generally speaking, the future values of a process $X(n)$ cannot be predicted in advance if the present value is known. The unpredictable nature of the process has two causes. Firstly, the time structure of a process realization depends on the outcome $\zeta_i$ of a random experiment.

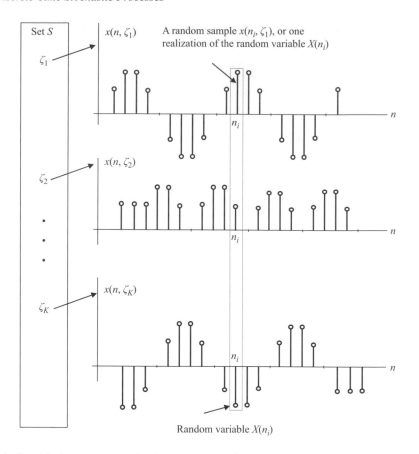

**Fig. 3.1.1** *Graphical presentation of a discrete-time stochastic process X(n).*

Secondly, there is not, generally, a functional description for all realizations of the stochastic process except for some special cases, when the process is called *predictable* or *deterministic* in the time domain. In these special cases, the realization of the stochastic process is expressed as a function of discrete time. If we cannot predict the future values of a stochastic process $X(n)$ in the time domain, the process is called a *regular process*. Most of the processes in real world are unpredictable, that is, regular processes.

As we said, we use $X(n)$ to denote the stochastic process that is defined for all values of $n$ and the random variables $X(n)$ defined at each time instant $n$. A single random value, defined at a particular time instant $n$, can be understood as a sample obtained by the ideal uniform sampling of a continuous-time stochastic process, which can have a continuous, mixed or discrete probability density function. In our analysis, we will assume that a realization of a stochastic process is represented by a discrete-time continuous-valued function that can have any real value, including infinity.

Examples of a predictable or deterministic process are samples of a sinusoidal process that are generated at the output of a sinusoidal generator in the case when the phase of the process is a random variable. This process can be expressed in the form $X(n) = A\cos(\Omega n + \theta)$, where the phase of process $\theta$ is a random variable uniformly distributed on the interval $[0, 2\pi]$. One realization of this stochastic process is a stochastic signal, expressed in the form $x_0(n) = A\cos(\Omega n + \theta_0)$, which is generated with a random phase $\theta_0$. In this example, if we know the stochastic signal sample value $x_0(n_1)$ at any time instant $n_1$, we can predict any value $x_0(n_2)$ at any future time instant $n_2$.

### 3.1.3 Mathematical Analysis of Stochastic Processes

Most of the stochastic processes of our interest can be represented as functions of time. Therefore, the simplest description of these processes is based on the analysis of these functions. This analysis results in some quantitative values that supply evidence about significant characteristics of the underlying process.

The *first* characteristic of discrete-time stochastic processes is the distribution of the random values of the process at the defined time instant $n$. Because these values are samples of a random variable $X(n)$, this characteristic can be expressed in terms of the first-order probability distribution or density function of these sample values. The *second* characteristic of the process is the degree of dependence between two sample values, which are realizations of random variables defined at those two time instants. This dependence is expressed in terms of the correlation that exists between them. The *third* characteristic of the process is its periodicity behaviour that can result in sharp peaks in its power spectrum and its autocorrelation function.

In particular, there is significant interest in the variability of some of the properties of a process, like its probability density function, mean value, variance, or spectral content. This variability can be analysed at the level of a limited number of sample values of the process $X(n)$, or on the whole process $X(n)$. In practice, these two analyses are interrelated and are used together. In the following section, these characteristics of a stochastic process will be mathematically formalized, defining the $k$th-order statistical description of the process.

At each time instant $n$, we define a random variable $X(n)$ that requires a first-order probability distribution function that is defined as

$$F_X(x_1; n_1) = P(X(n_1) \leq x_1), \tag{3.1.1}$$

and its related probability density function that is defined as a derivative of the distribution function, that is,

$$f_X(x_1; n_1) = \frac{dF_X(x_1; n_1)}{dx_1}. \tag{3.1.2}$$

for any real number $x_1$. Similarly, the random variables defined at two distinct instants $n_1$ and $n_2$, $X(n_1)$ and $X(n_2)$, are joint random variables that are statistically described by their *second-order joint probability distribution function* $F_X(x_{n1}, x_{n2}; n_1, n_2) = F_X(x_1, x_2; n_1, n_2)$:

$$F_X(x_1, x_2; n_1, n_2) = P(X(n_1) \leq x_1, X(n_2) \leq x_2), \tag{3.1.3}$$

and the corresponding second-order joint density function

$$f_X(x_1, x_2; n_1, n_2) = \frac{\partial^2 F_X(x_1, x_2; n_1, n_2)}{\partial x_1 \partial x_2}. \tag{3.1.4}$$

However, a stochastic process can contain a countable and finite number of such random variables. Therefore, they are to be statistically described by their *kth*-order joint distribution function

$$F_X(x_1, x_2, \ldots, x_k; n_1, n_1, \ldots, n_k)$$
$$= P(X(n_1) \le x_1, X(n_2) \le x_2, \ldots, X(n_k) \le x_k), \tag{3.1.5}$$

or the *k*th-order probability density function

$$f_X(x_1, x_2, \ldots, x_k; n_1, n_2, \ldots, n_k) = \frac{\partial^{2k} F_X(x_1, x_2, \ldots, x_k; n_1, n_1, \ldots, n_k)}{\partial x_{R1} \ldots \partial x_{Ik}}, \tag{3.1.6}$$

for any *k* value and assuming that the stochastic process can be complex, with both real (*R*) and imaginary (*I*) parts. This description of stochastic processes, based on the density and distribution functions, requires a lot of information that is difficult to obtain in practice. Fortunately, many properties of a stochastic process can be described in terms of averages associated with its first-order and second-order densities. In that context, it is important to define a special type of stochastic processes that can be more easily mathematically described as an *independent process*. This process is independent of the *k*th order if its joint probability density function is defined as the product of a related marginal probability density function, that is,

$$f_X(x_1, x_2, \ldots, x_k; n_1, n_2, \ldots, n_k)$$
$$= f_{X_1}(x_1; n_1) \cdot f_{X_2}(x_2; n_2) \cdot \ldots \cdot f_{X_k}(x_k; n_k), \tag{3.1.7}$$

for every *k* and $n_i$, where $i = 1, 2, \ldots, k$. In the case when all random variables have identical probability density functions $f(x)$, the process $X(n)$ is called an independent and identically distributed (iid) stochastic process.

---

**Example**

A white Gaussian noise process is an iid stochastic process $X(n)$ if each random variable $X(n)$, $-\infty < n < \infty$, is a zero-mean Gaussian random variable with the same variance $\sigma^2$. Find the *N*th-order joint probability density function of this process.

**Solution.** The probability density function of a random variable $X(n)$ which has the value $x(n)$ at time instant *n* is governed by the Gaussian distribution

$$f_X(x(n)) = f_X(x; n) = f_X(x) = \frac{1}{\sqrt{2\pi\sigma^2}} e^{-x^2/2\sigma^2}.$$

Here we can see three different notations for the density function. The last and most simplified notation will be used most of the time in this book. However, we need to keep in mind that the random variables $X(n)$ are defined at time instants $n$ and belong to the related stochastic process $X(n)$. Because $X(n)$ is an iid process, the joint ($N$th-order) density function is a product of marginal densities, that is,

$$f_X(x_1, x_2, \ldots, x_N; n_1, n_2, \ldots, n_N) = f_{X_1}(x_1; n_1) \cdot f_{X_2}(x_2; n_2) \cdot \ldots \cdot f_{X_N}(x_N; n_N)$$

$$= \prod_{n=1}^{N} f_{X_n}(x_n) = \frac{1}{\sqrt{2\pi\sigma^2}} e^{-x_1^2/2\sigma^2} \cdot \frac{1}{\sqrt{2\pi\sigma^2}} e^{-x_2^2/2\sigma^2} \cdots \frac{1}{\sqrt{2\pi\sigma^2}} e^{-x_N^2/2\sigma^2}$$

$$= \prod_{n=1}^{N} \frac{1}{\sqrt{2\pi\sigma^2}} e^{-x_n^2/2\sigma^2} = \frac{1}{\left(\sqrt{2\pi\sigma^2}\right)^N} e^{-\sum\limits_{n=1}^{N} x_n^2/2\sigma^2} = \frac{1}{(2\pi\sigma^2)^{N/2}} e^{-\frac{1}{2\sigma^2}\sum\limits_{n=1}^{N} x_n^2}.$$

## 3.2 Statistical Properties of Stochastic Processes

### 3.2.1 First-Order Statistics

As we said, for the sake of simplicity, we will sometimes use $X(n)$ to denote a random variable defined at the time instant $n$. One realization of this random variable at time instant $n$ will be denoted as $x(n)$ or just $x$. Then, for the sake of simplicity, the probability density function of this random variable is denoted as $f_X(x(n))$ or $f_X(x, n)$) or $f_X(x)$, and the probability distribution function as $F_X(x(n))$ or $F_X(x, n)$ or $F_X(x)$. The mean value of $X(n)$ is generally a function of $n$ and is expressed as

$$\eta_X(n) = E\{X(n)\} = \int_{-\infty}^{\infty} x(n) f_X(x(n)) \, dx. \tag{3.2.1}$$

The mean square value is

$$E\{X^2(n)\} = \int_{-\infty}^{\infty} x^2(n) f_X(x(n)) \, dx, \tag{3.2.2}$$

and the variance is

$$\sigma_X^2(n) = E\left\{[X(n) - \eta_X(n)]^2\right\} = E\{X^2(n)\} - \eta_X^2(n). \tag{3.2.3}$$

Generally speaking, the mean and the variance depend on the distribution function of the random variable $X(n)$ defined for each discrete-time instant $n$. Therefore, these values for the mean and the variance are the first-order statistics of the process and can be expressed as a function of $n$. Therefore, they can be represented as a sequence of values.

The mathematical expectation $E\{\}$ is a linear operator. The expectation of a function $Y(n)$ of one random variable, that is, $Y(n) = g(X(n))$, is defined as

$$\eta_Y(n) = E\{Y(n)\} = \int_{-\infty}^{\infty} g(x(n)) f_X(x(n)) \, dx. \tag{3.2.4}$$

The mean of a function of $k$ random variables $Y(\mathbf{X}) = g(X_1, X_2, \ldots, X_k)$ is

$$\eta_Y(n) = E\{Y(\mathbf{X})\} = E\{g(X_1, X_2, \ldots, X_k)\}$$
$$= \int_{-\infty}^{\infty} \cdots \int_{-\infty}^{\infty} g(x_1, x_2, \ldots, x_k) f_X(x_1, x_2, \ldots, x_k; n_1, n_2, \ldots, n_k) \, dx_1 \ldots dx_k. \tag{3.2.5}$$

Thus, we can find the expectation of a random function using the density function of the argument but not that of the function. The bold font is used to represent a vector of $k$ random variables, that is, $\mathbf{X}$ is a vector of $k$ variables $\mathbf{X} = (X_1, X_2, \ldots, X_k)$.

**Example**

Find the mean of a stochastic process that is defined as the sinusoidal process

$$X(n) = A\cos(\Omega n + \theta),$$

where the phase $\theta$ of the process is a random variable uniformly distributed in the interval $[0, 2\pi]$. Note that one realization of this process is a stochastic signal represented by a discrete-time series of deterministic values at the discrete-time instants $n$, as shown in Figure 3.2.1 for the first realization having the phase $\theta_0$.

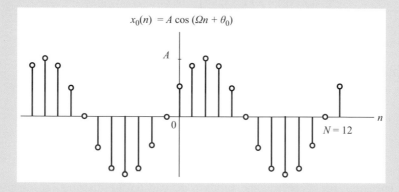

**Fig. 3.2.1**   *One realization of the stochastic process.*

**Solution.** In this case, the random variable $X(n)$, defined for any $n$, is a function of the random variable $\theta$, and its mean for a defined value of $n$ is

$$\eta_X(n) = \int_{-\infty}^{\infty} x(n) f_X(x(n)) \, dx = \int_{-\infty}^{\infty} x(n,\theta) f_\theta(\theta,n) \, d\theta$$

$$= \int_0^{2\pi} A\cos(\Omega n + \theta) \frac{1}{2\pi} d\theta$$

$$= \frac{A}{2\pi} \left[ \int_0^{2\pi} \cos\Omega n \cos\theta \, d\theta - \int_0^{2\pi} \sin\Omega n \sin\theta \, d\theta \right] = 0,$$

because the integrals of the sine and cosine functions of $\theta$ are both zero. Therefore, the solution is a discrete function of $n$ that has zero mean for every $n$.

**Example**

Find the mean of a stochastic process defined as a complex sinusoidal process expressed as

$$X(n) = Ae^{j(\Omega n + \theta)},$$

where the phase $\theta$ of the process is a random variable uniformly distributed in the interval $[0, 2\pi]$. Note that one realization of the process is composed of two deterministic discrete series of cosine and sine values which are the real and imaginary parts of the stochastic signal.

**Solution.** In this case, the random variable $X(n)$, defined for any $n$, is a function of the random variable $\theta$, and its mean, for a defined value of $n$, is

$$\eta_X(n) = \int_{-\infty}^{\infty} x(n) f_X(x(n)) \, dx = \int_{-\infty}^{\infty} x(n,\theta) f_\theta(\theta,n) \, d\theta$$

$$= \int_0^{2\pi} Ae^{j(\Omega n + \theta)} \frac{1}{2\pi} d\theta$$

$$= \frac{A}{2\pi} \left[ \int_0^{2\pi} \cos(\Omega n + \theta) \, d\theta + j \int_0^{2\pi} \sin(\Omega n + \theta) \, d\theta \right] = 0,$$

because the integrals of the sine and cosine functions of $\theta$ are both zero. Therefore, the mean value of this stochastic process is a discrete-time series of zeros for both the real and the imaginary parts.

## 3.2.2 Second-Order Statistics

In order to compare the similarities and dissimilarities of stochastic processes, the correlation (cross-correlation and autocorrelation) and covariance (cross-covariance and autocovariance) functions are used. In other words, these functions provide the degree of linear dependence inside a particular process or between two stochastic processes. In order to investigate the behaviour of the same stochastic process, the autocorrelation and autocovariance functions are used.

The *autocorrelation function* is defined as the mathematical expectation of the product of two complex random variables defined at time instants $m$ and $n$ of the same stochastic process $X(n)$, that is,

$$R_{XX}(m,n) = E\left\{X(m)X^*(n)\right\} = \int_{-\infty}^{\infty}\int_{-\infty}^{\infty} x_m x_n^* f_X\left(x_m, x_n^*\right) dx_m dx_n. \tag{3.2.6}$$

This function provides evidence about the statistical relations of the samples of a discrete-time stochastic process $X(n)$ at two different time instants $m$ and $n$. The function $f_X\left(x_m, x_n^*\right)$ is the joint probability density function of random variables $X(m)$ and $X^*(n)$.

For the process $X(n)$, we can also define the *autocovariance function* as

$$C_{XX}(m,n) = E\left\{\left[X(m) - \eta_{X(m)}\right]\left[X(n) - \eta_{X(n)}\right]^*\right\} = R_{XX}(m,n) - \eta_{X(m)}\eta_{X(n)}^*. \tag{3.2.7}$$

Generally speaking, the mean values for random variables at time instants $m$ and $n$ can be different, meaning that the mean value can be a function of time $n$.

The autocorrelation function can be represented as a function of two discrete variables $m$ and $n$, and expressed in the form of a symmetric $M \times M$ matrix with the values of the autocorrelation function obtained for any $n$ and $m$ in the interval of $M$ values, expressed in a transpose matrix form and the related complex conjugate matrix as

$$\mathbf{X} = [X(0) \; X(1) \; \ldots \; X(M-1)]^T,$$
$$\mathbf{X}^H = \left[ \; X^*(0) \quad X^*(1) \quad \ldots \quad X^*(M-1) \; \right]. \tag{3.2.8}$$

Keeping in mind the symmetry property of the autocorrelation, that is, $R_{XX}(l) = R_{XX}^*(-l)$, the autocorrelation can be calculated as

$$\mathbf{R}_{XX} = E\left\{\mathbf{X} \cdot \mathbf{X}^H\right\}$$

$$= \begin{bmatrix} X(0)X^*(0) & X(0)X^*(1) & \ldots & X(0)X^*(M-1) \\ X(1)X^*(0) & X(1)X^*(1) & \cdots & X(1)X^*(M-1) \\ \vdots & \vdots & \vdots & \vdots \\ X(M-1)X^*(0) & X(M-1)X^*(1) & \cdots & X(M-1)X^*(M-1) \end{bmatrix}$$

$$= \begin{bmatrix} R_{XX}(0) & R_{XX}(-1) & \ldots & R_{XX}(-M+1) \\ R_{XX}(1) & R_{XX}(0) & \cdots & R_{XX}(-M+2) \\ \vdots & \vdots & \vdots & \vdots \\ R_{XX}(M-1) & R_{XX}(M-2) & \cdots & R_{XX}(0) \end{bmatrix}$$

$$= \begin{bmatrix} R_{XX}(0) & R_{XX}^*(1) & \ldots & R_{XX}^*(M-1) \\ R_{XX}(1) & R_{XX}(0) & \cdots & R_{XX}^*(M-2) \\ \vdots & \vdots & \vdots & \vdots \\ R_{XX}(M-1) & R_{XX}(M-2) & \cdots & R_{XX}(0) \end{bmatrix}. \tag{3.2.9}$$

The matrix is called the *autocorrelation matrix of order M*. A symmetric $M \times M$ matrix with the values of the covariance function $\mathbf{C}_{XX}$ that can be obtained for any $n$ and $m$ is called the covariance matrix of order $M$. It can be calculated from autocorrelation matrices as

$$\mathbf{C}_{XX} = \mathbf{R}_{XX} - \boldsymbol{\eta}_{XX} \boldsymbol{\eta}_{XX}^H,$$ 

(3.2.10)

where $\boldsymbol{\eta}$ is a vector of $M$ mean values:

$$\boldsymbol{\eta}_X = [\eta_0 \quad \eta_1 \quad \ldots \quad \eta_{M-1}].$$ 

(3.2.11)

If a stochastic process $X(n)$ is represented as a vector $\mathbf{X}(n)$ of $M$ real random variables, the autocorrelation matrix can be expressed as

$$\mathbf{R}_{XX} = E\left\{\mathbf{X}(n) \cdot \mathbf{X}^T(n)\right\}.$$ 

(3.2.12)

The covariance function can be expressed in matrix form as

$$\begin{aligned} \mathbf{C}_{XX} &= \mathbf{R}_{XX} - \boldsymbol{\eta}_{XX} \boldsymbol{\eta}_{XX}^T = E\left\{\mathbf{X}(n) \cdot \mathbf{X}^T(n)\right\} - \boldsymbol{\eta}_{XX} \boldsymbol{\eta}_{XX}^T \\ &= E\left\{[\mathbf{X}(n) - \boldsymbol{\eta}_{XX}] \cdot [\mathbf{X}(n) - \boldsymbol{\eta}_{XX}]^T\right\}. \end{aligned}$$ 

(3.2.13)

In order to compare the similarities and dissimilarities of two stochastic processes, the cross-correlation and cross-covariance functions are used. In other words, these functions provide the degree of linear dependence between two stochastic processes. The *cross-correlation function* is defined for two stochastic processes $X(n)$ and $Y(n)$ as the mathematical expectation of the product of two different random variables $X(m)$ and $Y(n)$ expressed as

$$R_{XY}(m,n) = E\left\{X(m)Y^*(n)\right\} = \int_{-\infty}^{\infty} \int_{-\infty}^{\infty} x_m y_n^* f_{XY}\left(x_m, y_n^*\right) dx_m dy_n,$$ 

(3.2.14)

and represents the correlation between two different random variables of two discrete-time stochastic processes $X(m)$ and $Y(n)$. The function $f_{XY}( \ )$ is the joint probability density function of random variables $X(m)$ and $Y(n)$. The *cross-covariance function* is defined as

$$C_{XY}(m,n) = E\left\{[X(m) - \eta_X(m)][Y(n) - \eta_Y(n)]^*\right\} = R_{XY}(m,n) - \eta_X \eta_Y^*(n).$$ 

(3.2.15)

The first- and second-order statistics will not change if the process is shifted in the time domain. The autocorrelation, autocovariance, cross-corelation, and cross-covariance are two-dimensional sequences.

## Example

Find the autocorrelation function of a stochastic process that is defined as

$$X(n) = A\cos(\Omega\, n + \theta).$$

The phase $\theta$ is a random variable uniformly distributed in the interval from 0 to $2\pi$. This process is an ensemble of a discrete series of deterministic values with a random initial phase and a period $N$. The process is represented by an ensemble of its realizations as shown in Figure 3.2.2 for different random phases $\theta$. In this case, we say that the number of phases is infinite and uncountable.

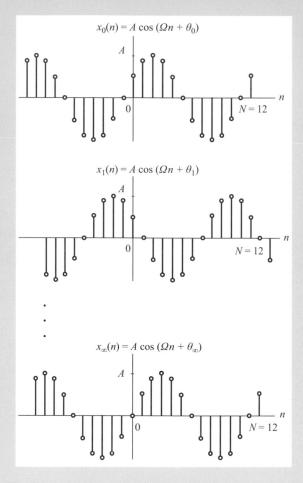

**Fig. 3.2.2**   *Ensemble of the realizations of the stochastic process.*

**Solution.** We may use the general expression (3.2.14) for the autocorrelation function calculation, that is,

$$
\begin{aligned}
R_{XX}(m,n) &= E\left\{X(m)X^*(n)\right\} \\
&= \int_{-\infty}^{\infty}\int_{-\infty}^{\infty} x(m,\theta)x^*(n,\theta)f_{\theta,\theta}(\theta,m,\theta,n)\,d\theta\,d\theta \\
&= \int_{0}^{2\pi}\int_{0}^{2\pi} A\cos(\Omega m+\theta)A\cos(\Omega n+\theta)\frac{1}{2\pi}\frac{1}{2\pi}\,d\theta\,d\theta \\
&= \frac{A^2}{8\pi^2}\int_{0}^{2\pi} 2\pi\cos\Omega\,(m-n)\,d\theta = \frac{A^2}{2}\cos\Omega\,(m-n) = \frac{A^2}{2}\cos\Omega l.
\end{aligned}
$$

In this problem, the random variable of the stochastic process is a phase angle, and the product inside the expectation operator $E$ can be treated as a function of the phase angle $\theta$. Therefore, the autocorrelation function can also be calculated as

$$
\begin{aligned}
R_{XX}(m,n) &= E\left\{X(m)X^*(n)\right\} = \int_{-\infty}^{\infty} x(m,\theta)x^*(n,\theta)f_\theta(\theta)\,d\theta \\
&= \int_{0}^{2\pi} A\cos(\Omega\,m+\theta)A\cos(\Omega\,n+\theta)\frac{1}{2\pi}\,d\theta \\
&= \frac{A^2}{2\pi}\int_{0}^{2\pi}\frac{1}{2}\left[\cos(\Omega\,m+\Omega\,n+2\theta)+\cos(\Omega\,m-\Omega\,n)\right]d\theta \\
&= \frac{A^2}{4\pi}\cos\Omega\,(m-n)\int_{0}^{2\pi} d\theta + \frac{A^2}{4\pi}\int_{0}^{2\pi}\cos(\Omega\,m+\Omega\,n+2\theta)\,d\theta \\
&= \frac{A^2}{4\pi}2\pi\cos\Omega\,(m-n)+0 = \frac{A^2}{2}\cos\Omega\,(m-n).
\end{aligned}
$$

The autocorrelation function depends only on the time difference $l=(m-n)$ between the samples of the stochastic process, that is,

$$
R_{XX}(l) = R_{XX}(m-n) = \frac{A^2}{2}\cos\Omega\,l.
$$

A close look at the presented stochastic process $X(n)$ in Figure 3.2.2. at the time instants $m$ and $n$, represented by infinite uncountable number of realizations (an infinite ensemble), reveals an imaginable infinite number of amplitudes that will, in the sum, result in zero value, which intuitively supports the theoretical calculation of the mean. If $m = n$, the autocorrelation function has a maximum value equal to the power of the process. This is intuitively clear and imaginable. Namely, in this case, all corresponding realizations accross the ensemble for each pair of the two random variables are the same. Therefore, their multiplications produce squared values, which, when added to each other, result in the power of the process. This positive maximum value of the correlation repeats after one period of the correlation function. It has a maximun negative value at half of the period of the correlation function because, in this case, the values of the magnitudes at both time instants are the same but of opposite sign. Thus, the autocorrelation attains again the maximum value, but a negative one.

Note that the random phases $\theta$ at time instants $m$ and $n$ are assumed to be identical and statistically independent random variables, that is, they are iid variables. The cosine stochastic process is widely used in communication systems. It represents the carrier of the system in the case when carrier phase has a random value. The calculated autocorrelation function is used to calculate the power spectral density of the process using the Wiener–Khintchine theorem.

## Example

In practice, a discrete cosine stochastic process is generated using a finite number of samples $N$ in one period. Therefore, the number of discrete phases different from zero is finite and equal to $N$. Following the presentation in Figure 3.2.2, the number of samples is $n = 12$ and, consequently, we can have 12 distinct phases. Following the definition of the related stochastic process, these 12 phases are the outcomes of the random experiment that generates these random phases. The possible phases form the set of elementary outcomes $S$. For this random experiment, find the autocorrelation function of the stochastic process that is defined as

$$X(n) = A\cos(\Omega\, n + \theta),$$

where the phases $\theta$ belong to a discrete random variable having the discrete uniform distribution defined in the interval from 0 to $2\pi$. This process itself then is an ensemble of discrete cosine functions with random discrete initial phases and a period of $N = 12$. The process can be represented by an ensemble of its realizations, as shown in Figure 3.2.2 for different discrete random phases $\theta$. In this case, we say that the number of phases is finite and countable. Assume that the uniform density is expressed in terms of the Dirac delta functions expressed as

$$
\begin{aligned}
f_d(\theta) &= \frac{1}{N} \sum_{n=-N/2}^{n=N/2-1} \delta(\theta - 2\pi n/N) \\
&= \frac{1}{N} [\delta(\theta + 2\pi N/2N) + \cdots + \delta(\theta) + \cdots + \delta(\theta - 2\pi(N/2-1)/N)] \\
&= \frac{1}{N} [\delta(\theta + \pi) + \cdots + \delta(\theta) + \cdots + \delta(\theta - \pi + 2\pi/N)],
\end{aligned}
$$

where the discrete phases are represented by the weights of the Dirac delta functions.

**Solution.** In this case, the process $X(n)$ is a function of the continuous random variable $\theta$, which is the phase of the process. The phases are different from zero only at 12 discrete phase instants defined by the delta functions and having the values $1/N$. The mean value of the process is zero because the sum of the cosine is zero, that is,

$$\eta(m) = E\{X(m,\theta)\} = \int_{-\infty}^{\infty} x(m,\theta) f_d(\theta)\, d\theta = \int_{-\infty}^{\infty} \left[ x(m,\theta) \sum_{n=-N/2}^{n=N/2-1} \frac{1}{N} \cdot \delta(\theta - 2\pi n/N) \right] d\theta$$

$$= \frac{A}{N} \sum_{n=-N/2}^{n=N/2-1} \int_{-\infty}^{\infty} \cos(\Omega m + \theta)\, \delta(\theta - 2\pi n/N)\, d\theta = \frac{A}{N} \sum_{n=-N/2}^{n=N/2-1} \cos(\Omega m + 2\pi n/N) = 0.$$

In the expression for the autocorrelation function, the phase angle and the product inside the expectation operator $E$ can be treated as a function of the phase angle $\theta$. Therefore, the autocorrelation function can also be calculated as

$$R_X(m,l) = E\{X(m,\theta) X(l,\theta)\} = \int_{-\infty}^{\infty} X(m,\theta) X(l,\theta) f_d(\theta)\, d\theta$$

$$= \int_{-\infty}^{\infty} \left[ A\cos(\Omega\, m + \theta) A\cos(\Omega\, l + \theta) \sum_{n=-N/2}^{n=N/2-1} \frac{1}{N} \cdot \delta(\theta - 2\pi n/N) \right] d\theta$$

$$= \frac{A^2}{2N} \sum_{n=-N/2}^{n=N/2-1} \int_{-\infty}^{\infty} \left[ \cos\Omega\, (m-l) + \cos[\Omega\, (m+l) + 2\theta] \right] \delta(\theta - 2\pi n/N)\, d\theta$$

$$= \frac{A^2}{2N} \cos\Omega\, (m-l) \sum_{n=-N/2}^{n=N/2-1} \int_{-\infty}^{\infty} \delta(\theta - 2\pi n/N)\, d\theta$$

$$+ \frac{A^2}{2N} \sum_{n=-N/2}^{n=N/2-1} \int_{-\infty}^{\infty} \cos[\Omega\, (m+l) + 2\theta] \delta(\theta - 2\pi n/N)\, d\theta$$

$$= \frac{A^2}{2N} \cos\Omega\, (m-l) \sum_{n=-N/2}^{n=N/2-1} 1 + \frac{A^2}{2N} \sum_{n=-N/2}^{n=N/2-1} \int_{-\infty}^{\infty} \cos[\Omega\, (m+l) + 4\pi n/N]$$

$$= \frac{A^2}{2} \cos\Omega\, (m-l).$$

**Example**

Find the autocorrelation function of a stochastic process that is defined as

$$X(n) = A\cos(\Omega\, n + \theta_n),$$

where the phases $\theta_n$ belong to a discrete random variable having the discrete uniform distribution defined in the interval from 0 to $2\pi$. Suppose the uniform discrete probability density function is expressed in terms of the Kronecker delta functions as

$$f_d(\theta_n) = \frac{1}{N} \sum_{n=-N/2}^{n=N/2-1} \delta(\theta_n - 2\pi n/N)$$

$$= \frac{1}{N}[\delta(\theta_n + 2\pi N/2N) + \cdots + \delta(\theta_n) + \cdots + \delta(\theta_n - 2\pi\,(N/2-1)/N)]$$

$$= \frac{1}{N}[\delta(\theta_n + \pi) + \cdots + \delta(\theta_n) + \cdots + \delta(\theta_n - \pi + 2\pi/N)].$$

**Solution.** For the uniform density expressed in terms of the Kronecker delta functions, the autocorrelation can be calculated as the sum

$$R_{XX}(m,k) = E\{X(m,\theta_n)X(k,\theta_n)\} = \sum_{n=-\infty}^{n=\infty} X(m,\theta_n)X(k,\theta_n)f_d(\theta_n)$$

$$= \frac{A^2}{2N} \sum_{n=-\infty}^{n=\infty} \left\{ [\cos\Omega\,(m-k) + \cos(\Omega\,(m+k)\,\theta_n)] \sum_{n=-N/2}^{n=N/2-1} \delta(\theta_n - 2\pi n/N) \right\}$$

$$= \frac{A^2}{2N} \sum_{n=-N/2}^{n=N/2-1} \sum_{n=-\infty}^{n=\infty} [\cos\Omega\,(m-k) + \cos(\Omega\,(m+k) + 2\theta_n)]\delta(\theta_n - 2\pi n/N)$$

$$= \frac{A^2}{2N}[\cos\Omega\,(m-k)] \sum_{n=-N/2}^{n=N/2-1} \sum_{n=-\infty}^{n=\infty} \delta(\theta_n - 2\pi n/N)$$

$$+ \frac{A^2}{2N} \sum_{n=-N/2}^{n=N/2-1} \sum_{n=-\infty}^{n=\infty} \cos(\Omega\,(m+k) + 2\theta_n)\,\delta(\theta_n - 2\pi n/N)$$

$$= \frac{A^2}{2N}[\cos\Omega\,(m-k)] \sum_{n=-N/2}^{n=N/2-1} 1 + \frac{A^2}{2N} \sum_{n=-N/2}^{n=N/2-1} \cos(\Omega\,(m+k) + 4\pi n/N)$$

$$= \frac{A^2}{2}[\cos\Omega\,(m-k)] + 0 = \frac{A^2}{2}\cos\Omega\,(m-k) = \frac{A^2}{2}\cos\Omega l = R_{XX}(l).$$

As expected from the previous solution, the autocorrelation function depends only on the time difference $l = (m - k)$ between the samples of the stochastic process.

## Example

In discrete communciation systems practice, the carrier is represented by the orthonormal bases functions that can be understood as the deterministic periodic signal

$$s_c(n) = \sqrt{2/N_b} \cos(\Omega_c n)$$

or as a stochastic cosine process with a discrete random phase

$$S_c(n) = \sqrt{2/N_b} \cos(\Omega_c n + \theta).$$

Following the results in previous examples, find the autocorrelation function of the carriers.

**Solution.** Considering the carrier as a periodic deterministic cosine function, we can calculate its autocorrelation function as

$$R_{sc}(l) = \frac{1}{N_b} \sum_{n=0}^{n=N_b-1} \left[ \sqrt{\frac{2}{N_b}} \cos \Omega_c (n+l) \cdot \sqrt{\frac{2}{N_b}} \cos \Omega_c n \right]$$

$$= \frac{1}{N_b} \frac{2}{N_b} \sum_{n=0}^{n=N_b-1} \frac{1}{2} [\cos \Omega_c l + \cos \Omega_c (2n+l)] = \frac{1}{N_b} \frac{1}{N_b} \sum_{n=0}^{n=N_b-1} \cos \Omega_c l + 0 = \frac{1}{N_b} \cos \Omega_c l.$$

The Fourier transform of the autocorrelation function will give us the power spectral density expressed as

$$S_{sc}(\Omega) = \frac{1}{N_b} \pi [\delta(\Omega - \Omega_c) + \delta(\Omega + \Omega_c)].$$

We can use directly any of the above obtained expression for the autocorrelation function expressed as $R_{XX}(l) = \left(A^2/2\right) \cos \Omega \, l$. By replacing the amplitude $A$ with the defined amplitude for the basis function expressed as $\sqrt{2/N_b}$, we can get the autocorrelation function of the stochastic carrier expressed as for the periodic carrier. If the constant term $1/N_b$ in front of the correlation integral is excluded, the obtained autocorrelation function is just a cosine function. The Fourier transform of this function will give us the energy spectral density expressed as

$$E_{sc}(\Omega) = \pi [\delta(\Omega - \Omega_c) + \delta(\Omega + \Omega_c)].$$

If we divide the energy spectral density by the number of samples $N_b$, we will get the power spectral density in the form in which it has already been calculated. Precisely speaking, if the carrier is treated as a periodic signal, then the Fourier transform of its autocorrelation function is the energy spectral density, which can be used to derive the related power spectral density function.

### 3.2.3 Higher-Order Statistics

*The third-order moments* of a stochastic process $X(n)$ can be defined about its mean as

$$C_{XX}^3(m,n,k) = E\{[X(m) - \eta_X(m)][X(n) - \eta_X(n)][X(k) - \eta_X(k)]\}, \qquad (3.2.16)$$

or about its origin as

$$R_{XX}^3(m,n,k) = E\{X(m)X(n)X(k)\}. \qquad (3.2.17)$$

These moments can be defined about its mean *for any order M* as

$$C_{XX}^M(n_1, n_2, \ldots, n_M)$$
$$= E\{[X(n_1) - \eta_X(n_1)][X(n_2) - \eta_X(n_2)] \cdots [X(n_M) - \eta_X(n_M)]\}, \qquad (3.2.18)$$

and about the origin as

$$R_{XX}^M(n_1, n_2, \ldots, n_M) = E\{[X(n_1)X(n_2) \cdots X(n_M)]\}. \qquad (3.2.19)$$

### 3.2.4 Types of Discrete-Time Stochastic Processes

We can define multiple random variables on the same sample set $S$ and then perform their theoretical analysis. This section is dedicated to the analysis of one or multiple processes.

**Analysis of one stochastic process.** A stochastic process $X(n)$ is an *independent process* if its joint probability density function is defined as

$$f_X(x_1, x_2, \ldots, x_k; n_1, n_1, \ldots, n_k) = f_1(x_1; n_1) \cdot f_2(x_2; n_2) \cdot \ldots \cdot f_k(x_k; n_k), \qquad (3.2.20)$$

for every $k$ and $n_i$, where $i = 1, 2, \ldots, k$. The mean value of an independent random process depends on $n$ and can be different from zero and expressed as $\eta_X(n) = E\{X(n)\}$. Furthermore, for an independent process, this condition holds for its autocovariance

$$
\begin{aligned}
C_{XX}(m,n) &= E\{X(m) - \eta_X(m)\} \cdot E\left\{X^*(n) - \eta_X^*(n)\right\} \\
&= E\{X(m)\} \cdot E\left\{X^*(n)\right\} - \eta_X(m)E\left\{X^*(n)\right\} - \eta_X^*(n)E\{X(m)\} + \eta_X(m)\eta_X^*(n) \\
&= E\{X(m)\} \cdot E\left\{X^*(n)\right\} - \eta_X(m)\eta_X^*(n) \\
&= \begin{cases} E\left\{X^2(m)\right\} - \eta_X^2(m) & m = n \\ E\{X(m)\} \cdot E\left\{X^*(n)\right\} - \eta_X(m)\eta_X^*(n) & m \neq n \end{cases} \\
&= \begin{cases} \sigma_X^2(m) & m = n \\ 0 & m \neq n \end{cases} = \sigma_X^2(m)\delta(m-n).
\end{aligned} \qquad (3.2.21)
$$

Therefore, the covariance is zero everywhere except at the point $n = m$. The autocorrelation of an independent process is

$$R_{XX}(m,n) = E\{X(m) \cdot X(n)\}$$

$$= \begin{cases} E\{X^2(m)\} & m = n \\ E\{X(m)\} \cdot E\{X(n)\} & m \neq n \end{cases} = \begin{cases} \sigma_X^2(m) + |\eta_X(m)|^2 & m = n \\ \eta_X(m)\eta_X^*(n) & m \neq n \end{cases}$$

$$= \begin{cases} \sigma_X^2(m) + |\eta_X(m)|^2 & m = n \\ 0 + \eta_X(m)\eta_X^*(n) & m \neq n \end{cases} = C_X(m,n) + \eta_X(m)\eta_X^*(n). \tag{3.2.22}$$

In the case when all random variables, defined for every $n$, have the same probability density function $f(x)$, with both the mean $\eta_X$ and the variance $\sigma_X^2$ constant, the process $X(n)$ is an iid process. For this process, we can find the first- and second-order statistics as follows. The mean value of an iid random process can be different from zero, does not depend on time $n$, and can be expressed as

$$\eta_X = E\{X(n)\}. \tag{3.2.23}$$

For an iid process, the following condition holds for its covariance:

$$C_{XX}(m,n) = E\{X(m) - \eta_X\} \cdot E\left\{X^*(n) - \eta_X^*\right\}$$

$$= E\{X(m)\} \cdot E\left\{X^*(n)\right\} - \eta_X E\left\{X^*(n)\right\} - \eta_X^* E\{X(m)\} + \eta_X \eta_X^*$$

$$= E\{X(m)\} \cdot E\left\{X^*(n)\right\} - \eta_X \eta_X^*$$

$$= \begin{cases} E\{X^2(m)\} - \eta_X^2 & m = n \\ E\{X(m)\} \cdot E\{X^*(n)\} - \eta_X \eta_X^* & m \neq n \end{cases}$$

$$= \begin{cases} \sigma_X^2(m) & m = n \\ 0 & m \neq n \end{cases} = \sigma_X^2(m)\delta(m-n). \tag{3.2.24}$$

The autocorrelation of an iid process is

$$R_{XX}(m,n) = E\left\{X(m) \cdot X^*(n)\right\}$$

$$= \begin{cases} E\{X^2(m)\} & m = n \\ E\{X(m)\} \cdot E\{X^*(n)\} & m \neq n \end{cases} = \begin{cases} \sigma_X^2 + |\eta_X|^2 & m = n \\ |\eta_X|^2 & m \neq n \end{cases}$$

$$= \begin{cases} \sigma_X^2 + |\eta_X|^2 & m = n \\ 0 + |\eta_X|^2 & m \neq n \end{cases} = C_X(m,n) + |\eta_X|^2. \tag{3.2.25}$$

A stochastic process $X(n)$ is an *uncorrelated stochastic process* if it is a sequence of uncorrelated random variables, that is, the following condition holds:

$$E\left\{X(m)X^*(n)\right\} = \begin{cases} 0 & \eta_X(m) = \eta_X^*(n) = 0 \\ \eta_X(m)\eta_X^*(n) & \eta_X(m) \neq 0, \eta_X^*(n) \neq 0 \end{cases}. \tag{3.2.26}$$

Therefore, its covariance is

$$C_{XX}(m,n) = \left\{ \begin{array}{cc} \sigma_X^2(m) & m=n \\ 0 & m \neq n \end{array} \right\} = \sigma_X^2(m)\delta(m-n), \qquad (3.2.27)$$

and the autocorrelation is

$$R_{XX}(m,n) = \left\{ \begin{array}{cc} \sigma_X^2(m) + |\eta_X(m)|^2 & m=n \\ \eta_X(m)\eta_X^*(m) & m \neq n \end{array} \right\}. \qquad (3.2.28)$$

The independent variables are also uncorrelated. The uncorrelated variables are not necessarily independent. However, this is true for Gaussian random variables because, if two Gaussian variables are uncorrelated, they are also independent.

A stochastic process $X(n)$ is said to be an *orthogonal stochastic process* if it is a sequence of orthogonal random variables, that is, its autocorrelation is

$$R_{XX}(m,n) = \left\{ \begin{array}{cc} \sigma_X^2(m) + |\eta_X(m)|^2 & m=n \\ 0 & m \neq n \end{array} \right\} = E\left\{|X(m)|^2\right\}\delta(m-n), \qquad (3.2.29)$$

which is the consequence of the orthogonality condition for two random variables, expressed as

$$E\{X(m)X(n)\} = 0. \qquad (3.2.30)$$

The theory of communication systems is based primarily on the orthogonality principle. That is the reason this book started with the orthogonal functions in Chapter 2 and returns in this chapter to the principle of orthogonality, which will be extensively used in the later chapters.

**Analysis of two stochastic processes.** Suppose two processes $X(n)$ and $Y(n)$ are defined on the same sample space $S$. We can apply the above definitions for these two joint stochastic processes. Two stochastic processes $X(n)$ and $Y(n)$ are *statistically independent* if, for all values of $m$ and $n$, their joint density function is

$$f_{XY}(x,y;m,n) = f_X(x;m) \cdot f_Y(y;n). \qquad (3.2.31)$$

Stochastic processes $X(n)$ and $Y(n)$ are *uncorrelated stochastic processes* if, for every $m$ and $n$, their covariance is

$$C_{XY}(m,n) = 0, \qquad (3.2.32)$$

and the autocorrelation function is

$$R_{XY}(m,n) = \eta_X(m)\eta_Y^*(n). \qquad (3.2.33)$$

Stochastic processes $X(n)$ and $Y(n)$ are *orthogonal* if, for every $m$ and $n$, their autocorrelation function is

$$R_{XY}(m,n) = 0. \tag{3.2.34}$$

**Power of a stochastic process $X(n)$.** For a deterministic aperiodic signal $x(n)$, the average power is defined as

$$P_X = \lim_{N \to \infty} \frac{1}{2N+1} \sum_{n=-N}^{N} x^2(n). \tag{3.2.35}$$

Similarly, taking into account the random nature of a stochastic process, the power of the process is defined as a limit of the sum of the mean square values, according to the following expression:

$$P_X = \lim_{N \to \infty} \frac{1}{2N+1} \sum_{n=-N}^{N} E\left\{|X(n)|^2\right\}. \tag{3.2.36}$$

In the case when these mean square values are constant, for each value of the time variable $n$, this expression becomes

$$P_X = \lim_{N \to \infty} \frac{1}{2N+1} \sum_{n=-N}^{N} E\left\{|X(n)|^2\right\} = E\left\{|X(n)|^2\right\}, \tag{3.2.37}$$

because there are $(2N + 1)$ terms inside the summation sign that are of equal value.

## 3.3 The Stationarity of Discrete-Time Stochastic Processes

### 3.3.1 The Stationarity of One Discrete-Time Stochastic Process

**Stationary processes.** A stochastic process $X(n)$ is stationary if its statistical characteristics are equal to the statistical characteristics of its time-shifted version $X(n + k)$, for every $k$. In other words, we say that the process $X(n)$ is stationary if its properties are invariant with respect to a time-shift. In particular, a stochastic process $X(n)$ is called *stationary of order N* if its joint probability density function fulfils the following condition:

$$\begin{aligned} &f_X(x_1, x_2, \ldots, x_N; n_1, n_2, \ldots, n_N) \\ &= f_X(x_{1+k}, x_{2+k}, \ldots, x_{N+k}; n_{1+k}, n_{2+k}, \ldots, n_{N+k}), \end{aligned} \tag{3.3.1}$$

for any value of time-shift $k$. In other words, for this stationarity, the two joint probability density functions are identical to each other.

**Strict-sense stationary processes.** A stochastic process $X(n)$ is called *strict-sense stationary* (or *strongly stationary*) if it is stationary for any order $N = 1, 2, 3, \ldots$ and any time-shift $k$.

An example of this process is an iid process. A process is first-order stationary (i.e. for $N = 1$) if the random variables defined at time instants $n = 1$ and $n = (1 + k)$, for any $k$, have the same density functions, that is, if this relation holds, $f_X(x_1; n_1) = f_X(x_{1+k}; n_{1+k})$.

The stationarity represents the time invariance of the statistical properties of a random process, that is, invariance of its statistical properties with respect to position in the time domain. Strict-sense stationary processes are restricted to those random processes that fulfil the above strict definition. However, in practical application, there are processes that can be defined by a more relaxed form of stationarity. These are termed wide-sense stationary processes, which are stationary up to the second order, or all their *first- and second-order moments* are finite and independent of sequence index $n$. They are defined as follows.

**Wide-sense stationary processes.** A stochastic process $X(n)$ is called *wide-sense stationary* (or *weak-sense stationary*) if its mean and its variance are constant and independent of $n$, and its autocorrelation function depends only on the time difference $l = (m - n)$, which is called the *correlation lag*, that is,

$$\eta_X = E\{X(n)\} = \text{constant}, \tag{3.3.2}$$

$$\sigma_X^2 = E\left\{[X(n) - \eta_X]^2\right\} = \text{constant}, \tag{3.3.3}$$

$$R_{XX}(m,n) = R_{XX}(m-n)\Big) = E\left\{X(n+l)X^*(n)\right\} = R_{XX}(l), \tag{3.3.4}$$

which depends on the lag. Consequently, the *autocovariance function* of a wide-sense stationary process depends only on the difference $l = m - n$:

$$C_{XX}(m,n) = C_{XX}(m-n) = R_{XX}(m-n) - \eta_X \eta_X^*$$
$$= R_{XX}(l) - |\eta_X|^2 = C_{XX}(l). \tag{3.3.5}$$

In summary, a wide-sense stationary process $X(n)$ is *first-order stationary* if its first-order statistics is independent of time, that is, the mean value and the variance are constants. In addition, the process is *second-order stationary* and its second-order statistics are independent of time, that is, the autocorrelation and the covariance are functions of the lag only. Understanding wide-sense stationary discrete processes is of essential importance for a deep understanding of the operation of discrete and digital communication systems.

**Jointly wide-sense stationary processes.** All stationary processes are wide-sense stationary since they satisfy conditions of wide-sense stationarity. Wide-sense stationarity is a weaker condition than second-order stationarity because the aforementioned conditions are defined on ensemble averages rather than on density functions. Two processes $X(n)$ and $Y(n)$ are *jointly wide-sense stationary* if each of them is wide-sense stationary and their cross-correlation function depends only on the time difference or lag between them, that is, the autocorrelation and autocovariance are functions of the lag:

$$R_{XY}(m,n) = E\left\{X(m)Y^*(n)\right\} = E\left\{X(m)Y^*(m-l)\right\} = C_{XY}(l) - \eta_X \eta_Y^*. \tag{3.3.6}$$

Consequently, wide-sense stationary processes have one-dimensional correlation and covariance functions. Due to this characteristic, their power spectral density can be nicely described.

All stationary processes are wide-sense stationary since they satisfy conditions of wide-sense stationarity. Generally speaking, if a stochastic process is strict-sense stationary, then it is also wide-sense stationary. However, the converse is not true, because the definition of wide-sense stationary processes takes into account the mean, the variance, and the covariance and does not include the properties of the joint probability density function. In addition, for a Gaussian stochastic process, wide-sense stationarity is equal to strict-sense stationarity, because the process is defined by its mean and its variance.

### Example

Suppose that $W(n)$ is a zero-mean, uncorrelated Gaussian stochastic process with unity variance $\sigma^2(n) = \sigma^2 = 1$.

    a. Analyse this process from a stationarity point of view.

    b. Define a new process $X(n) = W(n) + W(n-1)$ for any value of $n$. Characterize this process in terms of stationarity.

    c. Sketch the time series of the autocorrelation function $R_{XX}(n,m)$.

**Solution.**

    a.
$$R_{WW}(m,n) = \sigma^2 \delta(m-n) = R_{WW}(l),$$

because the mean of the process is zero. Therefore, this is a wide-sense stationary process, and the autocorrelation function can be represented as a one-dimensional function of the lag $l = m - n$, as shown in Figure 3.3.1.

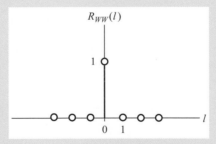

**Fig. 3.3.1** *The autocorrelation function of the Gaussian uncorrelated process W(n).*

The autocorrelation function can be expressed as a function of two variables $m$ and $n$, as shown in Figure 3.3.2, which visualizes the fact that the function has the positive values only when the lag is zero, that is, $m - n = 0$.

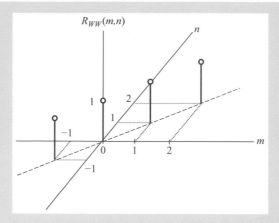

**Fig. 3.3.2**  *Two-dimensional presentation of the autocorrelation function of W(n).*

b.  The mean value of $X(n) = W(n) + W(n-1)$ is zero because the mean value of $W(n)$ is zero for every $n$. The autocorrelation of the process is

$$R_{XX}(m,n) = E\{X(m)X(n)\} = E\{[W(m) + W(m-1)][W(n) + W(n-1)]\}$$
$$= R_{WW}(m,n) + R_{WW}(m,n-1) + R_{WW}(m-1,n) + R_{WW}(m-1,n-1)$$
$$= \sigma^2\delta(m-n) + \sigma^2\delta(m-n+1) + \sigma^2\delta(m-n-1) + \sigma^2\delta(m-n)$$
$$= 2\sigma^2\delta(m-n) + \sigma^2\delta(m-n+1) + \sigma^2\delta(m-n-1).$$

For the unit variance, we may have

$$R_{XX}(m,n) = 2\delta(m-n) + \delta(m-n+1) + \delta(m-n-1).$$

c.  The sketch of the autocorrelation function $R_{XX}(n,m)$ is a time series shown in Figure 3.3.3. It is shown as one-dimensional function of the time difference $l = m - n$, that is,

$$R_{XX}(l) = 2\delta(l) + \delta(l+1) + \delta(l-1)$$

This function can be represented as a function of the two variables $m$ and $n$. In that case, the three components of the function are to be ploted for each value of $m$ and the corresponding values of $n$.

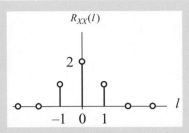

**Fig. 3.3.3**  *The autocorrelation function of X(n).*

Process $X(n)$ is a wide-sense stationary process. However, it is not an independent random process because succeeding realizations $X(n)$ and $X(n + 1)$ are not independent from each other but depend on $W(n)$ values. In contrast, the variables $X(n)$ and $X(n + 2)$ are not correlated because they do not depend on a common term $W$. That is the case when $l = 2$ and the autocorrelation is zero.

## Example

Suppose $W(n)$ is a process that can be defined as the tossing of a fair coin at each time instant $n$, for every $n$, defined as

$$W(n) = \left\{ \begin{array}{ll} +A & \text{with } P(\text{head}) = P(H) = 1/2 \\ -A & \text{with } P(\text{tail}) = P(T) = 1/2 \end{array} \right\},$$

for any $n$. Characterize statistically the defined process $W(n)$, and a *Wiener process*, which is defined as $X(n) = X(n-1) + W(n)$ for positive values of $n$.

**Solution.** The random variable $W(n)$, which is defined at a particular time instant $n$, can be considered a discrete random variable with two outcomes, $+A$ or $-A$, with equal corresponding probabilities equal to 0.5. Because the outcome $\zeta_i$ at time instant $(n + 1)$ does not depend on the outcome at the previous time instant $n$, the process $W(n)$ is an independent random process. The mean value of the discrete random variable, defined at time instant $n$, is

$$\eta_W = E\{W(n)\} = \sum_{\zeta_i = H \text{ or } T} w(n, \zeta_i) P(w = \zeta_i)$$

$$= w(n, H) P(w = H) + w(n, T) P(w = T) = A\frac{1}{2} - A\frac{1}{2} = 0$$

and the variance is

$$\sigma_W^2 = E\left\{W^2(n)\right\} = \sum_{\varsigma_i=H \ or \ T} w^2(n,\varsigma_i)\, P(w=\varsigma_i)$$

$$= w^2(n,H)\, P(w=H) + w^2(n,T)\, P(w=T) = A^2\frac{1}{2} + A^2\frac{1}{2} = A^2.$$

The Wiener process, defined as $X(n) = X(n-1) + W(n)$, at the time instant $n$, can be expressed as

$$X(n) = \sum_{i=1}^{n} W(i),$$

which is a running sum of independent increments. This is due to the following function expanssions:

$$X(1) = X(0) + W(1) = W(1)$$
$$X(2) = X(1) + W(2) = W(1) + W(2)$$
$$X(3) = X(2) + W(3) = W(1) + W(2) + W(3).$$

Therefore, because the outcomes at different time instants do not depend on each other, this is an independent-increment process, which is called a *discrete Wiener process* or a *random walk*. Because the mean value of each random variable $W(i)$ that is defined at time instant $i$ is zero, the mean value of the random process $X(n)$ is also zero for every $n$, as can be seen from the following expression:

$$\eta_X = E\{X(n)\} = E\left\{\sum_{i=1}^{n} W(i)\right\} = \sum_{i=1}^{n} E\{W(i)\} = 0.$$

The variance is

$$\sigma_{X(n)}^2 = E\left\{X^2(n)\right\} = E\left\{\sum_{i=1}^{n}\sum_{k=1}^{n} W(i)W(k)\right\} = \sum_{i=1}^{n}\sum_{k=1}^{n} E\{W(i)W(k)\}$$

$$= \sum_{i=1}^{n} [E\{W(i)W(1)\} + E\{W(i)W(2)\} + \cdots + E\{W(i)W(n)\}]$$

$$= [E\{W(1)W(1)\} + E\{W(2)W(2)\} + \cdots + E\{W(n)W(n)\}] + 0$$

$$= \sum_{i=1}^{n} E\left\{W^2(i)\right\} = nA^2,$$

which depends on $n$. Thus, this process is not stationary process. The variance is not a constant but increases linearly with $n$.

**Example**

Show that the variance of the above process can be calculated by direct development of the following expression:

$$\sigma^2_{X(n)} = E\left\{X^2(n)\right\} = E\left\{\left[\sum_{i=1}^{n} W(i)\right]^2\right\}.$$

## 3.3.2 Properties of the Autocorrelation Function

The autocorrelation function is widely used in practical discrete-time signal processing. This function, when calculated for a wide-sense stationary process, can be represented by a sequence of values that has the following properties:

a. For a real process, the autocorrelation function is an even function and, as such, it is symmetric in respect to the ordinate, that is,

$$R_{XX}(l) = R_{XX}(-l). \qquad (3.3.7)$$

For a complex wide-sense stationary process, this function is a conjugate function of the lag $l$, that is,

$$R_{XX}(l) = R^*_{XX}(-l), \qquad (3.3.8)$$

which is called the Hermitian symmetry.

b. The autocorrelation function value for the zero lag, which is equal to the mean square value of the process, is greater than or equal to the magnitude of the correlation values for any other lag value, that is,

$$E\left\{X^2(n)\right\} = R_{XX}(0) \geq |R_{XX}(l)|. \qquad (3.3.9)$$

This condition for the autocorrelation of a wide-sense stationary process also means that the function has its maximum at zero lag. The average power of a wide-sense stationary process is

$$R_{XX}(0) = \sigma^2_X + |\eta_X|^2 \geq 0. \qquad (3.3.10)$$

The mean value squared corresponds to the average direct current (DC) power of the stochastic process, while the variance corresponds to the average alternating current (AC) power of the process. Thus, the value of the autocorrelation at $l = 0$ corresponds to the total average power of the process.

c. The matrix of the autocorrelation function $R_{XX}(l)$ is non-negative definite, that is,

$$\mathbf{R}_{XX} = E\left\{\mathbf{X}(n) \cdot \mathbf{X}^T(n)\right\}. \tag{3.3.11}$$

This matrix has identical elements in each diagonal and is called a Hermitian Toeplitz matrix. Moreover, for a real-valued process, the correlation matrix is a symmetric Toeplitz matrix that has only $M$ different elements. Also, this matrix is *a positive semi-definite* matrix because it fulfils the condition

$$\alpha^T \mathbf{R}_{XX} \alpha \geq 0, \tag{3.3.12}$$

for any vector $\alpha$ of dimension $M$, that is, $\alpha = [\alpha_1\ \alpha_2\ \ldots\ \alpha_M]^T$ can be any $M \times 1$ vector.

The basic properties of the autocorrelation matrix can be summarized as follows:

1. The autocorrelation matrix of a wide-sense stationary process $X(n)$ is the Hermitian Toeplitz matrix

$$\mathbf{R}_{XX} = E\left\{\mathbf{X}(n) \cdot \mathbf{X}^T(n)\right\}. \tag{3.3.13}$$

2. The autocorrelation matrix of a wide-sense stationary process $X(n)$ is non-negative definite, that is,

$$\mathbf{R}_{XX} > 0. \tag{3.3.14}$$

3. The eigenvalues of the autocorrelation matrix of a wide-sense stationary process are real-valued and non-negative.

From the theory presented here, we can see that wide-sense stationary processes have one-dimensional correlation and covariance functions. Due to this characteristic, that is, their representation as a function of one variable, the power spectral density of these processes can be nicely described.

---

**Example**

A real-valued process is expressed as

$$X(n) = a \cdot X(n-1) + W(n),$$

for $n \geq 0$ and $X(-1) = 0$, where $W(n)$ is a wide-sense stationary process with the constant mean $\eta_W$ and the autocorrelation function $R_{WW}(l) = \sigma_W^2 \delta(l)$. Determine the mean of $X(n)$ and comment on its stationarity.

**Solution.** The general expression for the process $X(n)$ can be obtained from the following developments of the process:

$$X(0) = a \cdot X(-1) + W(0) = W(0)$$

$$X(1) = a \cdot X(0) + W(1) = a \cdot W(0) + W(1)$$

$$X(2) = a \cdot X(1) + W(2) = a^2 \cdot W(0) + a \cdot W(1) + W(2)$$

$$X(3) = a^3 \cdot W(0) + a^2 \cdot W(1) + a \cdot W(2) + W(3)$$

$$\vdots$$

$$X(n) = a^n W(0) + a^{n-1} W(1) + a^{n-2} W(2) + \cdots + a^0 W(n) = \sum_{k=0}^{n} a^k W(n-k).$$

Then, the mean value can be obtained as

$$\eta_X(n) = E\left\{ \sum_{k=0}^{n} a^k W(n-k) \right\} = \sum_{k=0}^{n} a^k E\{W(n-k)\} = \sum_{k=0}^{n} a^k \eta_W$$

$$= \eta_W \sum_{k=0}^{n} a^k = \eta_W \left\{ \begin{array}{ll} \frac{1-a^{n+1}}{1-a} & a \neq 1 \\ (n+1) & a = 1 \end{array} \right\}.$$

Because this value depends on $n$, the process $X(n)$ is not a stationary process. However, if $n$ tends to infinity, the mean value may be

$$\eta_X(\infty) = \eta_W \lim_{n \to \infty} \frac{1 - a^{n+1}}{1-a} = \eta_W \frac{1}{1-a}.$$

Because $W(n)$ is a stationary process, its mean is independent of $n$. Therefore, the mean value of $X(n)$, when $n$ tends to infinity, is also independent of $n$, that is, the process $X(n)$ is asymptotically stationary. The process $X(n)$ is known as a first-order autoregressive process, and also as a white noise process if the mean value is zero.

In practice, it is relatively simple to investigate the stationarity characteristics of a stochastic process. If this process is generated by a time-invariant system, then the process produced at the output of the system is a stationary process.

### 3.3.3    The Stationarity of Two Discrete-Time Stochastic Processes

**Joint strict-sense stationarity.** Two processes $X(n)$ and $Y(n)$ are jointly strict-sense stationary if the statistical properties of both processes, starting from any time instant $n$, are identical to the statistical properties of the processes starting at the shifted time instant $n + N$ for any $N$:

$$f_X(x_1, x_2; n_1, n_2) = f_X(x_{1+N}, x_{2+N}; n_{1+N}, n_{2+N}). \tag{3.3.15}$$

**Wide-sense stationarity.** Two processes $X(n)$ and $Y(n)$ are jointly wide-sense stationary if each of them is wide-sense stationary and their *cross-covariance function* depends only on the time difference between them, that is,

$$
\begin{aligned}
C_{XY}(m,n) &= E\left\{[X(m)-\eta_X][Y(n)-\eta_Y]^*\right\} \\
&= E\left\{X(m)Y^*(n)\right\} - E\left\{X(m)\eta_Y^*\right\} - E\left\{\eta_X Y^*(n)\right\} + E\left\{\eta_X\eta_Y^*\right\}. \\
&= R_{XY}(l) - \eta_X\eta_Y^*
\end{aligned}
\tag{3.3.16}
$$

**Properties of a cross-correlation function.** The cross-correlation function of two processes $X(n)$ and $Y(n)$ can be represented by a sequence of values that has the following properties:

1. The cross-correlation function is even and symmetric in respect to the origin, that is,

$$
R_{XY}(l) = R_{YX}(-l). \tag{3.3.17}
$$

2. The following relationship between the cross-correlation and autocorrelation functions holds:

$$
|R_{XY}(l)| \le [R_{XX}(0)R_{YY}(0)]^{1/2}. \tag{3.3.18}
$$

## 3.4 Ergodic Processes

### 3.4.1 Ensemble Averages and Time Averages

A stochastic process is defined by an ensemble of its realizations, which are called stochastic (random) discrete-time signals and are defined for all outcomes of a random experiment. For this process, the mathematical expectations, like the mean and the correlation, can be defined on the ensemble of realizations at the instants of time $n$. To calculate these statistical quantities for each time instant, we need to find the probability density function of the process, which is, most of the time, very hard or impossible to calculate and present in closed form. However, these expectations, calculated as the average values, are not known a priori for the stochastic system. On the other hand, we cannot find them easily, because we do not have all of the ensemble random sequences available to analyse in order to find the averages. Instead, we need to infer the properties of the underlying stochastic process by using a single realization of the process. This can be done for the so-called ergodic processes, for which we define time averages using the notion of limits of infinite sums. In communication system theory, we will have only one realization of the related stochastic process to be analyse.

A realization of an ergodic process can be defined as a power random signal, having a finite average power and infinite energy. Consequently, the analysis of this process will be similar to the analysis of power deterministic signals. However, in doing such an analysis, it is important to keep in mind the random nature of ergodic processes.

## 3.4.2    Ergodic Processes

A stochastic process that has statistical characteristics across the assemble of its realizations that are equivalent to statistical characteristics calculated on a single realization of the process is called an ergodic process. This process is of particular interest to communication theory and practice. Namely, stochastic (random) signals processed in transmitters and receivers can be understood as realizations of stochastic processes having a limited time duration that is usually equal to a message bit. As such, they are assumed to belong to ergodic processes and are processed and analysed as such. The mean value of these random signals, in principle, contains the power of the signal part of the transmitted message, while the variance contains the power of the noise part.

The foundation of the theoretical base of ergodic processes is the ergodic theorem: For a given stationary discrete-time stochastic process $X(n)$, having a finite mean value $\eta_X = E\{X(n)\}$, there exists with probability 1 a random variable $\overline{X}$ with the property

$$\overline{X} = \lim_{n \to \infty} \frac{1}{n} \sum_{n=0}^{\infty} X(n), \qquad (3.4.1)$$

that is, the limit of the average sum of random variables exists. If the process is also ergodic, then the process has only one realization, that is, $X(n) = x(n)$, and the sum is equal to the mean value of the process with probability 1, that is,

$$\lim_{n \to \infty} \frac{1}{n} \sum_{n=0}^{\infty} x(n) = \eta_X. \qquad (3.4.2)$$

This condition implies the convergence in the mean square that will be addressed later.

## 3.4.3    Estimate of the Mean across the Ensemble of Realizations of $X(n)$

Suppose a set of $I$ realizations $x_i(n)$, $i = 1, 2, \ldots, I$ of a stochastic process $X(n)$ is available. Then, the estimate of the mean function can be obtained at each time instant $n$ according to the expression

$$\overline{\eta}_X(n) = \frac{1}{I} \sum_{i=0}^{I} x_i(n). \qquad (3.4.3)$$

It is important to note that this is a function of $n$ because the mean value generally is a function of the discrete variable $n$. This estimation assumes that there is a 'sufficient' number $I$ of realizations. However, that is usually not the case. Namely, in practice, only one realization $x_1(n)$ of a stochastic process $X(n)$ is available.

The mean square value of the process, which is equal to the variance if the mean value is zero, can be estimated as

$$\left\langle |x(n)|^2 \right\rangle = \frac{1}{2N+1} \sum_{n=-N}^{N} |x(n)|^2. \tag{3.4.4}$$

If the process is stationary and ergodic, the mean values and variances of the process can be estimated from this single realization of the process.

### 3.4.4 Estimate of the Mean across a realization of $X(n)$

Suppose that one realization $x(n)$ of an ergodic process $X(n)$ is available. Then, the estimate of the mean function can be obtained by finding the mean value at each time instant $kN$, where $k$ is an integer and $N$ is the number of samples we are taking into account to estimate the mean. The mean is then calculated according to this expression (note a small subscript $x$ that is related to the ergodic processes):

$$\overline{\eta}_x(kN) = \frac{1}{N} \sum_{n=(k-1)N}^{kN} x(n). \tag{3.4.5}$$

This estimation assumes the existence of a 'sufficient' number of $N$ values of the ergodic process $X(n)$ inside one realization $x(n)$. If the process is stationary and ergodic, the mean values and variance of the process can be estimated from this single realization.

The last estimate can be considered a sample mean of the random variables $X(n)$ belonging to the samples taken $x(n)$. This sample mean can be expressed as follows (note a capital subscript $X$ related to the random process):

$$\overline{\eta}_X(kN) = \frac{1}{N} \sum_{n=(k-1)N}^{kN} X(n), \tag{3.4.6}$$

which can be considered a new discrete-time stochastic process. One realization of this process has the random values $x(kN)$ at time instants $kN$ belonging to the related random variables $X(kN)$. The expectation of the sample mean, which is a random variable, is the mean value of the process defined at time instants $kN$, that is,

$$E\left\{\overline{\eta}_X(kN)\right\} = \frac{1}{N} \sum_{n=(k-1)N}^{kN} E\{X(n)\} = E\{X(n)\} = \eta_X. \tag{3.4.7}$$

In summary, if the sample mean $\overline{\eta}_X(kN)$ converges in the mean square sense, that is,

$$\lim_{N \to \infty} E\left\{ \left| \overline{\eta}_X(kN) - \eta_X \right|^2 \right\} = 0, \tag{3.4.8}$$

the process is ergodic in the mean, which can be expressed in the form

$$\lim_{N \to \infty} \overline{\eta}_X(kN) = \lim_{N \to \infty} \frac{1}{N} \sum_{n=(k-1)N}^{kN} x(n) = \eta_X. \tag{3.4.9}$$

This limit provides the base for calculating the mean value of the ergodic process, when one realization of the process $x(n)$ is available, and allows the definition of the mean of ergodic processes that follows. In addition, the variance of the sample mean should tend to zero when $N$ tends to infinity. In a similar way, the necessary and sufficient condition for the ergodicity in autocorrelation can be deduced.

In communication systems theory and practice, in principle, we use this approach to find the mean and the variance of the signal and the noise in a bit interval. In addition, we extend the estimates of the mean inside a bit interval to multiple bit intervals when we are searching for a good estimate. For example, the number of bit intervals taken into account is very large when we are estimating the bit error probability in a communication system. If this probability is very low, we need an enormously large number of test bits, up to millions, to estimate the bit error probability.

## 3.4.5    Estimate of the Mean of an Ergodic Process $X(n)$

Following the above explanations, and the similar approach used to calculate the mean and average power of the power deterministic signals, here are the definitions of the mean, the variance, the autocorrelation, and the covariance of ergodic processes.

The mean value of an ergodic process $X(n)$ can be found as the limit of the average value of one realization $x(n)$ of the process, expressed as

$$\eta_x = \langle x(n) \rangle = \lim_{N \to \infty} \frac{1}{2N+1} \sum_{n=-N}^{N} x(n), \tag{3.4.10}$$

where the time average operator $\langle \rangle$ has the same properties as the ensemble average operator. The mean square value is

$$\left\langle |x(n)|^2 \right\rangle = \lim_{N \to \infty} \frac{1}{2N+1} \sum_{n=-N}^{N} |x(n)|^2. \tag{3.4.11}$$

The variance is

$$\sigma_x^2 = \left\langle |x(n) - \eta_x|^2 \right\rangle = \lim_{N \to \infty} \frac{1}{2N+1} \sum_{n=-N}^{N} |x(n) - \eta_x|^2. \tag{3.4.12}$$

An ergodic process is a wide-sense stationary process. Thus, the correlation and covariance functions depend on the lag *l*. However, a wide-sense stationary process is not necessarily ergodic. The autocorrelation of an ergodic process is

$$R_{xx} = \left\langle x(n)x^*(n-l) \right\rangle = \lim_{N \to \infty} \frac{1}{2N+1} \sum_{n=-N}^{N} x(n)x^*(n-l), \qquad (3.4.13)$$

and the autocovariance is

$$C_{xx} = \left\langle [x(n)-\eta_x][x(n-l)-\eta_x]^* \right\rangle$$

$$= \lim_{N \to \infty} \frac{1}{2N+1} \sum_{n=-N}^{N} [x(n)-\eta_x][x(n-l)-\eta_x]^*. \qquad (3.4.14)$$

For two ergodic processes, we can find the cross-correlation as

$$R_{xy} = \left\langle x(n)y^*(n-l) \right\rangle = \lim_{N \to \infty} \frac{1}{2N+1} \sum_{n=-N}^{N} x(n)y^*(n-l), \qquad (3.4.15)$$

and the cross-covariance is defined as

$$C_{xy} = \left\langle [x(n)-\eta_x]\left[y(n-l)-\eta_y\right]^* \right\rangle$$

$$= \lim_{N \to \infty} \frac{1}{2N+1} \sum_{n=-N}^{N} [x(n)-\eta_x]\left[y(n-l)-\eta_y\right]^*. \qquad (3.4.16)$$

These averages are very similar to the averages defined for deterministic power signals. However, the averages of ergodic processes are random variables, because they depend on the outcomes of a random experiment, as we mentioned above.

### 3.4.6  Summary of Ergodic Stochastic Processes

There are several degrees of ergodicity. A stochastic process $X(n)$ is *ergodic in the mean* if its average value on one realization $x(n, \zeta_i) = x(n)$ is equal to the mean value of the ensemble values $x(n)$ of $X(n)$, that is,

$$\eta_X = E\{X(n)\} = \langle x(n) \rangle = \lim_{N \to \infty} \frac{1}{2N+1} \sum_{n=-N}^{N} x(n). \qquad (3.4.17)$$

A stochastic process is *ergodic in the autocorrelation* if the following condition is satisfied:

$$R_{XX}(l) = E\left\{X(n)X^*(n-l)\right\}$$

$$= \left\langle x(n)x^*(n-l)\right\rangle = \lim_{N\to\infty} \frac{1}{2N+1} \sum_{n=-N}^{N} x(n)x^*(n-l) = R_{xx}(l). \tag{3.4.18}$$

If a process $X(n)$ is ergodic with respect to both the mean and the autocorrelation, the mean value $\langle x(n)\rangle$ of one realization is constant, and the autocorrelation $\langle x(n)x^*(n-l)\rangle$ depends on the time difference $l$, then this process is also wide-sense stationary. However, we cannot say that, if a process is wide-sense stationary, than the process is also an ergodic process, despite the fact that most real wide-sense stationary processes are also ergodic processes.

In practical applications, we cannot perform time averaging with an infinite number of samples of an ergodic process. Instead, we use a limited number of samples $N$ that allows us to achieve an in-advance-specified accuracy in the estimation. The new averaging expressions defined for this case are identical to the above ones when the *limits* are excluded and $N$ is a finite number. In this sense, the mean value is defined as the time average

$$\eta_x = \langle x(n)\rangle = \frac{1}{2N+1} \sum_{n=-N}^{N} x(n), \tag{3.4.19}$$

and the autocorrelation is

$$R_{XX}(l) = E\left\{X(n)X^*(n-l)\right\}$$

$$= \left\langle x(n)x^*(n-l)\right\rangle = \lim_{N\to\infty} \frac{1}{2N+1} \sum_{n=-N}^{N} x(n)x^*(n-l) = R_{xx}(l). \tag{3.4.20}$$

In the analysis of discrete communication systems, for example, we are processing the received bit that has $N$ discrete (usually interpolated) samples. Therefore, we assume that a realization of a stochastic process inside the bit interval is ergodic. This ergodicity is extended to a larger number of bits when more precise estimates of the mean value and the variance are required.

## 3.5 The Frequency-Domain Representation of Discrete-Time Stochastic Processes

Because a zero-mean stochastic stationary process $X(n)$ does not have finite energy and finite duration, its Fourier transform and its $z$-transform do not converge. On the other hand, a sample time function $x(n)$ of a stochastic process $X(n)$ is just one realization of the process taken from the ensemble of functions. Therefore, the basis functions for different realizations must be weighted with different weighting coefficients, and the whole process may have the averaged rate of change of ensemble coefficients. Thus, in a spectral analysis of these signals, we must take into consideration their random nature. For the sake of completeness, we will first present the Fourier transform pair for *continuous-time* stochastic signals.

### 3.5.1    Continuous-Time Stochastic Processes in the Frequency Domain

**Periodic process: Fourier series.** Because of the random nature of stochastic processes, we need to define their properties and values using random terms. We define the periodicity of a stochastic process in terms of the mean square error. A stochastic process $X(t)$ is the mean square periodic with period $T = 2\pi/\omega_0$ if

$$E\left\{|X(t+T)-X(t)|^2\right\} = 0,$$

for all $t$ values. Then, for a *periodic wide-sense stationary process* $X(t)$ with period $T$, we may define the sums that approximate Fourier series in the mean square sense as

$$\hat{X}(t) = \sum_{k=-\infty}^{\infty} \hbar_k e^{jk\omega_0 t},$$

and calculate the coefficients as

$$\hbar_k = \frac{1}{T}\int_0^T X(t)e^{-jk\omega_0 t}dt.$$

It can be proven that the calculated sum $\hat{X}(t)$ is equal to the signal $X(t)$ in the mean square sense, that is, $E\left\{\left|X(t)-\hat{X}(t)\right|^2\right\} = 0$, and the coefficients $\hbar_k$ are uncorrelated random variables with the finite mean for $k = 0$ and finite variance $E\{\hbar_k \cdot \hbar_l^*\}$ only for $k = l$. The Fourier series of a stochastic process is a special case of the Karhunen–Loève expansion of the process using a set of orthonormal functions.

A wide-sense stationary process is periodic in the mean square sense if its autocorrelation function is periodic with period $T$. Therefore, we can express the autocorrelation function as an infinite sum of exponentials as follows:

$$R_{XX}(\tau) = \sum_{k=-\infty}^{\infty} c_k e^{jk\omega_0 \tau},$$

where the coefficients of the sum are

$$c_k = \frac{1}{T}\int_0^T R_X(\tau)e^{-jk\omega_0 \tau}d\tau.$$

**Aperiodic signals: Fourier transform.** Due to the random behaviour of a stochastic process $X(t)$, its Fourier transform is again a stochastic process that can be expressed as

$$X(\omega) = \int_{-\infty}^{\infty} X(t)e^{-j\omega t}dt,$$

where the process is the inverse Fourier transform

$$X(t) = \frac{1}{2\pi} \int_{-\infty}^{\infty} X(\omega) e^{+j\omega t}d\omega.$$

These integrals are interpreted as the *mean square limits*. As we said before, the right-hand side of the second-to-last equation fails to converge in the usual sense for a stationary process $X(t)$, because this process is not square summable (with probability 1). The properties of the Fourier transform for random signals are the same as for deterministic signals.

## 3.5.2 Discrete-Time Stochastic Processes in the Frequency Domain

For the discrete-time process $X(n)$, we can find its discrete-time Fourier transform as

$$X\left(e^{j\Omega}\right) = \sum_{n=-\infty}^{\infty} X(n)e^{-jn\Omega}, \tag{3.5.1}$$

and its inverse transform as

$$X(n) = \frac{1}{2\pi} \int_{-\pi}^{\pi} X\left(e^{j\Omega}\right) e^{j\Omega\,n}d\Omega. \tag{3.5.2}$$

From the definition of the transform, it follows that the stochastic process in the frequency domain $X\left(e^{j\Omega}\right)$ is periodic with period $2\pi$.

As was shown in the previous sections, the correlation function of a wide-sense stationary process is a function of one variable. This property makes possible a relatively simple representation of discrete-time stochastic processes in the frequency domain and the $z$ domain. In this section, we will present a power spectral density (power spectrum) for a discrete-time stochastic process, using the discrete-time Fourier series and the discrete-time Fourier transform. It is important to note that some processes do not have the power spectral density function.

**The $z$-transform.** As was shown above, the correlation function of a wide-sense stationary process is a function of one variable. This function is deterministic and, in most cases, has finite energy, which implies that the correlation is absolutely summable. This property makes possible a relatively simple representation of the process in the frequency and the $z$ domains. The $z$-transform of the correlation function is

$$S_{XX}(z) = \sum_{l=-\infty}^{\infty} R_{XX}(l)z^{-l}, \tag{3.5.3}$$

which has some region of convergence. Because the correlation function is symmetric with respect to the origin, the following equation holds:

$$S_{XX}(z) = S_{XX}\left(z^{-1}\right). \tag{3.5.4}$$

**The discrete-time Fourier transform.** The $z$-transform developed on the unit circle $z = e^{j\Omega}$ is called the power spectral density or the power spectrum of the stochastic process $X(n)$.

**The aperiodic correlation function.** If the correlation function $R_{XX}(l)$ of a process $X(n)$ is aperiodic in $l$, then we can use the discrete-time Fourier transform to obtain the power spectral density. In this case, the correlation function $R_{XX}(l)$ should be absolutely summable, which guarantees the existence of the power spectral density, which means that $X(n)$ is a zero-mean process. The power spectral density is the discrete-time Fourier transform of the autocorrelation function $R_{XX}(l)$, which is expressed as

$$S_{XX}\left(e^{j\Omega}\right) = \sum_{l=-\infty}^{\infty} R_{XX}(l)e^{-j\Omega l}, \tag{3.5.5}$$

which is known as the *Wiener–Khintchine theorem*. Using the inverse discrete-time Fourier transform, we can find the autocorrelation function as

$$R_{XX}(l) = \frac{1}{2\pi}\int_{-\pi}^{\pi} S_{XX}\left(e^{j\Omega}\right)e^{j\Omega l}d\Omega. \tag{3.5.6}$$

**The periodic correlation function.** If the correlation function $R_{XX}(l)$ of a process $X(n)$ is periodic in $l$, which is the case when $X(n)$ is a wide-sense stationary periodic stochastic process, then we can use the discrete-time Fourier series to obtain the power spectrum (power spectral density) function. This density is represented by a line spectrum. Therefore, if the process $X(n)$ is periodic with a non-zero mean, or almost periodic, the power spectral density is expressed as

$$S_{XX}\left(e^{j\Omega}\right) = \sum_{i} 2\pi A_i \delta\left(\Omega - \Omega_i\right), \tag{3.5.7}$$

where $A_i$ are amplitudes of the correlation function $R_{XX}(l)$ at discrete frequencies $\Omega_i$ expressed in radians per sample.

### 3.5.3   Cross-Spectrum Functions

For two wide-sense stationary processes $X(n)$ and $Y(n)$, the cross-spectrum function is defined as

$$S_{XY}\left(e^{j\Omega}\right) = \sum_{l=-\infty}^{\infty} R_{XY}(l)e^{-j\Omega l}. \tag{3.5.8}$$

Because the cross-correlation is not an even function like the autocorrelation, the cross-spectrum function is a complex function. This function has the following properties:

$$S_{XY}\left(e^{j\Omega}\right) = S_{YX}^{*}\left(e^{j\Omega}\right),$$

(3.5.9)

and

$$S_{XY}^{*}\left(e^{j\Omega}\right) = S_{YX}\left(e^{j\Omega}\right).$$

(3.5.10)

**Proof.** Because $R_{XY}(l) = R_{YX}(-l)$, we may have

$$S_{XY}\left(e^{j\Omega}\right) = \sum_{l=-\infty}^{\infty} R_{XY}(l)e^{-j\Omega\, l} = \sum_{l=-\infty}^{\infty} R_{YX}(-l)\, e^{-j\Omega\, l},$$

or, if we change the index of summation, $l = -i$, we may have

$$S_{XY}\left(e^{j\Omega}\right) = \sum_{l=-\infty}^{\infty} R_{YX}(-l)\, e^{-j\Omega\, l} = \sum_{i=-\infty}^{\infty} R_{YX}(i)e^{+j\Omega\, i}$$

$$= \left[\sum_{l=-\infty}^{\infty} R_{YX}(l)e^{-j\Omega\, l}\right]^{*} = S_{YX}^{*}\left(e^{j\Omega}\right).$$

**Properties of the power spectral density function of wide-sense stationary processes.** There are three properties of the power spectral density function that follow from properties of the autocorrelation function and the characteristics of the discrete-time Fourier transform:

1.  The power spectral density function $S_{XX}(\Omega)$ of a process $X(n)$ is a real-valued periodic function of the normalized frequency $\Omega$ with period $2\pi$ for any real- or complex-valued process $X(n)$. If process $X(n)$ is real-valued, then the auto power spectral density $S_{XX}(\Omega)$ is an even function of frequency $\Omega$, that is, the following condition is fulfilled:

$$S_{XX}\left(e^{j\Omega}\right) = S_{XX}\left(e^{-j\Omega}\right).$$

(3.5.11)

   The proof for this follows from the characteristics of the autocorrelation function and the properties of the discrete-time Fourier transform.

2.  The auto power spectral density $S_{XX}(\Omega)$ is non-negative definite, that is,

$$S_{XX}\left(e^{j\Omega}\right) \geq 0.$$

(3.5.12)

   The proof for this follows from the non-negative definiteness of the autocorrelation function.

3. The integral value of the auto power spectral density function is non-negative and equal to the average power $P_{av}$ of the process $X(n)$. From the integral

$$R_{XX}(l) = \frac{1}{2\pi} \int_{-\pi}^{\pi} S_{XX}\left(e^{j\Omega}\right) e^{j\Omega l} d\Omega, \tag{3.5.13}$$

we can find the integral value of the power spectral density function as the average power

$$P_{av} = \frac{1}{2\pi} \int_{-\pi}^{\pi} S_{XX}\left(e^{j\Omega}\right) d\Omega = E\left\{X(m)X^*(n)\right\}$$

$$= E\left\{|X(n)|^2\right\} = R_{XX}(0). \tag{3.5.14}$$

## 3.6   Typical Stochastic Processes

### 3.6.1   Noise Processes

In this subsection, the following important discrete-time processes will be defined and analysed: purely random processes, strict white noise, white noise, wide-sense stationary white noise, and white Gaussian noise processes.

**Purely random processes.** A zero-mean, stationary discrete-time stochastic process that satisfies the condition

$$f_X(x_1, x_2, \ldots; n_1, n_2, \ldots) = f_X(x_1) \cdot f_X(x_2) \cdot \ldots, \tag{3.6.1}$$

where all random variables $X_i$ are identically distributed, is a *purely random process*. In this definition, the density function of a particular random variable is not defined, that is, it can have any distribution. We simply say that this process is defined by a series of iid random variables $X_i$. Thus, using this definition of the iid random variables, we say that these variables are also uncorrelated. However, the uncorrelated processes are not necessarily independent, or, better yet, the iid variables are subset of uncorrelated variables. This process is strictly stationary; therefore, it is also wide-sense stationary. The definition of a purely random process is used to define *noise processes*.

**White noise processes.** We say that a process $W(n)$, with uncorrelated variables but not independent, is a *white noise process* if and only if the mean is zero, $\eta_W(n) = E\{W(n)\} = 0$, and the autocovariance is

$$C_{WW}(n_1, n_2) = E\left\{W(n_1)W^*(n_2)\right\} = E\left\{W^2(n_1)\right\} = E\left\{W^2(n_2)\right\} = \sigma_W^2 \delta(n_1 - n_2) \tag{3.6.2}$$

**Strictly white noise processes.** The process $W(n)$ is a *strictly white noise process* if its variables are not only correlated but also independent. Therefore, this noise is a *purely random process*. We say that a wide-sense stationary process $W(n)$, with uncorrelated and independent variables, is a *wide-sense stationary white noise process* if and only if its mean is

$$\eta_W(n) = E\{W(n)\} = 0, \tag{3.6.3}$$

and autocorrelation, for the assumed zero mean, is

$$R_{WW}(l) = E\left\{W(n)W^*(n-l)\right\} = \sigma_W^2 \delta(l), \tag{3.6.4}$$

which implies that the power spectral density is constant and equal to the variance, that is,

$$S_{WW}\left(e^{j\Omega}\right) = \sum_{l=-\infty}^{\infty} R_{WW}(l)e^{-j\Omega\, l} = \sum_{l=-\infty}^{\infty} \sigma_W^2 \delta(l)e^{-j\Omega\, l} = \sigma_W^2, \tag{3.6.5}$$

for $-\pi \geq \Omega \geq \pi$. This noise is simply called *white noise*, due to the fact that all its frequency components are of equal power, as in the case of white light, which contains all possible colour components. The autocorrelation function and power spectral density function for this process is presented in Figure 3.6.1.

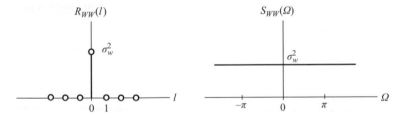

**Fig. 3.6.1** *Functions of the wide-sense stationary white noise process.*

Also, the correlation matrix of this process is a diagonal matrix defined by variances of random variables, that is, for $\eta_X(n) = 0$,

$$\mathbf{R}_{XX} = E\left\{\mathbf{X}(n)\cdot\mathbf{X}^T(n)\right\} = E\left\{\begin{bmatrix} X(0) \\ X(1) \\ \vdots \\ X(M-1) \end{bmatrix} [X(0)\ X(1)\ \cdots\ X(M-1)]\right\}$$

$$= \begin{bmatrix} \sigma_X^2 & 0 & \cdots & 0 \\ 0 & \sigma_X^2 & \cdots & 0 \\ \vdots & \vdots & \ddots & \vdots \\ 0 & 0 & \cdots & \sigma_X^2 \end{bmatrix} = \sigma_X^2 \mathbf{I}, \tag{3.6.6}$$

where **I** is an identity matrix. An iid process is always *strictly stationary*. It is also *weakly stationary* except for processes with unbiased moments. One example of this process is an iid process with a Cauchy distribution of random variables.

In the previous definition of a purely random process, the density function $f_X(n_i)$ of a particular random variable $X(n_i)$ is not defined, that is, it can have any distribution. We simply say that this process is defined by a series of iid random variables $X(n_i)$.

The *white noise process* defined above is the simplest random process, because it does not have any structure. However, it is widely used in practice as a model of a system or as a building block for more complicated signal-processing structures.

In the special case when a wide-sense stationary stochastic process $W(n)$ is a white noise process and has a Gaussian probability density function, the process is called a *white Gaussian noise process*. The variables of this process are uncorrelated and, being Gaussian, they are also independent. In the case when a stochastic process $W(n)$ is represented by a series of iid random variables, this process is called *strictly white Gaussian noise*. Consequently, as we said before, this noise is also a wide-sense stationary process. This noise, with zero mean and the same variances (iid), has the density function of each variable expressed as

$$f_W(w_n) = \frac{1}{\sqrt{2\pi\sigma^2}} e^{-w^2/2\sigma^2}, \tag{3.6.7}$$

and the joint density function of the process is

$$f_W(w_1, w_2, \ldots, w_N; n_1, n_2, \ldots, n_N) = f_W(w_1) \cdot f_W(w_2) \cdot \ldots \cdot f_W(w_N)$$

$$= \prod_{n=1}^{N} f_W(w_n) = \prod_{n=1}^{N} \frac{1}{\sqrt{2\pi\sigma^2}} e^{-w_n^2/2\sigma^2} = \frac{1}{(2\pi\sigma^2)^{N/2}} e^{-\frac{1}{2\sigma^2}\sum_{n=1}^{N} w_n^2}. \tag{3.6.8}$$

### 3.6.2    General Gaussian Noise Processes

A random process $X(n)$ is a general Gaussian stochastic process with different means and variances if its variables at different time instants $n$, $n + 1$, ..., $n + k$ are independent, but not identical, and jointly Gaussian random variables for all $k$ and any $n$. Therefore, a general Gaussian process has a joint Gaussian density function, which is a product of marginal density functions of different random variables that specify the stochastic process. The density function of the random variable $X(n)$ that is defined for a particular $n$ is

$$f_{X_n}(x_n) = \frac{1}{\sqrt{2\pi\sigma_{X_n}^2}} e^{[x_n - \eta_{X_n}]^2/2\sigma_{X_n}^2}. \tag{3.6.9}$$

Because $X(n)$ is an independent process, the joint ($M$th-order) density function is a product of marginal densities, that is,

$$f_{\mathbf{X}}(\mathbf{x}) = f_{\mathbf{X}}(x_0, x_1, \ldots, x_{M-1}; n_0, n_1, \ldots, n_{M-1})$$

$$= f_{X_0}(x_0) \cdot f_{X_1}(x_1) \cdot \ldots \cdot f_{X_{M-1}}(x_{M-1}) = \prod_{n=0}^{M-1} f_X(x_n)$$

$$= \frac{1}{\sqrt{2\pi\sigma_0^2}} e^{-[x_0 - \eta_0]^2 / 2\sigma_{X_0}^2} \cdot \frac{1}{\sqrt{2\pi\sigma_1^2}} e^{-[x_1 - \eta_1]^2 / \sigma_{X_1}^2} \cdots \frac{1}{\sqrt{2\pi\sigma_{M-1}^2}} e^{-[x_{M-1} - \eta_{M-1}]^2 / 2\sigma_{X_{M-1}}^2}$$

$$= \prod_{n=0}^{M-1} \frac{1}{\sqrt{2\pi\sigma_n^2}} e^{-[x_n - \eta_n]^2 / \sigma_{X_n}^2} = \frac{1}{(2\pi)^{M/2}\sqrt{\sigma_0^2\sigma_1^2 \cdots \sigma_{M-1}^2}} e^{-\sum_{n=0}^{M-1} [x_n - \eta_n]^2 / 2\sigma_{X_n}^2}$$

$$= \frac{1}{(2\pi)^{M/2}\Delta^{1/2}} e^{-\frac{1}{2}(\mathbf{X}-\boldsymbol{\eta})^T \mathbf{C}_{XX}^{-1}(\mathbf{X}-\boldsymbol{\eta})}.$$

Here we have the following definitions of matrices used in the final formula:

$$\mathbf{X} = \begin{bmatrix} X_0 - \eta_{X_0} & X_1 - \eta_{X_1} & \cdots & X_{M-1} - \eta_{X_{M-1}} \end{bmatrix}^T,$$

$$\boldsymbol{\eta} = \begin{bmatrix} \eta_{X_0} & \eta_{X_1} & \cdots & \eta_{X_{M-1}} \end{bmatrix}^T$$

$$\Delta = \sigma_{X_0}^2 \cdot \sigma_{X_1}^2 \cdots \sigma_{X_{M-1}}^2,$$

$$\mathbf{C}_{XX} = \mathbf{R}_{XX} - \boldsymbol{\eta}\boldsymbol{\eta}^T = E\left\{\mathbf{X}(n) \cdot \mathbf{X}^T(n)\right\} - \boldsymbol{\eta}\boldsymbol{\eta}^T = E\left\{[\mathbf{X}(n) - \boldsymbol{\eta}] \cdot [\mathbf{X}(n) - \boldsymbol{\eta}]^T\right\}$$

$$= \begin{bmatrix} \sigma_{X_0}^2 & 0 & \cdots & 0 \\ 0 & \sigma_{X_1}^2 & \cdots & 0 \\ \vdots & \vdots & \ddots & \vdots \\ 0 & 0 & \cdots & \sigma_{X_{M-1}}^2 \end{bmatrix}. \tag{3.6.10}$$

The inverse autocovariance function is

$$\mathbf{C}_{XX}^{-1} = \begin{bmatrix} \sigma_0^{-2} & 0 & \cdots & 0 \\ 0 & \sigma_1^{-2} & \cdots & 0 \\ \vdots & \vdots & \ddots & \vdots \\ 0 & 0 & \cdots & \sigma_{M-1}^{-2} \end{bmatrix}. \tag{3.6.11}$$

Let us present the key properties of the white Gaussian noise process:

1. We said that the independent random variables are uncorrelated. The converse is not necessarily true. However, the uncorrelation of Gaussian random variables implies their independence.

2. If a Gaussian process is transferred through a linear system, the output of the system will again be a Gaussian process.

3. A Gaussian distribution is completely described by its first two moments, the mean and the variance. Thus, its higher moments can be expressed as a function of these two moments, and the definitions of strict and weak stationarity for this process are equivalent.

The formulas derived above can be simplified for a strictly white Gaussian noise process, which is, in practice, just called white Gaussian noise, if we equate all mean values with zero and make all variances equal to each other. Then, the previously obtained formula will include a correlation matrix instead of a covariance matrix, and will take the form

$$f_X(x_0, x_1, \ldots, x_{M-1}; n_0, n_1, \ldots, n_{M-1}) = \frac{1}{(2\pi)^{M/2} \Delta^{1/2}} e^{-\frac{1}{2} \mathbf{X}^T \mathbf{R}_{XX}^{-1} \mathbf{X}}. \tag{3.6.12}$$

In the case when the mean value is zero, the inverse correlation function is

$$\mathbf{R}_{XX}^{-1} = \begin{bmatrix} \sigma_0^{-2} & 0 & \cdots & 0 \\ 0 & \sigma_1^{-2} & \cdots & 0 \\ \vdots & \vdots & \ddots & \vdots \\ 0 & 0 & \cdots & \sigma_{M-1}^{-2} \end{bmatrix} = \sigma_X^{-2} \begin{bmatrix} 1 & 0 & \cdots & 0 \\ 0 & 1 & \cdots & 0 \\ \vdots & \vdots & \ddots & \vdots \\ 0 & 0 & \cdots & 1 \end{bmatrix}, \tag{3.6.13}$$

and the determinant is given by

$$\Delta = \sigma_0^2 \cdot \sigma_1^2 \cdots \sigma_{M-1}^2 = \sigma_X^{2(M-1)}. \tag{3.6.14}$$

Therefore, the density function of the white Gaussian noise is

$$f_{\mathbf{X}}(\mathbf{x}) = f_{\mathbf{X}}(x_0, x_1, \ldots, x_{M-1}; n_0, n_1, \ldots, n_{M-1})$$

$$= \frac{1}{(2\pi)^{M/2} (\sigma_X^2)^{M/2}} e^{-\sum_{n=0}^{M-1} x_n^2 / 2\sigma_X^2}. \tag{3.6.15}$$

This expression is equivalent to the expression for the strictly Gaussian noise process derived in the previous section.

### 3.6.3   Harmonic Processes

A harmonic process is defined by the expression

$$X(n) = \sum_{k=1}^{M} A_k \cos(\Omega_k n + \theta_k), \tag{3.6.16}$$

where amplitudes and frequencies are constants, and phases $\theta_k$ are random variables uniformly distributed on the interval $[0, 2\pi]$. It can be shown that this is a stationary process with zero mean and an autocorrelation function defined as

$$R_{XX}(l) = \frac{1}{2} \sum_{k=1}^{M} A_k^2 \cos \Omega_k l, \qquad (3.6.17)$$

for $-\infty < l < \infty$. Let us comment on these harmonic processes. These processes are represented by a set of harmonic functions that are generated with a random initial phase and then oscillate in a predictable manner. In the case when the ratios $\Omega_k/2\pi$ are rational numbers, the autocorrelation function is periodic and can be transformed to the frequency domain using discrete-time Fourier series. The obtained Fourier coefficients represent the power spectrum $S_{XX}(k)$ of the random process $X(n)$. Because $R_{XX}(l)$ is a linear combination of cosine functions, it has a line spectrum with the frequency components $A^2{}_k/4$ positioned symmetrically with respect to the $y$-axis at the discrete points $\pm \Omega_k$.

In the case when the autocorrelation $R_{XX}(l)$ is *periodic*, the ratios $\Omega_k/2\pi$ are rational numbers, and the line components are equidistant. In the case when $R_{XX}(k)$ is *almost periodic*, that is, in the case when the ratios $\Omega_k/2\pi$ are not rational numbers, the line components in the frequency domain will be similar to the component of the periodic autocorrelation function. Hence, the power spectrum of a harmonic process can be generally expressed in the form

$$S_{WW}\left(e^{j\Omega}\right) = \sum_{l=-\infty}^{\infty} R_{WW}(l) e^{-j\Omega\,l} = \frac{1}{2} \sum_{k=1}^{M} A_k^2 \sum_{l=-\infty}^{\infty} \cos \Omega_k l e^{-j\Omega\,l}$$

$$= \frac{1}{2} \sum_{k=1}^{M} A_k^2 \pi \left[\delta\left(\Omega - \Omega_k\right) + \delta\left(\Omega + \Omega_k\right)\right]$$

$$= \frac{\pi}{2} \sum_{k=1}^{M} A_k^2 \left[\delta\left(\Omega - \Omega_k\right) + \delta\left(\Omega + \Omega_k\right)\right]. \qquad (3.6.18)$$

**Example**

For the harmonic process

$$X(n) = \cos\left(0.1\pi n + \theta_1\right) + 2\sin\left(1.5n + \theta_2\right),$$

where the phases are iid random variables uniformly distributed in the interval $[0, 2\pi]$, calculate:

a. the amplitude and period of each process component
b. the mean value of the process
c. the autocorrelation function, and
d. the components of the line spectrum.

**Solution.**

a. The periods of process componts are

$$\Omega_1 = 0.1\pi, A_1 = 1, N_1 = 2/0.1 = 20,$$
$$\Omega_2 = 1.5, A_2 = 2, N_2 = 2\pi/1.5 = 4.19.$$

The spectral components can be found from

$$X(n) = \cos(0.1\pi n + \theta_1) + 2\sin(1.5n + \theta_2)$$
$$= \frac{e^{j(0.1\pi n + \theta_1)} + e^{j(0.1\pi n + \theta_1)}}{2} + 2\frac{e^{j(1.5n + \theta_2)} - e^{j(1.5n + \theta_2)}}{2j}.$$

b. The mean value of the process is

$$\eta_X(n) = E\{\cos(0.1\pi n + \theta_1) + 2\sin(1.5n + \theta_2)\}$$
$$= \int_0^{2\pi} [\cos(0.1\pi n + \theta_1)]\frac{1}{2\pi}d\theta_1 + \int_0^{2\pi} [2\sin(1.5n + \theta_2)]\frac{1}{2\pi}d\theta_2 = 0.$$

c. The autocorrelation function is

$$R_{XX} = E\left\{x(n_1)x^*(n_2)\right\}$$
$$= E\{[\cos(0.1\pi n_1 + \theta_1) + 2\sin(1.5n_1 + \theta_2)][\cos(0.1\pi n_2 + \theta_1) + 2\sin(1.5n_2 + \theta_2)]\}$$
$$= E\{[\cos(0.1\pi n_1 + \theta_1)\cos(0.1\pi n_2 + \theta_1)] + [2\sin(1.5n_1 + \theta_2)2\sin(1.5n_2 + \theta_2)] + 0 + 0\}$$
$$= E\left\{\frac{1}{2}[\cos(0.1\pi(n_1 + n_2) + 2\theta_1) + \cos(0.1\pi(n_1 - n_2))]\right\}$$
$$+ \frac{1}{2}4[\cos(1.5n_1 + \theta_2 - 1.5n_2 - \theta_2) - \cos(1.5n_1 + \theta_2 + 1.5n_2 + \theta_2)]\}$$
$$= \frac{1}{2}\cos 0.1\pi(n_1 - n_2) + 2\cos 1.5(n_1 - n_2) - 2E\{\cos(1.5(n_1 + n_2) + 2\theta_2)\}$$
$$= \frac{1}{2}\cos 0.1\pi l + 2\cos 1.5l.$$

d. The calculated components of the line spectrum are

$$S_{X(n)} = \begin{cases} 1 & \Omega_1 = -1.5 \\ 1/4 & \Omega_2 = -0.1\pi \\ 1/4 & \Omega_3 = 0.1\pi \\ 1 & \Omega_4 = 1.5 \end{cases},$$

and the power spectrum is

$$S_{XX}\left(e^{j\Omega}\right) = 2\pi\delta\left(\Omega+1.5\right) + \frac{\pi}{2}\delta\left(\Omega+0.1\pi\right) + \frac{\pi}{2}\delta\left(\Omega-0.1\pi\right) + 2\pi\delta\left(\Omega-1.5\right).$$

### 3.6.4 Stochastic Binary Processes

Figure 3.6.2 presents one realization $x(n)$ of a discrete-time binary stochastic process $X(n)$ that is defined as a train of polar rectangular non-return-to-zero pulses of duration $N$. This process can be described by its probability density function and autocorrelation function.

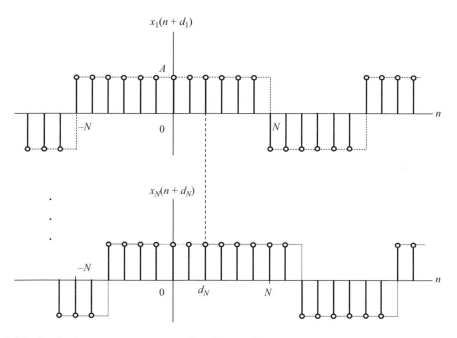

**Fig. 3.6.2** *A polar binary process presented in the time domain.*

We will assume that the probabilities of positive and negative binary-amplitude values $A$ are equal to 0.5. Therefore, at each time instant $n = n_k$, we have a discrete random variable $X = X(n_k)$ that takes the value of $A$ or $-A$, and is defined by its discrete probability density function expressed in terms of the Kronecker delta functions as

$$f_X(x) = \frac{1}{2}\delta\left(x-A\right) + \frac{1}{2}\delta\left(x+A\right). \tag{3.6.19}$$

In addition, it will be assumed that all realizations $\{x(n)\}$ of the stochastic process are asynchronously generated with random delays represented by a discrete random variable $D$ having the uniform discrete density function defined inside the interval $N$ as

$$f_D(d) = \left\{ \begin{array}{cc} 1/N & 0 \leq d \leq N \\ 0 & otherwise \end{array} \right\}. \tag{3.6.20}$$

Two realizations of the process are presented in Figure 3.6.2 for two delays $d_1 = 0$ and $d_N > 0$ that are realizations of the random variable $D$.

Since the amplitude levels of $x(n)$ inside the interval $N$ have the same magnitude values $A$, and occur with equal probabilities, the mean value of the process is zero, and the variance is $A^2$. Because the process is wide-sense stationary, having zero mean and constant variance, the autocorrelation function of the process will depend only on the time difference between two time instants defined as the lag $l$. If the lag is beyond the time interval $N$, the values of the autocorrelation function will be zero. If the lag is inside the time interval $N$, the values of the autocorrelation function will be different from zero and will decrease from zero lag to the lag of $N$. The autocorrelation function can be expressed in the general form

$$R_{XX}(l) = E\{X(n)X(n-l)\} = \sum_{-N}^{N} x(n,d)x(n-l,d)f_D(d)$$

$$= \frac{1}{N} \sum_{-N}^{N} x(n,d)x(n-l,d). \tag{3.6.21}$$

We can have two intervals of summation here related to the possible lags with respect to the referenced lag $l = 0$ when the autocorrelation function has the maximum value:

1. The lag $l < 0$ and the sum can be calculated as

$$R_{XX}(l) = \frac{1}{N} \sum_{-N}^{N-1} x(n,d)x(n-l,d) =_{l<0} \frac{1}{N} \sum_{0}^{N-1+l} A \cdot A$$

$$= A^2 \frac{1}{N}(N-1+l+1) = A^2 \frac{1}{N}(N+l). \tag{3.6.22}$$

2. The lag $l \geq 0$ and the sum can be calculated as

$$R_{XX}(l) = \frac{1}{N} \sum_{l}^{N-1} x(n,d)x(n-l,d) =_{l\geq0} \sum_{n=l}^{n=N-1} A \cdot A \frac{1}{N} =_{n-l=k} \sum_{k=0}^{N-1-l} A^2 \frac{1}{N}$$

$$= A^2 \frac{1}{N}(N-1-l+1) = A^2 \frac{1}{N}(N-l). \tag{3.6.23}$$

Therefore, the autocorrelation function can be expressed in the form

$$R_{XX}(l) = \left\{ \begin{array}{cc} A^2 \frac{1}{N}(N-|l|) & |l| \le N \\ 0 & \text{otherwise} \end{array} \right\}, \tag{3.6.24}$$

and represented by the triangular graph depicted in Figure 3.6.3 for the defined pulse duration $N = 6$, when the autocorrelation function is defined as

$$R_{XX}(l) = \left\{ \begin{array}{cc} A^2 \frac{1}{6}(6-|l|) & |l| \le 6 \\ 0 & \text{otherwise} \end{array} \right\}. \tag{3.6.25}$$

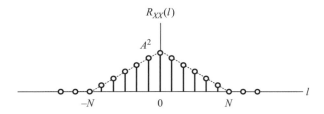

**Fig. 3.6.3** *The autocorrelation function of a binary stochastic process.*

This autocorrelation function is used in communication systems to calculate the power spectral density of a modulating discrete binary signal and the modulated discrete signal.

## 3.7   Linear Systems with Stationary Random Inputs

The theory of LTI systems for continuous-time stochastic processes is presented in Chapter 19. The same theory for continuous- and discrete-time deterministic signals is presented in Chapters 12 and 15, respectively. In this section, an analysis of an LTI system for discrete-time stochastic processes will be presented.

A discrete LTI system is represented by an input realization $x(n)$ of a stochastic process $X(n)$, an output realization $y(n)$ of a stochastic process $Y(n)$, and an LTI device defined by its impulse response $h(n)$, as shown in Figure 3.7.1. This system is analysed in detail in Chapters 14 and 15, which are dedicated to deterministic discrete-time domain processes. The fundamental result of this analysis is that *the output signal of any discrete-time LTI system is the convolution of the input signal and the system impulse response, that is,*

$$y(n) = x(n) * h(n). \tag{3.7.1}$$

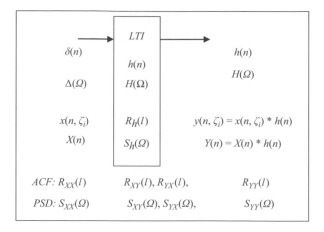

**Fig. 3.7.1** *A linear-time-invariant system; ACF, autocorrelation function; LTI, linear time invariant; PSD, power spectral density.*

### 3.7.1   An LTI System with Stationary Random Inputs in the Time Domain

In this section, we will analyse an LTI system that has a stationary stochastic process at its input. The analysis of input and output processes will be conducted in the time and frequency domains. Unlike the analysis of an LTI system that processes a deterministic signal, the analysis of an LTI system that has a stochastic process at its input must take into account the fact that the stochastic process is not just one discrete-time signal (sequence) but an ensemble of signals. Each realization $x(n, \zeta_i)$ of the input discrete stochastic process $X(n)$ produces an output discrete-time signal $y(n, \zeta_i)$ that can be treated as one realization of the output stochastic process $Y(n)$.

**Convergence issues.** Suppose an outcome $\zeta_i$ is generated with probability $p(\zeta_i)$, and the corresponding discrete-time random signal $x(n, \zeta_i)$, which represents one realization of the related stochastic process $X(n)$, is produced. This discrete-time process is an input of an LTI system, as shown in Figure 3.7.1. The output of the system is a new discrete-time random signal $y(n, \zeta_i)$ that is a convolution of the input discrete-time random signal and the impulse response of the system, that is,

$$y(n, \zeta_i) = \sum_{k=-\infty}^{k=\infty} h(k) x(n-k, \zeta_i). \tag{3.7.2}$$

The convolution expression (3.7.2) can be extended to the general case when the output stochastic process is defined as the convolution of the input stochastic process and the impulse response of the system, expressed as

$$Y(n) = \sum_{k=-\infty}^{k=\infty} h(k)X(n-k). \tag{3.7.3}$$

For this system we will define the input and output signals and related statistical functions as shown in Figure 3.7.1.

If the sum in the right-hand side of (3.7.2) exists for all $\zeta_i \in S$ and $p(S) = 1$, then the convergence exists with probability 1. This convergence is proven by the *Brockwell and Davis theorem*, which is as follows.

Suppose the discrete-time random signal $x(n, \zeta_i)$ is a realization of a stationary process with finite expected value $E\{|x(n, \zeta_i)|\} < \infty$. The discrete-time random signal is applied to the input of an LTI-bounded-input bounded-output (BIBO) system that is defined by $\sum_{n=-\infty}^{n=\infty} h(k) < \infty$. Then, the output discrete-time random signal of the system $y(n, \zeta_i)$ is stationary and converges absolutely with probability 1, or

$$y(n, \zeta_i) = \sum_{k=-\infty}^{k=\infty} h(k)x(n-k, \zeta_i), \tag{3.7.4}$$

for all $\zeta_i \in S$ and $p(S) = 1$. Furthermore, if $E\{|x(n, \zeta_i)|^2\} < \infty$, then $E\{|y(n, \zeta_i)|^2\} < \infty$ and $y(n, \zeta_i)$ is stationary and converges in the mean square to the same limit.

Also, the convergence condition can be expressed by a less restrictive condition of a finite power of the system impulse response, which guaranties the mean square existence of the output process, as expressed by the following theorem.

Suppose the input of an LTI system is a zero-mean stationary process that satisfies the condition $\sum_{l=-\infty}^{\infty} |R_{XX}(l)| < \infty$. If the system satisfies the condition

$$\sum_{n=-\infty}^{\infty} |h(n)|^2 = \frac{1}{2\pi} \int_{-\pi}^{\pi} |H(e^{j\Omega})|^2 d\Omega < \infty, \tag{3.7.5}$$

then the output $y(n, \zeta_i)$ is stationary and converges in the mean square sense.

If we used the expression for the output of an LTI system to find the output of an LTI system for a known input, we would have to calculate an infinite number of convolutions. Therefore, we want to look for an easier way to characterize the output of an LTI system. One alternative is to find the statistical properties of the output when we know both the statistical properties of the input and the characteristics of the system itself. This is of a great practical help in the case of Gaussian processes, where first- and second-order statistics are sufficient for the characterization of the system output.

**Statistics of the input process in the time and frequency domains.** The input of the system is a stochastic process $X(n)$ represented by its autocorrelation function

$$R_{XX}(l) = E\{X(n)X(n+l)\}.$$

The power spectral density function of the input process is

$$S_{XX}\left(e^{j\Omega}\right) = \sum_{l=-\infty}^{\infty} R_{XX}(l)e^{-j\Omega l}, \tag{3.7.6}$$

and the related autocorrelation function can be expressed as its inverse Fourier transform

$$R_{XX}(l) = \frac{1}{2\pi} \int_{-\pi}^{\pi} S_{XX}\left(e^{j\Omega}\right) e^{j\Omega l} d\Omega. \tag{3.7.7}$$

These two expressions form the Fourier transform pair, which is proven by the *Wiener–Khintchine theorem*.

If the impulse delta function $\delta(l)$ with its discrete-time Fourier transform $\Delta(\Omega)$ is applied to the input of an LTI, the impulse response $h(n)$ with its discrete-time Fourier transform $H(\Omega)$ is obtained at the output, and this characterizes the system in the time domain. Any input stochastic process $X(n)$ is represented in the time domain by its autocorrelation function $R_{XX}(l)$, and in the frequency domain by its power spectral density $S_{XX}(\Omega)$.

**Statistics of the system in the time domain.** An LTI system is defined in the time domain by its impulse response $h(n)$. The autocorrelation function of the impulse response is

$$R_h(l) = \sum_{n=-\infty}^{\infty} h(n)h^*(n-l) = h(l)*h^*(-l), \tag{3.7.8}$$

which is called the *system correlation function*. The *cross-correlation* of the input–output processes can be obtained as follows:

$$R_{XY}(l) = E\left\{x(n+l)y^*(l)\right\} = \sum_{k=-\infty}^{k=\infty} h^*(k)E\left\{x(n+l)x^*(n-k)\right\}$$

$$= \sum_{k=-\infty}^{k=\infty} h^*(k)E\left\{x(m)x^*\left(m-(l+k)\right)\right\} = \sum_{k=-\infty}^{k=\infty} h^*(k)R_{XX}(l+k)$$

$$= \sum_{m=-\infty}^{m=\infty} h^*(-m)R_{XX}(l-m) = h^*(-l)*R_{XX}(l). \tag{3.7.9}$$

Similarly, we may obtain the following expression:

$$R_{YX}(l) = h(l)* R_{XX}(l). \tag{3.7.10}$$

**Statistics of the output in the time and frequency domains.** The mean value of the output process is

$$\eta_Y = E\{Y(n)\} = \sum_{k=-\infty}^{k=\infty} h(k)E\{X(n-k)\} = \eta_X \sum_{\Omega=0 \, \& \, k=-\infty}^{k=\infty} h(k)e^{j\Omega k} = \eta_X H\left(e^{j0}\right), \tag{3.7.11}$$

where the constant $H(e^{j0})$ is the frequency-domain transfer function of the system, or the DC gain of the system. In other words, the mean of the output is equal to the input mean multiplied by the values of the frequency response of the LTI system at $\Omega = 0$. Because the mean value of the input process is constant, the mean value of the output is also a constant.

The autocorrelation function of the output can be obtained from the expression for the convolutional sum rearranged into the following form:

$$R_{YY}(l) = E\left\{y(n)y^*(n-l)\right\} = \sum_{k=-\infty}^{k=\infty} h(k)E\left\{x(n-k)y^*(n-l)\right\}$$

$$= \sum_{k=-\infty}^{k=\infty} h(k)R_{XY}(l-k) = h(l)* R_{XY}(l). \tag{3.7.12}$$

Using the relations for the cross-correlation, we obtain

$$R_{YY}(l) = h(l)* R_{XY}(l) = h(l)* h^*(-l) * R_{XX}(l) = R_h(l)* R_{XX}(l). \tag{3.7.13}$$

The output of a stable system $y(n)$ to a stationary input $x(n)$ is a stationary process, because its mean is constant and the autocorrelation function depends only on the lag $l$. Also, *the autocorrelation of the output process is equal to the convolution of the autocorrelation of the input process and the autocorrelation of the impulse response of the system itself*, as shown in Figure 3.7.2.

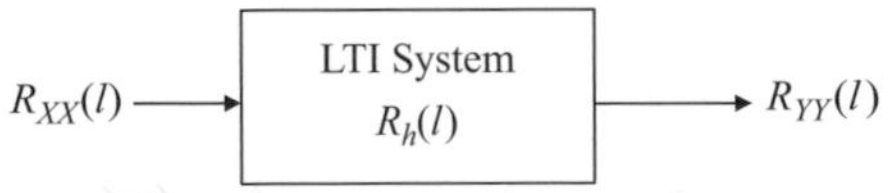

**Fig. 3.7.2** *Correlation characteristics of a linear-time-invariant system; LTI, linear time-invariant.*

**Output power.** The power of the output process is equal to the value of output autocorrelation at zero lag, that is,

$$P_Y = R_{YY}(0) = R_h(l) * R_{XX}(l)|_{l=0} = \sum_{k=-\infty}^{k=\infty} R_h(k)R_{XX}(0-k)$$

$$= \sum_{k=-\infty}^{\infty} \left[ h(k) * h^*(-k) \right] R_{XX}(k) = \sum_{k=-\infty}^{\infty} \sum_{m=-\infty}^{\infty} h(m) \cdot h^*(m-k) R_{XX}(k)$$

$$= \sum_{k=-\infty}^{\infty} R_{XX}(k) \sum_{m=-\infty}^{\infty} h(m) \cdot h^*(m-k)$$

$$= \sum_{k=-\infty}^{\infty} R_h(k)R_{XX}(k). \tag{3.7.14}$$

**Output probability density function.** Generally speaking, finding the probability density function of the output of an LTI system is a hard task. There are some cases when this problem is solvable. Firstly, if the input process is Gaussian, the output will be also Gaussian. Secondly, if the input is an iid process, the output is a weighted sum of independent random variables which has a density function which is a convolution of the densities of these individual variables.

### 3.7.2 Frequency-Domain Analysis of an LTI System

The $z$-transform of the real impulse response $h(n)$ of an LTI system is

$$z\{h(n)\} = H(z) = \sum_{n=-\infty}^{\infty} h(n)z^{-n}. \tag{3.7.15}$$

Similarly, we may have

$$z\left\{ h^*(-n) \right\} = H^*\left( \frac{1}{z^*} \right). \tag{3.7.16}$$

By using expression for the cross-correlation, we obtain the complex cross-power spectral densities

$$S_{XY}(z) = H^*\left( \frac{1}{z^*} \right) \cdot S_{XX}(z), \text{and } S_{YX}(z) = H(z) \cdot S_{XX}(z). \tag{3.7.17}$$

The output power spectral density is

$$S_{YY}(z) = H(z) \cdot H^*\left( \frac{1}{z^*} \right) \cdot S_{XX}(z). \tag{3.7.18}$$

In the case of a stable system, the unit circle lies within the receiver operating characteristics of $H(z)$ and $H(z^{-1})$. Therefore, we may derive the following relations for the power spectral densities of an LTI system:

$$S_{XY}\left(e^{j\Omega}\right) = H^*\left(\frac{1}{e^{-j\Omega}}\right) \cdot S_{XX}\left(e^{j\Omega}\right) = H^*\left(e^{j\Omega}\right) \cdot S_{XX}\left(e^{j\Omega}\right)$$

$$S_{YX}\left(e^{j\Omega}\right) = H\left(e^{j\Omega}\right) \cdot S_{XX}\left(e^{j\Omega}\right)$$

$$S_{YY}\left(e^{j\Omega}\right) = H\left(e^{j\Omega}\right) \cdot H^*\left(\frac{1}{e^{-j\Omega}}\right) \cdot S_{XX}\left(e^{j\Omega}\right) = \left|H\left(e^{j\Omega}\right)\right|^2 \cdot S_{XX}\left(e^{j\Omega}\right). \qquad (3.7.19)$$

From the last expression in (3.7.19) we can see that, if we have the autocorrelation or auto power spectral density of the input and output process of an LTI system, then we can determine the magnitude response of the system. However, we cannot determine the phase response of the system. To do this, we can use the cross-spectral densities. The relationship between the processes in the frequency domain is presented in Figure 3.7.3.

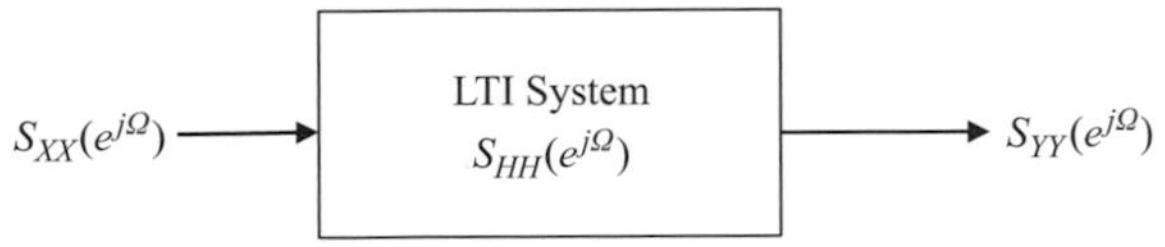

**Fig. 3.7.3** *Power spectral densities of a linear-time-invariant system; LTI, linear time-invariant.*

By using a frequency-domain representation of a random process, we can find the power of the output process in the following form:

$$E\left\{|Y(n)|^2\right\} = \frac{1}{2\pi}\int_{-\pi}^{\pi} S_{YY}\left(e^{j\Omega}\right)d\Omega = \frac{1}{2\pi}\int_{-\pi}^{\pi}\left|H\left(e^{j\Omega}\right)\right|^2 S_X\left(e^{j\Omega}\right)d\Omega$$

$$= R_{YY}(0) = \sum_{l=-\infty}^{\infty} R_{HH}(l)R_{XX}(l). \qquad (3.7.20)$$

**Example**

Suppose the impulse response $h(k)$ of an LTI system is

$$h(k) = \left\{\begin{array}{ll} 1/2 & k=0 \\ 1/2 & k=1 \\ 0 & otherwise \end{array}\right\},$$

and the input random signal $x(n)$ is a realization of a white Gaussian noise process $X(n)$ with variance $\sigma_X^2$.

a. Find the output of the system $y(n)$ in the time domain.

b. Find the statistics of the output process.

c. Find the transfer characteristics of the system in the $z$ domain.

d. Find the output of the system in the $z$ domain and the frequency domain.

e. Find the frequency-transfer characteristics of this system.

**Solution.**

a. The output of the system in the time domain is

$$y(n) = \sum_{k=-\infty}^{k=\infty} h(k)x(n-k) = \sum_{k=0}^{k=1} h(k)x(n-k)$$

$$= h(0)x(n-0) + h(1)x(n-1) = \frac{1}{2}x(n) + \frac{1}{2}x(n-1).$$

b. The mean value of the output process can be obtained from the time-domain analysis as

$$\eta_y = \sum_{k=-\infty}^{k=\infty} h(k)E\{X(n-k)\} = h(0)E\{X(n)\} + h(1)E\{X(n-1)\} = \eta_x[h(0) + h(1)].$$

The autocorrelation of the output can be obtained as follows:

$$R_{YY}(l) = E\left\{Y(n)Y^*(n-l)\right\} = \frac{1}{4}E\left\{[Y(n) + Y(n-1)][Y(n-l) + Y(n-l-1)]\right.$$

$$= \frac{1}{4}[E\{Y(n)Y(n-l) + E\{Y(n)Y(n-l-1) + E\{Y(n-1)Y(n-l)$$

$$+ E\{Y(n-1)Y(n-1-l)\}]$$

$$= \frac{1}{4}[R_{XX}(l) + R_{XX}(l+1) + R_{XX}(l-1) + R_{XX}(l)].$$

$$= \frac{1}{4}[2R_{XX}(l) + R_{XX}(l+1) + R_{XX}(l-1)].$$

The power of the output is

$$P_Y = R_{YY}(0) = \frac{1}{4}[2R_{XX}(0) + R_{XX}(0+1) + R_{XX}(0-1)] = \frac{1}{4}[2R_{XX}(0) + R_{XX}(1) + R_{XX}(-1)].$$

c. The *z*-transform of the real impulse response $h(n)$ of an LTI system is

$$z\{h(n)\} = H(z) = \sum_{n=-\infty}^{\infty} h(n)z^{-n} = \sum_{n=0}^{1} h(n)z^{-n} = \frac{1}{2} + \frac{1}{2}z^{-1} = \frac{z+1}{2z}.$$

d. The output of the system in the *z* domain is

$$z\{y(n)\} = Y(z) = X(z)H(z) = X(z)\frac{z+1}{2z} = \frac{1}{2}X(z)\left(1+z^{-1}\right),$$

and the output in the time domain can be obtained from

$$Y(z) = \frac{1}{2}X(z) + \frac{1}{2}X(z)z^{-1},$$

which corresponds to the following expression in the time domain:

$$y(n) = \frac{1}{2}x(n) + \frac{1}{2}x(n-1).$$

e. The output process in the frequency domain is

$$Y\left(e^{j\Omega}\right) = \frac{1}{2}X\left(e^{j\Omega}\right)\left(1+e^{-j\Omega}\right) = X\left(e^{j\Omega}\right)H\left(e^{j\Omega}\right),$$

where the transfer function alone is

$$H\left(e^{j\Omega}\right) = \frac{1}{2}\left(1+e^{-j\Omega}\right).$$

Its magnitude spectrum can be obtained as follows:

$$|H\left(e^{j\Omega}\right)| = \frac{1}{2}|(1+\cos\Omega - j\sin\Omega)| = \frac{1}{2}\sqrt{(1+\cos\Omega)^2 + \sin^2\Omega}$$
$$= \frac{1}{2}\sqrt{1+2\cos\Omega + \cos^2\Omega + \sin^2\Omega} = \frac{1}{2}\sqrt{2+2\cos\Omega}.$$

## 3.8   Summary

This chapter presents the fundamental theory and application of discrete-time stochastic processes. Definitions of stationary and ergodic processes, and their characterization in time and frequency domains, have been provided. The chapter also provided a detailed theoretical analysis of the mean, autocorrelation, and power spectral density functions of discrete-time stochastic processes and their application to the analysis of LTI systems.

.......................................................................................................................................

PROBLEMS

1.  Define a random process as the difference $X(n) = W(n) - W(n-1)$, where $W(n)$ is an iid random process with mean $\eta$ and variance $\sigma^2$.

    a.  Find the autocorrelation function of this process.
    b.  Sketch the time series of the autocorrelation function $R_{XX}(n, m)$.
    c.  Comment on the mean and the characteristics of the autocorrelation function of $X(n)$.

2.  Define a random process as the random DC process $X(n) = A$, where $A$ is a normal random variable with zero mean $\eta = 0$ and unity variance $\sigma^2 = 1$, that is, $N(0, 1)$.

    a.  Sketch a realization of this random process $X(n)$ if a generated sample of the random variable $x(n)$ is $A = 0.5$.
    b.  Find the mean value and autocorrelation function of this process.
    c.  Comment on the mean value of one realization $x(n)$ of the random process $X(n)$.

3.  Suppose the input of an LTI system is a wide-sense stationary $X(n)$ defined by its covariance and autocorrelation functions:

$$C_{XX}(m,n) = R_{XX}(m,n) = \sigma^2 \delta(m-n) = \sigma^2 \delta(l).$$

The output process $Y(n)$ is defined by its covariance function

$$C_{YY}(m,n) = \frac{1}{2}\sigma^2 \delta(m-n) + \frac{1}{4}\sigma^2 \delta(m-n+1) + \frac{1}{4}\sigma^2 \delta(m-n-1)$$

$$= \frac{1}{2}\sigma^2 \delta(l) + \frac{1}{4}\sigma^2 \delta(l+1) + \frac{1}{4}\sigma^2 \delta(l-1) = C_{YY}(l).$$

a. Find and sketch the graph of the power spectral density of the input process $X(n)$. Calculate the power of the process from its spectral characteristics.

b. Find the power spectral density of the output process $Y(n)$. Find the average power of this process.

c. Find the transfer function of the LTI system.

4. For the harmonic process

$$X(n) = \cos(0.1\pi n + \theta_1) + 2\sin(1.5n + \theta_2),$$

where the phases are iid random variables uniformly distributed in the interval $[0, 2\pi]$, do the following:

a. Find the periods of each process component.

b. Find the mean value of the process.

c. Find the autocorrelation function.

d. Calculate the components of the line spectrum.

5. For the exponential process

$$X(n) = Ae^{j\Omega_0 n} = |A| e^{j(\Omega_0 n + \theta)},$$

a. Find the mean value of the process.

b. Find the autocorrelation function.

c. Calculate the components of the power spectrum.

6. Suppose that two stochastic processes are $X(n) = W(n)$ and $Y(n) = W(n) + 2W(n-1)$, where $Y(n)$ is a general moving average process, and $W(n)$ is a white noise process with zero mean and variance $\sigma_W^2 = \sigma^2 = 1$.

a. Find the mean values of $X(n)$ and $Y(n)$. Find the autocorrelation and power spectral density of $X(n)$.

b. Sketch the graphs of the autocorrelation and power spectral density functions of $X(n)$, assuming that the variance is $\sigma^2 = 1$. Is $X(n)$ a stationary process? Prove your answer.

c. Find the autocorrelation of $Y(n)$. Sketch the autocorrelation sequence of $Y(n)$. Is $Y(n)$ a stationary process?

d. Find the cross-correlation function $R_{XY}(l)$ of $X(n)$ and $Y(n)$. Are $X(n)$ and $Y(n)$ jointly wide-sense stationary processes?

7. The autocorrelation function of a zero-mean wide-sense stationary process $X(n)$ is defined as

$$R_x(l) = \frac{\sigma^2}{1-a^2} a^{|l|},$$

where $-\infty < l < \infty$.

a. Prove that the expression of the power spectral density $S_{XX}(e^{j\Omega})$ for this wide-sense stationary process is

$$S_X(e^{j\Omega}) = \frac{\sigma^2}{1-2a\cos\Omega + a^2}.$$

b. What is the value of the average power of this process in the case when this condition is fulfilled $\sigma^2 = 1 - a^2$? How can we calculate this power by using the calculated power spectral density function of the process $X(n)$?

c. Sketch the graph of the power spectral density for this process. Is this graph a periodic function? What is the maximum value of the power spectral density in the case when $a = 0.5$ and $\sigma^2 = 1 - a^2$?

8. A stochastic process is defined as

$$X(n) = A\cos(\Omega n + \theta).$$

a. Find the autocorrelation function of this stochastic process.

b. For the calculated autocorrelation function, find a $4 \times 4$ autocorrelation matrix.

9. Suppose a random stochastic process $Y(n)$ is defined as a linear function of stochastic process $X(n)$, that is, $Y(n) = aX(n) - b$. Suppose $X(n)$ is defined by its iid random variables for any $n$ that have their mean $\eta$ and variance $\sigma^2$.

a. Find the mean of $Y(n)$.

b. Find the variance of $Y(n)$.

10. Suppose the process $X(n)$ is a purely random process, which means that the random variables defined for every $n$ are iid random variables. The process has a zero mean $\eta$ and variance $\sigma^2$. Define the moving average process of order $k$ as

$$Y(n) = a_0 X(n) + a_1 X(n-1) + a_2 X(n-2) + \cdots + a_k X(n-k).$$

a. Find the mean of $Y(n)$ and the variance of $Y(n)$.

b. Find the autocorrelation of $Y(n)$ for the leg $l = 1$ and then for any lag $l$.

c. Discuss the stationarity of the process $Y(n)$.

d. Discuss the stationarity of the process $Y(n)$, assuming that the process $X(n)$ is normal.

11. Suppose that the stochastic process $W(n)$ is defined as a series in the time domain of iid random variables, with mean $\eta$ and variance $\sigma^2$. Let us define the random walk process as $X(n) = X(n-1) + W(n)$ for $X(0) = 0$, which can be expressed as

$$X(n) = \sum_{i=1}^{n} W(i).$$

Find the mean value and the variance of this process.

12. A process $X(n)$ is called a purely random process if it is represented by a sequence of iid random variables.

a. Find the autocovariance and the autocorrelation function of this process.

b. Suppose that the mean value of the process is zero. Find the autocovariance and the autocorrelation function of this process.

c. Find the power spectral density for the process defined in b.

13. The autocorrelation function of a zero-mean wide-sense stationary process $X(n)$ is defined as

$$R_{XX}(l) = a^{|l|}, \text{where} -8 < l < 8 \text{ and } 0 < a < 1.$$

Prove that the expression of the power spectral density $S_{XX}(e^{j\Omega})$ for this wide-sense stationary process is

$$S_{XX}\left(e^{j\Omega}\right) = \frac{1-a^2}{1-2a\cos\Omega + a^2}.$$

Find the average power of the process. Sketch the graph of the power spectral density for this process. Is this graph a periodic function? What is the maximum value of the power spectral density in the case when $a = 0.5$?

14. The discrete-time Fourier pair for the autocorrelation function and the corresponding power spectral density are, respectively,

$$S_{XX}\left(e^{j\Omega}\right) = \sum_{l=-\infty}^{\infty} R_{XX}(l)e^{-j\Omega l}, \quad R_{XX}(l) = \frac{1}{2\pi}\int_{-\pi}^{\pi} S_{XX}\left(e^{j\Omega}\right) e^{j\Omega l} d\Omega.$$

Prove the expression for the autocorrelation function by evaluating the above integral.

15. Suppose $X(n)$ is a white Gaussian noise process that is defined by its zero mean and autocorrelation

$$R_{XX} = \sigma^2 \delta(l) = \delta(l).$$

   a. Find the power spectral density of $X(n)$.

   b. Form a new stochastic process $Y(n) = X(n) + aX(n-1)$. Find the autocorrelation function of this process.

   c. Find the power spectral density of $Y(n)$ and plot its graph.

16. The signal $Y(n)$ is the sum of the signal $X(n)$ and the noise $N(n)$, that is, $Y(n) = X(n) + N(n)$. The content of the signal $X(n)$ can be found by cross-correlating the signal $Y(n)$ with the signal $X(n)$.

   a. Find the cross-correlation $R_{XY}(n)$ and comment on the result, given the relationship of $X(n)$ and $N(n)$.

   b. Suppose the signal $X(n) = A$ has zero mean and variance $\sigma_X^2$, and $N(n)$ is white noise. Find the power spectral density of $Y(n)$.

17. A stochastic process is defined as

$$X(n) = A\cos^2(\Omega n + \theta),$$

   where the phase of signal $\theta$ is a random variable uniformly distributed in the interval $[0, 2\pi]$.

   a. Sketch this process in the time domain for a period of $T = 5$ and the zero phase $\theta$. Find its mean.

   b. Find the autocorrelation function of a stochastic process.

   c. Sketch the graph of this autocorrelation function. Find the power of the signal $X(n)$ from the calculated autocorrelation function?

   d. Confirm that the power found from the autocorrelation function in c. is equal to the power of the stochastic signal obtained by calculating the power of the stochastic process.

18. Prove the following expressions for cross-correlation functions:

$$S_{XY}\left(e^{j\Omega}\right) = S_{YX}^*\left(e^{j\Omega}\right) \text{ and } S_{XY}^*\left(e^{j\Omega}\right) = S_{YX}\left(e^{j\Omega}\right).$$

19. The stochastic process $Y(n)$ is a moving average process defined as

$$Y(n) = X(n) + a \cdot X(n-1),$$

where $X(n)$ is a white noise process defined by its zero mean, autocorrelation delta function $R_{XX}(l) = \sigma_X^2 \delta(l)$, and constant power spectral density $S_{XX}\left(e^{j\Omega}\right) = \sigma_X^2$.

a. Find the autocorrelation function of $Y(n)$.

b. Find the autocorrelation matrix of $Y(n)$.

c. Find the covariance matrix of the process from the autocorrelation matrix obtained in b.

d. Find the power spectral density $S_{YY}\left(e^{j\Omega}\right)$ of the stochastic process $Y(n)$. Plot the graph of this power spectral density function for $a = 1$. Calculate the average power $P$ of the process $Y(n)$.

20. The first-order Gauss–Markov process is defined by the expression

$$s(n) = as(n-1) + u(n) \quad n \geq 0,$$

where $u(n)$ is white Gaussian noise with variance $\sigma_u^2 = 2$ and constant $a < 1$, and $s(-1)$ is a Gaussian process defined by zero mean $\eta_s = 0$ and unit variance $\sigma_s^2 = 1$.

a. Find the mean and the variance of the process $s(n)$ in recursive form.

b. Prove that this process can be expressed in the general form

$$s(n) = a^{n+1} s(-1) + \sum_{k=0}^{n} a^k u(n-k).$$

Find the mean value of this process and sketch the graph showing the mean in the discrete-time domain.

c. Prove that the covariance function of $s(n)$, for $m \geq n$, is

$$C_{ss}(m,n) =_{m \geq n} a^{m+n+2}\sigma_s^2 + \sigma_u^2 a^{m-n} \sum_{k=0}^{n} a^{2k},$$

and the variance of the process is

$$\mathrm{var}\,(s(n)) = a^{2n+2}\sigma_s^2 + \sigma_u^2 \sum_{k=0}^{n} a^{2k}.$$

d. Using the results of the calculations from a. and b., find out if this process is wide-sense stationary and explain your answer.

e. Find the expressions of the mean and the variance of $s(n)$ when $n$ tends to infinity. Comment on the stationarity properties of this process.

21. A stochastic process is defined as $X(n) = A\cos(\Omega n + \theta) + B\cos(\Omega n + \theta)$, where $A$ and $B$ are two independent normal random variables with a zero mean and variance $\sigma^2$, and $\Omega$ and $\theta$ are constant values.

    a. Find the mean of $X(n)$ and present it in graphical form.

    b. Find the autocorrelation function of the stochastic process.

    c. Find the power of the signal $X(n)$ from the calculated autocorrelation function.

    d. Confirm that the power found from the autocorrelation function is equal to the power of the stochastic signal obtained by calculating the power of the stochastic process, using its expression in the time domain.

22. Suppose $X(n)$ is a white Gaussian noise process that is defined by its zero mean and autocorrelation $R_{XX} = \sigma^2\delta(l) = 4\delta(l)$.

    a. Find the power spectral density of $X(n)$. Sketch graphs of both the autocorrelation function and the power spectral density. Identify the period of the power spectral density on the graph representing the power spectral density. Investigate the stationarity of this process.

    b. Form a new stochastic process $Y(n) = X(n) + 2X(n-1) + 3X(n-2)$. Find the autocorrelation function of this process. Sketch graphs of this autocorrelation function. Is $Y(n)$ a stationary process? Prove your answer.

    c. Find the power spectral density of $Y(n)$. Investigate the periodicity of this function.

23. Suppose a reverse-link CDMA system uses a chaotic sequence to spread message symbols. Suppose the spreading sequences $S_n(n)$ are generated using a logistic map expressed by the following recursive formula:

$$c(n+1) = 2[c(n)]^2 - 1, -1 < c < +1.$$

The sequence of chaotic samples can be defined as a realization of a discrete-time stochastic process. At each time instant $n$, the samples are outcomes of a random variable $C(n)$ that has the density function

$$f_C(c) = \left\{ \begin{array}{ll} \frac{1}{\pi\sqrt{1-c^2}}, & -1 < c < +1 \\ 0, & otherwise \end{array} \right\}.$$

    a. Prove that the random variable $C(n)$ has zero mean and variance 0.5. How is this variance related to the power of the sequence $C(n)$?

    b. If we are using standard pseudorandom sequences, their average power is 1. Define a new random variable $Y(n)$, which is a function of $C(n)$, which will produce sequences that have an average power equal to 1.

c. Plot the graph of the probability density function of $Y(n)$, and find the mean value of random variable $Y(n)$. Prove that the variance of $Y(n)$ is 1.

d. The chaotic sequence is multiplied by itself at the receiver side. Then, at each point $n$, we can define a new random variable $Z$ that is the random variable $Y(n)$ squared, that is, $Z(n) = [Y(n)]^2$. Find the density function of random variables $Z(n)$. Plot the graph of the obtained density function.

24. Suppose the input of an LTI system is a white Gaussian process $X(n)$ defined by zero mean and variance $\sigma^2$ for every time instant $-\infty < n < \infty$. The output of the system is a moving average process expressed as

$$Y(n) = \frac{1}{2}[X(n) + X(n-1)].$$

a. Find the mean value $\eta_x$, variance $\sigma^2$, autocorrelation $R_{XX}$, and covariance $C_{XX}$ of the input process $X(n)$.

b. Find the mean value $\eta_y$ and variance $\sigma^2_y$ of the output process $Y(n)$.

c. Find the covariance $C_{YY}$ of the output process $Y(n)$.

d. Find and sketch the graph of the autocorrelation function $R_{YY}$ of the output process $Y(n)$ in the case when the variance of the input process is $\sigma^2 = 1$.

25. The impulse response $h(k)$ of an LTI system is defined as

$$h(k) = \begin{cases} 1/3 & k = 0 \\ 1/3 & k = 1 \\ 1/3 & k = 2 \\ 0 & otherwise \end{cases},$$

and the input process $X(n)$ is a white Gaussian noise process with variance $\sigma^2$.

a. Find the output of the system $y(n)$ in the time domain. Sketch seven samples of a hypothetical random process that best represent $X(n)$ from $n = -3$ to $n = 3$, and then sketch three samples of the output process $Y(n)$ at positions $n = 1$, $n = 2$, and $n = 3$.

b. Find the mean and the autocorrelation function of the output process $Y(n)$. Find the power of the output process.

c. Find the transfer characteristic of the LTI system in the $z$ domain. Find the output of the LTI system in the $z$ domain and the frequency domain.

d. Find the frequency transfer characteristic of this LTI system.

26. A first-order autoregressive process can be generated using the system shown in Figure 26. The output of system is defined as

$$Y(n) = b_1 Y(n-1) + X(n),$$

where the input $X(n)$ is a zero-mean white nose with variance $\sigma_X^2$.

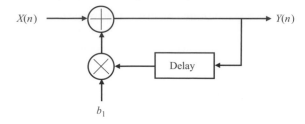

**Fig. 26** *An autoregressive process.*

   a.  Find the impulse response of the system in the time and frequency domains.

   b.  Find the power spectral density of the output signal of the system.

27.  Suppose the system that generates a first-order AR(1) process is defined by its transfer function

$$H(z) = \frac{1}{1 - b_1 z^{-1}},$$

where $|b_1| < 1$, meaning that the poles at $z = b_1$ are inside the unit circle, and the system is stable.

   a.  Find the difference equation specifying the output of the synthesis system.

   b.  What kind of sequence generates this system?

   c.  Calculate the autocorrelation function.

   d.  Calculate the power spectrum of the process.

# 4

# Noise Processes in Discrete Communication Systems

## 4.1 Gaussian Noise Processes in the Continuous-Time Domain

### 4.1.1 Continuous White Gaussian Noise Processes

There are a large number of noise sources in communication systems, including external sources existing around the systems, and internal sources existing inside the devices of communication systems. Therefore, the total noise in a system can be expressed as a sum of these independent noises. Precisely speaking, the noise can be represented by a random variable (or a random function) that is a sum of a large number of random variables representing the noises generated by independent noise sources. Following the central limit theorem, the noise random variable will have a Gaussian distribution defined by its zero mean and a variance that usually represents the noise power.

Because the noise is generated in the time domain, having at each time instant a random value, it is mathematically represented as a stochastic process, more precisely, a wide-sense stationary stochastic process as defined in Chapters 3 and 19. Because the random variables of the process, defined at each time instant, are Gaussian, the related process is Gaussian. Furthermore, in practical applications, the noise generator is designed to produce one realization of this stochastic process that contains complete statistical properties of the ensemble, that is, the noise will have the properties of an ergodic process.

In this chapter, we will start with an analysis of continuous-time noise processes and then continue with discrete-time noise processes. The notation in this chapter will be the same as the notation used in Chapter 3 and Chapter 19 for the theory of discrete- and continuous-time stochastic processes. It is advisable to read those chapters before starting with this chapter. One realization $w(t)$ of a wide-sense stochastic noise process $W(t)$ is shown by the dotted continuous-time graph in Figure 4.1.1. The noise is a Gaussian process if its amplitudes have the Gaussian probability density function expressed as

$$f_W(w) = \frac{1}{\sqrt{2\pi \sigma_W^2}} e^{-w^2/2\sigma_W^2},$$

(4.1.1)

*Discrete Communication Systems*. Stevan Berber, Oxford University Press. © Stevan Berber 2021.
DOI: 10.1093/oso/9780198860792.003.0004

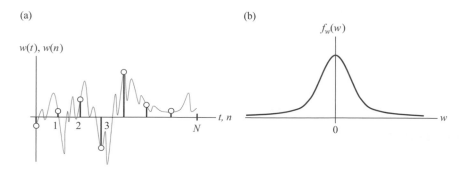

**Fig. 4.1.1** *(a) Samples of continuous-time and discrete-time white Gaussian noise processes W(t) and W(n). (b) The probability density function of the noise amplitudes.*

for the zero mean and finite variance $\sigma_W^2$. The noise samples are realizations of an uncorrelated, wide-sense stationary Gaussian process. Therefore, we say that the noise amplitudes are realizations of the identical and independent Gaussian random variables.

The autocorrelation function of the Gaussian process can be derived as

$$R_{WW}(\tau) = E\{W(t)W(t+\tau)\} = \begin{Bmatrix} E\{W^2(t)\} & \tau = 0 \\ 0 & \tau \neq 0 \end{Bmatrix} = \begin{Bmatrix} \sigma_W^2 & \tau = 0 \\ 0 & \tau \neq 0 \end{Bmatrix} = \sigma_W^2 \delta(\tau). \quad (4.1.2)$$

Note that $R_{WW}(\tau)$ is the Dirac delta function weighted by the factor $\sigma_W^2$ and occurring at $\tau = 0$. It is zero for all $\tau \neq 0$, meaning that any two different samples of the white Gaussian noise, no matter how closely together in time they are taken, are uncorrelated. Consequently, being Gaussian, the samples are independent.

The autocorrelation function has the infinite value at the origin. Therefore, for a finite variance $\sigma_W^2$, the process defined in this way has the infinite power defined by $R_{WW}(0)$ as

$$P_W = R_{WW}(0) = \sigma_W^2 \delta(0) \to \infty. \quad (4.1.3)$$

This statement can be supported by the following rationale. The noise amplitudes can take infinite values, and the power calculated in the time domain, either across the ensemble of the process or inside one realization of the process, will be infinite due to the existence of at least one amplitude of the infinite amplitude inside a sufficiently long interval of time. Following the Wiener–Khintchine theorem, this power calculated in the time domain needs to be the same as the power calculated in the frequency domain. Let us check the theorem. The power spectral density of the process is the Fourier transform of the autocorrelation function expressed as

$$S_{WW}(f) = \int_{-\infty}^{+\infty} R_W(\tau) e^{-j2\pi f\tau} \, d\tau = \sigma_W^2, \quad (4.1.4)$$

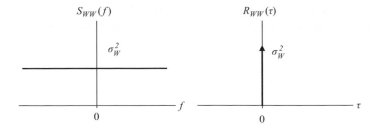

**Fig. 4.1.2** *Functions of a continuous-time wide-sense stationary white Gaussian noise process.*

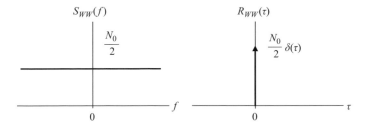

**Fig. 4.1.3** *Functions of a continuous-time wide-sense stationary white Gaussian noise process.*

for all frequencies $-\infty < f < \infty$, and can be represented by the two-sided power spectral density of the constant value variance expressed in watts per hertz. The noise is traditionally called white, because it is constant for all frequencies from minus to plus infinity.

**Example**

Prove eqn (4.1.2) using the strict definition of an autocorrelation function given in Chapter 19.

**Solution.** Slightly changing the notation for the sake of clarity, the autocorrelation function can be calculated as

$$R_{WW}(\tau) = R\{W(t_1)W(t_2)\} = E\{W_1 \cdot W_2\} = \int_{-\infty}^{\infty}\int_{-\infty}^{\infty} w_1 w_2 f_{W_1 W_2}(w_1, w_2) dw_1 w_2$$

$$= \int_{-\infty}^{\infty}\int_{-\infty}^{\infty} w_1 w_2 f_{W_1}(w_1) f_{W_2}(w_2) dw_1 dw_2 = \left\{ \begin{array}{l} \int_{-\infty}^{\infty} w_2^2 f_{W_2}(w_2) dw_2 \int_{-\infty}^{\infty} f_{W_1}(w_1) f_{W_1} \\ \int_{-\infty}^{\infty} w_2 f_{W_2}(w_2) dw_2 \int_{-\infty}^{\infty} f_{W_1}(w_1) f_{W_1} \end{array} \right\}$$

$$= \left\{ \begin{array}{ll} \sigma_W^2 & t_1 - t_2 = 0 \\ 0 & t_1 - t_2 = \tau \end{array} \right\} = \sigma_W^2 \delta(t_1 - t_2) = \sigma_W^2 \delta(\tau)$$

For white Gaussian noise in communication systems, the two-sided power spectral density $S_{WW}(f)$ of a constant value $\sigma_W^2 = N_0/2$ is defined and referenced to the input circuits of the receiver. According to the Wiener–Khintchine theorem, the autocorrelation function $R_{WW}(\tau)$ is the inverse Fourier transform of the power spectral density $S_{WW}(f)$. Both the autocorrelation function and the power spectral density function are presented in Figure 4.1.3.

Because the power spectral density has a constant value of $N_0/2$ for all frequencies, the white noise process has an infinite power, that is, the integral value of the power spectral density tends to infinity:

$$P_W = \int_{-\infty}^{+\infty} S_{WW}(f)df = \frac{N_0}{2} \int_{-\infty}^{+\infty} df \to \infty, \tag{4.1.5}$$

which complies with the previous discussion related to the power calculation from both the autocorrelation function and from the ensemble of the noise process in time domain. In practice, at the output of a noise generator, a random noise $w(t)$ needs to be generated in a limited time interval $T$. In simulating communication systems, for example, this interval corresponds to a bit duration. If the noise has the property of the above-mentioned white Gaussian noise, it cannot be realized, as can be seen from the following analysis. The power of noise, contained in the $k$th $T$ interval, will have a random value expressed as

$$p_{w,k} = \frac{1}{T} \int_{kT}^{(k+1)T} w^2(t)dt, \tag{4.1.6}$$

which can be considered to be one realization of a random variable $P_w$ defined as

$$P_w = \frac{1}{T} \int_{kT}^{(k+1)T} W^2(t)dt. \tag{4.1.7}$$

The expected power of the noise process in the time interval $T$ can be calculated by finding the expectation of the random variable $P_w$, which is the power of the stochastic process $P_W$ according to the expression

$$\begin{aligned}
E\{P_w\} = E\left\{\frac{1}{T} \int_{kT}^{(k+1)T} W^2(t)dt\right\} &= \frac{1}{T} \int_{kT}^{(k+1)T} E\left\{W^2(t)\right\} dt \\
&= \frac{1}{T} \int_{kT}^{(k+1)T} P_W \, dt = P_W \to \infty
\end{aligned} \tag{4.1.8}$$

where, in the case of a Gaussian noise process, the process $W(t)$ at various time instants $t$ is represented by iid random variables $W(t)$ defined for any time instant $t$ inside the interval $T$.

The infinite power and, consequently, the infinite energy of the noise value can also be confirmed in the time domain by calculating the total power of the ergodic process $w(t)$, which is a realization of the stochastic process $W(t)$. For this purpose, the realization of the noise process $w(t)$ will be subdivided into $N$ intervals, and the power in each of them will be calculated. If we let $N$ to tend to infinity, the average power can be found to be

$$\lim_{N\to\infty} \frac{1}{N} \sum_{k=0}^{N-1} \frac{1}{T} \int_{kT}^{(k+1)T} w^2(t)dt = \lim_{N\to\infty} \frac{1}{N} \sum_{k=0}^{N-1} p_{w,k} = \lim_{N\to\infty} P_W = P_W \to \infty. \tag{4.1.9}$$

This relation can be considered correct, because we can imagine that at least one interval $T$ contains a noise amplitude of an infinite value, which results in the infinite overall power. The noise energy is

$$\lim_{N\to\infty} \sum_{k=0}^{N-1} \int_{kT}^{(k+1)T} w^2(t)dt = \lim_{N\to\infty} \sum_{k=0}^{N-1} E_{w,k} = E_W \to \infty. \qquad (4.1.10)$$

The infinite energy results directly from the infinite power of the noise, that is,

$$\lim_{N\to\infty} \sum_{k=0}^{N-1} \frac{T}{T} \int_{kT}^{(k+1)T} w^2(t)dt = \lim_{N\to\infty} \sum_{k=0}^{N-1} Tp_{w,k} = T \lim_{N\to\infty} P_W \to \infty. \qquad (4.1.11)$$

White Gaussian noise is used to model channels in communication systems. Because this noise is added to the transmitter signal, the channel is called an additive white Gaussian noise channel. The theoretical model of an additive white Gaussian noise channel, which has noise of infinite power and infinite bandwidth with a constant power spectral density of $N_0/2$, has been traditionally used in theoretical analysis of communication systems, and also used in practice. Namely, due to the filtering at the input of the receiver, the receiver processing will be performed on the band-limited signal and band-limited noise, and the concept of a channel band-unlimited noise can be used in theoretical analysis.

The traditional assertion is that the term 'white' was taken from the name for the spectrum of visible light that, supposedly, has a constant value for all frequencies (wavelengths). This is not the case for at least two reasons. Firstly, the visible light spectrum is band limited. Secondly, the spectrum is not constant in the defined bandwidth and contains a limited power. Thus 'white' light is not similar to white noise.

Therefore, there is nearly no scientific foundation for calling white Gaussian noise 'white' in relation to the spectrum of visible light. We need to search for an additional measure of the noise's properties and include it in the definition of the white Gaussian noise process. The natural measure can be the entropy of the noise generator, which can be treated as a source of the noise process. The entropy, as an average information per noise sample, can give us additional insight into noise properties related to the noise amplitude distribution, as will be seen from the following section. Therefore, in our analysis, entropy will be used as an additional qualitative and quantitative measures of the noise process, in order make it easier to distinguish and define the controversial terms used to define the white Gaussian noise process, as noted by Lapidoth (2009, pp. 554–558).

In summary, the mathematical model of white Gaussian noise, as presented in the existing literature, is not able to define and explain precisely the controversial terms 'white' and 'Gaussian' in the case when the noise source generates the noise of an infinite power. The problem remains in the interpretation of the noise power, too. The noise power is not equal to the finite variance of the probability density function of the noise. In contrast, it is equal to infinity. This is one reason why the notion of the entropy of the noise source will be introduced and used to establish some additional qualitative and quantitative measures in order to more easily distinguish these controversial terms used to define the white Gaussian noise process.

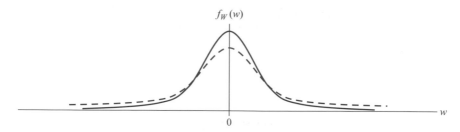

$f_W(w)$

0

w

**Fig. 4.1.4** *Probability density function of noise amplitudes for various variances.*

## 4.1.2   The Entropy of White Gaussian Noise Processes

The definition and basic mathematical presentation of entropy can be found in Chapter 19. In addition, the term entropy is defined from the theory-of-information point of view in Chapter 9. The analysis of the noise energy and power in the frequency domain should comply with the related analysis of the noise energy and power in the time domain. The variance of the defined white noise is constant, but the power is infinite. Therefore, any changes of variance will not change the power. Then, the question is, how does the variance affect the noise properties? For example, if the variance of noise increases, it will not affect its power, as it already has an infinite value! We can observe that an increase in the noise variance will flatten the noise probability density function, as shown in Figure 4.1.4, where the dashed graph has a larger variance.

To gain a better insight into this controversial statement related to the noise power and energy, let us pay attention to the changes in the noise amplitude distribution and the entropy of the noise process when the variance changes.

**Infinite entropy of the noise.** The behaviour of white noise can be related to the values of its entropy in the following way. The noise generated by a Gaussian source has a differential entropy that can be calculated following the information theory as

$$H(W) = \int_{-\infty}^{\infty} f_W(w) \log_2 \frac{1}{f_W(w)} dw = \frac{1}{2} \log_2 2\pi e \sigma_W^2. \tag{4.1.12}$$

When the noise variance increases towards infinity, the entropy increases towards infinity, too. Therefore, in that case, we can say that the probability of generating a sample of the noise of any amplitude is getting very small, tending to zero. Consequently, the information content in any noise sample tends to infinity, and the entropy tends to infinity, as we said. In other words, the average uncertainty of generating any value of the noise, measured by the entropy, is very high, tending to infinity, due to the fact the probability of generating that particular amplitude tends to zero. The definition of entropy as the uncertainty or average information in generating a sample from the source complies quite well with the definitions of entropy in statistical mechanics, where entropy is a measure of the disorder in a physical system, or the ignorance about the system, or irreversible changes in a system (as also used in thermodynamics; Sethna, 2006).

If we relate this analysis to an understanding of the power spectral density function of the noise, it appears that the following explanation will hold. If the variance approaches infinity, the Gaussian distribution tends to a uniform distribution, and the noise in the time domain will have amplitudes generated with nearly equal probability, no matter if they are small or large, still being concentrated in a rather large interval about zero. Due to their independence, the changes of noise values on neighbouring amplitudes in the time domain, no matter how close they are, can be extremely high. Thus, the spectrum of noise will contain all possible frequencies, making the power spectral density 'dense' and flat everywhere. The average power of the noise tends to infinity, and the possible energy content tends to infinity, too. The noise generated is a pure random process, having flat power spectral density and nearly flat probability density function. We say that the uncertainty in generating any noise value in the interval from minus to plus infinity is extremely high.

Of course, we cannot generate noise that is even close to the noise defined by this theoretical model. Namely, the amplitudes of noise in the model can include theoretically assumed infinite magnitudes leading to infinite noise power.

**Finite entropy of the noise.** We return to the initial case related to the noise of a finite variance. In that case, the entropy of the noise source was finite and depended on the variance value. The probability of generating noise samples in the defined interval was finite, and the process was uncorrelated and independent, preserving the delta autocorrelation and the flat power spectral density function.

We can summarize these findings as follows. Unlike the existing presentation of white Gaussian noise process and its interpretation using the concept of the autocorrelation function, the power spectral density, and the power of the noise, in this section, the definition of white Gaussian noise is extended by introducing the behaviour of the entropy of the noise source. In doing this, it is possible to understand the properties of the white Gaussian noise process and to find a meaningful explanation of its behaviour when using filtering. The introduction of entropy gives sense to the definition of white Gaussian noise and the related definition of filtered white Gaussian noise. It is shown that the difference in these definitions can be qualified and quantified by observing the behaviour of the related entropies.

The white Gaussian noise defined in this subsection has the following properties: a constant variance, infinite power, infinite energy, a power spectral density with a finite and constant value for all frequencies, and a finite entropy, due to the finite value of the variance. This noise is not physically realizable, which makes its theoretical model inapplicable to both the practical design of noise generators and the theoretical modelling of the discrete-time Gaussian noise process for application in discrete-time systems. Namely, due to the unlimited power and unlimited bandwidth, this noise does not fulfil the basic requirements for the sampling theorem, the bandwidth limitation, and practical realizability and, as such, cannot be sampled in time. For example, due to the infinite bandwidth, the sampling interval is zero, and possible amplitude samples in the time domain can be of infinite values, which makes the sampling unrealizable.

In the following subsection, a mathematical model of Gaussian noise is investigated, developed, and explained. The model is applicable to the practical design of noise generators, and is useful for the mathematical modelling of discrete-time noise processes and the design of related generators. A noise generator based on this model produces noise that has physically meaningful parameters: finite variance, finite power, infinite energy, finite entropy, and a constant value for the power spectral density function. The development of the model was motivated by our intention to answer the following questions:

1. Can continuous-time Gaussian noise have both infinite power and energy? Can white Gaussian noise be rigorously mathematically defined and interpreted?

2. Can the variance of Gaussian noise be equal to noise power? When the opposite happens, why does it happen?

3. Under which conditions is the variance of Gaussian noise finite and the power finite?

4. How are the noise power spectral density, the autocorrelation function, power, and energy then (in the above cases) related to each other?

5. Can the entropy of a continuous-time Gaussian noise process be used to define more precisely the noise process for the above cases, and, if so, how?

6. How would it be possible to define and model a continuous-time Gaussian noise process to use it for the design of a theoretical discrete-time Gaussian noise model and for the development of Gaussian noise generators operating in the discrete-time domain?

### 4.1.3   Truncated Gaussian Noise Processes

**Statistical characterization of a truncated Gaussian noise process.** The white Gaussian noise mentioned above, having infinite power and infinite energy, is not physically realizable. In order to generate the noise in a specified bandwidth and with a specified power, we need to process the white Gaussian noise and generate noise with a finite power. If we can relax the strict mathematical definition of white Gaussian noise, thus reducing mathematical rigor, we can define the noise to have its power spectral density constant over an arbitrary wide bandwidth $(-B_W, B_W)$. This noise is physically realizable and can be generated assuming also that the distribution of amplitudes is defined by the Gaussian truncated probability density function presented in Figure 4.1.5. By using density truncation, we will eliminate the possibility of having theoretically infinite amplitudes, limit the power of noise, and eliminate the very-high-frequency component in the power spectral density. Because amplitudes beyond $\pm A$ cannot exist, this noise process will have a limited average power, that is, its realization will be a random power signal.

The expression for a truncated probability density function is

$$f_W(w) = C(A)\frac{1}{\sqrt{2\pi\sigma^2}}e^{-w^2/2\sigma^2}, \tag{4.1.13}$$

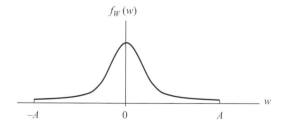

**Fig. 4.1.5** *A truncated continuous Gaussian density function.*

with a constant term $C(A)$ that will be obtained in Chapter 19 in eqn (19.2.22) and where $\sigma$ is the standard deviation of the un-truncated probability density function. The variance of the truncated density will depend on the width of the truncation interval according to an expression that will be derived in Chapter 19 in eqn (19.2.23). Fortunately, this variance can be calculated for the known variance of the un-truncated density $\sigma^2$ and the defined value of the truncation interval defined by $A$. When the variance $\sigma^2$ of the un-truncated probability density function increases, the truncated density tends to the uniform density. A similar scenario can occur when the truncation interval defined by $A$ reduces.

The assumption of limited amplitudes will support our requirement for a limited bandwidth for the noise. Namely, in spite of the fact that the noise samples are independent, their differences are not so large, and the case where infinite amplitude occurs is eliminated. Therefore, the high-frequency component of the noise will not be generated. We can assume that the noise bandwidth $B_W$ is sufficiently large, which is the case in practice, and defined as

$$S_{WW}(f) = \begin{cases} N_0/2 & |f| \le B_W \\ 0 & \text{otherwise} \end{cases}. \tag{4.1.14}$$

Following the Wiener–Khintchine theorem, the autocorrelation function is the Fourier transform of the power spectral density function, derived as

$$R_{WW}(\tau) = \int_{-B_W}^{B_W} S_{WW}(f)e^{j2\pi f\tau}\, df = \frac{N_0}{2}\int_{-B_W}^{B_W} e^{j2\pi f\tau}\, df = \frac{N_0}{2}\frac{e^{-j2\pi f\tau}}{-j2\pi\tau}\Big|_{-B_W}^{B_W} \tag{4.1.15}$$

$$= \frac{N_0}{2}\frac{e^{-j2\pi B_W\tau} - e^{j2\pi B_W\tau}}{-j2\pi\tau} = \frac{N_0}{2\pi\tau}\frac{e^{-j2\pi B_W\tau} - e^{j2\pi B_W\tau}}{-2j} = \frac{N_0}{2\pi\tau}\sin 2\pi B_W\tau$$

$$= N_0 B_W \frac{\sin 2\pi B_W\tau}{2\pi B_W\tau} = N_0 B_W \operatorname{sinc} 2\pi B_W\tau.$$

The graphs of these functions are presented in Figure 4.1.6, for an arbitrary value of the noise variance. We can now mathematically define the white Gaussian noise using limits: the noise bandwidth $B_W$ tends to infinity, and the truncation interval $A$ tends to infinity.

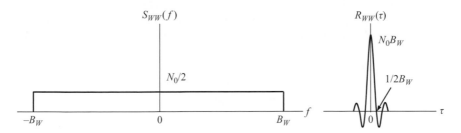

**Fig. 4.1.6** *The power spectral density function (left) and the autocorrelation function (right) of a continuous-time wide-sense stationary Gaussian noise process.*

**The power of the process.** We can calculate the noise power from the power spectral density and autocorrelation function as follows:

$$P_W = \int_{-\infty}^{+\infty} S_{WW}(f)df = \int_{-B_W}^{+B_W} \frac{N_0}{2}df = N_0 B_W = R_{WW}(0). \qquad (4.1.16)$$

If the bandwidth of the noise tends to infinity, which is the case of white Gaussian noise, the power will tend to infinity. This increase of the bandwidth can be achieved by allowing sudden changes of noise amplitudes and by increasing the interval $(-A, A)$ of the amplitude values which, in the final instance, returns back to the white noise generator.

**The energy of the process.** Since the average power of the process is finite and is represented by a constant variance in the time domain, any realization of the truncated white Gaussian noise stochastic process will be defined as a power signal having an infinite energy. If the estimate of the energy is found across the ensemble of all possible realizations of the noise process at the time instant $t$, we obtain

$$E_W = \lim_{T \to \infty} \int_{-A}^{+A} w^2(t)dt = \lim_{T \to \infty} T \frac{1}{T} \int_{-A}^{+A} w^2(t)dt \approx \lim_{T \to \infty} TP_W \to \infty. \qquad (4.1.17)$$

**The entropy.** Given the truncation of noise probability density function, the noise differential entropy can be is derived as in Chapter 19, eqns (19.2.24) and (19.2.25), and expressed as

$$H(W) = \frac{1}{2}\log_2 2\pi e\sigma^2 \left[ 1 - \frac{1}{1 - erfcA/\sqrt{2\sigma^2}} \frac{2A}{\sqrt{2\pi\sigma^2}} e^{-\frac{A^2}{2\sigma^2}} \right], \qquad (4.1.18)$$

where $\sigma^2$ is the variance of the un-truncated probability density function. In this case, the variance is finite, so the entropy will be also finite. That makes sense again. Namely, there is a high level of certainty that the amplitude contained in the truncation interval will be generated. There is zero certainty (infinite uncertainty) that amplitudes outside the truncation interval will be generated. Therefore, the uncertainty of generating noise samples that are in the truncation interval is finite. The average information per noise sample is defined by the differential entropy $H(W)$ and depends on the variance values and the width of the truncation interval.

It is interesting to know what happens when the truncation interval $(-A, A)$ increases and the magnitude value $|A|$ tends to infinity. In that limiting case, the variance of the noise tends to the variance of the un-truncated probability density function, that is,

$$\lim_{A \to \infty} \sigma_W^2 = \lim_{A \to \infty} \sigma^2 \left[ 1 - \frac{1}{1 - erfcA/\sqrt{2\sigma^2}} \frac{2A}{\sqrt{2\pi\sigma^2}} e^{-\frac{A^2}{2\sigma^2}} \right] = \sigma^2 \qquad (4.1.19)$$

and the entropy becomes equal to the entropy of the Gaussian white noise process, that is,

$$\lim_{A \to \infty} H(W) = \lim_{A \to \infty} \frac{1}{2} \log_2 2\pi e\sigma^2 \left[ 1 - \frac{1}{1 - erfcA/\sqrt{2\sigma^2}} \frac{2A}{\sqrt{2\pi\sigma^2}} e^{-\frac{A^2}{2\sigma^2}} \right].$$ (4.1.20)

$$= \frac{1}{2} \log_2 2\pi e\sigma^2$$

**An important independence observation.** We assumed that the samples of the noise are generated nearly independently. However, our certainty in generating and observing smaller samples rather than bigger ones is higher. This certainty will be measured by the probability of the samples' appearance at the output of the noise generator, and the uncertainty will be measured by their information content. Therefore, despite the fact that the samples are generated nearly independently (a narrow graph of the autocorrelation function), the shape of the probability density function is relevant for the development of our prior knowledge of the certainty of observing a particular sample at the noise generator output. Namely, our certainty in generating a particular noise sample depends on the shape of the probability density function. The smaller the variance of the truncated function is, the higher our certainty will be about generating noise samples with small amplitudes.

Finally, the noise samples defined by the time instants $k/2B_W$, for integer $k$ values, will be uncorrelated, as can be seen from the noise autocorrelation function in Figure 4.1.6. Being Gaussian, these samples will be independent. This fact will be used in constructing the noise generators.

## 4.1.4 Concluding Notes on Gaussian Noise Processes

**White Gaussian noise process as traditionally defined.** The white Gaussian noise process is traditionally defined as a stochastic process with a Gaussian distribution of amplitudes, constant power spectral density, finite variance, infinite power, and finite entropy, as presented in this section. This definition, excluding the entropy that is introduced in this book, complies with the definition that can be found nearly in all books on the theory of communications and the theory of stochastic processes. However, it appears that the use of the term 'white' to indicate the infinite bandwidth of the noise is not really appropriate. Also, the frequently used analogy between white light and white noise is problematic.

Suppose that a Gaussian noise with a finite variance is generated. Since there is a finite variance, no matter how large it is, the majority of the generated samples will be placed in the related $2\sigma$ interval of the Gaussian distribution, even though the probability of generating a sample of a very large magnitude does exist and the possibility of generating it is still present. The Gaussian density is preserved, too.

Due to the independence between the noise samples, the autocorrelation function will be a delta function with a weight equal to the variance measured in watts. If the power is estimated across the ensemble of the process, it will tend to infinity, due to the possible existence of at least one amplitude with infinite value. Of course, this infinite value could happen very rarely, almost never; however, it must be taken into account in a theoretical analysis.

The Fourier transform of the autocorrelation function will yield a power spectral density function that is constant and numerically equal to the variance for all frequencies from minus to plus infinity. Namely, the integration of the autocorrelation function results in a power spectral density magnitude that is equal to the variance and has units of watt-seconds or joules. The calculation of the power will result in an infinite value, which is expected. Consequently, the energy of the noise will be infinite, too.

However, in the case when the variance is constant, the entropy of the noise generator will have a constant value. Due to the Gaussian distribution, the information content of the samples will depend on their amplitude values, that is, the samples will be generated randomly and independently, but the level of uncertainty in the appearance of a particular sample value at the output of the generator will depend on the shape of the probability density function. At the average, for the whole process, the uncertainty is defined by the entropy, which depends exclusively on the variance value. However, for each sample, this uncertainty depends on the probability of its appearance at the generator output. As we said before, the larger the variance is, the higher the average uncertainty in generating the noise samples will be.

In summary, a white Gaussian noise with a finite variance (not necessarily power signal) is defined as a stochastic process that has infinite power and infinite energy but finite entropy and a constant power spectral density. The infinite energy is presented by the frequency components that have finite magnitudes, but they are spread out in an infinite interval, resulting in the infinite power. The infinite energy in the time domain can be imagined to be calculated as the infinite power times the infinite time. In the frequency domain, the power will be obtained as the product of the finite power spectral density and the infinite bandwidth. This noise does not comply with the definition of power signals. Of course, we are not able to generate this kind of stochastic processes.

**Gaussian noise processes with extremely large entropy.** In this, the most extreme, case, when the variance of the Gaussian noise process increases towards infinity, the power and energy remain infinite, and the entropy tends to infinity, the obtained noise is different from the defined white Gaussian noise. This noise can be called black Gaussian noise, because it has all of the power spectrum components spread from minus to plus infinity, and entropy that tends to infinity. This noise is the extreme case when all its parameters are maximized, due to the noise variance tending to infinity and, as such, is beyond our comprehension. However, it can be defined in terms of its power, energy, distribution of amplitudes, and entropy.

Furthermore, for this noise, due to the increase of variance towards infinity, the Gaussian probability density function 'flattens' and spreads the probability values to infinity on both sides by making its tails 'heavier', as shown in Figure 4.1.4. The amplitude probability distribution tends to a uniform distribution, having nearly the same infinitesimally small probability of any deferential event that can occur in an infinite interval of possible amplitude values. Consequently, the entropy tends to infinity, meaning that the uncertainty of generating any random value of the noise is extremely high, tending to infinity, too. In addition, the information content in any generated random amplitude is infinite. Therefore, any realization of the noise process, presented as a random signal generated in the time domain, will contain all amplitudes with nearly the same probabilities, and the time variations of noise samples, having independent amplitudes, no matter how close they are in the time domain, will be very large. Consequently, due to possible sudden changes of the noise amplitudes in the time domain, the number of different frequency components in the power spectral density will be enormously high, covering, theoretically, the continuous bandwidth from minus infinity to plus infinity.

In summary, black Gaussian noise is defined as a stochastic process that has infinite variance, infinite power, infinite entropy, infinite energy, and an infinite power spectral density. The energy in the time domain can be imagined to be the product of the averaged infinite power and the infinite time. In the frequency domain, the power will be the product of the infinite power spectral density times the infinite bandwidth. This noise does not comply with the definition of either the power or the energy signals. Of course, we are not able to generate this kind of stochastic process. Then, what can we realistically achieve? Let us discuss the following case.

***White Gaussian noise processes with truncated density and limited power.*** Having in mind the previous discussion, we cannot, in practice, generate white Gaussian noise that complies with the requirement of having infinite power, even if the noise has a limited variance. In practice, we are generating noise samples within a defined interval of possible values by limiting the maximum amplitudes of the noise samples and excluding samples of possible infinite amplitudes. In fact, we are generating noise samples defined by a finite variance and the truncated probability density function. We can still have a white noise process that has a finite power equal to the variance of the truncated probability density function. In this case, we need to be careful about the width of the truncation interval, to control the variance and, therefore, the power of the noise.

Defining white Gaussian noise in this way, we can apply the sampling theorem and generate noise amplitudes in the discrete-time domain. The time discretization and value quantization of the white Gaussian noise process will be based on the assumption that the continuous-time noise has limited power, finite variance, finite entropy, and a truncated Gaussian probability density function. In other words, the realization of the noise process will be random power signals.

## 4.2   Gaussian Noise Processes in the Discrete-Time Domain

### 4.2.1   White Gaussian Noise Processes with Discrete-Time and Continuous-Valued Samples

**Discrete noise with a truncated probability density function.** In this subsection, we will specify the key requirements for the Gaussian noise sampling. In discrete communication systems, the signals processed are represented as discrete-time discrete-valued signals. Therefore, the model of the continuous-time white Gaussian noise cannot be directly applied in discrete systems. The purpose of this section is to make a theoretical model of discrete-time noise by presenting it as a discrete-time stochastic process and analysing it in the frequency and time domains. In particular, the white Gaussian noise process, band-limited baseband Gaussian noise, and band-limited passband noise will be analysed.

All our efforts in the previous section had one simple aim: to make a theoretical model of a continuous-time Gaussian noise process that can be used to generate a discrete-time Gaussian noise with controllable statistical properties. This was achieved by making a model of a Gaussian noise process with a truncated probability density function that has the following important properties:

1. The process fulfils the requirement for physical realizability because its amplitudes have limited values and limited average power. Therefore, it fulfils the first requirement for the application of the sampling theorem.

2. The process is also band limited, having a power spectral density in the limited bandwidth $B_W$, which is an essential 'silent' requirement for the application of the sampling theorem.

3. An autocorrelation function is derived that allows the calculation of the time intervals $T_s$ for noise sampling that produces a set of independent Gaussian noise samples and specifies the sampling procedure that will result in uncorrelated noise samples. Therefore, the defined time intervals $T_s$ guarantee that the noise samples are uncorrelated and Gaussian.

We will analyse here the basic statistical characteristics of Gaussian noise that has a finite power and for which the probability density function of its amplitudes is defined by the truncated Gaussian distribution. Therefore, the power spectral density of the truncated discrete-time stochastic noise process $N_T(n)$, having the bandwidth defined by the cut-off normalized frequency $\Omega_T = 2\pi F_T$, can be expressed as

$$S_{NT}(\Omega) = \begin{cases} \frac{N_0}{2}, & -\Omega_T < \Omega < \Omega_T \\ 0, & \text{otherwise} \end{cases} = \begin{cases} \frac{N_0}{2}, & -F_T < F < F_T \\ 0, & \text{otherwise} \end{cases}. \tag{4.2.1}$$

Following the Wiener–Khintchine theorem, the autocorrelation function can be calculated as the inverse Fourier transform of the power spectral density, that is,

$$\begin{aligned} R_{NT}(l) &= \frac{1}{2\pi} \int_{-\pi}^{\pi} S_{NT}\left(e^{j\Omega}\right) e^{j\Omega l} d\Omega = \frac{N_0}{4\pi} \int_{-\Omega_T}^{\Omega_T} e^{j\Omega l} d\Omega = \frac{N_0}{4\pi} \frac{e^{j\Omega l}}{jl} \bigg|_{\Omega_T}^{\Omega_T} \\ &= \frac{N_0}{2\pi} \frac{\left(e^{j\Omega_T l} - e^{-j\Omega_T l}\right)}{2jl} = \frac{N_0}{2\pi} \frac{\sin(\Omega_T l)}{l} = \frac{N_0 \Omega_T}{2\pi} \operatorname{sinc}(\Omega_T l) \end{aligned} \tag{4.2.2}$$

This expression is equivalent to the expression presented in the book by Rice (2009, p. 207, eqn (4.75)). However, the difference is that, in this case, we present the discrete-time noise in terms of a discrete variable $n$ rather than relating it to the sampling interval $T_s$ as it was done in the aforementioned book. The autocorrelation function can be expressed in terms of the normalized frequency $F_T = \Omega_T/2\pi$, and the autocorrelation function will be

$$R_{NT}(l) = N_0 F_T \operatorname{sinc}(2\pi F_T l). \tag{4.2.3}$$

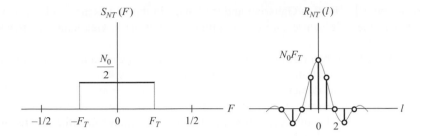

**Fig. 4.2.1** *The power spectral density (left) and the autocorrelation function (right) for N = 2.*

Because the correlation function $R_{NT}(l)$ of the process $N_T(n)$ is aperiodic in $l$, the discrete-time Fourier transfer can be used to obtain the power spectral density. In this case, the correlation function $R_{NT}(l)$ should be absolutely summable, which guarantees the existence of the power spectral density, which means that $N_T(n)$ is a zero-mean process.

The zero crossings of the autocorrelation function are defined by the condition $l = m/2F_T$. Examples of the power spectral density and the autocorrelation function of the process $N_T(n)$ are presented in Figure 4.2.1. The power spectral density is a periodic function with a period of $\Omega = 2\pi$ or $F = 1$, and the autocorrelation function is an aperiodic, even function, with its first zero crossing at $l = 2$, that specifies sampling intervals that produce uncorrelated noise samples. In that case, the discrete-time noise contains only iid Nyquist samples, and the autocorrelation function becomes the Kronecker delta function, because

$$R_{NT}(l) = N_0 F_T \operatorname{sinc}(\pi) = N_0 F_T \delta(l), \tag{4.2.4}$$

which gives zeros for every $l$ except for $l = 0$, according to the following condition: the autocorrelation is zero when $l\pi = m\pi$ or $l = m$, that is, for all $l$ except $l = 0$. For this case, the autocorrelation function is represented by the graph shown in Figure 4.2.1, showing that there is a strong correlation between the interpolated samples and strong independence between the Nyquist samples. This is the case mentioned in the book by Rice (2009, p. 207, eqn (4.76)) which becomes eqn (4.2.4) for $T = 1$, which implies $F_T = 1/2$.

For the sake of explanation, we have just presented the theory for a white Gaussian noise process defined by an truncated probability density function. In the next subsection, the theory for a noise process with an un-truncated Gaussian density function will be presented.

**White Gaussian noise processes with an un-truncated density function.** The discrete-time noise $W(n)$ can be obtained by sampling a continuous-time white Gaussian noise $W(t)$ at the sampling time instants $nT_s$ to get noise samples as a function of the discrete-time variable $n$, as depicted in Figure 4.1.1. The white Gaussian noise discrete-time sample function, denoted by $w(n)$, can be represented in the discrete-time domain as one realization of the discrete-time Gaussian stochastic process $W(n)$, as shown in Figure 4.1.1. Because the noise samples are realizations of iid random variables, the random variables at all discrete-time instants $n$ are also iid Gaussian variables. Each noise sample is independent of any other noise sample, or, in other words, if we observe any sample, we cannot predict the next one. Furthermore, because continuous-time noise $W(t)$ is Gaussian and stationary in the wide sense and, consequently, is stationary in the strict sense, the discrete-time noise $W(n)$ is Gaussian and strict-sense stationary.

Let us address the problem of the infinite power. At each time instant $n$, a random variable $W_n$ is defined for the discrete-time white Gaussian noise process $W(n)$ having zero mean and standard deviation $\sigma_W$, as depicted in Figure 4.1.1. The distribution of $W_n$ is defined by the un-truncated Gaussian probability density function, which is expressed by eqn (4.1.1) and shown in Figure 4.1.1. The autocorrelation function of this process depends on the time lag $l = m - n$, and can be calculated as

$$
\begin{aligned}
R_{WW}(m,n) &= E\{W(m)W(n)\} \\
&= \int_{-\infty}^{\infty} \int_{-\infty}^{\infty} w_m w_n f_{W_m W_n}(w_m, w_n)\, dw_m dw_n \\
&= \int_{-\infty}^{\infty} \int_{-\infty}^{\infty} w_m w_n f_{W_m}(w_m) f_{W_n}(w_n)\, dw_m dw_n \\
&= \left\{ \begin{array}{ll} \int_{-\infty}^{\infty} w_n^2 f_{W_n}(w_n)\, dw_n \int_{-\infty}^{\infty} f_{W_m}(w_m)\, dw_m & m = n \\ \int_{-\infty}^{\infty} w_n f_{W_n}(w_n)\, dw_n \int_{-\infty}^{\infty} w_m f_{W_m}(w_m)\, dw_m & m \neq n \end{array} \right\} \\
&= \left\{ \begin{array}{ll} \sigma_W^2 & m = n \\ 0 & m \neq n \end{array} \right\} = \sigma_W^2 \delta(m-n) = \frac{N_0}{2} \delta(l) = R_{WW}(l)
\end{aligned}
$$

(4.2.5)

where the joint probability density function $f_{W_m W_n}(w_m, w_n)$ is equal to the product of the marginal densities, due to the independence of the random variables $W_m$ and $W_n$, and the variance is equal to the noise two-sided power spectral density $N_0/2$. The autocorrelation function is expressed in terms of Kronecker delta functions, $\delta(l)$, *implying that the noise sampling was possible fulfilling conditions of the sampling theorem, that is, the power of the continuous-time noise samples is finite, and the power of the obtained discrete-time noise also has a finite value.*

According to the Wiener–Khintchine theorem, the power spectral density can be calculated as the discrete-time Fourier transform of the autocorrelation function, that is,

$$
S_{WW}(\Omega) = \sum_{l=-\infty}^{\infty} R_{WW}(l) e^{-j\Omega l} = \sigma_W^2 \sum_{l=-\infty}^{\infty} \delta(l) e^{-j\Omega l} = \sigma_W^2 = \frac{N_0}{2},
$$

(4.2.6)

for $-\pi \leq \Omega \leq \pi$. If we had assumed the possible existence of noise samples of infinite amplitudes, the relation (4.2.5) would have been expressed in terms of Dirac delta functions, and the power of the noise process would be of infinite value. However, that would conflict with the requirements of the sampling theorem, that is, that only physically realizable signals can be sampled. In our previous analysis, we defined noise to have a truncated probability density function and a finite variance. Therefore, we could sample this noise. In summary, the discrete noise has finite power, due to restrictions (assumed and used) on the continuous-time noise that we are sampling to get the discrete-time noise. We can now see the importance

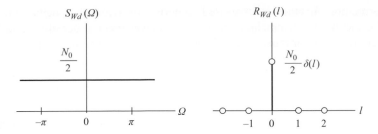

**Fig. 4.2.2** *The power spectral density (left) and the autocorrelation function (right) of a discrete white Gaussian noise process.*

of the definition of Gaussian noise process having a truncated Gaussian probability density function, because that noise can give discrete noise as a power signal with the autocorrelation function and power spectral density as presented in Figure 4.2.2.

Due to the importance of the theoretical presentation and practical generation of discrete noise, the following analysis will include the white Gaussian noise processes that fulfil the requirements of the sampling theorem and can be used to get discrete-time noise signals.

***Discrete truncated white Gaussian noise with finite power.*** In the process of sampling a continuous-time continuous-valued (analogue) deterministic signal, we assume that the obtained samples are of finite value. Therefore, we can say that the probability of the existence of samples beyond a certain limit $A$ is zero, and the power of the signal is finite, to fulfil the condition of the physical realizability of the signal. The second condition was that the signal has a limited bandwidth $B$. The Gaussian stochastic process needs to fulfil the same requirements. That is done by filtering (limiting the bandwidth to $B$) and generating a process that has a truncated Gaussian probability density function. The obtained sampled signal is a discrete-time continuous-valued noise having density, as presented in Figure 4.1.5 and expressed by eqn (4.1.13). In this case, the precise calculation of the autocorrelation function is

$$
\begin{aligned}
R_{WW}(m,n) &= E\{W(m)W(n)\} \\
&= \int_{-A}^{A}\int_{-A}^{A} w_m w_n f_{W_m W_n}(w_m, w_n)\, dw_m dw_n \\
&= \int_{-A}^{A}\int_{-A}^{A} w_m w_n f_{W_m}(w_m) f_{W_n}(w_n)\, dw_m dw_n \\
&= \begin{cases} \int_{-A}^{A} w_n^2 f_{W_n}(w_n)\, dw_n \int_{-\infty}^{\infty} f_{W_m}(w_m)\, dw_m & m = n \\ \int_{-A}^{A} w_n f_{W_n}(w_n)\, dw_n \int_{-\infty}^{\infty} w_m f_{W_m}(w_m)\, dw_m & m \neq n \end{cases} \\
&= \begin{cases} \sigma_W^2 & m = n \\ 0 & m \neq n \end{cases} = \sigma_W^2 \delta(m-n) = \sigma_W^2 \delta(l) = R_{WW}(l)
\end{aligned}
\tag{4.2.7}
$$

The autocorrelation function is expressed in terms of Kronecker delta functions, $\delta(l)$. According to the Wiener–Khintchine theorem, the power spectral density can be calculated as the discrete-time Fourier transform of the autocorrelation function, that is,

$$S_{WW}(\Omega) = \sum_{l=-\infty}^{\infty} R_{WW}(l) e^{-j\Omega l} = \sigma_W^2 \sum_{l=-\infty}^{\infty} \delta(l) e^{-j\Omega l} = \sigma_W^2, \qquad (4.2.8)$$

for $-\pi \leq \Omega \leq \pi$, where $\sigma_W^2$ is referenced to the input stage of the receiver of a communication system and is expressed in watts per hertz. It is worth saying that this function is periodic with a period of $2\pi$. Having in mind the definition of the delta function, we can derive the autocorrelation function as the inverse discrete-time Fourier transform of the power spectral density as

$$
\begin{aligned}
R_{WW}(l) &= \frac{1}{2\pi} \int_{-\pi}^{\pi} S_{WW}(\Omega) e^{j\Omega l} d\Omega = \frac{\sigma_W^2}{2\pi} \int_{-\pi}^{\pi} e^{j\Omega l} d\Omega \\
&= \frac{\sigma_W^2}{\pi l} \frac{e^{j l\pi} - e^{-j l\pi}}{2j} = \sigma_W^2 \frac{\sin(l\pi)}{\pi l} = \left\{ \begin{array}{cc} \sigma_W^2 & l = 0 \\ 0 & otherwise \end{array} \right\} = \sigma_W^2 \delta(l)
\end{aligned}
\qquad (4.2.9)
$$

where $\delta(n)$ is the Kronecker delta function. The autocorrelation function and the power spectral density for discrete-time additive white Gaussian noise are presented in Figure 4.2.3.

Note that $R_{WW}(l)$ is the Kronecker delta function weighted by the factor $\sigma_W^2$ and occurring at $l = 0$, and is zero for all $l \neq 0$, meaning that any two different samples $w_m$ and $w_n$ of the white noise $W(n)$ are two realizations of two uncorrelated random variables. In addition, if the noise is also Gaussian, then two samples are statistically independent, as we already said. In the discrete-time case, the power of the noise can be calculated from the power spectral density as

$$P_W = \frac{1}{2\pi} \int_{-\pi}^{\pi} S_{WW}(\Omega) d\Omega = \frac{\sigma_W^2}{2\pi} \int_{-\pi}^{\pi} d\Omega = \sigma_W^2 = R_W(0), \qquad (4.2.10)$$

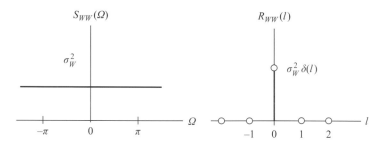

**Fig. 4.2.3** *The power spectral density (left) and autocorrelation function (right) of white Gaussian noise.*

The power is finite, as it was for continuous-time truncated noise. In this analysis, the discrete-time white noise process $w(n)$, which is, by definition, a wideband noise, is represented by its samples, which are realizations of iid random variables. The sampled noise is then analysed as a discrete-time stochastic process $W(n)$. It is assumed that the sampling rate is very high, and the time duration of the sampled noise realization $w(n)$ is sufficiently small to preserve the theoretical spectral characteristics of the white noise.

In practice, samples of noise are taken at finite sampling intervals and then represented as a stepwise continuous-time noise process. In that case, the bandwidth of the calculated power spectral density is finite and is defined by the sampling frequency. The higher the sampling frequency is, the wider the bandwidth is. Therefore, in practical applications where the noise is simulated and generated with a finite sampling frequency, we need to make this frequency much larger than the bandwidth of the signal affected by the wideband noise.

## 4.2.2 Discrete-Time White Gaussian Noise Processes with Discrete-Valued Samples

The realization of the discrete-time noise process presented in Figure 4.1.1 is a discrete-time continuous-valued function, that is, each sample $w(n)$ can theoretically have any real value. In real systems, where processing is done using digital technology, these samples are quantized, and discrete-time discrete-valued noise samples are generated. In this case, the statistical description of the process will be slightly different. Suppose that the quantization of the noise values is uniform. Therefore, a discrete-time noise sample value $w(n)$, taken at the discrete-time instant $n$, is represented by a quantized value $w_d(n)$ that is considered to be a realization of the discrete-time discrete-valued random variable $W_d(n)$. The sample obtained can be expressed as $w_d(n) = k\Delta w_d$, for the number of discrete levels $k = 1, 2, \ldots, k_{max}$ and for the quantization step $\Delta w_d$. Therefore, a series of quantized values $w_d(n)$, taking on the values in the interval $A = k_{max}\Delta w_d$, represents one realization of the discrete-time discrete-valued (quantized) stochastic process $W_d(n)$.

The statistical description of this process is slightly different from the previous description of the discrete-time continuous-valued process $W_d(n)$. If the quantized noise values $\Delta w_d$ are very small and, consequently, the number of quantization levels is very large, the probability density function of the random variable $W_d(n)$, defined for any $n$, will be represented by a discrete probability density function that preserves a Gaussian shape, as shown in Figure 4.2.4.

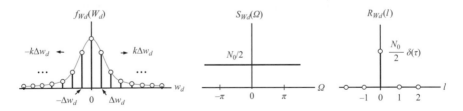

**Fig. 4.2.4** *A discrete truncated probability density function (left) and its related power spectral density (middle) and autocorrelation (right) functions.*

Because the discrete and quantized samples obtained exist in a limited interval of values, this function needs to be truncated. This condition makes the noise have a limited power, which is one condition for its sampling, that is, it becomes a physically realizable signal. Having in mind the statistical independence of the random variables at time instants $m$ and $n$ of the stochastic process $W_d(n)$, we can calculate the autocorrelation function as

$$
\begin{aligned}
R_{W_d}(m,n) = E\{W_d(m)W_d(n)\} &= \sum_{w_{md}=-A}^{A} \sum_{w_{nd}=-A}^{A} w_{md} w_{nd} f_{W_{md}W_{nd}}(w_{md}, w_{nd}) \\
&= \sum_{m=-A}^{A} w_{md} f_{W_{md}}(w_{md}) \sum_{n=-A}^{A} w_{nd} f_{W_{nd}}(w_{nd}) \\
&= \left\{ \begin{array}{ll} \sum_{m=-A}^{A} w_d^2 f_{W_{md}}(w_{md}) \sum_{n=-A}^{A} f_{W_{nd}}(w_{nd}) & m = n \\ \sum_{m=-A}^{A} w_{md} f_{W_{md}}(w_{md}) \sum_{n=-A}^{A} w_{nd} f_{W_{nd}}(w_{nd}) & m \neq n \end{array} \right. \\
&= \left\{ \begin{array}{ll} \sigma_{W_d}^2 & m = n \\ 0 & m \neq n \end{array} \right\} = \sigma_{W_d}^2 \delta(m-n) = R_{W_d}(l)
\end{aligned}
$$

(4.2.11)

If truncation is performed, the autocorrelation function is expressed in terms of Kronecker delta functions, $\delta(l)$. The power spectral density is identical to that derived in eqn (4.2.8), that is,

$$
S_{Wd}(\Omega) = \sum_{l=-\infty}^{\infty} R_{Wd}(l) e^{-j\Omega l} = \sigma_{Wd}^2 \sum_{l=-\infty}^{\infty} \delta(l) e^{-j\Omega l} = \sigma_{Wd}^2,
$$

(4.2.12)

for $-\pi \le \Omega \le \pi$. The graphs of these functions are presented in Figure 4.2.4.

It is important to say that discrete-time and quantized noise samples will give an approximate discrete probability density function due to the rounding of the sampled values. The discrete probability density function obtained will be statistically very close to the previously calculated density for continuous-valued noise samples.

### 4.2.3 White Gaussian Noise Processes with Quantized Samples in a Strictly Limited Interval

For the sake of exactness, we need to provide an additional explanation here. In the design of transceivers based on discrete-time and discrete-valued signal processing, the noise and signal samples processed will have values taken in a strictly limited (truncated) interval of possible values, without exception. Therefore, the related discrete probability density function should be replaced by a truncated discrete probability density function. The consequence of this truncation will be a change in the variance value, which should be taken into account in practical calculations. This change requires an additional theoretical development. For the sake of illustration, we can derive the autocorrelation function of the discrete truncated noise process with discrete values in a limited interval, as in eqn (4.2.13):

$$
R_{W_{dt}}(m,n) = E\{W_{dt}(m)W_{dt}(n)\}
$$

$$
= \sum_{w_{mdt}=-w_{mdt\,\max}}^{w_{mdt\,\max}} \sum_{w_{ndt}=-w_{ndt\,\max}}^{w_{ndt\,\max}} w_{mdt} w_{ndt} f_{W_{mdt}W_{ndt}}(w_{mdt}, w_{ndt})
$$

$$
= \sum_{w_{mdt}=-w_{mdt\,\max}}^{w_{mdt\,\max}} w_{mdt} f_{W_{mdt}}(w_{mdt}) \sum_{w_{ndt}=-w_{ndt\,\max}}^{w_{ndt\,\max}} w_{ndt} f_{W_{ndt}}(w_{ndt})
$$

$$
= \begin{cases} \displaystyle\sum_{w_{mdt}=-w_{mdt\,\max}}^{w_{mdt\,\max}} w_{dt}^2 f_{W_{mdt}}(w_{mdt}) \sum_{w_{ndt}=-w_{ndt\,\max}}^{w_{ndt\,\max}} f_{W_{ndt}}(w_{ndt}) & m = n \\ \displaystyle\sum_{w_{mdt}=-w_{mdt\,\max}}^{w_{mdt\,\max}} w_{mdt} f_{W_{mdt}}(w_{mdt}) \sum_{w_{ndt}=-w_{ndt\,\max}}^{w_{ndt\,\max}} w_{ndt} f_{W_{ndt}}(w_{ndt}) & m \neq n \end{cases}
$$

$$
= \begin{cases} \sigma_{W_{dt}}^2 & m = n \\ 0 & m \neq n \end{cases} = \sigma_{W_{dt}}^2 \delta(m-n) = \frac{N_0}{2}\delta(l) = R_{W_{dt}}(l)
$$

$$
(4.2.13)
$$

The autocorrelation function is expressed in terms of Kronecker delta functions, and the variance has a finite value. Therefore, despite the fact that the noise can have the values in a limited interval, it preserves the white nature. The power spectral density has the constant value $N_0/2$ in the whole interval of possible frequencies, as shown in Figure 4.2.4.

## 4.2.4 Band-Limited Continuous- and Discrete-Time Signals and Noise

**Baseband binary signal analysis:** In digital communication systems, the basic baseband signal is presented as a binary signal with bits of the magnitude $A$ and duration $T$. The bandwidth of this signal, containing most of the power, is inversely proportional to the bit duration, that is, $B = 1/T$. To present a bit in the discrete-time domain, we need to apply the Nyquist theorem, which will result in a sampling time that is half of the bit duration, that is, $T_s = 1/2B = T/2$. Therefore, in the discrete-time domain, a bit will be represented by at least $N = 2$ samples taken at $t = 0$ and $t = T/2$, as shown in Figure 4.2.5(a).

The procedure of discretization of a binary continuous-time signal is presented in Figure 4.2.5, where the dotted graph represents a continuous-time binary signal, and the bold samples represent discrete-time binary signals. Figure 4.2.5(b) and (c) presents the case when one sample is taken, up-sampled with $N$ samples, and then interpolated.

***Baseband M-ary signal analysis.*** If the baseband signal can have up to $M$ amplitude levels and the bit duration $T$ corresponds to an $M$-ary symbol duration, the number of samples for each symbol will again be $N = 2$, because the bandwidth of the symbol is defined by the same duration $T$. For the practical design of signal-processing blocks in discrete communication systems, having two samples is not enough to perform discrete-time signal processing. In order to preserve the bandwidth of bits and symbols, up-sampling and interpolation filtering will be used. In doing this, due to the constant values of amplitudes inside the bit or symbol interval, it is sufficient to represent each of them by the first sample and then interpolate them

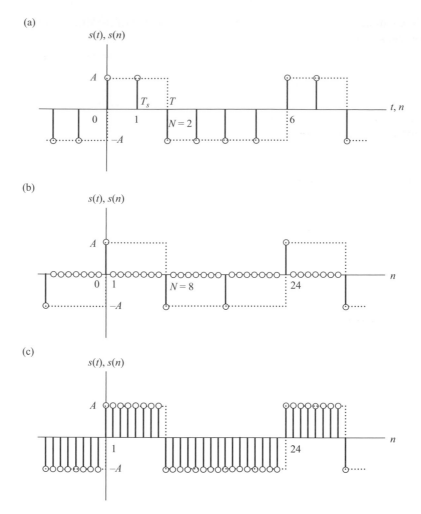

**Fig. 4.2.5** *(a) Continuous-time polar non-return-to-zero binary signal and its (b) interpolated and (c) filtered counterparts in the discrete-time domain.*

with the same values $N$ times, as presented in Figure 4.2.5(c) for $N = 8$ (as in the sample-and-hold procedure). This method of presenting bits and symbols will be used in most of the theoretical analyses in this book.

**Baseband noise analysis.** A similar analysis can be performed for baseband noise. Because this noise, after filtering in the receiver blocks, exists in the bandwidth of the signal, its bandwidth is $B$. Therefore, the required number of Gaussian independent noise samples is $N = 2$, which is generated inside the bit/symbol interval $T$. Likewise, for the signal, the noise will be up-sampled and interpolated by a sufficient number of samples which will not increase the bandwidth of the baseband noise and will preserve its Gaussian nature. A rationale similar to that for the application of the up-sampling procedure used for the discrete signal can be applied for the noise. Namely, due to the independence of the two Gaussian samples generated in one bit interval, it is sufficient to generate only one sample and repeat it $N$ times inside the up-sampler and interpolation filter to get the interpolated noise. The noise generated inside a sufficient number of bits, which is, in principle, a very high number when bit error probability is investigated, for example, will preserve its Gaussian nature, have limited bandwidth, and have certain level of correlation, as we will see in the subsequent sections.

In summary, in an analysis of a discrete communication system, the signal and the noise will be represented by discrete-time rectangular pulses with a duration of $N$ samples. The number of samples can be changed but will preserve the duration of pulses $T$ in the corresponding continuous-time domain via the use of up-sampling. The up-sampling will depend on the signal processing required in the building blocks of the discrete transmitters and receivers. The bandwidth of the original signals will be preserved during the discrete-time signal processing.

## 4.3 Operation of a Baseband Noise Generator

### 4.3.1 Band-Limited Continuous-Time Noise Generators

In the theoretical analysis, simulation, and emulation of communication systems, we need to investigate their characteristics in the presence of noise and fading. Therefore, noise with known and controllable power inside the required bandwidth need to be generated. In this section, generators of continuous Gaussian noise will be presented.

At the input of the baseband noise generator shown in Figure 4.3.1 is a continuous-time white Gaussian noise $w(t)$, which is a realization of a white Gaussian noise stochastic process

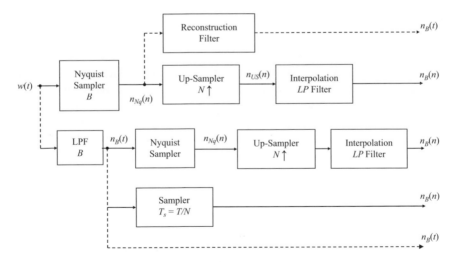

**Fig. 4.3.1** *Baseband noise generators; LP, low pass; LFP, low-pass filter.*

$W(t)$ with zero mean and standard deviation $\sigma_W$. To allow sampling according to the sampling theorem, we will assume that the distribution of noise amplitudes is defined by the Gaussian truncated probability density function. The question is how to process this wideband white Gaussian noise, which theoretically has nearly infinite bandwidth, to obtain the baseband Gaussian noise defined in its bandwidth $B$. The simplest method is to use the baseband filter with the cut-off frequency $B$ that produces at the output the baseband noise $n_B(t)$, as shown in Figure 4.3.1 in the lowest branch (dashed line). We can also take the noise samples and then use a reconstruction filter as shown in the top branch in Figure 4.3.1 (dashed line). Then, the Nyquist samples are passed through a reconstruction filter, and the continuous-time baseband noise $n_B(t)$ is generated.

### 4.3.2   Band-Limited Discrete-Time Noise Generators

We will present here three methods for generating discrete-time Gaussian noise $W(n)$. The implementations of these methods are presented as three block schematics in Figure 4.3.1. In order to have finite noise power, it is assumed that the distribution of the discrete noise amplitudes (sampled values) is specified by the Gaussian discrete truncated probability density function.

**The first method for generating discrete-time Gaussian noise.** Nyquist sampling is the simplest method that involves discrete-time noise processing, assuming that the standard deviation of the continuous-time noise process $W(t)$ is $\sigma_W$, and the values of the noise can be taken from a finite interval of possible values preserving a finite value of the average power. According to this method, implemented in the first branch in Figure 4.3.1, discrete-time samples of the continuous-time white Gaussian noise $w(t)$ are taken according to the Nyquist theorem for the defined bandwidth $B$, presented in Figure 4.3.2(a). Therefore, the bandwidth should be equal to the Nyquist frequency, that is, $B = f_{Nq}$, and the discrete-time samples of the noise need to be taken at the Nyquist sampling instants, $T_s = 1/2B$. The result of sampling is a discrete-time random signal $n_{Nq}(n)$ which is a realization of the discrete-time stochastic process $N_{Nq}(n)$. Because the continuous-time noise process $W(t)$ is Gaussian, the discrete-time process $N_{Nq}(n)$, containing Gaussian samples, is Gaussian, too. We can say that the discrete-time baseband noise samples obtained are realizations of the uncorrelated Gaussian random variables $N_{Nq}(n)$, which are defined for each $n$ and have zero mean and variance $N_0 B$. If these samples are transferred through an ideal low-pass reconstruction filter with bandwidth $B$, the output will be continuous-time correlated-baseband noise, as shown in the first branch in Figure 4.3.1.

The number of samples obtained using Nyquist criterion is not sufficient in practice because the signal, processed in the presence of noise, usually has many more samples in the same time interval. Therefore, in order to preserve the bandwidth $B$ and increase the number of noise samples, up-sampling and interpolation are used, as shown in the second branch of Figure 4.3.1. The up-sampling adds a sufficient number of zeros between the samples, and the interpolation filter generates discrete samples instead of zeros, preserving the bandwidth $B$ of the noise, as shown in Figure 4.3.2(b) and Figure 4.3.2(c).

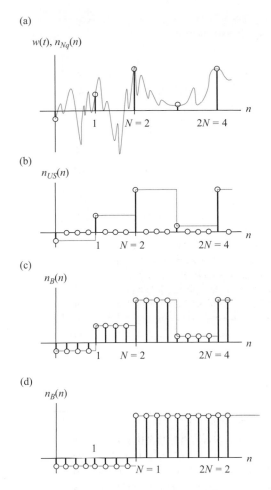

**Fig. 4.3.2** *Band-limited white Gaussian noise in continuous- and discrete-time domains, obtained by interpolation (sample and hold).*

In the example presented in Figure 4.3.2(c), the interpolation is performed by repeating the sample values three times inside the sampling interval. Likewise, in the example presented in Figure 4.3.2(d), the interpolation is performed by repeating the sample values seven times inside the two sampling intervals. If the symbol (bit) duration coincides with the two sampling intervals, the procedure in 4.3.2(d) will give both good noise statistics and an accurate signal-to-noise ratio for discrete communication system analysis. Instead of repeating the noise samples (similar to the sample-and-hold procedure), we can do exact interpolation, where the discrete samples follow the shape of the continuous-time noise. An example of this kind of precise interpolation is presented in Figure 4.3.3(d), where three noise samples are added and interpolated.

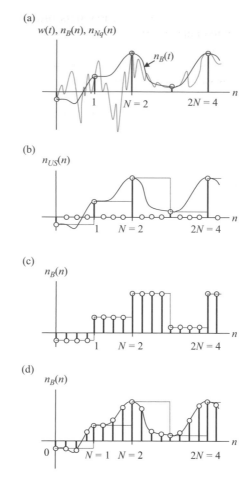

**Fig. 4.3.3** *Band-limited white Gaussian noise in continuous time and discrete time, obtained by precise interpolation.*

Both of these procedures preserve the bandwidth $B$ and make the processing of signal and noise when simulating or emulating discrete communication systems simpler. As we said before, if we are investigating communication system properties, like the bit error rate value, in the presence of noise, we need to transmit a sufficiently large number of bits to quantify these properties. In that case, each discrete bit can be affected by at least one independent noise sample, and the statistics of the noise on a large number of bits will remain Gaussian. When this procedure is followed, the discrete Gaussian noise will be as presented in Figure 4.3.2(d), where one noise sample is interpolated seven times.

**The second method for generating discrete-time Gaussian noise.** The second method for generating baseband noise involves transferring a white Gaussian noise $W(t)$ of zero mean and power spectral density $N_0/2$ through an ideal low-pass filter of bandwidth $B$ to obtain the continuous-time band-limited noise $n_B(t)$, as shown in Figure 4.3.1 (third branch). Then sampling is performed according to the Nyquist sampling theorem, and the discrete samples are up-sampled and interpolated as in the first method, producing the baseband noise, as shown in Figure 4.3.3(c). However, as we pointed out before, the samples can be interpolated using other interpolation methods (instead of sample and hold) that will yield discrete-time noise with amplitudes that precisely follow the shape of the baseband continuous-time noise, as shown in Figure 4.3.3(d).

**The third method for generating discrete-time Gaussian noise.** The third method for generating baseband discrete noise involves transferring a truncated white Gaussian noise $W(t)$ of zero mean and power spectral density $N_0/2$ through an ideal low-pass filter of bandwidth $B$ to obtain the continuous-time band-limited noise $n_B(t)$, as shown in Figure 4.3.3(a). As we said above, the noise can be sampled and the samples can be interpolated, as in the second method, which will give the discrete noise in Figure 4.3.3(c). However, the sampling can be performed according to the Nyquist sampling theorem but with a higher sampling rate, which is defined as $T_s = T/N = T_{Nyquist}/N$, as presented in the fourth branch in Figure 4.3.1. Because the baseband noise is sampled according to the Nyquist theorem but with a higher rate, the resulting discrete-time noise preserves the bandwidth $B$ and can be interpolated and represented by the graph in Figure 4.3.3(d). The samples of the noise are statistically the same as the samples obtained by precise interpolation, which are presented in Figure 4.3.3(a). The noise can be returned to the continuous-time domain by transferring these samples through a reconstruction low-pass filter.

### 4.3.3   Spectral Analysis of Continuous-Time Baseband Noise

The ideal baseband noise process $N_B(t)$ has a constant power spectral density $S_{NB}(f)$ in the bandwidth $B$, as presented in Figure 4.3.4. A realization of this noise process, denoted by $n_B(t)$, can be obtained at the output of an ideal low-pass filter of bandwidth $B$ and a magnitude response of 1 when the white Gaussian noise $w(t)$ is applied at the input of the filter. This procedure is shown in Figure 4.3.1 in the fourth branch, which generates the baseband waveforms $n_B(t)$ presented in Figure 4.3.3(a). This noise can also be obtained by transferring the discrete noise through a reconstruction filter as shown in Figure 4.3.1 (first branch).

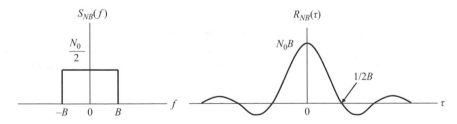

**Fig. 4.3.4** *Noise power spectral density (left) and autocorrelation (right) functions.*

It is important to observe the following behaviour of the noise: The samples of the noise taken at the sampling rate $2B$ are independent and uncorrelated, which corresponds to the zero crossings of the autocorrelation function, as shown in Figure 4.3.4. The distribution function of the noise samples remains Gaussian. That is obvious for the case when linear interpolation (sample and hold) is used for the noise samples, as shown in Figure 4.3.2(c) and (d). The samples taken at the lower sampling rate are not independent and are not uncorrelated, which is obvious from the shape of the autocorrelation function.

Therefore, we have here Gaussian noise that is band limited but not is 'white' according to the strict classical definition, due to the limited power and correlation. By specifying the procedure of generating noise, as presented in Figure 4.3.1, we can control the distribution, correlation, and independence of the samples. In addition, the definition of the white noise, as we used it, makes sense from both a theoretical and a practical point of view, which partially opposes the discussion and definition given in Lapidoth (2009, pp. 554–558). Namely, any limitation of the bandwidth of white Gaussian noise $W(t)$ will eliminate the term 'white', which is reserved for noise in an unlimited bandwidth. Theoretically, noise generated in a communication channel, for example, is modelled as additive white Gaussian noise. However, the noise processed in the receiver after the first filtering is no longer white noise but band-limited Gaussian noise with a certain level of correlation. It appears that, when the noise is in a limited bandwidth, the following analysis applies without any exception.

The autocorrelation function of the baseband noise may be obtained as the inverse Fourier transform of the baseband noise power spectral density $S_{NB}(f)$, as follows:

$$R_{NB}(\tau) = \int_{-B}^{B} \frac{N_0}{2} e^{j2\pi f \tau} df = N_0 B \operatorname{sinc}(2\pi B \tau) = N_0 B \frac{\sin(2\pi B \tau)}{2\pi B \tau}. \qquad (4.3.1)$$

The power spectral density and the autocorrelation function of this band-limited noise $N_B(t)$ are shown in Figure 4.3.4. Equation (4.3.1) was proven in the literature for the noise $n_B(t)$ generated using the first method. Following that method, a finite variance noise $W(t)$ is sampled at sampling instants $1/2B$, and then the samples are passed through a low-pass reconstruction filter with the bandwidth $B$. The output of the filter is the baseband noise having the bandwidth $B$ and the autocorrelation function defined by eqn (4.3.1).

The maximum value of the autocorrelation function is equal to the power of noise $N_0 B$. The zero crossings occur at $\tau = \pm k/2B$. For the defined power, when the bandwidth $B$ tends to infinity, the autocorrelation function tends to the delta function, the noise becomes white Gaussian noise, and the correlation between any two samples vanishes.

### 4.3.4   Spectral Analysis of Discrete-Time Baseband Noise

Suppose now that the continuous-time band-limited noise $N_B(t)$ is represented in the discrete-time domain, as shown in Figure 4.3.1 in the third branch. As we said before, the bandwidth should be related to the Nyquist frequency, that is, $2B = f_{Nq}$, and the discrete-time samples of the noise need to be taken at the Nyquist sampling instants, $T = 1/2B$. The generated noise samples $n_B(kT)$ are realizations of the discrete-time stochastic noise process $N_B(kT)$, where $k$ takes integer values. Because the continuous-time noise process $N_B(t)$ is Gaussian, the discrete-time process $N_{Nq}(n)$ generated from it is Gaussian, too. The random signal $n_{Nq}(n)$ is one realization of the discrete-time stochastic process $N_{Nq}(n)$. The discrete-time noise samples $n_{Nq}(n)$ obtained are realizations of the iid Gaussian random variable $N_{Nq}(n)$, which are defined for each $n$ and have zero mean and finite variance.

The discrete-time stochastic process $N_{Nq}(n)$ generated in this way will have the minimum number of samples per second defined by the Nyquist theorem. In this case, the power spectral density of the discrete noise will be defined by the bandwidth

$$\Omega_{\max} = 2\pi \frac{B}{f_s} = 2\pi \frac{B}{2B} = \pi, \tag{4.3.2}$$

where $f_s$ is the sampling frequency, and the maximum normalized frequency is $F_{max} = \Omega_T/2\pi = 1/2$. The power spectral density and the autocorrelation function for this noise are shown in Figure 4.2.1. The noise samples taken are realizations of the iid random variables representing the white Gaussian noise. Therefore, the random discrete-time signal is one realization of the white Gaussian discrete-time stochastic process that has bandwidth $\pm\pi$, and the autocorrelation function that is the Kronecker delta function. As such, the process $N_{Nq}(n)$ remains Gaussian, uncorrelated, and independent.

**Ideal filtered baseband discrete correlated Gaussian noise.** If we need both more samples per second and a preserved continuous-time noise bandwidth, the sampling of the band-limited noise $N_B(t)$ can be done at sampling instants $T_s$ that are smaller than the Nyquist sampling time $T$. Because the noise $W(t)$ is already filtered to get $N_B(t)$, this increase in the sampling rate will not change the noise bandwidth. Namely, if we pass the sampled noise $N_B(kT_s)$ though the Nyquist reconstruction filter, the output will be again the noise $N_B(t)$. For that purpose, we will use a sampler performing sampling on $N_B(t)$ at time instants $T_s = T/N$, which increases the sampling rate $N$ times, as shown in Figure 4.3.1, in the fourth branch. For a given $T_s$, the sampling frequency is $f_s = 1/T_s = N/T = 2NB$, which is $N$ times higher than the minimum Nyquist frequency $f_{Nq} = 2B$. The bandwidth $\Omega_B$ of the generated discrete-time stochastic noise process $N_B(n)$ can be expressed in terms of the normalized frequency $F_B$ as

$$\Omega_B = 2\pi \frac{B}{f_s} = 2\pi F_B = 2\pi \frac{B}{2NB} = \frac{\pi}{N}, \tag{4.3.3}$$

and the power spectral density of the discrete stochastic noise process is expressed as

$$S_{NB}(\Omega) = \begin{Bmatrix} \frac{N_0}{2}, & -\Omega_B < \Omega < \Omega_B \\ 0, & \text{otherwise} \end{Bmatrix} = \begin{Bmatrix} \frac{N_0}{2}, & -\pi/N < \Omega < \pi/N \\ 0, & \text{otherwise} \end{Bmatrix}. \tag{4.3.4}$$

Based on the Wiener–Khintchine theorem, the inverse Fourier transform of the power spectral density gives the autocorrelation function as

$$R_{NB}(l) = \frac{1}{2\pi} \int_{-\pi}^{\pi} S_N\left(e^{j\Omega}\right) e^{j\Omega l} d\Omega = \frac{N_0}{4\pi} \int_{-\Omega_B}^{\Omega_B} e^{j\Omega l} d\Omega = \frac{N_0}{4\pi} \frac{e^{j\Omega l}}{jl} \bigg|_{-\Omega_B}^{\Omega_B}$$

$$= \frac{N_0}{2\pi} \frac{\left(e^{j\Omega l} - e^{-j\Omega l}\right)}{2jl} = \frac{N_0}{2\pi} \frac{\sin(l\Omega_B)}{l} = \frac{N_0 \Omega_B}{2\pi} \frac{\sin(l\Omega_B)}{l\Omega_B}$$

$$= \frac{N_0 \Omega_B}{2\pi} \operatorname{sinc}(l\Omega_B). \tag{4.3.5}$$

This expression is equivalent to the expression presented in the book by Rice (2009, p. 207, expression (4.76)) for an assumed $l = 0$ and replacing $T$ by $N$. However, the difference is in that we needed to present discrete-time noise in terms of the discrete variable $n$ rather than relating it to the sampling interval $T$.

Because the correlation function $R_{NB}(l)$ of the process $N_B(n)$ is aperiodic in $l$, the discrete-time Fourier transform can be used to obtain the power spectral density. In this case, the correlation function $R_{NB}(l)$ should be absolutely summable, which guarantees the existence of the power spectral density, which means that $N_B(n)$ is a zero-mean process. The zero crossings of the autocorrelation function are found for these conditions: $F_T = \sin(l\Omega_B) = 0$ or $l\Omega_B = m\pi$, that is, $l = m\pi/\Omega_B$. Suppose $\Omega_B = \pi/2$; then the first crossing is at $l = \pi/\pi/2$. For the practical case when $\Omega_B = \pi/N$, we obtain

$$R_{NB}(l) = \frac{1}{2\pi} \int_{-\pi/N}^{\pi/N} S_{NB}\left(e^{j\Omega}\right) e^{j\Omega l} d\Omega = \frac{N_0}{4\pi} \int_{-\pi/N}^{\pi/N} e^{j\Omega l} d\Omega = \frac{N_0}{4\pi} \frac{e^{j\Omega l}}{jl} \bigg|_{-\pi/N}^{\pi/N}$$

$$= \frac{N_0}{2\pi} \frac{\left(e^{jl\pi/N} - e^{-jl\pi/N}\right)}{2jl} = \frac{N_0}{2\pi} \frac{\sin(l\pi/N)}{l} = \frac{N_0}{2N} \frac{\sin(l\pi/N)}{l\pi/N} \tag{4.3.6}$$

$$\frac{N_0}{2N} \operatorname{sinc}(l\pi/N)$$

The power of this noise can be directly seen from the autocorrelation function or calculated as

$$P_{NB} = \frac{1}{2\pi} \int_{-\pi/N}^{\pi/N} S_{NB}\left(e^{j\Omega}\right) d\Omega = \frac{N_0}{4\pi} \int_{-\pi/N}^{\pi/N} d\Omega = \frac{N_0}{4\pi} \frac{2\pi}{N} = \frac{N_0}{2N} = R_{NB}(0). \tag{4.3.7}$$

Zero crossings of the correlation function (4.3.6) occur for the condition $l\pi/N = m\pi$, that is, for the condition $l = mN$. The power spectral density and the autocorrelation function of the process $N_B(n)$ are presented in Figure 4.3.5, for the case when $N = 2$. The power spectral density is a periodic function with a period of $2\pi$.

In the case when the number of discrete samples is $N = 1$, the discrete-time noise contains only iid Nyquist samples, and the autocorrelation function becomes the Kronecker delta function, because

$$R_{NB}(l) = \frac{N_0}{2} \frac{\sin(l\pi)}{l\pi}, \tag{4.3.8}$$

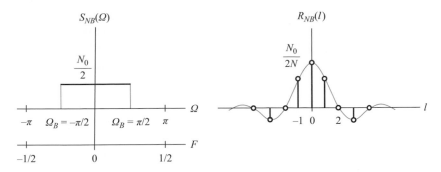

**Fig. 4.3.5** *The power spectral density (left) and autocorrelation (right) functions for N = 2.*

which gives zeros for every $l$ except for $l = 0$ according to the condition that the autocorrelation must be zero when $l\pi = m\pi$ or $l = m$, that is, for all $l$ except $l = 0$. For this case, the autocorrelation function is represented by the graph shown in Figure 4.2.1, showing that there is strong correlation between the interpolated samples, and strong independence between the Nyquist samples. This is the case mentioned in the book by Rice (2009, p. 207, eqn (4.75)) if we define the sampling interval to be $T = 1$.

**The practical design of a baseband discrete-time noise generator.** In practical applications, discrete-time band-limited Gaussian noise will be generated in a slightly different way. A block schematic of a discrete-time noise generator is shown in Figure 4.3.1, in the second branch. The truncated white Gaussian noise is sampled at the Nyquist rate. The samples obtained are realizations of iid random variables that have a truncated Gaussian probability density function with a finite variance. Then, depending on the number of samples, the discrete-time signal obtained is up-sampled at the rate $1/T_s$. After passing through the interpolation filter, the discrete-time signal $n_B(n)$ has the desired sampling rate and represents one realization of the discrete-time baseband noise process $N_B(n)$. The added interpolated samples do not change the bandwidth of the baseband continuous-time noise but increase the mutual correlation.

## 4.4 Operation of a Bandpass Noise Generator

### 4.4.1 Ideal Bandpass Continuous-Time Gaussian Noise

Bandpass noise can be generated using two methods: filtering and modulation. Firstly, we will explain the filtering method.

**The filtering method.** If a white Gaussian noise $W(t)$, with a truncated probability density function of zero mean and power spectral density $N_0/2$, is applied at the input of an ideal *bandpass* filter that has unity amplification, a mid-band frequency $\pm f_c$, and bandwidth $2B$, the output signal will be a *bandpass* noise process $N(t)$. The power spectral density of this noise is shown in Figure 4.4.1. Its power is $P_N = 2BN_0/2 + 2BN_0/2 = 2BN_0$. The autocorrelation function of this noise is

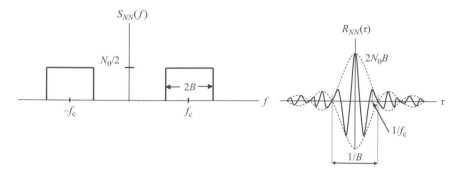

**Fig. 4.4.1** *The power spectral density function of bandpass noise (left), and its related autocorrelation function (right).*

$$R_{NN}(\tau) = \int_{-f_c-B}^{f_c+B} S_{NN}(f)e^{j2\pi f\tau} \, df = \int_{-f_c-B}^{f_c+B} \frac{N_0}{2}e^{j2\pi f\tau} \, df + \int_{-f_c-B}^{f_c+B} \frac{N_0}{2}e^{j2\pi f\tau} \, df$$

$$= N_0 B \operatorname{sinc}(2\pi B\tau) \left[ e^{-j2\pi f_c\tau} + e^{j2\pi f_c\tau} \right] = 2N_0 B \operatorname{sinc}(2\pi B\tau)\cos(2\pi f_c\tau)$$

(4.4.1)

Because the power spectral density of the noise is symmetric about the mid-band frequency $\pm f_c$, with power equal to the power of each component, the corresponding power spectral density of the baseband component is $2BN_0$. Namely, $2BN_0/2 + 2BN_0/2 = 2BN_0$, as we will see below. Its autocorrelation function is

$$R_{NB}(\tau) = 2N_0 B \operatorname{sinc}(2B\tau),$$

(4.4.2)

which is represented by the upper envelope (dashed line) in Figure 4.4.1. Thus, the noise power of this baseband noise may be expressed as

$$P_{NB} = \int_{-B}^{+B} N_0 \, df = 2N_0 B.$$

(4.4.3)

When the bandwidth $B$ tends to infinity, this noise becomes white Gaussian noise (tends to infinite bandwidth) and the correlation between any two samples becomes negligible. This can happen if we spread the power of the bandpass noise frequency components into frequencies beyond their bandwidth while preserving the average power at the same level, that is, $2BN_0$. Consequently, the autocorrelation function will tend to the Dirac delta function with a weight of $2BN_0$. If we extend the bandwidth and preserve the average power, the final value of the power spectral density $N_0/2$ becomes infinitesimally small. The noise in the time domain will be characterized by a wave shape that changes in value very fast, which could even have theoretically infinite values but preserves a finite variance, $2BN_0$. Therefore, we return to our definition of truncated Gaussian noise as noise with finite energy, finite variance, infinite power spectral density bandwidth (preserves classical definition of whiteness), finite entropy, and a Gaussian truncated probability density function.

**The modulation method.** Bandpass noise can also be generated by frequency shifting the spectrum of the baseband noise according to the modulation theorem. Suppose the baseband noise $N_B(t)$, having the ideal bandwidth defined in Figure 4.3.4, is multiplied by the carrier at the frequency $f_c$. The initial random noise amplitude and the carrier amplitude are statistically independent, which is achieved by the randomness of the noise amplitudes and the random uniform phase of the carrier $\theta$. Therefore, the bandpass noise process can be expressed as the product

$$N(t) = N_B(t) \cdot A \cos(2\pi f_c t + \theta), \tag{4.4.4}$$

where $A$ is the amplitude of the carrier, which will be used here as a proportionality factor, as will be discussed later. In order to find the power spectral density function of the noise, we need to find its autocorrelation function as

$$
\begin{aligned}
R_{NN}(\tau) &= E\{N(t)N(t+\tau)\} \\
&= E\{[N_B(t) \cdot A \cos(2\pi f_c t + \theta)][N_B(t+\tau) \cdot A \cos(2\pi f_c(t+\tau)+\theta)]\} \\
&= \frac{1}{2} A^2 E\{[N_B(t) \cdot N_B(t+\tau)][\cos(4\pi f_c t + 2\pi f_c \tau + 2\theta) + \cos(2\pi f_c \tau)]\}
\end{aligned} \tag{4.4.5}
$$

Bearing in mind that the expectation of a deterministic cosine function is zero, and due to the independence between the cosine carrier signal and the noise, we obtain

$$
\begin{aligned}
R_{NN}(\tau) &= \frac{1}{2} A^2 E\{[N_B(t) \cdot N_B(t+\tau)]\}[0 + E\{\cos(2\pi f_c \tau)\}] \\
&= \frac{1}{2} A^2 R_{NB}(\tau) \cdot \cos(2\pi f_c \tau)
\end{aligned} \tag{4.4.6}
$$

The related power spectral density is the Fourier transform of this autocorrelation function. To find the transform in this case, we will apply the modulation theorem, which will give us the power spectral density of the bandpass noise as the shifted power spectral density of the baseband noise. Due to the fact the autocorrelation of $N_B(t)$ is multiplied by the cosine function in (4.4.6), we obtain

$$
\begin{aligned}
S_{NN}(f) &= FT\{R_{NN}(\tau)\} = \frac{1}{2} A^2 FT\{R_{NB}(\tau) \cdot \cos(2\pi f_c \tau)\} \\
&= \frac{1}{4} A^2 [S_{NB}(f - f_c) + S_{NB}(f + f_c)]
\end{aligned} \tag{4.4.7}
$$

Because the baseband noise has constant value of $N_0/2$ in the $2B$ bandwidth, that is, $S_{NB} = N_0/2$, and assuming that the proportionality factor $A = 2$, we find that the power spectral density of bandpass noise is the same as the one obtained by filtering and presented in Figure 4.4.1, that is,

$$S_{NN}(f) = S_{NB}(f - f_c) + S_{NB}(f + f_c). \tag{4.4.8}$$

Also, for the defined autocorrelation of the baseband noise and the same proportionality factor $A = 2$, we find the autocorrelation function of the bandpass noise expressed as

$$
\begin{aligned}
R_{NN}(\tau) &= \frac{1}{2} 2^2 N_0 B \operatorname{sinc}(2\pi B\tau) \cdot \cos(2\pi f_c \tau) \\
&= 2 N_0 B \operatorname{sinc}(2\pi B\tau) \cdot \cos(2\pi f_c \tau)
\end{aligned}
\tag{4.4.9}
$$

as derived previously in (4.4.1). The important conclusion from these theoretical developments is that *the generator of passband noise can be realized by using a simple modulator that uses baseband noise to modulate the carrier.*

## 4.4.2 Ideal Bandpass Discrete Gaussian Noise

Discrete-time bandpass noise can also be obtained by either filtering or modulation. If a discrete white Gaussian noise $W(n)$ of zero mean and power spectral density $N_0/2$ is applied at the input of an ideal bandpass filter that has unity amplification, a mid-band frequency $\pm\Omega_c$ and bandwidth $2\Omega_B$, the output signal will be bandpass discrete noise. The power spectrum density of this noise is shown in Figure 4.4.2.

The autocorrelation function of this discrete noise is

$$
R_{NN}(l) = \frac{1}{2\pi} \int_{-\Omega_c-\Omega_B}^{\Omega_c+\Omega_B} S_{NN}(\Omega) e^{j\Omega l}\, d\Omega = \frac{1}{2\pi} \int_{-\Omega_c-\Omega_B}^{-\Omega_c+\Omega_B} \frac{N_0}{2} e^{j\Omega l}\, d\Omega + \frac{1}{2\pi} \int_{\Omega_c-\Omega_B}^{\Omega_c+\Omega_B} \frac{N_0}{2} e^{j\Omega l}\, d\Omega
\tag{4.4.10}
$$

$$
\begin{aligned}
&= \frac{1}{2\pi} \frac{N_0}{2} \frac{e^{j(-\Omega_c+\Omega_B)l} - e^{j(-\Omega_c-\Omega_B)l}}{jl} + \frac{1}{2\pi} \frac{N_0}{2} \frac{e^{j(\Omega_c+\Omega_B)l} - e^{j(\Omega_c-\Omega_B)l}}{jl} \\
&= \frac{1}{2\pi} \frac{N_0}{2} \frac{e^{j(-\Omega_c)l}\left[e^{j(\Omega_B)l} - e^{j(-\Omega_B)l}\right]}{jl} + \frac{1}{2\pi} \frac{N_0}{2} \frac{e^{j(\Omega_c)l}\left[e^{j(\Omega_B)l} - e^{j(-\Omega_B)l}\right]}{jl} \\
&= \frac{1}{2\pi} \frac{N_0}{2} \frac{\left[e^{j\Omega_B l} - e^{-j\Omega_B l}\right]}{jl}\left[e^{-j\Omega_c l} + e^{j\Omega_c l}\right] = \frac{1}{\pi} \frac{N_0}{l} \frac{\left[e^{j\Omega_B l} - e^{-j\Omega_B l}\right]}{2j} \frac{\left[e^{-j\Omega_c l} + e^{j\Omega_c l}\right]}{2} \\
&= \frac{\Omega_B N_0}{\pi} \frac{\sin(\Omega_B l)}{\Omega_B l} \cos\Omega_c l = P_N \operatorname{sinc}(\Omega_B l) \cos\Omega_c l
\end{aligned}
$$

where $P_N$ is the average power of the noise expressed as

$$
P_N = \frac{\Omega_B N_0}{\pi} = \frac{N_0 \pi / N}{\pi} = \frac{N_0}{N}.
\tag{4.4.11}
$$

Suppose that the baseband noise $N_B(n)$, having the ideal bandwidth defined in Figure 4.3.4, is multiplied by a carrier with frequency $\Omega_c$. The initial random noise amplitude and the carrier amplitude are statistically independent, which is represented by the randomness of the noise amplitude and the random uniform phase of the carrier $\theta$. Then, the bandpass noise can be expressed as the product

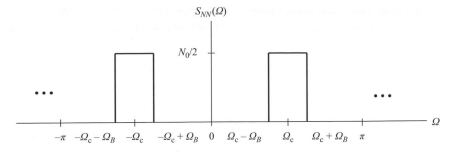

**Fig. 4.4.2** *The power spectral density function of discrete bandpass Gaussian noise.*

$$N(n) = N_B(n) \cdot A \cos(\Omega_c n + \theta), \tag{4.4.12}$$

where $A$ is the amplitude of the carrier, which will be used here as a proportionality factor. In order to find the power spectral density function of the noise, we need to find the autocorrelation function, which is expressed as

$$
\begin{aligned}
R_{NN}(l) &= E\{N_B(n) \cdot A \cos(\Omega_c n + \theta) N_B(n+l) \cdot A \cos(\Omega_c n + \Omega_c l + \theta)\} \\
&= \frac{1}{2} A^2 E\{[N_B(n) \cdot N_B(n+l) \cdot][\cos(2\Omega_c n + \Omega_c l + 2\theta) + \cos(\Omega_c l)]\}
\end{aligned}
$$

Due to the independence between the cosine carrier and the noise, we obtain

$$R_{NN}(l) = \frac{1}{2} A^2 E\{[N_B(n) \cdot N_B(n+l)]\} \cos(\Omega_c l) = \frac{1}{2} A^2 R_{NB}(l) \cdot \cos(\Omega_c l). \tag{4.4.13}$$

The power spectral density is the discrete-time Fourier transform of the autocorrelation function. To find the transform in this case, we will apply the modulation theorem, which will give us the power spectral density of the bandpass noise as the shifted power spectral density of the baseband noise, due to the fact the autocorrelation of $R_{NB}(l)$ is multiplied by the cosine function. Therefore, we obtain

$$
\begin{aligned}
S_{NN}(\Omega) &= FT\{R_N(l)\} = \frac{1}{2} A^2 FT\{R_{NB}(l) \cdot \cos(\Omega_c l)\} \\
&= \frac{1}{4} A^2 [S_{NB}(\Omega - \Omega_c) + S_{NB}(\Omega + \Omega_c)]
\end{aligned} \tag{4.4.14}
$$

Because the baseband noise has constant value of $N_0/2$ in the $2\Omega_B$ interval, that is, $S_{NB} = N_0/2$, and assuming that the proportionality factor $A = 2$, we find that the power spectral density of the bandpass noise is the same as that obtained by filtering and presented in Figure 4.4.2, that is,

$$S_{NN}(\Omega) = S_{NB}(\Omega - \Omega_c) + S_{NB}(\Omega + \Omega_c). \tag{4.4.15}$$

Also, for the defined autocorrelation function of the baseband noise $R_{NB}(l) = (N_0 \Omega_B / 2\pi) \, \text{sinc} \, (l \Omega_B)$ and $A = 2$, we may have the autocorrelation function of the bandpass noise expressed as

$$R_{NN}(l) = \frac{1}{2} 2^2 \frac{N_0 \Omega_B}{2\pi} \, \text{sinc} \, (l \Omega_B) \cdot \cos (\Omega_c l) = \frac{N_0 \Omega_B}{\pi} \, \text{sinc} \, (l \Omega_B) \cdot \cos (\Omega_c l). \qquad (4.4.16)$$

The important conclusion from these theoretical developments is that *the generator of the passband noise can be designed using a simple modulator that uses discrete baseband noise to modulate the discrete carrier.* Using this theory we will develop generators (modulators) and regenerators (demodulators) of bandpass discrete-time noise.

### 4.4.3 Modulators and Demodulators of Ideal Bandpass Discrete Gaussian Noise

In practical simulations or emulations of a designed communication system, we need to control the signal-to-noise ratio at the receiver side by changing the level of the noise in the simulated channel. Therefore, the noise generator needs to process the normally distributed Gaussian baseband noise to achieve the required noise power, and then modulate the orthonormal carrier by the baseband noise to obtain bandpass noise. The block schematics of the noise modulator and the noise demodulator, called the *j*th modulator and the *j*th demodulator, respectively, are shown in Figure 4.4.3. The block schemes are expressed in terms of mathematical operators that are used in the subsequent mathematical modelling. The inputs and outputs of the blocks are expressed in terms of the realizations of stochastic processes denoted by small letters. The related stochastic processes are indicated in capital letters beneath the block schematics. For the stochastic processes and their realization, we use standard notation that is used throughout this book. In addition, in order to simplify notation, we will use the term *noise energy*, which will be denoted as $E_N$, which is equivalent to the power spectral density of white Gaussian noise, that is, $E_N = N_0 / 2$.

**The noise modulator/generator.** At the input of the *j*th noise modulator is a baseband discrete-time Gaussian noise $n_{Gj}(n)$, having the ideal bandwidth defined in Figure 4.3.4. This noise is one realization of a zero-mean and unit-variance discrete-time Gaussian stochastic process $N_{Gj}(n)$, that is, the standard normal process. The noise can be generated according to one of the three methods explained in Section 4.3.1 and presented in Figure 4.3.1. Suppose the

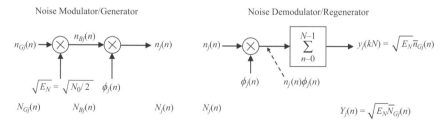

**Fig. 4.4.3** *The noise modulator and noise demodulator for a linearly interpolated noise sample.*

discrete noise samples $n_{Gj}(n)$ are obtained by interpolating the Nyquist samples as presented in Figure 4.3.2(c) by adding three samples and interpolating them. The interpolated noise samples $n_{Gj}(n)$ are multiplied by the square root of the noise energy $E_N = N_0/2$. By changing this value, we can generate the noise with the required power. Because the noise is a wide-sense stationary process, the samples of noise obtained are realizations of the baseband Gaussian random variable $N_{Bj}(n) = N_{Gj}(n) \cdot \sqrt{E_N}$ defined at each time instant $n$ and having zero mean and the standard deviation $\sigma_B$. In order to make the noise modulator compatible with the modulation procedure in discrete communications, which will be used in Chapters 5 and 7, the $j$th carrier will be defined as the orthonormal function, that is,

$$\phi_j(n) = \sqrt{\frac{2}{N}} \cos\left(\Omega_j n + \theta\right), \tag{4.4.17}$$

where $N$ is the total number of noise samples in a symbol interval $N$ after completion of the interpolation process. The baseband noise is multiplied by the carrier of the amplitude $A_N$, which is proportional to the square root of the noise energy $E_N$, which results in a bandpass noise at the output of the noise modulator that can be expressed as

$$
\begin{aligned}
n_j(n) &= n_{Gj}(n) \cdot \sqrt{E_N}\phi_j(n) = n_{Gj}(n) \cdot \sqrt{E_N}\sqrt{\frac{2}{N}} \cos\left(\Omega_j n + \theta\right), \\
&= n_{Gj}(n) \cdot A_N \cos\left(\Omega_j n + \theta\right)
\end{aligned}
\tag{4.4.18}
$$

which is a realization of the bandpass stochastic process $N_j(n)$, expressed as

$$N_j(n) = N_{Gj}(n) \cdot A_N \cos\left(\Omega_j n + \theta\right), \tag{4.4.19}$$

where $N_j(n)$ is the bandpass noise process presented as its realizations in Figure 4.4.4(c), 4.4.4(d), and 4.4.4(e), which are obtained for the three cases that include three methods of generating baseband noise $N_{Gj}(n)$, as mentioned in Section 4.3.2 and presented in Figures 4.3.2(c), 4.3.2(d), and 4.3.3(d), respectively. The process $N_j(n)$ is a wide-sense stationary process with the variance

$$
\begin{aligned}
\sigma_{Nj}^2 &= E\left\{N_j^2(n)\right\} = E\left\{\left[N_{Gj}(n) \cdot \sqrt{E_N}\phi_j(n)\right]^2\right\} \\
&= E\left\{N_{Gj}^2(n) \cdot E_N \frac{2}{N}\left(\frac{1}{2} + \frac{1}{2}\cos\left(2\Omega_j n + 2\theta\right)\right)\right\} \\
&= E_N \frac{1}{N}E\left\{N_{Gj}^2(n)\right\} + E_N \frac{1}{N}E\left\{N_{Gj}^2(n)\cos\left(2\Omega_j n + 2\theta\right)\right\} = E_N \frac{1}{N} = \frac{N_0}{2N}
\end{aligned}
\tag{4.4.20}
$$

where $N$ is the number of discrete random noise samples in a symbol interval. The baseband noise, shown in Figure 4.4.4(a), and the discrete carrier, shown in Figure 4.4.4(b), are statistically independent. The autocorrelation function of the bandpass noise can be calculated as

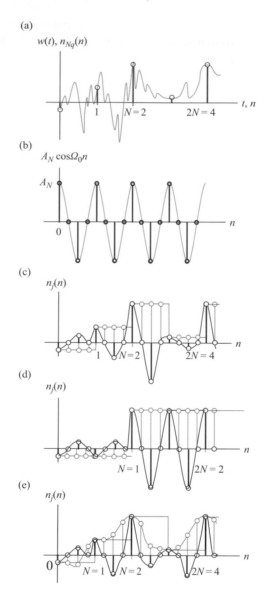

**Fig. 4.4.4** *Bandpass Gaussian noise in continuous- and discrete-time domains.*

$$R_{Nj}(l) = E\left\{N_{Gj}(n) \cdot A_N \cos\left(\Omega_c n + \theta\right) N_{Gj}(n+l) \cdot A_N \cos\left(\Omega_c n + \Omega_c l + \theta\right)\right\}$$
$$= \frac{1}{2} A_N^2 E\left\{\left[N_{Gj}(n) \cdot N_{Gj}(n+l)\right]\left[\cos\left(2\Omega_c n + \Omega_c l + 2\theta\right) + \cos\left(\Omega_c l\right)\right]\right\}.$$

(4.4.21)

Due to the independence between the cosine carrier and the noise, and having in mind that the expectation of a cosine function is zero, we obtain

$$R_{Nj}(l) = \frac{A_N^2}{2} E\left\{\left[N_{Gj}(n) \cdot N_{Gj}(n+l)\right] \cdot \cos\left(\Omega_c l\right)\right\} = \frac{1}{2} A_N^2 R_{Gj}(l) \cdot \cos\left(\Omega_c l\right).$$

(4.4.22)

The power spectral density is the discrete-time Fourier transform of the autocorrelation function, which can be found by applying the modulation theorem. The power spectral density of the bandpass noise is a shifted power spectral density of the baseband noise, due to the fact the autocorrelation of $R_{Gj}(l)$ is multiplied by the cosine function. Therefore, we obtain

$$S_{Nj}(\Omega) = FT\{R_{Nj}(l)\} = \frac{1}{2}A_N^2 FT\{R_{Gj}(l) \cdot \cos(\Omega_c l)\}$$
$$= \frac{1}{4}A_N^2[S_{Gj}(\Omega - \Omega_c) + S_{Gj}(\Omega + \Omega_c)] \tag{4.4.23}$$

Because the baseband noise has constant value of $N_0/2$ in the $2\Omega_B$ interval, that is, $S_{Gj}(\Omega) = N_0/2$, and assuming that the proportionality factor $A_N = 2$, we may have the power spectral density of the bandpass noise as that obtained by filtering and presented in Figure 4.4.2, that is,

$$S_{Nj}(\Omega) = S_{Gj}(\Omega - \Omega_c) + S_{Gj}(\Omega + \Omega_c). \tag{4.4.24}$$

Also, for the defined autocorrelation function of the baseband noise in (4.3.5), which is expressed as $R_{Gj}(l) = (N_0\Omega_B/2\pi)\,\text{sinc}(l\Omega_B)$, and $A_N = 2$, we may find the autocorrelation function of the bandpass noise using expression (4.4.22) in the form

$$R_{Nj}(l) = \frac{1}{2}2^2\frac{N_0\Omega_B}{2\pi}\,\text{sinc}(l\Omega_B) \cdot \cos(\Omega_c l) = \frac{N_0\Omega_B}{\pi}\,\text{sinc}(l\Omega_B) \cdot \cos(\Omega_c l), \tag{4.4.25}$$

as has been already derived in eqn (4.4.10). The following important conclusion comes from these theoretical developments: *the generator of the bandpass noise can be realized using a simple modulator that uses discrete baseband noise to modulate the discrete carrier.*

**A noise demodulator/regenerator.** Baseband noise can be demodulated by a correlator, as shown in Figure 4.4.3. for the *j*th correlator. A correlator operation will be precisely described in Chapters 5 and 7. The correlator output is

$$y_j(kN) = \sum_{n=0}^{N-1}\sqrt{E_N}n_{Gj}(n) \cdot \phi_j^2(n) = \sqrt{E_N}\frac{1}{N}\sum_{n=0}^{N-1}n_{Gj}(n) \cdot [1 + \cos 2(\Omega_j n + \theta)]$$
$$= \sqrt{E_N}\frac{1}{N}\sum_{n=0}^{N-1}n_{Gj}(n) = \sqrt{E_N}\bar{n}_{Gj}(n) \tag{4.4.26}$$

which is a realization of a discrete-time stochastic process $Y_j(kN)$ represented by the related random variables $Y_j(kN)$ calculated in the symbol intervals $kN$, $k = 1, 2, \ldots$. Therefore, for each $k$ value, we define a random variable $Y_j(kN)$ as the weighted sample mean $\bar{N}_{Gj}(kN)$ of $N$ Gaussian baseband noise variables $N_{Gj}(n)$, that is,

$$Y_j(kN) = \sqrt{E_N}\frac{1}{N}\sum_{n=(k-1)N}^{kN-1}N_{Gj}(kN) = \sqrt{E_N}\bar{N}_{Gj}(kN). \tag{4.4.27}$$

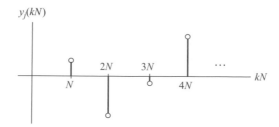

**Fig. 4.4.5** *One realization $y_j(kN)$ of the stochastic process $Y_j(kN)$.*

The sample mean $\overline{N}_{Gj}(kN)$ is defined as the sum of $N$ Gaussian random variables $N_{Gj}(kN)$ divided by $N$. The weighting factor is the square root of the noise energy $E_N$. If all $k$ values are taken into account, the discrete-time stochastic process $Y_j(kN)$ is completely defined. One realization of this stochastic process is the series of random discrete-time continuous-valued samples taken at time instants $kN$, as presented in Figure 4.4.5. for a hypothetical processing of noise in four symbol intervals.

The random variables $N_{Gj}(n)$, defined for $n = 1, 2, \ldots, N-1$, can be identical and strongly correlated with mean zero and unit variance, which follows from the explanation of noise generators in Section 4.3.1. The mean of $Y_j(kN)$, for the first symbol defined by $k = 1$, is

$$\eta_{Yj} = E\{Y_j\} = E\{Y_j(N)\} = \sqrt{E_N} \frac{1}{N} \sum_{n=0}^{N-1} E\{N_{Gj}(n)\} = 0, \qquad (4.4.28)$$

The standard Gaussian noise samples $n_{Gj}(n)$ can be generated by different methods, as explained in Section 4.3.1. Depending on the method, different variances of the noise process can be obtained. Recall the theorem that states that the variance of a sum of uncorrelated random variables is equal to the sum of their variances. In addition, we will include correlation coefficients in the case when the random variables are correlated. Calculations of the noise variances at time instants $kN$ can be performed for the baseband noise that is generated using previously demonstrated methods. Some of them will be presented in the following examples.

**Example**

Suppose a baseband noise is obtained by generating one Nyquist sample and interpolating it by adding $N$ identical samples. The obtained discrete-time noise modulates the carrier, as shown in Figure 4.4.4(d). The baseband noise samples are presented in Figure 4.3.2(d). Present the baseband noise process $N_B(n)$ as an ensemble of its realizations. Calculate the variance of the random variable $Y_j(kN)$.

**Solution.** The baseband noise process $N_B(n)$ is presented by an ensemble of its realizations in Figure 4.4.6. The mean is zero and the variance of the output of the correlator is

$$\sigma_{Y_j}^2 = E\left\{(Y_j)^2\right\} = E\left\{\left(\sqrt{E_N}\frac{1}{N}\sum_{n=0}^{N-1}N_{Gj}(n)\right)^2\right\} = \frac{E_N}{N^2}E\left\{\left(\sum_{n=0}^{N-1}N_{Gj}(n)\right)^2\right\}$$

$$= \frac{E_N}{N^2}E\left\{\begin{array}{l}N_{Gj}^2(0)+\ldots+N_{Gj}^2(N-1)\\+2N_{Gj}(0)N_{Gj}(1)+\ldots+2N_{Gj}(N-2)N_{Gj}(N-1)\end{array}\right\} \tag{4.4.29}$$

Due to the correlation between the random variables defined for every $n$ noise sample, the correlation coefficients are ones, and the mixed doubled terms will give the product of two identical variances, that is, we may have the following expectation for the first product:

$$E\left\{2N_{Gj}(0)N_{Gj}(1)\right\} = 2\rho_{N_{Gj}}\sigma_{N_{Gj}}\sigma_{N_{Gj}} = 2\sigma_{Gj}^2. \tag{4.4.30}$$

If we take into account all $N^2$ terms in the brackets of (4.4.29) for $\sigma_{Gj}^2 = 1$, the variance becomes

$$\sigma_{Y_j}^2 = \frac{E_N}{N^2}E\left\{N_{Gj}^2(0)+N_{Gj}^2(1)+\cdots+2N_{Gj}(0)N_{Gj}(1)+\ldots\right\} = \frac{E_N}{N^2}N^2\sigma_{Gj}^2 = E_N = \frac{N_0}{2}. \tag{4.4.31}$$

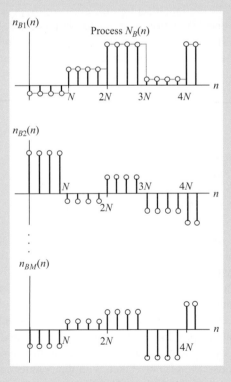

**Fig. 4.4.6** *A baseband noise process represented by its M realizations.*

## 4.5   Practical Design of a Band-Limited Discrete-Time Noise Modulator

A noise modulator is used to implement a generator of the discrete-time bandpass noise that can be used for the theoretical analysis, simulation and emulation of discrete communication systems. The demodulator/regenerator is added to investigate the noise properties after demodulation in a correlation receiver. The demodulator structure is equivalent to the structures defined in Chapters 5 and 7 and similar to that one mentioned in the book by Rice (2009, p. 346, Figure 6.5.1). In this section, we will describe the property of bandpass noise that is obtained by modulating a set of carriers by the independent baseband noise processes defined in the bandwidth $B$. The noise modulator and demodulator are presented in Figure 4.5.1. The block schemes are expressed in terms of mathematical operators. The inputs and outputs of the blocks are expressed in terms of the realizations of stochastic processes denoted by small letters. The related stochastic processes are listed in capital letters beneath the block schematics, and are followed by their vector forms in bold letters.

**Baseband noise processing.** At the inputs of the modulator branches are standard Gaussian noise samples $n_{Gj}(n)$, which are the realizations of Gaussian random variables $N_{Gj}(n)$, $j = 1, \ldots J$, defined for every $n$ in the symbol interval from $n = 0$ to $n = N - 1$, as explained in Section 4.3. The random discrete-time noise $n_{Gj}(n)$ generated in the time interval $N$ can be understood as one realization of a stochastic process $N_{Gj}(n)$ defined by its zero mean and unit variance. The notation here follows the notation for discrete-time stochastic processes presented in Chapter 3. Therefore, the input samples taken at any time instant $n$ are realizations of the related Gaussian random variables $N_{Gj}(n)$. These samples form a vector that can be represented in matrix form as

$$\mathbf{n_G} = [n_{G1} \ \ n_{G2} \ \ \cdots \ \ n_{GJ}]^T. \tag{4.5.1}$$

The elements of the matrix are realizations of $J$ random variables defined for any $n$. The random variables can also be expressed in matrix form as

$$\mathbf{N_G} = [N_{G1} \ \ N_{G2} \ \ \cdots \ \ N_{GJ}]^T. \tag{4.5.2}$$

We assume here that a symbol duration is $N$ and that all processing is done on both the discrete-time noise samples and the signal samples in that time frame. Because the noise is a wide-sense stationary process, all symbols have the same statistical properties, and the analysis of one symbol, as is done in this section, will be sufficient to explain the overall operation of both the modulator and the demodulator.

The $j$th random variables $N_{Gj}(n)$ are Gaussian with zero mean and unit standard deviation. Their samples are generated according to the principles explained for the baseband noise generation in Section 4.3 using the modulators presented in Figure 4.5.1. Firstly, the samples of the *truncated* white Gaussian noise are taken with the Nyquist rate to control the desired bandwidth $B$ of the noise. In order to modulate the carrier, the discrete-time noise is up-sampled and then interpolated to produce a sufficient number of samples for discrete-time

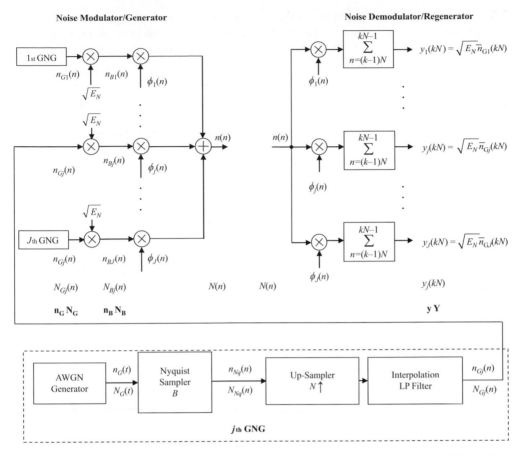

**Fig. 4.5.1** *A bandpass noise modulator/generator and demodulator/regenerator; AWGN, additive white Gaussian noise; GNG, Gaussian baseband noise generator; LP, low-pass.*

modulation. The obtained baseband discrete noise, denoted as $n_{Gj}(n)$ in Figure 4.5.1, modulates the carrier. This noise can be understood as a realization of the stochastic process $N_{Gj}(n)$. In principle, this procedure is to be used to generate baseband noise in each branch of the modulator. For the sake of explanation, only one branch in Figure 4.5.1. is presented in full. The mean value of the baseband noise is 0, and the variance is 1, that is,

$$E\left\{N_{Gj}^2\right\} = \sigma_{Gj}^2 = 1, \tag{4.5.3}$$

which is equal to the power of the noise calculated as

$$P_G = \lim_{N\to\infty} \frac{1}{2N+1} \sum_{n=-N}^{N} E\left\{|N_{Gj}(n)|^2\right\} = E\left\{|N_{Gj}(n)|^2\right\} = \sigma_{Gj}^2 = 1. \tag{4.5.4}$$

The noise samples are up-scaled by multiplying them by the square root of the noise energy, which is denoted by $\sqrt{E_N}$, and the baseband noise samples are generated as a realization of the stochastic process $N_{Bj}(n)$. These samples, for all branches and at the defined time instant $n$, can be expressed as the vector

$$\mathbf{n_B} = [n_{B1} \ n_{B2} \ \cdots \ n_{BJ}]^T. \tag{4.5.5}$$

The samples in the vector are realizations of $J$ random variables that form the vector expressed as

$$\mathbf{N_B} = [N_{B1} \ N_{B2} \ \cdots \ N_{BJ}]^T. \tag{4.5.6}$$

Thus, by specifying the value $\sqrt{E_N}$, baseband noise can be generated with controllable power, which is calculated as

$$
\begin{aligned}
P_B &= \lim_{N \to \infty} \frac{1}{2N+1} \sum_{n=-N}^{N} E\left\{ |N_{Bj}(n)|^2 \right\} = \lim_{N \to \infty} \frac{1}{2N+1} \sum_{n=-N}^{N} E\left\{ \left| \sqrt{E_N} N_{Gj}(n) \right|^2 \right\} \\
&= E_N \lim_{N \to \infty} \frac{1}{2N+1} \sum_{n=-N}^{N} E\left\{ |N_{Gj}(n)|^2 \right\} = E_N \sigma_{Gj}^2 = \sigma_B^2
\end{aligned}
\tag{4.5.7}
$$

Because there is signal transmission through the same channel, it is assumed that the generated noise samples have the same energy. The outputs of the first $J$ multipliers in Figure 4.5.1 are Gaussian noise samples with the zero mean, and variance defined by the noise energy $E_N$. In the case of a system design with multiple orthogonal carriers, the same samples of noise can be applied to the input of all of the branches of the modulator, due to the orthogonality of the carriers. In principle, the baseband modulating noise samples are independently generated to modulate the carriers.

**Bandpass noise processing.** The noise samples are multiplied by the carrier signals ($J$ basis functions) to generate the bandpass noise $n(n)$, which is a realization of the wide-sense stationary stochastic Gaussian noise process $N(n)$. Therefore, the output of the noise modulator is a bandpass random noise $n(n)$ expressed as

$$n(n) = \sum_{j=1}^{J} n_{Bj}(n)\phi_j(n) = \sqrt{E_N} \sum_{j=1}^{J} n_{Gj}(n)\phi_j(n). \tag{4.5.8}$$

The noise demodulators will reproduce the baseband noise components with the specified energy $E_N$. It is sufficient to analyse branch $j$ of the demodulator, because the same processing takes place in all $J$ branches. The bandpass noise samples are multiplied by the $j$th basis function representing the carrier, and then the calculated products for each $n = 0, 1, 2, \ldots, N-1$ are accumulated in the symbol interval, to obtain

$$y_j(kN) = \sum_{n=0}^{N-1} n(n)\phi_j(n) = \sum_{n=0}^{N-1} \left[ \sqrt{E_N} \sum_{j=1}^{J} n_{Gj}(n)\phi_j(n) \right] \phi_j(n).$$

$$= \sqrt{E_N} \sum_{n=0}^{N-1} \left[ n_{G1}(n)\phi_1(n)\phi_j(n) + \cdots + n_{Gj}(n)\phi_j^2(n) + \cdots + n_{GJ}(n)\phi_J(n)\phi_j(n) \right]. \quad (4.5.9)$$

$$= \sqrt{E_N} \sum_{n=0}^{N-1} n_{Gj}(n)\phi_j^2(n)$$

Similar processing can be found in the book by Rice (2009, p. 345). The basis function squared will produce a low-frequency term and a high-frequency term. The high-frequency term will be eliminated by the accumulator, and the bandpass noise component will be generated at the output of the correlation demodulator at the end of the symbol interval.

Suppose now the sinusoidal carriers are used, as presented in Section 4.4.3. The $j$th random output in Figure 4.5.1. is a realization $y_j(kN)$ of a discrete-time stochastic process $Y_j(kN)$ with values at the discrete-time instants $kN$, where $k$ is an integer and $N$ is the number of samples per symbol. One sample of the process $Y_j(kN)$ is

$$y_j(kN) = \sqrt{E_N} \frac{2}{N} \sum_{n=(k-1)N}^{kN-1} \left[ \begin{array}{l} n_{G1}(n)\cos\Omega_1 n\cos\Omega_j n + \cdots + n_{Gj}(n)\cos^2\Omega_j n + \ldots \\ + n_{GJ}(n)\cos\Omega_J n\cos\Omega_j n \end{array} \right]$$

$$= \sqrt{E_N} \frac{2}{N} \sum_{n=(k-1)N}^{kN-1} n_{Gj}(n)/2 \left[ 1 + \cos 2\Omega_j n \right] = \left( \sqrt{E_N}/N \right) \sum_{n=(k-1)N}^{N-1} n_{Gj}(n) \quad (4.5.10)$$

$$= \sqrt{E_N} \bar{n}_{Gj}(kN)$$

which can be calculated for each $k$ and interpreted as the weighted mean value of the Gaussian noise samples. Therefore, for each $k$ value, we can define a random variable $Y_j(kN)$ as the weighted sample mean value $\bar{N}Gj(n)$ of $N$ Gaussian noise variables $N_{Gj}(n)$, that is,

$$Y_j(kN) = \sqrt{E_N} \frac{1}{N} \sum_{n=(k-1)N}^{kN-1} N_{Gj}(n) = \sqrt{E_N} \bar{N}_{Gj}(kN). \quad (4.5.11)$$

The random variables at all correlator outputs, at the time instant $kN$, can be represented by the following vector of $J$ random variables,

$$\mathbf{Y} = [Y_1 \ Y_2 \ \cdots \ Y_J]^T, \quad (4.5.12)$$

and a realization of this vector at the time instant $kN$ can be represented by the corresponding vector of $J$ realizations of random variables in $\mathbf{Y}$, that is,

$$\mathbf{y} = [y_1 \ y_2 \ \cdots \ y_J]^T. \tag{4.5.13}$$

The mean value of any variable $Y_j$ is zero, that is,

$$\eta_{Yj} = E\left\{Y_j(kN)\right\} = \sqrt{E_N} \frac{1}{N} \sum_{n=(k-1)N}^{kN-1} E\left\{N_{Gj}(n)\right\} = 0, \tag{4.5.14}$$

and its variance is

$$\sigma_{Y_j}^2 = \sigma_B^2 = \frac{E_N}{N} = \frac{N_0}{2N}, \tag{4.5.15}$$

as derived in Section 4.4.3. Therefore, the output of the accumulator is the noise that has the same power as the baseband noise at the modulator side. By changing the noise energy $E_N$ at the modulator, we can easily control the noise power at the output of the demodulator. Therefore, if this noise modulator is used for the simulation of a discrete communication system, the bandpass noise process $N(n)$ of the defined power can be added to the transmitter signal. Then, the signal-to-noise ratio at the receiver correlator output can be controlled by changing the noise power at the noise modulator input.

## 4.6  Design of an Ordinary Band-Limited Discrete-Time Noise Modulator

In the analysis of modulation methods in digital communication systems, which are based on the signal and noise processing in the continuous-time domain, the bandpass noise is generated using the in-phase and quadrature noise components to comply with the generation of the in-phase and quadrature signal components, respectively. This noise modulator/generator, operating in the discrete-time domain, is presented in Figure 4.6.1. It is just a special case of the noise modulator/generator presented in Figure 4.5.1, when only two orthogonal carriers are used to produce the bandpass noise. Due to the fact this noise modulator is used in the theoretical and practical analysis of both the $M$-ary phase-shift keying and quadrature amplitude modulation schemes, we will present its operation in detail. The noise modulator and demodulator will be implemented using the in-phase and quadrature carriers, as will be done by implementing signal modulators and demodulators in Chapter 7.

**Baseband noise processing.** At the inputs of modulator branches are standard Gaussian noise samples $n_{Gj}(n)$, which are the realizations of random variables $N_{Gj}(n), j = 1, 2$, and are defined for every $n$ in the symbol interval $n = 0$ to $n = N - 1$ in the same way as for the general case presented in Figure 4.5.1.

**Bandpass noise processing.** The output of the noise modulator is a bandpass noise $n(n)$ expressed as

$$n(n) = \sum_{j=1}^{2} n_{Bj}(n)\phi_j(n) = \sqrt{E_N} \sum_{j=1}^{2} n_{Gj}(n)\phi_j(n), \tag{4.6.1}$$

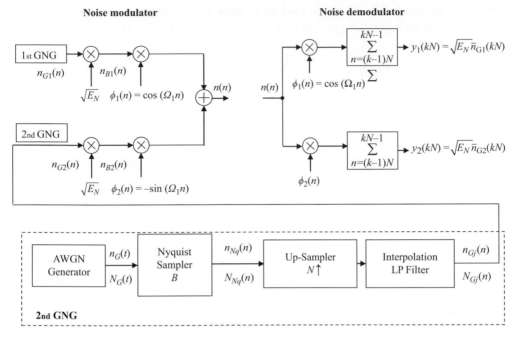

**Fig. 4.6.1** *A bandpass noise modulator and demodulator; AWGN, additive white Gaussian noise; GNG, Gaussian baseband noise generator; LP, low-pass.*

for the cosine and sine basis functions representing the carriers. The noise demodulator will reproduce the baseband noise components. The first output of the demodulator is

$$
y_1(kN) = \sum_{n=(k-1)N}^{kN-1} n(n)\phi_1(n) = \sum_{n=(k-1)N}^{kN-1} \left[ \sqrt{E_N} \sum_{j=1}^{2} n_{Gj}(n)\phi_j(n) \right] \phi_1(n)
$$

$$
= \sqrt{E_N} \sum_{n=(k-1)N}^{kN-1} [n_{G1}(n)\phi_1(n)\phi_1(n) + n_{G2}(n)\phi_1(n)\phi_2(n)] = \sqrt{E_N} \sum_{n=(k-1)N}^{kN-1} n_{G1}(n)\phi_1^2(n)
$$

$$(4.6.2)$$

In an ordinary noise modulator/demodulator, the first basis function is a cosine and the second one is a negative sine function (a basis cosine function shifted for $+90°$). Therefore, we have

$$
y_1(kN) = \sqrt{E_N} \frac{2}{N} \sum_{n=(k-1)N}^{kN-1} [n_{G1}(n)\cos\Omega_1 n \cdot \cos\Omega_1 n + n_{G2}(n)\cos\Omega_1 n \cdot \sin\Omega_1 n]
$$

$$
= \sqrt{E_N} \frac{2}{2N} \sum_{n=(k-1)N}^{kN-1} n_{Gj}(n)[1 + \cos 2\Omega_1 n] = \sqrt{E_N}/N \sum_{n=(k-1)N}^{kN-1} n_{G1}(n)
$$

$$(4.6.3)$$

$$
= \sqrt{E_N}\bar{n}_{G1}(kN)
$$

which is a realization of the related random variable $Y_1(kN)$ that can be expressed as the weighted mean value $\overline{N}_{G1}(kN)$ of a set of $N$ Gaussian noise variables $N_{G1}(n)$, that is,

$$Y_1(kN) = \sqrt{E_N} \frac{1}{N} \sum_{n=(k-1)N}^{KN-1} N_{G1}(n) = \sqrt{E_N} \overline{N}_{G1}(kN). \tag{4.6.4}$$

The mean value of any of $Y_1(kN)$ variables is zero, that is,

$$\eta_{Y_1} = E\{Y_1(kN)\} = \sqrt{E_N} \frac{1}{N} \sum_{n=(k-1)N}^{kN-1} E\{N_{G1}(n)\} = 0, \tag{4.6.5}$$

and the related variance is

$$\sigma_{Y_1}^2 = \sigma_B^2 = \frac{E_N}{N} = \frac{N_0}{2N}, \tag{4.6.6}$$

as derived in Section 4.3.3, relation (4.3.7). Therefore, the output of the accumulator is the noise that has the same power as the baseband noise at the modulator side. By changing the noise energy $E_N$ at the modulator, we can easily control the signal-to-noise ratio in the receiver for the known energy of the signal defined inside the symbol interval $N$.

........................................................................................................................................................

PROBLEMS

1. Suppose a random delay $\tau$ is represented by its continuous exponential density and distribution function expressed as

$$f_\tau(\tau) = \lambda e^{-\lambda \tau}, \quad F_\tau(\tau) = 1 e^{-\lambda \tau},$$

where parameter $\lambda$ defines the mean value $\eta = 1/\lambda$ and variance $\sigma^2 = 1/\lambda^2$. We can derive the discrete probability density function and then truncate it to exist in the interval from 1 to $S$, where $S$ is the number of truncated values with related probabilities. Therefore, the truncated discrete probability density function can be expressed using Dirac delta functions in the form

$$f_{\tau_t}(\tau) = \sum_{s=1}^{S} \frac{(e^\lambda - 1)}{1 - e^{-S\lambda}} e^{-s\lambda} \delta(\tau_t - s) = C \sum_{s=1}^{S} e^{-s\lambda} \delta(\tau_t - s).$$

a. What is the condition the given probability density function should fulfil to be a density function? Prove that the truncated density fulfils this condition.

b. Prove that the mean value of the truncated probability density function is

$$\eta_{t1} = \frac{1}{1-e^{-S\lambda}} \frac{Se^{-(S+1)\lambda} - (S+1)e^{-S\lambda} + 1}{1-e^{-\lambda}}.$$

c.  Compare the mean value calculated in b. with the corresponding mean value defined for the continuous exponential probability density function. Explain the difference between them and comment on the reason for this difference. You may do this without any calculations. Find the mean value for the trivial case when $S = 1$. Explain the meaning of the obtained value.

d.  For $S = 5$ and $\lambda = 0.5$, calculate the truncated exponential probability density function. Plot the graph of this function. Using the probability density function calculation, define on your own (invent) a stochastic process and present one realization of the process. Use at least ten realizations of the related random experiment for the assumed values of the outcomes of the random experiment.

2.  Suppose that a DC signal of amplitude $A$ is present in the white Gaussian noise $W(n)$ specified by the variance $\sigma^2$. Two hypotheses are defined as follows: the null hypothesis $H_0$ is that only the noise is present in the signal $X(n)$, and the alternative hypothesis $H_1$ is that both the DC signal of amplitude $A$ and the noise are present in the signal $X(n)$, which is expressed in the following form:

$$H_0 : X(n) = W(n) \qquad n = 1, 2, \ldots, N.$$
$$H_1 : X(n) = A + W(n) \quad n = 1, 2, \ldots, N$$

Under the null hypothesis $H_0$, we have the probability density function $f_X(X(n); H_0)$ for the signal $X(n)$, which is a Gaussian multivariate distribution defined as $(\mathbf{0}, \sigma^2\mathbf{I})$, where $\mathbf{I}$ is an identity matrix. Under the alternative hypothesis $H_1$, we have the density function $f_X(X(n); H_1)$ for the signal $X(n)$, which is a Gaussian multivariate distribution defined as $(\mathbf{A}, \sigma^2\mathbf{I})$ with the vector $\mathbf{A}$ containing all $A$s. If we apply the Neyman–Pearson theorem to maximize the probability of detection $P_{DE}$, we can get the test statistics $\overline{X}$ in the following form:

$$\overline{X} = \frac{1}{N} \sum_{n=1}^{N} X(n).$$

a.  Find the mean and variance of this test statistic $\overline{X}$ for the null hypothesis $H_0$. Find the probability density function of the test statistic $\overline{X}$ for this case. Find the expressions for these probability density functions defined for $N = 1$ and $N = 4$.

b.  Find the mean and variance of this test statistic $\overline{X}$ for the alternative hypothesis $H_1$. Find the probability density function of the test statistic for this case. On the same coordinate system, sketch the graphs for the probability density functions defined by $N = 1$ and $N = 9$, and indicate the value of standard deviation for each graph.

3. Bearing in mind that the *fai* function can be expressed in terms of the *erfc* functions as

$$\Phi(\tau) = \int_{-\infty}^{\tau} \frac{1}{\sqrt{2\pi}} e^{-x^2/2} dx = \frac{1}{2} erfc\left(-\tau/\sqrt{2}\right) = \frac{1}{2}\left(2 - erfc\left(\tau/\sqrt{2}\right)\right) = 1 - \frac{1}{2} erfc\left(\tau/\sqrt{2}\right),$$

derive the expression for the truncated Gaussian probability density function

$$f_{ct}(\tau) = \frac{\frac{1}{\sqrt{2\pi\sigma^2}} e^{-(\tau-\eta)^2/2\sigma^2}}{\Phi\left(\frac{b-\eta}{\sigma}\right) - \Phi\left(\frac{a-\eta}{\sigma}\right)},$$

and calculate its mean and variance in terms of the *erfc* functions.

4. For a continuous non-truncated Gaussian probability density function that has zero mean, that is, $\eta = 0$, assume that the corresponding truncated probability density function is asymmetric with respect to zero, that is, is defined in the interval $2T = (b - a)$ for $|a| \neq |b|$. Calculate expressions for the density, mean, and variance of the truncated density.

5. For a continuous symmetric truncated Gaussian density function:

   a. Derive expressions for the truncated probability density function, the mean, and the variance.

   b. Find the graph for the variances of the continuous density and the truncated density as a function of the variance of the non-truncated density, and then find the dependence of the truncated variance on the truncation interval $T = (0, 4)$.

   c. Plot graphs of both the unit-variance Gaussian probability density function $N(0, 1)$ inside the interval $2C = (-5, 5)$ and the truncated probability density function $N(0, 0.8796)$ inside the interval $2T = (-2, +2)$. Find the continuous probability density function and the related truncated probability density = function for the truncation interval $T$ as a parameter.

6. Find the variance of the noise $Y_j$ in the case when the filtered noise of the bandwidth $B$ is sampled at the Nyquist rate and interpolation is not performed.

7. A continuous-time white Gaussian noise is sampled two times in the interval $T$ with the Nyquist rate, and then each sample is repeated $N/2$ times. Find the variance of the noise in the interval of $N$ samples.

8. Find the variance of $Y_j$ in the case when the filtered noise of the bandwidth $B$ is sampled above the Nyquist rate and interpolation is not performed. Discuss this case from the correlation point of view.

9. Find the mean and the variance of the noise at the output of a noise demodulator/regenerator that is designed as a correlator, assuming that the noise samples are generated at the Nyquist rate at the input of the modulator. In this case, the output of the correlator can be expressed in the form

$$Y(kN) = \sqrt{E_N} \frac{1}{N} \sum_{n=(k-1)N}^{kN-1} N_G(n).$$

10. Suppose the amplitudes of the discrete Gaussian noise process $X(l)$ are distributed according to a discrete Gaussian probability density function that can have variable values in an infinite interval. The probability density function is expressed in terms of Dirac delta functions as

$$f_X(x) = \frac{1}{2} \sum_{n=-\infty}^{\infty} \left( erfc \frac{(2n-1)}{\sqrt{8\sigma^2}} - erfc \frac{(2n+1)}{\sqrt{8\sigma^2}} \right) \delta(x-n) = \frac{1}{2} \sum_{n=-\infty}^{\infty} Erfc(n)\delta(x-n).$$

Find the entropy of this discrete noise source.

11. Suppose the amplitudes of a quantized noise $X(l)$ are distributed according to a truncated discrete Gaussian probability density function that can have variable values in the truncation interval defined from $n = -S$ to $n = S$. The probability density function is expressed in terms of Dirac delta functions as

$$f_X(x) = f_d \left( x \| x | \leq S \right) = \frac{f_X(x)}{P(\tau \geq -S) - P(\tau > S)} = \frac{\frac{1}{2} \sum_{n=-S}^{n=S} \left( erfc \frac{(2n-1)}{\sqrt{8\sigma^2}} - erfc \frac{(2n+1)}{\sqrt{8\sigma^2}} \right) \delta(x-n)}{\frac{1}{2} \sum_{n=-S}^{n=S} \left( erfc \frac{(2n-1)}{\sqrt{8\sigma^2}} - erfc \frac{(2n+1)}{\sqrt{8\sigma^2}} \right)}$$

$$= P(S) \sum_{n=-S}^{n=S} Erfc(n) \cdot \delta(x-n).$$

Find the entropy of this noise process.

# 5

# Operation of a Discrete Communication System

## 5.1 Structure of a Discrete System

A system composed of a signal synthesizer and a signal analyser, as presented in Chapter 2, can be used to implement a multilevel *bandpass* discrete or digital communication system. In this system, the signal synthesizer is the modulator, and the signal analyser is the demodulator, as shown in Figure 5.1.1. The coefficients of the signal correspond to the modulating message signals that modulate the carriers represented by the basis functions. The signal analyser generates the coefficients of the signal at its output, which correspond to the received baseband message signal. In this sense, the coefficients generated in the time domain can be understood as one representation of modulating signals in the discrete communication system.

In practice, we can design a communication system by using a structure with a modulator as the signal synthesizer and a demodulator as the signal analyser. We test this structure by generating message signals and checking their expected properties at key points inside the communication system and at the output of the demodulator. When we are sure that the system works, we design a channel to investigate the system's properties, taking into account influences on the channel that can be due to noise, fading, or other impairments in the channel.

The structure of the basic discrete communication system, including the source of the message symbols, modulator, the transmission channel, and the demodulator, is represented by the scheme shown in Figure 5.1.1. The block scheme of the system is expressed in terms of mathematical operators. The inputs and outputs of the blocks are expressed in terms of signals which are denoted by small letters. These signals are, in principle, realizations of stochastic processes. The related vector representation of the signals is marked in bold letters beneath the block schematics. In this chapter, generic structures of discrete and digital communication systems will be developed and will be used to deduce various communication systems as special cases.

## 5.2 Operation of a Discrete Message Source

A message source is defined by its alphabet $A$, which is a set of $M$ elements, $\{a_m\}$, for $m = 1, 2, \ldots, M$, that is,

*Discrete Communication Systems.* Stevan Berber, Oxford University Press. © Stevan Berber 2021.
DOI: 10.1093/oso/9780198860792.003.0005

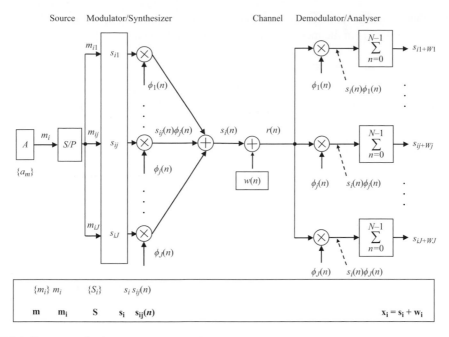

**Fig. 5.1.1** *Structure of a discrete communication system.*

$$A = \{a_m\} = (a_1, a_2, \ldots, a_m, \ldots, a_M).$$ (5.2.1)

The source takes the elements of $A$ and forms the messages at its output in the form of message symbols. The messages are obtained as the variations with repetition of the elements inside the alphabet $A$. Therefore, if a message symbol $m_i$ is composed of $J$ elements taken from the alphabet $A$, then we can generate $I = M^J$ different messages at the output of the message source.

The elements of the alphabet are generated randomly according to the associated probabilities

$$P(A) = (P(a_1), P(a_2), \ldots, P(a_m), \ldots, P(a_M)).$$ (5.2.2)

This is a discrete source of information that is theoretically described in Chapter 9, which is dedicated to information theory. The source generates a message symbol by randomly taking $J$ elements of the alphabet. The $i$th message symbol is expressed as the set $m_i = (m_{i1}, m_{i2}, \ldots, m_{ij}, \ldots, m_{iJ})$, where $m_{ij}$ are elements of the alphabet $A$. The symbol can be understood as a vector in $J$-dimensional space, called the message symbol vector $\mathbf{m_i}$, and expressed in matrix form as the one-column matrix

$$\mathbf{m_i} = [m_{i1} \quad m_{i2} \quad \cdots \quad m_{iJ}]^T.$$ (5.2.3)

All possible message symbols form the set $\{m_i\} = (m_1, m_2, \ldots, m_i, \ldots, m_I)$. They are also presented as columns in the message symbols matrix $\mathbf{m}$, that is,

$$\mathbf{m} = [\mathbf{m_1} \ \ \mathbf{m_2} \ \ \cdots \ \ \mathbf{m_i} \ \ \cdots \ \ \mathbf{m_{I-1}} \ \ \mathbf{m_I}]$$

$$= \begin{bmatrix} m_{11} & m_{21} & \cdots & m_{i1} & \cdots & m_{(I-1)1} & m_{I1} \\ m_{12} & m_{22} & \cdots & m_{i2} & \cdots & m_{(I-1)2} & m_{I2} \\ \vdots & \vdots & \ddots & \vdots & \ddots & \vdots & \vdots \\ m_{1J} & m_{2J} & \cdots & m_{iJ} & \cdots & m_{(I-1)J} & m_{IJ} \end{bmatrix}, \tag{5.2.4}$$

where the columns define all possible symbols that can be generated from the source. These message symbols are generated randomly defined by the matrix of their a priori probabilities expressed as

$$P(\mathbf{m}) = [P(\mathbf{m_1}) \ \ P(\mathbf{m_2}) \ \ \cdots \ \ P(\mathbf{m_i}) \ \ \cdots \ \ P(\mathbf{m_{I-1}}) \ \ P(\mathbf{m_I})]. \tag{5.2.5}$$

We will assume that these probabilities are equal to each other, meaning that the message symbols are generated by the source with the same probability expressed as

$$P(\mathbf{m_i}) = P(m_i) = \frac{1}{I}, \tag{5.2.6}$$

for $i = 1, 2, \ldots, I$. The source that generates the message symbols, defined in this way, can be analysed as a discrete memoryless source, as will be explained in Chapter 9. The probabilities of the message symbols can be related to the probabilities of the alphabet's elements.

Suppose the source generates a message symbol $m_i$ with the probability $P(m_i)$. The set of $J$ elements of $m_i$, taken from the alphabet $A$, are mapped into the set of $J$ signal coefficients forming the modulating signal $s_i$, as presented in Figure 5.1.1 in the vector form $\mathbf{s_i}$. The signal coefficients in $\mathbf{s_i}$ are also expressed in matrix form as the modulating signal vector

$$\mathbf{s_i} = [s_{i1} \ \ s_{i2} \ \ \cdots \ \ s_{iJ}]^T. \tag{5.2.7}$$

The signal coefficients define $J$ *modulating discrete-time signals*, usually their amplitudes, presented by the set $\{s_{ij}(n)\}$ in Figure 5.1.1 at the input of the transmitter modulator. The modulator is designed as a bank of $J$ individual modulators with their own carriers implemented as orthonormal functions. Each modulating signal is of duration $N$ in the discrete-time domain, which corresponds to the duration of a symbol $T = T_{Sy}$ in the continuous-time domain. The number of possible modulating signals is $I$. The modulating signals can be expressed in matrix form as

$$\mathbf{s} = [\mathbf{s_1} \ \ \mathbf{s_2} \ \ \cdots \ \ \mathbf{s_i} \ \ \cdots \ \ \mathbf{s_{I-1}} \ \ \mathbf{s_I}]$$

$$= \begin{bmatrix} s_{11} & s_{21} & \cdots & s_{i1} & \cdots & s_{(I-1)1} & s_{I1} \\ s_{12} & s_{22} & \cdots & s_{i2} & \cdots & s_{(I-1)2} & s_{I2} \\ \vdots & \vdots & \ddots & \vdots & \ddots & \vdots & \vdots \\ s_{1J} & s_{2J} & \cdots & s_{iJ} & \cdots & s_{(I-1)J} & s_{IJ} \end{bmatrix}. \tag{5.2.8}$$

Therefore, any set of modulating signals, defined by a column in the matrix **s,** can appear at the input of the modulator at the defined time instant $i$. The sets of message symbols are generated randomly at the output of the source. We will analyse one set of message symbols $m_i$ mapped into the set of modulating signals $s_i$ that modulate the carriers in the bank of $J$ individual modulators. The order of the message symbols generated at the source and transmitted through the system will be specified by the index $k$, $k = 1, 2, \ldots, s, \ldots, K$, where $K$ specifies the total number of message symbols transmitted.

**Example**

Suppose a discrete memoryless source is defined by an alphabet $A$ with $M = 2$ elements which are binary values +1 and -1, that is,

$$A = \{a_m\} = (a_1, a_2) = (-1, 1). \tag{5.2.9}$$

A message symbol, composed of $J = 3$ randomly generated elements of the alphabet $A$, is formed at the output of the source and mapped into the set of modulating signals. Assume that the modulating signals are rectangular pulses of duration $N = 8$.

Calculate the total number of message symbols that could be generated from the source. Form the matrix of the message symbols **m**. If the mapping is defined as the multiplication of the message symbols by a constant $\sqrt{E}$, find all $I$ possible modulating signals and present them in a matrix form. Present the discrete-time wave shape of the first modulating signal at the input of the first modulator for the last message symbol that can be generated from the source, denoted as $m_{81}$.

**Solution.** The number of message symbols is $I = M^J = 2^3 = 8$, which are defined by the matrix of message symbols

$$\mathbf{m} = [\mathbf{m}_1 \ \mathbf{m}_2 \ \cdots \ \mathbf{m}_i \ \cdots \ \mathbf{m}_8] = \begin{bmatrix} -1 & -1 & -1 & -1 & 1 & 1 & 1 & 1 \\ -1 & -1 & 1 & 1 & -1 & -1 & 1 & 1 \\ -1 & 1 & -1 & 1 & -1 & 1 & -1 & 1 \end{bmatrix}. \tag{5.2.10}$$

The message symbols need to be mapped into the modulating signals. In this case, all possible modulating signals are obtained by multiplying the matrix of message symbols by the specified constant $\sqrt{E}$, where $E$ can be considered to be the energy of any symbol. The signals are represented by eight columns in the following matrix of modulating signals:

$$\mathbf{s} = \sqrt{E}\mathbf{m} = \begin{bmatrix} -\sqrt{E} & -\sqrt{E} & -\sqrt{E} & -\sqrt{E} & \sqrt{E} & \sqrt{E} & \sqrt{E} & \sqrt{E} \\ -\sqrt{E} & -\sqrt{E} & \sqrt{E} & \sqrt{E} & -\sqrt{E} & -\sqrt{E} & \sqrt{E} & \sqrt{E} \\ -\sqrt{E} & \sqrt{E} & -\sqrt{E} & \sqrt{E} & -\sqrt{E} & \sqrt{E} & -\sqrt{E} & \sqrt{E} \end{bmatrix}. \tag{5.2.11}$$

The defined modulating signals are rectangular pulses with their amplitudes specified in the matrix **s**. The wave shape of the first discrete-time modulating signal, for the eighth message symbol, denoted as $s_{81}(n)$, is a rectangular pulse with duration $N$ and an amplitude equal to the $i$th coefficient, as presented in Figure 5.2.1.

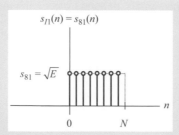

**Fig. 5.2.1** *A modulating rectangular pulse.*

This modulating signal is applied to the input of the first multiplier to modulate the first carrier $\phi_1(n)$, as presented in Figure 5.1.1 for defined $j = 1$. The number of samples of the modulating signal must be sufficiently large to accommodate the number of the related carrier oscillations. If we need at least two samples of the carrier per its oscillation and the number of oscillations inside the pulse is $N_c$, the number of samples inside the discrete pulse will be $N = 2N_c$. The sampling rate and interpolation procedure need to be controlled to preserve the signal properties in the time and frequency domains. In doing this, we need to follow principles of signal discretization that will be presented in Chapter 13, and principles of multi-rate signal processing that will be presented in Chapter 17. These principles were applied when we generated the modulated signal in Figure 5.3.1 for the modulating signal in Figure 5.2.1.

## 5.3   Operation of a Discrete Modulator

The *modulating signals* modulate $J$ carriers that form a set of discrete-time orthonormal basis functions as defined in Chapter 2. The modulator is composed of $J$ multipliers. All modulating signals, representing the message symbol, are samples with duration $N$. The amplitudes of the modulating signals are defined by the signal coefficients. The modulated signals obtained at the output of $J$ modulators are added in the adder at the output of the transmitter, as shown in Figure 5.1.1. Therefore, the transmitted modulated signal $s_i(n)$, expressed as the sum of modulated orthogonal carriers, is

$$s_i(n) = \sum_{j=1}^{J} s_{ij}\phi_j(n). \tag{5.3.1}$$

**Example**

As in the previous example, suppose a discrete memoryless source is defined by an alphabet $A$ with two elements that have the binary values +1 and -1. A message symbol, composed of $J = 3$ randomly generated elements from the alphabet $A$, is formed at the output of the source and mapped into the set of modulating signals. Assume that the modulating signals are rectangular pulses of duration $N = 8$. The mapping is defined as the multiplication of the message symbols by the constant $\sqrt{E}$. Present the discrete-time wave shape of the third modulating signal at the input of the third modulator for the last message symbol that can be generated from the source, $m_{83}$. Because the mapping is defined as the multiplication of the message symbols by the constant $\sqrt{E}$ and the signal coefficient, the message signal is defined as $m_{83} = \sqrt{E}$.

Present the discrete-time wave shape of the modulated signal at the output of the last modulator $J = 3$. Assume that the carrier is an orthonormal cosine function with two oscillations inside the discrete-time interval of $N = 8$ samples expressed as

$$\phi_3(n) = \sqrt{2/N} \cos (\Omega_3 n). \tag{5.3.2}$$

Plot the wave shapes of the carrier and the modulated signal for both the amplitude values and the number of samples.

**Solution.** The signal at the output of the adder is obtained by multiplying the modulating signal of the amplitudes $\sqrt{E}$ and the orthonormal carriers for $i = I = 8$.

$$s_I(n) = \sum_{j=1}^{J} s_{Ij}(n)\phi_j(n) = \sum_{j=1}^{3} s_{8j}(n)\phi_j(n) \tag{5.3.3}$$

Because the amplitudes of the modulating signals are identical and have constant values equal to $\sqrt{E}$, the modulated signal can be directly represented as a product of the coefficients and the carrier, that is,

$$s_I(n) = \sum_{j=1}^{J} s_{Ij}\phi_j(n) = \sum_{j=1}^{J} \sqrt{E}\phi_j(n),$$

which is identical to eqn (5.3.3). This is the case when the modulating signals are represented as ideal rectangular pulses. In practice, these modulating signals are filtered before modulation, using pulse shaping, which requires the modulated signal to be calculated according to relation (5.3.3). The pulse-shaping procedure is used to control the bandwidth of the modulated signal and, as such, is a separate topic in the theory of telecommunications. The third carrier and the modulated signal are presented in Figure 5.3.1. The period of the carrier is $N_c = 4$ and the duration of the pulse is $N = 8$. The third modulated signal is

$$s_{83}\phi_3(n) = \sqrt{E}\sqrt{2/N} \cos (\Omega_3 n) = A \cos (\Omega_3 n), \tag{5.3.4}$$

for $0 \le n < N$, and the amplitude $A = \sqrt{E}\sqrt{2/N}$.

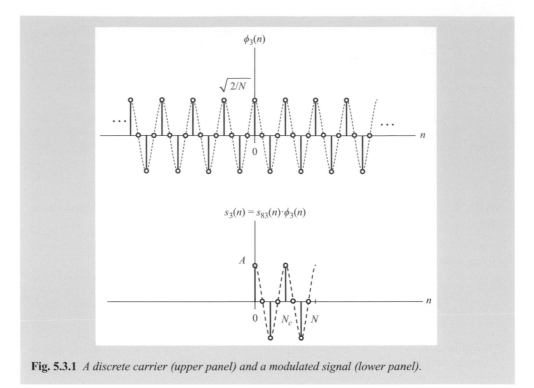

**Fig. 5.3.1** *A discrete carrier (upper panel) and a modulated signal (lower panel).*

In this chapter, we will also analyse a discrete communication system operation assuming that the noise is present in the channel. For that reason, the next section presents an elementary version of the theory for a noise process in a communication channel. The details of this theory can be found in Chapter 4.

## 5.4   Additive White Gaussian Noise Channels in a Discrete-Time Domain

The transmitter signal $s_i(n)$, expressed as the sum of $J$ modulated orthogonal carriers, is transferred through an additive white Gaussian noise channel. A detailed discussion related to the term 'white' for discrete-time noise can be found in Chapter 4. The discrete signal $r(n)$ that is received is the sum of the transmitted modulated signal $s_i(n)$ and the noise $w(n)$.

**White Gaussian noise processes.** The discrete-time noise process $W(n)$ can be obtained by sampling a continuous-time white Gaussian noise $W(t)$ at the sampling time instants $T_s$ to get noise samples as a function of the discrete-time variable $n$, as depicted in Chapter 4 in Figure 4.1.1. The function for the discrete-time sample of the white Gaussian noise, denoted by $w(n)$, can be represented in the discrete-time domain as one realization of the discrete-time Gaussian stochastic process $W(n)$. Because the samples of the noise are realizations of Gaussian random variables, the random variable at all discrete-time instants $n$ are also Gaussian variables. Furthermore, because the continuous-time noise process $W(t)$ is Gaussian, wide-sense stationary, and, consequently, stationary in the strict sense, the discrete-time noise process $W(n)$ is Gaussian and strict-sense stationary. The autocorrelation function of this process depends on the time lag $l = m - n$, and can be calculated as

$$
\begin{aligned}
R_{WW}(m,n) = E\{W(m)W(n)\} &= \int_{-\infty}^{\infty} \int_{-\infty}^{\infty} w_m w_n f_{W_m W_n}(w_m, w_n)\, dw_m dw_n \\
&= \int_{-\infty}^{\infty} \int_{-\infty}^{\infty} w_m w_n f_{W_m}(w_m) f_{W_n}(w_n)\, dw_m dw_n \\
&= \left\{ \begin{array}{ll} \int_{-\infty}^{\infty} w_n^2 f_{W_n}(w_n)\, dw_n \int_{-\infty}^{\infty} f_{W_m}(w_m)\, dw_m & m = n \\ \int_{-\infty}^{\infty} w_n f_{W_n}(w_n)\, dw_n \int_{-\infty}^{\infty} w_m f_{W_m}(w_m)\, dw_m & m \neq n \end{array} \right\}, \\
&= \left\{ \begin{array}{ll} \sigma_W^2 & m = n \\ 0 & m \neq n \end{array} \right\} = \sigma_W^2 \delta(m-n) = \frac{N_0}{2} \delta(l) = R_{WW}(l)
\end{aligned}
$$

$$(5.4.1)$$

where the joint probability density function $f_{W_m W_n}(w_m, w_n)$ is equal to the product of the marginal densities, due to the independence of the random variables $W_m$ and $W_n$, and the variance is equal to the two-sided power spectral density $N_0/2$ of the noise. This autocorrelation function is expressed in terms of Kronecker delta functions, $\delta(l)$, implying that the sampling was possible by fulfilling the conditions of the sampling theorem, that is, primarily, the condition of noise physical realizability. According to the Wiener–Khintchine theorem, the power spectral density can be calculated as the discrete-time Fourier transform of the autocorrelation function, that is,

$$
S_{WW}(\Omega) = \sum_{l=-\infty}^{\infty} R_{WW}(l) e^{-j\Omega l} = \sigma_W^2 \sum_{l=-\infty}^{\infty} \delta(l) e^{-j\Omega l} = \sigma_W^2 = \frac{N_0}{2}, \qquad (5.4.2)
$$

for $-\pi \leq \Omega \leq \pi$. If the possible existence of noise amplitudes of infinite values had been assumed, the power of this process would have been infinite and different from the calculated variance in eqn (5.4.1). Therefore, for that case, the process would not be physically realizable and, as such, it would not fulfil the conditions of the sampling theorem, which excludes the possible existence of amplitudes with infinite values. Furthermore, if an infinite signal power of noise is anticipated, then the definition and calculated value of the autocorrelation function in eqn (5.4.1) would require a Dirac delta function rather than a Kronecker delta function to fulfil that anticipation, which will define the infinite power at zero lag of the autocorrelation function, that is, the autocorrelation at zero lag will be $R_{WW}(0) = \infty$. However, in order to sample the noise, we need to make the signal realizable, that is, we must avoid the possibility of having amplitudes of infinite values. Therefore, the calculations presented for both the

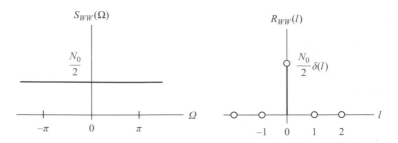

**Fig. 5.4.1** *The power spectral density function (left) and the autocorrelation function (right) for a discrete white Gaussian noise process.*

autocorrelation and the power spectral density functions assume, in fact, that the Gaussian noise density function is truncated in order to eliminate the possibility of having infinite values for the noise amplitudes. The graphs of these functions are shown in Figure 5.4.1.

The power of the noise is equal to the variance of the noise process, or to the numerical value of the power spectral density of the noise, which is expressed as

$$P_W = \frac{1}{2\pi} \int_{\pi}^{-\pi} S_{WW}(\Omega)\,d\Omega = \frac{1}{2\pi} \int_{-\pi}^{\pi} \sigma_W^2\,d\Omega = \sigma_W^2 = \frac{N_0}{2}. \tag{5.4.3}$$

The requirement of infinite power would further complicate the theoretical understanding of the energy and make it impossible to explain theoretically the sampling of noise and the generation of a discrete-time random noise waveform. For that reason, the following analysis will include white Gaussian noise processes that fulfil the requirements of the sampling theorem and can be used to get discrete-time noise waveforms.

## 5.5    Correlation Demodulators

### 5.5.1    Operation of a Correlator

The received signal is processed in a bank of $J$ correlators, and the estimates of signal coefficients are obtained at the output of the demodulator, as shown in Figure 5.1.1. In the case when no noise is present in the channel, these coefficients would be exactly the same as the coefficients at the transmitter side, and the transmitted message symbol can be ideally recovered from them. In the presence of noise, the estimates of coefficients will have a noise part added to the coefficient values that are specified at the transmitter side, which can cause some errors in signal transmission.

As we said in Chapter 2, a signal analyser contains a bank of $J$ correlators connected in a parallel manner. Each correlator calculates a coordinate of the received signal in the direction of the corresponding basis function. Similarly, in the case of signal transmission in a noisy channel, the receiver will contain a bank of parallel correlators. These correlators will calculate the coordinates of transmitted signal, and also the coordinates of channel noise which are added to the coordinates of each signal inside the channel.

Let $R(n)$ denote the discrete-time stochastic process at the output of the channel and at the input of the receiver. The received discrete-time random noisy signal $r(n)$ is a sample function or a realization of the discrete-time stochastic process $R(n)$. The stochastic process $R(n)$ is Gaussian, because it can be considered to be the sum of the deterministic signal $s_i(n)$ and the Gaussian stochastic process $W(n)$. The received random signal $r(n)$ is processed inside the bank of $J$ receiver correlators.

We did not use any indices for the stochastic processes $R(n)$ and $W(n)$ and the random signal $r(n)$ and noise $w(n)$, due to the following assumptions: firstly, we are talking about signal and noise received in an interval that can last for a number of message symbols (modulating signals) and, secondly, the noise $W(n)$ is defined as a discrete white Gaussian noise process. In addition, we will make the following important assumptions, which are valid for the operation of real communication systems:

1. The statistical characteristics of the noise in all symbol intervals are the same, that is, the random noise samples affecting succeeding symbols are statistically uncorrelated and, being Gaussian, independent.

2. The signal processing in all $J$ correlators is identical and the outputs of correlators are statistically independent, due to the orthogonality of the carriers and the statistical characteristics of noise that may affect the signals in different frequency ranges defined by the orthogonal carriers.

3. Due to the two previous assumptions, it is sufficient to analyse the transmission of the $j$th modulated signal and its processing in the $j$th correlator. Then, the obtained results can be generalized to all correlators. For this reason, the operation of a single correlator, the $j$th correlator that is presented in Figure 5.5.1, will be analysed.

Let the received discrete-time noisy signal $r(n)$, representing one realization of $R(n)$, be applied to the input of the $j$th correlator. The number of samples of the received discrete-time signal $r(n)$ is $N$, which defines a symbol interval. The output of the correlator is

$$x_j = \sum_{n=0}^{N-1} r(n)\phi_j(n) = \sum_{n=0}^{N-1} [s_i(n) + w(n)]\phi_j(n) = s_{ij} + w_j, \qquad (5.5.1)$$

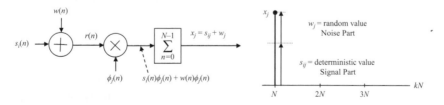

**Fig. 5.5.1** *An additive white Gaussian noise channel (left) with the first output of the correlation demodulator (right).*

where the particular first addend $s_{ij}$ is a deterministic quantity contributed by the transmitted signal $s_i(n)$, and the second addend, $w_j$, is a sample value of the random variable $W_j$ that arises in the symbol interval due to the channel noise $w(n)$. We said 'the *particular* first addend $s_{ij}$ is a deterministic quantity' because the coefficient $s_{ij}$, in fact, is a realization of a binary random variable which will be taken into account later. The value $s_{ij}$ is the signal coefficient that will be called the signal part, and the value $w_j$, as the realization of the noise random variable $W_j$, will be called the noise part. These parts are graphically presented in Figure 5.5.1 for the first received symbol at the time instant $N$.

The random variable $X_j$ is called the demodulation variable, and its sample value (one observation) is called the sample of demodulation $x_j$. This random variable belongs to the discrete-time stochastic process $X_j(kN)$ at the output of the $j$th correlator. One realization of this process will be a series of random values $x_j$ at equidistant instants of time $kN$. Therefore, if a series of message symbols $m_i$ is transmitted, the random variable $X_j$ will have its realizations (random values) $x_j$ at equidistant discrete-time instants $kN$, defined by the natural numbers $k$ and the duration of the discrete modulating signal $N$, as shown in Figure 5.5.1 for a single output of the correlator, that is, for $k = 1$.

The random variable $X_j$ is obtained by processing the Gaussian variables $W(n)$ and the deterministic signal $s_i(n)$ inside the correlator, and can be expressed as

$$X_j = s_{ij} + \sum_{n=0}^{N-1} W(n)\phi_j(n) = s_{ij} + W_j. \tag{5.5.2}$$

The random variable $W_j$ is represented by a sum of $N$ samples of white Gaussian noise multiplied by $N$ samples of the $j$th orthogonal carrier. Effectively, $W_j$ can be considered to be the sum of $N$ Gaussian random variables and, as such, it is a Gaussian variable. One realization of this random variable is expressed by the second addend in eqn (5.5.1). Random values $x_j = x_j(kN)$, for all $k$ values, form one realization of the discrete-time stochastic process $X_j(kN)$ at the $j$th output of the correlation demodulator.

For the sake of illustration, one realization of $X_j(kN)$ and the related Gaussian density function of the random variable $X_j$ are shown in Figure 5.5.2, assuming the existence of two values of the signal coefficients (signal parts). The example density function is presented only for the possible positive values of the signal part $s_{ij}$ inside the correlator output representing the coefficients of the demodulated symbol.

At this point, let us go back to the definition of the noise probability density function as a truncated Gaussian function. In the design and simulation or emulation of discrete-time communication systems, the sampling of the signal and the noise needs to comply with the sampling theorem. Therefore, the signal and the noise need to fulfil two conditions: they must be physically realizable and they must be sampled at least at the Nyquist rate. The modulated signal fulfils these conditions because it is defined as a band-limited energy signal. We will put aside the problem of the band-limited spectrum and assume that the noise probability density function is truncated to make the power of the noise finite. Therefore, in the theoretical modelling used in this chapter, the noise density function will be presented in its continuous truncated form. This presentation is also acceptable in the case when

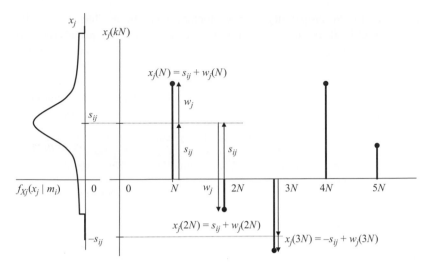

**Fig. 5.5.2** *Output of a correlation demodulator (right) and the related probability density function (left).*

the number of quantization levels is very high and finite or, in other words, when the quantization error is very small. That is the case when we are implementing transceiver blocks in digital technology, because we are dealing with discrete-time quantized samples expressed as processing bits and bytes.

It is important to note that the theory of discrete-time stochastic processes, which was presented in Chapter 3, will be used in this chapter. For the correlator operation, for example, the processes $X_j(kN)$ and $R(n)$ are equivalent to the process $X(n)$ as defined and analysed in Chapter 3. Therefore, the theory presented in that chapter can be considered to be a prerequisite for understanding the explanations in this chapter.

For the sake of simplicity, we will preserve the following notation: $x_j(kN) = x_j$, $X_j(kN) = X_j$, $W_j(kN) = W_j$, and $w_j(kN) = w_j$. Precise signal processing in the receiver correlator is presented in Figure 5.5.1. Of course, here we are analysing the noise part and the signal part separately, for the sake of explanation. However, in real systems, it is impossible to make such a separation because we cannot distinguish the signal part from the noise part at the receiver side. This impossibility is illustrated in the second sample $x_j(2N)$ in Figure 5.5.2, taken at the time instant $2N$, where the negative value of the noise is greater than the positive value of the signal; therefore, the sample's value is negative, even though the signal's value is positive.

It is necessary to comment here on the precise properties of the density function for the random variable $X_j$. Namely, the stochastic process $X_j(kN)$ is presented here as a discrete-time continuous valued process, which suggests that the related probability density function is continuous, as presented in Figure 5.5.2. However, if the implementation of the demodulator is in discrete digital technology, where quantized values are taken at the output of the correlator, as modern designs require, the stochastic process $X_j(kN)$ will be, in fact, a discrete-time discrete-valued stochastic process. Consequently, the probability density function of a discrete random variable defined at the discrete-time instants $kN$, with random values $X_j(kN)$, will be represented by a discrete probability density function. Even more, due to the finite number of possible quantized values, this function will be truncated and, therefore, represented as a discrete truncated probability density function.

## Example

Suppose we need to construct a discrete communication system that transmits bits, that is, the binary symbols 1 or 0. Suppose a bit 1 is transmitted with the carrier having phase $0°$, and the bit 0 is transmitted with the carrier having phase $180°$ (an inverted carrier). This system is called a binary phase-shift keying system.

    a. Based on the theory of correlation modulator operation and Figure 5.1.1, design a system that performs binary phase-shift keying signal transmission and reception. Include some additional signal processing blocks if necessary.

    b. Briefly explain the system operation, assuming that the message bit 1 is generated from the message source.

## Solution.

    a. The system is composed of baseband signal processing blocks including a binary source, a modulator, a white Gaussian noise communication channel, and a correlation receiver. The structure of the system, obtained by a simplification of the general scheme in Figure 5.1.1, is presented in Figure 5.5.3.

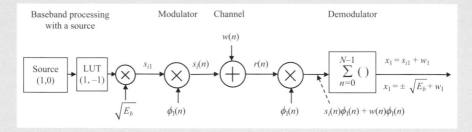

**Fig. 5.5.3** *Structure of a binary phase-shift keying system; LUT, look-up table.*

    b. The baseband processing block is composed of a binary source, a look-up table, and a multiplier. The source generates a binary digit 1 or 0 inside the time interval $T_b$ defined by $N$ discrete signal values. The look-up table converts these symbols into their polar form, assigning +1 to 1, and -1 to 0. The output of the look-up table is multiplied by a constant that is equal to the square root of the bit energy, $\sqrt{E_b}$. For a message bit 1, $\sqrt{+E_b}$ is generated $N$ times, forming a modulating signal which multiplies the carrier, which is a single discrete cosine basis function with phase 0, that is, $\phi_1(n) = \sqrt{2/N}\cos\Omega_c n$. The output of the modulator is the modulated signal $s_i(n)$, which is transmitted through a channel defined by the additive white Gaussian noise stochastic process $W(n)$. The samples of the received signal $s_i(n)$ plus the noise $w(n)$ are multiplied by the locally generated basis function $\phi_1(n) = \sqrt{2/N} \cos \Omega_c n$, and the products are accumulated, producing a sample $x_1$ of the random variable $X_1$. If the sample is positive, the receiver will decide in favour of +1. However, if the sample is negative, the receiver we will decide in favour of -1 and make an error.

## 5.5.2   Statistical Characterization of Correlator Output

It was anticipated that the process $X_j(kN)$ at the output of the $j$th correlator is a discrete-time wide-sense stationary Gaussian stochastic process. Therefore, any random variable $X_j$ is completely statistically characterized by its mean and variance. In this section, we will calculate the mean and variance of the process. Because the random variables defined for all discrete $k$ values inside the process $X_j(kN)$ are independent, it is sufficient to calculate the mean and variance of the random variable $X_j$ defined for $k = 1$. The noise part of $X_j$ is expressed as

$$W_j = \sum_{n=0}^{N-1} W(n)\phi_j(n). \tag{5.5.3}$$

The mean of $X_j$ is

$$\eta_{X_j} = E\{X_j\} = E\{s_{ij} + W_j\} = s_{ij} + E\{W_j\} = s_{ij}, \tag{5.5.4}$$

because the mean of $W_j$ is 0. It is important to note here that we are assuming that a particular bit, +1 or −1, is transmitted and that the signal part is a constant. The variance can be expressed as

$$\sigma_{X_j}^2 = E\{W_j^2\} = E\{W_j(k)W_j(l)\} = E\left\{\sum_{k=0}^{N-1} W(k)\phi_j(k) \sum_{l=0}^{N-1} W(l)\phi_j(l)\right\}$$
$$= E\left\{\sum_{k=0}^{N-1}\sum_{l=0}^{N-1} W(k)W(l)\phi_j(k)\phi_j(l)\right\} .$$

Due to the linearity of the expectation operator, the iid nature of the random variables $W(k)$ and $W(l)$ defined inside the symbol interval $N$, and the orthonormality of the basis functions, the variance of $X_j$ is equal to the variance of the white Gaussian noise in the channel $W(n)$, that is,

$$\sigma_{X_j}^2 = \sum_{k=0}^{N-1}\sum_{l=0}^{N-1} E\{W(k)W(l)\}\,\phi_j(k)\phi_j(l) = \sum_{k=0}^{N-1}\sum_{l=0}^{N-1} R_{WW}(k,l)\,\phi_j(k)\phi_j(l)$$

$$= \sum_{k=0}^{N-1}\sum_{l=0}^{N-1} \sigma_W^2\left[\delta(k-l)\,\phi_j(k)\right]\phi_j(l) = \sigma_W^2 \sum_{l=0}^{N-1}\left[\sum_{k=0}^{N-1}\left[\delta(k-l)\,\phi_j(k)\right]\phi_j(l)\right], \qquad (5.5.5)$$

$$= \sigma_W^2 \sum_{l=0}^{N-1} \phi_j(l)\phi_j(l) = \sigma_W^2 = \frac{N_0}{2}$$

for all $j = 1, 2, \ldots, J$. We can prove that any two random variables $X_j$ and $X_m$, defined at the outputs of two correlators, $j$ and $m$, for the message symbol $m_i$, are mutually uncorrelated. Due to the orthogonality of the basis functions, and the linearity of the expectation operator, the covariance can be calculated as

$$\text{cov}\{X_j X_m\} = E\left\{\left(X_j - \eta_{X_j}\right)\left(X_m - \eta_{X_m}\right)\right\} = E\{W_j W_m\}$$

$$= E\left\{\sum_{k=0}^{N-1} W(k)\phi_j(k) \sum_{l=0}^{N-1} W(l)\phi_m(l)\right\}$$

$$= \sum_{k=0}^{N-1}\sum_{l=0}^{N-1} E\{W(k)W(l)\}\,\phi_j(k)\phi_m(l)\right\} = \sum_{k=0}^{N-1}\sum_{l=0}^{N-1} R_{WW}(k,l)\,\phi_j(k)\phi_m(l) \qquad (5.5.6)$$

$$= \sum_{k=0}^{N-1}\sum_{l=0}^{N-1} \sigma_W^2\left[\delta(k-l)\,\phi_j(k)\right]\phi_m(l) = \sigma_W^2 \sum_{l=0}^{N-1}\left[\sum_{k=0}^{N-1}\delta(k-l)\,\phi_j(k)\right]\phi_m(l)$$

$$= \sigma_W^2 \sum_{k=0}^{N-1} \phi_j(l)\phi_m(l) = 0$$

for every $j \neq m$. Because the covariance is 0, the random variables $X_j$ and $X_m$ are uncorrelated. Because the two variables are Gaussian, they are also statistically independent. Therefore, the conditional probability density function of having $x_j$ at the $j$th correlator output, given that the message symbol $m_i$, which corresponds to the transmitted signal $s_i(n)$, is generated at the source, may be expressed as

$$f_{X_j}\left(x_j|m_i\right) = f_{X_j}\left(x_j|s_i(n)\right) = \frac{1}{\sqrt{\pi N_0}} e^{-(x_j - s_{ij})^2/N_0}, \qquad (5.5.7)$$

for any message symbol sent, $i = 1, 2, \ldots, I$, and at the output of any correlator $j = 1, 2, \ldots, J$. As we have seen previously, any signal $s_i(n)$ can be synthesized at the transmitter side, using $J$ basis functions as carriers. Consequently, we need $J$ correlators to analyse the received signal $r(n)$ that contains the signal transmitted $s_i(n)$ and noise $w(n)$ from the channel.

**Vector representation of the correlator output.** The structure of the correlation demodulator is presented in Figure 5.5.4. In the previous chapter, a *message symbol* and a *modulating signal* at the transmitter side are represented in the vector forms $\mathbf{m_i}$ and $\mathbf{s_i}$, respectively. Similarly,

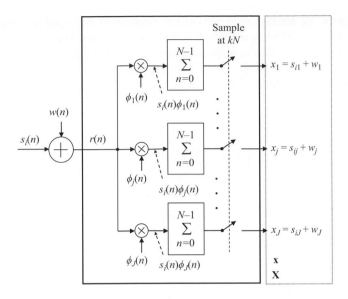

**Fig. 5.5.4** *Operation of a correlation demodulator.*

we may represent the correlation demodulator signals in vector forms. At the output of each correlator, we have a sample that corresponds to a coordinate of the signal received $r(n)$ (signal plus noise) in the direction of the corresponding basis function. For example, the random value $x_1$ is the coordinate of the signal $r(n)$ in the direction of the first basis function $\phi_1(n)$. The output samples of $J$ correlators, containing the signal values plus the noise values, are taken at time instants $kN$. These samples can be expressed in the form of random vectors, each defined by $J$ random values, that is,

$$\mathbf{x} = [x_1 \quad x_2 \quad \cdots \quad x_J]^T, \mathbf{s}_i = \begin{bmatrix} s_{i1} & s_{i2} & \cdots & s_{iJ} \end{bmatrix}^T, \mathbf{w} = \begin{bmatrix} w_1 & w_2 & \cdots & w_J \end{bmatrix}^T, \quad (5.5.8)$$

which are the coordinates of the demodulated received signal $r(n)$. The demodulation sample vector $\mathbf{x}$ is the sum of the signal vector $\mathbf{s_i}$ and the demodulation noise vector $\mathbf{w}$, that is,

$$\mathbf{x} = \mathbf{s_i} + \mathbf{w}. \quad (5.5.9)$$

The random coordinates of $\mathbf{x}$ can be considered the realizations of $J$ random variables. These variables are called the demodulation variables $X_j$ and presented in the form of a demodulation vector $\mathbf{X}$ as

$$\mathbf{X} = \begin{bmatrix} X_1 \\ X_2 \\ \vdots \\ X_J \end{bmatrix} = \begin{bmatrix} s_{i1} + W_1 \\ s_{i2} + W_2 \\ \vdots \\ s_{iJ} + W_J \end{bmatrix} = \mathbf{s_i} + \mathbf{W}, \quad (5.5.10)$$

which is the sum of the deterministic part, which is related to the signal, and the random part, which is related to the random noise, which is expressed as a vector of noise random variables. Random variables $X_j$ are Gaussian with means equal to $s_{ij}$ and variances equal to $N_0/2$. Since the variables in vector $\mathbf{X}$ are *independent*, their *joint conditional probability density function* is equal to the product of the marginal density functions of the individual random variables and is expressed as

$$f_{\mathbf{X}}(\mathbf{x}|m_i) \underset{iid}{=} \prod_{j=1}^{J} f_{X_j}(x_j|m_i) = (\pi N_0)^{-J/2} e^{-\frac{1}{N_0}\sum_{j=1}^{J}(x_j-s_{ij})^2}, \tag{5.5.11}$$

for $i = 1, 2, \ldots, I$, where the *demodulation sample vector* $\mathbf{x}$ is one realization of $\mathbf{X}$, and an element $x_j$ of the vector $\mathbf{x}$ is a sample of the random variable $X_j$. This joint density function of $\mathbf{X}$ specifies the conditional probability density function of having vector $\mathbf{x}$ at the correlator output, given that the message symbol $m_i$ was sent from the source.

The random noise parts $\mathbf{w}$ inside the demodulation vector $\mathbf{x}$ are obtained by multiplying the white noise and the different orthogonal basis functions and then accumulating the products inside a bit interval of length $N$, as shown in Figure 5.5.4. As such, these values are independent and represent one realization of a set of random variables defined by the vector $\mathbf{X}$.

Thus, the additive white Gaussian noise channel is equivalent to a $J$-dimensional vector channel represented by the demodulation noise vector $\mathbf{w}$. The dimension $J$ is the number of basis functions that formulate the *signal vector* $\mathbf{s_i}$.

**Practical simplifications.** In practice, we can use a highly simplified scheme of the discrete communication system presented in Figure 5.5.5. That scheme is of particular interest when we are investigating a baseband signal transmission or behaviour of various channel error-correcting codes.

In this case, we generate the baseband modulating signal $\mathbf{s_i}$ and add the white Gaussian noise $\mathbf{w}$ to get the received demodulation signal $\mathbf{x}$. Using the scheme in Figure 5.5.5, for example, the bit error rate curves for a binary phase-shift keying system, affected by the white Gaussian noise, can easily be obtained. In addition, if fading is present in the channel, the system can be analysed in the presence of the noise and the fading by adding one multiplier in front of the noise adder. For example, for a binary phase-shift keying system, the signal, the noise, and the demodulation vector will have the following matrix representations:

$$\mathbf{s}_i = \begin{bmatrix} s_{i1} \\ s_{i2} \end{bmatrix}, \quad \mathbf{w} = \begin{bmatrix} w_1 \\ w_2 \end{bmatrix}, \quad \mathbf{x} = \begin{bmatrix} x_1 \\ x_2 \end{bmatrix}. \tag{5.5.12}$$

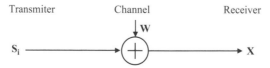

**Fig. 5.5.5** *The most simplified scheme of a discrete communication system.*

An example of the received signal is

$$x_1 = s_{i1} + w_1,$$     (5.5.13)

where the signal part is a constant, and the noise part is generated from the zero mean Gaussian process with the variance defined to fulfil the condition of the desired signal-to-noise ratio. For this example, the signal is sent as a single discrete value, which is usually a bit value in the polar form, say, +1 or -1. The noise generator produces the Gaussian noise sample $w_1$ with the defined variance. The received signal is the sum of the signal and the random noise. Due to the known variance of the noise, the signal-to-noise ratio can be directly calculated. In addition, by changing the variance of the noise, it is simple to change the signal-to-noise ratio in the system.

In the case when fading is present in the channel, the given scheme preserves its simplicity. Namely, the received signal, in the presence of Rayleigh fading in the channel, is

$$x_1 = \alpha \cdot s_{i1} + w_1,$$     (5.5.14)

where $\alpha$ represents fading coefficients. This simple and efficient scheme of the whole system is widely used in the research of communication systems, including, for example, cooperative communications, diversity systems, and direct-sequence CDMA systems. In a CDMA system, a small modification related to sequence spreading needs to be added. In the case when we investigate a bandpass signal transmission or the synchronization methods in communication systems, the finite duration of the signals needs to be taken into account to perform bandpass modulation or simulate the system delays. Therefore, the above-mentioned simplest scheme needs to be modified in these cases.

### 5.5.3     Signal Constellation

If noise is not present in the channel, the signal vector $\mathbf{s_i}$ generated at the transmitter can be represented at the receiver by the same *signal vector* $\mathbf{s_i}$ obtained at the output of a bank of correlators. Thus, the signal $\mathbf{s_i}$ itself can be represented as a point in a $J$-dimensional Euclidean space. Figure 5.5.6 presents a signal $\mathbf{s_i}$ in three-dimensional space. All possible signals $\{\mathbf{s_i}| i = 1, 2, \ldots, I\}$ can be visualized by defining a set of $I$ points in a $J$-dimensional Euclidean space, called the signal space or the constellation diagram of the signals $\mathbf{s_i}$. The corresponding set of modulated (transmitted) signals $\{s_i(n), i = 1, 2, \ldots, I\}$ corresponds to the set of message points $\{m_i, i = 1, 2, \ldots, I\}$.

In real communication systems, the noise $w(n)$ is added to the modulated signal. This noise can be represented by a Gaussian-distributed noise space presented as a sphere of noise points around the message point that has the centre at the modulated signal point (equivalent to the message point), as shown in Figure 5.5.6. One part of the noise $w(n)$, represented here by the *noise vector* $\mathbf{w}$, will affect the signal vector to form the *demodulation sample vector* $\mathbf{x}$. The remaining part of the noise is eliminated by bank of the correlators at the receiver side.

The received signal $r(n)$ is processed and expressed in a vector form by the demodulation sample vector $\mathbf{x}$, which is represented by a point in the Euclidian space. This point is called the

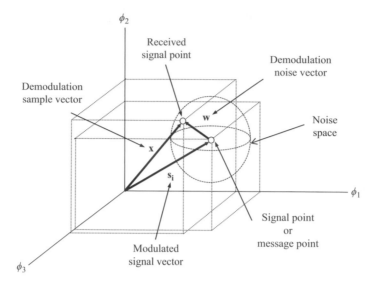

**Fig. 5.5.6** *Vector representation of the signal, the noise, and the received signal.*

received signal point. Because the orientation of the noise vector **w** randomly changes for each symbol transmitted, the received signal point may lie anywhere inside a Gaussian-distributed noise space, as can be seen from Figure 5.5.6.

If the signal vector $\mathbf{s_i}$ takes the place of any other signal vector $\mathbf{s_m}$ in the signal space, due to the position of the noise vector **w**, the receiver will decide in favour of $\mathbf{s_m}$ instead in favour of $\mathbf{s_i}$. In this case, the wrong decision will be made in the receiver.

## 5.6   Optimum Detectors

### 5.6.1   The Maximum Likelihood Estimator of a Transmitted Signal

The aim of the receiver is to detect the message symbols that are transmitted through the channel. At the output of the demodulator, we have a demodulation sample vector **x** containing the samples of the signal affected by the noise in the channel. The relationship between the demodulation sample vector **x** and the message symbols $m_i$ are expressed in eqn (5.5.11) by the joint conditional *probability* density function of **x** given that the message symbol $m_i$ was sent. The aim of the signal detection can be expressed as follows:

*For a given **x**, find the estimate $m_k$ of $m_i$ in such a way as to minimize the probability of symbol error.*

The outputs of $J$ correlators, taken at time instants $N$, $2N$, $3N$, ..., are represented by related demodulation vectors **X** of $J$ random variables $X_j$, where the density function of each $X_j$ is expressed by eqn (5.5.7), that is, their individual random variables are Gaussian with means equal to $s_{ij}$ and variances equal to $N_0/2$. Since these variables are independent (noise multiplied by different basis functions), the *joint conditional probability density function* of $J$ variables is equal to the product of marginal density functions of individual random variables, which is expressed in eqn (5.5.11). The vector **x** is a sample of **X**, and the demodulation element $x_j$ in **x** is a sample of the random variable $X_j$ in **X**. The density function (5.5.11) specifies the functional dependence on the demodulation sample vector **x**, given that the transmitted message symbol is $m_i$.

However, while the demodulation sample vector **x** is obtained by processing the received signal, the message symbol $m_i$ that was generated at the transmitter side has to be found. Precisely speaking, it is necessary that an estimate of the symbol $m_i$, denoted by $m_k$, that is closest in the probabilistic sense to the generated symbol $m_i$ be found. To achieve this, the method of maximum likelihood estimation will be used. For that purpose, the likelihood function $L(m_i)$ is defined as the joint conditional density function presented in eqn (5.5.11), that is,

$$L(m_i) = (\pi N_0)^{-J/2} e^{-\frac{1}{N_0} \sum_{j=1}^{J} (x_j - s_{ij})^2}. \tag{5.6.1}$$

Therefore, the maximum likelihood estimator of symbol $m_i$ is the symbol $m_k$ that maximizes the likelihood function $L(m_i)$, and the problem of message $m_i$ detection is reduced to the problem of maximizing the function $L(m_i)$.

In the following development, we will use the property of a logarithmic function as the monotonically increasing function of its argument. Therefore, if we observe the logarithm of a function, it preserves the position of the maximum value of the original function. The logarithm of $L(m_i)$, defined as a log-likelihood function $l(m_i)$, is given by

$$l(m_i) = \log L(m_i) = -(J/2) \log(\pi N_0) - \frac{1}{N_0} \sum_{j=1}^{J} (x_j - s_{ij})^2, \tag{5.6.2}$$

and can be used to estimate $m_i$. Namely, the likelihood function and the log-likelihood function have a common maximum. We can omit the first term in eqn (5.6.2) because, as it is constant due to the stationarity of the noise, it is not related to the message symbol $m_i$ defined by its coefficients $s_{ij}$ in eqn (5.6.2). Then, the log-likelihood function for an additive white Gaussian noise channel can be approximated by

$$l(m_i) \approx -\sum_{j=1}^{J} (x_j - s_{ij})^2, \tag{5.6.3}$$

for $i = 1, 2, \ldots, I$.

**The maximum likelihood decision rule.** Thus, for the received $x_j$, $j = 1, 2, \ldots, J$, the maximum likelihood decision rule may be expressed as set the estimate $m_k = m_i$ if $L(m_i) \triangleq f_X(x|m_i)$ is **maximum** for $i = k$, or

set the estimate $m_k = m_i$ if $l(m_i) \triangleq \log[f_X(x|m_i)]$ is **maximum** for $i = k$,

because $f_X(x|m_i)$ bears a one-to-one relationship to the log-likelihood function $l(m_i)$. Therefore, the maximum likelihood detector needs to compute the log-likelihood functions for all $I$ possible message symbols $m_i$, compare them, find a message symbol $m_i = m_k$ that gives the maximum value of the log-likelihood function, and then decide in favour of that $m_i$ as the estimate of the message bit sent, and set that $m_i = m_k$. The detector based on this rule is called the maximum likelihood detector.

The maximum likelihood decision rule can be graphically interpreted. Let denote the $J$-dimensional space of all possible demodulation vectors $x$ by N (not italic font), which is called the demodulation signal space. Because the number of message points is $I$, this space is partitioned into $I$ decision regions denoted by $N_1, N_2, \ldots, N_I$. The maximum likelihood rule may be restated as follows:

The demodulated vector $x$ lies in the region $N_k$ if $l(m_i)$ is **maximum** for $i = k$.

If the demodulation vector $x$ falls on the boundary between any two decision regions, the decision has to be made by flipping a fair coin, which has no effect on the ultimate value of the bit error probability in the communication system.

Let $I = 4$ signals be generated with equal probability and then transmitted through an additive white Gaussian noise channel. Suppose the signals have equal energy $E$ and are represented by $J = 2$ orthonormal basis functions. The four message points are on $\phi_1$ and $\phi_2$ axes. In this case, the signal space is partitioned into four decision regions, and the decision boundaries are two lines halving the quadrants and passing though the origin, as shown in Figure 5.6.1.

**The maximum a posteriori probability decision rule.** The maximum likelihood decision rule is a special case of the maximum a posteriori probability decision rule. Namely, to start the decision process, we need to have available the demodulation vector $x$ at the receiver. We are interested in knowing the probability that a particular message symbol $m_i$ is sent when we know $x$. This probability is the a posteriori probability $P(m_i|x)$, the probability that $m_i$ was sent at the transmitter, given that $x$ is received at the receiver.

The a posteriori probability can be expressed in terms of the a priori probabilities $P(m_i)$ of the transmitted symbols $m_i$ and the conditional probability that the demodulation signal vector is $x$ given that the message symbol $m_i$ is transmitted, $f_X(x|m_i)$. According to Bayes' theorem, the a posteriori conditional probability of the symbol $m_i$, given the demodulation vector $x$, $P(m_i|x)$, is

$$P(m_i|x) = \frac{P(m_i)}{f_X(x)} f_X(x|m_i) = \frac{P(m_i) f_X(x|m_i)}{\sum_{i=1}^{M} P(m_i) f_X(x|m_i)}, \tag{5.6.4}$$

where $f_X(x)$ is the unconditional probability density function of the random vector $X$. The maximum a posteriori probability decision rule can now be stated as follows:

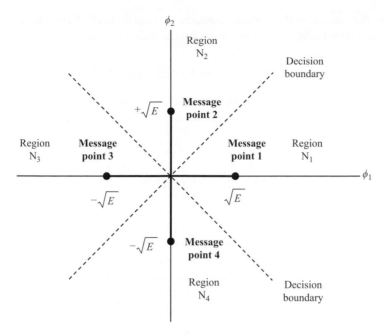

**Fig. 5.6.1** *Decision regions for a two-dimensional signal space and four message points.*

$$\text{Set } m_i = m_k \text{ if } P\left(m_i|\mathbf{x}\right) = \frac{P\left(m_i\right)}{f_{\mathbf{X}}\left(\mathbf{x}\right)} f_{\mathbf{X}}\left(\mathbf{x}|m_i\right) \text{ is } maximum \text{ for } k = i.$$

In the case when all a priori probabilities $P(m_i)$ are identical, given that the function $f_{\mathbf{X}}(\mathbf{x})$ is independent of the transmitted symbol, the maximum a posteriori probability rule can be expressed in terms of the conditional density $f_{\mathbf{X}}(\mathbf{x}|m_i)$. However, when the a priori probabilities $P(m_i)$ are not equal, the optimum maximum a posteriori probability detector makes a decision that is based on the a posteriori probabilities or, equivalently, on the metrics *PM* that includes the a priori probabilities, which is defined as follows:

$$PM = P\left(m_i|\mathbf{x}\right) f_X\left(\mathbf{x}\right) = P\left(m_i\right) f_{\mathbf{X}}\left(\mathbf{x}|m_i\right),$$

for $i = 1, 2, \ldots, I$. However, if the probabilities of the message symbols are the same, as we assumed in the maximum likelihood rule, the maximum a posteriori probability rule becomes equivalent to the maximum likelihood rule,

$$P\left(m_i|\mathbf{x}\right) = \frac{P\left(m_i\right)}{f_{\mathbf{X}}(x)} f_{\mathbf{X}}\left(\mathbf{x}|m_i\right) \approx f_X\left(\mathbf{x}|m_i\right) = \left(\pi N_0\right)^{-J/2} e^{-\frac{1}{N_0}\sum_{j=1}^{J}\left(x_j - s_{ij}\right)^2} = L\left(m_i\right). \tag{5.6.5}$$

Therefore, in the case when the message symbols are equally likely, the maximum likelihood detector is equivalent to the maximum a posteriori probability detector.

## 5.6.2 Application of the Maximum Likelihood Rule

**Numerical interpretation of the maximum likelihood decision rule.** The log-likelihood function (5.6.3) can be rearranged into a form that can be easily implemented in practice, that is,

$$l(m_i) \approx -\sum_{j=1}^{J} x_j^2 + 2\sum_{j=1}^{J} \left( x_j s_{ij} - \frac{s_{ij}^2}{2} \right).$$ (5.6.6)

This function needs to be calculated for every $i$, $i = 1, 2, \ldots, I$, with the aim of finding $m_i$ that maximizes the function. Because the first term does not depend on the index $i$, it can be ignored. Therefore, the decision rule can be expressed in terms of the sum in the second term containing the inner product of $\mathbf{x}$ and $\mathbf{s_i}$ and the coefficient squared that is equal to the energy $E_i$ of the transmitted signal $s_i(n)$ representing the $i$th symbol, that is,

$$l(m_i) \approx \sum_{j=1}^{J} x_j s_{ij} - \frac{1}{2}\sum_{j=1}^{J} s_{ij}^2 = \mathbf{x}^T \mathbf{s}_i - \frac{E_i}{2}.$$ (5.6.7)

For any $i$, the decision variable values, representing the estimates of all symbols $m_i$ and denoted by $Z_i$, can be obtained as follows. For each $i$ value, $i = 1, 2, \ldots, I$, the inner product needs to be calculated. Then half of the energy of each message symbol is to be subtracted. As a result, $I$ numerical random values (numbers) are calculated that are related to the received message symbols, denoted by $Z_i$, and expressed as

$$Z_i = \mathbf{x}^T \mathbf{s}_i - \frac{1}{2}E_i = \sum_{j=1}^{J} x_j s_{ij} - \frac{1}{2}E_i.$$ (5.6.8)

The calculated numbers $Z_i$ are decision variable values that are then compared, and the largest one is selected; say it is $Z_i = Z_k$. This largest number $Z_k$ corresponds to the *most likely transmitted message symbol* $m_k$. Therefore, the decision circuit will decide on the $Z_k$ value and generate at the output of the detector the related symbol $m_k$ taken from the set of all possible symbols $m_i$. Based on these results, primarily on expression (5.6.8), the optimum detector in the maximum likelihood sense can be constructed.

## 5.6.3 Design of an Optimum Detector

Based on the maximum likelihood decision rule, the optimum detector can be designed as a bank of $I$ detector correlators (composed of inner product calculators) and $I$ adders, followed

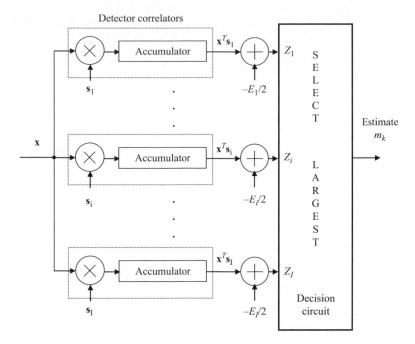

Detector correlators

**Fig. 5.6.2** *An optimum detector.*

by a decision circuit, as shown in Figure 5.6.2. The $I$ correlators calculate the cross-correlation of the demodulation sample vector $\mathbf{x}$ and all possible message signals.

The detector takes the demodulation sample vector $\mathbf{x}$ from the correlation demodulator, generates locally all possible message signal vectors $\mathbf{s}_1, \mathbf{s}_2, \ldots, \mathbf{s}_i, \ldots, \mathbf{s}_I$, and calculates the inner product for each of them. Firstly, $J$ elements of the matrix $\mathbf{x}$ are multiplied by $J$ elements of each of the $I$ possible signal vectors $\mathbf{s}_1, \mathbf{s}_2, \ldots, \mathbf{s}_i, \ldots, \mathbf{s}_I$. The accumulator in each branch of the correlator bank adds all these products. From the sum obtained, half of the signal energies is subtracted, as defined in eqn (5.6.7), and the $Z_i$ numerical random value is obtained for each $i = 1, 2, \ldots, I$. All $Z_i$ values are presented to the decision circuit.

The decision circuit compares all $Z_i$ values, selects the largest $Z_I = Z_k$, and selects $m_k$ as the transmitted message symbol. The estimated message symbol $m_k$ is generated at the output of the decision circuit.

---

### Example

Suppose the transmitter source generates a symbol $m_i$ taken from a binary alphabet defined as $A = (m_1, m_2) = (0, 1)$. Analyse the operation of the optimum detector for the case when one of two possible signals, $\mathbf{s}_1$ or $\mathbf{s}_2$, representing the message symbols, can be sent at the transmitter at a time. The duration of each message signal is $N$ samples.

a. Using the expression for the received signal $r(n) = s_1(n) + w(n)$, find the related output of the correlation demodulator $x_1$.

b. Make a scheme for the optimum detector. Analyse the output of the optimum detector, assuming that the first signal $s_1(n)$ is sent. Find the condition when the detector makes the wrong decision.

**Solution.**

a. Since two symbols can be generated from the source at the transmitter side, the sample of the received signal at the output of the first correlation demodulator is

$$x_1 = \sum_{n=0}^{N-1} r(n)\phi_j(n) = \sum_{n=0}^{N-1} [s_1(n) + w(n)]\phi_1(n)$$

$$= \sum_{n=0}^{N-1} s_1(n)\phi_1(n) + \sum_{n=0}^{N-1} w(n)\phi_1(n) = s_{11} + w_1. \qquad (5.6.9)$$

Therefore, the input of the optimum detector is $x = [x_1] = [s_{11} + w_1]$. Since two possible signals can be generated at the transmitter side, the block schematic of the optimum detector will include two correlators for both the anticipated signal $s_1$ and the second signal $s_2$, as shown in Figure 5.6.3.

b. Suppose that the first signal is sent, that is, $s_i(n) = s_1(n)$. The outputs of the detector are two decision variables that have random numerical values expressed as

$$Z_1 = x^T s_1 - \frac{1}{2}s_{11}^2 = [x_1][s_{11}] - \frac{1}{2}E_1 = [s_{11} + w_1][s_{11}] - \frac{1}{2}E_1$$

$$= \left(\sqrt{E_1}\right)^2 + w_1\sqrt{E_1} - \frac{1}{2}E_1 = \frac{1}{2}E_1 + w_1\sqrt{E_1}, \qquad (5.6.10)$$

and

$$Z_2 = x^T s_2 - \frac{1}{2}s_{21}^2 = [x_1][s_{21}] - \frac{1}{2}E_2 = s_{11}s_{21} + w_1 s_{21} - \frac{1}{2}E_2$$

$$= -\sqrt{E_1 E_2} - w_1\sqrt{E_2} - \frac{1}{2}E_2. \qquad (5.6.11)$$

These two random values need to be compared and the larger one selected. Assuming that the system is symmetric and the energies of both signals are the same, i.e. $E_2 = E_1$, the second decision variable value can be calculated as

$$Z_2 = -E_1 - w_1\sqrt{E_1} - \frac{1}{2}E_1 = -\frac{3}{2}E_1 - w_1\sqrt{E_1}. \qquad (5.6.12)$$

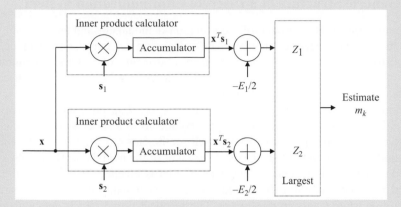

**Fig. 5.6.3** *A block schematic of an optimum detector.*

Now, we can compare $Z_2$ with $Z_1$ at the output of the first correlator to find which one is the largest, by checking this condition $Z_1 \geq Z_2$ as follows:

$$\frac{1}{2}E_1 + w_1\sqrt{E_1} \geq \frac{-3}{2}E_1 - w_1\sqrt{E_1}, \tag{5.6.13}$$

which results in

$$\sqrt{E_1} + w_1 \geq 0. \tag{5.6.14}$$

Thus, the decision rules can be reformulated as follows:

1. If $\sqrt{E_1} + w_1 \geq 0$, then $Z_1 \geq Z_2$, and the maximum likelihood symbol estimate $m_k = m_1$ should be generated at the output of the detector.

2. If $\sqrt{E_1} + w_1 < 0$, then $Z_1 < Z_2$, and the maximum likelihood symbol estimate $m_k = m_2$ should be generated at the output of the detector. In this case, assuming the symbol $s_1(n)$ is sent, the decision circuit will be forced to make a wrong decision in favour of $m_2$ instead of in favour of $m_1$.

In summary, if the first condition is fulfilled, the proper decision is made in favour of $m_1$, and, if the second condition is fulfilled, the wrong decision is made in favour of $m_2$. Finally, for this simple example of a communication system, it is not necessary to implement the optimum detector, because the proper decision can be made using just the demodulation sample $x_1$ at the output of the demodulator and make a decision as follows: if $x_1 \geq 0$, then $m_k = m_1$; or if $x_1 < 0$, then $m_k = m_2$.

### 5.6.4 Generic Structure of a Discrete Communication System

The optimum receiver is composed of a correlation demodulator and an optimum detector. Because the optimum receiver in this chapter is based on correlation demodulation, it is referred to as a correlation receiver. A generic structure of a discrete communication system is shown in Figure 5.6.4; we can use this structure to deduce various discrete-time communication systems as its special cases.

The source generates message symbols $m_i$, $i = 1, 2, \ldots, I$, taken from the source alphabet A (the number of possible symbols is $I = A^J$). By means of a series-to-parallel convertor, a message symbol $m_i$ is mapped into a set of signal coefficients $s_i$, $i = 1, 2, \ldots, I$. The coefficients define $J$ modulating discrete-time signals at the input of the modulator. For the $i$th symbol, a modulated signal $s_i(n)$ is generated using a set of $J$ basis functions that serve as the carriers. The signal is transmitted through the additive white Gaussian noise channel characterized by the additive white Gaussian noise $w(n)$. The received signal $r(n)$ is demodulated, and $J$ samples of the noisy signal are calculated using $J$ correlators forming a demodulation sample vector $\mathbf{x}$.

The $J$ calculated values of the demodulation sample vector $\mathbf{x}$ are applied to the input of the optimum detector that correlates the samples with all possible locally generated transmitted signals $s_i$, $i = 1, 2, \ldots, I$, expressed in vector forms $\mathbf{s_i}$. The result of the correlation is a set of random values defining the decision variables $Z_i$. A decision circuit is used to find the maximum value $Z_k$ in the set of decision variables $Z_i$ and generate the maximum likelihood estimated message symbol $m_k$ at its output.

## 5.7 Multilevel Systems with a Binary Source

A discrete communication system with a binary source that generates binary symbols is presented in Figure 5.7.1. This system can be considered to be a multilevel communication system because the number of possible levels in a message symbol is $I = 2^J$. For example, if $J = 2$, we will have a quaternary system with four possible levels defined as 00, 01, 10, and 11.

Discrete-time signal processing and modulation inside receivers and transmitters are, in principle, performed on an intermediate frequency. Therefore, a discrete communication system is composed of an intermediate-frequency transmitter, two radiofrequency blocks (one at the transmitter output, and one at the receiver input), a waveform additive white Gaussian noise channel, and an optimum intermediate-frequency receiver, as presented in Figure 5.7.1. In the practical design of a discrete communication system, orthonormal harmonic functions are usually used as the basis functions, which are called the sinusoidal carriers. The system transmitter, having a binary source and operating with symbols composed of $J$ bits at the same time, uses $J$ discrete orthonormal cosine carriers. The system receiver generates the same set of $J$ orthonormal carriers to demodulate the signals coherently, and the same set $I$ of possible message signals to detect the message symbol sent. Therefore, the system transmitter is composed of a set of modulators, and the receiver is composed of a set of demodulators and an optimum detector, as shown in Figure 5.7.1.

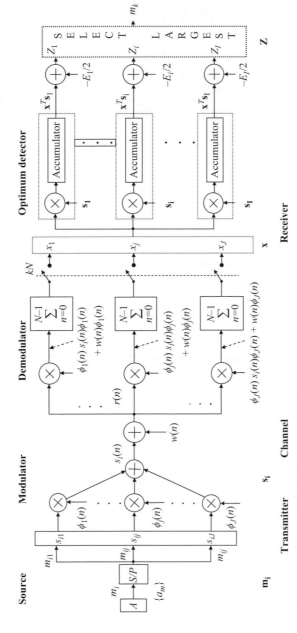

**Fig. 5.6.4** *A generic block schematic for a discrete communication system.*

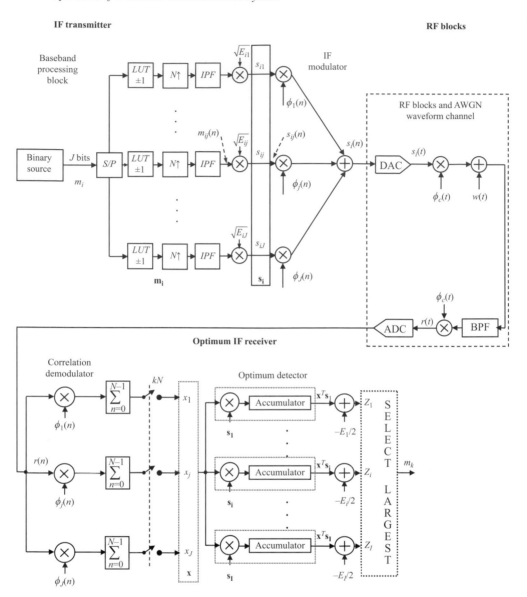

**Fig. 5.7.1** *A multilevel discrete communication system with a binary source; ADC, analogue-to-digital converter; BPF, bandpass filter; DAC, digital-to-analogue converter; IF, intermediate frequency; IPF, interpolation reconstruction filter; LUT, look-up table; RF, radiofrequency.*

The discrete-time signal processing, in general, takes place inside the baseband blocks and the intermediate-frequency blocks of the transmitter and the receiver. The radiofrequency blocks perform continuous-time signal processing. Their role is to up-convert the intermediate-frequency modulated signal to the radiofrequency carrier frequency suitable for transmission through the channel. For that purpose, the discrete-time intermediate-frequency modulated signal is converted into an analogue (continuous-time) signal which modulates the radiofrequency carrier. The conversion is performed by means of a digital-to-analogue converter. This up-conversion does not affect the content of the intermediate-frequency signal in the frequency domain.

Therefore, in the analysis of communication systems and the investigation of their properties, it is sufficient to investigate the intermediate-frequency part of the transmitter and the receiver. However, in that case, the communication channel properties at the intermediate-frequency level must be known. Furthermore, if the properties of intermediate-frequency blocks need to be investigated, a mathematical model of the intermediate-frequency channel that will serve as the basis for the simulation or emulation of a communication channel needs to be developed. Let us briefly explain the operation of the system. The explanation will be only presented for the $i$th symbol and the $j$th branch, because the same processing is undergone for any other branch of the system.

The binary source generates binary values from an alphabet $A = (0,1)$. A symbol $m_i$ of $J$ bits is converted to the parallel form and applied at the input of look-up tables in order to be converted into the polar form at each of $J$ transmitter branches. The polar bit is up-converted by a factor $N$, interpolated inside the interpolation filter and multiplied by a constant $\sqrt{E_{ij}} = \sqrt{E_{bj}}$ that specifies the amplitude of the carrier in the $j$th branch. The modulating signal that is obtained, $s_{ij}(n)$, which is defined by its coefficient $s_{ij}$, modulates the carrier specified by the $j$th basis function $\phi_j(n)$. The modulated signal is then added to the modulated signals from all the other branches to form the transmitted modulated signal $s_i(n)$.

## 5.7.1   Transmitter Operation

The transmitter is composed of two blocks: a baseband signal-processing block, and a discrete-time intermediate-frequency modulator. The source generates message symbols, which are mapped into the set of discrete-time modulating signals. The carriers are discrete-time orthonormal cosine signals. The modulation is performed in the discrete-time domain by multiplying the discrete modulating signals and the carriers.

**Baseband signal processing.** Baseband signal processing will be explained using Figure 5.7.1. At the output of the source, a message symbol $m_i$, composed of $J$ bits, is generated. The series of $J$ bits is converted into a parallel form by means of a serial-to-parallel convertor (a de-multiplexer). All the message bits are applied at the same time to the inputs of $J$ look-up tables. The look-up tables generate the polar binary values $(-1, +1)$ at their outputs to form a message symbol represented as a vector $\mathbf{m_i}$ of $J$ polar bits. These values are up-sampled in the bit interval and then linearly interpolated by repeating the bit sample $N$ times to get $J$ discrete binary signals in the form of rectangular pulses. The discrete binary signals are multiplied by the square root of the bit energy, $\sqrt{E_{ij}}$, and the baseband modulating signals thus obtained are applied to the input of the modulator multipliers to modulate the carriers. This procedure can also be understood as the magnitude scaling of the carrier by the coefficients $s_{ij}$ that are represented by positive or negative values of $\sqrt{E_{ij}}$. At the output of each multiplier, a binary phase-shift keying signal is obtained.

**Discrete modulation at an intermediate frequency.** At the output of the $j$th branch of the modulator, a discrete modulated signal is generated that can be expressed as

$$s_{ij}(n) = s_{ij}\phi_j(n) = s_{ij}\sqrt{2/N}\cos\Omega_j n, \tag{5.7.1}$$

with the average power

$$P_{ij} = \frac{\left(\sqrt{2E_{ij}/N}\right)^2}{2} = \frac{E_{ij}}{N} = \frac{s_{ij}^2}{N}. \tag{5.7.2}$$

Because the interpolated samples in our case are of equal amplitude, the modulation can be represented as the multiplication of the constant coefficients $s_{ij}$ by the carriers, where the coefficients can have one of the two possible values for the assumed binary phase-shift keying modulation, that is, $s_{ij} = \pm\sqrt{E_{ij}}$. Therefore, the modulated discrete-time intermediate-frequency signal at the output of the modulator is

$$s_i(n) = \sum_{j=1}^{J} s_{ij}(n) = \sum_{j=1}^{J} s_{ij}\phi_j(n) = \sum_{j=1}^{J} s_{ij}\sqrt{2/N}\cos\Omega_j n. \tag{5.7.3}$$

## 5.7.2    Radio Frequency Blocks and Additive White Gaussian Noise Waveform Channels

After digital-to-analogue conversion of the intermediate-frequency signal $s_i(n)$, the analogue signal obtained is up-converted to the radiofrequency frequency $\omega_c$ by means of an analogue up-converter, and transmitted to the receiver side through the channel characterized by an additive white Gaussian noise waveform channel defined by the continuous-time wide-sense-stationary stochastic process $W(t)$. The noisy radiofrequency signal received is filtered using a bandpass filter and then down-converted using a down-converter operating at frequency $\omega_c$. The intermediate-frequency analogue signal obtained, $r(t)$, is converted to the discrete-time intermediate-frequency signal $r(n)$ by means of an analogue-to-digital converter. Then the intermediate-frequency signal is processed in the discrete-time domain.

The noisy discrete-time signal received is band limited, due to filtering from a bandpass filter. Therefore, as we said before, the analysis of a discrete system can be done by analysing its intermediate-frequency part only, because the radiofrequency parts perform an up-conversion and a down-conversion of the intermediate-frequency signal, which has no influence on the intermediate-frequency signal processing inside the transmitter and the receiver. That is to say, these conversions do not affect the content of the intermediate-frequency signal, meaning that the intermediate-frequency signals before the up-conversion and after the down-conversion are the same. However, to investigate properties of an intermediate-frequency system in the presence of noise, an appropriate intermediate-frequency bandpass noise generator needs to be designed which is capable of generating, at the intermediate frequencies, a discrete noise that replaces the additive white Gaussian noise waveform process $W(t)$ in the radiofrequency channel.

### 5.7.3    Operation of a Bandpass Noise Generator

In Chapter 4, a noise generator was developed which produced the discrete-time bandpass intermediate-frequency noise $n(n)$ as a realization of the bandpass noise stochastic process $N(n)$, as shown in Figure 4.5.1. In that case, the whole intermediate-frequency communication system can be constructed by replacing the radiofrequency blocks with a bandpass discrete noise generator, as presented in Figure 5.7.2. For the sake of completeness and explanation, the intermediate-frequency transmitter and the optimum intermediate-frequency receiver from Figure 5.7.1 are presented as two compact blocks.

The bandpass Gaussian noise channel is designed as a bank of $J$ bandpass noise generators. At the input of the $j$th generator, a standard baseband Gaussian noise, defined by its zero mean and unit variance, is applied. The generated baseband noise $n_{Gj}(n)$, is multiplied by the required square root of noise energy $E_N$ to achieve the desired noise power. The baseband noise obtained, with the defined power and bandwidth, modulates the carrier to produce a passband noise $\sqrt{E_N}\phi_j(n)n_{Gj}(n)$ which has the specified power and occupies the bandwidth of the corresponding transmitter signal. The outputs of all $J$ generators are added to each other, and the wideband bandpass noise $N(n)$ is generated.

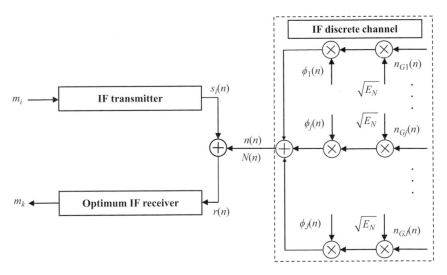

**Fig. 5.7.2** *A discrete intermediate-frequency communication system with a binary source and a bandpass noise generator; IF, intermediate frequency.*

## 5.7.4  Intermediate-Frequency Optimum Receivers

**Operation of a transmitter.** For the assumed binary source, the modulated discrete-time intermediate-frequency signal (5.7.3) at the output of the intermediate-frequency modulator can be expressed in the form

$$s_i(n) = \sum_{j=1}^{J} \sqrt{E_{ij}} \sqrt{2/N} \cos \Omega_j n, \qquad (5.7.4)$$

because the coefficients can have two values, $s_{ij} = \pm\sqrt{E_{ij}}$, for the defined energy of a bit $E_{ij}$ in the $j$th branch of the modulator. Because the signal processing is identical for all $J$ branches in the transmitter and receiver, the system operation will be explained and mathematically modelled for the $j$th signal transmission corresponding to the $j$th binary digit. For the bandpass intermediate-frequency noise process $N(n)$, the noisy intermediate-frequency discrete-time signal received is

$$r(n) = s_i(n) + n(n) = \sum_{j=1}^{J} s_{ij}\sqrt{2/N} \cos \Omega_j n + \sqrt{E_N} \sum_{j=1}^{J} n_G(n)\sqrt{2/N} \cos \Omega_j n, \qquad (5.7.5)$$

where $n(n)$ is a realization of the noise process $N(n)$, and $E_N$ is the energy of the bandpass noise and is expressed in terms of the power spectral density $N_0/2$, that is, $E_N = N_0/2$. The signal received, $r(n)$, is coherently demodulated using $J$ correlators inside the correlation demodulator.

**Operation of a correlation demodulator.** The intermediate-frequency optimum receiver is composed of a correlation demodulator and an optimum detector, as shown in Figure 5.7.1. The principle of operation of all correlators inside the correlation demodulator is the same. Therefore, we will analyse here the operation of the $j^{\text{th}}$ correlator that processes the received signal and generates a random sample at its output according to the expression

$$x_j = \sum_{n=0}^{N-1} r(n)\phi_j(n) = \sum_{n=0}^{N-1} \left[ \sum_{j=1}^{J} s_{ij}\sqrt{2/N} \cos \Omega_j n + \sqrt{E_N} \sum_{j=1}^{J} n_G(n)\sqrt{2/N} \cos \Omega_j n \right] \sqrt{2/N} \cos \Omega_j n$$

$$= \frac{2}{N} \sum_{n=0}^{N-1} \cos \Omega_j n \sum_{j=1}^{J} s_{ij} \cos \Omega_j n + \frac{2}{N} \sum_{n=0}^{N-1} \cos \Omega_j n \sum_{j=1}^{J} \sqrt{E_N} n_G(n) \cos \Omega_j n$$

Due to the orthonormality of the cosine carriers and the statistical independence of the noise samples, the output of the correlator is

$$x_j = \frac{2}{N} \sum_{n=0}^{N-1} s_{ij}\cos^2 \Omega_j n + \frac{2}{N}\sqrt{E_N} \sum_{n=0}^{N-1} n_{Gj}(n)\cos^2 \Omega_j n$$

$$= \frac{2}{N} N \frac{1}{2} s_{ij} + 0 + \frac{2}{N}\sqrt{E_N} \sum_{n=0}^{N-1} n_{Gj}(n)\left[\frac{1}{2}\left(1 + \cos 2\Omega_j n\right)\right], \tag{5.7.6}$$

$$= s_{ij} + \frac{1}{N}\sqrt{E_N} \sum_{n=0}^{N-1} n_{Gj}(n) = \sqrt{E_{ij}} + \bar{n}_{Gj}$$

which is one realization of the random variable $X_j$ defined as

$$X_j = s_{ij} + \sqrt{E_N}\frac{1}{N} \sum_{n=0}^{N-1} N_{Gj}(n) = \sqrt{E_{ij}} + \overline{N}_{Gj}, \tag{5.7.7}$$

where $\overline{N}_{Gj}$ is the sample mean of the normal noise process $N_{Gj}(n)$. The variable $X_j$ is the sum of a deterministic part $s_{ij}$, which belongs to the signal, and a random part $\overline{N}_{Gj}$, which belongs to the Gaussian noise. Therefore, the variable $X_j$ is Gaussian. Being Gaussian, it is completely characterized statistically by two parameters: its mean and its variance.

Let us analyse these parameters in the case when filtering and interpolation is performed and a baseband noise is generated, as discussed in Chapter 4 for baseband noise generators. The mean and the variance of the random variable at the output of the correlator can be calculated. The Gaussian noise samples $n_{Gj}(n)$ can be generated by different methods, as explained in Chapter 4. We will analyse here the case when one Nyquist sample is generated in a bit interval and interpolated by $N$ identical samples. The mean and the variance of the output of the correlator, for the defined, identical energy of a bit $E_{ij} = E_b$ in all branches, are, respectively,

$$\eta_{X_j} = E\left\{X_j\right\} = E\left\{s_{ij} + \sqrt{E_N}\frac{1}{N}N\cdot N_{Gj}(n)\right\} = \sqrt{E_{ij}} + \sqrt{E_N}E\left\{N_{Gj}(n)\right\} = \sqrt{E_b}, \tag{5.7.8}$$

and

$$\sigma_{X_j}^2 = E\left\{(X_j - s_{ij})^2\right\} = E\left\{\left(\sqrt{E_N}\frac{1}{N}\sum_{n=0}^{N-1} N_{Gj}(n)\right)^2\right\}$$

$$= \frac{E_N}{N^2}E\left\{(N\cdot N_{Gj}(n))^2\right\} = \frac{E_N}{N^2}N^2 E\left\{N_{Gj}^2(0)\right\} = E_N = \frac{N_0}{2} \tag{5.7.9}$$

**Statistics of the correlator output for a binary source.** The output of the correlator is a random variable that has Gaussian distribution, having its mean value and its variance defined by eqns (5.7.8) and (5.7.9) respectively, for all $j$, because the variance of the baseband noise is equal to its power spectral density. The probability density function of the $j$th variable $X_j$ is Gaussian and is expressed as

$$f_{X_j}\left(x_j|m_i\right) = \frac{1}{\sqrt{\pi N_0}} e^{-\left(x_j - s_{ij}\right)^2/N_0} = f_{X_j}\left(x_j|s_i(n)\right). \tag{5.7.10}$$

The bit error probability $p$ of the $j$th bit in the $i$th symbol can be calculated as

$$p = \frac{1}{2}\text{erfc}\left(\frac{2\sigma_{Xj}^2}{\eta_{Xj}^2}\right)^{-1/2} = \frac{1}{2}\text{erfc}\left(\frac{2\sigma_{Xj}^2}{s_{ij}^2}\right)^{-1/2} = \frac{1}{2}\text{erfc}\left(\frac{2E_N}{E_b}\right)^{-1/2}$$
$$= \frac{1}{2}\text{erfc}\left(\frac{2N_0/2}{E_b}\right)^{-1/2} = \frac{1}{2}\text{erfc}\sqrt{\frac{E_b}{N_0}} \tag{5.7.11}$$

which is identical to the expression for the bit error probability of the binary phase-shift keying system. At the output of each correlator, we have a sample that corresponds to the coordinate of the received signal $r(n)$ (signal plus noise) in the direction of the corresponding basis function. For example, the random value $x_1$ is the coordinate of the signal $r(n)$ in the direction of the first basis function $\phi_1(n)$. The outputs of $J$ correlators, sampled at time instants $kN$, $k = 1, 2, \ldots$, see Figure 5.7.1, can be represented by a vector of $J$ random values, that is,

$$\mathbf{x} = [x_1 \ x_2 \ \cdots \ x_J]^T, \tag{5.7.12}$$

which are the coordinates of the received signal $r(n)$. These coordinates are the sums of the coordinates of the signal vector $\mathbf{s_i}$ and the coordinates of the noise vector $\mathbf{w}$, that is,

$$\mathbf{x} = \mathbf{s_i} + \mathbf{w}. \tag{5.7.13}$$

The random outputs of $J$ correlators, sampled at any time instant $kN$, $k = 1, 2, \ldots$ (see Figure 5.7.2), can be represented by a vector of $J$ random variables, that is,

$$\mathbf{X} = [X_1 \ X_2 \ \cdots \ X_J]^T, \tag{5.7.14}$$

whose individual random variables are Gaussian with means equal to $s_{ij}$, and variances equal to $N_0/2$. Since these variables are independent (affected by different noises due to the frequency separation of the carriers, or by the same noise multiplied by different basis functions and then accumulated), the *joint conditional probability density function* of $J$ variables is equal to the product of density functions of individual random variables. This joint density function is derived in eqn (5.5.11) and presented here in its final form as

$$f_{\mathbf{X}}(\mathbf{x}|m_i) = (\pi N_0)^{-J/2} e^{-\frac{1}{N_0} \sum_{j=1}^{J} (x_j - s_{ij})^2}, \tag{5.7.15}$$

where the *demodulation sample vector* $\mathbf{x}$ is a sample of $\mathbf{X}$, and the correlator output $x_j$ is a sample of the random variable $X_j$. This density function specifies the joint probability density function of the demodulation sample vector $\mathbf{x}$, given that the transmitted message symbol was $m_i$. Therefore, $x_j$ is equal to the sum of the signal coefficient $s_{ij}$ and the noise sample $w_j$ at the $j$th correlator output.

### 5.7.5 Intermediate-Frequency Optimum Detectors

The log-likelihood function has been defined in eqn (5.6.7) followed by the expression (5.6.8) of the decision variable. Based on that expression, the optimum detector is developed as shown in Figure 5.7.1. The inputs of the detector are elements of the demodulation sample vector $\mathbf{x}$ and all possible $I$ message symbols $\mathbf{s}_1$ to $\mathbf{s}_I$. The detector is composed of $I$ multipliers and accumulators that perform the correlation of all possible message symbols with the sample vector $\mathbf{x}$. From the correlation values obtained inside the detector, $Z_i$, half of the symbol energies are subtracted. The decision circuit compares the correlation values $Z_i$, chooses the largest one, $Z_k$, and then presents it to the look-up-table, which generates the message symbol $m_k$ at its output.

## 5.8 Operation of a Digital Communication System

### 5.8.1 Digital versus Discrete Communication Systems

It is assumed that the reader of this book is familiar with the theory of digital communication systems which is based on continuous-time signal processing. For the sake of completeness and to understand the fundamentals of the operation of transmitters and optimum receivers, this section will give a brief review of the theory of digital communications and its relation to the theory of discrete communication systems as presented in this book. The digital communication systems theory in this section will be presented in author's way and will have a one-to-one correspondence with the discrete communication systems theory presented in previous sections. Finally, by reading this section, which is specifically structured to explain the operation of digital communication systems, based on Chapters 2, 3, and 4, the reader will be in a position to make a distinction between digital communication systems and discrete communication systems, as presented in this book. For these reasons, readers, even if they are familiar with digital modulation techniques, are encouraged to read this section before proceeding with the theory of discrete communication systems presented in previous sections.

The differences between digital and discrete communication systems can, in principle, be summarized as follows:

1. The continuous independent time variable $t$ in digital communications is replaced by the discrete-time variable $n$ in discrete communication systems, and all signal-processing operations are performed in the discrete-time domain.

2. The integrals in digital systems are replaced by sums in discrete systems.

3. Some of the blocks for baseband and radiofrequency signal processing are common to both digital and discrete systems.

4. The processing of signals in the frequency domain is based on the corresponding Fourier series and Fourier transforms for both continuous-time and discrete-time signals.

5. It is not necessary that the design of transceivers in discrete technology strictly follow strictly the theoretical structure presented in the book. However, the principle of the transceivers' operation is preserved in their design and follows the theoretical explanation presented in this book.

### 5.8.2 Generic Structure of a Digital Communication System

The discrete system structure presented in Figure 5.1.1 will be preserved in digital (continuous) systems, with the following main changes: the time variable will be $t$ instead of $n$, and the adders inside the demodulator will be replaced by integrators. The complete block schematic of the generic discrete system, presented in Figure 5.6.4, can be adapted to present a generic digital system, and vice versa, as shown in Figure 5.8.1. This figure will be used to explain the operation of a digital system.

**Operation of a continuous message source.** The structure of the message source and the generation of signal coefficients will be the same in the continuous-time system shown in Figure 5.8.1 as in the discrete system presented in Figure 5.1.1. The only difference is in the generation of a modulating signal. In the continuous-time domain, the signal depends on the

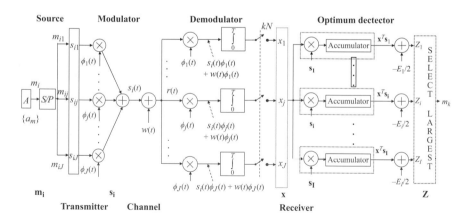

**Fig. 5.8.1** *A generic block schematic for a digital communication system.*

continuous-time variable $t$ instead of the discrete variable $n$. This difference can be observed in Figure 5.2.1, where the modulating signal will be changed to a continuous-time signal of duration $T$ instead of a discrete-time signal of duration $N$.

**Operation of a continuous modulator.** *Modulating signals* modulate $J$ carriers that form a set of the continuous-time orthonormal basis functions defined in Chapter 2, Section 2.4. All modulating signals, representing the message symbol, are of the continuous-time duration $T$, as shown in Figure 5.8.1. The $i$th transmitted signal $s_i(t)$, expressed as the sum of modulated orthogonal carriers $\phi_j(t)$, for $j = 1, 2, \ldots, J$, is

$$s_i(t) = \sum_{j=1}^{J} s_{ij}\phi_j(t), \tag{5.8.1}$$

where $s_{ij}$ are signal coefficients, as defined and presented in Chapter 2. To see the difference between digital and discrete systems, we can look at the signals presented in the discrete-time domain in Figure 5.3.1. These signals, when in the continuous-time domain, will be functions of the continuous variable $t$ instead of the discrete variable $n$, as presented by dashed graphs, and the bit duration will be the continuous interval $T$ instead of the discrete interval $N$.

**Additive white Gaussian noise channels in a continuous-time domain.** The transmitter signal $s_i(t)$, expressed as the sum of $J$ modulated orthonormal carriers, is transferred through the additive white Gaussian noise channel that is represented by the Gaussian white noise process $W(t)$, as shown in Figure 5.8.1. The analysis of this noise is presented in Chapter 4 alongside the presentation of discrete-time Gaussian noise. The theoretical base for the analysis of the continuous-time Gaussian noise process $W(t)$ is presented in Chapter 19.

**Operation of a continuous correlation demodulator.** The processing of signals in the continuous-time domain, as shown in Figure 5.8.1, is similar to their processing in the discrete-time domain, as presented in Figure 5.1.1 and Figure 5.6.4. The operation of a correlation demodulator in a digital system is based on continuous-time signal processing and will be explained using the block schematic of the system presented in Figure 5.8.1. The received signal can be expressed as

$$r(t) = s_i(t) + w(t), \tag{5.8.2}$$

for $0 \leq t \leq T$ and $i = 1, 2, \ldots, I$, where $w(t)$ is a sample function of the white Gaussian noise process $W(t)$ of zero mean and power spectral density $N_0/2$. The output of the $j$th correlator is

$$\begin{aligned} x_j &= \int_0^T r(t)\phi_j(t)dt = \int_0^T [s_i(t) + w(t)]\,\phi_j(t)dt \\ &= \int_0^T s_i(t)\phi_j(t)dt + \int_0^T w(t)\phi_j(t)dt = s_{ij} + w_j \end{aligned} \tag{5.8.3}$$

where the first component, denoted by $s_{ij}$, is a deterministic quantity contributed by the transmitted signal $s_i$ and is called the signal part, and the second component, $w_j$, is the sample value of the random variable $W_j$ that arises because of the channel noise $w(t)$, and is called the noise part.

The correlator random output $x_j$, $j = 1, 2, \ldots, J$, is a realization of the Gaussian random variable $X_j$, because it has been obtained by transferring a sum of the Gaussian process $W(t)$ and the deterministic signal $s_i(t)$ through a stable linear filter, and can be expressed as

$$X_j = s_{ij} + \int_0^T W(t)\phi_j(t)dt = s_{ij} + W_j. \tag{5.8.4}$$

In this case, the variable $X_j$, being Gaussian, is completely characterized by its mean and its variance, which can be derived via a process analogous to that used for discrete systems. Keeping in mind the continuous nature of noise process $W(t)$, we find that the mean value of $X_j$ is

$$\eta_{X_j} = E\{X_j\} = s_{ij} + E\{W_j\} = s_{ij}, \tag{5.8.5}$$

because the mean of $W_j$ is zero. The variance can be expressed as

$$\sigma_{X_j}^2 = E\left\{(X_j - s_{ij})^2\right\} = E\left\{W_j^2\right\}, \tag{5.8.6}$$

which gives the variance in the form

$$\sigma_{X_j}^2 = E\left\{W_j^2\right\} = E\{W(t_1)W(t_2)\} = E\{W(t)W(u)\} = E\left\{\int_0^T W(t)\phi_j(t)\,dt \int_0^T W(u)\phi_j(u)\,du\right\}$$
$$= E\left\{\int_0^T \int_0^T \phi_j(u)\phi_j(t)W(t)W(u)\,dt\,du\right\}.$$

Because the expectation $E$ is a linear operator, the order of integration and expectation can be changed, which results in

$$\sigma_{X_j}^2 = \int_0^T \int_0^T \phi_j(u)\,\phi_j(t)E\{W(t)W(u)\}\,dtdu = \int_0^T \int_0^T \phi_j(u)\phi_j(t)R_w(t,u)\,dt\,du,$$

where $R_{WW}(t, u)$ is the autocorrelation function of the white Gaussian noise process. Since the noise process $W(t)$ is stationary, its autocorrelation function $R_{WW}(t, u)$ depends only on the time difference $(t - u)$. Because the noise process $W(t)$ is white, its power spectral density is constant and equal to $N_0/2$. Thus, the autocorrelation function is

$$R_{WW}(t,u) = \frac{N_0}{2}\delta(t-u).$$

Taking into account the sifting property of the delta function (see Chapter 11), the variance can be derived as

$$\sigma_{X_j}^2 = \int_0^T \int_0^T \phi_j(u)\phi_j(t)\frac{N_0}{2}\delta(t-u) \ dtdu = \frac{N_0}{2}\int_0^T \phi_j(u)\int_0^T \phi_j(t)\delta(t-u) \ dtdu$$

$$= \frac{N_0}{2}\int_0^T \phi_j(u)\phi_j(u)du.$$

Because the basis functions have unit energy, the final expression for the variance is

$$\sigma_{X_j}^2 = \frac{N_0}{2}\int_0^T \phi_j^2(u)du = \frac{N_0}{2}\cdot 1 = \frac{N_0}{2}, \tag{5.8.7}$$

for all $j$. Because the basis functions form an orthogonal set, the random variable $X_j$ are mutually uncorrelated with a zero covariance function, as proved by

$$\text{cov}\{X_jX_k\} = E\{(X_j - \eta_{X_j})(X_k - \eta_{X_k})\} = E\{(X_j - s_{ij})(X_k - s_{ik})\} = E\{W_jW_k\}$$

$$= E\left\{\int_0^T W(t)\phi_j(t)dt \int_0^T W(u)\phi_k(u)du\right\} = E\left\{\int_0^T \int_0^T \phi_j(t)\phi_k(u)W(t)W(u)dtdu\right\}$$

$$= \int_0^T \int_0^T \phi_j(t)\phi_k(u)E\{W(t)W(u)\}dtdu = \int_0^T \int_0^T \phi_j(t)\phi_k(u)R_W(t,u) \ dtdu$$

$$= \frac{N_0}{2}\int_0^T \int_0^T \phi_j(t)\phi_k(u)\delta(t-u) \ dtdu = \frac{N_0}{2}\int_0^T \int_0^T \phi_j(u)\phi_k(u)du = 0$$

for every $j \neq k$. Since the variables $X_j$ are Gaussian, this proof implies that they are also statistically independent. The probability density function of the $j$th variable $X_j$ is

$$f_{X_j}(x_j|m_i) = \frac{1}{\sqrt{\pi N_0}}e^{-(x_j - s_{ij})^2/N_0}. \tag{5.8.8}$$

The processing of the random variable $X_j$ for digital continuous-time systems is equivalent to its processing in discrete systems, with one exception. Namely, when the samples $x_j$ of random variable $X_j$ are quantized and processed as discrete values, they will have a discrete Gaussian density function.

**Statistical characteristics of the optimum correlator.** At the output of each correlator, we have a sample that corresponds to a coordinate of the signal received $r(t)$ (signal plus noise) in the direction of the corresponding basis function. The outputs of $J$ correlators, sampled at time instants $t = T$, can be represented by vectors of $J$ elements, as presented by eqn (5.5.8):

$$\mathbf{x} = [x_1 \ x_2 \ \cdots \ x_J]^T, \mathbf{s}_i = [\ s_{i1} \ s_{i2} \ \cdots \ s_{iJ} \ ]^T, \mathbf{w} = [\ w_1 \ w_2 \ \cdots \ w_J \ ]^T, \tag{5.8.9}$$

where $\mathbf{x}$ represents the coordinates of the received signal $r(t)$, $\mathbf{s_i}$ are the signal coordinates, and $\mathbf{w}$ represents the channel noise coordinates. The coordinates of the received signal are obtained by summing the signal coordinates and the noise coordinates, that is,

$$\mathbf{x} = \mathbf{s_i} + \mathbf{w}. \tag{5.8.10}$$

The outputs of the $J$ correlators, sampled at time $t = kT$, $k = 1, 2, \ldots$ (see Figure 5.8.1), are realizations of $J$ random variables that can be represented as a vector $X$, that is,

$$\mathbf{X} = \begin{bmatrix} X_1 \\ X_2 \\ \vdots \\ X_J \end{bmatrix} = \begin{bmatrix} s_{i1} + W_1 \\ s_{i2} + W_2 \\ \vdots \\ s_{iJ} + W_J \end{bmatrix} = \mathbf{s_i} + \mathbf{W}, \qquad (5.8.11)$$

whose individual random variables are Gaussian with means equal to $s_{ij}$, and variances equal to $N_0/2$. A realization of $\mathbf{X}$ is $\mathbf{x}$. Since the variables in $\mathbf{X}$ are independent (noise multiplied by different basis functions), the joint conditional probability density function of $J$ variables is equal to the product of the density functions of individual random variables, as shown by

$$f_{\mathbf{X}}(\mathbf{x}) = \prod_{j=1}^{J} f_X\left(x_j | m_i\right) = \prod_{j=1}^{J} \frac{1}{\sqrt{\pi N_0}} e^{-(x_j - s_{ij})^2/N_0} = (\pi N_0)^{-J/2} e^{-\frac{1}{N_0} \sum_{j=1}^{J} (x_j - s_{ij})^2}, \qquad (5.8.12)$$

which is equivalent to the probability density function derived for a discrete communication system. Thus, the additive white Gaussian noise channel is equivalent to the $J$-dimensional vector channel described by the demodulation vector $\mathbf{x} = \mathbf{s_i} + \mathbf{w}$ defined for $i = 1, 2, \ldots, I$. The dimension $J$ is the number of basis functions that formulate the signal vector $\mathbf{s_i}$.

**Optimum detection.** For the derived joint density function of the random vector $X$ at the output of the optimum correlator (5.8.12), the maximum likelihood rule can be used as for the discrete optimum detector, as explained in Section 5.6. The log-likelihood function can be obtained as in eqn (5.6.7) and expressed as

$$l(m_i) \approx \mathbf{x}^T \mathbf{s_i} - \frac{E_i}{2}. \qquad (5.8.13)$$

Based on this expression, the decision variable values can be found and a decision made as for the discrete case, following the explanation in Section 5.6.2. Based on the expression for the log-likelihood function, the optimum detector can be constructed as presented in Figure 5.8.1. The explanation for the operation of the detector is equivalent to that for the operation of a discrete optimum detector, as presented in Section 5.6.3.

....................................................................................................................................................

### APPENDIX: OPERATION OF A CORRELATOR IN THE PRESENCE OF DISCRETE WHITE GAUSSIAN NOISE

In this case, noise samples are assumed to be realizations of a continuous-time white Gaussian noise process, taken at equidistant intervals of time. The sampling can be performed on physically realizable signals. Therefore, the noise is sampled under the assumption that its power has a finite value, which does not comply exactly with the definition of white noise. Details related to this issue can be found in Chapter 4. The autocorrelation function of the discrete noise process $W(n)$ may be calculated as

$$R_{WW}(k,l) = E\{W(k)\} \cdot E\{W(l)\} = \begin{Bmatrix} E\{W^2(k)\} & k=l \\ E\{W(k)\} \cdot E\{W(l)\} & k \neq l \end{Bmatrix}$$

$$= \begin{Bmatrix} \sigma_W^2 & k=l \\ 0 & k \neq l \end{Bmatrix} = \sigma_W^2 \delta(k-l) \tag{A1}$$

because the noise generated inside the channel is an uncorrelated discrete-time stochastic process. Since the noise process $W(n)$ is wide-sense stationary, its autocorrelation function $R_{WW}(k, l)$ depends only on the time difference $(k - l)$. The noise process $W(n)$ is white in the sense of discrete-time signal processing; thus, its power spectral density is constant and equal to $N_0/2$, that is,

$$S_{WW}\left(e^{j\Omega}\right) = \sum_{l=-\infty}^{\infty} R_{WW}(l)e^{-j\Omega l} = \sum_{l=-\infty}^{\infty} \sigma_W^2 \delta(l)e^{-j\Omega l} = \sigma_W^2 = \frac{N_0}{2}, \tag{A2}$$

which is defined for all frequencies $\Omega$ defined in the interval $-\pi < \Omega < \pi$. Inside a symbol interval, the random discrete noise is added to the discrete modulated signal, and the sum obtained is also an independent discrete-time stochastic process, $R(n)$. A realization $r(n)$ of the stochastic process $R(n)$ is processed in the bank of correlators. The output of the $j$th correlator is

$$x_j = s_{ij} + w_j, \tag{A3}$$

which is a realization of a random variable $X_j$ expressed as

$$X_j = s_{ij} + \sum_{n=0}^{N-1} W(n)\phi_j(n) = s_{ij} + W_j. \tag{A4}$$

The mean and variance of $X_j$ can be calculated as follows. The mean is

$$\eta_{X_j} = E\{X_j\} = E\{s_{ij} + W_j\} = s_{ij} + E\{W_j\} = s_{ij}, \tag{A5}$$

because the mean of $W$ is zero. The variance is

$$\sigma_{X_j}^2 = E\left\{(X_j - s_{ij})^2\right\} = E\{W_j^2\}, \tag{A6}$$

where $W_j$ is a sum of $N$ iid Gaussian random variables. Thus, the variance of $X_j$ can be calculated as

$$\sigma_{X_j}^2 = E\left\{W_j^2\right\} = E\left\{W_j(k)W_j(l)\right\} = E\left\{\sum_{k=0}^{N-1} W(k)\phi_j(k) \sum_{l=0}^{N-1} W(l)\phi_j(l)\right\}.$$

$$= E\left\{\sum_{k=0}^{N-1}\sum_{l=0}^{N-1} W(k)W(l)\phi_j(k)\phi_j(l)\right\} \tag{A7}$$

Due to the linearity of the expectation operator, and the iid nature of the random variables $W(k)$ and $W(l)$, the variance of $X_j$ is

$$\sigma_{X_j}^2 = \sum_{k=0}^{N-1}\sum_{l=0}^{N-1} E\{W(k)W(l)\}\,\phi_j(k)\phi_j(l) = \sum_{k=0}^{N-1}\sum_{l=0}^{N-1} R_{WW}(k,l)\,\phi_j(k)\phi_j(l)$$

$$= \sum_{k=0}^{N-1}\sum_{l=0}^{N-1} \sigma_W^2\left[\delta(k-l)\,\phi_j(k)\right]\phi_j(l) \tag{A8}$$

Taking into account the sifting property of the delta function, the variance is equal to the variance of the white Gaussian noise in the channel $W(n)$, that is,

$$\sigma_{X_j}^2 = \sigma_W^2 \sum_{l=0}^{N-1}\left[\sum_{k=0}^{N-1}\delta(k-l)\,\phi_j(k)\right]\phi_j(l) = \sigma_W^2 \sum_{k=0}^{N-1}\phi_j(l)\phi_j(l) = \sigma_W^2 = \frac{N_0}{2}, \tag{A9}$$

for all $j$. Because the basis functions form an orthogonal set, any pair of random variables $X_j$ and $X_m$ in the set $\mathbf{X}$ are mutually uncorrelated, as shown by calculating their covariance:

$$\text{cov}\left\{X_jX_m\right\} = E\left\{\left(X_j-\eta_{X_j}\right)\left(X_m-\eta_{X_m}\right)\right\} = E\left\{W_jW_m\right\}$$

$$= E\left\{\sum_{k=0}^{N-1} W(k)\phi_j(k)\sum_{l=0}^{N-1} W(l)\phi_m(l)\right\} = E\left\{\sum_{k=0}^{N-1}\sum_{l=0}^{N-1} W(k)W(l)\phi_j(k)\phi_m(l)\right\}$$

$$= \sum_{k=0}^{N-1}\sum_{l=0}^{N-1} R_{WW}(k,l)\,\phi_j(k)\phi_m(l) = \sum_{k=0}^{N-1}\sum_{l=0}^{N-1} \sigma_W^2\left[\delta(k-l)\,\phi_j(k)\right]\phi_m(l) \tag{A10}$$

$$= \sigma_W^2 \sum_{l=0}^{N-1}\left[\sum_{k=0}^{N-1}\delta(k-l)\,\phi_j(k)\right]\phi_m(l) = \sigma_W^2 \sum_{k=0}^{N-1}\phi_j(l)\phi_m(l) = 0$$

Because the covariance function is zero for every $j \neq m$, the variables $X_j$ and $X_m$ are uncorrelated. Because the variables $X_j$ and $X_m$ are Gaussian, this equation implies that they are also statistically independent. Therefore, the probability density function of the $j$th variable $X_j$ is Gaussian:

$$f_{X_j}\left(x_j|m_i\right) = \frac{1}{\sqrt{\pi N_0}}e^{-(x_j-s_{ij})^2/N_0}. \tag{A11}$$

# 6

# Digital Bandpass Modulation Methods

## 6.1 Introduction

Following the generic scheme for a digital communication system that was presented in Chapter 5 in Figure 5.8.1, Figure 6.1.1 shows a general block scheme of a system suitable for the presentation and explanation of modulation methods. For all blocks, the input and output signals are presented as realizations of related continuous-time or discrete-time stochastic processes, which are shown in the figure. In addition, the functions that are relevant for operation of each block are shown underneath each block and the related signals. We tried to unify the notation in this chapter with the notation in Chapters 5, 11–13, and 19, as we have in the whole book, at the level that could be done. However, due to the nature of the mathematical models used to analyse communication systems, which are based on different modulation techniques, that unification has not been always possible.

**Source and baseband signal processing.** A message source is defined by its alphabet $A$. The source generates the messages at its output in the form of message symbols $m_i$, which are mathematically defined as variations with repetition of the alphabet elements. Therefore, if a message symbol is composed of $J$ elements taken from the alphabet $A$ containing $M$ elements, then we can generate $I = M^J$ different messages at the output of the message source. This source can be considered to be a discrete source of information, which is theoretically explained and mathematically described in Chapter 9, which is dedicated to information theory. The $i$th message symbol is expressed as a set $m_i = (m_{i1}, m_{i2}, \ldots, m_{ij}, \ldots, m_{iJ})$, where $m_{ij}$ are elements of the alphabet $A$, and expressed in matrix form as a column matrix $\mathbf{m_i}$. Therefore, all possible message symbols form a set presented as a matrix $\mathbf{m}$, as shown in Figure 6.1.1. These message symbols are generated randomly as defined by the matrix of their a priori probabilities, which is expressed as

$$P(\mathbf{m}) = [P(\mathbf{m_1}) \quad P(\mathbf{m_2}) \quad \cdots \quad P(\mathbf{m_i}) \quad \cdots \quad P(\mathbf{m_{I-1}}) \quad P(\mathbf{m_I})]. \qquad (6.1.1)$$

Most of the time, we will assume that these probabilities are equal to each other, meaning that the message symbols are generated by the source with the same probability. The probabilities of the message symbols can be related to the probabilities of the alphabet's elements.

*Discrete Communication Systems.* Stevan Berber, Oxford University Press. © Stevan Berber 2021.
DOI: 10.1093/oso/9780198860792.003.0006

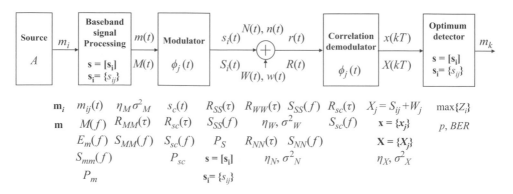

**Fig. 6.1.1** *Signals and functions in a digital communication system.*

The set of message symbol elements are mapped into the set of signal coefficients forming the modulating signal $s_i$, which is an element of the modulating signal vector $\mathbf{s_i}$. The signal coefficients define *modulating signals* $m(t)$ at the input of a transmitter modulator. The set of signal vectors $\mathbf{s_i}$ form the matrix of all possible signals at the output of the baseband signal processing block. There is a one-to-one correspondence between the message symbols $\mathbf{m_i}$ and the signals $\mathbf{s_i}$, and the modulating signal defined for the $i$th symbol can sometimes be expressed as $m_{ij}(t) = s_{ij}(t)$. A stream of these signals, presenting the message symbols, form the modulating signal $m(t)$ that is applied to the input of the modulator.

Due to the random nature of the source, the modulating signal $m(t)$ is defined as a realization of a continuous-time stochastic process $M(t)$. The analysis of this process is mainly based on the theory presented in Chapter 19. We consider it important to derive the autocorrelation function $R_{MM}(\tau)$ and the power spectral density $S_{MM}(f)$ of the process $M(t)$ and relate them to the analysis of the deterministic signals. Also, the probability density function of the process will be derived and the related mean $\eta_M$ and standard deviation $\sigma_M$ will be calculated. In addition, some of the properties of the modulating signal $m(t)$ can be investigated using the theory of deterministic signal process presented in Chapters 11 and 12. Using this theory, the energy spectral density $E_m(f)$ will be found and then related to the power spectral density of the signal. Using both stochastic and deterministic signal processing theory, the power $P_m$ and energy $E_m$ of the signal will be qualified and quantified.

**Modulation of the carriers.** The system modulator is designed, in principle, as a bank of multipliers. The input message signals $m(t)$ multiply the carriers, which are defined as a set of orthonormal basis function $\phi_j(t)$. The multiplied signals are added to each other, and the modulated signal $s_i(t)$ is obtained at the output of the modulator, which contains all the message signals applied at the input of the set of modulators. The modulated signal $s_i(t)$ is considered to be a realization of the modulated stochastic process $S_i(t)$, because the modulation is performed by the stochastic process $M(t)$ and a deterministic sinusoidal carrier $s_c(t)$. Alongside the analysis of the carrier in time and frequency domains, the derivation of its autocorrelation function $R_{sc}(\tau)$, power spectral density $S_{sc}(f)$, and power $P_{sc}$ are provided, to make calculations of the autocorrelation function $R_{SS}(\tau)$, the power spectral density $S_{SS}(f)$, and the power $P_S$ of the modulated signal easier, as shown in Figure 6.1.1.

Study of the power spectral density and the power of modulated signals is important for two reasons. Firstly, we have to know the expected occupancy of the channel bandwidth and, secondly, we are interested in knowing the level of cochannel interference in multiplex systems, which depends on the signal power spectral density. Following these requirements, we will derive the power spectral density of the baseband signal $m(t)$ and then find the power spectral density of the bandpass signal $s_i(t)$ for the main modulation schemes that are widely used in practice. Essential background theory, which is used for mathematical modelling and the analysis of various modulation methods and related communication systems, is presented in Chapters 11, 12, and 19.

In addition, such study is important for the calculation of bandwidth efficiency, which is generally defined as a ratio of the digital signal rate $R = 1/T$ and the required channel bandwidth $B_W$, that is,

$$\eta = \frac{R}{B_W}, \tag{6.1.2}$$

and expressed in bits per second per hertz for the defined duration of a symbol $T$. Multilevel encoding and spectral shaping can be used to increase bandwidth efficiency. Due to the limited size of this book, we will not present in detail the techniques of spectral shaping.

**Communication channels.** A modulated signal is amplified and transmitted through a channel. The output blocks of the transmitter and the input blocks of the receiver are excluded in the simplified presentation of a digital communication system in Figure 6.1.1, and the transmission of the modulated signal from the output of the modulator to the input of the demodulator is replaced by a communication channel. In our analysis, we will use two models of a communication channel.

The first model is the additive white Gaussian noise channel, which is traditionally used as a model for the theoretical analysis of communication systems. In the time domain, this channel is defined by a continuous-time wide-sense stationary Gaussian noise process with zero mean value, denoted by $W(t)$ in Figure 6.1.1. In theory, this noise will be presented by its autocorrelation function $R_{WW}(\tau)$, its power spectral density $S_{WW}(f)$ function, and the related probability density function of its amplitudes, as defined by its mean value $\eta_W$ and standard deviation $\sigma_W$. A detailed discussion of this channel can be found in Chapter 4, and the theory of continuous-time noise processes is presented in Chapter 19.

The second model is a bandpass channel represented by a continuous-time correlated Gaussian noise process $N(t)$ of zero mean and finite power. This noise will be presented by its autocorrelation function $R_{NN}(\tau)$ and power spectral density $S_{NN}(f)$ function, and the related probability density function of its amplitudes defined by its mean value $\eta_N$ and standard deviation $\sigma_N$. This channel model is analysed and presented in the investigated communication system structures for two reasons. Firstly, we need to use it for simulation and emulation. Namely, the generator of the noise $N(t)$ is physically realizable, unlike the white Gaussian noise generator, which is not physically realizable, due to the existence of infinite amplitudes which contribute to infinite noise power. In the frequency domain, $N(t)$ has values for power spectral density that are constant and are in a limited bandwidth that is equal to the bandwidth of the transmitted signal; this makes it possible to have simpler calculations and simulation of the signal-to-noise ratio in the communication system. Effectively, this noise has the same effect on the signal transmitted as the white Gaussian noise would have after signal filtering using a bandpass filter. Secondly, if we intend to analyse the transceiver operation at an intermediate frequency, this noise is a useful model for that analysis. As will be shown later, this noise model is important for the analysis of discrete communication systems because, at the present time, discrete-time signal processing inside transceivers take place at intermediate-frequency blocks, and the noise generated needs to be band limited.

**The correlation demodulator.** The received signal $r(t)$ is a realization of a continuous-time wide-sense-stationary process $R(t)$ which is the sum of the realizations of the modulated process $S_i(t)$ and the noise process $W(t)$ in the channel. The theoretical analysis of this process is not of particularly practical interest, unlike the signal processing that takes place inside the demodulator. In coherent demodulation, the received signal is correlated with locally generated synchronous carriers, resulting in a random correlation value at the output of each correlator. The output of each correlator is a discrete-time stochastic process $X(kT)$ with values at discrete instants of time defined by the duration of a symbol, $T$. One realization of this process is a series of random values $x(kT)$. The analysis of this process is based on the theory presented in Chapter 3. All outputs of the correlation demodulator at a particular time instant $kT$ are represented by a demodulation vector $\mathbf{x}$ containing the set of random values $\{x_j\}$ that are generated at the output of all correlators. Because these values are realizations of the related random variables defining the random outputs of all correlators $\{X_j\}$, we can define a vector $\mathbf{X} = [X_1 \ X_2 \ \dots \ X_J]^T$ which contains the random variables characterizing the outputs of all correlators at time instant $kT$ for integer values of $k$.

The discrete time stochastic process $X(kT)$ can be analysed by finding its probability density function. In the case of a Gaussian noise channel, this function will be Gaussian with zero mean and finite variance. At a particular time instant $kN$, the output $x_j$ of the $j$th correlator is a realization of the random variable $X_j$ which has the Gaussian distribution defined by its mean value $\eta_X$ and standard deviation $\sigma_X$. At a particular time instant $kN$, the outputs of all correlators form a set of random values $\{x_j\}$, which are realizations of the random variables $\{X_j\}$ in the vector $\mathbf{X}$. The random values in $\{x_j\}$ form the demodulation vector $\mathbf{x} = [x_1 \ x_2 \ \dots \ x_J]^T$.

The aim of the receiver operation is to estimate which message symbol $m_i$ was sent at the transmitter side, if a set of values defining the demodulation vector $\mathbf{x}$ is calculated by correlators.

**The optimum detector.** According to the general scheme of the optimum detector and its operation presented in Chapter 5, the detector correlates the incoming demodulation vector $\mathbf{x}$ with all possible signals sent $\mathbf{s} = \{\mathbf{s_i}\}$. The result is a set of numeric random values that define the decision variables $\{Z_i\}$. The decision is to be made according to the maximum likelihood decision rule, to find out which symbol $m_k$ is the best estimate of the symbol sent, $m_i$.

In the process of estimating the received symbol, the receiver can make a mistake or an error by estimating $m_i$ to be $m_k$. Due to statistical nature of the noise in communication channel, this error is quantitatively expressed by the conditional error probability $P(m_k \neq m_i \mid m_i)$, which states that the estimate $m_k$ differs from the transmitted symbol $m_i$, given that $m_i$ was sent. The basic requirement in designing the optimum receiver is to minimize the average probability of symbol

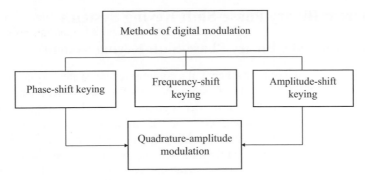

**Fig. 6.1.2** *Digital modulation methods.*

$$P_e = \sum_{i=1}^{I} P(m_i) \cdot P(m_k \neq m_i | m_i).$$ (6.1.3)

The receiver designed, based on this requirement, is said to be optimum in the minimum probability of symbol error sense. It is important to note here that the decision sometimes can be made at the output of the correlation demodulator.

Based on the general scheme of digital communication system presented in Figure 6.1.1, the implementation of basic digital modulation methods will be presented in the following sections. The methods under our investigation are presented in a block diagram in Figure 6.1.2. The main aim of this book is to present communication systems operating in the discrete-time domain. In order to use the space available in this book efficiently, we will limit our presentation of modulation methods to the basic ones. Following the same approach, it will be relatively easy to extend the mathematical models of modulation methods presented here to more complicated ones. In addition to that, in this chapter, we present a revised version of the theory of digital (not discrete) communication systems, for three reasons. Firstly, we do this to show that our presentation of the theory of digital communications systems is based on the general theory of the generic systems presented in Chapters 2 and 5, where a system using a particular modulation scheme is a special case. Secondly, we do this to show that the theory developed for digital systems is equivalent to the theory in discrete communication systems, with some specific features. Thirdly, we intend to show how our approach clearly differs from approaches previously used to explain the operation of communication systems.

In our analyses, we will extend the block schematic of communication systems presented in Figure 6.1.1, using defined signals and functions in time and frequency domains for both deterministic and stochastic signals. Then, we will explain the operation of various systems by using block schematics expressed in terms of mathematical operators, which will be followed by related mathematical models, as we did in Chapters 2 and 5. We start with communication systems that use phase-shift keying, primarily binary phase-shift keying and quaternary phase-shift keying. This will be followed by presenting a system that uses frequency-shift keying. Finally, quadrature amplitude modulation, which is a combination of amplitude-shift keying and phase-shift keying, will be presented.

## 6.2   Coherent Binary Phase-Shift Keying Systems

### 6.2.1   Operation of a Binary Phase-Shift Keying System

It is possible to simplify the generic scheme of a communication system, which was presented in Chapter 5 in Figure 5.8.1, to develop a binary phase-shift keying system, which is presented in Figure 6.2.1. In this system, the transmitter is composed of a baseband processing block, which generates the modulating signal $m(t) = m_i(t)$, and a multiplier, which functions as a modulator. The receiver is composed of a demodulation correlator and an optimum detector. The optimum detector can be excluded from the scheme, because the decision on the received message can be made at the output of the correlator, as we will explain later in this section. In the following sections, the mathematical model of the system and the related processing of signals will be presented in the continuous-time domain.

The message source generates one out of two symbols that belong to an alphabet $A$ of $I = 2$ elements: $m_1 = 1$ and $m_2 = 0$. The system is binary; therefore, a symbol generated at the output of the source is identical to a message bit. The message bit is applied to the input of a look-up table, which maps the input bits $(0, 1)$ into the corresponding polar output bits $(-1, +1)$. These bits are multiplied by the square root of the bit energy $\sqrt{E_b}$ in the bit interval $T_b$ to form the signal coefficient $s_{i1}$. The coefficient $s_{i1}$ defines the amplitude of a rectangular pulse that is the baseband modulating signal $m_i(t)$, as presented in Figure 6.2.1. The signal $m_i(t)$ is multiplied by the carrier, which is designed as a basis function, and the modulated binary phase-shift keying signal $s_i(t)$ is produced and transferred though an

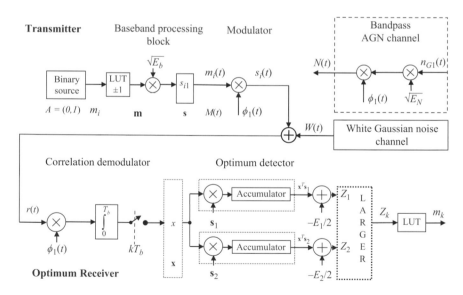

**Fig. 6.2.1** *Block schematic of a binary phase-shift keying system; AGN, additive Gaussian noise; LUT, look-up table.*

additive white Gaussian noise waveform channel represented by the wide-sense stationary Gaussian stochastic process $W(t)$. In Figure 6.2.1, alongside this channel, a generator of bandpass noise $N(t)$ is shown, which is a channel model that can be used for the simulation, emulation, and measurement of system characteristics. A detailed explanation of this noise, which is obtained by shifting the baseband noise to the carrier frequency, is presented in Chapter 4.

## 6.2.2   Transmitter Operation

### 6.2.2.1   *Modulating Signal Presentation*

**Signal-space or constellation diagram of the signal.** The number of possible message symbols is $I = A^J = 2^1 = 2$, which are defined by the matrix of message symbols

$$\mathbf{m} = [\mathbf{m}_1 \ \mathbf{m}_2] = \begin{bmatrix} 1 & -1 \end{bmatrix}. \tag{6.2.1}$$

These message symbols are mapped into the amplitudes of modulating signals by multiplying them by $\sqrt{E_b}$ to form a vector of modulating signal coefficients, which can be presented in one-dimensional signal space (see Section 6.2.2.2, Figure 6.2.6). The signal vector is represented by two columns in the matrix of signal coefficients, that is,

$$\mathbf{s} = [\mathbf{s}_1 \ \mathbf{s}_2] = \begin{bmatrix} s_{11} & s_{21} \end{bmatrix} = \sqrt{E_b}\mathbf{m} = \begin{bmatrix} \sqrt{E_b} & -\sqrt{E_b} \end{bmatrix}. \tag{6.2.2}$$

**Statistical characteristics of the modulating signal *m(t)* in the time domain.** The modulating signal $m(t)$ is a series of modulating rectangular pulses $m_i(t)$, each of which are of duration $T_b$. The modulating signal $m(t)$ is considered to be a realization of the random binary process $M(t)$, which randomly takes amplitude values specified by the vector defined in eqn (6.2.2), as shown in Figure 6.2.2. We will assume that the probabilities of positive and negative binary values are equal to 0.5. Therefore, at each time instant $t = t_k$ we have a discrete random variable $M = M(t_k)$ that takes the value of $\sqrt{E_b}$ or $\sqrt{-E_b}$, and is defined by its discrete probability density function expressed in terms of Dirac delta functions as

$$f_M(m) = \frac{1}{2}\delta\left(m - \sqrt{E_b}\right) + \frac{1}{2}\delta\left(m + \sqrt{E_b}\right), \tag{6.2.3}$$

as shown in Figure 6.2.2. The notation here follows the notation used in Chapter 19 for the theory of continuous-time stochastic processes.

Since the amplitude levels of $m(t)$ have the same magnitude values and occur with equal probabilities, the mean value of the process is zero, and the variance is $E_b$. Because the process is wide-sense stationary, with zero mean and constant variance, the autocorrelation function of the process depends only on the time difference between two time instants defined as the lag $\tau$. Although the function is derived in Chapter 19, in expression (19.2.10), it is also shown here in the form

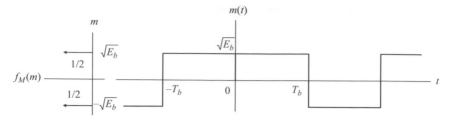

**Fig. 6.2.2** *A modulating polar binary signal presented in the time domain.*

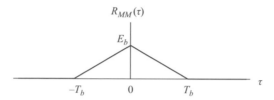

**Fig. 6.2.3** *Autocorrelation function of the modulating signal, represented as a binary stochastic process.*

$$R_{MM}(\tau) = \begin{cases} E_b\left(1 - \frac{|\tau|}{T_b}\right) & |\tau| < T_b \\ 0 & |\tau| \ge T_b \end{cases}, \qquad (6.2.4)$$

and represented by a triangular graph in Figure 6.2.3.

**Modulating signal presentation in the frequency domain.** Due to the Wiener–Khintchine theorem, the autocorrelation function and the power spectral density of a wide-sense stationary process form the Fourier transform pair. Therefore, the power spectral density of the modulating signal $m(t)$ can be calculated as follows:

$$S_{MM}(f) = \int_{-\infty}^{\infty} R_{MM}(\tau) e^{-j\omega t} d\tau = \int_{-T_b}^{T_b} E_b\left(1 - \frac{|\tau|}{T_b}\right) e^{-j\omega t} d\tau = E_b T_b \left[\frac{\sin(\pi f T_b)}{\pi f T_b}\right]^2 , \quad (6.2.5)$$
$$= E_b T_b \operatorname{sinc}^2(f T_b)$$

where the sinc function is defined as $\operatorname{sinc}(y) = \sin(\pi y)/\pi y$. The two-sided power spectral density normalized by $E_b T_b$ is

$$S_{MM}(f)/E_b T_b = \operatorname{sinc}^2(f T_b),$$

and presented Figure 6.2.4. The first zero crossing, which determines the first lobe of the spectrum, is defined by the frequency $f = 1/T_b$.

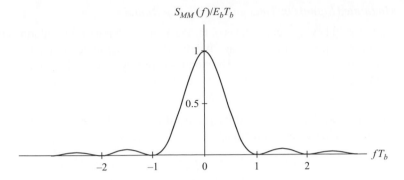

**Fig. 6.2.4** *Power spectrum density of the modulating signal, represented as a binary stochastic process.*

## Example

Calculate the mean and variance of the binary stochastic process $M(t)$ using the density function expressed by eqn (6.2.3). Then, calculate the power of the random signal $m(t)$ from its autocorrelation and power spectral density functions.

**Solution.** The mean and variance can be calculated as

$$
\eta_M = \int_{-\infty}^{\infty} m \cdot f_M(m) dm = \int_{-\infty}^{\infty} m \cdot \left[ \frac{1}{2}\delta\left(m - \sqrt{E_b}\right) + \frac{1}{2}\delta\left(m + \sqrt{E_b}\right) \right] dm ,
$$
$$
= \frac{1}{2}\sqrt{E_b} + \frac{1}{2}\left(-\sqrt{E_b}\right) = 0
$$

$$
\sigma_M^2 = \int_{-\infty}^{\infty} m^2 \cdot f_M(m) dm = \int_{-\infty}^{\infty} m^2 \cdot \left[ \frac{1}{2}\delta\left(m - \sqrt{E_b}\right) + \frac{1}{2}\delta\left(m + \sqrt{E_b}\right) \right] dm ,
$$
$$
= \frac{1}{2}E_b + \frac{1}{2}E_b = E_b
$$

respectively. The power is the value of the autocorrelation function at $\tau = 0$, which is numerically expressed as $E_m = R_{MM}(0) = E_b$. It is important to note that the amplitude of the rectangular pulse is expressed as the square root of the numerical value of the modulated signal energy $E_b$ to accommodate the use of sinusoidal carriers expressed in terms of orthonormal functions. The power calculated in the frequency domain is

$$
P_m = \int_{-\infty}^{\infty} S_{MM}(f) df = E_b T_b \int_{-\infty}^{\infty} \operatorname{sinc}^2 (f T_b) df = E_b T_b / T_b = E_b.
$$

The expression for the power spectral density of a single modulating baseband pulse, which is defined as an aperiodic deterministic signal in the interval $t = 0$ to $t = T_b$, as shown in Figure 6.2.2, can be calculated form its Fourier transform, as will be demonstrated in an example at the end of Section 6.2.2.

### 6.2.2.2    *Modulated Signals in Time and Frequency Domains*

The binary phase-shift keying signal $s_i(t)$ is obtained by multiplying the modulating signal and the carrier. The carrier is a deterministic signal defined by its amplitude, frequency, and phase. The modulating signal, as we said before, is a stochastic binary process. To find the modulated signal in the time and frequency domains, we need to find the carrier in both domains and then investigate the output of the modulator multiplier.

**Presentation of the carrier in time and frequency domains.** The carrier, as an orthonormal function, is expressed in the time domain as

$$s_c(t) = \phi_1(t) = \sqrt{\frac{2}{T_b}} \cos(\omega_c t). \tag{6.2.6}$$

If we consider the carrier to be a periodic function, we can calculate its autocorrelation function as

$$
\begin{aligned}
R_{sc}(\tau) &= \frac{1}{T_b} \int_{-\infty}^{\infty} s_c(t+\tau) s_c(t) dt = \int_{0}^{T_b} \sqrt{\frac{2}{T_b}} \cdot \cos 2\pi f_c(t+\tau) \sqrt{\frac{2}{T_b}} \cdot \cos 2\pi f_c t\, dt \\
&= \frac{1}{T_b} \frac{2}{T_b} \int_{0}^{T_b} \frac{1}{2} [\cos 2\pi f_c \tau + \cos(4\pi f_c t + 2\pi f_c \tau)] dt = \frac{1}{T_b} \cos 2\pi f_c \tau
\end{aligned} \tag{6.2.7}
$$

and the power spectral density as the Fourier transform of the autocorrelation function:

$$
\begin{aligned}
S_{sc}(f) &= \int_{-\infty}^{\infty} R_{sc}(\tau) e^{-j2\pi f \tau} d\tau = \frac{1}{T_b} \int_{-\infty}^{\infty} \cos 2\pi f_c \tau e^{-j2\pi f \tau} d\tau = \\
&= \frac{1}{2T_b} \int_{-\infty}^{\infty} \left( e^{j2\pi f_c \tau} + e^{-j2\pi f_c \tau} \right) e^{-j2\pi f \tau} d\tau = \frac{1}{2T_b} [\delta(f - f_c) + \delta(f + f_c)]
\end{aligned} \tag{6.2.8}
$$

Using the expressions for autocorrelation function and the power spectral density, we can calculate the power and the energy of the carrier as $P_{sc} = 1/T_b$, and $E_{sc} = 1$, respectively. The unit energy value is expected because the carrier is represented by an orthonormal function with unit energy. A precise theoretical analysis of the carrier in time and frequency domains can be found in Chapters 11 and 12.

**Modulated binary phase-shift keying signals in the time domain.** The modulated signal is represented by a pair of signals $s_1(t)$ and $s_2(t)$, generally expressed as

$$s_i(t) = \sqrt{E_b} \sqrt{\frac{2}{T_b}} \cos(\omega_c t + (i-1) \cdot \pi) \tag{6.2.9}$$

and defined as energy signals existing in the time interval $0 \le t < T_b$ with two phases determined by $i = 1$ or $i = 2$. The transmitted signal energy per bit is $E_b$. We assume that, for the source message bit 1, the $i$ value is 1 and, for the source bit 0, the $i$ value is 2. In addition, the carrier frequency is equal to $n/T_b$, for some constant $n$, in order to ensure that each transmitted signal $s_i(t)$ contains an integer number of the carrier cycles. This pair of sinusoidal signals, differing only in a relative phase-shift of 180°, are referred to as antipodal signals. They are expressed in terms of one basis function $\phi_1(t)$ and signal coefficients as

$$s_1(t) = \sqrt{\frac{2E_b}{T_b}} \cos(\omega_c t) = \sqrt{E_b}\sqrt{\frac{2}{T_b}} \cos(\omega_c t) = \sqrt{E_b}\phi_1(t) = s_{11}\phi_1(t), \qquad (6.2.10)$$

$$s_2(t) = \sqrt{\frac{2E_b}{T_b}} \cos(\omega_c t + \pi) = -\sqrt{E_b}\sqrt{\frac{2}{T_b}} \cos(\omega_c t) = -\sqrt{E_b}\phi_1(t) = s_{21}\phi_1(t), \qquad (6.2.11)$$

and presented in Figure 6.2.5 in the time domain. Obviously, these signals can be obtained by multiplying the periodic carrier by the inverted and the non-inverted rectangular modulating signals.

**Binary phase-shift keying signal constellation diagram.** From the expressions for the signal pair, it is obvious that there is only one basis function of unit energy, expressed as

$$\phi_1(t) = \sqrt{\frac{2}{T_b}} \cos(\omega_c t), \qquad (6.2.12)$$

and the related signal space is one dimensional ($M = 1$), with a signal constellation consisting of two message bit points having the coordinates

$$s_{11} = \int_0^{T_b} s_1(t)\phi_1(t)dt = \frac{2}{T_b}E_b \int_0^{T_b} \cos^2(\omega_c t)\ dt = +\sqrt{E_b}, \qquad (6.2.13)$$

$$s_{21} = \int_0^{T} s_2(t)\phi_1(t)dt = -\sqrt{E_b}, \qquad (6.2.14)$$

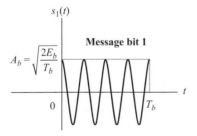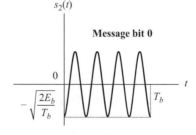

**Fig. 6.2.5** *Modulated signals in the time domain.*

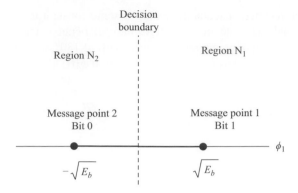

**Fig. 6.2.6** *Signal-space or constellation diagram for binary phase-shift keying signals.*

as shown in Figure 6.2.6. Message point 1 (for $m_1=1$ or, equivalently, $s_1(t)$) is located at $s_{11}$, and message point 2 (for $m_2 = 0$ or, equivalently, $s_2(t)$) is located at $s_{21}$. The constellation shown has a minimum energy.

The amplitude of this modulated signal is $A_b$ in the bit interval $T_b$, defining the symbol-shaping function as a rectangular pulse expressed as

$$s_{ss}(t) = \left\{ \begin{matrix} A_b = \sqrt{\frac{2E_b}{T_b}} & 0 \leq t \leq T_b \\ 0 & otherwise \end{matrix} \right\}, \tag{6.2.15}$$

which corresponds to a rectangular wave shape with an amplitude proportional to the amplitude of the modulating signal presented in Figure 6.2.2. In order to achieve better utilization of the signal spectrum, or for any other reason, the shape of this function can be changed. Most of the time, in this book, we will use rectangular pulses for the modulating signal. In this way, we will simplify the theoretical explanation and put the emphasis on understanding the theory of modulation. From the theoretical point of view, it is not so complicated to extend the modulation theory presented here to the cases when the pulse shape is different form a rectangle. The energy of the modulated signal representing a bit is calculated according to

$$E_b = \int_0^{T_b} [A_b \cos(\omega_c t)]^2 dt = A_b^2 \left[ \frac{1}{2}t + \frac{1}{4\omega_c} \sin(2\omega_c t) \right] \Big|_0^{T_b} = A_b^2 \frac{T_b}{2} = \frac{E_{bit}}{2}, \tag{6.2.16}$$

where $E_{bit}$ is the equivalent rectangular pulse energy that has the magnitude $A_b$.

**Modulated signals in the frequency domain.** As was stated in the introduction of this chapter, we are interested in knowing the spectral performance of the modulated signal. For that purpose, we need to calculate the power spectral density of the modulated signal and use it to calculate the power of the signal in a specified bandwidth.

   Due to the Wiener–Khintchine theorem, the autocorrelation function and the power spectral density of a wide-sense stationary process form a Fourier transform pair. We will present the operation of the modulator in terms of the autocorrelation function and the power spectral density. The baseband modulating signal $m(t)$, which has the power spectral density shown in Figure 6.2.4, is multiplied by the carrier, which operates at the frequency $f_c$. The initial amplitude of the random baseband signal and the carrier amplitude are statistically independent, due to the randomness of the signal amplitudes. Therefore, the bandpass modulated random signal $s(t)$, presented as a realization of a wide-sense stationary stochastic process $S(t)$, can be expressed as the product

$$s(t) = m(t) \cdot \sqrt{\frac{2}{T_b}} \cos(2\pi f_c t) = m(t) \cdot \phi_1(t), \tag{6.2.17}$$

where $m(t)$ is a realization of the baseband modulating stochastic process $M(t)$. The theory for these processes can be found in Chapter 19. In order to find the power spectral density of the modulated signal, we need to find the autocorrelation function of the process $S(t)$ as

$$R_{SS}(\tau) = E\{S(t) \cdot S(t+\tau)\} = E\{M(t)\phi_1(t) \cdot M(t+\tau)\phi_1(t+\tau)\}$$

$$= E\left\{M(t) \cdot M(t+\tau) \cdot \sqrt{\frac{2}{T_b}} \cos\omega_c t \sqrt{\frac{2}{T_b}} \cos\omega_c(t+\tau)\right\}$$

$$= \frac{2}{T_b}\frac{1}{2}E\{M(t) \cdot M(t+\tau) \cdot [\cos\omega_c \tau + \cos(2\omega_c t + \omega_c \tau)]\}$$

$$= \frac{1}{T_b}E\{M(t) \cdot M(t+\tau) \cdot \cos\omega_c \tau\} + \frac{1}{T_b}E\{M(t) \cdot M(t+\tau) \cdot \cos(2\omega_c t + \omega_c \tau)\}$$

Due to the independence between the cosine carrier signal and the baseband random signal, the expectation of the first term is the autocorrelation function of the modulating random signal $R_{MM}(\tau)$ multiplied by the cosine function. The expectation of the second term is zero, due to the high frequency of the carrier, which yields zero for the mean value. Therefore, the autocorrelation of the modulated signal is a product of the autocorrelation of the input modulating signal and the autocorrelation function of the carrier, expressed as

$$R_{SS}(\tau) = R_{MM}(\tau) \cdot \frac{1}{T_b}\cos\omega_c\tau + 0 = R_{MM}(\tau) \cdot R_{sc}(\tau). \tag{6.2.18}$$

With this autocorrelation function for the modulated signal, the power spectral density can be found, as the Fourier transform of this function. On the other hand, we can have this autocorrelation function expressed as the product of the autocorrelation function of the carrier and the autocorrelation function of the random binary process. Each of these autocorrelation functions is the inverse Fourier transform of the related power spectral density function. Therefore, we can directly obtain the power spectral density of the output random modulated signal as the convolution of the power spectral densities of the random modulating signal and

the carrier, calculated as

$$
\begin{aligned}
S_{SS}(f) &= \int_{-\infty}^{\infty} S_{MM}\,(f-\lambda)\,S_{sc}\,(\lambda)\,d\lambda \\
&= \int_{-\infty}^{\infty} E_b T_b \sin c^2\,(f-\lambda)\,T_b \cdot \frac{1}{2T_b}\,(\delta\,(\lambda-f_c)+\delta\,(\lambda+f_c))\,d\lambda. \\
&= \frac{E_b}{2}\sin c^2\,(f-f_c)\,T_b + \frac{E_b}{2}\sin c^2\,(f+f_c)\,T_b
\end{aligned}
\tag{6.2.19}
$$

The power spectral density can be normalized by $E_b$ and expressed as

$$
S_{SS}(f)/E_b = \frac{1}{2}\sin c^2\,(f-f_c)\,T_b + \frac{1}{2}\sin c^2\,(f+f_c)\,T_b.
$$

Therefore, the power spectral density of the binary phase-shift keying signal can be obtained as a shifted version of the baseband power spectral density, as shown in Figure 6.2.7.

The power spectral density is shown on a logarithmic scale in Figure 6.2.8, and is calculated in the following way. The normalized one-sided power spectral density, obtained by overlapping the amplitudes at negative frequencies with the positive frequencies in eqn (6.2.19), can be expressed as a function of the modified frequency $f_n = (f - f_c)T_b$ as

$$
S_{SS}\,(f_n) = S_{SS}(f)/E_b = \sin c^2 f_n = \left(\frac{\sin \pi f_n}{\pi f_n}\right)^2.
\tag{6.2.20}
$$

In practice, we need to see the level of signal spectrum in logarithmic units, and the bandwidth of the signal around the carrier frequency. Therefore, we may express the power spectral density in decibels as

$$
S_{SS}\,(f_n) = 10\log_{10}\left(\sin c^2 f_n\right) = 10\log_{10}\left(\frac{\sin \pi f_n}{\pi f_n}\right)^2 \quad dB,
\tag{6.2.21}
$$

and present it in graphical form as in Figure 6.2.8. For the sake of comparison, the power spectral density of signals modulated via quaternary phase-shift keying and frequency-shift

**Fig. 6.2.7** *Power spectral density of a binary phase-shift keying signal.*

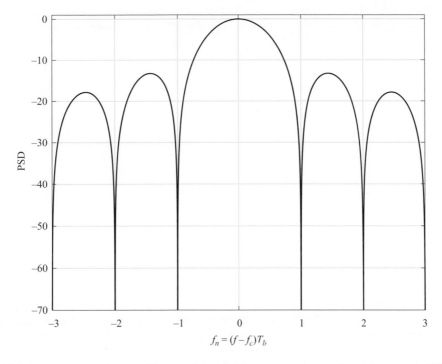

**Fig. 6.2.8** *Power spectral density of binary phase-shift keying signals, on a logarithmic scale; PSD, power spectral density.*

keying, alongside that in this graph, will be calculated in the next section and presented later on (see Section 6.4, Figure 6.4.10).

**Example**

Find the expression for the power spectral density of the modulating baseband pulse defined as an aperiodic deterministic signal inside the interval $t = 0$ to $t = T_b$, as shown in Figure 6.2.2.

**Solution.** The rectangular pulse of amplitude $E_b$ and duration $T_b$ has the Fourier transform

$$M(f) = \int_{-\infty}^{\infty} m(t)e^{-j\omega t}\,dt = \int_{0}^{T_b} \sqrt{E_b} \cdot e^{-j\omega t}\,dt = \frac{2\sqrt{E_b}}{\omega}\left(\frac{e^{-j\omega T_b/2} - e^{+j\omega T_b/2}}{-2j}\right)e^{-j\omega T_b/2}$$

$$= 2\sqrt{E_b}\frac{\sin\omega T_b/2}{\omega}e^{-j\omega T_b/2} = \sqrt{E_b}T_b\frac{\sin\pi f T_b}{\pi f T_b}e^{-j\omega T_b/2} = \sqrt{E_b}T_b e^{-j\pi f T_b}\,\mathrm{sinc}\,(f T_b)$$

*The energy spectral density* is the squared magnitude of the Fourier transform, that is,

$$E_m(f) = |M(f)|^2 = E_b T_b^2 \operatorname{sinc}^2 (f T_b).$$   (6.2.22)

Then, the power spectral density can be obtained if the energy spectral density is divided by the symbol duration $T_b$, as follows:

$$S_{mm}(f) = \frac{E(f)}{T_b} = \frac{E_b T_b^2 \operatorname{sinc}^2 (f T_b)}{T_b} = E_b T_b \operatorname{sinc}^2 (f T_b) = S_{MM}(f),$$   (6.2.23)

which is numerically equal to the power spectral density $SMM(f)$ calculated for the random binary modulating signal, which is expressed by eqn (6.2.5) and presented in Figure 6.2.4.

## Example

The power spectral density of a modulated signal can be found as the Fourier transform of its autocorrelation function. To find the transform in this case, we can apply the modulation theorem, which gives us the power spectral density of the bandpass binary phase-shift keying signal as a shifted power spectral density function of the baseband signal, due to the fact that the autocorrelation of the modulating process $M(t)$ is multiplied by the cosine function. Prove this statement. Find the power spectral density of the carrier and calculate its power.

**Solution.** The power spectral density of the modulated signal is the Fourier transform of its autocorrelation function, that is,

$$S_{SS}(f) = FT \left\{ R_{MM}(\tau) \cdot \frac{1}{T_b} \cos 2\pi f_c \tau \right\} = \frac{1}{T_b} FT \left\{ R_{MM}(\tau) \cdot \frac{e^{j2\pi f_c \tau} - e^{-j2\pi f_c \tau}}{2} \right\}.$$

$$= \frac{1}{2T_b} [S_{MM}(f + f_c) + S_{MM}(f - f_c)]$$

Because the power spectral density of the modulating signal is

$$S_{MM}(f) = E_b T_b \operatorname{sinc}^2 (f T_b),$$

the power spectral density of the modulated signal can be calculated as

$$S_{SS}(f) = \frac{1}{2T_b} [S_{MM}(f + f_c) + S_{MM}(f - f_c)]$$

$$= \frac{1}{2T_b} \left[ E_b T_b \operatorname{sinc}^2 (f + f_c) T_b + E_b T_b \operatorname{sinc}^2 (f - f_c) T_b \right]$$   (6.2.24)

$$= \frac{E_b}{2} \left[ \operatorname{sinc}^2 (f + f_c) T_b + \operatorname{sinc}^2 (f - f_c) T_b \right]$$

or

$$S_{SS}(f) = \frac{1}{4}\left[2E_b \operatorname{sinc}^2((f-f_c)T_b) + 2E_b \operatorname{sinc}^2((f+f_c)T_b)\right], \tag{6.2.25}$$

as derived in eqn (6.2.19). In the literature, the power spectral density of a carrier with amplitude $A_b$ is

$$S_{sc}(f) = \int_{\tau=-\infty}^{\infty} R_{sc}(\tau)e^{-2\pi f \tau}d\tau = \int_{-\infty}^{\infty} \frac{A_b^2}{2}\cos 2\pi f_c \tau\, e^{-2\pi f \tau}d\tau$$

$$= \frac{A_b^2}{2}\int_{-\infty}^{\infty}\left[\frac{1}{2}e^{j2\pi f_c \tau} + \frac{1}{2}e^{-j2\pi f_c \tau}\right]e^{-2\pi f \tau}d\tau, \tag{6.2.26}$$

$$= \frac{A_b^2}{2}\left[\frac{1}{2}\delta(f-f_c) + \frac{1}{2}\delta(f+f_c)\right] = \frac{A_b^2}{4}[\delta(f-f_c) + \delta(f+f_c)]$$

which can be used to find its power expressed in terms of the carrier amplitude, as

$$P_{sc} = \int_{t=-\infty}^{\infty} S_{sc}(f)df = \int_{t=-\infty}^{\infty} \frac{A_b^2}{2}\left[\frac{1}{2}\delta(f-f_c) + \frac{1}{2}\delta(f+f_c)\right]df = \frac{A_b^2}{2}. \tag{6.2.27}$$

In this book, for the sake of consistency, we use the term 'carrier' for the orthonormal basis function $\phi_j(t)$. Namely, by defining the carrier in that way, we can use the theoretical presentation and generic mathematical model for transmitters and receivers that was presented in Chapter 3 to deduce the transceiver structures for practical modulation schemes as the special cases of that generic model, as we did for the binary phase-shift keying modulation presented in Figure 6.2.1.

## 6.2.3  Receiver Operation

### 6.2.3.1  Correlation Demodulator Operation

A binary phase-shift keying receiver is composed of the correlation demodulator and the optimum detector, as presented in Figure 6.2.1. We will explain firstly the operation of the correlation demodulator. The received random signal, at the input of the demodulator, can be expressed as

$$r(t) = s_i(t) + w(t). \tag{6.2.28}$$

for $0 \le t \le T_b$, and $i = 1, 2$. The random noise $w(t)$ is a realization of the white Gaussian noise process $W(t)$ of zero mean and power spectral density $N_0/2$, and $r(t)$ is a realization of the received stochastic process $R(t)$. The correlator is composed of a multiplier, which multiplies the received noisy signal by a local coherent carrier, and an integrator, which integrates the product obtained in the bit interval $T_b$. The result is a random value expressed as

$$
\begin{aligned}
x &= \int_0^{T_b} r(t)\phi_1(t)dt = \int_0^{T_b} [s_i(t)+w(t)]\phi_1(t)dt \\
&= \int_0^{T_b} s_i(t)\phi_1(t)dt + \int_0^{T_b} w(t)\phi_1(t)dt = s_{i1} + w
\end{aligned}
\tag{6.2.29}
$$

where $x$ is considered to be a realization of a random variable $X(T_b) = X$ for the first transmitted bit, which has two parts. The first part, denoted by $s_{i1}$ and called the signal part, is a quantity contributed by the transmitted signals $s_i(t)$, which is calculated as the mean value of $X$. The second part, denoted by $w$ and called the noise part, is the sample value of the Gaussian random variable $W$ that arises because of the channel noise $w(t)$. It was proven in Chapter 5 that the density function of the discrete time stochastic process $X(kT_b)$, where $k$ has integer values, at the output of the correlator is Gaussian. We are interested in the statistical description of the random variable $X = X(T_b)$ at the output of the correlator at the time instant $T_b$, that is, at the end of the first received bit interval defined by $k = 1$. We will find the mean value and the variance of the random variable $X$ in the following section. Furthermore, because the received bits are statistically independent, the analysis of the first bit received is sufficient to statistically characterize any other bit received.

**Example**

Present the random value $x$ on the constellation diagram, assuming that the bit 1 was sent and the noise part $w$ is positively added to the signal part.

**Solution.** The constellation diagram is presented in Figure 6.2.9, with the noise part $w$ (dashed line) that is added to the signal part. The random value $x$ is positive, like the signal part.

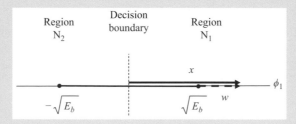

**Fig. 6.2.9**  *Constellation diagram.*

**Example**

Present the random value $x$ on the constellation diagram, assuming that the bit 1 was sent and the noise part $w$ has a negative value that is greater than the signal part.

**Solution.** The constellation diagram is presented in Figure 6.2.10, with the noise part $w$ (dashed line) that is added to the signal part. The random value $x$ is negative with respect to the signal part.

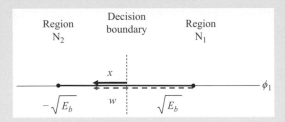

**Fig. 6.2.10** *Constellation diagram.*

### 6.2.3.2 *Operation of the Optimum Detector, and Structure of the Receiver*

As we said before, for this simple and important communication system based on binary phase-shift keying, we do not need an optimum detector. Let us briefly confirm this fact.

In a binary phase-shift keying system, two signals can be generated at the transmitter side, and the output of the correlation demodulator is defined by eqn (6.2.29) which is an input of the optimum detector expressed in vector form as $\mathbf{x} = [x_1] = [s_{11} + w]$ for the first binary symbol transmitted, or $\mathbf{x} = [x_2] = [s_{21} + w]$ for the second binary symbol transmitted. The transmitter signals can be expressed in vector form as $\mathbf{s}_1 = [s_{11}]$ and $\mathbf{s}_2 = [s_{21}]$. Each of the two signals, representing a message bit, is transmitted and processed in the same way inside the communication system. Furthermore, the noise affecting a bit in a stream of message bits is statistically independent of the noise affecting all other bits in the stream. Therefore, it is sufficient to analyse the transmission of one bit and then generalize the results obtained to all of the bits transmitted.

Let us assume that the first signal $\mathbf{s}_1$, which carries the message bit, is transmitted and the output of the correlator is a random value $x$. If we perform the mathematical operations defined inside the optimum detector presented in Figure 6.2.1, the outputs of detector can be calculated as

$$Z_1 = \mathbf{x}^T \mathbf{s}_1 - \frac{1}{2} E_1 = [x]\,[s_{11}] - \frac{1}{2} E_1 = \left[\sqrt{E_1} + w\right]\left[\sqrt{E_1}\right] - \frac{1}{2} E_1 = \frac{1}{2} E_1 + w\sqrt{E_1},$$

and

$$Z_2 = \mathbf{x}^T \mathbf{s}_2 - \frac{1}{2}E_2 = [x][s_{21}] - \frac{1}{2}E_2 \underset{E_2=E_1}{=} \left[\sqrt{E_1} + w\right]\left[-\sqrt{E_1}\right] - \frac{1}{2}E_1 = -\frac{3}{2}E_1 - w\sqrt{E_1}.$$

According to the optimum detector operation explained in Chapter 5, and the structure presented in Figure 6.2.1, these two numerical values need to be compared and then the largest one selected. The comparison can be performed by finding the difference of the output random values of the detector expressed as

$$Z_1 - Z_2 = \frac{1}{2}E_1 + w\sqrt{E_1} + \frac{3}{2}E_1 + w\sqrt{E_1} - \frac{1}{2}E_1 = 2E_1 + 2w\sqrt{E_1}.$$

The numerical decision value obtained, $y$, can be defined as

$$y = \frac{Z_1 - Z_2}{2\sqrt{E_1}} = \sqrt{E_1} + w,$$

because the term $2\sqrt{E_1}$, being a constant, does not influence our decision. Therefore, the decision rule can be defined as follows:

1. If $y = \sqrt{E_1} + w > 0$, meaning that $Z_1 > Z_2$, decide in favour of the symbol $m_k = m_1$.
2. If $y = \sqrt{E_1} + w < 0$, meaning that $Z_1 < Z_2$, decide in favour of the symbol $m_k = m_2$. Because we assumed that the message bit $m_1 = 1$ was sent, a wrong decision will be made here in favour of $m_2 = 0$ rather than in favour of $m_1 = 1$.
3. If $y = \sqrt{E_1} + w = 0$, meaning that $Z_1 = Z_2$, flip a coin to decide in favour of $Z_1$ or $Z_2$.

Likewise, let us assume that the second signal $\mathbf{s}_2$ is transmitted and the output of the correlator is a random value $x$. The following two random numerical values can be calculated at the output of the optimum detector:

$$Z_1 = \mathbf{x}^T \mathbf{s}_1 - \frac{1}{2}E_1 = [x][s_{11}] - \frac{1}{2}E_1 = [s_{21} + w][s_{11}] - \frac{1}{2}E_1 = -\frac{3}{2}E_1 + w\sqrt{E_1},$$

and, assuming the same power of both signals, we may have

$$Z_2 = \mathbf{x}^T \mathbf{s}_2 - \frac{1}{2}E_2 \underset{E_1=E_2}{=} [x][s_{21}] - \frac{1}{2}E_1 = [s_{21} + w][s_{21}] - \frac{1}{2}E_1$$

$$= E_1 - w\sqrt{E_1} - \frac{1}{2}E_1 = \frac{1}{2}E_1 - w\sqrt{E_1}$$

A comparison can be performed by finding the difference of the output random variables

$$y = Z_1 - Z_2 = -\frac{3}{2}E_1 + w\sqrt{E_1} - \left(\frac{1}{2}E_1 - w\sqrt{E_1}\right) = -2E_1 + 2w\sqrt{E_1}.$$

The numerical decision value can be defined as

$$y = \frac{Z_1 - Z_2}{2\sqrt{E_1}} = -\sqrt{E_1} + w.$$

Then we can decide as follows:

1. If $y = -\sqrt{E_1} + w > 0$, meaning that $Z_1 > Z_2$, decide in favour of the symbol $m_k = m_1$. Because we assumed that the message bit $m_2 = 0$ was sent, we decided in favour of $m_1 = 1$, instead in favour of $m_2 = 0$, which is an incorrect decision.

2. If $y = -\sqrt{E_1} + w < 0$, meaning that $Z_1 < Z_2$, decide in favour of the symbol $m_k = m_2$, which is the right decision.

3. If $y = -\sqrt{E_1} + w = 0$, meaning that $Z_1 = Z_2$, flip a coin and decide in favour of $Z_1$ or $Z_2$. The decision value $y$ is considered to be a realization of random variable defined as

$$Y = -\sqrt{E_1} + W,$$

which is the sum of the signal part $\pm\sqrt{E_1}$ and the noise part $W$ at the optimum detector output. The signal part depends on the message bit transmitted and is equal to the signal coefficient. Therefore, the decision variable $Y$ is identical to the random variable at the output of the correlation demodulator $X$ defined by eqn (6.2.29), that is,

$$Y = \pm\sqrt{E_1} + W = s_{i1} + W = X.$$

Because the random variables $X$ and $Y$ are identical, the decision can be made using the value obtained at the output of correlation demodulator and there is no need to use the optimum detector in the defined form, which will simplify the receiver structure. It is sufficient to take the output of the demodulator $x$ and apply the above-mentioned decision rules.

If we bear in mind the fact that the decision can be made using the output of the correlation demodulator, the block scheme of the binary phase-shift keying communication system can be simplified, as shown in Figure 6.2.11.

For the input binary values $m_i = 0$ or 1, a polar non-return-to-zero modulating signal $m(t)$ is generated that modulates a cosine carrier represented by a basis function $\phi_1(t)$. The modulating signal is a rectangular pulse with one of two possible amplitudes, $\pm\sqrt{E_b}$, depending on the value of a bit $m_i$ generated at the source output. The modulated signal is transmitted through the additive white Gaussian noise channel. The received signal $r(t)$ is processed inside the correlation demodulator. The output of the correlator $x$ is compared to

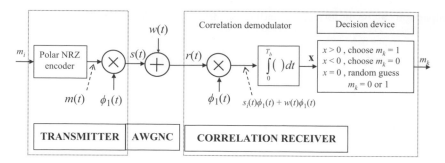

**Fig. 6.2.11** *Simplified scheme of the binary phase-shift keying communication system; AWGNC, additive white Gaussian noise channel; NRZ, non-return to zero.*

the threshold value, which is zero for a symmetric system, and the detected bit $m_k$ is generated at the output of the decision device.

**Example**

We can formulate a precise statistical processing of the correlator output. This output is a random variable $X$ that can be understood as the sum of the binary random variable $S_{i1}$ (representing the random generation of the message bits at the output of the source) and the noise variable $W$. Calculate the density function of $X$ and then the conditional densities of $X$, given $S_{i1} = s_{11}$ or $S_{i1} = s_{21}$. Suppose the probabilities of two possible signal values of the binary variable $S_{i1}$ are equal, that is, $P(s_{21}) = P(s_{11})$. Because a signal value at the correlator output corresponds to a message bit sent, we may have the following probabilities:

$$P(s_{11}) = P\left(+\sqrt{E_b}\right) = P(m_1) = P(1) = 1/2, \tag{6.2.30}$$

and

$$P(s_{21}) = P\left(-\sqrt{E_b}\right) = P(m_2) = P(0) = 1/2. \tag{6.2.31}$$

Find the density function of $X$ and the conditional densities of $X$ for the given messages $m_1 = 1$ and $m_2 = 0$ and present them in graphical form.

**Solution.** The density function of $S_{i1}$, which randomly takes the values $s_{11}$ and $s_{21}$, can be expressed in terms of Dirac delta functions as

$$f_{S_{i1}}(s_{i1}) = P(s_{11})\delta(s_{i1} - s_{11}) + P(s_{21})\delta(s_{i1} - s_{21}). \tag{6.2.32}$$

For the random variable $S_{i1}$, we can prove that the mean value is zero and the variance is $E_b$, as we did for the binary random variable $M$ in Section 6.2.2.1.

This function is equivalent to the function defined by eqn (6.2.3), as it has the same sample space. The noise variable $W$ is Gaussian, as presented in Chapter 5, and is expressed as

$$f_W(w) = \frac{1}{\sqrt{2\pi\sigma_W^2}} e^{-(w)^2/2\sigma_W^2}. \tag{6.2.33}$$

The variable $X$ is the sum of the two independent random variables $S_{i1}$ and $W$:

$$X = S_{i1} + W. \tag{6.2.34}$$

Therefore, its density function is the convolution of related densities calculated as

$$
\begin{aligned}
f_X(x) &= \int_{-\infty}^{\infty} f_{S_{i1}}(s_{i1}) f_W(x - s_{i1}) \, ds_{i1} \\
&= \int_{-\infty}^{\infty} [P(s_{11}) \delta(s_{i1} - s_{11}) + P(s_{21}) \delta(s_{i1} - s_{21})] \frac{1}{\sqrt{2\pi\sigma_W^2}} e^{-(x - s_{i1})^2/2\sigma_W^2} \, ds_{i1} \\
&= P(s_{11}) \frac{1}{\sqrt{2\pi\sigma_W^2}} e^{-(x - s_{11})^2/2\sigma_W^2} + P(s_{21}) \frac{1}{\sqrt{2\pi\sigma_W^2}} e^{-(x - s_{21})^2/2\sigma_W^2} \\
&= 0.5 \cdot f_X(x|S_{i1} = s_{11}) + 0.5 \cdot f_X(x|S_{i1} = s_{21})
\end{aligned}
\tag{6.2.35}
$$

The conditional density functions, expressed in terms of signals at the output of the correlator, or in terms of messages sent, are

$$
\begin{aligned}
f_X(x|S_{i1} = s_{11}) &= f_X\left(x|S_{i1} = \sqrt{E_b}\right) = \frac{1}{\sqrt{2\pi\sigma_W^2}} e^{-(x - s_{11})^2/2\sigma_W^2} \\
&= \frac{1}{\sqrt{2\pi\sigma_W^2}} e^{-\left(x - \sqrt{E_b}\right)^2/2\sigma_W^2} = f_X(x|m_i = 1)
\end{aligned}
\tag{6.2.36}
$$

and

$$
\begin{aligned}
f_X(x|S_{i1} = s_{21}) &= f_X\left(x|S_{i1} = -\sqrt{E_b}\right) = \frac{1}{\sqrt{2\pi\sigma_W^2}} e^{-(x - s_{21})^2/2\sigma_W^2} \\
&= \frac{1}{\sqrt{2\pi\sigma_W^2}} e^{-\left(x + \sqrt{E_b}\right)^2/2\sigma_W^2} = f_X(x|m_i = 0)
\end{aligned}
\tag{6.2.37}
$$

The random variable $X$ has the mean and the variance expressed as

$$\eta_X = E\{X\} = E\{S_{i1} + W\} = 0,$$

and

$$
\sigma_X^2 = E\{X^2\} = E\left\{(S_{i1} + W)^2\right\} = E\left\{(S_{i1})^2\right\} + 0 + E\left\{(W)^2\right\}
$$
$$
= E_b + \sigma_W^2 = E_b + \frac{N_0}{2}
$$
(6.2.38)

respectively. In a similar way, we can find the mean and the variance of the random variable defined by the conditional probabilities in eqns (6.2.36) and (6.2.37). The mean values will be $\sqrt{E_b}$ and $-\sqrt{E_b}$, respectively, and the variance for both will be $N_0/2$.

### 6.2.3.3   Bit Error Probability Calculation

The quality of the transmission of message bits depends on the signal-to-noise ratio at the correlator output. This ratio is directly related to the bit error probability for the given binary phase-shift keying system. Therefore, the aim of communication system analysis is to find the bit error probability as a function of $E_b/N_0$ in graphical form, that is, the system curve. In the case of binary phase-shift keying, we do not need to perform signal detection using the optimum detector presented in Figure 6.2.1. As we have already shown, it is sufficient to have the correlation demodulator output $x$ to make a decision. Therefore, we only added the optimum detector to the system in Figure 6.2.1 for the sake of explanation and confirmation that the general scheme for a correlation receiver, as presented in Chapter 5, Figure 5.8.1, can be simplified and applied to form a binary phase-shift keying system.

Following the derivatives related to the density function of random variable $X$, eqns (6.2.36) and (6.2.37), we may express the conditional probability density function of the correlator random variable $X$, given that $m_i$ was transmitted, as

$$
f_X(x|m_i) = f_X(x|s_{i1}) = \frac{1}{\sqrt{\pi N_0}} e^{-(x - s_{i1})^2/N_0}.
$$
(6.2.39)

In the case when the message bit $m_i = m_1 = 1$ is generated at the source and transmitted through the channel, the realization of the random variable $X$ is $x$, and its density function is Gaussian, defined by the mean value $s_{11} = \sqrt{E_b}$ and the variance $N_0/2$ and expressed as

$$
f_X(x|1) = \frac{1}{\sqrt{\pi N_0}} e^{-(x - \sqrt{E_b})^2/N_0}.
$$
(6.2.40)

In the case when $m_i = m_2 = 0$ is generated and transmitted, the realization of the random variable $X$ is $x$, and its conditional density function is Gaussian, defined by the mean value $s_{21} = -\sqrt{E_b}$ and the variance $N_0/2$ and expressed as

$$
f_X(x|0) = \frac{1}{\sqrt{\pi N_0}} e^{-(x + \sqrt{E_b})^2/N_0}.
$$
(6.2.41)

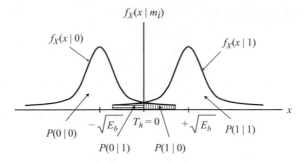

**Fig. 6.2.12** *Probability density functions of the conditional random variable X.*

These conditional densities are shown in Figure 6.2.12.

In order to explain the way we arrive at the decision we make at the correlation demodulator output, we will use both the constellation diagram in Figure 6.2.6 and the defined conditional density functions presented in Figure 6.2.12. Assuming that the threshold value is $T_h = 0$, defined for a symmetric system, the decision rules are as follows:

1. If the output of correlator $x$ falls in the region

$$N_1 : 0 < x < \infty, \tag{6.2.42}$$

then two probabilities can be identified. *Firstly*, the conditional probability that the message symbol $m_1 = 1$ is received, given that the message symbol $m_1 = 1$ (corresponding to the modulated signal $s_1(t)$) was transmitted, is expressed as

$$P(1|1) = \int_0^\infty f_X(x|1)\, dx = \int_0^\infty \frac{1}{\sqrt{\pi N_0}} e^{-(x-\sqrt{E_b})^2/N_0}\, dx. \tag{6.2.43}$$

This probability is large and corresponds to the area below the curve of density function $f_X(x|1)$ for $x > 0$. Thus, the right decision will be made in favour of signal $m_1 = 1$ with a high probability of $P(1|1)$. *Secondly*, the conditional probability that the message symbol $m_1 = 1$ is received, given that the message symbol $m_2 = 0$ (corresponding to the modulated signal $s_2(t)$) was transmitted, is expressed as

$$P(1|0) = \int_0^\infty f_X(x|0)\, dx = \int_0^\infty \frac{1}{\sqrt{\pi N_0}} e^{-(x+\sqrt{E_b})^2/N_0}\, dx. \tag{6.2.44}$$

This probability is small and corresponds to the area below the curve of density function $f_X(x|0)$ for $x > 0$. Thus, an erroneous decision will be made in favour of signal $m_1 = 1$ with a small probability $P(1|0)$.

2. If the output of correlator $x$ falls in the region

$$N_2 : -\infty < x < 0, \tag{6.2.45}$$

then two probabilities can be identified. *Firstly*, the conditional probability that the message symbol $m_2 = 0$ is received, given that the message symbol $m_2 = 0$ (corresponding to the modulated signal $s_2(t)$) was transmitted, is expressed as

$$P(0|0) = \int_0^\infty f_X(x|0)\,dx = \int_0^\infty \frac{1}{\sqrt{\pi N_0}} e^{-(x+\sqrt{E_b})^2/N_0}\,dx. \tag{6.2.46}$$

This probability is large and corresponds to the area below the curve of density function $f_X(x|0)$ for $x < 0$. Thus, the right decision will be made in favour of signal $m_2 = 0$ with a high probability $P(0|0)$. *Secondly*, the conditional probability that the message symbol $m_2 = 0$ is received, given that the message symbol $m_1 = 1$ (corresponding to the modulated signal $s_1(t)$ was transmitted, is expressed as

$$P(0|1) = \int_{-\infty}^0 f_X(x|1)\,dx = \int_{-\infty}^0 \frac{1}{\sqrt{\pi N_0}} e^{-(x-\sqrt{E_b})^2/N_0}\,dx. \tag{6.2.47}$$

This probability is small and corresponds to the area below the curve of density function $f_X(x|1)$ for $x < 0$. Thus, an erroneous decision will be made in favour of signal $m_2 = 0$ with a small probability $P(0|1)$.

The probabilities of erroneous decision (both of the second cases above) are equal due to the equality of the signal energies with respect to the threshold value $T_h = 0$. The conditional probability of the receiver deciding in favour of symbol $m_1 = 1$, given that symbol $m_2 = 0$ was transmitted, is equal to $P(1|0) = P(0|1)$, that is,

$$P(1|0) = \int_0^\infty f_X(x|0)\,dx = \int_0^\infty \frac{1}{\sqrt{\pi N_0}} e^{-(x+\sqrt{E_b})^2/N_0}\,dx. \tag{6.2.48}$$

Defining a new variable $z$ as

$$z = \left(x + \sqrt{E_b}\right)/\sqrt{N_0}, \tag{6.2.49}$$

and substituting the variable of integration $x$ to $z$, we obtain

$$P(1|0) = \int_0^\infty f_X(x|0)\,dx = \frac{1}{\sqrt{\pi}} \int_{\sqrt{E_b/N_0}}^\infty e^{-z^2}\,dz = \frac{1}{2}\,\mathrm{erfc}\left(\sqrt{\frac{E_b}{N_0}}\right), \tag{6.2.50}$$

where erfc($x$) is the complementary error function defined here as (see Figure 6.2.13)

$$erfc(x) = \frac{2}{\sqrt{\pi}} \int_x^\infty e^{-z^2} dz. \tag{6.2.51}$$

Taking into account the symmetry of distributions in Figure 6.2.12, it can also be proved that

$$P(0|1) = P(1|0) = \frac{1}{2} erfc\left(\sqrt{\frac{E_b}{N_0}}\right). \tag{6.2.52}$$

Assuming that the channel is symmetric, that is, that $P(0|1) = P(1|0)$, the probability of symbol error expression, generally defined as

$$P_e = \sum_{i=1}^{M} P(m_i) P(m_k \neq m_i | m_i), \tag{6.2.53}$$

can be applied here for $M = 2$ to calculate the bit error probability of the binary phase-shift keying system as

$$p = \sum_{i=1}^{2} P(m_i) P(m_k \neq m_i | m_i) = P(0)P(1|0) + P(1)P(0|1)$$

$$= P(1|0) = \frac{1}{2} erfc\left(\sqrt{\frac{E_b}{N_0}}\right) \tag{6.2.54}$$

where $m_k$ is the estimate of the transmitted bit value. This function is plotted in Figure 6.2.13, using linear scales for both coordinates. However, we cannot effectively use this scaling, due to the small expected values of the bit error probability $p$. Instead, we will present the bit error probability using logarithmic scales. In this case, the $x$-axis will present the signal-to-noise

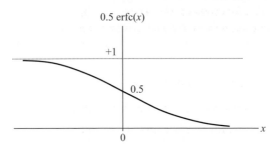

**Fig. 6.2.13** *The complementary error function.*

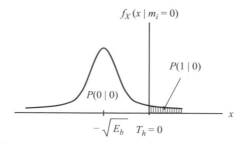

**Fig. 6.2.14** *A conditional probability density function.*

ratio in decibels, and the $y$-axis will present the bit error probability $p$ as the negative power of base 10, as shown in Figure 6.2.14.

**Example**

The bit error probability $p$, for a symmetric additive white Gaussian noise channel and for the same probability of generating binary values at the output of the source, can be calculated in a simpler way. In this case, the calculated $p$ will be the same whether the assumed bit transmitted is 1 or 0. Suppose that the bit transmitted is 1. Then, the decision variable is expressed as

$$X = +\sqrt{E_b} + W,$$

having Gaussian density functions expressed by eqn (6.2.40). Find the bit error probability in the system.

**Solution.** Assuming that bit 0 was sent, the graph of the probability density function of the decision variable $X = Y$ is presented in Figure 6.2.15.
A bit error will occur if $x$ values are greater than zero, which corresponds to the probability

$$p = \int_0^\infty f_X(x|0)\,dx = \int_0^\infty \frac{1}{\sqrt{\pi N_0}} e^{-(x+\sqrt{E_b})^2/N_0}\,dx.$$

Defining a new variable $z$ as $z = (x + \sqrt{E_b})/\sqrt{N_0}$, we may calculate $dz/dx = 1/\sqrt{N_0}$, and find the limits of integration as $x = 0$, $z = \sqrt{E_b N_0}$, $x = \infty$, $z = \infty$. Substituting the variable of integration $x$ to $z$, we obtain

$$p = \frac{1}{\sqrt{\pi N_0}} \int_{\sqrt{E_b/N_0}}^\infty e^{-z^2} \sqrt{N_0}\,dz = \frac{1}{\sqrt{\pi}} \int_{\sqrt{E_b/N_0}}^\infty e^{-z^2}\,dz = \frac{1}{2}\,\mathrm{erfc}\left(\sqrt{\frac{E_b}{N_0}}\right).$$

This expression for the bit error probability is equivalent to the previously derived expression for a general case and applied for this special symmetric case.

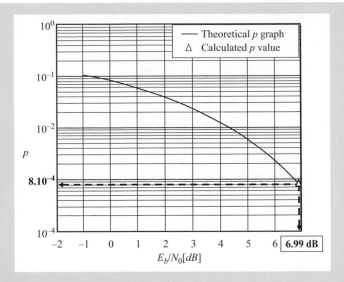

**Fig. 6.2.15** *Presentation of the numerical value for $p$ as a function of the related signal-to-noise ratio $E_b/N_0$.*

## Example

The bit error probability $p$, for a symmetric additive white Gaussian noise channel and for the same probability of generating binary values at the output of the sourse, can be calculated in a simpler way. In this case, the calculated $p$ will be the same whether the assumed transmitted bit is 1 or 0. Suppose the transmitted bit is 0. Then, the decision variable is expressd as

$$X = +\sqrt{E_b} + W.$$

Prove that the probability of error can be calculated as

$$p = \frac{1}{2}\operatorname{erfc}\left(\sqrt{\frac{\eta_X^2}{2\sigma_X^2}}\right), \tag{6.2.55}$$

for the defined values of the mean and the variance of the random variable $X$.

**Solution.** The mean and variance can be calculated as

$$\eta_X = E\left\{\sqrt{E_b} + W\right\} = \sqrt{E_b}$$

$$\sigma_X^2 = E\left\{\left(\sqrt{E_b} + W\right)^2\right\} - \eta_X^2 = \sigma_W^2.$$

The probability of error is

$$p = \frac{1}{2}\,\mathrm{erfc}\left(\sqrt{\frac{\eta_X^2}{2\sigma_X^2}}\right) = \frac{1}{2}\,\mathrm{erfc}\left(\sqrt{\frac{E_b}{2N_0/2}}\right) = \frac{1}{2}\,\mathrm{erfc}\left(\sqrt{\frac{E_b}{N_0}}\right).$$

## Example

A binary phase-shift keying system transmits a binary message signal with a bit rate of 1 Mbit/s. The modulated signal is transmitted through an additive white Gaussian noise channel. The amplitude of the signal is $A_b = 10$ mV. The power spectral density of the noise is $N_0 = 10^{-11}$ W/Hz. Calculate the bit error probability of the system. What kind of assumption did you make concerning the power and energy of the signal? Present the calculated value on a graph, defining the dependence of $p$ on $E_b/N_0$ in decibels.

**Solution.** Because the binary phase-shift keying signal can be expressed as

$$s(t) = s_{i1} \cdot \cos\left(2\pi f_c t\right) = \pm A_b \cos\left(2\pi f_c t\right),$$

the energy of a symbol is

$$E_b = \frac{A_b^2}{2}T_b,$$

and the period of the symbol stream is $T_b = 1/R_b$, the energy can be calculated according to

$$E_b = \frac{A_b^2}{2R_b} = \frac{10 \cdot 10^{-3}}{2 \cdot 10^6} = 5 \cdot 10^{-11}\,J.$$

The probability of error is

$$p = \frac{1}{2}erfc\left(\sqrt{\frac{E_b}{N_0}}\right) = \frac{1}{2}\,\mathrm{erfc}\left(\sqrt{\frac{5 \cdot 10^{-11}}{10^{-11}}}\right) = \frac{1}{2}\,\mathrm{erfc}\left(\sqrt{5}\right) = 0.5\,\mathrm{erfc}\,2.23 = 8 \cdot 10^{-4}.$$

We assume that the power and energy per symbol are normalized relative to a 1 Ohm load resistor. To present the bit error probability value $p$ on the theoretical graph, we need to calculate the signal-to-noise ratio in decibels as

$$\left.\frac{E_b}{N_0}\right|_{\text{in dB}} = 10\,\log_{10}(5) = 10 \cdot 0.69897 = 6.9897$$

for $p = 8 \cdot 10^{-4}$, as shown in Figure 6.2.15.

# 6.3 Quadriphase-Shift Keying

## 6.3.1 Operation of a Quadrature Phase-Shift Keying System

Based on the general scheme of a correlator-based communication system, shown in Chapter 5 in Figure 5.8.1, a quadrature phase-shift-keying system, or quadriphase-shift-keying system, can be obtained as a special case, as presented in Figure 6.3.1.

Suppose the message source can generate two binary values (or bits) that belong to an alphabet of $M = 2$ elements and are denoted by $m_1 = 1$ and $m_2 = 0$. A symbol generated at the output of the source is a pair of bits called a dibit. The dibit is converted into two parallel message bits by a serial-to-parallel converter (a de-multiplexer). Both message bits are applied at the same time to the inputs of two look-up tables. The look-up tables generate the polar

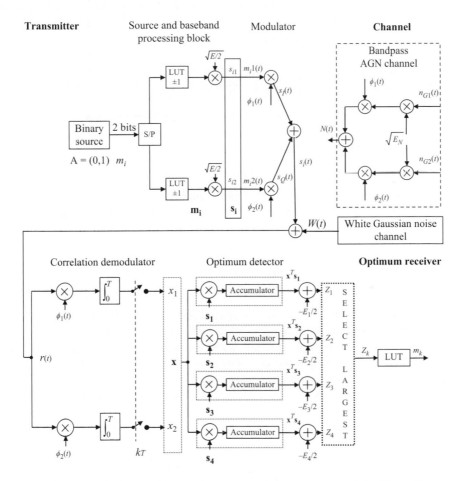

**Fig. 6.3.1** *A quadriphase-shift keying continuous communication system; AGN, additive Gaussian noise; LUT, look-up table; S/P, a serial-to-parallel converter.*

binary values $(-1, +1)$, which form message bits that are represented as the vector $\mathbf{m_i}$, which consists of two polar bits. These values are multiplied by the square root of the bit energy $\sqrt{E/2}$, and the signal coefficients obtained are used to generate two modulating signals, $m_{ij}(t)$: $m_{i1}(t)$ in the upper branch (the $I$ or in-phase branch) and $m_{i2}(t)$ in the lower branch (the $Q$ or quadrature branch) of the modulator. These signals are applied to the input of two modulator multipliers to modulate the carriers. This procedure can also be understood as magnitude scaling of the carrier by coefficients $s_{ij}$ that are represented by positive or negative values of $\sqrt{E/2}$, depending on the bit sign at the output of the look-up table. At the output of each multiplier, a binary phase-shift keying signal is obtained.

The signal received, $r(t)$, is coherently demodulated using two correlators inside the correlation demodulator. Because the principle of operation of both correlators is the same, it is sufficient to explain the operation of one of them. At the outputs of the correlators, random samples $x_1$ and $x_2$ are obtained at discrete-time instants $kT$; these samples can be presented to the optimum detector to correlate them with all possible symbols that could be generated at the transmitter side. The decision variables obtained are compared, the largest one is chosen, and an estimate of the transmitted dibit is made. However, in the quadrature phase-shift keying system, as in the binary phase-shift keying system, it is not necessary to use the detector. An equivalent decision can be made at the output of the correlation demodulator, as will be explained later.

## 6.3.2 Transmitter Operation

### 6.3.2.1 *Modulating Signals in Time and Frequency Domains*

**Signal-space or constellation diagram of the signal.** The number of possible message symbols is $I = A^J = 2^2 = 4$, which is defined by the matrix of possible message symbols at the output of the look-up tables as

$$\mathbf{m} = [\mathbf{m}_1 \quad \mathbf{m}_2 \quad \mathbf{m}_3 \quad \mathbf{m}_4] = \begin{bmatrix} 1 & -1 & -1 & 1 \\ -1 & -1 & 1 & 1 \end{bmatrix}. \tag{6.3.1}$$

Each dibit is assigned to one of four equally spaced phase values. These values correspond to a Gray-encoded set of dibits ordered as 10, 00, 01, and 11, assuring that only a single bit is changed inside the neighbouring dibits. These message symbols, presented in the form of dibits, need to be mapped into the coefficients of the modulating signals. In this case, all possible coefficients are obtained by multiplying the matrix of message symbols by the specified constant $\sqrt{E/2}$, where $E$ is the energy of any symbol. They are represented by four columns in the following matrix of modulating signals:

$$\begin{aligned} \mathbf{s} = [\mathbf{s}_1 \quad \mathbf{s}_2 \quad \mathbf{s}_3 \quad \mathbf{s}_4] &= \begin{bmatrix} s_{11} & s_{21} & s_{31} & s_{41} \\ s_{12} & s_{22} & s_{32} & s_{42} \end{bmatrix} \\ &= \mathbf{m}\sqrt{E/2} = \begin{bmatrix} \sqrt{E/2} & -\sqrt{E/2} & -\sqrt{E/2} & \sqrt{E/2} \\ -\sqrt{E/2} & -\sqrt{E/2} & \sqrt{E/2} & \sqrt{E/2} \end{bmatrix}. \end{aligned} \tag{6.3.2}$$

For each pair of coordinates, the signal space diagram for $I = 4$ is shown later on in this section (see Figure 6.3.7).

**Modulating signals in the time domain.** The defined coefficients in the matrix **s** are amplitudes of the modulating signals $m_{ij}(t)$, which are defined as rectangular pulses of duration $T$. The wave shape of the first modulating signal, for the fourth message symbol, $m_{41}(t)$, is a rectangular pulse of duration $T$ and amplitude equal to the $i$th coefficient, where $i = 4$, as shown in Figure 6.3.2.

**Statistical characteristics of modulating signals in the time domain.** Modulating signals $m_{ij}(t)$ in both branches are realizations of the random binary processes $M_{ij}(t)$, which randomly take the amplitude values specified by the vector defined in eqn (6.3.2) and generate two binary continuous-time modulating signals. We will analyse one of them, which is presented in Figure 6.3.3 and denoted by $m(t)$, which is a realization of the binary random process $M(t)$, because both have the same statistical characteristics. We will assume that the probabilities of both the positive and the negative binary values are equal to 0.5. Therefore, at each time instant $t = t_k$ we have a discrete random variable $M = M_{ij}(t_k)$ that takes the value of $\sqrt{E/2}$ or $-\sqrt{E/2}$, and is defined by its probability density function expressed in terms of a Dirac delta function as

$$f_M(m) = \frac{1}{2}\delta\left(m - \sqrt{E/2}\right) + \frac{1}{2}\delta\left(m + \sqrt{E/2}\right),\tag{6.3.3}$$

as shown in Figure 6.3.3. The notation here follows the notation used in Chapter 19 for the theory for continuous-time stochastic processes.

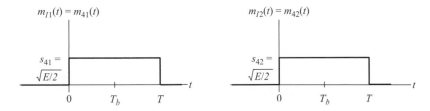

**Fig. 6.3.2** *Modulating rectangular pulses aligned in the time domain as they appear at the input of the modulator multipliers.*

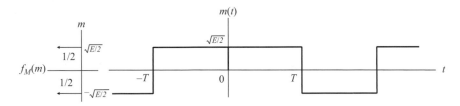

**Fig. 6.3.3** *A modulating polar binary signal presented in the time domain.*

Since the amplitude levels of $m(t)$ have the same magnitude values and occur with equal probabilities, the mean value of the process is zero, and the variance is $E/2$. The modulating signal is a realization of a wide-sense stationary process $M(t)$ with zero mean and constant variance, as has been proven for a binary phase-shift keying modulating signal. The autocorrelation function of the process in the upper ($I$) branch depends only on the time difference between two time instants, which is defined as the lag $\tau$, and is equivalent to the autocorrelation function in the lower ($Q$) branch. It is derived in Chapter 19, expression (19.2.10), repeated here in the form

$$R_{MI}(\tau) = R_{MQ}(\tau) = \begin{cases} \frac{E}{2}\left(1 - \frac{|\tau|}{T}\right) & |\tau| < T \\ 0 & |\tau| \ge T \end{cases}, \tag{6.3.4}$$

and represented by a triangular graph depicted in Figure 6.3.4 for this case.

**Presentation of a modulating signal in the frequency domain.** Due to the Wiener–Khintchine theorem, the power spectral density of the modulating signal $m(t)$ in the upper ($I$) branch can be calculated as the Fourier transform of its autocorrelation function, that is,

$$S_{MI}(f) = \int_{-\infty}^{\infty} R_{MI}(\tau)\, e^{-j\omega t}\, d\tau = \int_{-T}^{T} \frac{E}{2}\left(1 - \frac{|\tau|}{T}\right) e^{-j\omega t}\, d\tau = \frac{E}{2}T\left[\frac{\sin(\pi f\, T)}{\pi f\, T}\right]^2 = \frac{ET}{2}\operatorname{sinc}^2(f\, T),$$

where the sinc function is defined as $\operatorname{sinc}(y) = \sin(\pi y)/\pi y$. This power spectral density is equivalent to the power spectral density in the lower ($Q$) branch. For the defined dibit duration $T = 2T_b$, and the energy of a symbol $E = 2E_b$, we may have the power spectral density expressed as

$$S_{MI}(f) = S_{MQ}(f) = \frac{ET}{2}\operatorname{sinc}^2(f\, T) = 2E_b T_b \operatorname{sinc}^2(2f\, T_b). \tag{6.3.5}$$

The two-sided power spectral density, normalized by $E_b T_b$, is expressed as

$$S_{MI}(f)/E_b T_b = 2\operatorname{sinc}^2(2f\, T_b),$$

as shown in Figure 6.3.5.

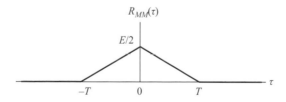

**Fig. 6.3.4** *The autocorrelation function for the modulating signal that is represented as a binary stochastic process.*

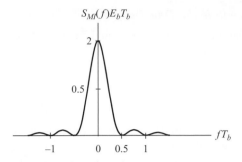

**Fig. 6.3.5** *Normalized power spectral density of the modulating signal, represented as a binary stochastic process.*

Due to the statistical independence between the modulating signals in the upper and lower branches, the power spectral density of the modulating signal is the sum of the individual power spectral density functions and is expressed as

$$S_{MM}(f) = S_{MI}(f) + S_{MQ}(f) = ET \operatorname{sinc}^2(fT) = 4E_b T_b \operatorname{sinc}^2(2fT_b). \qquad (6.3.6)$$

**Example**

Prove that the maximum value of the power spectral density function is $2E_b T_b$.

**Solution.** Applying L'Hopital's rule, we obtain

$$\lim_{f \to 0} S_{MI}(f) = 2E_b T_b \lim_{f \to 0} \frac{\sin^2(2\pi f T_b)}{(2\pi f T_b)^2} = 2E_b T_b \lim_{f \to 0} \frac{1}{2} \frac{1 - \cos(4\pi f T_b)}{(2\pi f T_b)^2}$$

$$= E_b T_b \lim_{f \to 0} \frac{0 + 4\pi T_b \sin(4\pi f T_b)}{2(2\pi f T_b) 2\pi T_b} = E_b T_b \lim_{f \to 0} \frac{\sin(4\pi f T_b)}{(2\pi f T_b)}$$

$$= E_b T_b \lim_{f \to 0} \frac{4\pi T_b \cos(4\pi f T_b)}{(2\pi T_b)} = 2E_b T_b \lim_{f \to 0} \cos(4\pi f T_b) = 2E_b T_b$$

or, for the energy $E$,

$$\lim_{f \to 0} S_{MI}(f) = \lim_{f \to 0} \frac{ET}{2} \operatorname{sinc}^2(fT) = \lim_{f \to 0} \frac{ET}{2} \frac{\sin^2(\pi f T)}{(\pi f T)^2} = \frac{ET}{2} \lim_{f \to 0} \frac{1}{2} \frac{1 - \cos(2\pi f T)}{(\pi f T)^2}$$

$$= \frac{ET}{4} \lim_{f \to 0} \frac{0 + 2\pi T \sin(2\pi f T)}{2(\pi f T)\pi T} = \frac{ET}{4} \lim_{f \to 0} \frac{\sin(2\pi f T)}{(\pi f T)} = \frac{ET}{4} \lim_{f \to 0} \frac{2\pi T \cos(2\pi f T)}{\pi T}.$$

$$= \frac{ET}{2} \lim_{f \to 0} \cos(2\pi f T) = \frac{ET}{2}$$

### 6.3.2.2    *Modulated Signals in the Time Domain*

If the phases of the carrier are $\pi/4$, $3\pi/4$, $5\pi/4$, and $7\pi/4$, the transmitted quadrature phase-shift keying signal is defined as

$$s_i(t) = \sqrt{\frac{2E}{T}} \cos\left[2\pi f_c t + (2i-1)\frac{\pi}{4}\right], \tag{6.3.7}$$

for $i = 1, 2, 3, 4$, where $0 \le t < T$ and $E$ is the transmitted signal energy per symbol (dibit). The carrier wave frequency is chosen to be equal to $n/T$, for the fixed integer $n$, in order to ensure that each transmitted dibit contains an integer number of cycles of the carrier. The wave shape of a quadrature phase-shift keying signal is shown in Figure 6.3.6. The first phase is $\pi/4$, which contains information about the symbol 10, and the message point is in the first quadrant. After the time interval $T$, which is the duration of a symbol, a new modulated signal is generated with phase $3\pi/4$, which corresponds to the transmitted symbol 00.

The quadrature phase-shift keying signal is generated according to the general structure shown in Figure 6.3.1 as a sum of the modulated signals in the $I$ and $Q$ branches, $s_I(t)$ and $s_Q(t)$. Bearing in mind the general expression of a quadrature phase-shift keying signal presented in eqn (6.3.7), we may express the quadrature phase-shift keying signal as

$$\begin{aligned}
s_i(t) = s_I(t) + s_Q(t) &= \sqrt{\frac{2E}{T}} \cos\left[(2i-1)\frac{\pi}{4}\right] \cos 2\pi f_c t - \sqrt{\frac{2E}{T}} \sin\left[(2i-1)\frac{\pi}{4}\right] \sin 2\pi f_c t \\
&= \sqrt{E} \cos\left[(2i-1)\frac{\pi}{4}\right] \phi_1(t) - \sqrt{E} \sin\left[(2i-1)\frac{\pi}{4}\right] \phi_2(t) = s_{i1}\phi_1(t) + s_{i2}\phi_2(t)
\end{aligned} \tag{6.3.8}$$

where the two orthonormal basis functions, representing a pair of quadrature carriers, are defined as

$$\phi_1(t) = \sqrt{\frac{2}{T}} \cos 2\pi f_c t, \qquad \phi_2(t) = \sqrt{\frac{2}{T}} \sin 2\pi f_c t, \tag{6.3.9}$$

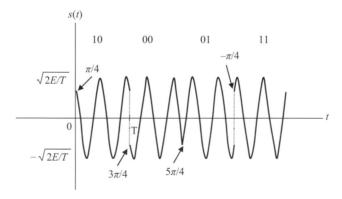

**Fig. 6.3.6** *Modulated signal obtained by direct changes in the symbols' phases.*

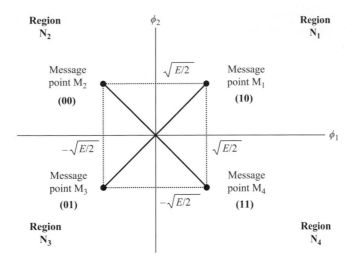

**Fig. 6.3.7** *Signal-space diagram and message constellation for a quadrature phase-shift keying system.*

where $0 \leq t < T$. Therefore, a signal space is two dimensional ($J = 2$), with a signal constellation diagram consisting of four message points ($M = 4$). The associated signal vectors are

$$\mathbf{s}_i = \begin{bmatrix} s_{i1} \\ s_{i2} \end{bmatrix} = \begin{bmatrix} \sqrt{E}\cos\left[(2i-1)\frac{\pi}{4}\right] \\ -\sqrt{E}\sin\left[(2i-1)\frac{\pi}{4}\right] \end{bmatrix}, \tag{6.3.10}$$

with coordinates $s_{i1}$ and $s_{i2}$, for $i = 1, 2, 3, 4$. All the coordinates and corresponding dibits are calculated and presented in matrix form in eqns (6.3.1) and (6.3.2). For each pair of coordinates, the signal-space diagram is shown in Figures 6.3.7. For example, message point 1 (for dibit $M_1 = m_1 m_2 = 10$ or, equivalently, for signal $s_1(t)$) is determined by the coordinates located at $s_{11}$ and $s_{12}$. The constellation diagram shown has the minimum energy.

**Example**

Form a quaternary phase-shift keying signal for the input binary sequence $\{m_i\} = \{0\ 1\ 1\ 0\ 1\ 1\ 0\ 0\}$, which corresponds to the sequence of four dibits or symbols $\{M_i\} = \{01\ 10\ 11\ 00\}$.

Divide the input binary sequence into two sequences, consisting of the first and the second bits of the dibit sequences $\{m_{ij}\}$.

For each bit inside a dibit or symbol, draw the related waveform. Add the waveforms for each bit in a dibit to obtain a quaternary phase-shift keying waveform for that dibit.

**Solution.** The sequence division is as follows:

$$\{m_{i1}\} = \{m_{11} \quad m_{21} \quad m_{31} \quad m_{41}\} = \{0 \quad 1 \quad 1 \quad 0\},$$
$$\{m_{i2}\} = \{m_{12} \quad m_{22} \quad m_{32} \quad m_{42}\} = \{1 \quad 0 \quad 1 \quad 0\}.$$

The quaternary phase-shift keying modulated signal is shown in Figure 6.3.8.

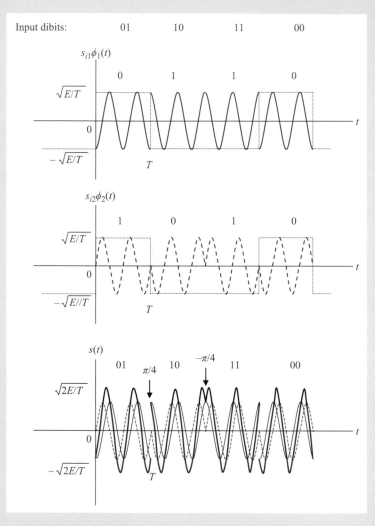

**Fig. 6.3.8**   *A quaternary phase-shift keying modulated signal.*

### 6.3.2.3 Modulated Signals in the Frequency Domain

A quadrature phase-shift keying signal is obtained by multiplying two modulating signals and their related carriers and then summing them up. The carriers are deterministic signals defined by their amplitudes, frequencies, and phases, and the modulating signals, as we said before, can be considered to be stochastic binary processes. Because the processing in both branches is identical, we will analyse only the upper or *I* branch signal processing. To find the modulated signal in time and frequency domains, we need to find the carrier in both domains and then investigate the output of the modulator multiplier.

**Presentation of carriers in time and frequency domains.** The carriers in both the *I* and the *Q* branches in the time domain are expressed as

$$s_{cI}(t) = \phi_1(t) = \sqrt{\frac{2}{T}} \cos 2\pi f_c t,$$

$$s_{cQ}(t) = \phi_2(t) = \sqrt{\frac{2}{T}} \sin 2\pi f_c t,$$

$$(6.3.11)$$

respectively. By considering the carrier to be a periodic cosine and sine power signal, we can calculate the autocorrelation function as for the binary phase-shift keying system and express it in the form

$$R_{sc}(\tau) = R_{sI}(\tau) = R_{sQ}(\tau) = \frac{1}{T} \cos 2\pi f_c \tau, \tag{6.3.12}$$

as shown in Section 12.5 and presented in Table 12.5.2. Likewise, the power spectral densities of both carriers are the Fourier transforms of the autocorrelation functions, expressed as

$$S_{sc}(f) = S_{sI}(f) = S_{sQ}(f) = \frac{1}{2T} (\delta(f - f_c) + \delta(f + f_c)]. \tag{6.3.13}$$

### 6.3.2.4 The Power Spectral Density of Signals in a Quadriphase-Shift Keying System

The baseband modulating signal $m_{i1}(t)$, having the power spectral density shown in Figure 6.3.5, is multiplied by the carrier operating at the frequency $f_c$. The initial amplitude of the random baseband signal $m_{i1}(t) = m(t)$ and the carrier $\phi_1(t)$ are statistically independent, due to the randomness of the signal amplitudes. Therefore, the bandpass modulated random signal $s_I(t)$ at the output of the first multiplier in Figure 6.3.1, represented as a realization of a wide-sense-stationary stochastic process $S_I(t)$, can be expressed as the product

$$s_I(t) = m(t) \cdot \sqrt{\frac{2}{T}} \cos(2\pi f_c t) = m(t) \cdot \phi_1(t), \tag{6.3.14}$$

where $m(t)$ is a realization of the baseband modulating process $M(t)$. In order to find the power spectral density function of the modulated signal, we need to find the autocorrelation function of $S_I(t)$, as we did for the binary phase-shift keying signal in eqn (6.2.18):

$$R_{SI}(\tau) = R_{MI}(\tau) \cdot \frac{1}{T} \cos \omega_c \tau = R_{MI}(\tau) \cdot R_{sc}(\tau). \tag{6.3.15}$$

Likewise, analogous to the derivative in eqn (6.2.19), we can express the power spectral density of the output random modulated signal in the upper ($I$) branch $S_{SI}$, bearing in mind that $E_b = E/2$ and $T_b = T/2$, as the convolution

$$
\begin{aligned}
S_{SI}(f) &= \int_{-\infty}^{\infty} S_{MI}(f-\lambda) S_{sc}(\lambda) d\lambda \\
&= \int_{-\infty}^{\infty} \frac{ET}{2} \operatorname{sinc}^2(f-\lambda) T \cdot \frac{1}{2T} (\delta(\lambda-f_c) + \delta(\lambda+f_c)) d\lambda \cdot \\
&= \frac{E_b}{2} \operatorname{sinc}^2 2(f-f_c) T_b + \frac{E_b}{2} \operatorname{sinc}^2 2(f+f_c) T_b
\end{aligned}
\tag{6.3.16}
$$

The signal in the lower ($Q$) branch $S_Q(f)$ has the identical power spectral density. Therefore, the power spectral density of the quadrature phase-shift keying modulated signal is the sum of the signals in both branches, expressed as

$$S_{SS}(f) = S_{SI}(f) + S_{SQ}(f) = E_b \operatorname{sinc}^2 2(f-f_c) T_b + E_b \operatorname{sinc}^2 2(f+f_c) T_b. \tag{6.3.17}$$

The normalized power spectral density by $E_b$ is

$$S_{SS}(f)/E_b = \operatorname{sinc}^2 2(f-f_c) T_b + \operatorname{sinc}^2 2(f+f_c) T_b,$$

which is presented in Figure 6.3.9. The amplitudes of the quadrature phase-shift keying power spectral density are two times larger than the amplitudes of the binary phase-shift keying spectrum; however, the first zero crossing of the quadrature phase-shift keying is two times smaller, meaning that the spectrum requirements for the quadrature phase-shift keying system are two times smaller than for the binary phase-shift keying system, which is a big advantage of the quadrature phase-shift keying system. The power spectral density of the quadrature phase-shift keying signal is a frequency-shifted version of the modulating signal power spectral density, as is shown in Figure 6.3.9.

The power spectral density can be calculated on a logarithmic scale in the following way. The one-sided power spectral density, obtained by overlapping the amplitudes at negative frequencies with the positive ones, is normalized by $2E_b$ and expressed as a function of frequency $f_n = (f - f_c) T_b$ in the form

$$S_{SS}(f)/2E_b = \operatorname{sinc}^2(f-f_c) 2T_b = \operatorname{sinc}^2 2f_n = \frac{\sin \pi 2f_n}{\pi 2f_n} = S_{SS}(f_n). \tag{6.3.18}$$

The normalization was done by $2E_b$ to have the value of power spectral density equal to 1 at zero frequency, to allow a comparison of the quadrature phase-shift keying signal spectrum with the spectra of other modulation schemes, primarily with the spectrum of the binary phase-shift keying signal. In practice, we need to see the level of the spectrum in logarithmic units.

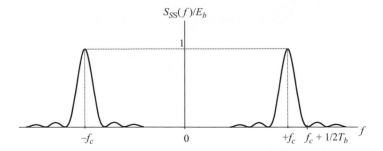

**Fig. 6.3.9** *Power spectral density of a quadrature phase-shift keying signal.*

Therefore, we can express the power spectral density as

$$S_{SS}(f_n) = 10 \log_{10}\left(\operatorname{sinc}^2 2f_n\right) = 10 \log_{10}\frac{\sin \pi 2f_n}{\pi 2f_n} \quad dB, \tag{6.3.19}$$

and present it in a graphical form (see Section 6.4, Figure 6.4.10), alongside with the power spectral density of a binary phase-shift keying signal and a binary frequency-shift keying modulated signal. This presentation of the power spectral density allows us to compare the bandwidths of binary phase-shift keying and quadrature phase-shift keying. Obviously, quadrature phase-shift keying is better than binary phase-shift keying, because the bandwidth required for it is two times narrower than that for binary phase-shift keying. As we will see later, the bit error probability curves for both schemes are identical.

**Exercise**

Calculate the power of a quadrature phase-shift keying signal, where the power spectral density is expressed as

$$S_{SS}(f) = 2E_b \operatorname{sinc}^2 (f - f_c) 2T_b.$$

**Solution.** For the function defined, bearing in mind the integral values

$$\int_{-\infty}^{\infty} \operatorname{sinc}^2 f df = \int_{-\infty}^{\infty} \operatorname{sinc} f df = \pi,$$

we may calculate the signal power in the frequency domain as

$$P_{SS} = \int_{-\infty}^{\infty} S_{SS}(f) df = 2E_b \int_{-\infty}^{\infty} \left( \operatorname{sinc}^2 (f - f_c) 2T_b \right) df.$$

If we substitute $x = \pi(f - f_c) 2T_b$, the integral value gives the power as

$$P_{SS} = 2E_b \int_{-\infty}^{\infty} \left( \operatorname{sinc}^2 x \right) \frac{dx}{2\pi T_b} = \frac{2E_b}{2\pi T_b} \pi = \frac{E_b}{T_b}.$$

## 6.3.3 Receiver Operation

### 6.3.3.1 *Operation of the Correlation Demodulator and the Optimum Detector*

The modulated signal is transmitted through the additive white Gaussian noise channel and received by a coherent quadrature phase-shift keying correlation receiver. The received signal can be expressed as

$$r(t) = s_i(t) + w(t), \tag{6.3.20}$$

for $0 \le t \le T$ and $i = 1, 2, 3, 4$, where $w(t)$ is a sample of the white Gaussian noise process $W(t)$ of zero mean and power spectral density $N_0/2$. At the output of the two correlators ($j = 1, 2$) are discrete-time random signals defined by their random values at the end of each symbol interval $T$ as

$$\begin{aligned} x_1 &= \int_0^T r(t)\phi_1(t)dt = \int_0^T s_i(t)\phi_1(t)dt + \int_0^T w(t)\phi_1(t)dt = s_{i1} + w_1 \\ &= \sqrt{E}\cos\left[(2i-1)\frac{\pi}{4}\right] + w_1 = \pm\sqrt{\frac{E}{2}} + w_1 \end{aligned} \tag{6.3.21}$$

and

$$\begin{aligned} x_2 &= \int_0^T r(t)\phi_2(t)dt = \int_0^T s_i(t)\phi_2(t)dt + \int_0^T w(t)\phi_2(t)dt = s_{i2} + w_2 \\ &= -\sqrt{E}\sin\left[(2i-1)\frac{\pi}{4}\right] + w_2 = \mp\sqrt{\frac{E}{2}} + w_2 \end{aligned} \tag{6.3.22}$$

where $s_{i1}$ and $s_{i2}$ can be considered to be deterministic quantities contributed by the transmitted signal $s_i$, and $w_1$ and $w_2$ are sample values of the random variables $W_1$ and $W_2$, respectively, that arise due to the channel noise process $W(t)$. The calculated random values are realizations of the random variables $X_1$ and $X_2$ and can be expressed in the form of a demodulation vector $\mathbf{x} = [x_1 \; x_2]$. Having this vector available, we can proceed with the processing of the signals inside the optimum detector. However, we can, as will be explained later, decide on the transmitted symbols by following the following decision rules:

1. If the output of the correlator, the demodulation vector **x**, falls in the region $N_1$, that is, the first quadrant of the coordinate system, then we conclude that the dibit 10 (corresponding to the transmitted signal $s_1(t)$) was generated with a high probability, and the decision will be made in favour of the dibit 10, as shown in Fig. 6.3.7.

2. If the output of the correlator **x** falls in the region $N_1$, that is, the first quadrant of the coordinate system, *given that* the dibit 00 (corresponding to the transmitted signal $s_2(t)$) was generated, then an erroneous decision will be made in favour of the dibit 10.

In order to calculate the probability of symbol error, a coherent quadrature phase-shift keying system can be treated as a parallel connection of two coherent binary phase-shift keying systems (sometimes called 'binary phase-shift keying channels'), because the quadrature phase-shift keying signal is defined by a pair of quadrature carriers. These two binary phase-shift keying systems are sometimes called *in-phase (I) systems* and *quadrature (Q) systems*, respectively. The correlator outputs $x_1$ and $x_2$ may be treated as the respective outputs of these two systems. As we said, they are realizations of the random variables $X_1$ and $X_2$. If these variables are conditioned to the possible bits sent, conditional probability density functions will be obtained that have the mean value $E/2$ or $-E/2$, respectively, and the variance $N_0/2$. The probability density functions of these random variables are derived in Appendix A.

### 6.3.3.2    Bit Error Probability Calculation

We proved that the bit error probability in each binary phase-shift keying system (or binary phase-shift keying channel) is

$$p = \frac{1}{2}\text{erfc}\left(\sqrt{\frac{E_b}{N_0}}\right),$$

(6.3.23)

where $E_b$ is the transmitted signal energy per bit. Because a bit duration in a binary phase-shift keying channel is two times larger than a bit duration at the output of the source, that is, $T = 2T_b$, the transmitted signal energy per bit of one binary phase-shift keying system inside the quadrature phase-shift keying system will be a half of the energy of a dibit, that is,

$$E_b = \frac{E}{2},$$

(6.3.24)

which results in the bit error probability in *each* binary phase-shift keying system of the quadrature phase-shift keying system expressed as

$$p_1 = p_2 = \frac{1}{2}\text{erfc}\left(\sqrt{\frac{E/2}{N_0}}\right).$$

(6.3.25)

The in-phase system makes a decision on one of the two bits inside the symbol (dibit) of the quadrature phase-shift keying signal, and the quadrature system makes the decision on the other bit inside the dibit. The errors in these decisions are statistically independent. Therefore, a correct decision will occur when the decisions in both the in-phase *and* the quadrature systems are correct. Thus, the probability of a correct decision for a symbol received in the quadrature phase-shift keying system $P_c$ is equal to the *product* of the probabilities of correct decisions for the bit in both the in-phase and the quadrature systems, $(1 - p_1)$ and $(1 - p_2)$, respectively, that is,

$$P_c = (1 - p_1)(1 - p_2) = \left[ 1 - \frac{1}{2}\mathrm{erfc}\left(\sqrt{\frac{E}{2N_0}}\right) \right]^2 = 1 - \mathrm{erfc}\left(\sqrt{\frac{E}{2N_0}}\right) + \frac{1}{4}\mathrm{erfc}^2\left(\sqrt{\frac{E}{2N_0}}\right),$$

and, consequently, the symbol error probability in a quadrature phase-shift keying system is

$$P_e = 1 - P_c = \mathrm{erfc}\left(\sqrt{\frac{E}{2N_0}}\right) - \frac{1}{4}\mathrm{erfc}^2\left(\sqrt{\frac{E}{2N_0}}\right). \tag{6.3.26}$$

In the case when $E/2N_0 \gg 1$, this formula for the symbol error probability can be approximated by

$$P_e \approx \mathrm{erfc}\left(\sqrt{\frac{E}{2N_0}}\right). \tag{6.3.27}$$

One symbol in a quadrature phase-shift keying system is composed of two bits (forming a dibit) having equal energy $E_b = E/2$. Thus, by inserting the energy of a dibit $E = 2E_b$ into the expression for the probability of symbol error, we obtain

$$P_e \approx \mathrm{erfc}\left(\sqrt{\frac{E_b}{N_0}}\right). \tag{6.3.28}$$

Gray encoding is used to specify the signal-space diagram. Therefore, an erroneous decision for a symbol will most likely result in an erroneous decision for one bit inside the symbol, because having an error for one bit in a symbol results in a move to the neighbouring region on the signal-space diagram (i.e. due to noise, the phase changes to the nearest value). Having an error for both bits corresponds to a move to the opposite region, which is the least likely. Thus, the bit error probability in a coherent quadrature phase-shift keying system is equal to the bit error probability of a coherent binary phase-shift keying system for the same bit rate $R_b$ and the same $E_b/N_0$, is exactly expressed as

$$p = \frac{1}{2}\mathrm{erfc}\left(\sqrt{\frac{E_b}{N_0}}\right) \tag{6.3.29}$$

(see Section 6.2.3.3, Figure 6.2.15). The basic advantage of quadrature phase-shift keying over binary phase-shift keying system is that, for the same bit error probability and the same bit rate, the quadrature phase-shift keying system requires only half of the channel bandwidth (see Section 6.4, Figure 6.4.10). Neglecting complexity, quadrature phase-shift keying systems are preferred to binary phase-shift keying systems.

### 6.3.3.3   *Signal Analysis and Transceiver Structure in a Quadrature Phase-Shift Keying System*

Expression (6.3.8) for the quadrature phase-shift keying signal may also be developed in terms of in-phase $I(t)$ and quadrature $Q(t)$ components as

$$s_i(t) = I(t)\phi_1(t) + Q(t)\phi_2(t), \qquad (6.3.30)$$

where the in-phase component is

$$I(t) = \sqrt{E}\cos\left[(2i-1)\frac{\pi}{4}\right] = \pm\sqrt{E}\frac{\sqrt{2}}{2} = \pm\sqrt{\frac{E}{2}} = \pm\sqrt{E_b}, \qquad (6.3.31)$$

and the quadrature component is

$$Q(t) = \sqrt{E}\sin\left[(2i-1)\frac{\pi}{4}\right] = \pm\sqrt{E}\frac{\sqrt{2}}{2} = \pm\sqrt{\frac{E}{2}} = \pm\sqrt{E_b}. \qquad (6.3.32)$$

The + or − sign is chosen depending on the value for $i$. Thus, the message bits 1's and 0's can be transformed directly into two levels, $+\sqrt{E_b}$ or $-\sqrt{E_b}$, respectively, using a polar non-return-to-zero encoder instead of the look-up tables and multipliers shown in Figure 6.3.1.

A de-multiplexer is used to divide the non-return-to-zero binary signal formed into two binary signals, $I(t)$ and $Q(t)$, which correspond to odd- and even-numbered input message bits, respectively, as shown in Figure 6.3.10. In this way, the first bit $m_k$ of an input dibit $m_k m_{k+1}$ goes to the $I$ system, and the second bit, $m_{k+1}$, goes to the $Q$ system, as shown in Figure 6.3.10. The duration of bits in both branches is $T = 2T_b$. The in-phase binary non-return-to-zero signal $I(t)$ modulates the carrier specified by the first orthonormal basis function $\phi_1(t)$. The quadrature binary non-return-to-zero signal $Q(t)$ modulates the carrier specified by the

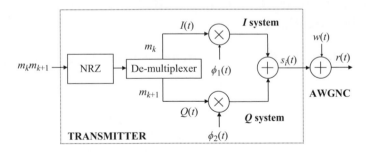

**Fig. 6.3.10** *Traditional scheme of a quadrature phase-shift keying transmitter; AWGNC, additive white Gaussian noise channel; NRZ, polar non-return-to-zero encoder.*

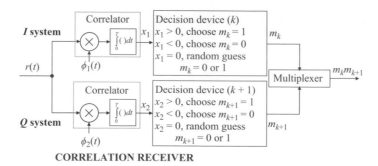

**CORRELATION RECEIVER**

**Fig. 6.3.11** *Traditional scheme for a quadrature phase-shift keying receiver.*

second orthonormal basis function $\phi_2(t)$. The resultant two binary phase-shift keying signals are added to form a quadrature phase-shift keying signal $s_i(t)$, as shown in Figure 6.3.10.

This signal is transmitted through an additive white Gaussian noise channel and then received using a correlation receiver, which is presented in Figure 6.3.11. The signal received is multiplied by the locally generated basis functions and then integrated in the symbol interval $T$. The results of these correlations are the random samples $x_1$ and $x_2$. These samples are compared with the threshold value of zero in the decision device, as depicted in Figure 6.3.11. The outputs of the decision device are two estimates, $m_k$ and $m_{k+1}$, for the $I$ system and the $Q$ system, respectively. These two estimates are multiplexed in the multiplexer, and an estimate of the transmitted symbol (dibit), denoted by $m_k \, m_{k+1}$, is generated at the output of the multiplexer.

The receiver in Figure 6.3.11 does not exactly follow the decision procedure used to derive the expression for the bit error probability in eqn (6.3.29), where the decision is made at the dibit level. Unlike that decision, the decision presented in Figure 6.3.11 is made in both the $I$ and the $Q$ systems at the bit level. This decision can be performed because the noise components affecting the bits in both the $I$ and the $Q$ branches are independent, and the bit error probability in both branches are the same and can be expressed by the relation (6.3.29).

## 6.4   Coherent Binary Frequency-Shift Keying with a Continuous Phase

### 6.4.1   Operation of a Binary Frequency-Shift Keying System

Binary phase-shift keying and quadrature phase-shift keying are examples of linear modulation techniques, in the sense that the spectrum of the baseband signal is shifted to the carrier frequency. Therefore, the frequencies of the baseband signal are preserved inside the bandpass signal, and no new frequency components are generated. In contrast, frequency-shift keying, which is studied in this section, is an example of a non-linear modulation technique.

Based on the general block scheme of a communication system presented in Chapter 5, in Figure 5.8.1, it is possible to develop a binary frequency-shift keying system as a special case,

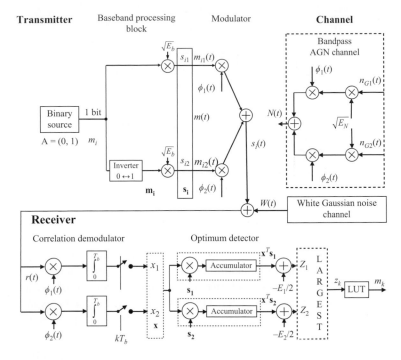

**Fig. 6.4.1** *Block schematic of a binary frequency-shift keying continuous communication system; AGN, additive Gaussian noise; LUT, look-up table.*

as shown in Figure 6.4.1. In this case, the transmitter is composed of a block for baseband signal processing to generate the modulating signal $m(t)$, and a multiplier, which functions as a modulator. The receiver is composed of a correlation demodulator and an optimum detector. The optimum detector can be excluded from the scheme because the decision for a received message can be made at the output of the correlator, as we did for a binary phase-shift keying system and a quadrature phase-shift keying system. This property will be explained later on in this section. The mathematical model of the system and the processing of signals will be presented in the continuous-time domain.

The message source generates one of two possible message symbols, which can have binary values denoted by $m_1 = 1$ and $m_2 = 0$ and belong to an alphabet A. Therefore, a symbol generated at the output of the source is identical to a message bit. Suppose the generated message bit is $m_1 = 1$. The bit is applied to the inputs of two modulation branches. In the upper branch, the binary value 1 is multiplied by the square root of the bit energy $\sqrt{E_b}$ in the bit interval $T_b$ to form the signal coefficient $s_{i1}$, which defines part of the baseband modulating signal $m(t) = m_{i1}(t)$, as presented in Figure 6.4.2. In the branch multiplier, the signal $m_{i1}(t)$ is multiplied by the first carrier defined by the basis function $\phi_1(t)$, which is designed as the first basis function, and the modulated binary frequency-shift keying signal $s_i(t)$ is produced and transferred though an additive white Gaussian noise waveform channel.

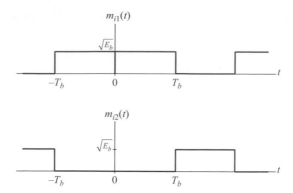

**Fig. 6.4.2** *Modulating binary signals presented in the time domain.*

The same message bit $m_1 = 1$ is also applied in the baseband processing block to the input of an inverter in the second modulation branch that inverts bit 1 into 0. Due to that, the zero multiplies $\sqrt{E_b}$ and then the second carrier. Therefore, the output of the second branch multiplier is 0. The modulated signal, defined in the bit interval $T_b$, can be expressed as

$$s_1(t) = s_{11}\phi_1(t) + s_{22}\phi_2(t) = s_{11}\phi_1(t) + 0 = \sqrt{E_b}\phi_1(t) = \sqrt{E_b}\sqrt{\frac{2}{T_b}}\cos{(2\pi f_1 t)}. \qquad (6.4.1)$$

In the case when the bit $m_2 = 0$ is generated at the output of the source, the modulated signal will be

$$s_2(t) = s_{11}\phi_1(t) + s_{22}\phi_2(t) = 0 + s_{22}\phi_2(t) = \sqrt{E_b}\phi_2(t) = \sqrt{E_b}\sqrt{\frac{2}{T_b}}\cos{(2\pi f_2 t)}. \qquad (6.4.2)$$

In summary, the message bits 1 and 0 generated at the source result in two distinct modulated signals that have the same amplitudes and phases but different frequencies. The frequencies are chosen in such a way to have orthogonal modulated signals.

## 6.4.2    Transmitter Operation

### 6.4.2.1    *Modulating Signals in Time and Frequency Domains*

Mathematical description of the frequency-shift keying transmitter and basic signal processing operations will be presented using its block schematic in Figure 6.4.1. The number of possible message symbols generated from the source is $I = 2$ and are defined by the matrix of message bits

$$\mathbf{m} = [\mathbf{m}_1 \quad \mathbf{m}_2] = \begin{bmatrix} 1 & 0 \end{bmatrix}. \qquad (6.4.3)$$

These message bits are mapped into the amplitudes of modulating signals by multiplying them by $\sqrt{E_b}$ to form two vectors of modulating signal coefficients represented by a two-column matrix of signal coefficients expressed as

$$\mathbf{s} = [\mathbf{s}_1 \;\; \mathbf{s}_2] = \begin{bmatrix} s_{11} & s_{21} \\ s_{12} & s_{22} \end{bmatrix} = \begin{bmatrix} \sqrt{E_b} & 0 \\ 0 & \sqrt{E_b} \end{bmatrix}. \tag{6.4.4}$$

If the source randomly generates a stream of bits, the modulating signal $m(t)$ will be composed of two rectangular trains $m_{i1}(t)$ and $m_{i2}(t)$ with mutually inverted amplitude values, as presented in Figure 6.4.2. These modulating signals multiply the carriers represented by two basis functions, as presented in the block schematic in Figure 6.4.1.

For the sake of simplicity, an analysis of this modulating signal in the frequency domain will be performed in the next section, where the modulated signal will be expressed in terms of the in-phase and quadrature components.

### 6.4.2.2 *Modulated Signals in the Time Domain and the Signal-Space Diagram*

The message source can generate two binary symbols (or bits) that belong to an alphabet $A$ of two symbols, denoted by $m_1 = 1$ and $m_2 = 0$. In binary frequency-shift keying, one of two sinusoidal signals with different frequencies is assigned to each symbol to obtain the modulated signal, according to the expression

$$s_i(t) = \left\{ \begin{array}{cc} \sqrt{\frac{2E_b}{T_b}} \cos\left(2\pi f_j t\right) & 0 \leq t \leq T_b \\ 0 & \text{elsewhere} \end{array} \right\}. \tag{6.4.5}$$

The frequencies of modulated signals are expressed as

$$f_j = \frac{n+i}{T_b}, \tag{6.4.6}$$

for the fixed integer values $n$ and $i = 1, 2$. The difference of these frequencies is a constant value that is equal to $1/T_b$. Defined in this way, binary frequency-shift keying preserves phase continuity, including inter-bit switching time, and, for that reason, is often called *continuous-phase binary frequency-shift keying*. An example of a binary frequency-shift keying signal is shown in Figure 6.4.3.

The functions $s_1(t)$ and $s_2(t)$ are orthogonal. If we divide them by $\sqrt{E_b}$, we obtain the orthonormal basis functions for the two carrier frequencies, which are expressed as

$$\phi_j(t) = \left\{ \begin{array}{cc} \sqrt{\frac{2}{T_b}} \cos\left(2\pi f_j t\right) & 0 \leq t \leq T_b \\ 0 & \text{elsewhere} \end{array} \right\}, \tag{6.4.7}$$

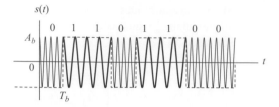

**Fig. 6.4.3** *A modulated signal in a binary frequency-shift keying system.*

for $j = 1, 2$. The signals can be presented in a vector form. The coefficients for the signal vectors can be calculated as

$$
\begin{aligned}
s_{ij} &= \int_0^{T_b} s_i(t)\phi_j(t)dt = \int_0^{T_b} \sqrt{\frac{2E_b}{T_b}} \cos 2\pi f_i t \sqrt{\frac{2}{T_b}} \cos 2\pi f_j t dt \\
&= \frac{2}{T_b}\sqrt{E_b} \int_0^{T_b} \cos 2\pi f_i t \cdot \cos 2\pi f_j t dt \\
&= \left\{ \begin{array}{ll} \sqrt{E_b}\frac{2}{T_b}\int_0^{T_b} \cos^2 2\pi f_i t dt = \sqrt{E_b} & i = j \\ 0 & i \neq j \end{array} \right\}
\end{aligned} \qquad (6.4.8)
$$

Unlike the case for coherent binary phase-shift keying, *the signal-space diagram for frequency-shift keying is two dimensional* ($J = 2$), with a signal constellation consisting of two message points defined for two signals in vector form as

$$
\mathbf{s}_1 = \left[ \begin{array}{c} s_{11} \\ s_{12} \end{array} \right] = \left[ \begin{array}{c} \sqrt{E_b} \\ 0 \end{array} \right], \quad \text{and} \quad \mathbf{s}_2 = \left[ \begin{array}{c} s_{21} \\ s_{22} \end{array} \right] = \left[ \begin{array}{c} 0 \\ \sqrt{E_b} \end{array} \right], \qquad (6.4.9)
$$

as has already been presented in eqn (6.4.4). For each pair of coordinates, the signal-space diagram is shown in Figure 6.4.4, and the schematic for a binary frequency-shift keying transceiver is the same as that in Figure 6.4.1. The possible transmitted signal waveforms corresponding to the two message points are presented in Figure 6.4.5.

The presence of noise in the channel can change the position of the message points. For the same energies of the modulated signals, we define the decision boundary as the line that separates the two-dimensional plane into two symmetric parts and lies under 45° with respect to the *x*-axis, as shown in Figure 6.4.4.

### 6.4.2.3  Modulating and Modulated Signals in Time and Frequency Domains

**In-phase and quadrature components in the time domain.** In order to find the power spectral density for both the modulating signal and the modulated signal, the modulated signal needs to be expressed in terms of the so-called low-frequency in-phase and quadrature components. Let us consider a continuous-phase binary frequency-shift keying system defined

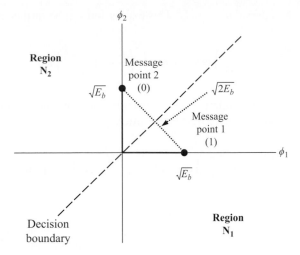

**Fig. 6.4.4** *Signal-space diagram and message constellation for a continuous-phase binary frequency-shift keying system.*

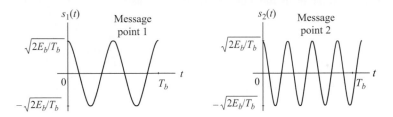

**Fig. 6.4.5** *Transmitted signal waveforms corresponding to two message points.*

by the two transmitted frequencies

$$
f_j = \begin{cases} f_c + \dfrac{1}{2T_b} & i = 2, m_i = 0 \\[2mm] f_c - \dfrac{1}{2T_b} & i = 1, m_i = 1 \end{cases} = f_c \pm \dfrac{1}{2T_b}, \tag{6.4.10}
$$

where $f_c$ is defined as the nominal carrier frequency equal to the arithmetic mean between the two carrier frequencies. Therefore, the modulated signal is represented by two signals with the frequencies $f_1$ and $f_2$ separated by the interval $1/T_b$, that is,

$$
s(t) = \sqrt{\frac{2E_b}{T_b}} \cos 2\pi \left( f_c \pm \frac{1}{2T_b} \right) t = \sqrt{\frac{2E_b}{T_b}} \cos \left( 2\pi f_c t \pm \frac{\pi t}{T_b} \right), \tag{6.4.11}
$$

which is defined in the interval $0 \leq t \leq T_b$. This expression can be further expended as

$$s(t) = \sqrt{\frac{2E_b}{T_b}} \cos\left(\pm\frac{\pi t}{T_b}\right) \cos 2\pi f_c t - \sqrt{\frac{2E_b}{T_b}} \sin\left(\pm\frac{\pi t}{T_b}\right) \sin 2\pi f_c t. \tag{6.4.12}$$

Taking into account the even property of the cosine function and the odd property of the sine function, we may express the binary frequency-shift keying signal as

$$s(t) = \sqrt{\frac{2E_b}{T_b}} \cos\left(\frac{\pi t}{T_b}\right) \cos 2\pi f_c t \mp \sqrt{\frac{2E_b}{T_b}} \sin\left(\frac{\pi t}{T_b}\right) \sin 2\pi f_c t, \tag{6.4.13}$$

where the '−' sign in front of the second addend corresponds to transmitting symbol 1, and the '+' sign corresponds to transmitting symbol 0. This signal can be expressed in terms of the defined basis cosine and sine functions. In our presentation of a communication system, the carriers are represented by basis functions which include the term $\sqrt{T_b/2}$ in front of the sine and the cosine functions. Therefore, the modulated signal is defined in terms of the orthonormal carrier, that is, in terms of the in-phase component $s_I(t)$, the quadrature component $s_Q(t)$, and the basis functions $\phi_1(t)$ and $\phi_2(t)$ as

$$\begin{aligned} s(t) &= \sqrt{E_b} \cos\left(\frac{\pi t}{T_b}\right) \sqrt{\frac{2}{T_b}} \cos 2\pi f_c t \mp \sqrt{E_b} \sin\left(\frac{\pi t}{T_b}\right) \sqrt{\frac{2}{T_b}} \sin 2\pi f_c t \\ &= \sqrt{E_b} \cos\left(\frac{\pi t}{T_b}\right) \phi_1(t) \mp \sqrt{E_b} \sin\left(\frac{\pi t}{T_b}\right) \phi_2(t) = s_I(t)\phi_1(t) \mp s_Q(t)\phi_2(t) \end{aligned} \tag{6.4.14}$$

The wave shapes of the in-phase and the quadrature components defined in eqn (6.4.13) are presented in Figures 6.4.6 and 6.4.7, assuming that the sequence of bits (0 1 1 0 1 1 0 0) is generated from the source and that the in-phase and quadrature components are defined for $T_b$, or their related periods equal to $T_I = T_Q = 2T_b$, which are expressed, respectively, as

$$s_I(t) = \sqrt{E_b} \cos\left(\frac{\pi t}{T_b}\right) \tag{6.4.15}$$

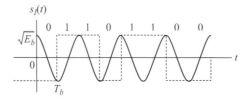

**Fig. 6.4.6** *Waveform of an in-phase signal.*

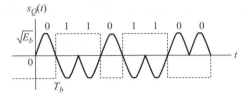

**Fig. 6.4.7** *Waveform of a quadrature signal.*

and

$$s_Q(t) = \sqrt{E_b} \sin\left(\frac{\pi t}{T_b}\right). \tag{6.4.16}$$

**In-phase and quadrature components in the frequency domain.** To find the modulating signal in the frequency domain in terms of its power spectral density, we will analyse the in-phase and the quadrature components separately. The in-phase component is a deterministic cosine signal that *does not* depend on the incoming random binary message signal $m(t)$. The value $A_b$ is the amplitude of the carrier as is traditionally defined. Since the in-phase component is a cosine function of time, its autocorrelation function is

$$R_{II}(\tau) - \frac{E_b}{2} \cos \frac{\pi}{T_b} \tau, \tag{6.4.17}$$

as defined in Chapter 12, Table 12.5.2, for the periodic power signal. The Fourier transform of the autocorrelation function will give the power spectral density function of the in-phase component, which consists of two delta functions weighted by $E_b/4$ and positioned at the two frequencies $f = \pm 1/2T_b$, that is,

$$S_{ii}(f) = \frac{E_b}{4} \left[\delta\left(f - \frac{1}{2T_b}\right) + \delta\left(f + \frac{1}{2T_b}\right)\right], \tag{6.4.18}$$

which can be expressed in the form

$$S_{ii}(f) = \frac{E_b T_b}{4} \left[\delta\left(fT_b - \frac{1}{2}\right) + \delta\left(fT_b + \frac{1}{2}\right)\right]. \tag{6.4.19}$$

The derivatives of these expressions can be found in Chapter 12, and their presentations in Table 12.5.2. The same expression for the autocorrelation function of the cosine carrier, which has a random phase, is derived in Chapter 19 in relation (19.2.9).

Unlike the in-phase component, the quadrature component is a random signal that *does depend* on the incoming random binary signal $m(t)$ and changes its sign in accordance with the sign of the incoming signal $m(t)$, as shown in Figure 6.4.7. This component can be understood as a bit (symbol) shaping function of the quadrature carrier and is expressed as

$$s_Q(t) = \mp\sqrt{E_b}\sin\left(\frac{\pi t}{T_b}\right).$$

(6.4.20)

Therefore, the quadrature component is a negative sine signal when bit 1 is transmitted, and a positive sine signal when symbol 0 is transmitted. Its energy spectral density is expressed as

$$E_{QQ}(f) = \frac{E_b T_b^2}{4}|\operatorname{sinc}T_b(f - 1/2T_b) + \operatorname{sinc}T_b(f + 1/2T_b)|^2$$

$$= \frac{4E_b T_b^2}{\pi^2}\left|\frac{\cos(\pi f T_b)}{4f^2 T_b^2 - 1}\right|^2,$$

(6.4.21)

as shown in Appendix D. Therefore, its power spectral density is

$$S_{QQ}(f) = \frac{E_{QQ}(f)}{T_b} = \frac{4E_b T_b \cos^2(\pi T_b f)}{\pi^2\left(4T_b^2 f^2 - 1\right)^2}.$$

(6.4.22)

Due to the independence of the binary message bits, the components $s_I(t)$ and $s_Q(t)$, which represent the modulating signal $m(t)$, are mutually independent of each other. Therefore, the power spectral density of the modulating signal components is equal to the sum of the power spectral densities of these two components, which is given by eqns (6.4.19) and (6.4.22), respectively, and expressed as

$$S_{MM}(f) = S_{ii}(f) + S_{QQ}(f)$$

$$= \frac{E_b T_b}{4}\left[\delta\left(fT_b - \frac{1}{2}\right) + \delta\left(fT_b + \frac{1}{2}\right)\right] + \frac{4E_b T_b\cos^2(\pi T_b f)}{\pi^2\left(4T_b^2 f^2 - 1\right)^2}.$$

(6.4.23)

In calculating the spectra of modulating signals for binary phase-shift keying systems and quadrature phase-shift keying systems, we used the carrier signals expressed in the form of orthonormal basis functions. To compare their spectra to the binary frequency-shift keying spectrum, it is necessary to calculate the power spectral density of the baseband binary frequency-shift keying signal for the case when the signal in the time domain is expressed in terms of orthonormal carriers and then calculate the power spectral density function. The power spectral density function needs to be normalized to have a unit value at zero frequency. The two-sided power spectral density, normalized by $E_b T_b$, is expressed as

$$S_{MM}(f)/E_b T_b = \frac{1}{4}\left[\delta\left(T_b f - \frac{1}{2}\right) + \delta\left(T_b f + \frac{1}{2}\right)\right] + \frac{4}{\pi^2}\frac{\cos^2(\pi T_b f)}{\left(4T_b^2 f^2 - 1\right)^2}$$

(6.4.24)

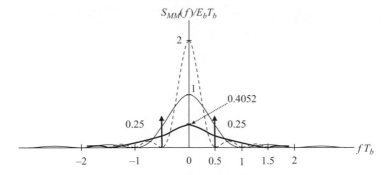

**Fig. 6.4.8** *Baseband spectra for a binary phase-shift keying signal (thin line), a quadrature phase-shift keying signal (dashed line), and a binary frequency-shift keying signal (thick line).*

and is shown in Figure 6.4.8. A value of the second component in eqn (6.4.24) for $fT_b = 0.5$ leads to a ratio of $0:0$ and L'Hopital's rule need to be applied, which gives the value 0.25. The normalized amplitude at $fT_b = 0$ is $4/\pi^2 = 0.4052$. The first and second zero crossings occur at $fT_b = 1.5$ and $fT_b = 2.5$, respectively.

The power spectral densities of the modulating signals for binary phase-shift keying (thin graph), quadrature phase-shift keying (dashed graph), and binary frequency-shift keying (thick graph) are presented in a normalized form in Figure 6.4.8. The power spectral densities for the binary phase-shift keying signal and the quadrature phase-shift keying signal, shown in Figures 6.2.4 and 6.3.5, respectively, are also presented on the same figure, for the sake of comparison, for the same normalization factor, $E_b T_b$.

**Example**

The power spectral density of the quadrature component of a binary frequency-shift keying signal is expressed by eqn (6.4.22) as

$$S_{QQ}(f) = \frac{4E_b T_b}{\pi^2} \frac{\cos^2(\pi T_b f)}{\left(4T_b^2 f^2 - 1\right)^2}.$$

Find the value of the function for $T_b f = 0.5$, and the positions of the zero crossings of this density function.

**Solution.** For $T_b f = 0.5$, the power spectral density function $S_{QQ}(f)$ has the undetermined value

$$\lim_{T_b f \to 0.5} S_{QQ}(f) = \frac{0}{0}. \tag{6.4.25}$$

Therefore, we need to apply L'Hopital's rule as follows:

$$\lim_{T_b f = x \to 0.5} S_{QQ}(f) = \frac{4E_b T_b}{\pi^2} \lim_{x \to 0.5} \frac{d/dx\left[\cos^2(\pi x)\right]}{d/dx\left[\left(4x^2 - 1\right)^2\right]}.$$

$$= \frac{4E_b T_b}{\pi^2} \lim_{Tx \to 0.5} \frac{\pi \sin(2\pi x)}{16x\left(4x^2 - 1\right)} = \frac{0}{0}$$

(6.4.26)

We need to repeat the rule to obtain

$$\frac{E_b T_b}{4\pi} \lim_{T_b f = x \to 0.5} \frac{d/dx\left[\sin(2\pi x)\right]}{d/dx\left[x\left(4x^2 - 1\right)\right]} = \frac{E_b T_b}{4\pi} \lim_{x \to 0.5} \frac{-2\pi \cos 2\pi x}{12x^2 - 1}.$$

$$= \frac{E_b T_b}{4\pi} \frac{2\pi}{2} = \frac{E_b T_b}{4}$$

(6.4.27)

The zero crossings can be calculated from the condition

$$\cos^2(\pi T_b f) = \frac{1}{2}[1 + \cos(2\pi T_b f)] = 0,$$

(6.4.28)

or $\cos(2\pi T_b f) = -1$, which results in the condition $2\pi T_b f = k\pi$. Therefore, the zero crossings occur for $T_b f = k/2$, where $k = 3, 5, 7, \ldots$ . The first crossing occurs for $k = 3$, and the second one for $k = 5$. This calculation makes sense, due to the fact that the quadrature component is composed of two sinc functions, as presented by expression (6.4.21), having maxima values at $T_b f = 1/2$.

### 6.4.2.4    Modulated Signals in the Frequency Domain

**Power spectral density of the modulated binary frequency-shift keying signal.** For the case when the carriers are expressed as a pair of orthonormal sine and cosine signals, having the same autocorrelation function and, consequently, the same power spectral density functions, their power spectral densities are the same and expressed in eqn (6.2.8) as

$$S_{sc}(f) = \frac{1}{2T_b}(\delta(f - f_c) + \delta(f + f_c)).$$

(6.4.29)

Then the power spectral density of the modulated signal can be obtained as a convolution of the power spectral density function of the modulating signal (6.4.24) and the power spectral density of carrier (6.4.29) as follows:

$$S_{SS}(f) = S_{MM}(f) * S_{sc}(f)$$

$$= \frac{E_b T_b}{2} \left\{ \frac{1}{4} \left[ \delta \left( T_b f - \frac{1}{2} \right) + \delta \left( T_b f + \frac{1}{2} \right) \right] + \frac{4 \cos^2 (\pi T_b f)}{\pi^2 \left( 4 T_b^2 f^2 - 1 \right)^2} \right\} * \left[ \delta (f - f_c) + \delta (f + f_c) \right]$$

$$= \left\{ \frac{E_b}{8} \left[ \delta \left( T_b (f - f_c) - \frac{1}{2} \right) + \delta \left( T_b (f - f_c) + \frac{1}{2} \right) \right] + \frac{2 E_b \cos^2 (\pi T_b (f - f_c))}{\pi^2 \left( 4 T_b^2 (f - f_c)^2 - 1 \right)^2} \right\}$$

$$+ \left\{ \frac{E_b}{8} \left[ \delta \left( T_b (f + f_c) - \frac{1}{2} \right) + \delta \left( T_b (f + f_c) + \frac{1}{2} \right) \right] + \frac{2 E_b \cos^2 (\pi T_b (f + f_c))}{\pi^2 \left( 4 T_b^2 (f + f_c)^2 - 1 \right)^2} \right\}$$

$$(6.4.30)$$

The power spectral density for a binary frequency-shift keying signal $S_{SS}(f)$, which is normalized by $E_b$, is presented in Figure 6.4.9.

The power spectral density on a logarithmic scale is shown in Figure 6.4.10. It is calculated in the following way. The one-sided power spectral density $S_{SS}(f_n)$, obtained by overlapping the amplitudes at negative frequencies with the positive frequencies in $S_{SS}(f)$ and expressed as a function of modified frequency $f_n = (f - f_c) T_b$, has the form

$$S_{SS}(f_n) = \frac{E_b}{4} \left[ \delta \left( f_n - \frac{1}{2} \right) + \delta \left( f_n + \frac{1}{2} \right) \right] + \frac{4 E_b}{\pi^2} \frac{\cos^2 (\pi f_n)}{\left( 4 f_n^2 - 1 \right)^2}. \qquad (6.4.31)$$

In order to compare this power spectral density with those for binary phase-shift keying and quadrature phase-shift keying, we need to express it in logarithmic units and normalize it to have unit amplitude at the modified frequency $f_n = 0$. This normalized power spectral density is expressed as

$$\frac{S_{SS}(f_n)}{4 E_b / \pi^2} = \frac{\pi^2}{16} \left[ \delta \left( f_n - \frac{1}{2} \right) + \delta \left( f_n + \frac{1}{2} \right) \right] + \frac{\cos^2 (\pi f_n)}{\left( 4 f_n^2 - 1 \right)^2}. \qquad (6.4.32)$$

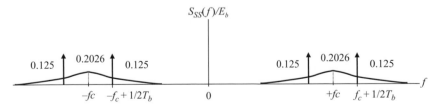

**Fig. 6.4.9** *Power spectral density for a binary frequency-shift keying signal.*

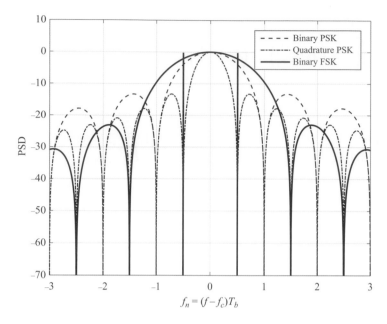

**Fig. 6.4.10** *Spectra for a modulated binary phase-shift keying signal, a binary frequency-shift keying signal, and a quadrature phase-shift keying signal; FSK, frequency-shift keying; PSD, power spectral density; PSK, phase-shift keying.*

In practice, we can express the power spectral density in logarithmic units as

$$\left.\frac{S_{SS}(f_n)}{4E_b/\pi^2}\right|_{dB} = \left(\log_{10}\frac{\pi^2}{16}\right) \cdot \left[\delta\left(f_n - \frac{1}{2}\right) + \delta\left(f_n + \frac{1}{2}\right)\right] + \log_{10}\frac{\cos^2(\pi f_n)}{\left(4f_n^2 - 1\right)^2}, \qquad (6.4.33)$$

and present it in graphical form as in Figure 6.4.10, alongside with the power spectral densities for binary phase-shift keying and quadrature phase-shift keying. The smoother the symbol shaping pulse $s_{ss}(t)$ is, the faster is the drop of the spectral tail. Since binary frequency-shift keying has a smoother pulse shape, it has lower side lobes than both phase-shift keying modulation schemes. However, the frequency-shift keying spectrum in its first lobe is wider than the corresponding spectra for binary phase-shift keying and quadrature phase-shift keying are.

## 6.4.3   Receiver Operation

### 6.4.3.1   *Operation of a Correlation Demodulator*

The received signal of a coherent binary frequency-shift keying can be expressed as

$$r(t) = s_i(t) + w(t). \qquad (6.4.34)$$

for $0 \leq t \leq T_b$, and $i = 1, 2$, where $w(t)$ is a sample function of the white Gaussian noise process $W(t)$ of zero mean and power spectral density $N_0/2$. At the output of the two correlators $(j = 1, 2)$ the calculated random values are

$$x_1 = \int_0^{T_b} r(t)\phi_1(t)dt = \int_0^{T_b} s_i(t)\phi_1(t)dt + \int_0^{T_b} w(t)\phi_1(t)dt = s_{i1} + w_1 \qquad (6.4.35)$$

and

$$x_2 = \int_0^{T_b} r(t)\phi_2(t)dt = \int_0^{T_b} s_i(t)\phi_2(t)dt + \int_0^{T_b} w(t)\phi_2(t)dt = s_{i2} + w_2, \qquad (6.4.36)$$

where $s_{i1}$ and $s_{i2}$ are treated as deterministic quantities contributed by the transmitted signal $s_i$, and $w_1$ and $w_2$ are the sample values of the random variables $W_1$ and $W_2$ that arise due to the channel noise $w(t)$. The demodulated signal vector $\mathbf{x}$ is defined by the pair $(x_1, x_2)$. If we apply the decision rule that the vector $\mathbf{x}$ lies in region $N_i$ if the Euclidean distance $||\mathbf{x} - \mathbf{s}_k||$ is minimum for $i = k$, then the decision boundary is the perpendicular bisector of the line joining two message points, as shown in Figure 6.4.4.

The outputs of the correlator, $x_1$ and $x_2$, are realizations of two random variables, $X_1$ and $X_2$. To simplify the decision procedure, we may define a new decision variable $X$ as the difference

$$X = X_1 - X_2. \qquad (6.4.37)$$

One realization of this variable is

$$x = x_1 - x_2 = (s_{i1} - s_{i2}) + (w_1 - w_2), \qquad (6.4.38)$$

which is used as the decision variable at the output of the correlator. This random value depends on the transmitted signal and the noise power in the channel. Keeping in mind the coefficient values, we find that, if bit 1 is sent, the decision value will be

$$x = (s_{i1} - s_{i2}) + (w_1 - w_2) = \left(\sqrt{E_b} - 0\right) + (w_1 - w_2) = \sqrt{E_b} + (w_1 - w_2) \qquad (6.4.39)$$

and, if bit 0 is sent, the decision value will be

$$x = (s_{i1} - s_{i2}) + (w_1 - w_2) = \left(0 - \sqrt{E_b}\right) + (w_1 - w_2) = -\sqrt{E_b} + (w_1 - w_2). \qquad (6.4.40)$$

This definition has the following logic. In any bit interval $T_b$, the transmitter generates a bit which can be 1 or 0. Consequently, the modulated signal is a sinusoidal pulse of frequency $f_1$ or $f_2$. The demodulator calculates two random values, $x_1$ and $x_2$, that define the elements of the demodulated vector $\mathbf{x}$. One of these two values contains the signal part, which depends on the power of the transmitted signal. The larger value will be the one which contains the signal part. Therefore, if we define the difference $x = x_1 - x_2$, it will contain evidence of where the signal part is. If the difference is positive, the most likely transmitted signal is $s_1$. If the difference is negative, the most likely transmitted signal is $s_2$. If either the calculated correlator outputs $x_1$ and $x_2$ or their difference $x = x_1 - x_2 = (s_{i1} - s_{i2}) + (w_1 - w_2)$ is available, the decision rules can be defined as follows:

1. If $x > 0$ or $x_1 > x_2$, then decide in favour of '1', because it is likely that the signal $s_1$ is transmitted, which implies that $s_{i2} = 0$. The demodulation vector $\mathbf{x}$ falls inside region $N_1$ on the constellation diagram in Figure 6.4.4.

2. If $x < 0$ or $x_2 > x_1$, then decide in favour of '0', because it is likely that the vector $\mathbf{x}$ is inside region $N_2$; we believe that the signal $s_2$ is transmitted, which implies that $s_{i1} = 0$.

3. If $x = 0$ or $x_2 = x_1$, then make a random guess in favour of '1' or '0'.

### 6.4.3.2    *Operation of an Optimum Detector*

As for a binary phase-shift keying system, we do not need the optimum detector to detect the message bits. Let us briefly confirm this fact by showing that the decision variable at the output of the optimum detector is equal to the decision variable $X$ at the output of the modulator.

In a binary frequency-shift keying system, two signals can be generated at the transmitter side, and the output of the correlation demodulator generates two calculated random values, $x_1 = s_{i1} + w_1$ and $x_2 = s_{i2} + w_2$, which are inputs for the optimum detector. These values are expressed in a transposed vector form as

$$\mathbf{x}^T = [x_1 \ x_2] = [s_{i1} + w_1 \quad s_{i2} + w_2]. \tag{6.4.41}$$

The transmitter signals can be expressed in the vector form presented in eqn (6.4.9). Each of the two signals is transmitted and processed in the same way inside the communication system. We will process the signals inside the optimum detector, as shown in Figure 6.4.1. Assuming the same energy for both possible symbols, that is, $E_1 = E_2 = E_b$, the following two random numerical values can be calculated at the output of the detector:

$$Z_1 = \mathbf{x}^T \mathbf{s}_1 - \frac{1}{2} E_b \underset{i=1}{=} [x_1 \ x_2] \begin{bmatrix} s_{11} \\ s_{12} \end{bmatrix} - \frac{1}{2} E_b = x_1 s_{11} + x_2 s_{12} - \frac{1}{2} E_b$$
$$\underset{s_{12}=0}{=} x_1 s_{11} - \frac{1}{2} E_b = (s_{i1} + w_1) s_{11} - \frac{1}{2} E_b \tag{6.4.42}$$

and

$$Z_2 = \mathbf{x}^T \mathbf{s}_2 - \frac{1}{2} E_b \underset{i=1}{=} [x_1 \ x_2] \begin{bmatrix} s_{21} \\ s_{22} \end{bmatrix} - \frac{1}{2} E_b = x_1 s_{21} + x_2 s_{22} - \frac{1}{2} E_b$$
$$\underset{s_{21}=0}{=} x_2 s_{22} - \frac{1}{2} E_b = (s_{i2} + w_2) s_{22} - \frac{1}{2} E_b \tag{6.4.43}$$

According to the operation of optimum detector for a generic communication system, as explained in Chapter 5, and its structure for the frequency-shift keying system, as presented in Figure 6.4.1, these two numerical values need to be compared and the largest one selected. Because the energies of the bits are equal, we obtain $s_{11} = s_{22} = E_b$. The comparison of calculated random values $Z_1$ and $Z_2$ can be performed by finding their difference:

$$Z_1 - Z_2 = (s_{i1} + w_1) s_{11} - \frac{1}{2} E_b - (s_{i2} + w_2) s_{22} + \frac{1}{2} E_b$$

$$= (s_{i1} + w_1) s_{11} - (s_{i2} + w_2) s_{22} = (s_{i1} s_{11} - s_{i2} s_{22}) + (w_1 s_{11} - w_2 s_{22})$$

$$(6.4.44)$$

Because the coefficients $s_{11} = s_{22}$ are constants, they have no influence on the decision process. Therefore, we can define the decision variable at the output of the optimum detector as

$$y = \frac{Z_1 - Z_2}{s_{11}} = (s_{i1} + s_{i2}) + (w_1 - w_2) = x, \qquad (6.4.45)$$

which is equal to the decision variable $x$ defined at the output of the correlator by eqn (6.4.38). Therefore, we do not need to implement the optimum detector for this modulation scheme, but just calculate the correlator output $x$ and make a decision, as explained in the previous section.

**Decision procedure.** The signals $\mathbf{s_1}$ and $\mathbf{s_2}$ are generated at the transmitter with the same probability, transmitted through the channel, and affected independently by the noise. Therefore, we can analyse one of the signals to find the bit error probability. Let us assume that the first signal $\mathbf{s_1}$ is transmitted and the signal coordinates are $s_{11} = \sqrt{E_b}$ and $s_{12} = 0$, as defined by eqn (6.4.4). The difference (6.4.44) of the optimum detector outputs in this case is

$$Z_1 - Z_2 = s_{11}^2 + (w_1 s_{11} - w_2 s_{22}) = E_b + (w_1 - w_2) \sqrt{E_b}. \qquad (6.4.46)$$

The numerical value of the decision variable can be defined as

$$y = \frac{Z_1 - Z_2}{\sqrt{E_b}} = \sqrt{E_b} + (w_1 - w_2). \qquad (6.4.47)$$

Therefore, the decision rule can be the same as for the random variable $X$ defined by eqn (6.4.39):

1. If $y > 0$, then decide in favour of '1', because $\mathbf{x}$ is inside $N_1$.
2. If $y < 0$, then decide in favour of '0', because $\mathbf{x}$ is inside $N_2$.
3. If $y = 0$, then make a random guess in favour of '1' or '0'.

Expression (6.4.47) is identical to the expression developed for the output of the correlation demodulator, eqn (6.4.39). Therefore, we can use this output to detect the bit. It is sufficient to process the correlator's outputs and make a decision, as explained in Section 6.4.3.1.

### 6.4.3.3 Calculation of the Bit Error Probability

The output of the correlation demodulator contains two random values that are realizations of the random variables $X_1$ and $X_2$. The decision variable is defined by eqn (6.4.38) as the difference of these two variables. This difference is defined as the realization of a new random decision variable $X$, which is defined by eqn (6.4.37) and expressed as

$$X = X_1 - X_2. \tag{6.4.48}$$

Then, the decision rules are defined in respect to the threshold value $x = 0$. Because the random variables $X_1$ and $X_2$ are Gaussian, the random variable $X$ is also Gaussian. Its mean value depends on which binary symbol has been transmitted, because the mean values of the individual variables $X_1$ and $X_2$ depend on the value of the binary symbol transmitted. The conditional mean of the random variables $X_1$ and $X_2$, given that bit 1 was transmitted, are

$$E\{X_1|1\} = \sqrt{E_b}, \; E\{X_2|1\} = 0, \tag{6.4.49}$$

respectively. Also, the conditional mean of the random variables $X_1$ and $X_2$, given that bit 0 was transmitted, are

$$E\{X_1|0\} = 0, \; E\{X_2|0\} = \sqrt{E_b}, \tag{6.4.50}$$

respectively. The conditional mean of the random variable $X$, given that bit 1 was transmitted, can be expressed as

$$E\{X|1\} = E\{X_1|1\} - E\{X_2|1\} = \sqrt{E_b} - 0 = \sqrt{E_b} \tag{6.4.51}$$

and, similarly, the conditional mean of the random variable $X$, given that bit 0 was transmitted, is

$$E\{X|0\} = E\{X_1|0\} - E\{X_2|0\} = 0 - \sqrt{E_b} = -\sqrt{E_b}. \tag{6.4.52}$$

Since $X_1$ and $X_2$ are statistically uncorrelated, Gaussian, and consequently independent, each with a variance equal to $N_0/2$, the variance of $X$ is equal to the sum of the variances for $X_1$ and $X_2$, that is,

$$\sigma_X^2 = 2\sigma_{X_i}^2 = 2\frac{N_0}{2} = N_0. \tag{6.4.53}$$

The conditional density function of $X$, given that bit 0 was transmitted, is then expressed as

$$f_X(x|0) = \frac{1}{\sqrt{2\pi N_0}} e^{-(x+\sqrt{E_b})^2/2N_0}, \tag{6.4.54}$$

and the conditional probability density function of $x$, given that bit 1 was transmitted, is

$$f_X(x|1) = \frac{1}{\sqrt{2\pi N_0}} e^{-(x-\sqrt{E_b})^2/2N_0}. \tag{6.4.55}$$

These densities are shown in Figure 6.4.11.

Suppose that bit 0 was sent. The receiver will make an error, that is, decide in favour of 1 instead of 0, with a small probability calculated as

$$P(1|0) = P(x > 0|0) = \int_0^\infty f_X(x|0)\, dx = \frac{1}{\sqrt{2\pi N_0}} \int_0^\infty e^{-(x+\sqrt{E_b})^2/2N_0}\, dx.$$

Substituting the variable of integration $x$ by $z = (x + \sqrt{E_b})/\sqrt{2N_0}$, we obtain

$$P(1|0) = \int_0^\infty f_X(x|0)\, dx = \frac{1}{\sqrt{\pi}} \int_{\sqrt{E_b/2N_0}}^\infty e^{-z^2}\, dz = \frac{1}{2} \operatorname{erfc} \sqrt{\frac{E_b}{2N_0}}. \tag{6.4.56}$$

If a transmission channel is symmetric, implying that $P(1|0) = P(0|1)$, the theoretical bit error probability in the binary frequency-shift keying system is

$$p = P(1)P(0|1) + P(0)P(1|0) = P(0|1) = \frac{1}{2}\operatorname{erfc}\sqrt{\frac{E_b}{2N_0}}. \tag{6.4.57}$$

This bit error probability as a function of $E_b/N_0$ is shown in Figure 6.4.12. alongside the probability of error for a binary phase-shift keying system and a quadrature phase-shift keying system. The ratio $E_b/N_0$ has to be doubled (3 dB) in the binary frequency-shift keying system to maintain the same probability of error as in a binary phase-shift keying system or a quadrature phase-shift keying system. This complies with the signal-space diagrams for a

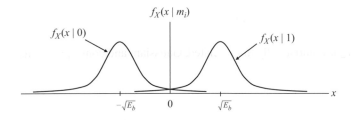

**Fig. 6.4.11** *Conditional probability density functions.*

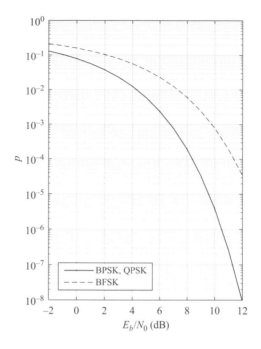

**Fig. 6.4.12** *Bit error probabilities for a binary phase-shift keying system, a binary frequency-shift keying system, and a quadrature phase-shift keying system; BFSK, binary frequency-shift keying system; BPSK, binary phase-shift keying; QPSK, quadrature phase-shift keying.*

binary phase-shift keying system and a binary frequency-shift keying systems because, for a prescribed bit energy $E_b$, the Euclidean distance in a binary phase-shift keying system is $\sqrt{2}$ times that in a binary frequency-shift keying system.

### 6.4.3.4   *Design of a Transceiver for a Binary Frequency-Shift Keying Signal*

**Transmitter operation.** Bearing in mind that the expression for a binary frequency-shift keying signal is

$$s(t) = \sqrt{\frac{2E_b}{T_b}} \cos 2\pi \left( f_c \mp \frac{1}{2T_b} \right) t = \begin{cases} \sqrt{E_b}\sqrt{\frac{2}{T_b}} \cos(2\pi f_1 t) & \text{for ` - 'sign} \\ \sqrt{E_b}\sqrt{\frac{2}{T_b}} \cos(2\pi f_2 t) & \text{for ` + 'sign} \end{cases},$$

the scheme of the coherent binary frequency-shift keying transmitter, shown in Figure 6.4.1, can be simplified as depicted in Figure 6.4.13. At the transmitter side, the message symbol 1's and 0's are transformed into two levels, $+\sqrt{E_b}$ and 0, respectively, using an on–off encoder. The non-return-to-zero unipolar binary signal $m(t)$ formed is directly applied to the input of a multiplier in the upper branch and, after inversion, the same signal is applied to the input of a multiplier in the lower branch. Thus, for the input bit 1, the oscillator in the upper branch is switched on and the incoming signal, with amplitude $+\sqrt{E_b}$, is multiplied by the generated basis function $\phi_1(t)$ defined by the frequency $f_1$. At the same time, the oscillator in the lower branch is switched off, because its input bit $m_{inv}(t)$, obtained from the inverter, is 0, and the output of the second multiplier is 0.

Similarly, in the case when the input bit is 1, the oscillator in the lower branch is switched on, and the incoming inverted signal with amplitude $+\sqrt{E_b}$ is effectively multiplied by the generated basis function $\phi_2(t)$ defined by its frequency $f_2$. At the same time, the multiplier output in the upper branch is 0, because its input is 0.

The following assumptions have been made:

1. The two oscillators in Figure 6.4.13 are synchronized, giving at their outputs two signals with frequencies $f_1$ and $f_2$. The two oscillators can be replaced by one voltage-controlled oscillator.
2. The frequency of the modulated signal is changed with continuous phase.

**Receiver operation.** We showed that the decision process inside the optimal binary frequency-shift keying receiver can take place at the output of the optimum detector or at the output of the correlation demodulator. For the sake of simplicity, we avoid the use of an optimum detector and design the receiver with the decision circuit at the output of demodulator, as shown in Figure 6.4.14.

To demodulate the received signal $r(t)$ and detect the message symbol, the receiver depicted in Figure 6.4.14 can be used. The receiver contains two correlators that calculate components of the correlation vector $\mathbf{x}$, defined by $x_1$ and $x_2$. The components are subtracted from each other to obtain the difference $x$ which is compared with a zero-level threshold. The decisions rules are simple, as shown in Figure 6.4.14, and defined as follows:

1. If $x > 0$, then decide in favour of '1'.
2. If $x < 0$, then decide in favour of '0'.
3. If $x = 0$, then make a random guess in favour of '1' or '0'.

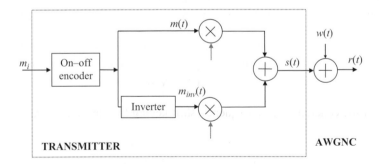

**Fig. 6.4.13** *A coherent binary frequency-shift keying transmitter; AWGNC, additive white Gaussian noise channel.*

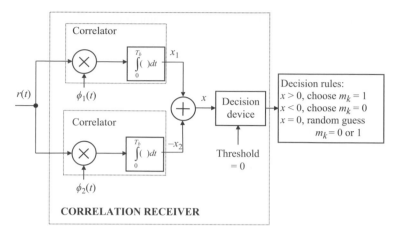

**Fig. 6.4.14** *A coherent binary frequency-shift keying receiver.*

## 6.5   *M*-ary Quadrature Amplitude Modulation

### 6.5.1   System Operation

An *M*-ary quadrature amplitude modulated signal can be expressed in terms of two quadrature carriers. Each of the carrier can be modulated with two independent amplitudes defined by $A_i$ and $B_i$ to obtain a quadrature amplitude modulation signal expressed as

$$s_i(t) = \sqrt{E}\sqrt{\frac{2}{T}}A_i \cos \omega_c t - \sqrt{E}\sqrt{\frac{2}{T}}B_i \sin \omega_c t, \qquad (6.5.1)$$

and defined for the *i*th symbol transmitted, where the constant $\sqrt{E}$ is the square root of energy of a symbol that has the minimum amplitude, and $T$ is the duration of a symbol. Based on the generic scheme of a correlator-based communication system, which is presented in Chapter 5 in Figure 5.8.1, an *M*-ary quadrature amplitude modulated system can be structured as presented in Figure 6.5.1.

Suppose the message source can generate two binary values (or bits) that belong to an alphabet A of two elements, where the messages are denoted by $m_1 = 1$ and $m_2 = 0$. A binary symbol $m_i$ generated at the output of the source contains *m* bits. Therefore, the number of message symbols is $M = 2^m$. By means of a serial-to-parallel converter, a message binary symbol $m_i$ is applied at the same time to the inputs of two look-up tables that are inside two parallel branches, the in-phase (*I*) system and the quadrature (*Q*) system. The look-up tables generate the amplitude values $A_i$ and $B_i$ at their outputs to form a message symbol represented as the pair vector $\mathbf{m_i}$, $i = 1, 2, \ldots, M$. The matrix $\mathbf{m}$ of all these vectors is presented in matrix form in eqn (6.5.2) for $M = 16$. For example, the symbol 1011 will be converted into the pair $\mathbf{m_1} = (A_1, B_1) = (-3, 3)$. The value $A_i$ will define the input of the *I* system, and $B_i$ will define

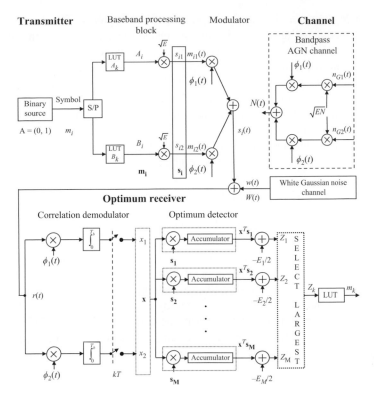

**Fig. 6.5.1** *A quadrature amplitude modulation continuous-time communication system; AGN, additive Gaussian noise; LUT, look-up table; S/P serial-to-parallel converter.*

the input of the $Q$ system. These values will be multiplied by $\sqrt{E}$ to generate the modulating signals in the two systems.

The pair vectors $\mathbf{m_i}$ are multiplied by the constant $\sqrt{E}$ to obtain signal coefficients, which are presented in matrix form by eqn (6.5.3).These coefficients define two modulating signals $m_{ij}(t)$: signal $m_{i1}(t)$ in the $I$ system, and $m_{i2}(t)$ in the $Q$ system. These signals are applied to the input of two modulator multipliers to modulate the orthogonal carriers. This procedure can also be understood as magnitude scaling of the carrier by the coefficients $s_{ij}$, which are represented by quadrature values $A_i\sqrt{E}$ and $B_i\sqrt{E}$. The level of scaling depends on the symbol structure at the input of the look-up table. At the output of each multiplier, an amplitude-shift keying signal is obtained. Two amplitude-shift keying signals are added to produce the quadrature amplitude modulation signal $s_i(t)$ at the output of the transmitter.

The modulated signal $s_i(t)$ is affected by the additive white Gaussian noise $w(t)$ in the communication channel, as shown in Figure 6.5.1, which is a realization of the wide-sense stationary noise process $W(t)$, which is presented theoretically in Chapter 19 and analysed in detail in Chapter 4. The communication channel in Figure 6.5.1 is also represented by the bandpass additive Gaussian noise $N(t)$, which is useful for system simulation and emulation, as can be seen in the theoretical analysis presented in Chapter 4. The received signal $r(t)$ is coherently demodulated using two correlators inside the correlation demodulator. At the outputs of correlators, the random samples $x_1$ and $x_2$ are obtained at the discrete-time instants $kT$. The samples are presented to the optimum detector to correlate them with all possible symbols that could be generated at the transmitter side, as was explained for a general correlation receiver in Chapter 5, Sections 5.5 and 5.6. However, in an $M$-ary quadrature amplitude modulation system, as in a quadrature phase-shift keying system and a binary phase-shift keying system, the decision about the symbol sent can be made at the output of the correlation demodulator by using the calculated random values $x_1$ and $x_2$ and applying the decision rules, as will be explained in Section 6.5.3.

## 6.5.2   Transmitter Operation

**Generation of the modulating signal.** The number of possible message symbols is $M = 2^m$, where $m$ is the number of bits in an $M$-ary symbol. We will analyse the case when an even number of bits form a symbol defining a square constellation diagram. It is possible to form a constellation diagram for an odd number of bits in a symbol. In that case, we define a cross constellation diagram. For the sake of explanation, we will analyse the case of a 16-ary quadrature amplitude modulation system defined by $M = 16$ message symbols presented in the constellation diagram in Figure 6.5.2. Each message symbol consists of $m = \log_2 M = 4$ message bits. The first two bits define the quadrant (in bold font) of a message symbol, and the last two bits define four distinct symbols that belong to the related quadrant.

For this constellation diagram, the pairs of coordinates $(A_i, B_i)$ specify all message symbols that can be expressed in the matrix form

$$\mathbf{m} = \begin{bmatrix} \mathbf{m}_1 & \mathbf{m}_2 & \mathbf{m}_3 & \mathbf{m}_4 \\ \mathbf{m}_5 & \mathbf{m}_6 & \mathbf{m}_7 & \mathbf{m}_8 \\ \mathbf{m}_9 & \mathbf{m}_{10} & \mathbf{m}_{11} & \mathbf{m}_{12} \\ \mathbf{m}_{13} & \mathbf{m}_{14} & \mathbf{m}_{15} & \mathbf{m}_{16} \end{bmatrix} \begin{bmatrix} -3,3 & -1,3 & 1,3 & 3,3 \\ -3,1 & -1,1 & 1,1 & 3,1 \\ -3,-1 & -1,-1 & 1,-1 & 3,-1 \\ -3,-3 & -1,-3 & 1,-3 & 3,-3 \end{bmatrix}. \tag{6.5.2}$$

These pairs of coordinates are multiplied by $\sqrt{E}$ to obtain a pair of amplitudes of modulating signals $s_{i1}$ and $s_{i2}$ that correspond to the generated symbols and can be expressed in the matrix form

$$\mathbf{s} = \mathbf{m}\sqrt{E} = \begin{bmatrix} -3\sqrt{E},3\sqrt{E} & -\sqrt{E},3\sqrt{E} & \sqrt{E},3\sqrt{E} & 3\sqrt{E},3\sqrt{E} \\ -3\sqrt{E},\sqrt{E} & -\sqrt{E},\sqrt{E} & \sqrt{E},\sqrt{E} & 3\sqrt{E},\sqrt{E} \\ -3\sqrt{E},-\sqrt{E} & -\sqrt{E},-\sqrt{E} & \sqrt{E},-\sqrt{E} & 3\sqrt{E},-\sqrt{E} \\ -3\sqrt{E},-3\sqrt{E} & -\sqrt{E},-3\sqrt{E} & \sqrt{E},-3\sqrt{E} & 3\sqrt{E},-3\sqrt{E} \end{bmatrix}. \tag{6.5.3}$$

The first row in the matrix corresponds to the message symbols specified by the top-most four points in the constellation diagram.

For the second message symbol $\mathbf{m}_2$, the wave shape of the first modulating signal $m_{21}(t)$ is a rectangular pulse of duration $T$ and amplitude equal to $-\sqrt{E}$ for the $i$th coefficient, where $i = 2$, as presented in Figure 6.5.3. In the same figure, the second modulating signal is shown

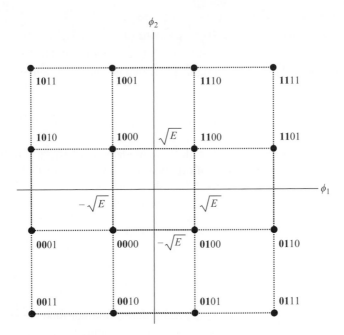

**Fig. 6.5.2** *Constellation diagram of a 16-ary quadrature amplitude modulation.*

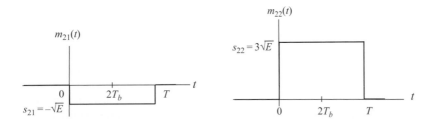

**Fig. 6.5.3** *Modulating rectangular pulses that appear at the input of modulator multipliers.*

with amplitude $3\sqrt{E}$. The duration of the modulating signal $T$ is four times longer than the duration of one bit, $T_b$.

**Generation of a modulated signal.** For the same example, we will explain the operation of the modulator. The $M$-ary quadrature amplitude modulation modulator uses two orthonormal basis functions defined as

$$\phi_1(t) = \sqrt{\frac{2}{T}} \cos \omega_c t, \qquad (6.5.4)$$

$$\phi_2(t) = -\sqrt{\frac{2}{T}} \sin \omega_c t, \tag{6.5.5}$$

which are used to obtain the modulated signal presented in eqn (6.5.1) and expressed in terms of these functions as

$$s_i(t) = \sqrt{E}A_i\phi_1(t) + \sqrt{E}B_i\phi_2(t) = s_{i1}(t)\phi_1(t) + s_{i2}(t)\phi_2(t). \tag{6.5.6}$$

The frequency of the carrier is much larger than the fundamental frequency of the rectangular pulse, that is, $f = 1/T << f_c$. For the sake of explanation, we will assume that the carrier makes four oscillations inside the symbol interval $T$, as shown in Figure 6.5.4.

For this particular example, the two modulated signals presented in Figure 6.5.4 are added to obtain a quadrature amplitude modulation signal at the transmitter output. The modulated signal is then transmitted through the channel. For our analysis of the receiver operation, we will assume that the signal is affected by the white Gaussian noise $w(t)$ in the channel. In Figure 6.5.1, a bandpass Gaussian noise $N(t)$ channel is also presented that can be designed and used in the simulation or emulation of communication systems, as explained in Chapter 4.

Bearing in mind the operation of an $M$-ary quadrature amplitude modulated system, we can simplify the scheme of its transmitter, as we did for the transmitter of a quadrature phase-shift keying system, as presented in Figure 6.5.5.

The operation of the transmitter will be explained with the following example. Suppose the second message symbol $m_2 = (1001)$ is generated at the output of the source, as shown in Figure 6.5.5. This binary symbol is transferred into a parallel form by a series-to-parallel converter and used as an address inside the look-up tables to generate the symbol pair $(A_i, B_i)$, which is multiplied by $\sqrt{E}$ to form the modulating rectangular pulses $m_{i1}(t) = -\sqrt{E}$ and $m_{i2}(t) = 3\sqrt{E}$ inside the interval $0 \le t \le T$, as shown in Figure 6.5.3. These are traditionally denoted the in-phase $I(t)$ component and the quadrature $Q(t)$ component, respectively, and are generated in the upper and lower branches of the transmitter, respectively, which are called the $I$ system and the $Q$-system, respectively. The modulating signals are filtered in

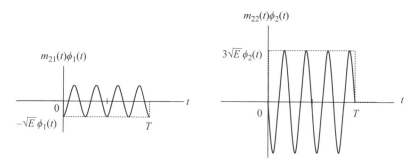

**Fig. 6.5.4** *In-phase and quadrature components of a modulated M-ary quadrature amplitude modulation signal.*

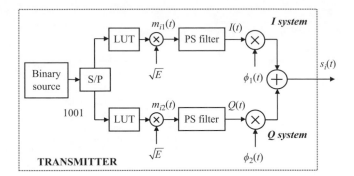

**Fig. 6.5.5** *A traditional scheme for an M-ary quadrature amplitude modulation transmitter; LUT, look-up table; PS, pulse-shaping; S/P, series-to-parallel converter.*

the pulse-shaping filters before the modulation takes place. For the sake of explanation, we will exclude the filtering procedure in our explanation and assume that the rectangular pulses are applied at the input of the multiplier performing modulation.

If the pulse shaping filter is used, the modulated signal can be expressed in terms of in-phase and quadrature components as

$$s_i(t) = I(t)\phi_1(t) + Q(t)\phi_2(t), \tag{6.5.7}$$

which is the standard expression usually used in practice. For the sake of explanation, we will exclude the operation of the pulse-shaping filter in the following analysis.

## 6.5.3   Receiver Operation

**The correlation demodulator.** A simplified receiver structure is presented in Figure 6.5.6. The modulated signal is transmitted through an additive white Gaussian noise channel and received by a coherent quadrature amplitude modulation correlation receiver. The received signal can be expressed as

$$r(t) = s_i(t) + w(t), \tag{6.5.8}$$

for $0 \leq t \leq T$, where $w(t)$ is a sample of a white Gaussian noise process $W(t)$ of zero mean and power spectral density $N_0/2$. Detailed explanation of this noise can be found in Chapter 4. At the output of the two correlators ($j = 1, 2$) are discrete-time random signals defined by their random values $x_1$ and $x_{2+}$ which are calculated at the end of each symbol interval $T$ as

$$x_1 = \int_0^T r(t)\phi_1(t)dt = \int_0^T s_i(t)\phi_1(t)dt + \int_0^T w(t)\phi_1(t)dt = \sqrt{E}A_i + w_1 = s_{i1} + w_1, \tag{6.5.9}$$

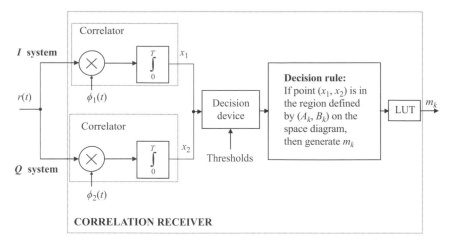

**Fig. 6.5.6** *Block schematic of an M-ary quadrature amplitude modulation receiver; LUT, look-up table.*

and

$$x_2 = \int_0^T r(t)\phi_2(t)dt = \sqrt{E}B_i + w_2 = s_{i2} + w_2, \tag{6.5.10}$$

Respectively, where $s_{i1}$ and $s_{i2}$ are deterministic quantities contributed by the transmitted signal $s_i$, and $w_1$ and $w_2$ are the sample values of the random variables $W_1$ and $W_2$ that arise due to the channel noise process $W(t)$.

The calculated random values $x_1$ and $x_2$ are realizations of the random variables $X_1$ and $X_2$ and can be expressed in the form of a demodulation vector $\mathbf{x} = [x_1 \, x_2]$. Having this vector, we can decide on the transmitted symbols, using the following decision rules:

1. If the output of the correlator, the demodulation vector $\mathbf{x}$, falls in the region of a particular symbol, then we conclude that symbol was sent. In our example, if the coordinates defined by vector $\mathbf{x}$ are in the region of the symbol $s_2$ defined by the pair $(A_2, B_2)$, we will rightly decide in favour of the binary message (1001).

2. If the output of the correlator $\mathbf{x}$ falls in a region of any other symbol, the wrong decision will be made in favour of a message symbol which is not (1001).

**The optimum detector.** The decision can be made by strictly using the procedure defined for the optimal detector. Suppose the signal $s_i(t) = s_2(t)$ was transmitted in an $M$-ary quadrature amplitude modulated system. For the sake of simplicity, we will investigate the operation of the optimum detector only for the signals $s_1(t)$, $s_2(t)$, and $s_7(t)$ by calculating the decision variables $Z_1$, $Z_2$, and $Z_7$ and then comparing them.

If the second signal $s_2(t)$ is sent, the decision variable will be calculated for the signal energy

$$E_2 = \left(3\sqrt{E}\right)^2 + \sqrt{E}^2 = 9E + E = 10E$$

and the related decision variable will be

$$Z_2 = \mathbf{x}^T \mathbf{s}_2 = \left[\sqrt{E}A_2 + w_1 \quad \sqrt{E}B_2 - w_2\right]\begin{bmatrix} A_2\sqrt{E} \\ B_2\sqrt{E} \end{bmatrix} - \frac{1}{2}E_2$$
$$= \sqrt{E}A_2\left(\sqrt{E}A_2 + w_1\right) + \sqrt{E}B_2\left(\sqrt{E}B_2 - w_2\right) - \frac{1}{2}10E$$
$$= EA_2^2 + EB_2^2 - 5E + \sqrt{E}A_2 w_1 - \sqrt{E}B_2 w_2 = E + 9E - 5E - \sqrt{E}w_1 + 3\sqrt{E}w_2$$
$$= 5E - \sqrt{E}\left(3w_2 - w_1\right)$$

The energy of the first signal is

$$E_1 = \left(3\sqrt{E}\right)^2 + \left(-3\sqrt{E}\right)^2 = 9E + 9E = 18E,$$

and the related energy, assuming that the second signal is transmitted, is

$$Z_1 = \mathbf{x}^T \mathbf{s}_1 = \left[\sqrt{E}A_2 + w_1 \quad \sqrt{E}B_2 - w_2\right]\begin{bmatrix} A_1\sqrt{E} \\ B_1\sqrt{E} \end{bmatrix} - \frac{1}{2}E_2$$
$$= \sqrt{E}A_1\left(\sqrt{E}A_2 + w_1\right) + \sqrt{E}B_1\left(\sqrt{E}B_2 - w_2\right) - \frac{1}{2}18E$$
$$= EA_1 A_2 + EB_1 B_2 - 9E + \sqrt{E}A_1 w_1 - \sqrt{E}B_1 w_2$$
$$= 3E + 9E - 9E - 3\sqrt{E}w_1 - 3\sqrt{E}w_2 = 3E - 3\sqrt{E}\left(w_2 - w_1\right)$$

The energy of the seventh signal is

$$E_7 = \left(\sqrt{E}\right)^2 + \left(\sqrt{E}\right)^2 = E + E = 2E$$

and the related energy, assuming that the second signal is transmitted, is

$$Z_7 = \mathbf{x}^T \mathbf{s}_7 = \left[\sqrt{E}A_2 + w_1 \quad \sqrt{E}B_2 - w_2\right]\begin{bmatrix} A_7\sqrt{E} \\ B_7\sqrt{E} \end{bmatrix} - \frac{1}{2}E_7$$
$$= \sqrt{E}A_7\left(\sqrt{E}A_2 + w_1\right) + \sqrt{E}B_7\left(\sqrt{E}B_2 - w_2\right) - \frac{1}{2}2E$$
$$= EA_7 A_2 + EB_7 B_2 - 9E + \sqrt{E}A_7 w_1 - \sqrt{E}B_7 w_2$$
$$= -3E + 3E - E + \sqrt{E}w_1 - \sqrt{E}w_2 = -E - \sqrt{E}\left(w_2 - w_1\right)$$

If we exclude the noise influence on the decision variables, we can see that the variable belonging to the supposedly sent symbol $s_2(t)$ is the largest. Therefore, the decision device would make the right decision if it decides on $m_i = m_2$ in that case. However, as often happens in reality, if the noise part that is present in all decision variables has a value that makes $Z_1$ or $Z_7$ greater than $Z_2$, the decision will be in favour of $Z_1$ or $Z_7$, causing an incorrect decision.

**Example**

Calculate the values of the demodulation vector **x** by expressing the basis orthogonal functions in their full form as in expressions (6.5.4) and (6.5.5). Comment on the minus sign in front of the second basis function.

**Solution.** If we insert eqns (6.5.4) and (6.5.5) into (6.5.9), we find that the first value of the demodulation vector is

$$
\begin{aligned}
x_1 &= \sqrt{E}\frac{2}{T}\int_0^T (A_i\cos\omega_c t - B_i\sin\omega_c t)(\cos\omega_c t)\,dt + \sqrt{\frac{2}{T}}\int_0^T w(t)\cos\omega_c t\,dt \\
&= \sqrt{E}A_i\frac{2}{T}\int_0^T \frac{1}{2}\,dt - 0 + w_1 = \sqrt{E}A_i + w_1 = s_{i1} + w_1
\end{aligned}
$$

and the second value can be obtained by inserting eqns (6.5.4) and (6.5.5) into eqn (6.5.9) to obtain

$$
\begin{aligned}
x_2 &= \sqrt{E}\frac{2}{T}\int_0^T (A_i\cos\omega_c t - B_i\sin\omega_c t)(-\sin\omega_c t)\,dt - \sqrt{\frac{2}{T}}\int_0^T w(t)\sin\omega_c t\,dt \\
&= 0 + \sqrt{E}B_i\frac{2}{T}\int_0^T \frac{1}{2}\,dt - 0 - w_2 = \sqrt{E}B_i - w_2 = s_{i2} - w_2
\end{aligned}
$$

The sign of the noise part $w_2$ does not statistically affet the decision because the density function of the noise is Gaussian with zero mean and constant variance. Therefore, the modulation and demodulation need to be performed with a sine function with a negative sign to fulfil the condition of the coherent demodulation. Namely, in that case, the second basis function $\phi_2(t)$ at the transmitter side will be synchronized in frequency, and the phase with the basis function $\phi_2(t)$ at the receiver side. The negative sign in front of the noise term $w_2$ will not change the mean value and the variance of its Gaussian density function and will not have any consequence on the statistical properties of the receiver operation. If the negative sign is not assigned to the amplitude of the quadrature carrier $\phi_2(t)$, which contains the coordinate of the message symbol, there would be an ambiguity in the decision process related to the position of the related message symbol in the constellation diagram.

## APPENDIX A: DENSITIES OF THE CORRELATION VARIABLES $X_1$ AND $X_2$ IN A QUADRATURE PHASE-SHIFT KEYING SYSTEM

For the correlation variables $X_1$ and $X_2$, defined at the output of a quadrature phase-shift keying correlation demodulator, the conditional probability density function of $X_1$, given that any $m_i$ was transmitted, is

$$f_{X_1}(x_1|m_i) = \frac{1}{\sqrt{\pi N_0}} e^{-(x_1-s_{i1})^2/N_0},$$

and the conditional probability density function of $X_2$, given that any $m_i$ was transmitted, is

$$f_{X_2}(x_2|m_i) = \frac{1}{\sqrt{\pi N_0}} e^{-(x_2-s_{i2})^2/N_0}.$$

The first density function can be expressed separately for $m_i = m_1 = 1$ and $m_i = m_2 = 0$, as follows:

$$f_{X_1}(x_1|1) = \frac{1}{\sqrt{\pi N_0}} e^{-(x_1-s_{11})^2/N_0}, \quad f_{X_1}(x_1|0) = \frac{1}{\sqrt{\pi N_0}} e^{-(x_1-s_{21})^2/N_0}.$$

These densities are shown in Figure A1.1.

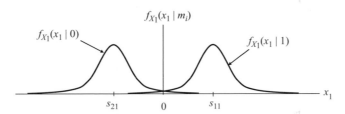

**Fig. A1.1** *Probability density functions.*

Likewise, the second density function for the random variable $X_2$ can be expressed separately for $m_i = m_1 = 1$ and $m_i = m_2 = 0$, as

$$f_{X_2}(x_2|1) = \frac{1}{\sqrt{\pi N_0}} e^{-(x_2-s_{12})^2/N_0}, \quad f_{X_2}(x_2|0) = \frac{1}{\sqrt{\pi N_0}} e^{-(x_2-s_{22})^2/N_0},$$

and presented analogously in graphical form as for $X_1$ in Figure A1.1.

.............................................................................................................................................

APPENDIX B: DERIVATIVES OF DENSITY FUNCTIONS FOR A BINARY
FREQUENCY-SHIFT KEYING SYSTEM

For the coherent binary frequency-shift keying receiver presented in Figure 6.4.1, we can
define the random variables $X_1$ and $X_2$ with realizations $x_1$ and $x_2$, respectively, as

$$x_1 = \int_0^{T_b} r(t)\phi_1(t)dt = \int_0^{T_b} s_i(t)\phi_1(t)dt + \int_0^{T_b} w(t)\phi_1(t)dt = s_{i1} + w_1$$

and

$$x_2 = \int_0^{T_b} r(t)\phi_2(t)dt = \int_0^{T_b} s_i(t)\phi_2(t)dt + \int_0^{T_b} w(t)\phi_2(t)dt = s_{i2} + w_2,$$

respectively. For the defined random variables $X_1$ and $X_2$, we can find the related probability
density functions. The random variable $X_1$ with its realization $x_1$ can be expressed in the form

$$X_1 = s_{i1} + W_1,$$

where the signal coefficients are defined by the matrix

$$\mathbf{s} = [\mathbf{s}_1 \ \ \mathbf{s}_2] = \begin{bmatrix} s_{11} & s_{21} \\ s_{12} & s_{22} \end{bmatrix} = \begin{bmatrix} \sqrt{E_b} & 0 \\ 0 & \sqrt{E_b} \end{bmatrix}.$$

Therefore, for the wide-sense stationary Gaussian noise process $W_1(t)$, the random variable
$X_1$ can be expressed in two ways: as

$$X_1 = s_{i1} + W_1 = s_{11} + W_1 = \sqrt{E_b} + W_1$$

if bit $m_1 = 1$ is transmitted, and as

$$X_1 = s_{i1} + W_1 = 0 + W_1 = W_1,$$

if bit $m_2 = 0$ is transmitted. Because the noise has Gaussian density, there are two conditional
density functions defined for the random variable $X_1$.

If bit $m_1 = 1$ is transmitted, the mean value and the variance of $X_1$ are defined as

$$E\{X_1|1\} = \sqrt{E_b},$$

$$\sigma_{X_1}^2 = E\left\{X_1^2|1\right\} - E^2\{X_1|1\} = E_b + \sigma_{W_1}^2 - E_b = \frac{N_0}{2},$$

respectively. If bit $m_2 = 0$ is transmitted, the mean value and the variance of $X_1$ are calculated as

$$E\{X_1|0\} = 0,$$

$$\sigma_{X_1}^2 = E\left\{X_1^2|1\right\} - E^2\{X_1|1\} = E\left\{W_1^2|1\right\} - E^2\{W_1|1\} = \sigma_{W_1}^2 - 0 = \frac{N_0}{2},$$

respectively. Therefore, the corresponding conditional density functions are

$$f_{X_1}(x_1|1) = \frac{1}{\sqrt{2\pi N_0/2}} e^{-(x_1-\sqrt{E_b})^2/2N_0/2} = \frac{1}{\sqrt{\pi N_0}} e^{-(x_1-\sqrt{E_b})^2/N_0}$$

and

$$f_{X_1}(x_1|0) = \frac{1}{\sqrt{2\pi N_0/2}} e^{-x_1^2/2N_0/2} = \frac{1}{\sqrt{\pi N_0}} e^{-x_1^2/N_0}.$$

Graphs of these density functions are presented in in Figure B1.1.

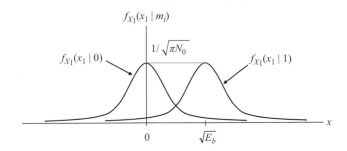

**Fig. B1.1** *Probability density functions.*

Analogously, the density function for the variable $X_2$ can be derived as follows. The random variable $X_2$ is

$$X_2 = s_{i2} + W_2.$$

If bit $m_1 = 1$ is transmitted, the variable is

$$X_2 = s_{12} + W_2 = 0 + W_2 = W_2$$

and, if bit $m_2 = 0$ is transmitted, the variable is

$$X_2 = s_{i2} + W_2 = s_{22} + W_2 = \sqrt{E_b} + W_2.$$

Because the noise has Gaussian density, there are two conditional density functions defined for the random variable $X_2$.

If bit $m_1 = 1$ is transmitted, the mean value and the variance of $X_2$ are

$$E\{X_2|1\} = 0,$$

$$\sigma_{X_2}^2 = E\left\{X_2^2|1\right\} - E^2\{X_2|1\} = E\left\{W_2^2|1\right\} - E^2\{W_2|1\} = \sigma_{W_2}^2 - 0 = \frac{N_0}{2},$$

respectively. If bit $m_2 = 0$ is transmitted, the mean value and the variance of $X_2$ are

$$E\{X_2|0\} = \sqrt{E_b},$$

$$\sigma_{X_2}^2 = E\left\{X_2^2|0\right\} - E^2\{X_2|0\} = E_b + \sigma_{W_2}^2 - E_b = \frac{N_0}{2},$$

respectively. The density function when bit $m_2 = 1$ is transmitted is

$$f_{X_2}(x_2|1) = \frac{1}{\sqrt{\pi N_0}} e^{-x_2^2/N_0},$$

and that when bit $m_2 = 0$ is transmitted is

$$f_{X_2}(x_2|0) = \frac{1}{\sqrt{\pi N_0}} e^{-\left(x_2 - \sqrt{E_b}\right)^2/N_0},$$

which are at the same positions as the graphs shown in Figure B1.1 for the random variable $X_1$.

..................................................................................................................................

## APPENDIX C: PRECISE DERIVATION OF THE BIT ERROR PROBABILITY FOR A BINARY FREQUENCY-SHIFT KEYING SYSTEM

The random variables at the output of a binary frequency-shift keying correlator, $X_1$ and $X_2$, are defined by their zero mean and variance $N_0/2$. The decision variable is defined as their difference, that is, $X = X_1 - X_2$. We can find the density function for these variables and derive the expression for the bit error probability in this binary frequency-shift keying system as follows.

The variables $X_1$ and $X_2$ are statistically independent Gaussian variables, and consequently uncorrelated, each with zero mean and variance equal to $N_0/2$. Following eqns (6.4.35) and (6.4.36), they can be expressed as

$$X_1 = s_{i1} + W_1$$

and

$$X_2 = s_{i2} + W_2.$$

If bit $m_1 = 1$ is transmitted, the variables are

$$X_1 = s_{i1} + W_1 = s_{11} + W_1 = \sqrt{E_b} + W_1,$$
$$X_2 = s_{12} + W_2 = 0 + W_2 = W_2,$$

and

$$X = X_1 - X_2 = \sqrt{E_b} + W_1 - W_2.$$

Because the noise is white and Gaussian, the mean value of $X$ is

$$E\{X|1\} = \sqrt{E_b} + 0 - 0 = \sqrt{E_b},$$

and the corresponding variance is

$$\sigma_X^2 = E\left\{X^2\right\} - E^2\{X\} = E\left\{\left(\sqrt{E_b} + W_1 - W_2\right)^2\right\} - E_b$$
$$= E_b + 0 + 0 + 0 + \sigma_{W1}^2 + \sigma_{W2}^2 - E_b = \frac{N_0}{2} + \frac{N_0}{2} = N_0.$$

The conditional density function of $X$, given that symbol $m_1 = 1$ is transmitted, is

$$f_X(x|1) = \frac{1}{\sqrt{2\pi N_0}} e^{-(x - \sqrt{E_b})^2 / 2N_0}.$$

If bit $m_2 = 0$ is transmitted, the variables are

$$X_1 = s_{i1} + W_1 = 0 + W_1 = W_1,$$
$$X_2 = s_{i2} + W_2 = s_{22} + W_2 = \sqrt{E_b} + W_2,$$

and

$$X = X_1 - X_2 = W_1 - \sqrt{E_b} - W_2.$$

Because the noise is white and Gaussian, the mean value and the variance of $X$ are

$$E\{X|0\} = -\sqrt{E_b} + 0 - 0 = -\sqrt{E_b},$$

and

$$\sigma_X^2 = E\left\{X^2\right\} - E^2\{X\} = E\left\{\left(-\sqrt{E_b} + W_1 - W_2\right)^2\right\} - E_b$$
$$= E_b + 0 + 0 + 0 + \sigma_{W1}^2 + \sigma_{W2}^2 - E_b = \frac{N_0}{2} + \frac{N_0}{2} = N_0,$$

respectively. The conditional density function of $X$, given that symbol $m_1 = 0$ is transmitted, is

$$f_X(x|0) = \frac{1}{\sqrt{2\pi N_0}} e^{-(x+\sqrt{E_b})^2/2N_0}.$$

These densities are shown in Figure C1.1.

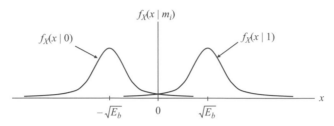

$f_X(x \mid m_i)$

$f_X(x \mid 0)$          $f_X(x \mid 1)$

$-\sqrt{E_b}$      $0$      $\sqrt{E_b}$      $x$

**Fig. C1.1** *Probability density functions.*

Suppose that symbol 0 was sent. The receiver will make an error, that is, decide in favour of 1 instead 0, with the small probability

$$P(1|0) = P(x > 0|0) = \int_0^\infty f_X(x|0)\,dx = \frac{1}{\sqrt{2\pi N_0}} \int_0^\infty e^{-(x+\sqrt{E_b})^2/2N_0}\,dx.$$

Substituting the variable $x$ with $z = (y + \sqrt{E_b})/\sqrt{2N_0}$, we obtain

$$P(1|0) = \int_0^\infty f_X(x|0)\,dx = \frac{1}{\sqrt{\pi}} \int_{\sqrt{E_b/2N_0}}^\infty e^{-z^2}\,dz = \frac{1}{2}\,\mathrm{erfc}\sqrt{\frac{E_b}{2N_0}}.$$

If a transmission channel is symmetric, the average probability of error in a binary frequency-shift keying system is

$$p = \frac{1}{2}\,\mathrm{erfc}\sqrt{\frac{E_b}{2N_0}}.$$

........................................................................................................................................

APPENDIX D: POWER SPECTRAL DENSITY OF A QUADRATURE COMPONENT IN A FREQUENCY-SHIFT KEYING SIGNAL

Prove that the power spectral density of a quadrature component of a binary frequency-shift keying signal is expressed by

$$S_{QQ}(f) = \frac{4E_b T_b}{\pi^2} \frac{\cos^2(\pi T_b f)}{\left(4T_b^2 f^2 - 1\right)^2}.$$

**Proof.** The power spectral density of a quadrature component of a binary frequency-shift keying signal can be expressed in terms of sinc functions as

$$S_{QQ}(f) = \frac{E_b T_b}{4} |\mathrm{sinc}\, T_b\, (f - 1/2T_b) + \mathrm{sinc}\, T_b\, (f + 1/2T_b)|^2.$$

Let us make a substitution to express the argument of the sinc function as

$$\pi T_b (f - 1/2T_b) = \pi T_b f - \pi/2 = \alpha - \beta.$$

We then may express the normalized power spectral density as

$$S_{QQ}(f)/E_b T_b/4 = |\mathrm{sinc}\, T_b\, (f - 1/2T_b) + \mathrm{sinc}\, T_b\, (f + 1/2T_b)|^2$$
$$= \left| \frac{\sin \pi T_b (f - 1/2T_b)}{\pi T_b (f - 1/2T_b)} + \frac{\sin \pi T_b (f + 1/2T_b)}{\pi T_b (f + 1/2T_b)} \right|^2 = \left| \frac{\sin (\alpha - \beta)}{(\alpha - \beta)} + \frac{\sin (\alpha + \beta)}{(\alpha + \beta)} \right|^2.$$

We can use trigonometric identities to rearrange the last expression as

$$S_{QQ}(f)/E_b T_b/4 = \left| \frac{\sin (\alpha - \beta)}{(\alpha - \beta)} + \frac{\sin (\alpha + \beta)}{(\alpha + \beta)} \right|^2$$

$$= \left| \frac{\sin \alpha \cos \beta - \cos \alpha \sin \beta}{(\alpha - \beta)} + \frac{\sin \alpha \cos \beta + \cos \alpha \sin \beta}{(\alpha + \beta)} \right|^2 \underset{\sin \beta = 1, \cos \beta = 0}{=} \left( \frac{-\cos \alpha}{(\alpha - \beta)} + \frac{\cos \alpha}{(\alpha + \beta)} \right)^2$$

$$= \cos^2 \alpha \left( \frac{1}{(\alpha - \beta)^2} - \frac{2}{(\alpha^2 - \beta^2)} + \frac{1}{(\alpha + \beta)^2} \right)$$

$$= \cos^2 \alpha \left( \frac{(\alpha^2 - \beta^2)(\alpha + \beta)^2 - 2(\alpha + \beta)^2(\alpha - \beta)^2 + (\alpha^2 - \beta^2)(\alpha - \beta)^2}{(\alpha - \beta)^2 (\alpha^2 - \beta^2)(\alpha + \beta)^2} \right)$$

$$= \cos^2 \alpha \left( \frac{(\alpha^2 - \beta^2)[(\alpha + \beta)^2 + (\alpha - \beta)^2] - 2(\alpha^2 - \beta^2)(\alpha^2 - \beta^2)}{(\alpha - \beta)^2 (\alpha^2 - \beta^2)(\alpha + \beta)^2} \right)$$

$$= \cos^2 \alpha \left( \frac{2(\alpha^2 - \beta^2)(\alpha^2 + \beta^2)] - 2(\alpha^2 - \beta^2)(\alpha^2 - \beta^2)}{(\alpha - \beta)^2 (\alpha^2 - \beta^2)(\alpha + \beta)^2} \right)$$

$$= \cos^2 \alpha \left( \frac{2(\alpha^2 - \beta^2)(\alpha^2 + \beta^2 - \alpha^2 + \beta^2)]}{(\alpha - \beta)^2 (\alpha^2 - \beta^2)(\alpha + \beta)^2} \right) = \cos^2 \alpha \left( \frac{4\beta^2}{(\alpha - \beta)^2(\alpha + \beta)^2} \right)$$

$$= \cos^2 \alpha \left( \frac{4\beta^2}{(\alpha^2 - \beta^2)^2} \right)$$

If we insert the values for $\alpha = \pi T_b f$ and $\beta = \pi/2$ into the last expression, we obtain

$$S_{QQ}(f)/E_b T_b/4 = \cos^2(\pi T_b f) \frac{16}{\pi^2 (4f^2 T_b^2 - 1)^2},$$

and the power spectral density is expressed as

$$S_{QQ}(f) = E_b T_b \frac{4\cos^2(\pi T_b f)}{\pi^2 (4f^2 T_b^2 - 1)^2}.$$

PROBLEMS

1. A binary phase-shift keying signal is applied to a correlator supplied with a phase reference that lies within $\varphi$ radians (phase error) of the exact carrier phase.

   a. Determine the effect of the phase error $\varphi$ on the average probability of error of the system.
   b. Find the values for the probability of error $P_e$ for the phase errors $\varphi = 0°$ and $\varphi = 90°$.
   c. Analyse the case when $\varphi = 180°$.

2. The signal component of a coherent phase-shift keying system is defined by

$$s(t) = A_c k \sin 2\pi f_c t \pm A_c \sqrt{1 - k^2} \cos 2\pi f_c t.$$

   where $0 \leq t \leq T_b$ and where the plus sign corresponds to symbol 1 and the minus sign corresponds to symbol 0. The first term represents a carrier component that is included for the purpose of synchronizing the receiver to the transmitter.

   a. Draw a signal-space diagram for the scheme described above. What observations can you make about this diagram?
   b. Show that, in the presence of an additive white Gaussian noise of zero mean and power spectral density $N_0/2$, the average bit error probability is

$$P_e = \frac{1}{2} \text{erfc} \left( \sqrt{\frac{E_b}{N_0} (1 - k^2)} \right)$$

   where

$$E_b = \frac{A_c^2 T_b}{2}.$$

   c. Suppose that 10% of the transmitted signal power is allocated to the carrier component. Determine the $E_b/N_0$ required to realize a bit error probability equal to $10^{-4}$.

   d. Compare this value of $E_b/N_0$ with that required for a conventional phase-shift keying
system with the same probability of error.

3. Find the bit error probability for a binary phase-shift keying system with a bit rate of
1 Mbit/s. The received signal is coherently demodulated with a correlator. The value of
amplitude $A_c$ is 10 mV. The one-sided noise power spectral density is $N_0 = 10^{-11}$ W/Hz,
and the signal power and energy per bit are normalized relative to a 1 Ohm load.

4. A binary phase-shift keying modulator is designed in such a way so as to have the phase
angle $\varphi$ between symbols, in order to provide a residual carrier, as shown in Figure 4. The
modulating signal has a bit rate of $R_b = 140$ Mbit/sec and is pulse shaped to constrain its
spectrum to the Nyquist bandwidth defined by $B = 1/T_b$, where $T_b$ is the duration of a
bit in the modulating signal. The received signal power is $P_b = 10$ mW, and the one-sided
noise power spectral density is $N_0 = 6 \cdot 10^{-12}$ W/Hz. The binary phase-shift keying signal
is demodulated by a correlator receiver.

   a. Suppose first that the phase angle is $\varphi = 180°$. For this case, find the power of the noise
$N$. Find the energy of the received signal and noise. Find the probability of error for
this case.

   b. Suppose now the phase angle is $\varphi = 165°$. Find the energy of the carrier component
and the energies of the message components.

   c. Find the probability of error for the case discussed in b., that is, for the case when the
phase angle is $\varphi = 165°$.

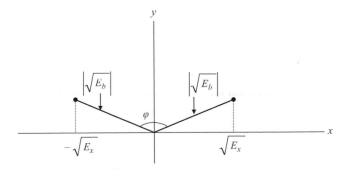

**Fig. 4** *State diagram of a binary phase-shift keying signal.*

5. A binary phase-shift keying system operates with a bit rate of $R_b = 1$ Mbit/s. The received
signal is coherently demodulated with a correlator. The carrier frequency is 10 MHz and
its amplitude $A_b$ is 10 mV. The one-sided noise power spectral density is $N_0 = 10^{-11}$ W/Hz.
The signal power and energy per bit are calculated on an $R = 1$ Ohm load. The modulating
binary signal $m(t)$ is +1 or −1 in a 1 bit interval.

a. Find the expression of this phase-shift keying signal in the time domain. Calculate the energy and the time duration of a bit. Calculate the number of carrier oscillations in a 1 bit interval.

b. Find the bit error probability for this binary phase-shift keying system.

c. Suppose the amplitude of the carrier is doubled, that is, the new amplitude is $A = 2A_b$. Calculate the bit error probability for this case. Comment on the difference between this bit error probability and that calculated in b.

d. Derive a general formula for the bit error probability if the amplitude of the carrier increases $M$ times. Sketch the graph for the bit error probability as a function of $E_b/N_0$, taking into account the calculated bit error probability values in b., c., and d..

6. A digital system is used to transmit digital signals at message bit rate of $R_b = 100$ Mb/sec. The modulated signal is represented as

$$s(t) = \pm \sqrt{\frac{2E_b}{T_b}} \cos\left(\frac{\pi t}{2T_b}\right) \cdot \cos\left(2\pi f_c t\right) \pm \sqrt{\frac{2E_b}{T_b}} \sin\left(\frac{\pi t}{2T_b}\right) \cdot \sin\left(2\pi f_c t\right), \text{for } 0 \leq t \leq 2T_b,$$

where the signs '$\pm$' depend on the information bits $m(t)$ transmitted, that is, on $m(t) \in (-1, +1)$. The message bits are grouped into pairs, which are called dibits and modulate the carrier according to the following procedure: the first bit inside a dibit modulates the in-phase carrier, and the second bit inside the same dibit modulates the quadrature carrier. Thus, the modulated signal is composed of both the in-phase and the quadrature components that last $2T_b$, as stated in the above relation.

a. Sketch a block diagram of the modulator that synthesizes the digital modulated signal $s(t)$. Find out the rate and duration of digital symbols entering the in-phase and quadrature branches of the modulator. Briefly explain the operation of the modulator.

b. Find the envelope $a(t)$ of the modulated signal $s(t)$. From the calculated value of envelope $a(t)$, state what kind of modulation is defined by the above relation. Support your statement with an explanation. Then, calculate the instantaneous angle $\theta(t)$ for the signal $s(t)$ and express the signal $s(t)$ in a closed form as a function of its envelope and instantaneous angle.

c. Express the signal $s(t)$ defined in the above relation as a function of two orthonormal functions $\phi_1(t)$ and $\phi_2(t)$, which are called a pair of quadrature carriers. In this case, the signal can be generally expressed as follows: $s(t) = s_1\phi_1(t) + s_2\phi_2(t)$, where $s_1$ and $s_2$ are constants. Find the expressions for $\phi_1(t)$ and $\phi_2(t)$. Prove that the energy of $\phi_1(t)$ is 1 J.

7. Phase-shift keying modulation is often used for reliable digital signal transmission.

a. Suppose that binary phase-shift keying is used to modulate a carrier in order to transmit a message sequence over an additive white Gaussian noise channel. The two-sided power spectral density of the channel is $N_0/2 = 10^{-10}$ W/Hz. Determine the carrier amplitude $A$ required to achieve a bit error probability of $10^{-6}$. Assume that the bit rate of the message sequence is $R = 1$ Mbits/s or $R = 10$ Mbits/s.

b. Derive the expression showing the dependence of the amplitude $A$ on the bit rate $R$. Deduce the general conclusion related to the relationship between the increase in message rate $R$ and the amplitude of the carrier $A$ for constant values of both the noise power and the bit error probability in the system. Explain in your own words the physical sense of the relation obtained between the amplitude of the carrier $A$ and the bit rate $R$ in the system.

c. Derive an expression showing the dependence of the carrier power $P$ on the bit rate $R$. Deduce the general conclusion related to the relationship between the increase in the message rate $R$ and the carrier power $P$ for constant values of both the noise power and the bit error probability in the system. Deduce the general conclusion related to the relationship between the increase in the required bandwidth of the message signal and the carrier power $P$ for constant values of both the noise power and the bit error probability in the system.

8. A modified coherent phase-shift keying signal is defined by

$$s(t) = A_c k \, \sin 2\pi f_c t \pm A_c \sqrt{1 - k^2} \cos 2\pi f_c t$$

inside the time interval $0 \leq t \leq T_b$. The first term represents an orthogonal carrier component included for the purpose of synchronizing the receiver to the transmitter. The second term is equivalent to the conventional binary phase-shift keying signal. The plus sign in front of the second term corresponds to the message symbol 1, and the minus sign corresponds to the message symbol 0.

a. The binary phase-shift keying terms are orthogonal to the first term (the orthogonal carrier component). Calculate the energy of the conventional binary phase-shift keying term, $E_1$, and then the energy of the orthogonal carrier, $E_2$. Calculate the sum of these energies, that is, the total energy $E_b$ consumed by the system. Prove that the orthogonal basis functions are

$$\phi_1(t) = \sqrt{2/T_b} \cos 2\pi f_c t$$

for the binary phase-shift keying terms, and

$$\phi_2(t) = \sqrt{2/T_b} \sin 2\pi f_c t$$

for the orthogonal carrier. Draw a signal-space diagram for the signal $s(t)$. Explain the difference between this signal-space diagram and the corresponding signal-space diagram of the conventional binary phase-shift keying signal.

a. Suppose that an additive white Gaussian noise of zero mean and two-sided power spectral density $N_0/2$ is present in the channel. Plot the scheme of a correlator receiver that demodulates the binary phase-shift keying term of the signal $s(t)$. Find the expression of the signal $x$ at the output of the correlator, and then the expression for the probability of error, $P_e$, in this system.

9. Suppose that a message source generates 0's and 1's which are transmitted through a binary phase-shift keying communications system. The modulated signal is expressed in the form

$$s_i(t) = \sqrt{\frac{2E_b}{T_b}} \cos(\omega_c t + (i-1) \cdot \pi), \quad i = 1 \text{ and } 2.$$

a. Sketch the block diagram of a binary phase-shift keying transmitter and explain its operation. Sketch the signal space diagram for this system. Plot the decision boundary, assuming that the source generates 0's and 1's with the same probability.

b. How is the frequency of the carrier chosen with respect to the duration of a bit $T_b$? Based on the expression for the modulated signal, define the basis function that is used to generate the binary phase-shift keying signal.

c. Sketch the block diagram of a binary phase-shift keying correlator receiver. Label the signals at the input of the correlator, at the output of the multiplier, and at the output of the integrator.

d. For a coherent binary phase-shift keying system, where the presence of an additive white Gaussian noise is assumed, prove that the probability of bit error, $P_e$, is given by

$$P_e = P(1|0) = \frac{1}{2}\text{erfc}\sqrt{\frac{E_b}{N_0}},$$

where $N_0/2$ is a two-sided power spectral density, and the complementary error function is defined as

$$\text{erfc}(x) = \int_x^\infty \frac{2}{\sqrt{\pi}} e^{-y^2} dy.$$

Keep in mind that the density function of the random variable $X$, which represents the input of the decision circuit, has a probability density function given by

$$f_X(x) = \frac{1}{\sqrt{\pi N_0}} e^{-\frac{1}{N_0}(x+\sqrt{E_b})^2},$$

in the case when the message signal $m_1 = +1$ is generated at the source and transmitted through the additive white Gaussian noise channel.

10. Set up a block diagram for the generation of a frequency-shift keying signal $s(t)$ expressed as

$$s(t) = \sqrt{\frac{2E_b}{T_b}} \cos\left(\frac{\pi t}{T_b}\right) \cos 2\pi f_c t \mp \sqrt{\frac{2E_b}{T_b}} \sin\left(\frac{\pi t}{T_b}\right) \sin 2\pi f_c t.$$

Calculate the envelope of the generated signal.

11. The signal vectors $s_1$ and $s_2$ are used to represent binary symbols 1 and 0, respectively, in a coherent binary frequency-shift keying system. The receiver decides in favour of symbol 1 when

$$\mathbf{x}^T \mathbf{s}_1 > \mathbf{x}^T \mathbf{s}_2,$$

where $\mathbf{x}^T \mathbf{s}_i$ is the inner product of the correlation vector $\mathbf{x}$ and the signal vector $\mathbf{s}_i$, where $i = 1, 2$. Show that this decision rule is equivalent to the condition $x_1 > x_2$, where $x_1$ and $x_2$ are the coordinates of the correlation vector $\mathbf{x}$. Assume that the signal vectors $s_1$ and $s_2$ have equal energy.

12. A frequency-shift keying system transmits binary data at a rate of $2.5 \times 10^6$ bits/s. During the course of transmission, a white Gaussian noise with zero mean and a power spectral density of $10^{-20}$ W/Hz is added to the signal. In the absence of noise, the amplitude of the sinusoidal wave received for digit 1 or 0 is 1 $\mu$V. Determine the average probability of symbol error for coherent binary frequency-shift keying.

13. Suppose the message source generates two binary symbols (or bits) that belong to an alphabet of $M = 2$ symbols, denoted by $m_1 = 1$ and $m_2 = 0$. In a binary frequency-shift keying system, these symbols are assigned sinusoidal signals with different frequencies, according to the expression

$$s_i(t) = \begin{cases} \sqrt{\frac{2E_b}{T_b}} \cos(2\pi f_i t) & 0 \le t \le T_b \\ 0 & \text{elsewhere} \end{cases},$$

for $i = 1, 2$. The difference of these frequencies is constant, expressed as $f_i = (n + i)/T_b$ for a fixed integer value of $n$ and $i = 1, 2$.

a. Prove that the signals $s_1(t)$ and $s_2(t)$ are orthogonal.

b. There are two orthonormal basis functions used for the modulation and demodulation of binary frequency-shift keying signals. Find them both and confirm that the first one calculated has unit energy.

c. Signals $s_1(t)$ and $s_2(t)$ can be represented in a vector form. Find the coordinates (coefficients) of the signal vectors and present them in matrix form. Plot the constellation diagram of this frequency-shift keying signal. Discuss the difference between this diagram and the related constellation diagram for binary phase-shift keying from the probability of error point of view.

d. Suppose the first carrier has the energy $E_b = 2$ $\mu$J, frequency $f_1 = 1$ MHz, and $n = 0$. Calculate the bit duration $T_b$ and then the second frequency $f_2$. Calculate the amplitude of the frequency-shift keying signal. For the signal parameters obtained, plot a frequency-shift keying signal for the message sequence $\mathbf{m} = 00111100$. Indicate the amplitude of the frequency-shift keying signal and the duration of a bit on the graph obtained.

e. Suppose the two-sided power spectral density of noise in the channel is $N_0/2 = 2.5 \times 10^{-7}$ W/Hz. Calculate the probability of error $P_e$ in the frequency-shift keying system that transmits the signal specified in d. Calculate the signal-to-noise ratio in decibels in this case.

14. An orthogonal frequency division multiplexing transmitter generates four subcarriers with the frequencies

$$f_i = \frac{i}{T_b},$$

for $i = 1, 2, 3$, and 4, where $T_b$ are time durations of subcarriers. All the subcarrier signals generated at the output of the transmitter have the same amplitude and can be expressed in the form

$$s_i(t) = \begin{cases} A_b \cos(2\pi f_i t) & 0 \leq t \leq T_b \\ 0 & \text{elsewhere} \end{cases}, \text{ for } i = 1, 2, 3, \text{and } 4.$$

a. Calculate the energy of signals $s_1(t)$ and $s_2(t)$. Prove that these signals are orthogonal. Are the energies of all four subcarriers equal to each other?

b. There are four orthonormal basis functions which can be used to form the orthogonal frequency division multiplexing signal. Calculate the basis functions for the first signal $s_1(t)$ and the second signal $s_2(t)$. Prove that these signals are orthogonal.

c. Suppose the first subcarrier has the energy $E_b = E_1 = 1$ $\mu$J, and frequency is $f_1 = 1$ MHz. Calculate the bit duration $T_b$ and then the second frequency $f_2$. Sketch graphs of all four signals in the time interval $T_b$. Calculate the amplitude of the first subcarrier.

d. Suppose the first subcarrier is modulated using binary phase-shift keying modulation. The energy of a signal $s_1(t)$ is $E_b = E_1 = 1$ $\mu$J. The power spectral density of noise in the channel is $N_0/2 = 2.5 \times 10^{-7}$ W/Hz. Calculate the probability of error $P_e$ at the output of the binary phase-shift keying demodulator of the first subcarrier. Calculate the signal-to-noise ratio in decibels in this case.

15. In orthogonal frequency division multiplexing systems, a set of orthogonal carriers is used to transmit a wideband digital message signal. Suppose the two carriers are the functions

$$\phi_i(t) = \sqrt{\frac{2}{T}} \cos(i\omega_c t) \text{ and } \phi_j(t) = \sqrt{\frac{2}{T}} \cos(j\omega_c t).$$

a. Prove that these two carriers are orthogonal. What conditions should $i$ and $j$ fulfil in that case? Suppose the functions represent the voltages on a unit resistor. Prove that the first function is normalized (with respect to energy). Suppose the basic carrier frequency is $f_c = 1$ MHz. Define a set of at least three carrier frequencies that satisfy the condition of the orthogonality of these two functions.

b. Suppose the carrier frequencies $\phi_1(t)$ and $\phi_2(t)$ are $f_{c1} = 1$ MHz and $f_{c2} = 2$ MHz. One symbol interval is equal to $T = 1/f_{c1}$. Plot the two carriers in the $T$ interval. Observe and explain their orthogonality, in the case that orthogonality exists. You can make this explanation without any calculations.

16. The frequencies of binary frequency-shift keying carriers are expressed by

$$f_i = \frac{n+i}{T_b} = \left\{ \begin{array}{ll} f_c - \frac{1}{2T_b} & i = 1, m_i = 1 \\ f_c + \frac{1}{2T_b} & i = 2, m_i = 0 \end{array} \right\} = f_c \mp \frac{1}{2T_b}.$$

Find the expression for $n$ that fulfils the conditions for both carriers. Confirm the values of the carrier frequencies.

17. The carriers of a binary frequency-shift keying signal are expressed as

$$s_1(t) = \sqrt{\frac{2E_b}{T_b}} \cos 2\pi \left( f_c - \frac{1}{2T_b} \right) t$$

$$s_2(t) = \sqrt{\frac{2E_b}{T_b}} \cos 2\pi \left( f_c + \frac{1}{2T_b} \right) t.$$

Confirm that they are orthogonal.

18. Find the power spectral density of both the in-phase and the quadrature components of the binary frequency-shift keying signal, assuming that the carrier is just expressed as sine and cosine functions and the baseband spectrum of the modulating signal is expressed by

$$S_{MM}(f) = S_I(f) + S_Q(f) = \frac{E_b}{2} \left[ \delta \left( f T_b - \frac{1}{2} \right) + \delta \left( f T_b + \frac{1}{2} \right) \right] + \frac{8 E_b \cos^2 (\pi T_b f)}{\pi^2 \left( 4 T_b^2 f^2 - 1 \right)^2}.$$

Calculate the power spectral density of the frequency-shift keying signal at the output of the modulator.

19. A binary frequency-shift keying signal has been expressed as

$$s(t) = \sqrt{\frac{2E_b}{T_b}} \cos\left(\frac{\pi t}{T_b}\right) \cos 2\pi f_c t \mp \sqrt{\frac{2E_b}{T_b}} \sin\left(\frac{\pi t}{T_b}\right) \sin 2\pi f_c t$$

$$= s_I(t) \cos 2\pi f_c t + s_Q(t) \sin 2\pi f_c t$$

Present the wave shapes of both the in-phase and the quadrature components, assuming that the sequence of bits (0 1 1 0 1 1 0 0) is generated from the source. Confirm that the in-phase and the quadrature low-frequency components are orthogonal.

20. The bit error probability $p$ of a frequency-shift keying system, for a symmetric additive white Gaussian noise channel and for the same probability of generation binary values at the output of the source, can be calculated in a simpler way. In this case, the calculated $p$ will be the same for the assumed bit sent regardless of whether it is 1 or 0. Suppose the sent bit is 1. Then, the decision variable, either at the correlator output or the detector output, is expressed as

$$X = \sqrt{E_b} + (W_1 - W_2).$$

Prove that the probability of error can be calculated as

$$p = \frac{1}{2} \operatorname{erfc}\left(\sqrt{\frac{\eta_X^2}{2\sigma_X^2}}\right).$$

21. The carrier is an orthogonal function expressed as

$$s_c(t) = \phi_1(t) = \sqrt{\frac{2}{T_b}} \cos(\omega_c t),$$

and defined in the interval $0 \le t < T$. The autocorrelation function and the power spectral density are expressed, respectively, as

$$R_{sc}(\tau) = \frac{1}{T_b} \cos 2\pi f_c \tau$$

and

$$S_{sc}(f) = \frac{1}{2T_b} [\delta(f - f_c) + \delta(f + f_c)].$$

Prove that this carrier is an orthonormal function. Calculate the energy and power of the carrier in the time and frequency domains.

# 7

# Discrete Bandpass Modulation Methods

## 7.1 Introduction

This is the central chapter of the book. It contains a condensed presentation of discrete-time communication systems presented in the form of block schematics followed by mathematical expressions for the input and output signals for each block. The operation of a block is explained using the theory of discrete-time stochastic processes and expressed in time and frequency domains, using related autocorrelation functions, power spectral densities, amplitude and phase spectral densities, and signal expressions in the time domain and in related vector forms. Whenever it is possible, the notation in the chapter follows the unique notation defined in related chapters containing the theory of deterministic and stochastic signal processing. The background for this chapter encompasses Chapters 2, 4, 5, and 6, which focus on communication systems theory, Chapters 3, 4 and 19, which present the Gaussian noise process and discrete- and continuous-time stochastic processes, and Chapters 13–18, which address the theory of discrete-time deterministic signal processing.

Following the generic scheme of a discrete communication system, presented in Chapter 5 in Figure 5.6.4, and, in particular a block scheme presenting the signals and functions in digital communication systems in Chapter 6, Figure 6.1.1, the block scheme of a discrete communication system, specifying the signals and functions, is shown in Figure 7.1.1.

For all blocks, input and output signals are presented as realizations of related discrete-time stochastic processes, as shown in Fig. 7.1.1. In addition, beneath each block, the functions that are relevant to the operation of that block are shown. We tried to unify the notation in this chapter with the notation in Chapters 3–6 and the related Chapters 13, 14, and 15, as we did in the whole book, at the level that could be done.

**Source and baseband signal processing.** The explanation of the source and the baseband signal processing is analogous to that for digital communication systems, which was explained in Chapter 6 and will be outlined here for the sake of completeness. A message source generates the messages at its output in the form of message symbols $m_i$, which are expressed as sets $m_i = (m_{i1}, m_{i2}, \ldots, m_{ij}, \ldots, m_{iJ})$, where $m_{ij}$ are elements of the alphabet A. Each message symbol is composed of $J$ elements taken from the alphabet A, which contains $A$ elements. Therefore, we can generate $I = A^J$ different messages at the output of the message source. This source can be considered to be a discrete source of information, as presented in

*Discrete Communication Systems.* Stevan Berber, Oxford University Press. © Stevan Berber 2021.
DOI: 10.1093/oso/9780198860792.003.0007

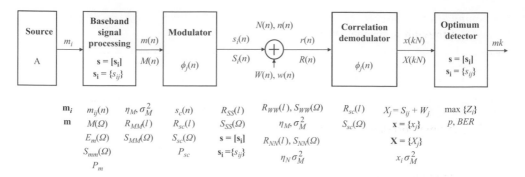

**Fig. 7.1.1** *Signals and functions in a discrete communication system.*

Chapter 9. All possible message symbols form a set which can be presented as the matrix **m**, and the related matrix of their a priori probabilities can be expressed as

$$P(\mathbf{m}) = [P(\mathbf{m}_1) \quad P(\mathbf{m}_2) \quad \cdots \quad P(\mathbf{m_i}) \quad \cdots \quad P(\mathbf{m_{I-1}}) \quad P(\mathbf{m_I})]. \tag{7.1.1}$$

Depending on the procedure used to form the message symbols, these probabilities can be related to the probabilities of the alphabet's elements.

The set of message symbol elements are mapped into the set of signal coefficients, which define *modulating signals m(n)* at the input of a transmitter modulator. There is a one-to-one correspondence between the message symbols $\mathbf{m_i}$ and the signals $\mathbf{s_i}$. The modulating signal defined for the *i*th symbol can be expressed as $m_{ij}(n) = s_{ij}(n)$. A stream of these signals, presenting the message symbols, form the modulating signal $m(n)$ that is defined as a realization of a discrete-time stochastic process $M(n)$. This signal is analysed in time and frequency domains by calculating the amplitude spectral density $M(\Omega)$, the energy spectral density $E_m(\Omega)$, and the power of the signal $P_m$. The autocorrelation function and the power spectral density of the signal $m(n)$ are calculated assuming that the signal is generated as a binary stochastic defined by its probability density function. The analysis of this process is based on the discrete-time stochastic processing theory presented in Chapter 3, and the theory of discrete-time deterministic signal processing, which is presented in Chapters 14 and 15.

**Modulation of the carriers.** The input message signals $m(n)$ are multiplied by the carriers, which are defined as a set of orthonormal basis functions $\phi_j(n)$. Then, the autocorrelation function $R_{sc}(l)$ and power spectral density function $S_{sc}(\Omega)$ of the carriers are calculated, in order to find the output signal of the modulator in the time and frequency domains. Next, the multiplied signals are added to each other, and the modulated signal $s_i(n)$ is obtained at the output of the modulator; it is considered to be a realization of the modulated stochastic process $S_i(n)$. The modulated signal is analysed in terms of its autocorrelation function $R_{SS}(l)$, the power spectral density $S_{SS}(\Omega)$, and the power of the modulated signal, as shown in Figure 7.1.1. The aforementioned autocorrelation function and power spectral density function are calculated in the following sections for the main modulation schemes that are widely used in practice. The background theory for this chapter is presented in Chapters 3, 14, and 15.

**Communication channels.** The discrete-time bandpass modulated signal is amplified and transmitted through a channel. In our analysis, we will use two models of a discrete-time channel. The first model is the additive white Gaussian noise channel, which is defined as a discrete-time wide-sense stationary Gaussian process which has zero mean value and is denoted by $W(n)$ in Figure 7.1.1. In theory, this noise will be mathematically described by its autocorrelation function $R_{WW}(l)$, the power spectral density function $S_{WW}(\Omega)$, and the probability density function. The second model is a passband channel that is represented by the discrete-time correlated Gaussian noise process $N(n)$ of zero mean and finite average power. This channel model is analysed and presented for two reasons. Firstly, we need to use this model for simulation and emulation. Namely, the generator of noise $N(n)$ is physically realizable. In the frequency domain, the noise $N(n)$ has constant power spectral density values in a limited bandwidth that is equal to the bandwidth of the transmitted signal, which allows simpler calculations and simulation of the signal-to-noise ratio in the communication system. Secondly, if we need to analyse the transceiver operation at an intermediate frequency, this is a useful model for that analysis. Detailed discussion of this channel can be found in Chapter 4, and the theory of the discrete-time noise process is presented in Chapter 3.

**The correlation demodulator.** The received signal $r(n)$ is a realization of a discrete-time wide-sense stationary process $R(n)$ which is a sum of the modulated signal and the noise in the channel. In coherent demodulation, which is the subject of our analysis, the received signal is correlated with locally generated synchronous carriers, resulting in a random correlation value $x(kN_s)$ at the output of the correlation demodulator, where $N_s$ is a symbol duration in samples. This value is a realization of the random variable $X(kN_s)$, which is defined by the integer $k$, which belongs to the stochastic process $X(kN_s)$. One realization of this process is a series of random values $x(kN_s)$, for integer $k$ values. The analysis of this process is based on the theory presented in Chapter 3. All outputs of the correlation demodulator at discrete-time instants $kN_s$ are represented by a demodulation vector $\mathbf{x}$ containing a set of the random values $\{x_j\}$ that are generated at the output of all correlators. These values are realizations of random variables that define a vector $\mathbf{X}$ that characterizes the outputs of all correlators at time instants $kN_s$.

**The optimum detector.** The general scheme of the optimum detector presented in Chapter 5 can be used to construct the optimum detector of a discrete-time correlation receiver. The detector correlates the incoming demodulation vector $\mathbf{x}$ with all possible signals sent, $\mathbf{s} = \{\mathbf{s_i}\}$, resulting in a set of numeric random values, $Z = \{Z_i\}$, that define the decision variables $Z_i$. The decision is to be made according to the maximum likelihood decision rule, as was explained in Chapter 5 and, for digital communication systems, in Chapter 6. The average probability of symbol error will be defined as for the digital communication system in eqn (6.1.3). A receiver, based on the requirement to minimize the probability of the symbol error, is said to be optimum in the sense of the minimum probability of symbol error.

Based on the general scheme for a digital communication system, which is presented in Figure 7.1.1, the implementation of basic discrete modulation methods will be presented in the following sections. The methods under our investigation are presented in a block diagram in Figure 7.1.2.

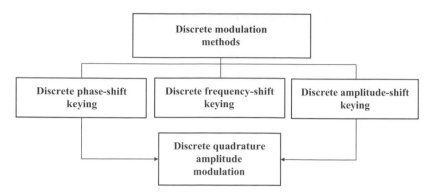

**Fig. 7.1.2** *Discrete modulation methods.*

The main aim of this book is to present basic communication systems operating in the discrete-time domain. To use the space available in the book efficiently, we will limit our presentation of modulation methods to the basic ones. Following the approach presented, it will be relatively easy to extend the mathematical models presented for these systems to those for more complicated ones.

As for digital communication systems, in our presentation, we will use precise block schematics based on the generic scheme in Figure 5.6.4, which will show the mathematical operators for the scheme, and follow these with the underlying mathematical model of the system. We will start by presenting a communication system that uses phase-shift keying, primarily binary phase-shift keying and quaternary phase-shift keying. This will be followed by a presentation of frequency-shift keying. Finally, quadrature amplitude modulation, which is a combination of amplitude-shift keying and phase-shift keying, will be presented.

## 7.2   Coherent Binary Phase-Shift Keying Systems

### 7.2.1   Operation of a Binary Phase-Shift Keying System

It is possible to simplify the generic scheme of a communication system presented in Figure 5.6.4, and its extension to a binary case in Figure 5.7.1, to develop the binary phase-shift keying system presented in Figure 7.2.1. In this case, the transmitter is composed of a baseband processing block that generates the modulating signal $m_i(n)$ and a multiplier that functions as a modulator. The receiver is composed of a demodulation correlator and an optimum detector. In the following sections, the mathematical model of the system and the related processing of signals will be presented in the discrete-time domain.

The message source generates one of two symbols that belong to an alphabet of two elements A = (1, 0), with the symbols being denoted by the message symbols $m_1 = 1$ and $m_2 = 0$. The system is binary, so a symbol generated at the output of the source is identical to a message bit. The message bit is applied to the input of a look-up table, which maps the input bits (0, 1) into the corresponding polar output bits (−1, +1). These bits are multiplied by the

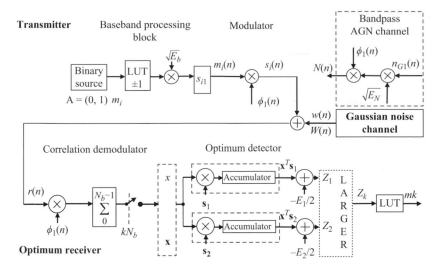

**Fig. 7.2.1** *Block schematic of a binary phase-shift keying system; AGN, additive Gaussian noise; LUT, look-up table.*

square root of the bit energy $\sqrt{E_b}$ in the bit interval $N_b$ to form the signal coefficient $s_{i1}$. The coefficient $s_{i1}$ defines the amplitude of a rectangular pulse that is the baseband modulating signal $m_i(n)$, as presented in Figure 7.2.1. The signal $m_i(n)$ is multiplied by the carrier, which is designed as a single basis function $\phi_1(n)$, and the modulated binary phase-shift keying signal $s_i(n)$ is produced and transferred though the discrete additive white Gaussian noise waveform channel represented by the discrete-time noise process $W(n)$. In the case of system simulation and emulation, the channel is represented by the bandpass discrete noise process $N(n)$. Both noise processes are precisely defined in Chapter 4.

## 7.2.2   Transmitter Operation

### 7.2.2.1   *Presentation of a Modulating Signal*

**Signal-space or constellation diagram of the signal.** The number of possible message symbols is $I = A^J = 2^1 = 2$, which are defined by the matrix of message symbols

$$\mathbf{m} = [\mathbf{m}_1 \quad \mathbf{m}_2] = \begin{bmatrix} 1 & -1 \end{bmatrix}. \tag{7.2.1}$$

These message symbols are mapped into the amplitudes of modulating signals by multiplying them by $\sqrt{E_b}$ to form a vector of modulating signal coefficients, which can be presented in one-dimensional signal space (see Section 7.2.2.2, Figure 7.2.6). The signal vector is represented by two columns in the matrix of signal coefficients, that is,

$$\mathbf{s} = [\mathbf{s}_1 \quad \mathbf{s}_2] = \begin{bmatrix} s_{11} & s_{21} \end{bmatrix} = \sqrt{E_b}\mathbf{m} = \begin{bmatrix} \sqrt{E_b} & -\sqrt{E_b} \end{bmatrix}. \tag{7.2.2}$$

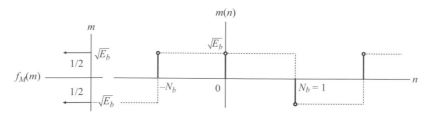

**Fig. 7.2.2** *A modulating discrete-time polar binary signal presented in the time domain.*

**Statistical characteristics of the modulating signal *m(n)* in the time domain.** Modulating signal $m(n)$ is a series of modulating rectangular pulses $m_i(n)$, each having $N_b$ samples. For the sake of simplicity and explanation, the pulses are represented by one sample, $N_b = 1$, in Figure 7.2.2. The modulating signal $m(n)$ is considered to be a realization of a discrete-time random binary process $M(n)$, which randomly takes the amplitude values specified by the vector defined in eqn (7.2.2), as presented in Figure 7.2.2. We will assume that the probabilities of positive and negative binary values are equal to 0.5. Therefore, at each time instant $n = n_k$, we have a discrete random variable $M = M(n_k)$ that takes the value of $\sqrt{E_b}$ or $-\sqrt{E_b}$, and is defined by its discrete probability density function expressed in terms of Kronecker (or Dirac) delta functions as

$$f_M(m) = \frac{1}{2}\delta\left(m - \sqrt{E_b}\right) + \frac{1}{2}\delta\left(m + \sqrt{E_b}\right), \qquad (7.2.3)$$

and shown in Figure 7.2.2. The notation here follows the notation in the theory of discrete-time stochastic processes presented in Chapter 3.

Since the amplitude levels of $m(n)$ have the same magnitude values and occur with equal probabilities, the mean value of the process is zero and the variance is $E_b$.

**Presentation of a modulating signal in the frequency domain.** Following its derivation in Chapter 15, the magnitude spectral density of a rectangular discrete-time pulse defined by $N_b$ samples is expressed as

$$|M(\Omega)| = \left\{ \begin{array}{ll} \sqrt{E_b}N_b & \Omega = \pm 2k\pi, k = 0,1,2,3,\ldots \\ \sqrt{E_b}\left|\frac{\sin \Omega N_b/2}{\sin \Omega/2}\right| & \text{otherwise} \end{array} \right\} \qquad (7.2.4)$$

and shown in Figure 7.2.3 for $N_b = 32$.

Detailed analysis of the magnitude spectral density and the importance of the sampling frequency can be found in the solution to Problem 14 below. The energy spectral density of the pulse is

$$E_m(\Omega) = |M(\Omega)|^2 = E_b\left|\frac{\sin \Omega N_b/2}{\sin \Omega/2}\right|^2,$$

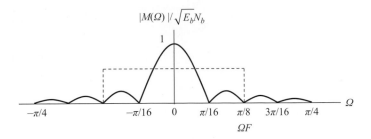

**Fig. 7.2.3** *The magnitude spectral density of a discrete-time rectangular pulse.*

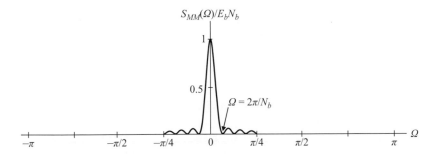

**Fig. 7.2.4** *Power spectral density of a modulating discrete-time binary signal.*

and the power spectral density is

$$S_{MM}(\Omega) = \frac{|M(\Omega)|^2}{N_b} = \frac{E_b}{N_b}\left|\frac{\sin \Omega N_b/2}{\sin \Omega/2}\right|^2 = E_b N_b \frac{1}{N_b^2}\left|\frac{\sin \Omega N_b/2}{\sin \Omega/2}\right|^2. \tag{7.2.5}$$

Zero crossings can be calculated for the condition $\sin \Omega N_b/2 = 0$, $\Omega N_b/2 = k\pi$. Therefore, the crossings are defined by $\Omega = 2k\pi/N_b$, for $k = 1, 2, 3, \ldots$. For the graph in Figure 7.2.4, the number of samples is obtained from the condition expressed as $\Omega = 2\pi/N_b = \pi/16$, which results in $N_b = 32$.

**Example**

Calculate the mean and variance of the binary stochastic process $M(n)$ presented in Figure 7.2.2.

**Solution.** The mean and variance are

$$\eta_M = \int_{-\infty}^{\infty} m \cdot f_M(m)dm = \int_{-\infty}^{\infty} m \cdot \left[\frac{1}{2}\delta\left(m - \sqrt{E_b}\right) + \frac{1}{2}\delta\left(m + \sqrt{E_b}\right)\right]dm$$

$$= \frac{1}{2}\sqrt{E_b} + \frac{1}{2}\left(-\sqrt{E_b}\right) = 0$$

and

$$\sigma_M^2 = \sum_{m=-\sqrt{E_b}}^{m=-\sqrt{E_b}} m^2 \cdot f_M(m) = \sum_{m=-\sqrt{E_b}}^{m=-\sqrt{E_b}} m^2 \cdot \left[ \frac{1}{2}\delta\left(m - \sqrt{E_b}\right) + \frac{1}{2}\delta\left(m + \sqrt{E_b}\right) \right]$$

$$= \frac{1}{2}E_b + \frac{1}{2}E_b = E_b,$$

respectively. The variance corresponds to the power of the signal, as has been already calculated.

**Example**

Express the power spectral density of $m(n)$ in terms of the normalized frequency $F = \Omega/2\pi$ and confirm the calculated power. Calculate the power of the modulating pulse in the frequency domain.

**Solution.** We can find the magnitude spectral density, eqn (7.2.4), as a function of $F$ as

$$\sqrt{E_b}\left|\frac{\sin \Omega N_b/2}{\sin \Omega/2}\right| = \sqrt{E_b}\left|\frac{\sin 2\pi FN_b/2}{\sin 2\pi F/2}\right| = \sqrt{E_b}\left|\frac{\sin \pi FN_b}{\sin \pi F}\right| = |M(F)|,$$

with zero crossings for $\pi FN_b = k\pi$ or $F = k/N_b = k/32$. The related power spectral density is

$$S_{MM}(F) = \frac{|M(F)|^2}{N_b} = E_b N_b \frac{1}{N_b^2}\left|\frac{\sin \pi FN_b}{\sin \pi F}\right|^2.$$

Using this expression and the integral value $\int_{-\infty}^{\infty}\left|\frac{\sin \Omega N/2}{\sin \Omega/2}\right|^2 d\Omega = 2N\pi$, we can calculate the numerical value of the power as

$$P_m = \int_{-1/2}^{1/2} S_{MM}(F)dF = \frac{E_b}{N_b}\int_{-1/2}^{1/2}\left|\frac{\sin 2\pi FN_b/2}{\sin 2\pi F/2}\right|^2 dF \underset{2\pi F=x}{=} \frac{E_b}{N_b}\int_{-\pi}^{\pi}\left|\frac{\sin xN_b/2}{\sin x/2}\right|^2 \frac{dx}{2\pi} = \frac{E_b}{2\pi N_b}\cdot 2N_b\pi = E_b.$$

### 7.2.2.2 *Modulated Signals in Time and Frequency Domains*

The binary phase-shift keying modulated signal $s_i(n)$ is obtained by multiplying the modulating signal and the carrier. The carrier is a deterministic signal defined by its amplitude, frequency, and phase. It is important to notice that the carrier can be considered to be a realization of a sinusoidal stochastic process with a random phase. The modulating signal, as we said before, is a stochastic binary process. To find the modulated signal in time and frequency domains, we need to find the carrier in both domains and then investigate the output of the modulator multiplier.

**Presentation of the carrier in time and frequency domains.** The carrier is expressed in the time domain as

$$s_c(n) = \phi_1(n) = \sqrt{\frac{2}{N_b}} \cos \Omega_c n. \tag{7.2.6}$$

Based on the theory presented in Chapter 15, and considering the carrier as a periodic function, we can calculate its autocorrelation function as

$$
\begin{aligned}
R_{sc}(l) &= \frac{1}{N_b} \sum_{n=0}^{n=N_b-1} \left[ \sqrt{\frac{2}{N_b}} \cos \Omega_c (n+l) \cdot \sqrt{\frac{2}{N_b}} \cos \Omega_c n \right] \\
&= \frac{1}{N_b} \frac{2}{N_b} \sum_{n=0}^{n=N_b-1} \frac{1}{2} [\cos \Omega_c l + \cos \Omega_c (2n+l)] = \frac{1}{N_b} \frac{1}{N_b} \sum_{n=0}^{n=N_b-1} \cos \Omega_c l + 0 \\
&= \frac{1}{N_b} \cos \Omega_c l = \frac{1}{N_b} \cos \Omega_c l.
\end{aligned}
\tag{7.2.7}
$$

Then the power spectral density function can be calculated as the Fourier transform of the autocorrelation function, which is a cosine function expressed as

$$S_{sc}(\Omega) = \frac{1}{N_b} \pi [\delta(\Omega - \Omega_c) + \delta(\Omega + \Omega_c)], \tag{7.2.8}$$

where the carrier frequencies are related as $F_c = \Omega_c/2\pi = 1/N_c$.

**Example**

Calculate the energy and power of carriers in the time and frequency domains. Suppose the power is calculated in one period defined by $N_b$ samples of the signal.

**Solution.**
The energy and power can be calculated from the autocorrelation function and the related power spectral density and energy spectral density functions as

$$E_{sc} = N_b R_{sc}(0) = 1, \tag{7.2.9}$$

$$
\begin{aligned}
E_{sc} &= \frac{1}{2\pi} \int_{-\pi}^{\pi} E_{sc}(\Omega) d\Omega = \frac{1}{2\pi} \int_{-\pi}^{\pi} \pi [\delta(\Omega - \Omega_c) + \delta(\Omega + \Omega_c)] d\Omega \\
&= \frac{1}{2\pi} 2\pi = 1
\end{aligned}
\tag{7.2.10}
$$

$$
\begin{aligned}
P_{sc} &= \frac{1}{2\pi} \int_{-\pi}^{\pi} S_{sc}(\Omega) d\Omega = \frac{1}{2\pi} \int_{-\pi}^{\pi} \frac{\pi}{N_b} [\delta(\Omega - \Omega_c) + \delta(\Omega + \Omega_c)] d\Omega \\
&= 2 \frac{1}{2\pi} \frac{\pi}{N_b} = \frac{1}{N_b}
\end{aligned}
\tag{7.2.11}
$$

and

$$P_{sc} = \frac{E_{sc}}{N_b} = R_{sc}(0) = \frac{1}{N_b}, \tag{7.2.12}$$

respectively.

In the time domain, the energy and power are, respectively,

$$E_{sc} = \sum_{n=0}^{N_b-1} s_{sc}^2(n) = \sum_{n=0}^{N_b-1} \left(\sqrt{2/N_b}\cos\Omega_c n\right)^2 = \frac{2}{N_b}\sum_{n=0}^{N_b-1}\frac{1}{2}(1+\cos 2\Omega_c n) = 1, \tag{7.2.13}$$

$$P_{sc} = \frac{E_{sc}}{N_b} = \frac{1}{N_b}. \tag{7.2.14}$$

**Modulated binary phase-shift keying signals in the time domain.** The modulated signal is represented by a pair of signals $s_1(n)$ and $s_2(n)$, generally expressed as

$$s_i(n) = \sqrt{E_b}\sqrt{\frac{2}{N_b}}\cos\left(\Omega_c n + (i-1)\cdot\pi\right). \tag{7.2.15}$$

Each signal is defined as an energy signal existing in the time interval $0 \le n < N_b$, with two phases determined by $i = 1$ or $i = 2$. The transmitted signal energy per bit is $E_b$. We assume that, for the source message bit 1, the value of $i$ is 1 and, for the source bit 0, the value of $i$ is 2. In addition, the carrier frequency is equal to $k/N_b$, for some constant integer $k$, in order to ensure that each transmitted signal $s_i(n)$ contains an integer number of carrier cycles. This pair of sinusoidal signals, differing only in a relative phase shift of $180°$, are referred to as antipodal signals. They are expressed in terms of one basis function $\phi_1(n)$ as

$$s_1(n) = \sqrt{E_b}\sqrt{\frac{2}{N_b}}\cos\left(\Omega_c n\right) = \sqrt{E_b}\phi_1(n) = s_{11}\phi_1(n), \tag{7.2.16}$$

$$s_2(n) = \sqrt{E_b}\sqrt{\frac{2}{N_b}}\cos\left(\Omega_c n + \pi\right) = -\sqrt{E_b}\sqrt{\frac{2}{N_b}}\cos\Omega_c n = -\sqrt{E_b}\phi_1(n) = s_{21}\phi_1(n), \tag{7.2.17}$$

where $\phi_1(n)$ is an orthonormal signal, and $s_{11}$ and $s_{21}$ are coefficients of signals $s_1(n)$ and $s_2(n)$, which are presented in Figure 7.2.5 in the time domain. Both signals may be defined as vectors in one-dimensional signal space, following the procedure in Chapter 2. Therefore, for the given signals $s_1(n)$ and $s_2(n)$, the basis function $\phi_1(n)$ can be calculated using the Gram–Schmidt orthogonalization procedure. Obviously, these signals can be obtained by multiplication of the periodic carrier and the non-inverted and inverted discrete rectangular modulating pulses, as shown in Figure 7.2.1.

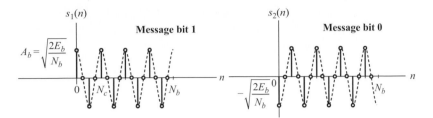

**Fig. 7.2.5** *Modulated signals in the discrete-time domain.*

**Constellation diagram of a binary phase-shift keying signal.** From the expressions for the signal pair (7.2.15), it is obvious that there is only one basis function of unit energy, expressed as

$$\phi_1(n) = \sqrt{\frac{2}{N_b}} \cos(\Omega_c n), \qquad (7.2.18)$$

and the related signal space is one dimensional ($M = 1$), with a signal constellation consisting of two message bit points with coordinates that can be calculated using the procedure presented in Chapter 2 as follows:

$$s_{11} = \sum_{n=0}^{N_b-1} s_1(n)\phi_1(n) = \sum_{n=0}^{N_b-1} \sqrt{E_b}\frac{2}{N_b}\cos^2 \Omega_c n = \sqrt{E_b}\frac{2}{N_b}\sum_{n=0}^{N_b-1} \cos^2 \Omega_c n = \sqrt{E_b}, \qquad (7.2.19)$$

$$s_{21} = \sum_{n=0}^{N_b-1} s_2(n)\phi_1(n) = -\sqrt{E_b}. \qquad (7.2.20)$$

These coefficients are coordinate of two vectors defining the signals $s_1(n)$ and $s_2(n)$, expressed as $\mathbf{s}_1 = [s_{11}]$ and $\mathbf{s}_2 = [s_{21}]$, respectively, as shown in Figure 7.2.6. Message point 1 (for $m_1 = 1$ or, equivalently, $s_1(n)$) is located at $s_{11}$, and message point 2 (for $m_2 = 0$ or, equivalently, $s_2(n)$) is located at $s_{21}$. The constellation diagram shown has the minimum energy.

The amplitude of the modulated signal presented is $A_b$ in the bit interval $N_b$, defining the symbol-shaping function as a rectangular pulse expressed as

$$s_{ss}(n) = \begin{cases} A_b = \sqrt{\frac{2E_b}{N_b}} & 0 \leq n < N_b \\ 0 & otherwise \end{cases}, \qquad (7.2.21)$$

which corresponds to a rectangular wave shape with amplitude proportional to the amplitude of the modulating signal presented in Figure 7.2.2. In order to achieve better utilization of the signal spectrum, or for some other reason, the shape of this function can be changed. Most of the time, in this book, we will use rectangular pulses for the modulating signal. In this way, we will simplify the theoretical explanation and put the emphasis on understanding the

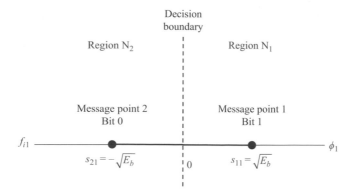

**Fig. 7.2.6** *The signal-space or constellation diagram for binary phase-shift keying signals.*

theory of modulation. From the theoretical point of view, it is not so complicated to extend the modulation theory to the cases when the pulse shape is not rectangular. The energy of the modulated signal representing a bit is calculated according to the expression

$$E = \sum_{n=0}^{N_b-1} (A_b \cos \Omega_c n)^2 = \sum_{n=0}^{N_b-1} \frac{A_b^2}{2} (1 + \cos 2\Omega_c n) = \frac{A_b^2 N_b}{2} = \frac{E_{bit}}{2}, \tag{7.2.22}$$

where $E_{bit}$ is equivalent to the energy of a rectangular pulse with magnitude $A_b$.

### 7.2.2.3   *The Power Spectral Density of Binary Phase-Shift Keying Modulated Signals*

Due to the Wiener–Khintchine theorem, the autocorrelation function and the power spectral density of a wide-sense stationary process form a Fourier transform pair. We will present the operation of the modulator in terms of the autocorrelation functions and power spectral densities. The baseband modulating signal $m(n)$, which has the power spectral density shown in Figure 7.2.4, is multiplied by the carrier operating at the frequency $\Omega_c$, for $\Omega_c \gg \Omega_B$, where $\Omega_B$ is the bandwidth of the modulating signal. The initial amplitude of the random baseband signal and the carrier amplitude are statistically independent, due to the randomness of the signal amplitudes. Therefore, the bandpass modulated random signal $s_i(n)$, presented as a realization of a wide-sense stationary stochastic process $S_i(n)$, can be expressed as the product

$$s_i(n) = m(n) \cdot \sqrt{\frac{2}{N_b}} \cos(\Omega_c n) = m(n) \cdot \phi_1(n), \tag{7.2.23}$$

defined in the interval $0 \le n < N_B$, where $m(n)$ is a realization of the baseband modulating process $M(n)$, or a series of randomly generated signals $m_i(n)$. In order to find the power spectral density function of the modulated signal, we need to find the autocorrelation function $R_{SS}(l)$ of the process $S_i(n)$ and use the Wiener–Khintchine theorem to find the power spectral density of the signal. The autocorrelation function is

$$R_{SS}(l) = E\{S(n) \cdot S(n+l)\} = E\{M(n)\phi_1(n) \cdot M(n+l)\phi_1(n+l)\}$$

$$= E\left\{M(n) \cdot M(n+l) \cdot \sqrt{\frac{2}{N_b}} \cos \Omega_c n \sqrt{\frac{2}{N_b}} \cos \Omega_c(n+l)\right\}$$

$$= \frac{2}{N_b}\frac{1}{2}E\{M(n) \cdot M(n+l) \cdot [\cos \Omega_c l + \cos(2\Omega_c n + \Omega_c l)]\} \qquad (7.2.24)$$

$$= \frac{1}{N_b}E\{M(n) \cdot M(n+l) \cdot \cos \Omega_c l\}$$

$$+ \frac{1}{N_b}E\{M(n) \cdot M(n+l) \cdot \cos(2\Omega_c n + \Omega_c l)\}.$$

Due to the independence between the cosine carrier signal and the baseband random signal, the expectation of the first addend is the autocorrelation function of the modulating random signal $R_{MM}(l)$ that is multiplied by the cosine function. The expectation of the second addend is 0, due to the high frequency of the carrier, which yields a mean value equal to 0. Therefore, the autocorrelation of the modulated signal is a product of the autocorrelation of the input modulating signal and the autocorrelation function of the carrier, expressed as

$$R_{SS}(l) = R_{MM}(l) \cdot \frac{1}{N_b} \cos \Omega_c l + 0 = R_{MM}(l) \cdot R_{sc}(l). \qquad (7.2.25)$$

Having available the autocorrelation function of the modulated signal, the power spectral density can be found as the Fourier transform of this function. On the other hand, we have this autocorrelation function expressed as a product of the autocorrelation function of the carrier and the autocorrelation function of the random binary process. These autocorrelation functions are the inverse Fourier transforms of the related power spectral density functions. Therefore, we can obtain the power spectral density of the output random modulated signal as the convolution of the power spectral densities of the random modulating signal calculated in eqn (7.2.5) and the carrier calculated in eqn (7.2.8). Bearing in mind the convolution property of the Fourier transform that will be presented in Section 15.3, we obtain

$$S_{SS}(\Omega) = \frac{1}{2\pi}S_{MM}(\Omega) * S_{sc}(\Omega) = \frac{1}{2\pi}\int_{-\pi}^{\pi} S_{MM}(\Omega - \lambda) S_{sc}(\lambda) d\lambda$$

$$= \frac{1}{2\pi}\int_{-\pi}^{\pi} \frac{E_b}{N_b}\left|\frac{\sin(\Omega - \lambda)N_b/2}{\sin(\Omega - \lambda)/2}\right|^2 \cdot \frac{1}{N_b}\pi[\delta(\lambda - \Omega_c) + \delta(\lambda + \Omega_c)]d\lambda$$

$$= \frac{E_b}{2N_b^2}\left|\frac{\sin(\Omega - \Omega_c)N_b/2}{\sin(\Omega - \Omega_c)/2}\right|^2 + \frac{E_b}{2N_b^2}\left|\frac{\sin(\Omega + \Omega_c)N_b/2}{\sin(\Omega + \Omega_c)/2}\right|^2$$

$$= \frac{E_b}{2}\frac{1}{N_b^2}\left|\frac{\sin(\Omega - \Omega_c)N_b/2}{\sin(\Omega - \Omega_c)/2}\right|^2 + \frac{E_b}{2}\frac{1}{N_b^2}\left|\frac{\sin(\Omega + \Omega_c)N_b/2}{\sin(\Omega + \Omega_c)/2}\right|^2$$

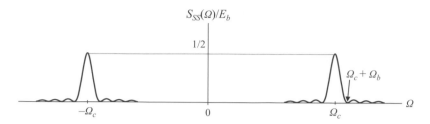

**Fig. 7.2.7** *Power spectral density of a discrete-time binary phase-shift keying signal.*

and its normalized expression

$$S_{SS}(\Omega)/E_b = \frac{1}{2N_b^2}\left|\frac{\sin(\Omega-\Omega_c)N_b/2}{\sin(\Omega-\Omega_c)/2}\right|^2 + \frac{1}{2N_b^2}\left|\frac{\sin(\Omega+\Omega_c)N_b/2}{\sin(\Omega+\Omega_c)/2}\right|^2 \qquad (7.2.26)$$

Therefore, the power spectral density of a binary phase-shift keying signal is obtained as a shifted version of the baseband power spectral density, as shown in Figure 7.2.7. The *k*th zero crossing can be calculated as $\sin(\Omega-\Omega_c)N_b/2 = 0$ or $\Omega = \Omega_c + 2k\pi/N_b$. The first zero crossing is at the frequency $\Omega = \Omega_c + \Omega_b = 2\pi/N_c + 2\pi/N_b$ or expressed in terms of the frequency $F$ as $F = \Omega/2\pi = (1/N_c + 1/N_b) = (F_c + F_b)$, where $N_b$ and $N_c$ are the number of samples inside the modulating pulse and inside a period of the carrier, respectively. The normalized frequencies of the carrier and the pulse are $F_c$ and $F_b$, respectively.

The power spectral density in the logarithmic scale is calculated in the following way. The one-sided power spectral density, obtained by overlapping the amplitudes at negative frequencies with the positive frequencies, can be normalized to the unit amplitude as in eqn (7.2.26) and defined for a zero middle frequency by substituting $\Omega_n = \Omega - \Omega_c$ and expressing eqn (7.2.26) as

$$S_{SS}(\Omega_n) = 2S_{SS}(\Omega_n)/E_b = \frac{1}{N_b^2}\left|\frac{\sin\Omega_n N_b/2}{\sin\Omega_n/2}\right|^2. \qquad (7.2.27)$$

In practice, we need to see the level of the signal spectrum in logarithmic units. Therefore, we may express the power spectral density as

$$S_{SS}(\Omega_n)\big|_{dB} = 10\log_{10}\frac{1}{N_b^2}\left|\frac{\sin\Omega_n N_b/2}{\sin\Omega_n/2}\right|^2 \qquad (7.2.28)$$

and present it in graphical form, as in Figure 7.2.8, as a function of frequency $\Omega_n$.

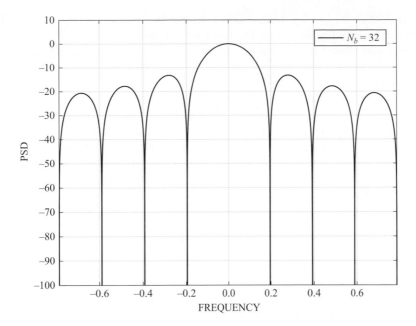

**Fig. 7.2.8** *Power spectral densities of binary phase-shift keying signals in the logarithmic scale; PSD, power spectral density.*

## 7.2.3 Receiver Operation

### 7.2.3.1 Operation of a Correlation Demodulator

A block schematic of a binary phase-shift keying system is presented in Figure 7.2.1. Operation of the transmitter was explained in Section 7.2.2. The receiver is composed of a correlation demodulator and an optimum detector. We will explain firstly the operation of the correlation demodulator. The received noisy modulated signal, at the input of demodulator, can be expressed as

$$r(n) = s_i(n) + w(n),$$  (7.2.29)

for $0 \leq n \leq N_b$, and $i = 1, 2$. The random noise $w(n)$ is a realization of a white Gaussian noise process $W(n)$ of zero mean and power spectral density $N_0/2$, and $r(n)$ is a realization of the received discrete-time stochastic process $R(n)$. The correlator is composed of a multiplier, which multiplies the received noisy signal by a local coherent carrier, and an adder that accumulates the obtained products in the bit interval $N_b$. The result is a random value expressed as

$$
\begin{aligned}
x = x_d(N_b) &= \sum_{n=0}^{N_b-1} r(n)\phi_1(n) = \sum_{n=0}^{N_b-1} [s_i(n) + w(n)]\phi_1(n) \\
&= \sum_{n=0}^{N_b-1} s_i(n)\phi_1(n) + \sum_{n=0}^{N_b-1} w(n)\phi_1(n) = s_{i1} + w,
\end{aligned}
$$  (7.2.30)

where $x$ is considered to be a realization of the random variable $X(N_b) = X$ for the first transmitted bit, which has two parts. The first part, denoted by $s_{i1}$ and called the signal part, is a quantity contributed by the transmitted signals $s_i(n)$, which is calculated as the mean value of $X$. The second part, denoted by $w$ and called the noise part, is the sample value of the Gaussian random variable $W$ that arises because of the channel noise $w(n)$. It was proven in Chapter 5 that the density function of the discrete-time stochastic process $W(kN_b)$ at the output of the correlator is Gaussian, as presented in Figure 5.5.2 in Section 5.5. The integer value $k$ indicates the order of transmitted bits in the time domain. We are interested in the statistical description of the random variable $X = X(N_b)$ at the output of the correlator at the time instant $N_b$, that is, at the end of the first received bit interval. We will find the mean value and variance of the random variable $X$ in the following subsection. Furthermore, because the received bits are statistically independent, the analysis of the first bit received is sufficient to statistically characterize any received bit.

**Example**

Present the random value $x$ on a constellation diagram, assuming that the bit 1 was sent and the noise part $w$ is positively added to the signal part.

**Solution.** The solution is shown in Figure 7.2.9.

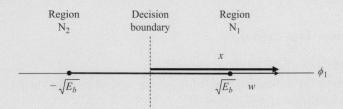

**Fig. 7.2.9**  *Constellation diagram.*

**Example**

Present the random value $x$ on a constellation diagram, assuming that the bit 1 was sent and the noise part $w$ has a negative value that is greater than the signal part.

**Solution.** The solution is shown in Figure 7.2.10.

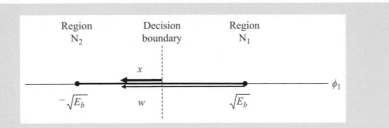

**Fig. 7.2.10** *Constellation diagram.*

### 7.2.3.2 *Operation of an Optimum Detector, and Structure of a Receiver*

As we said before, for this simple and important communication system based on binary phase-shift keying, we can decide at the output of the correlator instead of at the output of the optimum detector. Let us briefly confirm this fact.

In a binary phase-shift keying system, two signals can be generated at the transmitter side, and the output of the correlation demodulator is expressed by eqn (7.2.30), which is an input of the optimum detector expressed in vector form as $\mathbf{x} = [x_1] = [s_{11} + w]$ for the first binary symbol transmitted, or $\mathbf{x} = [x_2] = [s_{21} + w]$ for the second binary symbol transmitted. The transmitter signals can be expressed in a vector form as $\mathbf{s}_1 = [s_{11}]$ and $\mathbf{s}_2 = [s_{21}]$. Each of the two signals, which represent message bits, is transmitted and processed in the same way inside the communication system. Furthermore, the noise affecting a bit in a message bit stream is statistically independent of the noise affecting all other bits in the stream. Therefore, it is sufficient to analyse the transmission of one bit and generalize the obtained results to all transmitted bits.

Let us assume that the first signal $\mathbf{s}_1$ carrying the first message bit is transmitted, and the calculated output of the correlator is the random value $x$. At the output of the optimum detector, the following two random numerical values can be calculated as

$$Z_1 = \mathbf{x}^T \mathbf{s}_1 - \frac{1}{2}E_1 = [x][s_{11}] - \frac{1}{2}E_1 = \left[\sqrt{E_1} + w\right]\left[\sqrt{E_1}\right] - \frac{1}{2}E_1 = \frac{1}{2}E_1 + w\sqrt{E_1}, \quad (7.2.31)$$

and

$$Z_2 = \mathbf{x}^T \mathbf{s}_2 - \frac{1}{2}E_2 = [x][s_{21}] - \frac{1}{2}E_2 \underset{E_2 = E_1}{=} \left[\sqrt{E_1} + w\right]\left[-\sqrt{E_1}\right] - \frac{1}{2}E_1 = -\frac{3}{2}E_1 - w\sqrt{E_1}. \quad (7.2.32)$$

According to the explanation for the operation of an optimum detector, as explained in Chapter 5, and the structure presented in Figure 7.2.1, these two numerical values need to be compared and the largest one selected. The comparison can be performed by finding the difference of the output random values of the detector; this is expressed as

$$Z_1 - Z_2 = \frac{1}{2}E_1 + w\sqrt{E_1} + \frac{3}{2}E_1 + w\sqrt{E_1} - \frac{1}{2}E_1 = 2E_1 + 2w\sqrt{E_1}. \quad (7.2.33)$$

The numerical decision value obtained can be defined as

$$y = \frac{Z_1 - Z_2}{2\sqrt{E_1}} = \sqrt{E_1} + w, \tag{7.2.34}$$

because the term $2\sqrt{E_1}$, being a constant, does not influence our decision. Therefore, the decision rules can be defined as follows:

1. If $y = \sqrt{E_1} + w > 0$, meaning that $Z_1 > Z_2$, decide in favour of the symbol $m_k = m_1$.
2. If $y = \sqrt{E_1} + w < 0$, meaning that $Z_1 < Z_2$, decide in favour of the symbol $m_k = m_2$. Because we assumed that the message bit $m_1 = 1$ was sent, an incorrect decision will be made here in favour of $m_2 = 0$ instead of in favour of $m_1 = 1$.
3. If $y = \sqrt{E_1} + w = 0$, meaning that $Z_1 = Z_2$, flip a coin and decide in favour of $Z_1$ or $Z_2$.

Likewise, let us assume that the second signal $s_2$ is transmitted, and the output of the correlator is the random value $x$. Then the following two random numerical values can be calculated at the output of the optimum detector:

$$Z_1 = \mathbf{x}^T \mathbf{s}_1 - \frac{1}{2}E_1 = [x][s_{11}] - \frac{1}{2}E_1 = [s_{21} + w][s_{11}] - \frac{1}{2}E_1 = -\frac{3}{2}E_1 + w\sqrt{E_1}. \tag{7.2.35}$$

Assuming the same power for both signals, we obtain

$$Z_2 = \mathbf{x}^T \mathbf{s}_2 - \frac{1}{2}E_2 \underset{E_1 = E_2}{=} [x][s_{21}] - \frac{1}{2}E_1 = [s_{21} + w][s_{21}] - \frac{1}{2}E_1$$

$$= E_1 - w\sqrt{E_1} - \frac{1}{2}E_1 = \frac{1}{2}E_1 - w\sqrt{E_1}. \tag{7.2.36}$$

The comparison can be performed by finding the difference of the output random variables:

$$y = Z_1 - Z_2 = -\frac{3}{2}E_1 + w\sqrt{E_1} - \left(\frac{1}{2}E_1 - w\sqrt{E_1}\right) = -2E_1 + 2w\sqrt{E_1}. \tag{7.2.37}$$

The numerical decision value can be defined as

$$y = \frac{Z_1 - Z_2}{2\sqrt{E_1}} = -\sqrt{E_1} + w. \tag{7.2.38}$$

1. If $y = -\sqrt{E_1} + w > 0$, meaning that $Z_1 > Z_2$, decide in favour of the symbol $m_k = m_1$. Because we assumed that the message bit $m_2 = 0$ was sent, an incorrect decision will be made in favour of $m_1 = 1$ instead of in favour of $m_2 = 0$.
2. If $y = -\sqrt{E_1} + w < 0$, meaning that $Z_1 < Z_2$, decide in favour of the symbol $m_k = m_2$, which is the right decision.
3. If $y = -\sqrt{E_1} + w = 0$, meaning that $Z_1 = Z_2$, flip a coin and decide in favour of $Z_1$ or $Z_2$.

The decision value $y$ is considered to be a realization of a random variable defined as

$$Y = \pm\sqrt{E_1} + W, \qquad (7.2.39)$$

which is the sum of the signal part $\pm\sqrt{E_1}$ and the noise part $W$ at the optimum detector output. The signal part depends on the message bit transmitted and is equal to the signal coefficient. Therefore, the decision variable is identical to the random variable at the output of the correlation demodulator $X$ defined by eqn (7.2.30), that is,

$$Y = \pm\sqrt{E_1} + W = s_{i1} + W = X. \qquad (7.2.40)$$

Because the random variables $X$ and $Y$ are identical, the decision can be made using the value obtained at the output of the correlation demodulator. It is sufficient to take the output of the demodulator $x$ and apply the above-mentioned decision rules.

Bearing in mind the fact that the decision can be made using the output of the correlation demodulator, the block scheme of the binary phase-shift keying communication system can be simplified as presented in Figure 7.2.11. This is structurally the traditional presentation of a binary phase-shift keying system. The difference between our presentation of the system operation and the traditional system is in the definition of the carriers and the related mathematical model. The carriers we used at the transmitter and the receiver side are defined strictly by basis functions, as specified in Chapter 2. In that way, we made the mathematical expressions in this chapter consistent with those in both previous chapters and all succeeding chapters, excluding ambiguity in the presentation of the signals and avoiding the use of meaningless coefficients. The same approach is used throughout the book, including for the presentation of noise and the design of noise generators, thus making it possible to obtain a simplified procedure for calculating the signal-to-noise ratio and the related bit error probability in systems.

For the input binary values $m_i = 1$ or 0, a polar non-return-to-zero modulating signal $m(n)$ is generated that modulates a cosine carrier represented by a basis function $\phi_1(n)$. The modulating signal $m(n)$ is a rectangular discrete-time pulse having one of two possible

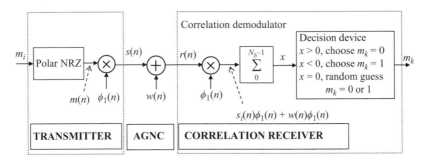

**Fig. 7.2.11** *Simplified scheme of a binary phase-shift keying communication system; AGNC, additive Gaussian noise correlator; NRZ, polar non-return-to-zero encoder.*

amplitudes $\pm\sqrt{E_b}$, depending on the value of a bit $m_i$ generated at the source output. The modulated signal $s(n)$ is transmitted through the additive Gaussian noise channel. The received signal $r(n)$ is processed inside the correlation demodulator. The output of the correlator, $x$, is compared with the threshold value, which is 0 for a symmetric system, and the detected bit $m_k$ is generated at the output of the decision device.

**Example**

We can formulate a precise statistical processing of the correlator output. This output is a random variable $X$ that can be understood as a sum of the binary random variable $S_{i1}$ (representing the random generation of the message bits at the output of the source) and the noise variable $W$. We may calculate the density function of $X$ and then the conditional densities of $X$ given $S_{i1} = s_{11}$ or $S_{i1} = s_{21}$. Suppose the probabilities of two possible signal values of the binary variable $S_{i1}$ are equal, that is, $P(s_{11}) = P(s_{21}) = 1/2$. Because the signal values at the correlator output correspond to the message bits sent, we may have the following probabilities:

$$P(s_{11}) = P\left(+\sqrt{E_b}\right) = P(m_1) = P(1) = 1/2, \tag{7.2.41}$$

and

$$P(s_{21}) = P\left(-\sqrt{E_b}\right) = P(m_2) = P(0) = 1/2. \tag{7.2.42}$$

Find the density function of $X$ and the conditional densities of $X$ for the given message bits $m_1 = 1$ and $m_2 = 0$, and present them in graphical form.

**Solution.** The probability density function of $S_{i1}$, which randomly takes the values $s_{11}$ and $s_{21}$, can be expressed in terms of Kronecker delta functions as

$$f_{S_{i1}}(s_{i1}) = P(s_{11})\delta(s_{i1} - s_{11}) + P(s_{21})\delta(s_{i1} - s_{21}). \tag{7.2.43}$$

This function is equivalent to the function defined by eqn (7.2.3), as the sample space is the same. The noise variable $W$ is Gaussian, as presented in Chapter 5, and is expressed as

$$f_W(w) = \frac{1}{\sqrt{2\pi\sigma_W^2}} e^{-w^2/2\sigma_W^2}. \tag{7.2.44}$$

The variable $X$ can be expressed as the sum of the two independent random variables $S_{i1}$ and $W$:

$$X = S_{i1} + W. \tag{7.2.45}$$

Therefore, its density function is the convolution of related densities that can be calculated as

$$
\begin{aligned}
f_X(x) &= \int_{-\infty}^{\infty} f_{S_{i1}}(s_{i1}) f_W(x - s_{i1}) \, ds_{i1} \\
&= \int_{-\infty}^{\infty} [P(s_{11}) \delta(s_{i1} - s_{11}) + P(s_{21}) \delta(s_{i1} - s_{21})] \frac{1}{\sqrt{2\pi \sigma_W^2}} e^{-(x - s_{i1})^2 / 2\sigma_W^2} \, ds_{i1} \\
&= P(s_{11}) \frac{1}{\sqrt{2\pi \sigma_W^2}} e^{-(x - s_{11})^2 / 2\sigma_W^2} + P(s_{21}) \frac{1}{\sqrt{2\pi \sigma_W^2}} e^{-(x - s_{21})^2 / 2\sigma_W^2} \\
&= 0.5 \cdot f_X(x | S_{i1} = s_{11}) + 0.5 \cdot f_X(x | S_{i1} = s_{21}).
\end{aligned}
\tag{7.2.46}
$$

The conditional density functions, expressed either in terms of signals at the output of the correlator or in terms of message bits sent, are

$$
\begin{aligned}
f_X(x | S_{i1} = s_{11}) = f_X\left(x | S_{i1} = \sqrt{E_b}\right) &= \frac{1}{\sqrt{2\pi \sigma_W^2}} e^{-(x - s_{11})^2 / 2\sigma_W^2} \\
&= \frac{1}{\sqrt{2\pi \sigma_W^2}} e^{-(x - \sqrt{E_b})^2 / 2\sigma_W^2} = f_X(x | m_i = 1),
\end{aligned}
\tag{7.2.47}
$$

and

$$
\begin{aligned}
f_X(x | S_{i1} = s_{21}) = f_X(x | s_{i1}) = -\sqrt{E_b}) &= \frac{1}{\sqrt{2\pi \sigma_W^2}} e^{-(x - s_{21})^2 / 2\sigma_W^2} \\
&= \frac{1}{\sqrt{2\pi \sigma_W^2}} e^{-(x + \sqrt{E_b})^2 / 2\sigma_W^2} = f_X(x | m_i = 0).
\end{aligned}
\tag{7.2.48}
$$

The random variable $X$ has its mean and variance expressed as

$$
\eta_X = E\{X\} = E\{S_{i1} + W\} = 0
$$

and

$$
\begin{aligned}
\sigma_X^2 = E\left\{X^2\right\} = E\left\{(S_{i1} + W)^2\right\} &= E\left\{(S_{i1})^2\right\} + 0 + E\left\{(W)^2\right\} \\
&= E_b + \sigma_W^2 = E_b + \frac{N_0}{2},
\end{aligned}
\tag{7.2.49}
$$

respectively. In a similar way, we can find the mean and variance of the conditional probabilities defined by eqns (7.2.47) and (7.2.48). The mean values will be $\sqrt{E_b}$ and $-\sqrt{E_b}$, respectively, and the variance will be $N_0/2$.

### 7.2.3.3 Calculation of the Bit Error Probability

The quality of message bit transmission depends on the signal-to-noise ratio at the correlator output. This ratio is directly related to the bit error probability in the given binary phase-shift keying system. Therefore, the aim of communication system analysis is to find the bit error probability as a function of $E_b/N_0$, which is called the system property and is presented by a related graph called the system curve. In the case of binary phase-shift keying, it is sufficient to have the correlation demodulator output $x$ to make a decision.

Following the derivations related to the density function of the random variable $X$, eqns (7.2.47) and (7.2.48), we may express the conditional probability density function of the correlator random variable $X$, given that a message bit $m_i = 0$ or 1 is transmitted, and present them as shown in Figure 7.2.12.

In order to explain the procedure of decision that we may make at the correlation demodulator output, we will use both the constellation diagram in Figure 7.2.6 and the defined conditional density functions presented in Figure 7.2.12. The decision rules, assuming that the threshold value is $T_h = 0$ defined for a symmetric system, are as follows:

1. If the output of the correlator, $x$, falls in the region

$$N1 : 0 < x < \infty, \tag{7.2.50}$$

then two probabilities can be identified. *Firstly*, the conditional probability that the message symbol $m_1 = 1$ is received, given that the message symbol $m_1 = 1$ (corresponding to the modulated signal $s_1(n)$) was transmitted, is expressed as

$$P(1|1) = \int_0^\infty f_X(x|1)\, dx = \int_0^\infty \frac{1}{\sqrt{\pi N_0}} e^{-\frac{1}{N_0}\left(x-\sqrt{E_b}\right)^2} dx. \tag{7.2.51}$$

This probability is large and corresponds to the area below the curve of density function $f_X(x|1)$ for $x > 0$. Thus, the right decision will be made in favour of signal $m_1 = 1$ with

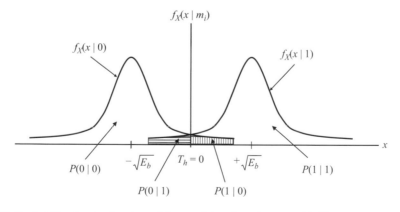

**Fig. 7.2.12** *Probability density function of the conditional random variable X.*

a high probability of $P(1|1)$. *Secondly,* the conditional probability that the message symbol $m_1 = 1$ is received, given that the message symbol $m_2 = 0$ (corresponding to the modulated signal $s_2(n)$) was transmitted, is expressed as

$$P(1|0) = \int_0^\infty f_X(x|0)\,dx = \int_0^\infty \frac{1}{\sqrt{\pi N_0}} e^{-\frac{1}{N_0}\left(x + \sqrt{E_b}\right)^2} dx. \qquad (7.2.52)$$

This probability is small and corresponds to the area below the curve of density function $f_X(x|0)$ for $x > 0$. Thus, an erroneous decision will be made in favour of signal $m_1 = 1$ with a small probability $P(1|0)$.

2. If the output of the correlator, $x$, falls in the region

$$N2 : -\infty < x < 0, \qquad (7.2.53)$$

then two probabilities can be identified. *Firstly,* the conditional probability that the message symbol $m_2 = 0$ is received, given that the message symbol $m_2 = 0$ (corresponding to the modulated signal $s_2(n)$) was transmitted, is expressed as

$$P(0|0) = \int_{-\infty}^0 f_X(x|0)\,dx = \int_{-\infty}^0 \frac{1}{\sqrt{\pi N_0}} e^{-\frac{1}{N_0}\left(x + \sqrt{E_b}\right)^2} dx. \qquad (7.2.54)$$

This probability is large and corresponds to the area below the curve of density function $f_X(x|0)$ for $x < 0$. Thus, the right decision will be made in favour of signal $m_2 = 0$ with a high probability $P(0|0)$. *Secondly,* the conditional probability that the message symbol $m_2 = 0$ is received, given that the message symbol $m_1 = 1$ (corresponding to the modulated signal $s_1(n)$) was transmitted, is expressed as

$$P(0|1) = \int_{-\infty}^0 f_X(x|1)\,dx = \int_{-\infty}^0 \frac{1}{\sqrt{\pi N_0}} e^{-\frac{1}{N_0}\left(x - \sqrt{E_b}\right)^2} dx. \qquad (7.2.55)$$

This probability is small and corresponds to the area below the curve of density function $f_X(x|1)$ for $x < 0$. Thus, an erroneous decision will be made in favour of signal $m_2 = 0$ with a small probability $P(0|1)$. The probabilities of erroneous decisions (both of the second cases above) are equal, due to the equality of the signal energies with respect to the threshold value $T_h = 0$.

For a symmetric channel, the conditional probabilities of erroneous decision are identical. The conditional probability of the receiver deciding in favour of symbol $m_1 = 1$, given that symbol $m_2 = 0$ was transmitted, is equal to $P(1|0) = P(0|1)$, that is,

$$P(1|0) = \int_0^\infty f_X(x|0)\,dx = \int_0^\infty \frac{1}{\sqrt{\pi N_0}} e^{-\frac{1}{N_0}\left(x + \sqrt{E_b}\right)^2} dx. \qquad (7.2.56)$$

Defining a new variable $z$ as

$$z = \frac{1}{\sqrt{N_0}} \left( x + \sqrt{E_b} \right),$$

(7.2.57)

and substituting the variable of integration $x$ to $z$, we obtain

$$P(1|0) = \int_0^\infty f_X(x|0)\, dx = \frac{1}{\sqrt{\pi}} \int_{\sqrt{E_b/N_0}}^\infty e^{-z^2}\, dz = \frac{1}{2} \mathrm{erfc}\left( \sqrt{\frac{E_b}{N_0}} \right),$$

(7.2.58)

where $\mathrm{erfc}(x)$ is the complementary error function defined here as

$$\mathrm{erfc}(x) = \frac{2}{\sqrt{\pi}} \int_x^\infty e^{-z^2}\, dz.$$

(7.2.59)

Taking into account the symmetry of the distributions in Figure 7.2.12, it can also be proved that

$$P(0|1) = P(1|0) = \frac{1}{2} \mathrm{erfc}\left( \sqrt{\frac{E_b}{N_0}} \right).$$

(7.2.60)

Assuming that the channel is symmetric, that is, $P(0|1) = P(1|0)$, the probability of symbol error expression, generally defined in eqn (6.1.3), can be applied here for $M = 2$ to calculate the bit error probability of the binary phase-shift keying system as

$$p = \sum_{i=1}^{2} P(m_i) P(m_k \neq m_i | m_i) = P(0)P(1|0) + P(1)P(0|1)$$

$$= P(1|0) = \frac{1}{2} \mathrm{erfc}\left( \sqrt{\frac{E_b}{N_0}} \right),$$

(7.2.61)

where $m_k$ is the estimate of the transmitted bit value. This function is usually plotted using logarithmic scales. In this case, the $x$-axis will present the signal-to-noise ratio $E_b/N_0$ in decibels and the $y$-axis will present the bit error probability $p$ as the negative power of the base 10, as presented in Chapter 6 in Figure 6.2.15.

## 7.3   Quadriphase-Shift Keying

### 7.3.1   System Operation

Based on the general scheme of a correlator-based communication system, which is presented in Figure 5.6.4 and in Figure 5.7.1 for the binary source, a quadriphase-shift keying system can be obtained as shown in Figure 7.3.1.

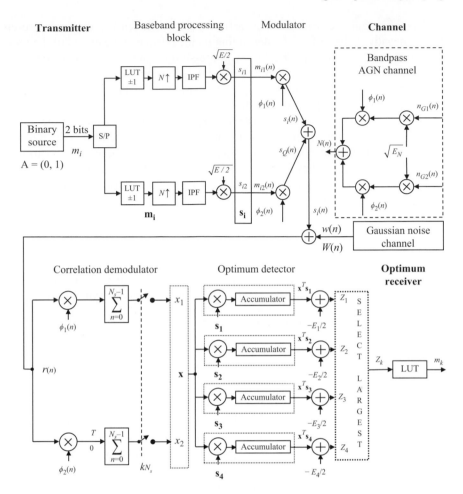

**Fig. 7.3.1** *A quadriphase-shift keying discrete-time communication system; AGN, additive Gaussian noise; IPF, interpolation filter; LUT, look-up table; S/P, serial-to-parallel convertor.*

Suppose the message source can generate two binary values (or bits) that belong to an alphabet A of $A = 2$ elements and are denoted by $m_1 = 1$ and $m_2 = 0$. A symbol generated at the output of the source is a pair of bits called a dibit. The dibit is converted into two parallel message bits by means of a serial-to-parallel convertor (de-multiplexer). Both message bits are applied at the same time to the inputs of two look-up-tables. The look-up tables generate polar binary values at their outputs, taken from the set $(-1, +1)$, that form message bits represented as a vector $\mathbf{m_i}$ consisting of two polar bits. These values are multiplied by the square root of a bit energy, $\sqrt{E/2}$, and the signal coefficient obtained is used to generate two modulating signals $m_{ij}(n)$: $m_{i1}(n)$ in the upper branch (which is also called the *I* system) and $m_{i2}(n)$ in the lower branch (which is also called the *Q* system) of the modulator. These signals are applied to the input of two modulator multipliers to modulate the carriers. This procedure can also be understood as the magnitude scaling of the carrier by the coefficients $s_{ij}$ that are represented by positive or negative values of $\sqrt{E/2}$, depending on the bit sign at the output of the look-up table. At the output of each multiplier, a binary phase-shift keying signal is obtained.

The received signal $r(n)$ is coherently demodulated using two correlators inside the correlation demodulator. Because the principle of operation of both correlators is the same, it is sufficient to explain the operation of one of them. At their outputs, random samples $x_1$ and $x_2$ are obtained at the discrete-time instants $kN_s$ that can be presented to the optimum detector to correlate them with all possible symbols that could be generated at the transmitter side. The decision variables obtained are compared, the largest is chosen, and an estimate of the transmitted dibit is made. However, in a quadriphase-shift keying system, as in a binary phase-shift keying system, the equivalent decision can be made at the output of the correlation demodulator, as will be explained later.

## 7.3.2 Transmitter Operation

### 7.3.2.1 Modulating Signals in Time and Frequency Domains

**Signal-space or constellation diagram of the signal.** The number of possible message symbols is $I = A^J = 2^2 = 4$, which are defined by the matrix of message symbols at the output of the look-up tables as

$$\mathbf{m} = [\mathbf{m}_1 \ \mathbf{m}_2 \ \mathbf{m}_3 \ \mathbf{m}_4] = \begin{bmatrix} 1 & -1 & -1 & 1 \\ -1 & -1 & 1 & 1 \end{bmatrix}. \tag{7.3.1}$$

Each dibit is assigned by one of four equally spaced phase values. These values correspond to the Gray-encoded set of dibits ordered as 10, 00, 01, and 11, ensuring that only a single bit is changed inside the neighbouring dibits. These message symbols, presented in the form of dibits, need to be mapped into the coefficients of the modulating signals. In this case, all possible coefficients are obtained by multiplying the matrix of message symbols by the specified constant $\sqrt{E/2}$, where $E$ is the energy of any symbol. They are represented by four columns in the following matrix of modulating signals:

$$\begin{aligned}
\mathbf{s} = [\mathbf{s}_1 \ \mathbf{s}_2 \ \mathbf{s}_3 \ \mathbf{s}_4] &= \begin{bmatrix} s_{11} & s_{21} & s_{31} & s_{41} \\ s_{12} & s_{22} & s_{32} & s_{42} \end{bmatrix} \\
&= \mathbf{m}\sqrt{E/2} = \begin{bmatrix} \sqrt{E/2} & -\sqrt{E/2} & -\sqrt{E/2} & \sqrt{E/2} \\ -\sqrt{E/2} & -\sqrt{E/2} & \sqrt{E/2} & \sqrt{E/2} \end{bmatrix}.
\end{aligned} \tag{7.3.2}$$

Later on, we will show the signal-space diagram for each pair of coordinates, for $I = 2$ (see Section 7.3.2.2, Figure 7.3.6).

**Modulating signals in the time domain.** The defined coefficients in matrix $\mathbf{s}$ are amplitudes of the modulating signals $m_{ij}(n)$ that are defined as rectangular pulses of symbol duration $N_s$. The wave shape of the first modulating signal, for the fourth message symbol, $m_{41}(n)$, is a rectangular pulse of duration $N_s$ and amplitude equal to the $i$th coefficient, where $i = 4$, as presented in Figure 7.3.2 for the case when $N_s = 2$.

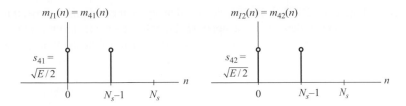

**Fig. 7.3.2** *Modulating rectangular pulses aligned in the time domain as they appear at the input of the modulator multipliers.*

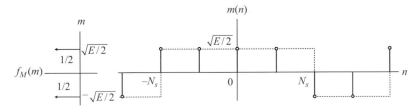

**Fig. 7.3.3** *A modulating polar binary signal presented in the time domain.*

**Statistical characteristics of the modulating signals in the time domain.** Modulating signals $m_{ij}(n)$ in both branches are realizations of the random binary *processes* $M_{ij}(n)$, which takes randomly the amplitude values specified by the vector defined in eqn (7.3.2) and generates two binary discrete-time modulating signals. We will analyse one of them which is presented in Figure 7.3.3 and denoted by $m(n)$, which is a realization of the binary random process $M(n)$, because they both have the same statistical characteristics. We will assume that the probabilities of both the positive and the negative binary values are equal to 0.5. Therefore, at each time instant $t = t_k$, we have a discrete random variable $M = M_{ij}(t_k)$ that takes the value of $\sqrt{E/2}$ or $-\sqrt{E/2}$ and is defined by its probability density function expressed in terms of Dirac delta function as

$$f_M(m) = \frac{1}{2}\delta\left(m - \sqrt{E/2}\right) + \frac{1}{2}\delta\left(m + \sqrt{E/2}\right), \tag{7.3.3}$$

as shown in Figure 7.3.3. The notation here follows the notation used in the theory of continuous-time stochastic processes, as presented in Chapter 19.

Since the amplitude levels of $m(n)$ have the same magnitude values and occur with equal probabilities, the mean value of the *process* is zero and the variance is $E/2$. This signal is statistically the same in both the upper and the lower branch of the modulator.

**Presentation of the modulating signal in the frequency domain.** Following the derivation in Chapter 15 and relation (7.2.4), the magnitude spectral density of a rectangular discrete-time pulse is

$$|M(\Omega)| = \begin{cases} \sqrt{E/2}N_s & \Omega = \pm 2k\pi, k = 0, 1, 2, 3, \ldots \\ \sqrt{E/2}\left|\frac{\sin \Omega N_s/2}{\sin \Omega/2}\right| & otherwise \end{cases}, \tag{7.3.4}$$

as presented in Figure 7.2.3. The energy spectral density is the magnitude spectral density squared. The power spectral density in the upper ($I$) branch is equivalent to the power spectral density in the lower ($Q$) branch, and can be expressed it in terms of the symbols $N_s$ and $E$ and bit parameters $N_b = N_s/2$ and $E_b = E/2$, respectively, as

$$S_{MI}(\Omega) = S_{MQ}(\Omega) = \frac{|M(\Omega)|^2}{N_s} = \frac{E}{2N_s}\left|\frac{\sin \Omega N_s/2}{\sin \Omega/2}\right|^2 = \frac{E_b N_b}{2}\frac{1}{N_b^2}\left|\frac{\sin 2\Omega N_b/2}{\sin \Omega/2}\right|^2. \quad (7.3.5)$$

Due to the random nature of the modulating signal, its power spectral density function can be calculated using its autocorrelation function, as shown in Appendix A. The maximum value of this function is $S_{MI}(\Omega)\max = 2E_b N_b$. Because the baseband signal components in the modulator branches are independent, the baseband power spectral density can be numerically expressed as the sum of individual densities as

$$S_{MM}(\Omega) = S_{MI}(\Omega) + S_{MQ}(\Omega) = E_b N_b \frac{1}{N_b^2}\left|\frac{\sin \Omega N_b}{\sin \Omega/2}\right|^2.$$

The maximum value of this function is $S_{MM}(\Omega)\max = 4E_b N_b$. The power spectral density function in the upper branch, $S_{MI}(\Omega)$ is normalized by $E_b N_b$, and its graph is presented in Figure 7.3.4, for values of frequency up to $\pi/4$, even though the spectrum exists in the whole interval from $-\pi$ to $+\pi$. Zero crossings can be calculated for this condition as $\sin 2\Omega N_b/2 = 0$ or $2\Omega N_b/2 = k\pi$. Therefore, the crossings are defined by $\Omega = k\pi/N_b$, for $k = 1, 2, 3, \ldots$. For the graph in Figure 7.2.4, the number of samples is obtained from the condition expressed as $\Omega = 2\pi/N_b = \pi/16$, for the number of samples $N_b = 32$ or $N_s = 64$. The maximum value of the quadratic term in eqn (7.3.5) is $4N_b^2$. Therefore, the maximum value of the normalized power spectral density in one branch is 2, as shown in Figure 7.3.4.

The modulating signals $m(n)$ in the upper and lower branches are realizations of the statistically independent and identical stochastic processes $M(n)$. Therefore, the power spectral density of the modulating signal is expressed as the sum of power spectral densities in both branches. Furthermore, to compare the power spectral densities and the powers of various modulation schemes, the power spectral density is expressed as a function of common parameters: the bit duration $N_b$ and the bit energy $E_b$.

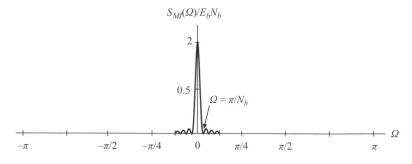

**Fig. 7.3.4** *Power spectral density of a modulating discrete binary signal.*

### 7.3.2.2 Modulated Signals in the Time Domain

If the phases of carrier are $\pi/4$, $3\pi/4$, $5\pi/4$, and $7\pi/4$, the transmitted discrete-time quadriphase-shift keying signal is defined as

$$s_i(n) = \sqrt{\frac{2E}{N_s}} \cos\left[\Omega_c n + (2i-1)\frac{\pi}{4}\right] = \sqrt{\frac{2E_b}{N_b}} \cos\left[\Omega_c n + (2i-1)\frac{\pi}{4}\right], \tag{7.3.6}$$

for $i = 1, 2, 3, 4$, where $0 \leq n < N_s$, and $E$ is the transmitted signal energy per symbol (dibit). The period of the carrier $N_c$ and the modulation pulse duration $N_s$ are chosen to fulfil the condition $N_s/N_c = k$, for a fixed integer $k$, in order to ensure that each transmitted dibit contains an integer number of cycles of the carrier. In that case, expression (7.3.6) for the modulated signal can be expressed as

$$s_i(n) = \sqrt{\frac{2E}{N_s}} \cos\left[\frac{2\pi}{N_c}n + (2i-1)\frac{\pi}{4}\right], \tag{7.3.7}$$

where $N_c$ is the period of the discrete-time carrier. In addition, the number of samples inside the period of the carrier should be at least $N_c = 8$ to accommodate generation of the symbols with four possible phases, as presented in Figure 7.3.5. These conditions are controlled by the up-sampling procedure inside the transmitter, which is performed by the up-sampler and the interpolation filter, as shown in Figure 7.3.1, according to the theory presented in Chapter 17. For the example, in the wave shape of a quadriphase-shift keying signal shown in Figure 7.3.5, the first phase is $\pi/4$, which contains information about the dibit 10, which corresponds to the message point in the first quadrant of the related constellation diagram. After the discrete-time interval defined by $N_s = 8$, which is the duration of a symbol, a new modulated signal is generated with phase $3\pi/4$, which corresponds to the transmitted dibit 00.

The discrete-time quadriphase-shift keying signal is generated according to general transmitter structure shown in Figure 7.3.1. Bearing in mind the general expression of a quadriphase-shift keying signal presented in eqn (7.3.6), we may also express the quadriphase-shift keying signal as

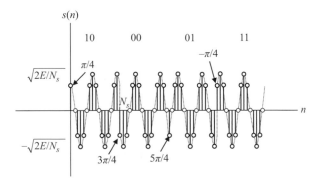

**Fig. 7.3.5** *The discrete-time modulated signal obtained by direct changes of symbols' phases.*

$$s_i(n) = \sqrt{\frac{2E}{N_s}} \cos\left[(2i-1)\frac{\pi}{4}\right]\cos\Omega_c n - \sqrt{\frac{2E}{N_s}} \sin\left[(2i-1)\frac{\pi}{4}\right]\sin\Omega_c n$$

$$= \sqrt{E}\cos\left[(2i-1)\frac{\pi}{4}\right]\phi_1(n) - \sqrt{E}\sin\left[(2i-1)\frac{\pi}{4}\right]\phi_2(n) = s_{i1}\phi_1(n) + s_{i2}\phi_2(n), \tag{7.3.8}$$

where the two orthonormal basis functions, which represent a pair of quadrature carriers called the in-phase (*I*) carrier and the quadrature (*Q*) carrier, respectively, are defined as

$$s_{cI}(n) = \phi_1(n) = \sqrt{\frac{2}{N_s}}\cos\Omega_c n, \quad s_{cQ}(n) = \phi_2(n) = \sqrt{\frac{2}{N_s}}\sin\Omega_c n, \tag{7.3.9}$$

where $0 \le n < N_s$. Therefore, a signal space is two dimensional ($J = 2$), with a signal constellation diagram consisting of four message points ($I = 4$). The associated signal vectors are

$$\mathbf{s}_i = \begin{bmatrix} s_{i1} \\ s_{i2} \end{bmatrix} = \begin{bmatrix} \sqrt{E}\cos\left[(2i-1)\frac{\pi}{4}\right] \\ -\sqrt{E}\sin\left[(2i-1)\frac{\pi}{4}\right] \end{bmatrix}, \tag{7.3.10}$$

with coordinates $s_{i1}$ and $s_{i2}$, for $i = 1, 2, 3, 4$. All the coordinates and corresponding dibits are calculated and presented in matrix form in eqns (7.3.1) and (7.3.2). For each pair of coordinates, the signal-space diagram is shown in Figure 7.3.6. For example, message point 1 (for dibit $M_1 = m_1 m_2 = 10$ or, equivalently, for signal $s_1(n)$) is determined by the coordinates located at $s_{11}$ and $s_{12}$. The constellation shown has a minimum energy.

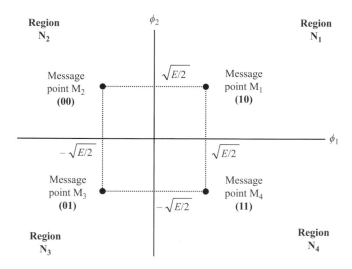

**Fig. 7.3.6** *Signal-space diagram and message constellation for quadriphase-shift keying.*

## Example

Form a quadriphase-shift keying signal for the source binary sequence $\{m_i\} = \{0\ 1\ 1\ 0\ 1\ 1\ 0\ 0\}$, which corresponds to the sequence of four dibits $\{M_i\} = \{01\ 10\ 11\ 00\}$.

  a. Divide the input binary sequence into two sequences, consisting of the first and second bits of the dibit sequences $\{m_{ij}\}$.
  b. For each bit inside a dibit, draw the waveform. Add the waveforms for each bit in a dibit to obtain a quadriphase-shift keying waveform for that dibit.

## Solution.

  a. The sequence division is as follows:

  $$\{m_{i1}\} = \{m_{11}m_{21}m_{31}m_{41}\} = \{0\quad 1\quad 1\quad 0\},$$
  $$\{m_{i2}\} = \{m_{12}m_{22}m_{32}m_{42}\} = \{1\quad 0\quad 1\quad 0\},$$

  b. The quadriphase-shift keying modulated signal is shown in Figure 7.3.7.

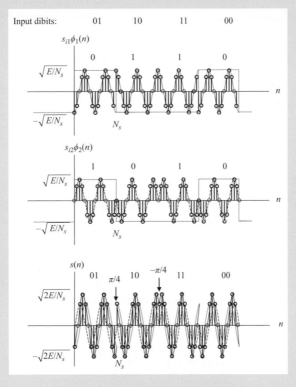

**Fig. 7.3.7** *Generation of a quadriphase-shift keying modulated signal.*

### 7.3.2.3  *Modulated Signals in the Frequency Domain*

A quadriphase-shift keying modulated signal is obtained by multiplying two modulating signals with their related carriers and then summing them up, as shown in Figure 7.3.1. The carriers can be considered to be deterministic signals defined by their amplitudes, frequencies, and phases, and the modulating signals, as we said before, are stochastic binary processes. Because the processing methods in both branches are identical, we will analyse only the processing in the upper branch signal. To find the modulated signal in time and frequency domains, we need to find the carrier in both domains and then investigate the output of the individual modulator multiplier.

**Presentation of the carrier in time and frequency domains.** The in-phase and quadrature carriers are defined in eqn (7.3.9). The autocorrelation function of the cosine carrier is calculated in Chapter 3, Section 3.2.2, and expressed in the form

$$R_{sI}(l) = \frac{1}{N_s} \cos \Omega_c l. \tag{7.3.11}$$

In addition, the autocorrelation function of the sine signal representing the quadrature carrier is the same as the autocorrelation function of the cosine carrier. Therefore, the autocorrelation function for both carriers is

$$R_{sc}(l) = R_{cI}(l) = R_{cQ}(l) = \frac{1}{N_s} \cos \Omega_c l. \tag{7.3.12}$$

Likewise, the power spectral density of each carrier is the Fourier transform of its autocorrelation function, given by

$$S_{sc}(\Omega) = S_{sI}(\Omega) = S_{sQ}(\Omega) = \frac{1}{N_s} \pi \left[ \delta(\Omega - \Omega_c) + \delta(\Omega + \Omega_c) \right], \tag{7.3.13}$$

as was shown in Chapter 3, Section 3.6.3, for a harmonic stochastic process.

**The power spectral density of quadriphase-shift keying modulated signals.** The baseband modulating signal $m_{i1}(n)$, which has the power spectral density as shown in Figure 7.3.4, is multiplied by the carrier operating at the frequency $\Omega_c$. The initial amplitudes of the random baseband signals $m_{i1}(n)$ forming the modulating signal $m(n)$, and the in-phase carrier $\phi_1(n)$, are statistically independent due to the randomness of the signal amplitudes. Therefore, the bandpass modulated random signal at the output of the in-phase multiplier $s_I(n)$, represented as a realization of a wide-sense stationary stochastic process $S_I(n)$ as shown in Figure 7.3.1, can be expressed as the product

$$s_I(n) = m(n) \cdot \phi_1(n) = m(n) \cdot \sqrt{\frac{2}{N_s}} \cos(\Omega_c n), \tag{7.3.14}$$

where $m(n)$ is a realization of the baseband modulating stochastic process $M(n)$, and $s_I(n)$ is one realization of the discrete-time stochastic process $S_I(n)$. In order to find the power spectral density function of the modulated signal, we need to find the autocorrelation function of the stochastic process $S_I(n)$, as we did for the binary phase-shift keying signal in eqn (7.2.25), and express it as

$$R_{SI}(l) = R_{MI}(l) \cdot R_{sc}(l) = R_{MI}(l) \cdot \frac{1}{N_s} \cos \Omega_c l. \tag{7.3.15}$$

Likewise, like the derivative in eqn (7.2.26), the power spectral density of the output random modulated signal can be expressed as the convolution

$$
\begin{aligned}
S_{SI}(\Omega) &= \frac{1}{2\pi} \int_{-\pi}^{\pi} S_{MI}(\Omega - \lambda) S_{sc}(\lambda) d\lambda \\
&= \frac{1}{2\pi} \int_{-\pi}^{\pi} \frac{E}{2N_s} \left| \frac{\sin(\Omega - \lambda) N_s/2}{\sin(\Omega - \lambda)/2} \right|^2 \cdot \frac{1}{N_s} \pi \left[ \delta(\lambda - \Omega_c) + \delta(\lambda + \Omega_c) \right] d\lambda \\
&= \frac{E}{4} \frac{1}{N_s^2} \left( \left| \frac{\sin(\Omega - \Omega_c) N_s/2}{\sin(\Omega - \Omega_c)/2} \right|^2 + \left| \frac{\sin(\Omega + \Omega_c) N_s/2}{\sin(\Omega + \Omega_c)/2} \right|^2 \right) = S_{SQ}(\Omega),
\end{aligned}
$$

which is equivalent to the power spectral density in the $Q$ branch. Bearing in mind the relationship between a bit and a symbol, defined by $E_b = E/2$ and $N_b = N_s/2$, we obtain

$$S_{SI}(\Omega) = S_{SQ}(\Omega) = \frac{E_b}{8N_b^2} \left( \left| \frac{\sin(\Omega - \Omega_c) N_b}{\sin(\Omega - \Omega_c)/2} \right|^2 + \left| \frac{\sin(\Omega + \Omega_c) N_b}{\sin(\Omega + \Omega_c)/2} \right|^2 \right) \tag{7.3.16}$$

This is the power spectral density of the modulated signal in the upper branch. The signal in the lower branch has the same power spectral density. The signals in the two branches are independent from each other. Therefore, the power spectral density of the quadriphase-shift keying modulated signal is the sum of power spectral densities of both branches; it can be calculated as

$$S_{SS}(\Omega) = S_{SI}(\Omega) + S_{SQ}(\Omega) = \frac{E_b}{4N_b^2} \left( \left| \frac{\sin(\Omega - \Omega_c) N_b}{\sin(\Omega - \Omega_c)/2} \right|^2 + \left| \frac{\sin(\Omega + \Omega_c) N_b}{\sin(\Omega + \Omega_c)/2} \right|^2 \right), \tag{7.3.17}$$

and its normalized function is

$$S_{SS}(\Omega)/E_b = \frac{1}{4N_b^2} \left( \left| \frac{\sin(\Omega - \Omega_c) N_b}{\sin(\Omega - \Omega_c)/2} \right|^2 + \left| \frac{\sin(\Omega + \Omega_c) N_b}{\sin(\Omega + \Omega_c)/2} \right|^2 \right),$$

which is presented in Figure 7.3.8. This expression is formally equivalent to the power spectral density of a binary phase-shift keying signal, as expressed by eqn (7.2.26). The normalized power spectral density of a quadriphase-shift keying modulated signal is presented graphically in Figure 7.3.8. The amplitudes of the quadriphase-shift keying power spectral density are two times larger than the amplitudes of the binary phase-shift keying spectrum. However,

**Fig. 7.3.8** *Power spectral density of a discrete quadriphase-shift keying modulated signal.*

the first zero crossing of the quadriphase-shift keying is two times smaller, meaning that the spectrum requirements for the quadriphase-shift keying system are two times smaller than for the binary phase-shift keying system, which is a big advantage of the quadriphase-shift keying system. Because the maximum value of the quadratic terms in eqn (7.3.17) is $4N_b^2$, the power spectral density of the quadriphase-shift keying signal normalized by $2E_b$ is a frequency-shifted version of the power spectral density of the related modulating signal.

Figure 7.4.11 shows the power spectral density in the logarithmic scale, which is calculated in the following way. The one-sided power spectral density, obtained by overlapping the amplitudes at negative frequencies with the positive ones, can be normalized to the unit amplitude and zero middle frequency and expressed as a function of frequency $\Omega_n = \Omega - \Omega_c$ in the form

$$S_{SS}(\Omega_n) = 2\frac{E_b}{4N_b^2}\left|\frac{\sin \Omega_n N_b}{\sin \Omega_n/2}\right|^2 = \frac{E_b}{2N_b^2}\left|\frac{\sin \Omega_n N_b}{\sin \Omega_n/2}\right|^2. \tag{7.3.18}$$

In practice, we need to see the level of the spectrum in logarithmic units. Therefore, we will normalize expression (7.3.18) and find the logarithm of the normalized power spectral density as

$$S_{SS}(\Omega_n)\big|_{dB} = 10\log_{10}\frac{S_{SS}(\Omega_n)}{2E_b} = 10\log_{10}\frac{1}{4N_b^2}\left|\frac{\sin \Omega_n N_b/2}{\sin \Omega_n/2}\right|^2 \tag{7.3.19}$$

(see Section 7.4, Figure 7.4.11, to see this expression in graphical form together with the power spectral density functions of a binary phase-shift keying signal and a binary frequency-shift keying signal, for the sake of comparison).

### 7.3.3  Receiver Operation

#### 7.3.3.1  *Operation of the Correlation Demodulator and the Optimum Detector*

The modulated signal is transmitted through the additive white Gaussian noise channel defined in Chapter 4, and received by a coherent quadriphase-shift keying correlation receiver. The received signal can be expressed as

$$r(n) = s_i(n) + w(n), \tag{7.3.20}$$

for $0 \leq n \leq N_s$ and $i = 1, 2, 3, 4$, where $w(n)$ is a sample of a white Gaussian noise process $W(n)$ of zero mean and power spectral density $N_0/2$. At the output of the two correlators $(j = 1, 2)$ are the discrete-time random signals, which are defined by their random values at the end of each symbol interval $N_s$ as

$$x_1 = \sum_{n=0}^{N_s-1} x(n)\phi_1(n) = \sum_{n=0}^{N_s-1} s_i(n)\phi_1(n) + \sum_{n=0}^{N_s-1} w(n)\phi_1(n) = s_{i1} + w_1$$

$$= \sqrt{E}\cos\left[(2i-1)\frac{\pi}{4}\right] + w_1 = \pm\sqrt{\frac{E}{2}} + w_1$$

(7.3.21)

and

$$x_2 = \sum_{n=0}^{N_s-1} x(n)\phi_2(n) = \sum_{n=0}^{N_s-1} s_i(n)\phi_2(n) + \sum_{n=0}^{N_s-1} w(n)\phi_2(n) = s_{i2} + w_2$$

$$= -\sqrt{E}\sin\left[(2i-1)\frac{\pi}{4}\right] + w_2 = \mp\sqrt{\frac{E}{2}} + w_2,$$

(7.3.22)

where $s_{i1}$ and $s_{i2}$ are deterministic quantities contributed by the transmitted signal $s_i$, and $w_1$ and $w_2$ are the sample values of the random variables $W_1$ and $W_2$ that arise due to the channel noise process $W(n)$. The calculated random values are realizations of the random variables $X_1$ and $X_2$ and can be expressed in the form of a demodulation vector $\mathbf{x} = [x_1 \ x_2]$. With this vector, we can proceed with processing the signals inside the optimum detector. However, as will be explained later, we can decide on the transmitted symbols by using the following decision rules:

1.  If the output of the correlator, the demodulation vector $\mathbf{x}$, falls in the region $N_1$, that is, the first quadrant of the coordinate system, then we conclude that the dibit 10 (corresponding to the transmitted signal $s_1(n)$) was generated with a high probability, and the decision will be made in favour of the dibit 10.

2.  However, if the output of the correlator $\mathbf{x}$ falls in the region $N_1$, that is, the first quadrant of the coordinate system, *given that* the dibit 00 (corresponding to the transmitted signal $s_2(n)$) was generated, then the erroneous decision will be made in favour of the dibit 10.

In order to calculate the probability of symbol error, a coherent quadriphase-shift keying system can be treated as a parallel connection of two coherent binary phase-shift keying systems (sometimes called binary phase-shift keying channels), because the quadriphase-shift keying signal is defined by a pair of quadrature carrier components. These two binary phase-shift keying systems are called the *in-phase (I) and the quadrature (Q) systems*. The correlator outputs $x_1$ and $x_2$ may be treated as the outputs of these two systems. As we said, they are realizations of the random variables $X_1$ and $X_2$. If these variables are conditioned to the possible bits sent, conditional probability density functions will be obtained that have as their mean either $E/2$ or $-E/2$ and the variance $N_0/2$.

### 7.3.3.2 Calculation of the Bit Error Probability

Previously, we proved that the bit error probability in each binary phase-shift keying system (or binary phase-shift keying channel) is

$$p = \frac{1}{2}\mathrm{erfc}\left(\sqrt{\frac{E_b}{N_0}}\right), \tag{7.3.23}$$

where $E_b$ is the transmitted signal energy per bit. Because a bit duration in a binary phase-shift keying channel is two times larger than a bit duration at the output of the source, that is, $N_s = 2N_b$, the transmitted signal energy per bit of one binary phase-shift keying system inside the quadriphase-shift keying system will be half of the energy of a dibit, that is,

$$E_b = \frac{E}{2}, \tag{7.3.24}$$

which results in the bit error probability in *each* binary phase-shift keying system of the quadriphase-shift keying system expressed as

$$p_1 = p_2 = \frac{1}{2}\mathrm{erfc}\left(\sqrt{\frac{E/2}{N_0}}\right). \tag{7.3.25}$$

The in-phase ($I$) system makes a decision on one of the two bits inside the symbol (dibit) of the quadriphase-shift keying signal, and the quadrature ($Q$) system makes the decision on the other bit inside the dibit. The errors in these decisions are statistically independent. Therefore, a correct decision will occur when the decisions in both the in-phase *and* the quadrature systems are correct. Thus, the probability of a correct decision on a symbol received in the quadriphase-shift keying system, denoted by $P_c$, is equal to *the product* of the probabilities of correct decisions on the bit in both the in-phase and quadrature systems, expressed as $(1 - p_1)$ and $(1 - p_2)$, respectively, that is,

$$P_c = (1 - p_1)(1 - p_2) = \left[1 - \frac{1}{2}\mathrm{erfc}\left(\sqrt{\frac{E}{2N_0}}\right)\right]^2 = 1 - \mathrm{erfc}\left(\sqrt{\frac{E}{2N_0}}\right) + \frac{1}{4}\mathrm{erfc}^2\left(\sqrt{\frac{E}{2N_0}}\right),$$

and, consequently, the symbol error probability in the quadriphase-shift keying system is

$$P_e = 1 - P_c = \mathrm{erfc}\left(\sqrt{\frac{E}{2N_0}}\right) - \frac{1}{4}\mathrm{erfc}^2\left(\sqrt{\frac{E}{2N_0}}\right). \tag{7.3.26}$$

In the case when $E/2N_0 \gg 1$, this formula can be approximated by

$$P_e \approx \text{erfc}\left(\sqrt{\frac{E}{2N_0}}\right). \tag{7.3.27}$$

One symbol in a quadriphase-shift keying system is composed of two message bits (forming a dibit) with equal energy, $E_b = E/2$. Thus, inserting the energy of a dibit, $E = 2E_b$, into the expression for the probability of symbol error, we obtain

$$P_e \approx \text{erfc}\left(\sqrt{\frac{E_b}{N_0}}\right). \tag{7.3.28}$$

The Gray encoding is used to specify the signal-space diagram. Therefore, an erroneous decision on a symbol will most likely result in an erroneous decision on one bit inside the symbol, because the error occurrence on one bit in a symbol results in a move to the neighbouring region on the signal-space diagram (i.e. due to noise, the phase changes to the nearest value). The error occurrence on both bits corresponds to the move to the opposite region, which is the least likely. Thus, the bit error probability in a coherent quadriphase-shift keying system is equal to the bit error probability of a coherent binary phase-shift keying system for the same bit rate $R_b$ and the same $E_b/N_0$, and is exactly expressed as

$$p = \frac{1}{2}\text{erfc}\left(\sqrt{\frac{E_b}{N_0}}\right), \tag{7.3.29}$$

and graphically presented in Figure 6.2.15 in logarithmic scale. The basic advantage of a quadriphase-shift keying over a binary phase-shift keying system is that, for the same bit error probability and the same bit rate, the quadriphase-shift keying system requires only half of the channel bandwidth (see Section 7.4, Figure 7.4.11). Neglecting complexity, quadriphase-shift keying systems are preferred to binary phase-shift keying systems.

### 7.3.3.3 *Signal Analysis and Structure of the Transceiver in a Quadriphase-Shift Keying System*

Expression (7.3.8) for the quadriphase-shift keying signal may also be developed in terms of the in-phase and quadrature components as

$$s_i(n) = \sqrt{E}\cos\left[(2i-1)\frac{\pi}{4}\right]\phi_1(n) - \sqrt{E}\sin\left[(2i-1)\frac{\pi}{4}\right]\phi_2(n)$$
$$= I(n)\phi_1(n) + Q(n)\phi_2(n), \tag{7.3.30}$$

where the in-phase component is

$$I(n) = \sqrt{E}\cos\left[(2i-1)\frac{\pi}{4}\right] = \pm\sqrt{E}\frac{\sqrt{2}}{2} = \pm\sqrt{\frac{E}{2}} = \pm\sqrt{E_b}, \tag{7.3.31}$$

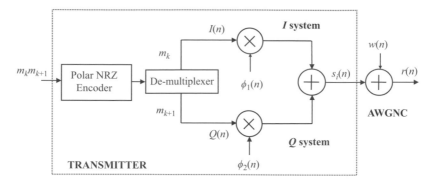

**Fig. 7.3.9** *Traditional scheme of a quadriphase-shift keying transmitter; AWGNC, additive white Gaussian noise channel; NRZ, non-return-to-zero.*

and the quadrature component is

$$Q(n) = \sqrt{E}\sin\left[(2i-1)\frac{\pi}{4}\right] = \pm\sqrt{E}\frac{\sqrt{2}}{2} = \pm\sqrt{\frac{E}{2}} = \pm\sqrt{E_b}. \tag{7.3.32}$$

The + or − sign is chosen depending on the value for *i*. Thus, the message bits 1 and 0 can be transformed directly into the two levels $+\sqrt{E_b}$ and $-\sqrt{E_b}$, respectively, using a polar non-return-to-zero encoder instead of the look-up tables and multipliers presented in Figure 7.3.1. The non-return-to-zero binary signal formed is divided into two binary signals using a de-multiplexer, denoted by $I(n)$ and $Q(n)$, corresponding to the odd- and even-numbered input message bits, as shown in Figure 7.3.9. In this way, the first bit $m_k$ of an input dibit, $m_k m_{k+1}$, goes to the *I* system, and the second bit, $m_{k+1}$, goes to the *Q* system. The duration of a bit in both branches is $N_s = 2N_b$. The in-phase binary non-return-to-zero signal, $I(n)$, modulates the carrier specified by the first orthonormal basis function, $\phi_1(n)$. The quadrature binary non-return-to-zero signal, $Q(n)$, modulates the carrier specified by the second orthonormal basis function, $\phi_2(n)$. The resultant two binary phase-shift keying signals are added to form a quadriphase-shift keying signal $s_i(n)$, as shown in Figure 7.3.9.

This signal is transmitted through the additive white Gaussian noise channel and then received by a correlation receiver, which is presented in Figure 7.3.10. The received signal is multiplied by the locally generated basis functions, and the results are then accumulated in the symbol interval $N_s$. The outputs of these correlators are the random samples $x_1$ and $x_2$. These samples are compared with the threshold value of 0 in the decision device, as shown in Figure 7.3.10. The outputs of the decision device are two estimates, $m_k$ and $m_{k+1}$, for the *I* system and the *Q* system, respectively. These two estimates are multiplexed in the multiplexer, and then an estimate of the transmitted symbol (dibit), denoted by $m_k m_{k+1}$, is generated at the output of the multiplexer.

The receiver in Figure 7.3.10 does not exactly follow the decision procedure used to derive the expression for the bit error probability in eqn (7.3.29), where the decision is made at the dibit level. Unlike that decision, the decision presented in Figure 7.3.10 is made in both

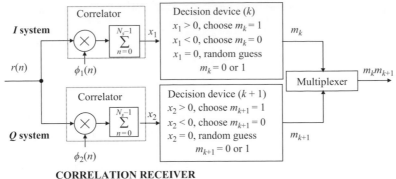

**CORRELATION RECEIVER**

**Fig. 7.3.10** *Traditional scheme of a quadriphase-shift keying receiver.*

the $I$ and the $Q$ systems at the bit level. This decision can be performed because the noise components affecting the bits in the $I$ and the $Q$ branches are independent, and the bit error probabilities in both branches are the same and can be expressed by relation (7.3.29).

Figures 7.3.9 and 7.3.10 are traditional presentations of the digital quadriphase-shift keying transmitter and receiver system that is applied here for the discrete quadriphase-shift keying system. The difference is in the definition of carriers and the related mathematical model, as discussed for the binary phase-shift keying system presented in Figure 7.2.11.

## 7.4 Coherent Binary Frequency-Shift Keying with Continuous Phase

### 7.4.1 Operation of a Binary Frequency-Shift Keying System

Binary phase-shift keying and quadriphase-shift keying are examples of linear modulation techniques in the sense that the spectrum of a baseband signal is shifted to the carrier frequency. Therefore, the frequencies of the baseband signal are preserved inside the passband signal and no new frequency components are generated. In contrast, frequency-shift keying, which is studied in this section, is an example of a non-linear modulation technique.

Based on the generic block scheme of a discrete communication system presented in Figure 5.6.4 and in Figure 5.7.1, it is possible to simplify it to develop the binary frequency-shift keying system presented in Figure 7.4.1. In this case, the transmitter is composed of a block for baseband signal processing, to generate the modulating signal $m(n)$, and two multipliers which function as a modulator. The receiver is composed of a correlation demodulator and an optimum detector. The optimum detector can be replaced by a decision device, because the decision on a received message can be made using the signal at the output of the correlator, as we did for a binary phase-shift keying system and a quadriphase-shift keying system. This property will be explained later on in this section. The mathematical model of the system and the signal processing will be presented in the discrete-time domain.

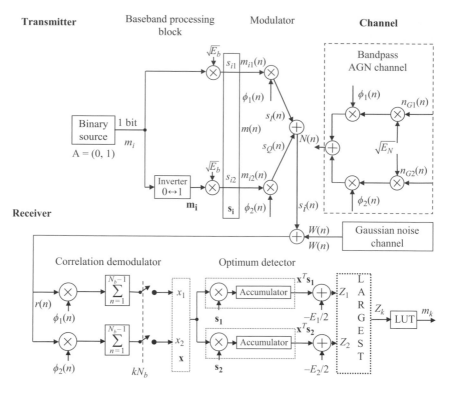

**Fig. 7.4.1** *Block schematic of a binary frequency-shift keying discrete-time communication system; AGN, additive Gaussian noise; LUT, look-up table.*

The message source generates one of two possible message symbols, which can have binary values denoted by $m_1 = 1$ and $m_2 = 0$ and belong to an alphabet A. Therefore, a symbol generated at the output of source is identical to a message bit. Suppose the message bit generated is $m_1 = 1$. The bit is applied to the inputs of two modulation branches. In the upper branch, the binary value 1 is multiplied by the square root of the bit energy $\sqrt{E_b}$ in the bit interval of $N_b$ samples to form a signal coefficient $s_{i1}$ that defines part of the baseband modulating signal $m(n) = m_{i1}(n)$, which is presented in Figure 7.4.1. In the branch multiplier, the signal $m_{i1}(n)$ is multiplied by the first carrier defined by the function $\phi_1(n)$, which is designed as the first basis function.

The same message bit, $m_1 = 1$, is also applied, in the same baseband processing block, to the input of an inverter in the second modulation branch that inverts the bit 1 into 0. Due to that, the constant $\sqrt{E_b}$ and then the second carrier is multiplied by 0. Therefore, the output of the second branch multiplier is 0, which is added to the modulated signal from the first branch. In this case, the modulated binary frequency-shift keying signal $s_i(n)$ is equal to the signal from the first branch, which is generated at the output of the transmitter and transferred though the additive white Gaussian noise waveform channel. In this case, the modulated signal, defined in the bit interval $N_b$, can be expressed as

$$s_1(n) = s_{11}\phi_1(n) + s_{22}\phi_2(n) = s_{11}\phi_1(n) + 0 = \sqrt{E_b}\phi_1(n)$$

$$= \sqrt{E_b}\sqrt{\frac{2}{N_b}}\cos(\Omega_1 n) = \sqrt{E_b}\sqrt{\frac{2}{N_b}}\cos(2\pi F_1 n). \tag{7.4.1}$$

In the case when bit $m_2 = 0$ is generated at the output of the source, the modulated signal is formed in the second branch and expressed as

$$s_2(n) = s_{11}\phi_1(n) + s_{22}\phi_2(n) = 0 + s_{22}\phi_2(n) = \sqrt{E_b}\phi_2(n)$$

$$= \sqrt{E_b}\sqrt{\frac{2}{N_b}}\cos(\Omega_2 n) = \sqrt{E_b}\sqrt{\frac{2}{N_b}}\cos(2\pi F_2 n). \tag{7.4.2}$$

We are using here our standard notation for the frequencies of the discrete-time signals, that is, the Greek letter $\Omega = 2\pi F = 2\pi/N$ is the radian normalized frequency presenting the number of radians per sample. The capital letter $F$ denotes the normalized frequency of the discrete-time signal which is equal to the continuous-time signal of the frequency $f$ normalized by the sampling frequency $f_s$, that is, $F = f/f_s = 1/N$, where $N$ is the number of samples. In summary, the message bits 1 or 0 at the output of the information source result in two distinct modulated signals that have the same amplitudes and phases but different frequencies. The frequencies are chosen in such a way to have two orthogonal modulated signals.

## 7.4.2 Transmitter Operation

### 7.4.2.1 Modulating Signals in Time and Frequency Domains

The mathematical description of the frequency-shift keying transmitter and basic signal processing operations will be presented using the block schematic in Figure 7.4.1. The number of possible message symbols generated from the source is $I = A^{J-1} = 2^1 = 2$, which is defined by the matrix of message bits

$$\mathbf{m} = [\mathbf{m}_1 \ \ \mathbf{m}_2] = \begin{bmatrix} 1 & 0 \end{bmatrix}. \tag{7.4.3}$$

These message bits are mapped into the amplitudes of modulating signals by multiplying them by $\sqrt{E_b}$ to form two vectors of modulating signal coefficients represented by a two-column matrix of signal coefficients expressed as

$$\mathbf{s} = [\mathbf{s}_1 \ \ \mathbf{s}_2] = \begin{bmatrix} s_{11} & s_{21} \\ s_{12} & s_{22} \end{bmatrix} = \begin{bmatrix} \sqrt{E_b} & 0 \\ 0 & \sqrt{E_b} \end{bmatrix}, \tag{7.4.4}$$

which can be calculated using the procedure presented in Chapter 2. If the source randomly generates a stream of bits, the modulating signal $m(n)$ will be defined by two rectangular

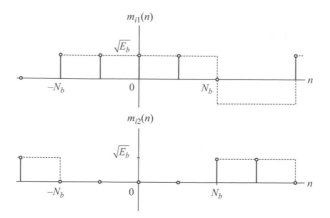

**Fig. 7.4.2** *Modulating discrete binary signals presented in the time domain.*

trains $m_{i1}(n)$ and $m_{i2}(n)$ having inverted amplitude values, as presented in Figure 7.4.2. These modulating signals are multiplying the carriers represented by two basis functions, as presented in Figure 7.4.1. For the sake of simplicity, the analysis of these modulating signals in the frequency domain will be performed in the next subsection, where the modulated signal will be expressed in terms of the in-phase and quadrature components.

### 7.4.2.2 Modulated Signal Analysis in the Time Domain and a Signal-Space Diagram

The message source generates two binary symbols (or bits) that belong to an alphabet A of $I = 2$ symbols, denoted by $m_1 = 1$ and $m_2 = 0$. In a binary frequency-shift keying system, one of two sinusoidal carrier signals having different frequencies is assigned to each symbol to obtain the modulated signal defined in eqns (7.4.1) and (7.4.2), which can be presented in a common expression as

$$s_i(n) = \left\{ \begin{array}{ll} \sqrt{\frac{2E_b}{N_b}} \cos\left(\Omega_j n\right) & 0 \le n < N_b \\ 0 & \text{elsewhere} \end{array} \right\}. \tag{7.4.5}$$

The frequencies of modulated signals are expressed as

$$\Omega_j = 2\pi \frac{n+i}{N_b}, \text{ or } F_j = \Omega_i/2\pi = \frac{n+i}{N_b}, \tag{7.4.6}$$

for a fixed integer values $n$ and $i = 1, 2$. The difference of these frequencies is a constant value that is equal to $1/N_b$. Defined in this way, binary frequency-shift keying preserves phase continuity, including the inter-bit switching time and, for that reason, is called continuous-phase binary frequency-shift keying. An example of a continuous-phase binary frequency-shift keying signal is shown in Figure 7.4.3 for $N_b = 16$.

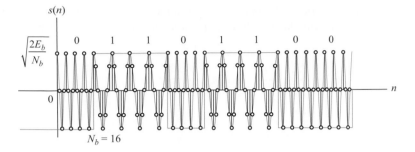

**Fig. 7.4.3** *A discrete binary frequency-shift keying modulated signal.*

**Example**

For the signal in Figure 7.4.3, calculate the frequencies of the carriers, their expressions in the time domain, and the ratio of their frequencies.

**Solution.** The number of samples per period of the two carrier signals are $N_1 = 8$ and $N_2 = 4$. Therefore, the frequencies of signals can be calculated according to expression (7.4.6) for $n = 0$ as

$$\Omega_1 = 2\pi \frac{n+1}{8} = 2\pi \frac{0+1}{8} = 2\pi \frac{1}{8}$$

$$\Omega_2 = 2\pi \frac{n+2}{8} = 2\pi \frac{0+2}{8} = 2\pi \frac{1}{4},$$

with the ratio $\Omega_1/\Omega_2 = 1/2$ for $n = 1$. The time domain expressions of the signals are

$$s_1(n) = \sqrt{\frac{2E_b}{N_b}} \cos(\Omega_1 n) = \sqrt{\frac{2E_b}{N_b}} \cos(\pi/4)\, n,$$

and

$$s_2(n) = \sqrt{\frac{2E_b}{N_b}} \cos(\Omega_2 n) = \sqrt{\frac{2E_b}{N_b}} \cos(\pi/2)\, n,$$

respectively. The signals presented in Figure 7.4.3 have a different number of oscillations, which does not change their orthogonality.

The functions $s_1(n)$ and $s_2(n)$ are orthogonal. If we divide them by $\sqrt{E_b}$ we obtain the orthonormal basis functions for the two carrier frequencies expressed as

$$\phi_j(n) = \left\{ \begin{array}{ll} \sqrt{\frac{2}{N_b}} \cos(\Omega_j n) & 0 \le n < N_b \\ 0 & \text{elsewhere} \end{array} \right\}, \qquad (7.4.7)$$

for $j = 1, 2$. The signals $s_1(n)$ and $s_2(n)$ can be presented in vector form. The coefficients for the signal vectors can be calculated following the procedure presented in Chapter 2 as

$$
\begin{aligned}
s_{ij} &= \sum_{n=0}^{N_b-1} s_i(n)\phi_j(n) = \sum_{n=0}^{N_b-1} \sqrt{\frac{2E_b}{N_b}} \cos \Omega_i n \sqrt{\frac{2}{N_b}} \cos \Omega_j n \\
&= \frac{2}{N_b} \sqrt{E_b} \sum_{n=0}^{N_b-1} \cos \Omega_i n \cos \Omega_j n = \left\{ \begin{array}{ll} \sqrt{E_b} \frac{2}{N_b} \sum_{n=0}^{N_b-1} \cos^2 \Omega_j n = \sqrt{E_b} & i = j \\ 0 & i \neq j \end{array} \right\}.
\end{aligned}
\tag{7.4.8}
$$

In contrast to the coherent binary phase-shift keying, the signal-space diagram of the frequency-shift keying is two-dimensional ($J = 2$), with a signal constellation consisting of two message points defined for two signals in a vector form as

$$
\mathbf{s}_1 = \begin{bmatrix} s_{11} \\ s_{12} \end{bmatrix} = \begin{bmatrix} \sqrt{E_b} \\ 0 \end{bmatrix}, \quad \text{and} \quad \mathbf{s}_2 = \begin{bmatrix} s_{21} \\ s_{22} \end{bmatrix} = \begin{bmatrix} 0 \\ \sqrt{E_b} \end{bmatrix},
\tag{7.4.9}
$$

as was shown in eqn (7.4.4). For each pair of coordinates, the signal-space diagram is shown in Figure 7.4.4, and a schematic of the binary frequency-shift keying transceiver is presented in Figure 7.4.1. The possible transmitted signal waveforms, corresponding to the two message points, are presented in Figure 7.4.5 for the number of samples $N_b = 16$.

The presence of noise in the channel can change the position of the message points. For the same energies of the modulated signals, we define the decision boundary as the line that separates the two-dimensional plane into two symmetric parts and lies under $45°$ with respect to the $x$-axis, as shown in Figure 7.4.4.

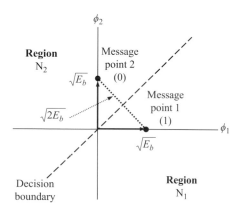

**Fig. 7.4.4** *Signal-space diagram and message constellation for continuous-phase binary frequency-shift keying.*

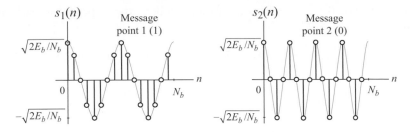

**Fig. 7.4.5** *Transmitted discrete-time signal waveforms corresponding to the two message points.*

### 7.4.2.3 Modulated Signal Analysis in Time and Frequency Domains

**In-phase and quadrature components in the time domain.** In order to find the power spectral density of the modulated signal, the modulated signal needs to be expressed in terms of the so-called low-frequency in-phase and quadrature components. Let us consider a binary continuous-phase frequency-shift keying signal specified by two normalized angular frequencies defined by $i = 1$ or 2 as

$$\Omega_j = \begin{cases} \Omega_c + 2\pi/2N_b & i = 2, m_i = 0 \\ \Omega_c - 2\pi/2N_b & i = 1, m_i = 1 \end{cases} = \Omega_c \pm \pi/N_b = \Omega_c \pm \Omega_b/2, \tag{7.4.10}$$

or, in terms of the normalized cycle frequency $F_j = \Omega_j/2\pi$, as

$$F_j = \begin{cases} F_c + 1/2N_b & i = 2, m_i = 0 \\ F_c - 1/2N_b & i = 1, m_i = 1 \end{cases} = F_c \pm 1/2N_b = F_c \pm F_b/2,$$

where $\Omega_c = 2\pi F_c$ is defined as the nominal carrier frequency equal to the arithmetic mean of the two carrier frequencies, which guarantees the orthogonality of the two frequency-shift keying carriers and the related baseband normalized frequencies $\Omega_b = 2\pi/N_b = 2\pi F_b$, that is, $F_b = 1/N_b$. Therefore, the modulated signal can be represented by two signals with the frequencies $F_1$ and $F_2$ separated by the interval $1/N_b$, that is,

$$s_i(n) = \sqrt{\frac{2E_b}{N_b}} \cos\left(\Omega_c \pm \frac{\Omega_b}{2}\right) n = \sqrt{\frac{2E_b}{N_b}} \cos\left(\Omega_c \pm \frac{\pi}{N_b}\right) n,$$

or

$$s_i(n) = \sqrt{\frac{2E_b}{N_b}} \cos 2\pi \left(F_c \pm \frac{F_b}{2}\right) n = \sqrt{\frac{2E_b}{N_b}} \cos 2\pi \left(F_c \pm \frac{1}{2N_b}\right) n, \tag{7.4.11}$$

defined in the interval $0 \leq n < N_b$. This expression can be further expended as the function

$$
\begin{aligned}
s_i(n) &= \sqrt{\frac{2E_b}{N_b}} \cos\left(\pm\frac{\Omega_b}{2}n\right)\cos\Omega_c n - \sqrt{\frac{2E_b}{N_b}}\sin\left(\pm\frac{\Omega_b}{2}n\right)\sin\Omega_c n \\
&= \sqrt{\frac{2E_b}{N_b}}\cos\left(\pm\frac{\pi n}{N_b}\right)\cos\Omega_c n - \sqrt{\frac{2E_b}{N_b}}\sin\left(\pm\frac{\pi n}{N_b}\right)\sin\Omega_c n.
\end{aligned}
\tag{7.4.12}
$$

for $\Omega_b/2 = 2\pi/2N_b = (2\pi/N_b)/2 = 2\pi F_b/2$. Taking into account the even property of the cosine function, and the odd property of the sine function, we may express the binary frequency-shift keying signal in terms of the in-phase ($I$) component $s_I(n)$ and the quadrature ($Q$) component $s_Q(n)$ as

$$
\begin{aligned}
s_i(n) &= \sqrt{E_b}\cos\left(\frac{\Omega_b}{2}n\right)\sqrt{\frac{2}{N_b}}\cos\Omega_c n \mp \sqrt{E_b}\sin\left(\frac{\Omega_b}{2}n\right)\sqrt{\frac{2}{N_b}}\sin\Omega_c n \\
&= \sqrt{E_b}\cos\left(\frac{\pi n}{N_b}\right)\sqrt{\frac{2}{N_b}}\cos\Omega_c n \mp \sqrt{E_b}\sin\left(\frac{\pi n}{N_b}\right)\sqrt{\frac{2}{N_b}}\sin\Omega_c n \\
&= s_I(n)\sqrt{\frac{2}{N_b}}\cos\Omega_c n + s_Q(n)\sqrt{\frac{2}{N_b}}\sin\Omega_c n = s_I(n)\phi_1(n) + s_Q(n)\phi_2(n),
\end{aligned}
\tag{7.4.13}
$$

where the '$-$' sign in the second addend corresponds to transmitting symbol 1, and the '$+$' sign corresponds to transmitting symbol 0. The wave shapes of the in-phase and quadrature components defined in eqn (7.4.13) are presented in Figure 7.4.6 and Figure 7.4.7, respectively, assuming that the message sequence of bits (0 1 1 0 1 1 0 0) is generated from the source, and the in-phase and quadrature components are defined for $N_b = 4$, or their related periods equal to $N_I = N_Q = 2N_b = 8$ and frequency $\Omega_I = 2\pi/N_I = \Omega_b/2$, which are expressed as functions of $N_I$ in the forms

$$
s_I(n) \underset{N_I=2N_b}{=} \sqrt{E_b}\cos\left(\frac{2\pi n}{N_I}\right) = \sqrt{E_b}\cos(\Omega_I n) = \sqrt{E_b}\cos\left(\frac{\pi n}{4}\right),
\tag{7.4.14}
$$

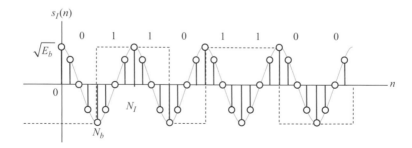

**Fig. 7.4.6** *Waveform of the in-phase signal.*

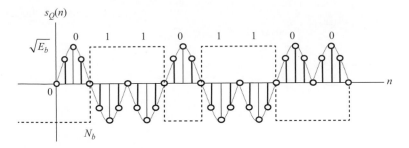

**Fig. 7.4.7** *Waveform of the quadrature signal.*

and

$$s_Q(n) \Big|_{N_Q=2N_b} = \mp\sqrt{E_b}\sin\left(\frac{2\pi n}{N_Q}\right) = \mp\sqrt{E_b}\sin\left(\Omega_Q n\right) = \mp\sqrt{E_b}\sin\left(\frac{\pi n}{4}\right). \qquad (7.4.15)$$

**In-phase and quadrature components in the frequency domain.** To find the modulated signal in the frequency domain in terms of its power spectral density, we will analyse the in-phase and the quadrature components separately. The in-phase component is a deterministic cosine signal that *does not* depend on the incoming random binary message signal $m(n)$. The value $A_b$ is the amplitude of the carrier, as traditionally defined. In our notation, this amplitude is proportional to the related energy of a symbol. Since the in-phase component is a cosine periodic function of time with a period of $2N_b$, its autocorrelation function can be calculated following the theory presented in Chapter 15, using Table 15.6.1, as

$$
\begin{aligned}
R_I(l) &= \frac{1}{2N_b} \sum_{n=0}^{2N_b-1} \sqrt{E_b}\cos\frac{\Omega_b}{2}(n+l)\sqrt{E_b}\cos\frac{\Omega_b}{2}n \\
&= \frac{E_b}{4N_b} \sum_{n=0}^{2N_b-1} \cos\frac{\Omega_b}{2}l = \frac{E_b}{4N_b}2N_b\cos\frac{\Omega_b}{2}l = \frac{E_b}{2}\cos\frac{\Omega_b}{2}l.
\end{aligned}
\qquad (7.4.16)
$$

**Example**

Calculate the autocorrelation function of the in-phase component as a function of the period $N_I = 2N_b$.

**Solution.** The normalized frequency of the in-phase component is $\Omega_I = 2\pi/N_I = 2\pi/2N_b$, the signal in the time domain is expressed by eqn (7.4.14), and the related autocorrelation function can be calculated as

$$R_I(l) = \frac{1}{N_I} \sum_{n=0}^{n=N_I-1} \sqrt{E_b} \cos \Omega_I (n+l) \sqrt{E_b} \cos \Omega_I n$$

$$= \frac{E_b}{N_I} \sum_{n=0}^{N_I-1} \frac{1}{2} [\cos \Omega_I l + \cos \Omega_I (2n+l)] = \frac{E_b}{2N_I} \sum_{n=0}^{N_I-1} \cos \Omega_I l + 0 = \frac{E_b}{2} \cos \Omega_I l$$

$$= \frac{E_b}{2} \cos \frac{2\pi l}{N_I} \underset{N_I=2N_b}{=} \frac{E_b}{2} \cos \frac{2\pi l}{2N_b} = \frac{E_b}{2} \cos \frac{\Omega_b}{2} l = \frac{E_b}{2} \cos 2\pi \frac{F_b}{2} l.$$

The Fourier transform of the autocorrelation function will give a power spectral density function of the in-phase component that consists of two Dirac delta functions weighted by $E_b \pi / 2$ and positioned at two frequencies, $\Omega_b / 2 = \pm \pi / N_b$. This power spectral density can be calculated following the theory presented in Chapter 15, using Table 15.6.2, which results in the expression

$$S_I(\Omega) = \frac{E_b}{2} \pi [\delta(\Omega - \Omega_I) + \delta(\Omega + \Omega_I)] = \frac{E_b}{4} [\delta(F - F_I) + \delta(F + F_I)].$$

Because $\Omega_b = 2\pi / N_b = 2\Omega_I$ and $F_b = 1/N_b = 2/N_I = 2F_I$, we also obtain

$$S_I(\Omega) = \frac{E_b}{2} \pi \left[ \delta\left( \Omega - \frac{\Omega_b}{2} \right) + \delta\left( \Omega + \frac{\Omega_b}{2} \right) \right] = \frac{E_b}{4} \left[ \delta\left( F - \frac{F_b}{2} \right) + \delta\left( F + \frac{F_b}{2} \right) \right]. \quad (7.4.17)$$

However, the quadrature component is a random signal that *does* depend on the incoming random binary signal $m(n)$, and changes its sign in accordance with the sign changes of the incoming symbols $m_{ij}(n)$, as shown in Figure 7.4.7. This component can be understood as a bit-shaping (symbol-shaping) function of the quadrature carrier and expressed as

$$s_Q(n) = \mp \sqrt{E_b} \sin \left( \frac{\pi n}{N_b} \right) = \mp \sqrt{E_b} \sin \left( \frac{2\pi n}{2N_b} \right) = \mp \sqrt{E_b} \sin \left( \frac{\Omega_b}{2} n \right). \quad (7.4.18)$$

Therefore, the quadrature component is a negative sine signal when message bit 1 is transmitted, and a positive sine signal when message bit 0 is transmitted. Its amplitude spectral density is defined as the discrete-time Fourier transform presented in Chapter 15, which can be calculated for the quadrature component as

$$s_Q(\Omega) = \sum_{n=0}^{N_b-1} \sqrt{E_b} \sin \left( \frac{\Omega_b}{2} n \right) e^{-j\Omega n} = \sum_{n=0}^{N_b-1} \frac{\sqrt{E_b}}{2j} \left( e^{j\Omega_b n/2} - e^{-j\Omega_b n/2} \right) e^{-j\Omega n}$$

$$= \frac{\sqrt{E_b}}{2j} \sum_{n=0}^{N_b-1} e^{-j(\Omega - \Omega_b/2)n} - \frac{\sqrt{E_b}}{2j} \sum_{n=0}^{N_b-1} e^{-j(\Omega + \Omega_b/2)n}. \quad (7.4.19)$$

The first term is

$$
\begin{aligned}
s_{Q1}(\Omega) &= \frac{\sqrt{E_b}}{2j} \sum_{n=0}^{N_b-1} \left( e^{-j(\Omega-\Omega_b/2)} \right)^n = \frac{\sqrt{E_b}}{2j} \frac{1 - e^{-jN_b(\Omega-\Omega_b/2)}}{1 - e^{+j(\Omega-\Omega_b/2)}} \\
&= \frac{\sqrt{E_b}}{2} e^{-j\pi/2} \frac{e^{-jN_b(\Omega-\Omega_b/2)/2}}{e^{-j(\Omega-\Omega_b/2)/2}} \frac{e^{+jN_b(\Omega-\Omega_b/2)/2} - e^{-jN_b(\Omega-\Omega_b/2)/2}}{e^{j(\Omega-\Omega_b/2)/2} - e^{-j(\Omega-\Omega_b/2)/2}} \\
&= e^{-j(N_b-1)(\Omega-\Omega_b/2)/2 - j\pi/2} \frac{\sqrt{E_b}}{2} \frac{\sin N_b(\Omega-\Omega_b/2)/2}{\sin(\Omega-\Omega_b/2)/2}.
\end{aligned}
\tag{7.4.20}
$$

Analogously, the second term is calculated as

$$
s_{Q2}(\Omega) = -\frac{\sqrt{E_b}}{2j} \sum_{n=0}^{N_b-1} e^{-j(\Omega+\Omega_b/2)n} = e^{-j(N_b-1)(\Omega+\Omega_b/2)/2 + j\pi/2} \frac{\sqrt{E_b}}{2} \frac{\sin N_b(\Omega+\Omega_b/2)/2}{\sin(\Omega+\Omega_b/2)/2}.
$$

The energy spectral densities of the first term and the related second term are

$$
\begin{aligned}
E_{Q1}(\Omega) &= \left| \frac{\sqrt{E_b}}{2} \frac{\sin N_b(\Omega-\Omega_b/2)/2}{\sin(\Omega-\Omega_b/2)/2} \right|^2 \\
E_{Q2}(\Omega) &= \left| \frac{\sqrt{E_b}}{2} \frac{\sin N_b(\Omega+\Omega_b/2)/2}{\sin(\Omega+\Omega_b/2)/2} \right|^2,
\end{aligned}
\tag{7.4.21}
$$

the energy spectral density of the quadrature component is the sum of the energy spectral densities of these components, expressed as

$$
\begin{aligned}
E_{QQ}(\Omega) &= E_{Q1}(\Omega) + E_{Q2}(\Omega) \\
&= \frac{E_b}{4} \left| \frac{\sin N_b(\Omega-\Omega_b/2)/2}{\sin(\Omega-\Omega_b/2)/2} \right|^2 + \frac{E_b}{4} \left| \frac{\sin N_b(\Omega+\Omega_b/2)/2}{\sin(\Omega+\Omega_b/2)/2} \right|^2,
\end{aligned}
\tag{7.4.22}
$$

and its power spectral density is calculated for $A_b^2 = 2E_b/N_b$ as

$$
S_{QQ}(\Omega) = \frac{E_{QQ}(\Omega)}{N_b} = \frac{E_b}{4N_b} \left| \frac{\sin N_b(\Omega-\Omega_b/2)/2}{\sin(\Omega-\Omega_b/2)/2} \right|^2 + \frac{E_b}{4N_b} \left| \frac{\sin N_b(\Omega+\Omega_b/2)/2}{\sin(\Omega+\Omega_b/2)/2} \right|^2.
\tag{7.4.23}
$$

Due to the independence between the binary message bits, the components $s_I(n)$ and $s_Q(n)$, which represent the modulating signal $m(n)$, are mutually independent of each other. Therefore, the power spectral density of the modulating signal is equal to the sum of the power spectral densities of these two components, which are given by eqns (7.4.17) and (7.4.23), expressed as

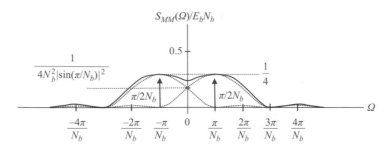

**Fig. 7.4.8** *Baseband spectra of the discrete-time binary frequency-shift keying signal.*

$$
S_{MM}(\Omega) = S_I(\Omega) + S_{QQ}(\Omega) = \frac{E_b N_b}{2} \frac{\pi}{N_b} \left[ \delta\left(\Omega - \frac{\Omega_b}{2}\right) + \delta\left(\Omega + \frac{\Omega_b}{2}\right) \right]
$$
$$
\frac{E_b N_b}{4} \frac{1}{N_b^2} \left( \left| \frac{\sin N_b (\Omega - \Omega_b/2)/2}{\sin (\Omega - \Omega_b/2)/2} \right|^2 + \left| \frac{\sin N_b (\Omega + \Omega_b/2)/2}{\sin (\Omega + \Omega_b/2)/2} \right|^2 \right),
$$

(7.4.24)

and shown in Figure 7.4.8 as a function of $\Omega$ with normalized amplitudes. The first zero crossings of both spectral components can be calculated as follows. For the condition $\sin N_b (\Omega + \Omega_b/2)/2 = 0$, we can get the $k$th zero crossing for the left spectral component if $N_b (\Omega + \Omega_b/2)/2 = k\pi$, or for the frequency

$$
\Omega = \frac{2k\pi}{N_b} - \Omega_b/2 = \frac{2k\pi}{N_b} - \frac{\pi}{N_b} = \frac{2k\pi - \pi}{N_b} = \frac{(2k-1)\pi}{N_b}.
$$

(7.4.25)

For the condition $\sin N_b (\Omega - \Omega_b/2)/2 = 0$, we can get the $k$th zero crossing for the right spectral component if $N_b (\Omega - \Omega_b/2)/2 = k\pi$, or for the frequency

$$
\Omega = \frac{2k\pi}{N_b} + \Omega_b/2 = \frac{2k\pi}{N_b} + \frac{\pi}{N_b} = \frac{2k\pi + \pi}{N_b} = \frac{(2k+1)\pi}{N_b}.
$$

(7.4.26)

The maximum value of the spectral component on the right-hand side, normalized by $E_b N_b$, is

$$
\frac{1}{4N_b^2} \lim_{\Omega \to \Omega_b/2} \left| \frac{\sin N_b (\Omega - \Omega_b/2)/2}{\sin (\Omega - \Omega_b/2)/2} \right|^2 = \frac{1}{4}.
$$

(7.4.27)

The same maximum value can be calculated for the spectral component on the left-hand side at the frequency $\Omega = -\pi/N_b$.

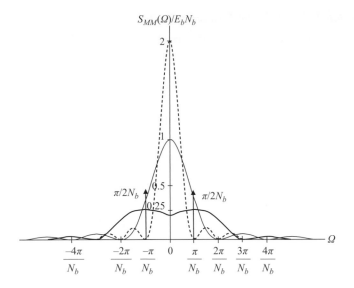

**Fig. 7.4.9** *Baseband spectra for a discrete-time binary phase-shift keying signal, a quadriphase-shift keying signal, and a binary frequency-shift keying signal.*

The normalized power spectral density can be calculated using eqn (7.4.24) in the form

$$\frac{S_{MM}(\Omega)}{E_b N_b} = \frac{\pi}{2N_b}\left[\delta\left(\Omega - \frac{\Omega_b}{2}\right) + \delta\left(\Omega + \frac{\Omega_b}{2}\right)\right]$$
$$+ \frac{1}{4N_b^2}\left|\frac{\sin N_b\,(\Omega - \Omega_b/2)\,/2}{\sin\,(\Omega - \Omega_b/2)\,/2}\right|^2 + \frac{1}{4N_b^2}\left|\frac{\sin N_b\,(\Omega + \Omega_b/2)\,/2}{\sin\,(\Omega + \Omega_b/2)\,/2}\right|^2. \tag{7.4.28}$$

In calculating the spectra of the modulating signals for a binary phase-shift keying system and a quadriphase-shift keying system, we assumed that the carrier signals would be expressed as orthonormal basis functions. For the sake of comparison, Figure 7.4.9 shows the power spectral densities of the modulating signals for a binary phase-shift keying signal (thin line), a quadriphase-shift keying signal (only the in-phase component $S_{MI}(\Omega)$; dashed line), and a binary frequency-shift keying signal (thick line).

When the number of samples increases, the signal in the time domain tends to the continuous-time function, and the spectrum becomes narrower, but the maxima values remain the same. The relative weights of the delta functions representing the cosine term in the signal decrease with the number of samples $N_b$. The number of side lobes increases when the number of samples increases, that is, only one lobe exists for $N_b = 8$. However, we should keep in mind that the energy of the signal remains the same, as will be discussed in Chapter 15, Section 15.3.

### 7.4.2.4   *Modulated Signals in the Frequency Domain*

**The power spectral density of a modulated binary frequency-shift keying signal**. For the case when the carriers are expressed as a pair of orthonormal sine and cosine signals, with the same autocorrelation function and, consequently, the same power spectral density functions expressed in eqn (7.2.8), the power spectral density of the modulated signal can be obtained as a convolution of the power spectral density function of the modulating signal in eqn (7.4.24) and the power spectral density of the carrier in eqn (7.2.8), as follows:

$$S_{SS}(\Omega) = \frac{1}{2\pi} S_{MM}(\Omega) * S_{sc}(\Omega) = \frac{1}{2\pi} \left[ S_I(\Omega) + S_{QQ}(\Omega) \right] * \frac{1}{N_b} \pi \left[ \delta(\Omega - \Omega_c) + \delta(\Omega + \Omega_c) \right]$$

$$= \frac{1}{2N_b} \left\{ \begin{array}{l} \frac{E_b}{2} \pi \left[ \delta(\Omega - \Omega_b/2) + \delta(\Omega + \Omega_b/2) \right] \\[8pt] + \frac{E_b N_b}{4} \frac{1}{N_b^2} \left| \frac{\sin N_b(\Omega - \Omega_b/2)/2}{\sin(\Omega - \Omega_b/2)/2} \right|^2 \\[8pt] + \frac{E_b N_b}{4} \frac{1}{N_b^2} \left| \frac{\sin N_b(\Omega + \Omega_b/2)/2}{\sin(\Omega + \Omega_b/2)/2} \right|^2 \end{array} \right\} * \left[ \delta(\Omega - \Omega_c) + \delta(\Omega + \Omega_c) \right],$$

$$(7.4.29)$$

which results in

$$S_{SS}(\Omega) = \frac{E_b}{4N_b} \pi \left[ \delta((\Omega - \Omega_c) - \Omega_b/2) + \delta((\Omega - \Omega_c) + \Omega_b/2) \right]$$

$$+ \frac{E_b}{4N_b} \pi \left[ \delta((\Omega + \Omega_c) - \Omega_b/2) + \delta((\Omega + \Omega_c) + \Omega_b/2) \right]$$

$$+ \frac{E_b}{8} \frac{1}{N_b^2} \left[ \left| \frac{\sin N_b((\Omega - \Omega_c) - \Omega_b/2)/2}{\sin((\Omega - \Omega_c) - \Omega_b/2)/2} \right|^2 + \left| \frac{\sin N_b((\Omega - \Omega_c) + \Omega_b/2)/2}{\sin((\Omega - \Omega_c) + \Omega_b/2)/2} \right|^2 \right]$$

$$+ \frac{E_b}{8} \frac{1}{N_b^2} \left[ \left| \frac{\sin N_b((\Omega + \Omega_c) - \Omega_b/2)/2}{\sin((\Omega + \Omega_c) - \Omega_b/2)/2} \right|^2 + \left| \frac{\sin N_b((\Omega + \Omega_c) + \Omega_b/2)/2}{\sin((\Omega + \Omega_c) + \Omega_b/2)/2} \right|^2 \right].$$

$$(7.4.30)$$

The power spectral density of a continuous-phase binary frequency-shift keying signal is presented in Figure 7.4.10.

Figure 7.4.11 shows the power spectral density in the logarithmic scale, which is calculated in the following way. The one-sided power spectral density, obtained by overlapping the

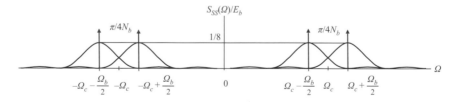

**Fig. 7.4.10** *Power spectral density of a discrete-time binary frequency-shift keying signal.*

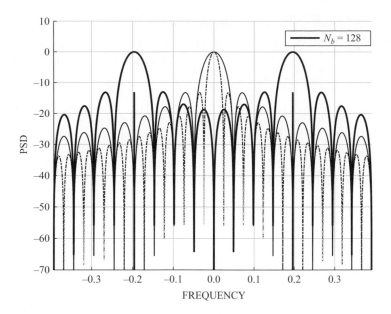

**Fig. 7.4.11** *Spectra for a modulated binary phase-shift keying signal, a binary frequency-shift keying signal, and a quadriphase-shift keying signal; PSD, power spectral density.*

amplitudes at negative frequencies with the positive ones and defining the common frequency $\Omega_n = (\Omega - \Omega_c)$, can be normalized to the unit amplitude and expressed as

$$
S_{SS}(\Omega_n) = S_{SS}(\Omega)/E_b/4 = 2\frac{\pi}{N_b}\left[\delta(\Omega_n - \Omega_b/2) + \delta((\Omega - \Omega_c) + \Omega_b/2)\right]
$$
$$
+ \frac{1}{N_b^2}\left[\left|\frac{\sin N_b(\Omega_n - \Omega_b/2)/2}{\sin(\Omega_n - \Omega_b/2)/2}\right|^2 + \left|\frac{\sin N_b(\Omega_n + \Omega_b/2)/2}{\sin(\Omega_n + \Omega_b/2)/2}\right|^2\right] \tag{7.4.31}
$$

In practice, we need to see the level of the spectrum in logarithmic units. Therefore, we express the power spectral density as

$$
S_{SS}(\Omega_n)|_{dB} = 10\log_{10}\left(2\frac{\pi}{N_b}\right)\cdot\left[\delta(\Omega_n - \Omega_b/2) + \delta((\Omega - \Omega_c) + \Omega_b/2)\right]
$$
$$
+ 10\log_{10}\frac{1}{N_b^2}\left|\frac{\sin N_b(\Omega_n - \Omega_b/2)/2}{\sin(\Omega_n - \Omega_b/2)/2}\right|^2 + 10\log_{10}\frac{1}{N_b^2}\left|\frac{\sin N_b(\Omega_n + \Omega_b/2)/2}{\sin(\Omega_n + \Omega_b/2)/2}\right|^2, \tag{7.4.32}
$$

and present it in graphical form, as in Figure 7.4.11. The position of the delta functions remains the same with the associated weights. The value of the remaining two terms in eqn (7.4.28) at $\Omega_n = 0$ will depend on the number of samples, as can be seen from the expression

$$S_{SS}(0) = \frac{1}{N_b^2}\left[\left|\frac{\sin N_b\,(-\Omega_b/4)}{\sin\,(-\Omega_b/4)}\right|^2 + \left|\frac{\sin N_b\,(\Omega_b/4)}{\sin\,(\Omega_b/4)}\right|^2\right] = \frac{2}{N_b^2}\left|\frac{\sin N_b\,(\Omega_b/4)}{\sin\,(\Omega_b/4)}\right|^2$$

$$= \frac{2}{N_b^2}\left|\frac{\sin N_b\,(2\pi/4N_b)}{\sin\,(2\pi/4N_b)}\right|^2 = \frac{2}{N_b^2}\left|\frac{\sin\,(\pi/2)}{\sin\,(\pi/2N_b)}\right|^2 = \frac{2}{N_b^2}\left|\frac{1}{\sin\,(\pi/2N_b)}\right|^2 \tag{7.4.33}$$

We can observe that the smoother the symbol-shaping pulse $s_i(n)$ is, the faster is the drop of the spectral tail. Since binary frequency-shift keying has a smoother pulse shape than phase-shift keying, it has lower side lobes.

## 7.4.3 Receiver Operation

### 7.4.3.1 *Operation of the Correlation Demodulator*

The received signal of a coherent binary frequency-shift keying can be expressed as

$$r(n) = s_i(n) + w(n). \tag{7.4.34}$$

for $0 \leq n \leq N_b$, and $i = 1, 2$, where $w(n)$ is a sample function of a discrete white Gaussian noise process $W(n)$ of zero mean and power spectral density $N_0/2$. At the output of the two correlators ($j = 1, 2$) the calculated random values are, respectively,

$$x_1 = \sum_{n=0}^{N_b-1} r(n)\phi_1(n) = \sum_{n=0}^{N_b-1} s_i(n)\phi_1(n) + \sum_{n=0}^{N_b-1} w(n)\phi_1(n) = s_{i1} + w_1, \tag{7.4.35}$$

and

$$x_2 = \sum_{n=0}^{N_b-1} r(n)\phi_2(n) = \sum_{n=0}^{N_b-1} s_i(n)\phi_2(n) + \sum_{n=0}^{N_b-1} w(n)\phi_2(n) = s_{i2} + w_2, \tag{7.4.36}$$

where $s_{i1}$ and $s_{i2}$ are deterministic quantities contributed by the transmitted signal $s_i$, and $w_1$ and $w_2$ are the sample values of the random variables $W_1$ and $W_2$, respectively, that arise due to the channel noise $w(n)$. The demodulated signal vector $\mathbf{x}$ is defined by the pair $(x_1, x_2)$. If we apply the *decision rule* that the vector $\mathbf{x}$ lies in region $N_i$ if the Euclidean distance $||\mathbf{x} - \mathbf{s}_k||$ is minimum for $i = k$, then the decision boundary is the perpendicular bisector of the line joining the two message points, as shown in Figure 7.4.4.

The outputs of the correlator, $x_1$ and $x_2$, are realizations of the two random variables $X_1$ and $X_2$. To simplify the decision procedure, we may define a new decision variable $X$ as the difference of these variables, expressing it in the form

$$X = X_1 - X_2. \tag{7.4.37}$$

One realization of this variable is

$$x = x_1 - x_2 = (s_{i1} - s_{i2}) + (w_1 - w_2), \tag{7.4.38}$$

which is used as the decision variable at the output of the correlator. This random value depends on the signal sent and the noise power in the channel. Bearing in mind the values of the coefficients, if bit 1 is sent, the decision value will be

$$x = (s_{i1} - s_{i2}) + (w_1 - w_2) = \left(\sqrt{E_b} - 0\right) + (w_1 - w_2) = \sqrt{E_b} + (w_1 - w_2) \qquad (7.4.39)$$

and, if bit 0 is sent, the decision value is

$$x = (s_{i1} - s_{i2}) + (w_1 - w_2) = \left(0 - \sqrt{E_b}\right) + (w_1 - w_2) = -\sqrt{E_b} + (w_1 - w_2). \qquad (7.4.40)$$

This definition has the following sense. The transmitter generates in any bit interval of length $N_b$ a bit which can be 1 or 0. Consequently, the modulated signal is a sinusoidal pulse of frequency $\Omega_1$ or $\Omega_2$. The demodulator calculates two random values, $x_1$ and $x_2$, that define the elements of the demodulated vector **x**. One of these two values contains the signal part, which depends on the power of the transmitted signal. The larger value will be the one which contains the signal part. Therefore, if we define a difference $x = x_1 - x_2$, it will contain the evidence for where the signal part is. If the difference is positive, the most likely transmitted signal is $s_1$. If the difference is negative, the most likely transmitted signal is $s_2$. Having available either the calculated correlator outputs $x_1$ and $x_2$ or their difference $x = x_1 - x_2 = (s_{i1} - s_{i2}) + (w_1 - w_2)$, the decision rules can be defined as follows:

1. If $x > 0$ or $x_1 > x_2$, then decide in favour of '1', because it is likely that the signal $s_1$ is transmitted, which implies that $s_{i2} = 0$. The demodulation vector **x** falls inside region $N_1$ on the constellation diagram in Figure 7.4.4.

2. If $x < 0$ or $x_2 > x_1$, then decide in favour of '0', because it is likely that the vector **x** is inside region $N_2$; we believe that the signal $s_2$ is transmitted, which implies that $s_{i1} = 0$.

3. If $x = 0$ or $x_2 = x_1$, then make a random guess in favour of '1' or '0'.

### 7.4.3.2 Operation of the Optimum Detector

As for a binary phase-shift keying system, we do not need a block with an optimum detector to detect the message bits. Let us briefly confirm this fact by showing that the decision variable at the output of the optimum detector is equal to the decision variable $X$ at the output of the modulator.

In a binary frequency-shift keying system, two signals can be generated at the transmitter side, and the output of the correlation demodulator generates two calculated random values in eqns (7.4.35) and (7.4.36) that are inputs to the optimum detector. These values are expressed in a transposed vector form as

$$\mathbf{x}^T = [x_1 \ x_2]^T = [s_{i1} + w_1 \ \ s_{i2} + w_2]^T. \qquad (7.4.41)$$

The transmitter signals can be expressed in the vector form presented in eqn (7.4.9). Each of the two signals is transmitted and processed in the same way inside the communication system. We will process the signals inside the optimum detector as shown in Figure 7.4.1. Assuming the same energy for both possible symbols, that is, $E_1 = E_2 = E_b$, the following two random numerical values can be calculated at the output of the detector:

$$
z_1 = \mathbf{x}^T \mathbf{s}_1 - \frac{1}{2} E_b \underset{i=1}{=} [x_1 \ x_2] \begin{bmatrix} s_{11} \\ s_{12} \end{bmatrix} - \frac{1}{2} E_b = x_1 s_{11} + x_2 s_{12} - \frac{1}{2} E_b
$$

$$
\underset{s_{12}=0}{=} x_1 s_{11} - \frac{1}{2} E_b = (s_{i1} + w_1) s_{11} - \frac{1}{2} E_b
$$

(7.4.42)

and

$$
z_2 = \mathbf{x}^T \mathbf{s}_2 - \frac{1}{2} E_b \underset{i=1}{=} [x_1 \ x_2] \begin{bmatrix} s_{21} \\ s_{22} \end{bmatrix} - \frac{1}{2} E_b = x_1 s_{21} + x_2 s_{22} - \frac{1}{2} E_b
$$

$$
\underset{s_{21}=0}{=} x_2 s_{22} - \frac{1}{2} E_b = (s_{i2} + w_2) s_{22} - \frac{1}{2} E_b.
$$

(7.4.43)

According to the operation of an optimum detector as explained in Chapter 5, and the structure presented in Figure 7.4.1, these two numerical values need to be compared and the larger one selected. Because the energies of the bits are equal, we obtain $s_{11} = s_{22} = E_b$. The calculated random values $z_1$ and $z_2$ can be compared by finding their difference:

$$
z_1 - z_2 = (s_{i1} + w_1) s_{11} - \frac{1}{2} E_b - (s_{i2} + w_2) s_{22} + \frac{1}{2} E_b
$$

$$
= (s_{i1} + w_1) s_{11} - (s_{i2} + w_2) s_{22} = (s_{i1} s_{11} - s_{i2} s_{22}) + (w_1 s_{11} - w_2 s_{22})
$$

(7.4.44)

Because the coefficients $s_{11} = s_{22}$ are constants, they have no influence on the decision process. Therefore, we can define the decision variable at the output of the optimum detector as

$$
y = \frac{z_1 - z_2}{s_{11}} = (s_{i1} + s_{i2}) + (w_1 - w_2) = x,
$$

(7.4.45)

which is equal to the decision variable $x$ defined at the output of the correlator by expression (7.4.38). Therefore, we can calculate the correlator output $x$ and make a decision, as explained in the previous section.

**The decision procedure.** The signals $\mathbf{s}_1$ and $\mathbf{s}_2$ are generated at the transmitter with the same probability, transmitted through the channel, and affected independently by the noise. Therefore, we can analyse one of the signals transmitted to find the bit error probability. Let us assume that the first signal $\mathbf{s}_1$ is transmitted and the signal coordinates are $s_{11} = \sqrt{E_b}$ and $s_{12} = 0$, as defined by eqn (7.4.4). The difference (7.4.44) of the optimum detector outputs in this case is

$$
z_1 - z_2 = s_{11}^2 + (w_1 s_{11} - w_2 s_{22}) = E_b + (w_1 - w_2) \sqrt{E_b}
$$

(7.4.46)

The value of the numerical decision variable obtained can be defined as

$$y = \frac{z_1 - z_2}{\sqrt{E_b}} = \sqrt{E_b} + (w_1 - w_2), \tag{7.4.47}$$

because the constant value $\sqrt{E_b}$ does not influence our decision. Therefore, the decision rules can be the same as for the random variable $X$ and are defined as follows:

1.  If $y > 0$, then decide in favour of '1', because **x** is inside the region $N_1$.
2.  If $y < 0$, then decide in favour of '0', because **x** is inside the region $N_2$.
3.  If $y = 0$, then make a random guess in favour of '1' or '0'.

Expression (7.4.47) is identical to the expression developed for the output of the correlation demodulator, eqn (7.4.39). Therefore, we do not need to use an optimum detector to detect the bit. It is sufficient to process the outputs of the correlation demodulator outputs and make a decision, as explained in Section 7.4.3.1.

### 7.4.3.3  Calculation of the Bit Error Probability

The output of the correlation demodulator contains two random values, $x_1$ and $x_2$, which are realizations of the random variables $X_1$ and $X_2$. The decision variable is defined by eqn (7.4.37) as their difference. This difference is defined as a new random decision variable $X$, which is defined by eqn (7.4.37) and expressed as

$$X = X_1 - X_2. \tag{7.4.48}$$

Then, the decision rules are defined with respect to the threshold value $x = 0$. Because the random variables $X_1$ and $X_2$ are Gaussian, the random variable $X$ is also Gaussian. Its mean value depends on which binary symbol has been transmitted, because the mean values of the individual variables $X_1$ and $X_2$ depend on the values of the binary symbols transmitted. The conditional means of the random variables $X_1$ and $X_2$, given that bit 1 was transmitted, are

$$E\{X_1|1\} = \sqrt{E_b}, \quad E\{X_2|1\} = 0, \tag{7.4.49}$$

respectively. Also, the conditional means of the random variables $X_1$ and $X_2$, given that bit 0 was transmitted, are

$$E\{X_1|0\} = 0, \quad E\{X_2|0\} = \sqrt{E_b}, \tag{7.4.50}$$

respectively. The conditional mean of the random variable $X$, given that bit 1 was transmitted, can be expressed as

$$E\{X|1\} = E\{X_1|1\} - E\{X_2|1\} = \sqrt{E_b} - 0 = \sqrt{E_b}, \tag{7.4.51}$$

and, similarly, the conditional mean of the random variable $X$, given that bit 0 was transmitted, is

$$E\{X|0\} = E\{X_1|0\} - E\{X_2|0\} = 0 - \sqrt{E_b} = -\sqrt{E_b}. \qquad (7.4.52)$$

Since $X_1$ and $X_2$ are statistically uncorrelated and Gaussian, and, consequently, can be considered independent, each with a variance equal to $N_0/2$, the variance of $X$ is equal to the sum of the variances for $X_1$ and $X_2$, that is,

$$\sigma_X^2 = 2\sigma_{X_i}^2 = 2\frac{N_0}{2} = N_0. \qquad (7.4.53)$$

The conditional density function of $X$, given that bit 0 was transmitted, is then expressed as

$$f_X(x|0) = \frac{1}{\sqrt{2\pi N_0}} e^{-(x+\sqrt{E_b})^2/2N_0}, \qquad (7.4.54)$$

and the conditional probability density function of $x$, given that bit 1 was transmitted, is

$$f_X(x|1) = \frac{1}{\sqrt{2\pi N_0}} e^{-(x-\sqrt{E_b})^2/2N_0}. \qquad (7.4.55)$$

These densities are shown in Figure 7.4.12.

Due to the statistical independence between the received bits, the probability of error can be calculated for a bit 0 or a bit 1 sent. Suppose that a particular bit 0 was sent. The receiver will make an error, that is, decide in favour of 1 instead 0, with a small probability calculated as

$$P(1|0) = P(x>0|0) = \int_0^\infty f_X(x|0)\, dx = \frac{1}{\sqrt{2\pi N_0}} \int_0^\infty e^{-(x+\sqrt{E_b})^2/2N_0} dx.$$

Substituting $z = (x+\sqrt{E_b})/\sqrt{2N_0}$ for the variable of integration $x$, we obtain

$$P(1|0) = \int_0^\infty f_X(x|0)\, dx = \frac{1}{\sqrt{\pi}} \int_{\sqrt{E_b/2N_0}}^\infty e^{-z^2} dz = \frac{1}{2}\mathrm{erfc}\sqrt{\frac{E_b}{2N_0}}. \qquad (7.4.56)$$

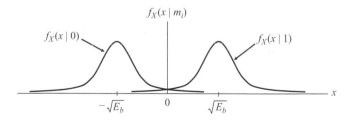

**Fig. 7.4.12** *Conditional probability density functions.*

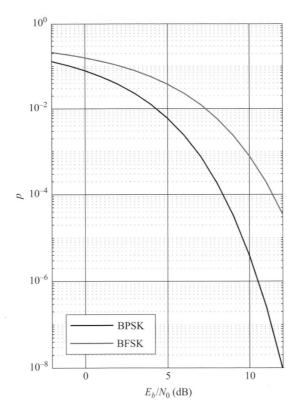

**Fig. 7.4.13** *Bit error probabilities for a binary phase-shift keying system, a binary frequency-shift keying system, and a quadriphase-shift keying system; BFSK, binary frequency-shift keying system; BPSK, binary phase-shift keying system; p, bit error probability.*

If a transmission channel is symmetric, implying that $P(1|0) = P(0|1)$, and the probability of generating 0 or 1 at the source is 0.5, the theoretical bit error probability in a binary frequency-shift keying system is

$$p = P(1)P(0|1) + P(0)P(1|0) = P(0|1) = \frac{1}{2}\text{erfc}\sqrt{\frac{E_b}{2N_0}}. \tag{7.4.57}$$

This bit error probability, as a function of $E_b/N_0$, is shown in Figure 7.4.13, along with the probability of error for a binary phase-shift keying system. The ratio $E_b/N_0$ has to be doubled (3 dB) in a binary frequency-shift keying system to maintain the same probability of error as in a binary phase-shift keying system. This complies with the signal-space diagrams for a binary phase-shift keying system and a binary frequency-shift keying system because, for a prescribed bit energy $E_b$, the Euclidean distance in a binary phase-shift keying system is $\sqrt{2}$ times greater than that in a binary frequency-shift keying system. The bit error probability of a quadriphase-shift keying system is equal to the probability of error in a binary phase-shift keying system.

### 7.4.3.4 Transceiver Design for a Binary Frequency-Shift Keying Signal

**Transmitter operation.** Bearing in mind that the expression for a binary frequency-shift keying signal is

$$s(n) = \sqrt{\frac{2E_b}{N_b}} \cos 2\pi \left( F_c \mp \frac{1}{2T_b} \right) t = \begin{cases} \sqrt{E_b}\sqrt{\frac{2}{N_b}} \cos \left( 2\pi F_1 n \right) & \text{for '} - \text{' sign} \\ \sqrt{E_b}\sqrt{\frac{2}{N_b}} \cos \left( 2\pi F_2 n \right) & \text{for '} + \text{' sign} \end{cases}, \quad (7.4.58)$$

the block schematic of coherent binary frequency-shift keying transmitter, shown in Figure 7.4.1, can be simplified as depicted in Figure 7.4.14.

At the transmitter side, the message symbols 1's and 0's are transformed into two levels, $+\sqrt{E_b}$ and 0, using an on–off encoder. The non-return-to-zero, unipolar binary signal $m(n)$ formed is directly applied to the input of a multiplier in the upper branch and, after inversion, the same signal is applied to the input of a multiplier in the lower branch. Thus, for the input bit 1, the oscillator in the upper branch is switched on and the incoming signal, with amplitude $+\sqrt{E_b}$, is multiplied by the generated basis function $\phi_1(n)$ defined by the frequency $F_1$. At the same time, the oscillator in the lower branch is switched off, because its input bit $m_{inv}(n)$, obtained from the inverter, is 0, and the output of the second multiplier is 0.

Similarly, in the case the input bit is 0, the oscillator in the lower branch is switched on, and the incoming inverted signal with amplitude $+\sqrt{E_b}$ is effectively multiplied by the generated basis function $\phi_2(n)$ defined by its frequency $F_2$. At the same time, the multiplier output in the upper branch is 0, because its input is 0.

The following assumptions are made:

1. The two oscillators in Figure 7.4.14 are synchronized, giving at their outputs two signals with frequencies $F_1$ and $F_2$.

2. The frequency of the modulated signal is changed with a continuous phase.

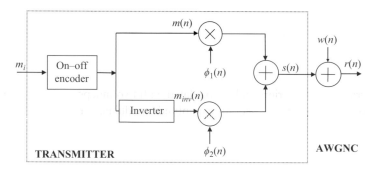

**Fig. 7.4.14** *A coherent binary frequency-shift keying transmitter; AWGNC; additive white Gaussian noise channel.*

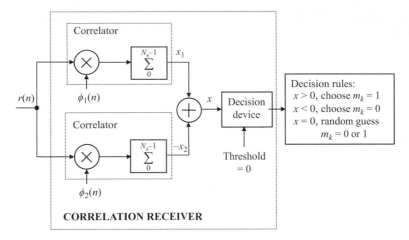

**Fig. 7.4.15** *A coherent binary frequency-shift keying receiver.*

**Receiver Operation.** We showed that the decision process inside an optimal binary frequency-shift keying receiver can take place at the output of the optimum detector or at the output of the correlation demodulator. For the sake of simplicity, we avoid the optimum detector and design the receiver with the decision circuit at the output of the demodulator, as shown in Figure 7.4.15. The receiver contains two correlators that calculate components of the correlation vector **x**, which are defined as $x_1$ and $x_2$. The components are subtracted from each other to obtain the difference $x$, which is compared with a zero-level threshold value. The decision rules are simple, as shown in Figure 7.4.15, and defined as follows:

1. If $x > 0$, then decide in favour of '1'.
2. If $x < 0$, then decide in favour of '0'.
3. If $x = 0$, then make a random guess in favour of '1' or '0'.

## 7.5 *M*-ary Discrete Quadrature Amplitude Modulation

### 7.5.1 Operation of a Discrete *M*-ary Quadrature Amplitude Modulation System

An *M*-ary quadrature amplitude modulated signal can be expressed in terms of two discrete quadrature carriers. Each carrier can be modulated with two independent amplitudes defined by $A_i$ and $B_i$ to obtain a discrete quadrature amplitude modulated signal expressed as

$$s_i(n) = \sqrt{E}\sqrt{\frac{2}{N}}A_i \cos \Omega_c n - \sqrt{E}\sqrt{\frac{2}{N}}B_i \sin \Omega_c n, \tag{7.5.1}$$

which is defined for the *i*th symbol transmitted, where the constant $\sqrt{E}$ is the square root of the energy of the symbol that has the minimum amplitude, and $N$ is the duration of a

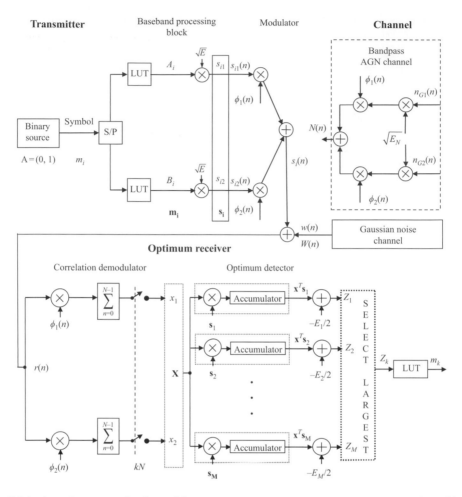

**Fig. 7.5.1** *A quadrature amplitude modulation continuous-time communication system; AGN, additive Gaussian noise; LUT, look-up table; S/P, serial-to-parallel converter.*

symbol expressed in discrete samples. Based on the generic scheme of a correlator-based communication system, which was presented in Chapter 5, Figure 5.6.4, for the general case, and in Figure 5.7.1 for a binary source, an $M$-ary quadrature amplitude modulation system can be structured as presented in Figure 7.5.1.

Suppose the message source can generate two binary values (or bits) that belong to an alphabet A of $A = 2$ elements, denoted by $m_1 = 1$ and $m_2 = 0$. A symbol generated at the output of the source contains $m$ bits. Therefore, the number of message symbols is $M = A^m$. A message symbol is applied at the inputs of two parallel branches, called the in-phase ($I$ system) and the quadrature system ($Q$ system), by means of a serial-to-parallel converter. The look-up-tables in the $I$ system and the $Q$ system generate the amplitude values

$A_i$ and $B_i$ at their outputs to form message symbols which are represented as the pair vector $\mathbf{m_i}$, where $i = 1, 2, \ldots, M$. The matrix of these vectors is expressed in matrix form in eqn (7.5.2) for $M = 16$.

The pair vectors $\mathbf{m_i}$ are multiplied by the constant $\sqrt{E}$ to obtain the signal coefficients, which are presented in matrix form in eqn (7.5.3). These coefficients define two modulating signals $s_{ij}(n)$: signal $s_{i1}(n)$, in the $I$ system of the modulator, and $s_{i2}(n)$, in the $Q$ system of the modulator. These signals are applied to the input of two modulator multipliers to modulate the carriers. This procedure can also be understood as magnitude scaling of the carrier by the coefficients $s_{ij}$, which are represented by the quadrature values $A_i\sqrt{E}$ and $B_i\sqrt{E}$. The level of scaling depends on the symbol structure at the output of the look-up table, as can be seen from the constellation diagram in Figure 7.5.2. At the output of each multiplier, an amplitude-shift keying signal is obtained. The two amplitude-shift keying modulated signals are added to produce the quadrature amplitude modulated signal $s_i(n)$ at the output of the transmitter.

The modulated discrete-time signal $s_i(n)$ is affected by additive white Gaussian noise. The discrete communication systems are related to the theoretical modelling and design of an intermediate-frequency block inside the transmitter that is followed by a digital-to-analogue converter and then transferred through a continuous-time channel, as explained in Chapter 5 and presented in Figure 5.7.1. This continuous-time channel is characterized by the additive white Gaussian noise $w(t)$, which is a realization of the wide-sense stationary noise process $W(t)$ presented theoretically in Chapter 19 and analysed in detail in Chapter 4.

When we model, design, and test the properties of a discrete-time system operating in the presence of noise, the noise affecting the transmitted signal needs to be presented in the discrete-time domain because the intermediate signal in the system is presented in the discrete-time domain. This noise is represented by the random discrete signal $w(n)$, which is a realization of the wide-sense stationary noise process $W(n)$ that is analysed theoretically in Chapter 3, and in detail in Chapter 4. The noise is added to the discrete-time transmitted signal, as shown in Figure 7.5.1. The difficult problem of getting the discrete Gaussian noise $W(n)$ from its counterpart continuous noise $W(t)$ is elaborated in Chapter 4. The communication channel in Figure 7.5.1 is also represented by the passband additive Gaussian noise $N(n)$, which is a useful model for system simulation and emulation, as can be seen in the theoretical analysis presented in Chapter 4.

The received signal $r(n)$ is coherently demodulated using two correlators inside the correlation demodulator. At the outputs of the correlators, the two random samples $x_1$ and $x_2$ are obtained at the discrete-time instants $kN$. The samples are presented to the optimum detector to correlate them with all possible discrete-time symbols that could be generated at the transmitter side, as was explained for the general optimum detector presented in Chapter 5, Section 5.6.

## 7.5.2 Transmitter Operation

**Generation of a discrete-time modulating signal.** The number of possible message symbols is $M = 2^m$, where $m$ is the number of bits in an $M$-ary symbol at the output of the source. We will analyse the case when an even number of bits form a symbol defining a square constellation diagram. It is possible to form a constellation diagram for an odd number of bits

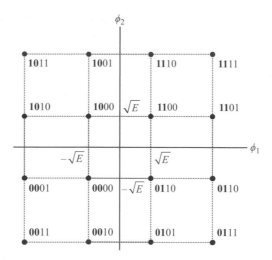

**Fig. 7.5.2** *Constellation diagram for a 16-ary quadrature amplitude modulation system.*

in a symbol. In that case, we define a cross-constellation diagram. For the sake of explanation, we will analyse the case of a 16-ary quadrature amplitude modulation system defined by $M = 16$ message symbols presented in the constellation diagram in Figure 7.5.2. Each message symbol consists of $m = \log_2 M = 4$ message bits. The first two bits (in bold) define the quadrant of a message symbol, and the last two bits define four distinct symbols that belong to the related quadrant.

For this constellation diagram, pairs of coordinates $(A_i, B_i)$ specify all message symbols that can be expressed in the matrix form

$$
\mathbf{m} =
\begin{bmatrix}
\mathbf{m}_1 & \mathbf{m}_2 & \mathbf{m}_3 & \mathbf{m}_4 \\
\mathbf{m}_5 & \mathbf{m}_6 & \mathbf{m}_7 & \mathbf{m}_8 \\
\mathbf{m}_9 & \mathbf{m}_{10} & \mathbf{m}_{11} & \mathbf{m}_{12} \\
\mathbf{m}_{13} & \mathbf{m}_{14} & \mathbf{m}_{15} & \mathbf{m}_{16}
\end{bmatrix}
\begin{bmatrix}
-3,3 & -1,3 & 1,3 & 3,3 \\
-3,1 & -1,1 & 1,1 & 3,1 \\
-3,-1 & -1,-1 & 1,-1 & 3,-1 \\
-3,-3 & -1,-3 & 1,-3 & 3,-3
\end{bmatrix}.
\tag{7.5.2}
$$

These pairs of coordinates are multiplied by $\sqrt{E}$ to obtain pairs of amplitudes of modulating signals $s_{i1}$ and $s_{i2}$ that correspond to the generated symbols and can be expressed in the matrix form

$$
\mathbf{s} = \mathbf{m}\sqrt{E} =
\begin{bmatrix}
-3\sqrt{E},3\sqrt{E} & -\sqrt{E},3\sqrt{E} & \sqrt{E},3\sqrt{E} & 3\sqrt{E},3\sqrt{E} \\
3\sqrt{E},\sqrt{E} & -\sqrt{E},\sqrt{E} & \sqrt{E},\sqrt{E} & 3\sqrt{E},\sqrt{E} \\
-3\sqrt{E},-\sqrt{E} & -\sqrt{E},-\sqrt{E} & \sqrt{E},-\sqrt{E} & 3\sqrt{E},-\sqrt{E} \\
-3\sqrt{E},-3\sqrt{E} & -\sqrt{E},-3\sqrt{E} & \sqrt{E},-3\sqrt{E} & 3\sqrt{E},-3\sqrt{E}
\end{bmatrix}.
\tag{7.5.3}
$$

The first row in the matrix corresponds to the message symbols specified by the topmost four points in the constellation diagram. For each message point in the constellation diagram, we

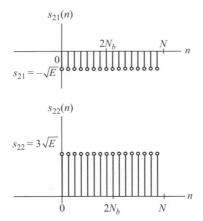

**Fig. 7.5.3** *Modulating discrete-time rectangular pulses aligned in the time domain as they appear at the input of the modulator multipliers.*

can calculate the energy of the related modulated signal and make a matrix of these energies in the same way we made the **m** and **s** matrices. These energies are to be calculated following the principles presented in Chapter 2, which is dedicated to orthonormal functions and the orthogonalization procedure.

For all message symbols, we can find the related wave shapes. For example, for the second message symbol $m_2 = (1001)$, we can define the wave shapes of the first and second modulating signals, which are $s_{21}(n)$ and $s_{22}(n)$, respectively. The first modulating signal, $s_{21}(n)$, is a rectangular pulse of duration $N$ and amplitude equal to $-\sqrt{E}$ defined by the second signal coefficient $s_{21}$, as presented in Figure 7.5.3. On the same figure, the second modulating signal, $s_{22}(n)$, is shown to have the amplitude $3\sqrt{E}$.

**Generation of the modulated signal.** For the same example, message symbol $m_2 = (1001)$, we will explain the operation of the modulator. The $M$-ary quadrature amplitude modulator uses two orthonormal basis functions, which are defined as

$$\phi_1(n) = \sqrt{\frac{2}{N}} \cos \Omega_c n \qquad (7.5.4)$$

and

$$\phi_2(n) = -\sqrt{\frac{2}{N}} \sin \Omega_c n, \qquad (7.5.5)$$

to obtain the modulated signal presented in eqn (7.5.1) and expressed in terms of these functions as

$$s_i(n) = \sqrt{E}A_i\phi_1(n) + \sqrt{E}B_i\phi_2(n) = s_{i1}(n)\phi_1(n) + s_{i2}(t)\phi_2(n). \qquad (7.5.6)$$

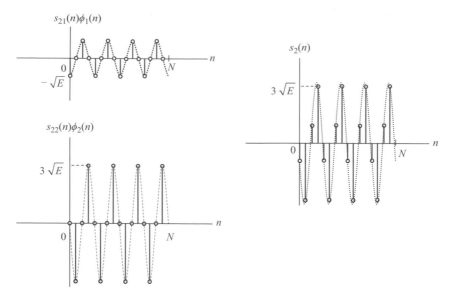

**Fig. 7.5.4** *In-phase and quadrature components at the inputs of the transmitter adder, and the modulated 16-ary quadrature amplitude modulation signal $s_2(n)$ at the output of the transmitter adder.*

The normalized frequency of the carrier is much larger than the fundamental frequency of the rectangular pulse, that is, $F = 1/N \ll F_c = 1/N_c$. For the sake of explanation, we will assume that the carrier makes four oscillations inside the symbol interval $N$, as shown in Figure 7.5.4.

For this particular example, the two modulated signals in the $I$ system and the $Q$ system presented in Figure 7.5.4 are added to obtain the quadrature amplitude modulated signal at the transmitter output, as shown in the figure on the right-hand side. The modulated signal $s_2(n)$ is transmitted through the channel. For our analysis of the receiver operation, we will assume that the modulated signal is affected by the discrete white Gaussian noise $w(n)$ in the channel. The channel used in Figure 7.5.1 is also represented by passband noise $N(n)$. Discrete-time noise is analysed in detail in Chapters 3 and 4, due to its importance in simulating or emulating discrete communication systems.

Bearing in mind the operation of an $M$-ary quadrature amplitude modulation system, we can simplify the schemes of both the transmitter and the receiver, as we did for a quadriphase-shift keying transmitter, as presented in Figures 7.5.5 and 7.5.6. We use these schemes for two reasons. Firstly, these simplified schemes are traditionally used in most of the books on the theory of digital communication systems. Secondly, similar structures will be used to explain the design of discrete communication systems in Chapter 10.

The transmitter operation will be explained for message symbol $m_2 = (1001)$, as an example. Suppose this message symbol is generated at the output of the source, as shown in Figure 7.5.5. This binary symbol is transferred into parallel form by a series-to-parallel converter and used as an address inside look-up tables to generate the symbol pair $(A_i, B_i)$, which is multiplied by $\sqrt{E}$ to form the modulating rectangular pulses $s_{21}(n) = -\sqrt{E}$ and

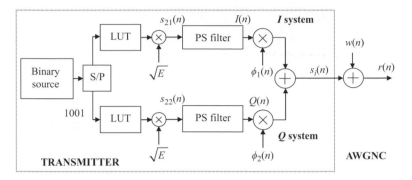

**Fig. 7.5.5** *Traditional scheme of an M-ary quadrature amplitude modulation discrete transmitter; AWGNC, additive white Gaussian noise channel; LUT, look-up table; PS, pulse-shaping; S/P, serial-to-parallel converter.*

$s_{22}(n) = 3\sqrt{E}$ inside the interval $0 \leq n \leq N$, as shown in Figure 7.5.5. The modulating rectangular signals presented in Figure 7.5.3 are filtered in pulse-shaping filters before modulation takes place. The output of the filters in the $I$ system and the $Q$ system are traditionally denoted as the in-phase $I(n)$ and quadrature $Q(n)$, respectively, modulating signal components.

If a pulse-shaping filter is used, then the modulated signal can be expressed in terms of its in-phase and quadrature components as

$$s_i(n) = I(n)\phi_1(n) + Q(n)\phi_2(n), \tag{7.5.7}$$

which is the standard expression usually used in practice. For the sake of explanation, we will exclude the filtering procedure in our explanation of the transmitter operation and assume that the rectangular pulses are applied at the input of the carrier multipliers performing modulation. For that case, relation (7.5.7) will be equivalent to relation (7.5.6).

## 7.5.3   Operation of a Correlation Demodulator

A simplified receiver structure is presented in Figure 7.5.6. The modulated signal is transmitted through the additive white Gaussian noise channel and received by a coherent quadrature amplitude modulation correlation receiver. The received signal can be expressed as

$$r(n) = s_i(n) + w(n), \tag{7.5.8}$$

defined inside the interval $0 \leq n < N$. The random discrete signal $w(n)$ is a realization of the discrete-time white Gaussian noise process $W(n)$ of zero mean and power spectral density $N_0/2$. A detailed explanation of this noise can be found in Chapter 4. Therefore, at the output of the two correlators ($j = 1, 2$) are discrete-time random signals defined by their random

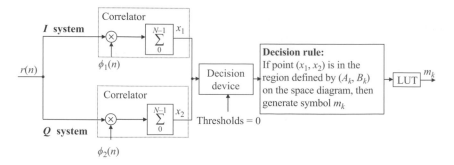

**Fig. 7.5.6** *Block schematic of an M-ary quadrature amplitude modulation correlation receiver LUT, look-up table.*

values $x_1$ and $x_2$, which are calculated at the end of each symbol interval $N$ as

$$x_1 = \sum_{n=0}^{N-1} r(n)\phi_1(n) = \sum_{n=0}^{N-1} s_i(n)\phi_1(n) + \sum_{n=0}^{N-1} w(n)\phi_1(n) = \sqrt{E}A_i + w_1 = s_{i1} + w_1 \quad (7.5.9)$$

and

$$x_2 = \sum_{n=0}^{N-1} r(n)\phi_2(n) = \sqrt{E}B_i + w_2 = s_{i2} + w_2, \quad (7.5.10)$$

where $s_{i1}$ and $s_{i2}$ are deterministic quantities contributed by the transmitted signal $s_i$. In other words, they are the coefficients of the transmitted signals, as defined in the theory of orthogonal signals, which was presented in Chapter 2. The random values $w_1$ and $w_2$ are the sample values of the random variables $W_1$ and $W_2$, which arise due to the channel noise process $W(n)$.

The random values $x_1$ and $x_2$ calculated are realizations of the Gaussian random variables $X_1$ and $X_2$ and can be expressed as the demodulation vector $\mathbf{x} = [x_1\ x_2]$. With this vector, we can decide on the symbols transmitted, using the following decision rules:

1. If the output of the correlators, defined as the demodulation vector $\mathbf{x}$, falls in the region of a particular symbol, then we conclude that symbol was sent. In our example, if the coordinates defined by vector $\mathbf{x}$ are in the region of symbol $s_2$ defined by the pair $(-1, 3)$, we will rightly decide in favour of the binary message (1001).

2. However, if the output of the correlator, $\mathbf{x}$, falls in a region of any other symbol, an incorrect decision will be made in favour of a message symbol which is not (1001).

## Example

Calculate the values of the demodulation vector **x** by expressing the basis orthogonal functions in their full form as in expressions (7.5.4) and (7.5.5). Comment on the minus sign in front of the second basis function. Can we use a cosine function for the second basis function? Show how that can be done.

**Solution.** If we insert eqns (7.5.1) and (7.5.4) into eqn (7.5.9), we obtain the first value of the demodulation vector as

$$
\begin{aligned}
x_1 &= \sum_{n=0}^{N-1} s_i(n)\phi_1(n) + \sum_{n=0}^{N-1} w(n)\phi_1(n) \\
&= \sqrt{E}\frac{2}{N}\sum_{n=0}^{N-1}(A_i\cos\Omega_c n - B_i\sin\Omega_c n)\cos\Omega_c n + \sqrt{\frac{2}{N}}\sum_{n=0}^{N-1}w(n)\cos\Omega_c n \\
&= \sqrt{E}A_i + w_1 = s_{i1} + w_1
\end{aligned}
\tag{7.5.11}
$$

The second value can be found by inserting eqns (7.5.1) and (7.5.5) into eqn (7.5.10) to obtain

$$
\begin{aligned}
x_2 &= \sum_{n=0}^{N-1} s_i(n)\phi_2(n) + \sum_{n=0}^{N-1} w(n)\phi_2(n) \\
&= \sqrt{E}\frac{2}{N}\sum_{n=0}^{N-1}(A_i\cos\Omega_c n - B_i\sin\Omega_c n)(-\sin\Omega_c n) - \sqrt{\frac{2}{N}}\sum_{n=0}^{N-1}w(n)\sin\Omega_c n \\
&= \sqrt{E}B_i - w_2 = s_{i2} - w_2.
\end{aligned}
\tag{7.5.12}
$$

The sign of the noise part $w_2$ does not statistically affect the operation of the decision device because the density function of this noise is Gaussian with the zero mean and constant variance. Therefore, the modulation and the demodulation need to be performed with a sine function with a negative sign to fulfil the condition of coherent demodulation. Namely, in that case, the second basis function $\phi_2(n)$ at the transmitter side will be synchronized both in frequency and in phase with the basis function $\phi_2(n)$ at the receiver side. The negative sign in front of the noise term $w_2$ will not change the mean value or the variance of its Gaussian density function and will not have any consequence on the statistical properties of the operation of the receiver.

In contrast, if a negative sign is not assigned to the amplitude of the quadrature carrier $\phi_2(n)$ at the receiver side, which contains the coordinate of the message symbol, there would be an ambiguity in the decision process with respect to the position of the related message symbol in the constellation diagram. In addition, in that case, the carrier at the receiver side would not be phase synchronized with its counterpart at the transmitter side, that is, they will be 180° out of phase. The second orthogonal carrier can be expressed in terms of a cosine function via the simple trigonometric identity

$$
\phi_2(n) = -\sqrt{\frac{2}{N}}\sin\Omega_c n = \sqrt{\frac{2}{N}}\cos\left(\Omega_c n + \frac{\pi}{2}\right).
\tag{7.5.13}
$$

### 7.5.4   Operation of an Optimum Detector

The structure of an optimum detector was presented in Figure 7.5.1. We will explain and demonstrate its operation with a simple example. We will assume that the signal $s_2(n)$ was transmitted in a 16-ary quadrature amplitude modulation system, and then investigate the operation of the optimum detector, demonstrating how it works for signals $s_1(n)$ and $s_7(n)$ by calculating the decision variables $Z_2$, $Z_1$, and $Z_7$ and comparing them. The analysis presented here can then be extended to the remaining 13 signals.

If the second signal is sent, its energy can be calculated using the signal constellation diagram in Figure 7.5.2 or expression (7.5.3) as

$$E_2 = \left(-\sqrt{E}\right)^2 + \left(3\sqrt{E}\right)^2 = 10E. \tag{7.5.14}$$

Using eqns (7.5.11) and (7.5.12) where $i = 2$, the related decision variable can be calculated as

$$
\begin{aligned}
Z_2 = \mathbf{x}^T \mathbf{s}_2 &= \left[\sqrt{E}A_2 + w_1 \quad \sqrt{E}B_2 - w_2\right] \begin{bmatrix} A_2\sqrt{E} \\ B_2\sqrt{E} \end{bmatrix} - \frac{1}{2}E_2 \\
&= \sqrt{E}A_2\left(\sqrt{E}A_2 + w_1\right) + \sqrt{E}B_2\left(\sqrt{E}B_2 - w_2\right) - 5E \\
&= EA_2^2 + EB_2^2 - 5E + \sqrt{E}A_2 w_1 - \sqrt{E}B_2 w_2 = E + 9E - 5E - \sqrt{E}w_1 - 3\sqrt{E}w_2 \\
&= 5E - \sqrt{E}(3w_2 + w_1).
\end{aligned}
\tag{7.5.15}
$$

For the sake of explanation, we will compare the calculated variable $Z_2$ with the two variables $Z_1$ and $Z_7$. The energy of the first signal is

$$E_1 = \left(-3\sqrt{E}\right)^2 + \left(3\sqrt{E}\right)^2 = 18E$$

and the related decision variable, assuming that the second signal is transmitted, is

$$
\begin{aligned}
Z_1 = \mathbf{x}^T \mathbf{s}_1 &= \left[\sqrt{E}A_2 + w_1 \quad \sqrt{E}B_2 - w_2\right] \begin{bmatrix} A_1\sqrt{E} \\ B_1\sqrt{E} \end{bmatrix} - \frac{1}{2}E_1 \\
&= \sqrt{E}A_1\left(\sqrt{E}A_2 + w_1\right) + \sqrt{E}B_1\left(\sqrt{E}B_2 - w_2\right) - 9E \\
&= EA_1 A_2 + EB_1 B_2 - 9E + \sqrt{E}A_1 w_1 - \sqrt{E}B_1 w_2 \\
&\quad - 3E + 9E - 9E - 3\sqrt{E}w_1 - 3\sqrt{E}w_2 = 3E - 3\sqrt{E}(w_2 + w_1).
\end{aligned}
\tag{7.5.16}
$$

Likewise, the energy of the seventh signal is

$$E_7 = \left(\sqrt{E}\right)^2 + \left(\sqrt{E}\right)^2 = 2E \tag{7.5.17}$$

and the related energy, assuming that the second signal is transmitted, is

$$
\begin{aligned}
Z_7 = \mathbf{x}^T \mathbf{s}_7 &= \left[ \sqrt{E}A_2 + w_1 \quad \sqrt{E}B_2 - w_2 \right] \left[ \begin{array}{c} A_7\sqrt{E} \\ B_7\sqrt{E} \end{array} \right] - \frac{1}{2}E_7 \\
&= \sqrt{E}A_7 \left( \sqrt{E}A_2 + w_1 \right) + \sqrt{E}B_7 \left( \sqrt{E}B_2 - w_2 \right) - E \\
&= EA_7A_2 + EB_7B_2 - E + \sqrt{E}A_7w_1 - \sqrt{E}B_7w_2 \\
&= -E + 3E - E + \sqrt{E}w_1 - \sqrt{E}w_2 = E - \sqrt{E}(w_2 - w_1).
\end{aligned}
\tag{7.5.18}
$$

If we exclude the influence of the noise on the decision variables, we can see that the variable belonging to the supposedly sent symbol, $Z_2$ in this case, is the largest. Therefore, deciding in favour of message $m_k = m_2$ would be the right decision in that case. However, if noise is present, as occurs in reality, the values of the decision variable can be changed, causing an incorrect decision.

........................................................................................................................................................

## APPENDIX A: POWER SPECTRAL DENSITY OF A QUADRIPHASE-SHIFT KEYING MODULATING SIGNAL

Calculate the power spectral density of the modulating signal of a quadriphase-shift keying signal, using its autocorrelation function. Calculate the maximum value of the power spectral density. Plot graphs of the power spectral density using logarithmic scales.

The autocorrelation function for a discrete-time binary stochastic process, which is defined as a train of polar rectangular pulses like the modulating signal, was calculated in Chapter 3, Section 3.6.4, and expressed in eqn (3.6.24). For our modulating signal, in both branches, that expression will be adapted for the new parameters $N_b = N_s/2$ and $E_b = E/2$ as

$$
R_{MI}(l) = R_{MQ}(l) = \left\{ \begin{array}{ll} \frac{E}{2}\left(1 - \frac{|l|}{N_s}\right) & |l| \le N_s \\ 0 & \text{otherwise} \end{array} \right\} = \left\{ \begin{array}{ll} E_b\left(1 - \frac{|l|}{2N_b}\right) & |l| \le N_b \\ 0 & \text{otherwise} \end{array} \right\}.
\tag{A.1}
$$

The Fourier transform of the autocorrelation function can be calculated as

$$
FT\{R_{MI}(l)\} = \sum_{l=-N_s}^{N_s} R_{MI}(l)e^{-j\Omega l} = \frac{E}{2} \frac{\sin^2 \Omega N_s/2}{\sin^2 \Omega/2} \Bigg|_{E=2E_b, N_s=2N_b} = E_b \frac{\sin^2 2\Omega N_b/2}{\sin^2 \Omega/2}.
$$

Due to the definition of the modulating pulse amplitude, the power spectral density of the upper branch is

$$
S_{MI}(\Omega) = \frac{E}{2N_s} \frac{\sin^2 \Omega N_s/2}{\sin^2 \Omega/2} \Bigg|_{E=2E_b, N_s=2N_b} = \frac{E_b}{2N_b} \frac{\sin^2 \Omega N_b}{\sin^2 \Omega/2}.
\tag{A.2}
$$

Because the signals in both branches are identical and independent from each other, the power spectral density of the modulating signal is the sum of the power spectral densities in the $I$ and $Q$ branches, expressed as

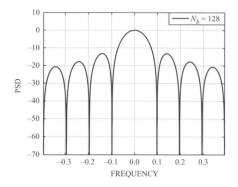

**Fig. A1** *Power spectral density function in linear logarithmic units; PSD, power spectral density.*

$$S_{MM}(\Omega) = S_{MI}(\Omega) + S_{MQ}(\Omega) = \frac{E}{N_s}\frac{\sin^2 \Omega N_s/2}{\sin^2 \Omega/2}\Bigg|_{E=2E_b, N_s=2N_b} = \frac{E_b}{N_b}\frac{\sin^2 \Omega N_b}{\sin^2 \Omega/2},$$

as was calculated using the expression for the energy spectral density of a rectangular discrete pulse. The maximum value can be calculated by applying L'Hopital's rule:

$$
\begin{aligned}
S_{MI}(\Omega)/(E_b N_b/2)_{\max} &= \frac{1}{N_b^2} \lim_{\Omega \to 0} \left| \frac{\sin N_b \Omega}{\sin \Omega/2} \right|^2 = \frac{1}{N_b^2} \lim_{\Omega \to 0} \frac{1 - \cos 2N_b \Omega}{1 - \cos 2\Omega/2} \\
&= \frac{1}{N_b^2} \lim_{\Omega \to 0} \frac{d\left(1 - \cos 2N_b \Omega\right)/d\Omega}{d\left(1 - \cos \Omega\right)/d\Omega} = \frac{1}{N_b^2} \lim_{\Omega \to 0} \frac{2N_b \sin 2N_b \Omega}{\sin \Omega} \\
&= \frac{1}{N_b^2} 2N_b \lim_{\Omega \to 0} \frac{2N_b \cos 2N_b \Omega}{\cos \Omega} = \frac{1}{N_b^2} 2N_b 2N_b = 4.
\end{aligned}
\tag{A.3}
$$

The graph of the normalized power spectral density function $S_{MI}(\Omega)$ is presented in Figure A1 on both linear and logarithmic scales.

...................................................................................................................................

## APPENDIX B: PROBABILITY DENSITY FUNCTIONS FOR A QUADRIPHASE-SHIFT KEYING SYSTEM

The conditional probability density function of $X_1$, given that any $m_i$ was transmitted, is

$$f_{X_1}(x_1|m_i) = \frac{1}{\sqrt{\pi N_0}} e^{-(x_1 - s_{i1})^2/N_0}, \tag{B.1}$$

and the conditional probability density function of $X_2$, given that any $m_i$ was transmitted, is

$$f_{X_2}(x_2|m_i) = \frac{1}{\sqrt{\pi N_0}}e^{-(x_2-s_{i2})^2/N_0}. \tag{B.2}$$

The first density function can be expressed separately for $m_i = m_1 = 1$ and $m_i = m_2 = 0$, respectively, as follows:

$$f_{X_1}(x_1|1) = \frac{1}{\sqrt{\pi N_0}}e^{-(x_1-s_{11})^2/N_0}, \quad f_{X_1}(x_1|0) = \frac{1}{\sqrt{\pi N_0}}e^{-(x_1-s_{21})^2/N_0}. \tag{B.3}$$

These densities are shown in Figure B1.

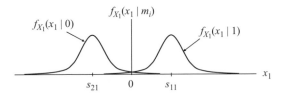

**Fig. B1** *Probability density functions.*

In the same way, we can find the conditional probability density functions of the random variable $X_2$:

$$f_{X_2}(x_2|1) = \frac{1}{\sqrt{\pi N_0}}e^{-(x_2-s_{12})^2/N_0}, \quad f_{X_2}(x_2|1) = \frac{1}{\sqrt{\pi N_0}}e^{-(x_2-s_{22})^2/N_0}. \tag{B.4}$$

APPENDIX C: DENSITY FUNCTIONS FOR $X_1$ AND $X_2$ IN A FREQUENCY-SHIFT KEYING SYSTEM

For the coherent binary frequency-shift keying receiver presented in Figure 7.4.1, we can define the random variables $X_1$ and $X_2$ with the realizations $x_1$ and $x_2$, respectively, as

$$x_1 = \sum_{n=0}^{N_b-1} r(n)\phi_1(n) = \sum_{n=0}^{N_b-1} s_i(n)\phi_1(n) + \sum_{n=0}^{N_b-1} w(n)\phi_1(n) = s_{i1} + w_1,$$

and

$$x_2 = \sum_{n=0}^{N_b-1} r(n)\phi_2(n) = \sum_{n=0}^{N_b-1} s_i(n)\phi_2(n) + \sum_{n=0}^{N_b-1} w(n)\phi_2(n) = s_{i2} + w_2.$$

For the defined random variables $X_1$ and $X_2$, we can find the probability density functions as follows. The random variable $X_1$ with the realization $x_1$ can be expressed in the form

$$X_1 = s_{i1} + W_1,$$

where the signal coefficients are defined by the matrix

$$\mathbf{s} = [\mathbf{s}_1 \ \mathbf{s}_2] = \begin{bmatrix} s_{11} & s_{21} \\ s_{12} & s_{22} \end{bmatrix} = \begin{bmatrix} \sqrt{E_b} & 0 \\ 0 & \sqrt{E_b} \end{bmatrix}.$$

Therefore, for the wide-sense stationary Gaussian noise process $W_1(n)$, the random variable $W_1$ can be expressed in two ways: as

$$X_1 = s_{i1} + W_1 = s_{11} + W_1 = \sqrt{E_b} + W_1,$$

if bit $m_1 = 1$ is transmitted, and as

$$X_1 = s_{i1} + W_1 = 0 + W_1 = W_1,$$

if bit $m_2 = 0$ is transmitted. Because the noise has Gaussian density, there are two conditional density functions defined for the random variable $X_1$.

If bit $m_1 = 1$ is transmitted, the mean value and the variance of $X_1$ are defined by eqn (7.4.49) as

$$E\{X_1|1\} = \sqrt{E_b}$$

and

$$\sigma_{X_1}^2 = E\left\{X_1^2|1\right\} - E^2\{X_1|1\} = \left(E_b + \sigma_{W_1}^2\right) - E_b = \frac{N_0}{2},$$

respectively. If bit $m_2 = 0$ is transmitted, the mean value and the variance of $X_1$ are calculated as

$$E\{X_1|0\} = 0$$

and

$$\sigma_{X_1}^2 = E\left\{X_1^2|1\right\} - E^2\{X_1|1\} = E\left\{W_1^2|1\right\} - E^2\{W_1|1\} = \sigma_{W_1}^2 - 0 = \frac{N_0}{2},$$

respectively. Therefore, the corresponding conditional density functions are

$$f_{X_1}(x_1|1) = \frac{1}{\sqrt{2\pi N_0/2}} e^{-(x_1 - \sqrt{E_b})^2 / 2N_0/2} = \frac{1}{\sqrt{\pi N_0}} e^{-(x_1 - \sqrt{E_b})^2 / N_0}$$

and

$$f_{X_1}(x_1|0) = \frac{1}{\sqrt{2\pi N_0/2}} e^{-x_1^2/2N_0/2} = \frac{1}{\sqrt{\pi N_0}} e^{-x_1^2/N_0},$$

respectively. Graphs of these density functions are presented in Figure C1.

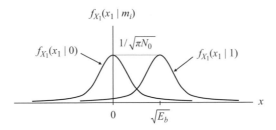

**Fig. C1** *Probability density functions.*

Analogously, the density function for the variable $X_2$ can be derived as follows. The random variable $X_2$ is

$$X_2 = s_{i2} + W_2.$$

If bit $m_1 = 1$ is transmitted, the variable is

$$X_2 = s_{12} + W_2 = 0 + W_2 = W_2,$$

and, if bit $m_2 = 0$ is transmitted, the variable is

$$X_2 = s_{i2} + W_2 = s_{22} + W_2 = \sqrt{E_b} + W_2.$$

Because the noise has Gaussian density, there are two conditional density functions defined for the random variable $X_2$. If bit $m_1 - 1$ is transmitted, the mean value and the variance of $X_2$ are

$$E\{X_2|1\} = 0$$

and

$$\sigma_{X_2}^2 = E\left\{X_2^2|1\right\} - E^2\{X_2|1\} = E\left\{W_2^2|1\right\} - E^2\{W_2|1\} = \sigma_{W_2}^2 - 0 = \frac{N_0}{2},$$

respectively. If bit $m_2 = 0$ is transmitted, the mean value and the variance of $X_2$ are defined as

$$E\{X_2|0\} = \sqrt{E_b}$$

and

$$\sigma_{X_2}^2 = E\left\{X_2^2|0\right\} - E^2\{X_2|0\} = E_b + \sigma_{W_2}^2 - E_b = \frac{N_0}{2},$$

respectively. The density function for bit $m_2 = 1$ transmitted is

$$f_{X_2}(x_2|1) = \frac{1}{\sqrt{\pi N_0}} e^{-x_2^2/N_0}$$

and for bit $m_2 = 0$ transmitted is

$$f_{X_2}(x_2|0) = \frac{1}{\sqrt{\pi N_0}} e^{-(x_2-\sqrt{E_b})^2/N_0},$$

which are at the same positions as the graphs shown in Figure A1 for the random variable $X_1$.

...............................................................................................................................

## APPENDIX D: STATISTICS OF THE DECISION VARIABLE $X = X_1 - X_2$

The random variables at the output of the binary frequency-shift keying correlator $X_1$ and $X_2$ are defined by their zero mean and variance $N_0/2$. The decision variable is defined as their difference, that is, $X = X_1 - X_2$. We can find the density function for these variables and derive the expression for the bit error probability in this binary frequency-shift keying system.

The variables $X_1$ and $X_2$ are statistically independent Gaussian variables and, consequently, are uncorrelated, each with zero mean and variance equal to $N_0/2$. Following eqns (7.4.35) and (7.4.36), they can be expressed as

$$X_1 = s_{i1} + W_1,$$

and

$$X_2 = s_{i2} + W_2$$

If bit $m_1 = 1$ is transmitted, the variables are

$$X_1 = s_{i1} + W_1 = s_{11} + W_1 = \sqrt{E_b} + W_1$$

and

$$X_2 = s_{12} + W_2 = 0 + W_2 = W_2,$$

and

$$X = X_1 - X_2 = \sqrt{E_b} + W_1 - W_2.$$

Because the noise is white and Gaussian, the mean value of the variable $X$ is

$$E\{X\} = \sqrt{E_b} + 0 - 0 = \sqrt{E_b},$$

and the corresponding variance is

$$\sigma_X^2 = E\left\{X^2\right\} - E^2\{X\} = E\left\{\left(\sqrt{E_b} + W_1 - W_2\right)^2\right\} - E_b$$

$$= E_b + 0 + 0 + 0 + \sigma_{W1}^2 + \sigma_{W2}^2 - E_b = \frac{N_0}{2} + \frac{N_0}{2} = N_0.$$

The conditional density function of $X$, given that symbol $m_1 = 1$ is transmitted, is

$$f_X(x|1) = \frac{1}{\sqrt{2\pi N_0}} e^{-(x - \sqrt{E_b})^2 / 2N_0}.$$

If bit $m_2 = 0$ is transmitted, the variables are

$$X_1 = s_{i1} + W_1 = 0 + W_1 = W_1$$

and

$$X_2 = s_{i2} + W_2 = s_{22} + W_2 = \sqrt{E_b} + W_2,$$

and

$$X = X_1 - X_2 = W_1 - \sqrt{E_b} - W_2.$$

Because the noise is white and Gaussian, the mean value and the variance of $X$ are

$$E\{X\} = -\sqrt{E_b} + 0 - 0 = -\sqrt{E_b},$$

and

$$\sigma_X^2 = E\left\{X^2\right\} - E^2\{X\} = E\left\{\left(-\sqrt{E_b} + W_1 - W_2\right)^2\right\} - E_b$$

$$= E_b + 0 + 0 + 0 + \sigma_{W1}^2 + \sigma_{W2}^2 - E_b = \frac{N_0}{2} + \frac{N_0}{2} = N_0,$$

respectively. The conditional density function of $X$, given that symbol $m_1 = 0$ is transmitted, is

$$f_X(x|0) = \frac{1}{\sqrt{2\pi N_0}} e^{-(x+\sqrt{E_b})^2/2N_0}.$$

These densities are shown in Figure D1.

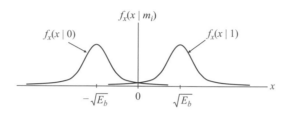

**Fig. D1** *Probability density functions.*

Suppose that symbol 0 was sent. The receiver will make an error, that is, decide in favour of 1 instead of 0, with the small probability

$$P(1|0) = P(x > 0|0) = \int_0^\infty f_X(x|0)\,dx = \frac{1}{\sqrt{2\pi N_0}} \int_0^\infty e^{-(x+\sqrt{E_b})^2/2N_0}\,dx.$$

Substituting variable $x$ with $z = \left(y + \sqrt{E_b}\right)/\sqrt{2N_0}$, we obtain

$$P(1|0) = \int_0^\infty f_X(x|0)\,dx = \frac{1}{\sqrt{\pi}} \int_{\sqrt{E_b/2N_0}}^\infty e^{-z^2}\,dz = \frac{1}{2}\mathrm{erfc}\sqrt{\frac{E_b}{2N_0}}.$$

If a transmission channel is symmetric, the average probability of error in a binary frequency-shift keying system is

$$p = \frac{1}{2}\mathrm{erfc}\sqrt{\frac{E_b}{2N_0}}.$$

...........................................................................................................................................

PROBLEMS

1. Calculate the power spectral density of the modulating signal $m(n)$ presented in Figure 1, using its autocorrelation function. Prove that the maximum normalized value of the power spectral density is 1. Plot the graph of the normalized power spectral density on a linear scale. Calculate the power of the signal in the frequency and time domains.

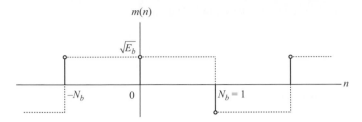

**Fig. 1** *A modulating polar binary signal presented in the time domain.*

2. Suppose the carrier is defined as the orthonormal basis function

$$s_c(n) = \phi_1(n) = \sqrt{\frac{2}{N_b}} \cos \Omega_c n \big).$$

Calculate the autocorrelation function of the carrier, taking into account the period of the carrier. Find the power spectral density and energy spectral density of the carrier as a function of the normalized frequency $F$. Calculate the energy and the power of the carrier in both the time and the frequency domains.

3. Using eqns (7.2.5) and (7.2.8), find the power spectral density of the modulated signal expressed in terms of the normalized frequency $F$. Then, find the power spectral density of the binary phase-shift keying modulated signal as the convolution of the power spectral densities of the carrier and the modulating signal.

4. Find the autocorrelation function of the carrier, treating it as a power deterministic signal. Calculate the power spectral density of the signal.

5. The power spectral density of a binary phase-shift keying modulated signal can be found as the Fourier transform of its autocorrelation function. To find the Fourier transform in this case, we can apply the modulation theorem that will give us the power spectral density of the bandpass binary phase-shift keying signal as a shifted power spectral density of the baseband signal, due to the fact that the autocorrelation function of the modulating process $M(n)$ is multiplied by the cosine function. Prove this statement.

6. The bit error probability $p$ of a binary phase-shift keying system, for a symmetric additive white Gaussian noise channel and for the same probability of generating binary values 1 and 0 at the output of the source, can be calculated by taking into account the distance between the message points in the constellation diagram. In this case, the calculated $p$ will be the same whether the assumed transmitted bit is 1 or 0. Suppose the sent bit is 0. Then, the decision variable is expressed as a realization of the random variable

$$X = +\sqrt{E_b} + W.$$

Prove that the bit error probability can be calculated as

$$p = \frac{1}{2}\text{erfc}\left(\sqrt{\frac{\eta_X^2}{2\sigma_X^2}}\right).$$

7. A component of the normalized power spectral density for a frequency-shift keying signal is expressed as

$$S_{MMNorm}(\Omega) = \frac{1}{4N_b^2}\left|\frac{\sin N_b(\Omega + \Omega_b/2)/2}{\sin(\Omega + \Omega_b/2)/2}\right|^2.$$

Find the expression for this normalized value at zero frequency. Calculate the values for $N_b = 2$, 4, and 6. Calculate the value of the spectrum $S_{MM}(\Omega)$ at the point $\Omega = \Omega_b/2$. Plot the graph of the power spectral density for $N_b = 8$, 16, 32, and 64.

8. Calculate the power of both the in-phase and the quadrature components given by expressions (7.4.17) and (7.4.18). Confirm the calculated power by calculating it in the time domain. Calculate the total power of the signal.

9. Suppose the carrier frequencies are defined as $(n + i)/N_b$. Find the values of $n$ that fulfil the condition expressed by

$$\Omega_j = \begin{cases} \Omega_c + 2\pi/2N_b & i = 2, m_i = 0 \\ \Omega_c - 2\pi/2N_b & i = 1, m_i = 1 \end{cases} = \Omega_c \pm \pi/N_b = \Omega_c \pm \Omega_b/2.$$

Find the lowest frequency of the carrier that fulfils the condition.

10. The power spectral density of the binary frequency-shift keying modulating signal is

$$\frac{S_{MM}(\Omega)}{E_b N_b} = \frac{\pi}{2N_b}\left[\delta\left(\Omega - \frac{\Omega_b}{2}\right) + \delta\left(\Omega + \frac{\Omega_b}{2}\right)\right] + \frac{1}{4N_b^2}\left|\frac{\sin N_b(\Omega - \Omega_b/2)/2}{\sin(\Omega - \Omega_b/2)/2}\right|^2$$
$$+ \frac{1}{4N_b^2}\left|\frac{\sin N_b(\Omega + \Omega_b/2)/2}{\sin(\Omega + \Omega_b/2)/2}\right|^2.$$

Express the power spectral density in terms of the normalized frequency $F$.

11. The one-sided power spectral density $S_{SS}(\Omega_n)$ of a modulated binary frequency-shift keying signal is expressed by eqn (7.4.31) as

$$S_{SS}(\Omega_n) = 2\frac{\pi}{N_b}[\delta(\Omega_n - \Omega_b/2) + \delta((\Omega - \Omega_c) + \Omega_b/2)]$$
$$+ \frac{1}{N_b^2}\left[\left|\frac{\sin N_b(\Omega_n - \Omega_b/2)/2}{\sin(\Omega_n - \Omega_b/2)/2}\right|^2 + \left|\frac{\sin N_b(\Omega_n + \Omega_b/2)/2}{\sin(\Omega_n + \Omega_b/2)/2}\right|^2\right].$$

Express the power spectral density in terms of the signal frequency $F_n$. Express the power spectral density function in the form

$$S_{SS}(N_b(F-F_c)) = \frac{E_b}{8} \frac{1}{N_b^2} \left| \frac{\sin \pi (N_b(F-F_c)+1/2)}{\sin \pi (N_b(F-F_c)+1/2)/N_b} \right|^2$$

and find its maximum value. Plot the graph of the power spectral density in the linear and logarithmic scales.

12. Find the power of the modulated frequency-shift keying signal expressed as

$$S_{SS}(F) = \frac{E_b}{8} [\delta (N_b(F-F_c)-1/2)+\delta (N_b(F-F_c)+1/2)]$$

$$+ \frac{E_b}{8} \left[\delta ((N_b(F+F_c)-1/2)+\delta (N_b(F+F_c)+1/2)\right]$$

$$+ \frac{E_b}{8} \frac{1}{N_b^2} \left[\left| \frac{\sin \pi (N_b(F-F_c)-1/2)}{\sin \pi (N_b(F-F_c)-1/2)/N_b} \right|^2 + \left| \frac{\sin \pi (N_b(F-F_c)+1/2)}{\sin \pi (N_b(F-F_c)+1/2)/N_b} \right|^2\right]$$

$$+ \frac{E_b}{8} \frac{1}{N_b^2} \left[\left| \frac{\sin \pi (N_b(F+F_c)-1/2)}{\sin \pi (N_b(F+F_c)-1/2)/N_b} \right|^2 + \left| \frac{\sin \pi (N_b(F+F_c)+1/2)}{\sin \pi (N_b(F+F_c)+1/2)/N_b} \right|^2\right].$$

13. The power spectral density of the modulated signal expressed by eqn (7.3.16) is calculated using the convolution of the power spectral density functions of the modulating signal and the carrier. The same power spectral density can be calculated as the Fourier transform of the autocorrelation function of the modulated signal. Derive the expression for the autocorrelation function of the modulated signal in one branch of the modulator, calculate the signal power, and present the procedure of this calculation and the basic expressions involved in this calculation.

14. Plot the graph of the magnitude spectral density of a discrete rectangular pulse with amplitude $\sqrt{E_b}$ and number of samples $N_b$, as shown in Figure 14.

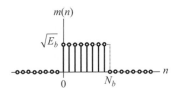

**Fig. 14** *A discrete rectangular pulse.*

a. Calculate the energy of the pulse in time and frequency domains for the pulse duration $N_b = 8$ and amplitude $\sqrt{E_b} = 1$. Calculate the total energy, the energy in one arcade, and the energy in two arcades.

b. Analyse the case when $N_b$ tends to infinity. Relate that case to the sampling interval $T_s = T/N_b$.

# 8

# Orthogonal Frequency Division Multiplexing and Code Division Multiple Access Systems

## 8.1 Introduction

Chapters 5, 6, and 7 contain the basic theory related to the mathematical description, modelling, analysis, and synthesis of digital and discrete communication systems. The systems are dedicated, in principle, to the transmission of a single-user message signal that is generated at the output of the message source. The single-user systems analysed in Chapters 6 and 7 are obtained by the simplification of the generic communication system schemes presented in Chapter 5, Figure 5.6.4, for discrete systems, and Figure 5.8.1, for digital systems. However, due to the generality of the theory presented, the same generic schemes can be used to design multicarrier or multi-user systems.

**The multicarrier system.** A discrete system, based on the theory in Chapters 2 and 5 and presented in Figure 5.6.4, can be used as a discrete multicarrier system, as will be explained using the system's block schematic in Figure 8.1.1. The system uses $J$ orthonormal carriers to transmit the source messages for a single user. The system transmitter transmits $J$ message symbols at the same time and effectively extends the duration of each message symbol by $J$ times. The message bits $\{m_i\}$ are generated from the source and transferred into the parallel form $\{m_{ij}\}$ by means of serial-to-parallel circuits. The bits form a set of modulating signals $\{s_{ij}\}$ that modulate $J$ carriers in this multicarrier system. The received signals can be processed using $J$ correlation demodulators, which will generate the estimates of the signals transmitted at their outputs $\{s_{ij} + w_j\}$. The system in Figure 8.1.1 presents a baseband multicarrier communication system block operating in discrete time, including a noise generator that can be designed to represent the communication channel in discrete time following the theoretical principles presented in Chapter 4. The same structure can be used to design a baseband system operating in the continuous-time domain.

A multicarrier system is advantageous with respect to the robustness to fading and intersymbol interference because the effect of the channel dispersion will be signficantly reduced. The bandwidth of the transmitted signal will increase with the increase in the number

*Discrete Communication Systems*. Stevan Berber, Oxford University Press. © Stevan Berber 2021.
DOI: 10.1093/oso/9780198860792.003.0008

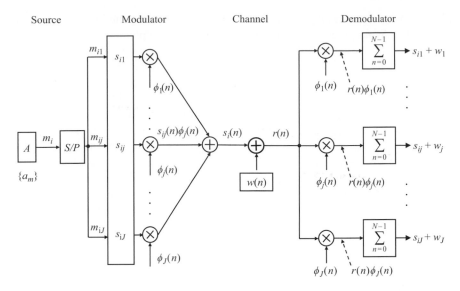

**Fig. 8.1.1** *Structure of a baseband discrete multicarrier communication system S/P, serial-to-parallel converter.*

of carriers. The carriers in a multicarrier system are called subcarriers, and related frequency bands belonging to them are called subbands. The total bandwidth of the multicarrier signal is equivalent to the sum of the subbands. This system can be designed using two main methods: *frequency division multiplexing and orthogonal frequency division multiplexing.*

In frequency division multiplexing systems, the message symbols sent in a particular subband will be demodulated and detected at the receiver side, using filters tuned to the subband carrier frequencies. In this case, the subbands cannot overlap and filters with steep transition characteristics should be used. The systems based on frequency division multiplexing principles, with some variations, were widely used in the early days of the development of analogue frequency division multiplexing multichannel telephone communication systems.

In digital orthogonal frequency division multiplexing systems, the orthogonal carriers are separated by a frequency $1/T_b$, where $T_b$ is the duration of a symbol, as we used in the simple binary frequency-shift keying system presented in Chapter 6, Section 6.4. In discrete orthogonal frequency division multiplexing systems, the orthogonal carriers are separated by a frequency $1/N_b$, where $N_b$ is the time duration of a discrete symbol, as we used in the simple discrete binary frequency-shift keying system presented in Chapter 7, Section 7.4. Therefore, the adjacent subbands can overlap and their separation at the receiver is performed using the orthogonality properties of the subcarriers, eliminating any need to use filters to detect the symbols. The multicarrier system presented in Figure 8.1.1, operating as a discrete orthogonal frequency division multiplexing system, can be used for the transmission of a single-user or a multi-user signal.

In this chapter, we will present three types of orthogonal frequency division multiplexing systems. The first type is traditionally called analogue orthogonal frequency division multiplexing, where processing of the modulating signal is presented using concepts from continuous-time signal processing. The second type can be implemented using the structure shown in Figure 8.1.1, with all the signals expressed as functions of the continuous-time variable $t$ rather than the discrete variable $n$. These systems can be termed 'digital' when the modulating signal is a binary or a multilevel digital signal that uses principles of continuous-time signal processing, as presented in Chapter 7. The third type is discrete orthogonal frequency division multiplexing, which is implemented using mainly the concepts of discrete-time signal processing, as presented in Chapter 7.

An orthogonal frequency division multiplexing system can be designed for baseband or bandpass signal transmission. The system presented in Figure 8.1.1 can be termed a baseband orthogonal frequency division multiplexing system. If the signal at the output of the baseband system is up-converted to the carrier frequency to be transmitted though a wireless channel, for example, the system will be called a bandpass orthogonal frequency division multiplexing system. This chapter will present both baseband and bandpass systems.

Another communication technology which is widely used for multi-user signal transmission is so-called code division multiple access technology. In such systems, the orthogonality principle is used at the level of message-symbol processing. The bits of a user message signal are spread using so-called orthogonal spreading sequences. The spreading is performed by multiplying a user message bit by the user-assigned orthogonal sequence of chips. The number of chips generated in a bit interval defines the processing gain of the spreading procedure. The result of single-user message bit spreading is a spreaded sequence of chips. The spreaded sequences of multiple users are added inside the transmitter baseband block, producing a modulating code division multiple access signal that modulates the carrier by using one of modulation methods mentioned in Chapters 6 or 7. The system can be implemented as a code division multiple access multicarrier system if the signal obtained modulates multiple orthogonal carriers. In this case, we may have a hybrid system that incorporates both code division multiple access and orthogonal frequency division multiplexing.

In this chapter, both orthogonal frequency division multiplexing systems and code division multiple access systems will be analysed and mathematical models for analogue, digital, and discrete signal transmission in these systems will be discussed.

## 8.2 Digital Orthogonal Frequency Division Multiplexing Systems

### 8.2.1 Introduction

It is possible to use the generic scheme of a digital communication system presented in Figure 5.8.1, or adopt the block schematic of a discrete communication system with a binary source shown in Figure 8.1.1, to obtain a digital orthogonal frequency division multiplexing system operating on continuous-time signals, as shown in Figure 8.2.1.

For the sake of explanation, the submodulators are implemented as multipliers, which is equivalent to the way binary phase-shift keying modulators are presented in Chapter 6, Section 6.2. Such modulators can be replaced by any of the modulators presented in Chapter 6. Likewise, the correlation subband demodulators and optimum detectors in

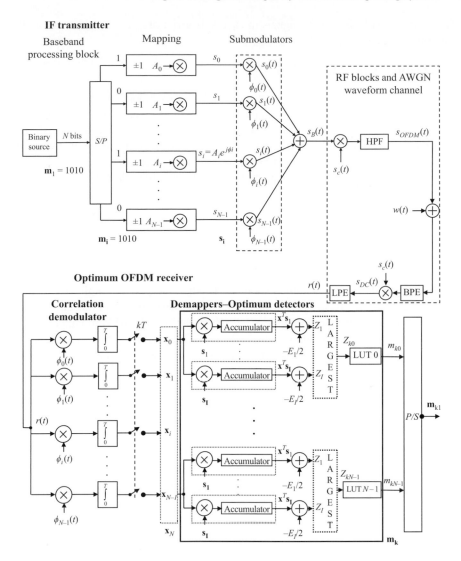

**Fig. 8.2.1** *Digital orthogonal frequency multiplexing system with a binary source, AWGN, additive white Gaussian noise; BPF, bandpass filter; HPF, high-pass filter; IF, intermediate frequency; LPF, lowpass filter; LUT, look-up table; OFDM, orthogonal frequency division multiple access; P/S, parallel-to-series converter; RF, radiofrequency; S/P, serial-to-parallel converter.*

Figure 8.2.1 are presented for binary phase-shift keying signals, even though they can be implemented using any of the modulation methods presented in Chapter 6. In this section, the operation of a digital orthogonal frequency division multiplexing system will be explained and the related signal processing operations in the time and frequency domains will be presented in mathematical form. The general explanation will be supported by an example of how binary phase-shift keying signals in each subcarrier are processed, assuming that the system uses four subcarriers.

## 8.2.2   Transmitter Operation

The transmitter is composed of a baseband signal processing block that includes the source of message bits and a serial-to-parallel converter, a mapper device, and a set of $N$ submodulators. Generally, the source can generate a stream of message bits $\{m_i\}$ that are transferred into $N$ substreams of $L$ bits by means of a serial-to-parallel converter. Each substream forms a message subsymbol that will be used at the final instance to modulate the corresponding subcarrier. For example, if $N = 4$ binary phase-shift keying modulators are used to form an orthogonal frequency division multiplexing system, the stream of four bits $\{m_i\} = (1010)$ will be generated from the source and transferred into $N$ parallel streams composed of $L = 1$ bits that can be 1 or 0. Each bit at the input of the corresponding mapper is a message subsymbol. This example is presented in Figure 8.2.1 and will be used in the analysis of the system.

It is important to notice that all the structures of the modulators in Figure 8.2.1 are made for a binary phase-shift keying system. However, in principle, they represent all the modulators mentioned in Chapter 6, or combinations thereof, which can all be implemented inside an orthogonal frequency division multiplexing transmitter structure. A similar rationale applies for the structure of the receiver in Figure 8.2.1, where the correlation demodulators are based on binary phase-shift keying demodulation. By replacing the demodulator multipliers with demodulators from other possibly used modulation techniques, we can see that the scheme has a general structure that is applicable for any modulation method that can be used in an orthogonal frequency division multiplexing system. If we implement any of the modulation methods mentioned in Chapter 6 to generate subband signals, we need to take care about the indices $i$ and $j$ that are used in the modelling of related modulators and demodulators in Chapters 6.

**The mapping procedure.** Generally speaking, submessages that are composed of $L$ bits are applied to the input of $N$ mappers. At the output of each mapper is a complex symbol that will be used to modulate the subcarrier, which is defined by the constellation diagram of the applied modulation method. Therefore, the $i$th complex symbol is expressed as

$$s_i = A_i e^{j\phi_i} = A_i \cos\phi_i + jA_i \sin\phi_i = I_i + Q_i, \qquad (8.2.1)$$

for the index $i = 0, 1, 2, \ldots, N - 1$, and is used to generate the modulating signal that modulates the $i$th subcarrier. Each $s_i$ in this notation corresponds to a vector of signal coefficients $\mathbf{s_i}$ used in Chapter 5, 6, and 7, and care must be taken about the notation in this chapter and the previous chapters. The modulation of each subcarrier can be performed using the methods presented in Chapters 6 and 7. By specifying amplitudes $A_i$ and phases $\phi_i$, various modulation methods can be implemented inside the orthogonal frequency division multiplexing submodulators. The parameters $A_i$ and $\phi_i$ specify a particular symbol and are related to the modulation method applied. If phase-shift keying is used, $A_i$ is a constant, and $\phi_i$ is specified by the $i$th message symbol. Likewise, if quadrature amplitude modulation is used, then both parameters $A_i$ and $\phi_i$ are specified by a complex symbol.

For example, if we implement binary phase-shift keying, the parameters inside a complex symbol will be specified and related to the message symbols as follows:

$$s_i = A_i e^{j\phi_i} = A \cos \phi_i = \left\{ \begin{array}{ll} A & \phi_i = 0 \text{ for } m_i = 1 \\ -A & \phi_i = \pi \text{ for } m_i = 0 \end{array} \right\}, \tag{8.2.2}$$

where $A$ is proportional to the energy of a bit, that is, $A = \sqrt{E_b}$ is the amplitude of the rectangular modulating pulse of duration $T = T_b$, as can be seen in Chapter 6, Section 6.2. Furthermore, the modulating signals at the inputs of the submodulators are random binary signals $m(t)$ presented in Chapter 7 in Figure 7.2.2, having magnitude and power spectral densities defined by eqns (7.2.4) and (7.2.5) and presented in Figures 7.2.3 and 7.2.4, respectively. It is important to note that the complex symbol defines the modulated signal coefficients and, consequently the constellation diagram of the related modulation method. For binary phase-shift keying, these coefficients will be specified using eqn (8.2.2) as

$$s_i = \left\{ \begin{array}{ll} A & \phi_i = 0 \\ -A & \phi_i = \pi \end{array} \right\} = \left\{ \begin{array}{ll} s_{11} = \sqrt{E_b} & \phi_i = 0 \\ s_{21} = -\sqrt{E_b} & \phi_i = \pi \end{array} \right\}. \tag{8.2.3}$$

The analogous relations between the complex symbols $s_i$ and the modulating signals can be established for other modulation methods presented in Chapter 6. It is important for the readers of this book to make an effort to establish these relations.

We can make a general summary of the mapping procedure of the orthogonal frequency division multiplexing system by referring to Figure 8.2.1. The set of complex symbols makes the message vector $\mathbf{m_i}$. The mappers map the message bits in $\mathbf{m_i}$ into the complex symbols $s_i$ in a symbol vector $\mathbf{s_i}$. The complex symbols in $\mathbf{s_i}$ define the modulating signals $s_i$ that modulate the subcarriers. After signal transmission and demodulation, the correlation values contained in vector $\mathbf{X_N}$ that correspond to the sent complex symbols $s_i$ are calculated. The correlation values inside $\mathbf{X_N}$ are then demapped to the message bits $\mathbf{m_k}$. Due to the noise in the channel, the content of $\mathbf{X_N}$ may be mapped into $\mathbf{m_k}$, which is different from $\mathbf{m_i}$. In that case, we say that bit errors have occurred in the signal transmission.

**Submodulation.** The complex symbol is defined in a subsymbol interval $T$, and the modulation is defined as a multiplication of the modulating symbol and the orthonormal subcarriers inside the $T$ interval; $N$ subsymbols are used to modulate $N$ subcarriers, which are defined by a set of orthonormal sinusoidal functions with frequencies defined by

$$f_i = \frac{i}{T}, \tag{8.2.4}$$

where $T$ is a symbol duration, and $i = 0, 1, 2, \ldots, N - 1$. The orthogonal subcarriers are expressed in their real and complex forms as

$$\phi_i(t) = \sqrt{2/T} \cos 2\pi \frac{i}{T} t = \operatorname{Re} \left\{ \sqrt{2/T} e^{j2\pi \frac{i}{T} t} \right\}. \tag{8.2.5}$$

The subcarrier frequencies are defined as carrier frequencies for the frequency-shift keying modulation method presented in Chapter 6. For the defined frequencies in eqn (8.2.4), the power spectral density of the signal starts at zero frequency. If we want the symbol spectra to start at a fixed frequency $f_{sc}$, the subcarrier frequencies will be defined by

$$f_i = f_{sc} + \frac{i}{T}, \tag{8.2.6}$$

where $i = 1, 2, \ldots, N - 1$, and $sc$ stands for subcarriers. The complex symbols define the modulating signals $s_i$ that modulate the subcarriers producing the modulated signals at the outputs of all of the submodulators. The modulated signals are added to each other in the time domain, as shown in Figure 8.2.1, and the signal is obtained at the output of the modulator, which is expressed in the time domain as

$$
\begin{aligned}
s_B(t) &= \operatorname{Re}\left\{\sum_{i=0}^{N-1} s_i \cdot \phi_i(t)\right\} = \operatorname{Re}\left\{\sqrt{\frac{2}{T}} \sum_{i=0}^{N-1} A_i e^{j\phi_i} \cdot e^{j2\pi \frac{i}{T}t}\right\} \\
&= \operatorname{Re}\left\{\sqrt{\frac{2}{T}} \sum_{i=0}^{N-1} A_i \cdot e^{j\left(2\pi \frac{i}{T}t + \phi_i\right)}\right\} = \sqrt{\frac{2}{T}} \sum_{i=0}^{N-1} A_i \cdot \cos\left(2\pi \frac{i}{T}t + \phi_i\right)
\end{aligned}
\tag{8.2.7}
$$

For the sake of consistency, we define the subcarriers as orthonormal functions. The amplitudes and phases inside the subcarriers are expressed in a generic form defined by the index $i$. However, these values are dependent on the modulation used (i.e. amplitude-shift keying, $M$-ary phase-shift keying, or quadrature amplitude modulation) as we will demonstrate for the example of binary phase-shift keying. The modulated subcarrier signals for various modulation schemes can be found in Chapter 6. Because the signals at the output of the submodulators are orthogonal, the power spectral density of the baseband orthogonal frequency division multiplexing signal is equal to the sum of the power spectral densities of all the submodulators. In the case of binary phase-shift keying, bearing in mind expression (6.2.19) in Section 6.2.2, the power spectral density of a baseband orthogonal frequency division multiplexing modulated signal, based on binary phase-shift keying modulation, can be expressed as

$$S_B(f) = \sum_{i=0}^{N-1} S_i(f) = \frac{E_b}{2} \sum_{i=0}^{N-1} \left(\operatorname{sinc}^2 (f - f_i) T_b + \operatorname{sinc}^2 (f + f_i) T_b\right), \tag{8.2.8}$$

for the complex symbol defined in the interval $T_b$ by eqn (8.2.3). The one-sided normalized power spectral density of baseband orthogonal frequency division multiplexing is shown in Figure 8.2.2, for $N = 4$. In the case of other modulation schemes, the parameters $E_b$ and $T_b$ and the number of addends for each subcarrier will be changed, as can be seen by looking at the expressions for the power spectral densities for various modulation methods presented in Chapter 6. The bandwidth of a baseband orthogonal frequency division multiplexing signal is $B = N/T_b$, that is, it increases linearly with the number of subcarriers. If we need to discretize the obtained signal, the Nyquist sampling frequency can be calculated as

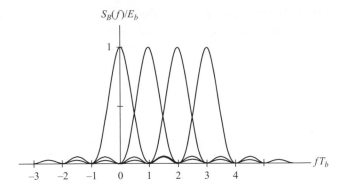

**Fig. 8.2.2** *Normalized power spectral density of an orthogonal frequency division multiplexing baseband signal.*

$$f_{Ny} = 2B = 2\frac{N}{T_b}.$$
(8.2.9)

The one-sided power spectral density of the baseband orthogonal frequency division multiplexing signal can be expressed in the general form

$$S_B(f) = \sum_{i=0}^{N-1} S_i(f) = E \cdot \sum_{i=0}^{N-1} \text{sinc}^2 (f - f_i) T_b,$$

defined for positive frequencies $f \geq 0$, where $E$ is the proportionality factor that depends on the modulation method used.

**Operation of a radiofrequency block.** The baseband signal obtained in an orthogonal frequency division multiplexing system is sometimes transmitted in the form presented in eqn (8.2.7) through the communication channel. If radio transmission is needed, the bandwidth of the modulated signal will be shifted to the carrier radiofrequency $f_{RF}$ by means of a mixer and a bandpass filter, as shown in Figure 8.2.1. This shift is called up-conversion. Sometimes the up-conversion of the signal is performed on an intermediate frequency, and then the spectrum of the obtained signal is shifted to a radio frequency. In the following analysis, we will exclude this up-conversion and discuss only direct radiofrequency up-conversion. The radiofrequency carrier is expressed in the form

$$s_c(t) = A_c \cos (2\pi f_{RF} t),$$
(8.2.10)

where the carrier frequency is defined as

$$f_{RF} = f_c - \frac{N-1}{2T},$$
(8.2.11)

where RF stands for radiofrequency; this guarantees that the subbands of the signal are distributed symmetrically around the nominal carrier frequency $f_c$. Therefore, the signal at the output of the radiofrequency modulator, which is implemented as an up-converter, is

$$s_{RF}(t) = s_B(t)s_c(t) = \text{Re}\left\{ A_c\sqrt{2/T} \sum_{i=0}^{N-1} A_i \cdot e^{j\left(2\pi \frac{i}{T}t+\phi_i\right)} e^{j2\pi f_{RF}t} \right\}$$

$$= \sum_{i=0}^{N-1} A_i\sqrt{2/T} \cos\left( 2\pi\frac{i}{T}t+\phi_i \right) \cdot A_c \cos(2\pi f_{RF}t)$$

$$= \sum_{i=0}^{N-1} \frac{A_c}{2} A_i\sqrt{2/T}\left[ \cos\left( 2\pi\left(f_{RF}+\frac{i}{T}\right)t+\phi_i \right) + \cos\left( 2\pi\left(f_{RF}-\frac{i}{T}\right)t-\phi_i \right) \right]$$

$$(8.2.12)$$

Suppose the amplitude of the radiofrequency carrier is $A_c = 2$, and the signal is filtered using a bandpass filter that eliminates the lower sideband of the modulated signal. Then, at the filter output, a signal is produced that is expressed as

$$s_{OFDM}(t) = \sum_{i=0}^{N-1} A_i\sqrt{2/T} \cos\left( 2\pi\left(f_{RF}+\frac{i}{T}\right)+\phi_i \right), \qquad (8.2.13)$$

where OFDM stands for orthogonal frequency division multiplexing signal. Bearing in mind the definition of radiofrequency, and the frequencies of the subcarriers, the position of the subbands with respect to the nominal frequency $f_c$ can be calculated according to the expression of the sideband subcarrier frequencies as

$$f_{iRF} = f_{RF} + \frac{i}{T} = f_c - \frac{N-1}{T} + \frac{i}{T}. \qquad (8.2.14)$$

In the case when the binary phase-shift keying is used, the signal at the output of the transmitter can be expressed as

$$s_{OFDM}(t) = \sum_{i=0}^{N-1} \sqrt{\frac{2E_b}{T_b}} \cos(2\pi f_{iRF}t+\phi_i) = \text{Re}\left\{ \sum_{i=0}^{N-1} \sqrt{\frac{2E_b}{T_b}} e^{j(2\pi f_{iRF}t+\phi_i)} \right\}. \qquad (8.2.15)$$

The normalized power spectral density of the modulated signal, corresponding to the power spectral density of the baseband signal defined by eqn (8.2.8) but for a one-sided spectrum, is presented in Figure 8.2.3. The frequencies of the subbands, for this practical case, are calculated using eqn (8.2.14) for $N = 4$.

**The two-sided power spectral density of a modulated orthogonal frequency division multiplexing signal.** The power spectral density of the carrier (8.2.10) can be calculated from its autocorrelation function, which is expressed as

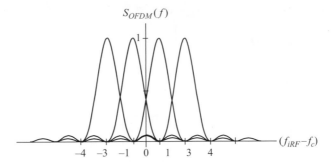

**Fig. 8.2.3** *One-sided normalized power spectral density of an orthogonal frequency division multiplexing modulated signal.*

$$
\begin{aligned}
R_{sc}(\tau) &= \frac{1}{T}\int_{-\infty}^{\infty} s_c(t+\tau)\,s_c(t)dt = \int_0^T A_c^2 \cdot \cos 2\pi f_{RF}(t+\tau) \cdot \cos 2\pi f_{RF}t\,dt \\
&= \frac{1}{T}A_c^2 \int_0^T \frac{1}{2}\left[\cos 2\pi f_{RF}\tau + \cos\left(4\pi f_{RF}' + 2\pi f_{RF}\tau\right)\right]dt = \frac{A_c^2}{2}\cos 2\pi f_{RF}\tau
\end{aligned}
\qquad (8.2.16)
$$

The Fourier transform of the autocorrelation function of the signal is its power spectral density, which can be expressed in the form

$$
\begin{aligned}
S_{sc}(f) &= \int_{-\infty}^{\infty} R_{sc}(\tau)\,e^{-j2\pi f\tau}\,d\tau = \frac{A_c^2}{2}\int_{-\infty}^{\infty}\cos 2\pi f_{RF}\tau \cdot e^{-j2\pi f\tau}\,d\tau \\
&= \frac{A_c^2}{4}\int_{-\infty}^{\infty}\left(e^{j2\pi f_{RF}\tau} + e^{-j2\pi f_{RF}\tau}\right)e^{-j2\pi f\tau}\,d\tau = \frac{A_c^2}{4}\left[\delta\left(f - f_{RF}\right) + \delta\left(f + f_{RF}\right)\right]
\end{aligned}
\qquad (8.2.17)
$$

which is the result obtained in Chapter 12, presented in Table 12.5.2, and used in Chapters 6 and 7. Because the autocorrelation function at the output of the radiofrequency modulator is the product of the autocorrelation function of the input baseband signal and the autocorrelation function of the carrier, the power spectral density of the modulated signal is the convolution of the power spectral densities of the input baseband signal expressed in eqn (8.2.8), and the power spectral density of the carrier. Therefore, the power spectral density of the modulated signal can be expressed as

$$
\begin{aligned}
S_{RF}(f) &= \int_{-\infty}^{\infty} S_B(f-\lambda)\,S_{sc}(\lambda)\,d\lambda \\
&= \int_{-\infty}^{\infty} \frac{E_b}{2}\sum_{i=0}^{N-1}\left(\operatorname{sinc}^2(f-\lambda-f_i)\,T_b + \operatorname{sinc}^2(f-\lambda+f_i)\,T_b\right)\cdot\frac{A_c^2}{4}\left[\delta(\lambda-f_{RF}) + \delta(\lambda+f_{RF})\right]d\lambda \\
&= \sum_{i=0}^{N-1}\frac{E_b}{2}\cdot\frac{A_c^2}{4}\int_{-\infty}^{\infty}\left(\operatorname{sinc}^2(f-\lambda-f_i)\,T_b + \operatorname{sinc}^2(f-\lambda+f_i)\,T_b\right)\left[\delta(\lambda-f_{RF}) + \delta(\lambda+f_{RF})\right]d\lambda
\end{aligned}
$$

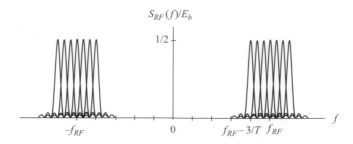

**Fig. 8.2.4** *Normalized two-sided power spectral density of an orthogonal frequency division multiplexing radio-frequency signal.*

Using the property of the integral of the Dirac delta function, the power spectral density can be obtained in the form

$$S_{RF}(f) = \frac{E_b}{2} \cdot \frac{A_c^2}{4} \sum_{i=0}^{N-1} \frac{\left(\text{sinc}^2\,(f - f_{RF} - f_i)\,T_b + \text{sinc}^2\,(f - f_{RF} + f_i)\,T_b\right)}{+ \left(\text{sinc}^2\,(f + f_{RF} - f_i)\,T_b + \text{sinc}^2\,(f + f_{RF} + f_i)\,T_b\right)}. \tag{8.2.18}$$

For the amplitude $A_c = 2$, and $N = 4$, the expression for the power spectral density of the two-sided radiofrequency signal can be simplified and presented in graphical form in Figure 8.2.4.

## 8.2.3   Receiver Operation

**Operation of the receiver when there is no noise in the channel.** The signal is transmitted through an additive white Gaussian noise channel and received by a correlation receiver. In order to explain the operation of the receiver, we will exclude for a while the influence of noise on the signal transmitted. Therefore, the signal received is passed through a bandpass filter, which would be important to eliminate noise and other impairments in a real system, and then mixed (multiplied) by a locally generated carrier to obtain the signal expressed in the form

$$r_{OFDM}(t) = s_c(t)s_{OFDM}(t) = A_c \cos\,(2\pi f_{RF} t) \sum_{i=0}^{N-1} A_i \sqrt{2/T} \cos\,(2\pi\,(f_{RF} + i/T)\,t + \phi_i)$$

$$= \sum_{i=0}^{N-1} \sqrt{2A_c^2/T} A_i \cos\,((2\pi i/T)\,t + \phi_i) + \sum_{i=0}^{N-1} \sqrt{2A_c^2/T} A_i \cos\,(4\pi\,(f_{RF} + i/T)\,t + \phi_i)$$

$$\tag{8.2.19}$$

which is passed through a lowpass filter to eliminate the second addend and produce the signal that will be demodulated in the correlation demodulator. For a carrier amplitude $A_c = 1$, we obtain the following signal at the inputs of the correlation demodulators:

$$r(t) = \sum_{i=0}^{N-1} A_i \sqrt{2/T} \cos\,((2\pi i/T)\,t + \phi_i). \tag{8.2.20}$$

Note that, while the structures of the correlation demodulators in Figure 8.2.1 are for a binary phase-shift keying system, they would be the same for any of the orthogonal carriers of the other modulation methods mentioned in Chapter 6. If we implemented any of these modulation methods in this system, we need to use the indices $i$ and $j$, as in Chapter 6. Therefore, the correlators in Figure 8.2.1 would need to be replaced by the demodulators defined for the modulation methods used.

The outputs of the demodulator correlators are the integral values taken at the time instants defined by the subsymbol duration $T$ and presented as vectors $\mathbf{x}_0, \ldots, \mathbf{x}_{N-1}$. At any discrete-time instant $kT$, where $k$ is a positive integer, the outputs of all of the correlators comprise one element of the vector $\mathbf{X_N}$ that contains all the vectors $\mathbf{x}_0, \ldots, \mathbf{x}_{N-1}$. The structure of these vectors, of course, depends on the modulation methods used to form the signal. In the case when binary phase-shift keying is used, each vector contains a single sample representing the related message bit. Assuming that binary phase-shift keying modulation has been used in the first subband, the demodulation of the first subsymbol defined by $i = 0$ results in

$$x_0 = \int_0^T \sqrt{2/T} \cos(2\pi f_0 t) \sum_{i=0}^{N-1} A_i \sqrt{2/T} \cos(2\pi f_i t + \phi_i) \, dt$$

$$= \frac{2}{T} \int_0^T (A_0 \cos(2\pi f_0 t) \cos(2\pi f_0 t + \phi_0)) \, dt + \frac{2}{T} \int_0^T \cos(2\pi f_0 t) \sum_{i=1}^{N-1} A_i \cos(2\pi f_i t + \phi_i) \, dt$$

The second integral of the sum will be 0 due to the orthogonality of the cosine terms. The first integral will have the value

$$x_0 = (2A_0/T) \int_0^T \cos(2\pi f_0 t) \cos(2\pi f_0 t + \phi_0) \, dt$$

$$= (2A_0/T) \int_0^T \frac{1}{2} \cos(\phi_0) + \cos(4\pi f_0 t + \phi_0) \, dt. \qquad (8.2.21)$$

$$= (A_0/T) \int_0^T \cos(\phi_0) \, dt = A_0 \cos(\phi_0)$$

According to the example of the message symbols (1010) and the binary phase-shift keying modulation presented in Figure 8.2.1, the first symbol sent was 1, which means that the phase of the related complex symbol would be $\phi_0 = 0$, and the amplitude would be $A_0 = \sqrt{E_b}$. Therefore, the calculated symbol value $s_0$ would be equal to the calculated demodulator value $x_0$, as expected, that is

$$s_0 = A_0 \cos\phi_0 = \sqrt{E_b} \cos 0 = \sqrt{E_b}. \qquad (8.2.22)$$

Note that we assumed that there is no noise in the channel. Consequently, the received signal is the same as the transmitted signal, and the decision device will generate the binary value 1 at the output. This is result expected, which confirms the proper operation of the transmitter and the receiver.

**Demappers: Optimum detectors.** The vector $\mathbf{X_N}$ defines all possible outputs of correlation demodulators, which is, in the general case, defined by the subvectors $\mathbf{x_0}$ to $\mathbf{x_{N-1}}$. If there is no noise in the channel, the elements of the vector $\mathbf{X_N}$, obtained at the output of the correlators, are identical to the complex symbols in the symbol vector $\mathbf{s_i}$ at the transmitter side, and the implementation of the receiver demappers is simple in this case. A demapper can be a look-up-table that maps the elements of $\mathbf{X_N}$ into a set of message bits $\mathbf{m_k}$. If there is no noise in the channel, the set $\mathbf{m_k}$ will be equal to the set $\mathbf{m_i}$ that is defined at the inputs of the mappers at the transmitter side.

This procedure can be used in practice to check the operation of a transceiver in its design, simulation, and implementation. However, in practical systems, noise and fading are present and so need to be taken into account. In the presence of noise, the elements of the matrix $\mathbf{X_N}$, obtained at certain time instant $kT$, are not identical to the corresponding elements inside the symbol vector $\mathbf{s_i}$. Therefore, the elements of $\mathbf{s_i}$ need to be estimated on the basis of the demodulated elements of $\mathbf{X_N}$ according to the procedures presented in Chapter 6. For this estimation, we may use optimum detectors, as presented in Figure 8.2.1.

**Operation of a receiver when noise is present in the channel.** For each subband signal, we need one detector. For example, in the $i$th subband, the incoming demodulated value $\mathbf{x_i}$ will be correlated with all possible signals that could be sent in that subband. For example, the first detector correlates the first demodulated values inside $\mathbf{x_0}$ with $\mathbf{I}$ possibly sent signals in the first subband, which are denoted by $\mathbf{s_1}$ to $\mathbf{s_I}$. The number $\mathbf{I}$ depends on the modulation method used in the subband, as can be seen in Chapter 6, which explains the operation of an optimum detector for various modulation methods. For example, the index is $\mathbf{I} = 2$ for binary phase-shift keying and has only one element $x_0$ inside $\mathbf{x_0}$. In this case, a single correlation value is applied to the input of the first detector, whereas, for quadriphase-shift keying, the index $\mathbf{I} = 4$ and two random values exist at the input of the related detector.

If we follow the definition of a generic scheme of a communication system with a correlation receiver in Chapter 5, we can see that the detectors in an orthogonal frequency division multiplexing system will use the principle of maximum likelihood estimation. The maximum likelihood estimates $Z_{ki}$, $i = 0, 1, 2, \ldots, N - 1$, are generated at the outputs of all of the detectors, following the principles of applied modulation techniques in a given subband, as shown in Figure 8.2.1. Using a set of $N$ look-up-tables, the corresponding message bits from the source are assigned to each estimate. Then, by using a parallel-to-series procedure, a stream of estimated bits is generated at the output of the receiver, $\mathbf{m_{k1}}$, which represents the best estimate of the stream of bits sent at the transmitter side, $\mathbf{m_1}$. In the case when no errors occur due to noise in the channel, the received and the sent streams of bits are equal to each other, that is, $\mathbf{m_{k1}} = \mathbf{m_1}$.

**Analog versus digital versus discrete orthogonal frequency division multiplexing systems.** The system presented in this section is sometimes called an analogue orthogonal frequency division multiplexing system. However, due to the digital nature of the modulating signal, which is represented as a random series of rectangular pulses in our analysis, we will call it a 'digital orthogonal frequency division multiplexing system'. The system will be termed an analogue orthogonal frequency division multiplexing system if the modulating signal is represented by a set of analogue signals (continuous-time signals), which is not the subject of our analysis. In contrast, if an orthogonal frequency division multiplexing system has modulating and baseband modulated signals processed in the discrete-time domain, we will refer to it as a discrete orthogonal frequency division multiplexing system. With this nomenclature, we will be able to distinguish analogue, digital, and discrete systems clearly.

In the next section, we will address issues involved in the design and mathematical modelling of a discrete orthogonal frequency division multiplexing system. In this book, for the sake of consistency, we use the term 'carriers' for the orthonormal basis function $\phi_j(t)$. By doing this, we can use the generic models of the transmitter and receiver from Chapter 5 to derive the structures of transceivers in practical modulation schemes; for example, the discrete binary phase-shift keying system presented in Chapter 7, Figure 7.2.1, was derived from the generic model presented in Chapter 5, Figure 5.6.4.

## 8.2.4 Operation of a Receiver in the Presence of Noise

In real orthogonal frequency division multiplexing communication systems the modulated signal is transmitted through an additive white Gaussian noise channel and then received by a correlation receiver. Both the signal and the noise present are passed through a bandpass filter, which eliminates noise beyond the bandwidth of the signal, and then are down-converted, as shown in Figure 8.2.1. The output of the down-converter can be expressed as

$$r_{DC}(t) = s_c(t) \cdot [s_{OFDM}(t) + w_{BP}(t)], \tag{8.2.23}$$

where $w_{BP}(t)$ is a sample function of the bandpass white Gaussian noise process $W_{BP}(t)$ of zero mean and power spectral density $N_0/2$, which has properties as presented in Chapter 4. The filtered signal plus noise is then multiplied by the locally generated carrier according to expression (8.2.19) to obtain the signal (8.2.23) in the form

$$r_{DC}(t) = s_c(t) \cdot s_{OFDM}(t) + s_c(t) \cdot w_{BP}(t)$$

$$= \left( \frac{A_c}{2} \sqrt{\frac{2}{T}} \sum_{i=0}^{N-1} A_i \cos\left((2\pi i/T)t + \phi_i\right) + \sum_{i=0}^{N-1} A_i \cos\left(4\pi f_{RF}t + it/T + \phi_i\right) \right) + s_c(t)w_{BP}(t)^,$$

$$\tag{8.2.24}$$

This signal is filtered through a lowpass filter to produce the received signal $r(t)$ that contains the baseband signal and the baseband noise $w_B(t)$. For a carrier amplitude $A_c = 2$, we may obtain the baseband received signal at the inputs of correlation demodulators as

$$r(t) = \sum_{i=0}^{N-1} A_i \sqrt{2/T} \cos\left((2\pi i/T)t + \phi_i\right) + w_B(t). \tag{8.2.25}$$

For the sake of explanation, let us assume that the structure of all of the correlation demodulators in Figure 8.2.1 is made for a system employing binary phase-shift keying. Then, the output of the $i$th correlator is

$$x_i = \int_0^T \sqrt{2/T} \cos{(2\pi f_i t)} \left( \sum_{i=0}^{N-1} A_i \sqrt{2/T} \cos{(2\pi f_i t + \phi_i)} + w_B(t) \right) dt$$

$$= \frac{2A_i}{T} \int_0^T \cos{(2\pi f_i t)} \left[ \cos{(2\pi f_0 t + \phi_0)} + w_B(t) \right] dt + \dots$$

$$+ \frac{2A_i}{T} \int_0^T \cos{(2\pi f_i t)} \left[ \cos{(2\pi f_i t + \phi_i)} + w_B(t) \right] dt$$

$$+ \dots + \frac{2}{T} \int_0^T \cos{(2\pi f_i t)} \left[ A_{N-1} \cos{(2\pi f_{N-1} t + \phi_{N-1})} + w_B(t) \right] dt$$

Due to the orthogonality of the subcarriers, the integral values are 0 for all terms except the *i*th one, which can be calculated as

$$x_i = \frac{2}{T} \int_0^T \cos{(2\pi f_i t)} \left[ A_i \cos{(2\pi f_i t + \phi_i)} + w_B(t) \right] dt$$

$$= \frac{2}{T} \int_0^T \cos{(2\pi f_i t)} \left[ A_i \cos{(2\pi f_i t + \phi_i)} + w_B(t) \right] dt$$

$$= \frac{2A_i}{2T} \int_0^T \left[ \cos{(\phi_i)} + \cos{(4\pi f_i t + \phi_i)} \right] dt + \frac{2}{A_i T} \int_0^T w_B(t) \cos{(2\pi f_i t)} dt$$

$$= A_i \cos{(\phi_i)} + w_i$$

(8.2.26)

where the first component is a deterministic quantity contributed by the transmitted signal $s_i$, called the signal part, and the second component, $w_i$, is the sample value of the random variable $W_j$ that arises because of the baseband noise $w_B(t)$, called the noise part. Generally, for any modulation method, the outputs of all demodulator correlators are the integral values taken at the time instants defined by the subsymbol duration $T$ and presented as vectors $\mathbf{x_0}, \dots,$ $\mathbf{x_{N-1}}$. At any discrete-time instant $kT$, where $k$ is a positive integer, the outputs of all of the correlators comprise one element of the vector $\mathbf{X_N}$ that contains all vectors $\mathbf{x_0}, \dots, \mathbf{x_{N-1}}$. The content of these vectors, of course, depends on the modulation methods used to form the signal. In the case when binary phase-shift keying is used, each of the vectors $\mathbf{x_0}, \dots, \mathbf{x_{N-1}}$ contains a single random sample representing the related message bit, which can be 1 or 0. For the example presented in Figure 8.2.1, the first symbol sent was 1, followed by 0, 1, and 0, which means that the phase of the first complex symbol was $\phi_0 = 0$, and the amplitude was $A_0 = \sqrt{E_b}$. Therefore, the calculated symbol value $x_0$ at the output of the first correlator is equal to the calculated demodulator symbol value $s_0$ plus the noise $w_0$. The outcomes of all four integrators are as follows:

$$x_0 = A_0 \cos{(\phi_0)} + w_0 = A_0 + w_0 = \sqrt{E_b} + w_0,$$

$$x_1 = A_1 \cos{\pi} + w_1 = -A_1 + w_1 = -\sqrt{E_b} + w_1,$$

$$x_2 = A_2 \cos{0} + w_2 = A_2 + w_2 = \sqrt{E_b} + w_2,$$

$$x_3 = A_3 \cos{\pi} + w_3 = -A_3 + w_3 = -\sqrt{E_b} + w_3.$$

(8.2.27)

The correlation values calculated in eqn (8.2.27) are statistically independent, because the symbols are transmitted inside different subbands and the modulation is performed on orthogonal carriers. Therefore, if we calculate the bit error probability $p$ for one subband, it will be the same as the bit error probability for any other subband. The integral value $x_0$ at the output of the first correlator is a realization of the Gaussian random variable $X_0$, because the integral of the noise in the subband will give the random value $w_0$, which is a realization of the Gaussian random variable $W_0$ defining the discrete-time Gaussian stochastic process $W_0(kT)$ at the output of the first correlator. The background theory related to the definitions of these variables and related stochastic processes can be found in Chapters 3, 4, 5, 6, and 19. To calculate the bit error probability in the first subchannel, we need to find the mean and the variance of the variable $X_0$. The mean is

$$\eta_0 = E\{X_0\} = \sqrt{E_b} + E\{W_0\} = \sqrt{E_b}, \qquad (8.2.28)$$

and the variance is

$$\sigma_{X_0}^2 = E\left\{X_0^2\right\} - \eta_0^2 = E\left\{W_0^2\right\} = \frac{N_0}{2}. \qquad (8.2.29)$$

Bearing in mind the explanation given in Chapter 6, Section 6.2.3, the bit error probability in the subchannel can be calculated as

$$p = \frac{1}{2}\operatorname{erfc}\left(\sqrt{\frac{\eta_0^2}{2\sigma_{X_0}^2}}\right) = \frac{1}{2}\operatorname{erfc}\left(\sqrt{\frac{E_b}{2N_0/2}}\right) = \frac{1}{2}\operatorname{erfc}\left(\sqrt{\frac{E_b}{N_0}}\right), \qquad (8.2.30)$$

where $E_b$ is the energy of a transmitted symbol, and $N_0/2$ is the two-sided power spectral density of the white Gaussian noise as defined in Chapter 4. The calculated probability in this subchannel is the same as the bit error probability in any of the subchannels. Due to the statistical independence between the symbols transmitted in all the subchannels, the bit error probability of the system will be equal to the bit error probability in any of the subchannels and can be expressed as eqn (8.2.30).

## 8.3 Discrete Orthogonal Frequency Division Multiple Access Systems

### 8.3.1 Principles of Discrete Signal Processing in an Orthogonal Frequency Division Multiple Access System

An orthogonal frequency division multiple access system can be implemented in the discrete-time domain via two approaches. In the first approach, the system can be implemented by adapting the basic digital system presented in Section 8.2.1 to a system based on discrete-time signal processing, using the principles presented in Chapter 7. Namely, in each subband, discrete-time signal processing will be performed according to the chosen modulation method. The message bits will then be processed to generate the discrete-time modulating signals which will modulate the subcarriers via multiplication in the discrete-time domain.

The baseband signal obtained will be modulated using an intermediate-frequency carrier. Then, the intermediate-frequency signal will be converted into an analogue intermediate-frequency signal by means of a digital-to-analogue converter. If necessary, the intermediate-frequency signal will be up-converted in the frequency domain to the analogue carrier frequency via a mixer. The received signal will then be down-converted to the intermediate frequency. The analogue intermediate-frequency signal will be converted to the discrete-time domain by means of an analogue-to-digital converter and then processed in a correlation demodulator and an optimal detector in the discrete-time domain, using the procedures presented in Chapter 7.

Implementation of an orthogonal frequency division multiple access system in the discrete-time domain assumes that the signal processing is performed using multipliers, oscillators, correlators, and detectors. Due to the complexity of the system based on this kind of processing, this implementation became obsolete, and a new system, based on the use of discrete Fourier transform, was developed; this will be the subject of this section.

**Basic model of a discrete orthogonal frequency division multiple access system.** The bandpass signal at the output of the transmitter in an orthogonal frequency division multiple access system is defined by eqn (8.2.12), which can be expressed in its complex form as

$$s_{RF}(t) = \text{Re}\left\{ \sum_{i=0}^{N-1} A_i e^{j\phi_i} \cdot \sqrt{\frac{2}{T}} e^{j\left(2\pi \frac{i}{T} t\right)} A_c e^{j2\pi f_{RF} t} \right\}. \tag{8.3.1}$$

The terms in the sum can be understood as a product of the signal envelope and the carrier signal at the transmitter mixer operating at radio frequency $f_{RF}$. The baseband signal in its complex form is equal to the envelope of the radiofrequency signal and can be expressed as

$$s_B(t) = \sum_{i=0}^{N-1} A_i e^{j\phi_i} \cdot \sqrt{\frac{2}{T}} e^{j\left(2\pi \frac{i}{T} t\right)} = \sum_{i=0}^{N-1} s_i \cdot \sqrt{\frac{2}{T}} e^{j\left(2\pi \frac{i}{T} t\right)}, \tag{8.3.2}$$

for the defined complex symbols $s_i$ and the defined symbol duration $T$. The real part of the complex envelope is the baseband signal expressed as

$$s_B(t) = \text{Re}\left\{ \sum_{i=0}^{N-1} A_i e^{j\phi_i} \cdot \sqrt{\frac{2}{T}} e^{j\left(2\pi \frac{i}{T} t\right)} \right\} = \text{Re}\left\{ \sum_{i=0}^{N-1} s_i \cdot \sqrt{\frac{2}{T}} e^{j\left(2\pi \frac{i}{T} t\right)} \right\}. \tag{8.3.3}$$

A continuous-time baseband signal in an orthogonal frequency division multiple access system, in the complex form (8.2.3), can be sampled with the uniform sampling time instants $T_s = T/N$ defining the sampling frequency $f_s = N/T$, following the principles that will be presented in Chapter 13. The number of samples $N$ is defined here as the number of subcarrier signals. This definition will be addressed later on in this section.

To transfer the signal into its discrete-time form, we may express the continuous-time variable $t$ as $t = nT_s = nT/N$ and normalize the amplitudes of the complex baseband signal by $N$. The complex signal then has the form

$$s_B(n) = \frac{1}{N} \sum_{i=0}^{N-1} A_i e^{j\phi_i} \cdot \sqrt{\frac{2}{N}} e^{j\left(2\pi \frac{i}{N}n\right)} = \frac{1}{N} \sum_{i=0}^{N-1} s_i \cdot \sqrt{\frac{2}{N}} e^{j\left(2\pi \frac{i}{N}n\right)}, \tag{8.3.4}$$

for $n = 0, 1, 2, \ldots, N - 1$. If we compare this expression with expression (15.4.24) in Chapter 15, we can see that it is equivalent to the inverse Fourier transform of a signal $s(n)$ which has its discrete Fourier transform $s_i$ that can be expressed using the first relation in eqn (15.4.24) in the form

$$s_i = \sum_{n=0}^{N-1} s_B(n) \cdot \sqrt{\frac{2}{N}} e^{-j\left(2\pi \frac{i}{N}n\right)}, \tag{8.3.5}$$

for $i = 0, 1, 2, \ldots, N - 1$. Therefore, we can design a discrete-time orthogonal frequency division multiple access system by simply implementing the inverse discrete Fourier transform at the transmitter to get the discrete-time orthogonal frequency division multiple access signal $s_B(n)$, and then performing the discrete Fourier transform at the receiver to obtain the complex symbols $s_i$. We can speed up the processing in these two transforms by using one of the fast Fourier algorithms, primarily the fast Fourier transform algorithm presented in Chapter 15 in Section 15.5.3. In this chapter, we will use the terms 'discrete Fourier transform' and 'inverse discrete Fourier transform' instead of 'fast Fourier transform' and 'inverse fast Fourier transform' because the processing in an orthogonal frequency division multiple access system is not limited to fast Fourier algorithms, as can be seen from the solutions of the problems presented in this chapter.

**The issue with sampling frequency.** According to the expression for an orthogonal frequency division multiple access signal, the number of samples for the signal inside its interval $T$ is $N$, resulting in the sampling frequency $T_s = N/T$. However, the Nyquist sampling frequency, which guarantees the reconstruction procedure without distortion, is calculated by eqn (8.2.9) to be $T_{Ny} = 2N/T$, for $T = T_b$. Two problems have to be resolved here: Firstly, we should design a system that will be able to generate a discrete-time orthogonal frequency division multiple access signal with at least the minimum number of samples defined by the Nyquist frequency and directly allow the processing and reconstruction of signals without any distortion. Secondly, if we need to do modulation and demodulation at an intermediate frequency, as we suggested in Chapter 7, the number of samples inside the modulating signal before the carrier modulation should comply with the number of samples required in a carrier period, which is achieved by the up-sampling and interpolation procedure. Therefore, the number of samples inside a discrete-time orthogonal frequency division multiple access symbol should comply with the Nyquist theorem to allow its up-sampling and reconstruction without any distortion.

These problems can be solved by adding $N$ zero-valued complex symbols that will, in turn, increase the number of samples inside the signal in the discrete-time domain. By adding $N$ complex symbols, we ensure that the Nyquist criterion will be fulfilled and the number of samples inside a symbol interval $N_s = 2N$ will fulfil the condition of the Nyquist sampling frequency that becomes $T_s = 2N/T = N_s/T = T_{Ny}$.

## 8.3.2    A Discrete Baseband Orthogonal Frequency Division Multiple Access System Based on Binary Phase-Shift Keying

For the sake of explanation, we will analysis two different cases. In the first case, the system has $N = 4$ subcarriers which are modulated via binary phase-shift keying, where the source message is $m_1 = (1010)$, as presented in Figure 8.2.1, and the duration of a symbol in real time is $T = 1$ sec. In the second case, the number of complex symbols is extended by adding $N$ zero-valued symbols to form a signal with $N_s = 2N$ complex symbols. Consequently, the baseband signal will be defined by $N_s$ samples in the discrete-time domain.

**First case.** In this case, the number of subcarriers is equivalent to the number of symbol samples in the time domain. Using eqn (8.2.2) as the expression for the complex symbol, we can calculate complex symbols with amplitudes $A_b$ as

$$s_i = A_i e^{j\phi_i} = A_b \cos \phi_i = \left\{ \begin{array}{ll} A_b & \phi_i = 0 \text{ for } m_i = 1 \\ -A_b & \phi_i = \pi \text{ for } m_i = 0 \end{array} \right\}. \tag{8.3.6}$$

Then, we can find the symbols defined for $m_1 = (1010)$ and for equal amplitudes of all symbols and express them as

$$s_0 = A_0 e^{j0} = A_b, s_1 = A_1 e^{j\pi} = -A_b, s_2 = A_2 e^{j0} = A_b, \text{ and } s_3 = A_3 e^{j\pi} = -A_b. \tag{8.3.7}$$

These complex symbols define the corresponding symbols of the orthogonal frequency division multiple access signal expressed in a general form by eqn (8.2.7) and for the discrete case in eqn (8.3.4), given here as

$$s_B(n) = \frac{1}{N} \sum_{i=0}^{N-1} s_i \cdot \sqrt{\frac{2}{N}} \cos \left( 2\pi \frac{i}{N} n \right), \tag{8.3.8}$$

which can be represented as the sum of the signals in all the subbands as

$$s_B(n) = \frac{1}{N} \left( A_b - A_b \cdot \cos \left( 2\pi \frac{1}{N} n \right) + A_b \cdot \cos \left( 2\pi \frac{2}{N} n \right) - A_b \cdot \cos \left( 2\pi \frac{3}{N} n \right) \right), \tag{8.3.9}$$

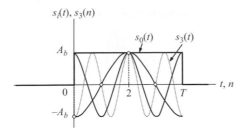

**Fig. 8.3.1** *Signal components in the continuous time domain of an orthogonal frequency division multiplexing signal.*

where amplitudes $A_b$ are related to the symbol energy, which can be calculated as

$$A_b = \sqrt{\frac{2E_b}{N}}. \tag{8.3.10}$$

The graphs of all the normalized signals in the continuous-time domain are presented in Figure 8.3.1. If we sample all the signals with the number of samples $N = 4$, as was done in Figure 8.3.1, visual inspection can prove that the Nyquist criterion is not fulfilled for all of the signals. The problematic signal is $s_3(t)$ (the thinnest graph), shown in Figure 8.3.1. For this signal, we presented the four sampling instants and the four sampled amplitude values by the dotted lines in the same figure. Obviously, the signal cannot be reconstructed with these four samples. We say that the aliasing occurs in the system, which is theoretically explained in Chapter 13. The Nyquist sampling frequency for this signal should be $f_s = 2f_3 = 6/T$, that is, we need to have at least $N_s = 6$ samples per a symbol of duration $T$. The number of samples $N = 4$ would be sufficient for symbols $s_0(t)$, $s_1(t)$, and $s_2(t)$, as can be seen in Figure 8.3.1. For this example, the normalized baseband signal in the discrete-time domain is defined by the discrete frequencies $F_i = i/2N$, or $\Omega_i = 2\pi F_i$, for $i = 0, 1, 2, 3$, and expressed as

$$s_B(n) \cdot 4 = A_b - A_b \cdot \cos(\pi n/2) + A_b \cdot \cos(\pi n) - A_b \cdot \cos(3\pi n/2). \tag{8.3.11}$$

**Second case.** In this case, the number of subcarriers is doubled, and the corresponding number of symbol samples in the time domain is doubled. If we extend the number of complex symbols to $N_s = 2N = 8$ by adding $N$ additional zero-valued symbols, the symbols in the time domain can be calculated as in eqn (8.3.4). Bearing in mind relations (8.3.6) and (8.3.8), the signals in the time domain can be calculated as

$$s_B(n) = \frac{1}{N_s} \sum_{i=0}^{N_s-1} s_i \cdot \sqrt{\frac{2}{N_s}} \cos\left(2\pi \frac{i}{N_s} n\right) = \frac{1}{N_s} \sum_{i=0}^{7} A_i e^{j\phi_i} \cdot \sqrt{\frac{2}{8}} \cos\left(2\pi \frac{i}{8} n\right)$$

$$= \frac{1}{N_s} \left(\frac{A_0}{2} \cos\left(2\pi \frac{0}{8} n\right) - \frac{A_1}{2} \cos\left(2\pi \frac{1}{8} n\right) + \frac{A_2}{2} \cos\left(2\pi \frac{2}{8} n\right) - \frac{A_3}{2} \cos\left(2\pi \frac{3}{8} n\right)\right).$$

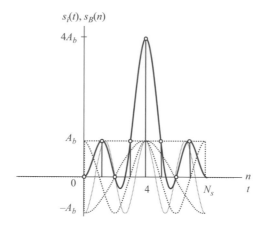

**Fig. 8.3.2** *Signal components in the discrete time domain of an orthogonal frequency division multiplexing signal.*

Because the amplitude of the discrete-time subcarriers have the same amplitudes $A_b$, the normalized baseband signal can be expressed as

$$s_B(n) \cdot N_s = A_b - A_b \cos\left(\pi \frac{1}{4}n\right) + A_b \cos\left(\pi \frac{1}{2}n\right) - A_b \cos\left(\pi \frac{3}{4}n\right). \qquad (8.3.12)$$

The samples of the signal in the discrete-time domain can be calculated as $s_B(0) = 0$, $s_B(1) = s_B(3) = s_B(5) = s_B(7) = A_b$, and $s_B(4) = 4A_b$. The discrete-time baseband signal is presented via its samples in Figure 8.3.2. In the same figure, the continuous-time baseband signal (solid line) is presented alongside with the related subband signals (dotted lines).

Of course, we may find all the possible baseband signals that could be generated at the input of the intermediate-frequency block. In this particular example, for the $N = 4$ bits at the output of the serial-to-parallel converter, the number of signals is equal to the number of variations, with repetition calculated as $2^4 = 16$. Following this example, and our discussion related to the number of samples, we can say that an orthogonal frequency division multiple access system can be implemented by adding $N$ zero-valued complex symbols, which, in turn, produce the same number of samples inside the baseband signal that now fulfils the Nyquist criterion. Having this procedure in mind, we will explain the operation of an orthogonal frequency division multiple access system in the next section.

### 8.3.3 Structure and Operation of a Discrete Orthogonal Frequency Division Multiple Access System

Figure 8.3.3 shows the structure of a discrete orthogonal frequency division multiple access system. In this system, the transmitter is composed of a baseband signal processing block, mappers, an inverse discrete Fourier transform block, intermediate-frequency modulators, and radiofrequency blocks followed by the communication channel.

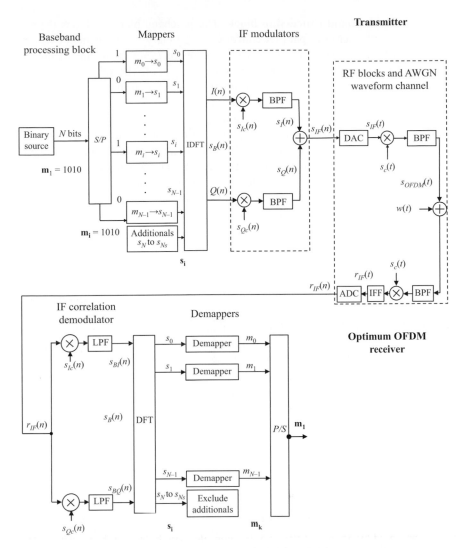

**Fig. 8.3.3** *A discrete orthogonal frequency division multiplexing system with a binary source; ADC, analogue-to-digital converter; AWGN, additive white Gaussian noise; BPF, bandpass filter; DAC, digital-to-analogue converter; DFT, discrete Fourier transform block; HPF, high-pass filter; IDFT, inverse discrete Fourier transform block; IF, intermediate frequency; IFF, intermediate-frequency filter; LPF, lowpass filter; LUT, look-up table; OFDM, orthogonal frequency division multiple access; P/S, parallel-to-series converter; RF, radiofrequency; S/P, serial-to-parallel converter.*

**Operation of the baseband processing block.** The baseband block includes the source of the message bits and a serial-to-parallel converter. The binary source generates a stream of message bits $\{m_i\}$ that is transferred into $N$ substreams of $L$ bits by means of the serial-to-parallel converter. Each substream forms a message subsymbol. The subsymbols are mapped into complex symbols using $N$ mappers that operate according to the constellation diagram of the related modulation methods implemented inside the subbands. The complex symbols at the output of the mappers are applied to the input of an inverse discrete Fourier transform block. At the same time, $N$ zero-valued complex symbols (called 'additionals') are added to the input of the inverse discrete Fourier transform block and are used to increase the number of samples in the baseband signal to avoid aliasing, as we have already explained.

For example, if the system operates with $N = 4$ subcarriers, and binary phase-shift keying modulators are used to form the system, a stream containing the four bits $\{m_i\}$, $i = 0, 1, 2, 3$, will be generated from the source and transferred into $N$ parallel streams composed of $L = 1$ bit that can be 1's or 0's, as shown in Figure 8.3.3. Each bit at the input of the corresponding mapper is a message subsymbol. These bits are mapped into a set of $N = 4$ complex symbols $s_i$, as defined by expression (8.2.2). In addition, at the input of the inverse discrete Fourier transform block, an additional set of $N = 4$ zero-valued symbols will be added, as we have already mentioned. This example is illustrated in Figure 8.2.3 for the message steam $\{m_i\} = (1010)$.

The example given in Figure 8.3.3 assumes that binary phase-shift keying is used in the system. However, any of the modulation schemes mentioned in Chapter 7 can be used here to generate the signal. The inverse discrete Fourier transform block uses $N_s = 2N$ symbols from its input and produces a discrete-time baseband signal at its output that is expressed in complex form by eqn (8.3.4). Because the number of complex symbols is increased to $N_s$, the discrete-time signal at the output of the inverse discrete Fourier transform block is

$$s_B(n) = \frac{1}{N_s} \sum_{i=0}^{N_s-1} s_i \cdot e^{j(2\pi i n / N_s)}. \tag{8.3.13}$$

Generally speaking, submessages composed of $L$ bits are applied to the inputs of $N$ mappers. At the output of each mapper is a complex symbol that will be used to modulate the subcarrier. As in eqn (8.2.1), the $i$th complex symbol is expressed as

$$s_i = A_i e^{j\phi_i} = A_i \cos \phi_i + j A_i \sin \phi_i, \tag{8.3.14}$$

for the new index values $i = 0, 1, 2, \ldots, N_s - 1$, and is used to generate the modulating signal that modulates the $i$th subcarrier. Each $s_i$ in this notation is defined by the amplitude of the subcarrier $A_i$ and the related phases $\phi_i$. Modulation of each subcarrier can be performed using the methods mentioned in Chapter 7, which will specify the amplitudes $A_i$ and phases $\phi_i$.

Separating the real and the imaginary parts in eqn (8.3.13), we obtain the discrete-time signal expressed in terms of in-phase and quadrature components as

$$s_B(n) = \frac{1}{N_s} \sum_{i=0}^{N_s-1} A_i e^{j\phi_i} \cdot e^{j(2\pi in/N_s)}$$

$$= \frac{1}{N_s} \sum_{i=0}^{N_s-1} A_i \cos(\Omega_i n + \phi_i) + j \frac{1}{N_s} \sum_{i=0}^{N_s-1} A_i \sin(\Omega_i n + \phi_i), \quad (8.3.15)$$

$$= \mathrm{Re}\{s_B(n)\} + j\,\mathrm{Im}\{s_B(n)\} = I(n) + jQ(n)$$

where $\Omega_i = 2\pi i/Ns = 2\pi F_i i$ is the frequency of both the in-phase component, $I(n)$, and the quadrature component, $Q(n)$, of the signal. There are two ways to process the baseband signal. Firstly, the signal can be transferred through the channel, as we explained in detail in Section 8.2.1 for a digital orthogonal frequency division multiple access system, where we directly modulate the radiofrequency carrier with the baseband signal. Secondly, as we mentioned before, the baseband signal $s_B(n)$ can modulate the carrier at the discrete-time intermediate frequency $\Omega_{IF}$, and then the intermediate-frequency signal obtained can be shifted to a radiofrequency signal for transmission through the channel by mixing the intermediate-frequency signal with a continuous-time radiofrequency carrier with frequency $f_{RF}$. In our analysis, bearing in mind the reasons behind the design of the discrete system presented in Chapter 7, and the one that will be presented in Chapter 10, we will present the operation of an orthogonal frequency division multiple access system that includes modulation on an intermediate frequency.

**Operation of the intermediate-frequency block.** The intermediate-frequency carriers in the in-phase and quadrature branches are expressed in the forms

$$s_{Ic}(n) = A_{IF} \cos \Omega_{IF} n,$$

$$s_{Qc}(n) = -A_{IF} \sin \Omega_{IF} n, \quad (8.3.16)$$

where the frequency of the radiofrequency carrier is defined in terms of its sampling rate according to the expression $\Omega_{IF} = 2\pi/N_{IF} = 2\pi F_{IF}$. Based on the theory that will be presented in Chapter 15, and considering the carrier to be a deterministic periodic function, we can calculate its autocorrelation function as

$$R_{if}(l) = \frac{1}{N_s} \sum_{n=0}^{N_s-1} \left[ A_{IF}^2 \cos \Omega_{IF}(n+l) \cdot \cos \Omega_{IF} n \right] = \frac{A_{IF}^2}{N_s} \sum_{n=0}^{N_s-1} \frac{1}{2} [\cos \Omega_{IF} l + \cos \Omega_{IF}(2n+l)]$$

$$= \frac{A_{IF}^2}{N_s} \sum_{n=0}^{N_s-1} \cos \Omega_{IF} l = \frac{A_{IF}^2}{2} \cos \Omega_{IF} l$$

Then, the power spectral density can be calculated as the Fourier transform of the autocorrelation function as

$$S_{IF}(\Omega) = \frac{A_{IF}^2}{2} \pi [\delta(\Omega - \Omega_{IF}) + \delta(\Omega + \Omega_{IF})] = \frac{A_{IF}^2}{4} [\delta(F - F_{IF}) + \delta(F + F_{IF})],$$

where the carrier frequencies are related as $F_{IF} = \Omega_{IF}/2\pi = 1/N_{IF}$, where $N_{IF}$ is the number of samples in the intermediate-frequency carrier.

The theory that will be presented in Chapter 15, for deterministic signals, and the related theory presented in Chapter 3, for stochastic signals, are essential for understanding the operation of a discrete communication system. In addition, the theoretical principles related to the design of discrete transceivers and how to apply these principles in practice, avoiding possible serious errors in the design, need to be understood in depth.

The complex baseband signal given in eqn (8.3.15) contains both real and imaginary parts that need to be modulated separately. The real part will modulate the in-phase carrier as

$$
s_{Ic}(n) \cdot \mathrm{Re}\{s_B(n)\} = \frac{1}{N_s}\sum_{i=0}^{N_s-1} A_i A_{IF} \cos\Omega_{IF} n \cos(\Omega_i n + \phi_i)
$$

$$
= \frac{A_{IF}}{2N_s}\sum_{i=0}^{N_s-1} A_i[\cos((\Omega_{IF}+\Omega_i)n+\phi_i) + \cos((\Omega_{IF}-\Omega_i)n-\phi_i)]
$$

and the signal obtained at the output of the in-phase bandpass filter is

$$
s_I(n) = \frac{A_{IF}}{2N_s}\sum_{i=0}^{N_s-1} A_i \cos((\Omega_{IF}+\Omega_i)n+\phi_i). \tag{8.3.17}
$$

The imaginary part will modulate the quadrature carrier defined in eqn (8.3.16) as

$$
s_{Qc}(n) \cdot \mathrm{Im}\{s_B(n)\} = \frac{1}{N_s}\sum_{i=0}^{N_s-1} A_i A_{IF} \sin(\Omega_i n + \phi_i)(-\sin\Omega_{IF} n)
$$

$$
= \frac{A_{IF}}{2N_s}\sum_{i=0}^{N_s-1} A_i[-\cos((\Omega_{IF}-\Omega_i)n-\phi_i) + \cos((\Omega_{IF}+\Omega_i)n+\phi_i)]
$$

and the signal obtained at the output of the quadrature bandpass filter is

$$
s_Q(n) = \frac{A_{IF}}{2N_s}\sum_{i=0}^{N_s-1} A_i \cos((\Omega_{IF}+\Omega_i)n+\phi_i). \tag{8.3.18}
$$

The signal at the output of the intermediate-frequency modulator is the sum of the in-phase and the quadrature components and is expressed as

$$
s_{IF}(n) = s_I(n) + s_Q(n) = \frac{A_{IF}}{N_s}\sum_{i=0}^{N_s-1} A_i \cos((\Omega_{IF}+\Omega_i)n+\phi_i). \tag{8.3.19}
$$

**Operation of the radiofrequency block.** The discrete-time intermediate-frequency signal is converted into its continuous-time domain via a digital-to-analogue converter, as shown in Figure 8.3.3. The theoretical procedure for transferring the discrete-time variable into a continuous-time variable can be found in Chapter 13, Section 13.5. Following that procedure, we may express the continuous variable $t$ as $t = nT_s = nT/N$ for the assumed sampling interval $T_s$ and the duration of a symbol $N$. Then, for the defined frequencies $\Omega_{IF} = 2\pi/N_{IF} = 2\pi F_{IF}$ and $\Omega_i = 2\pi i/N_s = 2\pi F_i i$, we may calculate the continuous-time frequencies as

$$\Omega_i n = 2\pi i n/N_s = 2\pi i t/T_s = 2\pi f_s i t = 2\pi f_i t = \omega_i t$$
$$\Omega_{IF} n = 2\pi n/N_{IF} = 2\pi t/T_{IF} = 2\pi f_{IF} t = \omega_{IF} t$$

(8.3.20)

respectively, and then calculate the continuous-time intermediate-frequency signal as

$$s_{IF}(t) = \frac{A_{IF}}{N_s} \sum_{i=0}^{N_s-1} A_i \cos\left((\omega_{IF} + \omega_i)t + \phi_i\right).$$

(8.3.21)

The intermediate-frequency signal is then up-converted to the carrier frequency via a mixer, as shown in Figure 8.3.3. To shift the signal to the carrier frequency, the mixer carrier signal is defined as

$$s_c(t) = A_c \cos\left(\omega_c - \omega_{IF}\right)t,$$

(8.3.22)

which is multiplied by the incoming intermediate-frequency signal to obtain

$$s_c(t)s_{IF}(t) = \frac{A_{IF}A_c}{N_s} \sum_{i=0}^{N_s-1} A_i \cos\left((\omega_{IF} + \omega_i)t + \phi_i\right) A_c \cos\left(\omega_c - \omega_{IF}\right)t$$
$$= \frac{A_{IF}A_c}{2N_s} \sum_{i=0}^{N_s-1} A_i \left[\cos\left((\omega_c + \omega_i)t + \phi_i\right) + \cos\left((\omega_c - 2\omega_{IF} - \omega_i)t - \phi_i\right)\right]$$

The mixer output is filtered via a bandpass filter to obtain the bandpass signal at the output of the transmitter; this signal is expressed as

$$s_{OFDM}(t) = \frac{A_{IF}A_c}{2N_s} \sum_{i=0}^{N_s-1} A_i \cos\left((\omega_c + \omega_i)t + \phi_i\right).$$

(8.3.23)

The signal is then transmitted through an additive white Gaussian noise channel, which is mathematically described in detail in Chapter 4.

## 8.3.4 Operation of the Receiver in an Orthogonal Frequency Division Multiple Access System

We will analyse the operation of the receiver in the absence of noise in the communication channel. At the input of the receiver, the signal received is mixed with a carrier with the frequency $(\omega_c - \omega_{IF})$ to obtain the following signal, which is expressed as

$$
s_c(t)s_{OFDM}(t) = A_c \cos{(\omega_c - \omega_{IF})} t \frac{A_{IF}A_c}{2N_s} \sum_{i=0}^{N_s-1} A_i \cos{((\omega_c + \omega_i)t + \phi_i)}
$$

$$
= \frac{A_{IF}A_c^2}{4N_s} \sum_{i=0}^{N_s-1} A_i [\cos{((2\omega_c - \omega_{IF} + \omega_i)t + \phi_i)} + \cos{((\omega_{IF} + \omega_i)t + \phi_i)}]
$$

$(8.3.24)$

The signal is then filtered via an intermediate-frequency filter to obtain an intermediate-frequency continuous-time signal, which is expressed as

$$
r_{IF}(t) = \frac{A_{IF}A_c^2}{4N_s} \sum_{i=0}^{N_s-1} A_i \cos{((\omega_{IF} + \omega_i)t + \phi_i)}.
$$

$(8.3.25)$

Using the relations in eqn (8.3.20), the signal can be converted back to the discrete-time domain via an analogue-to-digital converter to give the discrete intermediate-frequency signal, which is expressed as

$$
r_{IF}(n) = \frac{A_{IF}A_c^2}{4N_s} \sum_{i=0}^{N_s-1} A_i \cos{((\Omega_{IF} + \Omega_i)n + \phi_i)}.
$$

$(8.3.26)$

If we use the carrier amplitude $A_c = 2$, the receiver intermediate-frequency signal is equal to the intermediate-frequency signal obtained at the transmitter side and expressed by eqn (8.3.19), that is,

$$
r_{IF}(n) = s_{IF}(n) = \frac{A_{IF}}{N_s} \sum_{i=0}^{N_s-1} A_i \cos{((\Omega_{IF} + \Omega_i)n + \phi_i)}.
$$

$(8.3.27)$

**The intermediate-frequency correlation demodulator block.** The intermediate-frequency signal is demodulated in an intermediate-frequency demodulator block that is implemented as a correlation demodulator. The signal is multiplied by the locally generated synchronous intermediate-frequency carrier, as defined in eqn (8.3.16), to obtain the signal in the in-phase branch, which is expressed as

$$r_{IF}(n)s_{Ic}(n) = \frac{A_{IF}^2}{N_s} \sum_{i=0}^{N_s-1} A_i \cos\left((\Omega_{IF} + \Omega_i)n + \phi_i\right) \cos\left(\Omega_{IF}n\right)$$

$$= \frac{A_{IF}^2}{2N_s} \sum_{i=0}^{N_s-1} A_i \left[\cos\left(\Omega_i n + \phi_i\right) + \cos\left((2\Omega_{IF} + \Omega_i)n + \phi_i\right)\right]$$

This signal is passed through a lowpass filter to get the baseband in-phase signal

$$s_{BI}(n) = \frac{A_{IF}^2}{2N_s} \sum_{i=0}^{N_s-1} A_i \cos\left(\Omega_i n + \phi_i\right). \tag{8.3.28}$$

Likewise, using relation (8.3.16), the intermediate-frequency signal is demodulated in the quadrature branch according to the expression

$$r_{IF}(n)s_{Qc}(n) = \frac{A_{IF}^2}{N_s} \sum_{i=0}^{N_s-1} A_i \cos\left((\Omega_{IF} + \Omega_i)n + \phi_i\right) \left(-\sin\Omega_{IF}n\right)$$

$$= \frac{A_{IF}^2}{2N_s} \sum_{i=0}^{N_s-1} A_i \left[\sin\left(\Omega_i n + \phi_i\right) - \sin\left((2\Omega_{IF} + \Omega_i)n + \phi_i\right)\right]$$

The signal obtained is passed through a quadrature lowpass filter to obtain the baseband signal, which is expressed as

$$s_{BQ}(n) = \frac{A_{IF}^2}{2N_s} \sum_{i=0}^{N_s-1} A_i \sin\left(\Omega_i n + \phi_i\right), \tag{8.3.29}$$

as shown in Figure 8.3.3. If the intermediate-frequency amplitude is $A_{IF}^2/2 = \sqrt{2/N_s}$, the signal obtained in the in-phase branch can be expressed in a complex form that is equivalent to the baseband signal expressed in eqn (8.3.15) and proved here as

$$s_{BI}(n) = \frac{1}{N_s} \sum_{i=0}^{N_s-1} A_i \sqrt{\frac{2}{N_s}} \cos\left(2\pi \frac{i}{N_s}n + \phi_i\right) = \mathrm{Re}\left(\frac{1}{N_s} \sum_{i=0}^{N_s-1} A_i e^{j\phi_i} \sqrt{\frac{2}{N_s}} e^{j\left(2\pi \frac{i}{N_s}n\right)}\right)$$

$$= \mathrm{Re}\left(\frac{1}{N_s} \sum_{i=0}^{N_s-1} s_i \cdot \sqrt{\frac{2}{N_s}} e^{j\left(2\pi \frac{i}{N_s}n\right)}\right) = I(n)$$

Likewise, the signal in the quadrature branch can be calculated as

$$s_{BQ}(n) = \frac{A_{IF}}{2N_s} \sum_{i=0}^{N_s-1} A_i \sin(\Omega_i n + \phi_i) = \frac{1}{N_s} \sum_{i=0}^{N_s-1} A_i \sqrt{\frac{2}{N_s}} \sin\left(2\pi \frac{i}{N_s} n + \phi_i\right)$$

$$= \mathrm{Im}\left(\frac{1}{N_s} \sum_{i=0}^{N_s-1} A_i e^{j\phi_i} \sqrt{\frac{2}{N_s}} e^{j\left(2\pi \frac{i}{N_s} n\right)}\right) = \mathrm{Im}\left(\frac{1}{N_s} \sum_{i=0}^{N_s-1} s_i \cdot \sqrt{\frac{2}{N_s}} e^{j\left(2\pi \frac{i}{N_s} n\right)}\right) = Q(n)$$

for $n = 0, 1, 2, \ldots, N_s - 1$. Having available both the real and the imaginary parts of the baseband signal in the discrete-time domain, we may obtain the signal at the output of the correlation demodulator in its complex form, which is defined as

$$s_{BI}(n) + j s_{BQ}(n) = \frac{1}{N_s} \sum_{i=0}^{N_s-1} s_i \cdot \sqrt{\frac{2}{N_s}} e^{j\left(2\pi \frac{i}{N_s} n\right)} = s_B(n). \tag{8.3.30}$$

**The discrete Fourier transform block.** For a demodulated discrete-time baseband orthogonal frequency division multiple access system, we can use the discrete Fourier transform to calculate the complex symbols. As shown by eqn (8.3.5), the discrete Fourier transform of $s_B(n)$, expressed by eqn (8.3.30), will give us the complex symbols $s_i$ of the baseband signal according to the expression

$$s_i = \sum_{n=0}^{N_s-1} s_B(n) \cdot \sqrt{\frac{2}{N_s}} e^{-j\left(2\pi \frac{i}{N_s} n\right)}, \tag{8.3.31}$$

for $i = 1, 2, 3, \ldots, N_s - 1$. The calculated values of the discrete-time signal $s_B(n)$ are then processed in the discrete Fourier transform block to calculate the complex symbols $s_i$, as shown in Figure 8.3.3.

**Demappers and the parallel-to-series block.** Demappers are used to map the complex symbols $s_i$ into message bits for all the subcarriers, $m_0$ to $m_{N-1}$. The additional $N$ symbols are eliminated from the set of $N_s$ symbols. The first $N$ complex symbols are mapped into the set of parallel message symbols at $N$ demapper outputs. If there is noise in the channel, the set of bits received, $\mathbf{m_k}$, will be different from the set of bits sent, $\mathbf{m_j}$. In our analysis, noise was excluded, so the bits received will be identical to the bits sent.

In simple words, we can say the following: by means of a parallel-to-series block, parallel message symbols $s_i$ are converted into a series of message bits $\mathbf{m_k}$. If there have been no errors in transmission, the series of bits received, $\mathbf{m_k}$, will be identical to the stream of bits generated from the source, $\mathbf{m_j}$.

In this book, for the sake of consistency, we use the term 'carrier' both for the orthonormal basis function $\phi_j(t)$ in the continuous-time domain and for the basis functions $\phi_j(n)$ in the discrete-time domain. By defining the term in this way, we can use the theoretical presentation and generic mathematical model of the transmitter and receiver presented in Chapter 5 to deduce the transceiver structures for practical modulation methods. The same approach was used in Chapters 6 and 7 to derive the binary phase-shift keying system, the quadriphase-shift keying system, the frequency-shift keying system, and the quadrature amplitude modulation, using the generic model developed in Chapter 5.

## 8.3.5   Operation of the Receiver in the Presence of Noise

The signal received after transmission through a noisy channel can be expressed as the sum of the signal expressed in eqn (8.3.23) and the additive white Gaussian noise $w(t)$, in the form

$$s_{OFDM}(t) = \frac{A_{IF}A_c}{2N_s} \sum_{i=0}^{N_s-1} A_i \cos((\omega_c + \omega_i)t + \phi_i) + w(t). \tag{8.3.32}$$

The received signal is passed through a bandpass filter to eliminate noise beyond the bandwidth of the signal, and then is mixed with a carrier with the frequency $(\omega_c - \omega_{IF})$ to obtain the signal, which is expressed as

$$s_c(t) \cdot s_{OFDM}(t) = A_c \cos(\omega_c - \omega_{IF})t \left[ \frac{A_{IF}A_c}{2N_s} \sum_{i=0}^{N_s-1} A_i \cos((\omega_c + \omega_i)t + \phi_i) + w(t) \right]$$

$$= \frac{A_{IF}A_c^2}{4N_s} \sum_{i=0}^{N_s-1} A_i [\cos((2\omega_c - \omega_{IF} + \omega_i)t + \phi_i) + \cos((\omega_{IF} + \omega_i)t + \phi_i)] + w(t)A_c \cos(\omega_c - \omega_{IF})t$$

The signal and the noise are then filtered via an intermediate-frequency filter. The noise $w(t)$ is defined in an infinite bandwidth. Multiplying the additive white Gaussian noise by a carrier with the frequency $(\omega_c - \omega_{IF})$ preserves the bandwidth defined by the intermediate-frequency filter and affects the bandwidth of each of the subcarrier signals. If we had modelled the noise $w(t)$ as a bandpass noise at an intermediate frequency, as explained in detail in Chapter 4, and used that model for the simulation of the system, we would have obtained the same result. Therefore, the output of the bandpass intermediate-frequency filter is an intermediate-frequency continuous-time signal plus the noise. The noise affects each baseband signal component independently, so that

$$r_{IF}(t) = \frac{A_{IF}A_c^2}{4N_s} \sum_{i=0}^{N_s-1} [A_i \cos((\omega_{IF} + \omega_i)t + \phi_i) + w(t)A_N \cos(\omega_{IF} + \omega_i)t], \tag{8.3.33}$$

where $A_N$ defines the level of noise with respect to the subband frequency components. Using the relations in (8.3.20), the intermediate-frequency noisy signal can be converted back to a discrete-time signal via an analogue-to-digital converter to obtain the discrete received intermediate-frequency signal, which is expressed as

$$r_{IF}(n) = \frac{A_{IF}A_c^2}{4N_s} \sum_{i=0}^{N_s-1} A_i \cos((\Omega_{IF} + \Omega_i)n + \phi_i) + w(n)A_N \cos(\Omega_{IF} + \Omega_i)n. \tag{8.3.34}$$

If we use the carrier amplitude $A_c = 2$, the received intermediate-frequency signal is equal to the intermediate-frequency signal obtained at the transmitter side, as expressed in eqn (8.3.19), plus the noise in the signal bandwidth, and is expressed as

$$r_{IF}(n) = \frac{A_{IF}}{N_s} \sum_{i=0}^{N_s-1} [A_i \cos((\Omega_{IF} + \Omega_i) n + \phi_i) + w(n) A_N \cos(\Omega_{IF} + \Omega_i) n]. \qquad (8.3.35)$$

**The intermediate-frequency demodulator block.** The intermediate-frequency signal is multiplied by the locally generated synchronous intermediate-frequency carrier and then filtered in a lowpass filter (see Figure 8.3.3) to obtain the signal in the in-phase branch, as in eqn (8.3.28), with an added noise component:

$$s_{BI}(n) = \frac{A_{IF}^2}{2N_s} \sum_{i=0}^{N_s-1} [A_i \cos(\Omega_i n + \phi_i) + w(n) A_N \cos \Omega_i n]. \qquad (8.3.36)$$

Likewise, using the carriers defined by relation (8.3.16), the signal in the quadrature branch at the output of the lowpass filter may be obtained as

$$s_{BQ}(n) = \frac{A_{IF}^2}{2N_s} \sum_{i=0}^{N_s-1} [A_i \sin(\Omega_i n + \phi_i) + w(n) A_N \sin \Omega_i n], \qquad (8.3.37)$$

as shown in Figure 8.3.3. If the intermediate-frequency amplitude is $A_{IF}^2/2 = \sqrt{2/N_s}$, the signal obtained in the in-phase branch can be expressed in a complex form that is equivalent to that for the baseband signal expressed in eqn (8.3.4) with noise added. The in-phase signal can be derived as

$$
\begin{aligned}
s_{BI}(n) &= \frac{1}{N_s} \sum_{i=0}^{N_s-1} \left[ A_i \sqrt{\frac{2}{N_s}} \cos(2\pi i n/N_s + \phi_i) + w_i \sqrt{\frac{2}{N_s}} \cos(2\pi i n/N_s) \right] \\
&= \mathrm{Re} \left\{ \frac{1}{N_s} \sum_{i=0}^{N_s-1} \left[ A_i e^{j\phi_i} \sqrt{\frac{2}{N_s}} e^{j(2\pi i n/N_s)} + w_i \sqrt{\frac{2}{N_s}} e^{j(2\pi i n/N_s)} \right] \right\} \\
&= \mathrm{Re} \left( \frac{1}{N_s} \sum_{i=0}^{N_s-1} (s_i + w_i) \cdot \sqrt{\frac{2}{N_s}} e^{j(2\pi i n/N_s)} \right) = \mathrm{Re}\{s_B(n)\}
\end{aligned}
$$

and, by following the same procedure, the signal in the quadrature branch can be derived as

$$s_{BQ}(n) = \frac{1}{N_s} \left[ \sum_{i=0}^{N_s-1} A_i \sqrt{\frac{2}{N_s}} \sin\left(2\pi in/N_s + \phi_i\right) + w_i \sqrt{\frac{2}{N_s}} \sin\left(2\pi in/N_s\right) \right]$$

$$= \text{Im} \left( \frac{1}{N_s} \sum_{i=0}^{N_s-1} A_i e^{j\phi_i} \sqrt{\frac{2}{N_s}} e^{j(2\pi in/N_s)} + w_i \sqrt{\frac{2}{N_s}} e^{j(2\pi in/N_s)} \right) \quad ,$$

$$= \text{Im} \left( \frac{1}{N_s} \sum_{i=0}^{N_s-1} (s_i + w_i) \cdot \sqrt{\frac{2}{N_s}} e^{j\left(2\pi \frac{i}{N_s} n\right)} \right) = \text{Im}\{s_B(n)\}$$

for $n = 0, 1, 2, \ldots, N_s - 1$. Using the real and the imaginary parts of the baseband signal $s_B(n)$ in the discrete-time domain, we may obtain the signal at the output of correlation demodulator in its complex form:

$$s_{BI}(n) + js_{BQ}(n) = \frac{1}{N_s} \sum_{i=0}^{N_s-1} (s_i + w_i) \cdot \sqrt{\frac{2}{N_s}} e^{j(2\pi in/N_s)} = s_B(n). \tag{8.3.38}$$

**A discrete Fourier transform block.** Using the demodulated discrete-time baseband signals $s_B(n)$, we can calculate the complex symbols with noise, $(s_i + w_i)$, using the discrete Fourier transform as follows:

$$s_i + w_i = \sum_{n=0}^{N_s-1} s_B(n) \cdot \sqrt{\frac{2}{N_s}} e^{-j\left(2\pi \frac{i}{N_s} n\right)}, \tag{8.3.39}$$

for $i = 0, 1, 2, \ldots, N_s - 1$. The calculated values of the discrete-time signal $s_B(n)$ are processed in a discrete Fourier transform block, and the noisy complex symbols $(s_i + w_i)$ are calculated, as shown in Figure 8.3.3 for the signal parts.

**Demappers and the parallel-to-series block.** Demappers map the noisy complex symbols $(s_i + w_i)$ into the message bits for all the subcarriers, $m_0$ to $m_{N-1}$. The additional $N$ noisy symbols are eliminated from the set of $N_s$ symbols calculated, and the first $N$ complex symbols are mapped into the set of parallel message symbols at $N$ demapper outputs. Due to the noise in the symbols, this mapping can be wrong, that is, a message symbol that has not been generated from the source can be assigned to the noisy symbols. In this case, we know that noise is present in the channel and that the set of bits received, $\mathbf{m_k}$, may be different from the set of bits sent, $\mathbf{m_i}$.

In simple words, we can say the following: by means of a parallel-to-series block, the parallel noisy symbols $(s_i + w_i)$ are converted into the series of message bits $\mathbf{m_k}$. If there have been no errors in transmission, the series of bits received that is generated, $\mathbf{m_k}$, is identical to the stream of bits generated from the source, $\mathbf{m_i}$. However, if there have been errors in transmission, $\mathbf{m_k}$ will be different from $\mathbf{m_i}$.

**Note on the bit error probability.** The discrete Fourier transform is a linear operation. Therefore, the discrete Fourier transform block preserves the power relationship between the signal part and the noise part. Consequently, the bit error probability of an orthogonal frequency division multiple access system will be equivalent to the value calculated for the modulation technique used in the related subbands, if the same modulation method is implemented in all the subbands.

## 8.4   Discrete Code Division Multiple Access Systems

### 8.4.1   Principles of Operation of a Discrete Code Division Multiple Access System

In ordinary direct-sequence code division multiple access communication systems, all users transmit their message bits using their own orthogonal signals. These signals can be the Walsh functions and Walsh sequences, chaotic sequences, or other random or semi-random sequences. The signals can spread the spectrum of the transmitted signals and enhance the security and anti-jamming capabilities of a communication system. For the sake of explanation, we will present the theory of operation of a so-called forward-link code division multiple access system that is used in mobile communications, as shown in Figure 8.4.1.

In this system, a number of user signals are combined at the base station transmitter and transmitted through the channel as a multi-user signal. That signal is received in each mobile receiver. Using the orthogonal sequence assigned to each user, the message content belonging to a user is extracted by the procedure of dispreading. For the sake of explanation, all receivers are presented in the same figure, even though they belong to users who are scattered in space around the base station.

The system in Figure 8.4.1 can be deduced from the generic scheme presented for the discrete system in shown in Chapter 5 in Figure 5.6.4. Because the system has multiple users,

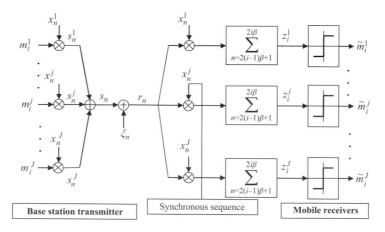

**Fig. 8.4.1** *Code division multiple access system structure and signal processing.*

we will replace the serial-to-parallel converter in Figure 5.6.4 by $J$ independent sources of message bits belonging to $J$ users. The orthogonal basis functions $\phi_j(n)$ will be replaced by the orthogonal spreading sequences $x_n^j$. The receiver will have only a correlation demodulator, with the orthogonal spreading sequences $x_n^j$ replacing the orthogonal basis functions $\phi_j(n)$.

**Transmitter operation.** At the input of the transmitter are sequences of message bits from $J$ users. The $i$th bit of the $j$th user is denoted by $m_i^j$, which can have one of two values from the set $\{-1, +1\}$. The users' sequences are multiplied by $J$ orthogonal spreading sequences $x_n^j$.

The spreaded signals of all the users are added to each other, resulting in the transmitter signal

$$s_n = \sum_{j=1}^{J} s_n^j = \sum_{j=1}^{J} m_i^j x_n^j, \tag{8.4.1}$$

where the $j$th user sample is expressed as

$$s_n^j = m_i^j x_n^j = \begin{cases} -1 \cdot x_n^j & m_i^j = -1 \\ +1 \cdot x_n^j & m_i^j = +1 \end{cases}, \tag{8.4.2}$$

and the spreading sequence $x_n^j = \left( x_1^j, x_2^j, \ldots, x_{2\beta}^j \right)$ is a series of $2\beta$ samples called chips. The number of chips per message bit is $2\beta$, which defines the processing gain of the system. For example, if the spreading sequence is the discrete-time Walsh sequence, as presented in Chapter 2, Figure 2.5.1, the processing gain will be $2\beta = 8$. The spreaded sequence of the $j$th user, for the two message bits $m_1^j = 1$ and $m_2^j = -1$ from the same user, will be as presented in Figure 8.4.2. Therefore, for the positive message bit ($i = 1$), we just generate the Walsh sequence as it is. For a negative message bit ($i = 2$), we invert the Walsh sequence. To multiply

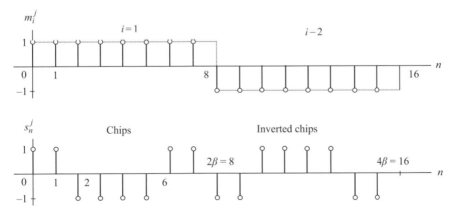

**Fig. 8.4.2** *Interpolated discrete bits and corresponding spread sequence.*

the message sequence by the spreading chip sequence in this example, we need to interpolate each message bit with seven additional samples.

**Modulation and demodulation.** In a practical wireless communication system, the signal obtained, eqn (8.4.1), modulates the carrier. The modulated signal is transmitted through the white Gaussian noise channel, and then demodulated at the receiver side. That process shifts the baseband signal spectrum at the carrier frequency and returns it back to the baseband at the output of the receiver demodulator. To explain the operation of a code division multiple access system, we will exclude the modulation and demodulation in our theoretical analysis. Therefore, without losing generality in our explanation, we will analyse only the baseband system presented in Figure 8.4.1.

**Characterization of the channel.** Because the system under consideration operates via discrete-time baseband signals, the baseband channel is represented by the discrete noise samples $\xi_n$, which can be generated using one of the methods explained in Chapter 4. These discrete-time noise samples are added to the baseband discrete signal, as shown in Figure 8.4.1. In the case of system simulation or emulation, the channel simulator will be a noise generator that produces discrete noise samples, as explained in Chapter 4. The signal samples at the output of the channel and at the input of the receivers can be expressed as

$$r_n = \sum_{j=1}^{J} s_n^j + \xi_n = \sum_{j=1}^{J} m_i^j x_n^j + \xi_n. \tag{8.4.3}$$

**Receiver operation.** Processing of the signal received is identical for each user. For that reason, we will demonstrate the operation of the receiver only for the $j$th user. Samples of the multi-user received signal $r_n$, taken at time instants $n$, will be correlated by a locally generated synchronous spreading sequence defined for the $j$th user. The correlation is performed by multiplying the multi-user signal by a locally generated spreading sequence and then summing all $2\beta$ products inside a bit interval. The result of these operations is a correlation sample for the $i$th bit at the output of the $j$th correlator; this is expressed as

$$\begin{aligned}
z_i^j &= \sum_{n=2\beta(i-1)+1}^{2\beta i} r_n^j x_n^j = \sum_{n=2\beta(i-1)+1}^{2\beta i} m_i^j \left(x_n^j\right)^2 + \sum_{n=2\beta(i-1)+1}^{2\beta i} \sum_{g=1, g\neq j}^{J} m_i^g x_n^g x_n^j + \sum_{n=2\beta(i-1)+1}^{2\beta i} x_n^j \xi_n \\
&= \sum_{n=2\beta(i-1)+1}^{2\beta i} m_i^j \left(x_n^j\right)^2 + \sum_{g=1, g\neq j}^{J} m_i^g \sum_{n=2\beta(i-1)+1}^{2\beta i} x_n^g x_n^j + \sum_{n=2\beta(i-1)+1}^{2\beta i} x_n^j \xi_n
\end{aligned} \tag{8.4.4}$$

This correlation operation can be considered to be a dispreading procedure. The $2\beta$ chips inside a received bit are affected in the channel by the statistically independent noise samples. Therefore, it is sufficient to analyse the transmission and reception of one user bit and then generalize the result to all received bits. For that reason, let us analyse the procedure for the demodulation and detection of the first bit sent by the $j$th user, in the bit interval denoted by $i = 1$. From eqn (8.4.4) the first correlation sample of the $j$th user can be expressed as

$$z_1^j = \sum_{n=1}^{2\beta} m_1^j \left(x_n^j\right)^2 + \sum_{g=1, g\neq j}^{J} m_1^g \sum_{n=1}^{2\beta} x_n^g x_n^j + \sum_{n=1}^{2\beta} x_n^j \xi_n = z_A + z_B + z_C. \qquad (8.4.5)$$

In this expression, $z_A$, $z_B$, and $z_C$ are realizations of the random variables $Z_A$, $Z_B$, and $Z_C$. Generally, it is better to say that they are realizations of the random functions $Z_A$, $Z_B$, and $Z_C$, which sometimes depend on a large number of random variables. The random value in eqn (8.4.5) is calculated at the end of the first bit interval defined by the spreading factor $2\beta$. Let us consider the case when binary phase-shift keying is used in the system. In this case, the calculated correlation value is the decision variable that is presented to the decision circuit, as shown in Figure 8.4.1. The decision about the message bit sent is made according to the following rule:

$$\tilde{m}_1^j = \left\{ \begin{array}{ll} +1 & z_1^j \geq 0 \\ -1 & z_1^j < 0 \end{array} \right\}. \qquad (8.4.6)$$

The random correlation sample $z_1^j$ in eqn (8.4.5) is a realization of the random variable $Z_1^j$, as defined for the binary phase-shift keying system in Chapter 7. A set of random values $z_i^j$ for $i = 1, 2, 3, \ldots$ can be considered as one realization of the discrete-time stochastic process $Z_i^j$, which is equivalent to the process $X_j(kN)$ at the output of the correlation receiver presented in Chapter 5 in Figure 5.5.2.

Furthermore, in this case, when a multi-user signal is transmitted through the channel, the random variable $Z_1^j$ will be a random function of a large number of random variables, as will be seen from the following analysis. Firstly, the random function $Z_1^j$ has three components. The first component, $Z_A$, is the part of the message signal that contains the $j$th user's message bit information. It is calculated as the autocorrelation value of the $j$th user's spreading sequence, which has $2\beta$ chips multiplied by the message coefficient. Generally, this variable can be the sum of $2\beta$ identically distributed and independent random variables in the case when the stream of chips is not a deterministic sequence. Therefore, due to the central limit theorem, the variable $Z_A$ can be approximated by a Gaussian random variable.

The second component, $Z_B$, is the inter-user interference part. It is calculated as the cross-correlation value of the $j$th user's spreading sequence by the spreading sequences of other users. If we use the Walsh sequences, this value will be 0, due to the zero value of the calculated cross-correlations, as defined by eqn (2.5.2.) in Chapter 2, Section 2.5. However, if we use chaotic spreading sequences, this calculated cross-correlation will have a finite random value that depends on the correlation characteristics of the chaotic sequence implemented. We are interested in knowing what the distribution function of these products are. The product of two random independent chip values is, again, a random value. All chips are identically distributed, as presented in Chapter 2, relation (2.5.6). Therefore, all the chip products have the same distribution, too. The variable $Z_B$ is, in fact, the sum of $2\beta(J - 1)$ identical and independent random variables. Therefore, due to the central limit theorem, $Z_B$ is the Gaussian random variable.

The third component is the noise part, which is the sum of $2\beta$ terms obtained by multiplying the chip samples by the noise samples. Because the chip sequence has a mean of 0, the calculated products are realizations of the zero-mean Gaussian random variables. Consequently, the third variable, $Z_C$, being the sum of $2\beta$ Gaussian variables, is the Gaussian random variable.

Therefore, the decision variable $z_1^j$ is a realization of the decision random variable $Z_1^j$, which is approximately Gaussian, because it is expressed as the sum of three independent random variables that are approximately Gaussian, that is,

$$Z_1^j = Z_A + Z_B + Z_C. \tag{8.4.7}$$

Now we can use the same procedure to derive the bit error probability in a code division multiple access system by using the theory presented in Chapter 7 for a binary phase-shift keying system. To that end, we should find the mean and the variance of the decision random variable, which is the subject of the analysis in the next section.

## 8.4.2   Derivation of the Probability of Error

The message bits of any user are affected independently by the noise in the channel. Therefore, if an expression for the bit error probability is derived for a particular bit, it will be valid for all of the transmitted bits. Suppose the bit $+1$ is transmitted, and the decision is made at the output of the correlator according to expression (8.4.6). The theoretical bit error probability for the transmission of the $j$th user's message bit can be calculated using the simplified expression from Chapter 6, eqn (6.2.55), expressed as

$$p^j = \frac{1}{2}\mathrm{erfc}\sqrt{\frac{\eta_{Z_1^j}^2}{2\cdot\sigma_{Z_1^j}^2}} = \frac{1}{2}\mathrm{erfc}\left(\frac{2\cdot\sigma_{Z_1^j}^2}{\eta_{Z_1^j}^2}\right)^{-1/2}, \tag{8.4.8}$$

where the error complimentary function, erfc, is defined in Chapter 6, Section 6.2.3. The mean value of the decision random variable can be calculated as

$$\eta_{Z_1}^j = E\{Z_A + Z_B + Z_C\} = E\{Z_A\} = \sum_{n=1}^{2\beta} E\left\{\left(x_n^j\right)^2\right\} = 2\beta\left(\sigma^j\right)^2 = 2\beta P_c \tag{8.4.9}$$

because the mean values of the second variable, $Z_B$, and the third variable, $Z_C$, are both 0. The variance of the sum in eqn (8.4.7) is the sum of the individual variances, due to the independence of the addends. Also, the variance of each addend in eqn (8.4.5) is the sum of the variances of its component. Therefore, the variance of the decision variable is

$$\sigma_{Z_1}^j = 2\beta\cdot\mathrm{var}\left(\left(x_n^j\right)^2\right) + 2\beta P_c\,(J-1)\,P_c + 2\beta P_c\sigma^2, \tag{8.4.10}$$

where $P_c$ is the average power of a random chip of the $j$th user. The probability of error, eqn (8.4.8), can then be expressed as

$$p^j = \frac{1}{2}\text{erfc}\left(\frac{2 \cdot \sigma^2_{z_1^j}}{\eta^2_{z_1^j}}\right)^{-1/2} = \frac{1}{2}\text{erfc}\left(\frac{4\beta \cdot \text{var}\left(\left(x_n^j\right)^2\right) + 4\beta P_c\,(J-1)\,P_c + 4\beta P_c\sigma^2}{4\beta^2\,P_c^2}\right)^{-1/2} .$$

$$= \frac{1}{2}\text{erfc}\left(\frac{\text{var}\left(\left(x_n^j\right)^2\right)}{\beta\,P_c^2} + \frac{(J-1)}{\beta} + \frac{\sigma^2}{\beta P_c}\right)^{-1/2}$$

$$(8.4.11)$$

Bearing in mind that the variance of the channel noise is equal to the noise power spectral density, that is, $\sigma^2 = N_0/2$, and the energy of a bit is $E_b = 2\beta \cdot P_c$, the probability of error in this system can be expressed as

$$p^j = \frac{1}{2}\text{erfc}\left(\frac{\psi}{\beta} + \frac{(J-1)}{\beta} + \left(\frac{E_b}{N_0}\right)^{-1}\right)^{-1/2} .$$

$$(8.4.12)$$

The parameter $\psi$ depends on the statistical characteristics of the user's spreading sequence. Obviously, the bit error probability depends on the signal-to-noise ratio $E_b/N_0$ and an additional three parameters, $\psi$, $J$, and $2\beta$. If the number of users, $J$, increases, the probability of error increases, due to the increase in the inter-user interference in the system. If the statistical properties of the spreading sequence cause an increase in the statistical parameter $\psi$, the bit error probability will increase. In contrast, if the spreading factor increases, the bit error probability will decrease. Finally, if the signal-to-noise ratio increases, the probability of error decreases. In addition, if the number of users is $J = 1$, and the spreading sequences are binary, like Walsh functions, the bit error probability is equal to the bit error probability in the binary phase-shift keying system defined in Chapter 7 in eqn (7.2.62), that is,

$$p^j = \frac{1}{2}\text{erfc}\sqrt{\frac{E_b}{N_0}} .$$

$$(8.4.13)$$

In this case, the statistical parameter $\psi$ is 0, because there are no variations in the positive or negative amplitude values of the message bit. This is a single-user code division multiple access system, which is equivalent to a direct-sequence spread-spectrum system.

........................................................................................................................................

PROBLEMS

1. A real signal is expressed as the sum

$$s(t) = \sum_{i=0}^{N-1} s_i(t) = \sum_{i=0}^{N-1} A_i \cdot \cos\left(2\pi \frac{i}{T}t + \phi_i\right),$$

which is defined in the time interval $0 \leq t \leq T$, for $N = 4$, $A_i = 10 = $ constant, and $\phi_i = i \cdot \pi$, where $T = 10^{-3}$ seconds is the duration of all signal components that define the frequencies of all frequency components, that is, $f_i = i/T$.

a. Calculate the frequencies and phases of the signal components. Develop the expression for $s(t)$ in terms of its components, with defined numerical values for the parameters for all four components.

b. Plot the graphs of all four components in the time domain.

c. Find the amplitude spectral density of all the signal components, using the Fourier transform. (Hint: When you find the amplitude spectral density of the first component for $i = 0$, you can apply the modulation theorem to find the spectra of the other components; or you can apply the Fourier transform directly on the sinusoidal pulses for $i = 1, 2,$ and $3$.)

d. Plot the magnitude spectral density of the signal.

2. A baseband orthogonal frequency division multiple access signal is expressed in the form

$$s_B(t) = \sum_{i=0}^{N-1} A_i \cdot \cos\left(2\pi f_i t + \phi_i\right),$$

for $i = 0, 1, 2,$ and $3$, $N = 4$, and $fi = i/T$, where $T$ is the duration of the symbols.

a. Calculate the power spectral densities of the first and second signal components. Then, using these results, determine the power spectral density of the whole signal.

b. Calculate the powers of the first and second signal components. Then, using these results, determine the power of the whole signal.

3. Prove that the neighbouring subband signals inside a baseband signal in an orthogonal frequency division multiple access system are orthogonal, where the subband signals are defined in the interval $0 \leq t \leq T$ and expressed as $A_i \cdot \cos\left(2\pi f_i t + \phi_i\right)$.

4. A subband symbol of a baseband signal in an orthogonal frequency division multiple access system has the amplitude spectral density expressed as

$$S(f) = -\frac{1}{2} e^{-j\pi\left(f - \frac{1}{T}\right)T} \cdot AT \cdot \text{sinc}\left(f - \frac{1}{T}\right)T - \frac{1}{2} e^{-j\pi\left(f - \frac{1}{T}\right)T} \cdot AT \cdot \text{sinc}\left(f + \frac{1}{T}\right)T,$$

where $T$ is the duration of the symbol, and $A$ is its amplitude. Prove that the power spectral density of this signal is equal to the sum of the power spectral densities obtained for the signal components defined for positive and negative frequencies.

5. Derive the autocorrelation function expression of a radiofrequency carrier, considering it as a power signal expressed as

$$s_c(t) = A_c \cos(2\pi f_{RF} t),$$

where the frequency of the radiofrequency carrier is defined as $f_{RF} = f_c - (N-1)/2T$. Calculate the power spectral density of the carrier. Then, calculate the power of the carrier from its autocorrelation function and power spectral density function.

6. Suppose an orthogonal frequency division multiple access system is based on application of binary phase-shift keying modulation in each subband. The symbols in all $N$ subbands are of equal energy and are affected by additive white Gaussian in the channel. The value of the demodulated symbol can be expressed as

$$x_i = A_i \cos(\phi_i) + w_i.$$

Calculate the probability of error in the system. Then, confirm that the bit error probability in the system is equal to the bit error probability in any of the subbands.

7. Suppose that the demodulated $I$ and $Q$ components of an intermediate-frequency signal in an orthogonal frequency division multiple access system are

$$s_{BI}(n) = \frac{A_{IF}^2}{2N_s} \sum_{i=0}^{N_s-1} [A_i \cos(\Omega_i n + \phi_i) + w(n)A_N \cos\Omega_i n]$$

and

$$s_{BQ}(n) = \frac{A_{IF}^2}{2N_s} \sum_{i=0}^{N_s-1} [A_i \sin(\Omega_i n + \phi_i) + w(n)A_N \sin\Omega_i n],$$

respectively. Define the amplitude of the intermediate-frequency carrier at the modulator and the demodulator of the transceiver as $A_{IF} = \sqrt{2}$. Find the expression for the demodulated baseband signal. Define the signal-plus-noise term in the baseband signal.

8. Suppose the baseband demodulated noisy signal in an orthogonal frequency division multiple access system is expressed as

$$s_B(n) = \frac{1}{N_s} \sum_{i=0}^{N_s-1} (s_i + w_i) \cdot e^{j(2\pi i n/N_s)},$$

where the complex symbol received is $s_i = A_i e^{j\phi_i}$, and $w_i$ is a baseband noise sample with Gaussian distribution. Prove that the expression given is the sum of the signal part and the noise part.

9. The noisy complex symbol at the output of the receiver is expressed in the form

$$s_{noisy} = s_i + w_i = s_i + \sum_{n=0}^{N_s-1} w(n) \cdot e^{-j(2\pi i n/N_s)}.$$

Calculate the expression for the bit error probability, assuming that the noise is Gaussian with the two-sided power spectral density $N_0/2$. Assume that the quadriphase-shift keying used in the system is defined by four points in its constellation diagram as $s_i = (\pm 1 \pm j) A \sqrt{2}/2$.

10. Suppose the transmitter uses $N = 4$ subbands to transmit four signals and uses a discrete Fourier transform defined as a linear transform presented in the form of matrices. The modulation is performed using quadriphase-shift keying, with the defined constellation diagram of the signals presented in the matrix form as

$$\mathbf{s}_i = A \frac{\sqrt{2}}{2} \begin{bmatrix} 1-j \\ -1+j \\ 1+j \\ -1-j \end{bmatrix}.$$

a. Firstly, calculate the modulated baseband signal expressed in matrix form as $\mathbf{s_B}$ with the components $I(n)$ and $Q(n)$. Next, present the components in the time domain in terms of Kronecker delta functions. Finally, plot the block schematic of the receiver.

b. Process the received baseband signal $\mathbf{r_B}$ inside the receiver blocks, assuming that there is no noise in the channel.

11. Suppose that the baseband signal in an orthogonal frequency division multiple access system, to be transmitted in $N = 4$ subbands using quadriphase-shift keying modulation, is

$$\mathbf{s}_B = \begin{bmatrix} s(0) \\ s(1) \\ s(2) \\ s(3) \end{bmatrix} = A \frac{\sqrt{2}}{8} \begin{bmatrix} 0 \\ -2-2j \\ 4 \\ -2-j2 \end{bmatrix} = I + jQ.$$

The noise present in each subchannel is added to the signal to form a noisy signal at the input of the receiver; this signal can be expressed as

$$\mathbf{r}_{BW} = \mathbf{s}_B + \mathbf{w}_B = \begin{bmatrix} s(0) \\ s(1) \\ s(2) \\ s(3) \end{bmatrix} + \begin{bmatrix} w(0) \\ w(1) \\ w(2) \\ w(3) \end{bmatrix},$$

or as the function of the discrete variable $n$ as

$$r_{BW}(n) = [I(n) + I_W(n)] + j[Q(n) + Q_W(n)].$$

Firstly, make a block schematic of the receiver. Next, process the received signal in the receiver blocks to get estimates of the message bits transmitted. Finally, derive the expression for the bit error probability in the system.

# 9

# Information Theory and Channel Coding

The purpose of a communication system is to transfer *information*. The information theory in communication systems is a theoretical discipline for the mathematical modelling of a communication system. It can be considered to be part of probability theory, the theory of random variables, and the theory of stochastic processes. It is used for the mathematical characterization of a message (information) source, a communication channel, and a communication system as a whole.

The communication system shown in Figure 9.1.1 will be used to explain the concepts of information generation, information transfer, and communication channel capacity. At the output of the source, a symbol $m$ is generated, encoded in a source encoder, encoded in a channel encoder to produce a code word $c$, which is interleaved (resulting in a transmitter signal $x$), and then transmitted through a continuous waveform or a discrete communication channel. The received signal $y$ is processed in the reverse order to obtain an estimate $m'$ of the source message symbol $m$.

This is a highly simplified scheme of communication system, but it is sufficient to explain the main concepts of the theory of information. Two types of communication channels are presented in Figure 9.1.1. A discrete channel is presented inside the inner rectangle. This channel is represented by an adder which adds an error sequence $e$ to the transmitted discrete signal $x$ to obtain the received signal $y$ at the output of the channel. In the binary case, when the input signal $x$ is a sequence of 0's and 1's, the error sequence is also a binary sequence, having 1' at the positions of channel errors, and 0's at other positions. Because the channel in this case is designed as a modulo-2 adder, the binary sequence $y$ generated at its output contains the received sequence of message bits with errors at the positions determined by the error sequence. In the case when interleavers and de-interleavers are not used, the coded sequence is denoted by $c$ and applied to the input of the channel, and the received sequence is $r$, which is decoded to produce the estimated sequence $c'$ and then the estimated message symbol $m'$. This notation will be used in the coding part of this chapter.

In the case when input signal $x$ is a continuous-time waveform, the channel is represented by a continuous-time additive white Gaussian noise $w$. This channel is presented in Figure 9.1.1 in the outer rectangle. The noise waveform $w$ is added to the transmitted signal to obtain the received continuous-time signal $y$.

The analysis of the system, from the theory of information point of view, will start with the characterization of the discrete source of information. The main terms, like information

*Discrete Communication Systems*. Stevan Berber, Oxford University Press. © Stevan Berber 2021.
DOI: 10.1093/oso/9780198860792.003.0009

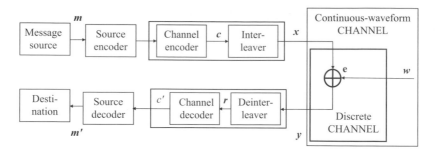

**Fig. 9.1.1** *The structure of a communication system from an information-theory point of view.*

and entropy, will be defined. Then, a discrete memoryless channel will be analysed and the expression for the channel capacity will be derived. A similar analysis of a channel with a continuous-time input signal $x$ and a waveform channel will be performed, to derive the expression for the channel capacity and to define the theoretical limits of the capacity of the communication system. Finally, the coding theorem will be presented, to get some ideas related to the theoretical communication limits that can be achieved by the application of channel encoding. Several sections are dedicated to channel coding, covering the theory of block, convolutional, and iterative coding.

## 9.1   Characterization of a Discrete Source

Suppose the output of a discrete source is represented by a finite set of $K$ symbols which is called an alphabet $A$, expressed as

$$A = \{m_0, m_1, \ldots, m_k, \ldots, m_{K-1}\}, \tag{9.1.1}$$

for $k = 0, 1, \ldots, K - 1$, and defined by the corresponding probabilities

$$P = \{p_0, p_1, \ldots, p_k, \ldots, p_{K-1}\}, \tag{9.1.2}$$

which satisfy the axioms of probability theory. These probabilities are called a priori probabilities. This source is called a *discrete memoryless source* in the case when there is no statistical dependence between any two symbols generated from the source. That is, the occurrence of a message symbol is statistically independent from the occurrence of all past and future symbols.

**Information in a symbol.** Each symbol from the alphabet $A$ is a carrier of information. This *symbol information* can be defined as *an uncertainty in generating a particular symbol*. The higher the probability of generating a symbol is, the smaller is the uncertainty in its appearance at the output of the source (before its generation), and the smaller is the information content in that symbol (after its generation). Thus, *the amount of information* in a symbol $m_k$ can be expressed as the reciprocal of the probability of its occurrence, which is defined as the logarithmic function

$$I(m_k) = \log\left(\frac{1}{p_k}\right) = -\log(p_k),\tag{9.1.3}$$

which is sometimes called the *self-information* of the message $m_k$. It is standard practice to use log base 2 for this function, that is,

$$I(m_k) = \log_2\left(\frac{1}{p_k}\right) = -\log_2(p_k),\tag{9.1.4}$$

and express the information in *bits*. When the base of the logarithm is $e$, the unit of information is the natural unit *nat* and, for log base 10, the unit is *dit*. The amount of information defined in this way is an objective measure of information that satisfies the conditions of being real valued, monotonic, and additive for statistically independent events.

Consider the following three somewhat extreme but illustrative cases. If the appearance of a symbol at the output of a source is *certain*, that is to say, it happens with probability 1, then that symbol contains *zero information*. If a symbol appears with *zero probability*, it contains *infinite information*. If a source can generate two symbols which are *equiprobable*, then the information content in one symbol is one bit of information. Furthermore, in this case, we say that one binary digit is required to transmit the knowledge of its occurrence.

The basic unit of information is a *bit*. This should not be confused with the name for a binary digit or a bit in communication theory, which is sometimes also called a *binit*. In this book, we will call bits used in information theory, when necessary, information bits, to distinguish them from message bits in communication theory.

**The entropy of the source.** The amount of information in all $K$ source symbols can be represented by a set of $K$ real numbers greater than 0, as calculated by eqn (9.1.4), for all $k = 0, 1, 2, \ldots, K - 1$. Therefore, the source of information can be represented by a *discrete random variable* $I(m_k)$ that takes on the values from the set of discrete values $\{I(m_0), I(m_1), \ldots, I(m_{K-1})\}$, with the corresponding probabilities $\{p_0, p_1, \ldots, p_{K-1}\}$. The mean value of this random variable is called the *entropy* of a discrete memoryless source, denoted by $H$. Thus, the entropy, calculated as

$$H = E\{I(m_k)\} = \sum_{k=0}^{K-1} I(m_k)p_k = \sum_{k=0}^{K-1} p_k \log_2\left(\frac{1}{p_k}\right) = -\sum_{k=0}^{K-1} p_k \log_2 p_k,\tag{9.1.5}$$

is defined as the average information per message symbol. In other words, entropy is *the average information contained in one symbol* generated by the source, which depends only on the probability distribution of the symbols in the alphabet $A$ of the source.

## Example

The four symbols $m_0$, $m_1$, $m_2$, and $m_3$ occur with probabilities 1/2, 1/4, 1/8, and 1/8, respectively.

  a. Compute the information in a three-symbol message defined as $m = m_0 m_0 m_1$, assuming that the symbols are statistically independent.
  b. Assume that the symbols occur with the same probability. What is now the information in the three-symbol message $m = m_0 m_0 m_1$?

## Solution.

  a. Because the symbols are independent, the probability of their occurrence is equal to the product of their a priori probabilities, that is,

$$P(m) = P(m_0) P(m_0) P(m_1).$$

The information of the message $X$ is

$$I(m) = \log_2 \frac{1}{P(m)} = \log_2 \frac{1}{P(m_0) P(m_0) P(m_1)}$$
$$= \log_2 \frac{1}{P(m_0)} + \log_2 \frac{1}{P(m_0)} + \log_2 \frac{1}{P(m_1)}$$
$$= \log_2(2) + \log_2(2) + \log_2(4) = 1 + 1 + 2 = 4 \; bits$$

  b. In this case, the information content in the message is

$$I(m) = \log_2 \frac{1}{P(m)} = \log_2 \frac{1}{P(m_0)} + \log_2 \frac{1}{P(m_0)} + \log_2 \frac{1}{P(m_1)} = \log_2(4) + \log_2(4) + \log_2(4) = 6 \; bits.$$

The information content in this case is greater than the information content in the first case, even though the number of the transmitted messages is the same, because the probability of occurrence of the first two symbols is decreased in respect to case a., so their information content is increased.

## Example

Suppose eight symbols occur with the same probability of 1/8. Compute the information in one- and two-symbol messages, assuming that the symbols are statistically independent.

**Solution.** Because the symbols are independent, the measure of information is additive, so the information content in one symbol is

$$I = \log_2(8) = 3 \ bits.$$

The information contained in two symbols is

$$I = \log_2(8) + \log_2(8) = 3 + 3 = 6 \ bits.$$

We can then extend this result to say that *four* equiprobable messages require four distinct binary pulse patterns or *two* bits, and *eight* equiprobable messages require eight distinct binary pulse patterns or *three* bits. Generally, any one of $n$ equiprobable messages contains $\log_2(n)$ bits of information. In this case the probability of any message occurrence is $p_k = 1/n$, resulting in the information content

$$I(m_k) = \log_2 \left( \frac{1}{p_k} \right) = \log_2(n) \ \text{bits}.$$

## Example

A discrete memoryless source that generates two symbols is called a binary source. It is characterized by an alphabet with two symbols, $A = (m_0, m_1)$. The symbols are generated by two respective probabilities, $P = (p_0, p_1)$. Suppose that the symbols are 0 and 1 and generated with probabilities $P(0) = p_0$ and $P(1) = p_1$. Find and characterize the entropy of this source.

**Solution.** The source entropy is

$$H = -\sum_{k=0}^{1} p_k \log_2 p_k = -p_0 \log_2 p_0 - p_1 \log_2 p_1, \tag{9.1.6}$$

which is expressed in bits. Obviously, the average information per symbol depends on the probabilities of the symbols. When the probability of, say, symbol 0 is equal to 0, the average information is 0, because there is no uncertainty in the symbol occurrence at the output of the source, that is, we are sure that 1 will be generated at the output of the source, with probability 1. A similar conclusion will be drawn in the case when the probability of 1 is 0. The average information generated at the output of the source is maximum when the probabilities of both symbols are equal. In that case, the entropy, or the average information content, is 1 bit.

In order to find the entropy of the source for any value of the probability of 0, we can define the entropy function as a function of one probability by setting $p_1 = (1 - p_0)$:

$$H(p_0) = -p_0 \log_2 p_0 - (1 - p_0) \log_2 (1 - p_0), \tag{9.1.7}$$

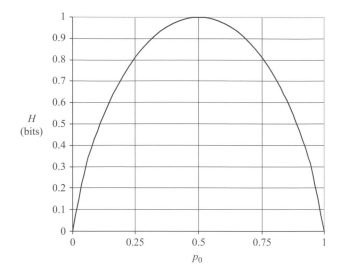

**Fig. 9.1.2** *Entropy of a binary source.*

which is plotted in Figure 9.1.2. The graph complies with the previous analysis. If, for example, the probability of generating 0 from the source is 0 or 1, the information content of the source is 0, because we are certain what the source will generate at any time instant. In other words, if the probability of generating 0 is 0, then we are certain that the source will always generate 1; alternatively, if the probability of generating 0 is 1, we are certain that the source will always generate 0. We use the term 'uncertainty' to express the information content or the entropy of the source. In this context, the certainty increases when the probability of the event increases (the information decreases) and the certainty decreases when the probability of the event decreases (the information increases).

## 9.2 Characterization of a Discrete Channel

### 9.2.1 A Discrete Memoryless Channel

A discrete channel is a statistical model of a real channel that is defined by its input discrete random variable $X$, its output discrete random variable $Y$, and a transition probability matrix $\mathbf{P_{YX}}$ of a finite size, which relates a finite number of outputs to the finite number of inputs in a probabilistic sense. This discrete channel is memoryless if its output $y_k$ depends only on the corresponding input $x_j$ and does not depend on any previous value of $X$. A highly simplified structure of a communication system that includes a discrete memoryless channel is presented in Figure 9.2.1. This scheme is even more simplified than the scheme in Figure 9.1.1, because the symbols at the input of the channel can be different from those ones generated by the source. In the most simplified case, the channel input symbols correspond to the output message symbols generated by the source. The signals appear in the discrete channel as a

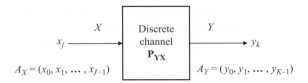

**Fig. 9.2.1** *Simplified structure of a communication channel.*

series of random message symbols which are taken from a finite alphabet and represented as realizations of the input random variable $X$. After the symbols are transmitted through the discrete channel, they are received at the receiver side and represented as realizations of the output random variable $Y$.

The *input $X$* and *output $Y$* of the discrete channel are random variables that take their values (symbols) from their alphabets. The input alphabet is a set of $J$ independent symbols

$$A_X = \{x_0, x_1, \ldots, x_j, \ldots, m_{J-1}\},$$
(9.2.1)

and corresponding a priori probabilities of the input symbols

$$p(x_j) = p(X = x_j)$$
(9.2.2)

define the probability distribution of the input symbols for $j = 0, 1, \ldots, J-1$. The output alphabet is a set of $K$ independent symbols

$$A_Y = \{y_0, y_1, \ldots, y_k, \ldots, y_{K-1}\},$$
(9.2.3)

and the corresponding probabilities of the output symbols are

$$p(y_k) = p(Y = y_k),$$
(9.2.4)

which define the probability distribution of the output symbols for $k = 0, 1, \ldots, K-1$. The input and output alphabets can, in general, have different numbers of symbols. The channel is called discrete because the number of messages in both alphabets is finite.

The *channel* itself is defined by its transition probabilities, which are expressed in the form of conditional probabilities for all $j$ and $k$ as

$$p(y_k|x_j) = p(Y = y_k|X = x_j),$$
(9.2.5)

which is the probability that $y_k$ is received given that $x_j$ is transmitted, or

$$p(x_j|y_k) = p(X = x_j|Y = y_k),$$
(9.2.6)

which is the probability that $x_j$ is transmitted, given that $y_k$ is received. These probabilities can be also expressed in matrix form as a *channel matrix* or a *transition matrix*. The probabilities of all channel outputs, given the inputs, is presented in the following matrix:

$$\mathbf{P_{YX}} = \begin{bmatrix} p(y_0|x_0) & p(y_1|x_0) & \cdots & p(y_{K-1}|x_0) \\ p(y_0|x_1) & p(y_1|x_1) & \cdots & p(y_{K-1}|x_1) \\ \vdots & \vdots & \ddots & \vdots \\ p(y_0|x_{J-1}) & p(y_1|x_{J-1}) & \cdots & p(y_{K-1}|x_{J-1}) \end{bmatrix}. \tag{9.2.7}$$

The probabilities of all channel inputs, given the outputs, are presented in the following matrix:

$$\mathbf{P_{XY}} = \begin{bmatrix} p(x_0|y_0) & p(x_1|y_0) & \cdots & p(x_{J-1}|y_0) \\ p(x_0|y_1) & p(x_1|y_1) & \cdots & p(x_{J-1}|y_1) \\ \vdots & \vdots & \ddots & \vdots \\ p(x_0|y_{K-1}) & p(x_1|y_{K-1}) & \cdots & p(x_{J-1}|y_{K-1}) \end{bmatrix}. \tag{9.2.8}$$

Having defined the *a priori probabilities* $p(x_j)$ and the conditional probabilities, it is possible to calculate the joint probabilities of the random variables $X$ and $Y$ as

$$p(x_j, y_k) = p(X = x_j, Y = y_k) = p(x_j) p(y_k|x_j),$$
$$p(y_k, x_j) = p(Y = y_k, X = x_j) = p(y_k) p(x_j|y_k), \tag{9.2.9}$$

which can be understood as the probabilities of the occurrence of the input–output pairs. The output *marginal probabilities* are defined as the total probability of the occurrence of the output symbol for the given a priori and transition probabilities, that is,

$$p(y_k) = \sum_{j=0}^{J-1} p(x_j) p(y_k|x_j), \text{ for every } k = 0, 1, \ldots, K-1. \tag{9.2.10}$$

## 9.2.2 Discrete Binary Channels with and without Memory

### 9.2.2.1 Discrete Binary Channels

This section presents a theoretical model of a binary discrete channel. This channel is obtained when additive white noise and other impairments are present in the waveform channel, and a hard decision is made in the receiver detector/decision block. If the input of the channel is a binary sequence, then the output is also a binary sequence obtained at the output of the decision device. Because this channel has both binary inputs and binary outputs, it is called a binary digital channel. The channel itself can be characterized by a binary signal, which is called the 'error sequence' and is denoted by the symbol **e** in Figure 9.2.2. The error sequence specifies the relationship between the channel inputs and outputs.

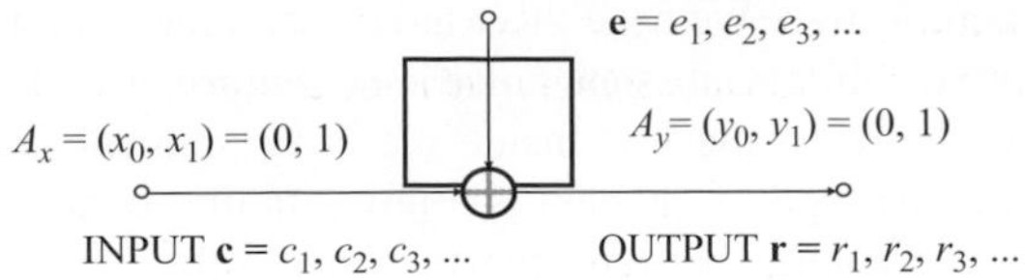

**e** $= e_1, e_2, e_3, \ldots$

$A_x = (x_0, x_1) = (0, 1)$     $A_y = (y_0, y_1) = (0, 1)$

INPUT **c** $= c_1, c_2, c_3, \ldots$     OUTPUT **r** $= r_1, r_2, r_3, \ldots$

**Fig. 9.2.2** *Simplified scheme of a discrete binary channel.*

**Inputs and outputs.** The input $X$ and output $Y$ of the binary channel are random variables that take their values from their binary alphabets, which represent the message symbols called *message bits*. It is important to note that these bits correspond to message symbols and are not the information bits defined in information theory. In this context, the input alphabet is a set of *two* bits, that is,

$$A_X = \{x_0, x_1\} = \{0, 1\}, \tag{9.2.11}$$

which are generated independently with corresponding a priori probabilities defined as

$$p(x_j) = p(X = x_j), \tag{9.2.12}$$

which define the probability distribution of the input bits for $j = 0, 1$. The output alphabet is a set of *two* bits,

$$A_Y = \{y_0, y_1\} = \{0, 1\}, \tag{9.2.13}$$

which appear at the output of the channel independently, with the corresponding probabilities defined as

$$p(y_k) = p(Y = y_k). \tag{9.2.14}$$

The channel is called 'discrete' because the number of message bits in both alphabets is finite.

**The channel.** The channel is specified by the error sequence and is represented as the modulo-2 addition of the input sequence of message bits **c** and the error sequence **e**, that is, the output sequence **r** is

$$\mathbf{r} = \mathbf{c} \oplus \mathbf{e}, \tag{9.2.15}$$

where **c** is a sequence taken randomly from the input alphabet $A_X$, **e** is an error sequence that specifies the statistical properties of the channel, and **r** is a sequence at the output of the channel.

The error sequence is represented by a binary signal. The statistical structure of the error sequence determines the properties of the channel. If there is no statistical dependence between the bits inside the error sequence, then the channel is without memory; in other words, the channel is memoryless. We have this kind of channel when the waveform channel is represented by additive white Gaussian noise (see Figure 9.1.1). In contrast, if there is statistical dependence between the bits inside the error sequence, then the channel has memory. An example of this type of channel is a noisy fading channel, because the fading introduces statistical dependencies between the bits inside the error sequence, due to the multipath effects in the channel.

Strictly speaking, mathematically, an error sequence is a realization of a binary stochastic process and can be analysed using the theory of discrete-time stochastic processes (see Chapter 3). Furthermore, inside an error sequence, we can design various structures and construct new processes that are useful for the statistical characterization of binary channels with memory. For example, if the transmission of blocks of bits is observed, that would be a block process.

**Example of a binary channel.** One example of an error sequence is presented in Figure 9.2.3. Two processes are defined in the channel error sequence: the error and the error gap process.

The error process generates 1's and 0's with a specified procedure that depends on the statistical nature of the channel. If the channel is memoryless, then the 1's and 0's are generated independently and 1's appear with a fixed probability $p$, that is, the structure of the error sequence depends on a single parameter, $p$. If the channel has memory, then the generation of 1's in the error sequence will depend on more than one parameter.

A gap is defined as a series of 0's finishing with 1 in the error sequence. In Figure 9.2.3, the gap sequence is a series of integers, that is, **gap** = {... , 4, 4, 1, 10, 4, 1, 9, ... }. The use of gap statistics is important in the analysis of channels with memory.

### 9.2.2.2   Discrete Binary Memoryless Channels

A channel without memory has often been used as a good model for a real digital channel. This channel is usually the basic part of a channel with memory. The channel is memoryless if there is a transitional probability assignment $p(y_k \mid x_j)$, which was defined in Section 9.2.1 as the fixed conditional probability of the received bit $y_k$, given that the transmitted bit is $x_j$. These probabilities are presented in matrix form and relate all possible outputs to all possible inputs of the channel and characterize entirely a digital signal transmission through the channel. In the case of a binary channel, there are four conditional (transitional) probabilities that completely describe the channel, defined in matrix form as

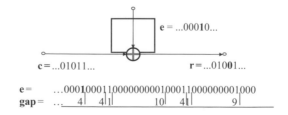

**Fig. 9.2.3**  *Channel representation by an error process and an error gap process.*

$$\mathbf{P_{YX}} = \begin{bmatrix} p(y_0|x_0) & p(y_1|x_0) \\ p(y_0|x_1) & p(y_1|x_1) \end{bmatrix} = \begin{bmatrix} p(0|0) & p(1|0) \\ p(0|1) & p(1|1) \end{bmatrix}, \qquad (9.2.16)$$

and

$$\mathbf{P_{XY}} = \begin{bmatrix} p(x_0|y_0) & p(x_1|y_0) \\ p(x_0|y_1) & p(x_1|y_1) \end{bmatrix}.$$

The transition probabilities are marked on the model of the channel in Figure 9.2.4. The input to the channel consists of message bits $x_j$ that form the input message sequence **c**. The bits inside the sequence **c** will be transmitted either correctly or incorrectly, depending on the transition probabilities defined for the channel itself.

In this channel, the relationship between input and output sequences of bits, which can be represented as the input and output vectors (arrays) **c** and **r**, respectively, is defined by the conditional probabilities of each input–output pair of bits in the form

$$P(\mathbf{r}|\mathbf{c}) = P(r_1, r_2, \ldots, r_n | c_1, c_2, \ldots, c_n) = \prod_{i=1}^{n} P(r_i|c_i), \qquad (9.2.17)$$

because the error sequence affects each input bit independently.

The *bit error probability* is the basic measure of the quality (and reliability) of signal transmission in digital communication systems. The bit error probability in a memoryless channel, denoted by $p$, depends on the conditional probabilities $P(0 \mid 1)$ and $P(1 \mid 0)$, as well as on the a priori probabilities of generating 0 and 1 at the input of the channel, which are designated by $P(0)$ and $P(1)$; it can be expressed in the form

$$p = P(1)P(0|1) + P(0)P(1|0).$$

For a symmetric channel, where the probabilities of transition of 0 to 1, and vice versa, in the channel are equal, the bit error probability is equal to the transition probabilities, that is,

$$p = P(0|1) = P(1|0). \qquad (9.2.18)$$

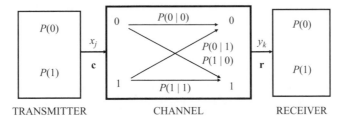

**Fig. 9.2.4** *Transitional probability assignments in a binary memoryless channel.*

In conclusion, the basic and single parameter that characterizes channels without memory is the bit error probability $p$.

---

**Example**

Find the probability of reception of a sequence of 10 bits in the case when an error has occurred only on the tenth bit with bit error probability $p$. Assume that the channel is memoryless and symmetric and the bit error probability on each bit is $p$. If the input sequence has nine 1's, what would be the tenth bit in the received sequence? Find the probability that the received sequence $\mathbf{r}$ has an error at any position among 10 received bits.

**Solution.** The solution is as follows:

$$
\begin{aligned}
P(\mathbf{r}|\mathbf{c}) = P(r_1, r_2, \ldots, r_{10}|c_1, c_2, \ldots, c_{10}) &= \prod_{i=1}^{10} P(r_i|c_i) \\
&= P(r_1|c_1) \cdot P(r_2|c_2) \ldots P(r_9|c_9) \cdot P(r_{10}|c_{10}) \\
&= (1-p) \cdot (1-p) \ldots (1-p) \cdot p = (1-p)^9 p
\end{aligned}
\tag{9.2.19}
$$

The tenth bit is 1 or 0. This is equal to the probability that a single error occurs in a sequence of $\mathbf{r}$ bits at particular bit. If a single error can happen on any bit, the probability is

$$
P(\mathbf{r}|\mathbf{c}) = P(r_1, r_2, \ldots, r_{10}|c_1, c_2, \ldots, c_{10}) = 10 \prod_{i=1}^{10} P(r_i|c_i) = 10(1-p)^9 p.
\tag{9.2.20}
$$

---

### 9.2.2.3   *Discrete Binary Channels with Memory*

In channels with memory, there is statistical dependence between the bits in the error sequence. Thus, simple transitional probability assignments, which were sufficient to mathematically represent channels without memory, are not applicable for channels with memory. In channels with memory, the error sequence consists of bursts of errors that are separated by long error-free intervals, meaning that the independence in generating errors is partially lost. A burst is defined as an interval inside the error sequence where errors are predominantly generated. Therefore, the memory of a channel increases its capacity. Furthermore, it follows that a binary symmetric channel without memory has the least capacity.

Models for channels with memory can be divided into two broad classes: *generative models* and *descriptive models*. A generative model is a Markov chain consisting of either a finite or an infinite number of states with a defined set of the transition probabilities that produce error sequences. Generative models can be divided into three broad groups: two-state models, finite-state models, and models of higher complexity. Descriptive channel models describe the statistical structure of error sequences by using known standard statistical probability distribution functions. For the sake of illustration, we will present here a two-state generative model. The theory of binary channels with and without memory is a broad area in communication systems theory. Here we have just touched on the main issue related to channel modelling, at the level that was necessary to follow the content of this book.

The simplest generative two-state model, with a bad state (state B) and a good state (state G), is based on the theory of Markov chains. The state transition diagram and the corresponding probabilities for this model are shown in Figure 9.2.5. We can assume that the bit error probabilities for state B and state G are $p_B \neq 0$ and $p_G = 0$, respectively. In order to simulate bursts, both of the states must tend to persist. Thus, the conditional probabilities of state G given state B, $p' = P(G \mid B)$, and the probability of state B given state G, $P = P(B \mid G)$, ought to be small, while the conditional probabilities of remaining in the $G$ or the $B$ state, defined as $Q = P(G \mid G)$ and $q' = P(B \mid B)$, ought to be large. When the channel is in state $G$, long runs of 0's will be produced, while, in state $B$, errors (1's in the related error sequence) will be produced with probability $P(1 \mid B) = p_B$. When the channel is in one of the states, it behaves like a binary symmetric channel without memory, as shown in Figure 9.2.5.

A *generalized two-state model* is obtained when errors are also allowed to occur in state G. As shown in Figure 9.2.5, the model is completely defined by error probabilities for both states, where $e_n$ are the corresponding values of the error sequence,

$$
\begin{aligned}
P\,(e_n = 1 | G) &= p_G & P\,(e_n = 1 | B) &= p_B \\
P\,(e_n = 0 | G) &= 1 - p_G & P\,(e_n = 0 | B) &= 1 - p_B
\end{aligned}
\tag{9.2.21}
$$

The state transition probabilities are defined as the conditional probabilities

$$
P\,(B|G) = P, P\,(G|B) = p', P\,(B|B) = 1 - p' = q', \text{and } P\,(G|G) = 1 - P,
\tag{9.2.22}
$$

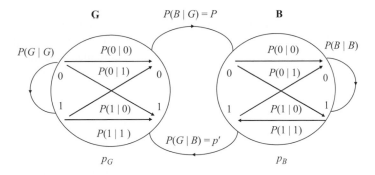

**Fig. 9.2.5** *Transitional diagram of a generalized two-state model.*

and the bit error probability is

$$p = \frac{p_B P + p_G p'}{p' + P}.$$ (9.2.23)

Therefore, the statistical properties of this generative two-state channel with memory cannot be described by a single parameter, the bit error probability $p$, as was possible for a memoryless channel. The channel is specified by a set of parameters. The number of model parameters increases with the number of channel states and with the complexity of the channel.

## 9.2.3   Capacity of a Discrete Memoryless Channel

### 9.2.3.1   *Capacity of a Discrete Channel*

We will here analyse separately the input of the channel, the channel itself, and the output of the channel for the specified input and the channel.

**The channel input.** The amount of information in each input symbol $x_j$, which is a realization of the input random variable $X$, can be represented by a real number greater than 0, as calculated by the expression

$$I\left(x_j\right) = \log_2\left(\frac{1}{p_j}\right) = -\log_2\left(p_j\right).$$ (9.2.24)

The information content in a symbol $x_j$ is the measure of *a priori uncertainty* about the symbol, in contrast to the probability of the symbol, which is the measure of a priori certainty about the symbol. The higher the probability of a symbol's appearance is, the higher the certainty is; the higher the information is, the higher is the uncertainty.

The input of the channel can be generally represented by a discrete random variable $I$ that takes on its values $I(x_j)$ from the set of discrete values $I = \{I(x_0), I(x_1), \dots, I(x_{J-1})\}$. Therefore, these values are assigned to all messages of the channel input alphabet $A_x$, which occur with the corresponding probabilities $\{p_0, p_1, \dots, p_{J-1}\}$, as shown in Figure 9.2.1. We mapped the input alphabet into the set of information values and assigned them the corresponding probabilities. Therefore, the random variable $I$ is completely described by its probability density and distribution function and complies with all requirements defined in the theory of random variables. The information, defined as a random variable taking real values, can also be analysed in terms of its parameters, as we would do for any random variable. The information of the channel input can be represented as a column vector of real numbers:

$$I(X) = [I(x_0) \quad I(x_1) \quad \cdots \quad I(x_{J-1})]^T,$$ (9.2.25)

where $T$ denotes the transpose matrix. The information in the matrix is inversely *log-proportional* to the probabilities of the symbols at the input of the channel. The higher the probability of a symbol's occurrence is, the smaller is the information content or, in other words, the smaller is the uncertainty of the symbol's appearance at the input of the channel.

The *mean value of the information I* is called the entropy of the channel input symbols for a discrete memoryless channel, denoted by $H$. It can be calculated as

$$H(X) = E\{I\} = \sum_{j=0}^{J-1} I(x_j) p_j = \sum_{j=0}^{J-1} p_j \log_2 \left( \frac{1}{p_j} \right) = -\sum_{j=0}^{J-1} p_j \log_2 p_j. \tag{9.2.26}$$

This entropy represents the average information contained in one input symbol that can be generated at the input of the channel. We can also say that entropy is the measure of the average uncertainty about input $X$ *before* the channel output is observed. This entropy depends only on the probability values of symbols in the channel input alphabet $A_X = \{x_0, x_1, \ldots, x_{J-1}\}$.

**The channel.** In reality, the symbols taken from alphabet $A_X$ are transmitted through the discrete memoryless channel and appear at the channel output as the elements of alphabet $A_y$. From the system designer's point of view, it is important to know what the information content of the channel input $X$ is which can be taken from the symbols observed from the channel output $Y$. This can be obtained from the conditional information, which is defined as

$$I(x_j|y_k) = \log_2 \left( \frac{1}{p(x_j|y_k)} \right) = -\log_2 p(x_j|y_k), \tag{9.2.27}$$

which can be calculated for every input symbol $x_j$. The conditional information of symbol $x_j$ given that the symbol $y_k$ is received is the information content about the input symbol $x_j$ for the observed symbol $y_k$ at the output of the channel, and can also be interpreted as the measure of uncertainty that *remains* about input symbol $x_j$ when $y_k$ is received.

The amount of information in each of the $J$ symbols in $X$, given the output $y_k$, can be represented by a set of $J$ real numbers greater than 0, as calculated by the above expression. Therefore, the channel can be represented by a random variable $I(X|y_k)$ that takes on the values from the set of discrete values $\{I(x_0 \mid y_k), I(x_1 \mid y_k), \ldots, I(x_{J-1} \mid y_k)\}$ with the corresponding probabilities $\{p(x_0 \mid y_k), p(x_1 \mid y_k), \ldots, p(x_{J-1} \mid y_k)\}$. The conditional information can be calculated for every $k = 0, 1, \ldots, K - 1$ and presented as a matrix of information expressed as

$$I_{XY} = \begin{bmatrix} I(x_0|y_0) & I(x_1|y_0) & \cdots & I(x_{J-1}|y_0) \\ I(x_0|y_1) & I(x_1|y_1) & \cdots & I(x_{J-1}|y_1) \\ \vdots & \vdots & \ddots & \vdots \\ I(x_0|y_{K-1}) & I(x_1|y_{K-1}) & \cdots & I(x_{J-1}|y_{K-1}) \end{bmatrix}. \tag{9.2.28}$$

Therefore, the matrix rows define the random variables $I(X \mid y_k)$ for $k = 0, 1, \ldots, K - 1$. The mean values of these random variables are called the *conditional entropies* of a discrete memoryless channel input $X$, given that the outputs are $Y = y_k$, for all $k$ values.

These conditional entropies, denoted by $H(X \mid y_k)$, may be calculated as the mathematical expectations

$$H(X|y_k) = E\{I(X|y_k)\} = \sum_{j=0}^{J-1} I(X|y_k)p(x_j|y_k) = \sum_{j=0}^{J-1} p(x_j|y_k)\log_2 \frac{1}{p(x_j|y_k)}, \qquad (9.2.29)$$

for all $k$ values, and define the average uncertainty about symbols at the input when one output symbol is received. This entropy can be considered to be a realization of a new random variable $H(X \mid y)$. For each $k$, the values $H(X \mid y_k)$ of the random variable $H(X \mid y)$ can be calculated and then assigned by a related conditional probability from the set $\{p(X \mid y_0), p(X \mid y_1), \ldots, p(X \mid y_{K-1})\}$. The calculated entropies can be presented in matrix form as

$$H(X|y_k) = [H(X|y_0) \ H(X|y_1) \ \cdots \ H(X|y_{K-1})]^T. \qquad (9.2.30)$$

The entropies in the matrix occur with the probabilities $\{p(y_0), p(y_1), \ldots, p(y_{K-1})\}$. Then, we can find the mean of the random variable $H(X \mid y_k)$, which represents the average amount of information about the input $X$ when symbols $y_k$, $k = 0, 1, \ldots, K - 1$, are received. This entropy is expressed as

$$
\begin{aligned}
H(X|Y) = E\{H(X|y_k)\} &= \sum_{k=0}^{K-1} H(X|y_k)p(y_k) \\
&= \sum_{k=0}^{K-1} \left[ \sum_{j=0}^{J-1} p(x_j|y_k)\log_2 \frac{1}{p(x_j|y_k)} \right] p(y_k) \\
&= \sum_{k=0}^{K-1}\sum_{j=0}^{J-1} p(y_k)p(x_j|y_k)\log_2 \frac{1}{p(x_j|y_k)} \\
&= \sum_{k=0}^{K-1}\sum_{j=0}^{J-1} p(x_j,y_k)\log_2 \frac{1}{p(x_j|y_k)}
\end{aligned}
\qquad (9.2.31)
$$

This conditional entropy is the average amount of information about the channel input when the channel outputs are observed. In other words, this is the mean value of information that *remains* about the channel input $X$ after all the channel outputs $y_k$ have been observed. Therefore, in terms of uncertainty, the channel entropy $H(X \mid Y)$ represents the uncertainty that remains about the channel input $X$ *after* the channel output $Y$ is observed and is called *the conditional entropy* of the channel input $X$ given the channel output $Y$.

**The output of the channel.** The information *provided* about the input symbol $x_j$ by the occurrence (reception) of the symbol $y_k$ can be expressed as the random variable

$$I\left(x_j;y_k\right) = I\left(x_j\right) - I\left(x_j|y_k\right) = \log_2\frac{p\left(x_j|y_k\right)}{p\left(x_j\right)} = -\log_2\frac{p\left(x_j\right)}{p\left(x_j|y_k\right)}, \qquad (9.2.32)$$

which is called the *mutual information* between the input symbol $x_j$ and the output symbol $y_k$. We can now find its mean value and express it as

$$I\left(X;Y\right) = H(X) - H\left(X|Y\right). \qquad (9.2.33)$$

As we said before, the entropy $H(X)$ represents the mean value of information (uncertainty) of the input of the channel $X$ *before* the channel output $Y$ has been observed. The conditional entropy $H(X \mid Y)$ represents the mean value of information (uncertainty) of the channel input $X$ that *remains* after the channel output $Y$ has been observed. Thus, their difference, $I(X; Y)$, represents the mean value of information (uncertainty) provided about the input $X$ *that is resolved* by observing the channel output $Y$.

The information $I(X; Y)$ can be expressed in terms of channel probabilities. Because the sum of conditional probabilities $p(y_k \mid x_j)$ for all $k$ values from 0 to $K - 1$ is 1, that is,

$$\sum_{k=0}^{K-1} p\left(y_k|x_j\right) = 1, \qquad (9.2.34)$$

the entropy of the channel input can be expressed as

$$
\begin{aligned}
H(X) &= \left[\sum_{j=0}^{J-1} p\left(x_j\right)\log_2\frac{1}{p\left(x_j\right)}\right] = \left[\sum_{j=0}^{J-1} p\left(x_j\right)\log_2\frac{1}{p\left(x_j\right)}\right] \cdot \sum_{k=0}^{K-1} p\left(y_k|x_j\right) \\
&= \sum_{j=0}^{J-1}\sum_{k=0}^{K-1} p\left(x_j\right)p\left(y_k|x_j\right)\log_2\frac{1}{p\left(x_j\right)} = \sum_{j=0}^{J-1}\sum_{k=0}^{K-1} p\left(x_j,y_k\right)\log_2\frac{1}{p\left(x_j\right)}
\end{aligned}
\qquad (9.2.35)
$$

Now we can calculate the mutual information as

$$
\begin{aligned}
I\left(X;Y\right) &= H(X) - H\left(X|Y\right) \\
&= \sum_{j=0}^{J-1}\sum_{k=0}^{K-1} p\left(x_j\right)p\left(y_k \mid x_j\right)\log_2\frac{1}{p\left(x_j\right)} - \sum_{k=0}^{K-1}\sum_{j=0}^{J-1} p\left(x_j\right)p\left(y_k \mid x_j\right)\log_2\frac{1}{p\left(x_j \mid y_k\right)} \\
&= \sum_{j=0}^{J-1}\sum_{k=0}^{K-1} p\left(x_j\right)p\left(y_k \mid x_j\right)\log_2\frac{p\left(x_j \mid y_k\right)}{p\left(x_j\right)} = \sum_{j=0}^{J-1}\sum_{k=0}^{K-1} p\left(x_j,y_k\right)\log_2\frac{p\left(x_j \mid y_k\right)}{p\left(x_j\right)}
\end{aligned}
$$

$$(9.2.36)$$

The mutual information has the following properties:

1. The mutual information of a discrete memoryless channel is symmetric, that is,

$$I(X;Y) = I(Y;X) = H(Y) - H(Y|X).$$ (9.2.37)

2. This information is always non-negative, that is,

$$I(X;Y) \geq 0.$$ (9.2.38)

3. This information can be expressed as

$$I(X;Y) = H(X) + H(Y) - H(X,Y),$$ (9.2.39)

where the joint entropy $H(X, Y)$ is defined as the mean value of the joint information $I(x_j, y_k)$, that is,

$$
\begin{aligned}
H(X,Y) = E\left\{I\left(x_j, y_k\right)\right\} &= \sum_{k=0}^{K-1}\sum_{j=0}^{J-1} I\left(x_j, y_k\right) \cdot p\left(x_j, y_k\right) \\
&= \sum_{k=0}^{K-1}\sum_{j=0}^{J-1} \log_2 \frac{1}{p\left(x_j, y_k\right)} \cdot p\left(x_j, y_k\right) = H(X) + H(Y|X)
\end{aligned}
$$ (9.2.40)

**The channel capacity.** In practice, we want to resolve the information content of the input, given the output available. Thus, we want to maximize the mutual information, which leads us to the definition of the channel capacity. *The capacity of a discrete memoryless channel is defined as the maximum of mutual information in any single use of the channel, taken over all possible input probability distributions $p(x_j)$ in the channel input alphabet $A_x$, which is expressed as*

$$C = \max_{p(x_j)} I(X;Y) = \max_{p(x_j)} \sum_{j=0}^{J-1}\sum_{k=0}^{K-1} p\left(x_j\right) p\left(y_k|x_j\right) \log_2 \frac{p\left(x_j|y_k\right)}{p\left(x_j\right)}.$$ (9.2.41)

In other words, the channel capacity represents the maximum mean value of information (uncertainty) *that is resolved* about the input $X$ by observing the channel output $Y$. It is measured in bits per channel use, or bits per transmission. The capacity depends only on the conditional probabilities $p(y_k \mid x_j)$ because the time maximization should be done with respect to all possible a priori probabilities $p(x_j)$.

### 9.2.3.2 *Example of the Capacity of a Binary Memoryless Channel*

A simplified structure of a binary communication system with a binary memoryless channel is presented in Figure 9.2.6. A discrete memoryless source that generates two symbols is called

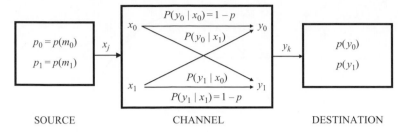

SOURCE                    CHANNEL                    DESTINATION

**Fig. 9.2.6** *Definition of a binary channel in a probabilistic sense.*

a binary source. It is characterized by an alphabet of two symbols, $A = (m_0, m_1)$, where the symbols are generated by the respective probabilities $P = (p_0, p_1)$. Assume that these symbols are 0 and 1. They are generated with probabilities $P(0) = p_0$ and $P(1) = p_1$, respectively, and applied to the channel input. The symbols at the input of the channel, $x_0$ and $x_1$, are taken directly from the source and transmitted through a binary symmetric channel without memory and defined by the bit error probability $p$. We want to know the capacity of the channel in this communication system.

The entropy of the source can be expressed as

$$H = -\sum_{k=0}^{1} p_k \log_2 p_k = -p_0 \log p_0 - p_1 \log p_1. \tag{9.2.42}$$

This entropy defines the average number of information bits per generated symbol. In this case, a symbol is a message bit. In order to find the entropy of the source for any value of the probability of a symbol being 0, $p_0$, we can define the entropy function as

$$H(p_0) = -p_0 \log p_0 - (1 - p_0) \log (1 - p_0), \tag{9.2.43}$$

which is plotted in Figure 9.2.7.

One of two source messages can appear at the channel input as the input symbol $x_j$. The input of the channel can have two binary values, 0 and 1, with probabilities that are the same as those for the corresponding source symbols. In this problem, we will assume that the message symbols and the input to the channel are the same and can have the binary values 0 and 1. Therefore, the entropy of the input is equal to the entropy of the source, that is,

$$\begin{aligned} H(X) &= -\sum_{j=0}^{1} p(x_j) \log_2 p(x_j) = -p_0 \log p_0 - p_1 \log p_1 \\ &= -p_0 \log p_0 - (1 - p_0) \log (1 - p_0) \end{aligned} \tag{9.2.44}$$

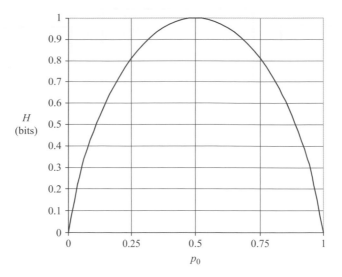

**Fig. 9.2.7** *Entropy of a binary source as a function of bit probability.*

The channel capacity is defined as the maximum of the mutual information taken across all possible distributions of the input symbols of the channel, that is,

$$C = \max_{p(x_j)} I(X;Y) = \max_{p(x_j)} [H(X) - H(X|Y)]. \qquad (9.2.45)$$

Using the expression for the conditional entropy, and keeping in mind the relation for the joint probability $p(x_j, y_k) = p(x_j)p(y_k \mid x_j)$, the conditional entropy can be derived in the form

$$
\begin{aligned}
H(X|Y) &= \sum_{k=0}^{K-1} \sum_{J-1}^{j=0} p(x_j, y_k) \log_2 \frac{1}{p(x_j|y_k)} \\
&= p(x_0, y_0) \log_2 \frac{1}{p(x_0|y_0)} + p(x_1, y_1) \log_2 \frac{1}{p(x_1|y_1)} \\
&\quad + p(x_0, y_1) \log_2 \frac{1}{p(x_0|y_1)} + p(x_1, y_0) \log_2 \frac{1}{p(x_1|y_0)} \\
&= -p_0 (1-p) \log_2 (1-p) - (1-p_0)(1-p) \log_2 (1-p) \\
&\quad - p_0 p \log_2 p - (1-p_0) p \log_2 p
\end{aligned}
\qquad (9.2.46)
$$

The conditional entropy depends on the bit error probability $p$ in the channel. However, it also depends on the probability distribution of the input symbols, like the entropy of the input $X$. Therefore, the capacity in eqn (9.2.46) depends on the probability of the symbol 0, and the maximization of the mutual information will be performed in respect to this probability, that is,

$$C = \max_{p_0} I\,(X;Y) = \max_{p_0} [H(X) - H\,(X|Y)]. \tag{9.2.47}$$

The first derivative of $H(X \mid Y)$ with respect to $p_0$ is a constant. Therefore, the maximum of the mutual information is where the maximum of $H(X)$ is, and that is proven to be for $p_0 = 0.5$. Therefore, for a binary symmetric channel without memory, the capacity of the channel is

$$C = \max_{p_0} [H(X) - H\,(X|Y)]\Big|_{p_0=1/2} = \left[ -\frac{1}{2} \log \frac{1}{2} - \frac{1}{2} \log \frac{1}{2} \right.$$
$$+ \frac{1}{2}(1-p)\log_2(1-p) + \left(1 - \frac{1}{2}\right)(1-p)\log_2(1-p) + \frac{1}{2}p\log_2 p + \left(1 - \frac{1}{2}\right)p\log_2 p \Big].$$
$$= \left[ 1 + (1-p)\log_2(1-p) + p\log_2 p \right] = 1 - \left[ -(1-p)\log_2(1-p) - p\log_2 p \right] \tag{9.2.48}$$

For the bit error probability $p = 0.5$, the capacity of the channel is 0, meaning that the resolved uncertainty about the input when the output is observed is 0. When the bit error probability is different from 0.5, this resolution becomes greater than 0. There is a graph symmetry with respect to $p = 0.5$ that can be related to the transmission of a binary symbol and its inverted version (see Figure 9.2.8).

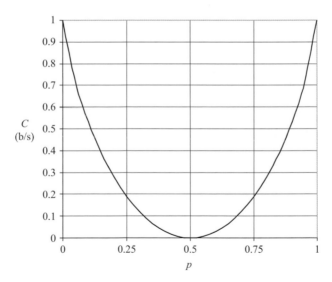

**Fig. 9.2.8** *Capacity of a binary channel as a function of the bit error probability; b/s, bits per second.*

## 9.3 Characterization of Continuous Channels

The source can generate a continuous-time signal that carries information about the phenomena it represents. In that case, the input signal of the channel can be represented by a random signal $x(t)$ that is a realization of a stochastic continuous-time process $X(t)$. The theory for this kind of process can be found in Chapter 19. We follow here the notation introduced in that chapter. At each time instant $t$ of the stochastic process $X(t)$, we define a random variable $X = X(t)$. Due to the continuum of the time variable $t$, we have a continuum of random variables $X = X(t)$. A block schematic of the channel, including the inputs, the outputs, and the channel itself, is shown in Figure 9.3.1.

If we sample the input process, we can get a discrete-time stochastic process, which is theoretically analysed in Chapter 3. In this case, the input and the output of the channel can be represented by the vectors $X$ and $Y$, respectively, which contain the random variables of the input and output stochastic processes, respectively, defined at discrete-time instants, as shown in Figure 9.3.1. The output discrete process is expressed as the sum of the input process and the noise process $N$ in the channel: $Y = X + N$. One realization of these discrete-time processes is a series of random numbers presented in a vector form. For example, one realization of the input stochastic process $X$ is $x = [x_1 \, x_2 \ldots x_M]$. At a particular time instant $t = m$, we define a random variable $X_m$ and its realization $x_m$ belonging to the process $X$.

### 9.3.1 Differential Entropy

Assume that a random variable $X$ can have continuous-valued realizations. Therefore, it can take an uncountable, infinite number of values in a limited or unlimited interval of possible values. In the case when two random variables $X$ and $Y$, representing the input and output of a communication channel, respectively, are continuous random variables, the *mutual information* can be obtained by analogy with the discrete case analysed in previous sections (see relation (9.2.36)), as

$$I(X;Y) = h(X) - h(X|Y) = \int_{-\infty}^{\infty} \int_{-\infty}^{\infty} f_{X,Y}(x,y) \log_2 \frac{f_X(x|y)}{f_X(x)} dx dy, \qquad (9.3.1)$$

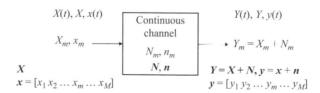

**Fig. 9.3.1** *Simplified structure of a continuous communication system.*

where $h(X)$ is the *differential entropy* of continuous random variable $X$ defined as

$$h(X) = \int_{-\infty}^{\infty} f_X(x) \log_2 \frac{1}{f_X(x)} dx, \tag{9.3.2}$$

and the *conditional differential entropy* of $X$ given $Y$ is defined as

$$h(X|Y) = \int_{-\infty}^{\infty} \int_{-\infty}^{\infty} f_{X,Y}(x,y) \log_2 \frac{1}{f_X(x|y)} dxdy. \tag{9.3.3}$$

Suppose a random vector $X$ consists of $M$ random variables $X_1, X_2, \ldots, X_M$. Its differential entropy is defined as the $M$-fold integral

$$h(X) = \int_x \cdots \int_{M-fold} f_X(x) \log_2 \frac{1}{f_X(x)} dx, \tag{9.3.4}$$

where $f_X(x)$ is the joint probability density function of the random vector $X$. Following eqn (9.3.1), the mutual information for two random vectors $X$ and $Y$ is given by the expression

$$I(X;Y) = \int_x \cdots \int_{M-fold} \int_y \cdots \int_{M-fold} f_X(x) f_{XY}(y|x) \log_2 \frac{f_{XY}(y|x)}{f_Y(y)} dxdy. \tag{9.3.5}$$

The input of the channel can be understood as a discrete-time stochastic process $X(m)$ with the random samples $x_1, x_2, \ldots, x_M$ of the related random variables $X_1, X_2, \ldots, X_M$ at the discrete-time instants. Likewise, the output of channel can be also understood as a discrete-time stochastic process $Y(m)$ with the random samples $y_1, y_2, \ldots, y_M$ of the related random variables $Y_1, Y_2, \ldots, Y_M$ at discrete-time instants. The theory for discrete-time stochastic processes that was presented in Chapter 3 can be applied in the analysis of this channel.

**Example**

Suppose $X_m = X$ is a Gaussian random variable defined by its density:

$$f_X(x) = \frac{1}{\sqrt{2\pi}\sigma_X} e^{-(x-\eta_X)^2/2\sigma_X^2}.$$

Find its differential entropy.

**Solution.** Keeping in mind the identity $\log_2 x = \log_e x \log_2 e$, we find that the entropy is

$$
\begin{aligned}
h(X) &= \int_{-\infty}^{\infty} f_X(x) \log_2 \sqrt{2\pi}\, \sigma_X e^{(x-\eta)^2/2\sigma_X^2} dx \\
&= \log_2 e \int_{-\infty}^{\infty} f_X(x) \left[ \frac{(x-\eta)^2}{2\sigma_X^2} + \log_e \sqrt{2\pi}\, \sigma_X \right] dx \\
&= \frac{1}{2\sigma_X^2} \log_2 e \int_{-\infty}^{\infty} (x-\eta)^2 f_X(x) dx + \log_2 e \log_e \sqrt{2\pi}\, \sigma_X \int_{-\infty}^{\infty} f_X(x) dx. \\
&= \frac{1}{2\sigma_X^2} \sigma_X^2 \log_2 e + \frac{1}{2} \log_2 e \log_e 2\pi \sigma_X^2 \\
&= \frac{1}{2} \log_2 e + \frac{1}{2} \log_2 2\pi \sigma_X^2 = \frac{1}{2} \log_2 2\pi e \sigma_X^2
\end{aligned}
\tag{9.3.6}
$$

## 9.3.2 Channel Information for Random Vectors

In the case when $X$ and $Y$ are continuous random variables, their mutual information is derived and expressed in terms of differential entropies in eqn (9.3.2). This expression is restricted to the transmission of one symbol through the channel, which is the transmission of one random amplitude value of the process $X(m)$ through the channel. However, in practice, we can say that the amplitudes are transmitted continuously through the channel until the last one is received. In the following analysis, we will present an analysis of the system for this realistic communication system, develop a related mathematical model of the channel, and calculate the channel capacity.

Suppose a time series of symbols is transmitted through the channel. Therefore, the input of the channel is a series of symbols $x = [x_1 \ x_2 \ldots x_m \ldots x_M]$, and the output of the channel is a corresponding series of the output symbols $y = [y_1 \ y_2 \ldots y_m \ldots y_M]$. The symbols can be continuous valued, as defined in Chapter 1. The input symbols are taken from the input alphabet, and the output symbols are taken from the output alphabet.

In order to analyse the properties of the channel, we will represent the input series $x$ as one realization of a random vector $X$ that consists of $M$ random variables $X_1, X_2, \ldots, X_M$. Likewise, the output series $y$ is represented by one realization of a random vector $Y$ that consists of $M$ random variables $Y_1, Y_2, \ldots, Y_M$. In this context, the series of random input samples $x = [x_1 \ x_2 \ldots x_m \ldots x_M]$ is one realization of the random vector $X$, and the series of random output samples $y = [y_1 \ y_2 \ldots y_m \ldots y_M]$ is one realization of the random vector $Y$.

The term *symbol*, which is used for discrete systems, is replaced here with the term *sample*, which represents the input or output process at discrete-time instants. These random samples are realizations of continuous-valued random variables contained in vectors $X$ and $Y$. Therefore, in their analysis from the information-theory point of view, will use the notion of differential entropy, as defined in classical information theory. The differential entropy of the channel input can be defined as the $M$-fold integral

$$h(X) = \int_x \cdots \int_{M\text{-}fold} f_X(x) \log_2 \frac{1}{f_X(x)} dx,$$　(9.3.7)

where $f_X(x)$ is the joint probability density function of the random variables defined inside the random vector $X$. The mutual information for the two random vectors $X$ and $Y$ is given by the expression

$$I(X;Y) = \int_x \cdots \int_{M\text{-}fold} \int_y \cdots \int_{M\text{-}fold} f_X(x) f_{XY}(y|x) \log_2 \frac{f_{XY}(y|x)}{f_Y(y)} dx dy.$$　(9.3.8)

## 9.3.3　Definition of the Capacity of a Continuous Channel

For a discrete channel, where the source generates symbol values, it was sufficient to characterize the source and the communication channel in terms of the related symbol probabilities and derive the expression for the channel capacity. For that case, we derived and expressed the mutual information as

$$I(X;Y) = H(X) - H(X|Y)$$
$$= \sum_{j=0}^{J-1} \sum_{k=0}^{K-1} p(x_j, y_k) \log_2 \frac{p(x_j|y_k)}{p(x_j)} = \sum_{j=0}^{J-1} \sum_{k=0}^{K-1} p(x_j, y_k) \log_2 \frac{p(y_k|x_j)}{p(y_k)}.$$　(9.3.9)

Thus, this information depends *not* only on the characteristics of the channel, which are defined by the conditional probabilities, *but also* on the a priori probabilities $p(x_j)$, which do not depend on the channel characteristics. Using the expression for the mutual information, the channel capacity can be expressed as

$$C = \max_{p(x_j)} \sum_{j=0}^{J-1} \sum_{k=0}^{K-1} p(x_j) p(y_k|x_j) \log_2 \frac{p(x_j|y_k)}{p(x_j)}.$$　(9.3.10)

In other words, the channel capacity is measured in information bits per channel use, or information bits per transmission. The capacity will depend only on the transition probabilities $p(y_k \mid x_j)$, after maximizing with respect to all possible a priori probabilities $p(x_j)$. The rate of information transmission for a continuous channel, expressed in information bits per symbol, is

$$R = H(X) - H(X|Y) \equiv h(X) - h(X|Y) = I(X;Y),$$　(9.3.11)

which is analogous to the way a discrete channel rate is defined. Then, the channel capacity may be defined as the maximum of this rate $R$ when the channel input is varied over all possible distributions of the channel input $f(x)$ and for the case when the duration of a symbol $T$ (in seconds per symbol) tends to infinity, that is,

$$C = \lim_{T \to \infty} \max_{f(x)} \frac{1}{T} R = \lim_{T \to \infty} \max_{f(x)} \frac{1}{T} I\left(X;Y\right) = \lim_{T \to \infty} \max_{f(x)} \frac{1}{T} \left[H(X) - H\left(X|Y\right)\right]$$

$$= \lim_{T \to \infty} \max_{f(x)} \frac{1}{T} \int_{-\infty}^{-\infty} \int_{-\infty}^{-\infty} f(x) f\left(y|x\right) \log_2 \frac{f(y|x)}{f(y)} \, dx dy \tag{9.3.12}$$

where $T$ is the duration of a symbol in seconds per symbol. For a channel with a limited bandwidth, it is possible to use the following method to calculate the capacity, starting with the following representation of the channel input, the channel itself, and the channel output.

### 9.3.4 Proof of the Channel Capacity Theorem

**The channel input.** Suppose the input of the continuous channel is a realization of the zero-mean stationary stochastic process $X(t)$ that is band limited to $B$ Hz (see Figure 9.3.1). This signal is uniformly sampled at the Nyquist rate of $2B$ samples per second, and an $M$-dimensional sample vector $x = [x_1 \ x_2 \ ... \ x_m \ ... \ x_M]$ is produced. Vector $x$ represents one realization or a sample of the random vector $X$ defined by $M$ random variables, that is, $X = [X_1 \ X_2 \ ... \ X_m \ ... \ X_M]$. The sample $x$ is transmitted in the time interval $T$ over the channel. Thus, the number of transmitted samples is

$$M = 2BT = T/T_s, \tag{9.3.13}$$

for the Nyquist sampling interval $T_s = 1/2B$. The power of the input signal is

$$P = E\left[X_m^2\right] = \sigma_x^2, \tag{9.3.14}$$

where $X_m$ is a random variable of the zero-mean stationary stochastic process $X(m)$ defined at the time instant $m$, that is, $X_m = X(m)$. The average power of the input signal can be expressed as

$$P_{av} = \frac{1}{T} \int_0^T E\left[X^2(t)\right] dt = \frac{1}{T} \sum_{m=1}^{M} E\left[X_m^2\right] = \frac{M\sigma_x^2}{T}, \tag{9.3.15}$$

assuming that the input of the channel is represented by the stochastic process $X(t)$. For a unit sampling instant $T_s = 1$, expressions (9.3.14) and (9.3.15) are equivalent, as expected. The sampling of the input signal must comply with all conditions required by the sampling theorem, as is pointed out in Chapters 3 and 13. Firstly, the sampled signal should be band limited and physically realizable, as defined in Chapter 1. The process defined here fulfils the required conditions. The distribution of the input $X = X_m$ that maximizes the rate of transmission is Gaussian, that is,

$$f_{X_m}(x_m) = \frac{1}{\sqrt{2\pi \sigma_x^2}} e^{-(x_m)^2/2\sigma_x^2}. \tag{9.3.16}$$

One random input signal $x(t)$, as one realization of stochastic process $X(t)$ among the ensemble of its realizations, can be represented by a sum of sinc functions expressed as

$$\hat{x}(t) = \sum_{m=-\infty}^{\infty} x_m \frac{\sin \pi (2Bt - m)}{\pi (2Bt - m)} = \sum_{k=-\infty}^{\infty} x_m \, \text{sinc} \, \pi (2Bt - m), \qquad (9.3.17)$$

as defined in Chapter 13. Here, $x_m$ are the samples of the signal $x(t)$ taken according to the sampling theorem at time instants $m/2B$, where $m = \ldots, -1, 0, 1, 2, \ldots$, and is defined on an unlimited interval. However, in real systems, we may assume that the value of $x_m$ is 0 beyond the point $t = T$ and that the signal is band limited in $B$. Then, the function $x(t)$ is represented by $M = 2TB$ points in a $2TB$-dimensional space. Of course, this is an approximation, because we cannot limit a signal in both the time and the frequency domain. For the sake of illustration, the channel is shown in Figure 9.3.2. The input signal, in both continuous time and discrete time, and the discrete-time continuous-valued output signal are shown in Figure 9.3.3.

**The channel.** The communication channel is characterized by an additive white Gaussian noise of zero mean, and a power spectral density of $N_0/2$, which is band limited to $B$ Hz, like the input signal. The precise definition of this channel can be found in Chapter 4. The noise sample is one realization $\mathbf{n} = [n_1 \; n_2 \; \ldots \; n_m \; \ldots \; n_M]$ of a random noise process presented in vector form as $\mathbf{N} = [N_1 \; N_2 \; \ldots \; N_m \; \ldots \; N_M]$. Each noise random variable $N_m$ is Gaussian, that is,

$$f_{N_m}(n_m) = \frac{1}{\sqrt{2\pi}\sigma_n} e^{-n_m^2/2\sigma_n^2}, \qquad (9.3.18)$$

with zero mean, and variance expressed as $\sigma_n^2 = N_0/2$.

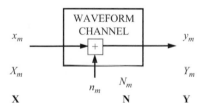

**Fig. 9.3.2** *A discrete-time channel with specified inputs and outputs.*

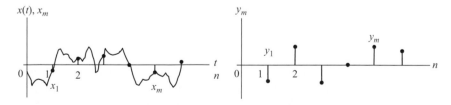

**Fig. 9.3.3** *The input signal in the continuous- and discrete-time domains, and the output signal in the discrete-time domain, representing the sum of the input signal and the discrete-time noise.*

**The output of the channel.** The output of the channel can be expressed as a new random variable, in fact, as a function of two random variables, expressed as

$$Y_m = X_m + N_m, \qquad (9.3.19)$$

where $m = 1, 2, \ldots, M$, or in matrix form as the sum of random functions of $M$ random variables, expressed as

$$\boldsymbol{Y} = \boldsymbol{X} + \boldsymbol{N} = [X_1 \ X_2 \ \cdots \ X_M] + [N_1 \ \ N_2 \ \ \cdots \ \ N_M], \qquad (9.3.20)$$

where one realization of the vector $\boldsymbol{Y}$ is a vector of random values $\boldsymbol{y} = [y_1 \ y_2 \ldots y_M]$, which can be expressed as

$$\begin{aligned} \boldsymbol{y} = \boldsymbol{x} + \boldsymbol{n} &= [x_1 \ x_2 \ \cdots \ x_M] + [n_1 \ \ n_2 \ \ \cdots \ \ n_M] \\ &= [x_1 + n_1 \ \ x_2 + n_2 \ \ \cdots \ \ x_M + n_M] \end{aligned}. \qquad (9.3.21)$$

Because the output is the sum of two uncorrelated and independent Gaussian random variables, its density is

$$f_{Y_m}(y_m) = \frac{1}{\sqrt{2\pi \sigma_y^2}} e^{-(y_m)^2/2\sigma_y^2} = \frac{1}{\sqrt{2\pi \left(\sigma_x^2 + \sigma_n^2\right)}} e^{-(y_m)^2/2\left(\sigma_x^2 + \sigma_n^2\right)}. \qquad (9.3.22)$$

Due to the statistical independence of the input $\boldsymbol{X}$ and the noise $\boldsymbol{N}$, the conditional probability density function of the output for the given input is defined as

$$\begin{aligned} f_{X_m Y_m}(y_m \mid x_m) &= \frac{1}{\sqrt{2\pi \sigma_n^2}} e^{-(y_m - x_m)^2/2\sigma_n^2} = \frac{1}{\sqrt{\pi N_0}} e^{-(y_m - x_m)^2/N_0} \\ &= f_{N_m}(y_m - x_m) \end{aligned}, \qquad (9.3.23)$$

because the variance of the Gaussian noise distribution is $\sigma_{N_m}^2 = N_0/2$.

**Channel capacity.** The capacity of the channel can be obtained from the mutual information for random vectors $\boldsymbol{X}$ and $\boldsymbol{Y}$ that is given by expression (9.3.8). As we said, the maximum occurs when the random variables in the vector $\boldsymbol{X} = [X_1 \ X_2 \ldots X_M]$ are statistically independent and zero-mean Gaussian random variables with the probability density function defined in eqn (9.3.16). Therefore, we first need to find this mutual information and then maximize it for Gaussian input. Due to the statistical independence of the variables inside vectors $\boldsymbol{X}$ and $\boldsymbol{Y}$, we may develop the $2 \times M$-fold integral into two-fold integrals as follows:

$$I(X;Y) = \int_x \cdots \int_{M\text{-}fold} \int_y \cdots \int_{M\text{-}fold} f_X(x_1,\ldots,x_M) f_{XY}(y_1,\ldots,y_M|x_1,\ldots,x_M)$$
$$\cdot \log_2 \frac{f_{XY}(y_1,\ldots,y_M|x_1,\ldots,x_M)}{f_y(y_1,\ldots,y_M)} dx_1 \ldots dx_M dy_1 \ldots dy_M$$
$$= \int_x \cdots \int_{M\text{-}fold} \int_y \cdots \int_{M\text{-}fold} \prod_{m=1}^{M} f_{X_m}(x_m) \prod_{m=1}^{M} f_{X_m Y_m}(y_m|x_m) \cdot \sum_{m=1}^{M} \log_2 \frac{f_{X_m Y_m}(y_m|x_m)}{f_{Ym}(y_m)} dx_m dy_m$$

$$(9.3.24)$$

The first term inside the integrals will contain the first term in the integrand sum and finish with the $M$th term in the integrand sum, that is,

$$I(X;Y) = \int_{-\infty}^{\infty} \int_{-\infty}^{\infty} f_{X_1}(x_1) f_{X_1 Y_1}(y_1|x_1) \log_2 \frac{f_{X_1 Y_1}(y_1|x_1)}{f_{Y_1}(y_1)} dx_1 dy_1$$
$$\cdot \int_{-\infty}^{\infty} \int_{-\infty}^{\infty} f_{X_2}(x_2) f_{X_2 Y_2}(y_2|x_2) dx_2 dy_2 \quad \cdots \quad \int_{-\infty}^{\infty} \int_{-\infty}^{\infty} f_{X_M}(x_M) f_{X_M Y_M}(y_M|x_M) dx_M dy_M$$
$$+ \cdots + \int_{-\infty}^{\infty} \int_{-\infty}^{\infty} f_{X_1}(x_1) f_{X_1 Y_1}(y_1|x_1) dx_1 dy_1 \cdot \int_{-\infty}^{\infty} \int_{-\infty}^{\infty} f_{X_2}(x_2) f_{X_2 Y_2}(y_2|x_2) dx_2 dy_2$$
$$\cdots \int_{-\infty}^{\infty} \int_{-\infty}^{\infty} f_{X_M}(x_M) f_{X_M Y_M}(y_M|x_M) \log_2 \frac{f_{X_M Y_M}(y_M|x_M)}{f_{Y_M}(y_M)} dx_M dy_M$$

Due to the statistical independence of random variables in the random vectors, most of the integral terms will be equal to 1, that is,

$$\int_{-\infty}^{\infty} \int_{-\infty}^{\infty} f_{X_m}(x_m) f_{X_m Y_m}(y_m|x_m) dx_m dy_m = 1.$$

Therefore, the expression for mutual information can be simplified as follows:

$$I(X;Y) = \int_{-\infty}^{\infty} \int_{-\infty}^{\infty} f_{X_1}(x_1) f_{X_1 Y_1}(y_1|x_1) \log_2 \frac{f_{X_1 Y_1}(y_1|x_1)}{f_{Y_1}(y_1)} dx_1 dy_1 + \cdots$$
$$+ \int_{-\infty}^{\infty} \int_{-\infty}^{\infty} f_{X_M}(x_M) f_{X_M Y_M}(y_M|x_M) \log_2 \frac{f_{X_M Y_M}(y_M|x_M)}{f_{Y_M}(y_M)} dx_M dy_M. \quad (9.3.25)$$
$$= \sum_{m=1}^{M} \int_{-\infty}^{\infty} \int_{-\infty}^{\infty} f_{X_m}(x_m) f_{X_m Y_m}(y_m|x_m) \log_2 \frac{f_{X_m Y_m}(y_m|x_m)}{f_{Y_m}(y_m)} dx_m dy_m$$

Because, as we said in eqn (9.3.23), the conditional probability density function of the output for a given input is

$$f_{X_m Y_m}(y_m|x_m) = f_{N_m}(y_m - x_m), \quad (9.3.26)$$

we obtain

$$I(X;Y) = \sum_{m=1}^{M} \int_{-\infty}^{\infty} \int_{-\infty}^{\infty} f_{X_m}(x_m) f_{N_m}(y_m - x_m) \log_2 \frac{f_{N_m}(y_m - x_m)}{f_{Y_m}(y_m)} dx_m dy_m. \qquad (9.3.27)$$

Then, the Gaussian distribution of the input, and, consequently, the Gaussian output that is a sum of two Gaussian variables, maximize this sum, that is,

$$
\begin{aligned}
\max_{f(x)} I(X;Y) &= \max_{f(x)} \sum_{m=1}^{M} \int_{-\infty}^{\infty} \int_{-\infty}^{\infty} f_{X_m}(x_m) f_{N_m}(y_m - x_m) \log_2 \frac{f_{N_m}(y_m - x_m)}{f_{Y_m}(y_m)} dx_m dy_m \\
&= \sum_{m=1}^{M} \int_{-\infty}^{\infty} \int_{-\infty}^{\infty} f_{X_m}(x_m) f_{N_m}(y_m - x_m) \log_2 \frac{\sqrt{2\pi\sigma_y^2} e^{-(y_m - x_m)^2/N_0}}{\sqrt{2\pi N_0/2} e^{-(y_m)^2/2\sigma_y^2}} dx_m dy_m \\
&= \sum_{m=1}^{M} \int_{-\infty}^{\infty} \int_{-\infty}^{\infty} f_{X_m}(x_m) f_{N_m}(y_m - x_m) \log_2 \sqrt{2\sigma_y^2/N_0} e^{-(y_m - x_m)^2/N_0 + (y_m)^2/2\sigma_y^2} dx_m dy_m
\end{aligned}
$$

Bearing in mind the identity $\log_2 e^x = \log_e e^x / \log_e 2$, we obtain

$$
\begin{aligned}
\max_{f(x)} I(X;Y) &= \log_2 \sqrt{2\sigma_y^2/N_0} \sum_{m=1}^{M} \int_{-\infty}^{\infty} f_{X_m}(x_m) \int_{-\infty}^{\infty} f_{N_m}(y_m - x_m) dy_m dx_m \\
&\quad + \frac{1}{\log_e 2} \sum_{m=1}^{M} \int_{-\infty}^{\infty} \int_{-\infty}^{\infty} f_{X_m}(x_m) f_{N_m}(y_m - x_m) \left[ -(y_m - x_m)^2/N_0 + (y_m)^2/2\sigma_y^2 \right] dy_m dx_m \\
&= \log_2 \sqrt{2\sigma_y^2/N_0} \sum_{m=1}^{M} \int_{-\infty}^{\infty} f_{X_m}(x_m) \int_{-\infty}^{\infty} f_{N_m}(y_m - x_m) dy_m dx_m \\
&\quad + \frac{1}{\log_e 2} \sum_{m=1}^{M} \int_{-\infty}^{\infty} \int_{-\infty}^{\infty} \left[ -(y_m - x_m)^2/N_0 \right] f_{X_m}(x_m) f_{N_m}(y_m - x_m) dy_m dx_m \\
&\quad + \frac{1}{\log_e 2} \sum_{m=1}^{M} \int_{-\infty}^{\infty} \int_{-\infty}^{\infty} \left[ (y_m)^2/2\sigma_y^2 \right] f_{X_m}(x_m) f_{N_m}(y_m - x_m) dy_m dx_m \\
&= I1 + I2 + I3
\end{aligned}
$$

$$(9.3.28)$$

The solution of the first integral gives the value $MK$, and the sum of the second and the third integrals is 0, where $K$ is

$$K = \log_2 \sqrt{2\sigma_y^2/N_0} = \log_2 \sqrt{2\left(\sigma_x^2 + N_0/2\right)/N_0} = \frac{1}{2}\log_2\left(\frac{2\sigma_x^2}{N_0} + 1\right), \qquad (9.3.29)$$

for $\sigma_y^2 = \sigma_x^2 + N_0/2$. The complete proof is in Appendix A. Therefore, the maximum value of the mutual information is

$$\max_{f_X(x)} I\,(X;Y) = MK. \tag{9.3.30}$$

*Thus, the capacity of this channel is the maximum of the mutual information in any single use of the channel, taken over all possible input probability densities* $f_{Xn}(x_n)$ *in the input signal.* The maximum of this mutual information, which is obtained when the random variables in the vector $X = [X_1\,X_2\,\dots\,X_N]$ are statistically independent and zero-mean Gaussian random variables with the same variance $\sigma_x{}^2$. This maximum can be obtained from eqns (9.3.29) and (9.3.30), bearing in mind that $M = 2BT$, in the following form:

$$\max_{f_X(x)} I\,(X,Y) = MK = \frac{1}{2}M\log_2\left(1 + \frac{2\sigma_x^2}{N_0}\right) = BT\log_2\left(1 + \frac{2\sigma_x^2}{N_0}\right). \tag{9.3.31}$$

From the average power, expressed by eqn (9.3.15), we can find the variance of the input. Inserting that variance into the expression for the maximum of the mutual information, eqn (9.3.31), we can get the capacity of the channel as *the maximum of the mutual information in any single use of the channel, taken over all possible input probability densities* $f_{Xm}(x_m)$ *in the input signal*, and express it in the form

$$C = \lim_{T \to \infty} \max_{f_X(x)} \frac{1}{T} I\,(X;Y) = \lim_{T \to \infty} \max_{f_X(x)} \frac{1}{T} BT\log_2\left(\frac{2\sigma_x^2}{N_0} + 1\right) = B\log_2\left(1 + \frac{P_{av}}{BN_0}\right), \tag{9.3.32}$$

because $B = M/2T$ is a constant. This expression is the famous Hartley–Shannon theorem, which can be expressed as

$$C = B\log_2\left(1 + \frac{S}{N}\right), \tag{9.3.33}$$

where $S/N$ is the mean-square signal-to-noise ratio. Therefore, the capacity of a channel affected by additive white Gaussian noise is linearly proportional to the channel bandwidth $B$ and logarithmically depends on the signal-to-noise ratio in the channel.

**Example**

For the conditional density function

$$f_{YX}\,(y|x) = \frac{1}{\sqrt{2\pi}\,\sigma_n} e^{-(y-x)^2/2\sigma_n^2},$$

calculate its differential entropy and find the mutual information.

**Solution.** The differential entropy is

$$
\begin{aligned}
h(Y|X) &= \int_{-\infty}^{\infty} \int_{-\infty}^{\infty} f_{XY}(x,y) \log_2 \frac{1}{f_{XY}(y|x)} \, dx \, dy \\
&= \int_{-\infty}^{\infty} \int_{-\infty}^{\infty} f_X(x) f_{XY}(y|x) \log_2 \left[ \sqrt{2\pi}\sigma_n e^{(y-x)^2/2\sigma_n^2} \right] dx \, dy \\
&= \int_{-\infty}^{\infty} f_X(x) \int_{-\infty}^{\infty} f_{XY}(y|x) \left[ \log_2 \sqrt{2\pi}\sigma_n + \frac{(y-x)^2}{2\sigma_n^2} \log_2 e \right] dy \, dx \\
&= \int_{-\infty}^{\infty} f_X(x) \left[ \int_{-\infty}^{\infty} \log_2 \sqrt{2\pi}\sigma_n f_{XY}(y|x) \, dy + \frac{\log_2 e}{2\sigma_n^2} \int_{-\infty}^{\infty} (y-x)^2 f_{XY}(y|x) \, dy \right] dx \\
&= \int_{-\infty}^{\infty} f_X(x) \left[ \log_2 \sqrt{2\pi}\sigma_n + \frac{\log_2 e}{2\sigma_n^2} \sigma_n^2 \right] dx \\
&= \int_{-\infty}^{\infty} f_X(x) \left[ \frac{1}{2}\log_2 2\pi\sigma_n^2 + \frac{\log_2 e}{2} \right] dx = \frac{1}{2}\log_2 2\pi\sigma_n^2 + \frac{\log_2 e}{2} = \frac{1}{2}\log_2 2\pi e\sigma_n^2
\end{aligned}
$$

Therefore, the mutual information is

$$
\begin{aligned}
I(X;Y) &= h(Y) - h(Y|X) \\
&= \frac{1}{2}\log_2 2\pi e \left(\sigma_x^2 + \sigma_n^2\right) - \frac{1}{2}\log_2 2\pi e\sigma_n^2 \\
&= \frac{1}{2}\log_2 \frac{2\pi e \left(\sigma_x^2 + \sigma_n^2\right)}{2\pi e\sigma_n^2} = \frac{1}{2}\log_2 \left(1 + \frac{\sigma_x^2}{\sigma_n^2}\right)
\end{aligned}
$$

**Special case: A noiseless channel.** When variance $\sigma_n^2$ approaches 0 (no noise), the output $Y$ approximates the input $X$ more and more exactly, and mutual information approaches infinity. This is expected because the self-information of any given sample value of $X$ should be infinity, that is,

$$
I(X;Y) = \lim_{\sigma_n^2 \to 0} \frac{1}{2}\log_2 \left(1 + \frac{\sigma_x^2}{\sigma_n^2}\right) \to \infty. \tag{9.3.34}
$$

In this case, the entropy of the output is equal to the entropy of the input, that is,

$$
h(Y) = \lim_{\sigma_n^2 \to 0} \frac{1}{2}\log_2 2\pi e \left(\sigma_x^2 + \sigma_n^2\right) \to \frac{1}{2}\log_2 2\pi e \left(\sigma_x^2\right) = h(X), \tag{9.3.35}
$$

meaning that, by receiving $Y$, we can completely resolve what was $X$ at the input of the channel.

**Special case: Infinite noise power.** In this case, the variance $\sigma_n{}^2$ approaches infinity (a noisy channel), the output $Y$ cannot resolve what was at the channel input $X$, and the mutual information approaches 0:

$$I(X;Y) = \lim_{\sigma_n^2 \to \infty} \frac{1}{2} \log_2 \left(1 + \frac{\sigma_x^2}{\sigma_n^2}\right) \to 0. \tag{9.3.36}$$

In this case, the entropy of the output tends to the entropy of the noise source, and nothing can be resolved about the input $X$ after receiving $Y$, that is,

$$h(Y) = \lim_{\sigma_n^2 \to \infty} \frac{1}{2} \log_2 2\pi e \left(\sigma_x^2 + \sigma_n^2\right) \to \frac{1}{2} \log_2 2\pi e \sigma_n^2 = h(N). \tag{9.3.37}$$

## 9.4 Capacity Limits and the Coding Theorem

### 9.4.1 Capacity Limits

Capacity can be expressed as

$$C = B \log_2 \left(1 + \frac{S}{N}\right) = B \log_2 \left(1 + \frac{P_{av}}{BN_0}\right). \tag{9.4.1}$$

If the bandwidth tends to infinity, the capacity limit is

$$\begin{aligned} C_\infty &= \lim_{B \to \infty} C = \lim_{B \to \infty} B \log_2 \left(1 + \frac{P_{av}}{BN_0}\right) = \frac{1}{\ln 2} \lim_{B \to \infty} B \ln \left(1 + \frac{P_{av}}{BN_0}\right) \\ &= \frac{1}{\ln 2} \lim_{B \to \infty} B \cdot \left(\frac{P_{av}}{BN_0} - \frac{1}{2}\left(\frac{P_{av}}{BN_0}\right)^2 + \frac{1}{3}\left(\frac{P_{av}}{BN_0}\right)^3 + \ldots\right) = \frac{P_{av}}{N_0} \frac{1}{\ln 2}. \end{aligned} \tag{9.4.2}$$

The energy of a bit is $E_b = P_{av} T_b = P_{av}/R$, and the capacity limit is

$$C_\infty = \frac{P_{av}}{N_0} \frac{1}{\ln 2} = \frac{E_b R}{N_0} \frac{1}{\ln 2} = \frac{E_b R}{N_0} \log_2 e. \tag{9.4.3}$$

If we apply the channel-coding theorem expressed by the condition

$$\frac{C_\infty}{R} > 1, \tag{9.4.4}$$

we obtain

$$\frac{C_\infty}{R} = \frac{E_b}{N_0} \frac{1}{\ln 2} > 1 \tag{9.4.5}$$

which gives the following limit: *in any reliable communication system with a Gaussian noise channel, the ratio $E_b/N_0$ cannot be less than the fundamental Shannon limit, which is calculated as*

$$\frac{E_b}{N_0} > \ln 2 = 0.693 = -1.6 \; dB. \tag{9.4.6}$$

This limit is shown in Figure 9.4.1 with a bold vertical line. Also, as long as $E_b/N_0$ exceeds this limit, Shannon's channel-coding theorem guarantees the existence of a system for reliable communication over the channel.

**Possible coding gain in digital communications.** The quality of signal transmission from the modulator input to the demodulator output is expressed by the bit error probability $p$. For example, if the system uses binary phase shift keying, the bit error probability can be expressed as a function of the signal-to-noise ratio:

$$p = \frac{1}{2} \mathrm{erfc} \sqrt{\frac{E_b}{N_0}}, \tag{9.4.7}$$

where $E_b$ is the energy of the bit transmitted, and $N_0/2$ is the two-sided power spectral density of the additive white Gaussian noise in the channel. An example of bit error probability as a function of the signal-to-noise ratio $E_b/N_0$ is presented in Figure 9.4.1, along with the

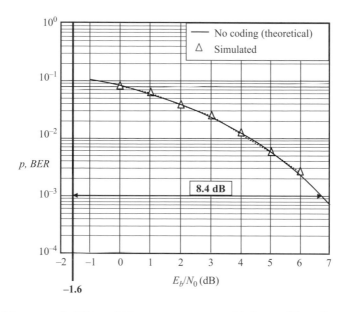

**Fig. 9.4.1** *The bit error probability and bit error rate curves, with the possible coding gain indicated; p, bit error probability BER, bit error rate.*

simulated bit error rate (indicated by triangles). We can see that, with a bit error probability of $10^{-3}$, an additive white Gaussian noise channel will require $E_b/N_0$ to be at least 6.8 dB. Therefore, for this $E_b/N_0$, the potential coding gain is 8.4 dB.

It is important to distinguish the terms bit error probability and bit error rate. According to the notation used in this book, the bit error probability $p$ is a theoretically derived expression, and the bit error rate is an estimate of the bit error probability obtained by simulation, emulation, or measurements for a defined signal-to-noise ratio. In other words, bit error rate values are estimates of the corresponding theoretically expected $p$ values for each value of the signal-to-noise ratio.

## 9.4.2   The Coding Theorem and Coding Channel Capacity

**Rate of information.** The channel capacity is related to the rate of information $R$, which is defined for the source output and the input of a channel, and expressed as

$$R = R_{Sym}H, \tag{9.4.8}$$

which is expressed in bits per second, where $R_{Sym}$ is the symbol rate, and $H$ is the average number of information bits per source symbol, or the entropy of the source. The limiting rate $R$ of information transmission through the channel is the *channel capacity C*, that is,

$$R \le C. \tag{9.4.9}$$

**Shannon's channel-coding theorem:** If a discrete memoryless channel has capacity $C$, and the source generates information at a rate $R < C$, then there exists a coding technique such that the information can be transmitted over the channel with an *arbitrarily low probability* of symbol error.

We can also say that it is not possible to transmit messages without error if $R > C$. In that sense, the channel capacity is sometimes expressed as the maximum rate of reliable transmission through the channel.

Discovering coding techniques which let $R = C$ and allow an arbitrarily low bit error probability can be considered a major engineering challenge. *Turbo codes* and *low-density parity-check codes* nearly achieved this theoretical limit. In practice, we often use the ratio of the information rate and the channel capacity,

$$\mu = \frac{R}{C}, \tag{9.4.10}$$

as a measure of the *efficiency* of a communication system.

**Capacity in communication systems with channel coding.** A general scheme of a coding communication system is presented in Figure 9.4.2, which is a modification of Figure 9.1.1 that includes a modulator and a demodulator. This modification is done to present in a better way the capacity of a channel when both a channel encoder and a channel decoder are used in the communication system.

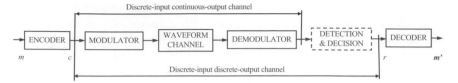

**Fig. 9.4.2** *Definition of a channel for a coding communication system.*

The transmitter is composed of a channel encoder and a modulator. The receiver includes a demodulator with a detector/decision device and a channel decoder. We will assume that the encoder inputs are binary message symbols (message bits) *m* and that its outputs are encoded bits *c*. These encoded bits modulate a carrier, and a modulated signal is transmitted through a communication channel characterized by additive white Gaussian noise and other impairments. The encoded bits are represented by two discrete values: 0 and 1.

The demodulator demodulates the received signal and produces a continuous-valued output that is the sum of the discrete encoded term and the channel random term, as explained in Chapter 5 for the correlation demodulator. From the coding theory point of view, we define a *discrete-input continuous-output channel* as a channel that incorporates a modulator, a waveform channel, and a demodulator, as shown in Figure 9.4.2. The output of this channel is a continuous-valued discrete-time process, as explained in Chapter 5.

If the output of the demodulator is used to determine a discrete binary value of the encoded bit, the channel will also include a detector/decision block. The channel that includes this block on top of a discrete-input continuous-output channel is called a *discrete-input discrete-output channel*. In the case when these discrete-output values are expressed in digital form, the channel is called a *digital channel*. In our case of binary signal transmission, these encoder output values and decision-device output discrete values are represented by the members of the additive group (0, 1) and sometimes by members of the multiplicative group (1, −1). Because this channel has both binary inputs and binary outputs, it is called a *binary digital channel*, as was discussed in Section 9.2.2.

In the following section, the basic operation of the system is presented, with a focus on the calculation of the capacity of the binary channel.

**Transmitter operation.** The input of the encoder is a binary message generated from the source with the bit rate $R_b = 1/T_b$. This message is encoded by adding a number of redundancy bits. The encoder is defined by its code rate $R = k/n$, where *k* is the number of input bits *m* needed to produce *n* encoded bits *c*. Thus, the data rate at the output of the encoder is increased with respect to its input to $R_c = R_b/R = nR_b/k$, and the duration of an encoded bit is reduced with respect to its input value to $T_c = 1/R_c = k/nR_b = T_b k/n$. In this way, we have introduced redundancy into the message, which results in an increase in the channel bandwidth.

The modulator takes a set of encoded bits and produces a corresponding waveform that is suitable for transmission in a waveform channel. In the general case, the modulator takes *l* encoded bits and generates one waveform. This waveform is taken from the set of $M = 2^l$ possible waveforms, which are called symbols. The duration of a symbol corresponds to the duration of an *l* bit-encoded bit sequence. Thus, the rate of these symbols is $R_S = R_c/l$, and the duration is *l* times greater than the duration of an encoded bit, that is,

$$T_S = lT_c = lT_b k/n = lkT_b/n, \text{ and } T_b = nT_S/lk. \tag{9.4.11}$$

**Receiver operation.** The demodulator receives the symbols' waveforms at the symbol rate $R_S$, demodulates them, and then generates estimates of *l* encoded bits. These bits appear at the input of the decoder with the encoded bit rate $R_c$. There are two possibilities here. If detection and decision are not performed, the received bits have continuous values that are applied directly to the inputs of the decoder. If detection and decision are performed, the received bits have discrete values at the output of the detector and are applied in binary form to the input of the decoder. After being decoded, the estimates of the message bits, $m'$, are generated at the output of the decoder with the message rate $R_b$.

In our explanation, we did not mention the delay due to bit processing. This is a separate topic which is outside the main stream of our explanation.

**Calculation of capacity and its limits.** Suppose that the major limitations for signal transmission are thermal noise and finite bandwidth *B*, where the bandwidth is directly related to the data rate of the signal in the channel. Firstly, the bandwidth limitation of the channel is quantitatively expressed in the number of bits transmitted per second per hertz of the bandwidth, that is,

$$\eta = \frac{R_b}{B}, \tag{9.4.12}$$

which is expressed in bits per second per hertz. This can be expressed as a function of the symbol rate $R_S$ in the waveform channel as

$$\eta = \frac{R_b}{B} = \frac{1}{BT_b} = Rl\frac{R_S}{B} \tag{9.4.13}$$

for $T_b = nT_S/lk$. The minimum bandwidth *B* for the modulated signal is equal to the symbol rate $R_S$. Thus, the maximum spectral efficiency is

$$\eta_{max} = Rl\frac{R_S}{B} = Rl\frac{R_S}{R_S} = Rl. \tag{9.4.14}$$

Secondly, the Gaussian noise present in the channel is measured by its two-sided power spectral density $N_0/2$. It causes errors in signal transmission, which are theoretically expressed in terms of the bit error probability *p*, and in terms of the bit error rate in the case of system simulation or measurement. The influence of noise can be reduced by increasing the power of the signal transmitted. Thus, the power efficiency of the system is determined by the ratio of the bit energy and the noise power spectral density, $E_b/N_0$, that is required in a channel to achieve a specified bit error probability *p*.

Due to coding and modulation, the channel bandwidth required for signal transmission changes. Thus, for the bit error probability $p$ required, we define the bit signal-to-noise power ratio, $P_b/P_N$, as a function of the required signal-to-noise energy ratio, $E_b/N_0$, and the encoder–modulator parameters, that is,

$$\frac{S}{N} = \frac{P_b}{P_N} = \frac{P_b T_b}{P_N T_b} = \frac{kl}{n} \frac{E_b}{P_N T_s} = l\frac{k}{n} \frac{E_b}{P_N/(1/T_s)} = IR \frac{E_b}{P_N/B_s} = IR \frac{E_b}{N_0} = \frac{R_b}{R_s} \frac{E_b}{N_0}, \qquad (9.4.15)$$

where $B_S$ is the bandwidth of the transmitted symbols, and $IR = R_b/R_S$. The maximum rate at which a message can be transmitted over a noisy channel is specified by the Shannon–Hartley theorem,

$$C = B \log_2 \left( 1 + \frac{S}{N} \right) = B \log_2 \left( 1 + IR \frac{E_b}{N_0} \right). \qquad (9.4.16)$$

The Shannon coding theorem states that there are coding schemes that guarantee information transmission with an arbitrarily small bit error probability if the information rate $R_b$ is less than the capacity of the channel, that is,

$$R_b \leq C. \qquad (9.4.17)$$

The maximum value of the defined spectral efficiency is $\eta_{max} = Rl$. This maximum occurs when the information rate is equal to the capacity

$$R_b = C = B \log_2 \left( 1 + \frac{S}{N} \right), \qquad (9.4.18)$$

which gives the maximum spectral efficiency as

$$\eta_{max} = \frac{R_b}{B} = \frac{C}{B} = \log_2 \left( 1 + \frac{S}{N} \right) = \log_2 \left( 1 + IR \frac{E_b}{N_0} \right) = \log_2 \left( 1 + \eta_{max} \frac{E_b}{N_0} \right). \qquad (9.4.19)$$

From this expression, the $E_b/N_0$ required for transmission without errors is

$$\frac{E_b}{N_0} = \frac{2^{\eta_{max}} - 1}{\eta_{max}}.$$

When $B$ tends to infinity, the maximum spectral efficiency tends to 0, and the minimum required $E_b/N_0$ is $-1.59$ dB, due to

$$\lim_{B \to \infty} \frac{E_b}{N_0} = \lim_{B \to \infty} \frac{2^{\eta_{max}} - 1}{\eta_{max}} = \lim_{B \to \infty} \frac{2^{C/B} - 1}{C/B} \Bigg|_{\text{substitute } C/B = x}$$
$$= \lim_{x \to 0} \frac{2^x - 1}{x} = \lim_{x \to 0} \frac{2^x \ln 2}{1} = \ln 2 = 0.693 = -1.6 \, dB \qquad (9.4.20)$$

Therefore, in a reliable coding communication system with a Gaussian noise channel, the ratio $E_b/N_0$ cannot be less than the fundamental Shannon limit of $-1.6$ dB. As long as the signal-to-noise-ratio exceeds this limit, there exists a system for reliable communication over the channel.

**Example**

A black-and-white TV picture frame may be considered to be composed of approximately 300,000 picture elements. Assume that each picture element is equiprobable among ten distinguishable brightness levels. Assume that 30 picture frames are transmitted each second. Calculate the minimum bandwidth required to transmit the video signal, assuming that a 30 dB signal-to-noise ratio is necessary for satisfactory picture reception.

**Solution.** The information per picture element $m_k$ is

$$I(m_k) = \log_2\left(\frac{1}{p_k}\right) = \log_2\left(\frac{1}{1/10}\right) = 3.32 \text{ bits/picture elements.}$$

The information per frame for the equiprobable picture elements is

$$I = 300000 \cdot I(m_k) = 300000 \cdot 3.32 = 996,000 \text{ bits/frame.}$$

Because 30 picture frames are transmitted per second, the total information rate is

$$R = 30 \cdot I = 30 \cdot 996,000 = 29.88 \left(\text{ frames/ sec}\right) / \times \text{bits/frame}\right) = \text{bits/ sec.}$$

Let the rate be equal to the capacity of channel. Then the bandwidth can be calculated according to the Hartley–Shannon theorem as

$$B_{\min} = \frac{C}{\log_2\left[1 + \frac{S}{N}\right]} = \frac{29.88 \times 10^6}{(3.32)(3.0004)} = 2.9996 \text{ MHz.}$$

This value is close to that of a commercial TV, which actually uses a bandwidth of about 4 MHz.

## 9.5   Information and Entropy of Uniform Density Functions

The entropy of a source generating a stochastic process can be analysed from an information-theory point of view. To this end, a stochastic process generator can be treated as a source of information in an information-theory sense. If the source has a countable (finite or infinite) number of outcomes, the information content of the outcomes can be found and the entropy, defined as the average information per outcome, can be quantified according to the principles of information theory. Likewise, if the source produces a continuous-time process (with an infinite, uncountable number of possible outcomes), the differential entropy can be calculated as the mean value of the information content of the process. The details related to entropy for different sources of information can be found in the previous sections.

The entropy of a continuous-time stochastic process $X(t)$ cannot be calculated directly, because the number of elementary events is uncountably large, so we cannot assign finite probabilities to them. The entropy can be obtained by discretizing the density function, finding the entropy of the discrete one, and then finding the limit when the discretization interval tends to 0, as Popoulis and Pillai presented in their book (2002, p. 654). Then, the entropy can be calculated according to the expression

$$H = -\int_{-\infty}^{\infty} f_X(x)\log f_X(x)dx,$$

where the logarithm is assumed to be base 2. In this section, the information and entropy will be calculated for both uniform and Gaussian stochastic processes.

### 9.5.1   Continuous Uniform Density Functions

**Continuous uniform untruncated density.** The density function of a uniform continuous random variable $x$ can be expressed as

$$f_c(x) = \begin{cases} \frac{1}{2T_c} & -T_c \leq x < T_c \\ 0 & \text{otherwise} \end{cases} = \begin{cases} \frac{1}{2\sigma_c\sqrt{3}} & -T_c \leq x < T_c \\ 0 & \text{otherwise} \end{cases} \tag{9.5.1}$$

and is graphically presented in Figure 9.5.1, for a mean value equal to 0 and the variance $\sigma_c^2 = T_c^2/3$.

The *differential entropy* of the function is

$$h(X) = \int_{-\infty}^{\infty} f_c(x)\log_2 f_c(x)dx = -\int_{-T_c}^{T_c} \frac{1}{2T_c}\log_2 \frac{1}{2T_c}dx = -\frac{x}{2T_c}\log_2 \frac{1}{2T_c}\Big|_{-T_c}^{T_c}.$$

$$= -\frac{2T_c}{2T_c}\log_2 \frac{1}{2T_c} = \log_2 2T_c = \log_2 2\sigma_c\sqrt{3} \tag{9.5.2}$$

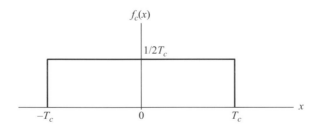

$f_c(x)$

$1/2T_c$

$-T_c$      $0$      $T_c$      $x$

**Fig. 9.5.1** *A continuous uniform density function.*

The *information content* is

$$I(X) = -\log_2 f_X(x) = -\log_2 \frac{1}{2T_c} = \begin{cases} \log_2 2T_c & -T_c \leq x < T_c \\ 0 & \text{otherwise} \end{cases}$$

$$= \begin{cases} \log_2 2\sigma_c\sqrt{3} & -T_c \leq x < T_c \\ 0 & \text{otherwise} \end{cases}.$$

(9.5.3)

---

**Example**

Calculate the information content $I(X)$ for $2T_C = 1$, $2T_C < 1$, and $2T_C > 1$.

**Solution.** The solution is as follows:

$$2T_c = 1 \Rightarrow T_c = 1/2, I(X) = \begin{cases} \log_2 2T_c = 0 & -T_c \leq x < T_c \\ 0 & \text{otherwise} \end{cases}$$

$$2T_c > 1 \Rightarrow T_c > 1/2, I(X) = \begin{cases} \log_2 2T_c > 0 & -T_c \leq x < T_c \\ 0 & \text{otherwise} \end{cases}.$$

$$2T_c < 1 \Rightarrow T_c < 1/2, I(X) = \begin{cases} \log_2 2T_c < 0 & -T_c \leq x < T_c \\ 0 & \text{otherwise} \end{cases}$$

---

**Continuous uniform truncated density.** A truncated uniform density function can be derived from eqn (9.5.1) and expressed as

$$f_{ct}(x) = \begin{cases} \frac{1}{2(T_c - A)} = \frac{1}{2\sigma_{ct}\sqrt{3}} & -T_c + A \leq x < T_c - A, A \geq 0 \\ 0 & \textit{otherwise} \end{cases},$$

(9.5.4)

where $A$ is the truncation factor. The function is presented in Figure 9.5.2. The mean value is 0, and the variance is

$$\sigma_{ct}^2 = \frac{(T_c - A)^2}{3} = \frac{T_c^2}{3}(1 \quad A/T_c)^2 - \sigma_c^2(1 \quad A/T_c)^2.$$

(9.5.5)

The *differential entropy* of the function is

$$h(X) = \int_{-\infty}^{\infty} f_X(x)\log_2 f_X(x)dx = -\int_{-T_c+A}^{T_c-A} \frac{1}{2(T_c - A)} \log_2 \frac{1}{2(T_c - A)} dx$$

$$= -\frac{x}{2(T_c - A)} \log_2 \frac{1}{2(T_c - A)} \Big|_{-T_c+A}^{T_c-A} = -\frac{T_c - A + T_c - A}{2(T_c - A)} \log_2 \frac{1}{2(T_c - A)}.$$

(9.5.6)

$$= -\log_2 \frac{1}{2(T_c - A)} = \log_2 2(T_c - A) = \log_2 2\sqrt{3}\frac{(T_c - A)}{\sqrt{3}} = \log_2 2\sqrt{3}\sigma_{ct}$$

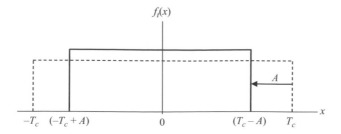

**Fig. 9.5.2** *A continuous truncated uniform density function.*

The *information content* is

$$
\begin{aligned}
I(X) = -\log_2 f_X(x) &= -\log_2 \frac{1}{2\,(T_c - A)} \\
&= \begin{cases} \log_2 2\,(T_c - A) & -T_c + A \le x < T_c - A \\ 0 & \text{otherwise} \end{cases}. \\
&= \begin{cases} \log_2 2\sigma_{ct}\sqrt{3} & -T_c + A \le x < T_c - A \\ 0 & \text{otherwise} \end{cases}
\end{aligned}
\tag{9.5.7}
$$

---

**Example**

Calculate the information content $I(X)$ for $2T_c - 2A = 1$, $2T_c - 2A < 1$, and $2T_c - 2A > 1$.

**Solution.** The solution is as follows:

$$
2T_c - 2A = 1 \Rightarrow T_c = (1 + 2A)\,/2, I(X) = \begin{cases} \log_2 2\,(T_c - A) = 0 & -T_c + A \le x < T_c - A \\ 0 & \text{otherwise} \end{cases}
$$

$$
2T_c - 2A > 1 \Rightarrow T_c > (1 + 2A)\,/2, I(X) = \begin{cases} \log_2 2\,(T_c - A) > 0 & -T_c + A \le x < T_c - A \\ 0 & \text{otherwise} \end{cases}.
$$

$$
2T_c - 2A < 1 \Rightarrow T_c < (1 + 2A)\,/2, I(X) = \begin{cases} \log_2 2\,(T_c - A) < 0 & -T_c + A \le x < T_c - A_c \\ 0 & \text{otherwise} \end{cases}
$$

---

### 9.5.2   Discrete Uniform Density Functions

**Discrete uniform untruncated density.** The density function of a uniform symmetric discrete random variable $X$ can be expressed in terms of Dirac delta functions as

$$
f_d(x) = \sum_{n=-S}^{n=S} \frac{1}{2S + 1} \cdot \delta\,(x - n),
\tag{9.5.8}
$$

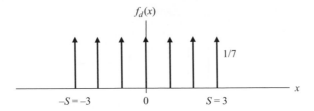

**Fig. 9.5.3** *A discrete uniform density function presented using Dirac delta functions for S = 3.*

and graphically presented, as shown in Figure 9.5.3 for $S = 3$, for a mean value equal to 0, and the variance $\sigma_d^2 = S(S+1)/3$.

The *information content* of the function for each weight of the delta function is

$$I(X) = -\log_2 \sum_{n=-S}^{n=S} \frac{1}{2S+1} \delta(x-n) = -\log_2 \frac{1}{2S+1} = \log_2 (2S+1) = I(n), \qquad (9.5.9)$$

for all $n = (-S, \ldots, 0, \ldots, S)$.

**Example**

Calculate the information content $I(X)$ for $2S + 1 = 1$, $2S + 1 < 1$, and $2S + 1 > 1$.

**Solution.** The solution is as follows:

$$2S+1 = 1 \Rightarrow S = 0, I(X) = \begin{cases} \log_2 (2S+1) = 0 & -S \leq n < S \\ 0 & \text{otherwise} \end{cases}$$

$$2S+1 > 1 \Rightarrow S > 0, I(X) = \begin{cases} \log_2 (2S+1) > 0 & -S \leq n < S \\ 0 & \text{otherwise} \end{cases}.$$

$$2S+1 < 1 \Rightarrow S < 0, I(X) \text{ does not exist}$$

The *entropy* of the function has a constant value for all $n$ that can be calculated as

$$\begin{aligned} H(X) &= -\int_{-\infty}^{\infty} f_d(x) \log_2 f_d(x) dx \\ &= -\int_{-S}^{S} \sum_{n=-S}^{n=S} \frac{1}{2S+1} \cdot \delta(x-n) \log_2 \left[ \sum_{n=-S}^{n=S} \frac{1}{2S+1} \cdot \delta(x-n) \right] dx \\ &= -\frac{1}{2S+1} \cdot \sum_{n=-S}^{n=S} \int_{-S}^{S} \delta(x-n) \log_2 \sum_{n=-S}^{n=S} \frac{1}{2S+1} \cdot \delta(x-n) dx \\ &= -\frac{1}{2S+1} \cdot \sum_{n=-S}^{n=S} \log_2 \frac{1}{2S+1} = -\frac{1}{2S+1} \cdot (2S+1) \log_2 \frac{1}{2S+1} \\ &= \log_2 (2S+1) \end{aligned} \qquad (9.5.10)$$

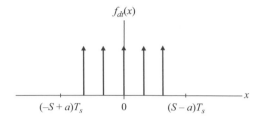

**Fig. 9.5.4** *A discrete truncated uniform density function presented using Dirac delta functions.*

**Discrete uniform truncated density.** In practical applications, the discrete values of a random variable take values in a limited interval defined as the truncated interval $(-S + a, S - a)$, where $a \leq S$ is a positive whole number called the 'truncation factor'. Therefore, the function is truncated and has the values in the truncated interval, as shown in Figure 9.5.4.

A truncated discrete density function is given in its closed form by the expression

$$f_{dt}(x) = \frac{1}{2(S-a)+1} \sum_{n=-S+a}^{n=S-a} \delta(x-nT_s) \underset{T_s=1}{=} \frac{1}{2(S-a)+1} \sum_{n=-S+a}^{S-a} \delta(x-n), \qquad (9.5.11)$$

and the variance is

$$\sigma_{d\,t}^2 = E\{x^2\} - \eta_{dt}^2 = T_s^2 \frac{(S-a)(S-a+1)}{3} \underset{T_s=1}{=} \frac{(S-a)(S-a+1)}{3}. \qquad (9.5.12)$$

The *information* of the function for every $n$ is

$$I(X) = -\log_2 \frac{1}{2(S-a)+1} \sum_{n=-S+a}^{S-a} \delta(x-n) = -\log_2 \frac{1}{2(S-a)+1},$$
$$= \log_2[2(S-a)+1] = I(n) \qquad (9.5.13)$$

with a constant value for each $n = (-S+a, \ldots, 0, \ldots, S-a)$.

**Example**

Calculate the information content $I(X)$ for $2(S-a)+1 = 1$, $2(S-a)+1 < 1$, and $2(S-a)+1 > 1$.

**Solution.** The solution is as follows:

$$2(S-a)+1 = 1 \Rightarrow S = a, I(X) = \begin{cases} \log_2(2(S-a)+1) = 0 & -S \leq n < S \\ 0 & \text{otherwise} \end{cases}$$

$$2(S-a)+1 > 1 \Rightarrow S > a, I(X) = \begin{cases} \log_2(2(S-a)+1) > 0 & -S \leq n < S \\ 0 & \text{otherwise} \end{cases}.$$

$$2(S-a)+1 < 1 \Rightarrow S < a, I(X) = \{\text{never, because } a < S\}$$

The *entropy* of the function is

$$H(X) = -\int_{-\infty}^{\infty} f_{dt}(x) \log_2 f_{dt}(x) dx$$

$$= -\int_{-S+a}^{S-a} \left[ \frac{1}{2(S-a)+1} \sum_{n=-S+a}^{S-a} \delta(x-n) \right] \cdot \log_2 \frac{1}{2(S-a)+1} \sum_{n=-S+a}^{S-a} \delta(x-n)\, dx. \quad (9.5.14)$$

$$= -\frac{1}{2(S-a)+1} \sum_{n=-S+a}^{S-a} \log_2 \frac{1}{2(S-a)+1}$$

## 9.6   Information and Entropy of Gaussian Density Functions

Suppose $X(t)$ is defined as an independent, identically distributed Gaussian stochastic process with zero mean and variance $\sigma^2$. We are interested in knowing how the entropy changes when the variance tends to infinity. Following the definition of entropy in information theory, log base 2 will be used and the differential entropy calculated.

### 9.6.1   Continuous Gaussian Density Functions

**Gaussian continuous untruncated density.** The probability density function of a Gaussian continuous random variable $X$ can be expressed as

$$f_c(x) = \frac{1}{\sqrt{2\pi\sigma^2}} e^{-(x-\eta)^2/2\sigma^2}, \quad (9.6.1)$$

for the mean value $\eta$ and the finite variance $\sigma^2$. The information content of the function is

$$I = \log_2 \sqrt{2\pi}\,\sigma\, e^{(x-\eta)^2/2\sigma^2} = \log_2 e \left[ \frac{(x-\eta)^2}{2\sigma^2} + \log_e \sqrt{2\pi}\,\sigma \right], \quad (9.6.2)$$

$$= \frac{(x-\eta)^2}{2\sigma^2} \log_2 e + \log_2 e \log_e \sqrt{2\pi}\,\sigma$$

and the entropy is

$$h(X) = \int_{-\infty}^{\infty} f_X(x) \log_2 \sqrt{2\pi}\,\sigma\, e^{(x-\eta)^2/2\sigma^2} dx = \frac{1}{2} \log_2 2\pi e\sigma^2, \quad (9.6.3)$$

as calculated in relation (9.3.6).

**Gaussian continuous truncated density.** In the case when the random variable has zero mean and a symmetric density function defined in the interval $(a, b)$ with width $(b - a) =$

$T - (-T) = 2T$, we can derive the related density function and its parameters. The density function can be expressed as

$$f_{ct}(x) = C(T) \frac{1}{\sqrt{2\pi\sigma^2}} e^{-(x-\eta)^2/2\sigma^2} = C(T) \frac{1}{\sqrt{2\pi\sigma^2}} e^{-x^2/2\sigma^2}, \tag{9.6.4}$$

where $a \leq x < b$, and the constant $C(T)$ is

$$C(T) = \frac{2}{\mathrm{erfc}\dfrac{a}{\sqrt{2\sigma^2}} - \mathrm{erfc}\dfrac{b}{\sqrt{2\sigma^2}}} = \frac{1}{1 - \mathrm{erfc}\,T/\sqrt{2\sigma^2}}. \tag{9.6.5}$$

Figure 9.6.1 presents a continuous density function (indicated by the thick line) and its truncated densities for the truncation interval $T$ as a parameter (indicated by the thin lines).

The variance of a continuous truncated density function is a function of the variance of the untruncated function and the truncation interval and is expressed as

$$\sigma_t^2 = \sigma^2 \left[ 1 - \frac{1}{1 - \mathrm{erfc}\,T/\sqrt{2\sigma^2}} \frac{2T}{\sqrt{2\pi\sigma^2}} e^{-\frac{T^2}{2\sigma^2}} \right]. \tag{9.6.6}$$

In Figure 9.6.2, the variance of the truncated density function is presented as a function of the truncation interval $T$ for a fixed variance of the continuous density function $\sigma^2$. When $T$ increases, the variance of the truncated function increases and goes to saturation for a

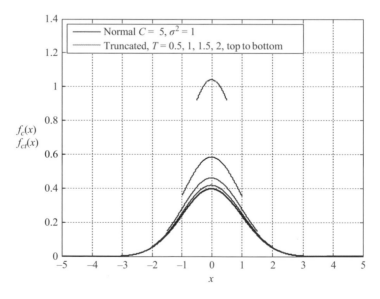

**Fig. 9.6.1** *A continuous density function and its truncated density for the truncation interval T as a parameter.*

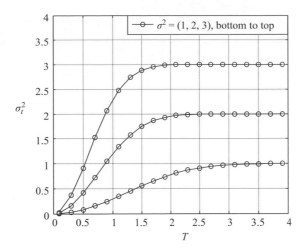

**Fig. 9.6.2** *Dependence of the truncated variance on the truncation interval T.*

reasonably high $T$ value and tends to the variance of the continuous variable when $T$ tends to infinity. That makes sense because, for an extremely high $T$, the truncated density function is very close to the untruncated continuous density function.

## 9.6.2   Discrete Gaussian Density Functions

**Discrete Gaussian untruncated density.** A discrete Gaussian density function can be expressed in terms of Dirac delta functions as

$$f_X(x) = \frac{1}{2} \sum_{n=-\infty}^{\infty} \left( \text{erfc}\frac{(2n-1)}{\sqrt{8\sigma^2}} - \text{erfc}\frac{(2n+1)}{\sqrt{8\sigma^2}} \right) \delta(x-n) = \frac{1}{2} \sum_{n=-\infty}^{\infty} Erfc(n)\delta(x-n), \quad (9.6.7)$$

where the function $Erfc(n)$ is defined as

$$Erfc(n) = \text{erfc}\frac{(2n-1)}{\sqrt{8\sigma^2}} - \text{erfc}\frac{(2n-1)}{\sqrt{8\sigma^2}}. \quad (9.6.8)$$

The variance of the discrete variable is equal to the variance of the continuous variable $\sigma$. The discrete function is presented in Figure 9.6.3 along with the corresponding continuous function.

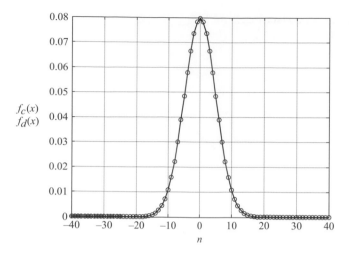

**Fig. 9.6.3** *Discrete and continuous Gaussian density functions.*

The information content of the discrete density function for every $n$ is

$$
I(X) = -\log_2 f_X(x) = -\log_2\left[\frac{1}{2}\sum_{n=-S}^{n=S} Erfc(n)\cdot \delta\,(x-n)\right]
$$

$$
= \cdots -\log_2\left[\frac{1}{2}Erfc\,(-1)\right] -\log_2\left[\frac{1}{2}Erfc(0)\right] -\log_2\left[\frac{1}{2}Erfc(1)\right] -\cdots \qquad (9.6.9)
$$

$$
= -\log_2\left[\frac{1}{2}Erfc(n)\right] = -\log_2 2^{-1} -\log_2 Erfc(n) = 1 -\log_2 Erfc(n)
$$

Following the definition of entropy in information theory, log base 2 can be used and the differential entropy can be calculated as

$$
H(X) = -\int_{-\infty}^{\infty} f_X(x)\log_2 f_X(x)dx
$$

$$
= -\int_{-\infty}^{\infty}\left[\frac{1}{2}\sum_{n=-\infty}^{\infty} Erfc(n)\cdot \delta\,(x-n)\right]\log_2\left[\frac{1}{2}\sum_{n=-\infty}^{\infty} Erfc(n)\cdot \delta\,(x-n)\right]dx
$$

$$
= -\frac{1}{2}\sum_{n=-\infty}^{\infty}\int_{x=-\infty}^{\infty} Erfc(n)\cdot \delta\,(x-n)\log_2\left[\frac{1}{2}\sum_{n=-\infty}^{\infty} Erfc(n)\cdot \delta\,(x-n)\right]dx \quad . \qquad (9.6.10)
$$

$$
= \cdots +\frac{1}{2}[Erfc(0)]\log_2\left[\frac{1}{2}Erfc(0)\right] +\frac{1}{2}[Erfc(1)]\log_2\left[\frac{1}{2}Erfc(1)\right] +\cdots
$$

$$
= -\frac{1}{2}\sum_{n=-S}^{S} Erfc(n)\cdot \log_2 Erfc(n)
$$

**Discrete truncated Gaussian density.** In practical applications, discrete random variable values are taken from a limited interval of, say, $S$ possible values. For example, the function which describes the delay in a discrete communication system has a density function that is truncated to $S$, as shown in Figure 9.6.4.

This function is defined as

$$f_{dt}(x) = f_d\left(x \,\middle|\, |x| \le S\right) = \frac{f_d(x)}{P(x \ge -S) - P(x > S)} = P(S) \sum_{n=-S}^{n=S} Erfc(n) \cdot \delta(x-n) \qquad (9.6.11)$$

for the unit sampling interval $T_s = 1$ and the constant $P(S)$ expressed as

$$\frac{1}{P(S)} = \sum_{n=-S}^{n=S} \left( \text{erfc} \frac{(2n-1)}{\sqrt{8\sigma^2}} - \text{erfc} \frac{(2n+1)}{\sqrt{8\sigma^2}} \right). \qquad (9.6.12)$$

The variance of the discrete truncated density is

$$\sigma_{dt}^2 = \sigma^2 \left[ 1 - \frac{1}{1 - \text{erfc}\,(2S+1)/\sqrt{8\sigma^2}} \frac{(2S+1)}{\sqrt{2\pi\sigma^2}} e^{-\frac{(2S+1)^2}{8\sigma^2}} \right], \qquad (9.6.13)$$

and can be approximated by the variance of the continuous truncated density expressed in eqn (9.6.6). The related information function is

$$I(X) = -\log_2 f_X(x) = -\log_2 \left[ P(S) \sum_{n=-S}^{n=S} Erfc(n) \cdot \delta(x-n) \right]$$

$$= \cdots - \log_2 [P(S)Erfc(-1)] - \log_2 [P(S)Erfc(0)] - \log_2 [P(S)Erfc(1)] - \cdots$$

$$= -\log_2 [P(S)Erfc(x)] \qquad (9.6.14)$$

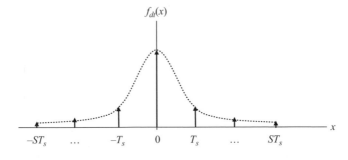

**Fig. 9.6.4** *A discrete truncated Gaussian density function presented using Dirac delta functions.*

for $x = (-S, \ldots, 0, \ldots, S)$. The truncated density function and related information functions are presented in Figure 9.6.5 for $S = 20$ with standard deviation as a parameter. The same functions are presented in Figure 9.6.6 for a fixed value of the standard deviation equal to 10 and for $S = 2, 3, 5, 10$, and 20. The entropy calculated for $\sigma = 10$ is 5.1597.

Graphs of the discrete truncated Gaussian density function, for variance as a parameter, are presented in Figure 9.6.5. It is worth noting that, if the variance of the untruncated density function increases, the truncated density function becomes 'flatter' and tends to the uniform density function. In addition, if the variance of the underlying untruncated function increases, the truncated discrete function tends faster to the uniform density. More importantly, we can see from the panel on the right-hand side of the figure that, when the values of the probability density function decrease, the information content at related values of the random variable increases. The gradient of this increase depends on the variance of the probability density function.

Figure 9.6.6 presents the behaviour of the truncated Gaussian probability density function and the related information ($I$) function when the truncation interval $S$ changes. When we reduce the truncation interval $S$, the probability density function becomes narrower and the

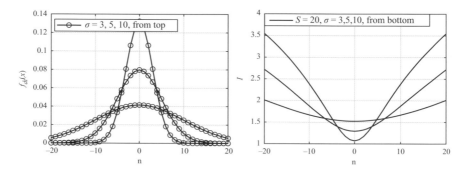

**Fig. 9.6.5** *A discrete truncated density function (left) and the related information function for variance as a parameter (right).*

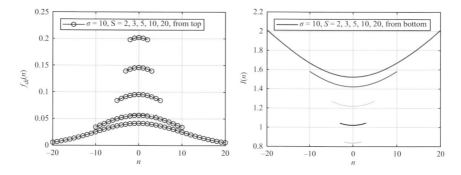

**Fig. 9.6.6** *A discrete truncated density function (left) and the related information function for the truncation factor S as a parameter (right).*

information content inside the interval becomes smaller, and vice versa. In other words, when we truncate the probability density function, we reduce the uncertainty around the occurrence of differential events inside the truncation interval.

The entropy is

$$H(X) = -\int_{-\infty}^{\infty} f_X(x)\log_2 f_X(x)dx = -P(S)\sum_{n=-S}^{S} Erfc(n) \cdot \log_2 P(S)Erfc(n), \qquad (9.6.15)$$

as presented in detail in Appendix B.

## 9.7 Block Error Control Codes

### 9.7.1 Theoretical Basis and Definitions of Block Code Terms

There are two main categories of channel codes: block codes and convolutional codes. The basic structure of a coding communication system, presented in Figure 9.1.1, will be used to investigate the characteristics of the block and convolution coding techniques. The interleaver and de-interleaver operations will be excluded from our analysis of encoding and decoding message bits. Therefore, the encoded bits **c** will be transmitted through a channel characterized by the error sequence **e**, received as the encoded noisy bits **r**, and then decoded inside the decoder to obtain the estimate of the bits transmitted, **c′**. In the case of block code analysis, the channel will be represented exclusively by the error sequence **e**. In the case of convolutional code analysis, more sophisticated channel structures will be used. In this section, we will present basic theory related to block codes. In Section 9.8, the convolutional codes will be analysed and, in Section 9.9, an introduction to iterative and turbo coding will be presented.

The mathematical basis for block codes is field theory, in particular the Galois field GF(2), for linear binary block codes. We will present the basics of this theory, which is necessary to understand the procedure of coding and decoding using block codes.

**The Galois field GF(2).** The finite Galois field GF(2) is defined by two elements in the set {0, 1} and two operations:

1.  addition: $0 \pm 0 = 0$, $0 \pm 1 = 1 \pm 0 = 1$, and $1 \pm 1 = 0$ (modulo-2 addition), and
2.  multiplication: $0 \times 0 = 0$, $0 \times 1 = 1 \times 0 = 0$, and $1 \times 1 = 1$.

**Vector space in GF(2).** We can form a vector space **V** over GF(2) that contains vectors of $n$ elements. For example, if $n = 4$, the vector space contains $2^n = 2^4$ vectors defined as $\mathbf{V} = \{v_1, v_2, \ldots, v_{16}\} = \{0000, 0001, \ldots, 1111\}$.

**The basis B.** The basis of a vector space **V** is a set of vectors **B** defined in such a way that their linear combinations are vectors in the vector space **V**, that is, $av_i + \ldots + bv_j \in \mathbf{V}$, for any $v_i, \ldots, v_j \in \mathbf{B}$, and the coefficients $a$ and $b$ are taken from the Galois field, that is, $a, b \in \{0, 1\}$. The number of vectors in the basis is called the base dimension. For our example, a base of **V** is $\mathbf{B} = \{1000, 0100, 0010, 0001\}$ with the dimension $\dim(\mathbf{B}) = 4$.

**Vector subspace.** The set of vectors $\mathbf{S}$ is a vector subspace of vector space $\mathbf{V}$ if and only if $av_i + bv_j \in \mathbf{S}$, for any $v_i$, $v_j \in \mathbf{S}$ and $a$, $b \in GF(2)$. For example, a vector subspace of the example $\mathbf{V}$ is $\mathbf{S} = \{0000, 0101, 0001, 0100\}$.

**Dual space.** A subspace $\mathbf{S}^{\perp}$ is a dual space of $\mathbf{S}$ if the product of their two vectors is 0, that is, $v_i \cdot v_j = 0$ for all $v_i \in \mathbf{S}$ and $v_j \in \mathbf{S}^{\perp}$. For example, a dual space of space $\mathbf{S}$ is a vector subspace $\mathbf{S}^{\perp} = \{0000, 1010, 1000, 0010\}$.

We will use the following standard notation in presenting the block codes:

$\mathbf{m} = (m_0, m_1, \ldots, m_{k-1})$ is a message word (represented as a vector) composed of a stream of $k$ symbols. If the symbols in the message word can take any value in $GF(2)$, then the collection of all $k$-tuples forms a vector space $\mathbf{C}$ over $GF(2)$. For a message word of $k$ bits, we can form, in total, $M = 2^k$ possible $k$-bit message words that are vectors on the vector space $\mathbf{C}$.

$\mathbf{C} = \{\mathbf{c}_0, \mathbf{c}_1, \ldots, \mathbf{c}_{M-1}\}$ is a set of $M$ valid code words that define a block error control code.

$\mathbf{c} = (c_0, c_1, \ldots, c_{n-1})$ is a code word composed of $n$ elements. If symbols $c_i \in GF(2)$, then the code $\mathbf{C}$ is said to be binary. In this case, we may calculate all possible $n$-tuples $N$ over $GF(2)$, $N = 2^n$, which form a vector space $\mathbf{N}$ containing $2^n$ vectors. The number of valid code words $M = 2^k$ is smaller than the number of possible $n$-bit code words $N = 2^n$. Therefore, $(N - M)$ $n$-bit words are not associated with message words and are thus not valid code words. The difference $r$ of the number of bits in a code word and a message word is called the redundancy of the code $\mathbf{C}$, that is, $r = n - \log_2 M = n - k$. These $r$ bits are called redundancy bits or parity bits. The ratio of the number of bits in the message word and that in the code word is called the code rate of a block code and is expressed as $R = k/n$.

$\mathbf{e} = (e_0, e_1, \ldots, e_{n-1})$ is an error sequence that characterizes a binary channel; it is defined as a binary stream of $n$ bits which forms $E = 2^n$ possible $n$-bit error vectors. Binary channels are defined in Section 9.2.2.

$w(\mathbf{c})$ is the weight of a code word, defined as the number of non-zero coordinates in the code word $\mathbf{c}$ or in the error sequence $\mathbf{e}$. For example, for the code word $\mathbf{c} = (1, 0, 0, 1, 1, 0, 1, 1, 1, 1)$, the weight is $w(\mathbf{c}) = 7$.

$\mathbf{r} = \mathbf{c} + \mathbf{e} = (r_0, r_1, \ldots, r_{n-1}) = (c_0 + e_0, c_1 + e_1, \ldots, c_{n-1} + e_{n-1})$ is the received binary code word for the defined binary code word $\mathbf{c}$ and the error sequence $\mathbf{e}$, which are added to each other via modulo-2 addition.

$\mathbf{m}' = \left(m'_0, m'_1, \ldots, m'_{k-1}\right)$ is a decoded message word.

**Error detection.** Error detection is a procedure for determining whether errors are present in a received code word $\mathbf{r}$. Undetectable error patterns exist if and only if the received code word is a valid code word other than the one transmitted. Thus, there are $M - 1$ undetectable error patterns. There are three main decoding strategies. Namely, the decoder can react to a detected error in three ways:

1. Request a retransmission of the code word by using the automatic repeat request protocol.

2. Tag the received code word as being incorrect and pass it along to the message sink by using the MUTING procedure.

3. Attempt to correct the errors in the received code word by using the structure of the code words via the forward error correction procedure.

**Definition: Hamming distance.** The Hamming distance between the two code words (vectors) $\mathbf{c_1}$ and $\mathbf{c_2}$ is the number of coordinates in which the two code words differ, that is,

$$d_{Hamming}\,(\mathbf{c_1},\mathbf{c_2}) = d\,(\mathbf{c_1},\mathbf{c_2}) = \{i | c_{1i} \neq c_{2i}, i = 0, 1, 2, \ldots, n-1\}. \qquad (9.7.1)$$

**Definition: Minimum distance.** The minimum distance $d_{min}$ of a block code $\mathbf{C}$ is the minimum Hamming distance between all distinct pairs of code words in $\mathbf{C}$. The following rules apply:

A code $\mathbf{C}$ with $d_{min}$ can *detect* all error patterns of weight less than or equal to $(d_{min} - 1)$ because, for an error pattern weight equal to $d_{min}$, we get another code word; therefore, this is an undetectable error pattern.

A code $\mathbf{C}$ with $d_{min}$ can *correct* all error patterns of weight less than or equal to *the largest integer less than or equal to* $(d_{min} - 1)/2$.

The decoding is, in principle based on the theory of maximum likelihood estimation. An error in a maximum likelihood decoder occurs whenever the received code word is closer to an incorrect word than to the correct word. Decoder errors are possible only if the weight of the error pattern induced by the channel is $\geq d_{min}/2$.

## 9.7.2 Coding Procedure Using a Generator Matrix

A block code $\mathbf{C} = \{\mathbf{c_0}, \mathbf{c_1}, \ldots, \mathbf{c_{M-1}}\}$ consisting of $n$-tuples $\mathbf{c} = (c_0, c_1, \ldots, c_{n-1})$ of symbols from GF(2) is a binary linear code if and only if $\mathbf{C}$ forms a vector subspace over GF(2). The dimension of this linear code is the dimension of the corresponding vector space $\mathbf{C}$. A binary $(n, k)$ code is a linear code of length $n$ and dimension $k$ with symbols in GF(2) having $M = 2^k$ code words of length $n$. The linear codes have the following properties:

1. The linear combination of any set of code words is a code word (consequently, linear codes always contain the all-zero vector).

2. The minimum distance of a linear code is equal to the weight of the lowest-weight non-zero code word.

3. The undetectable error patterns for a linear code are independent of the code word transmitted and always consist of the set of all non-zero code words.

Following the definition for a basis in a vector space, we can define a basis for a linear binary code that can be understood as the vector space $\mathbf{C}$. Let $\{\mathbf{g}_0, \mathbf{g}_1, \ldots, \mathbf{g}_{k-1}\}$ be a basis of code words for the $(n, k)$ binary code $\mathbf{C}$. There exists a unique representation of code words that is a linear combination of the basis vectors, expressed as $\mathbf{c} = m_0 \mathbf{g}_0 + m_1 \mathbf{g}_1, \ldots, + m_{k-1} \mathbf{g}_{k-1}$ for every code word $\mathbf{c} \in \mathbf{C}$. Then, we can define for this code a *generator matrix* that contains the vectors of the basis in its rows and is expressed in matrix form as

$$\mathbf{G} = \begin{bmatrix} \mathbf{g}_0 \\ \mathbf{g}_1 \\ \vdots \\ \mathbf{g}_{k-1} \end{bmatrix} = \begin{bmatrix} g_{0,0} & g_{0,1} & \cdots & g_{0,n-1} \\ g_{1,0} & g_{1,1} & \cdots & g_{1,n-1} \\ \vdots & \vdots & \ddots & \vdots \\ g_{k-1,0} & g_{k-1,1} & \cdots & g_{k-1,n-1} \end{bmatrix}. \tag{9.7.2}$$

The $n \times k$ matrix $\mathbf{G}$ can be used to directly encode $k$-bit message words. Let the message word $\mathbf{m} = (m_0, m_1, \ldots, m_{k-1})$ be a binary block of uncoded message bits. Then $k$-bit blocks can be directly encoded by the multiplication of the message word $\mathbf{m}$ by the generator matrix in the following manner:

$$\mathbf{c} = \mathbf{mG} = [m_0 \; m_1 \ldots m_{k-1}] \begin{bmatrix} \mathbf{g}_0 \\ \mathbf{g}_1 \\ \vdots \\ \mathbf{g}_{k-1} \end{bmatrix} = [m_0 \; m_1 \ldots m_{k-1}] \begin{bmatrix} g_{0,0} & g_{0,1} & \cdots & g_{0,n-1} \\ g_{1,0} & g_{1,1} & \cdots & g_{1,n-1} \\ \vdots & \vdots & \ddots & \vdots \\ g_{k-1,0} & g_{k-1,1} & \cdots & g_{k-1,n-1} \end{bmatrix}. \tag{9.7.3}$$

A binary $(n, k)$ code $\mathbf{C}$ can be understood as a vector subspace formed within the space $\mathbf{N}$ of all $n$-tuples over $GF(2)$.

---

### Example

Let $\mathbf{C}$ be a $(7, 3)$ binary linear code with the basis $\{1000111, 0101011, 0011101\}$. Find the generator matrix $\mathbf{G}$ and the coded words for the message words $\mathbf{m} = 001$ and $\mathbf{m} = 111$. Find the code words $\mathbf{c}$ for all possible message words $\mathbf{m}$.

**Solution.** The number of message words is $M = 2^3 = 8$, defined as $\mathbf{M} = \{000, 001, 010, 011, 100, 101, 110, 111\}$. The basis elements are $\{\mathbf{g}_0, \mathbf{g}_1, \mathbf{g}_2\} = \{1000111, 0101011, 0011101\}$. Therefore, the generator matrix is

$$\mathbf{G} = \begin{bmatrix} \mathbf{g}_0 \\ \mathbf{g}_1 \\ \mathbf{g}_2 \end{bmatrix} = \begin{bmatrix} g_{0,0} & g_{0,1} & g_{0,2} & g_{0,3} & g_{0,4} & g_{0,5} & g_{0,6} \\ g_{1,0} & g_{1,1} & \cdots & & & & g_{1,6} \\ g_{2,0} & & \cdots & & & & g_{2,6} \end{bmatrix} = \begin{bmatrix} 1 & 0 & 0 & 0 & 1 & 1 & 1 \\ 0 & 1 & 0 & 1 & 0 & 1 & 1 \\ 0 & 0 & 1 & 1 & 1 & 0 & 1 \end{bmatrix}.$$

The generator matrix $\mathbf{G}$, which has $n \times k = 7 \times 3 = 21$ elements, can be used to directly encode three-symbol data blocks as follows. The message word $\mathbf{m} = (m_0, m_1, m_2) = (0, 0, 1)$ can be directly encoded into the code word **c** in the following manner:

$$\mathbf{c} = \mathbf{mG} = [m_0\ m_1\ m_2] \begin{bmatrix} g_0 \\ g_1 \\ g_2 \end{bmatrix} = [0\ 0\ 1] \begin{bmatrix} 1 & 0 & 0 & 0 & 1 & 1 & 1 \\ 0 & 1 & 0 & 1 & 0 & 1 & 1 \\ 0 & 0 & 1 & 1 & 1 & 0 & 1 \end{bmatrix} = [0011101].$$

Likewise, if the message word is **m** = (1, 1, 1), the code word can be obtained by direct matrix multiplication as follows:

$$\mathbf{c} = [m_0\ m_1\ m_2] \begin{bmatrix} g_0 \\ g_1 \\ g_2 \end{bmatrix} = [1\ 1\ 1] \begin{bmatrix} 1 & 0 & 0 & 0 & 1 & 1 & 1 \\ 0 & 1 & 0 & 1 & 0 & 1 & 1 \\ 0 & 0 & 1 & 1 & 1 & 0 & 1 \end{bmatrix} = [1\ 1\ 1\ 0\ 0\ 0\ 1\ ].$$

For any **m**, we can have all the code words **c**:

$$\mathbf{c} = \begin{bmatrix} m_0^1\ m_1^1\ m_2^1 \\ m_0^2\ m_1^2\ m_2^2 \\ m_0^3\ m_1^3\ m_2^3 \\ m_0^4\ m_1^4\ m_2^4 \\ m_0^5\ m_1^5\ m_2^5 \\ m_0^6\ m_1^6\ m_2^6 \\ m_0^7\ m_1^7\ m_2^7 \\ m_0^8\ m_1^8\ m_2^8 \end{bmatrix} \begin{bmatrix} g_0 \\ g_1 \\ g_2 \end{bmatrix} = \begin{bmatrix} 000 \\ \mathbf{001} \\ 010 \\ 011 \\ 100 \\ 101 \\ 110 \\ \mathbf{111} \end{bmatrix} \begin{bmatrix} 1 & 0 & 0 & 0 & 1 & 1 & 1 \\ 0 & 1 & 0 & 1 & 0 & 1 & 1 \\ 0 & 0 & 1 & 1 & 1 & 0 & 1 \end{bmatrix} = \begin{bmatrix} 0000000 \\ \mathbf{0011101} \\ 0101011 \\ 0110110 \\ 1000111 \\ 1011010 \\ 1101100 \\ \mathbf{1110001} \end{bmatrix} = \begin{bmatrix} c_0 \\ c_1 \\ c_2 \\ c_3 \\ c_4 \\ c_5 \\ c_6 \\ c_7 \end{bmatrix}.$$

The previously obtained code words and corresponding message words inside the above matrices are indicated in bold.

## 9.7.3   Error Detection Using a Parity Check Matrix

Block code can be either *systematic* or *non-systematic*. A code is non-systematic if the order of the basis vectors in generator matrix **G** is arbitrary. If the vectors in the generator matrix form an identity submatrix, we say that the encoder is systematic. The decoding procedure will be presented for both systematic and non-systematic codes, using the notion of a parity check matrix **H** that will be defined as follows.

Let $\mathbf{C}^\perp$ be the dual space of subspace **C** within the vector space **N**, having dimension $r = (n - k)$. Then, it follows that the basis $\{\mathbf{h}_i\}$ of $\mathbf{C}^\perp$, which has the basis elements $\{\mathbf{h}_0, \mathbf{h}_1, \ldots, \mathbf{h}_{n-k-1}\}$, can be found and used to construct an $n \times (n - k)$ parity check matrix **H** expressed as

$$\mathbf{H} = \begin{bmatrix} \mathbf{h}_0 \\ \mathbf{h}_1 \\ \vdots \\ \mathbf{h}_{n-k-1} \end{bmatrix} = \begin{bmatrix} h_{0,0} & h_{0,1} & \cdots & h_{0,n-1} \\ h_{1,0} & h_{1,1} & \cdots & h_{1,n-1} \\ \vdots & \vdots & \ddots & \vdots \\ h_{n-k-1,0} & h_{n-k-1,1} & \cdots & h_{n-k-1,n-1} \end{bmatrix}. \tag{9.7.4}$$

The error detection procedure for a general non-systematic case is based on the parity check theorem, which states the following:

1. A code vector $\mathbf{c}$ is a code word in $\mathbf{C}$ if and only if

$$\mathbf{cH}^T = \mathbf{0}. \tag{9.7.5}$$

2. The minimum distance $d_{min}$ for an $(n, k)$ code is bounded by $d_{min} \leq n - k + 1$.

The error detection algorithm is based on the matrix condition in eqn (9.7.5). We need to find $\mathbf{cH}^T$ and then make a decision based on the following:

If $\mathbf{cH}^T = \mathbf{0}$, then no errors occurred in the code word $\mathbf{c}$.

If $\mathbf{cH}^T \neq \mathbf{0}$, there are errors present in the code word $\mathbf{c}$.

We can conclude the following: the matrices $\mathbf{G}$ and $\mathbf{H}$ for linear block codes greatly simplify *encoding* and *error detection* procedures, which are reduced to the defined matrix multiplication.

In the non-systematic case, the code word $\mathbf{c}$ is obtained by using the generator matrix $\mathbf{G}$ that contains the message word bits spread inside the code word. This spreading of bits complicates the procedure of coding and decoding. In order to simplify the decoding procedure, systematic encoding can be used. Using Gaussian elimination change (which does not alter the code word set for the associated code) and column reordering (which may generate code words that are not present in the original code), it is always possible to obtain a *systematic generator matrix* $\mathbf{G}$ in the form

$$\mathbf{G} = [\mathbf{P}|\mathbf{I}_k] = \begin{bmatrix} p_{0,0} & p_{0,1} & \cdots & p_{0,n-k-1} & 1 & 0 & \cdots & 0 \\ p_{1,0} & p_{1,1} & \cdots & p_{1,n-k-1} & 0 & 1 & \cdots & 0 \\ \vdots & \vdots & \ddots & \vdots & \vdots & \vdots & \ddots & \vdots \\ p_{k-1,0} & p_{k-1,1} & \cdots & p_{k-1,n-k-1} & 0 & 0 & \cdots & 1 \end{bmatrix}, \tag{9.7.6}$$

where $\mathbf{P}$ is the parity portion of $\mathbf{G}$, and $\mathbf{I}_k$ is an identity matrix. Using the systematic matrix $\mathbf{G}$, message bits can be embedded without modification in the last $k$ coordinates of the resulting code word, that is,

$$\mathbf{c} = \mathbf{mG} = \mathbf{m} \cdot [\mathbf{P}|\mathbf{I}_k]$$

$$= [m_0 \; m_1 \ldots m_{k-1}] \begin{bmatrix} p_{0,0} & p_{0,1} & \cdots & p_{0,n-k-1} & 1 & 0 & \cdots & 0 \\ p_{1,0} & p_{1,1} & \cdots & p_{1,n-k-1} & 0 & 1 & \cdots & 0 \\ \vdots & \vdots & \ddots & \vdots & \vdots & \vdots & \ddots & \vdots \\ p_{k-1,0} & p_{k-1,1} & \cdots & p_{k-1,n-k-1} & 0 & 0 & \cdots & 1 \end{bmatrix}. \tag{9.7.7}$$

$$= [c_0 \; c_1 \ldots c_{n-k-1} | m_0 \; m_1 \ldots m_{k-1}]$$

The corresponding systematic parity check matrix can be obtained in the form

$$\mathbf{H} = \left[ \mathbf{I}_{n-k} | \mathbf{P}^T \right] = \begin{bmatrix} 1 & 0 & \cdots & 0 & p_{0,0} & p_{1,0} & \cdots & p_{k-1,0} \\ 0 & 1 & \cdots & 0 & p_{0,1} & p_{1,1} & \cdots & p_{k-1,1} \\ \vdots & \vdots & \ddots & \vdots & \vdots & \vdots & \ddots & \vdots \\ 0 & 0 & \cdots & 1 & p_{0,n-k-1} & p_{1,n-k-1} & \cdots & p_{k-1,n-k-1} \end{bmatrix}. \tag{9.7.8}$$

**Procedures for encoding and error detection.** The procedures for encoding and error detection are equivalent to those in the non-systematic case, that is, we need to find the matrices $\mathbf{c} = \mathbf{mG}$ and the condition $\mathbf{rH}^T = 0$, and proceed as for the non-systematic case.

**Example**

Let $\mathbf{C}$ be a $(7, 3)$ binary linear code with the basis $\{1000111, 0101011, 0011101\}$. Find the systematic generator matrix $\mathbf{G}$ and the coded word for the message word $\mathbf{m} = 001$. Demonstrate the procedure for error detection, assuming that the received code words are $\mathbf{r} = 1010001$ and $\mathbf{r} = 1110001$.

**Solution.** Using Gaussian elimination, the systematic generator matrix $\mathbf{G}$ can be obtained in the form

$$\mathbf{G} = \begin{bmatrix} p_{0,0} & p_{0,1} & p_{0,2} & p_{0,3} & 1 & 0 & 0 \\ p_{1,0} & p_{1,1} & \cdots & & 0 & 1 & 0 \\ p_{2,0} & & \cdots & & 0 & 0 & 1 \end{bmatrix} = \begin{bmatrix} 1 & 1 & 0 & 1 & 1 & 0 & 0 \\ 1 & 0 & 1 & 1 & 0 & 1 & 0 \\ 1 & 1 & 1 & 0 & 0 & 0 & 1 \end{bmatrix} = [\mathbf{P} \ \mathbf{I}_k].$$

For the message word $\mathbf{m} = 001$, the code word is

$$\mathbf{c} = \mathbf{mG} = [0 \ 0 \ 1] \begin{bmatrix} 1 & 1 & 0 & 1 & 1 & 0 & 0 \\ 1 & 0 & 1 & 1 & 0 & 1 & 0 \\ 1 & 1 & 1 & 0 & 0 & 0 & 1 \end{bmatrix} = [1 \ 1 \ 1 \ 0 \ 0 \ 0 \ 1].$$

The corresponding systematic parity check matrix $\mathbf{H}$ can be obtained in the form

$$\mathbf{H} = \left[ \mathbf{I}_{n-k} | \mathbf{P}^T \right] = \begin{bmatrix} 1 & 0 & 0 & 0 & 1 & 1 & 1 \\ 0 & 1 & 0 & 0 & 1 & 0 & 1 \\ 0 & 0 & 1 & 0 & 0 & 1 & 1 \\ 0 & 0 & 0 & 1 & 1 & 1 & 0 \end{bmatrix},$$

and the transpose matrix of $\mathbf{H}$ is

$$\mathbf{H}^T = \begin{bmatrix} \mathbf{I}_{n-k} \\ \mathbf{P}^T \end{bmatrix} = \begin{bmatrix} 1 & 0 & 0 & 0 \\ 0 & 1 & 0 & 0 \\ 0 & 0 & 1 & 0 \\ 0 & 0 & 0 & 1 \\ 1 & 1 & 0 & 1 \\ 1 & 0 & 1 & 1 \\ 1 & 1 & 1 & 0 \end{bmatrix}.$$

For the received code word $\mathbf{r} = 1010001$, we can detect possible errors by investigating the matrix

$$\mathbf{rH}^T = [1\ 0\ 1\ 0\ 0\ 0\ 1] \begin{bmatrix} 1 & 0 & 0 & 0 \\ 0 & 1 & 0 & 0 \\ 0 & 0 & 1 & 0 \\ 0 & 0 & 0 & 1 \\ 1 & 1 & 0 & 1 \\ 1 & 0 & 1 & 1 \\ 1 & 1 & 1 & 0 \end{bmatrix} = [0\ 1\ 0\ 0] \neq \mathbf{0}.$$

Since, for the received word $\mathbf{r} = 1010001$, we have

$$\mathbf{rH}^T \neq \mathbf{0},$$

the conclusion is that the error(s) occurred in the received code word $\mathbf{r}$, because it does not belong to the set of valid code words. However, in the case when the received code word is $\mathbf{r} = 1110001$, we have

$$\mathbf{rH}^T = [1\ 1\ 1\ 0\ 0\ 0\ 1] \begin{bmatrix} 1 & 0 & 0 & 0 \\ 0 & 1 & 0 & 0 \\ 0 & 0 & 1 & 0 \\ 0 & 0 & 0 & 1 \\ 1 & 1 & 0 & 1 \\ 1 & 0 & 1 & 1 \\ 1 & 1 & 1 & 0 \end{bmatrix} = [0\ 0\ 0\ 0].$$

The matrix product is 0; therefore, the received code word is the valid code word and has no errors. Since, in this case, we are dealing with a systematic code, we can directly find the detected message word $\mathbf{m} = 001$ from the received code word $\mathbf{r}$.

### 9.7.4   Standard Array Decoding

Decoding of block codes can be performed using a so-called standard array, which will be presented in this section. A standard array, shown in Figure 9.7.1, is an array that contains all $n$-tuples obtained in the following manner (without replacement):

1. Select, from the set of all $V_{n/2} = 2^n$ $n$-tuples over GF(2), the code words belonging to the code $\mathbf{C}$. Write down the code words in the first row, starting with the all-zero code word (this word must be in the code $\mathbf{C}$).

2. Select from the remaining $V_{n/2}$ a total of $(2^{n-k} - 1)$ patterns of lowest weight, and place them into the first column under the all-zero code word. These are the error words containing correctable errors.

3. Now add the patterns taken from the first column to each of the other code words, and write down the resulting sum at the place that corresponds to the code word and the error pattern (in the sense of the arrows in Figure 9.7.1).

$$\begin{vmatrix} \mathbf{e}_0 + \mathbf{c}_0 = \mathbf{c}_0 & \mathbf{c}_1 & \cdots & \mathbf{c}_{2^k} \\ \mathbf{e}_1 \longrightarrow & \mathbf{e}_1 + \mathbf{c}_1 & \cdots & \mathbf{e}_1 + \mathbf{c}_{2^k} \\ \vdots & \vdots & \ddots & \vdots \\ \mathbf{e}_{2^{n-k}} & \mathbf{e}_{2^{n-k}} + \mathbf{c}_1 & \cdots & \mathbf{e}_{2^{n-k}} + \mathbf{c}_{2^k} \end{vmatrix}$$

**Fig. 9.7.1** *A standard array.*

The rows are called *cosets*, and the first term in a coset is the *coset leader*. The coset leaders correspond to the error patterns that can be corrected or, more precisely, the coset leaders are the error patterns that are most likely to have corrupted the corresponding code word (which is in the first row of the array).

Once you have this standard array, you should use the following three-step procedure for decoding:

1. Find the received code word inside the standard array (say this is $\mathbf{r}_1 = \mathbf{e}_1 + \mathbf{c}_1$).
2. Find the corresponding coset leader $\mathbf{e}_1$.
3. Calculate the corrected code word (via modulo-2 addition) for the assumed case $\mathbf{c}_1 = \mathbf{e}_1 + \mathbf{r}_1$.

Note the following for the specification of error sequences and the applicability of the standard array. All of the weight-1 error patterns were used as row headers because it is most likely that we will have one error inside a particular code word. In addition, while some of the weight-2 and weight-3 patterns were used, in the above case, we cannot guarantee that all of the patterns with weight 2 or higher will be corrected. Generally, for a *perfect code*, we have to use the error patterns that have the minimum weights.

The problem with using a standard array is that such arrays can be extremely large and so require large amounts of memory to process them. For example, a (64, 56) Reed–Solomon code contains $1.4 \times 10^{101}$ code words, which is more than the number of atomic particles in the entire universe. Consequently, the processing time for decoding can be prohibitive. However, the time required for processing can be reduced by calculating the syndrome values of the code $\mathbf{C}$; this is the subject of analysis in the following section.

## 9.7.5　Syndrome Table Decoding

Consider the received code vector $\mathbf{r} = \mathbf{c} + \mathbf{e}$, where $\mathbf{c}$ is a valid code word, and $\mathbf{e}$ is an error pattern induced by the channel. The parity check matrix $\mathbf{H}$ is a linear transformation whose null space is $\mathbf{C}$, because $\mathbf{cH}^T = \mathbf{0}$, and, generally, non-null space is the space of received code words, that is, $\mathbf{rH}^T \neq \mathbf{0}$, which defines the syndrome vector $\mathbf{s}$. The syndrome vector $\mathbf{s}$ for the received vector $\mathbf{r}$ is thus defined as

$$\mathbf{s} = \mathbf{rH}^T = (\mathbf{c} + \mathbf{e})\,\mathbf{H}^T = \mathbf{cH}^T + \mathbf{eH}^T = 0 + \mathbf{eH}^T = \mathbf{eH}^T. \tag{9.7.9}$$

The syndrome vector **s** is thus solely a function of **e** and $\mathbf{H}^{\mathrm{T}}$ and is independent of the transmitted code word **c**, which can be used for the syndrome decoding of block codes via the following procedure:

1. Compute the syndrome values $\mathbf{s} = \mathbf{e}\mathbf{H}^{\mathrm{T}}$ $(2^{n-k})$ for all error vectors **e** and *store them in the look-up table* $|\mathbf{s}, \mathbf{e}|$.
2. Compute the syndrome of a received vector **r**, $\mathbf{s_r} = \mathbf{r}\mathbf{H}^{\mathrm{T}}$.
3. Using the look-up table, find the error pattern **e** that corresponds to the computed syndrome.
4. Add the error pattern to the received code word **r**. The result is the maximum likelihood code word $\mathbf{c}'$ associated with **r**.

**Example**

Let **C** be a (7, 3) binary linear code with the basis {1000111, 0101011, 0011101}. Demonstrate the procedures of error detection and correction, using syndrome table decoding, assuming that the received code word is **r** = 0011011.

**Solution.** The syndrome values can be calculated as $\mathbf{s} = \mathbf{e}\mathbf{H}^{\mathrm{T}}$:

$$
\mathbf{s} = \mathbf{e}\mathbf{H}^T =
\begin{bmatrix}
0000000 \\
0000001 \\
0000010 \\
0000100 \\
0001000 \\
0010000 \\
0100000 \\
1000000 \\
0000011 \\
0000110 \\
0001100 \\
0011000 \\
0001010 \\
0010100 \\
0010010 \\
0111000
\end{bmatrix}
\begin{bmatrix}
1 & 0 & 0 & 0 \\
0 & 1 & 0 & 0 \\
0 & 0 & 1 & 0 \\
0 & 0 & 0 & 1 \\
1 & 1 & 0 & 1 \\
1 & 0 & 1 & 1 \\
1 & 1 & 1 & 0
\end{bmatrix}
=
\begin{bmatrix}
0 & 0 & 0 & 0 \\
1 & 1 & 1 & 0 \\
1 & 0 & 1 & 1 \\
1 & 1 & 0 & 1 \\
0 & 0 & 0 & 1 \\
0 & 0 & 1 & 0 \\
0 & 1 & 0 & 0 \\
1 & 0 & 0 & 0 \\
0 & 1 & 0 & 1 \\
0 & 1 & 1 & 0 \\
1 & 1 & 0 & 0 \\
0 & 0 & 1 & 1 \\
1 & 0 & 1 & 0 \\
1 & 1 & 1 & 1 \\
1 & 0 & 0 & 1 \\
0 & 1 & 1 & 1
\end{bmatrix}.
$$

These values can then be used to construct a table containing the error words **e** and the corresponding syndrome words **s** (see Figure 9.7.2).

| SYNDROME VALUES | | ERROR WORDS |
|---|---|---|
| 0 0 0 0 | | 0000000 |
| 1 1 1 0 | | 0000001 |
| 1 0 1 1 | | 0000010 |
| 1 1 0 1 | | 0000100 |
| 0 0 0 1 | | 0001000 |
| 0 0 1 0 | | 0010000 |
| 0 1 0 0 | | 0100000 |
| 1 0 0 0 | ⇔ | 1000000 |
| 0 1 0 1 | | 0000011 |
| **0 1 1 0** | | **0000110** |
| 1 1 0 0 | | 0001100 |
| 0 0 1 1 | | 0011000 |
| 1 0 1 0 | | 0001010 |
| 1 1 1 1 | | 0010100 |
| 1 0 0 1 | | 0010010 |
| 0 1 1 1 | | 0111000 |

**Fig. 9.7.2** *Syndromes and error sequences.*

Then the procedure for decoding is as follows:

1. Calculate the syndrome value for the received code word $\mathbf{r} = 0011011$:

$$\mathbf{s_r} = \mathbf{rH}^T = [0\ 0\ 1\ 1\ 0\ 1\ 1] \begin{bmatrix} 1 & 0 & 0 & 0 \\ 0 & 1 & 0 & 0 \\ 0 & 0 & 1 & 0 \\ 0 & 0 & 0 & 1 \\ 1 & 1 & 0 & 1 \\ 1 & 0 & 1 & 1 \\ 1 & 1 & 1 & 0 \end{bmatrix} = \begin{bmatrix} 0 & 1 & 1 & 0 \end{bmatrix}.$$

2. Looking up this syndrome in the table, we can find that the corresponding error pattern (error word) is $\mathbf{e} = 0000110$ (the word highlighted in bold in Figure 9.7.2).

We can then find that the maximum likelihood transmitted code word $\mathbf{c}'$ is $\mathbf{c} = \mathbf{r} - \mathbf{e} = (0011011) - (0000110) = 0011101$, because the modulo-2 addition corresponds to the modulo-2 subtraction.

## 9.8   Convolutional Codes

As we said in the previous section, a linear block code is described by two integers, $n$ and $k$, and a generator matrix. The integer $k$ is the number of message bits that form an input to a block encoder. The integer $n$ is the total number of bits in the associated code word out of the encoder. A characteristic of linear block codes is that each code word's $n$-tuple is uniquely determined by the input message $k$-tuple. The ratio $k/n$ is a measure of the amount of added redundancy and is called the rate of the code.

A convolutional code is described by three integers: $n$, $k$, and $L$. The ratio $k/n$ has the same significance that it had for block codes: it specifies the amount of added redundancy or additional information per coded bit. However, the integer $n$ does not define a block of the code word's length, as it does for block codes. The integer $L$ is the constraint length parameter, which represents the number of $k$-tuple stages in the coding shift register. Convolutional codes differ from block codes in that a convolutional encoder has memory, that is, an $n$-tuple generated by a convolutional encoder is not only a function of an input $k$-tuple but also a function of the previous $L - 1$ input $k$-tuple in the coding shift register.

The constraint length $L$ of a convolutional code is the maximum number of bits in a single output sequence that can be affected by any input bit, and can be expressed as the maximum number of taps in the shift register of the encoder. The memory of an encoder is defined to be the total number of memory elements in the encoder. A convolutional code word is considered to be complete when the contents of the encoder memory elements have been 'emptied' and the output encoded word is completely generated. Thus, due to the encoder bit processing, there will be some additional bits of redundancy, which will reduce the effective rate of the code. This reduction is expressed in terms of a fractional rate loss. In practice, the input stream is very long with respect to the total encoder memory, which makes the fractional rate negligible.

### 9.8.1   Linear Convolutional Codes

Traditionally, the procedure of convolutional coding can be represented by using shift-registers, a generator matrix, a state diagram, a $D$-transform, and a trellis diagram. Such representation will be the subject of our analysis.

**Shift-register design for convolutional encoders.** A convolutional encoder can be designed using a shift-register circuit. A typical 1/2-rate linear convolutional encoder is presented in Figure 9.8.1. The input binary message sequence

$$\mathbf{m} = (m_0, m_1, m_2, \dots) \tag{9.8.1}$$

is fed into the encoder (which is composed of a shift-register circuit consisting of a series of memory elements). It is assumed that the shift-register content is initialized to all 0's before the coding process begins. With each successive input to the shift register, the values of the memory elements are tapped off and added according to a fixed pattern, creating a pair of output coded sequences. The encoder output sequence can be called a convolutional code word

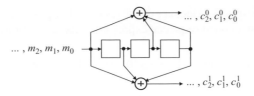

**Fig. 9.8.1** *A convolutional rate-1/2 encoder.*

if it is defined for a finite number of bits in a corresponding message sequence or message word that is also defined for a finite number of message bits.

At the first and second outputs of the encoder, we obtain two binary coded sequences expressed as

$$c^0 = \left(c_0^0, c_1^0, c_2^0, \ldots\right) \text{ and } c^1 = \left(c_0^1, c_1^1, c_2^1, \ldots\right), \tag{9.8.2}$$

respectively, which can be multiplexed to create the single coded sequence

$$c = \left(c_0^0 c_0^1, c_1^0\, c_1^1, c_2^0 c_2^1, \ldots\right). \tag{9.8.3}$$

This encoder can be generalized for multiple input sequences defined by $k$, and multiple output sequences defined by $n$. Then, the input and output bit sequences can be expressed in interleaved forms as

$$\mathbf{m} = \left(m_0^0 m_0^1 \ldots m_0^{k-1}, m_1^0 m_1^1 \ldots m_1^{k-1}, m_2^0 m_2^1 \ldots m_2^{k-1}, \ldots\right) \tag{9.8.4}$$

and

$$\mathbf{c} = \left(c_0^0 c_0^1 \ldots c_0^{n-1}, c_1^0 c_1^1 \ldots c_1^{n-1}, c_2^0 c_2^1 \ldots c_2^{n-1}, \ldots\right), \tag{9.8.5}$$

respectively. Each element of $\mathbf{c}$ is a linear combination of the elements in the input sequences $\mathbf{m}^0$, $\mathbf{m}^1$, ..., $\mathbf{m}^{k-1}$. For example, the output stream in Figure 9.8.1 is calculated from a single input stream $\mathbf{m} = \mathbf{m}^0$. Assuming that the initial state of the shift register is (000), the outputs are $c_0^1 = m_0^0 + 0 + 0$, $c_1^1 = m_1^0 + m_0^0 + 0$, $c_2^1 = m_2^0 + m_1^0 + 0$, and $c_3^1 = m_3^0 + m_2^0 + m_0^0$. Therefore, the output can be expressed in the following general form:

$$c_i^1 = m_i^0 + m_{i-1}^0 + m_{i-3}^0,$$
$$c_i^j = m_i g_0^j + m_{i-1} g_1^j + m_{i-2} g_2^j + m_{i-3} g_3^j,$$

where $g$ specifies the positions of the taps of the encoder. For $\mathbf{g}^1 = \left(g_0^1, g_1^1, g_2^1, g_3^1\right) = (1101)$, we may have a second output coded sequence expressed in its general form as

$$c_i^1 = m_i \cdot 1 + m_{i-1} \cdot 1 + m_{i-2} \cdot 0 + m_{i-3} \cdot 1 = m_i + m_{i-1} + m_{i-3}. \tag{9.8.6}$$

For $\mathbf{g}^0 = \left(g_0^1, g_1^1, g_2^1, g_3^1\right) = (1011)$, we may have the first output sequence expressed in its general form as

$$c_i^0 = m_i + m_{i-2} + m_{i-3}. \tag{9.8.7}$$

The convolutional encoder in Figure 9.8.1 is a linear system as defined in Chapters 11, 12, 14, and 15. Therefore, if $\mathbf{c}_1$ and $\mathbf{c}_2$ correspond to inputs $\mathbf{m}_1$ and $\mathbf{m}_2$, respectively, then $(\mathbf{c}_1 + \mathbf{c}_2)$ is the code word corresponding to the input message $(\mathbf{m}_1 + \mathbf{m}_2)$.

**Example**

For the convolutional encoder in Figure 9.8.1, find the coded sequence $\mathbf{c}$ for the input message sequence $\mathbf{m} = (1\ 0\ 1\ 0)$. Due to the finite lengths of $\mathbf{m}$ and $\mathbf{c}$, we can also refer to them as 'the message word' and 'the coded word', respectively.

**Solution.** There are two ways to solve this problem.

1. The coded word can be obtained by using the block schematic of the encoder in Figure 9.8.1 and then applying the procedure of shift-register operation for its initial 000 state. Let us present this procedure, using Table 9.8.1.
   The output coded sequence is obtained by multiplexing the two $\mathbf{c}$ sequences from the table and expressed as $\mathbf{c} = \left(c_0^0 c_0^1, c_1^0 c_1^1, \ldots, c_5^0 c_5^1\right) = (11\ 01\ 01\ 10\ 10\ 11\ 00)$.

2. The output coded sequences $\mathbf{c}^0$ and $\mathbf{c}^1$ can be obtained from the general formulas (9.8.6) and (9.8.7). For example, the encoded bits at the second output $\mathbf{c}^1$ can be calculated as

$$c_0^1 = m_0^0 + 0 + 0 = 1 + 0 + 0 = 1,$$

$$c_1^1 = m_1^0 + m_0^0 + 0 = 0 + 1 + 0 = 1,$$

$$c_2^1 = m_2^0 + m_1^0 + 0 = 1 + 0 + 0 = 1,$$

$$c_3^1 = m_3^0 + m_2^0 + m_0^0 = 0 + 1 + 1 = 0,$$

$$c_4^1 = m_4^0 + m_3^0 + m_1^0 = 0 + 0 + 0 = 0,$$

$$c_5^1 = m_5^0 + m_4^0 + m_2^0 = 0 + 0 + 1 = 1,$$

$$c_6^1 = m_6^0 + m_5^0 + m_3^0 = 0 + 0 + 0 = 0.$$

**Table 9.8.1** *Coded word calculations.*

| m | Shifted-register states | $c^0 \, c^1$ |
|---|---|---|
| – | 0 0 0 | – – |
| $m_0 = 1$ | 1 0 0 | 1 1 |
| $m_1 = 0$ | 0 1 0 | 0 1 |
| $m_2 = 1$ | 1 0 1 | 0 1 |
| $m_3 = 0$ | 0 1 0 | 1 0 |
| $m_4 = \mathbf{0}$ | 0 0 1 | **1 0** |
| $m_5 = \mathbf{0}$ | **0 0 0** | **1 1** |
| $m_6 = \mathbf{0}$ | **0 0 0** | **0 0** |

**The generator matrix.** As we have already said, a convolutional encoder can be represented by a linear time-invariant system. Therefore, if we find its impulse response, then, for any input message sequence, we can find an encoded sequence that is a convolution of the input sequence and the encoder's impulse response. Let us demonstrate the procedure for finding the impulse response of the encoder.

The unit impulse here will be defined as it was in Chapter 14, that is, as a sequence of 0's that starts with a single 1. Therefore, for a multiple input and multiple output encoder, when a single 1 is applied at the *j*th input of the encoder and then followed by a string of 0's and, at the same time, strings of 0's are applied to all other inputs, then the impulse response $g_j^i$

is obtained for the *i*th output of the encoder. The *j*th input bit sequence $m_i^j = \delta = 10000\ldots$, which can be understood as a Kronecker delta function, serves the same role as the Dirac delta function $\delta(t)$ in the analysis of continuous systems. Because the encoder is a linear-time-invariant system, the output coded sequence is the convolution of the input of the encoder and its impulse response, and eqns (9.8.6) and (9.8.7) can be re-expressed in the general form

$$c_i^j = \sum_{l=0}^{k-1} m_{i-l} g_l^j, \tag{9.8.8}$$

which is the discrete convolution of a pair of sequences, hence the term 'convolutional code'. The output coded sequence $c^j$ of a 1/*n*-rate code is the convolution of the input sequence **m** and the impulse response sequence $g^j$ and can be expressed in the form

$$c^j = m * g^j. \tag{9.8.9}$$

If an encoder has $k$ inputs, the discrete convolution formula becomes

$$c_i^j = \sum_{t=0}^{k-1} \left( \sum_{l=0}^{m} m_{i-l}^t g_{t,l}^i \right),$$

(9.8.10)

which can be expressed in matrix form as

$$\mathbf{c}^i = \sum_{t=0}^{k-1} \mathbf{m}^t * \mathbf{g}_t^i.$$

(9.8.11)

Therefore, the output sequence can be expressed as a matrix multiplication defined by the convolution, thus providing a generator matrix similar to that developed for block codes. Because a message sequence at the input of an encoder can have unlimited number of bits, or the message sequence of an encoder is not necessarily bounded in length, both the generator matrix and the parity check matrix are semi-infinite for convolutional codes. For the 1/2-rate encoder given as an example, the generator matrix is formed by interleaving the two generator sequences $g^0$ and $g^1$ as follows:

$$\mathbf{G} = \begin{bmatrix} g_0^{(0)} g_0^{(1)} & g_1^{(0)} g_1^{(1)} & \cdots & g_m^{(0)} g_m^{(1)} & & \mathbf{0} \\ & g_0^{(0)} g_0^{(1)} & g_1^{(0)} g_1^{(1)} & \cdots & g_m^{(0)} g_m^{(1)} & \\ & & g_0^{(0)} g_0^{(1)} & g_1^{(0)} g_1^{(1)} & \cdots & g_m^{(0)} g_m^{(1)} \\ \mathbf{0} & & \ddots & \ddots & \ddots & \ddots \end{bmatrix}.$$

(9.8.12)

Therefore, a code word $\mathbf{c}$ corresponding to a message word $\mathbf{m}$ can be obtained via the following matrix multiplication:

$$\mathbf{c} = \mathbf{m} \cdot \mathbf{G}.$$

(9.8.13)

**Example**

For the convolutional encoder in Figure 9.8.1, find the output code word $\mathbf{c}$ for the input message word $\mathbf{m} = (1\ 0\ 1\ 0)$.

**Solution.** The coded word can be obtained with the generator sequences obtained using the block schematic of the encoder in Figure 9.8.1. The generator sequences are obtained as the impulse response of the encoder, according to the procedure presented in Table 9.8.2.

For the generator sequences $g^0$ and $g^1$, we may calculate the coded word using matrix multiplication expressed as

$$\mathbf{c} = \mathbf{mG} = [m_0 \ m_1 \ m_2 \ m_3] \begin{bmatrix} g_0^0 g_0^1 & g_1^0 g_1^1 & g_2^0 g_2^1 & g_3^0 g_3^1 & 00 & 00 & 00 \\ 00 & g_0^0 g_0^1 & g_1^0 g_1^1 & g_2^0 g_2^1 & g_3^0 g_3^1 & 00 & 00 \\ 00 & 00 & g_0^0 g_0^1 & g_1^0 g_1^1 & g_2^0 g_2^1 & g_3^0 g_3^1 & 00 \\ 00 & 00 & 00 & g_0^0 g_0^1 & g_1^0 g_1^1 & g_2^0 g_2^1 & g_3^0 g_3^1 \end{bmatrix}$$

$$= [1\ 0\ 1\ 0] \begin{bmatrix} 11 & 01 & 10 & 11 & 00 & 00 & 00 \\ 00 & 11 & 01 & 10 & 11 & 00 & 00 \\ 00 & 00 & 11 & 01 & 10 & 11 & 00 \\ 00 & 00 & 00 & 11 & 01 & 10 & 11 \end{bmatrix} = [11 \quad 01 \quad 01 \quad 10 \quad 10 \quad 11 \quad 00]$$

$$= \begin{bmatrix} c_0^0 c_0^1 & c_1^0 c_1^1 & c_2^0 c_2^1 & c_3^0 c_3^1 & c_4^0 c_4^1 & c_4^0 c_4^1 & c_5^0 c_5^1 & c_6^0 c_6^1 \end{bmatrix}$$

**Table 9.8.2** *Calculation of the encoder impulse response.*

| x | Shifted-register status | $g^0 g^1$ |
|---|---|---|
| – | 0 0 0 | – – |
| $x_0 = 1$ | 1 0 0 | 1 1 |
| $x_1 = 0$ | 0 1 0 | 0 1 |
| $x_2 = 0$ | 0 0 1 | 1 0 |
| $x_3 = 0$ | 0 0 0 | 1 1 |
| $x_4 = 0$ | 0 0 0 | 0 0 |
| $x_5 = 0$ | 0 0 0 | 0 0 |
| $x_6 = 0$ | 0 0 0 | 0 0 |

**The *D*-transform.** The procedure of convolutional coding can be described as a discrete convolution operation where the output is a convolution of the input and the impulse response. Using the *D*-transform, a sequence of bits can be expressed in terms of a unit delay operator *D* that represents a time interval between the succeeding bits. The message sequence, coded sequence, and impulse response sequence, expressed by the *D*-transform, are

$$\mathbf{m}^i = \left(m_0^i, m_1^i, m_2^i, \ldots\right) \leftrightarrow \mathbf{M}^i(D) = m_0^i D^0 + m_1^i D^1 + m_2^i D^2 \ldots, \tag{9.8.14}$$

$$\mathbf{c}^i = \left(c_0^i, c_1^i, c_2^i, \ldots\right) \leftrightarrow \mathbf{C}^i(D) = c_0^i D^0 + c_1^i D^1 + c_2^i D^2 \ldots, \tag{9.8.15}$$

and

$$\mathbf{g}_j^i = \left(g_{j0}^i, g_{j1}^i, g_{j2}^i, \dots\right) \leftrightarrow \mathbf{G}_j^i(D) = g_{j0}^i D^0 + g_{j1}^i D^1 + g_{j2}^i D^2 \dots, \tag{9.8.16}$$

respectively, where the superscript $i$ defines the order of the inputs and outputs of the encoder. The indeterminate $D$ is a unit delay operator. The exponent of $D$ denotes the number of time units that a particular coefficient is delayed with respect to the coefficient of the $D^0$ term. The $D$-transform is similar to the $z$-transform that will be presented in Chapter 15, but there is no corresponding interpretation of $D$ as a number in the complex plane. For a single input encoder, the coding operation can be represented as follows:

$$\mathbf{C}^i(D) = \mathbf{M}^i(D)\mathbf{G}^i(D). \tag{9.8.17}$$

For a $k$-input encoder, the coding operation can be represented in matrix form as

$$\mathbf{C}^i(D) = \sum_{j=0}^{k-1} \mathbf{M}^j(D)\mathbf{G}_j^i(D). \tag{9.8.18}$$

This equation can be represented as a matrix multiplication operation. If we define

$$\mathbf{C}^i(D) = \begin{bmatrix} \mathbf{M}^0(D) & \mathbf{M}^1(D) & \cdots & \mathbf{M}^{k-1}(D) \end{bmatrix} \begin{bmatrix} \mathbf{G}_0^i(D) \\ \mathbf{G}_1^i(D) \\ \vdots \\ \mathbf{G}_{k-1}^i(D) \end{bmatrix} \tag{9.8.19}$$

then we can obtain the output sequence by multiplying the matrix of the input sequence by the transfer-function generator matrix $\mathbf{G}(D)$ as follows:

$$\mathbf{C}(D) = \mathbf{M}(D)\mathbf{G}(D)$$

$$= \begin{bmatrix} \mathbf{M}^{(0)}(D) & \mathbf{M}^{(1)}(D) & \cdots & \mathbf{M}^{(k-1)}(D) \end{bmatrix} \begin{bmatrix} \mathbf{G}_0^{(0)}(D) & \mathbf{G}_0^{(1)}(D) & \cdots & \mathbf{G}_0^{(n-1)}(D) \\ \mathbf{G}_1^{(0)}(D) & \mathbf{G}_1^{(1)}(D) & \cdots & \mathbf{G}_1^{(n-1)}(D) \\ \vdots & \vdots & \ddots & \vdots \\ \mathbf{G}_{k-1}^{(0)}(D) & \mathbf{G}_{k-1}^{(1)}(D) & \cdots & \mathbf{G}_{k-1}^{(n-1)}(D) \end{bmatrix}. \tag{9.8.20}$$

## Example

The message at the input of the encoder in Figure 9.8.2 is $\mathbf{m} = (11, 10, 11)$. Find the output code word $\mathbf{c}$ by applying the $D$-transform.

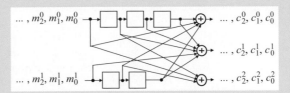

**Fig. 9.8.2** *A 2/3-rate convolutional encoder.*

**Solution.** The matrix representation of the input signal is

$$\mathbf{m} = (11,10,11) \leftrightarrow \left[\mathbf{M}^0(D)\mathbf{M}^1(D)\right] = \left[D^0 + D^1 + D^2 \ \ D^0 + D^2\right] = \left[1 + D + D^2 \ \ 1 + D^2\right].$$

To get the transfer-function matrix, the generator sequences need to be calculated as

$$g_j^i = \left(g_{j0}^i, g_{j1}^i, g_{j2}^i, \ldots, \right) \leftrightarrow \mathbf{G}_j^i(D) = g_{j0}^i D^0 + g_{j1}^i D^1 + g_{j2}^i D^2 + \ldots.$$

The first input ($j = 0$) to the first output ($i = 0$) gives

$$g_0^0 = \left(g_{00}^0, g_{01}^0, g_{02}^0, g_{03}^0\right) = (1001) \leftrightarrow \mathbf{G}_0^0(D) = 1D^0 + 0D^1 + 0D^2 + 1D^3 = 1 + D^3.$$

The second input ($j = 1$) to the first output ($i = 0$) gives

$$\mathbf{g}_1^0 = \left(g_{10}^0, g_{11}^0, g_{12}^0, g_{13}^0\right) = (0110) \leftrightarrow \mathbf{G}_1^0(D) = 0D^0 + 1D^1 + 1D^2 + 0D^3 = D + D^2.$$

The first and second inputs to the second output ($i = 1$) give, respectively,

$$g_0^1 = \left(g_{00}^1, g_{01}^1, g_{02}^1, g_{03}^1\right) = (0111) \leftrightarrow \mathbf{G}_0^1(D) = 0D^0 + 1D^1 + 1D^2 + 1D^3 = D + D^2 + D^3,$$
$$\mathbf{g}_1^1 = \left(g_{10}^1, g_{11}^1, g_{12}^1, g_{13}^1\right) = (1010) \leftrightarrow \mathbf{G}_1^1(D) = 1D^0 + 0D^1 + 1D^2 + 0D^3 = 1 + D^2.$$

The first and second inputs to the third output ($i = 2$) give, respectively,

$$\mathbf{g}_0^2 = \left(g_{00}^2, g_{01}^2, g_{02}^2, g_{03}^2\right) = (1100) \leftrightarrow \mathbf{G}_0^2(D) = 1D^0 + 1D^1 + 0D^2 + 0D^3 = 1 + D,$$
$$\mathbf{g}_1^2 = \left(g_{10}^2, g_{11}^2, g_{12}^2, g_{13}^2\right) = (0100) \leftrightarrow \mathbf{G}_1^2(D) = 0D^0 + 1D^1 + 0D^2 + 0D^3 = D.$$

Therefore, the transfer-function matrix for the encoder is

$$\mathbf{G}(D) = \begin{bmatrix} G_0^{(0)} & G_0^{(1)} & G_0^{(2)} \\ G_1^{(0)} & G_1^{(1)} & G_1^{(2)} \end{bmatrix} = \begin{bmatrix} 1+D^3 & D+D^2+D^3 & 1+D \\ D+D^2 & 1+D^2 & D \end{bmatrix}.$$

The $D$-transform of the output bit sequence is then computed as follows:

$$\mathbf{C}(D) = \mathbf{M}(D)\mathbf{G}(D) = \begin{bmatrix} \mathbf{M}^{(0)}(D) & \mathbf{M}^{(1)}(D) \end{bmatrix} \begin{bmatrix} G_0^{(0)} & G_0^{(1)} & G_0^{(2)} \\ G_1^{(0)} & G_1^{(1)} & G_1^{(2)} \end{bmatrix}$$

$$= \begin{bmatrix} 1+D+D^2 & 1+D^2 \end{bmatrix} \begin{bmatrix} 1+D^3 & D+D^2+D^3 & 1+D \\ D+D^2 & 1+D^2 & D \end{bmatrix}$$

$$= \begin{bmatrix} 1+D^5 & 1+D+D^3+D^4+D^5 & 1+D \end{bmatrix} = \begin{bmatrix} \mathbf{C}^{(0)}(D) & \mathbf{C}^{(1)}(D) & \mathbf{C}^{(2)}(D) \end{bmatrix}$$

The output sequences can be obtained from

$$c^i = \left( c_0^i, c_1^i, c_2^i, \ldots, \right) \leftrightarrow \mathrm{C}i(D) = c_0^i D0 + c_1^i D1 + c_2^i D2 + \ldots,$$

as

$$\mathbf{c}^0 = \left( c_0^0, c_1^0, c_2^0, c_3^0, c_4^0, c_5^0 \right) = (100001) \leftrightarrow \mathbf{C}^0(D) = 1+D^5,$$

$$\mathbf{c}^1 = \left( c_0^1, c_1^1, c_2^1, c_3^1, c_4^1, c_5^1 \right) = (110111) \leftrightarrow \mathbf{C}^1(D) = 1+D+D^3+D^4+D^5,$$

$$\mathbf{c}^2 = \left( c_0^2, c_1^2, c_2^2, c_3^2, c_4^2, c_5^2 \right) = (110000) \leftrightarrow \mathbf{C}^2(D) = 1+D.$$

The output code word is then

$$\mathbf{c} = \left( c_0^0 c_0^1 c_0^2, c_1^0 c_1^1 c_1^2, \ldots, c_5^0 c_5^1 c_5^2 \right) - (\mathbf{111}, 011, 000, 010, 010, 110).$$

**The state diagram.** A convolutional encoder can be treated as a finite-state machine. It contains memory elements whose content (called the encoder state) determines a mapping between the input and output bits of an encoder. The encoder in Figure 9.8.3, for example, has two memory elements. Thus, the number of states the encoder can take is 4. If we designate these states as $\{S_0, S_1, S_2, S_3, S_4\}$ and associate them with the contents of the memory elements, we obtain $S_0 \leftrightarrow (00)$, $S_1 \leftrightarrow (10)$, $S_2 \leftrightarrow (01)$, and $S_3 \leftrightarrow (11)$. The state diagram is an oriented graph that consists of states and directed lines that represent the sense of state transitions. These lines are labelled with the input/output sequence pairs. Therefore, the label $a/bcd$ means that the input bit $a$ causes a state transition, and $bcd$ is the corresponding set of the encoded bits. Using the circuit diagram of the encoder in Figure 9.8.3, these transitions are calculated and presented in Figure 9.8.4 on the diagram for the encoder state.

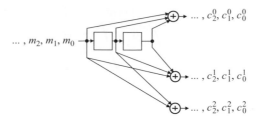

**Fig. 9.8.3** *Diagram of a 1/3-rate encoder circuit.*

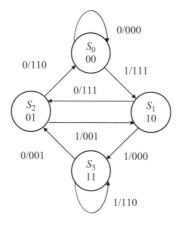

**Fig. 9.8.4** *State diagram of a 1/3-rate encoder.*

**Trellis diagrams.** A trellis diagram explicitly shows the passage time of an encoder, possible transitions between the states, and possible generated encoded bits for the given inputs. Thus, it is an extension of a convolutional-code state diagram. We will assume that, after the input sequence has been entered into the encoder, $m$ state transitions are necessary to return the encoder to state $S_0$. The characteristics of the trellis diagram will be demonstrated for the example $(n, k) = (3, 1)$. For this code, every code word is associated with a unique path that starts and stops at state $S_0$. In Figure 9.8.5, the path indicated in bold corresponds to the input message bits $\mathbf{m} = (01100)$, or $\mathbf{c} = (000, 111, 000, 001, 110)$ encoder output bits.

The first three bits in $\mathbf{m}$ are message bits. The number of nodes at each unit time increment is $2^M = 4$, for a total number of memory elements $M = 2$. Any two neighbouring nodes are connected by a branch. A number of concatenated branches forms a partial path. The number of branches leaving each node is $2^k = 2$, where $k$ is the number of the encoder's inputs. The number of branches entering each node after $m = 2$ time increments is 2. If a sequence of $kL = 3$ message bits has been entered into the encoder, then the trellis diagram must have $(L + m) = 5$ stages. There are $2^{kL} = 8$ distinct paths through the trellis diagram, each corresponding to a convolutional code word of length $n(L + m) = 15$.

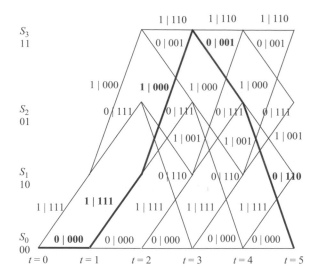

**Fig. 9.8.5** *Trellis diagram for a 1/3-rate encoder.*

## 9.8.2 Operation of a Coding Communication System

Traditionally, convolutional-code decoding has been performed using the Viterbi algorithm, which is asymptotically optimum in the maximum likelihood sense. Due to the importance of the Viterbi algorithm, a concise explanation of it is presented in this section.

Figure 9.8.6 shows a communication system which incorporates a rate-$1/n$ convolutional encoder that generates a set of $n$ coded bits for each message bit at the input to the encoder, at the discrete-time instant $t$. It is important to note here that we are using $t$ instead of $n$ to denote a discrete-time variable. The reason is that, in coding theory, the letter $n$ is usually used to denote the number of encoder outputs, and we wanted to preserve that meaning of $n$. We will perform the investigation and testing of encoders and decoders by designing a communication system composed of an encoder, a communication channel, and a decoder. For the sake of completeness, we will analyse the system and define the channel structures relevant for the testing of encoder and decoder operation.

**Encoder operation.** Suppose a convolutional encoder is defined by its constraint length $L$, a set $M$ of shift-register stages (encoder cells), and adders, which define the mapping of input bits into code words at the encoder output. In this case, at the discrete-time instant $t$, a set of $L$ bits from all encoder cells defined as $B(t) = [m(t), m(t-1), \ldots, m(t-L+1)]$ is combined in a separate block of adders to form at its output a set of $n$ coded bits denoted as $c(B(t)) = [c_0(B(t)), c_1(B(t)), \ldots, c_i(B(t)), \ldots, c_{n-1}(B(t))]$ or, expressed in a simplified way, $c_t = [c_{t,0}, c_{t,1}, \ldots, c_{t,i}, \ldots, c_{t,n-1}]$, which will be used later. Here we express bits as a function of time $t$, which can be interpreted in two ways. If the encoder is implemented using a shift register, each bit will be represented by a binary signal with a unit duration equivalent to the delay between succeeding message bits. However, if the encoder is designed in discrete

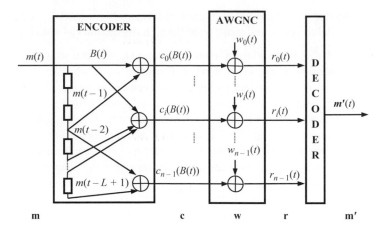

**Fig. 9.8.6** *Communication system with a 1/n-rate convolutional encoder and decoder; AWGNC, additive white Gaussian noise channel.*

technology, or if the encoder operation is simulated on a computer, the message bits and the corresponding coded bits will be presented as the discrete values 1 and 0 at the discrete-time instants $t$.

Because the output set of $n$ bits depends on the values of $L$ bits generated inside the encoder structure, the coding procedure can be represented as a mapping of an $L$-dimensional message vector $B(t)$ into an $n$-dimensional code vector $c(B(t))$, and represented by a matrix of impulse responses $\mathbf{g}_L$ expressed in the form

$$\mathbf{g}_L = \begin{bmatrix} g_{01} & g_{02} & \cdots & g_{0L} \\ g_{11} & g_{12} & \cdots & g_{1L} \\ \vdots & \vdots & \ddots & \vdots \\ g_{n-11} & g_{n-12} & \cdots & g_{n-1L} \end{bmatrix}. \tag{9.8.21}$$

The elements of this matrix can be 1's and 0's that specify the connections between the shift-register cells and the modulo-2 adders inside the encoder structure. Using the structure of the adders in the encoder, the $i$th encoded bit may be represented as a modulo-2 sum of bits at the output of the shift-register stages of the encoder, that is,

$$c_j(B(t)) = \sum_{i=1}^{L} \oplus m(t-i+1) \cdot g_{j,i}. \tag{9.8.22}$$

The number of additions depends on the number of connections between the shift-register stages and the encoder adders. At each output of the encoder, one such bit is generated and a set of coded bits $c(B(t)$ is obtained, which can be viewed as a vector in an $n$-dimensional Euclidean space, with each component $c_i(B(t)$ being a coordinate in the $i$th dimension.

The set $\mathbf{c}$ of encoded bits is presented in parallel form in Figure 9.8.6. However, they are converted from parallel to series and transferred through the channel as a series of bits.

**Characterization of the transmission channel.** Suppose the transmission channel is a waveform channel defined as an additive white Gaussian noise channel. For the set of $n$ encoded bits generated at time instant $t$, this channel can be represented by a sequence of $n$ noise samples $w(t) = [w_0(t), w_1(t), \ldots, w_i(t), \ldots, w_{n-1}(t)]$, which are realizations of a white Gaussian noise process as defined in Chapter 4 and are added to the set of $n$ encoded bits. The encoded bits $\mathbf{c}$ and the corresponding noise samples $\mathbf{w}$ are represented in parallel formation in Figure 9.8.6. However, in a real situation, they would be generated as a serial sequence of encoded bits and a serial sequence of noise samples generated in the channel. Because of that, the noise samples $w_i(t)$ can be generally represented as a series of realizations of independent and identically distributed Gaussian random variables $W_i(t)$, $i = 0,1,\ldots, n - 1$, as defined in Chapters 3 and 4. If the channel is designed in a discrete form, then the noise is represented by the error sequence $\mathbf{e}$ defined in Section 9.2.2 and presented in Figures 9.1.1 and 9.2.2. In this case, the error sequence $\mathbf{e}$ is added modulo-2 at each time instant $t$ to the encoded $\mathbf{c}$ sequence to obtain the received encoded sequence $\mathbf{r}$.

**Operation of the decoder.** At the receiver, we have an $n$-dimensional received encoded sequence $r(t) = [r_0(t), r_1(t), \ldots, r_i(t), \ldots, r_{n-1}(t)]$ for each message bit generated at the discrete-time instants $t = \{0, 1, \ldots\}$. This sequence is the sum of the encoded sequence at the output of the encoder and the sequence of noise samples, that is,

$$r(t) = c(t) + w(t). \tag{9.8.23}$$

If the transmission channel is represented by additive white Gaussian noise, we can say that the channel is memoryless, as defined in Section 9.2.2, and theoretically analysed in Chapter 3. Thus, the noise affecting a given encoded bit is independent of the noise affecting any proceeding or succeeding encoded bit.

**Practical considerations.** In practice, a sequence of message bits can also be represented by a series of bits or as a message vector $\mathbf{m}$. A sequence of encoded bits produced at the output of the encoder can be denoted as a code vector $\mathbf{c}$, which is a series of encoded bits usually presented in polar form. Each of the received bits is corrupted by the noise inside the coding channel. The noise is represented by a series of noise samples denoted by $\mathbf{w}$ and called a noise vector. The noise vector is added to the code vector to form the received vector $\mathbf{r}$, according to the expression

$$\mathbf{r} = \mathbf{c} + \mathbf{w}. \tag{9.8.24}$$

This relation was defined in Chapter 5, and demonstrated in Figure 5.5.5. Additional details related to the processing of the encoded sequence before its transmission through the channel will be presented later on in this section.

**Application of channel interleavers.** In the case when the channel has memory, bursts and clusters of errors can occur. In order to randomize the errors and make the channel behave

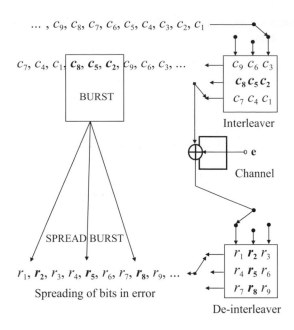

**Fig. 9.8.7** *Operation of block interleaver/de-interleaver blocks.*

like a memoryless channel, channel interleavers and de-interleavers can be used. These are placed after the encoder and in front of the decoder, respectively, as shown in Figure 9.1.1. The operation of a simple block interleaver is shown in Figure 9.8.7, which presents as an example an $(n \times n) = 3 \times 3$ block interleaver. The interleaver is designed as a memory matrix that writes in the encoded three-bit symbols column-wise and then reads them out row-wise. Suppose a series of errors (called a burst) occurred in the channel, affecting the succeeding bits $c_2$, $c_5$, and $c_8$. The received bits are stored in the receiver memory column-wise and read out row-wise, returning the order of the received bits as it was at the transmitter side. Therefore, the bits affected by the channel noise, in this case $r_2$, $r_5$, and $r_8$, are spread across the received encoded sequence, causing each symbol to have one error.

The errors are effectively randomized, and the bits from any given encoded sequence are well separated during transmission through the channel, which can increase the efficiency of the decoding algorithm. As we will see later, interleavers and de-interleavers are also the building blocks of turbo encoders and decoders.

## 9.8.3 Operation of a Decoder

The problem in decoding is finding the encoded sequence **c** which is the closest in *some sense* to the received noisy encoded sequence **r**, or, more precisely, finding an estimate of the received message vector **m′** which is, in some sense, the closest to the transmitted message vector **m**. Various techniques and algorithms have been developed to solve this problem. In principle, we aim to minimize the probability of error in the encoded words in the communication system.

**Minimization of word error probability.** Assuming that the message sequence $\mathbf{m}$ is transmitted in a limited time interval from $t = 1$ to $t = T$, the sequence error probability, also called the word error probability, due to its finite duration, can be expressed in the form

$$P_T = 1 - \int_{\mathbf{r}} p(\mathbf{m}, \mathbf{r}) \, d\mathbf{r} = 1 - \int_{\mathbf{r}} p(\mathbf{m}|\mathbf{r}) p(\mathbf{r}) \, d\mathbf{r} \tag{9.8.25}$$

for a received encoded sequence $\mathbf{r}$. The decoding problem can be formulated as follows: for a given received sequence $\mathbf{r}$, we want to find the message sequence $\mathbf{m}$ (or encoded sequence $\mathbf{c}$) which minimizes the word error probability $P_T$. Because there is one-to-one correspondence between the message sequence $\mathbf{m}$ and the encoded sequence $\mathbf{c}$, the above relation is equivalent to the relation

$$P_T = 1 - \int_{\mathbf{r}} p(\mathbf{c}, \mathbf{r}) \, d\mathbf{r} = 1 - \int_{\mathbf{r}} p(\mathbf{c}|\mathbf{r}) p(\mathbf{r}) \, d\mathbf{r}. \tag{9.8.26}$$

In addition, we can say that the joint probabilities of sequences $\mathbf{m}$ and $\mathbf{r}$ are equivalent to the joint probability of sequences $\mathbf{c}$ and $\mathbf{r}$, due to the direct functional correspondence between the sequence $\mathbf{m}$ and $\mathbf{c}$ as defined by the encoder structure. If we apply the Bayesian rule

$$p(\mathbf{c}|\mathbf{r}) p(\mathbf{r}) = p(\mathbf{r}|\mathbf{c}) p(\mathbf{c}), \tag{9.8.27}$$

we can express this word error probability as

$$P_T = 1 - \int_{\mathbf{r}} p(\mathbf{r}, \mathbf{c}) \, d\mathbf{r} = 1 - \int_{\mathbf{r}} p(\mathbf{r}|\mathbf{c}) p(\mathbf{c}) \, d\mathbf{r}, \tag{9.8.28}$$

and the decoding process can be redefined as follows: we want to find the sequence of bits $\mathbf{c}$ that maximizes the conditional probability $p(\mathbf{r} \mid \mathbf{c})$.

**The maximum likelihood estimator.** The maximum likelihood algorithm is based on the maximization of the corresponding likelihood function. Therefore, to decode a convolutionally encoded sequence, we have to find a likelihood function that relates the encoded sequence (word) $\mathbf{c}$ to the received noise sequence $\mathbf{r}$. Let us analyse the general case when the input of the coding channel is a sequence of discrete values $\mathbf{c}$, and the outputs are continuous-valued random samples represented by the received vector $\mathbf{r}$. Let us also assume that the noise affecting a given bit is independent of the noise affecting any proceeding or succeeding bit, that is, that the channel is memoryless. Furthermore, assume that the output of the encoder can have only two discrete values. Therefore, the coding channel can be represented as a binary discrete-input continuous-output channel, as shown in Figure 9.8.8.

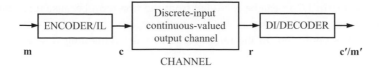

**Fig. 9.8.8** *A coding communication system with interleavers and de-interleavers; DI, de-interleaver; IL, interleaver.*

For this channel, we can express the conditional density function of the received encoded bit $r = r_i$ for a given encoded bit $c = c_i$ as

$$p_{R_i}(r = r_i | c = c_i) = \frac{1}{\sqrt{2\pi\sigma^2}} e^{-(r_i - c_i)^2 / 2\sigma^2}, \qquad (9.8.29)$$

where $r_i$ is a realization of a Gaussian random variable $R_i$, and $\sigma^2$ is the variance of the channel noise. In our case, the joint conditional probabilities calculated for each time instant defined by $t$ can be found. Bearing in mind that the channel is symmetric and without memory, the joint conditional probability density function $p(r(t) | c(t))$, at every time instant $t$, is

$$p(r(t)|c(t)) = \prod_{i=0}^{n-1} \frac{1}{\sqrt{2\pi\sigma^2}} e^{-(r_i(t) - c_i(t))^2 / 2\sigma^2}, \qquad (9.8.30)$$

where $i = 0, 1, 2, \dots, n - 1$, and $n$ specifies the number of bits generated at the output of the encoder for each message bit. The index $i$ shows the order of a serial transmission of $n$ encoded bits generated in parallel at the encoder output, as shown in Figure 9.8.6, and transmitted through the channel. This transmission of $n$ bits occurs at each time instant $t$. The conditional probability $p(\mathbf{r} | \mathbf{c})$ for all time instants $t$, where $t = 1, \dots, T$, can be expressed as

$$p(\mathbf{r}|\mathbf{c}) = \prod_{t=1}^{T} \prod_{i=1}^{n-1} \frac{1}{\sqrt{2\pi\sigma^2}} e^{-(r_i(t) - c_i(t))^2 / 2\sigma^2}. \qquad (9.8.31)$$

*This conditional probability density function corresponds to the likelihood function, and its maximization is equivalent to the minimization of the word (sequence) error probability defined by eqn (9.8.28).* The maximum of this function is equivalent to the maximum of the logarithm of the same function, expressed as

$$\ln p(\mathbf{r}|\mathbf{c}) = -\left[ \frac{1}{2\sigma^2} \sum_{t=1}^{T} \sum_{i=0}^{n-1} (r_i(t) - c_i(t))^2 + Tn \ln \sqrt{2\pi\sigma^2} \right]. \qquad (9.8.32)$$

Due to the assumption that the variance of the noise is the same for each received bit (due to the stationarity of the stochastic noise process), the last term in eqn (9.8.32) is a constant, and the multiplication term $1/2\sigma^2$ can be excluded from the first term because it has no influence on the minimization procedure whatsoever. Due to the negative sign of the likelihood function, its maximization is equivalent to the minimization of the remaining function $D$, which is expressed as

$$D = \sum_{t=1}^{T} \sum_{i=0}^{n-1} (r_i(t) - c_i(t))^2. \tag{9.8.33}$$

Because this function specifies in some sense the 'distances' between the received sequence and all possible transmitted sequences, we will call it the *distance function*. In addition, this function is sometimes called the *noise energy function*, because it specifies the energy of the noise present in the received sequence with respect to the encoded sequence.

Because the function $D$ is derived from the likelihood function, the decoder can be defined as a maximum likelihood estimator, which calculates the sequence $\mathbf{c}'$ that gives the minimum distance $D$ with respect to the receiver's sequence, $\mathbf{r}$.

**Definition of a coding channel.** The received sequence $\mathbf{r}$ can be taken at the output of the correlation demodulator, as defined in Chapters 4, 5, and 6. In the case when an encoder is used at the transmitter side in front of the modulator, the vector $\mathbf{r}$ will be the received encoded sequence that can be used to estimate the encoded bits $\mathbf{c}$. The sequence $\mathbf{r}$, in this sense, can be understood as a realization of a discrete-time continuous-valued Gaussian stochastic process. This process is defined in Chapter 6 for digital binary phase-shift keying (see Figure 6.2.1), and in Chapter 7 for discrete binary phase-shift keying (see Figure 7.2.1).

If we can detect the received samples at the output of the demodulator, we can form a sequence of bits $\mathbf{r}$ at the output of the decision device. In this case, the communication system is characterized by a discrete-input discrete-output channel (a digital channel), as shown in Figure 9.8.9, and can be represented by a binary error sequence $\mathbf{e}$, as defined in Section 9.2.2.1 and presented in Figure 9.2.2. It is always important to understand the placement of the encoders and decoders inside communication systems with respect to the related modulators and demodulators.

## 9.8.4   Decoding Algorithms

**Direct or brute force decoding.** The simplest way, and perhaps the most reliable algorithm for finding the maximum likelihood sequence (maximum likelihood estimator), is to calculate the distances $D$ for all possibly sent sequences defined by the trellis diagram of a particular code. This is called *direct decoding*, which involves comparing all of the encoded sequences

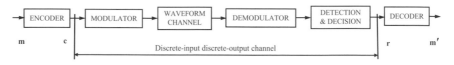

**Fig. 9.8.9**  *A discrete-input discrete-output channel.*

that could be sent with the received sequence $\mathbf{r}$, selecting the sequence $\mathbf{c}'$ that has the smallest calculated distance $D$, and then calculating the message sequence $\mathbf{m}$ from $\mathbf{c}'$. However, the number of encoded sequences increases exponentially with the length of the message sequence, so the number of calculated distances can be enormously large; hence, the process of decoding can be very time consuming and expensive. For this reason, direct decoding is not an attractive approach in practice.

**Using an asymptotic maximum likelihood estimator: The Viterbi algorithm.** Instead of using direct decoding, Viterbi proposed a method that eliminates less probable encoded sequences, which correspond to the less probable paths in the trellis diagram, in favour of more probable encoded sequences. For this purpose, the distances at the level of the *bit*, the *branch*, the *partial path*, and the *path* (called metrics) need to be defined and calculated. In order to explain this algorithm, let us define these metrics first.

The *bit metric* is defined as the 'distance' between the $i$th received bit and the $i$th transmitted bit at the time instant $t$, expressed as

$$d_t^i = \|r_i(t) - c_i(t)\|. \tag{9.8.34}$$

**Example**

Suppose that we use the $L1$ norm to define the distance, that is, we take absolute value of the difference between the $i$th received bit and the $i$th transmitted bit, and that the binary decision performed at the receiver side produces the received bits in the binary forms 0 and 1. Calculate the bit metrics if the received bits are $(1, 0, 1)$ assuming that the bits sent are $(1, 1, 1)$, respectively.

**Solution.** The bit metric is the absolute value of the difference, that is,

$$d_t^i = |r_i(t) - c_i(t)|, \tag{9.8.35}$$

which corresponds to the Hamming distance between the $i$th received bit and the $i$th bit transmitted at the time instant $t$. The smaller the absolute value is, the closer the received bit is to the transmitted one. The bit metrics represented by these distances are $d_1 = |1 - 1| = 0$, $d_1 = |0 - 1| - 1$, and $d_1 = |1 - 1| = 0$, for $i = 0$, 1, and 2, respectively. Also, this bit metric corresponds to the modulo-2 addition of binary numbers representing the $j$th transmitted and received bits, that is, $d_1 = 1 \oplus 1 = 0$, $d_1 = 0 \oplus 1 = 1$, and $d_1 = 1 \oplus 1 = 0$, for $i = 0$, 1, and 2, respectively, which is usually used in practice. If we define the bit metric as the difference between the transmitted bit and its demodulated value, taken at the decision point in front of the receiver decision circuit, where theoretically it can be any real number, then the bit metric is usually defined as the Euclidean distance between the two values, and is expressed as that distance squared, that is,

$$d_t^i = (r_i(t) - c_i(t))^2. \tag{9.8.36}$$

For the sake of simplicity, the $L1$ norm, or the sum of the Hamming distances between the bits sent and the bits received, will be used to explain the operation of the basic form of the Viterbi algorithm.

**The branch metric.** Suppose that the encoder generates a symbol of $n$ encoded bits for a given message input bit, and then changes its state accordingly. As we said before, this transition from the previous state to a new state is represented in a trellis diagram by a branch starting at the time instant $t$. The branch metric is defined as the sum of bit metrics that belong to that branch, and is expressed in the form

$$d_t^x = \sum_{i=0}^{n-1} \|r_i(t) - c_i(t)\|, \tag{9.8.37}$$

assuming that the branch belongs to a path $x$. Again, depending on whether this metric is expressed as the absolute value or the difference squared, we may represent it by the Hamming distance or the Euclidean distance, respectively. For our example in Figure 9.8.10, where the received sequence is supposed to be (1 0 1), the branch metric at node $A$ is the distance $d_1 = |1 - 1| + |0 - 1| + |1 - 1| = 1$.

**The partial path metric.** The partial path metric is the metric that represents the difference between the received encoded sequence and the transmitted encoded sequence that includes $\tau$ branches, and is expressed as

$$D_\tau^x = \sum_{t=1}^{\tau} \sum_{i=0}^{n-1} \|r_i(t) - c_i(t)\|. \tag{9.8.38}$$

**The path metric.** If the encoder starts with the all-zero state and returns back to that state after time $T$ through the addition of the required reset dummy bits, we can define the path metric as

$$D_T^x = \sum_{t=1}^{T} \sum_{i=0}^{n-1} \|r_i(t) - c_i(t)\|. \tag{9.8.39}$$

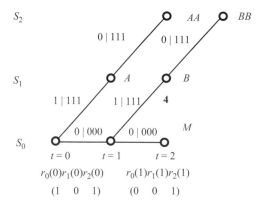

**Fig. 9.8.10** *Initial part of a trellis diagram.*

The decoding procedure now can be defined as a minimization procedure which searches for a path defined in a time interval $T$ that has the minimum path distance defined by eqn (9.8.39).

**Implementation of the Viterbi Algorithm.** The Viterbi algorithm is an 'asymptotically optimum' algorithm for decoding of convolutional codes. We will explain this algorithm for the case when the channel is a discrete-input discrete-output memoryless binary channel that can be represented by an error sequence $\mathbf{e}$, as shown in Figure 9.2.2. The encoding procedure can be represented by a trellis diagram, where each path represents a possibly generated encoded sequence, or code vector $\mathbf{c}$. Therefore, the procedure for decoding convolutional code can be reduced to finding a path in the trellis diagram, where the path represents an encoded sequence $\mathbf{c}$ that is closest to the received sequence $\mathbf{r}$, in the maximum likelihood sense. The algorithm will be explained using the trellis diagram shown in Figure 9.8.10.

**Initial procedure.** The branches of the trellis are labelled with the output $n$-bit sequences of the encoder for a particular input message bit, that is, 1 | 111 means that the output of the encoder at the time instant $t$ is the sequence 111 if the input bit is 1. Suppose the received successive $n$-bit sequences are written underneath the trellis diagram, as shown in Figure 9.8.10 for a 1/3-rate code.

Each node in the trellis is assigned a number which corresponds to the metric of the partial path that starts at the initial state $S_0$ at time $t = 0$ and terminates at that node. For example, this number, denoted by $B = 4$ for the node $S_1$ at the time instant $t = 2$, is

$$D_{t=2}^{B} = \sum_{t=1}^{2}\sum_{i=0}^{2} \|r_i(t) - c_i(t)\| = \sum_{i=0}^{2} \|r_i(1) - c_i(1)\| + \sum_{i=0}^{2} \|r_i(2) - c_i(2)\|,$$

$$= (|1-0| + |0-0| + |1-0|) + (|0-1| + |0-1| + |1-1|) = 4$$

(9.8.40)

which represents the sum of two branch metrics. In the same way, we may calculate the metric of the partial path finishing at $BB$.

**The steady-state decision procedure: The survivor partial path.** When the number of encoded bits $\mathbf{c}$ is large, then the number of paths can be extremely high. In order to eliminate the less probable paths, Viterbi used the properties of trellis diagrams. Namely, in a trellis, after the initial transition, there is a steady-state part that is represented by a fixed structure of the branches between neighbouring stages. In this case, more than one branch merges at a particular node, as shown in Figure 9.8.11. When this happens, the best partial path, determined by the lowest value of the partial path metric and called the 'survivor', should be chosen. The other paths entering the node are called non-survivors and will be disregarded. If the best metric is shared by more than one partial path, the survivor is selected at random.

**The decision procedure.** Consider the hypothetical middle part of the trellis diagram shown in Figure 9.8.11, where the partial paths merge at nodes $Y$ and $Z$. Because the maximum

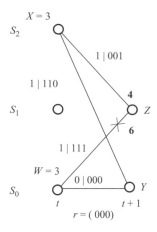

**Fig. 9.8.11** *Steady state of a trellis diagram.*

likelihood sequence corresponds to the partial path with the minimum metric, the decision at the time instant $(t + 1)$ and at node $S_1$ will be made according to the expression

$$D_{t+1}^Z = \min\left\{D_{t+1}^{XZ}, D_{t+1}^{WZ}\right\}, \tag{9.8.41}$$

which means that we will choose the partial path with the minimum distance between the two partial paths that pass through points $X$ and $W$ and merge at point $Z$.

**Example**

Find the survivor at node $Z$ in Figure 9.8.11 if the received code word for the stage shown is $\mathbf{r} = (000)$, and the node metrics are $X = 3$ and $W = 3$.

**Solution.** The solution is as follows:

$$D_{t+1}^Z = \min\left\{D_{t+1}^{XZ}, D_{t+1}^{WZ}\right\} = \min\{X + 1, W + 3\} = \min\{4, 6\} = 4.$$

Because the path coming from $X$ is the survivor, the path from $W$ is excluded and so is crossed out in Figure 9.8.11.

**Termination of the algorithm.** The algorithm is terminated when all of the nodes on the maximum likelihood path are labelled and the entering survivors are determined. The last node in the trellis corresponds to the state $S_0$ and is reached at time $T$. From this node, we have to trace back through the trellis to determine the estimated sequence of message bits. Because each node has a single survivor, this trace-back operation gives a unique path called

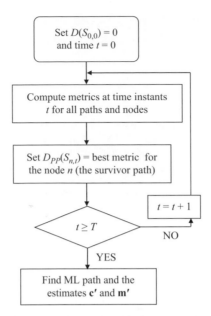

**Fig. 9.8.12** *Flow diagram of the Viterbi algorithm; ML, maximum likelihood.Because the maximum*

the *maximum likelihood path*, which specifies the maximum likelihood estimate $\mathbf{c}'$ of the code word $\mathbf{c}$ and, consequently, the asymptotically maximum likelihood estimate $\mathbf{m}'$ of the message sequence $\mathbf{m}$.

If the nodes corresponding to the states $S_n$ at time $t$ are denoted by $S_{n,t}$, and the assigned survivor partial path metric values for these nodes are $D_{PP}(S_{n,t})$, then the Viterbi algorithm can be described using a flow diagram, as shown in Figure 9.8.12. The procedure starts with the initial state $S_{0,0} = S_0$. The partial path metrics are calculated and the survivors determined for each node. Non-survivor paths are eliminated. The procedure lasts until the condition $t \geq T$ is fulfilled, that is, until the time instant when the whole received sequence is processed. At that moment, the survivor path in the trellis corresponds to the asymptotically maximum likelihood path, and the sequence of bits $\mathbf{m}'$ on that path corresponds to the maximum likelihood estimator of the message sequence $\mathbf{m}$.

The Viterbi algorithm is an asymptotically optimum algorithm, in the maximum likelihood sense, for decoding convolutional codes. The algorithm is based on the use of the trellis diagram to compute the branch, partial path, and path metrics and find the maximum likelihood path. The branches are labelled with $n$-bit encoded sets of the convolutional code for a particular input of $k$-bit message set. Each node in the trellis is assigned a number, which corresponds to the metric of a partial path that starts at the first state $S_0$ and terminates at that node. The aim of the algorithm is to find the path with the smallest Hamming distance, that is, the maximum likelihood path. The algorithm works according to the rules outlined in the example below.

**Example**

Suppose the message sequence is

$$\mathbf{m} = (10101\mathbf{100}),$$

where the bits in bold are dummy bits used to reset the decoder. For the encoder defined by its circuit and the state diagram, as shown in Figure 9.8.13, find the expected code word $\mathbf{c}$. Assume that the received encoded sequence is $\mathbf{r} = (11\ 10\ 10\ 10\ 00\ 01\ 01\ 11)$. Find the decoded sequence $\mathbf{c}'$ by using the Viterbi algorithm. Compare the decoded sequence $\mathbf{c}'$ with the expected code word $\mathbf{c}$.

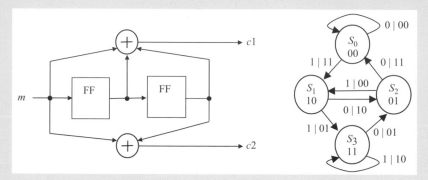

**Fig. 9.8.13** *A 1/2-rate encoder (left) and its state diagram (right); FF, flip-flop.*

**Solution.** Using the state diagram, it is simple to find, for the given message sequence $\mathbf{m} = (10101100)$, the expected code sequence expressed as $\mathbf{c} = (11\ 10\ 00\ 10\ 00\ 01\ 01\ 11)$, which makes a path inside the trellis diagram.

To find the decoded sequence $\mathbf{c}'$ we can apply the Viterbi algorithm, which can be executed in the following four steps.

**First step.** Make a trellis diagram for the encoder. The four states are placed vertically on the trellis diagram, as in Figure 9.8.14. In this case, the number of stages is 8, which is equal to the number of pairs in the given received sequence $\mathbf{r}$, which corresponds to the number of message bits $\mathbf{m}$ that includes two dummy bits for the resetting encoder. The stages are placed horizontally on the trellis diagram. The input/output values are presented on the trellis diagram next to the branches. For example, the first two branches starting at state $S_0$ on the left-hand side of the trellis are 0 | 00 and 1 | 11. This procedure for defining the branches can be continued, and the whole trellis diagram can be populated until $T = 8$, as shown in Figure 9.8.14.

**Second step.** Find the branch metrics of the trellis diagram for the time instants $t = 0$ to $t = T = 8$. A branch metric is the Hamming distance between the received pair of bits and the possible generated bits by the encoder, presented at the corresponding branches inside each stage. For example, for the first stage, we can calculate two distances: one for the first upper branch, using modulo-2 addition, as $(11 \oplus 00) = 2$, and the second one for the corresponding lower branch, as $(00 \oplus 00) = 0$. In the same way, we may calculate all of the branch metrics up to $t = 8$, and then assign them as numbers to all of the branches, as shown in Figure 9.8.14.

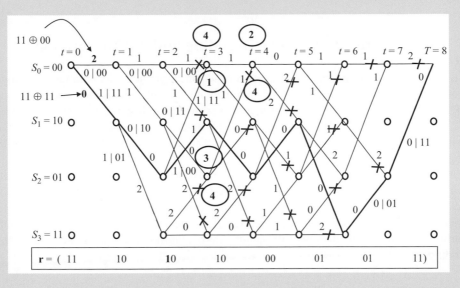

**Fig. 9.8.14** *Calculation of an asymptotic maximum likelihood path.*

**Third step.** Find the partial path metrics up to stage $t = 3$, where pairs of branches merge at four nodes, and a decision is to be made about the survivor and non-survivor partial paths. For example, the partial path metrics for the topmost node at $t = 3$ are 4 and 1, shown as the circled numbers above and below the node. Therefore, it is more probable that the path $S_0$–$S_1$–$S_2$–$S_0$ would have distance 1, so we proclaim that one the 'survivor'. The merging path $S_0$–$S_0$–$S_0$–$S_0$ has distance 4, so it is less probable and, as such, is a non-survivor and will be eliminated from further consideration (on Figure 9.8.14, it is crossed out at the merging node). Because it is possible for us to make a mistake at this decision point, we say that the algorithm is not optimal in the maximum likelihood sense; rather, it is suboptimal.

In the case of a tie at a merging node, we will arbitrarily choose either the upper or the lower path. This should be done throughout the execution of the algorithm.

**Fourth step.** Repeat the third step until $t = 8$, as shown in Figure 9.8.14, and all of the non-survivor paths are crossed out. The path that remains, the survivor, is the maximum likelihood path that defines the maximum likelihood sequence define by the node transitions $S_0$–$S_1$–$S_2$–$S_1$–$S_2$–$S_1$–$S_3$–$S_2$–$S_0$, the maximum likelihood code sequence is $\mathbf{c}' = (11\ 10\ 00\ 10\ 00\ 01\ 01\ 11)$, and the corresponding message sequence is $\mathbf{m}' = 10101100$. The decoded sequence $\mathbf{c}'$ differs from the received sequence $\mathbf{r}$ in that it has 0 instead of 1 at the fifth position. Therefore, we can say that the decoder has corrected one error in the received code sequence and presented the sent message sequence $\mathbf{m}$ at its output.

In a real system, we would not know what the message sequence **m** was. Therefore, upon finding **c′** and **m′**, we would state that the decoded sequence **c′** and the related message sequence **m′** are the sequences most likely to have been transmitted in the system.

## 9.9   Introduction to Iterative Decoding and Turbo Decoding

As we said in the previous section, Viterbi proposed an 'asymptotically optimum' method for decoding convolutional codes used in a communication system with the memoryless channels, a method which minimizes the word (sequence) error probability. We presented the simplest case, which assumes that a hard decision was made at the optimum detector when it produced estimates of the encoded bits, which can be 1's or 0's, and then applied the Viterbi algorithm, using the Hamming distances as a metric.

If we use Euclidean distance as the metric to estimate coded sequences, the algorithm is called a hard-output Viterbi algorithm. However, it is also possible to develop a soft Viterbi algorithm that produces soft outputs of the message bits. Using these soft values, we will be able to develop more sophisticated algorithms, like the bidirectional soft-output Viterbi algorithm and the iterative soft-output Viterbi algorithm, which will be used for turbo decoding.

### 9.9.1   Coding Models for Communication Systems

In this section, we will present the hard-output Viterbi algorithm first. This algorithm will be explained using the basic structure of a system for digital signal transmission, as shown in Figure 9.9.1, which is slightly different from the system shown in Figure 9.8.8, due to the introduction of the modulator block at the transmitter side. For that reason, the notation in this section will be slightly different from that used in Section 9.8. The theory will be presented for an encoder that has $k$ inputs and $n$ outputs, that is, it is a $k/n$-rate encoder.

**Transmitter operation.** A binary message sequence **m**, containing $T$ message sets, each of which has $k$ bits, is applied at the input of the encoder and expressed as

$$\mathbf{m} = (m_1, m_2, \ldots, m_t, \ldots, m_T), \tag{9.9.1}$$

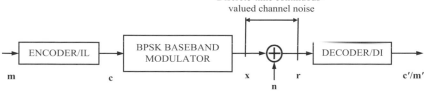

**Fig. 9.9.1** *Coding model for a digital communication system; BPSK, binary phase-shift keying; DI, de-interleaver; IL, interleaver.*

where each message symbol $m_t$ of $k$ bits, which are denoted as $m_{t,i}$, is generated at the output of the encoder at the discrete-time instant $t$ and expressed as

$$m_t = \left(m_{t,0}, m_{t,1}, \ldots, m_{t,i}, \ldots, m_{t,k-1}\right), \tag{9.9.2}$$

for $i = 0, 1, \ldots, k - 1$. The message sequence $\mathbf{m}$ is encoded to form the encoded sequence $\mathbf{c}$, that is,

$$\mathbf{c} = (\mathbf{c}_1, \mathbf{c}_2, \ldots, \mathbf{c}_t, \ldots, \mathbf{c}_T), \tag{9.9.3}$$

which contains $T$ encoded symbols, each of which has $n$ encoded bits, that is, one symbol can be expressed in the form

$$c_t = \left(c_{t,0}, c_{t,1}, \ldots, c_{t,i}, \ldots, c_{t,n-1}\right), \tag{9.9.4}$$

for $t = 1, \ldots, T$. Because, for a number $k$ of input message bits in a symbol, the encoder generates $n$ encoded bits in an encoded symbol, this is a $k/n$ encoder. The encoded sequence $\mathbf{c}$ is then modulated, resulting in a modulated encoded sequence composed of $T$ modulated symbols, that is,

$$\mathbf{x} = (x_1, x_2, \ldots, x_t, \ldots, x_T), \tag{9.9.5}$$

where a modulated symbol contains $n$ modulated encoded bits, that is,

$$x_t = \left(x_{t,0}, x_{t,1}, \ldots x_{t,i}, \ldots, x_{t,n-1}\right), \tag{9.9.6}$$

defined for $t = 1, \ldots, T$. Suppose the encoded bits belong to the additive group $(1, 0)$. If the baseband encoded bits are modulated via binary phase shift keying, the output of the modulator can be expressed as

$$x_{t,i} = 2 \cdot c_{t,i} - 1, \tag{9.9.7}$$

for $i = 0, 1, \ldots, n - 1$, and $t = 1, \ldots, T$. Suppose the noise in the channel is additive white Gaussian noise, which is described in Chapters 3 and 4, with a wide-sense stationary discrete-time stochastic process. Then, the received encoded sequence $\mathbf{r}$ is the sum of the channel noise sequence and the input modulated encoded sequence, that is,

$$\begin{aligned} \mathbf{r} = \mathbf{x} + \mathbf{n} &= (x_1, x_2, \ldots, x_t, \ldots, x_T) + (n_1, n_2, \ldots, n_t, \ldots, n_T) \\ &= (r_1, r_2, \ldots, r_t, \ldots, r_T) \end{aligned}, \tag{9.9.8}$$

where the received encoded symbol at the time instant $t$ is

$$r_t = \left(r_{t,0}, r_{t,1}, \ldots, r_{t,i}, \ldots, r_{t,n-1}\right), \tag{9.9.9}$$

and a received encoded bit is

$$r_{t,i} = x_{t,i} + n_{t,i}, \tag{9.9.10}$$

defined for any $i = 0, 1, \ldots, n - 1$, and any $t = 1, \ldots, T$. The channel noise sequence $\mathbf{n}$ is composed of noise samples affecting each of $T$ modulated symbols, expressed as

$$\mathbf{n} = (n_1, n_2, \ldots, n_t, \ldots, n_T), \tag{9.9.11}$$

where the noise samples affecting the $t$th modulated symbol are

$$n_t = \left(n_{t,0}, n_{t,1}, \ldots, n_{t,i}, \ldots, n_{t,n-1}\right) \tag{9.9.12}$$

is a realization of the discrete-time Gaussian noise process $N_{t,i}$ as defined in Chapters 3 and 4. Each sample $n_{t,i}$ is a realization of a zero-mean Gaussian independent random variable $N_{t,i}$ with variance $\sigma^2$ that is defined at the time instant determined by $t$ and $i$. Therefore, the received encoded symbol $r_t$ can be understood as a series of samples $x$ at the output of the correlation receiver shown in Chapter 7, Figure 7.2.1, if we assume that the signal part of each received bit at the correlation demodulator output is normalized to have a unit absolute value, and the noise added has the variance $\sigma^2$. That series of samples is a realization of the wide-sense stationary discrete-time process, with random values at the time instants $kN_b$, as shown in Chapter 7, Figure 7.2.1, that correspond to the time instants defined here by $t$ and $i$. If we want to simulate this coding system, we can exclude the operation of the bandpass modulator and the demodulator presented in Figure 7.2.1 and replace the waveform channel with the discrete baseband Gaussian channel defined in Chapter 4.

The received encoded sequence $\mathbf{r}$ is processed in the decoder to obtain the estimated decoded sequence $\mathbf{c}'$ is obtained, which is mapped into the estimates of the message bits $\mathbf{m}'$, that is, the decoder generates the decoded message sequence

$$\mathbf{m}' = \left(m'_1, m'_2, \ldots, m'_t, \ldots, m'_T\right). \tag{9.9.13}$$

An interleaver and a de-interleaver can be used in the transmission of the signals, as shown in Figure 9.9.1.

## 9.9.2 The Hard-Output Viterbi Algorithm

Due to the definition of Gaussian noise samples, the channel is considered to be memoryless. Therefore, the channel noise affecting a given encoded bit is statistically independent of the noise affecting any proceeding or succeeding bit. Consequently, the channel noise affecting a given message bit is independent of the noise affecting any proceeding or succeeding message bit. Let us assume that the message sequence (also called 'message word') $\mathbf{m}$ is transmitted in the form of message symbols in the time interval defined by $t = 0$ to $t = T$. For the received encoded sequence (code word) $\mathbf{r}$, the word error probability is defined in eqn

(9.8.25). We can minimize the code word error probability by maximizing the integral value in eqn (9.8.25). Due to the independence of the noise samples, the probability of the received encoded sequence, $p(\mathbf{r})$, is independent of the message sequence $\mathbf{m}$. Therefore, minimization of the word error probability in eqn (9.8.25) is equivalent to maximizing the a posteriori probability $p(\mathbf{m} \mid \mathbf{r})$, expressed as

$$p(\mathbf{m}|\mathbf{r}) = \frac{p(\mathbf{m})p(\mathbf{r}|\mathbf{m})}{p(\mathbf{r})}, \qquad (9.9.14)$$

as defined by Bayes' rule. The algorithm that maximizes this probability is called the maximum a posteriori probability algorithm. If the message and the received encoded symbols are equally likely, then the maximum a posteriori probability $p(\mathbf{m} \mid \mathbf{r})$ is equivalent to the maximum of the conditional probability $p(\mathbf{r} \mid \mathbf{m})$ that defines the likelihood function. Because there is direct correspondence between the message sequence $\mathbf{m}$ and the modulated code sequence $\mathbf{x}$, the Bayesian rule can be expressed in the form

$$p(\mathbf{x}|\mathbf{r}) = \frac{p(\mathbf{x})p(\mathbf{r}|\mathbf{x})}{p(\mathbf{r})}. \qquad (9.9.15)$$

The maximum likelihood algorithm is based on the maximization of the maximum likelihood function $p(\mathbf{r} \mid \mathbf{m})$ or the corresponding function, $p(\mathbf{r} \mid \mathbf{x})$. Assuming that the channel is a binary symmetric channel without memory, as it is in our case, due to the definition of the Gaussian noise, the conditional probability of receiving $\mathbf{r}$, given that the modulated encoded sequence $\mathbf{x}$ is transmitted, can be expressed as

$$p(\mathbf{r}|\mathbf{x}) = p\left(\mathbf{r}^T|\mathbf{x}^T\right) = \prod_{t=1}^{T} p(r_t|x_t) = \prod_{t=1}^{T}\prod_{i=0}^{n-1} p\left(r_{t,i}|x_{t,i}\right)$$
$$= \prod_{t=1}^{T}\prod_{i=0}^{n-1} \frac{1}{\sqrt{2\pi\sigma^2}} e^{-(r_{t,i}-x_{t,i})^2/2\sigma^2} \qquad (9.9.16)$$

where the Gaussian density function is analogous to that defined in eqn (9.8.29). Because the maximum of a function is equivalent to the maximum of the function's logarithm, the maximum likelihood estimator of $\mathbf{r}$ will be found by maximizing the logarithmic function

$$\log p(\mathbf{r}|\mathbf{x}) = \sum_{t=1}^{T}\left[\sum_{i=0}^{n-1} \ln \frac{1}{\sqrt{2\pi\sigma^2}} e^{-(r_{t,i}-x_{t,i})^2/2\sigma^2}\right] = \sum_{t=1}^{T}\left\{\sum_{i=0}^{n-1}\left[-\frac{(r_{t,i}-x_{t,i})^2}{2\sigma^2} - \ln\sqrt{2\pi\sigma^2}\right]\right\}$$
$$= -\sum_{t=1}^{T}\sum_{i=0}^{n-1} \frac{(r_{t,i}-x_{t,i})^2}{2\sigma^2} - Tn\ln\sqrt{2\pi\sigma^2} = -\sum_{t=1}^{T}\sum_{i=0}^{n-1} \frac{(r_{t,i}-x_{t,i})^2}{2\sigma^2} - \frac{Tn}{2}\ln 2\pi - Tn\ln\sigma$$

$$(9.9.17)$$

which is equivalent to expression (9.8.32). Due to the negative sign of this function, the *maximization* of this logarithm is equivalent to the *minimization* of the Euclidean distance

$$\sum_{t=1}^{T}\sum_{i=0}^{n-1}\left(r_{t,i}-x_{t,i}\right)^{2} \qquad (9.9.18)$$

between the received encoded sequence **r** and the modulated encoded sequence **x**, which is equivalent to the distance presented by eqn (9.8.33). Thus, at each time instant $t$, we assign the *branch metric* to each branch of the related trellis diagram on the path $x$; this is expressed as

$$d_{t}^{x} = \sum_{i=0}^{n-1}\left(r_{t,i}-x_{t,i}\right)^{2}. \qquad (9.9.19)$$

The *path metric* of the path $x$ in the trellis diagram, from $t' = 1$ to $t' = t$, includes $t$ branch metrics. It can be expressed in a recursive form as

$$D_{t}^{x} = \sum_{t'=1}^{t} d_{t'}^{x} = D_{t-1}^{x} + d_{t}^{x}. \qquad (9.9.20)$$

Having defined the metrics in this way, we can estimate the message sequence by using the received encoded sequence and applying the Viterbi algorithm, as explained in the previous section. In this case, however, we use the Euclidean distances between the received samples and the transmitted bits, instead of the Hamming distances.

**Example**

Suppose that a 1/2-rate encoder is defined by its scheme, state, and trellis diagram, as shown in Figure 9.9.2. Find the maximum likelihood estimate of the code sequence **c** for the case when the received encoded sequence is $\mathbf{r} = [(1, -1), (0.8, 1), (-1, 1), (-1, 1), (1, -1)]$.

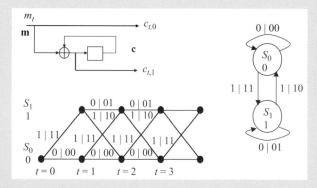

**Fig. 9.9.2** *Encoder specifications.*

**Solution.** The trellis diagram in Figure 9.9.2 is drawn for the case when $(0, 1)$ binary logic is assumed. In our case, the signal at the output of the encoder is modulated according to the expression

$$x_{t,i} = 2 \cdot c_{t,i} - 1,$$

for $i = 0, 1$, which results in the trellis diagram shown in Figure 9.9.3. In this diagram, the previous binary logic, $(0, 1)$, is converted to the binary logic $(-1, 1)$. The received encoded sequence **r**, which represents a code word, is written underneath the trellis diagram.

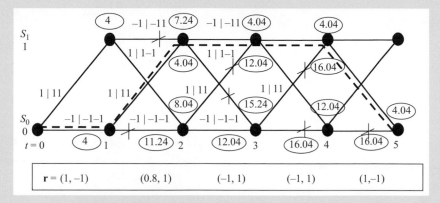

**Fig. 9.9.3** *The Viterbi algorithm.*

The branch metrics are calculated according to eqn (9.9.19). The first branch metric for the $x = 00$ path is

$$d_{t=1}^{x=00} = \sum_{i=0}^{1} (r_{t,i} - x_{t,i})^2 = (r_{1,0} - x_{1,0})^2 + (r_{1,1} - x_{1,1})^2 = (1 - (-1))^2 + (-1 - (-1))^2 = 4 + 0 = 4.$$

and the second branch metric for the 01 path is

$$d_{t=1}^{x=01} = \sum_{i=0}^{1} (r_{t,i} - x_{t,i})^2 = (r_{1,0} - x_{1,0})^2 + (r_{1,1} - x_{1,1})^2 = (1 - 1)^2 + (-1 - 1)^2 = 0 + 4 = 4.$$

At this point in time, the branch metrics correspond to the partial path metrics being equal to 4, that is,

$$D_{t=1}^{x=00} = \sum_{t'=1}^{t=1} d_{t'}^{x} = D_{t-1}^{x} + d_t^{x} = D_{1-1}^{x} + d_1^{x} = 0 + 4 = 4 = D_{t=1}^{x=01}.$$

The values of the path metrics are encircled and presented next to each node in Figure 9.9.3. The branch metrics at time $t = 2$ are

$$d_{t=2}^{x=00} = \sum_{i=0}^{1} \left(r_{2,i} - x_{2,i}\right)^2 = \left(r_{2,0} - x_{2,0}\right)^2 + \left(r_{2,1} - x_{2,1}\right)^2 = (0.8+1)^2 + (1+1)^2 = 3.24 + 4 = 7.24,$$

$$d_{t=2}^{x=01} = \sum_{i=0}^{1} \left(r_{2,i} - x_{2,i}\right)^2 = (0.8-1)^2 + (1-1)^2 = 0.04 + 0 = 0.04,$$

$$d_{t=2}^{x=11} = \sum_{i=0}^{1} \left(r_{2,i} - x_{2,i}\right)^2 = (0.8-(-1))^2 + (1-1)^2 = 3.24 + 0 = 3.24,$$

$$d_{t=2}^{x=10} = \sum_{i=0}^{1} \left(r_{2,i} - x_{2,i}\right)^2 = (0.8-1)^2 + (1-(-1))^2 = 0.04 + 4 = 4.04.$$

The path metrics are

$$D_{t=2}^{x=000} = D_{t-1}^{x} + d_t^{x} = D_1^{00} + d_2^{00} = 4 + 7.24 = 11.24,$$
$$D_{t=2}^{x=001} = D_{t-1}^{x} + d_t^{x} = D_1^{00} + d_2^{01} = 4 + 0.04 = 4.04,$$
$$D_{t=2}^{x=011} = D_{t-1}^{x} + d_t^{x} = D_1^{01} + d_2^{11} = 4 + 3.24 = 7.24,$$

and

$$D_{t=2}^{x=010} = D_{t-1}^{x} + d_t^{x} = D_1^{01} + d_2^{10} = 4 + 4.04 = 8.04.$$

These metrics, as well as the rest of them, are calculated and encircled at each node. At each node starting from $t = 2$, the non-survivor branch is crossed out. The survivor path is presented by a dashed bold line. Following this survivor path, we can read out the maximum likelihood estimate of the message sequence and express it in terms of elements of the set $(-1, 1)$ as

$$\mathbf{m}' = [-1, 1, -1, -1, 1],$$

which corresponds to the estimate of the message sequence expressed in $(0, 1)$ logic as

$$\mathbf{m}' = [0, 1, 0, 0, 1].$$

### 9.9.3 Iterative Algorithms and Turbo Coding

Based on the theory presented above, it is possible to develop the soft-output Viterbi algorithm and the bidirectional soft-output Viterbi algorithm, which can be used for the development of turbo coding algorithms. Due to constraints on the size of this book, the soft-output Viterbi algorithm and the related turbo coding algorithms will not be presented in this chapter.

However, in order to give the reader a chance to learn about and develop design skills in this important subject, we will present the main algorithms here, and Project 5 in the Supplementary Material will contain topics and tasks related to these algorithms. In that way, instructors and practising engineers will be in a position to acquire the theoretical knowledge needed for these algorithms. In addition, based on that knowledge, they will have the opportunity to develop simulations of sophisticated coding communication systems and investigate their properties.

..................................................................................................................................

## APPENDIX A: DERIVATION OF MUTUAL INFORMATION

The expression for the maximum entropy in eqn (9.3.28) has three integrals. The first integral is

$$
I1 = \log_2 \sqrt{2\sigma_y^2/N_0} \sum_{m=1}^{M} \int_{-\infty}^{\infty} f_{X_m}(x_m) \int_{-\infty}^{\infty} f_{N_m}(y_m - x_m)\, dy_m dx_m.
$$

If we perform the substitutions $(y_m - x_m) = n_m, dn_m = dy_m$, the integral is simplified to

$$
I1 = \log_2 \sqrt{2\sigma_y^2/N_0} \sum_{m=1}^{M} \int_{-\infty}^{\infty} f_{X_m}(x_m) \int_{-\infty}^{\infty} f_{N_m}(n_m)\, dn_m dx_m = \log_2 \sqrt{2\sigma_y^2/N_0} \sum_{m=1}^{M} 1
$$

$$
= \log_2 \sqrt{2\sigma_y^2/N_0} \cdot M = K \cdot M
$$

where

$$
K = \log_2 \sqrt{2\sigma_y^2/N_0} = \log_2 \sqrt{2\left(\sigma_x^2 + N_0/2\right)/N_0} = \frac{1}{2}\log_2\left(\frac{2\sigma_x^2}{N_0} + 1\right).
$$

The following density functions will be needed to calculate the remaining integrals:

$$
f_{X_m}(x_m) = \frac{1}{\sqrt{2\pi\sigma_x^2}} e^{-(x_m)^2/2\sigma_x^2},
$$

$$
f_{Y_m}(y_m) = \frac{1}{\sqrt{2\pi\sigma_y^2}} e^{-(y_m)^2/2\sigma_y^2} = \frac{1}{\sqrt{2\pi\left(\sigma_x^2 + \sigma_n^2\right)}} e^{-(y_m)^2/2\left(\sigma_x^2 + \sigma_n^2\right)},
$$

$$
f_{N_n}(n_n) = \frac{1}{\sqrt{2\pi\sigma_{n_n}^2}} e^{-n_n^2/2\sigma_{n_n}^2},
$$

$$
f_{X_n Y_n}(y_n|x_n) = \frac{1}{\sqrt{2\pi N_0/2}} e^{-(y_n - x_n)^2/2N_0/2} = \frac{1}{\sqrt{\pi N_0}} e^{-(y_n - x_n)^2/N_0}.
$$

The second integral is

$$I2 = \frac{1}{\log_e 2} \sum_{m=1}^{M} \int_{-\infty}^{\infty} \int_{-\infty}^{\infty} \left[-(y_m - x_m)^2/N_0\right] f_{X_m}(x_m) f_{N_m}(y_m - x_m)\, dy_m dx_m$$

$$= \frac{1}{\log_e 2} \sum_{m=1}^{M} \int_{-\infty}^{\infty} \int_{-\infty}^{\infty} \left[-(y_m - x_m)^2/N_0\right] \frac{1}{\sqrt{2\pi \sigma_x^2}} e^{-(x_m)^2/2\sigma_x^2} \frac{1}{\sqrt{\pi N_0}} e^{-(y_n - x_n)^2/N_0}\, dy_m dx_m$$

$$= \frac{-1}{N_0 \log_e 2} \sum_{m=1}^{M} \int_{-\infty}^{\infty} \frac{1}{\sqrt{2\pi \sigma_x^2}} e^{-(x_m)^2/2\sigma_x^2} \int_{-\infty}^{\infty} (y_m - x_m)^2 \frac{1}{\sqrt{\pi N_0}} e^{-(y_n - x_n)^2/N_0}\, dy_m dx_m$$

$$= \frac{-1}{N_0 \log_e 2} \sum_{m=1}^{M} \int_{-\infty}^{\infty} \frac{1}{\sqrt{2\pi \sigma_x^2}} e^{-(x_m)^2/2\sigma_x^2} \int_{-\infty}^{\infty} (n_m)^2 \frac{1}{\sqrt{\pi N_0}} e^{-(n_m)^2/N_0}\, dn_m dx_m$$

$$= \frac{-1}{N_0 \log_e 2} \sum_{m=1}^{M} \int_{-\infty}^{\infty} \frac{1}{\sqrt{2\pi \sigma_x^2}} e^{-(x_m)^2/2\sigma_x^2} \frac{N_0}{2}\, dx_m = \frac{-1}{N_0 \log_e 2} \sum_{m=1}^{M} \frac{N_0}{2} = \frac{-M}{2 \log_e 2}$$

The third integral is

$$I3 = \frac{1}{\log_e 2} \sum_{m=1}^{M} \int_{-\infty}^{\infty} \int_{-\infty}^{\infty} (y_m)^2 f_{X_m}(x_m) f_{N_m}(y_m - x_m)\, dy_m dx_m \big|_{\text{substitution } y_m - x_m = n_m}.$$

$$= \frac{1}{2\sigma_y^2 \log_e 2} \sum_{m=1}^{M} \int_{-\infty}^{\infty} \int_{-\infty}^{\infty} (x_m + n_m)^2 f_{X_m}(x_m) f_{N_m}(n_m)\, dn_m dx_m$$

$$= \frac{1}{2\sigma_y^2 \log_e 2} \sum_{m=1}^{M} \int_{-\infty}^{\infty} f_{X_m}(x_m) \int_{-\infty}^{\infty} \left[(x_m)^2 + 2x_m n_m + (n_m)^2\right] f_{N_m}(n_m)\, dn_m dx_m$$

$$= \frac{1}{2\sigma_y^2 \log_e 2} \sum_{m=1}^{M} \int_{-\infty}^{\infty} f_{X_m}(x_m)$$

$$\left[\int_{-\infty}^{\infty} (x_m)^2 f_{N_m}(n_m)\, dn_m + \int_{-\infty}^{\infty} (n_m)^2 f_{N_m}(n_m)\, dn_m + 2x_m \int_{-\infty}^{\infty} n_m f_{N_m}(n_m)\, dn_m\right] dx_m$$

Solving the integrals in the sum results in

$$I3 = \frac{1}{2\sigma_y^2 \log_e 2} \sum_{m=1}^{M} \int_{-\infty}^{\infty} f_{X_m}(x_m)$$

$$\left[ \int_{-\infty}^{\infty} (x_m)^2 f_{N_m}(n_m) \, dn_m + \int_{-\infty}^{\infty} (n_m)^2 f_{N_m}(n_m) \, dn_m + 2x_m \int_{-\infty}^{\infty} n_m f_{N_m}(n_m) \, dn_m \right] dx_m$$

$$= \frac{1}{2\sigma_y^2 \log_e 2} \sum_{m=1}^{M} \int_{-\infty}^{\infty} f_{X_m}(x_m) \left[ \sigma_x^2 + \sigma_n^2 + 0 \right] dx_m$$

$$= \frac{1}{2\sigma_y^2 \log_e 2} \sum_{m=1}^{M} \left( \sigma_x^2 + \sigma_n^2 \right) = \frac{1}{2\sigma_y^2 \log_e 2} \sum_{m=1}^{M} \sigma_y^2 = \frac{M}{2 \log_e 2}$$

By inserting the integral values obtained, we find the solution for the maximum mutual information in the simple form

$$\max_{f(x)} I(X; Y) = I1 + I2 + I3 = K \cdot M + \frac{-M}{2 \log_e 2} + \frac{M}{2 \log_e 2} = K \cdot M.$$

........................................................................................................................

## APPENDIX B: ENTROPY OF A TRUNCATED DISCRETE GAUSSIAN DENSITY FUNCTION

Because

$$f_X(x) = f_d\left(x \,\Big\|\, |x| \le S\right) = \frac{f_X(x)}{P(\tau \ge -S) - P(\tau > S)}$$

$$= \frac{\frac{1}{2} \sum_{n=-S}^{n=S} \left( \mathrm{erfc} \frac{(2n-1)}{\sqrt{8\sigma^2}} - \mathrm{erfc} \frac{(2n+1)}{\sqrt{8\sigma^2}} \right) \delta(x-n)}{\frac{1}{2} \sum_{n=-S}^{n=S} \left( \mathrm{erfc} \frac{(2n-1)}{\sqrt{8\sigma^2}} - \mathrm{erfc} \frac{(2n+1)}{\sqrt{8\sigma^2}} \right)} = P(S) \sum_{n=-S}^{n=S} Erfc(n) \cdot \delta(x-n),$$

where

$$1/P(S) = \sum_{n=-S}^{n=S} \left( \mathrm{erfc} \frac{(2n-1)}{\sqrt{8\sigma^2}} - \mathrm{erfc} \frac{(2n+1)}{\sqrt{8\sigma^2}} \right)$$

and

$$Erfc(n) = \mathrm{erfc} \frac{(2n-1)}{\sqrt{8\sigma^2}} - \mathrm{erfc} \frac{(2n+1)}{\sqrt{8\sigma^2}},$$

the entropy is

$$H(X) = -\int_{-\infty}^{\infty} f_X(x)\log_2 f_X(x)dx$$

$$= -\int_{-\infty}^{\infty}\left[P(S)\sum_{n=-S}^{n=S} Erfc(n)\cdot\delta(x-n)\right]\log_2\left[P(S)\sum_{n=-S}^{n=S} Erfc(n)\cdot\delta(x-n)\right]dx$$

$$= \cdots -\int_{-\infty}^{\infty}\left[P(S)\sum_{n=-1}^{n=-1} Erfc(-1)\cdot\delta(x+1)\right]\log_2\left[P(S)\sum_{n=-S}^{n=S} Erfc(n)\cdot\delta(x-n)\right]dx$$

$$-\int_{-\infty}^{\infty}\left[P(S)\sum_{n=0}^{n=0} Erfc(0)\cdot\delta(x+0)\right]\log_2\left[P(S)\sum_{n=-S}^{n=S} Erfc(n)\cdot\delta(x-n)\right]dx$$

$$-\int_{-\infty}^{\infty}\left[P(S)\sum_{n=1}^{n=1} Erfc(1)\cdot\delta(x-1)\right]\log_2\left[P(S)\sum_{n=-S}^{n=S} Erfc(n)\cdot\delta(x-n)\right]dx - \cdots$$

$$= \cdots -\int_{-\infty}^{\infty} P(S)Erfc(-1)\cdot\delta(x+1)\log_2\left[P(S)\sum_{n=-S}^{n=S} Erfc(n)\cdot\delta(x-n)\right]dx$$

$$-\int_{-\infty}^{\infty}[P(S)Erfc(0)\cdot\delta(x-0)]\log_2\left[P(S)\sum_{n=-S}^{n=S} Erfc(n)\cdot\delta(x-n)\right]dx$$

$$-\int_{-\infty}^{\infty}[P(S)Erfc(1)\cdot\delta(x-1)]\log_2\left[P(S)\sum_{n=-S}^{n=S} Erfc(n)\cdot\delta(x-n)\right]dx$$

$$= \cdots - P(S)Erfc(-1)\log_2 P(S)Erfc(-1) - P(S)Erfc(0)\log_2 P(S)Erfc(0)$$
$$- P(S)Erfc(1)\log_2 P(S)Erfc(1) - \cdots$$

$$= -P(S)\sum_{n=-S}^{S} Erfc(n)\cdot\log_2 P(S)Erfc(n)$$

---

## PROBLEMS

1. A set of four symbols, $m_0$, $m_1$, $m_2$, and $m_3$, is generated by the source with probabilities 1/2, 1/4, 1/8, and 1/8, respectively.

   a. What is the entropy of this source? What is the capacity of the channel for the symbol rate of the source of $r = 100$ symbols/second?

b. Use the code $C$ that encodes symbols in the following way: $C(m_0) = 0$, $C(m_1) = 10$, $C(m_2) = 110$, and $C(m_3) = 111$. Find the average length of the source code $L(C)$ and compare it with the calculated entropy of the source.

c. What would be the average length if we used simple binary encoding, that is, if we encode each symbol with two binary symbols (bits)?

d. If the symbol rate is $r = 100$ symbols/second, find the bit rate in the channel for case b..

e. What is the symbol rate achievable in case c.?

2. A message source generates a binary signal. Assume that the signal is transmitted through an additive white Gaussian channel. Express the average probability of symbol errors for the following cases:

a. The message sequence is unbalanced, that is, the number of generated 1's is different from the number of generated 0's.

b. The channel is symmetric.

c. The message sequence is balanced, that is, the number of 1's is equal to the number of 0's.

3. Find the maximum likelihood estimator of the mean value $s_{ij}$ if the log-likelihood function is defined by

$$l(m_i) = \log L(m_i) = -(J/2)\log(\pi N_0) - \frac{1}{N_0}\sum_{j-1}^{J}(x_j - s_{ij})^2,$$

and the energies of the signals are equal, that is, $s_{ij}$ are equal for all $j = 1, 2, \ldots, J$.

4. A source generates a black-and-white picture to be transmitted through a channel with a bandwidth of 10 kHz. Suppose that the picture consists of picture elements (pixels) that can be black or white. The pixels are black with a probability of 40%.

a. What is the amount of information in each of these pixels? Sketch the probability density function of the source information. What is the entropy of the source?

b. Suppose there is no statistical dependence for the appearance of the pixels at the output of the source. What is the information content of the picture that contains 1,000,000 pixels?

c. What would be the information content of an arbitrarily number of pixels $n$? How would this content depend on $n$ in a real communication system?

d. Assume that the black and white pixels are generated at the output of the source with equal probabilities. Compare the characteristics of this source with those of the source mentioned in a., b., and c..

5. Suppose we use a binary symmetric channel without memory, as shown in Figure 5. The average bit error probability is $p = 10^{-1}$.

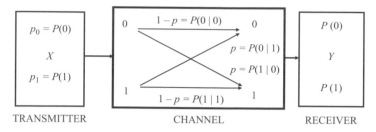

TRANSMITTER                CHANNEL                RECEIVER

**Fig. 5** *A binary symmetric memoryless channel.*

a. Find the entropy of the input of the channel $H(X)$. Sketch the graph of this entropy as a function of the probability $p_0$. Explain the meaning of the maximum of this function. The amount of information in the input symbol 1 at the input of the channel is defined as

$$I(x_1) = \log_2 \frac{1}{p(x_1)} = \log_2 \frac{1}{p_1},$$

where $p_1$ is the probability of 1 being generated at the input of the channel. Sketch the graph and list three properties of this function.

b. The channel capacity $C$ is equal to the maximum of the mutual information $I(X; Y)$. For the channel in Figure 5, this maximum occurs when the input binary symbols are generated with the probability 1 or 0. Find the capacity $C$ of the channel.

c. Sketch a graph of the channel capacity $C$ as a function of the bit error probability $p$. Explain the meaning of the minimum point of this function. Calculate the value of the capacity of this channel for $p = 0.1$.

6. Suppose that a convolutional 2/3-rate encoder is defined by the scheme shown in Figure 6. A message sequence **m** is composed of pairs of message binary digits (message bits). The encoded sequence **c** contains three coded bits for each pair of message bits. Each message bit is taken from its alphabet $A = \{0, 1\}$.

a. Suppose the message sequence **m** is randomly generated with equal probabilities for generating 1's and 0's. Consider the encoder to be a source of information that generates sequence of symbols **c** composed of three bits each, as shown in Figure 6. What is the probability of generating the symbol $c_k$ at the output of the encoder? Calculate the value of the entropy of the source (i.e. the output of the encoder).

b. Following the explanation in a., suppose now that the probability of a symbol at the output of the encoder is proportional to the number of 0's in that symbol. Find the probability of each symbol. What would be the entropy of this encoder, which is considered to be a source of information?

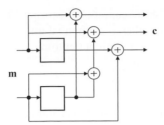

**Fig. 6** *A convolutional 2/3-rate encoder.*

7. A robot equipped with a monochrome TV camera has been used to take a picture and send the information to a distant receiver. Each digitized picture is composed of $n = 120,000$ pixels, and each pixel has 16 possible and equally probable brightness levels. Suppose the power generated by the transmitter is $P_t = 20$ W. The power attenuation of the signal inside the radio link is $a = 186$ dB. The transmission bandwidth is $B = 90$ kHz. The noise power spectral density at the receiver is $N_0 = 8 \times 10^{-22}$ W/Hz.

   a. Calculate the signal power $S$ at the receiver. Calculate the noise power $N$ at the receiver.

   b. Calculate the average value of information per pixel. Calculate the total information content $I$ in one picture. Find the relationship between the required information rate $R$, the time $T$ needed to transmit one picture, and the total information content $I$.

   c. Calculate the capacity $C$ of the radio link and the time $T$ needed to transmit the picture.

   d. Suppose the transmission bandwidth tends to infinity. Find the capacity of the channel and the time $T_\infty$ needed to transmit the picture in this case.

8. Suppose the noise $N$ in a digital memoryless channel is characterized by a zero-mean Gaussian probability density function defined by its standard deviation, $\sigma_n$, and the corresponding noise power spectral density, $N_0$. The input of the channel is a sample $x$ of a zero-mean Gaussian random variable $X$, defined by its standard deviation $\sigma_x$. Each sample $x$ is taken at the Nyquist sampling rate $R_s = 2B$, where $B$ is the bandwidth of the channel. The output sample $y$ of the channel is the sum of the input sample $x$ and the noise sample $n$, expressed as $y = x + n$, where $x$ and $n$ are realizations of the independent random variables $X$ and $N$, respectively. Assume that the variance of the output $Y$ is the sum of the variance of the input $X$ and the variance of the channel noise $N$.

   a. Express, in closed form, the probability density functions of both the input signal and the channel noise. Find the differential entropy $h(X)$ of the signal at the input of the channel.

   b. Find the differential entropy of the output $Y$, $h(Y)$.

c. Derive the expression for the conditional entropy $H(Y/X)$. Find the mutual information $I(X; Y)$. What is the value of the mutual information if the channel is noiseless? What is relationship between the input entropy $h(X)$ and the output entropy $h(Y)$ of the noiseless channel, and why?

d. Find the capacity $C$ of this channel.

9. A binary source generates a polar binary signal defined by two possible discrete symbols, $A$ and $-A$. Suppose that the probability of generating symbol $A$ is $N$ times greater than the probability of generating symbol $-A$.

a. Prove that the probability of generating symbol $A$ is $P(A) = N/(N + 1)$, and the probability of generating symbol $-A$ is $P(-A) = 1/(N + 1)$. What is the amount of information in each of these symbols? What is the unit for the information? Plot the probability density function of the information $I$, which is defined as a random variable.

b. Calculate the entropy of the source. What is the unit for this entropy? What is the value of $N$ that will maximize this entropy? Find the values of $N$ that will minimize the entropy and explain the physical meaning of this case.

c. Suppose there is no statistical dependence for the appearance of different symbols at the output of the source. For any value of $N$, find the average information content of a message that contains 100 symbols. Suppose $N = 1$; what then is the information content?

d. Suppose that the probability of generating symbol $A$ is four times greater than the probability of generating symbol $-A$. What is the amount of information in each of these symbols in this case? Calculate the entropy of the source in this case.

10. In a source node of a sensor network, four threshold values of the temperature are sensed and transmitted through a channel with a signal-to-noise ratio of 25 dB at the receiver. Suppose that 400,000 threshold values are collected and stored in the memory of the source node before transmission. Suppose that the probabilities of these threshold values are $Th_1 = 30\%$, $Th_2 = 15\%$, $Th_3 = 15\%$, and $Th_4 = 40\%$.

a. What is the entropy of the source node? What does this entropy represent?

b. The information of the symbols is a random variable. How many components does the information probability density function have? Try to sketch the graph of this density function. How much information is contained in the memory of a source where 400,000 threshold values are stored?

c. What is the minimum bandwidth required to transmit the memory content through the channel in a time interval of 20 seconds?

11. A binary communication system is defined by its inputs, a memoryless binary channel, and its outputs. The inputs and their corresponding probabilities are shown in Figure 11.

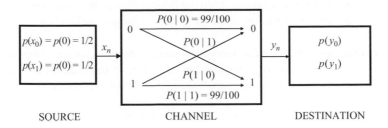

**Fig. 11** *A binary communication system.*

   a. Using Figure 11, specify the transitional probability matrix $\mathbf{P_{YX}}$ of the channel.
   b. Find the entropy of the input of the channel. What does this entropy represent?
   c. Calculate the probabilities of the output symbols $p(y_k)$. Did you get the probabilities you expected? Explain the reasons for your expectation.
   d. Find the joint and conditional probabilities of the channel. Using Bayes' formulas, determine the matrix of the transition probabilities $P(X \mid Y)$.
   e. Calculate the entropy $H(X \mid Y)$ and then the mutual information $I(X; Y)$. What will happen to the mutual information if the bit error probability increases?

12. A discrete memoryless source is characterized by an alphabet of two symbols, $M = (0, 1)$, where the symbols are generated by the two probabilities $P = (p_0, p_1)$, where $P(0) = p_0$, and $P(1) = p_1$.

   a. Find the entropy of this source, $H$, and express it as a function of $p_0$. What is the physical meaning and the unit of this entropy?
   b. For the values $p_0 = (0, 0.1, 0.25, 0.5, 0.75, 0.9, 1)$, calculate the entropy and sketch the graph of the entropy function $H$ as a function of $p_0$.
   c. Using the graph obtained in b., explain the meaning of the entropy values for $p_0 = 0$, $p_0 = 0.5$, and $p_0 = 1$.
   d. Express the entropy $H$ as a function of probability $p_1$ and sketch the graph of this function. Can we draw similar conclusions as in c., using this graph?

13. An information memoryless source generates four possible symbols defined by the alphabet $M = \{m_0, m_1, m_2, m_3\}$ with probabilities 1/8, 1/2, ¼, and 1/8, respectively.

   a. Find the information of each symbol. What is the entropy of the source? What does this entropy represent? The information of the symbols is a random variable.

Sketch the graph of the density function of this random variable, bearing in mind the calculated information of all the symbols.

b. Suppose the source generates two messages composed of four symbols: $X = m_0 m_0 m_0 m_0$ and $Y = m_1 m_1 m_1 m_1$. Find the information of these two messages. Explain the reasons why the information for these two messages are different.

14. Suppose that the channel of a communication system is represented by a binary memoryless channel, and suppose that the input bits to the channel are coming from a channel encoder that generates symbols 0 and 1 with equal probability.

a. Find the bit error probability in this channel in the case when the channel is asymmetric and specified by the transitional probabilities $P(0 \mid 1) = 0.1$ and $P(1 \mid 0) = 0.2$.

b. Sketch a graph that shows the dependence of the bit error probability $p$ on the transition probability $P(1 \mid 0)$ for a constant value of the transition probability $P(0 \mid 1) = 0.1$. What is the maximum possible bit error probability in the channel for this case?

c. What is the condition that should be fulfilled if this memoryless channel is symmetric? Find the expression for the bit error probability for this symmetric memoryless channel. What is the bit error probability in the symmetric memoryless channel if the transition probability is $P(0 \mid 1) = 0.1$? Suppose the sequence $\mathbf{c} = 1111101111$ is generated by the encoder and applied to the input of the channel. What is the probability that the received sequence is $\mathbf{r} = 1010101011$ if the bit error probability is $p = 10^{-1}$?

d. A channel with memory can be represented by a two-state Markov chain. Sketch the model of this channel and specify its states and the related probabilities. Derive the expression for the bit error probability in this channel.

15. A source generates two symbols with amplitudes of 4 V and 6 V, respectively. There is no statistical dependence between the symbols generated. The probability of their generation is proportional to their amplitudes. Suppose a message signal contains 10,000 symbols that should be transmitted in 1 second. The channel bandwidth is 5 kHz. What is the entropy of the source? What is the minimum value of the signal-to-noise ratio in the channel that is needed to transmit this message signal?

16. A source generates a black-and-white picture to be transmitted through a channel with a bandwidth of $B = 10$ kHz. Suppose that the picture consists of 2,000,000 picture elements (pixels) that can be black or white. The pixels are black with probability 0.4. What is the information content of the picture? Suppose the source needs 2 minutes to generate the picture. What is the minimum value of the signal-to-noise ratio in the channel to transmit this picture?

17. Suppose that a discrete binary symmetric memoryless channel is defined by its input alphabet, channel transition matrix, and output alphabet as follows. The input alphabet is

$$A_X = \{x_0, x_1\} = \{0, 1\},$$

defined by the corresponding a priori probabilities $p(x_0) = p(X = x_0 = 0) = 0.5$, and $p(x_1) = p(X = x_1 = 1) = 0.5$. The output alphabet is

$$A_Y = \{y_0, y_1\} = \{0, 1\},$$

and the corresponding a priori probabilities are $p(y_0) = p(y = y_0 = 0) = p(0) = 0.5$, and $p(y_1) = p(Y = y_1 = 1) = p(1) = 0.5$. The channel itself is defined by its transition probabilities in the form of the conditional probabilities expressed as $p(y_k \mid x_j) = p(Y = y_k \mid X = x_j)$, which are $p(0 \mid 0) = 9/10$, $p(1 \mid 1) = 9/10$, $p(0 \mid 1) = 1/10$, and $p(1 \mid 0) = 1/10$, and expressed in matrix form as

$$\mathbf{P} = \begin{bmatrix} p(y_0|x_0) = p(0|0) & p(1|0) \\ p(0|1) & p(1|1) \end{bmatrix} = \begin{bmatrix} 9/10 & 1/10 \\ 1/10 & 9/10 \end{bmatrix}.$$

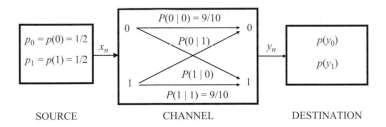

**Fig. 17** *Traditional representation of a binary symmetric memoryless channel.*

   a. Calculate all of the probabilities that define this channel.
   b. Calculate the information content of the channel.
   c. Calculate the conditional entropy and the mutual information of the channel

18. Suppose that a discrete memoryless channel is defined as in Problem 17, but with the following matrix of transitional probabilities:

$$\mathbf{P} = \begin{bmatrix} p(y_0|x_0) = p(0|0) & p(1|0) \\ p(0|1) & p(1|1) \end{bmatrix} = \begin{bmatrix} 0 & 1 \\ 1 & 0 \end{bmatrix}.$$

This is the case when the probability of correct transmission is 0, and the bit error probability is 1. Such a channel is considered to be completely contaminated by noise and so is useless.

a. Calculate the joint probabilities $p(x_j, y_k)$ of this channel, the marginal probabilities $p(y_k)$, and the related information content for each output symbol.

b. Calculate the entropy of the input and the conditional entropy $H(X \mid Y)$ of this discrete memoryless channel.

c. Calculate the mutual information of the channel $I(X; Y)$. What does this mutual information represent?

19. Suppose that a discrete memoryless channel is defined as in Problem 18, but with the following matrix of transitional probabilities:

$$\mathbf{P} = \begin{bmatrix} p(y_0|x_0) = p(0|0) & p(1|0) \\ p(0|1) & p(1|1) \end{bmatrix} = \begin{bmatrix} 1 & 0 \\ 0 & 1 \end{bmatrix}.$$

This is the case when the probability of correct transmission is 1, and the bit error probability is 0. Such a channel is considered to be noiseless.

a. Calculate the joint probabilities $p(x_j, y_k)$ of this channel, the marginal probabilities $p(y_k)$, and the related information content for each output symbol.

b. Calculate the entropy of the input and the conditional entropy $H(X \mid Y)$ of this channel.

c. Calculate the mutual information of the channel, $I(X; Y)$. What does this mutual information represent?

20. Suppose that a discrete memoryless channel is defined as in Problem 18 but with the following matrix of transitional probabilities:

$$\mathbf{P} = \begin{bmatrix} p(y_0|x_0) = p(0|0) & p(1|0) \\ p(0|1) & p(1|1) \end{bmatrix} = \begin{bmatrix} 0.5 & 0.5 \\ 0.5 & 0.5 \end{bmatrix}.$$

This is the case when both the probability of correct transmission and the bit error probability are 0.5. Such a channel is considered to be useless.

a. Calculate the joint probabilities $p(x_j, y_k)$ of this channel, the marginal probabilities $p(y_k)$, and the related information content for each output symbol. Comment on the result.

b. Calculate the entropy of the input and the conditional entropy $H(X \mid Y)$ of this channel. Comment on the result.

c. Calculate the mutual information of the channel, $I(X; Y)$. What does this mutual information represent?

21. A communication system can transmit four different symbols that are coded for this purpose into pulses of varying duration. The width of each coded pulse is inversely proportional to the probability of the symbol that it represents. The transmitted pulses have durations of $T_i = T_0, T_1, T_2,$ and $T_3$, or 1, 2, 3, and 4 ms, respectively.

   a. Find the average rate of information transmitted by this system.
   b. If the four symbols are coded into two-digit binary words, find the digit rate needed to maintain the same average transmitted information rate. Comment on your answer with respect to that for part a..

22. Solve the following problems:

   a. A telephone line has a signal-to-noise ratio of 1000 and a bandwidth of 4000 kHz. What is the maximum data rate supported by this line?
   b. The performance of a telephone line was measured. The bandwidth of the line was 4000 Hz. The measured voltage of the signal was 10 V and the noise voltage was 5 mV. What is the maximum data rate supported by this line?

23. Suppose that the input of a white Gaussian noise channel is a chaotic signal defined by the probability density function

$$f_{X_n}(x_n) = \frac{1}{\pi\sqrt{2-x_n^2}}.$$

Find the capacity of the channel. Firstly, make an assumption that the output of the channel is approximated by a Gaussian variable. Secondly, find the output density function for the input chaotic signal and the Gaussian noise in the channel and then find the capacity of the channel. Compare the capacities obtained by these two calculations.

24. A zero-mean Gaussian stationary process $X(t)$ is band limited to $B$ Hz. This signal is sampled at the rate of $2B$ chips per second, and one $2\beta$-dimensional sample vector $x = [x_1 \, x_2 \, \ldots \, x_{2\beta}]$ is produced. The sample vector is used as the spreading sequence in a direct-sequence spread-spectrum system. The vector can be understood as a realization of a random vector $X$ defined by $2\beta$ Gaussian random variables, that is, $X = [X_1 \, X_2 \, \ldots \, X_{2\beta}]$. The sample $x$ spreads a bit of duration $T$. The bit is transmitted through a zero-mean white Gaussian noise channel. Find the capacity of the channel for this input sequence. What is the value of the capacity when the bandwidth of the channel tends to infinity?

25. Prove that the entropy of a binary memoryless source is maximum for $p_0 = 0.5$. Prove that the capacity of the binary memoryless source is minimum for the bit error probability $p = 0.5$.

26. A source generates a black-and-white picture to be transmitted through a binary symmetric channel without memory. Suppose that the picture consists of picture elements (pixels) that can be black or white. The pixels are black with probability $P(B) = 0.4$. Suppose that a black pixel is transmitted through the channel without memory with the transitional probability $P(B \mid B) = 0.9$.

   a. Calculate the a priori probability of a white pixel, $P(W)$, and then all the transitional (conditional) probabilities of this channel. Form the matrix **P** of transitional probabilities.

   b. Sketch the scheme of the channel and indicate the transitional probabilities. Calculate the probability $p$ that a pixel is transmitted with error.

   c. Calculate the probability $P_R(B)$ that a black pixel is received at the receiver side. Repeat this calculation for a white pixel.

   d. Calculate the values of joint probabilities of the channel. We say this channel is memoryless. Explain the meaning of the term 'memoryless', that is, define a memoryless channel.

27. Let the binary random variables $X$ and $Y$ represent the input and output, respectively, of the memoryless channel presented in Figure 27. The a priori probabilities of generating 0 and 1 at the source are $P_X(0) = 0.5$, and $P_X(1) = 0.5$, respectively. Suppose the channel is specified by the transition probabilities $P(0 \mid 1) = P_0$, and $P(1 \mid 0) = P_1$.

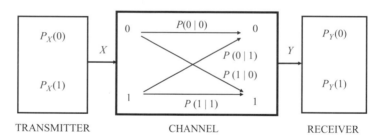

**Fig. 27** *A binary communication system.*

   a. Find the expressions for the probabilities of 0 and 1 at the output of the channel, that is, $P_Y(0)$ and $P_Y(1)$ as functions of $P_0$ and $P_1$, respectively. Then, calculate their values for the transition probabilities $P(0 \mid 1) = P_0 = 0.1$, and $P(1 \mid 0) = P_1 = 0.2$. Prove that your calculated values are correct.

   b. For the transition probabilities $P(0 \mid 1) = P_0 = 0.1$, and $P(1 \mid 0) = P_1 = 0.2$, calculate the transition probabilities $P(1 \mid 1)$ and $P(0 \mid 0)$. Prove that your calculated values are correct. Calculate the probability of bit error, $p$, in this system.

   c. The channel in Figure 27 is binary and asymmetric. What is the main condition for a channel to be symmetric? The channel in Figure 27 is memoryless.

Explain the meaning of the term 'memoryless', that is, define a memoryless channel. How many parameters do we need to describe this channel? How can we express the 'memoryless' property in mathematical terms, using the joint probability $P(\mathbf{m}'|\mathbf{m}) = P(m_1', m_2', \ldots, m_i'|m_1, m_2, \ldots, m_n)$, where $\mathbf{m}'$ is the estimated message binary sequence, and $\mathbf{m}$ is the sent message binary sequence.

28. Suppose that the communication system presented in Figure 27 has a binary memoryless channel, and suppose that the source generates symbols 0 and 1 with equal probability.

   a. Find the bit error probability in this channel in the case when the channel is asymmetric and specified by the transitional probabilities $P(0 \mid 1) = 0.3$, and $P(1 \mid 0) = 0.2$.

   b. Sketch a graph that shows the dependence of the bit error probability $p$ on the transition probability $P(1 \mid 0)$ for a constant value of the transition probability $P(0 \mid 1) = 0.3$. What is the maximum possible bit error probability in the channel for this case?

   c. What is the condition that should be fulfilled if this memoryless channel is symmetric? Find the expression in its general form for the bit error probability for this symmetric memoryless channel. What is the bit error probability in the channel if the transition probability is $P(0 \mid 1) = 0.3$?

   d. Suppose the sequence $X = \mathbf{c} = 1111101111$ is generated by the source and transmitted through a binary symmetric channel without memory. What is the probability that the received sequence is $Y = \mathbf{r} = 1010101011$ if the bit error probability is $p = 10^{-1}$?

29. Find the Hamming distance between the following pairs of binary code words: $\mathbf{c}_0 = (0, 0, 0, 0)$ and $\mathbf{c}_0' = (0, 1, 0, 1)$; $\mathbf{c}_1 = (0, 1, 1, 1, 0)$ and $\mathbf{c}_1' = (1, 1, 1, 0, 0)$; $\mathbf{c}_2 = (0, 1, 0, 1, 0, 1)$ and $\mathbf{c}_2' = (1, 0, 1, 0, 0, 1)$; $\mathbf{c}_3 = (1, 1, 1, 0, 1, 1, 1)$ and $\mathbf{c}_3' = (1, 1, 0, 1, 0, 1, 1)$.

30. A code $\mathbf{C} = \{\mathbf{c}_0, \mathbf{c}_1, \ldots, \mathbf{c}_8\}$ consists of a set of the following code words: $\mathbf{c}_0 = (0, 0, 0, 0, 0, 0, 0)$, $\mathbf{c}_1 = (1, 0, 0, 0, 1, 1, 1)$, $\mathbf{c}_3 = (0, 1, 0, 1, 0, 1, 1)$, $\mathbf{c}_4 = (0, 0, 1, 1, 1, 0, 1)$, $\mathbf{c}_5 = (1, 1, 0, 1, 1, 0, 0)$, $\mathbf{c}_6 = (1, 0, 1, 1, 0, 1, 0)$, $\mathbf{c}_7 = (0, 1, 1, 0, 1, 1, 0)$, $\mathbf{c}_8 = (1, 1, 1, 0, 0, 0, 1)$. Find the minimum Hamming distance and determine the error correction and error detection capabilities of this code.

31. Let $\mathbf{C}$ be a (7, 3) binary linear code with the basis $\{1101100, 1110001, 1011010\}$. Demonstrate the procedure for error detection, assuming that the received code words are $\mathbf{r} = 1010001$ and $\mathbf{r} = 1110001$.

   a. Find the systematic generator matrix $\mathbf{G}$. Indicate the parity bits in this matrix. Find the code word $\mathbf{c}$ for the message word $\mathbf{m} = 101$. Find the corresponding systematic parity check matrix $\mathbf{H}$ and its transpose matrix.

   b. Demonstrate the procedure for error detection, assuming that the received code words are $\mathbf{r} = 0011101$ and $\mathbf{r} = 1110101$, and indicate the message sequence in the case of correct transmission. If you were using a standard array to detect and correct errors in the received code words, what would be the first nine error sequences in your standard array for this code?

32. Let **C** be a (7, 3) binary linear code with the generator matrix

$$\mathbf{G} = \begin{bmatrix} 1 & 1 & 0 & 1 & 1 & 0 & 0 \\ 1 & 1 & 1 & 0 & 0 & 0 & 1 \\ 1 & 0 & 1 & 1 & 0 & 1 & 0 \end{bmatrix}.$$

  a. How many message words can be encoded using this generator matrix? Choose the message word **m** with the maximum weight and find the corresponding code word **c**.

  b. Find the corresponding systematic generator matrix $\mathbf{G_s}$ and the code word $\mathbf{c_s}$ for your message word **m**. What is the difference between the code word calculated in a. and the code word calculated here? Explain the reason for this difference.

  c. Demonstrate the procedure for error detection, assuming that the received code word is **r** = 1111111. Can we indicate the message bits in this code word?

  d. What is the purpose of using a standard array? What would be the number of error patterns (error sequences) in the standard array for the code analysed above? What is the guaranteed number of errors you can correct with this standard array?

33. Solve the following problems:

  a. Suppose that a block code has the following parity check matrix:

$$\mathbf{H} = \begin{bmatrix} 1 & 0 & 0 & 1 & 0 & 1 \\ 0 & 1 & 0 & 1 & 1 & 0 \\ 0 & 0 & 1 & 0 & 1 & 1 \end{bmatrix}.$$

Determine the code word for the message 1001. Find the syndrome table for error sequences that contain at most one error. If the code word 0011010 is received, determine if any error has been made and correct the error if necessary.

  b. Suppose now a block code is a simple repetition code. Then the parity check matrix is

$$\mathbf{H} = \begin{bmatrix} 1 & 0 & 0 & 0 & 1 \\ 0 & 1 & 0 & 0 & 1 \\ 0 & 0 & 1 & 0 & 1 \\ 0 & 0 & 0 & 1 & 1 \end{bmatrix}.$$

Determine the corresponding generator matrix **G**. Find the code words for the input message sequence **m**=1011. Suppose the received code words are (10011 10101 00110 10011 11010). Determine the most likely message sequence by using a majority logic decoder.

c. Suppose that the bits at the input of the block encoder are equally likely and that the code words are transmitted through an additive white Gaussian noise channel. Find the possible symbols that could be applied to the input of the channel. What is the entropy of the source? What is the entropy of the input of the channel?

34. Solve the following problems:

a. Suppose that a parity check matrix $H$ for a $(6, 3)$ block code **C** is

$$H = \begin{bmatrix} 1 & 0 & 0 & 1 & 0 & 1 \\ 0 & 1 & 0 & 1 & 1 & 0 \\ 0 & 0 & 1 & 0 & 1 & 1 \end{bmatrix}.$$

Suppose we perform systematic encoding with the message bits at the last three positions of the code word. Find the code word that carries the message word **m** = 101.

b. Find the general expressions that are conditions sufficient for calculating parity check bits for any code word **m** = $m_1$ $m_2$ $m_3$. Find the generator matrix **G** of code **C**. For the message word **m** = 101, use **G** to confirm that the code word **c** is equivalent to the one obtained in a.. Find the dual code $\mathbf{C}^{\perp}$ of the code **C** by defining its generator matrix $\mathbf{G_D}$ and parity check matrix $\mathbf{H_D}$. What is the rate of this code?

35. Suppose that a generator matrix **G** for a $(5, 1)$ block code is **G** = [1 1 1 1 1].

a. Determine the code words for a message sequence **m** that is composed of three message words, that is, **m** = 101. What kind of code is this? What is the code rate of this code?

b. Find the parity check matrix of this code. Suppose the received code word is **r** = 10111. Prove that some errors occurred in the transmitted code word. Suppose the sequence of received code words is (10001 10111 01110 10111 10010) and find the most likely message sequence by using a majority logic decoder.

c. Suppose the parity check matrix **H** calculated in b. is a generator matrix $\mathbf{G_D}$ of a dual code. Present $\mathbf{G_D}$ and find the parity check matrix $\mathbf{H_D}$ for this code.

36. Prove that the convolutional encoder in Figure 36 generates a linear code. Find the impulse response, find the transfer-function matrix, and determine the code word associated with the input sequence **x** = (11101), using the transfer-function matrix.

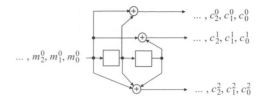

**Fig. 36** *A rate-1/3 encoder.*

37. Construct an encoder for the convolutional code represented by the following transfer-function matrix:

$$\mathbf{G}(D) = \begin{bmatrix} 1+D+D^2 & 1+D+D^3 & 1+D^2+D^3 \\ 1+D^3 & 1+D^2 & 1+D+D^3 \end{bmatrix}.$$

38. Draw the first five stages of the trellis diagram corresponding to the encoder in Figure 38.

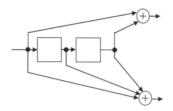

**Fig. 38** *A convolutional encoder.*

39. A rate-1/3 convolutional code is described by the following impulse responses:

$$\mathbf{g}^{(0)} = \left( g_0^{(0)}, g_1^{(0)}, g_2^{(0)} \right) = (110),$$
$$\mathbf{g}^{(1)} = \left( g_0^{(1)}, g_1^{(1)}, g_2^{(1)} \right) = (101),$$

and

$$\mathbf{g}^{(2)} = \left( g_0^{(2)}, g_1^{(2)}, g_2^{(2)} \right) = (111),$$

which are, respectively, the first, second, and third outputs of the encoder.

a. Draw an encoder corresponding to this code. Draw the state-transition diagram for this code. Find the generator matrix **G** of the encoder. What is the size of this matrix?

b. Find the encoded sequence **c** for the input message sequence **m** = (1001). What is now the size of generator matrix **G**?

c. Draw the trellis diagram for this encoder. Assume that four message bits, followed by two 0 bits, have been encoded and sent via a binary symmetric memoryless channel with an error probability of $p = 10^{-3}$. Suppose the received sequence is $\mathbf{r} = (111\ 111\ 111\ 111\ 111\ 111)$. Use the Viterbi algorithm to find the most likely message sequence. In the case of a tie, delete the upper branch.

40. The encoder for the 1/3 convolutional code is shown in Figure 40. Find the impulse response of the encoder. Find the generator matrix of the encoder. Find the encoded sequence $\mathbf{c}$ if the input message sequence of the encoder is $\mathbf{m} = (11101)$.

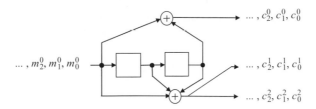

**Fig. 40** *A rate-1/3 convolutional encoder.*

41. Draw the state diagram for the encoder in Figure 40. Draw the trellis diagram for the encoder and use the Viterbi algorithm for the given received encoded sequence $\mathbf{r} = (101001011110111)$. Assume that the encoder returns to the 00 state after encoding each message sequence.

42. A 1/2-rate encoder for a convolutional code is shown in Figure 42.

a. Find the impulse response and the generator matrix of the encoder. What is the size of this matrix?

b. Find the encoded sequence $\mathbf{c}$ if the input message sequence of the encoder is $\mathbf{m} = (11001)$. What is the size of the generator matrix in this case, and why?

c. Draw the state-transition diagram for this encoder for its states, which are defined as $S_0 = 00$, $S_1 = 01$, $S_2 = 10$, and $S_3 = 11$.

d. Assume that the bit error probability in the transmission channel is $p = 10^{-6}$. Suppose that the received encoded sequence is $\mathbf{r} = (01\ 10\ 11\ 11\ 01\ 00\ 01)$. Using the Viterbi algorithm, decode the received sequence $\mathbf{r}$. What is the estimate $\mathbf{m'}$ of the input message sequence $\mathbf{m}$? Specify the path that corresponds to this sequence. Assume that the encoder returns to the $S_0 = 00$ state after encoding each message sequence. In the case of a tie, delete the upper branch.

43. Suppose that a convolutional 2/3-rate encoder is defined by the scheme in Problem 6. A message sequence $\mathbf{m}$ is composed of pairs of message bits. The encoded sequence $\mathbf{c}$ contains three bits for each pair of message bits. Each bit is defined on its alphabet $\{0, 1\}$.

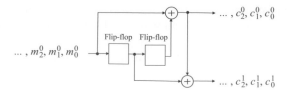

**Fig. 42** *A rate-1/2 encoder for a convolutional code.*

Suppose that two message pairs (four bits) followed by a two-0 dummy pair (two 0 bits) are encoded and the binary sequence $r = (010\ 111\ 000)$ is received.

a. Find the impulse response in the binary and $D$ forms of this encoder. Form the transfer-function matrix $\mathbf{G}(D)$, where $D$ is a unit delay operator. Using $\mathbf{G}(D)$, find the output code word $\mathbf{c}^1$ for the input message sequence $\mathbf{m}^1 = (11\ 10\ 11)$.

b. Plot and then complete the state diagram of this encoder. Plot the path on the state diagram that corresponds to the message sequence $\mathbf{m} = (00\ 01\ 11\ 00\ 10\ 01)$.

44. Suppose that a convolutional 2/3-rate encoder is defined by the scheme given in Problem 6.

a. Suppose that a message sequence $\mathbf{m}$ is encoded, producing a sequence $\mathbf{c}$ that is transmitted through the channel, and the sequence $\mathbf{r} = (010\ 111\ 000)$ is received. Using the hard-decision Viterbi algorithm, determine estimates of the message $\mathbf{m}$ and the encoded sequence $\mathbf{c}$, denoted by $\mathbf{m}'$ and $\mathbf{c}'$, respectively. Suppose that the encoder starts from the $S_0 = 00$ state and finishes at the $S_0 = 00$ state.

   We applied the Viterbi algorithm here. What is the basic assumption related to the nature of the transmission channel when we apply this algorithm? Is the Viterbi algorithm optimal or asymptotically optimum, in the maximum likelihood sense? Explain the reason for your answer.

45. Figure 45 shows a convolutional encoder with a rate of 1/3. The convolutional code is used for digital signal transmission on a binary symmetric memoryless channel with an error probability of $p = 10^{-4}$.

a. Find the impulse responses of the encoder. How many bits inside the impulse response are there at each output of the encoder? Why is there that number of bits?

b. Draw the state-transition diagram for this encoder for the states defined as $S_0 = 00$, $S_1 = 01$, $S_2 = 10$, and $S_3 = 11$.

c. Suppose the received sequence is $\mathbf{r} = (110\ 110\ 110\ 111\ 010\ 101\ 101)$. Draw the trellis diagram for this encoder. Find the transmitted message sequence $\mathbf{m}'$, using the Viterbi algorithm. Assume that the encoder returns to the $S_0 = 00$ state after encoding each message sequence. In the case of a tie, delete the upper branch. You can use the trellis diagram previously developed.

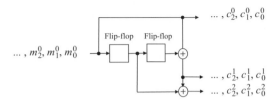

**Fig. 45** *A rate-1/3 convolutional encoder.*

46. A convolutional encoder is defined by the convolutional encoding matrix $\mathbf{G}(D) = [1 + D + D^2 \ 1 + D^2]$, has rate ½, and is composed of $m = 2$ memory elements. This convolutional encoder is used in a system for digital signal transmission over a binary symmetric memoryless channel with an error probability of $p = 10^{-3}$.

   a. Find the impulse response of the encoder in binary form. Plot the block schematic of the encoder with specified inputs, outputs, memory elements, adders, and their interconnections. Define the constraint length of the encoder and find its value for the defined encoder structure.

   b. Draw the state-transition diagram for this encoder for states defined as $S_0 = 00$, $S_1 = 01$, $S_2 = 10$, and $S_3 = 11$. Find the path in this state diagram that corresponds to the message bit sequence $\mathbf{m} = 1110$. Find the encoded sequence $\mathbf{c}$ for this message sequence. Find the path inside the state diagram for the case when dummy bits are added to force the encoder to go into state $S_0 = 00$.

47. Suppose that a convolutional encoder is defined by the scheme shown in Figure 47. A binary message sequence $\mathbf{m}$, which is defined on its alphabet (0, 1), is applied to the encoder input. The encoded sequence $\mathbf{c}$ contains the message and encoded redundancy bits. For each message sequence $\mathbf{m}$, one bit is added to return the encoder into the 0 state.

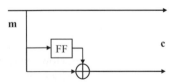

**Fig. 47** *One memory-element encoder; FF, flip-flop.*

   a. Find the state diagram of this encoder. Assume that the initial and final states of the encoder are $S_0 = 0$. Plot the path on the state diagram that corresponds to the message sequence $\mathbf{m} = (0\ 1\ 0\ 1)$ and find the corresponding encoded sequence $\mathbf{c}$. What is the value of the dummy bit that is needed to reset the encoder?

b.  Plot the trellis diagram of the encoder. Find the path in the trellis for the message sequence $\mathbf{m} = (0\ 1\ 1)$. Add one bit to $\mathbf{m}$ to force the trellis diagram to go to the $S_0 = 0$ state. Find the encoder output sequence $\mathbf{c}$ in this case.

c.  Suppose that the encoded sequence $\mathbf{c}$ is transmitted through an additive white Gaussian noise channel, and the received sequence is $\mathbf{r} = (1\ 1\ 0\ 1\ 1\ 0\ 1\ 0)$. Using the hard-decision Viterbi algorithm, determine the estimates of message $\mathbf{m}$ and encoded sequence $\mathbf{c}$. Suppose that the encoder starts from the $S_0 = 0$ state and finishes at the $S_0 = 0$ state. How many errors occurred in the channel?

d.  Explain the procedure for obtaining the impulse response of an encoder. Find the impulse response of this encoder.

48. Suppose that a recursive systematic convolutional encoder is defined by the scheme shown in Figure 48. A binary message sequence $\mathbf{m}$, which is defined on its alphabet $(0, 1)$, is applied to its input. The encoded sequence $\mathbf{c}$ contains the systematic and parity bits.

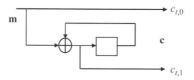

**Fig. 48** *A recursive systematic convolutional encoder.*

a.  Find the state diagram of this encoder. Plot the path on the state diagram that corresponds to the message sequence $\mathbf{m} = (0\ 1\ 0\ 1)$ and find the corresponding encoded sequence $\mathbf{c}$.

b.  Plot the trellis diagram of the encoder. Find the path in the trellis for the message sequence $\mathbf{m} = (0\ 1\ 0\ 1)$.

c.  Suppose that the encoded sequence $\mathbf{c}$ is transmitted through an additive white Gaussian noise channel, and the received sequence is $\mathbf{r} = (1\ 1\ 0\ 1\ 1\ 0\ 1\ 0)$. Using the hard-decision Viterbi algorithm, determine the estimates of the message $\mathbf{m}$ and the encoded sequence $\mathbf{c}$. Suppose that the encoder starts from state $S_0 = 0$ and finishes at state $S_0 = 0$. How many errors occurred in the channel?

d.  How could you reset this encoder and force it to return to the initial, say, $S_0 = 0$ state?

49. A 1/3-rate turbo encoder is specified by two $(2, 1, 4)$ recursive systematic convolutional encoders, which are shown in Figure 49. Find the output sequence, assuming that the input sequence is $\mathbf{m} = (1011001)$ and the interleaver permutes the input sequence $\mathbf{m}$ to $\mathbf{m}' = (1011001)$.

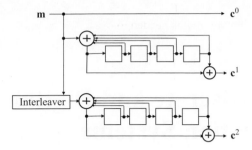

**Fig. 49** *A rate-1/3 turbo encoder.*

# 10

# Designing Discrete and Digital Communication Systems

## 10.1 Introduction

The design of discrete communication systems, including their mathematical modelling, simulation, emulation, and implementation in digital technology, is, in some aspects, different from the design of digital communication systems operating in the continuous-time domain. In this chapter, we will present issues related to the digital design of discrete communication systems. Due to the very close connection between digital and discrete communication systems, we will also discuss the design of digital systems. This chapter will use the principles of discrete-system modelling that were presented in previous chapters, primarily Chapters 5, 6, 7, 8, and 9, as well as concepts related to the design of discrete systems, including digital filters, multi-rate signal processing, and multi-rate filter banks, which will be presented in Chapters 16, 17, and 18. Therefore, the information presented in these chapters is essential for reading this chapter.

Starting with Chapter 5, and extending our theoretical analysis in Chapters 6, 7, 8, and 9, we investigated the operation of discrete communication systems and the application of various modulation methods. We rightly implied that discrete-time signal processing in these systems would be implemented in system transceivers for the baseband and passband signal processing. Furthermore, a significant part of Chapter 4, which was dedicated to discrete-time noise analysis and synthesis, was dedicated to the mathematical modelling of discrete-time passband noise, and the procedure for generating such noise. In doing this, the intention was to develop a theoretical base for the development of discrete-noise simulators and emulators for the testing of discrete communication systems operating in both the baseband and the passband.

In our explanations, we will use the terms 'digital technology' and 'digital design'. The term 'digital technology' refers to modern digital devices like digital signal processors, field-programmable gate arrays, and custom-made devices with discrete-time signal-processing functions. The term 'digital design' refers to the implementation of transceiver function in one of the digital technologies or their combination. The explanation in this chapter will be at the level of the block schematics of the transceivers. Block schematics of real designs will

*Discrete Communication Systems*. Stevan Berber, Oxford University Press. © Stevan Berber 2021.
DOI: 10.1093/oso/9780198860792.003.0010

not necessarily follow these block schematics, even though the theoretically defined functions presented in the previous chapters and in this chapter will be implemented.

An example of the design of a transceiver in field-programmable gate array technology will be presented as a project in the Complementary Material for this book. It will be shown that the blocks of the transceiver are designed in VHDL code and implemented on a field-programmable gate array board. The project content will present the phases in the practical development of modern digital design in a communication system: mathematical modelling, simulation, emulation on a digital signal-processor board, and testing. This project was added to the book because it was impossible for the author to cover the basic theory in discrete communication systems, which was the main aim of the book, and at the same time present the issues related to the modern design of the communication devices inside the book. The design of modern transceivers in digital technology can be considered to be a separate topic that would be well worth publishing in a separate book.

In this chapter, for the sake of explanation and due to our intention to present the design of discrete transceivers, we will use the simplified block schematics of transmitters and receivers separately and exclude communication channels. In this approach, we will use the block schematics of transceivers presented in Chapter 7, which contain a precise description and exact mathematical presentation of functions that are performed inside discrete transceiver signal-processing blocks. How to implement these functions and how to deal with the issues of multi-rate signal processing and the application of digital filters and filter banks will, in part, be the subjects of this chapter.

The discrete-time communication system in Chapter 7 is mathematically modelled and analysed assuming that discrete-time signal processing is performed on intermediate frequencies. Therefore, a digital-to-analogue converter was implemented at the output of the intermediate-frequency modulator block, and an analogue-to-digital converter was implemented at the input of the intermediate-frequency modulator block. We would like to add an important comment at this point.

Generally, we want to make most of the processing inside a transceiver structure in the discrete-time domain. Therefore, the allowable sampling rate in a particular digital processing device is the most important parameter that determines the implementation of our transceiver. The higher the sampling rate is, the more flexibility we have in our design. For the design of our transceiver, the ideal case would be if all signal processing could be done in the discrete time, up to or, if possible, including up-conversion to the carrier radiofrequency. In that case, we would be able to perform all of the signal processing in the discrete-time domain, use an digital-to-analogue converter to make an analogue (continuous-time domain) radiofrequency signal, filter the signal, amplify it, and transmit it through the channel. The limiting factor in this is the possible maximum sampling rate that is achievable in digital technology, as it is usually below the carrier frequencies of modern transceivers.

This chapter will address the design issues of discrete communication systems based on phase-shift keying and quadrature amplitude modulation. Most of the material presented will be on systems based on quadrature amplitude modulation, which encompass phase-shift keying modulation as a special case. In particular, we will discuss the problem of using analogue-to-digital and digital-to-analogue converters, and their place and operation inside the transceiver structure.

## 10.2 Designing Quadriphase-Shift Keying Transceivers

### 10.2.1 Quadriphase-Shift Keying Systems with Digital-to-Analogue and Analogue-to-Digital Conversion of Modulating Signals

In Chapter 7, we analysed basic modulation methods, assuming that the signal processing was performed in the discrete-time domain. The analysis started with binary and quaternary discrete-time phase-shift keying systems. It is quite simple to extend these systems to $M$-ary phase-shift keying systems. For the sake of explanation, we will focus on the design of a quadriphase-shift keying system. A discrete communication system based on quadriphase-shift keying modulation is shown in Chapter 7, Figure 7.3.1. We will analyse the design of the transmitters and receivers of this system, pointing out, in particular, the places and roles of analogue-to-digital and digital-to-analogue converters.

#### 10.2.1.1 *Quadriphase-Shift Keying Transmitters with Baseband Discrete-Time Signal Processing*

The structure of a quadriphase-shift keying transmitter was given in Chapter 7, based on both Figure 7.3.1 and the simplified form presented in Figure 7.3.9. In the version presented in this section, analogue-to-digital converters are placed inside the in-phase ($I$) and quadrature ($Q$) branches of the transmitter, as presented in Figure 10.2.1.

The operation of the baseband block up to the input of the digital-to-analogue converter is explained in Section 7.3.1. If we analyse the general case, when the input of the digital-to-analogue converter is a discrete-time rectangular pulse, the output of the converter will be a continuous-time rectangular pulse. Therefore, the modulation procedure will follow the explanation presented in Chapter 6, Section 6.3, where the operation of a digital

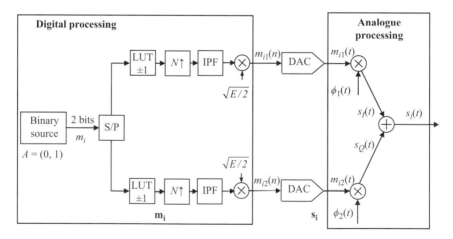

**Fig. 10.2.1** *A quadriphase-shift keying transmitter with baseband digital-to-analogue converters; DAC, digital-to-analogue converter; IPF, interpolation reconstruction filter; LUT, look-up table; S/P, serial-to-parallel converter.*

quadriphase-shift keying transmitter is explained, and thus the explanation will not be repeated in this section. A realistic structure of the transmitter will be directly obtained if the interpolation reconstruction filter performs signal processing according to the procedure defined for the appropriate pulse-shaping filter widely used in practice. We will look here at the properties of the transmitter from the design point of view.

The digital-to-analogue converter is implemented in both the $I$ and the $Q$ branches of the modulator, where the low-frequency baseband signal is processed. The advantage of this design is in the low sampling rate required to generate the discrete-time rectangular pulses. The theoretical minimum rate is 2, as presented in Chapter 7, Figure 7.3.2, which simplifies the design of the converter. The modulating continuous-time signals in both the $I$ and the $Q$ branches modulate the carriers, which are defined by orthonormal cosine and sine functions, according to the modulation procedure presented in Chapter 6, Section 6.3.2. However, the continuous-time modulation used in this system is a disadvantage. Namely, it is necessary to use two carriers that are in quadrature, which is hard to achieve in practice. Furthermore, we need to have two perfectly balanced multipliers with the same gain, which is also hard to achieve using analogue technology. The consequence of these disadvantages is in the increase in the signal bandwidth and the deterioration of the bit error probability in the system. Therefore, due to the development of digital-to-analogue converters and digital circuits with a higher clock rate, this design became obsolete, and the design of modulators was possible in the discrete-time domain without processing in the continuous-time domain.

### *10.2.1.2 Designing a Quadriphase-Shift Keying Receiver*

In this case, the received signal is presented in the continuous-time domain. It can be demodulated using the correlation demodulator and an optimum detector, as presented in Chapter 6, Figure 6.3.1, and also shown in Figure 10.2.2 for the sake of completeness and explanation. The procedures for demodulation and decision-making were explained in

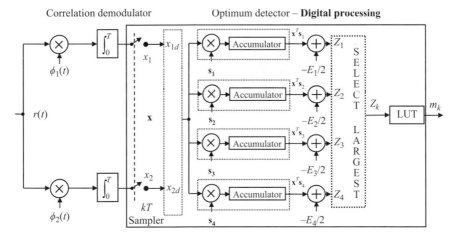

**Fig. 10.2.2** *A quadriphase-shift keying receiver with baseband analogue-to-digital converters; LUT, look-up table; S/P, serial-to-parallel converter.*

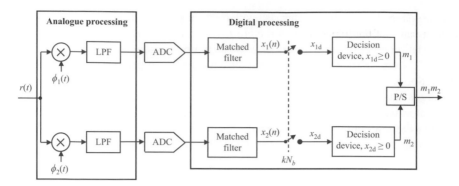

**Fig. 10.2.3** *A first-generation quadriphase-shift keying receiver with baseband analogue-to-digital converters; DAC, digital-to-analogue converter; LPF, lowpass filter; P/S, parallel-to-serial converter.*

Chapter 6, Section 6.3, and so will not be repeated in this section. In brief, the received signal is demodulated using a correlation demodulator that produces the demodulated vector **x**, which has two elements, in this case, $x_1$ and $x_2$, which are used inside the optimum detector to obtain the message symbol $m_k$, which is the optimal estimate of the message symbol $m_i$ that has been sent from the transmitter source.

The elements $x_1$ and $x_2$ of the demodulated vector **x** can be understood as samples of a discrete-time continuous-valued signal taken by a sampler at the time instants $kT$, where $T$ is the symbol duration, and index $k$ defines the order of symbols in the time domain. The continuous-valued samples can be discretized to obtain the discrete values $x_{1d}$ and $x_{2d}$ and then digitalized, as shown in Figure 10.2.2, and further processed using digital technology to obtain the estimates $m_k$ of the symbols sent. The processing can be simplified if the decision is made in both the $I$ and the $Q$ branches, as discussed in Chapter 6. However, if we extend the number of modulation levels to $M$ and use $M$-ary phase-shift keying methods, simple processing cannot be applied.

The demodulation and processing of the received signal in its baseband can be performed by including discrete-time signal processing. In this case, the receiver may be designed as presented in Figure 10.2.3. Firstly, the received signal $r(t)$ is multiplied by the locally generated carriers, and then the output signals are filtered using lowpass filters, which eliminate the double-carrier frequency component and allow the baseband signal components to pass through. To obtain the discrete-time baseband signal, the demodulated continuous-time signal is applied at the input of the analogue-to-digital converter, which generates a discrete-time baseband signal at its output. To obtain the demodulated sample values of the received signal, $x_1$ and $x_2$, we can use a discrete-time matched filter that is matched to the input modulating signal. The output of the matched filter is a discrete-time signal with a random value determined by the time instants defined by the duration of the bits inside the $I$ and the $Q$ branches.

Due to the signal processing in the baseband, the analogue-to-digital converter can operate at a low sampling rate, as was the case with the digital-to-analogue converter at the transmitter side, which reduces the complexity and the price of the converter. Also, the digital matched filter, which operates in the discrete-time domain, is simple to design. The output of the filter is a sample value of the transmitted bits that are affected by the noise in the channel, like the output of the correlators, as explained in Chapter 7.

The decision on the samples $x_1$ and $x_2$ can be made in each branch, as explained in Chapter 7, Section 7.3, and presented in Figure 7.3.10. This procedure is also shown in Figure 10.2.3, where the 1's are generated at the output of the decision circuits if the discrete-valued samples $x_{1d}$ and $x_{2d}$ are greater than 0. The detected bits, denoted $m_1$ and $m_2$, are generated in series at the output of the receiver by means of a parallel-to-series block.

The disadvantages of this design are similar to the ones mentioned for the design of the corresponding transmitter. At the input of the receiver, we need to generate orthonormal carriers and achieve the same gain in both branches, which is hard to achieve in practice. The possible imbalances between the $I$ and the $Q$ branches can cause distortion that will change the constellation diagram and increase the bit error probability in the system.

An additional problem is related to the operation of the symmetric $I$ and $Q$ multipliers, where the local oscillator (signal) of one branch 'leaks' into the input port of the other branch, producing a constant term that causes a distortion called 'DC offset'. This distortion is transferred through the related branches and is added to the sample values $x_{1d}$ and $x_{2d}$. Therefore, this offset changes the position of the message points inside the constellation diagram and influences the degradation of the bit error probability.

Another problem with the DC offset is the possible leakage of the local oscillator signals that return back to and are radiated by the receiving antenna. The radiated sinusoidal signal can be considered to be interference for the receivers that are in the vicinity of our receiver.

### *10.2.1.3 Practical Design of a Quadriphase-Shift Keying Receiver*

A quadriphase-shift keying transceiver that uses analogue-to-digital and digital-to-analogue baseband signals can be considered to be a first-generation discrete transceivers. Because the design of the transmitter can be considered to be self-explanatory, we will present and explain the issues related to the design of the receiver. The block schematics of the receiver, including the positions of the circuit blocks for clock and carrier generation, is presented in Figure 10.2.4.

The system includes front-end receiver processing in the continuous-time domain, with the related disadvantages mentioned before. The signal is transferred in the discrete-time domain for digital processing that includes a decision device implemented in the digital signal processing block. This block also implements functions related to the synchronization in the system, and control and communication functions with other blocks in the receiver. As shown in Figure 10.2.4, this block generates the information needed for the phase-locked loop in the discrete-time domain that is transferred into the continuous-time domain to regulate the voltage-controlled oscillator and synchronize the frequency and phase of the locally generated carriers with the carrier of the incoming radiofrequency signal. The synchronization of the carriers is performed in the continuous-time domain. In addition, the digital signal processor block controls the timing phase-locked loop used to adjust the frequency and phase of the sampling instants in the system. Due to the size of this book and the initial intention to cover the theory of discrete communication system operation, the problems of the system's synchronizations are not covered and can be, in the author's opinion, a good subject for a separate book.

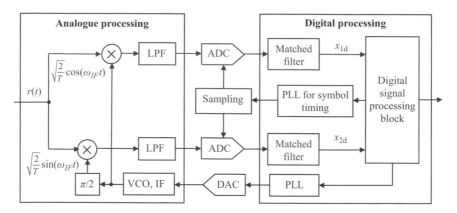

**Fig. 10.2.4** *Design of a first-generation quadriphase-shift keying receiver with baseband analogue-to-digital converters; ADC, analogue-to-digital converter; DAC, digital-to-analogue converter; IF, intermediate-frequency modulator block; LPF, lowpass filter; PLL, phase-locked loop; VCO, voltage-controlled oscillator.*

## 10.2.2 Quadriphase-Shift Keying Systems with Digital-to-Analogue and Analogue-to-Digital Conversion of Intermediate-Frequency Signals

### 10.2.2.1 *Designing a Digital Quadriphase-Shift Keying Transmitter at Intermediate Frequency*

For the sake of completeness, in the previous section, we presented the design of quadriphase-shift keying transceivers with digital-to-analogue and analogue-to-digital converters inside the baseband processing block. However, discrete-time processing of transceiver signals is usually performed in both the baseband block and the intermediate-frequency block. Thus, digital-to-analogue and analogue-to-digital converters are placed at the output of the intermediate-frequency transmitter block and at the input of the intermediate-frequency receiver block. The design of a transceiver that performs discrete-signal processing in the discrete-time domain in all blocks, including intermediate-frequency modulators and demodulators, is the subject of our analysis in this section.

The issues related to signal processing in discrete communication systems, together with the placement of analogue-to-digital and digital-to-analogue converters, was already raised in Chapter 5, and a block schematic of the system was presented in Figure 5.7.1 and followed by the intermediate-frequency system structure shown in Figure 5.7.2, which incorporates the intermediate-frequency passband noise generator recommended for system testing in Chapter 4. Based on these figures, a quadriphase-shift keying system structure was presented in Chapter 7, Figure 7.3.1, with a related mathematical model and a detailed theoretical explanation in Section 7.3.

In this section, we will refer to Chapter 5, Figure 5.7.1, which was adapted for the quadriphase-shift keying system and addresses the structure of the transmitter and the receiver from the design point of view. Detailed explanations of their operation, together with related

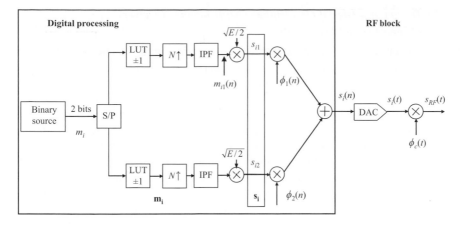

**Fig. 10.2.5** *A quadriphase-shift keying transmitter with an intermediate-frequency digital-to-analogue converter; DAC, digital-to-analogue converter; IPF, interpolation reconstruction filter; LUT, look-up table; RF, radiofrequency; S/P, serial-to-parallel converter.*

mathematical descriptions, can be found in Chapter 5, Section 5.7. The structure of the transmitter is presented in Figure 10.2.5.

Both baseband and intermediate-frequency signal processing are performed in the discrete-time domain in digital technology. Therefore, all of the disadvantages of the receiver design presented in Figure 10.2.1, where digital-to-analogue conversion is performed on a baseband signal, and intermediate-frequency modulation is implemented in the continuous-time domain, are eliminated. For a sufficiently high clock rate and, in turn, a sufficiently high sampling rate, the desired precision in representing the amplitudes and phases of the carriers can be achieved. In addition, the balance of gain for mixers will be preserved with the accuracy dictated by the sampling rate. That is to say, the design of discrete systems is directly connected to and depends on the sampling rate. That is the reason why we are asking what the highest achievable clock rate is, or what the highest sampling rate is, when we are talking about the design capability of digital circuits based on modern digital technology.

The problem of the sampling rate is addressed in Chapter 5, Section 5.2, from the theoretical point of view, and one example, which demonstrates one procedure for generating a modulated signal for a given modulating discrete-time pulse and a discrete-time sinusoidal carrier, is presented in Chapter 5, Section 5.3.

The intermediate-frequency discrete-time signal obtained at the output of the modulator is converted into a continuous-time signal that is applied to the input of the mixer operating at the radiofrequency of the transmitter. This mixer would have been excluded if the frequency of the intermediate-frequency carrier had been equal to that of the radiofrequency carrier. In that case, we would say that we had performed a direct discrete modulation of the radiofrequency carrier. However, for that kind of design, the sampling rate of the digital circuitry has to be sufficiently higher than the frequency of the radiofrequency carrier.

### 10.2.2.2   Design of a Digital Quadriphase-Shift Keying Receiver at Intermediate Frequency

The block schematic of a receiver that is designed in digital technology at the level of intermediate-frequency discrete-time processing is presented in Figure 10.2.6.

If the signal is transmitted at the radiofrequency frequency $\phi_c(t)$, then, at the receiver input, there is a mixer that down-converts the signal to an intermediate frequency that is filtered by an intermediate0frequency filter. Such processing is performed in analogue technology. Then, the analogue intermediate-frequency signal is converted to a discrete-time intermediate-frequency signal that will be processed in an intermediate-frequency demodulator by means of an intermediate-frequency analogue-to-digital converter. The intermediate-frequency demodulator is implemented in digital technology, and the discrete amplitude correlation samples $x_{1d}$ and $x_{2d}$ are processed in the decision devise to generate the estimates of the bits sent.

The decision device can be designed as an optimum detector, as presented in Figure 10.2.6, and the signal processing that leads to the estimation of the message bits can be performed as explained in Chapter 7, Section 7.3. The intermediate-frequency demodulator is implemented in digital technology, and the correlation samples obtained are processed in the decision devise. The digital signal processing is done on the discrete-time and discrete-valued signals.

Because the signal-processing procedures in the $I$ and $Q$ branches are identical, and related signals are statistically independent, the decision circuits can be placed in each branch, as was presented in Chapter 7, Section 7.3, Figure 7.3.10. The decision can be made in each branch following the decision procedure for binary phase-shift keying signals, as presented in Chapter 7, Section 7.2.3. The receiver based on this decision procedure is presented in Figure 10.2.7.

With this design, most of the disadvantages related to the design using analogue-to-digital converters in $I$ and $Q$ branches (presented in Figure 10.2.3) are eliminated. The generation of

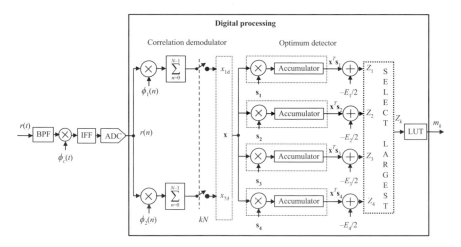

**Fig. 10.2.6** *A quadriphase-shift keying receiver with an intermediate-frequency analogue-to-digital converter and an optimum detector; ADC, analogue-to-digital converter; BPF, bandpass filter; IFF, intermediate-frequency filter; LUT, look-up table.*

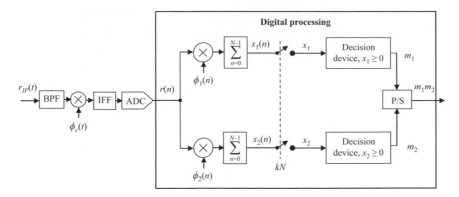

**Fig. 10.2.7** *A quadriphase-shift keying receiver with an intermediate-frequency analogue-to-digital converter, and a decision device in the I and Q branches; ADC, analogue-to-digital converter; BPF, bandpass filter; IFF, intermediate-frequency filter; LUT, look-up table. P/S, parallel-to-serial converter.*

intermediate-frequency carriers and the demodulation procedure are executed in the discrete-time domain using digital technology. Therefore, it is easy to achieve the same gain in both the $I$ and the $Q$ branches and maintain the phase angle of $\pi$ between the orthonormal local carriers, which can be achieved with the desired precision following the principles of signal processing presented in Chapters 1, 13, and 15. The demodulators are implemented using arithmetic multipliers with the same gain in both branches, which alleviates the problem with the analogue demodulators presented in Figure 10.2.3. The problem of the $I$ and $Q$ imbalances is eliminated.

There are some disadvantages in the discrete-time design of the intermediate-frequency blocks of a transceiver. First of all, discrete-time processing at a higher clock rate increases the processing power and the cost of the devices. However, if we reduce the clock rate, we risk reducing the precision in quantizing the signals, which will increase the bit error probability in the system. At the same time, this reduction will degrade the dynamic range of the signals.

### 10.2.2.3 Practical Design of a Discrete Quadriphase-Shift Keying Receiver at Intermediate Frequency

The advancement of digital technology allowed faster sampling rates and the possibility of discrete-time signal processing inside transceiver blocks at intermediate frequencies, subsequently leading to the second generation of receiver designs. A block scheme of a receiver with discrete-time signal processing in digital technology is presented in Figure 10.2.8. Analogue-to-digital conversion is performed by means of an analogue-to-digital converter, which converts the intermediate-frequency continuous-time signal $r(t)$ to a discrete-time signal. The discrete signal is applied to the inputs of two correlation demodulators to produce the discrete quantized samples $x_1$ and $x_2$ that contain information about the message bits sent from the source. A demodulation correlator is implemented as a multiplier and an accumulator. The quantized samples are transferred to the digital signal processor block to perform detection and estimation of the bits sent. From a theoretical point of view, these signals should be treated as discrete-time discrete-valued signals and processed according to the procedure used for noise processes in discrete-time systems in Chapter 4.

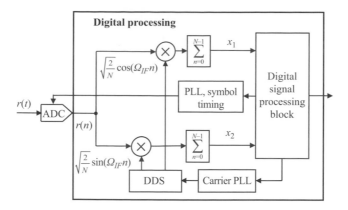

**Fig. 10.2.8** *A second-generation quadriphase-shift keying receiver with intermediate-frequency analogue-to-digital converters; ADC, analogue-to-digital converter; DDS, direct digital synthesizer; PLL, phase-locked loop.*

The sampling rate is controlled from a phase-locked symbol-timing loop. The discrete-time carrier, which is synchronized in frequency and phase with the incoming carrier, is generated at the output of the carrier phase-locked loop, as shown in Figure 10.2.8.

If the receiver is a part of a communication network, the detected bits are sent for further processing at the higher layers of the network. The additional processing is implemented in the digital signal processor block, which performs control functions needed to communicate data in the network.

## 10.3 Designing Quadrature Amplitude Modulation Transceivers

Practically speaking, the structure of a quadrature amplitude modulation transceiver is not so different from the structure of an $M$-ary phase-shift keying transceiver. As we have said repeatedly throughout the book, $M$-ary phase-shift keying modulation methods can be considered to be simplified quadrature amplitude modulation methods where the variability of the phases is preserved and the variability of amplitudes is excluded. For that reason, we will not spend too much time on the design of quadrature amplitude modulation transceivers but will mention their specific properties with respect to their quadriphase-shift keying counterparts.

### 10.3.1 Quadrature Amplitude Modulation Systems with Digital-to-Analogue and Analogue-to-Digital Conversion of Modulating Signals

In Chapter 7, we presented the block schematic of a discrete quadrature amplitude modulation communication system and precise mathematical models for the transmitter and the receiver. A discrete communication system based on quadrature amplitude modulation and the related discrete-time signal processing performed at intermediate frequencies is shown in Chapter 7,

Figure 7.5.1. In this section, we will analyse the design of the transmitters and receivers of this system, pointing out, in particular, the place and roles of the analogue-to-digital and digital-to-analogue converters, as we did for quadriphase-shift keying systems. In this context, we will analyse the case when analogue-to-digital and digital-to-analogue conversions are performed on baseband signals and then the case when these conversions are performed in the intermediate-frequency modulation block. The block schematics of quadrature amplitude modulation systems and quadriphase-shift keying systems are very similar, and some of the issues noted for the operation of quadriphase-shift keying systems will be the same as for that of quadrature amplitude modulation systems and, as such, will be just noted without detailed explanation.

### 10.3.1.1   *Quadrature Amplitude Modulation Transmitters with Baseband Discrete-Time Signal Processing*

The structure of a quadrature amplitude modulation transmitter presented here is based on both the diagram in Chapter 7, Figure 7.5.1, and its simplification presented in Figure 7.5.5. We added some new blocks that it is important to mention in the design of quadrature amplitude modulation transmitters. Analogue-to-digital converters are placed inside the in-phase ($I$) and quadrature ($Q$) branches of the transmitter, as presented in Figure 10.3.1.

The basic operation of the baseband block up to the input of the digital-to-analogue converter is explained in Chapter 7, Section 7.5.1. We added here two blocks: the up-sampling block and the interpolation filter block. The amplitudes of the modulating signals in both the $I$ and the $Q$ branches will depend on the level of amplitudes $A_i$ and $B_i$ generated at the output of the look-up table blocks. If we analyse a general case when the input of a digital-to-analogue converter is a discrete-time rectangular pulse, the output of the converter will be a continuous-time rectangular pulse. Therefore, the modulation procedure will follow the

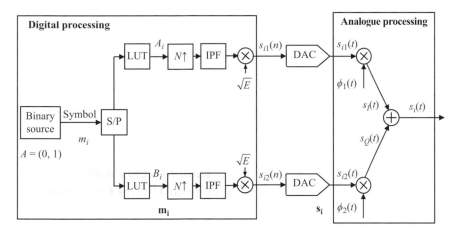

**Fig. 10.3.1** *A quadrature amplitude modulation transmitter with baseband digital-to-analogue converters; DAC, digital-to-analogue converter; IPF, interpolation reconstruction filter; LUT, look-up table; S/P, serial-to-parallel converter.*

explanation presented in Chapter 6, Section 6.5, where the operation of a digital quadrature amplitude modulation transmitter is explained.

Digital-to-analogue converters are implemented in both the *I* and the *Q* branches of the transmitter where the low-frequency baseband signal is processed. The advantages and disadvantages of the design are equivalent to their counterparts for the quadriphase-shift keying systems presented in Section 10.2 and will not be repeated here. Furthermore, as we said for quadriphase-shift keying systems, due to the development of digital technology, that is, the development of digital-to-analogue and analogue-to-digital converters with a higher clock rate, this particular design became obsolete and an intermediate-frequency digital design became possible in the discrete-time domain.

### *10.3.1.2  Designing a Quadrature Amplitude Modulation Receiver*

In the design of a baseband digital transmitter, the received modulated signal is presented in the continuous-time domain. It can be demodulated using the correlation demodulator and optimum detector implemented in digital technology, as presented in Chapter 6, Figure 6.5.1. As can be seen in the figure, the received signal is demodulated using two correlation demodulators that produce the demodulated vector **x**, which has two elements, in this case, the samples $x_1$ and $x_2$. The amplitudes of these samples are discretized using a sampler, converted into digital form, and processed in an optimum detector to obtain the message symbol $m_k$ that is the optimal estimate of the message symbol $m_i$ sent from the transmitter source. The number of branches in the optimum detector depends on the number of possible symbols generated in either the *I* or the *Q* branch of the modulator.

In addition to the receiver operation discussed above, the demodulation and processing of the received signal in its baseband can be performed via discrete-time signal processing. In this case, the receiver may be designed as presented in Figure 10.3.2. The received signal $r(t)$ is processed as for the quadriphase-shift keying receiver case; as already explained, to obtain the demodulated sample values of the received signal, $x_1$ and $x_2$, we can use a discrete-time matched filter that is matched to the input modulating signal. The output of the matched filter

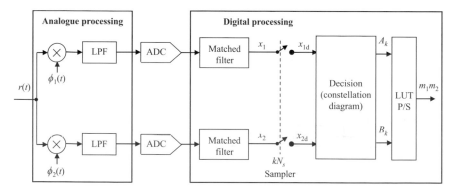

**Fig. 10.3.2** *A quadriphase-shift keying receiver with baseband analogue-to-digital converters; ADC, analogue-to-digital converter; LPF, lowpass filter; LUT, look-up table; P/S, parallel-to-serial converter.*

is a discrete-time signal with random value as at the time instants defined by the duration of the bits inside the $I$ and the $Q$ branches.

The analogue-to-digital converter and the digital matched filter operate as in the quadriphase-shift keying system explained in Section 10.2. The output of each filter is a sample value of the symbols transmitted affected by the noise in the channel. The decision for the digitalized samples $x_{1d}$ and $x_{2d}$ is performed using the constellation diagram to get the estimates $A_k$ and $B_k$, which are generated at the output of the decision circuit that operates using the constellation diagram. The detected symbols, denoted $m_1$ and $m_2$, are generated in parallel using a look-up table and then transferred to a series of bits at the output of the receiver by means of parallel-to-series processing.

The disadvantages of this design are equivalent to the ones mentioned in the design of the quadriphase-shift keying receiver and will not be repeated here.

### 10.3.1.3 Practical Design of a Quadrature Amplitude Modulation Receiver

The design presented here for a quadrature amplitude modulation transceiver is based on analogue-to-digital and digital-to-analogue conversions of the baseband signals and belongs to the first generation of discrete transceivers. The block schematic of the receiver is similar to the design of the quadriphase-shift keying receiver, including the positions of the blocks for the system clock and carrier generation, as presented in Figure 10.3.3.

The system includes a front-end receive processing circuit for the processing of the received signal $r(t)$ in the continuous-time domain, and has the disadvantages mentioned before for quadriphase-shift keying systema. After demodulation, the signal is transferred in the discrete-time domain by means of analogue-to-digital conversion to perform digital processing that includes a decision device in the digital signal processing block. This block also implements functions related to the synchronization of the system, and control and communications functions with the other blocks of the receiver. As shown in Figure 10.3.3,

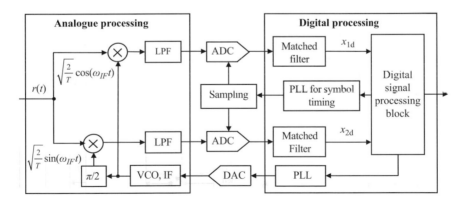

**Fig. 10.3.3** *Design of a first-generation quadrature amplitude modulation receiver with baseband analogue-to-digital converters; ADC, analogue-to-digital converter; DAC, digital-to-analogue converter; IF, intermediate-frequency modulator block; LPF, lowpass filter; PLL, phase-locked loop; VCO, voltage-controlled oscillator.*

this block generates the control signals necessary for the phase-locked loop in the discrete-time domain to be transferred into the continuous-time domain to control the voltage-controlled oscillator and synchronize the frequency and phase of locally generated carriers with the carriers of the incoming intermediate-frequency signal. The synchronization of the carriers is performed in the continuous-time domain. In addition, the digital signal processing block controls the symbol-timing phase-locked loop used to adjust the frequency and phase of the sampling instants in the system.

## 10.3.2 Quadrature Amplitude Modulation Systems with Digital-to-Analogue and Analogue-to-Digital Conversion of Intermediate-Frequency Signals

### 10.3.2.1 Digital Design of a Quadrature Amplitude Modulation Transmitter at Intermediate Frequency

We will present here a quadrature amplitude modulation system in which discrete-time processing is performed in both the baseband block and the intermediate-frequency block. Therefore, the digital-to-analogue and analogue-to-digital converters are placed at the output the input, respectively, of the intermediate-frequency block.

A quadrature amplitude modulation discrete communication system was already studied in Chapter 7, Section 7.5, assuming that discrete-time processing is performed inside both the baseband blocks and the intermediate-frequency modulator and demodulator blocks, as presented in Chapter 7, Figure 7.5.1. In this section, we will refer to Figure 7.5.1, and address the structures of the transmitter and the receiver from the design point of view.

The structure of the transmitter, which has a digital-to-analogue converter at the output of the modulator, is presented in Figure 10.3.4. As in the equivalent quadriphase-shift

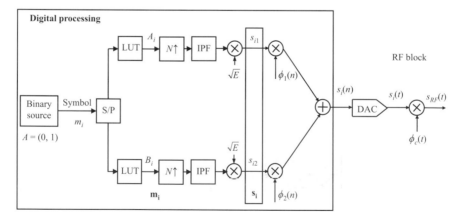

**Fig. 10.3.4** *A quadrature amplitude modulation transmitter with an intermediate-frequency digital-to-analogue converter; DAC, digital-to-analogue converter; IPF, interpolation reconstruction filter; LUT, look-up table; RF, radiofrequency; S/P, serial-to-parallel converter.*

keying transmitter, baseband and intermediate-frequency signal processing is performed in the discrete-time domain and implemented in digital technology.

Therefore, all of the disadvantages of the design of the receiver presented in Figure 10.3.1, where digital-to-analogue conversion is performed on the baseband signal, and intermediate-frequency modulation is implemented in the continuous-time domain, are eliminated, as for the case of quadriphase-shift keying systems. As in quadriphase-shift keying systems, the intermediate-frequency discrete-time signal $s_i(n)$ is converted into the continuous-time signal $s_i(t)$ that is applied to the input of a mixer operating at the radiofrequency $\phi_c(t)$ of the transmitter.

### 10.3.2.2 *Digital Design of a Quadrature Amplitude Modulation Receiver at Intermediate Frequency*

The block schematic of a receiver that is designed in digital technology at the level of intermediate-frequency discrete-time processing is presented in Figure 10.3.5. If the signal is transmitted at the radiofrequency frequency $\phi_c(t)$, then, at the input of the receiver, there is a mixer that down-converts the signal to an intermediate frequency that is filtered by an intermediate-frequency filter. This processing is performed in analogue technology. Then, the analogue intermediate-frequency signal is converted to a discrete-time intermediate-frequency signal that is processed in an intermediate-frequency demodulator by means of an analogue-to-digital converter. An intermediate-frequency correlation demodulator is implemented in digital technology, and the discrete amplitude correlation samples $x_{1d}$ and $x_{2d}$ are processed in the decision device to generate estimates of the bits sent.

The decision device can be designed as an optimum detector, as presented in Figure 10.3.5, and the signal processing that leads to the estimation of the message bits can be performed as explained in Chapter 7, Section 7.5.

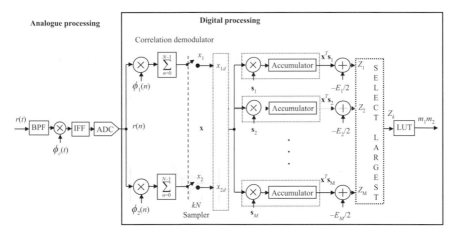

**Fig. 10.3.5** *A quadrature amplitude modulation receiver with an intermediate-frequency analogue-to-digital converter and an optimum detector; ADC, analogue-to-digital converter; BPF, bandpass filter; IFF, intermediate-frequency filter; LUT, look-up table.*

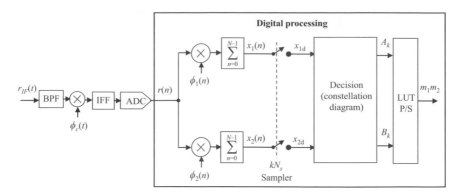

**Fig. 10.3.6** *A quadrature amplitude modulation receiver with an intermediate-frequency analogue-to-digital converter ADC, analogue-to-digital converter; BPF, bandpass filter; IFF, intermediate-frequency filter; LUT, look-up table; P/S, parallel-to-serial converter.*

The intermediate-frequency demodulator is implemented in digital technology, and the correlation samples obtained, $x_{1d}$ and $x_{2d}$, are processed in the decision device to generate estimates of the received bits. Then, digital signal processing is done on the discrete-time and discrete-valued signals. The receiver using this decision procedure is presented in Figure 10.3.6.

By using this design for the receiver, most of the disadvantages related to the design with analogue-to-digital converters in the $I$ and $Q$ branches are eliminated, as was the case with the quadriphase-shift keying system, as presented in Section 10.2.2.

### 10.3.2.3  *Practical Design of a Discrete Quadrature Amplitude Modulation Receiver at Intermediate Frequency*

As we pointed out for quadriphase-shift keying systems, the advancement of digital technology allowed faster sampling rates and the possibility of discrete-time signal processing inside the transceiver blocks at intermediate frequencies. A block scheme of a receiver, with discrete-time signal processing in digital technology, is presented in Figure 10.3.7. Analogue-to-digital conversion is performed by means of an analogue-to-digital converter that converts the intermediate-frequency continuous-time signal $r(t)$ to the discrete-time signal $r(n)$. The discrete signal is applied to the inputs of two correlation demodulators, in the $I$ and $Q$ branches, to produce the discrete quantized samples $x_{1d}$ and $x_{2d}$ that contain the information about the message symbols sent from the source. Each demodulation correlator is implemented as a multiplier and an accumulator. The quantized samples are transferred to the digital signal processor block to perform the detection and estimation of the received bits. From the theoretical point of view, the signals need to be treated as discrete-time discrete-valued stochastic signals and processed according to the procedure mentioned in Chapters 3 and 4.

The sampling rate of the analogue-to-digital converter is controlled from the phase-locked loop used for symbol timing. The discrete-time carrier, which is synchronized in frequency

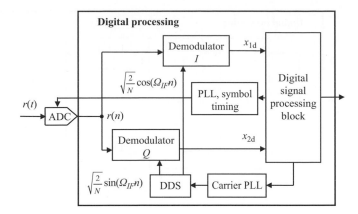

**Fig. 10.3.7** *A second-generation quadrature amplitude modulation receiver with intermediate-frequency analogue-to-digital converters; ADC, analogue-to-digital converter; DDS, direct digital synthesizer; PLL, phase-locked loop.*

and phase with the incoming carrier, is controlled from the carrier phase-locked loop. It is generated at the output of the direct digital synthesizer, as shown in Figure 10.3.7.

If the receiver is a part of a communication network, the detected bits are sent for further processing in the higher layers of the network. The additional processing is implemented in the digital signal processor block, which performs control functions that are necessary for communicating data in the network.

## 10.4 Overview of Discrete Transceiver Design

### 10.4.1 Introduction

When designing a discrete communication system, it is important to note that its structure in digital technology is different from the theoretical structure of the corresponding digital or discrete communication system. However, a good understanding of the theoretical model and its mathematical presentation remains a prerequisite for a good design. In addition, a good understanding of the digital technology, its capabilities, its design requirements, and the development tools required for prototyping are essential to optimize transceiver production. Therefore, to manufacture a transceiver for modern digital technology, it is necessary to have both a theoretical understanding of the modelling and simulation of discrete systems and the theoretical knowledge and practical skills needed for the design and prototyping of communication devices in modern technology.

A modern design for a discrete-time receiver implemented in digital technology can have the structure presented in Figure 10.4.1. The design is normally based on implementing processing functions in digital signal processors and/or field-programmable gate arrays, which depend on the system requirements, primarily, the sampling rate (related to speed) of the digital devices used. In principle, functions and algorithms that require a high rate of digital signal processing are used in devices using field -programmable gate array devices,

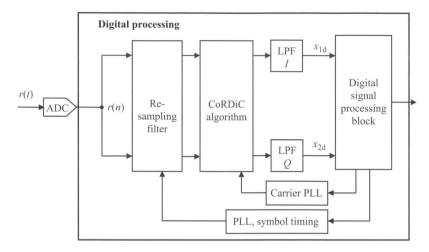

**Fig. 10.4.1** *Design of a third-generation quadrature amplitude modulation receiver with complete discrete signal processing of intermediate-frequency digital signals; ADC, analogue-to-digital converter; LPF, lowpass filter; PLL, phase-locked loop.*

while those with lower-rate functions and algorithms are implemented in devices using digital signal processors.

Furthermore, if it is necessary to develop transceivers for digital application-specific integrated circuits, such development will increase the overall price of the product. These designs are acceptable if the number of products needed is extremely high, to pay off the investment in all phases of their development, including their production. During the research, design, and development phase, we develop digital platforms, such as field -programmable gate arrays and digital signal processors, which will give us key evidence about vital properties of the devices under development and, even more, confirmation of the validity of our theoretical, simulation, and emulation results.

To illustrate the procedure of designing a system, an example of prototyping a code division multiple access communication system, based on the application of chaotic spreading sequences, is proposed as Project 5 in the Supplementary Material of this book. The prototyping is based on the use of Verilog and VHDL as register transfer languages, and an Altera DE2-115 board is used to verify the designs. In terms of software, Quartus is used to synthesize the designs, and ModelSim and Active-HDL are used for the simulation of the designs. The language Python is used to generate the data for various test benches for the simulation. Also, the Anaconda package is used, as it includes libraries required for generating the test benches.

## 10.4.2   Designing Quadrature Amplitude Modulation Systems

As we have already said, block schematics for modern transceivers do not follow the structure of the transmitter and the receiver used for their mathematical modelling and simulation. For example, the design in Figure 10.3.7 still partially follows the structure of the theoretical model and related mathematical description because the order of signal processing, starting with the demodulators and finishing with the baseband processing, is preserved in the block schematic of the digital design presented. This makes sense, because the principles of operation in the designed transceiver follow the mathematical model. However, the designed blocks and their structure do not necessarily strictly follow the mathematical model.

An advanced structure of a quadrature amplitude modulation system, designed using digital technology for third-generation systems, is presented in Figure 10.4.1. The system globally follows the structure of the discrete system presented in Figure 10.3.7, as will be seen from the explanation of its operation. The received analogue signal $r(t)$ is converted into the discrete-time domain by means of an analogue-to-digital converter. The converter allows the running clock to sample the signal and is followed by a resampling filter that is controlled by a symbol-timing phase-locked loop that, in turn, is controlled by the digital signal processor block.

The demodulators in Figure 10.3.7 are two new types of blocks: a resampling filter, and a CoRDiC block instead of a direct digital synthesizer block. Although using a direct digital synthesizer is a good way of generating sine and cosine signals with sufficient precision, in order to avoid the high complexity associated with it and its memory requirements, the CoRDiC algorithm can be used, as it simplifies the generation of orthogonal carriers and allows for the necessary multiplication of sine and cosine inside the correlator demodulators.

The digital signal processing block implements all of the baseband processing using the outputs of lowpass filters operating in the discrete-time domain and implemented in digital technology. In addition, this block controls the symbol-timing and the carrier phase-locked loops. The output of this receiver is composed of message bits that are generated in series for further processing in higher layers if the receiver is a part of a communication network.

# 11

# Deterministic Continuous-Time Signals and Systems

A *continuous-time deterministic* signal $x(t)$ is defined as a function that takes on any value in the continuous interval of its values $(x_a, x_b)$ for all values of the *continuous* independent variable $t$ taken in the interval $(a, b)$, where $a$ may be $-\infty$ and $b = \infty$. This does not mean that the signal $x(t)$ is represented by a continuous function of time or that this definition states only that the continuous-time function is defined on a continuum of time. Following this definition, the mathematical terms 'continuous function' and 'continuous-time function' mean different things.

## 11.1  Basic Continuous-Time Signals

The first set of continuous-time signals that will be analysed in this section are aperiodic signals, or singularity functions. These are functions that do not have continuous derivatives. The second set of signals that will be analysed in this section are complex exponentials and sinusoidal signals.

**Unit step function.** The unit step function is defined as

$$u(t) = \begin{cases} 0 & t < 0 \\ \text{not defined} & t = 0 \\ 1 & t > 0 \end{cases} \tag{11.1.1}$$

and shown in Figure 11.1.1.

The value of $x(t)$ at $t = 0$ is not defined, as there is discontinuity at this point. This is to say that this value is finite. It is irrelevant what the value of the function is at this point. Namely, any two signals which have finite values everywhere and differ in value only at a finite number of isolated points are equivalent in their effect on any physical system. Such signals are used for signal representation because they can mathematically represent the behaviour of real physical systems such as switching systems. For example, a step function is used to switch 12 V source if the input–output relationship is defined as

*Discrete Communication Systems.* Stevan Berber, Oxford University Press. © Stevan Berber 2021.
DOI: 10.1093/oso/9780198860792.003.0011

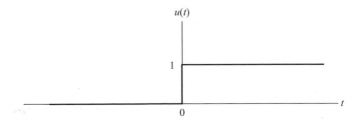

**Figure 11.1.1** *Unit step function.*

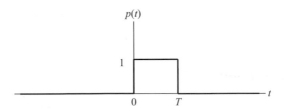

**Fig. 11.1.2** *Rectangular pulse.*

$$y(t) = 12 \cdot u(t) = \begin{cases} 0 \; V & t < 0 \\ \text{not defined} & t = 0 \\ 12 \; V & t > 0 \end{cases}.$$

**Rectangular pulse.** A *rectangular pulse* can be expressed as the difference of two unit step functions of the form

$$p(t) = u(t) - u(t - T), \tag{11.1.2}$$

where

$$u(t - T) = \begin{cases} 0, & t - T < 0, \text{or } t < T \\ 1, & t - T > 0, \text{or } t > T \end{cases}, \tag{11.1.3}$$

as shown in Figure 11.1.2. This is also defined as the switching function.

**Unit impulse function, or Dirac delta function.** Consider the continuous signal $u_\Delta(t)$ shown in Figure 11.1.3(a). The unit step function $u(t)$ is the limiting form of this function as $\Delta$ tends to 0, that is, $u(t) = \lim_{\Delta \to 0} u_\Delta(t)$. Define the delta function $\delta_\Delta(t)$ to be the derivative of the unit step function $u_\Delta(t)$, as shown in Figure 11.1.3(b):

$$\delta_\Delta(t) = \frac{du_\Delta(t)}{dt} = \begin{cases} \frac{1}{\Delta}, & 0 \leq t \leq \Delta \\ 0, & \text{otherwise} \end{cases}. \tag{11.1.4}$$

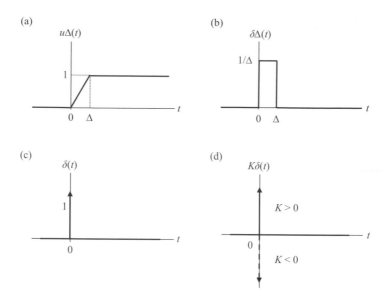

**Fig. 11.1.3** *Derivation of the delta function.*

The unit impulse (or delta, or Dirac delta) function is defined as the limiting function

$$\delta(t) = \lim_{\Delta \to 0} \delta_\Delta(t) = \frac{du(t)}{dt}. \tag{11.1.5}$$

However, when $\Delta$ tends to 0, the value of the delta function $\delta_\Delta(t)$ tends to infinity, and the area under this function is constant and equal to 1, as shown in Figure 11.1.3(c). Because the delta function is defined as a derivative of the unit step function $u(t)$, the unit step function $u(t)$ can be expressed as the running integral

$$u(t) = \int_{-\infty}^{t} \delta(\tau)\, d\tau, \tag{11.1.6}$$

which is the same as the relationship between the unit ramp and the unit step function.

Impulse delta functions are of great importance in the analysis of linear-time-invariant systems and their applications in the communication theory and practice. Delta functions will be widely used in the analysis and synthesis of communication systems in both the time and the frequency domains. For that reason, we will present some of their properties, which will be used in the theoretical analysis of linear-time-invariant systems.

1.  Magnitude scaling: if we multiply both sides of expression (11.1.5) by $K$, we obtain

$$K\delta(t) = K\frac{du(t)}{dt}, \tag{11.1.7}$$

which can be interpreted as the scaled impulse function, which is equal to the derivative of the scaled unit step function, as is shown in Figure 11.1.3(d) for $K > 0$ and $K < 0$.

2. Time scaling: time scaling by $a$ of the impulse function, that is, $\delta(at)$, is equal to the scaling of magnitude by $1/|a|$, that is,

$$\delta(at) = \frac{1}{|a|}\delta(t). \tag{11.1.8}$$

3. A more formal mathematical definition of the impulse function proceeds by defining it in terms of the *functional*

$$\int_{-\infty}^{\infty} x(t)\delta(t)dt = x(0), \tag{11.1.9}$$

or

$$\int_{-\infty}^{\infty} x(t)\delta(t-t_0)\,dt = x(t_0), \tag{11.1.10}$$

where $x(t)$ is continuous at $t = 0$, or $t = t_0$. Thus, the impulse function assigns the value $x(0)$ to any function $x(t)$ that is continuous at the origin, and the value $x(t_0)$ to any function $x(t)$ that is continuous at the time instant $t = t_0$.

4. Scaling by a value of the function, or *the equivalence property*: it can be proved that

$$x(t)\delta(t) = x(0)\delta(t),$$

or

$$x(t)\delta(t-\lambda) = x(\lambda)\,\delta(t-\lambda); \tag{11.1.11}$$

that is to say, the shifted delta function is simply scaled by the value of the function at $t = \lambda$. This characteristic is sometimes called *the equivalence property* of the unit impulse function.

5. The shifting property of the impulse function is expressed by the relation

$$\int_{-\infty}^{\infty} x(t)\delta(t-\lambda)\,dt \Rightarrow \int_{-\infty}^{\infty} x(\lambda)\,\delta(t-\lambda)\,dt = x(\lambda) \int_{-\infty}^{\infty} \delta(t-\lambda)\,dt = x(\lambda). \tag{11.1.12}$$

In this sense, an integral with a finite limit can be considered to be a special case by defining $x(t)$ to be 0 outside a certain interval, say $a < t < b$, which yields

$$\int_{a}^{b} x(t)\delta(t-\lambda)\,dt = \begin{cases} x(\lambda) & a < \lambda < b \\ 0 & \text{otherwise} \end{cases}.$$

Unlike the other functions, the delta function cannot be graphically represented, because its amplitude is undefined when its argument is 0. Due to this difficulty, the usual practice is to represent the impulse by a vertical arrow, as we did in Figure 11.1.3(c). The area under the impulse is called the 'strength' or the 'weight' of the impulse. In some way, we must add this strength to the graphical representation. It is commonly done in one of the following two ways. The first way is to write the strength beside the arrow representing the impulse, as can be seen in Figure 11.1.3(c) and (d). The second way is to draw a vertical arrow whose height corresponds to the weight of the impulse. Both ways are used in this book.

**Unit rectangular pulse function.** A unit rectangular pulse function is defined as

$$\Pi(t) = u\left(t + \frac{1}{2}\right) - u\left(t - \frac{1}{2}\right) = \begin{cases} 1, & |t| \leq \frac{1}{2} \\ 0, & |t| > \frac{1}{2} \end{cases} \tag{11.1.13}$$

and shown in Figure 11.1.4. We use the word 'unit' in its name because its width, height, and area are all equal to 1. This function represents a very common type of signal in physical systems and, in particular, in communication system. A scaled version of this signal is used as a modulating signal in digital communication systems. This function can also control a system that has to switch on at some time and then back off at a later time. In this respect, this signal can be thought as a gate function.

**Unit ramp function.** A unit ramp function is defined as

$$r(t) = \begin{cases} 0, & t < 0 \\ t, & t \geq 0 \end{cases}, \tag{11.1.14}$$

as shown in Figure 11.1.5. This function is used to represent a signal in real systems that is switched on at some time and then changes linearly after that time.

This function can be defined as an integral of the unit step function. It is called the unit ramp function because its slope is equal to 1, as can be seen from the following definition:

$$r(t) = \int_{-\infty}^{t} u(z)\, dz. \tag{11.1.15}$$

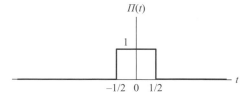

**Fig. 11.1.4** *A unit rectangular pulse.*

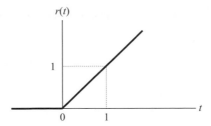

**Fig. 11.1.5** *A unit ramp function.*

Therefore, a unit step function can be considered to be the derivative of a unit ramp function.

**Complex exponential functions.** Complex exponential functions are widely used in communication systems as the carriers of message signals. The *complex exponential function w* of a complex variable $z$ is defined by

$$w(z) = Ae^z, \tag{11.1.16}$$

where $A$ and $z$ are generally complex numbers. For both $A$ and $z$ real, this function becomes the familiar real exponential function. As a function of time $t$, it can be expressed as the real-valued function

$$w(t) = Ae^{\alpha t}. \tag{11.1.17}$$

Then, if $\alpha > 0$, the function increases exponentially with time $t$, and is said to be a growing exponential. If $\alpha < 0$, then the function decreases exponentially with time $t$, and is said to be a decaying exponential. If $\alpha = 0$, the function is simply a constant.

A set of *harmonically related complex exponential signals* can be expressed in the form

$$\left\{ e^{jk\omega_0 t}, k = 0, \pm 1, \dots \right\} = \left\{ \dots, e^{-j\omega_0 t}, 1, e^{j\omega_0 t}, e^{j2\omega_0 t}, \dots \right\},$$

where $k$ takes on increasing values from $-\infty$ to $\infty$, and represents the *order* of the harmonics.

**Sinusoidal signals in the time domain.** A complex sinusoid is an exponential function, expressed in the form

$$s(t) = Ae^{j\omega_0 t} = A(\cos \omega_0 t + j \sin \omega_0 t), \tag{11.1.18}$$

and defined for real values of $\omega_0$. This function is periodic, which follows from Euler's formula

$$e^{j\omega_0 t} = \cos \omega_0 t + j \sin \omega_0 t. \tag{11.1.19}$$

A *complex sinusoid*, which has a complex-valued $A$ in the form $|A|e^{j\phi}$ and is expressed as

$$s(t) = Ae^{j\omega_0 t} = |A|e^{j\phi}e^{j\omega_0 t} = |A|e^{j(\omega_0 t+\phi)} = |A|[\cos(\omega_0 t+\phi)+j\sin(\omega_0 t+\phi)] = s(t+T), \tag{11.1.20}$$

is a periodic signal for all values of $t$ and some constant $T$. The smallest positive (non-zero) value of $T$ is said to be a (fundamental) period of the signal. The proof of the periodicity of the signal $s(t)$ can be left as an exercise.

A *real sinusoid* can be expressed in terms of the complex sinusoid as its real part,

$$s(t) = \text{Re}\left\{|A|e^{j(\omega_0 t+\phi)}\right\} = \text{Re}\{|A|\cos(\omega_0 t+\phi)+j|A|\sin(\omega_0 t+\phi)\} = |A|\cos(\omega_0 t+\phi), \tag{11.1.21}$$

or its imaginary part,

$$s(t) = \text{Im}\left\{|A|e^{j(\omega_0 t+\phi)}\right\} = \text{Im}\{|A|\cos(\omega_0 t+\phi)+j|A|\sin(\omega_0 t+\phi)\} = |A|\sin(\omega_0 t+\phi). \tag{11.1.22}$$

Two cosine function are shown in Figure 11.1.6 for two different values of the phase: $\phi = 0$, and $0 < \phi < \pi/2$. The ellipses on both sides of the graph indicate that the signal exists in an infinite interval $(-\infty, +\infty)$, which strictly follows the definition of a periodic function. In practice, we omitting the ellipses when we plot the graphs of periodic functions, but we anticipate the existence of the signal from $-\infty$ to $+\infty$. The above expressions for the signal are referred to as the *time-domain representations*.

The parameter $\omega_0$ is the *radian frequency* of the sinusoidal signal, which is expressed in radians per second, while the phase $\phi$ is simply expressed in radians. The parameter $\omega_0$ is also written as $2\pi f_0$, where $f_0$ is the *circular frequency* of the sinusoidal signal in hertz. The relationship between these parameters and the period $T$ of the signal is given by

$$\omega_0 = 2\pi f_0 = 2\pi/T. \tag{11.1.23}$$

**Phasor representation of sinusoidal signals.** The *complex sinusoid* (11.1.20) can be represented by one phasor in the complex plane. The real sinusoid defined in eqn (11.1.21) can also be expressed as the sum of two complex sinusoids in the form

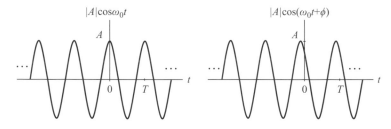

**Fig. 11.1.6** *Sinusoidal signals in the time domain.*

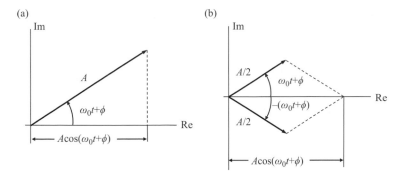

**Fig. 11.1.7** *Phasor representation of sinusoidal signals.*

$$s(t) = \frac{|A|}{2} \left[ e^{j(\omega_0 t + \phi)} + e^{-j(\omega_0 t + \phi)} \right], \tag{11.1.24}$$

which is a representation in terms of two complex conjugate, oppositely rotating phasors. These two presentations of the signal are illustrated schematically in Figure 11.1.7. The phasors are plotted for the assumed initial time instant $t = 0$. Physical systems always interact with real signals. However, the representation of real signals in terms of complex quantities is often mathematically convenient and necessary. Thus, the complex amplitude $A$ can be generally expressed as the phasor quantity

$$A = |A| e^{j\phi} = |A| \angle \phi. \tag{11.1.25}$$

The second representation assumes the existence of two conjugate, oppositely rotating phasors. This can be thought of as the sum of positive-frequency and negative-frequency rotating phasors. In the literature, there is an argument that states that negative frequencies do not physically exist but that the concept of negative frequencies is a convenient mathematical abstraction with a wide application in signal analysis. We cannot exactly agree with this. Namely, we can imagine a scenario where there are two sources of energy (generators) consisting of two oppositely rotating machines which constructively support each other. The question then is, which one of these machines operates with a positive frequency, and which one is a kind of mathematical construction? The fact, though, is that the machines must work synchronously, which is not a problem to implement. Furthermore, we can define the negative sense of phasors rotation and present the spectrum of an underlying signal only for the negative frequencies. Then, in this case, someone can claim that the positive frequencies represent the mathematical construction. It appears that the presentation of a signal using positive, negative, and both positive and negative frequencies is equally valid and has a basis in real systems.

**Representation of sinusoidal signals in the frequency domain.** In addition to representing sinusoidal signals in the time domain or with phasors, it is also possible to represent them in the frequency domain. Because the signal in the time domain is completely specified by $A, f_0$, and $\phi$, it can be completely specified by two relations: one showing the dependence of

(a) (b)

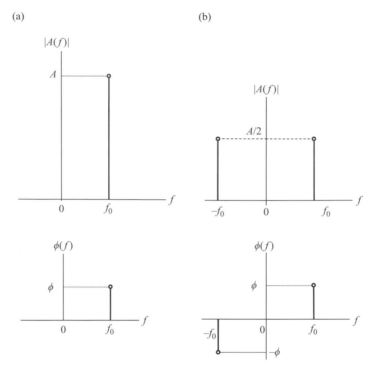

**Fig. 11.1.8** *Magnitude and phase spectra of sinusoidal signals: (a) one-sided spectra; (b) two-sided spectra.*

absolute values of amplitudes (or modulus or magnitudes) $|A|$ on the frequency $f = f_0$ (called the *magnitude spectrum*), and one showing the phases $\phi$ against the frequency (called the *phase spectrum*). This is called a *frequency domain representation*. The frequency domain representation of the phasors in Figure 11.1.7 is shown in Figure 11.1.8.

In the case when the signal is represented by one phasor, as shown in Figure 11.1.7(a), the magnitude spectrum depends only on a single frequency $f_0$, and the resulting plot consists of a single point or 'line' at $f = f_0$, as illustrated in Figure 11.1.8(a). The plot of magnitude versus frequency for this case is referred as a *one-sided magnitude spectrum* because it has points or lines only on positive frequencies. The phase spectrum is represented by one value at the discrete frequency $f_0$.

In the case when the signal is represented by two phasors that rotate in opposite directions, as shown in Figure 11.1.7(b), the corresponding plot consists of two points or 'lines' at $f = f_0$ and at $f = -f_0$, as illustrated in Figure 11.1.8(b). The plot of magnitude versus frequency for this case is referred as a *two-sided magnitude spectrum* because it has points or lines for both positive and negative frequencies. The plot of phase versus frequency is referred as a *two-sided phase spectrum*, because there are two points or lines for both positive and negative frequencies.

**Important note.** It is important to point out that the values of the phases cannot be arbitrarily changed to preserve symmetry with respect to the *y*-axis. If the positive rotation of the phasors is defined, then the phase angle for the positive frequencies should have a positive value defining the initial position of the phasor rotating in the positive, that is, the counter-clockwise direction. Consequently, the sign of the phases is strictly related to the assumed positive rotation of the phasors if we want to keep consistency between the definitions of frequency, phase, and amplitude. Furthermore, the 'negative' and 'positive' frequencies of a generator are equally 'natural'. In this context, the oppositely rotating phasors can be physically realized as two synchronous generators rotating in opposite directions with the same angular frequency and opposite initial phases, which produce the same voltages that are synchronously added. This point is essential for understanding magnitude and phase spectra for both continuous-time and discrete-time signals.

In the discussion above, we used a single sinusoidal signal. Had a sum of sinusoidal components with different frequencies and phases been presented, the spectral plot would have consisted of multiple lines. Thus, the signal spectrum would be in a defined band of frequencies, which would contain multiple lines, and the number of lines would correspond to the number of signals in the sum.

A two-sided magnitude spectrum has even symmetry about the origin, and a phase spectrum has odd symmetry. This symmetry is a consequence of the fact that the signal is real, which implies that the conjugate rotating phasors must be added to obtain a real quantity.

Precisely speaking, if the amplitudes *A* are exclusively shown as a function of the frequency, the spectrum is called an 'amplitude spectrum'. In contrast, the dependence of the magnitudes, defined as the absolute values (modulus) of amplitudes $|A|$, is called the 'magnitude spectrum'. In signal processing, we use the amplitude spectrum whenever possible. A typical case is when spectral components have $\pm\pi$ phase angles, that is, when they can have positive and negative values and we can plot the amplitude spectrum. However, generally speaking, the amplitude spectrum is expressed as a complex function. Therefore, we need to find expressions for both the magnitude spectra and the phase spectra in order to present the signal in the frequency domain.

## 11.2   Modification of the Independent Variable

Many simple continuous-time signal processing operations can be described mathematically via the modification of the independent variable *t*. This assumes that the functional of the function $x(t)$ does not change as the independent variable *t* is modified. However, the resulting functions are different. The modification can be time reversal, time scaling, time shifting, or combinations of these, as we will present here.

**Time reversal.** A function $y(t)$ is a time reversal of $x(t)$ if

$$y(t) = x(-t), \tag{11.2.1}$$

as depicted in Figure 11.2.1. Time reversal means flipping the signal $x(t)$ with the $x(t)$ axis as the rotation axis of the flip. The circle and the rectangle in Figure 11.2.1 indicate the

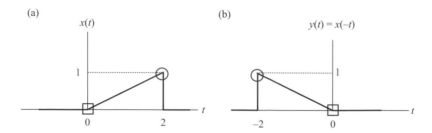

**Fig. 11.2.1**  *Time reversal.*

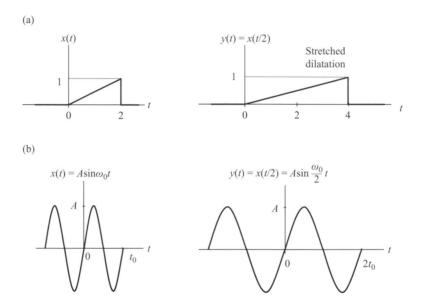

**Fig. 11.2.2**  *Time scaling: (a) illustration of time scaling; (b) time scaling of a sinusoidal signal.*

positions of the original signal and its modified version, respectively. Such a modification can be practically implemented by recording $x(t)$ on a tape recorder and then playing the tape backward.

**Time scaling.** A function $y(t)$ is a time-scaled version of $x(t)$ by a factor $\alpha$ if

$$y(t) = x(\alpha t). \tag{11.2.2}$$

If $\alpha > 1$, the function $y(t)$ is a time-compressed or speeded-up version of $x(t)$. If $\alpha < 1$, say, $\alpha = 1/2$, as in Figure 11.2.2(a), the function $y(t)$ is a time-stretched or slowed-down version of $x(t)$. The two characteristic points, in this case, can be found as

$$y(t) = \begin{cases} x(t/2) = 0 & t/2 = 0 \Rightarrow & t = 0 \\ x(t/2) = 1 & t/2 = 2 \Rightarrow & t = 4 \end{cases}.$$

Time scaling for a sinusoidal signal is, in fact, a change of its frequency, as shown in Figure 11.2.2(b).

## Example

Let $x(t)$ be the periodic signal shown in Figure 11.2.3. It is applied to the input of a system that does time scaling and produces the signal

$$y(t) = x(t/2).$$

Plot the graph of the signal at the output of the system.

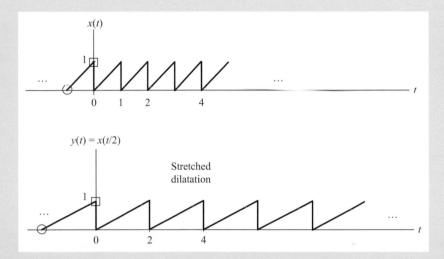

**Figure 11.2.3** *An example of time scaling.*

**Solution.** The input signal is a periodic signal. The output signal is supposed to be a periodic signal with a different period. The two characteristic points in this case, indicated respectively by the circle and the rectangle, can be found for $t = -1$ and 0 as

$$y(t) = \begin{cases} x(t/2) = 0 & t/2 = -1 \Rightarrow & t = -2 \\ x(t/2) = 1 & t/2 = 0 \Rightarrow & t = 0 \end{cases}.$$

For another two points, say, $t = 1$ and $t = 2$, we obtain

$$y(t) = \begin{cases} x(t/2) = 0 & t/2 = 1 \Rightarrow & t = 2 \\ x(t/2) = 1 & t/2 = 2 \Rightarrow & t = 4 \end{cases}.$$

With these two points, we can reconstruct the whole periodic signal $y(t)$, as shown in Figure 11.2.3.

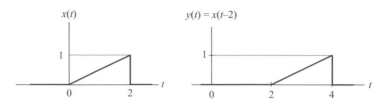

**Fig. 11.2.4** *Time shifting.*

**Time shifting.** A function $y(t)$ is a time-shifted version of $x(t)$, shifted by a factor $\beta$, if

$$y(t) = x(t + \beta). \tag{11.2.3}$$

If $\beta > 0$, the function $y(t)$ is a left-shifted version of $x(t)$. If $\beta < 0$, the function $y(t)$ is a right-shifted version of $x(t)$, as shown in Figure 11.2.4 for $\beta = -2$. However, shifting does not necessarily have to occur in the time domain. If the independent variable is for space or frequency, we will have space shifting or frequency shifting, respectively.

For two characteristic points of the signal $x(t)$, we may find the two corresponding values for signal $y(t)$ as follows:

$$y(t) = \begin{cases} x(t-2) = 0 & t-2 = 0 \Rightarrow & t = 2 \\ x(t-2) = 1 & t-2 = 2 \Rightarrow & t = 4 \end{cases}.$$

**Magnitude scaling.** The function $y(t)$ is the magnitude-scaled version of $x(t)$, scaled by a factor $A$, if

$$y(t) = Ax(t). \tag{11.2.4}$$

For an arbitrary instant of time $t$, this function multiplies the returned value $x(t)$ by the factor $A$. If $A > 1$, the function $y(t)$ is an amplified version of $x(t)$, represented by the dashed line in Figure 11.2.5. If $A < 1$, the function $y(t)$ is an attenuated version of $x(t)$, represented by the dot-dashed line in Figure 11.2.5. For $A < 0$, the function $y(t)$ is a 180° phase-shifted version of $x(t)$, represented by the dotted line in Figure 11.2.5.

To see clearly that the phase shift for $A < 0$ is $\pi$, we can express the signal in this form $y(t) = Ax(t) = |A|e^{j\pi}x(t)$. Therefore, the negative scaling factor $A$ flips the function $x(t)$ with respect to the $t$-axis. In this case, the scaled function can be expressed in the form

$$y(t) = Ax(t) = |A| [-x(t)],$$

which can be interpreted as flipping $x(t)$ with respect to the abscissa and then scaling it by the factor $|A|$. In this context, this modification can be interpreted as a modification of the dependent variable $x$.

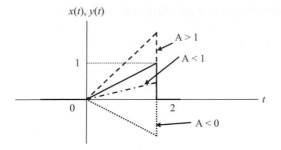

**Fig. 11.2.5** *Magnitude scaling.*

## 11.3   Combined Modifications of the Independent Variable

Generally speaking, the independent variable $t$ is modified using a linear transformation. Thus, the general expression, containing all the above modifications, is

$$y(t) = Ax(\alpha t + \beta), \tag{11.3.1}$$

which can also be expressed as

$$y(t) = Ax[\alpha(t + \beta/\alpha)], \tag{11.3.2}$$

where

$A$ represents the amplitude scaling, where

$A > 1$ is amplification,
$A < 1$ is attenuation, and
$A < 0$ is phase reversal;

$\alpha$ represents time scaling, where

$\alpha > 1$ is compression or speeding up,
$\alpha < 1$ is dilatation or stretching, and
$\alpha < 0$ is time reversal; and

$\beta/\alpha$ represents time shifting, where

$\beta/\alpha > 0$ is shift to the left, and
$\beta/\alpha < 0$ is a shift to the right.

## 11.4 Cross-Correlation and Autocorrelation

### 11.4.1 The Cross-Correlation Function

In the theory and practice of signal processing in communication systems, we often compare different signals or a given signal by itself under specified conditions. We are looking at signals' similarities or self-similarities. However, these similarities should have a meaningful measure, a kind of quantitative expression that can be compared and mathematically manipulated. The correlation function is a 'measure' of the similarity of two signals, or a measure of the linear dependence of one function on another. For example, in modulation theory, a signal received in a receiver is correlated with the locally generated signal to find their similarities. In this section, we will address the problem of correlation, present a definition of the cross-correlation function, and demonstrate the procedure of its calculation.

The cross-correlation of the energy signals $x(t)$ and $y(t)$, as the measure of their similarity, is defined as

$$R_{xy}(\tau) = \int_{-\infty}^{\infty} x(t+\tau) y(t) dt, \text{ or, equivalently, } R_{xy}(\tau) = \int_{-\infty}^{\infty} x(t-\tau) y(t) dt, \qquad (11.4.1)$$

because we are comparing them in the whole interval of possible values from $-\infty$ to $+\infty$. This function can be obtained in the following way:

1. Shift $x(t)$ by $\tau$ to obtain $x(t + \tau)$.
2. Multiply the results by $y(t)$ to obtain $x(t + \tau) y(t)$.
3. Integrate the product as shown in eqn (11.4.1) and then make a conclusion based on following rationale:

   a. If the integral value is small (or 0) for a particular value of $\tau$, then the waveforms of $x(t)$ and $y(t)$ are quite dissimilar from that for $\tau$.
   b. If the integral value is large for a particular value of $\tau$, then the waveforms of $x(t)$ and $y(t)$ are quite similar to that for $\tau$. In other words, in this case, the intervals of partial or substantial overlapping of $x(t)$ and $y(t)$ are significant.

### 11.4.2 The Autocorrelation Function

A signal in real systems can have various wave shapes that can change substantially in a long time interval. The question is what the similarity is between this waveform and its advanced or delayed version. This similarity can be measured using the autocorrelation function. The autocorrelation function is a special case of the correlation function. Therefore, we say that the autocorrelation function is a 'measure' of the self-similarity of a signal. For an energy signal $x(t)$, the autocorrelation is defined by the integral

$$R_x(\tau) = \int_{-\infty}^{\infty} x(t+\tau) x(t) dt, \qquad (11.4.2)$$

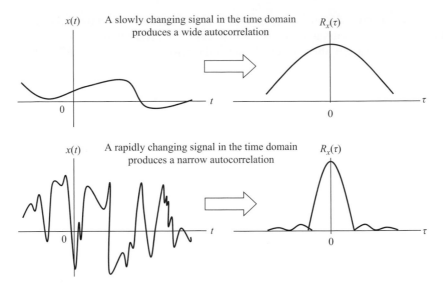

**Figure 11.4.1** *Dependence of the autocorrelation function on the shape of the signal in the time domain.*

or, equivalently,

$$R_x(\tau) = \int_{-\infty}^{\infty} x(t-\tau)\,x(t)dt. \tag{11.4.3}$$

Therefore, the autocorrelation function is obtained when a signal is cross-correlated by itself. If a signal is 'smooth' in the time domain, then its autocorrelation function is 'wide'. However, if the signal is 'rough' in the time domain, then its autocorrelation is 'narrow'. Both cases are demonstrated in Figure 11.4.1. Thus, using the autocorrelation function, we can make conclusions about the frequency contents of the signal in the time domain.

**Properties of the autocorrelation function.** The value $R_x(0)$ is equal to the total energy of the signal:

$$R_x(0) = \int_{-\infty}^{\infty} |x(t)|^2 dt = E_x. \tag{11.4.4}$$

The maximum value of the autocorrelation function is $R_x(0)$, that is,

$$R_x(0) \geq R_x(\tau). \tag{11.4.5}$$

The function is even, that is,

$$R_x(-\tau) = R_x(\tau). \tag{11.4.6}$$

This can be easily proved by using the definition of the autocorrelation function, changing the variable $t' = t + \tau$, and solving the correlation integral with respect to $x$.

For a power signal $x(t)$, the autocorrelation at the zero time shift is defined by the limit

$$R_x(0) = \lim_{a \to \infty} \frac{1}{2a} \int_{-a}^{a} x(t)^2 dt = P_x \tag{11.4.7}$$

that is equal to the average power of the signal.

**Example**

Compute the autocorrelation function for the signal

$$x(t) = A \cdot u(t + T/2) - A \cdot u(t - T/2), \tag{11.4.8}$$

which is shown in Figure 11.4.2.

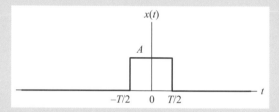

**Fig. 11.4.2**  *Presentation of the signal in the time domain.*

**Solution.** The general approach for finding the correlation function is to distinguish the intervals of $\tau$ values where the integrand is 0 from the intervals where the integrand is different from 0. In this case, we have two intervals:

- $\tau < -T$, and $\tau > T$, where the correlation integral is 0, and
- $-T < \tau < T$, where the integrand is different from 0 and the function depends on $\tau$.

For positive $\tau$ in the interval $-T < \tau < T$, there can be two cases, as is shown in Figure 11.4.3.

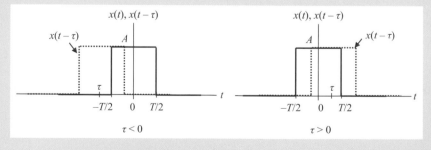

**Fig. 11.4.3**  *The signal and its shifted versions.*

1. The first case is when $\tau < 0$, as shown on the left-hand side of Figure 11.4.3, and we may calculate the correlation integral as

$$R_x(\tau) = \int_{-\infty}^{\infty} x(t-\tau)\, x(t) dt = \int_{-T/2}^{\tau+T/2} A^2 dt = A^2(T+\tau). \tag{11.4.9}$$

2. The second case is when $\tau > 0$, that is, $0 < \tau < T$, as shown on the right-hand side of Figure 11.4.3, so we obtain

$$R_x(\tau) = \int_{-\infty}^{\infty} x(t-\tau)\, x(t) dt = \int_{\tau-T/2}^{T/2} A^2 dt = A^2(T-\tau). \tag{11.4.10}$$

Therefore, the autocorrelation function can be expressed in the form

$$R_x(\tau) = \begin{cases} A^2(T-\tau) & \tau > 0 \\ A^2(T+\tau) & \tau < 0 \end{cases} = A^2 T\left(1 - \frac{|\tau|}{T}\right). \tag{11.4.11}$$

and presented graphically, as shown in Figure 11.4.4. This autocorrelation function will be widely used in the theory of digital communication systems for the calculation of the energy spectral density and the design of the correlation demodulators.

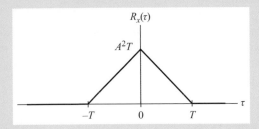

**Fig. 11.4.4**  *The autocorrelation function.*

## 11.5   System Classification

**Systems with memory.** Let the input and output signals of a system be $x(t)$ and $y(t)$, respectively. A system is said to *have memory* if its output $y(t) = y(t_0)$, at an arbitrary time $t = t_0$, depends on an input value other than, or in addition to, $x(t_0)$. The additional input value needed to compute $y(t_0)$ may be in the time intervals $t < t_0$ or $t > t_0$. This system is sometimes called a *dynamic system*. A system is said to be *memoryless* if $y(t_0)$ depends only on $x(t_0)$. Such a system is sometimes called a *static system* or a *zero-memory system*. We deal with systems with memory and memoryless systems in both modulation and coding theory.

**Causal systems.** A system is causal or non-anticipatory if its response $y(t)$, at an arbitrary time instant $t = t_0$, say, $y(t) = y(t_0)$, depends only on the input value $x(t)$ for $t \leq t_0$. Thus, all physical systems in the real world are causal, due to fact they cannot anticipate the future.

All memoryless systems are causal, but not vice versa, that is, a causal system can (and usually does) have memory.

**Linear systems.** A system is said to be linear if the superposition condition is fulfilled: Given that input $x_1(t)$ produces output $y_1(t)$, and input $x_2(t)$ produces output $y_2(t)$, then the linear combination of inputs

$$x(t) = ax_1(t) + bx_2(t) \tag{11.5.1}$$

must produce the linearly combined output

$$y(t) = ay_1(t) + by_2(t), \tag{11.5.2}$$

for the arbitrary complex scaling constants $a$ and $b$. The superposition contains two requirements: *additivity* and *homogeneity*, or *scaling*, which are expressed, respectively, as follows:

1. Given that input $x_1(t)$ produces output $y_1(t)$, and input $x_2(t)$ produces output $y_2(t)$, the input $x_1(t) + x_2(t)$ produces output $y_1(t) + y_2(t)$ for arbitrary $x_1(t)$ and $x_2(t)$.
2. Given that input $x(t)$ produces output $y(t)$, the scaled input $ax(t)$ must produce the scaled output $ay(t)$ for arbitrary $x(t)$ and complex constant $a$.

**Time-invariant systems.** A system is said to be time invariant or fixed if a shift (delay or advance) in the input signal $x(t)$ causes a similar shift in the output signal. In other words, if $x(t)$ produces $y(t)$, then $x(t - t_0)$ produces $y(t - t_0)$ for any value of $t_0$. The procedure for testing time invariance is shown schematically in Figure 11.5.1.

Decisions are made as follows:

- If $y_{t0}(t) = y(t - t_0)$, then the system is time invariant.
- If $y_{t0}(t) \neq y(t - t_0)$, then the system is *not* time invariant.

In other words, we are saying that the system is time invariant if its input–output relationship does not change with time. Otherwise, the system is time varying.

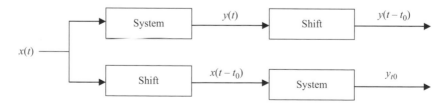

**Fig. 11.5.1** *Schematic for calculating time-invariant systems.*

**Stable systems.** One of the most important considerations in any system is the question of its stability. The stability can be defined in various ways. We will use the definition of bounded-input/bounded-output stability. An input signal $x(t)$ and output signal $y(t)$ are bounded if $|x(t)| \leq A$, and $|y(t)| \leq B$. A system is stable in bounded-input/bounded-output stability if any bounded input $|x(t)| \leq A$, for some finite constant $A$, produces a bounded output $y(t)$ satisfying $|y(t)| \leq B$ for some finite constant $B$.

## 11.6 Continuous-Time Linear-Time-Invariant Systems

### 11.6.1 Modelling of Systems in the Time Domain

In communication systems, we model and analyse a transceiver's signal processing blocks as systems. A communication channel itself can be understood as a system. The majority of transceiver blocks can be treated as time-invariant systems that are subject of the analysis in this section. The analysis will be conducted here in the time domain; it will extended into the frequency domain in Chapter 12.

In the time domain, a system can be modelled in three ways:

1. as an instantaneous input–output relationship,

2. via constant coefficient ordinary differential equations, and

3. via a convolution integral, which is of a particular importance in communication theory.

In this section we will present a method for describing a *linear-time-invariant* system, using the convolution integral and the concept of the impulse response. Let a linear-time-invariant system be represented by its input signal $x(t)$, the output signal $y(t)$, and its impulse response $h(t)$, as shown in Figure 11.6.1.

The problem of interest in this section is how to describe a linear-time-invariant device in order to find the output signal as a function of the input signal. For this purpose, the concept of the impulse response of the linear-time-invariant system will be used. In this context, the input signal will be represented using the impulse delta function, and the output signal will be represented using the response to the impulse delta function.

### 11.6.2 Representation of an Input Signal

A continuous-time input signal $x(t)$ can be represented as a continuum of impulse delta functions. In other words, we can decompose the signal $x(t)$ into an infinite uncountable set of delta functions using the following procedure. For the defined unit-area pulse

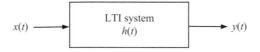

**Fig. 11.6.1** *Basic representation of a linear-time-invariant system; LTI, linear-time-invariant.*

$$\delta_\Delta(t) = \frac{1}{\Delta}\left[u(t) - u\,(t - \Delta)\right], \tag{11.6.1}$$

a unit-amplitude pulse is

$$\delta_\Delta(t) \cdot \Delta = u(t) - u\,(t - \Delta).$$

The factor $\Delta$ is included in the expression in order for the delta function to normalize the height of the pulse $\delta_\Delta(t)\Delta$ to unity, so that the *unit* rectangular pulse function can be defined inside the interval $\Delta$ that starts at the time instant $k\Delta$ as $\delta_\Delta(t - k\Delta)\Delta$.

Suppose now that the input signal $x(t)$ is sampled at equidistant points defined by $\Delta$, which generates samples denoted by $x(k\Delta)$, where $k = 0, \pm1, \pm2, \ldots$ . Now, the signal $x(t)$ in the $k$th time interval of duration $\Delta$ can be approximated by a constant value expressed as

$$x_{k\Delta}(t) = x\,(k\Delta)\,\Delta\delta_\Delta\,(t - k\Delta), \tag{11.6.2}$$

for $k\Delta < t < (k + 1)\Delta$, where $x(k\Delta)$ is the sample of the function $x(t)$ at the time $t = k\Delta$. Thus, the *weighted* rectangular pulse function is defined inside the $k$th $\Delta$ interval, as shown in Figure 11.6.2(a).

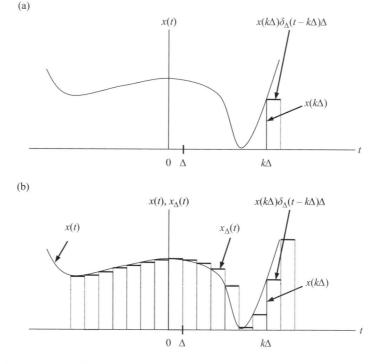

**Fig. 11.6.2** *Approximation of a continuous-time signal by the staircase function.*

For all values of time $t$, the 'staircase' approximation of $x(t)$ can be obtained and expressed as the sum of shifted rectangular pulses

$$x_\Delta(t) = \sum_{k=-\infty}^{k=\infty} x(k\Delta)\delta_\Delta(t-k\Delta)\Delta,$$

as shown in Figure 11.6.2(b). Then, if $\Delta \to 0$, the approximation $x_\Delta(t) \to x(t)$, and thus

$$x(t) = \lim_{\Delta \to 0} x_\Delta(t) = \lim_{\Delta \to 0} \sum_{k=-\infty}^{k=\infty} x(k\Delta)\delta_\Delta(t-k\Delta)\Delta. \tag{11.6.3}$$

Also, as $\Delta \to 0$, this summation approaches an integral, and $\delta_\Delta(t-k\Delta)$ approaches the impulse function. Letting $\Delta \to d\tau$, and $k\Delta \to \tau$, the last equation becomes

$$x(t) = \int_{-\infty}^{\infty} x(\tau)\delta(t-\tau)d\tau. \tag{11.6.4}$$

Therefore, no matter how complicated the function $x(t)$ may be, the integral of the product $x(\tau)\delta(t-\tau)$ over all $\tau$ values simply equals the value of $x(t)$ at the point $t = \tau$. This is known as the *sifting* property of the impulse function. In other words, the input signal is 'decomposed' into a sum of shifted and weighted impulse $\delta$-functions.

## 11.6.3 Basic Representation of a Linear-Time-Invariant System

Let a linear-time-invariant system be represented by its impulse response $h(t)$. The impulse response of a linear-time-invariant system is defined as its output signal for the case when the input signal is the impulse signal $\delta(t)$. The system can be represented by the scheme in Figure 11.6.3, where $L$ is a linear operator. Then, the output can be obtained from the input via a linear operation, that is,

$$h(t) = L[\delta(t)]. \tag{11.6.5}$$

In the general case, for an arbitrary input signal $x(t)$, the output of the linear-time-invariant system $y(t)$ can be obtained as a linear transformation $L$ of the input signal, according to the expression

$$y(t) = L[x(t)]. \tag{11.6.6}$$

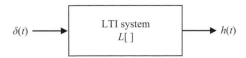

**Fig. 11.6.3** *Impulse response of a linear-time-invariant system; LTI, linear-time-invariant.*

## 11.6.4   Representation of an Output Signal

Let $h_\Delta(t)$ be the output of a linear-time-invariant system for the input $\delta_\Delta(t)$. Then, due to the *linearity* of the system, the product $\Delta\delta_\Delta(t)$ gives the response $\Delta h_\Delta(t)$. Also, due to the *time invariance* of the system, the shifted response $h_\Delta(t - k\Delta)$ is the response of the shifted input signal $\delta_\Delta(t - k\Delta)$. In summary, the input–output relationship is

$$\Delta\delta_\Delta(t - k\Delta) \underset{\text{LTI}}{\rightarrow} \Delta h_\Delta(t - k\Delta).$$

Furthermore, due to the linearity of the system, the input multiplied by a scalar $x_\Delta(t) = x(k\Delta)$ gives the output multiplied by the same scalar (homogeneity or scaling), that is, we obtain

$$x(k\Delta) \cdot \Delta\delta_\Delta(t - k\Delta) \underset{\text{LTI}}{\rightarrow} x(k\Delta) \cdot \Delta h_\Delta(t - k\Delta).$$

Also, due to the additivity property of the system, the sum of the inputs gives the sum of the outputs, as expressed by the relation

$$\sum_{k=-\infty}^{k=\infty} x(k\Delta)\,\Delta\delta_\Delta(t - k\Delta) \underset{\text{LTI}}{\rightarrow} \sum_{k=-\infty}^{k=\infty} x(k\Delta)\,\Delta h_\Delta(t - k\Delta). \tag{11.6.7}$$

Also, as $\Delta \to 0$, these summations approach integrals. Letting $k\Delta \to \tau$, and thus $\Delta \to d\tau$, $\delta_\Delta(t - k\Delta)$ approaches the impulse function $\delta(t - \tau)$, $h_\Delta(t - k\Delta)$ approaches the impulse response $h(t - \tau)$, and eqn (11.6.7) becomes

$$\int_{\tau=-\infty}^{\infty} x(\tau)\,\delta(t - \tau)\,d\tau \underset{\text{LTI}}{\rightarrow} \int_{\tau=-\infty}^{\infty} x(\tau)\,h(t - \tau)\,d\tau. \tag{11.6.8}$$

Therefore, for the input $x(t)$, the output of the linear-time-invariant system, represented by the impulse response $h(t)$, is

$$y(t) = \int_{\tau=-\infty}^{\infty} x(\tau)\,h(t - \tau)\,d\tau, \tag{11.6.9}$$

where $h(t)$ represents the impulse response of the linear-time-invariant system. The output is obtained using the convolution integral of the input and the impulse response of the system, that is,

$$y(t) = x(t) * h(t).$$

The fundamental result can be expressed as follows: *the output of any continuous-time linear-time-invariant system is the convolution of the input signal and the system impulse response.*

The output signal can also be found in the following way, which is sometimes used in this book. Due to the linearity of the system, the system can be represented by a linear operator $L$.

Then, the output of the system for any input $x(t)$ is

$$y(t) = L[x(t)] = L\left[\int_{-\infty}^{\infty} x(\tau)\,\delta(t-\tau)\,d\tau\right] = \int_{-\infty}^{\infty} x(\tau)\,L[\delta(t-\tau)]\,d\tau.$$

Because the system is time invariant, which is equivalent to the condition $L[\delta(t-\tau)] = h(t-\tau)$, the output is equal to eqn (11.6.9). Using the *commutativity* property of convolution, the convolutional integral can also be written as

$$y(t) = \int_{\tau=-\infty}^{\infty} h(\tau)x(t-\tau)\,d\tau.$$

A comprehensive presentation of a linear-time-invariant system in the time and frequency domains, and its operation expressed in terms of the impulse response, autocorrelation function, and power spectral density, will be presented in Chapter 12, Section 12.5.3.

## 11.6.5    Properties of Convolution

Let us express convolution in the following form:

$$y(t) = x(t)*h(t) = x*h = \int_{\tau=-\infty}^{\infty} x(\tau)h(t-\tau)d\tau. \tag{11.6.10}$$

The convolution is characterized by the following properties: the *commutativity* property is defined by the expression

$$x*h = h*x = \int_{\tau-\infty}^{\infty} x(\tau)h(t-\tau)d\tau = \int_{\tau=-\infty}^{\infty} h(\tau)x(t-\tau)d\tau. \tag{11.6.11}$$

The *associativity* property is expressed as

$$x*(h_1*h_2) = (x*h_1)*h_2, \tag{11.6.12}$$

and the impulse response of the whole system is

$$h(t) = \left[\int_{\tau=-\infty}^{\infty} h_1(\tau)h_2(t-\tau)d\tau\right] = (h_1*h_2), \tag{11.6.13}$$

as shown in Figure 11.6.4 for two concatenated systems.

The *distributivity property* is defined as

$$x*(h_1+h_2) = (x*h_1)+(x*h_2),$$

which is related to the parallel connection of two systems, as shown in Figure 11.6.5.

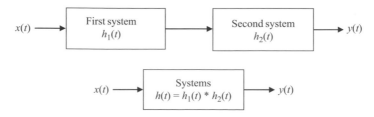

**Fig. 11.6.4** *Illustration of the associativity property.*

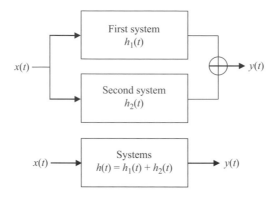

**Fig. 11.6.5** *Illustration of the distributivity property.*

## 11.7   Properties of Linear-Time-Invariant Systems

The characteristics of a linear-time-invariant system are completely determined by its impulse response $h(t)$. Therefore, other attributes of the system, such as causality, stability, and memory, must be expressible in terms of conditions imposed on $h(t)$.

**Systems with memory.** As was mentioned before, a system is said to be memoryless if the output at certain time instant $t_0$ depends only on the input at the same time instant $t_0$, that is, $y(t_0)$ depends only on $x(t_0)$. Then, if the system is linear and time variant, this relationship can only be of the form

$$y(t) = Kx(t), \tag{11.7.1}$$

where $K$ is the constant gain. The corresponding impulse response is

$$h(t) = K\delta(t). \tag{11.7.2}$$

If $h(t_0) \neq 0$ for any index value other than $t_0 = 0$, the system is said to have memory.

**Causal systems.** As mentioned previously, a system is causal if $y(t)$, at an arbitrary time $t = t_0$, expressed as $y(t) = y(t_0)$, depends only on the input value $x(t)$ for $t \leq t_0$. In other words, the response of an event beginning at $t = t_0$ is non-zero only for $t \geq t_0$. Thus, the impulse response for the $\delta(t)$ of the causal system can be non-zero only for $t \geq 0$, that is,

$$h(t) = 0, \tag{11.7.3}$$

for $t < 0$, which gives the convolution integral in the form

$$y(t) = \int_{\tau=0}^{\infty} h(\tau) x(t-\tau) d\tau, \tag{11.7.4}$$

where the integral values for every $\tau < 0$ are zero. Additional conclusions can be drawn if the alternative form of the convolution integral is used. Thus, the output of the system is

$$y(t) = \int_{\tau=-\infty}^{t} x(\tau) h(t-\tau) d\tau. \tag{11.7.5}$$

This expression clearly shows that the only values of $x(\tau)$ that can be used to find $y(t)$ are those for $\tau < t$.

**Stable systems.** Assume that the input signal $x(t)$ is bounded, that is, $|x(t)| \leq A$ for all $t$ values. It is known that the magnitude of an integral is less than or equal to the integral of the magnitude (recall Schwarz's inequality), and the magnitude of a product is equal to the product of the magnitudes. Hence, for

$$y(t) = \int_{\tau=-\infty}^{\infty} x(\tau) h(t-\tau) d\tau = \int_{\tau=-\infty}^{\infty} h(\tau) x(t-\tau) d\tau, \tag{11.7.6}$$

we find that

$$|y(t)| \leq \int_{\tau=-\infty}^{\infty} |h(\tau) x(t-\tau)| d\tau = \int_{\tau=-\infty}^{\infty} |h(\tau)| \cdot |x(t-\tau)| d\tau \leq \int_{\tau=-\infty}^{\infty} |h(\tau)| \cdot A d\tau$$
$$= A \int_{\tau=-\infty}^{\infty} |h(\tau)| d\tau$$

because of the assumption $|x(t)| \leq A$. Therefore, if the impulse response is absolutely integrable, that is, if

$$\int_{\tau=-\infty}^{\infty} |h(\tau)| d\tau = C < \infty, \tag{11.7.7}$$

the inequality (11.7.7) provides a bound on the output of the form

$$|y(t)| \leq AC = B, \tag{11.7.8}$$

for a finite constant $B$, and the system is stable in terms of bounded-input/bounded-output stability. In conclusion, we can state the following:

- A sufficient condition for a continuous-time system to be stable is that its impulse response should be absolutely integrable.
- The absolute integrability of the impulse response is also a necessary condition for stability, and thus the system is stable if and only if the relation for $C$ is satisfied.

PROBLEMS

1. Suppose the signal $x(t)$ is as shown in Figure 1.1.

   a. Write $x(t)$ in terms of singularity functions.
   b. Is $x(t)$ an energy signal or a power signal? Support your answer with proof.
   c. Let the signal $z(t)$ be defined as a periodic signal obtained by repeating the signal $x(t)$ every time interval $T_p = 5$, as shown in Figure 1.2. Is the signal $z(t)$ an energy signal or a power signal? Support your answer with an explanation (do not make any calculations).

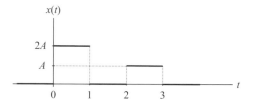

**Fig. 1.1** *A signal in the time domain.*

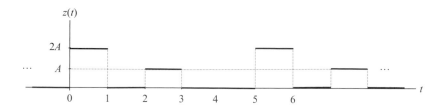

**Fig. 1.2** *A periodic signal in the time domain.*

2. Find the even and odd parts of each of the following signals:

   a. $x(t) = \cos(t) + \sin(t) + \sin(t)\cos(t)$,
   b. $x(t) = 1 + t + 3t^2 + 5t^3 + 9t^4$,

c. $x(t) = 1 + t\cos(t) + t^2\sin(t) + t^3\sin(t)\cos(t)$,

d. $x(t) = (1 + t^3)\cos^3(10t)$.

3. Determine whether the following signals are periodic; if they are periodic, find the fundamental period:

a. $x(t) = \cos^2(2\pi t)$.

b. $x(t) = \displaystyle\sum_{k=-5}^{5} w\,(t - 2k)$, where $w(t)$ is depicted in Figure 3.

c. $x(t) = \displaystyle\sum_{k=-\infty}^{\infty} w\,(t - 3k)$, where $w(t)$ is depicted in Figure 3.

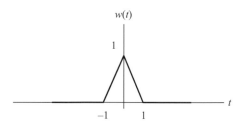

**Fig. 3** *A signal in the time domain.*

4. Solve the following problems:

a. Show that the time reversal of $x(t)$ in Figure 4.1(a) gives the signal $y(t)$ in Figure 4.1(b).

b. Let $x(t)$ be the signal shown in Figure 4.2. Express the signal in terms of singularity functions. Draw the modified signal expressed as $y(t) = 2x(1 + t/2)$.

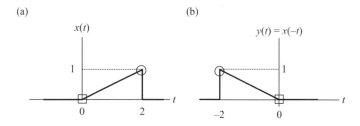

**Fig. 4.1** *Time reversal.*

5. Evaluate the following integrals:

a. $\int_0^5 \cos 2\pi t \cdot \delta\,(t - 0.5)\ dt$,

b. $\int_{-\infty}^{\infty} (t - 2)^2 \cdot \delta\,(t - 2)\ dt$.

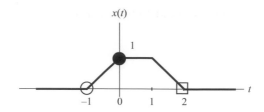

**Fig. 4.2** *The original signal in the time domain.*

6.  Sketch the following signals:

    a.  $x(t) = r(t+2) - 2r(t) + r(t-2)$,
    b.  $x(t) = 2u(t) + \delta(t-2)$,
    c.  $x(t) = 2u(t) \, \delta(t-2)$.

7.  Write the expression for the signal shown in Figure 7 in terms of singularity functions.

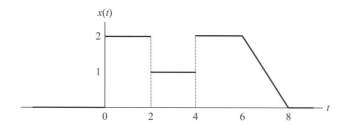

**Fig. 7** *A signal in the time domain.*

8.  A triangle pulse signal $x(t)$ is equivalent to $w(t)$ depicted in Figure 3. Sketch each of the following signals derived from $x(t)$:

    a.  $x(3t)$,
    b.  $x(3t + 2)$, and
    c.  $x(-2t - 1)$.

9.  Let a system be defined by its input–output relationship:

    a.  $y(t) = \cos x(t)$, and
    b.  $y(t) = \int_{-\infty}^{t/2} x(\tau) \, d\tau$.

    Determine whether each of them is memoryless, stable, causal, linear, and time invariant.

10. Passive electrical components can be treated as systems.

    a. In what kind of systems is a capacitor an ideal resistor, and is an ideal resistor in series with a capacitor?

    b. Prove that a capacitor is a time-invariant system.

    c. Is a resistor a stable system?

    d. Is a capacitor a stable system?

11. Solve the following problems:

    a. A time-reversal system is defined as

    $$y(t) = x(-t).$$

    Is this system a time-variant system?

    b. The impulse response of a system is a unit step signal $h(t) = u(t)$. Does this system have memory?

    c. Let the impulse response be the exponential signal

    $$h(t) = e^{-at}u(t).$$

    Is this system stable?

12. Find the impulse response for a differentiator and an integrator.

13. Let the input of a linear-time-invariant system be the signal $x(t) = 2\delta(t + 1) - \delta(t - 1)$. Find the response of the system to $x(t)$, for its impulse response shown in Figure 13. and expressed as

$$h(t) = \begin{cases} t+1 & -1 \le t \le 0 \\ 1-t & 0 \le t \le 1 \end{cases}.$$

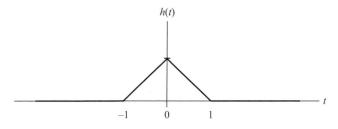

**Fig. 13** *Impulse response of the system.*

14. Find the expression for the impulse response relating the input $x(t)$ to the output $y(t)$ in terms of the impulse response of each subsystem for the complex linear-time-invariant system depicted in Figure 14.

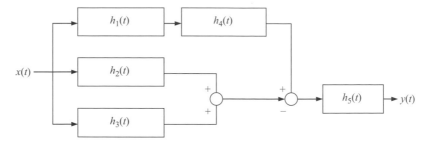

**Fig. 14** *A complex linear-time-invariant system.*

15. Consider the cascaded system shown in Figure 15, which processes the signal $x(t) = u(t) - 2u(t-1) + 2u(t-2)$. The impulse responses of the subsystems are as follows:

$$h_1(t) = u(t)$$

and

$$h_2(t) = \sum_{k=0}^{3} (0.5)^k \delta(t - 4k).$$

a. Sketch the signal $x(t)$.
b. Describe the type of signal processing done by the first subsystem, and sketch $y_1(t)$.
c. Find an expression for the function $y(t)$ in terms of $y_1(t)$.
d. Determine and sketch $y(t)$.
e. Determine the impulse response $h_0(t)$ of the overall cascaded system.

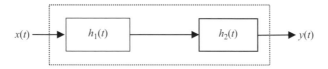

**Fig. 15** *A complex linear-time-invariant system.*

16. For the impulse response $h(t) = e^{2t}u(t-1)$, determine whether the corresponding system is memoryless, stable, and causal.

17. Suppose the input signal $x(t)$ of a correlator is an exponential truncated function specified by

$$x(t) = e^{-2t}$$

on the interval $0 \leq t < 2$, as depicted in Figure 17.

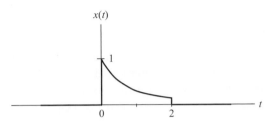

**Fig. 17** *An exponential truncated signal.*

a. Find out the output autocorrelation function $R(\tau)$ of the correlator using the autocorrelation integral.

b. Sketch the graph of the autocorrelation function $R(\tau)$. Find the energy of the signal $x(t)$ by using the expression for the autocorrelation function.

c. Calculate the energy of the signal $x(t)$ from its expression in the time domain. Is signal $x(t)$ an energy signal or a power signal?

d. From the expression for the autocorrelation function, is it possible to find the frequency-domain characteristics of the signal $x(t)$? If so, explain how this can be done.

18. An aperiodic signal $x(t)$ is expressed as $x(t) = e^{-2t}$, on the interval $0 \leq t < 2$, as depicted in Figure 18. Consider the case when the impulse response of a linear-time-invariant system is the repeated rectangular pulse $h(t)$ shown in Figure 18. Find the response $y(t)$ of the system for this case.

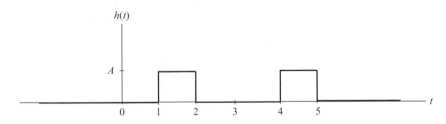

**Fig. 18** *Impulse response of the system.*

19. The impulse response of a linear-time-invariant system is

$$h(t) = \frac{1}{T} e^{-t/T} u(t).$$

The input signal $x(t)$ is a rectangular pulse specified by

$$x(t) = \begin{cases} 1/T_0 & 0 \leq t \leq T_0 \\ 0 & \text{otherwise} \end{cases}.$$

Find the output signal $y(t)$ of the system.

a. Assume that $T = T_0$. Suppose the maximum value of the output signal $y(t)$ is proportional to the energy of the input signal $x(t)$. Find the percentage of the total energy of the input signal $x(t)$ that corresponds to the maximum value of $y(t)$.

b. Is it possible to increase the percentage of the energy calculated in (b) to 100%? Support your answer with proof.

c. Suppose the impulse response of the system is a delta function, that is, $h(t) = \delta(t)$. Draw the wave shape of the system response $y(t)$ to the input signal $x(t)$.

20. The impulse response of a linear-time-invariant system is

$$h(t) = \begin{cases} 2 & 0 \leq t \leq 2 \\ 0 & \text{otherwise} \end{cases}.$$

a. Suppose the input signal $x(t)$ is a rectangular pulse specified by

$$x(t) = \begin{cases} 1 & 0 \leq t \leq 2 \\ 0 & \text{otherwise} \end{cases}.$$

Find out and sketch the output signal $y(t)$ of the system, using the convolution integral.

b. Suppose the input signal $x(t)$ is a magnitude-scaled delta function, that is, $x(t) = 2\delta(t)$. Sketch the wave shape of the system response $y(t)$.

c. Suppose the input signal $x(t)$ is a magnitude-scaled and time-shifted delta function, that is, $x(t) = 2\delta(t - a)$, where $a$ is a constant. Sketch the system response $y(t)$ for $|a| = 2$.

21. Given the signal

$$x(t) = \sin^2 (7\pi t - \pi/6) + \cos(3\pi t - \pi/3),$$

a. sketch its one-sided magnitude and phase spectra,

b. sketch its two-sided magnitude and phase spectra, and

c. sketch its phasor diagram after writing it as the sum of complex conjugate rotating phasors.

22. Suppose that the signal $x(t)$ shown in Figure 22 is applied to the input of a matched filter. Find and sketch the output signal $y(t)$ of the matched filter for $t$ values in the time interval

$-\infty < t \leq 8$. It is important to keep in mind that, if the input signal $x(t)$ is of duration $T = 3$, then the impulse response of its matched filter is $h(t) = x(T - t) = x(3 - t)$.

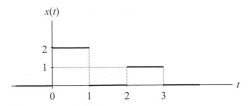

**Fig. 22** *Input signal of a matched filter.*

23. Let the input of an linear-time-invariant system be the unit step signal $x(t) = u(t)$. Find the response of the system if its impulse response is

$$h(t) = e^{-at}u(t),$$

as shown in Figure 23.

**Fig. 23** *Input signal and linear-time-invariant impulse response.*

24. The impulse response of an linear-time-invariant system is a causal exponential signal $h(t)$ defined as

$$h(t) = e^{-2t}u(t).$$

Let the input signal of the system be the exponential signal $x(t)$ defined as

$$x(t) = e^{-t}u(t).$$

a. Find the response of the system $y(t)$ for the input signal $x(t)$.

b. Draw the graph of the response $y(t)$ and find the maximum value of the signal $y(t)$.

c. Find the response of the linear-time-invariant system for the input signal $x(t) = \delta(t - T)$.

Consider the signal $x(t) = e^{-2|t|}$ shown in Figure 24. Is this signal a power signal or an energy signal? Support your answer with proof.

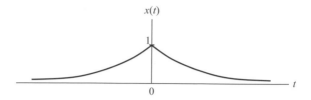

**Fig. 24** *Input signal of a linear-time-invariant system.*

25. The impulse response of a linear-time-invariant system is

$$h(t) = \begin{Bmatrix} A & 0 \leq t \leq 2 \\ 0 & \text{otherwise} \end{Bmatrix}.$$

Suppose the input signal of the system $x(t)$ is an exponential function specified by

$$x(t) = e^{-2t},$$

on the interval $0 \leq t < T = 2$, as depicted in Figure 25.

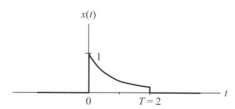

**Fig. 25** *Input signal.*

a. Find the output signal $y(t)$ of the system. Sketch the graph of $y(t)$.

b. Find the output of the system in the case when the duration of the input signal is defined by the same exponential function that exists inside the interval $0 \leq t < T = 1$.

26. Suppose the input signal $x(t)$ of a correlator is an exponential function specified by $x(t) = e^{-2t}$ on the interval $0 \leq t < 2$.

a. Find the output autocorrelation function $R(\tau)$ of the correlator by using the autocorrelation integral.

b. Sketch the graph of the autocorrelation function $R(\tau)$. Find the energy of the signal $x(t)$ by using the expression for the autocorrelation function.

c. From the expression for the autocorrelation function, it is possible to find the frequency-domain characteristics of the signal $x(t)$. Explain how this can be done.

27. Suppose the continuous-time signal shown in Figure 27 represents the voltage applied across a resistor $R$. Is this signal a power signal or an energy signal?

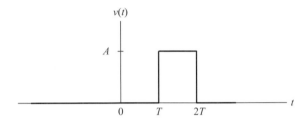

**Fig. 27** *A continuous-time signal.*

28. A sinusoidal signal with period $T$ is shown in Figure 28. Is this signal a power signal or an energy signal?

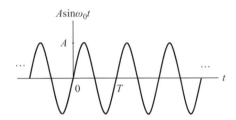

**Fig. 28** *A continuous-time sinusoidal signal.*

29. Consider the signal $x(t) = e^{-\varepsilon t}u(t)$, where $A$ and $\varepsilon$ are constants. Is it a power signal?

30. Categorize each of the following signals as an energy signal or a power signal, and find the energy or power for each of them:

a. $x(t) = \begin{cases} t & 0 \le t \le 1 \\ 2-t & 1 \le t \le 2 \\ 0 & \text{otherwise} \end{cases}$, and

b. $x(t) = \begin{cases} 5\cos(\pi t) & -0.5 \le t \le 0.5 \\ 0 & \text{otherwise} \end{cases}$.

31. Compute the cross-correlation function for the signals

$$x(t) = u(t) - u(t - T)$$

and

$$y(t) = u(t+T) - u(t),$$

which are both shown in Figure 31.

**Fig. 31** *Correlated signals in the time domain.*

32. Calculate the autocorrelation function for the sine signal $x(t) = A \sin(2f_c t)$ for two cases. The first case is when the signal is treated as a periodic signal. The second case is when the signal is treated as a power signal. Investigate its power and energy.

# 12

# Transforms of Deterministic Continuous-Time Signals

## 12.1 Introduction

In Chapter 11, all of the equations, operations, and properties related to signals and systems analysis were presented in the time domain. The signals were defined and operated upon using functions that had time as the independent variable.

However, these signals can be converted into other domains that can be more convenient for analysing, synthesizing, and/or processing the signals. Also, better insight into the nature and properties of the signals and systems can be gained by presenting the signals in various domains. In this chapter, the Fourier series and the Fourier transform for *continuous-time signals* will be presented. Using the Fourier series and the Fourier transform, the amplitudes of a signal can be expressed as a function of frequency. Thus, they specify the frequency contents or the spectrum of that signal. The procedure of obtaining the spectrum of a signal is called *frequency analysis*, or *spectral analysis*.

By analysing a linear-time-invariant system, it was found that any signal $x(t)$ can be 'decomposed' into an infinite sum of elementary functions represented by the shifted impulse delta functions; this sum is expressed as

$$x(t) = \int_{-\infty}^{\infty} x(\tau) \, \delta(t-\tau) \, d\tau. \tag{12.1.1}$$

Thus, no matter how complicated the function $x(t)$ may be, the integral of the product $x(\tau) \, \delta(t - \tau)$ over all $\tau$ simply equals the value of $x(t)$ at the point $t$. Then, the response of the linear-time-invariant system can be found by finding the response to each of these delta functions individually and then adding all the responses to get the overall response. By following this procedure, it was proved that the output of any continuous-time linear-time-invariant system is the *convolution* of the input signal and the system impulse response, that is,

$$y(t) = \int_{\tau=-\infty}^{\infty} x(\tau) h(t-\tau) \, d\tau, \tag{12.1.2}$$

*Discrete Communication Systems*. Stevan Berber, Oxford University Press. © Stevan Berber 2021.
DOI: 10.1093/oso/9780198860792.003.0012

where $h(t)$ represents the impulse response of the linear-time-invariant system. In this chapter, we will show that a continuous-time signal can be decomposed into the sum of real or complex sinusoids that are special cases of the complex exponentials. Because the response of a linear-time-invariant system to a sinusoid is also a sinusoid of the same frequency but, in principle, different magnitude and phase, the response to a sum of sinusoids will be a sum of sinusoids of the same frequencies. Therefore, the main problem is how to decompose the input continuous-time signal into a sum of complex exponentials or complex sinusoids, instead of a sum of the delta functions.

The solution to this problem will lead to the frequency-domain representation of the time-domain signals, one based on the Fourier series and the Fourier transform, as will be presented in this chapter. It will also be shown that any periodic signal can be expressed as a linear combination of complex sinusoidal signals, or, precisely speaking, as an infinite sum of elementary sinusoids. This finding will be extended to the case of aperiodic signals by applying the Fourier transform. Using this transform, any aperiodic signal will be represented as an integral of the elementary sinusoidal signals, because they will fulfil all the orthogonality conditions. For these signals, the power and energy will be calculated assuming that the load is a 1 ohm resistor, that is, $R = 1$ ohm. In this way, the values calculated for power will be expressed in watts, and the values calculated for energy will be calculated in joules.

## 12.2   The Fourier Series

### 12.2.1   The Fourier Series in Trigonometric Form

A periodic signal $x(t) = x(t + T)$ (not necessarily a continuous-valued signal) may be represented as a trigonometric series in the form

$$x(t) = a_0 + \sum_{k=1}^{\infty} a_k \cos k\omega_0 t + \sum_{k=1}^{\infty} b_k \sin k\omega_0 t, \tag{12.2.1}$$

where the coefficients are

$$a_0 = \frac{1}{T} \int_{-T/2}^{T/2} x(t)dt, \tag{12.2.2}$$

$$a_k = \frac{2}{T} \int_{-T/2}^{T/2} x(t) \cos k\omega_0 t dt, \tag{12.2.3}$$

and

$$b_k = \frac{2}{T} \int_{-T/2}^{T/2} x(t) \sin k\omega_0 t dt. \tag{12.2.4}$$

The component of the series(12.2.1) for $k = 1$ (at $\omega = \omega_0 = 2\pi f_0 = 2\pi/T$) is called the *fundamental component* of the series, while those for $k \geq 2$ are called *harmonic components* (or simply *harmonics*). Generally, the range of integration can start anywhere, providing that it spans exactly one period. In addition, the condition of integral convergence must be fulfilled.

**Example**

Find the components of the Fourier series for the rectangular pulse train $x(t)$ shown in Figure 12.2.1.

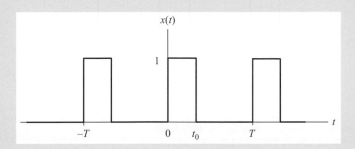

**Fig. 12.2.1** *Rectangular pulse train.*

**Solution.** This signal can be described in the first period by the expression

$$x(t) = \begin{cases} 1 & 0 < t < t_0 \\ 0 & t_0 < t < T \end{cases}.$$ 
(12.2.5)

Then, the coefficients of the Fourier series can be calculated as

$$a_0 = \frac{1}{T} \int_{-T/2}^{T/2} x(t)dt = \frac{1}{T} \int_0^{t_0} 1 \cdot dt = \frac{t_0}{T},$$
(12.2.6)

$$a_k = \frac{2}{T} \int_{-T/2}^{T/2} x(t)\cos k\omega_0 t\, dt = \frac{2}{T} \int_0^{t_0} 1 \cdot \cos k\omega_0 t\, dt = \frac{1}{k\pi} \sin k\omega_0 t_0,$$
(12.2.7)

$$b_k = \frac{2}{T} \int_{-T/2}^{T/2} x(t)\sin k\omega_0 t\, dt = \frac{2}{T} \int_0^{t_0} 1 \cdot \sin k\omega_0 t\, dt = \frac{1}{k\pi} (1 - \cos k\omega_0 t_0).$$
(12.2.8)

Thus, the Fourier series of $x(t)$ is given by

$$x(t) = \frac{t_0}{T} + \sum_{k=1}^{\infty} \left( \frac{1}{k\pi} \sin k\omega_0 t_0 \right) \cos k\omega_0 t + \sum_{k=1}^{\infty} \left[ \frac{1}{k\pi} (1 - \cos k\omega_0 t_0) \right] \sin k\omega_0 t.$$
(12.2.9)

Therefore, the magnitudes of coefficients are inversely proportional to their order $k$.

## Example

Find the components of the Fourier series for the symmetric rectangular pulse train $x(t)$ shown in Figure 12.2.2.

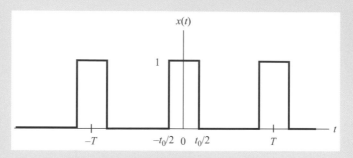

**Fig. 12.2.2** *Symmetric rectangular pulse train.*

**Solution.** This signal can be defined by the expression

$$x(t) = \begin{cases} 1 & -t_0/2 < t < t_0/2 \\ 0 & \text{for all other points in the interval} \quad -T/2 \text{ to } T/2 \end{cases}. \tag{12.2.10}$$

The coefficients of the Fourier series are calculated as

$$a_0 = \frac{1}{T} \int_{-t_0/2}^{t_0/2} 1 \cdot dt = \frac{t_0}{T},$$

$$a_k = \frac{2}{T} \int_{-T/2}^{T/2} x(t) \cos k\omega_0 t \, dt = \frac{2}{T} \int_{-t_0/2}^{t_0/2} 1 \cdot \cos k\omega_0 t \, dt = \frac{2}{k\pi} \sin k\pi \frac{t_0}{T},$$

$$b_k = \frac{2}{T} \int_{-T/2}^{T/2} x(t) \sin k\omega_0 t \, dt = \frac{2}{T} \int_{-t_0/2}^{t_0/2} 1 \cdot \sin k\omega_0 t \, dt = 0.$$

For $x(t)$ even, all of the coefficients $b_k$ are 0, and the coefficients of the series can be expressed as

$$a_k = \begin{cases} t_0/T & k = 0 \\ 2\dfrac{\sin k\pi t_0/T}{k\pi} & k \neq 0 \end{cases}. \tag{12.2.11}$$

Using the Fourier coefficients, the Fourier series of the signal $x(t)$ in the time domain is given by

$$x(t) = \frac{t_0}{T} + \sum_{k=1}^{\infty} \left( 2 \frac{\sin k\pi t_0/T}{k\pi} \right) \cos k\omega_0 t \tag{12.2.12}$$

## 12.2.2  An Example of Periodic Signal Analysis

It is important to understand the conventions related to the time-domain, frequency-domain, and phasor-domain presentations of a periodic signal $x(t)$. In particular, it is important to properly calculate and present the magnitude spectrum and the phase spectrum of the signal. To demonstrate this analysis, a special rectangular pulse train defined by eqn (12.2.10) for the redefined pulse duration $t_0 = T/2$, as presented in Figure 12.2.3, will be analysed.

In this case, the Fourier coefficients presented in eqn (12.2.11) are recalculated to be

$$a_k = \left\{ \begin{array}{ll} 1/2 & k = 0 \\ \frac{\sin k\pi/2}{k\pi/2} = \frac{(-1)^{(k-1)/2}}{k\pi/2} & k \text{ odd} \\ 0 & k \text{ even} \end{array} \right\}, \tag{12.2.13}$$

and the signal is expressed as the Fourier series in the form

$$\begin{aligned} x(t) &= \frac{t_0}{T} + \sum_{k=1}^{\infty} \frac{(-1)^{(k-1)/2}}{k\pi/2} \cos k\omega_0 t \\ &= \frac{1}{2} + \frac{2}{\pi} \left( \cos \omega_0 t - \frac{\cos 3\omega_0 t}{3} + \frac{\cos 5\omega_0 t}{5} - \frac{\cos 7\omega_0 t}{7} + \dots \right) \end{aligned} \tag{12.2.14}$$

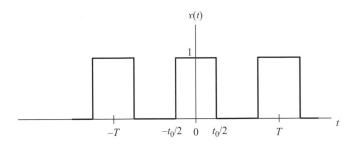

**Fig. 12.2.3** *A symmetric rectangular pulse train with pulse duration equal to half of the value of the train's period.*

Only cosine terms appear in the series; therefore, the series is in a compact form, except that the amplitudes of the alternating harmonics are negative. In this case, the numerical values of coefficients are real numbers having the values $a_0 = 0.5$, $a_1 = 0.637$, $a_3 = -0.212$, $a_5 = 0.127$, and $a_7 = -0.091$. These coefficients can be expressed as a function of the harmonic order $k$ or as a function of the harmonic frequencies $k\omega_0$, which defines the *amplitude spectrum* of the signal $x(t)$, as presented in Figure 12.2.4. The negative signs of the cosine terms carry information about the phases of the related spectral components. Due to the presentation of the spectrum in terms of lines, this spectrum is traditionally called the *line spectrum*.

The shape of the spectrum is represented by a dotted curve. The zero crossings of this graph occur if the condition $\sin k\pi/2 = 0$ is fulfilled, or for $k\pi/2 = l\pi$, which results in the condition $k = 2l$, for $l = 1, 2, 3, \dots$ . Therefore, the existing spectral components are at the odd values of $k$, and the zero-valued components are at the even values of $k$.

**Representation of the signal $x(t)$ in the frequency domain.** Using the identities $e^{j\pi} = e^{-j\pi} = -1$, the Fourier coefficients can be expressed in terms of their magnitudes and phases as

$$a_k = \begin{cases} 1/2 & k = 0 \\ (2/\pi, -2/3\pi, 2/5\pi, -2/7\pi \dots) & \\ = (2/\pi, e^{j\pi}2/3\pi, 2/5\pi, e^{j\pi}2/7\pi \dots) & k \text{ odd} \\ 0 & k \text{ even} \end{cases} \qquad (12.2.15)$$

where the magnitudes are

$$|a_k| = \begin{cases} 1/2 & k = 0 \\ 2/k\pi & k \text{ odd} \\ 0 & k \text{ even} \end{cases}, \qquad (12.2.16)$$

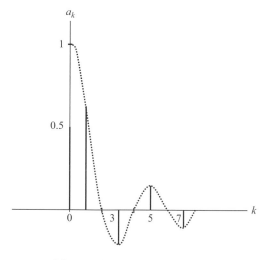

**Fig. 12.2.4** *Amplitude spectrum of the signal.*

and the corresponding phases are

$$\phi_k = \begin{cases} \tan^{-1}\left(\text{Im}\,(a_k)/\text{Re}\,(a_k)\right) = \tan^{-1}\left(\sin\pi/\cos\pi\right) = \pi & \text{for} \quad k = 3,7,11,15,\ldots \\ 0 & \text{otherwise} \end{cases}.$$

(12.2.17)

By using the magnitudes and phases of the signal, it is possible to present the signal in the frequency domain as the magnitude spectrum and the phase spectrum, as shown in Figures 12.2.5 and 12.2.6, respectively.

**Representation of the signal $x(t)$ in the phasor domain.** To present the signal in the phasor domain, we need, firstly, to express the cosine functions as the real part of the corresponding phasors defined in the exponential form by their frequencies and phases as follows:

$$x(t) = \text{Re}\left\{\frac{1}{2} + \sum_{k=1}^{\infty} a_k e^{jk\omega_0 t}\right\} = \frac{1}{2} + \text{Re}\left\{\sum_{k=1}^{\infty} |a_k| e^{j\phi_k} e^{jk\omega_0 t}\right\} = \frac{1}{2} + \text{Re}\left\{\sum_{k=1}^{\infty} |a_k| e^{j(k\omega_0 t + \phi_k)}\right\}.$$

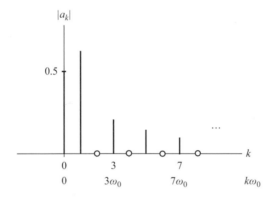

**Fig. 12.2.5** *Magnitude spectrum.*

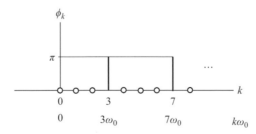

**Fig. 12.2.6** *Phase spectrum.*

Taking into account the magnitudes and phases of the Fourier coefficients in our example, the phasors $X_k$, defined as the rotating vectors in the complex plane, can be expressed as

$$X_k = a_0 + \sum_{k=1}^{\infty} |a_k| e^{j(k\omega_0 t + \phi_k)}$$

$$= \frac{1}{2} + |a_1| e^{j(\omega_0 t + 0)} + |a_3| e^{j(3\omega_0 t + \pi)} + |a_5| e^{j(5\omega_0 t + 0)} + |a_7| e^{j(5\omega_0 t + \pi)} + \dots, \quad (12.2.18)$$

$$= \frac{1}{2} + \frac{1}{\pi} e^{j(\omega_0 t + 0)} + \frac{1}{3\pi} e^{j(3\omega_0 t + \pi)} + \frac{1}{5\pi} e^{j(5\omega_0 t + 0)} + \frac{1}{7\pi} e^{j(7\omega_0 t + \pi)} + \dots$$

and presented in the phasor diagram in Figure 12.2.7. It is important to note the following: the positive sense of a phasor rotation is defined to be the counter-clockwise direction, that is, by a positive fundamental angular frequency $\omega_0 > 0$, and the initial positions of the phasors are determined by the initial phases of phasors defined by the time instant $t = 0$. The coefficient $a_0$ represents a DC component, having zero frequency. The fundamental harmonic has zero initial phase and rotates with the fundamental frequency $\omega_0$. The third and seventh harmonics have phases equal to $\pi$, and rotate counter-clockwise with frequencies $3\omega_0$ and $7\omega_0$, respectively.

**Precise representation of the signal in the time domain.** The signal presented in the time domain can be simply obtained as the real part of the signal presented in the phasor domain, that is,

$$x(t) = \operatorname{Re}\{X_k\}$$

$$= \operatorname{Re}\left\{\frac{1}{2} + \frac{1}{\pi} e^{j(\omega_0 t + 0)} + \frac{1}{3\pi} e^{j(\omega_0 t + \pi)} + \frac{1}{5\pi} e^{j(\omega_0 t + 0)} + \frac{1}{7\pi} e^{j(\omega_0 t + \pi)} + \dots\right\},$$

$$= \frac{1}{2} + \frac{1}{\pi} \cos(\omega_0 t) + \frac{1}{3\pi} \cos(3\omega_0 t + \pi) + \frac{1}{5\pi} \cos(5\omega_0 t) + \frac{1}{7\pi} \cos(7\omega_0 t + \pi) + \dots$$

$$(12.2.19)$$

expressed as the Fourier series.

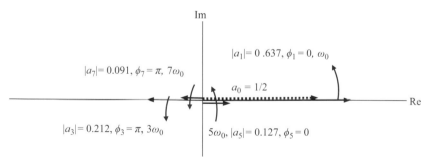

**Fig. 12.2.7** *Phasor diagram for the signal x(t).*

**Important notes.** In our analysis, we assumed the positive sense of phasor rotation to be counter-clockwise, which, in turn, specified the corresponding positive angular frequency. The initial phases presented in the phase diagram are defined by their positive values. Therefore, when the phasors start to rotate, this initial phase angle increases in the positive direction. In this way, everything is consistent with the expected behaviour of the signal in the time, frequency, and phasor domains. We can state that this consistency will be preserved if the angular frequency is negative, that is, if we assume that the phasors are rotating counter-clockwise. In that case, the initial phases will be negative, the components of the magnitude spectrum will be defined at negative frequencies, and the magnitudes of the signals will be the same as for the positive frequencies. We will briefly analyse this case. Due to the identity $\cos(x) = \cos(-x)$, we may present the signal in the time domain as

$$
\begin{aligned}
x(t) &= \mathrm{Re}\,\{X_k\} \\
&= \frac{1}{2} + \frac{1}{\pi}\cos\left(-\omega_0 t\right) + \frac{1}{3\pi}\cos\left(-3\omega_0 t - \pi\right) + \frac{1}{5\pi}\cos\left(-5\omega_0 t\right) + \dots
\end{aligned}
\tag{12.2.20}
$$

and in the phasor domain as

$$
\begin{aligned}
X_k &= a_0 + \sum_{k=1}^{\infty} \mid a_k \mid e^{j(-k\omega_0 t - \phi_k)} = a_0 + \sum_{k=1}^{\infty} \mid a_k \mid e^{-j\phi_k} e^{j(-k\omega_0 t)} \\
&= \frac{1}{2} + \mid a_1 \mid e^{j(-\omega_0 t)} + \mid a_3 \mid e^{-j\pi} e^{j(-3\omega_0 t)} + \mid a_5 \mid e^{j(-5\omega_0 t)} + \mid a_7 \mid e^{-j\pi} e^{j(-5\omega_0 t)} + \dots \\
&= \frac{1}{2} + \frac{1}{\pi} e^{j(-\omega_0 t)} + \frac{1}{3\pi} e^{-j\pi} e^{j(-3\omega_0 t)} + \frac{1}{5\pi} e^{j(-5\omega_0 t)} + \frac{1}{7\pi} e^{-j\pi} e^{j(-5\omega_0 t)} + \dots
\end{aligned}
\tag{12.2.21}
$$

The corresponding phasor diagram is presented in Figure 12.2.8. The components of the Fourier series rotate in the opposite direction with respect to the previous case, because the sense of rotation is assumed to be clockwise. The real parts of the phasors are added and give the same shape and values of the time-domain signal as in the case when the sense of the phasor rotation was counter-clockwise.

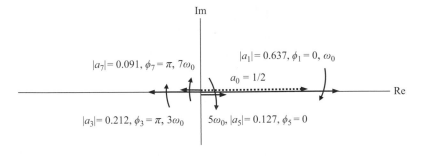

**Fig. 12.2.8** *Phasor diagram of the signal x(t), for negative frequencies.*

The magnitude of the signal can be calculated as

$$
\mid a_k \mid = \begin{cases} 1/2 & k = 0 \\ \mid 2/k\pi \mid & k \text{ odd} \\ 0 & k \text{ even} \end{cases}, \tag{12.2.22}
$$

and the corresponding phases are

$$
\phi_k = \begin{cases} \tan^{-1}\frac{\text{Im}(a_k)}{\text{Re}(a_k)} = \tan^{-1}\frac{-\sin\pi}{\cos\pi} = -\pi & \text{for} \quad k = -3, -7, -11, \ldots \\ 0 & \text{otherwise} \end{cases}. \tag{12.2.23}
$$

The magnitude spectrum and the phase spectrum are shown in Figure 12.2.9.

The magnitude spectrum has the same values of the magnitudes at the corresponding negative frequencies. The initial phases, defining the phase spectrum, have all negative values, due to the clockwise rotation of the phasors. We can say that the presentations of the signal in all the domains are the same because the information content about the spectral component (magnitudes, phases, and frequencies) can be recognized in all three domains. Knowing the signal in one domain is sufficient to find the signal in all other domains, as was the case when we assumed positive frequencies. The final conclusion is that negative frequencies are equally natural for the presentation of the signal as positive frequencies are. We can imagine voltage generators rotating in opposite directions and producing synchronously the same voltage at their outputs in the time domain. This conclusion will be further elaborated and confirmed for signals that are represented by the Fourier series in complex exponential forms, in the next section.

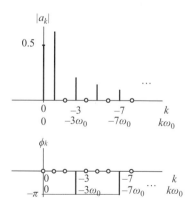

**Fig. 12.2.9** *Magnitude spectrum (upper panel) and phase spectrum (lower panel) of the signal x(t), for negative frequencies.*

## 12.2.3   The Fourier Series in Complex Exponential Form

Using Euler's formulas, the trigonometric terms in the Fourier series can be expressed in the form

$$\cos k\omega_0 t = \frac{1}{2}e^{jk\omega_0 t} + \frac{1}{2}e^{-jk\omega_0 t} = \frac{e^{jk\omega_0 t} + e^{-jk\omega_0 t}}{2},\qquad(12.2.24)$$

$$\sin k\omega_0 t = \frac{1}{j2}e^{jk\omega_0 t} - \frac{1}{j2}e^{-jk\omega_0 t} = \frac{e^{jk\omega_0 t} - e^{-jk\omega_0 t}}{j2}.\qquad(12.2.25)$$

Then, the periodic signal $x(t) = x(t + T)$ is expressed as a sum of exponential terms, in the form

$$x(t) = \sum_{k=-\infty}^{\infty} c_k e^{+jk\omega_0 t},\qquad(12.2.26)$$

where Fourier coefficients can be calculated as

$$c_k = \frac{1}{T}\int_{-T/2}^{T/2} x(t)e^{-jk\omega_0 t}dt.\qquad(12.2.27)$$

The last two equations are called the *Fourier series pair*. A plot of coefficients $c_k$ as a function of $k$, or $k\omega_0$, is called the *amplitude spectrum of the signal*. The inter-relation of these coefficients and the coefficients obtained from eqns (12.2.2), (12.2.3), and (12.2.4) can be expressed as

$$c_0 = a_0, \text{and } c_k = \frac{1}{2}(a_k - jb_k).\qquad(12.2.28)$$

Therefore, the signal $x(t)$ in the time domain can be reconstructed from its Fourier series obtained and expressed in terms of harmonic components or in terms of complex exponentials. The decomposition of a periodic signal into complex exponentials needs to give the same results as the decomposition of the signal in terms of harmonic components. Of course, the decompositions will have formally different mathematical forms; however, at the final instant, the correspondence between the time, frequency, and phasor domains will be preserved. To demonstrate this simple fact, we will re-analyse the special rectangular pulse train and show that the two aforementioned Fourier series are strictly related.

**Continuation of the example in Section 12.2.2.** Analysis of the same periodic signal in the frequency domain can be done in two ways, as presented in Section 12.2.2 and here in Section 12.2.3, resulting in essentially the same (qualitatively and quantitatively) signal content. To confirm that statement, let us analyse the signal presented in Section 12.2.2, using the Fourier series in complex form. In this context, let us firstly find the components of the Fourier series for the symmetric (with respect to the $y$-axis) rectangular pulse train $x(t)$ shown in Figure 12.2.2. The signal in the time domain is expressed by eqn (12.2.10). The coefficients of the Fourier series are calculated as

$$c_k = \frac{1}{T}\int_{-T/2}^{T/2} x(t)e^{-jk\omega_0 t}\,dt = \frac{1}{T}\int_{-t_0/2}^{t_0/2} 1\cdot e^{-jk\omega_0 t}\,dt = -\frac{1}{jTk\omega_0}e^{-jk\omega_0 t}\Big|_{-t_0/2}^{t_0/2} \tag{12.2.29}$$

$$= \frac{2\sin k\omega_0 t_0/2}{Tk\omega_0} = \frac{\sin k\pi t_0/T}{k\pi} = \frac{t_0}{T}\frac{\sin k\pi t_0/T}{k\pi t_0/T}.$$

For $k = 0$, the indeterminate case is in place. This case is readily resolved by noting that $\sin(x) \to 0$ when $x \to 0$. Thus, $\sin(x)/x \to 1$ when $\sin(x) \to 0$ and $x \to 0$. Therefore, the coefficients are

$$c_k = \begin{cases} t_0/T & k=0 \\ \frac{t_0}{T}\frac{\sin k\pi t_0/T}{k\pi t_0/T} & k\neq 0 \end{cases}. \tag{12.2.30}$$

The first zero crossing occurs for the condition $\sin k\pi t_0/T = 0$, that is, for $k\pi t_0/T = \pi$ or at $k = T/t_0$. The Fourier series of $x(t)$ is expressed as an infinite sum of exponential basis functions, given by

$$x(t) = \sum_{k=-\infty}^{\infty}\left(\frac{t_0}{T}\frac{\sin k\pi t_0/T}{k\pi t_0/T}\right)e^{k2\pi f_0 t} = \sum_{k=-\infty}^{\infty}\left(\frac{t_0}{T}\frac{\sin k\pi t_0/T}{k\pi t_0/T}\right)\cos k\omega_0 t. \tag{12.2.31}$$

Analyse a special case when the duration of the pulse is half of the period $T$, that is, $t_0 = T/2$, as defined in Section 12.2.2 and shown in Figure 12.2.3. We will find the amplitude spectrum, the magnitude spectrum, and the phase spectrum, and then find the power spectrum and calculate the power of the signal in the frequency domain. Following the procedure in the previous example, for this special case, the Fourier series of $x(t)$ can be expressed as the infinite sum of complex exponential functions, that is,

$$x(t) = \sum_{k=-\infty}^{\infty} c_k e^{jk2\pi f_0 t} = \sum_{k=-\infty}^{\infty}\left(\frac{1}{2}\frac{\sin k\pi/2}{k\pi/2}\right)\cos k\omega_0 t, \tag{12.2.32}$$

where the coefficients are

$$c_k = \begin{cases} 1/2 & k=0 \\ \frac{1}{2}\frac{\sin k\pi/2}{k\pi/2} = \frac{(-1)^{(k-1)/2}}{k\pi} & k\ \text{odd},\ -\infty < k < \infty \\ 0 & k\ \text{even} \end{cases}. \tag{12.2.33}$$

The plot of the first five coefficients ($c_0 = 0.5$, $c_{\pm 1} = 0.3185$, $c_{\pm 3} = -0.106$, $c_{\pm 5} = 0.0635$, $c_{\pm 7} = -0.0455$), representing the amplitude spectrum, is shown in Figure 12.2.10.

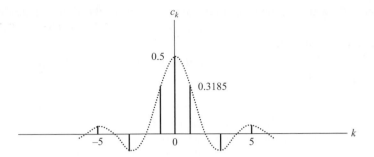

**Fig. 12.2.10** *A two-sided amplitude line spectrum.*

Because $\cos(x) = \cos(-x)$, we may present the signal in the time domain as

$$x(t) = \sum_{k=-\infty}^{\infty} c_k e^{jk2\pi f_0 t} = \sum_{k=-\infty}^{\infty} \left( \frac{1}{2} \frac{\sin k\pi/2}{k\pi/2} \right) \cos k\omega_0 t = \sum_{k=-\infty}^{\infty} \frac{(-1)^{(k-1)/2}}{k\pi} \cos k\omega_0 t$$

$$= \frac{1}{2} + \frac{1}{\pi} \cos(-\omega_0 t) - \frac{1}{3\pi} \cos(-3\omega_0 t) + \frac{1}{5\pi} \cos(-5\omega_0 t) - \frac{1}{7\pi} \cos(-7\omega_0 t) + \dots \quad (12.2.34)$$

$$+ \frac{1}{\pi} \cos \omega_0 t - \frac{1}{3\pi} \cos 3\omega_0 t + \frac{1}{5\pi} \cos 5\omega_0 t - \frac{1}{7\pi} \cos 7\omega_0 t + \dots$$

Because $-\cos(-x) = \cos(-x - \pi)$ and $-\cos(x) = \cos(x + \pi)$, we may present the signal in the time domain as

$$x(t) = \frac{1}{2} + \frac{1}{\pi} \cos(-\omega_0 t) + \frac{1}{3\pi} \cos(-3\omega_0 t - \pi) + \frac{1}{5\pi} \cos(-5\omega_0 t) + \frac{1}{7\pi} \cos(-7\omega_0 t - \pi) + \dots$$

$$+ \frac{1}{\pi} \cos \omega_0 t + \frac{1}{3\pi} \cos(3\omega_0 t + \pi) + \frac{1}{5\pi} \cos(5\omega_0 t) + \frac{1}{7\pi} \cos(7\omega_0 t + \pi) + \dots$$

$$(12.2.35)$$

where the signal in the phasor domain is

$$X_k = c_0 + \sum_{k=1}^{\infty} |c_k| e^{j(k\omega_0 t + \phi_k)} = c_0 + \sum_{k=-1}^{-\infty} |c_k| e^{j\phi_k} e^{j(k\omega_0 t)} + \sum_{k=1}^{\infty} |c_k| e^{j\phi_k} e^{j(k\omega_0 t)}$$

$$= \frac{1}{2} + |c_{-1}| e^{j(-\omega_0 t)} + |c_{-3}| e^{-j\pi} e^{j(-3\omega_0 t)} + |a_{-5}| e^{j(-5\omega_0 t)} + |a_{-7}| e^{-j\pi} e^{j(-7\omega_0 t)} + \dots$$

$$+ |c_1| e^{j(\omega_0 t)} + |c_3| e^{j\pi} e^{j(3\omega_0 t)} + |a_5| e^{j(5\omega_0 t)} + |a_7| e^{j\pi} e^{j(7\omega_0 t)} + \dots$$

$$= \frac{1}{2} + \frac{1}{\pi} e^{j(-\omega_0 t)} + \frac{1}{3\pi} e^{-j\pi} e^{j(-3\omega_0 t)} + \frac{1}{5\pi} e^{j(-5\omega_0 t)} + \frac{1}{7\pi} e^{-j\pi} e^{j(-7\omega_0 t)} + \dots$$

$$+ \frac{1}{\pi} e^{j(\omega_0 t)} + \frac{1}{3\pi} e^{j\pi} e^{j(3\omega_0 t)} + \frac{1}{5\pi} e^{j(5\omega_0 t)} + \frac{1}{7\pi} e^{j\pi} e^{j(7\omega_0 t)} + \dots$$

$$(12.2.36)$$

The real part of the signal in the phasor domain, $X_k$, gives us the signal in the time domain, $x(t)$.

## 12.2.4 Amplitude Spectra, Magnitude Spectra, and Phase Spectra of Periodic Signals

If a periodic signal $x(t) = x(t + T)$ is plotted versus time, it can be seen how the signal amplitudes are distributed with time. The complex Fourier coefficients $c_k(k\omega_0)$ and $c_k(kf_0)$ contain all the information about the signal $x(t)$. That is to say, if we know these coefficients, the time-domain signal $x(t)$ can be entirely reconstructed from an infinite sum of the complex exponential components. Since $c_k(kf_0)$ represents a sequence of complex quantities, expressed as

$$c_k = | c_k | \, e^{j\phi_k}, \tag{12.2.37}$$

the information about $x(t)$ can be obtained if the Fourier coefficients are plotted versus multiples of the fundamental frequency $\omega_0$. In the general case, the coefficients are presented in complex form and cannot be graphically presented as the amplitude spectrum (although they can be in some special cases). However, the magnitudes and phases of the coefficients can be calculated, and their distribution in the frequency domain can be found.

**Definitions.** In this analysis, we will use the following definitions:

- The plot of Fourier coefficients $c_k$ versus $k$ (or frequencies $k\omega_0$ or $kf_0$) is called the *amplitude spectrum* of the signal or its *amplitude response*.

- The plot of the absolute values $|c_k|$ (or $\sqrt{a_k^2 + b_k^2}/2$) versus $k$ (or frequencies $k\omega_0$ or $kf_0$) is called the *magnitude spectrum* of the signal or the *magnitude response*.

- The plot of the values of arguments or phases $\phi_k$ versus $k$ (or frequencies $k\omega_0$ or $kf_0$) is called the *phase spectrum* or the *phase response* of the signal.

These spectra are analysed for the rectangular pulse train and discussed in detail in previous examples for the calculated Fourier coefficients $a_k$. The notion of negative frequency was introduced, and the presentation of the same signal in the time, frequency, and phasor domains was discussed. In this section, the presentation of the same signal in all three domains will be discussed, using the Fourier coefficients $c_k$, which are calculated using the Fourier integral.

Since the frequency is a discrete variable, the magnitude spectrum and the phase spectrum of the periodic signal are represented by vertical lines. For this reason, the spectra of periodic functions are traditionally called *line spectra*.

**Continuation of the example from Sections 12.2.2 and 12.2.3.** For the signal presented by its amplitude spectrum in Figure 12.2.10, and its phasor expression, we will here find the phasor diagrams, the magnitude spectrum, and the phase spectrum.

The *phasor diagram* can be obtained from the presentation of the signal in the phasor domain derived in eqn (12.2.36). The phasors are composed of a DC component $c_0$, terms that rotate with negative frequencies defined by $k < 0$, and terms that rotate with positive frequencies defined by $k > 0$. There is a clear distinction between phasors rotating in opposite directions. For the phasors rotating with positive frequency (counter-clockwise), the initial angles are 0 or $\pi$, while the initial angles of the oppositely rotating phasors (those rotating clockwise) are 0 or $-\pi$. For the sake of explanation and clarity, the phasor diagram is presented in two figures: Figure 12.2.11 for positive frequencies, including the DC component, and Figure 12.2.12 for negative frequencies. Of course, the complete phasor diagram of the signal is easily obtained when these two figures overlap.

The *magnitudes* of the spectral components can be obtained directly as the modulus of the phasors and calculated as

$$| c_k | = \begin{cases} 0.5 & k = 0 \\ (1, 1/3\pi, 1/5\pi, 1/7\pi, \ldots,) & k \text{ odd, pozitive} \\ (1, 1/3\pi, 1/5\pi, 1/7\pi, \ldots,) & k \text{ odd, negative} \end{cases}. \qquad (12.2.38)$$

For $k < 0$, the corresponding phases are calculated as

$$\phi_k = \tan^{-1} \frac{\mathrm{Im}\,(c_k)}{\mathrm{Re}\,(c_k)} = \tan^{-1} \frac{\sin(-\pi)}{\cos(-\pi)} = \begin{cases} -\pi & k = -3, -7, -11, -15, \ldots \\ 0 & \text{otherwise} \end{cases} \qquad (12.2.39)$$

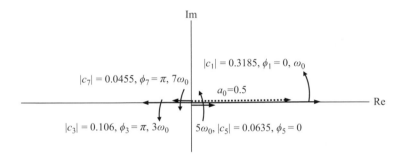

**Fig. 12.2.11** *Phasor diagram for positive frequencies of the signal.*

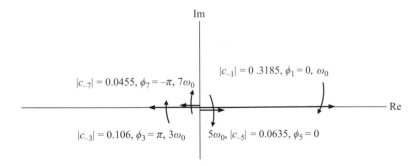

**Fig. 12.2.12** *Phasor diagram for negative frequencies of the signal.*

and, for $k > 0$, the phases are

$$\phi_k = \tan^{-1} \frac{\text{Im}(c_k)}{\text{Re}(c_k)} = \tan^{-1} \frac{\sin(\pi)}{\cos(\pi)} = \begin{cases} \pi & k = 3, 7, 11, 15, \ldots \\ 0 & \text{otherwise} \end{cases}. \qquad (12.2.40)$$

The magnitude spectrum and the phase spectrum are presented in Figure 12.2.13. The magnitudes and phases can be easily obtained from the calculated values of the Fourier coefficients: $c_0 = 0.5$, $c_{\pm 1} = 0.3185$, $c_{\pm 3} = -0.106$, $c_{\pm 5} = 0.0635$, $c_{\pm 7} = -0.0455$.

It is important to note that the initial angles for negatively rotating phasors are presented by their negative values. In the literature, some authors mix these frequencies, which can lead to controversial repercussions. If an initial angle for a positively rotating phasor is, for example, $3\pi/2$, we can only replace it by $-\pi/2$ in calculations; however, we need to keep in mind that the phasor actually rotates counter-clockwise. For that reason, to avoid any ambiguity, we presented a strict definition of the phases for the positive and negative frequencies. Some authors say the phases in Figure 12.2.13 can change their sign while preserving the symmetry, having positive values for negative frequencies, and negative values for positive frequencies. However, according to our observations and mathematical expressions, that would be incorrect. This kind of mistake is even more serious if we are analysing more complicated signals, like those in communication systems.

This issue raises the question posed in a number of books about the nature of negative frequencies. We can argue that the negative frequencies and positive frequencies are equally 'natural' because both of them contain complete information about the signal they represent. Namely, the sum of all of the real parts of all the phasors give us the signal in the time domain,

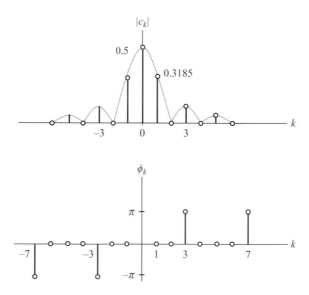

**Fig. 12.2.13** *Two-sided magnitude spectrum (upper panel) and phase spectrum (lower panel).*

and the sum of all of the magnitudes with their related phase angles gives us the same signal in the time domain. Furthermore, if the magnitude spectrum with its related phases is flipped to the left or to the right with respect to the $y$-axis, the newly obtained frequency components summed together will give the same signal. Therefore, the notion of negative frequencies has the same meaning as the notion of positive frequencies, and the voltages obtained by two phasors rotating in opposite directions can be considered to be a combination of the voltages at the output of two identical generators rotating synchronously in opposite directions. Both generators are equally 'natural'.

## 12.2.5   The Power and Energy of Signals, and Parseval's Theorem

In practice, we need to know the power and energy of a signal. It has been shown that a signal can be entirely presented in the time and frequency domains and complete information about the signal can be taken in any of these two domains. Therefore, we may expect that the signal's power can be found in either of these domains. The method to calculate the power of a periodic deterministic complex signal is specified by *Parseval's theorem*, which states that, for a periodic signal $x(t) = x(t + T)$, with period $T$, the average normalized power over one period is defined as

$$P_{av} = \frac{1}{T} \int_{-T/2}^{T/2} |x(t)|^2 dt, \qquad (12.2.41)$$

which is equivalent to the power calculated in the frequency domain as

$$P_{av} = \sum_{k=-\infty}^{\infty} |c_k|^2 = c_0^2 + 2 \sum_{k=1}^{\infty} |c_k|^2. \qquad (12.2.42)$$

**Proof.** For a complex-valued signal $x(t)$, we obtain

$$\begin{aligned} P_{av} &= \frac{1}{T} \int_{-T/2}^{T/2} x(t) x^*(t) dt = \frac{1}{T} \int_{-T/2}^{T/2} x(t) \sum_{k=-\infty}^{\infty} c_k^* e^{-jk\omega_0 t} dt \\ &= \sum_{k=-\infty}^{\infty} c_k^* \frac{1}{T} \int_{-T/2}^{T/2} x(t) e^{-jk\omega_0 t} dt = \sum_{k=-\infty}^{\infty} c_k^* c_k = \sum_{k=-\infty}^{\infty} |c_k|^2 = c_0^2 + 2 \sum_{k=1}^{\infty} |c_k|^2 \end{aligned}, \qquad (12.2.43)$$

which completes our proof. Similarly, the average power can be calculated using the Fourier coefficients $a_k$ and $b_k$. Firstly, from

$$c_k = \frac{1}{2} (a_k - jb_k), \qquad (12.2.44)$$

and

$$c_k^* = \frac{1}{2}\left(a_k + jb_k\right),$$
(12.2.45)

we can obtain

$$|c_k|^2 = c_k^* c_k = \frac{1}{4}\left(a_k^2 + b_k^2\right),$$
(12.2.46)

and then express the power in the form

$$P_{av} = \sum_{k=-\infty}^{\infty} \frac{1}{4}\left(a_k^2 + b_k^2\right) = a_0^2 + \sum_{k=1}^{\infty} \frac{1}{2}\left(a_k^2 + b_k^2\right).$$
(12.2.47)

In summary, the power of the signal calculated in the time domain is equal to the power of the signal calculated in the frequency domain, that is,

$$P_{av} = \frac{1}{T}\int_{-T/2}^{T/2} |x(t)|^2 dt = c_0^2 + 2\sum_{k=1}^{\infty} |c_k|^2 = a_0^2 + \sum_{k=1}^{\infty} \frac{1}{2}\left(a_k^2 + b_k^2\right).$$
(12.2.48)

The average power calculated in one period of the signal has a finite value. However, by definition, the signal exists in the interval ranging $-\infty$ to $+\infty$. Therefore, the energy of the signal can be calculated in one period with a finite value

$$E = P_{av}T = \int_{-T/2}^{T/2} |x(t)|^2 dt,$$
(12.2.49)

or in an infinite period with infinite value. In practice, we deal with signals of finite duration and finite energy.

**The power spectrum.** Using the power spectral components defined by eqn (12.2.48), we can find the power spectrum of the signal $x(t)$. The power spectrum is defined as a plot of $|c_k|^2$ or $\left(a_k^2 + b_k^2\right)/4$ versus frequency defined by the discrete variable $k$ (or $k\omega_0$ or $kf_0$) ranging from $-\infty$ to $+\infty$. This spectrum can be *one-sided* or *two-sided*, depending on the expression for power. If the power values are expressed in terms of the Fourier coefficients $a_k$ and $b_k$, the spectrum is called a *one-sided power spectrum*. If the power values are expressed in terms of the Fourier coefficient $c_k$, the spectrum is called a *two-sided power spectrum*.

**The root mean square value of the signal.** Traditionally, the root mean square value of a signal is used in electrical engineering to calculate the power of the signal. This value can be expressed in terms of Fourier coefficients as

$$x_{RMS} = \sqrt{\frac{1}{T}\int_{-T/2}^{T/2}|x(t)|^2 dt} = \sqrt{a_0^2 + \sum_{k=1}^{\infty}\frac{1}{2}(a_k^2 + b_k^2)} = \sqrt{c_0^2 + 2\sum_{k=1}^{\infty}|c_k|^2}, \qquad (12.2.50)$$

where the subscript *RMS* stands for 'root mean square'. Because the root mean square value of the *k*th harmonic is

$$x_{RMS}(kf_o) = \sqrt{\frac{1}{2}(a_k^2 + b_k^2)} = \sqrt{2|c_k|^2}, \qquad (12.2.51)$$

the root mean square value of the signal, which contains a DC component and AC components, is

$$x_{RMS} = \sqrt{(DC)^2 + \sum_{k=1}^{\infty}x_{RMS}^2(kf_o)}. \qquad (12.2.52)$$

**Calculation of power for the example presented in Sections 12.2.2 and 12.2.3.** In this part, we will analyse a signal with respect to its phase and energy, using the following approach: we will find the components of the Fourier series for the symmetric rectangular pulse train $x(t)$ shown in Figure 12.2.2 and defined by eqn (12.2.10), assuming that the amplitude of the pulse is $A$, and the duration of the pulse is equal to half of the signal period, that is, the condition $t_0 = T/2$ is fulfilled; next, we will calculate the average power of the signal and the corresponding energy in the case when the amplitude of the pulse is $A = 5$ V and the period of the train is $T = 1$ ms; then, we find the power in the first arcade (main lobe), defined by the bandwidth $BW = 2f_0$ and denoted as $P_{BW}$; finally, we will calculate the percentage of the average power of the signal $P_{BW}\%$, which is contained in the first arcade.

Following the procedure resulting in eqn (12.2.29) and bearing in mind that the amplitude of the pulse is $A$, the coefficients of the Fourier series can be calculated as

$$c_k = \frac{At_0}{T}\frac{\sin k\pi t_0/T}{k\pi t_0/T}. \qquad (12.2.53)$$

For $k = 0$, the indeterminate case is in place. This case is readily resolved by noting that $\sin(x) \to 0$ when $x \to 0$. Thus, $\sin(x)/x \to 1$ when $\sin(x) \to 0$ and $x \to 0$. Therefore, the coefficients are given by eqn (12.2.30). When $t_0 = T/2$ and the amplitude $A$, the Fourier series coefficients can be expressed as

$$c_k = \begin{cases} A/2 & k = 0 \\ \frac{A}{2}\frac{\sin k\pi/2}{k\pi/2} = A\frac{(-1)^{(k-1)/2}}{k\pi} & k \text{ odd} \\ 0 & k \text{ even} \end{cases}, \qquad (12.2.54)$$

with the values $c_0 = 0.5A$, $c_{\pm 1} = 0.3185A$, $c_{\pm 3} = -0.106A$, $c_{\pm 5} = 0.0635A$, and $c_{\pm 7} = -0.0455A$. The coefficients, as a function of $k$, constitute the amplitude line spectrum of the signal that is presented in Figure 12.2.14. The dotted line presents the shape of the discrete frequency spectrum and has no other meaning. That line crosses the $x$-axis at the points specified by $\sin k\pi/2 = \pm l\pi$ or $k = \pm 2l$, for $l = 1, 2, 3, \ldots$. The first zero crossing defines the first arcade of the spectrum, where most of the signal power is concentrated.

**Calculation of the power of the signal.** The average power can be calculated in both the frequency domain and the time domain. In the frequency domain, power is expressed as

$$
\begin{aligned}
P_{av} &= \sum_{k=-\infty}^{\infty} |c_k|^2 = \sum_{k=-\infty}^{\infty} \left( \frac{A \sin k\pi/2}{2k\pi/2} \right)^2 = \frac{A^2}{4} + 2\frac{A^2}{\pi^2} \sum_{k=1}^{\infty} \left( \frac{\sin k\pi/2}{k} \right)^2 \\
&= \frac{A^2}{4} + 2\frac{A^2}{\pi^2} \sum_{k=1, k=\text{odd}}^{\infty} \frac{1}{k^2}
\end{aligned}
\tag{12.2.55}
$$

which shows the values of all of the spectral components. These values are inversely proportional to $k$ squared. The two-sided power spectrum is shown in Figure 12.2.15.

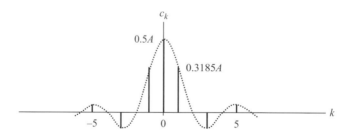

**Fig. 12.2.14** *Two-sided amplitude spectrum.*

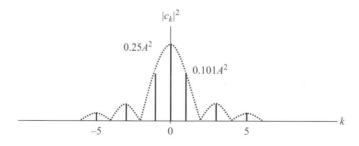

**Fig. 12.2.15** *Two-sided power spectrum.*

Therefore, the average power of the signal is

$$P_{av} = \frac{A^2}{4} + 2\frac{A^2}{\pi^2}\left(1 + 3^{-2} + 5^{-2} + \dots\right) = \frac{A^2}{4} + \frac{2A^2}{\pi^2}\pi^2/8 = \frac{A^2}{2}, \tag{12.2.56}$$

where the sum of the series components in the brackets is $\pi^2/8$. We can find the average power in the time domain as

$$P_{av} = \frac{1}{T}\int_{-T/2}^{T/2}|x(t)|^2\,dt = \frac{1}{T}\int_{-t_0/2}^{t_0/2}|A|^2\,dt = \frac{A^2}{T}t_0 = \frac{A^2}{2} = \frac{25}{2} = 12.5\text{ W},$$

which is the same as the power in the frequency domain, as we expected, given Parseval's theorem. The root mean square value of the signal can be calculated as

$$x_{RMS} = \sqrt{P_{av}} = \sqrt{\frac{1}{2}A^2} = \frac{\sqrt{2}}{2}A.$$

The energy dissipated in one period $T = 1$ ms of the signal across a resistor of unit resistance, $R = 1$ ohm, is

$$E = P_{av}T = \frac{A^2}{2R}T = 12.5 \cdot 10^{-3}\text{ joules.}$$

**The power in the main lobe.** The power in the first arcade is the sum of the power of the three power components inside the arcade and is expressed as

$$P_{BW} = 2 \cdot 0.101A^2 + 0.25A^2 = 0.452A^2,$$

which is expressed as the percentage of the average signal power as

$$P_{BW}\% = \frac{0.452A^2}{A^2/2}100\% = 90.4\%.$$

Therefore, the power contained in the first arcade is over 90% of the average signal power.

## 12.2.6　Existence of Fourier Series

A Fourier series exists if the following three Dirichlet conditions are satisfied:

1. absolute integrability: the integral of the absolute signal value is finite, that is,

$$\frac{1}{T} \int_{-T/2}^{T/2} | x(t) | \, dt < \infty \qquad (12.2.57)$$

2. bounded variation of the signal: the signal $x(t)$ has a finite number of maxima or minima in its period,

3. finite number of discontinuities: the signal $x(t)$ has a finite number of discontinuities in its period, and the values of the signal at all discontinuities have finite values.

If these three conditions are satisfied, then the signal $x(t)$ can be perfectly reconstructed from their Fourier coefficients.

## 12.2.7   Orthogonality Characteristics of the Fourier Series

Orthogonal signals, as defined in Chapter 2, are widely used in communication systems, as can be seen from the first ten chapters of this book. We will investigate the orthogonality of the components inside the Fourier series. As we previously said, using Fourier series, a periodic signal can be represented by an infinite sum of exponential functions or sinusoidal functions expressed as

$$x(t) = \sum_{k=-\infty}^{\infty} c_k e^{jk\omega_0 t} = \sum_{k=-\infty}^{\infty} c_k \left( \cos k\omega_0 t + j \sin k\omega_0 t \right). \qquad (12.2.58)$$

We can also say that the signal is decomposed into an infinite sum of orthogonal harmonic components, where orthogonality is defined as follows: any two harmonic components in the series, having frequency $k_1 \omega_0$ and $k_2 \omega_0$, are orthogonal if the integral of their product, calculated in the fundamental period $T$, is 0.

We can demonstrate the orthogonality principle by analysing a series of harmonic components with real coefficients that is expressed as

$$x(t) = \sum_{k=-\infty}^{\infty} c_k \cos k\omega_0 t. \qquad (12.2.59)$$

We will find the integral of the given components in the fundamental period $T$ expressed as

$$I = \int_0^T c_{k1} c_{k2} \cos(k_1 \omega_0 t) \cos(k_2 \omega_0 t) \, dt = \frac{c_{k1} c_{k2}}{2} \int_0^T \left( \cos(k_1 + k_2) \omega_0 t + \cos(k_1 - k_2) \omega_0 t \right) dt$$

$$= \frac{c_{k1} c_{k2}}{2} \int_0^T \left( \cos(k_1 + k_2) \omega_0 t \right) dt + \frac{c_{k1} c_{k2}}{2} \int_0^T \left( \cos(k_1 - k_2) \omega_0 t \right) dt = 0$$

Because the frequencies of harmonics are defined by the whole numbers $k_1$ and $k_2$, their sum and difference contain the whole number of their oscillations in the fundamental period.

Therefore, the value of both integrals is 0, and their sum $I$ is 0, which proves that the harmonic components are orthogonal.

## 12.2.8  Table of the Fourier Series

The basic relations, related to the Fourier series, are presented in Table 12.2.1. The signals are expressed in voltages and the power is expressed in watts, assuming that the load resistor is 1 ohm.

**Table 12.2.1** *Fourier series.*

| Signals/Theorems | Time Domain | Frequency Domain |
|---|---|---|
| Trigonometric form | $x(t) = a_0 + \sum\limits_{k=1}^{\infty} a_k \cos k\omega_0 t$ $+ \sum\limits_{k=1}^{\infty} b_k \sin k\omega_0 t$ | $a_0 = \frac{1}{T} \int_{-T/2}^{T/2} x(t) dt$ $a_k = \frac{2}{T} \int_{-T/2}^{T/2} x(t) \cos k\omega_0 t\, dt$ $b_k = \frac{2}{T} \int_{-T/2}^{T/2} x(t) \sin k\omega_0 t\, dt$ |
| Complex (exponential) form | $x(t) = \sum\limits_{k=-\infty}^{\infty} c_k e^{+jk\omega_0 t}$ | $c_k = \frac{1}{T} \int_{-T/2}^{T/2} x(t) e^{-jk\omega_0 t} dt$ |
| Cosine function | $A \cos \omega_0 t = \frac{A}{2}\left(e^{+j\omega_0 t} + e^{-j\omega_0 t}\right)$ | $c_1 = c_{-1} = \frac{A}{2},$ $c_k = 0, \text{for } k \neq \pm 1$ |
| Sine function | $A \sin \omega_0 t = \frac{A}{2j}\left(e^{+j\omega_0 t} - e^{-j\omega_0 t}\right)$ | $c_1 = \frac{A}{2j}, c_{-1} = -\frac{A}{2j}$ $c_k = 0, \text{for } k \neq \pm 1$ |
| Rectangular symmetric pulse of amplitude $A$, duration $t_0$, and period $T$ | $x(t) = A \text{ for } |t| \leq \frac{t_0}{2}$ | $c_k = \begin{cases} A t_0/T & k = 0 \\ \frac{t_0}{T} \frac{\sin k\pi t_0/T}{k\pi t_0/T} & k \neq 0 \end{cases}$ |
| Triangular symmetric pulse train, duration $t_0$ inside the period $T$ | $x(t) = A\left(2 - 2\frac{|t|}{t_0}\right), -\frac{t_0}{2} \leq t < \frac{t_0}{2}$ | $c_k = \frac{A t_0}{2T} \text{sinc}^2\left(\pi k t_0/2T\right)$ |
| Parseval's theorem | $P = \frac{1}{T} \int_{-T/2}^{T/2} |x(t)|^2 dt$ | $P = \sum\limits_{k=-\infty}^{\infty} \frac{1}{4}\left(a_k^2 + b_k^2\right)$ $= a_0^2 + \frac{1}{2} \sum\limits_{k=1}^{\infty} \left(a_k^2 + b_k^2\right)$ $P = \sum\limits_{k=-\infty}^{\infty} |c_k|^2 = c_0^2 + 2 \sum\limits_{k=1}^{\infty} |c_k|^2$ |
| Autocorrelation function | $R_x(\tau) = \frac{1}{T} \int_{-T/2}^{T/2} x(t-\tau) x(t) dt$ | $S_x(f) = \int_{-\infty}^{\infty} R_x(\tau) e^{-j2\pi f \tau} d\tau$ |

## 12.3 Fourier Transform of Continuous Signals

The theory of the Fourier series is limited in application to periodic signals. Using that theory, it is possible to describe any periodic signal as a linear combination of orthogonal sinusoidal signals. However, it cannot be used to analyse aperiodic signals.

In this section, the Fourier transform, which is used for the analysis of aperiodic signals, will be presented. It will be shown that the Fourier series is a special case of the Fourier transform. The amplitude spectrum obtained by the Fourier series will be replaced by the amplitude spectral density that is obtained by the Fourier transform. Based on the amplitude spectral density, the expression for the energy spectral density will be derived.

In order to relate this energy spectral density to the autocorrelation function of a signal, the correlation and cross-correlation properties of deterministic signals will be defined and analysed. Finally, the relationship between the energy spectral density and the autocorrelation function of a signal, as defined by the Wiener–Khintchine theorem, will be established. It will be shown that the energy spectral density and the autocorrelation function form the Fourier transform pair.

Periodic signals can be considered to be special cases of aperiodic signals. Consequently, the presentation of periodic signals in the frequency domain can be obtained from their Fourier transform instead of the Fourier series and expressed in terms of Dirac delta functions. In contrast to the discrete frequency representation of the magnitude spectrum and phase spectrum obtained by the Fourier series, the magnitude spectral density and phase spectral density of periodic signals obtained by the Fourier transform will be expressed as functions of continuous frequency. The Fourier transform of periodic signals will be separately investigated in this section, due to its importance in communication theory and practice.

### 12.3.1 Derivative and Application of Fourier Transform Pairs

**Derivation of Fourier transform pairs.** An aperiodic signal can be considered to be a periodic signal with infinite period $T$. As previously shown, a periodic signal can be represented by the Fourier series (12.2.26), where Fourier coefficients are calculated using eqn (12.2.27). If a measure of 'spectral amplitude per unit frequency' is defined as

$$X(kf_0) = \frac{c_k}{f_0}, \tag{12.3.1}$$

the signal $x(t)$ in eqn (12.2.26) can be expressed as

$$x(t) = \sum_{k=-\infty}^{\infty} X(kf_0)\, e^{jk\omega_0 t} f_0, \tag{12.3.2}$$

and, using eqn (12.2.27), the corresponding amplitudes per unit frequency as

$$X(kf_0) = \frac{c_k}{f_0} = \frac{1}{f_0 T} \int_{-T/2}^{T/2} x(t) e^{-jk2\pi f_0 t} dt. \tag{12.3.3}$$

If the signal $x(t)$ is aperiodic, which is equivalent to letting $T \to \infty$, then $f_0$ approaches the differential $df$, that is, $f_0 = 1/T \to df$, the product $Tf_0 \to 1$, and $kf_0$ becomes a continuous variable $f$, that is, $kf_0 \to f$. Then, the function of the discrete frequency $X(kf_0)$ approaches a function $X(f)$ of the continuous frequency $f$, that is,

$$X(f) = \lim_{T \to \infty} X(kf_0) = \lim_{T \to \infty} \frac{1}{f_0 T} \int_{-T/2}^{T/2} x(t) e^{-jk2\pi f_0 t} \, dt = \int_{-\infty}^{\infty} x(t) e^{-j2\pi ft} \, dt, \qquad (12.3.4)$$

which is called the *Fourier transform* of the aperiodic continuous-time signal $x(t)$. The time-domain signal $x(t)$ can be expressed as a limiting case of eqn (12.3.2) for $T \to \infty$ in the form

$$x(t) = \lim_{T \to \infty} \sum_{k=-\infty}^{\infty} X(kf_0) \, e^{jk2\pi f_0 t} f_0 = \int_{-\infty}^{\infty} X(f) e^{j2\pi ft} \, df, \qquad (12.3.5)$$

and is called the *inverse Fourier transform* of $x(t)$. Expressions (12.3.4) and (12.3.5) form the *Fourier transform pair*. These equations are also called a *symmetric transform pair* because of their similarity. Namely, the Fourier transform and its inverse are almost identical, being different only in the sign of the exponent and the variable of integration. Let us note the following:

- The signal $x(t)$ can be interpreted as a continuous sum of the complex exponentials $e^{j2\pi ft}$, each with the infinitesimal magnitude $|X(f)| df$. These complex exponentials oscillate for all time; even the signal $x(t)$ is a real signal that can be time limited. Thus, by choosing the weighted function $|X(f)|$ in the proper way, the complex exponentials will be able to cancel each other outside the time interval where the signal $x(t)$ exists. At the same time, they will be combined in the interval where the signal $x(t)$ exists in such a way as to make it possible to reconstruct the signal entirely.

- The spectral magnitude $|X(f)| df$ is infinitesimally small, tending to 0, because, for finite $|X(f)|$ and $df \to 0$, we have $|X(f)| df \to 0$.

- An additional note related to the notion of negative frequencies is as follows: the spectrum of the signal calculated using the Fourier transform will have positive and negative frequency components of the same magnitude but of the opposite phases. When calculating the magnitude and phase spectral densities, we need to bear in mind the previous discussions related to the spectrum of the periodic signals. Namely, for positive continuous frequencies, we need to preserve the positive initial values of the phases and, consistently, for the negative frequencies to preserve negative initial phases. Of course, in doing this, we need to take care that we are dealing here with the magnitude spectral density function and the phase spectral density function. This fact and the related procedures for presenting the magnitude spectral density and the phase spectral density will be demonstrated later on in some of the examples in this section.

In the case when the radian (angular) frequency is used, when $\omega = 2\pi f$, the *asymmetric transform pair* can be obtained as

$$X(\omega) = \int_{-\infty}^{\infty} x(t) e^{-j\omega t} dt, \tag{12.3.6}$$

which is the analysis equation, and

$$x(t) = \frac{1}{2\pi} \int_{-\infty}^{\infty} X(\omega) e^{j\omega t} d\omega, \tag{12.3.7}$$

which is the synthesis equation. A plot of $|X(\omega)|$ versus $\omega$ (or $|X(f)|$ vs $f$) gives the spectral magnitude per unit frequency or the *magnitude spectral density*. A plot of argument of $X(\omega)$ versus $\omega$ (or argument of $X(f)$ vs $f$) gives the spectral phase per unit frequency or the *phase spectral density*. Sometimes we can present the graph of the amplitude spectral density $X(\omega)$ as a function of the angular frequency $\omega$. When that is possible, the graphs are called the *amplitude spectral density*. We present the Fourier transform as $X(\omega)$ or $X(j\omega)$.

Generally speaking, the Fourier transform of an aperiodic signal is a complex function with both real and imaginary parts. For this function, we can define its magnitudes as

$$|X(j\omega)| = \sqrt{X(j\omega) X^*(j\omega)} = \sqrt{\mathrm{Re}^2 X(j\omega) + \mathrm{Im}^2 X(j\omega)} \tag{12.3.8}$$

and its phases as the argument of a complex function calculated as

$$\phi(\omega) = \arg(X(j\omega)) = \tan^{-1}(\mathrm{Re}\{X(j\omega)\} / \mathrm{Im}\{X(j\omega)\}). \tag{12.3.9}$$

The magnitude and phase spectral densities vary as the dependence of the magnitudes and phases on the frequency. Sometimes the Fourier transform can be used to directly plot the amplitude spectral density $X(\omega)$ that contains the magnitude spectral density and the phase spectral density (see 'Application of the Fourier transform', Figure 12.3.2, which shows the Fourier transform calculated for the rectangular pulse and expressed by eqn (12.3.10)).

**Application of the Fourier transform.** To demonstrate the application of the Fourier transform, we will analyse, in the time and frequency domains, the rectangular pulse shown in Figure 12.3.1.

The Fourier transform is

$$X(j\omega) = \int_{-\infty}^{\infty} x(t) e^{-j\omega t} dt = \int_{-t_0/2}^{t_0/2} A \cdot e^{-j\omega t} dt = \frac{A}{-j\omega} \left( e^{-j\omega t_0/2} - e^{j\omega t_0/2} \right),$$

$$= A t_0 \frac{\sin \omega t_0/2}{\omega t_0/2} = A t_0 \operatorname{sinc} \omega t_0/2 \tag{12.3.10}$$

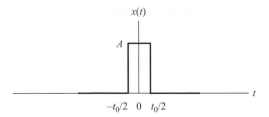

**Fig. 12.3.1** *Rectangular pulse.*

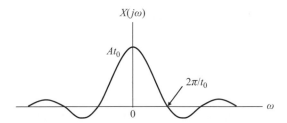

**Fig. 12.3.2** *Fourier transform of a rectangular pulse.*

where the sinc function is defined as $\text{sinc}\, y = (\sin y)/y$. The transform is shown in Figure 12.3.2. Due to the expression $\lim_{y \to 0} (\sin y/y) = 1$, the maximum value of the transform is $At_0$. The zero crossing occurs when $\omega t_0/2 = \pm\, l\pi$ or $\omega = \pm\, 2l\pi/t_0$ for $l = 1, 2, 3, \dots$ . The first zero crossing occurs at $\omega = \pm\, 2\pi/t_0$. Phases for the spectrum can be found from the conditions: $\sin \omega t_0/2 = \pm 1$, $\omega t_0/2 = \pm l\pi/2$, $\omega = \pm l\pi/t_0$. For example, the first zero crossing is at the frequency $\omega = \pm 2\pi/t_0$, and the first phase change is also at this frequency.

The position of the first zero crossing depends on the duration of the rectangular signal $t_0$. Generally speaking, the relationship between the signal duration in the time domain and the first zero crossing can be expressed by the following rule: *the duration of a signal in the time domain is inversely proportional to the bandwidth of the function in the frequency domain*. This is one form of the Heisenberg uncertainty principle. In our example, as the pulse duration $t_0$ decreases, the main lobe of the amplitude density function becomes wider, and vice versa. When the width of the pulse approaches 0, the amplitude spectral density approaches a constant value for all frequencies. In conclusion, we can say that, if a signal has a finite duration in the time domain, its amplitude spectrum has infinite bandwidth, and vice versa. This conclusion is of paramount importance in communication systems. It tells us that we must sacrifice bandwidth to reduce the signal in the time domain, and vice versa.

Using the amplitude spectral density, we may calculate the magnitude spectral density as

$$|X(j\omega)| = \sqrt{X(j\omega)X^*(j\omega)} = \sqrt{|At_0 \,\text{sinc}\, \omega t_0/2|^2} = |\,At_0 \,\text{sinc}\, \omega t_0/2\,| \qquad (12.3.11)$$

because the phases of $\pi$ inside the amplitude spectral density and its conjugate cancel each other out. These phases can be calculated as

$$\phi(\omega) = \arg(X(j\omega)) = \arg(At_0) + \arg(\text{sinc } \omega t_0/2). \qquad (12.3.12)$$

In this case, the amplitude $A$ and duration of the pulse $t_0$ are positive values, and the phase of the first term is 0, that is, $\arg(At_0) = 0$. The second term has the values $+\pi$ or $-\pi$, depending on the changes in the sign of the amplitude spectrum shown in Figure 12.3.2, and can be calculated for the arcades starting with the defined frequencies and expressed as

$$\phi(\omega) = \begin{cases} \pm\pi & \omega = \pm 1 \cdot 2\pi/t_0, \quad \pm 3 \times 2\pi/t_0, \quad \pm 5 \times 2\pi/t_0, \quad \pm 7 \times 2\pi/t_0 \ldots \\ 0 & \text{otherwise} \end{cases}. \qquad (12.3.13)$$

The first phase can be calculated for the arcade that starts with positive $\omega = +2\pi/t_0$ and ends with the frequency $\omega = +2 \cdot 2\pi/t_0$, where the amplitude spectrum is negative, meaning that the phase shifts of all of the frequencies are $+\pi$, due to the counter-clockwise rotation of any infinitesimally small spectral component inside the arcade's bandwidth. In this bandwidth, the amplitude spectrum can be expressed in terms of the magnitudes and phases as follows:

$$X(j\omega) = |X(j\omega)| \, e^{j\phi} = -|At_0 \text{ sinc } \omega t_0/2| = |At_0 \text{ sinc } \omega t_0/2| \, e^{j\pi}, \qquad (12.3.14)$$

where the magnitude obviously is $|At_0 \text{ sinc } \omega t_0/2|$, and the phase is precisely calculated as

$$\phi(\omega) = \arg(X(j\omega)) = \tan^{-1}(\sin\pi/\cos\pi) = \pi. \qquad (12.3.15)$$

The corresponding arcade of the spectral density for negative frequencies will have the same magnitudes and the phase angle $-\pi$. The magnitudes and phases are calculated from the Fourier transform as

$$X(j\omega) = |X(j\omega)| \, e^{j\phi} = -|At_0 \text{ sinc } \omega t_0/2| = |At_0 \text{ sinc } \omega t_0/2| \, e^{-j\pi}. \qquad (12.3.16)$$

The magnitude spectral density and phase spectral density, expressed as the dependence of the magnitudes and the phases on the frequency, respectively are presented in Figures 12.3.3 and 12.3.4, respectively. The phase changes occur at the same frequencies as the zero crossings.

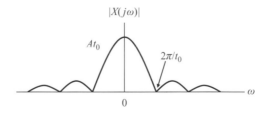

**Fig. 12.3.3** *Magnitude spectral density of a rectangular pulse.*

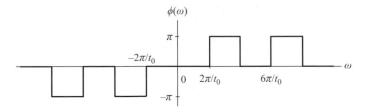

**Fig. 12.3.4** *Phase spectral density of a rectangular pulse.*

## 12.3.2 Convergence Conditions

For a given signal $x(t)$, it is sufficient to fulfil the following Dirichlet conditions in order to guarantee the existence of the Fourier transform:

Condition 1: The signal $x(t)$ has a finite number of finite discontinuities.

Condition 2: The signal $x(t)$ has a finite number of maxima and minima in the time domain.

Condition 3: The function $x(t)$ is absolutely integrable (except in some pathological cases that have no obvious engineering use), that is,

$$\int_{-\infty}^{\infty} |x(t)| \, dt \leq \infty. \tag{12.3.17}$$

The Fourier transform will exist in the case when the signal $x(t)$ is square integrable (or when it is an energy signal), that is,

$$\int_{-\infty}^{\infty} |x(t)|^2 dt \leq \infty. \tag{12.3.18}$$

If a signal $x(t)$ is absolutely integrable, it is also square integrable. Therefore, the Dirichlet conditions are sufficient but not necessary for the existence of the Fourier transform.

## 12.3.3 The Rayleigh Theorem and the Energy of Signals

This theorem is an equivalent to Parseval's theorem but for the Fourier transform. The theorem states that the energy of a signal in the time domain is equal to the energy of that signal in the frequency domain. Let us prove this theorem.

The energy of a complex signal $x(t)$ is

$$E = \int_{-\infty}^{\infty} |x(t)|^2 dt = \int_{-\infty}^{\infty} x(t) x^*(t) dt \tag{12.3.19}$$

where

$$x^*(t) = \int_{-\infty}^{\infty} X^*(f) e^{-j2\pi ft} df. \tag{12.3.20}$$

Thus,

$$E = \int_{-\infty}^{\infty} x(t) \int_{-\infty}^{\infty} X^*(f) e^{-j2\pi ft} df dt = \int_{-\infty}^{\infty} X^*(f) \left[ \int_{-\infty}^{\infty} x(t) e^{-j2\pi ft} dt \right] df$$

$$= \int_{-\infty}^{\infty} X(f) X^*(f) df = \int_{-\infty}^{\infty} |X(f)|^2 df. \tag{12.3.21}$$

Finally,

$$E = \int_{-\infty}^{\infty} |x(t)|^2 dt = \int_{-\infty}^{\infty} |X(f)|^2 df, \tag{12.3.22}$$

which means that the energy in the time domain is equal to the energy in the frequency domain, or the Fourier transform is a lossless transform. Since the energy $E$ can be calculated by integrating the magnitude spectral density squared, $|X(f)|^2$, the spectrum $|X(f)|^2$ is referred to as the *energy spectral density* or *energy-density spectrum*, and denoted by

$$S_x(f) = |X(f)|^2. \tag{12.3.23}$$

Similarly, for $\omega = 2\pi f$, the energy is

$$E = \int_{-\infty}^{\infty} |x(t)|^2 dt = \int_{-\infty}^{\infty} |X(\omega)|^2 \frac{d\omega}{2\pi} = \frac{1}{2\pi} \int_{-\infty}^{\infty} S_x(\omega) d\omega, \tag{12.3.24}$$

where the energy spectral density is expressed in the form

$$S_x(\omega) = |X(\omega)|^2. \tag{12.3.25}$$

### 12.3.4    Properties of the Fourier Transform

The properties of the Fourier transform are of particular interest in communication systems, because they simplify understanding signal-processing procedures in transceiver blocks and show the relationship between the presentation of the signal in the time domain and that in the frequency domain. For that reason, we will present the most important properties in this section.

**The time-domain case: The duality of convolution and multiplication.** The output of a continuous-time linear-time-invariant system with impulse response $h(t)$ to an arbitrary input $x(t)$ is given by the convolution

$$y(t) = x(t) * h(t) = \int_{-\infty}^{\infty} x(\tau) h(t-\tau) d\tau. \tag{12.3.26}$$

We then state the following: *convolution in the time domain corresponds to multiplication in the frequency domain.*

**Proof.** This can be proved by finding the corresponding Fourier transform as

$$Y(j\omega) = \int_{-\infty}^{\infty} y(t)e^{-j\omega t}dt = \int_{-\infty}^{\infty}\int_{-\infty}^{\infty} x(\tau)h(t-\tau)\,d\tau\,e^{-j\omega t}dt$$

$$= \int_{-\infty}^{\infty} x(\tau)\left[\int_{-\infty}^{\infty} h(t-\tau)\,e^{-j\omega t}dt\right]d\tau = \int_{-\infty}^{\infty} x(\tau)H(j\omega)e^{-j\omega\tau}d\tau. \qquad (12.3.27)$$

$$= H(j\omega)\int_{-\infty}^{\infty} x(\tau)e^{-j\omega\tau}d\tau = H(j\omega)\cdot X(j\omega)$$

Thus, *convolution* in the time domain corresponds to *multiplication* in the frequency domain, that is,

$$x(t)*h(t) \iff X(j\omega)\cdot H(j\omega), \qquad (12.3.28)$$

which is a property of great importance in the analysis of linear-time-invariant systems. Because the convolution in the time domain implies the product in the frequency domain, then, from the duality principle, it can be expected that the product in the time domain implies the convolution in the frequency domain. Let us show this is indeed the case. Let signal $y(t)$ be a product of two signals:

$$y(t) = x(t)\cdot h(t). \qquad (12.3.29)$$

Its Fourier transform is

$$Y(j\omega) = \int_{-\infty}^{\infty} y(t)e^{-j\omega t}dt = \int_{-\infty}^{\infty} x(t)\cdot h(t)e^{-j\omega t}dt = \int_{-\infty}^{\infty} x(t)\left[\frac{1}{2\pi}\int_{-\infty}^{\infty} H(j\lambda)e^{+j\lambda t}d\lambda\right]e^{-j\omega t}dt$$

$$= \frac{1}{2\pi}\int_{-\infty}^{\infty} H(j\lambda)\,d\lambda\int_{-\infty}^{\infty} x(t)e^{-j(\omega-\lambda)t}dt = \frac{1}{2\pi}\int_{-\infty}^{\infty} H(j\lambda)X[j(\omega-\lambda)]\,d\lambda$$

$$= \frac{1}{2\pi}H(j\omega)*X(j\omega).$$

$$(12.3.30)$$

Therefore, multiplication in the time domain corresponds to convolution in the frequency domain, that is,

$$x(t)\cdot h(t) \iff X(f)*H(f). \qquad (12.3.31)$$

**The frequency-domain case: The duality of multiplication and convolution.** Let signal $Y(f)$ be the product of two signals in the frequency domain:

$$Y(f) = X(f)\cdot H(f). \qquad (12.3.32)$$

Its Fourier transform is

$$
\begin{aligned}
y(t) &= \int_{-\infty}^{\infty} Y(f)e^{j2\pi ft}\,df = \int_{-\infty}^{\infty} H(f)X(f)e^{j2\pi ft}\,df \\
&= \int_{-\infty}^{\infty} H(f)\int_{-\infty}^{\infty} x(\tau)\,e^{-j2\pi f\tau}\,e^{j2\pi ft}\,d\tau\,df = \int_{-\infty}^{\infty} H(f)\int_{-\infty}^{\infty} x(\tau)\,e^{-j2\pi f(t-\tau)}\,d\tau\,df. \\
&= \int_{-\infty}^{\infty} x(\tau)\int_{-\infty}^{\infty} H(f)e^{j2\pi f(t-\tau)}\,d\tau\,df = \int_{-\infty}^{\infty} x(\tau)\,h(t-\tau)\;d\tau = x(t)*h(t)
\end{aligned}
$$

$$(12.3.33)$$

Therefore, multiplication in the frequency domain corresponds to convolution in the time domain. In summary, we find that

$$
x(t)*h(t) \iff X(f)\cdot H(f), \quad \text{or} \quad x(t)*h(t) \iff X(j\omega)\cdot H(j\omega),
$$

$$
x(t)\cdot h(t) \iff X(f)*H(f), \quad \text{or} \quad x(t)\cdot h(t) \iff \frac{1}{2\pi}X(j\omega)*H(j\omega),
$$

and

$$
X(f)\cdot H(f) \iff x(t)*h(t), \quad \text{or} \quad X(j\omega)\cdot H(j\omega) \iff x(t)*h(t) \tag{12.3.34}
$$

(also see Section 12.3.6., Tables 12.3.1 and 12.3.2, which present the basic properties of the Fourier transform).

**The modulation or shifting property.** The Fourier transform of the product of a signal $x(t)$ and a cosine signal produces two frequency components in the frequency domain; this is expressed as

$$
x(t)\cdot \cos\omega_0 t \iff X(f)*\frac{1}{2}\{\delta(f+f_0)+\delta(f-f_0)\} \tag{12.3.35}
$$

or

$$
x(t)\cdot \cos\omega_0 t \iff \frac{1}{2}[X(f+f_0)+X(f-f_0)]. \tag{12.3.36}
$$

In other words, the product of a signal and a cosine function in the time domain has a Fourier transform that contains the amplitude spectral densities of the signal, which are shifted to the negative and positive carrier frequencies and normalized by a factor of 2.

## 12.3.5   Important Problems and Solutions

In communication systems, we will extensively use Dirac delta functions and harmonic functions, which will be analysed in this section.

## Example

Find the Fourier transform of the delta function $x(t) = \delta(t)$.

**Solution.** The impulse function $x(t) = \delta(t)$ is not square integrable, since its square is not defined. However, it is absolutely integrable. It also satisfies the other Dirichlet requirements. Hence, its Fourier transform exists and is given by

$$X(j\omega) = \int_{-\infty}^{\infty} \delta(t) e^{-j\omega t} dt = e^{-j\omega 0} = 1. \qquad (12.3.37)$$

Thus, the magnitude and phase spectral densities are

$$|X(j\omega)| = 1, \arg X(j\omega) = 0, \qquad (12.3.38)$$

respectively, as shown in Figure 12.3.5.

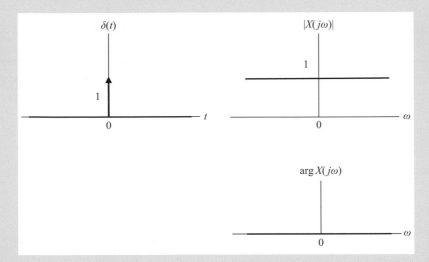

**Fig. 12.3.5** *Delta function (left) with its magnitude spectral density (upper right) and phase spectral density (lower right).*

With these graphs, we can conclude the following:

1.  The frequency components are equal in magnitude and phase for all $f = -\infty$ to $+\infty$, because

$$x(t) = \int_{-\infty}^{\infty} \left( \frac{1}{2\pi} X(j\omega) d\omega \right) e^{+j\omega t}. \qquad (12.3.39)$$

2.  The Heisenberg uncertainty principle is satisfied.

**Example**

Find the Fourier transform of the time-shifted impulse function $x(t) = \delta(t - t_0)$.

**Solution.** The Fourier transform of the time-shifted impulse function is

$$X(j\omega) = \int_{-\infty}^{\infty} \delta(t - t_0) e^{-j\omega t} dt = e^{-j\omega t_0} = \cos(\omega t_0) - j\sin(\omega t_0). \tag{12.3.40}$$

The magnitude and phase spectral densities are

$$|X(j\omega)| = 1, \tag{12.3.41}$$

and

$$\phi(\omega) = \arg X(j\omega) = \arctan \frac{-\sin(\omega t_0)}{\cos(\omega t_0)} = -\arctan \frac{\sin(\omega t_0)}{\cos(\omega t_0)} = -\omega t_0, \tag{12.3.42}$$

respectively. They are shown in Figure 12.3.6. The time shift causes the phase change.

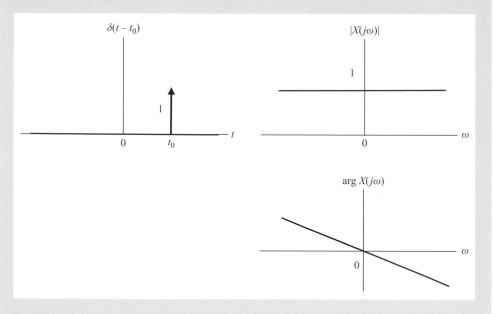

**Fig. 12.3.6** *Time-shifted impulse function (left) and its spectra (right).*

## Example

Find the inverse Fourier transform of a power spectral density defined by the shifted delta functions $X(f) = \delta(f - f_0)$, and $X(f) = \delta(f + f_0)$.

**Solution.** For the amplitude spectral density defined by

$$X(f) = \delta(f - f_0), \tag{12.3.43}$$

the inverse Fourier transform exists and is given by

$$x(t) = \int_{-\infty}^{\infty} \delta(f - f_0) e^{j2\pi ft} df = e^{+j2\pi f_0 t} = \cos 2\pi f_0 t + j \sin 2\pi f_0 t. \tag{12.3.44}$$

This signal is complex and periodic. It is composed of two harmonic components, which are shown in Figure 12.3.7.

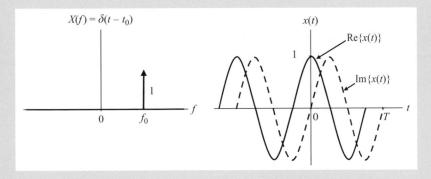

**Fig. 12.3.7** *Signal (left) and inverse Fourier transform (right), in the time domain.*

The magnitude of the signal is 1 because it is composed of two harmonic components with a phase angle of $90°$. If the frequency $f_0$ drops down towards 0, the period of these components increases, tending to infinity. At $f_0 = 0$, the sine term vanishes and the cosine term becomes a constant and equal to unity (DC).

Let the spectrum be defined by $X(f) = \delta(f + f_0)$. The inverse Fourier transform exists and is given by

$$x(t) = \int_{-\infty}^{\infty} \delta(f + f_0) e^{j2\pi ft} df = e^{-j2\pi f_0 t} = \cos 2\pi f_0 t - j \sin 2\pi f_0 t. \tag{12.3.45}$$

As in the previous case, this signal is composed of two harmonic components; however, in this case, the imaginary part is inverted.

**Example**

Find the spectra of the sine and cosine functions if the functions are expressed as functions of frequency $f$.

**Solution.** From the previous examples, we have

$$\delta(f-f_0) \xrightarrow{IFT} x(t) = e^{+j2\pi f_0 t} = \cos 2\pi f_0 t + j \sin 2\pi f_0 t \qquad (12.3.46)$$

and

$$\delta(f+f_0) \xrightarrow{IFT} x(t) = e^{-j2\pi f_0 t} = \cos 2\pi f_0 t - j \sin 2\pi f_0 t, \qquad (12.3.47)$$

where IFT stands for inverse Fourier transform. Then, from eqns (12.3.46) and (12.3.47), we obtain

$$\cos(2\pi f_0 t) = \frac{1}{2} e^{j2\pi f_0 t} + \frac{1}{2} e^{-j2\pi f_0 t}. \qquad (12.3.48)$$

Using the same expressions, we can get the Fourier transform of the cosine signal as

$$\cos(2\pi f_0 t) \xrightarrow{FT} \frac{1}{2}\delta(f-f_0) + \frac{1}{2}\delta(f+f_0), \qquad (12.3.49)$$

where FT stands for Fourier transform. Also, using eqns (12.3.46) and (12.3.47), we can deduce the sine signal

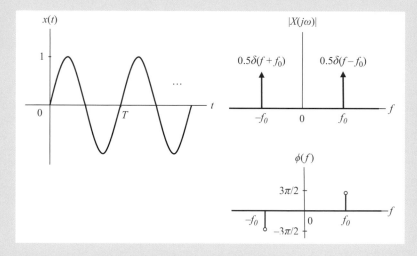

**Figure 12.3.8** *Sine signal (left) and its Fourier transform (right).*

$$\sin(2\pi f_0 t) = \frac{1}{2j}e^{j2\pi f_0 t} - \frac{1}{2j}e^{-j2\pi f_0 t},$$

and then, using the same expressions, we can deduce the Fourier transform of the sine signal as

$$\sin(2\pi f_0 t) \xrightarrow{FT} \frac{-j}{2}\delta(f-f_0) + \frac{j}{2}\delta(f+f_0). \tag{12.3.50}$$

The sine signal and its magnitude and phase spectral densities are presented in Figure 12.3.8. The precise value of the phases for the positive frequency $f = f_0$ is $3\pi/2$, counter-clockwise sense, and, for $f = -f_0$, it is $-3\pi/2$.

## Example

Find the signal in time domain for the weighted impulse train with frequency spacing $f_0$, which is expressed as

$$X(f) = \sum_{k=-\infty}^{k=\infty} c_k \delta(f-kf_0). \tag{12.3.51}$$

**Solution.** The Fourier transform is

$$
\begin{aligned}
x(t) &= \int_{-\infty}^{\infty} X(f)e^{j2\pi ft}\,df = \int_{-\infty}^{\infty} \sum_{k=-\infty}^{k=\infty} c_k \delta(f-kf_0)\ e^{j2\pi ft}\,df \\
&= \sum_{k=-\infty}^{k=\infty} c_k \int_{-\infty}^{\infty} \delta(f-kf_0)\,e^{j2\pi ft}\,df = \sum_{k=-\infty}^{k=\infty} c_k e^{j2\pi kf_0 t}
\end{aligned} \tag{12.3.52}
$$

which is the Fourier series representation of an arbitrary periodic signal with period $T$. Thus, the Fourier transform for a periodic signal can be represented as the weighted impulse train with impulse weights equal to the Fourier series coefficients $c_k$.

## Example

Find the Fourier transform of the uniform impulse train defined as

$$s(t) = \sum_{k=-\infty}^{k=\infty} \delta(t-kT). \tag{12.3.53}$$

Calculate the Fourier series of this periodic signal. For the calculated Fourier coefficients, find the expression for the signal $x(t)$ in the time domain and then find its Fourier transform $X(j\omega)$.

**Solution.** The graph of the signal is presented in Figure 12.3.9. It is a continuous-time periodic signal with the period $T$. Using the shifting property of the delta function, and integrating in one period where delta function exists, we can find the Fourier series of the signal as

$$c_k = \frac{1}{T} \int_{-T/2}^{T/2} s(t) e^{-jk\omega_s t}\, dt = \frac{1}{T} \int_{-T/2}^{T/2} \sum_{k=-\infty}^{\infty} \delta(t-kT)\, e^{-jk\omega_s t}\, dt = \frac{1}{T} \int_{-T/2}^{T/2} \delta(t)\, e^{-jk\omega_s t}\, dt = \frac{1}{T},$$

(12.3.54)

which is a series of coefficients $c_k$ defined for every $k$, as shown in Figure 12.3.9.

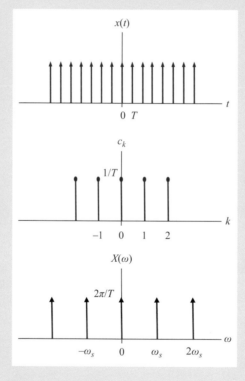

**Fig. 12.3.9** *Transforms of the impulse train.*

The inverse Fourier series of the signal is

$$x(t) = \sum_{k=-\infty}^{\infty} c_k e^{+jk\omega_s t} = \frac{1}{T} \sum_{k=-\infty}^{\infty} e^{+jk\omega_s t}, \tag{12.3.55}$$

where the impulse train is expressed in terms of periodic exponential functions. The amplitude spectral density can be calculated using the Fourier transform, that is,

$$\begin{aligned} X(\omega) &= \int_{-\infty}^{\infty} x(t) e^{-j\omega t} dt = \int_{-\infty}^{\infty} \frac{1}{T} \sum_{k=-\infty}^{k=\infty} e^{jk\omega_s t} e^{-j\omega t} dt \\ &= \sum_{k=-\infty}^{k=\infty} \frac{1}{T} 2\pi \int_{-\infty}^{\infty} \frac{1}{2\pi} e^{-j(\omega - k\omega_s)t} dt = \frac{2\pi}{T} \sum_{k=-\infty}^{k=\infty} \delta(\omega - k\omega_s) \end{aligned} \tag{12.3.56}$$

This amplitude spectral density is presented in Figure 12.3.9.

## Example

In the time domain, find the signal for the weighted impulse train with frequency spacing $f_0$, which is expressed as

$$X(f) = \sum_{k=-\infty}^{k=\infty} c_k \delta(f - kf_0). \tag{12.3.57}$$

**Solution.** The inverse Fourier transform is

$$\begin{aligned} x(t) &= \int_{-\infty}^{\infty} X(f) e^{j2\pi ft} df = \int_{-\infty}^{\infty} \sum_{k=-\infty}^{k=\infty} c_k \delta(f - kf_0) e^{j2\pi ft} df \\ &= \sum_{k=-\infty}^{k=\infty} c_k \int_{-\infty}^{\infty} \delta(f - kf_0) e^{j2\pi ft} df = \sum_{k=-\infty}^{k=\infty} c_k e^{j2\pi kf_0 t} \end{aligned} \tag{12.3.58}$$

which is the Fourier series representation of the given periodic signal. Thus, the Fourier transform for a periodic signal can be represented as the weighted impulse train with impulse weights equal to the Fourier series coefficients $c_k$.

## Example

In the time domain, find the signal for the weighted impulse train with frequency spacing $\omega_0$, which is expressed as

$$X(\omega) = \sum_{k=-\infty}^{k=\infty} c_k \delta(\omega - k\omega_0). \tag{12.3.59}$$

**Solution.** The inverse Fourier transform is

$$x(t) = \frac{1}{2\pi} \int_{-\infty}^{\infty} X(\omega) e^{j\omega t} d\omega = \frac{1}{2\pi} \int_{-\infty}^{\infty} \sum_{k=-\infty}^{k=\infty} c_k \delta(\omega - k\omega_0) e^{j\omega t} d\omega$$

$$= \frac{1}{2\pi} \sum_{k=-\infty}^{k=\infty} c_k \int_{-\infty}^{\infty} \delta(\omega - k\omega_0) e^{j\omega t} d\omega = \frac{1}{2\pi} \sum_{k=-\infty}^{k=\infty} c_k e^{jk\omega_0 t} \tag{12.3.60}$$

which is the Fourier series representation of the given periodic signal. Thus, the Fourier transform for a periodic signal can be represented as the weighted impulse train with impulse weights equal to the Fourier series coefficients $c_k$.

## Example

Calculate the Fourier transform, magnitude spectral density, energy spectral density, and power spectral density of a symmetric half-period sine pulse of amplitude $A_b$ and duration $T_b$.

**Solution.** The amplitude spectral density is

$$S_Q(f) = \int_{-\infty}^{\infty} s_Q(t) e^{-j\omega t} dt = \int_{-T_b/2}^{T_b/2} A_b \sin(\pi t/T_b) e^{-j\omega t} dt = \frac{A_b}{2j} \int_{-T_b/2}^{T_b/2} \left( e^{j2\pi t/2T_b} - e^{-j2\pi t/2T_b} \right) e^{-j2\pi f t} dt$$

$$= \frac{A_b}{2j} \int_{-T_b/2}^{T_b/2} \left( e^{-j2\pi(f-1/2T_b)t} - e^{-j2\pi(f+1/2T_b)t} \right) dt$$

$$= \frac{A_b}{2j} \left( \frac{e^{-j2\pi(f-1/2T_b)t}}{-j2\pi(f-1/2T_b)} - \frac{e^{-j2\pi(f+1/2T_b)t}}{-j2\pi(f+1/2T_b)} \right) \Bigg|_{-T_b/2}^{T_b/2}$$

$$= \frac{A_b}{2j} \left( \frac{e^{-j2\pi(f-1/2T_b)T_b/2} - e^{j2\pi(f-1/2T_b)T_b/2}}{-j2\pi(f-1/2T_b)} - \frac{e^{-j2\pi(f+1/2T_b)T_b/2} - e^{j2\pi(f+1/2T_b)T_b/2}}{-j2\pi(f+1/2T_b)} \right)$$

$$= \frac{A_b}{2j} \left( \frac{e^{j\pi(f-1/2T_b)T_b} - e^{-j\pi(f-1/2T_b)T_b}}{j2\pi(f-1/2T_b)} - \frac{e^{j\pi(f+1/2T_b)T_b} - e^{-j\pi(f+1/2T_b)T_b}}{j2\pi(f+1/2T_b)} \right)$$

$$= \frac{A_b T_b}{2j} \left( \frac{\sin(\pi(f-1/2T_b)T_b)}{\pi(f-1/2T_b)T_b} - \frac{\sin(\pi(f+1/2T_b)T_b)}{\pi(f+1/2T_b)T_b} \right)$$

$$= \frac{A_b T_b}{2j} \left( \text{sinc}\ (\pi(f-1/2T_b)T_b) - \text{sinc}\ (\pi(f+1/2T_b)T_b) \right)$$

The magnitude spectral density is

$$|S_Q(f)| = \left| \frac{A_b T_b}{2j} \left( \text{sinc}\ (\pi(f-1/2T_b)T_b) - \text{sinc}\ (\pi(f+1/2T_b)T_b) \right) \right|.$$

Due to the orthogonality characteristics of the sinc functions the ESD is

$$E_{QQ}(f) = \frac{A_b^2 T_b^2}{4} \left( \text{sinc}^2\ (f-1/2T_b)\ T_b + \text{sinc}^2\ (f+1/2T_b)\ T_b \right),$$

and the power spectral density is

$$P_{QQ}(f) = \frac{E_{QQ}(f)}{T_b} = \frac{A_b^2 T_b}{4} \left( \text{sinc}^2\ (f-1/2T_b)\ T_b + \text{sinc}^2\ (f+1/2T_b)\ T_b \right).$$

### Example

Find the expression in the time domain, and the Fourier transform, of the triangular pulse presented in Figure 12.3.10.

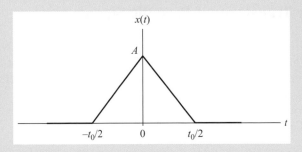

**Fig. 12.3.10** *Triangular pulse signal.*

**Solution.** The time-domain expression for the pulse is

$$x(t) = \begin{cases} A(1-2|t|/t_0) & -t_0/2 \le t < t_0/2 \\ 0 & \text{otherwise} \end{cases}. \tag{12.3.61}$$

The Fourier transform of the pulse is

$$X(f) = \int_{-\infty}^{\infty} x(t)e^{-j2\pi f\,t}dt = \int_{-t_0/2}^{t_0/2} A\left(1-2\frac{|t|}{t_0}\right)\cdot e^{-j2\pi f\,t}dt = \frac{2A}{t_0}\left(\frac{\sin\pi f t_0/2}{\pi f}\right)^2$$

$$= \frac{2A(2t_0/4)^2}{t_0}\left(\frac{\sin 2\pi f t_0/4}{2\pi f t_0/4}\right)^2 = \frac{At_0}{2}\left(\frac{\sin 2\pi f t_0/4}{2\pi f t_0/4}\right)^2 \underset{\omega=2\pi f}{=} \frac{At_0}{2}\left(\frac{\sin\omega\,t_0/4}{\omega\,t_0/4}\right)^2 = X(j\omega)$$

or, in terms of the sinc function,

$$X(f) = \frac{At_0}{2}\left(\frac{\sin 2\pi f t_0/4}{2\pi f t_0/4}\right)^2 = \frac{At_0}{2}\left(\frac{\sin\pi f t_0/2}{\pi f t_0/2}\right)^2 = \frac{At_0}{2}\text{sinc}^2\left(f t_0/2\right), \tag{12.3.62}$$

where the sinc function is defined as $\text{sinc}(x) = (\sin\pi x)/\pi x$.

### 12.3.6   Tables of the Fourier Transform

The basic properties of the Fourier transform are listed in Table 12.3.1. These properties can be confirmed using the basic Fourier transform pairs. In the table, the signals $x(t)$ and $y(t)$ have their Fourier transforms $X(f)$ and $Y(f)$, or the transforms $X(\omega)$ and $Y(\omega)$, respectively. The Fourier transforms, which are relevant for the mathematical modelling and analysis of communication systems, are presented in Table 12.3.2.

## 12.4   Fourier Transform of Periodic Signals

Spectral characteristics of a periodic signal $x(t)$ with period $T$ can be analysed by using their Fourier transform instead of the Fourier series. This analysis is of great importance in the theory of digital communication systems, where the amplitude spectral density and the power spectral density of carrier signals, for example, are represented by weighted delta functions.

To find the Fourier transform of a periodic signal, we need firstly to calculate the Fourier coefficients using Fourier series, that is,

$$c_k = \frac{1}{T}\int_{-T/2}^{T/2} x(t)e^{-jk\omega_0 t}dt, \tag{12.4.1}$$

**Table 12.3.1** *Basic properties of the Fourier transform.*

| Theorem | Time Domain | Frequency Domain |
|---|---|---|
| Linearity ($a$ and $b$, arbitrary constants) | $ax(t)+by(t)$ | $aX(f)+bY(f)$ <br> $aX(j\omega)+bY(j\omega)$ |
| Time delay | $x(t-\tau)$ | $X(f)e^{-j2\pi f\tau}$ <br> $X(j\omega)e^{-j\omega\tau}$ |
| Axis scale change | $x(at)$ | $|a|^{-1}X(f/a)$ <br> $|a|^{-1}X(\omega/a)$ |
| Frequency translation | $x(t)e^{+j\omega_0 t}$ | $X(f-f_0)$ <br> $X(\omega-\omega_0)$ |
| Duality | $X(jt)$ | $x(-f)$ <br> $2\pi \cdot x(-\omega)$ |
| Time reversal | $x(-t)$ | $X(-f)=X^*(f)$ |
| Modulation | $x(t)\cos\omega_0 t$ | $[X(f-f_0)+X(f+f_0)]/2$ <br> $\Big[[X(\omega-\omega_0)]+X(\omega+\omega_0)\Big]/2$ |
| Differentiation | $dx(t)/dt$ | $(j2\pi f)^n \cdot X(f),\ (j\omega)^n \cdot X(j\omega)$ |
| Integration | $\int\limits_{-\infty}^{t} x(\tau)\,d\tau$ | $(j2\pi f)^{-1}X(f)+[X(0)/2]\delta(f)$ <br> $(j\omega)^{-1}X(j\omega)+\pi X(0)\delta(\omega)$ |
| Convolution | $\int_{-\infty}^{\infty} x(\tau)y(t-\tau)\,d\tau = x*y$ | $X(f)\cdot Y(f),\ X(j\omega)\cdot Y(j\omega)$ |
| Multiplication | $x(t)\cdot y(t)$ | $\int_{-\infty}^{\infty} X(\lambda)Y(f-\lambda)\,d\lambda = X(f)*Y(f)$ <br> $\frac{1}{2\pi}X(j\omega)*Y(j\omega)$ |
| Conjugate | $x^*(t)$ | $X^*(-f),\ X^*(-j\omega)$ |
| Finite energy signals, Rayleigh theorem | $E=\int_{-\infty}^{\infty} x^2(t)dt$, real signal <br> $E=\int_{-\infty}^{\infty} |x(t)|^2 dt$, complex | $E=\int_{-\infty}^{\infty} |X(f)|^2 df$ <br> $E=\frac{1}{2\pi}\int_{-\infty}^{\infty} |X(\omega)|^2 d\omega$ |

and then find the expression of the signal in the time domain $x(t)$ as a function of the Fourier coefficients, that is,

$$x(t) = \sum_{k=-\infty}^{\infty} c_k e^{+jk\omega_0 t}. \tag{12.4.2}$$

**Table 12.3.2** *Fourier transforms.*

| Signal | Time Domain | Frequency Domain |
|---|---|---|
| Sine and cosine | $A \cdot \cos \omega_0 t$ <br> $A \cdot \sin \omega_0 t$ | $A \frac{\delta(f-f_0)+\delta(f+f_0)}{2} A\pi \left[\delta(\omega - \omega_0) + \delta(\omega + \omega_0)\right],$ <br> $A \frac{\delta(f-f_0)-\delta(f+f_0)}{2j} jA\pi \left[\delta(\omega + \omega_0) + \delta(\omega - \omega_0)\right],$ |
| Dirac delta | $\delta(t)$ | $X(f) = 1, -\infty < f < \infty, X(j\omega) = 1, -\infty < \omega < \infty$ |
| Delayed impulse | $\delta(t - t_0)$ | $e^{-j\omega t_0}$ |
| Exponential function | $e^{+j2\pi f_0 t}$ | $\delta(f - f_0)$ <br> $2\pi \cdot \delta(\omega - \omega_0)$ |
| Constant | $A, -\infty < t < \infty$ | $A \cdot \delta(f), 2\pi A \cdot \delta(\omega)$ |
| Uniform impulse train | $\sum\limits_{k=-\infty}^{k=\infty} \delta(t - kT)$ | $\frac{1}{T} \sum\limits_{k=-\infty}^{k=\infty} \delta(f - k/T)$ <br> $\omega_s \sum\limits_{k=-\infty}^{k=\infty} \delta(\omega - k\omega_s), \omega_s = 2\pi/T$ |
| Symmetric rectangular pulse of amplitude $A$ and duration $t_0$ | $x(t) = A$ <br> for $\quad |t| \le \frac{t_0}{2}$ | $X(j\omega) = At_0 \frac{\sin \omega t_0 / 2}{\omega t_0 / 2}, \quad$ for $\operatorname{sinc}(ft_0) = \frac{\sin(\pi f t_0)}{\pi f t_0}$ <br> $X(f) = At_0 \operatorname{sinc}(ft_0)$ |
| Asymmetric rectangular pulse | $x(t) = A, 0 \le t \le t_0$ | $X(f) = At_0 \cdot e^{-j\pi f t_0} \cdot \operatorname{sinc}(ft_0)$ |
| Symmetric triangular pulse of amplitude $A$ and duration $t_0$ | $x(t) = A \left(1 - \frac{2|t|}{t_0}\right)$ <br> for $|t| \le \frac{t_0}{2}$ | $X(j\omega) = \frac{At_0}{2} \left(\frac{\sin \omega t_0 / 4}{\omega t_0 / 4}\right)^2$ <br> $X(f) = (At_0/2) \operatorname{sinc}^2(ft_0/2)$ |
| Symmetric half-period sine pulse of amplitude $A$ and duration $T$ | $A \sin \left(\frac{\pi t}{T}\right)$ | $\frac{AT}{2j} \left( \begin{array}{l} \operatorname{sinc}(\pi (f - 1/2T) T) \\ - \operatorname{sinc}(\pi (f + 1/2T) T) \end{array} \right)$ |
| Radiofrequency pulse | $A \cos \omega_c t$, for <br> $-\tau/2 \le t \le \tau/2$ | $\frac{A\tau}{2} \left[\operatorname{sinc} \frac{(\omega + \omega_c)\tau}{2} + \operatorname{sinc} \frac{(\omega - \omega_c)\tau}{2}\right]$ |
| Half-period sine signal | $A_b \sin(\pi t/T_b)$ <br> $|t| \le T_b/2$ | $\frac{A_b T_b}{2j} \left[\operatorname{sinc} \pi \left(f - \frac{1}{2T_b}\right) T_b - \operatorname{sinc} \pi \left(f + \frac{1}{2T_b}\right) T_b\right]$ |

The Fourier transform then can be applied to the obtained signal $x(t)$. Bearing in mind that the Fourier transform of an exponential function can be expressed in terms of Dirac delta function, that is,

$$FT\left\{e^{+jk\omega_0 t}\right\} = \int_{-\infty}^{\infty} e^{+jk\omega_0 t} e^{+j\omega t} dt = \int_{-\infty}^{\infty} e^{-j(\omega-k\omega_0)t} dt = 2\pi\delta(\omega-k\omega_0), \qquad (12.4.3)$$

where FT stands for Fourier transform, and the inverse Fourier transform of the delta function is

$$IFT\{\delta(\omega-k\omega_0)\} = \frac{1}{2\pi}\int_{-\infty}^{\infty} X(\omega) e^{j\omega t} d\omega = \frac{1}{2\pi}\int_{-\infty}^{\infty} \delta(\omega-k\omega_0) e^{j\omega t} d\omega = \frac{1}{2\pi}e^{jk\omega_0 t}.$$
$$(12.4.4)$$

where IFT stands for inverse Fourier transform. We then obtain

$$X(j\omega) = \int_{-\infty}^{\infty} \sum_{k=-\infty}^{\infty} c_k e^{+jk\omega_0 t} e^{+j\omega t} dt = \sum_{k=-\infty}^{\infty} c_k \int_{-\infty}^{\infty} e^{-j(\omega-k\omega_0)t} dt$$
$$= \sum_{k=-\infty}^{\infty} c_k 2\pi \frac{1}{2\pi}\int_{-\infty}^{\infty} e^{-j(\omega-k\omega_0)t} dt = \sum_{k=-\infty}^{\infty} 2\pi c_k \delta(\omega-k\omega_0) \qquad (12.4.5)$$

which gives the amplitudes of the spectrum as a function of the continuous angular frequency $\omega$. Using the scaling property of the delta function, the amplitude spectral density can also be expressed as the function of the continuous cyclic frequency $f$ as

$$X(f) = \sum_{k=-\infty}^{\infty} 2\pi c_k \delta(2\pi f - 2\pi k f_0) = \sum_{k=-\infty}^{\infty} 2\pi c_k \frac{1}{2\pi}\delta(f-kf_0) = \sum_{k=-\infty}^{\infty} c_k \delta(f-kf_0). \quad (12.4.6)$$

Therefore, using the Fourier transform instead of the Fourier series for a periodic signal, the line spectrum is replaced with the spectrum expressed in terms of Dirac delta functions. It is important to note that the obtained amplitude spectrum is the function of the continuous frequency $f$ as the independent variable.

## Example

Find the components of the Fourier series for the rectangular pulse train $x(t)$ shown in Figure 12.2.3, for the case when $t_0 = T/2$. Then find the Fourier transform of the same periodic signal.

**Solution.** The coefficients of the Fourier series are calculated as

$$c_k = \frac{1}{T} \int_{-T/2}^{T/2} x(t) e^{-jk\omega_0 t} \, dt = \frac{t_0}{T} \frac{\sin k\pi t_0/T}{k\pi t_0/T} = \begin{cases} 1/2 & k=0 \\ \frac{1}{2} \frac{\sin k\pi/2}{k\pi/2} & k \text{ odd} \\ 0 & k \text{ even} \end{cases}. \tag{2.4.7}$$

The Fourier series of $x(t)$ is given by

$$x(t) = \sum_{k=-\infty}^{\infty} c_k e^{jk\omega_0 t} = \sum_{k=-\infty}^{\infty} \left( \frac{1}{2} \frac{\sin k\pi/2}{k\pi/2} \right) e^{jk\omega_0 t}. \tag{2.4.8}$$

For the signal $x(t)$ obtained in the time domain, we obtain its Fourier transform as the function of continuous cyclic frequency $f$ as

$$X(f) = \sum_{k=-\infty}^{\infty} c_k \delta (f - kf_0) = \sum_{k=-\infty}^{\infty} \left( \frac{1}{2} \frac{\sin k\pi/2}{k\pi/2} \right) \delta (f - kf_0). \tag{2.4.9}$$

The coefficients are expressed in terms of Dirac delta functions with weight corresponding to the values of the Fourier coefficients. In this case, the weights are: $c_0 = 0.5$, $c_{\pm 1} = 0.3185$, $c_{\pm 3} = -0.106$, $c_{\pm 5} = 0.0635$, and $c_{\pm 7} = -0.0455$. If we present the delta functions as arrows, with the same signs and weights as the coefficients, the two-sided amplitude spectrum can be obtained as shown in Figure 12.4.1.

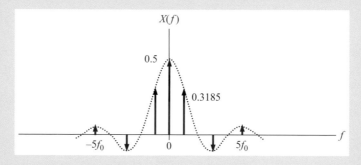

**Fig. 12.4.1** *Two-sided amplitude spectrum.*

The magnitude and phase spectra can be obtained from the amplitude spectrum and expressed in terms of delta functions, as shown in Figure 12.4.2.

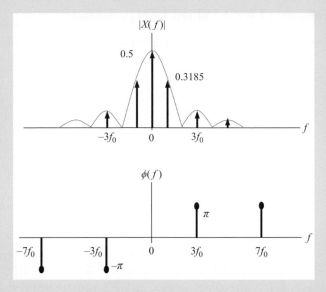

**Fig. 12.4.2** *Two-sided magnitude spectrum (upper panel) and phase spectrum (lower panel).*

The amplitude spectral density is defined by the following expression, which includes the magnitudes and phases of all of the spectral components:

$$X(f) = \sum_{k=-\infty}^{\infty} c_k \delta(f - kf_0) = \sum_{k=-\infty}^{\infty} \left( \frac{1}{2} \frac{\sin k\pi/2}{k\pi/2} \right) \delta(f - kf_0)$$

$$= \sum_{\substack{k=2n+1 \\ n=1,2,..}}^{\infty} e^{j\pi} \left( \frac{1}{2} \left| \frac{\sin k\pi/2}{k\pi/2} \right| \right) \delta(f - kf_0) + \sum_{\substack{k=2n-1 \\ n=-1,-2,..}}^{\infty} e^{-j\pi} \left( \frac{1}{2} \left| \frac{\sin k\pi/2}{k\pi/2} \right| \right) \delta(f - kf_0).$$

$$+ \sum_{\substack{other\ k\ values}}^{\infty} \frac{1}{2} \left| \frac{\sin k\pi/2}{k\pi/2} \right| \delta(f - kf_0)$$

(12.4.10)

## 12.5    Correlation Functions, Power Spectral Densities, and Linear-Time-Invariant Systems

### 12.5.1    Correlation of Real-Valued Energy Signals

The Wiener–Khintchine theorem relates the correlation function of signals to their energy spectral density. If the signal $x(t)$ is an *energy signal*, its autocorrelation function can be expressed as

$$R_x(\tau) = \int_{-\infty}^{\infty} x(t+\tau)x(t)dt, \tag{12.5.1}$$

or, equivalently,

$$R_x(\tau) = \int_{-\infty}^{\infty} x(t-\tau)x(t)dt. \tag{12.5.2}$$

Both relations will be used in this book. The Wiener–Khintchine autocorrelation theorem states that the Fourier transform of the autocorrelation function of a signal $x(t)$ is equal to its energy spectral density $E_x(f)$, sometimes denoted as $S_x(f)$, and can be proved as follows:

$$
\begin{aligned}
E_x(f) &= \int_{-\infty}^{\infty} R_x(\tau) e^{-j2\pi f \tau} d\tau = \int_{-\infty}^{\infty} \left( \int_{-\infty}^{\infty} x(t+\tau)x(t)dt \right) e^{-j2\pi f \tau} d\tau \\
&= \int_{-\infty}^{\infty} x(t) \left[ \int_{-\infty}^{\infty} x(t+\tau)\, e^{-j2\pi f \tau} d\tau \right] dt = \int_{-\infty}^{\infty} x(t)X(f)e^{+j2\pi f t} dt. \\
&= X(f) \int_{-\infty}^{\infty} x(t)\, e^{+j2\pi f t} dt = X(f)X^*(f) = |X(f)|^2
\end{aligned}
\tag{12.5.3}
$$

Since $R_x(\tau)$ is real and even, the energy spectral density $E_x(f)$ is purely real. Therefore, we say the autocorrelation function $R_x(\tau)$ and the energy spectral density $E_x(f)$ form the *Fourier transform pair*,

$$E_x(f) = \int_{-\infty}^{\infty} R_x(\tau) e^{-j2\pi f \tau} d\tau, \tag{12.5.4}$$

$$R_x(\tau) = \int_{-\infty}^{\infty} E_x(f) e^{+j2\pi f \tau} df. \tag{12.5.5}$$

Thus, the theory of Fourier transform can be applied to relate various autocorrelation and power spectral density functions $R_x(\tau)$ and $E_x(f)$. Likewise, the Fourier transform of the cross-correlation function of signals $x(t)$ and $y(t)$ expressed as

$$R_{xy}(\tau) = \int_{-\infty}^{\infty} x(t+\tau) y(t)dt \text{ or } R_{xy}(\tau) = \int_{-\infty}^{\infty} x(t-\tau) y(t)dt \qquad (12.5.6)$$

is equal to their cross energy spectral density $E_{xy}(f)$, which is proved as follows:

$$
\begin{aligned}
E_{xy}(f) &= \int_{-\infty}^{\infty} R_{xy}(\tau) e^{-j2\pi f\tau} d\tau = \int_{-\infty}^{\infty} \left[ \int_{-\infty}^{\infty} x(t+\tau) y(t)dt \right] e^{-j2\pi f\tau} d\tau \\
&= \int_{-\infty}^{\infty} y(t) \left[ \int_{-\infty}^{\infty} x(t+\tau) e^{-j2\pi f\tau} d\tau \right] dt = \int_{-\infty}^{\infty} y(t) X(f) e^{+j2\pi ft} dt. \qquad (12.5.7) \\
&= X(f) \int_{-\infty}^{\infty} y(t) e^{+j2\pi ft} dt = X(f) Y^*(f)
\end{aligned}
$$

Generally, $E_{xy}(f)$ is a function of complex variables. Furthermore, a change in the signals changes the cross-correlation function, that is, $E_{xy}(f) \neq E_{yx}(f)$. Instead, we obtain

$$E_{xy}(f) = E_{yx}^*(f). \qquad (12.5.8)$$

**Example**

The autocorrelation function for the rectangular pulse (11.4.8) is calculated in Section 11.4.2, expressed by eqn (11.4.11) as

$$R_x(\tau) = A^2 T\left(1 - \frac{|\tau|}{T}\right), \qquad (12.5.9)$$

and shown in Figure 12.5.1.

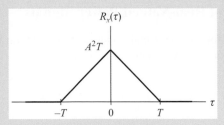

**Fig. 12.5.1** *Correlation function.*

Calculate the energy spectral density of the rectangular pulse based on this autocorrelation function.

**Solution.** The energy spectral density is

$$E_x(f) = \int_{-\infty}^{\infty} R_x(\tau) e^{-j\omega t} d\tau = \int_{-T}^{T} A^2 T \left(1 - \frac{|\tau|}{T}\right) e^{-j\omega t} d\tau ,$$

(12.5.10)

$$= A^2 T^2 \left[\frac{\sin(\pi f T)}{\pi f T}\right]^2 = A^2 T^2 \text{sinc}^2(f T)$$

where the sinc function is defined as sinc$(y) = \sin(\pi y)/\pi y$. The two-sided power spectral density, normalized by $A^2 T^2$, is shown in Figure 12.5.2.

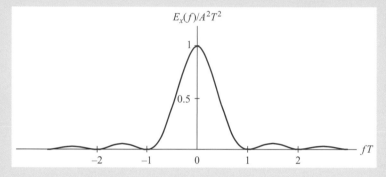

**Fig. 12.5.2** *Power spectrum density of the modulating signal.*

The signal energy is

$$E_x = \int_{-\infty}^{\infty} S_{xx}(f) df = \int_{-T}^{T} A^2 T^2 \text{sinc}^2(f T) df = A^2 T.$$

(12.5.11)

## 12.5.2 Correlation of Real-Valued Power Signals

The foregoing derivations for cross-correlation and autocorrelation functions assume that the $x(t)$ and $y(t)$ signals are energy signals. Thus, it is required that

$$R_x(0) = \int_{-\infty}^{\infty} |x(t)|^2 dt = E < \infty.$$

(12.5.12)

For power signals, which have infinite energy, the definitions of correlation and cross-correlation functions should be changed to

$$R_{xx}(\tau) = \lim_{a \to \infty} \frac{1}{2a} \int_{-a}^{a} x(t+\tau) x(t) dt,$$

(12.5.13)

and

$$R_{xy}(\tau) = \lim_{a \to \infty} \frac{1}{2a} \int_{-a}^{a} x(t+\tau)y(t)dt, \qquad (12.5.14)$$

respectively. If the power signals $x(t)$ and $y(t)$ are also *periodic*, with period $T$, then the limits in these equations may be dropped to give

$$R_{xx}(\tau) = \frac{1}{T} \int_{-T/2}^{T/2} x(t+\tau)x(t)dt, \qquad (12.5.15)$$

and

$$R_{xy}(\tau) = \frac{1}{T} \int_{-T/2}^{T/2} x(t+\tau)y(t)dt, \qquad (12.5.16)$$

respectively.

## Example

Calculate the autocorrelation function for the signals $x(t) = A \cdot \sin(2\pi ft)$ and $y(t) = A \cdot \cos(2\pi ft)$.

**Solution.** Because the sine signal is periodic, with period $T = 1/f$, the autocorrelation function is

$$R_{xx}(\tau) = \frac{1}{T} \int_{-T/2}^{T/2} x(t+\tau)x(t)dt = \frac{1}{T} \int_{-T/2}^{T/2} A^2 \sin 2\pi f(t+\tau) \cdot \sin 2\pi ft\, dt$$

$$= \frac{A^2}{2T} \int_{-T/2}^{T/2} [\cos 2\pi f\tau - \cos(4\pi ft + 2\pi f\tau)]dt = \frac{A^2}{2} \cos 2\pi f\tau$$

The cosine signal is a periodic power signal, like the sine signal. For the sake of demonstration, we will use the second equivalent autocorrelation function expression for the power signals to find its autocorrelation function as

$$R_{xx}(\tau) = \lim_{a \to \infty} \frac{1}{2a} \int_{-a}^{a} y(t)y(t-\tau)dt = \lim_{a \to \infty} \frac{1}{2a} \int_{-a}^{a} A^2 \cdot \cos(2\pi ft)\cos(2\pi f(t-\tau))dt$$

$$= \lim_{a \to \infty} \frac{A^2}{4a} \int_{-a}^{a} \cos(2\pi f\tau)dt = \frac{A^2}{2} \cos(2\pi f\tau)$$

As we expected, both calculations give the same results, due to the periodicity of the signals and the definition of the power signals.

## 12.5.3 Comprehensive Analysis of Linear-Time-Invariant Systems

### 12.5.3.1 *System Presentation*

The following standard notation, accepted in this book for the deterministic power and energy signals, will be used for the analysis of a deterministic linear-time-invariant system. The system itself will be defined by the following:

> $h(t)$: the impulse response in the time domain
>
> $H(f)$: the impulse response in the frequency domain, amplitude spectral density
>
> $R_h(\tau)$: the system autocorrelation function of the impulse response
>
> $E_h(f)$: the energy spectral density of the impulse response

A linear-time-invariant system with deterministic energy signals at its input (one letter for the subscripts) will be defined by the following:

> $R_x(\tau)$, $R_y(\tau)$: the autocorrelation functions of the input and output signals, respectively
>
> $R_{xy}(\tau)$, $R_{yx}(\tau)$: the cross-correlation functions
>
> $E_x(f) = |X(f)|^2$: the energy spectral density of the input signal
>
> $E_y(f) = |Y(f)|^2$: the energy spectral density of the output signal
>
> $S_{xy}(f)$: the cross energy spectral density

A linear-time-invariant system with deterministic power signals at its input (two letters for the subscripts) will be defined by the following:

> $R_{xx}(\tau)$: the autocorrelation function of the input signal
>
> $R_{xy}(\tau)$, $R_{yx}(\tau)$: the cross-correlation functions
>
> $S_{xx}(f)$: the power spectral density of the input signal
>
> $S_{yy}(f)$: the power spectral density of the output signal
>
> $S_{xy}(f)$: the cross power spectral density

A linear-time-invariant system for processing a *deterministic signal* is defined by its impulse response in the time and frequency domains, $h(t)$ and $H(f)$, respectively, as shown in Figure 12.5.3. Then, it is possible, for any input signal $x(t)$, to find the output signal in the time domain, $y(t)$, as the convolution of the input signal and the impulse response, or to find the output signal in frequency domain $Y(f)$ as the multiplication of the input signal $X(f)$ and the impulse response $H(f)$. The power and energy of all signals in the system can be found in the time and frequency domain using the expressions of the instantaneous power and energy spectral density.

This concept of signal processing in a linear-time-invariant system is directly applicable for *energy signals* because they fulfil the condition for existence of the Fourier transform. Furthermore, it is relatively simple to find the autocorrelation function and related energy

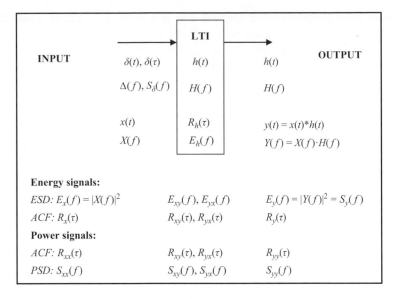

**Fig. 12.5.3** *Specification of a linear-time-invariant system; ACF, autocorrelation function; ESD, energy spectral density; LTI, linear-time-invariant system; PSD, power spectral density.*

spectral density function of the system and analyse its behaviour by using this representation. The relationship of these two functions is defined by their Fourier transform pair.

For *power signals*, which can have infinite energy, the Fourier transform cannot be always performed. The linear-time-invariant system is still defined by its impulse response, but the analysis of signal can be done using the autocorrelation function and the power spectral density function. In this case, the autocorrelation function is defined on a limited interval of time and then averaged over the whole interval. The power spectral density function can then be derived as the Fourier transform of the autocorrelation function. If the power signal is periodic, the autocorrelation function is calculated inside one period of the signal. The energy and power of the signal can be calculated in the time and frequency domains.

A block schematic of a linear-time-invariant system is presented in Figure 12.5.3, where all relevant signals and functions are indicated, including time-domain signals, autocorrelation functions, and energy and power spectral densities.

At the input of the system, two signals are defined in the time and frequency domains via a unique notation: Dirac delta functions and the arbitrary deterministic input signal. The Dirac delta function is used to define the system itself and is represented in its time- and frequency-domain forms. It is used to test the system to find the system's impulse response. The input signal can be an energy signal or a power signal. As an energy signal, it is defined by its autocorrelation function, the amplitude spectral density, and the energy spectral density, with a single lowercase letter used for the indices, to unify notation. If the input is a power signal, it is defined by its autocorrelation function, the amplitude power spectral density, and the power spectral density, with two lowercase letters used for the indices, to unify notation.

The system itself is defined by its impulse response in the time and frequency domains, together with the system correlation function and the related power spectral density. To characterize arbitrary signal transmission through the system, the cross power spectral density and cross-correlation functions are also defined.

The output of the system is expressed by its impulse response in the time and frequency domains. The output of the system is defined in the time domain as a convolution of the input and the impulse response and, in the frequency domain, as a product of the input and the system's transfer function. The output for an arbitrary energy input signal is defined by its energy spectral density and its autocorrelation function, while, for an input power signal, it is defined by the power spectral density and the autocorrelation function. The notation used is as for the input of the system. In terms of autocorrelation functions, the output autocorrelation function is expressed as a product of the input autocorrelation function and the system autocorrelation function, while the output power spectral density is the convolution of the input power spectral density and the system power spectral density. Understanding this section is vital for understanding the operation of communication systems.

In the time domain, for any input signal $x(t)$ of a linear-time-invariant system, the output signal $y(t)$ is the convolution of the input signal $x(t)$ and the system impulse response $h(t)$, that is,

$$y(t) = \int_{\tau=-\infty}^{\infty} x(\tau)\,h(t-\tau)\,d\tau = x(t)*h(t), \tag{12.5.17}$$

as was proven in Chapter 11. In the frequency domain, for any input signal $X(f)$ of a linear-time-invariant system, the output signal $Y(f)$ is the product of the input signal $X(f)$ and the system impulse response $H(f)$, that is,

$$Y(f) = X(f) \cdot H(f), \quad \text{or} \quad Y(\omega) = X(\omega) \cdot H(\omega). \tag{12.5.18}$$

**The delta signal as the test input of the linear-time-invariant system.** The impulse response of the system is obtained for the input signal defined by the Dirac delta function. If we relax our rigorous mathematical procedures to some extent, we can process the Dirac impulse delta function in the time and frequency domains. The Fourier transform of the impulse signal is

$$\Delta(\omega) = \int_{-\infty}^{\infty} \delta(t) e^{-j\omega t}\,dt = e^{-j\omega 0} = 1, \quad |\omega| \leq \infty. \tag{12.5.19}$$

The delta signal is presented in the time and frequency domains in Figure 12.5.4. Therefore, the amplitude spectral density of the delta function has a unit value for all frequencies from $-\infty$ to $+\infty$. The energy of the signal can be calculated from its energy spectral density as

$$E_\delta = \int_{-\infty}^{\infty} |\Delta(f)|^2 df = \int_{-\infty}^{\infty} 1 \cdot df \to \infty, \tag{12.5.20}$$

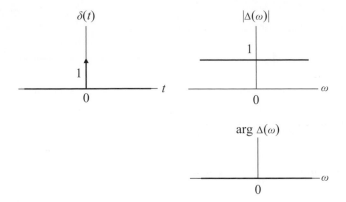

**Fig. 12.5.4** *Amplitude spectral density and phase spectral density of a delta impulse.*

or, from the time-domain representation of the signal, bearing in mind the shifting property of the Dirac delta function, as

$$E_\delta = \int_{-\infty}^{\infty} \delta^2(t)dt = \int_{-\infty}^{\infty} \delta(t)\delta(t)dt = \delta(0) \rightarrow \infty. \tag{12.5.21}$$

In other words, the energy is infinity if the following relation holds:

$$\int_{-\infty}^{\infty} dt \Delta_\varepsilon^2(t) \underset{\varepsilon \to \infty}{=} \frac{1}{\varepsilon}\varepsilon \cdot \varepsilon \to \infty. \tag{12.5.22}$$

In order to analyse this system in terms of its impulse response, the correlation and spectral density functions for the power and energy signals will be defined and derived. For the input delta impulse, we can find the autocorrelation function as follows:

$$R_\delta(\tau) = \int_{-\infty}^{\infty} x(t+\tau)x(t)dt = \int_{-\infty}^{\infty} \delta(t+\tau)\,\delta(t)dt = \delta(\tau), \tag{12.5.23}$$

and the energy spectral density as

$$E_\delta(f) = \int_{\tau=-\infty}^{\infty} R_\delta(\tau)e^{-j2\pi f\tau}d\tau = \int_{-\infty}^{\infty} \delta(\tau)e^{-j2\pi f\tau}d\tau = 1, \tag{12.5.24}$$

defined for all frequencies $-\infty < f < \infty$ and shown in Figure 12.5.5. The energy of the delta impulse is infinite, as follows form its calculation in the time domain,

$$E_\delta = R_\delta(0) = \int_{-\infty}^{\infty} \delta(t)\delta(t)dt = \delta(0) \rightarrow \infty, \tag{12.5.25}$$

**Fig. 12.5.5** *Autocorrelation function (left) and energy spectral density (right) of a Dirac delta signal.*

and in the frequency domain as

$$E_\delta = \int\limits_{f=-\infty}^{\infty} S_\delta(f)df = \int_{-\infty}^{\infty} df \to \infty, \tag{12.5.26}$$

which confirm the results obtained by calculations in the time and frequency domains.

We cannot say too much about the power of this impulse. Because the duration of the impulse is infinitesimally small and the amplitude is of infinite value, the energy can be understood to be of infinite value. Due to the short pulse duration, this impulse can be considered to have infinite power. Therefore, it does not fulfil the definition of either an energy signal or a power signal. If we avoid rigorous mathematics, we may calculate the power of the impulse function as

$$P_\delta = \lim_{a\to\infty} \frac{1}{2a}\int_{-a}^{a} \delta(t)\delta(t)dt = \lim_{a\to\infty}\frac{1}{2a}2a\delta(0) = \delta(0) \to \infty. \tag{12.5.27}$$

This is not the case with the Kronecker delta function, as we will see in Chapters 14 and 15. In discrete systems, the discrete amplitudes of a signal are obtained via the sampling theorem, which requires that the signal be physically realizable, so the signal is assumed to have finite energy.

The autocorrelation function of white Gaussian noise is a Dirac delta function with weight equal to the variance of the noise. The noise has infinite power; however, this is not due to the finite variance of the noise but to the possible (though very unlikely) infinite amplitude values of the noise generated. Therefore, a strict definition of white Gaussian noise implies that the power and energy of the noise are of infinite value. However, in reality, where the noise generated fulfils the condition of physical realizability, its power is finite because the expected noise amplitudes are finite. We say that the noise preserves Gaussian density, which is now truncated, and the noise generator can generate only the samples of finite amplitudes. This problem is addressed in detail in Chapter 4.

### 12.5.3.2   *Correlation and Energy Spectral Density of Complex Energy Signals*

The properties of a linear-time-invariant system, defined by its impulse response, can be investigated using the concepts of correlation, energy spectral density, and power spectral density. These functions will be defined and derived for the input and output signals and for the impulse response. They also can relate the output of the system to the input of the system, if we bear in mind the autocorrelation function of the impulse response. For that reason, the relationship between the correlation functions of the input and output signals will be established. The correlation functions of the energy and power signals are treated separately, due to their differences in definition and meaning. Namely, the correlation function of the energy signals is related to their energy spectral density, while the correlation function of the power signals is related to their power spectral density.

In this section, the important procedure of calculating the power and energy of the system's signals is also demonstrated. For the sake of precision and clarity, energy signals and power signals will be presented separately in this section, due to slight but important differences in their analysis and presentation.

**Input for a linear-time-invariant system.** For the input signal $x(t)$, we can find the autocorrelation function, power spectral density function, and related power and energy. If the signal $x(t)$ is a complex *energy signal*, its autocorrelation can be expressed as

$$R_x(\tau) = \int_{-\infty}^{\infty} x(t+\tau)x^*(t)dt. \tag{12.5.28}$$

The value of the autocorrelation function at zero lag is the energy of the signal that fulfils the condition

$$R_x(0) = \int_{-\infty}^{\infty} |x(t)|^2 dt = E < \infty. \tag{12.5.29}$$

The energy of the signal calculated in the time domain is equal to the energy of the signal calculated in the frequency domain, as may be proved by

$$E_x = \int_{-\infty}^{\infty} |x(t)|^2 dt = \int_{-\infty}^{\infty} x(t)x^*(t)dt = \int_{-\infty}^{\infty} x(t) \int_{-\infty}^{\infty} X^*(f)e^{-j2\pi ft} dfdt$$
$$= \int_{-\infty}^{\infty} X^*(f) \int_{-\infty}^{\infty} x(t)e^{j2\pi ft} dtdf = \int_{-\infty}^{\infty} X^*(f)X(f)df \underset{\omega=2\pi f}{=} \frac{1}{2\pi} \int_{-\infty}^{\infty} |X(\omega)|^2 d\omega \tag{12.5.30}$$

The energy spectral density is defined as the magnitude spectral density squared, that is,

$$E_x(f) = |X(f)|^2, E_x(\omega) = |X(\omega)|^2. \tag{12.5.31}$$

**The autocorrelation theorem.** The Wiener–Khintchine theorem relates the correlation function of an energy signal to its energy spectral density. The Fourier transform of the autocorrelation function of a signal $x(t)$ is equal to its energy spectral density $E_x(f)$. If we introduce the Fourier transform operator as $FT\{.\}$, we may prove the theorem as follows:

$$
\begin{aligned}
FT\{R_x(\tau)\} &= \int_{\tau=-\infty}^{\infty} R_x(\tau)\, e^{-j2\pi f\tau}\, d\tau = \int_{-\infty}^{\infty}\int_{-\infty}^{\infty} x(t+\tau)x^*(t)dt\, e^{-j2\pi f\tau}\, d\tau \\
&= \int_{-\infty}^{\infty} x^*(t)\left[\int_{-\infty}^{\infty} x(t+\tau)\, e^{-j2\pi f\tau}\, d\tau\right]dt = \int_{-\infty}^{\infty} x^*(t)X(f)e^{j2\pi ft}\, dt. \quad (12.5.32) \\
&= X(f)\int_{-\infty}^{\infty} x^*(t)e^{j2\pi ft}\, dt = X(f)X^*(f) = X^2(f) = E_x(f)
\end{aligned}
$$

Since $R_x(\tau)$ is Hermitian, the energy spectral density $E_x(f)$ is purely real. The autocorrelation function $R_x(\tau)$ and the energy spectral density $E_x(f)$ form *the Fourier transform pair,*

$$
E_x(f) = \int_{\tau=-\infty}^{\infty} R_x(\tau)\, e^{-j2\pi f\tau}\, d\tau, \tag{12.5.33}
$$

$$
R_x(\tau) = \int_{-\infty}^{\infty} E_x(f)e^{j2\pi f\tau}\, df. \tag{12.5.34}
$$

Thus, the theory of Fourier transforms can be applied for various functions of $R_x(\tau)$ and $S_x(f)$ of an energy input signal $x(t)$.

**Characterization of a linear-time-invariant system.** The energy spectral density of the impulse response $h(t)$ is the energy density or the system's magnitude response squared, expressed as

$$
E_h(f) = H(f)\cdot H^*(f) = |H(f)|^2. \tag{12.5.35}
$$

**The cross-correlation theorem.** The cross-correlation function is defined in eqn (12.5.6) as the correlation between two different signals. The Fourier transform of the cross-correlation function of signals $x(t)$ and $y(t)$ is equal to their cross energy spectral density $E_{xy}(f)$, that is,

$$
\begin{aligned}
FT\{R_{xy}(\tau)\} &= \int_{\tau=-\infty}^{\infty} R_{xy}(\tau)\, e^{-j2\pi f\tau}\, d\tau = \int_{-\infty}^{\infty}\int_{-\infty}^{\infty} x(t+\tau)y^*(t)dt\, e^{-j2\pi f\tau}\, d\tau \\
&= \int_{-\infty}^{\infty} y^*(t)\left[\int_{-\infty}^{\infty} x(t+\tau)\, e^{-j2\pi f\tau}\, d\tau\right]dt = \int_{-\infty}^{\infty} y^*(t)X(f)e^{j2\pi ft}\, dt. \quad (12.5.36) \\
&= X(f)\int_{-\infty}^{\infty} y^*(t)e^{j2\pi ft}\, dt = X(f)Y^*(f) = E_{xy}(f)
\end{aligned}
$$

The Fourier transform of the cross-correlation function of signals $y(t)$ and $x(t)$ is equal to their cross energy spectral density $E_{yx}(f)$ calculated as

$$FT\{R_{yx}(\tau)\} = \int_{\tau=-\infty}^{\infty} R_{yx}(\tau) e^{-j2\pi f\tau} d\tau = \int_{-\infty}^{\infty}\int_{-\infty}^{\infty} y(t+\tau)x^*(t)dt \; e^{-j2\pi f\tau} d\tau$$

$$= \int_{-\infty}^{\infty} x^*(t)\left[\int_{-\infty}^{\infty} y(t+\tau)e^{-j2\pi f\tau} d\tau\right] dt \Bigg|_{\substack{\text{substitution:}\\ z=t+\tau,\; d\tau=dz}}$$

$$= \int_{-\infty}^{\infty} x^*(t)\left[\int_{-\infty}^{\infty} y(z)e^{-j2\pi fz} e^{+j2\pi ft} dz\right] dt = \int_{-\infty}^{\infty} x^*(t)e^{+j2\pi ft}\left[\int_{-\infty}^{\infty} y(z)e^{-j2\pi fz} dz\right] dt$$

$$= X^*(f)Y(f) = E_{yx}(f)$$

$$(12.5.37)$$

Generally, $E_{xy}(f)$ and $E_{yx}(f)$ are functions of complex variables. Furthermore, they are not the same, that is, $E_{xy}(f) \neq E_{yx}(f)$. Instead, we have

$$E_{xy}(f) = E_{yx}^*(f). \qquad (12.5.38)$$

In summary, we can derive the energy spectral densities and cross energy spectral densities as

$$E_x(f) = X(f) \cdot X^*(f) = |X(f)|^2,$$
$$E_h(f) = H(f) \cdot H^*(f) = |H(f)|^2,$$
$$E_y(f) = Y(f) \cdot Y^*(f) = |Y(f)|^2,$$
$$E_y(f) = |H(f)|^2 S_x(f),$$
$$E_{xy}(f) = X(f)Y^*(f),$$
$$E_{yx}(f) = X^*(f)Y(f).$$

$$(12.5.39)$$

It is worth noting the following: for an energy signal, the value of the autocorrelation function at the origin, $R_x(0)$, represents the signal's energy. In contrast, for a power signal, the value $R_{xx}(0)$ corresponds to the power of the signal. In the case of stochastic processes, the value of the autocorrelation function at the origin is always the power of the process, as can be seen in Chapters 3 and 19.

### 12.5.3.3 *Correlation and Power Spectral Density of Complex Power Signals*

The foregoing derivations for the cross-correlation and autocorrelation functions are based on the assumption that $x(t)$ and $y(t)$ are *energy signals*. Thus, it is required that the signal energy be less than infinity. The energy of power signals can be infinite, implying that the duration of the signals can be infinite. Therefore, the definition and calculation of the correlation function need to include an infinite time interval.

**Aperiodic power signals.** For aperiodic power signals, the correlation and cross-correlation functions are defined for complex signals $x(t)$ and $y(t)$ as was done for real signals in eqns

(12.5.13) and (12.5.14), respectively, that is,

$$R_{xx}(\tau) = \lim_{a \to \infty} \frac{1}{2a} \int_{-a}^{a} x(t+\tau) x^*(t) dt \qquad (12.5.40)$$

and

$$R_{xy}(\tau) = \lim_{a \to \infty} \frac{1}{2a} \int_{-a}^{a} x(t+\tau) y^*(t) dt, \qquad (12.5.41)$$

for the complex conjugate signals $x^*(t)$ and $y^*(t)$, respectively. The power spectral density is defined as the Fourier transform of the autocorrelation function:

$$S_{xx}(f) = \int_{-\infty}^{\infty} R_{xx}(\tau) e^{-j2\pi f \tau} d\tau. \qquad (12.5.42)$$

The power of the signal can be calculated from the time-domain or frequency-domain representation of the signal as

$$P_x = \int_{f=-\infty}^{\infty} S_{xx}(f) df = \frac{1}{2\pi} \int_{\omega=-\infty}^{\infty} S_{xx}(\omega) d\omega = \lim_{a \to \infty} \frac{1}{2a} \int_{\omega=-\infty}^{\infty} |x(t)|^2 dt = R_{xx}(0). \qquad (12.5.43)$$

**Periodic power signals.** If the complex-valued signals $x(t)$ and $y(t)$ are periodic power signals, their correlation and cross-correlation functions are defined as was done for the real signals $x(t)$ and $y(t)$ in eqns (12.5.15) and (12.5.16), respectively, that is,

$$R_{xx}(\tau) = \frac{1}{T} \int_{t=-T/2}^{T/2} x(t+\tau) x^*(t) dt, \qquad (12.5.44)$$

and the cross-correlation function is

$$R_{xy}(\tau) = \frac{1}{T} \int_{t=-T/2}^{T/2} x(t+\tau) y^*(t) dt. \qquad (12.5.45)$$

It is important to note that the Fourier transform of an energy signal gives the energy spectral density, and the Fourier transform of the power and periodic signals gives the power spectral density. Furthermore, the Fourier transform of a power signal may not exist and the power spectral density of the signal can be calculated as the Fourier transform of its autocorrelation function.

We may make the following notes on the definition and interpretation of the power spectral density. By definition, power signals have infinite energy, and the Fourier transform and energy spectral density may not exist. Therefore, the power signals need alternate spectral functions with similar properties as the energy spectral density. In practice, the power spectral density can be obtained for a power signal $x(t)$ that is time windowed with window size $2a$. The power spectral density can be defined as the normalized limit of the energy spectral density for the windowed signal $x_T(t)$ as

$$S_{xx}(f) = \lim_{a \to \infty} \frac{1}{2a} E_{xT}(f) = \lim_{a \to \infty} \frac{1}{2a} |X_T(f)|^2. \tag{12.5.46}$$

The power spectral density represents the distribution of signal power in the frequency domain, and the power of the signal can be calculated as

$$P_x = \lim_{a \to \infty} \frac{1}{2a} \int_{-a}^{a} |x(t)|^2 dt = \int_{f=-\infty}^{\infty} S_{xx}(f) df. \tag{12.5.47}$$

The following calculations are performed for harmonic signals in two different ways, which give the same results, of course. They are instructive cases that will allow the reader to understand both the behaviour of the signal and the definitions of power signals and their related autocorrelation and power spectral density functions. In addition, the expressions that will be obtained are necessary for the analysis of digital communication systems.

**Example**

Calculate the autocorrelation function and the power spectral density of the cosine signal $x(t) = A \cos(2\pi f_c t)$ and investigate its power and energy in the time and frequency domains.

   a.　Solve the problem by processing the signal as a periodic function.

   b.　Solve the problem by processing the signal as a power signal.

   c.　Find the energy spectral density and calculate the energy and power of the signal.

   d.　Calculate the energy and power of the signal in the time domain.

**Solution.** The solution is as follows:

a. Because the signal is periodic with the period $T$, the autocorrelation function is

$$R_{xx}(\tau) = \frac{1}{T} \int_{t=-T/2}^{T/2} x(t+\tau)x(t)dt = \frac{1}{T} \int_{-T/2}^{T/2} A\cos 2\pi f_c(t+\tau) \cdot A\cos 2\pi f_c t\, dt$$

$$= \frac{A^2}{2T} \int_{-T/2}^{T/2} [\cos 2\pi f_c \tau + \cos(4\pi f_c t + 2\pi f_c \tau)]\, dt = \frac{A^2}{2T} \cos 2\pi f_c \tau \int_{-T/2}^{T/2} dt + 0.$$

$$= \frac{A^2}{2T} T \cos 2\pi f_c \tau = \frac{A^2}{2} \cos 2\pi f_c \tau$$

The power of the signal is

$$R_{xx}(0) = \frac{A^2}{2}.$$

The power spectral density is

$$S_{xx}(f) = \int_{\tau=-\infty}^{\infty} R_{xx}(\tau) e^{-j2\pi f \tau}\, d\tau = \int_{-\infty}^{\infty} \frac{A^2}{2} \cos(2\pi f_c \tau) e^{-j2\pi f \tau}\, d\tau$$

$$= \frac{A^2}{2} \int_{-\infty}^{\infty} \left[\frac{1}{2} e^{j2\pi f_c \tau} + \frac{1}{2} e^{-j2\pi f_c \tau}\right] e^{-j2\pi f \tau}\, d\tau \quad,$$

$$= \frac{A^2}{2} \left[\frac{1}{2}\delta(f-f_c) + \frac{1}{2}\delta(f+f_c)\right] = \frac{A^2}{4} [\delta(f-f_c) + \delta(f+f_c)]$$

which can be used to find the signal power

$$P_x = \int_{f=-\infty}^{\infty} S_{xx}(f)df = \int_{f=-\infty}^{\infty} \frac{A^2}{4} [\delta(f-f_c) + \delta(f+f_c)]df$$

$$= \frac{A^2}{4} \int_{f=-\infty}^{\infty} [\delta(f-f_c) + \delta(f+f_c)]df = \frac{A^2}{4} 2 = \frac{A^2}{2}$$

b. Because the signal is a power signal with the period $T$, the autocorrelation function is

$$R_{xx}(\tau) = \lim_{a \to \infty} \frac{1}{2a} \int_{-a}^{a} x(t+\tau)x(t)dt = \lim_{a \to \infty} \frac{1}{2a} \int_{-a}^{a} A \cos 2\pi f_c(t+\tau) \cdot A \cos 2\pi f_c t \, dt$$

$$= \lim_{a \to \infty} \frac{1}{2a} \int_{-a}^{a} \frac{A^2}{2} [\cos 2\pi f_c \tau + \cos(4\pi f_c t + 2\pi f_c \tau)] dt$$

$$= \lim_{a \to \infty} \frac{1}{2a} \int_{-a}^{a} \frac{A^2}{2} [\cos 2\pi f_c \tau] dt + 0 = \lim_{a \to \infty} \frac{1}{2a} 2a \frac{A^2}{2} \cos 2\pi f_c \tau = \frac{A^2}{2} \cos 2\pi f_c \tau$$

The power is

$$P_x = R_{xx}(0) = \frac{A^2}{2}.$$

The power spectral density and powers are as calculated in a..

c. The signal in the time domain is $x(t) = A \cos(2\pi f_c t)$. Its Fourier transform is

$$X(f) = \frac{1}{2}\delta(f - f_0) + \frac{1}{2}\delta(f + f_0).$$

Following the definition of the absolute values squared, the energy spectral density may be expressed as

$$E_x(f) = |X(f)|^2 = \left| \frac{1}{2}A\delta(f - f_c) + \frac{1}{2}A\delta(f + f_c) \right|^2 = \left( \frac{1}{2}A\delta(f - f_c) + \frac{1}{2}A\delta(f + f_c) \right)^2$$

$$\underset{\substack{\text{product of deltas is 0} \\ \text{if defined at 2 points}}}{=} \frac{1}{4}A^2\delta^2(f - f_c) + 0 + \frac{1}{4}A^2\delta^2(f + f_c) = \frac{1}{4}A^2\delta^2(f - f_c) + \frac{1}{4}A^2\delta^2(f + f_c)$$

and the energy can be calculated as

$$E_x = \int_{f=-\infty}^{\infty} E_x(f)df = \frac{A^2}{4} \int_{f=-\infty}^{\infty} \left( \delta^2(f - f_c) + \delta^2(f + f_c) \right) df = \frac{A^2}{4} \lim_{\varepsilon \to \infty} \frac{1}{\varepsilon}(\varepsilon\varepsilon + \varepsilon\varepsilon) \to \infty.$$

The delta function is not square integrable; in other words, its square diverges and tends to infinity.

c.  The calculations for power and energy in the time domain are

$$P_x = \lim_{a \to \infty} \frac{1}{2a} \int_{-a}^{a} |x(t)|^2 dt = \lim_{a \to \infty} \frac{1}{2a} \int_{-a}^{a} A^2 \cos^2 2\pi f_c t \, dt = \lim_{a \to \infty} \frac{1}{2a} \int_{-a}^{a} \frac{A^2}{2} (1 + \cos 4\pi f_c t) \, dt ,$$

$$= \lim_{a \to \infty} \frac{1}{2a} 2a \frac{A^2}{2} + 0 = \frac{A^2}{2}$$

and

$$E_x = \lim_{a \to \infty} \int_{-a}^{a} |x(t)|^2 dt = \lim_{a \to \infty} \int_{-a}^{a} A^2 \cos^2 (2\pi f_c t) \, dt = \lim_{a \to \infty} 2a \frac{A^2}{2} \to \infty,$$

respectively. These calculations comply with the previous calculations for power and energy in the frequency domain.

### 12.5.3.4  *Analysis of a Linear-Time-Invariant System with Deterministic Energy Signals*

We will now calculate the functions for a linear-time-invariant system for the case when an *energy signal* is applied at the system input. A block schematic of a linear-time-invariant system specified by the transfer function $H(\omega)$ or the impulse response $h(t)$ was presented in Figure 12.5.3. The output signals of the system can be found using the convolution of the input signal and the impulse response. The analysis of the system can also be conducted using signal representation in the form of the autocorrelation function and the power spectral density function. In this case, the impulse response of the system will be replaced by the *system correlation function* $R_h(\tau)$, which is defined as the autocorrelation function of the system impulse response. In this analysis, the input, the system itself, and the output will be represented by the correlation functions and the related spectral densities.

**The input signal in the time and frequency domains.** The input signal of the system $x(t)$ can be represented in the time domain by its autocorrelation function as

$$R_x(\tau) = \frac{1}{2\pi} \int_{-\infty}^{+\infty} E_x(\omega) e^{j\omega\tau} \, d\omega, \tag{12.5.48}$$

which was expressed in terms of the frequency $f$ in eqn (12.5.34), and in the frequency domain by the related power or energy spectral density function that is the Fourier transform of the autocorrelation function

$$E_x(\omega) = \int_{-\infty}^{+\infty} R_x(\tau) e^{-j\omega\tau} \, d\tau, \tag{12.5.49}$$

which is analogous to expression (12.5.33). These two functions constitute a Fourier transform pair, as proven by the Wiener–Khintchine theorem.

**Representation of the system in the time and frequency domains.** The system is defined by its finite energy impulse response $h(t)$, which is assumed to be a real-valued signal in this section. The autocorrelation function of the impulse response is

$$R_h(\tau) = \int_{-\infty}^{\infty} h(t)h(t+\tau)\,dt, \tag{12.5.50}$$

which is called the *system correlation function*. The energy spectral density of the impulse response is its Fourier transform:

$$E_h(\omega) = \int_{-\infty}^{+\infty} R_h(\tau)e^{-j\omega\tau}\,d\tau. \tag{12.5.51}$$

These two functions define the linear-time-invariant system as shown in Figure 12.5.3 and can be used in the system analysis in the same way that the impulse response functions $h(t)$ and $H(f)$ are used for the analysis of linear-time-invariant systems in the time and frequency domains. Namely, it will relate the autocorrelation function of the system output to the autocorrelation function of the system input according to the following explanation. The input–output cross-correlation function can be obtained as follows:

$$
\begin{aligned}
R_{xy}(\tau) &= \int_{-\infty}^{\infty} x^*(t)y(t+\tau)\,dt \underset{conv}{=} \int_{-\infty}^{\infty} x^*(t)\left(\int_{-\infty}^{\infty} h(\alpha)x(t+\tau-\alpha)\,d\alpha\right)dt \\
&= \int_{-\infty}^{\infty} h(\alpha)\left(\int_{-\infty}^{\infty} x^*(t)x(t+\tau-\alpha)\,dt\right)d\alpha = \int_{-\infty}^{\infty} h(\alpha)R_x(\tau-\alpha)\,d\alpha. \\
&= R_x(\tau)*h(\tau)
\end{aligned}
\tag{12.5.52}
$$

Consequently, we obtain the following expression for the output–input cross-correlation:

$$R_{yx}(\tau) = h(-\tau)*R_x(\tau). \tag{12.5.53}$$

The cross energy spectral density of the two signals $x(t)$ and $y(t)$ is the Fourier transform of the cross-correlation, that is,

$$E_{xy}(\omega) = \int_{-\infty}^{+\infty} R_{xy}(\tau)e^{-j\omega\tau}\,d\tau = E_{yx}^*(\omega), \tag{12.5.54}$$

and the cross-corelation is

$$R_{xy}(\tau) = \frac{1}{2\pi} \int_{-\infty}^{+\infty} E_{xy}(\omega) e^{j\omega\tau} d\omega = R_{yx}^*(-\tau). \tag{12.5.55}$$

**The system output in the time and frequency domains.** The *autocorrelation* of the output can be obtained from the expression for the convolution integral rearranged to the form

$$\begin{aligned}
R_y(\tau) &= \int_{-\infty}^{\infty} y^*(t) y(t+\tau) \, dt \underset{conv}{=} \int_{-\infty}^{\infty} y^*(t) \left( \int_{-\infty}^{\infty} h(\alpha) x(t+\tau-\alpha) \, d\alpha \right) dt \\
&= \int_{-\infty}^{\infty} h(\alpha) \left( \int_{-\infty}^{\infty} y^*(t) x(t+\tau-\alpha) \, dt \right) d\alpha = \int_{-\infty}^{\infty} h(\alpha) R_{yx}(\tau-\alpha) \, d\alpha. \\
&= h(\tau) * R_{yx}(\tau)
\end{aligned} \tag{12.5.56}$$

Using relation (12.5.53) for the cross-correlation, we may express eqn (12.5.56) as

$$R_y(\tau) = h(\tau) * h(-\tau) * R_x(\tau). \tag{12.5.57}$$

Because the autocorrelation of the deterministic impulse response, called the *system correlation function*, is

$$R_h(\tau) = \int_{-\infty}^{\infty} h(\alpha) h(\tau+\alpha) \, d\alpha = h(\tau) * h(-\tau), \tag{12.5.58}$$

the autocorrelation of the output signal is

$$R_y(\tau) = R_h(\tau) * R_x(\tau). \tag{12.5.59}$$

Therefore, the autocorrelation of the output signal is equal to the convolution of the autocorrelation of the input signal, and the autocorrelation of the impulse response of the system itself. The output of the linear-time-invariant system is calculated from the twofold convolution of the input autocorrelation function and the system's impulse response.

If we have the autocorrelation or autospectral density of the input and output signals of a linear-time-invariant system, then we can determine the magnitude response of the system. However, we cannot determine the phase response of the system. For this purpose, we can use the cross-spectral densities. In the analysis of linear-time-invariant systems, we will follow these steps:

1. Find the system correlation function $R_h(\tau)$ defined as the autocorrelation function of the system impulse response $h(t)$.

2. For any input signal $x(t)$, defined by its autocorrelation function, we can then calculate the autocorrelation function for the output signal $y(t)$ as the convolution of the input and the impulse response from eqn (12.5.59).

3. The energy spectral density of the output signal $y(t)$ will be calculated as the product of the system magnitude response squared and the input energy spectral density, that is,

$$E_y(\omega) = |H(\omega)|^2 E_x(\omega).$$

4. For the calculated autocorrelation function and energy spectral density, the energy of the output signal will be calculated as

$$E_y = \int_{-\infty}^{\infty} |y(t)|^2 dt = \int_{-\infty}^{\infty} |Y(f)|^2 df \underset{\omega=2\pi f}{=} \frac{1}{2\pi} \int_{-\infty}^{\infty} |Y(\omega)|^2 d\omega = R_y(0).$$

**Example**

Prove that the expression for the cross-correlation function of a linear-time-invariant system is

$$R_{yx}(\tau) = h(-\tau) * R_x(\tau).$$

**Solution.** The solution is as follows:

$$R_{yx}(\tau) = \int_{-\infty}^{\infty} y^*(t)x(t+\tau)\, dt \underset{t+\tau=z}{=} \int_{-\infty}^{\infty} y^*(z-\tau)x(z)dt = \int_{-\infty}^{\infty} x(z)\left(\int_{-\infty}^{\infty} h(\alpha)x^*(z-\tau-\alpha)\, d\alpha\right) dz$$

$$= \int_{-\infty}^{\infty} h(\alpha)\left(\int_{-\infty}^{\infty} x(z)x^*(z-\tau-\alpha)\, dz\right) d\alpha = \int_{-\infty}^{\infty} h(\alpha)R_x(-\tau-\alpha)\, d\alpha.$$

$$\underset{R(-x)=R(x)}{=} \int_{-\infty}^{\infty} h(\alpha)R_x(\tau+\alpha)\, d\alpha \underset{\substack{\alpha=-v \\ +\infty}}{=} \int_{+\infty}^{-\infty} h(-v)R_x(\tau-v)(-dv) = \int_{-\infty}^{\infty} h(-v)R_x(\tau-v)\, dv$$

$$= h(-\tau) * R_x(\tau)$$

## 12.5.4  Tables of Correlation Functions and Related Spectral Densities

The relationships between autocorrelation and cross-correlation functions, power spectral densities, and the energy of periodic and aperiodic signals are presented in Table 12.5.1.

**Table 12.5.1** *Correlation functions and spectral densities of energy signals and power signals.*

| | **Correlation Functions** | **Spectral Densities** |
|---|---|---|
| **Energy signals** | | |
| Correlation and energy spectral density, Wiener–Khintchine theorem | $R_x(\tau) = \int_{-\infty}^{\infty} x(t)x(t+\tau)\,dt$ <br> $= \int_{-\infty}^{\infty} E_x(f)e^{j2\pi f\tau}\,df$ | $E_x(f) = \int_{-\infty}^{\infty} R_x(\tau)e^{-j2\pi f\tau}\,d\tau$ <br> $= |X(f)|^2$ |
| Cross-correlation and energy spectral density, Wiener–Khintchine theorem | $R_{xy}(\tau) = \int_{-\infty}^{\infty} x(t)y(t+\tau)\,dt$ <br> $= \int_{-\infty}^{\infty} E_{xy}(f)e^{j2\pi f\tau}\,df$ | $E_{xy}(f) = \int_{-\infty}^{\infty} R_{xy}(\tau)e^{-j2\pi f\tau}\,d\tau$ <br> $= X(f)Y^*(f) = E_{xy}(f)$ |
| Rayleigh theorem for finite energy signals | $E = \int_{-\infty}^{\infty} |x(t)|^2\,dt$ <br> $E = R_x(0)$ | $E = \int_{-\infty}^{\infty} |X(f)|^2\,df = \int_{-\infty}^{\infty} E_x(f)\,df$ <br> $E = \frac{1}{2\pi}\int_{-\infty}^{\infty} |X(\omega)|^2\,d\omega$ <br> $= \frac{1}{2\pi}\int_{-\infty}^{\infty} E_x(\omega)\,d\omega$ |
| **Power signals** | | |
| Aperiodic power signals | $R_{xx}(\tau) = \lim_{a\to\infty} \frac{1}{2a}\int_{-a}^{a} x(t+\tau)x(t)\,dt$ <br> $R_{xy}(\tau) = \lim_{a\to\infty} \frac{1}{2a}\int_{-a}^{a} x(t+\tau)y(t)\,dt$ | $S_{xx}(f) = \int_{\tau=-\infty}^{\infty} R_{xx}(\tau)e^{-2\pi f\tau}\,d\tau$ <br> $S_{xy}(f) = \int_{\tau=-\infty}^{\infty} R_{xy}(\tau)e^{-2\pi f\tau}\,d\tau$ |
| Periodic power signal | $R_{xx}(\tau) = \frac{1}{T}\int_{-T/2}^{T/2} x(t+\tau)x(t)\,dt$ <br> $R_{xy}(\tau) = \frac{1}{T}\int_{-T/2}^{T/2} x(t+\tau)y(t)\,dt$ | $S_{xx}(f) = \int_{-\infty}^{\infty} R_{xx}(\tau)e^{-j2\pi f\tau}\,d\tau$ <br> $S_{xy}(f) = \int_{-\infty}^{\infty} R_{xy}(\tau)e^{-j2\pi f\tau}\,d\tau$ |
| Power calculations | $P = \frac{1}{T}\int_{-T/2}^{T/2} |x(t)|^2\,dt$ <br> $P = R_{xx}(0)$ | $P = \int_{-\infty}^{\infty} S_{xx}(f)\,df$ |

**Table 12.5.2** *Examples of correlation functions and their power spectral densities.*

| Signal $x(t)$ | Correlation | Spectral Density, Power, Energy |
|---|---|---|
| $A \cdot \sin(2\pi f t)$ $A \cdot \cos(2\pi f t)$ | $R_{xx}(\tau) = \frac{A^2}{2}\cos 2\pi f \tau$ $R_{xx}(\tau) = \frac{A^2}{2}\cos 2\pi f \tau$ $P = R_{xx}(0) = \frac{A^2}{2}$ | $A^2 \frac{\delta(f-f_0)+\delta(f+f_0)}{4}$ $A^2 \frac{\pi[\delta(\omega-\omega_0)+\delta(\omega+\omega_0)]}{2}$ $A^2 \frac{\delta(f-f_0)-\delta(f+f_0)}{4j}$ $jA^2\pi \frac{[\delta(\omega+\omega_0)+\delta(\omega-\omega_0)]}{2}$ |
| $A \cdot e^{-at}u(t)$ | $R_x(\tau) = \frac{A^2}{2a}e^{-a|\tau|}$ | $S_x(\omega) = \frac{A^2}{\omega^2+a^2},$ $E = R_x(0) = \frac{A^2}{2a}$ |
| Deterministic rectangular pulse $Au\left(t+\frac{T}{2}\right)-Au\left(t-\frac{T}{2}\right)$ | $\begin{cases} A^2 T \left(1-\frac{|\tau|}{T}\right) & |\tau|<T \\ 0 & |\tau|\geq T \end{cases}$ | $S_{xx}(f) = A^2 T^2 \cdot \text{sinc}^2(fT) = E_x(f)$ $E = R_x(0) = A^2 T$ |
| $\delta(t)$ | $\delta(\tau)$ | $S_\delta(f) = 1 = E_x(f),$ for $-\infty < f < \infty$ $E_\delta \rightarrow \infty$ |

---

PROBLEMS

1. Find the magnitude spectrum, the phase spectrum, and the power spectrum of the periodic sawtooth signal shown in Figure 1. Calculate the power of the signal in the time and frequency domains.

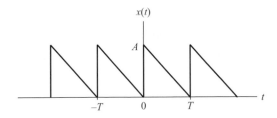

**Fig. 1** *Sawtooth signal.*

2. Find the Fourier transform for the rectangular pulse shown in Figure 2, using the shifting property of the Fourier transform. Calculate and plot the graphs of the magnitude spectrum and the phase spectrum of the signal. Calculate the power and energy of the signal.

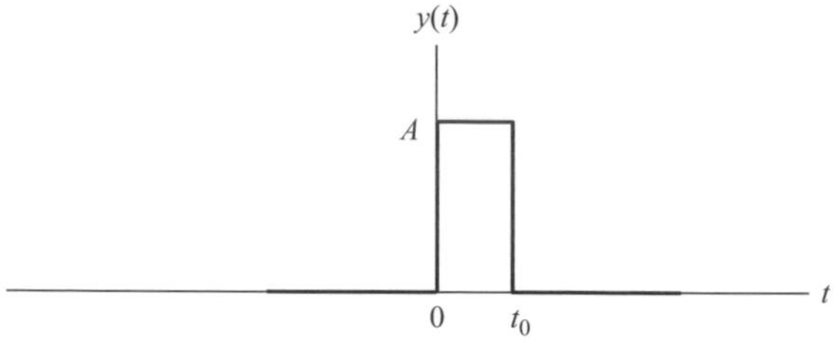

$y(t)$

$A$

$0$    $t_0$

**Fig. 2** *Rectangular pulse.*

3. An aperiodic signal $x(t)$ is expressed as $x(t) = e^{-2t}$, on the interval $0 \le t < 2$, as depicted in Figure 3. Calculate the Fourier transform $X(j\omega)$ of $x(t)$.

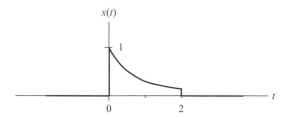

$x(t)$

$1$

$0$    $2$

**Fig. 3** *Exponential truncated function.*

4. A real signal has an amplitude spectral density defined as

$$S(f) = \begin{cases} S & -B < f < B \\ 0 & \text{otherwise} \end{cases}.$$

a. Plot the graphs of the magnitude spectral density and the phase spectral density of the signal.

b. Find the signal in the time domain.

c. Plot the signal in the time domain and find the zero-crossing point(s).

d. What is the maximum value of the signal in the time domain?

5. Suppose the impulse response of the linear-time-invariant system in the time domain is

$$h(t) = \begin{cases} A & 0 \le t \le 2 \\ 0 & \text{otherwise} \end{cases},$$

and the Fourier transform of the input signal $x(t)$ is

$$X(j\omega) = \frac{\sin \omega}{\omega} e^{-j\omega}.$$

a. Find out the Fourier transform $Y(j\omega)$ of the output signal $y(t)$.

b. Sketch the graph for the magnitude spectral density of $x(t)$. Find the zero-crossing point of this graph.

c. Sketch the magnitude spectral density and the phase spectral density of $y(t)$.

d. Sketch the graph of the signal $x(t)$. (Hint: Compare the signals $h(t)$ and $x(t)$ in the frequency domain.)

6. Derive the expressions for the asymmetric Fourier transform pair, using the expressions for the Fourier series.

7. Use the expression for the Fourier series coefficients of a rectangular pulse train,

$$c_k = \begin{cases} A\frac{t_0}{T} & k = 0 \\ \frac{A\sin k\pi t_0/T}{k\pi} & k \neq 0 \end{cases},$$

to derive the Fourier transform of a single rectangular pulse.

8. Find the amplitude spectral densities of a unit step function.

9. Find the inverse Fourier transform of a spectrum defined by the shifted delta functions of the angular frequency $\omega$, that is,

a. $X(j\omega) = \delta(\omega - \omega_0)$, and

b. $X(j\omega) = \delta(\omega + \omega_0)$.

10. Find the Fourier transform of a DC signal $x(t) = 1$, which is constant for any $t$ value.

11. Find the signal $x(t)$ in the time domain if its Fourier transform is defined as $X(\omega) = 1$ for any $\omega$ value.

12. Find the spectra of the cosine and sine functions for the case when they are expressed as functions of the angular frequency $\omega$.

13. The causal real exponential signal $h(t)$ is defined as

$$h(t) = e^{-t/T}u(t).$$

a. Calculate and sketch the magnitude spectral density and the phase spectral density of this signal.

b. Suppose that $1/T$ tends to 0. Sketch the magnitude spectrum and the phase spectrum of $h(t)$ for this case.

c. Suppose $1/T < 0$. Is it possible to obtain the Fourier transform of $h(t)$ for this case? Support your answer with proof.

d. Suppose the impulse response of the linear-time-invariant system is $h(t)$, and the Fourier transform of the input signal $x(t)$ is

$$X(j\omega) = \frac{\sin \omega T_0/2}{\omega T_0/2} e^{-j\omega T_0/2}.$$

Find and sketch the magnitude spectral density of $y(t)$.

14. Consider the case when the input signal of a linear-time-invariant system is a rectangular pulse $x(t)$ centred at time $t = 0$ with the total width $T$, as shown in Figure 14. The impulse response of the system is $h(t) = x(-t)$.

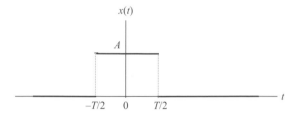

**Fig. 14** *Input signal for a linear-time-invariant system.*

a. Calculate the amplitude spectral density $X(j\omega)$ of the input signal $x(t)$, using the Fourier transform. Calculate the amplitude spectral density $H(j\omega)$ of the impulse response $h(t)$, using the Fourier transform.

b. Calculate the amplitude spectral density $Y(j\omega)$ of the output signal $y(t)$, using the Fourier transforms of $x(t)$ and $h(t)$.

c. Plot the magnitude spectral density and the phase spectral density of $x(t)$, $h(t)$, and $y(t)$. What are the differences between the magnitude spectral densities and the phase spectral densities of $x(t)$ and $h(t)$? What are the differences between $x(t)$ and $h(t)$ in the time domain? Explain.

d. Show how to calculate the time-domain expression for the output signal $y(t)$ by using its inverse Fourier transform. Set up the formula without further calculations. Sketch the output signal $y(t)$ by convolving $x(t)$ and $h(t)$.

15. A periodic signal $x(t)$ is expressed as $x(t) = e^{-2t}$ on the interval $0 \le t < T$, where $T = 2$ is the period of the signal, as depicted in Figure 15.

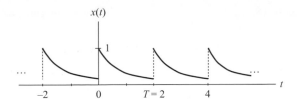

**Fig. 15** *Periodic signal.*

a. Calculate and sketch the magnitude spectrum of $x(t)$. Suppose the signal $x(t)$ represents the voltage applied to the input of an ideal lowpass filter. The cut-off frequency $f_c$ of the filter is specified as $f_c = 1.5 \cdot f_0$, where $f_0$ is the frequency of the fundamental component $c_{+1}$ in the magnitude spectrum of the signal $x(t)$.

b. Calculate the power of the signal at the output of this lowpass filter.

c. What per cent of the average power of the signal $x(t)$ is transferred to the output of the filter?

16. Find the autocorrelation function of a sine signal $x(t) = \sin(2\pi t)$. Analyse graphically the special case when $\tau = 0.5$, $\tau = 1$, and $\tau = 1/4$.

17. The signal is shown in Figure 17. Find its autocorrelation function. Plot the graph of the autocorrelation function. Calculate the energy of the signal by using the autocorrelation function and the signal expression in the time domain. Calculate the energy spectral density of the signal.

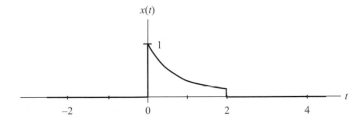

**Fig. 17** *Signal in the time domain.*

18. Compute the cross-correlation for the signals

$$x(t) = u(t) - u(t - T)$$

and

$$y(t) = u(t + T) - u(t),$$

as shown in Figure 18.

**Fig. 18** *Signals in the time domain.*

19. Calculate the autocorrelation function and the power spectral density of the sine signal $x(t) = A \sin(2\pi f_c t)$ and investigate its power and energy in the time and frequency domains.

    a. Solve the problem by processing the signal as a periodic function.

    b. Solve the problem by processing the signal as a power signal.

    c. Find the energy spectral density and calculate the energy and power of the signal.

20. Calculate the autocorrelation function for the unit step signal $u(t)$ in Figure 20, drawn for $\tau > 0$.

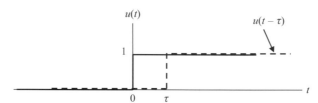

**Fig. 20** *Unit step signal.*

21. Calculate the autocorrelation function of the real-valued signal $x(t) = A \cdot e^{-at} u(t)$, for parameters $A$ and $a$ greater than 1. For the obtained autocorrelation function, find the energy spectral density and energy of this signal.

22. Prove the following expressions for the cross-correlation functions and power spectral densities for a linear-time-invariant system:

    a. $R_{yx}(\tau) = h(-\tau) * R_x(\tau)$,

    b. $R_h(\tau) = \int_{-\infty}^{\infty} h(\alpha) h(\tau + \alpha) d\alpha = h(\tau) * h(-\tau)$,

    c. $S_{xy}(\omega) = H(\omega) S_x(\omega)$, and $S_{yx}(\omega) = H^*(\omega) S_x(\omega) = S_{xy}^*(\omega)$.

23. Find the Fourier transform of the signal $x(t) = 10\cos(200\pi t)$. Present the amplitude spectral density function in terms of delta functions. Calculate the Nyquist sampling rate and sampling interval for this function.

24. An amplitude modulated signal $s(t)$ is obtained by modulating a carrier with amplitude $A_c = 20$ V and frequency $10^5$ Hz, with a tone signal having amplitude $A_m = 2$ V and frequency 1500 Hz. The modulated signal is expressed in the form

$$s(t) = [20 + 2\cos 2\pi 1500\ t]\cos\left(2\pi 10^5 t\right).$$

Find the spectral components of the signal $s(t)$ by using the Fourier series and Fourier transform.

25. The input of a linear-time-invariant system is an energy signal $x(t)$. Prove that the expressions for the system correlation function, the cross power spectral densities and the cross-correlation function are

$$R_h(\tau) = \int_{-\infty}^{\infty} h(\alpha)h(\tau+\alpha)\,d\alpha = h(\tau) * h(-\tau),$$
$$E_{xy}(\omega) = H(\omega)E_x(\omega),$$
$$E_{yx}(\omega) = H^*(\omega)E_x(\omega) = E_{xy}^*(\omega),$$
$$R_{yx}(\tau) = h(-\tau) * R_x(\tau).$$

26. Find the expression in the time domain and the Fourier series of the triangular pulse train presented in Figure 26. Find the average power of the signal defined in the time and frequency domains.

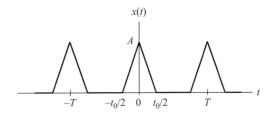

**Fig. 26** *Periodic triangular pulse train.*

# 13

# Sampling and Reconstruction of Continuous-Time Signals

## 13.1  Introduction

In the signal processing of the continuous-time signals of interest to us, all of the equations, operations, and properties related to signals and systems analysis are presented under the assumption that time is a variable and, furthermore, that time is a continuous variable denoted by $t$. These signals have been defined and operated upon in the form of functions. In addition, these signals are converted into the frequency domain, which is sometimes more convenient for signal analysis, synthesis, and processing. However, it has been shown that some better insight into the nature and properties of the signals and systems can be provided using presentation of the signals in various domains.

Nonetheless, some signals are, by their nature, functions of time as a discrete variable. In addition, it is usual practice to obtain a discrete-time signal $x(nT_s)$ by taking the values of an analogue signal $x(t)$, which is effectively a continuous-time signal, via sampling at discrete instants of time $nT_s$. In principle, by using this technique, we can obtain a discrete signal $x(nT_s)$ which is composed of samples of the continuous-time signal $x(t)$ that are taken at equidistant points of time defined as $nT_s$. This procedure is sometimes called the *discretization of signals in the time domain*, or *time-domain discretization*. The values of the obtained signal samples are rounded or truncated to have integer values, using the procedure of *discretization of signal values*, or *amplitude discretization*. In this way, an analogue signal is represented by a series of numbers that can be transferred to binary form and processed using digital signal processors. This procedure is traditionally known as an analogue-to-digital conversion, which was explained briefly in Chapter 1, Section 1.3.1. In the current chapter, the basic theoretical issues related to the theory of signal sampling and reconstruction will be presented.

If we had a discrete-time signal $x(nT_s)$, which expressed as $x(n)$ for a unified sampling interval $T_s$, the question would be how to find its representation in the frequency domain. To this end, the aim would be to determine the relationship between the Fourier transform for the continuous-time signals $x(t)$, and the discrete-time Fourier transform for the corresponding discrete-time signals $x(n)$. Also, in the case when the discrete-time signal is defined on a finite interval $N$, $n = 1, 2, 3, \ldots, N$, the discrete Fourier transform should be defined. Instead of the discrete-time Fourier transform, the discrete Fourier transform should be used for both the

*Discrete Communication Systems*. Stevan Berber, Oxford University Press. © Stevan Berber 2021.
DOI: 10.1093/oso/9780198860792.003.0013

time-domain and the frequency-domain analysis of signals on digital computers, because the signal spectrum obtained by the discrete-time Fourier transform is a function of the continuous normalized frequency and, as such, is not suitable for digital processing. Namely, the signal, being defined on a continuum of frequencies, requires a theoretically infinite number of frequency components to be processed, while the discrete Fourier transform only requires a limited number of these components.

To solve the problem of the spectral analysis of discrete signals using a limited set of spectral components, or a limited number of signal samples, fast Fourier transform algorithms were developed to compute discrete Fourier transforms efficiently. How to process both deterministic and stochastic signals from time and frequency domains in the discrete-time domain, and how to apply the developed theory in practice, are subjects of interest in several chapters of this book.

In this chapter, we will address basic issues related to the theory of signal discretization, primarily the time and amplitude discretization of continuous-time signals. The procedures of analogue-to-digital and digital-to-analogue conversions will be briefly explained in the context of understanding the place and importance of signal discretization in the time domain (sampling), and discretization of signal amplitude (quantization).

## 13.2   Sampling of Continuous-Time Signals

Sampling is the key operation for converting a signal from a continuous-time form to a discrete-time form. Signal reconstruction is the inverse operation of converting the sampled signal into its continuous-time (analogue) form. Both operations are part of the analogue-to-digital and digital-to-analogue conversion of signals.

The sampling operation must comply with the sampling (Nyquist) theorem, which states that any physically realizable signal $x(t)$, which has a limited spectrum in the bandwidth $B$, can be uniformly sampled at the sampling instants of time defined as $T_s \leq 1/2B$, and then perfectly reconstructed by passing the samples through an ideal lowpass filter with the bandwidth $B$.

The period of sampling, $T_s$, is called the sampling period, and its reciprocal value $1/T_s = f_s$ is called the sampling frequency, or the sampling rate $R_s$, and is expressed in the number of samples per second. The maximum value of the sampling period is $T_s = 1/2B$, which is called the Nyquist interval, and the value $1/T_s = f_s$ is called the Nyquist frequency, or the Nyquist rate. The reconstructed continuous-time signal $\hat{x}(t)$ must comply in the time and frequency domains with the original analogue signal $x(t)$ that was applied at the input of the sampler. As we mention at several places in the book, the signal $x(t)$ should be physically realizable, meaning that the power of the signal is limited. Furthermore, the condition of the signal's physical realizability implies that it is an energy signal. We had to bear this in mind when we explained the procedure of sampling white Gaussian noise, which, by definition, has infinite power. Details about this noise sampling and related issues can be found in Chapter 4.

Therefore, to prove the sampling theorem, all the signals involved in both the sampling and the reconstruction procedure need to be analysed in the time and frequency domains, as will be presented in the following explanation.

Sampling of an analogue band-limited signal $x(t)$ can be mathematically modelled as the multiplication of that signal by the periodic train of Dirac delta impulses $s(t)$. The schematic of a sampler, which is designed as a multiplier with the signals involved presented in the time and frequency domains, is shown in Figure 13.2.1. An example of the hypothetical signals involved in the sampling procedure is shown in Figure 13.2.2. A band-limited analogue (continuous-time and continuous-valued) signal $x(t)$ is applied to the input of the sampler. The signal can be expressed in the time and frequency domains by its Fourier transform pair. The amplitude spectral density of the signal is expressed as

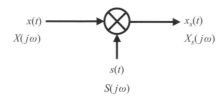

**Fig. 13.2.1**  *Implementation of a sample as a multiplier.*

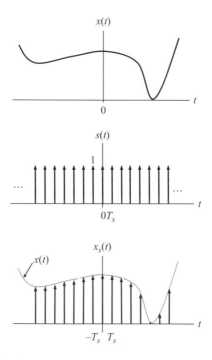

**Fig. 13.2.2**  *Sampling procedure for a continuous-time (analogue) signal x(t).*

$$X(\omega) = \int_{-\infty}^{\infty} x(t) e^{-j\omega t} dt, \tag{13.2.1}$$

and its inverse Fourier transform gives the signal in the time domain:

$$x(t) = \frac{1}{2\pi} \int_{-\infty}^{\infty} X(\omega) e^{j\omega t} d\omega. \tag{13.2.2}$$

This analogue signal will be sampled using the sampling signal $s(t)$, which has the spectrum $X_s(j\omega)$. Namely, we need to perform the sampling in such a way so as to preserve the signal $x(t)$ unchanged in both the time domain and the frequency domain after the reconstruction operation. Therefore, it is important to find the sampling $s(t)$ and sampled signal $x_s(t)$ in frequency domain by calculating their amplitude spectral densities, and then analyse their relationship from the sampling-procedure point of view.

The sampling signal $s(t)$ is expressed in the form

$$s(t) = \sum_{k=-\infty}^{\infty} \delta(t - kT_s). \tag{13.2.3}$$

This signal multiplies the input analogue signal, resulting in the output sampled signal $x_s(t)$, which can be represented in the time domain as the weighted sum of the time-shifted impulse delta functions:

$$x_s(t) = x(t)s(t) = \sum_{k=-\infty}^{\infty} x(t)\delta(t - kT_s) = \sum_{k=-\infty}^{\infty} x(kT_s)\delta(t - kT_s). \tag{13.2.4}$$

The input analogue signal, the sampling signal, and the output sampled signal are presented in Figure 13.2.2, for an arbitrarily chosen sampling period $T_s$.

Let us, firstly, find the spectrum of the sampling signal $s(t)$. Because this signal is an impulse train, and therefore a periodic signal with the period $T_s$, we can apply the Fourier series to calculate the Fourier coefficients as

$$c_k = \frac{1}{T_s} \int_{-T_s/2}^{T_s/2} s(t) e^{-jk\omega_s t} dt = \frac{1}{T_s}, \tag{13.2.5}$$

which have constant values and are defined for all $k$, as derived in Chapter 12, Section 12.3.5, and shown in Figure 12.3.9. Using the Fourier coefficients, the sampling signal in the time domain can be expressed as an infinite sum of complex exponentials, that is,

$$s(t) = \sum_{k=-\infty}^{\infty} c_k e^{+jk\omega_s t} = \frac{1}{T_s} \sum_{k=-\infty}^{\infty} e^{+jk\omega_s t}. \tag{13.2.6}$$

The signal obtained is a periodic signal presented in the time domain. Its Fourier transform can be obtained as presented in Section 12.3.5, expression (12.3.56). For the sampling frequency $\omega_s = 2\pi/T_s$ , we find

$$S(j\omega) = \int_{-\infty}^{\infty} \sum_{k=-\infty}^{\infty} c_k e^{+jk\omega_s t} e^{+j\omega t} dt = \sum_{k=-\infty}^{\infty} 2\pi c_k \frac{1}{2\pi} \int_{-\infty}^{\infty} e^{-j(\omega-k\omega_s)t} dt. \qquad (13.2.7)$$

Due to the known transform of an exponential function expressed as

$$FT\left\{ \frac{1}{2\pi} e^{+jk\omega_s t} \right\} = \frac{1}{2\pi} \int_{-\infty}^{\infty} e^{-j(\omega-k\omega_s)t} dt = \delta(\omega - k\omega_s), \qquad (13.2.8)$$

where FT stands for Fourier transform, we may solve the integral in eqn (13.2.7) as follows:

$$S(j\omega) = 2\pi \frac{1}{T_s} \sum_{k=-\infty}^{\infty} \delta(\omega - k\omega_s) = \omega_s \sum_{k=-\infty}^{\infty} \delta(\omega - k\omega_s), \qquad (13.2.9)$$

Because the sampled signal is the product of two signals in the time domain, $x_s(t) = x(t) \cdot s(t)$, its amplitude spectral density will be a convolution of the Fourier transforms of multipliers and can be expressed as

$$X_s(j\omega) = \frac{1}{2\pi} X(j\omega) * S(j\omega). \qquad (13.2.10)$$

Substituting eqn (13.2.9) into eqn (13.2.10), we may calculate the amplitude spectral density of the sampled signal as

$$X_s(j\omega) = \frac{1}{2\pi} X(j\omega) * S(j\omega) = \frac{1}{T_s} \sum_{k=-\infty}^{\infty} X(j\omega) * \delta[j(\omega - k\omega_s)]$$

$$= \frac{1}{T_s} \sum_{k=-\infty}^{\infty} \int_{-\infty}^{\infty} X(j\lambda) \cdot \delta[j(\omega - k\omega_s - \lambda)] d\lambda = \frac{1}{T_s} \sum_{k=-\infty}^{\infty} X[j(\omega - k\omega_s)] \qquad (13.2.11)$$

The spectrum of the sampled signal contains the spectrum of the analogue input signal that repeats periodically with a period equal to the sampling frequency $\omega_s = 2\pi/T_s$. The amplitudes of these repeated spectral components are normalized by the sampling period $T_s$. The amplitude spectral densities of the analogue signal, the sampling, and the sampled signal are presented in Figure 13.2.3.

From the expressions of the signals in the frequency domain, eqns (13.2.1), (13.2.9), and (13.2.11), and their graphical presentations in Figure 13.2.3, the following can be concluded:

- $X_s(j\omega)$ is a periodic signal containing an infinite number of components that are all images of the input signal $X(j\omega)$.

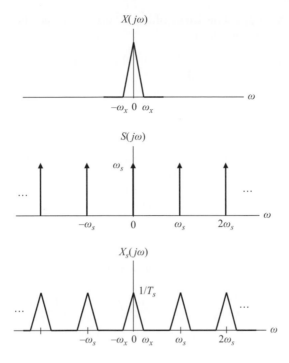

**Fig. 13.2.3** *The case when replicas do not overlap.*

- The entire spectrum of the input signal $X(j\omega)$ is contained inside one component of the spectrum of the sampled signal.
- Depending on the relationship between the input-signal bandwidth $\omega_x$ and the sampling frequency $\omega_s$, it may be possible to reconstruct the input signal $x(t)$ from the sampled signal $x_s(t)$ by using an image taken from the spectrum of the sampled signal.

Therefore, it may be possible to reconstruct the input signal from the sampled signal by using a baseband filter that will extract the frequency band containing the input signal. How to design the filter and fulfil the conditions required to reconstruct the input signal is the subject of the next section.

## 13.3 Reconstruction of Analogue Signals

Depending on the bandwidth of the analogue signal $X(j\omega)$, there can be some overlapping of its replicas inside the spectrum of the sampled signal $X_s(j\omega)$, so that *aliasing* occurs in the frequency domain; alternatively, when there is sufficient frequency separation between the replicas, overlapping is avoided, so aliasing in the frequency domain cannot occur. Therefore, for an analogue input signal with bandwidth $\omega_x$, two cases, specified by the relationship between $\omega_s$ and $\omega_x$, can be considered.

**Signal reconstruction without distortion: No aliasing in the frequency domain.** Figure 13.3.1 presents the amplitude spectral density of the sampled signal that is composed of a series of spectral replicas of the input analogue signal. If the following inequality is fulfilled,

$$\omega_s \geq 2\omega_x, \tag{13.3.1}$$

than the spectral replicas do not overlap and there is no aliasing, as depicted in the right-hand side of Figure 13.3.1. This inequality is also known as the *Nyquist condition* or *Nyquist criterion*. The frequency $\omega_s$ is often referred to as the *folding frequency* or *Nyquist frequency*. In order to avoid aliasing, the following statement, a rewording of the sampling theorem, must hold:

> *The sampling frequency must exceed twice the analogue signal bandwidth in order for the original analogue signal x(t) to be recoverable from the sampled signal $x_s$(t).*

Therefore, it is possible to define the *reconstruction procedure* of the analogue signal $x(t)$ from the sampled signal $x_s(t)$. The first replica of the analogue signal $X(j\omega)$ lies inside the spectrum of the sampled signal $X_s(j\omega)$ and is centred at the zero frequency. Therefore, it is sufficient to use an ideal lowpass filter with the cut-off frequency $\omega_c = \omega_s/2$ and the gain $T_s$ in order to reconstruct the analogue signal $x(t)$ from the sampled signal $x_s(t)$ without any distortion. This cut-off frequency should fulfil the condition $\omega_c \geq \omega_x \leq \omega_s/2$. Thus, the transfer characteristic of the ideal lowpass filter required should be as shown in the right-hand side of Figure 13.3.1.

If such a lowpass filter is used, the baseband analogue signal can be reconstructed without any distortion, as shown in the left-hand side of Figure 13.3.1 because all of the replicas in the sampled signal spectrum beyond the bandwidth of the analogue signal will be eliminated by the filter.

**Signal reconstruction with distortion: Aliasing in the frequency domain.** If the following inequality is fulfilled,

$$\omega_s \leq 2\omega_x, \tag{13.3.2}$$

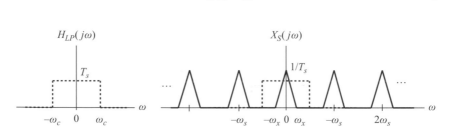

**Fig. 13.3.1** *Transfer characteristics of an ideal lowpass filter (left), and use of a lowpass filter to reconstruct an analogue signal (right).*

the spectral replicas of the input signal $x(t)$ contained in the spectrum of the sampled signal $x_s(t)$ will irreversibly overlap. This irreversible overlap of the spectral replicas is known as *aliasing in the frequency domain*. Figure 13.3.2 symbolically presents the amplitude spectral density of the analogue signal in the case when there is overlapping of replicas in the frequency domain. In this case, the original signal $x(t)$ cannot be recovered from the signal $x_s(t)$ that is obtained using sampling procedure, because of the distortion caused by the replicas. Consequently, it is not sufficient to use a lowpass filter with a cut-off frequency $\omega_c$ equal to $\omega_s/2$ and with a gain of $T_s$, as shown in the right-hand side of Figure 13.3.1, to recover the analogue signal $x(t)$ from its sampled signal $x_s(t)$. In this case, the bandwidth of the analogue signal is wide, so we need to use a higher sampling frequency than that used in Figure 13.3.2, to overcome the aliasing in the frequency domain.

Here, the sampling theorem was demonstrated for the case when the samples of an analogue signal are taken uniformly and periodically. However, the samples can be taken at arbitrary time instants while still fulfilling the condition of having a sufficient number of samples per second defined by the average sampling rate $f_s = 1/2B$.

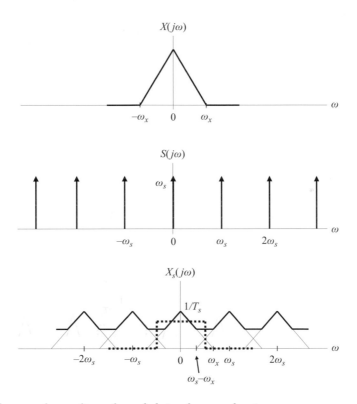

**Fig. 13.3.2** *The case when replicas of sampled signal are overlapping.*

## 13.4 Operation of a Lowpass Reconstruction Filter

If the Nyquist criterion is fulfilled, the continuous-time signal $x(t)$ can be fully recovered from its samples $x_s(t)$ obtained by the sampling procedure. For this purpose, the samples $x_s(t)$ should be passed through a lowpass filter with a cut-off frequency $\omega_c$ that fulfils the condition

$$\omega_x < \omega_c < \omega_s - \omega_x, \tag{13.4.1}$$

and has a transfer function defined as

$$H_{LP}(j\omega) = \begin{cases} T_s & |\omega| \leq \omega_c \\ 0 & |\omega| > \omega_c \end{cases}. \tag{13.4.2}$$

where LP stands for lowpass. The impulse response of the lowpass filter in the frequency domain $H(j\omega)$ is equivalent to its transfer function $H_{LP}(j\omega)$, which is shown in Figure 13.4.1. Therefore, the impulse response of this filter in the time domain, $h(t)$, is equal to the inverse Fourier transform of its frequency response, that is,

$$h(t) = \frac{1}{2\pi} \int_{-\infty}^{\infty} H(j\omega) e^{+j\omega t} dt = \frac{1}{2\pi} \int_{-\omega_c}^{\omega_c} T_s \cdot e^{+j\omega t} dt = \frac{1}{2\pi/T_s} \frac{\sin \omega_c t}{t/2} = \frac{\sin \omega_c t}{\omega_s t/2}, \tag{13.4.3}$$

which is defined in an open interval specified by $-\infty \leq t \leq +\infty$. This impulse response can also be expressed in terms of a sinc function as

$$h(t) = \frac{\omega_c t}{\omega_s t/2} \frac{\sin \omega_c t}{\omega_c t} = 2 \frac{\omega_c}{\omega_s} \operatorname{sinc}(\omega_c t) = 2 \frac{T_s}{T_c} \operatorname{sinc}(\omega_c t). \tag{13.4.4}$$

The graph of the impulse response in the time domain $h(t)$ is presented in Figure 13.4.1. The first zero crossing occurs when $\omega_c t = \pi$, or $t = \pi/2\pi f_c = 1/2f_c = T_c/2$. The sampled signal is represented by the weighted sum of the time-shifted impulse functions and expressed by eqn (13.2.4). If we pass this sampled signal through a lowpass filter with a transfer function equal to $H_{LP}(j\omega)$, as specified by eqn (13.4.2), the analogue signal $x(t)$ can be reconstructed from

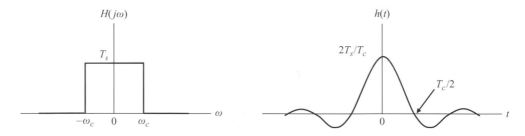

**Fig. 13.4.1** *Lowpass filter impulse response in the frequency and time domains.*

this sampled signal, according to the block schematic in Figure 13.4.2. This lowpass filter is called an *analogue reconstruction filter*.

It is important to note that, for the input impulse Dirac delta signal $\delta(t)$, the output is the impulse response $h(t)$ and, for the input shifted-impulse delta signal $\delta(t - kT_s)$ the output is the time-shifted impulse response $h(t - kT_s)$, due to the time invariance of the system. Therefore, due to the linearity of the lowpass filter, its output $\hat{x}(t)$ can be obtained as the convolution of the input sampled signal $x_s(t)$ and the impulse response of the filter, $h(t)$. Furthermore, because of the linearity and time invariance of the lowpass filter, the output signal can be obtained by simply applying the linear operation $L$ on the input signal $x_s(t)$, which results in the following expression for the output reconstructed analogue signal:

$$\hat{x}(t) = L\{x_s(t)\} = \sum_{k=-\infty}^{\infty} x(kT_s) L\{\delta(t - kT_s)\} = \sum_{k=-\infty}^{\infty} x(kT_s) h(t - kT_s). \qquad (13.4.5)$$

By inserting expression (13.4.3) for the impulse response of the filter, $h(t)$, into this expression, we can have the reconstructed analogue signal expressed in terms of sinc functions as

$$\hat{x}(t) = \sum_{k=-\infty}^{\infty} x(kT_s) \frac{\sin\omega_c(t - kT_s)}{\omega_s(t - kT_s)/2} = \sum_{k=-\infty}^{\infty} x(kT_s) 2 \frac{\omega_c}{\omega_s} \frac{\sin\omega_c(t - kT_s)}{\omega_c(t - kT_s)}$$
$$= \sum_{k=-\infty}^{\infty} x(kT_s) 2 \frac{T_s}{T_c} \operatorname{sinc}\omega_c(t - kT_s) \qquad (13.4.6)$$

If we assume that the cut-off frequency has the minimum possible value for distortionless reconstruction, that is, $\omega_c = \omega_x = \omega_s/2 = \pi/T_s$, which is specified by the Nyquist sampling frequency $f_s = \omega_x/\pi = 2\pi B/\pi = 2B$, we obtain

$$\hat{x}(t) = \sum_{k=-\infty}^{\infty} x(kT_s) \frac{\sin\pi(t - kT_s)/T_s}{\pi(t - kT_s)/T_s} = \sum_{k=-\infty}^{\infty} x(kT_s) \operatorname{sinc}[(t - kT_s)\pi/T_s]. \qquad (13.4.7)$$

The output of the reconstruction filter, for the sampling interval expressed as $T_s = 1/2B$, can be expressed in terms of sinc functions as

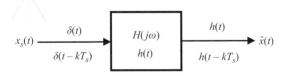

**Fig. 13.4.2** *Analogue reconstruction filter.*

$$\hat{x}(t) = \sum_{k=-\infty}^{\infty} x_k \frac{\sin \pi \, (2Bt - k)}{\pi \, (2Bt - k)} = \sum_{k=-\infty}^{\infty} x_k \operatorname{sinc} \pi \, (2Bt - k), \tag{13.4.8}$$

where $x_k$ are the samples at the time instant defined by the sampling interval. In summary, the reconstructed analogue signal $\hat{x}(t)$ is obtained by generating the weighted impulse responses of the lowpass filter at the positions defined by $kT_s$, for $-\infty \le k \le +\infty$, and then summing all the weighted responses. The weights of the delta impulses are defined by the sampled values $x(kT_s)$ of the original analogue signal $x(t)$. According to the sampling theorem, the reconstructed signal $\hat{x}(t)$ will be the ideal replica of the original analogue signal $x(t)$. That will be the case if the sampling procedure is carried out on an analogue signal that is of sufficient duration in time, ideally if the procedure is performed in the interval that tends to infinity.

A block schematic of the sampler and reconstruction lowpass-filter blocks and related input–output signals is shown in Figure 13.4.3. The wave shapes of the signals presented in the block schematic are shown in Figure 13.4.4.

The zero crossings for the $k$th response occur when the following condition is fulfilled: $(t - kT_s) \pi / T_s = \pm l\pi$, or $t = \pm lT_s + kT_s$, where $l$ is a positive integer. The zero crossings for the response defined by $k = 0$ occur at the time instants $t = \pm lT_s$. The sampled analogue signal $x_S(t)$ and the related impulse responses and their sum representing the reconstructed signal, for the analogue signal $x(t)$, are shown in Figure 13.4.4.

**Anti-aliasing filters and aliasing**. Let us state that the sampling theorem is valid for a *physically realizable* and *band-limited* signal. Therefore, an input continuous-time (analogue) and band-unlimited signal $x_a(t)$ has to be filtered to be in the bandwidth $B$ before the sampling procedure is performed. However, when the signal is band limited, it will be extended in the time domain, because we cannot limit the signal in the time and frequency domains at the same time. In practice, we are sacrificing the extension of the signal in the time domain, and adding an anti-aliasing filter with the transfer characteristic $H_{LP}(\omega)$ with a cut-off frequency specified by the Nyquist theorem, as shown in Figure 13.4.3. This filter has a double role. Firstly, it limits the bandwidth of the signal to comply with the sampling theorem and, secondly, it eliminates a part of the noise $w(t)$ that can exist beyond the bandwidth of the input signal. In this way, the input of the sampler is a band-limited signal $x(t)$ that is processed as explained in the previous sections of this chapter.

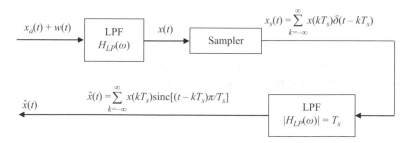

**Fig. 13.4.3** *Sampler and reconstruction lowpass-filter operations; LPF, lowpass filter.*

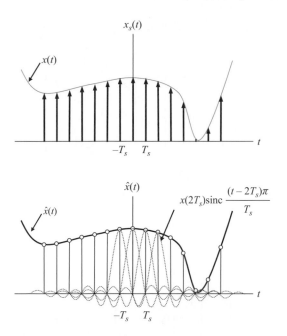

**Fig. 13.4.4** *Sampling procedure (upper panel) and reconstruction of an analogue signal (lower panel).*

The inclusion of the anti-aliasing filter causes a loss of the tail containing the high-frequency components of the signal. Otherwise, if the signal is not limited in the frequency domain via the anti-aliasing filter, then the signal $x_s(t)$, passing through the lowpass recon-struction filter, will have a distorted bandwidth, due to two reasons. Firstly, the signal will lose the tail of its higher frequencies that are beyond the Nyquist bandwidth. Secondly, the signal bandwidth will contain some frequency components which can be considered to comprise its lost tail folded back with respect to the folding frequency onto the signal bandwidth. In spite of that, if the anti-aliasing filter is not used, the spectrum of the reconstructed signal will still be band limited. Of course, in practice, we cannot construct that anti-aliasing filter with ideal rectangular characteristics. Instead, we design the filter with a steep transfer characteristic that achieves the required attenuation of the frequencies that are beyond the Nyquist frequency.

## 13.5   Generation of Discrete-Time Signals

The values of a discrete-time signal $x(n)$ are generated by periodically sampling a continuous-time signal $x(t)$ at uniform time intervals $t_n = nT_s$, which can be expressed as

$$x(n) = x(t) \mid_{t=t_n=nT_s} = x(nT_s), \tag{13.5.1}$$

for the following values of the discrete independent variable $n$: $n = \ldots, -2, -1, 0, 1, 2, \ldots$. The procedure of sampling is shown in Figure 13.5.1, and the discrete-time signal is shown in Figure 13.5.2. As we said before, the time interval between two consecutive samples $T_s$

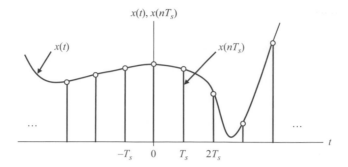

**Fig. 13.5.1** *Sampling procedure for a continuous-time signal.*

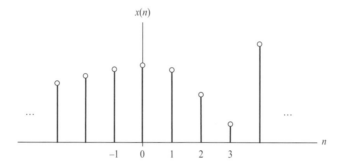

**Fig. 13.5.2** *Discrete-time signal.*

is called the *sampling interval*, or the *sampling period*, and its reciprocal value $f_s = 1/T_s$ is called the *sampling frequency*. Generally speaking, the continuous-time variable $t$ is related to the discrete-time variable $n$ according to the expression

$$t = t_n = nT_s = \frac{n}{f_s} = \frac{2\pi}{\omega_s} n \Rightarrow n = \frac{t}{T_s}, \tag{13.5.2}$$

where $\omega_s = 2\pi f_s$ denotes the *sampling angular frequency*, and the discrete-time variable can take any integer value, including 0, that is, $n = \dots, -2, -1, 0, 1, 2, \dots$.

**Example of sampling a sinusoidal signal.** Suppose a continuous-time sinusoidal signal is defined as

$$x_c(t) = A \cos(\omega_c t + \phi) = A \cos(2\pi f_c t + \phi), \tag{13.5.3}$$

with period $T_c = 1/f_c$ and frequency $f_c$. Let us find the discrete-time signal $x(n)$ obtained by sampling signal $x_c(t)$ at the time instants $T_s = 1/f_s$, where $f_s$ is the sampling frequency. If the number of samples in one period is $N$, we are interested in finding the expression for the discrete-time signal in terms of $N$.

The discrete-time sinusoidal signal is

$$x(n) = x_c(t) \,|_{t=nT_s} = A\cos(\omega_c nT_s + \phi) = A\cos(2\pi f_c nT_s + \phi). \tag{13.5.4}$$

Since the sampling interval is

$$T_s = \frac{1}{f_s} = \frac{2\pi}{\omega_s}, \tag{13.5.5}$$

we obtain the following expression for the discrete-time sinusoidal signal:

$$x(n) = A\cos\left(\omega_c n \frac{2\pi}{\omega_s} + \phi\right) = A\cos\left(\frac{2\pi\omega_c}{\omega_s} n + \phi\right) = A\cos(\Omega_c n + \phi), \tag{13.5.6}$$

where $\Omega_c$ is the *normalized discrete angular frequency* of $x(n)$, expressed as

$$\Omega_c = 2\pi \frac{\omega_c}{\omega_s} = 2\pi \frac{f_c}{f_s} = \frac{2\pi}{T_c/T_s} = \frac{2\pi}{N}. \tag{13.5.7}$$

The angular frequency $\omega_c$ of a continuous-time signal is expressed in radians per second. The normalized digital angular frequency $\Omega_c$ is expressed in radians per sample, because the ratio $N = T_c/T_s$ corresponds to the number of samples per period $T_c$ of the continuous-time sinusoidal signal $x_c(t)$. Therefore, if the number of samples per period is $N$, we may express the discrete-time signal in terms of the number of samples $N$ as

$$x(n) = A\cos(\Omega_c n + \phi) = A\cos\left(\frac{2\pi}{N} n + \phi\right). \tag{13.5.8}$$

**Example of reconstructing a sinusoidal signal.** A continuous-time signal can be obtained from a discrete-time signal via the following reversed process:

$$x(t) = A\cos\left(\frac{2\pi\omega_c}{\omega_s} n + \phi\right)\bigg|_{n=t/T_s} = A\cos\left(\frac{2\pi\omega_c}{2\pi/T_s} \frac{t}{T_s} + \phi\right) = A\cos(\omega_c t + \phi). \tag{13.5.9}$$

### Example

Suppose a cosine signal $x(t)$ has a period of $T_c = 2$ seconds and amplitude $A = 1$. Sample this signal using the Nyquist sampling frequency. Plot the graphs of both the discrete-time and continuous-time cosine signals and indicate the sampling interval and the period of the signal. Find the expression of the signal in the discrete-time domain and the continuous-time domain.

Next, increase the sampling frequency to be $f_s = 6$ Hz. Find the number of samples in one period, and the expression of the signal in the discrete-time domain and the continuous-time domain. Plot the graphs of the signals for this case.

**Solution.** The frequency of the first sinusoid is $f_c = 1/T_c = 0.5$ Hz. The Nyquist sampling interval is $T_s = 1/2f_c = 1$ second, the sampling frequency is $f_s = 1$ Hz, and the number of samples is $N = T_c/T_s = 2/1 = 2$ samples. The signal in the discrete-time domain is expressed as

$$x(n) = A\cos(\Omega_c n) = A\cos\left(\frac{2\pi}{N}n\right) = A\cos(\pi n).$$

The graph of the signal was shown in Chapter 7, Figure 7.4.3. The discrete-time signal when $N = 2$ is indicated in Figure 13.5.3 by the two thick, bold samples inside each period. The graph of the corresponding continuous-time sinusoidal signal is indicated by a dotted line.

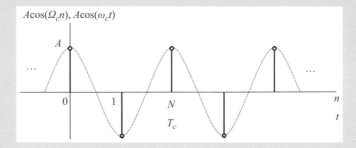

**Fig. 13.5.3** *Cosine signal in the continuous-time domain and the discrete-time domain, for $f_s = 1$ Hz.*

If the sampling frequency is $f_s = 6$ Hz, the sampling interval is $T_s = 1/6$, the number of samples is $N = T_c/T_s = f_s/f_c = 6/0.5 = 12$, and the signal in the discrete-time domain is

$$x(n) = A\cos(\Omega_c n) = A\cos\left(\frac{2\pi}{N}n\right) = A\cos\left(\frac{\pi}{6}n\right).$$

The graph of the signals for $N = 12$ is shown in Figure 13.5.4.

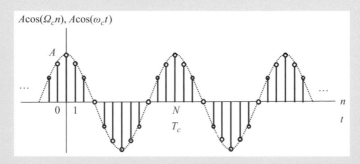

**Fig. 13.5.4** *Cosine signal in the continuous-time domain and the discrete-time domain, for $f_s = 1$ Hz.*

It is important to note that the signal in the continuous-time domain is of the same shape and duration for both values of $N$.

For $N = 2$, the signal in the continuous-time domain is expressed as

$$x(t) = A \cos\left(\frac{2\pi \omega_c}{\omega_s} n\right)\Bigg|_{n=t/T_s} = A \cos\left(\frac{2\pi f_c}{f_s} \frac{t}{T_s}\right) = A \cos\left(\frac{2\pi/2}{1} \frac{t}{1/1}\right) = A \cos(\pi t)$$

and, for $N = 12$, the signal is

$$x(t) = A \cos\left(\frac{2\pi \omega_c}{\omega_s} n\right)\Bigg|_{n=t/T_s} = A \cos\left(\frac{2\pi f_c}{f_s} \frac{t}{T_s}\right) = A \cos\left(\frac{2\pi/2}{6} \frac{t}{1/6}\right) = A \cos(\pi t).$$

The period of the continuous-time signal can be checked by using the following simple calculation: $\pi t = \omega_c t = 2\pi f_c t$, where $f_c = 0.5$ *Hz*, and $T_c = 2$ seconds.

# 14

# Deterministic Discrete-Time Signals and Systems

Discrete-time signal processing is an area of science and engineering that started to develop in the 1960's as a consequence of the advancement in the development of digital integrated circuits and digital signal processing technology. This point is to emphasize that the development of cheaper but smaller and faster digital processors and complex and programmable digital circuits allowed the construction of highly sophisticated digital systems. Due to these characteristics, discrete-time and related digital systems replaced the corresponding analogue systems in many (although not all) areas of application.

There is one more advantage of digital systems. Namely, the essential parts of these systems are programmed signal processing functions that can be performed by hardware. Thus, it is possible to modify these functions in software to be performed by the same hardware. This chapter introduces basic discrete-time signals and their characteristics, including a detailed explanation of the modification of independent variables.

## 14.1 Discrete-Time Signals

To say that a signal is a discrete-time signal is not to say that it is a discrete function of time $t$, but only that the time $t$ is a discrete variable. In discrete signal analysis, we will denote discrete time by the variable $n$ instead of $t$. In this chapter, a set of discrete signals represented by aperiodic functions will be analysed and then a set of exponential and sinusoidal signals, represented by corresponding exponential and sinusoidal functions, will be presented.

### 14.1.1 Elementary Discrete-Time Signals

**Discrete-time unit step signals.** A discrete-time unit step signal is defined as

$$u(n) = \begin{Bmatrix} 0, & n < 0 \\ 1, & n \geq 0 \end{Bmatrix}, \tag{14.1.1}$$

and is shown in Figure 14.1.1. The value of $u(n)$ at $n = 0$ is defined to be 1, unlike a continuous-time step signal, where this value is defined as an unknown finite value.

*Discrete Communication Systems*. Stevan Berber, Oxford University Press. © Stevan Berber 2021.
DOI: 10.1093/oso/9780198860792.003.0014

**Fig. 14.1.1** *Two presentations of a discrete-time unit step signal.*

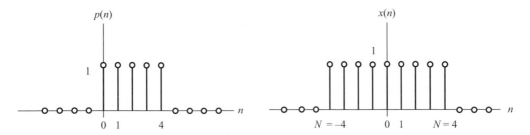

**Fig. 14.1.2** *Discrete rectangular pulse.*

Traditionally, discrete-time signals are represented as bars that show the amplitudes of the samples, as in the left-hand side of Figure 14.1.1. However, the precise mathematical presentation of a discrete-time signal uses dots in a coordinate system, as in the right-hand side of Figure 14.1.1.

**A discrete rectangular pulse.** A *rectangular asymmetric pulse* can be expressed as the difference of a unit step signal and its delayed version:

$$p(n) = u(n) - u(n-N-1) = \begin{cases} 0, & n < 0 \text{ and } n > N \\ 1, & -N \le n \le N \end{cases}. \tag{14.1.2}$$

A rectangular symmetric pulse can be expressed as

$$x(n) = u(n+N) - u(n-N-1) = \begin{cases} 0, & n < -N \text{ and } n > N \\ 1, & -N \le n \le N \end{cases}.$$

Both pulses are shown in Figure 14.1.2, for $N = 4$.

**A unit impulse signal, or delta signal.** A discrete unit impulse signal, or Kronecker delta function $\delta(n)$, is defined as

$$\delta(n) = \begin{cases} 0, & n \ne 0 \\ 1, & n = 0 \end{cases}, \tag{14.1.3}$$

and is shown in Figure 14.1.3. It is a signal that is 0 everywhere except at the origin, where its value is unity. The relationship between the unit step signal and the unit impulse is given by the first difference

$$\delta(n) = u(n) - u(n-1),\tag{14.1.4}$$

and in the form of the running sum

$$u(n) = \sum_{k=0}^{n} \delta(k).\tag{14.1.5}$$

The product of the delayed impulse with an arbitrary signal is

$$x(n)\,\delta(n-n_0) = x(n_0)\,\delta(n-n_0),\tag{14.1.6}$$

as is shown in Figure 14.1.4. The discrete-time signal $x(n)$ is multiplied by the impulse function, which is defined to be 1 at the point $n = n_0$, and 0 everywhere else. The result is a delta function with a weight that is equal to the value of the signal at the point $n = n_0$, and 0 everywhere else.

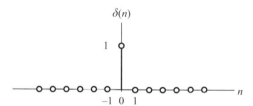

**Fig. 14.1.3** *Kronecker delta function.*

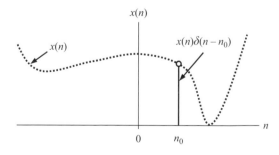

**Fig. 14.1.4** *Product of the delta signal and an arbitrary signal.*

**Discrete-time complex exponential signals.** A discrete-time complex exponential signal is defined as

$$x(n) = Aa^n,$$

where $A$ and $a$ are generally complex numbers. For the real-valued $A$ and $a$, this function becomes the familiar real-valued exponential function, as in the continuous-time case. For the discrete case, there are two boundary cases: one when $a = 1$, resulting in $x(n) = A = $ constant (DC signal), and one where $a = -1$, resulting in $x(n) = \pm A$ (a signal with alternating signs), as shown in Figure 14.1.5.

However, four distinct general cases can be identified, depending on the $a$ values. These cases are shown in Figure 14.1.6.

**Discrete sinusoidal signals.** A *complex sinusoid* can be produced from the exponential signal, defined as

$$x(n) = Aa^n, \tag{14.1.7}$$

by taking

$$a = e^{j\Omega_0}, \tag{14.1.8}$$

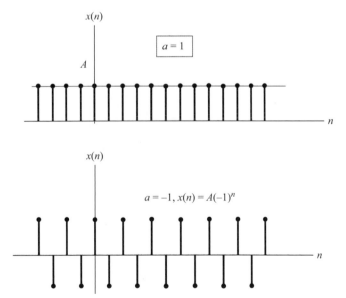

**Fig. 14.1.5** *Limit cases of discrete exponential functions.*

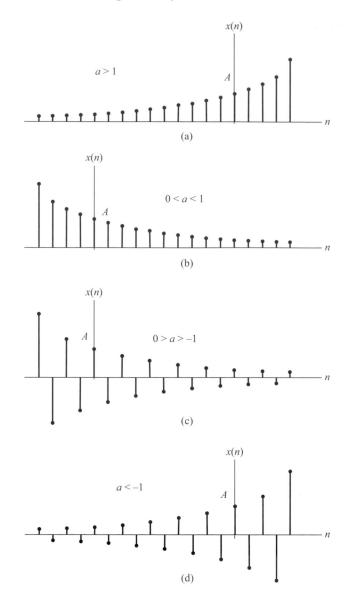

**Fig. 14.1.6** *A variety of discrete exponential functions.*

and a complex-valued $A$ being

$$A = |A|e^{j\phi}, \tag{14.1.9}$$

where the angular frequency $\Omega_0$ is real valued, which gives the complex-valued signal

$$x(n) = Ae^{j\Omega_0 n} = |A|e^{j(\Omega_0 n + \phi)} = |A|\cos(\Omega_0 n + \phi) + j|A|\sin(\Omega_0 n + \phi). \tag{14.1.10}$$

The *real sinusoid* $c(n)$ can be expressed in terms of the complex exponential (sinusoid) as its real part

$$\begin{aligned} c(n) &= \operatorname{Re}\left\{|A|e^{j(\Omega_0 n + \phi)}\right\} = \operatorname{Re}\{|A|\cos(\Omega_0 n + \phi) + j|A|\sin(\Omega_0 n + \phi)\} \\ &= |A|\cos(\Omega_0 n + \phi) \end{aligned} \tag{14.1.11}$$

or as its imaginary part

$$\begin{aligned} s(n) &= \operatorname{Im}\left\{|A|e^{j(\Omega_0 n + \phi)}\right\} = \operatorname{Im}\{|A|\cos(\Omega_0 n + \phi) + j|A|\sin(\Omega_0 n + \phi)\} \\ &= |A|\sin(\Omega_0 n + \phi) \end{aligned} \tag{14.1.12}$$

These expressions are referred to as the *discrete-time-domain representations of sinusoidal signals*. The graph of the discrete sine signal is shown in Figure 14.1.7 for $\phi = 0$, and $\Omega_0 = 2\pi f_0 = 2\pi/N = 2\pi/10$, that is, $N = 10$.

If the normalized frequency $f_0 = \Omega_0/2\pi$ is a rational number $k/N$, the complex sinusoid is periodic, that is,

$$x(n) = x(n + N), \tag{14.1.13}$$

for all values of $n$ and some period $N > 0$. We can prove this periodicity as follows:

$$\begin{aligned} |A|\sin\Omega_0(n+N) &= |A|\sin(2\pi k/N)(n+N) = |A|\sin 2\pi k/N(n+N) \\ &= |A|\sin(2\pi kn/N + 2\pi k) = |A|\sin(2\pi k/N)n = |A|\sin\Omega_0 n \end{aligned}$$

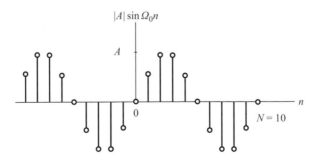

**Fig. 14.1.7** *A discrete sinusoid.*

In the example above, the ratio is $f_0 = \Omega_0/2\pi = k/N = 1/10$, that is, $k = 1$. If the normalized frequency is not a rational number, then the values of the samples will never repeat, no matter how long the sampling period is.

Unlike the case for continuous-time complex sinusoidal signals, discrete complex sinusoids with frequencies $\Omega_0$ and $(\Omega_0 + 2\pi)$ are indistinguishable. Namely, for all integer values of $n$, the following holds:

$$e^{j(\Omega_0+2\pi)n} = e^{j\Omega_0 n}e^{j2\pi n} = e^{j\Omega_0 n},$$

because $e^{j2\pi n} = 1$. This is also true for frequencies separated by multiples of $2\pi$. The consequence of this proof is that, when discrete-time sinusoidal signals are defined and used, it is sufficient to consider a frequency interval of length equal to $2\pi$, say, $(0 \leq \Omega_0 < 2\pi)$ or $(-\pi \leq \Omega_0 < \pi)$. For the *complex sinusoid*

$$x(n) = |A|e^{j(\Omega_0 n+\phi)},$$

the *real sinusoid* can be obtained as

$$c(n) = \mathrm{Re}\left\{|A|e^{j(\Omega_0 n+\phi)}\right\},$$

which can also be expressed as a sum of two complex components in the form

$$c(n) = \frac{|A|}{2}\left[e^{j(\Omega_0 n+\phi)} + e^{-j(\Omega_0 n+\phi)}\right]. \tag{14.1.14}$$

**A discrete unit ramp signal.** A *unit ramp signal* is defined as

$$r(n) = \begin{cases} n, & \text{for } n \geq 0 \\ 0, & \text{for } n < 0 \end{cases}, \tag{14.1.15}$$

as is shown in Figure 14.1.8.

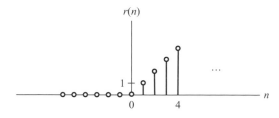

**Fig. 14.1.8** *Discrete unit ramp signal.*

## 14.1.2   Modification of Independent Variables

As for continuous-time signal processing, many simple signal processing operations can be described mathematically by a modification of the independent variable $n$. This assumes that the functional $x(\ )$ of the function $x(n)$ does not change as the independent variable $n$ is modified. However, the resulting functions are different. These basic modifications are time shift, time reversal, time scaling, and combinations thereof.

**Time reversal.** The signal $y(n)$ is the time reversal of $x(n)$ if

$$y(n) = x(-n),\qquad\qquad(14.1.16)$$

as depicted in Figure 14.1.9. This modification can be implemented by storing $x(n)$ in ascending order of $n$ and then reading it out in descending order of $n$. Time reversal is sometimes called a *folding* or *reflection* of the signal about the time origin defined by $n = 0$.

**Time shifting.** The signal $y(n)$ is a version of $x(n)$ that is time shifted by a factor $\beta$ if

$$y(n) = x(n + \beta).\qquad\qquad(14.1.17)$$

If $\beta > 0$, the signal $y(n)$ is a version of $x(n)$ that is shifted to the left. If $\beta < 0$, the signal $y(n)$ is a of $x(n)$ that is shifted to the right. It is always possible to insert a delay into the discrete values of a signal. Also, it is impossible to insert an advance into the discrete values of a real time signal. However, in the case when a signal is stored in memory, it is simple to modify the signal by adding an advance or a delay. In Figure 14.1.10, the original signal $x(n)$ and the delayed version $x(n-2)$ are shown. The delayed signal is shifted to the right for two time units.

If we want to fold and shift a signal, *we must first fold it, and then shift it*. This operation is not commutative. For example, the time-reversed (folded) and shifted signal $y(n) = x(-n-2) = x(-(n+2))$ is shown in Figure 14.1.10.

**Time scaling, or down-sampling.** The discrete-time signal $y(n)$ is a version of $x(n)$ that is time scaled by a factor $\alpha$ if

$$y(n) = x(\alpha n),\qquad\qquad(14.1.18)$$

where $\alpha$ is an integer.

**Fig. 14.1.9** *Example of time reversal.*

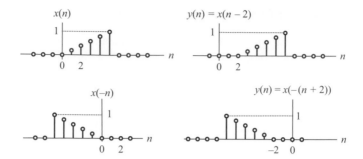

**Fig. 14.1.10**  *Time shift and both time reversal and time shift.*

**Example**

Draw the graph of $y(n)$, which is a version of $x(n)$ that is time scaled by a factor of 1/2, that is, $y(n) = x(n/2)$.

**Solution.** For the case when $\alpha = 1/2$, the scaled signal $y(n)$ is stretched or down-sampled. This type of *signal compression in the time domain* is called the *down-sampling* of the signal $x(n)$. The graphs of the signals are presented in Figure 14.1.11.

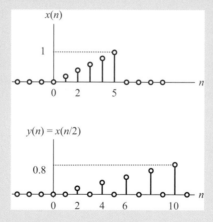

**Fig. 14.1.11**  *Example of time scaling, or down-sampling.*

**Magnitude scaling.** The signal $y(n)$ is a version of $x(n)$ that is magnitude (amplitude) scaled by a factor $A$ if every sample of $x(n)$ is multiplied by $A$, that is,

$$y(n) = Ax(n). \tag{14.1.19}$$

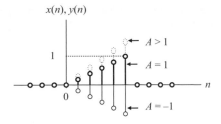

**Fig. 14.1.12** *Magnitude scaling.*

If $A > 1$, the signal $y(n)$ is an amplified version of $x(n)$, as depicted by the dashed line in Figure 14.1.12. If $A < 1$, the signal $y(n)$ is an attenuated version of $x(n)$. For $A = -1$, the signal $y(n)$ is a 180° phase-shifted version of $x(n)$, as shown by the thin lines in Figure 14.1.12.

## Example

For a given signal $x(n)$, draw the signal defined by following linear interpolation:

$$y(n) = \begin{cases} x(n/2) & n \text{ even} \\ \dfrac{v(n)}{3} = \dfrac{x[(n-1)] + x(n) + x[(n+1)]}{3} & n \text{ odd} \end{cases}.$$

**Solution.** The signal $y(n)$ in Figure 14.1.13 is the signal obtained by interpolating the two signals.

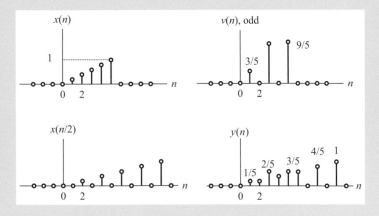

**Fig. 14.1.13** *Graphs of linearly interpolated signals.*

**Combined modification of an independent variable.** Generally speaking, an independent variable $n$ is modified using a *linear* transformation. Thus, its expression, containing all the above modifications, is

$$y(n) = Ax(\alpha n + \beta),\qquad\qquad(14.1.20)$$

which can also be expressed as

$$y(n) = Ax[\alpha(n + \beta/\alpha)],\qquad\qquad(14.1.21)$$

where

> $A$ represents amplitude scaling: $A > 1$ is amplification, $A < 1$ is attenuation, and $A < 0$ is phase reversal;
>
> $\alpha$ represents time scaling: $\alpha > 1$ is compression, $\alpha < 1$ is dilatation or stretching, and $\alpha < 0$ is time reversal; and
>
> $\beta/\alpha$ represents time shifting: $\beta/\alpha > 0$ represents a shift to the left, and $\beta/\alpha < 0$ represents a shift to the right.

### 14.1.3  Cross-Correlation and Autocorrelation Functions

In this section, we will present the basic theory related to the calculation of correlation functions for energy signals and give several typical examples that are relevant for the theory of discrete-time communication systems. A detailed analysis of correlation functions and their relationship to the energy and power spectral densities is presented in Chapter 15, Section 15.6. The cross-correlation of two *energy signals* $x(n)$ and $y(n)$, as a measure of their similarity, is defined as

$$R_{xy}(l) = \sum_{n=-\infty}^{\infty} x(n)y(n-l),\qquad\qquad(14.1.22)$$

for $l = 0, \pm 1, \pm 2, \ldots$, because we are comparing them in the whole interval of possible values $-\infty < n < +\infty$. For an *energy signal* $x(n)$, the autocorrelation is defined by the sum

$$R_x(l) = \sum_{n=-\infty}^{\infty} x(n)x(n-l).\qquad\qquad(14.1.23)$$

Therefore, the autocorrelation function is obtained when a signal is cross-correlated by itself. As is the case for the autocorrelation function for continuous-time signals, the following is true: if a discrete-time signal is 'smooth' in the time domain, then its autocorrelation function will be 'wide'. However, if the signal is 'rough' in the time domain, then its autocorrelation will be 'narrow'. Thus, using the autocorrelation function, we can make conclusions about the frequency contents of a signal in the discrete-time domain. Details related to this behaviour of the autocorrelation function will be presented in Chapter 15, Section 15.6.

**Properties of the autocorrelation function.** The value $R_x(0)$ is equal to the total energy of the signal, that is,

$$E_x = R_x(0) = \sum_{n=-\infty}^{\infty} |x(n)|^2. \tag{14.1.24}$$

The maximum value of an autocorrelation function is $R_x(0)$, that is,

$$R_x(0) \geq R_x(l). \tag{14.1.25}$$

The function is even, that is,

$$R_x(-l) = R_x(l). \tag{14.1.26}$$

This can be easily proven by using the definition of the autocorrelation function, changing the variable $k = n - l$, and solving the correlation integral with respect to $x$.

**Example**

Compute the autocorrelation function for a discrete rectangular pulse expressed in terms of step functions as

$$x(n) = A \cdot u(n+N) - A \cdot u(n-N-1). \tag{14.1.27}$$

**Solution.** The pulse is presented in Figure 14.1.14. For $l \leq 0$, the pulse is shifted to the left for the $l$ value. The overlapping of the pulses starts from $n = 0$ to $n = l + N$, and the correlation sum is

$$R_x(l) = \sum_{n=-\infty}^{\infty} x(n-l)x(n) = \sum_{n=0}^{l+N} A^2 = (N+l)A^2. \tag{14.1.28}$$

For $l \geq 0$, we obtain

$$R_x(l) = \sum_{n=-\infty}^{\infty} x(n-l)x(n) = \sum_{n=l}^{N} A^2 = (N-l)A^2. \tag{14.1.29}$$

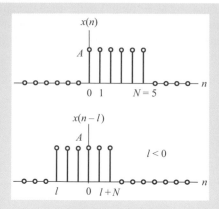

**Fig. 14.1.14**  *Presentation of the signal in the time domain.*

For all values of $l$, we obtain

$$R_x(l) = \begin{cases} (N+l)A^2 & N \le l \le 0 \\ (N-l)A^2 & N \ge l \ge 0 \\ 0 & \text{otherwise} \end{cases} = (N+|l|)A^2, \text{for} - N \le l \le N,$$

which is shown in Figure 14.1.15.

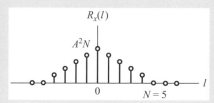

**Fig. 14.1.15**  *Autocorrelation function.*

## 14.2   Discrete-Time Systems

### 14.2.1   Systems Classification

The attributes defined for continuous-time systems have discrete-time counterparts. Therefore, the underlying concepts of continuous-time systems are the same for discrete-time systems. These system attributes have physical significance because they allow the categorization and description of the systems' behaviour.

**Linear systems.** A system is said to be linear if the following *superposition condition* is fulfilled: given that the input $x_1(n)$ produces the output $y_1(n)$, and the input $x_2(n)$ produces the output $y_2(n)$, then the linear combination of inputs

$$x(n) = ax_1(n) + bx_2(n) \tag{14.2.1}$$

must produce the linear combination of outputs

$$y(n) = ay_1(n) + by_2(n), \tag{14.2.2}$$

for the arbitrary complex scaling constants $a$ and $b$.

The superposition contains two requirements: *additivity* and *homogeneity* (or scaling), expressed respectively as follows:

1. Given that the input $x_1(n)$ produces the output $y_1(n)$, and the input $x_2(n)$ produces the output $y_2(n)$, the input $x_1(n) + x_2(n)$ produces the output $y_1(n) + y_2(n)$ for arbitrary $x_1(n)$ and $x_2(n)$.

2. Given that the input $x(n)$ produces the output $y(n)$, then the scaled input $ax(n)$ must produce the scaled output $ay(n)$ for arbitrary $x(n)$ and the complex constant $a$.

The superposition condition can be generalized. Then, for an input of the form

$$x(n) = \sum_k a_k x_k(n), \tag{14.2.3}$$

the output is

$$y(n) = \sum_k a_k y_k(n), \tag{14.2.4}$$

for the arbitrary constants $a_k$. If $a_k = 0$, then a zero input for all times yields a zero output. A system is non-linear if it does not satisfy the superposition principle.

**Time-invariant systems.** A system is said to be *time invariant* if a shift (delay or advance) in the input signal $x(n)$ causes a like shift in the output signal $y(n)$. That is, if $x(n)$ produces $y(n)$, then $x(n - n_0)$ produces $y(n - n_0)$ for any (integer) value of $n_0$.

The procedure for testing time invariance is shown schematically in Figure 14.2.1. Let us assume that, for the given $x(n)$, the system produces $y(n)$. If the system is linear-time invariant, then the shifted input $x(n - n_0)$ produces the shifted output $y(n - n_0)$. Suppose the input of the system is shifted, that is, $x(n - n_0)$, which produces the output $y_{n0}(n)$. Then, if the following condition is fulfilled,

$$y_{n0}(n) = y(n - n_0), \tag{14.2.5}$$

the system is said to be time invariant; otherwise, it is not. Therefore, we will use the scheme in Figure 14.2.1 to check if the system is linear-time invariant.

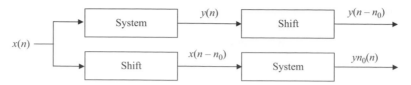

**Fig. 14.2.1** *Scheme for investigating the time invariance of linear-time-invariant systems.*

**Systems with memory.** A system is said to have *memory* if its output $y(n) = y(n_0)$, at an arbitrary time $n = n_0$, depends on input values other than, or in addition to, $x(n_0)$. The additional input value needed to compute $y(n_0)$ may be in the past ($n < n_0$) or in the future ($n > n_0$).

A system is said to be *memoryless* if $y(n_0)$ depends only on $x(n_0)$. For example, an accumulator is a system that has memory, because its *output* at an arbitrary instant $n = n_0$ depends on the *input values* starting from the infinite past, that is, on all $n \leq n_0$. Also, a simple *unit delay* $y(n) = x(n-1)$, and a simple *unit advance* $y(n) = x(n+1)$, are systems with memory, because the value at the instant of time $n = n_0$ depends on past or future values, respectively. In contrast, a system described by the function $y(n) = x^3(n)$ is memoryless, because its *output* at an arbitrary instant $n = n_0$ depends only on the input value at $n = n_0$.

**Causal systems.** A system is causal if $y(n)$, at an arbitrary time instant $n = n_0$, say, $y(n) = y(n_0)$, depends only on the input value $x(n)$ for $n \leq n_0$. In other words, the output at the present time $n_0$ depends only on present and past values ($n = n_0$, and $n < n_0$), or on present or past values ($n = n_0$, or $n < n_0$), but not on future values. Thus, all physical systems in the real world are causal, due to the fact they cannot anticipate the future. For example, a simple *unit advance* $y(n) = x(n-1)$, is a non-causal system, because the value of the output $y(n)$ at the instant of time $n = n_0$, $y(n_0)$, depends on the future input value $x(n_0 - 1)$. Similarly, the time reversal $y(n) = x(-n)$ is a non-causal system. Namely, for negative values of $n = n_0$, the values $-n_0$ are greater than $n_0$, and so are in the future. A system defined by

$$y(n_0) = \sum_{n=n_0-2}^{n_0+2} x(n),$$

is a non-causal system, because the present output values depend on the future values of the input. For example, for $n_0 = 0$, the outputs depend on the values $n = -2, -1, 0, 1, 2$, which includes the future for $n = 1, 2$.

All memoryless systems are causal, but not vice versa; that is, a causal system can (and usually does) have memory.

**Stable systems.** An input signal $x(n)$ and output signal $y(n)$ are bounded if $|x(n)| \leq A$, and $y(n)| \leq B$. A system is stable in terms of bounded-input/bounded-output stability if any bounded input $|x(n)| \leq A$, for some finite constant $A$, produces a bounded output $y(n)$ satisfying $|y(n)| \leq B$ for some finite constant $B$.

## 14.2.2 Discrete-Time Linear-Time-Invariant Systems

Let a discrete linear-time-invariant system be represented by an input signal $x(n)$, an output signal $y(n)$, and a linear-time-invariant device defined by its impulse response $h(n)$, as shown in Figure 14.2.2.

The problem of interest in this section is how to describe the linear-time-invariant device needed to find the output signal as a function of the input signal. For this purpose, the concept of the impulse response will be used. As in continuous linear-time-invariant systems, the fundamental input–output relationship for a discrete-time linear-time-invariant system is defined by the convolution of the input signal and the impulse response of the system. Thus, the impulse response completely characterizes the linear-time-invariant system.

**Representation of the input signal.** A discrete-time signal $x(n)$ is shown in Figure 14.2.3. If we keep in mind the definition of a discrete-time impulse signal, we see that each value of $x(n)$ at the time instant $k$ is equivalent to an impulse signal multiplied (weighted) by the value of the signal $x(n)$ at that instant $k$, that is,

$$x(n)\delta(n-k) \equiv x(k)\delta(n-k). \tag{14.2.6}$$

Thus, the whole signal $x(n)$ can be represented by the sum of weighted impulse signals in the form

$$x(n) = \sum_{k=-\infty}^{k=\infty} x(k)\delta(n-k). \tag{14.2.7}$$

That is to say, no matter how complicated the signal $x(n)$ may be, the sum of the products $x(k)\delta(n-k)$ over all $k$ simply equals the value of $x(n)$ at any point $n=k$. By analogy with the continuous-time case, this is called the *shifting* property of the discrete-time impulse signal. In other words, the input signal is 'decomposed' into the sum of shifted discrete-time impulse $\delta$-functions with weights equal to the signal $x(k)$.

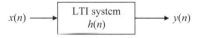

**Fig. 14.2.2** *General scheme of a linear-time-invariant system; LTI, linear-time-invariant.*

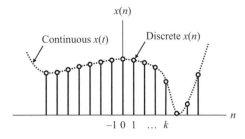

**Fig. 14.2.3** *Input discrete-time signal.*

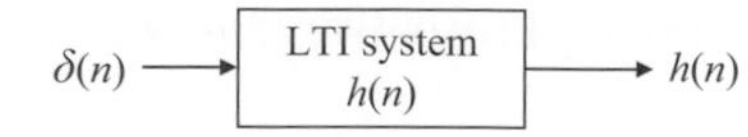

$$\delta(n) \longrightarrow \boxed{\begin{array}{c} \text{LTI system} \\ h(n) \end{array}} \longrightarrow h(n)$$

**Fig. 14.2.4** *Definition of the system, based on the system's impulse response; LTI, linear-time-invariant.*

**Representation of a linear-time-invariant system.** Let a linear-time-invariant system be represented by its impulse response $h(n)$. The impulse response of an linear-time-invariant system is defined as its output signal $h(n)$ for the case when the input signal is the impulse signal $\delta(n)$. The system can be represented by the scheme in Figure 14.2.4.

**Representation of the output signal.** Let $h(n)$ be the output of a linear-time-invariant system for the input $\delta(n)$. Also, due to the time invariance of the system, the shifted response $h(n-k)$ is the response to the shifted-input delta impulse $\delta(n-k)$. In summary, the input–output relationship is

$$\delta(n-k) \underset{\text{LTI}}{\rightarrow} h(n-k), \tag{14.2.8}$$

where LTI stands for linear-time-invariant system. Furthermore, due to the linearity of the system, the input multiplied by a scalar $x(k)$ gives the output multiplied by the same scalar (homogeneity), that is,

$$x(k)\delta(n-k) \underset{\text{LTI}}{\rightarrow} x(k)h(n-k), \tag{14.2.9}$$

and the sum of the inputs gives the sum of the outputs (additivity):

$$\sum_k x(k)\delta(n-k) \underset{\text{LTI}}{\rightarrow} \sum_k x(k)h(n-k). \tag{14.2.10}$$

Therefore, for the input $x(n)$ given as

$$x(n) = \sum_{k=-\infty}^{k=\infty} x(k)\delta(n-k), \tag{14.2.11}$$

the output of the linear-time-invariant system, represented by the impulse response $h(n)$, is

$$y(n) = \sum_{k=-\infty}^{k=\infty} x(k)h(n-k), \tag{14.2.12}$$

which is the *convolution sum* of the input and the impulse response of the system:

$$y(n) = x(n)* h(n). \tag{14.2.13}$$

The fundamental result can be expressed as follow: *the output signal of any discrete-time linear-time-invariant system is the convolution of its input signal and the system impulse response.*

The output signal can also be found in the following way. Due to the linearity of the system, it can be represented by a linear operator $L$. Therefore, the output of the system is

$$y(n) = L[x(n)] = L\left[\sum_{k=-\infty}^{k=\infty} x(k)\delta(n-k)\right] = \sum_{k=-\infty}^{k=\infty} x(k)L[\delta(n-k)].$$

Because the system is time invariant, the expression $L[\delta(n-k)] = h(n-k)$ holds, and the output becomes

$$y(n) = \sum_{k=-\infty}^{k=\infty} x(k)h(n-k).$$

Using the commutativity property of convolution, the convolutional sum can also be written as

$$y(n) = \sum_{k=-\infty}^{k=\infty} h(k)x(n-k).$$

## 14.3   Properties of Linear-Time-Invariant Systems

The input–output characteristics of a discrete-time linear-time-invariant system in the time domain are completely determined by its impulse response $h(n)$. Therefore, system attributes, such as stability, causality, and memory, have to be expressed in terms of conditions imposed on the impulse response $h(n)$.

**Systems with memory.** As was mentioned before, a discrete-time system is said to be *memoryless* if $y(n)$ for $n = n_0$ depends only on the present input $x(n)$ for $n = n_0$. Furthermore, if the system is also linear and time variant, this relationship can only be of the general form

$$y(n) = Kx(n), \tag{14.3.1}$$

where $K$ is the constant gain. The corresponding impulse response is thus simply

$$h(n) = K\delta(n). \tag{14.3.2}$$

Therefore, if $h(n_0) \neq 0$ for any index value other than $n_0 = 0$, the system is said to *have memory.*

**Causal systems.** A system can be said to be causal if $y(n)$, at an arbitrary time $n = n_0$, depends only on the input value $x(n)$ for $n \leq n_0$. This is to say that the response of an event beginning at $n = n_0$ is non-zero only for $n \geq n_0$. Thus, the impulse response of the causal system $h(n)$ can be non-zero only for $n \geq 0$, that is,

$$h(n) = 0, n < 0, \tag{14.3.3}$$

because the input different from 0 exists only at $n = 0$. Then the convolution sums are expressed in the form

$$y(n) = \sum_{k=0}^{k=\infty} h(k)x(n-k) \tag{14.3.4}$$

and

$$y(n) = \sum_{k=-\infty}^{k=n} x(k)h(n-k). \tag{14.3.5}$$

The first form clearly shows that the only values of $x(n)$ used to compute $y(n)$ are those for $k \leq n$. Generally, a signal $x(n)$ is causal if

$$x(n) = 0, \text{for } n < 0, \tag{14.3.6}$$

whether that signal is actually an impulse response or not.

**Stable systems.** Assume that the input signal $x(n)$ is bounded by a finite constant $A$, that is,

$$|x(n)| \leq A, \text{for all } n. \tag{14.3.7}$$

If the system is stable, the output can also be bounded. In order to find the conditions that must be imposed on the impulse response $h(n)$ to have a stable system, let write the magnitude of the output in terms of the convolution sum as

$$|y(n)| = \left| \sum_{k=-\infty}^{k=\infty} h(k)x(n-k) \right|. \tag{14.3.8}$$

Since the magnitude of the sum is less than or equal to the sum of magnitudes, and the magnitude of a product is equal to the product of the magnitudes, the following holds:

$$|y(n)| = \left| \sum_{k=-\infty}^{k=\infty} h(k)x(n-k) \right| \leq \sum_{k=-\infty}^{k=\infty} \left| h(k) \right| \left| x(n-k) \right| \leq A \sum_{k=-\infty}^{k=\infty} |h(k)|, \tag{14.3.9}$$

because of the assumption $|x(n)| \leq A$. Therefore, if the impulse response is absolutely summable, that is, if

$$\sum_{k=-\infty}^{k=\infty} |h(k)| = C < \infty, \tag{14.3.10}$$

the inequality for $|y(n)|$ provides a bound on the output of the form

$$|y(n)| \leq AC = B, \tag{14.3.11}$$

for a finite constant $B$, and the system is said to be in bounded-input/bounded-output stability, or *BIBO stable*.

In conclusion, the *necessary and sufficient condition* for a discrete-time system to be stable is that its impulse response must be absolutely summable. In other words, the system is stable *if and only if* the condition for $C$ in eqn (14.3.10) is satisfied.

**Cascade interconnection.** If the output of one system is connected to the input of another system, as depicted in Figure 14.3.1, the impulse response of this cascaded system is simply the convolution of the impulse responses of the individual systems, that is,

$$h = h_1 * h_2, \tag{14.3.12}$$

as shown in Figure 14.3.1.

**Parallel interconnection.** If the outputs of two elementary systems are summed to produce a single output for a common input, the systems are said to be *parallel*. The impulse response of the parallel interconnection is

$$h = h_1 + h_2, \tag{14.3.13}$$

as shown in Figure 14.3.2. This result can be generalized to the parallel interconnection of $N$ elementary systems.

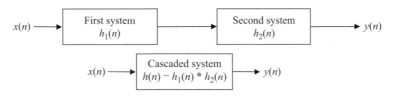

**Fig. 14.3.1** *Cascade interconnection.*

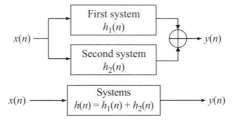

**Fig. 14.3.2** *Parallel system interconnection.*

## 14.4 Analysis of Linear-Time-Invariant Systems in Time and Frequency Domains

A linear-time-invariant system is completely described in the discrete-time domain by its impulse response $h(n)$. Further operation of the system can be described by representing the input–output signals and the impulse response in the frequency domain, which is the subject of work presented in Chapter 15. The signals can also be represented in terms of their autocorrelation functions. Then, the output of an linear-time-invariant system can be expressed as the convolution of the input autocorrelation function and the system autocorrelation function. The corresponding operations of the system in the frequency domain can be expressed in terms of the energy and power spectral density functions, depending on the nature of the system's input and output signals, which can be power signals or energy signals. For these reasons, a detailed analysis of the autocorrelation functions and the power and energy spectral density functions is presented in Chapter 15, Section 15.6.

Finally, the theory for the operation of linear-time-invariant systems is of vital importance for the analysis of systems with stochastic inputs, because practically the same procedure for explaining the operation of an linear-time-invariant system for deterministic signals should be used when the input signal of the system is a stochastic process. For that reason, the description of a linear-time-invariant system with continuous-time stochastic input is presented in Chapter 19, Section 19.4, and a related description of the system with discrete-time stochastic input was presented in Chapter 3, particularly, Section 3.7.

The theory of the operation of linear-time-invariant systems for deterministic and stochastic processes is of key importance for understanding the operation of digital and discrete communication systems, where the input modulating signal carrying a message is, by nature, a stochastic discrete- or continuous-time process, and all the signal processing procedures inside the transmitter, the channel, and the receiver can be explained using the concepts of statistical signal processing.

PROBLEMS

1. Draw the signal $y(n) = u(m-n)$ for general values of $m$ and then for $m = 3$.

2. Let $x(n)$ be the signal in Figure 2. Draw the transformed signal $y(n) = x(2-2n)$.

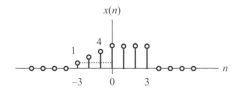

**Fig. 2** *Time-domain signal.*

3. The input $x(n)$ and the response $h(n)$ of an linear-time-invariant system are the step functions shown in Figure 3. Find the output $y(n)$ of the system.

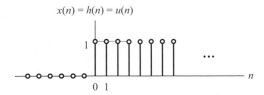

**Fig. 3** *Signal x(n) and h(n) in the time domain.*

4. Present the signal $x(n) = (n-5)[u(n-3) - u(n-8)]$ in graphical form and express it as a scaled sum of $\delta$-functions.

5. Compute and then plot the convolutions $y = x * h$ and then $z = h * x$ for the pairs of signals shown in Figure 5.

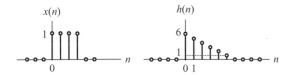

**Fig. 5** *Signals in the time domain.*

6. A discrete-time signal $x(n) = \{\underline{1}, -2, 3\}$ is applied to the input of a linear-time-invariant system, where the underlined number corresponds to the signal magnitude defined by $n = 0$, as presented in Figure 6. The impulse response of the linear-time-invariant system is $h(n) = \{\underline{0}, 0, 1, 1, 1, 1\}$.

   a. Prove the following rule: if the convolution of the input signal $x(n)$ and the impulse response $h(n)$ is $y(n) = \sum_{k=-\infty}^{\infty} h(k)x(n-k)$, then the equality $\sum_n y(n) = \sum_k h(k) \sum_n x(n)$ holds.

   b. Plot the graphs of $x(k)$ and $h(n-k)$. Find the time-domain output $y(n)$ of the linear-time-invariant system for the input signal $x(n)$, using the convolution expression.

   c. Confirm that the rule proved in a. applies for the response calculated in b.

7. Determine the convolution $y(n)$ of the following signals:

$$x(n) = \begin{cases} \alpha^n, & -3 \le n \le 5 \\ 0, & othervise \end{cases} \text{ and } h(n) = \begin{cases} 1, & 0 \le n \le 3 \\ 0, & othervise \end{cases}.$$

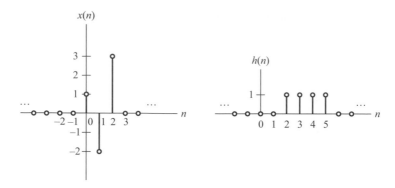

**Fig. 6** *Signals in the time domain.*

8. The impulse response of the linear-time-invariant system is $h(n) = (\underline{1}, 1, 0, 0)$. Find the time-domain response $y(n)$ of the linear-time-invariant system for the input signal $x(n) = (3, 2, 3)$.

9. Express the signal

$$x(n) = (n-5)\left[u(n-3) - u(n-8)\right]$$

as a scaled sum of $\delta$-functions.

10. Draw the graph of $y(n)$ which is a scaled version of $x(n)$ by a factor 2, that is,

$$y(n) = x(2n),$$

and then plot the graph of the following time-shifted signal:

$$z(n) = x(2n+1).$$

11. The time-reversal system is defined by $y(n) = x(-n)$. Prove that the system is not a time variant.

12. The input $x(n)$ and the response $h(n)$ of a linear-time-invariant system are the step functions shown in Figure 12. Find the output $y(n)$ of the system.

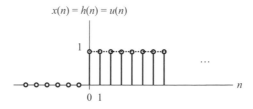

**Fig. 12** *Input signal of the system.*

13. Solve the following problems:

   a. Let the impulse response of an linear-time-invariant system be the unit impulse delay signal $h(n) = \delta(n - 1)$. Does this system have memory?

   b. Let the impulse response of a system be the unit step signal $h(n) = u(n)$. Does this system have memory?

   c. Let the impulse response be the exponential signal $h(n) = a^n u(n)$. Is this system stable?

# 15
# Deterministic Discrete-Time Signal Transforms

## 15.1 Introduction

The Fourier series of a continuous-time periodic signal consists of an infinite sum of sinusoidal or exponential spectral components ranging from $-\infty$ to $+\infty$. These components are separated by $1/T$, where $T$ is the fundamental period of the signal, and cover the frequency range from $-\infty$ to $+\infty$. However, as we will see, the frequency range of a discrete-time periodic signal is in the interval from $-\pi$ to $+\pi$ or from 0 to $2\pi$. The density of these components depends on the value of the fundamental period $N$. Namely, the components are separated by $2\pi/N$ radians or $1/N$ cycles. Thus, the number of frequency components is equal to $N$.

The Fourier transform or, in precise terminology, the continuous-time Fourier transform, is used to represent aperiodic discrete signals in the frequency domain, where frequency is a continuous variable. The inverse Fourier transform is used to find the signal in the time domain when the Fourier transform is available. Therefore, the transformation from the time domain to the frequency domain, and vice versa, can be performed by using the Fourier transform pair.

This chapter starts by presenting theoretical characteristics of the discrete-time Fourier series, the discrete-time Fourier transform, and the discrete Fourier transform. At the end of the chapter, the algorithms for the discrete Fourier transform will be presented. These algorithms are used in discrete communication systems to transform the signals in the signal processing blocks of the system transmitter and receiver.

## 15.2 The Discrete-Time Fourier Series

### 15.2.1 Continuous-Time Fourier Series and Transforms

Continuous-time Fourier series and transforms are presented in detail in Chapter 12. For the sake of explanation and completeness, the main findings from that chapter will be briefly summarized in this chapter. A continuous-time periodic signal $x(t) = x(t + T)$ is expressed as a sum of exponential terms of the form

*Discrete Communication Systems*. Stevan Berber, Oxford University Press. © Stevan Berber 2021.
DOI: 10.1093/oso/9780198860792.003.0015

$$x(t) = \sum_{k=-\infty}^{\infty} c_k e^{+jk\omega_0 t}, \qquad (15.2.1)$$

where the coefficients of the terms are calculated as

$$c_k = \frac{1}{T} \int_{-T/2}^{T/2} x(t) e^{-jk\omega_0 t} dt. \qquad (15.2.2)$$

For an aperiodic continuous-time signal the Fourier transform pair is defined as

$$X(f) = \int_{-\infty}^{\infty} x(t) e^{-j2\pi ft} dt, \qquad (15.2.3)$$

and

$$x(t) = \int_{-\infty}^{\infty} X(f) e^{j2\pi ft} df. \qquad (15.2.4)$$

These equations are also called the *symmetric transform pair* because of their similarity. In the case when the radian frequency is used, that is, when $f = 2\pi/\omega$, the *asymmetric transform pair* can be derived in the form

$$X(\omega) = \int_{-\infty}^{\infty} x(t) e^{-j\omega t} dt, \qquad (15.2.5)$$

and

$$x(t) = \frac{1}{2\pi} \int_{-\infty}^{\infty} X(\omega) e^{j\omega t} d\omega. \qquad (15.2.6)$$

It is important to note that the Fourier transform can be applied to a periodic signal, as presented in Chapter 12. In that case, the discrete frequency spectrum was presented in terms of shifted Dirac delta functions with weights corresponding to the magnitudes of the spectral components. This fact is extensively used for signal analysis and synthesis in the communication systems presented in this book.

## 15.2.2  The Discrete-Time Fourier Series

A discrete-time Fourier series is used to transform a discrete-time periodic signal into the frequency domain. Let a discrete-time periodic signal $x(n) = x(n + N)$ be expressed as a sum of exponential terms of the form

$$x(n) = \sum_{k=0}^{N-1} c_k e^{+jk2\pi n/N}, \tag{15.2.7}$$

where $k = 0, 1, 2, \ldots, N-1$, where $c_k$ are the Fourier coefficients that can be found in the following way. Because

$$\sum_{n=0}^{N-1} a^n = \left\{ \begin{array}{ll} N & a=1 \\ \frac{1-a^N}{1-a} & a \neq 1 \end{array} \right\}, \tag{15.2.8}$$

we obtain

$$\sum_{n=0}^{N-1} \left( e^{+jk2\pi/N} \right)^n = \left\{ \begin{array}{ll} N & k = 0, \pm N, \pm 2N, \ldots \\ \frac{1-e^{+jk2\pi}}{1-e^{+jk2\pi/N}} = \frac{1-\cos 2k\pi - j\sin 2k\pi}{1-e^{+jk2\pi/N}} = 0 & \text{otherwise} \end{array} \right\}.$$

Multiplying expression (15.2.7) for $x(n)$ by the exponential and summing the obtained product from $n = 0$ to $n = N-1$, we find

$$\sum_{n=0}^{N-1} x(n) e^{-jl2\pi n/N} = \sum_{n=0}^{N-1} e^{-jl2\pi n/N} \sum_{k=0}^{N-1} c_k e^{+jk2\pi n/N}$$

$$= \sum_{n=0}^{N-1} \sum_{k=0}^{N-1} c_k e^{+j(k-l)2\pi n/N} = \sum_{k=0}^{N-1} c_l N \tag{15.2.9}$$

Bearing in mind that

$$\sum_{n=0}^{N-1} e^{+j(k-l)2\pi n/N} = \left\{ \begin{array}{ll} N & k-l = 0, \pm N, \pm 2N, \ldots \\ 0 & \text{otherwise} \end{array} \right\}, \tag{15.2.10}$$

we obtain (note that the sum is equal to $N$ only for $k = l$)

$$\sum_{n=0}^{N-1} x(n) e^{-jl2\pi n/N} = c_l \cdot N, \tag{15.2.11}$$

or the Fourier coefficients expressed in terms of the signal $x(n)$ as

$$c_l = \frac{1}{N} \sum_{n=0}^{N-1} x(n) e^{-jl2\pi n/N}, \tag{15.2.12}$$

for $l = 0, 1, 2, \ldots, N - 1$. Therefore, the synthesis and analysis equations, expressed as

$$x(n) = \sum_{k=0}^{N-1} c_k e^{+jk2\pi n/N} \qquad (15.2.13)$$

and

$$c_k = \frac{1}{N} \sum_{n=0}^{N-1} x(n) e^{-jk2\pi n/N}, \qquad (15.2.14)$$

respectively, represent the discrete-time Fourier series pair. The sum representing the time-domain signal $x(n)$ is called the discrete-time Fourier series. The second expression gives a set of Fourier coefficients $\{c_k\}$, representing the amplitude spectrum of the signal that describes the signal $x(n)$ in the frequency domain. These coefficients contain all of the information about the magnitude spectrum and the phase spectrum of the signal. Because

$$c_k = c_{k+N}, \qquad (15.2.15)$$

we say that the sequence $\{c_k\}$ is periodic with the fundamental period $N$, or that the spectrum of a periodic signal $x(n)$ with period $N$ is *a periodic sequence $\{c_k\}$ with period $N$*. In practice, we are dealing with a single period ranging from $k = 0$ to $k = N - 1$, because it corresponds to the fundamental range of frequencies expressed as

$$0 \leq \Omega_k = k2\pi/N \leq 2\pi, \qquad (15.2.16)$$

where $\Omega_1 = 2\pi/N$ is the fundamental frequency expressed in radians per sample, and $\Omega_k = 2\pi k/N$ are signal harmonics. Therefore, any set of $N$ consecutive values of the signal $x(n)$, or any set of its $N$ spectrum components, completely describes the signal in the time or frequency domains.

**Important remark on notation.** In this book, $\Omega = 2\pi F$ will be used to denote the angular frequency of a discrete-time signal, as opposed to $\omega = 2\pi f$, which is used to denote the angular frequency of continuous-time signals. Here the frequency $F$ is the circular frequency $f$ normalized by the sampling frequency, that is, $F = f/f_s = \Omega/2\pi$. In the case of Fourier series, we define $\Omega_k$ as the normalized angular frequency, $\Omega_k = 2\pi k/N$, expressed in radians per sample, and $F_k$ as the normalized frequency, $F_k = k/N$, and $k$ is the order of the frequency components. The distinction between these frequencies is shown in Figure 15.2.1, which presents the amplitude spectrum of a sinusoidal signal. For the sake of simplicity, we will use just the word 'frequency' for both $\Omega$ and $F$, because they will be easily distinguishable in the text and formulas via this unique notation.

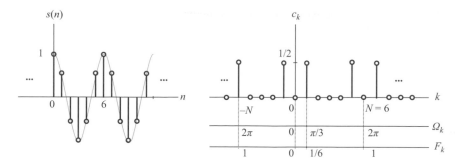

**Fig. 15.2.1** *Signal in the time domain (left) and the frequency domain (right).*

### 15.2.3  Fourier Series Examples Important for Communication Systems

**Cosine signal.** A discrete-time cosine signal is usually used as a carrier in discrete communication systems. Suppose a cosine signal is defined as $x(n) = \cos(2\pi n/N)$, where $N$ is the number of samples in the fundamental period of the signal. We are interested in presenting this signal in time and frequency domains, and in calculating the power and energy of the signal from its time- and frequency-domain presentations.

A continuous-time sinusoidal signal is a periodic signal, regardless of the value of its frequency. However, this is not the case for a discrete-time sinusoidal signal, which is periodic if the ratio $\Omega/2\pi$ is a rational number.

We will analyse the signal for $N = 6$. In this case, from $2\pi n/N = \pi n/3 = 2\pi F_0 n = \Omega n$, we can have $F_0 = 1/6$, or $\Omega = 2\pi/6$. Because $F_0$ is a rational number, which means that the signal is periodic with the fundamental period $N = 6$, the discrete values of the signal are calculated in one period and presented as a function of discrete time $n = 1, 2, \ldots, 5$, as shown in Figure 15.2.1. The amplitude values are 1, +1/2, −1/2, −1, −1/2, and +1/2. The signal is real with power

$$P_x = \frac{1}{N}\sum_{n=0}^{N-1}x^2(n) = \frac{1}{6}\sum_{n=0}^{5}x^2(n) = \frac{1}{6}(1 + 2/4 + 1 + 2/4) = \frac{1}{2},$$

and energy

$$E_x = \sum_{n=0}^{N-1}x^2(n) = \sum_{n=0}^{5}x^2(n) = (1 + 2/4 + 1 + 2/4) = 3.$$

The Fourier coefficients can be calculated from the analysis equation as

$$c_k = \frac{1}{N}\sum_{n=0}^{N-1}x(n)e^{-jk2\pi n/N} = \frac{1}{6}\sum_{n=0}^{5}x(n)e^{-jk2\pi n/6},$$

for $k = 0, 1, 2, \ldots, 5$. These calculations can be demanding. However, the coefficients can be obtained directly from the expression for the signal $x(n)$ in the time domain by using the synthesis equation expressed in exponential form as

$$x(n) = \cos(2\pi n/6) = \frac{1}{2}e^{j2\pi n/6} + \frac{1}{2}e^{-j2\pi n/6},$$

which shows that there are two spectral components at $k = 1$ and at $k = -1$, respectively, both with magnitude 0.5. These components are shown in Figure 15.2.1. We can find the spectral components in the interval of one period, from $k = 0$ to $k = 5$, by rearranging the previous signal expression to the form

$$x(n) = \frac{1}{2}e^{j2\pi n/6} + \frac{1}{2}e^{+j2\pi n(5-6)/6} = \frac{1}{2}e^{j2\pi n/6} + \frac{1}{2}e^{+j2\pi n5/6},$$

which shows that there are two components in the fundamental period, at the positions $k = 1$ and $k = 5$, respectively. The amplitude spectrum of the signal is shown in Figure 15.2.1 for two abscises, with one specifying the coefficient $k$, and one specifying the frequency. In this way, the relationship between the coefficients and the discrete frequencies can be clearly seen.

The energy of the signal can be calculated from its spectrum as

$$E_x = N \sum_{k=0}^{N-1} c_k^2 = 6(1/4 + 1/4) = 3,$$

and the power as

$$P_x = E_x/N = 1/2.$$

The calculated power and energy values comply with the corresponding values calculated for the signal in the time domain.

**Periodic rectangular pulses: The asymmetric case.** We are interested in finding the discrete-time Fourier series, and deriving expressions for the magnitude spectrum and the phase spectrum of the asymmetric, periodic, discrete rectangular pulse train that is defined as a series of the discrete-time rectangular pulses shown in Figure 15.2.2. We want to analyse the case when the duration of a pulse in the train is $L = 5$ and the period is $N = 10$.

According to eqn (15.2.14), we can find the coefficients

$$c_k = \frac{1}{N} \sum_{n=0}^{N-1} x(n)e^{-jk2\pi n/N} = \frac{A}{N} \sum_{n=0}^{L-1} e^{-jk2\pi n/N}, \tag{15.2.17}$$

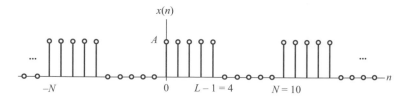

**Fig. 15.2.2** *Periodic asymmetric train of discrete pulses.*

for $k = 0, 1, \ldots, N - 1$, which is the sum of the first $L$ terms of the given geometric series, which can be expressed as

$$
c_k = \begin{cases} \dfrac{AL}{N} & k = 0, \pm N, \pm 2N, \ldots \\ \dfrac{A}{N}\dfrac{1 - e^{-jk2\pi L/N}}{1 - e^{-jk2\pi/N}} = \dfrac{A}{N} e^{-jk\pi(L-1)/N} \dfrac{\sin(k\pi L/N)}{\sin(k\pi/N)} & otherwise \end{cases}, \tag{15.2.18}
$$

because

$$
\frac{A}{N}\frac{1 - e^{-jk2\pi L/N}}{1 - e^{-jk2\pi/N}} = \frac{A}{N}\frac{e^{-jk\pi L/N}}{e^{-jk\pi/N}}\frac{e^{jk\pi L/N} - e^{-jk\pi L/N}}{e^{jk\pi/N} - e^{-jk\pi/N}} = \frac{A}{N} e^{-jk\pi(L-1)/N}\frac{\sin(k\pi L/N)}{\sin(k\pi/N)}.
$$

Generally, the coefficients are complex numbers with real and imaginary parts. The moduli of the coefficients are their magnitudes, and the arguments of the coefficients are their phases. They can be calculated as follows. Firstly, the coefficients can be expressed as

$$
c_k = \frac{A}{N}\frac{\sin(k\pi L/N)}{\sin(k\pi/N)} e^{-jk\pi(L-1)/N} = \frac{A}{N}\frac{\sin(k\pi L/N)}{\sin(k\pi/N)}\left(\cos k\pi\,(L-1)/N - j\sin k\pi\,(L-1)/N\right).
$$

The magnitudes (moduli or absolute values) of the coefficients are

$$
|c_k| = \sqrt{(\mathrm{Re}\,(c_k))^2 + (\mathrm{Im}\,(c_k))^2} = \left|\frac{A}{N}\frac{\sin(k\pi L/N)}{\sin(k\pi/N)}\right|.
$$

Due to the rule of the product of arguments, $\phi_k = \arg(z_1 z_2 z_3) = \arg(z_1) + \arg(z_2) + \arg(z_3)$, and for the given positive amplitude, $A > 0$, we obtain

$$
\begin{aligned}
\phi_k = \arg(c_k) &= \arg\left(\frac{A}{N}\frac{\sin(k\pi L/N)}{\sin(k\pi/N)} e^{-jk\pi(L-1)/N}\right) = \arg\left(\frac{1}{N}\frac{\sin(k\pi L/N)}{\sin(k\pi/N)}\right) + \arg\left(e^{-jk\pi(L-1)/N}\right) \\
&= \phi_{kSymm} + \arctan\left(\frac{-\sin(k\pi\,(L-1)/N)}{\cos(k\pi\,(L-1)/N)}\right) = \phi_{kSymm} + \arctan\left[-\tan(k\pi\,(L-1)/N)\right] \\
&= \phi_{kSymm} - \phi_{kAsymm} = \phi_{kSymm} - k\pi\,(L-1)/N
\end{aligned}
$$

Now, the magnitude spectrum and the phase spectrum are

$$|c_k| = \begin{cases} AL/N & k=0,\pm N,\pm 2N,\ldots \\ \frac{A}{N}\left|\frac{\sin(k\pi L/N)}{\sin(k\pi/N)}\right| & \text{otherwise} \end{cases}, \qquad (15.2.19)$$

$$\phi_k = \begin{cases} 0 & k=0,\pm N,\pm 2N,\ldots \\ \phi_{kSymm} - k\pi\,(L-1)\,/N & \text{otherwise} \end{cases}, \qquad (15.2.20)$$

respectively, whereas the phase $\phi_{kSymm}$ is the phase spectrum of a periodic pulse stream that is symmetric with respect to the $y$-axis. The power spectrum of the signal is its magnitude squared:

$$|c_k|^2 = \begin{cases} (AL/N)^2 & k=0,\pm N,\pm 2N,\ldots \\ \left(\frac{A}{N}\right)^2\left[\frac{\sin(k\pi L/N)}{\sin(k\pi/N)}\right]^2 & \textit{otherwise} \end{cases}. \qquad (15.2.21)$$

For the particular case when $N = 10$ and $L = 5$, we can calculate the spectra and present them in graphical form. The power and energy can be calculated from the signal presentation in the time domain. The power can be calculated in one signal period according to

$$P_x = \frac{1}{10}\sum_{n=0}^{N-1} x^2(n) = \frac{1}{10}\sum_{n=0}^{4} x^2(n) = \frac{5}{10}A^2 = \frac{A^2}{2}, \qquad (15.2.22)$$

and the energy over a single period is

$$E_x = \sum_{n=0}^{N-1}|x(n)|^2 = NP_x = 5A^2. \qquad (15.2.23)$$

We can find the expressions for the magnitude spectrum and the phase spectrum of the asymmetric periodic signal shown in Figure 15.2.2. Using the previously derived expression (15.2.18), the coefficients are calculated and expressed in complex form as

$$c_k = \frac{A\sin(2k\pi/5)}{N\sin(k\pi/10)}(\cos 2k\pi/5) - j\frac{A\sin(2k\pi/5)}{N\sin(k\pi/10)}(j\sin 2k\pi/5). \qquad (15.2.24)$$

The magnitude (modulus) spectrum components are calculated from eqn (15.2.19) as

$$|c_k| = \sqrt{(\text{Re}(c_k))^2 + (\text{Im}(c_k))^2} = \left|\frac{A\sin(2k\pi/5)}{N\sin(k\pi/10)}\right|. \qquad (15.2.25)$$

Via eqn (15.2.20), the phase spectrum of the asymmetric signal, for the given amplitude $A > 0$, can be calculated as the sum of the phase for the symmetric pulse stream and the added phase caused by the shift of the asymmetric stream to the right, that is,

$$\phi_k = \phi_{kSymm} - \phi_{kAsymm} = \phi_{kSymm} - k\pi 5/10. \tag{15.2.26}$$

The phases for the symmetric case have values of $\pi$ when the sine function is negative, that is, when $\sin(2k\pi/5) = -1$, or $k\pi/2 = 3\pi/2$ and $k\pi/2 = 7\pi/2$. Therefore, the phases are $\phi_{ksymm} = \pi$ for $k = 3$ and 7. The phases, due to the signal asymmetry, can be calculated inside the fundamental period as $\phi_{kAsymm} = -k\pi (L-1)/N = -2k\pi/5$, resulting in $\phi_{kAsymm} = 0, -2\pi/5, -4\pi/5, -6\pi/5, -8\pi/5, -2\pi, -12\pi/5, -14\pi/5, -16\pi/5$, and $-18\pi/5$, for $k = 0, 1, \ldots, 9$, respectively. The magnitudes for the asymmetric pulse are shown in Figure 15.2.3.

**Periodic rectangular pulse trains: The symmetric case.** A symmetric periodic pulse train is shown in Figure 15.2.4. For the duration of a pulse $L = 5$ with period $N = 10$, we will find the inverse Fourier transform to demonstrate the presentation of the spectral components in the discrete-time domain. It will be shown that the sum of the components can reconstruct the original signal, as is theoretically expected.

According to the analysis equation, we can find the coefficients for any chosen interval of $N$ signal values. In this case, we will use the expression

$$c_k = \frac{1}{N} \sum_{n=-N/2}^{N/2-1} x(n)e^{-jk2\pi n/N} = \frac{A}{N} \sum_{n=-L/2}^{L/2} e^{(-jk2\pi/N)n}, \tag{15.2.27}$$

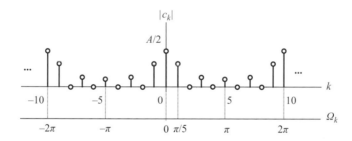

**Fig. 15.2.3** *Magnitude spectrum of an asymmetric pulse.*

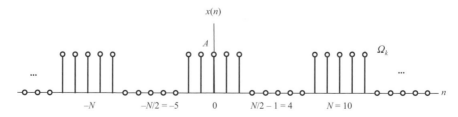

**Fig. 15.2.4** *Periodic symmetric series of discrete pulses.*

for $k = 0, 1, \ldots, N-1$, which is the sum of the $L$ terms of the given geometric series that can be expressed as

$$c_k = \begin{cases} \frac{AL}{N} & k = 0, \pm N, \pm 2N, \ldots \\ \frac{A}{N} \frac{\sin(k\pi L/N)}{\sin(k\pi/N)} & \text{otherwise} \end{cases}. \tag{15.2.28}$$

The magnitude spectrum and the phase spectrum can be calculated as

$$|c_k| = \begin{cases} \frac{AL}{N} & k = 0, \pm N, \pm 2N, \ldots \\ \frac{A}{N} \left| \frac{\sin(k\pi L/N)}{\sin(k\pi/N)} \right| & \text{otherwise} \end{cases}, \tag{15.2.29}$$

and

$$\phi_k = \arg(c_k) = \arctan \frac{\text{Im}(c_k)}{\text{Re}(c_k)}, \tag{15.2.30}$$

respectively. The power spectrum of this periodic signal can be calculated using eqn (15.2.21). For the given positive amplitude $A$, pulse width $L = 5$, and signal period $N = 10$, we may calculate the values for the power and the energy of the signal in the time domain as presented in eqns (15.2.22) and (15.2.23). For $L = 5$ and $N = 10$, we may use eqn (15.2.29) to calculate the magnitude spectral components. Then, for all $k$ values, we can find the numerical values of the Fourier coefficients, or the amplitudes of the spectral components, as $c_1 = 0.32A$, $c_2 = 0$, $c_3 = -0.12A$, $c_4 = 0$, $c_5 = 0.1A$, $c_6 = 0$, $c_7 = -0.12A$, $c_8 = 0$, and $c_9 = 0.32A$. The amplitude spectrum can be presented in graphical form because the phase changes can be only $+\pi$ or $-\pi$, as presented in the upper graph in Figure 15.2.5. The spectrum is a function of periodic discrete frequency.

The magnitude spectra of the symmetric and asymmetric pulses are the same because the time shift does not change the amplitudes of the spectral components. The power and energy can be calculated from the magnitude spectrum of the signal as

$$P_x = \left(\frac{A}{2}\right)^2 + 2\left(\frac{A}{3.1}\right)^2 + \left(\frac{A}{10}\right)^2 + 2\left(\frac{A}{8.1}\right)^2 = 0.50A^2$$

and

$$E_x = N \sum_{k=0}^{N-1} c_k^2 = N \sum_{k=0}^{9} P_k = 10 \cdot 0.5A^2 = 5A^2,$$

respectively, which are equivalent to the power and energy values calculated in the time domain and expressed by eqns (15.2.22) and (15.2.23), respectively.

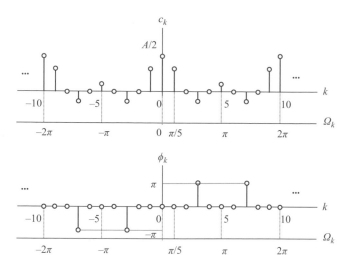

**Fig. 15.2.5** *Amplitude spectrum (upper panel) and phase spectrum (lower panel) of a symmetric rectangular periodic discrete pulse.*

The signal in the time domain can be obtained from its presentation in the frequency domain by applying the inverse discrete-time Fourier series. In this case, the signal in the time domain contains $N = 10$ discrete-time sinusoidal components having the already calculated amplitudes. Bearing in mind that $c_1 = c_9$, and $c_3 = c_7$, we can calculate the signal in the time domain as

$$
\begin{aligned}
x(n) &= \sum_{k=0}^{N-1} c_k e^{+jk2\pi n/N} = \sum_{k=0}^{9} c_k e^{+jk\pi n/5} \\
&= c_0 + c_1 e^{+j\pi n/5} + c_3 e^{+j3\pi n/5} + c_5 e^{+j\pi n} + c_7 e^{+j7\pi n/5} + c_9 e^{+j9\pi n/5} \\
&= c_0 + c_1 \left( e^{+j\pi n/5} + e^{+j9\pi n/5} \right) + c_3 \left( e^{+j3\pi n/5} + e^{+j7\pi n/5} \right) + c_5 e^{+j\pi n} \\
&= c_0 + c_1 \left( \cos\pi n/5 + j\sin\pi n/5 + \cos 9\pi n/5 + j\sin 9\pi n/5 \right) \\
&\quad + c_3 \left( \cos 3\pi n/5 + j\sin 3\pi n/5 + \cos 7\pi n/5 + j\sin 7\pi n/5 \right) + c_5 \left( \cos\pi n + j\sin\pi n \right) \\
&= c_0 + 2c_1 \cos\pi n/5 + 2c_3 \cos 3\pi n/5 + c_5 \cos\pi n \\
&= 0.5A + 0.64A \cos\pi n/5 - 0.24A \cos 3\pi n/5 + 0.1A \cos\pi n
\end{aligned}
$$

The signal in the time domain will be obtained by calculating the amplitude values of all of the components for every $n$, which results, with limited accuracy, in $x(0) = A$, $x(1) = 0.9919A$, $x(-1) = x(1)$, $x(2) = 0.9917A$, $x(3) = 0.0081A$, and $x(10) = A$.

For the sake of presentation and understanding, the presented spectral components are plotted in the time domain as continuous functions. In fact, it is necessary to add only the amplitude values of $N = 10$ components at the discrete-time instants $n$. The values of

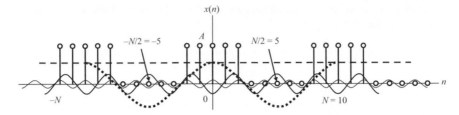

**Fig. 15.2.6** *Periodic series of discrete pulses as a superposition of discrete spectral components.*

these components are presented in Figure 15.2.6. They are calculated for $n = 0$ to 10, as follows. The first sinusoidal component with phase 0 has the following amplitudes multiplied by $A$ (indicated by the thick dotted line): [0.6400 0.5178 0.1978 –0.1978 –0.5178 –0.64 –0.5178 –0.1978 0.1978 0.5178 0.64]. The second sinusoidal component has phase $\pi$ and the following amplitudes multiplied by $A$ (indicated by the solid thick line): [–0.24 0.0742 0.1942 –0.1942 –0.0742 0.24 –0.0742 –0.1942 0.1942 0.0742 –0.24]. The third sinusoidal component has phase 0 and the following amplitudes multiplied by $A$ (indicated by the solid thin line): [0.1000 –0.1000 0.1000 –0.1000 0.1000 –0.1000 0.1000 –0.1000 0.1000 –0.1000 0.1000]. For each discrete $n$ value, we can get the sum of the amplitudes for the sinusoidal components. By adding the DC value of $0.5A$ (indicated by the horizontal dashed line in Figure 15.2.6), we can reassemble the discrete-time periodic signal.

We can see that, by adding the frequency components for every $n$, we can reconstruct the rectangular pulses, even though the addition is performed at the discrete-time instants.

## 15.3 The Discrete-Time Fourier Transform

### 15.3.1 Derivation of the Discrete-Time Fourier Transform Pair

The amplitude spectral density, or discrete-time Fourier transform, $X\left(e^{j\Omega}\right)$, is presented as a function of the complex argument $e^{j\Omega}$, to remind the reader that the transform is, in principle, a function of complex variables. For the sake of simplicity, $X\left(e^{j\Omega}\right)$ will sometimes be replaced by $X(\Omega)$.

The discrete-time Fourier transform can be treated as a simplification of the $z$-transform. Namely, for the $z$-transform defined as

$$X(z) = \sum_{n=-\infty}^{\infty} x(n)z^{-n}, \qquad (15.3.1)$$

if the region of convergence includes the unit circle, then the discrete-time Fourier transform equals the $z$-transform evaluated at that unit circle, that is,

$$X\left(e^{j\Omega}\right) = \sum_{n=-\infty}^{\infty} x(n)e^{-j\Omega n}. \qquad (15.3.2)$$

The inverse discrete-time Fourier transform can be obtained from the following expression for the inverse *z-transform*:

$$x(n) = \frac{1}{2\pi j} \int_l X(z) z^{n-1} dz, \tag{15.3.3}$$

where *l* is a counter-clockwise contour of integration enclosing the origin defined for the unit circle. In this case, the contour is

$$l = \left\{ \; z : z = e^{j\Omega}, \Omega \in (-\pi, \pi) \right\}, \tag{15.3.4}$$

so that $dz = je^{j\Omega} d\Omega$, which results in the Fourier integral, or the synthesis equation of the discrete-time Fourier transform, expressed as

$$
\begin{aligned}
x(n) &= \frac{1}{2\pi j} \int_l X(z) z^{n-1} dz = \frac{1}{2\pi j} \int_{-\pi}^{\pi} X\left(e^{j\Omega}\right) e^{j\Omega n} e^{-j\Omega} je^{j\Omega} d\Omega \\
&= \frac{1}{2\pi} \int_{-\pi}^{\pi} X\left(e^{j\Omega}\right) e^{j\Omega n} d\Omega
\end{aligned}
\tag{15.3.5}
$$

Thus, the *analysis* and *synthesis* equations, forming the discrete-time Fourier transform pair, are

$$X\left(e^{j\Omega}\right) = \sum_{n=-\infty}^{\infty} x(n) e^{-j\Omega n} \tag{15.3.6}$$

and

$$x(n) = \frac{1}{2\pi} \int_{-\pi}^{\pi} X\left(e^{j\Omega}\right) e^{j\Omega n} d\Omega, \tag{15.3.7}$$

respectively. The discrete-time Fourier transform $X(e^{j\Omega})$ is a complex function of the continuous angular frequency $\Omega$. The modulus of the function represents the *magnitude spectrum* of the signal $x(n)$, while its phase (argument) represents the *phase spectrum* of the signal $x(n)$. These equations can be expressed in terms of the normalized cycle frequency $F = \Omega/2\pi$, as

$$X(F) = \sum_{n=-\infty}^{\infty} x(n) e^{-j2\pi F n} \tag{15.3.8}$$

and

$$x(n) = \int_{-0.5}^{0.5} X(F) e^{j2\pi F n} dF, \tag{15.3.9}$$

respectively. The discrete-time Fourier transform pair corresponds to the equations for the Fourier series of a periodic continuous-time signal, where the signal in the time domain is represented by a sum of complex exponential basis functions with discrete coefficients, as in eqn (15.2.1), and the coefficients are calculated using the integral transform (15.2.2). Therefore, eqn (15.3.6) can be interpreted as a Fourier series expansion of the periodic function $X\left(e^{j\Omega}\right)$ with coefficients $x(n)$. Thus, the theory concerning the continuous-time Fourier series applies to the discrete-time Fourier transform, but with the time and frequency domains reversed. From the expression for $c_k$, the inverse discrete-time Fourier transform for $x(n)$ directly follows.

The expression for $X\left(e^{j\Omega}\right)$ is the *spectrum* of $x(n)$, because the signal $x(n)$ is represented as the (limiting) sum of complex exponential (or sinusoidal) components of the form

$$\left[\frac{1}{2\pi}X\left(e^{j\Omega}\right)d\Omega\right]e^{j\Omega n}. \tag{15.3.10}$$

Since $e^{j\Omega}$ is a periodic function of the continuous angular frequency $\Omega$ with the period $2\pi$, $X\left(e^{j\Omega}\right)$ is also periodic in $\Omega$ with period $2\pi$.

## 15.3.2 The Problem of Convergence

There are two conditions that guarantee the existence of the discrete-time Fourier transform:

1. The discrete-time Fourier transform exists if $x(n)$ is absolutely summable, that is,

$$\sum_{n=-\infty}^{\infty} |x(n)| < \infty. \tag{15.3.11}$$

Then, in mathematical terms, the function $X\left(e^{j\Omega}\right)$ is absolutely convergent, and it converges uniformly to a continuous function of $\Omega$.

2. The function $X\left(e^{j\Omega}\right)$ also exists for all discrete-time signals $x(n)$ having finite energy, that is, for the signals that fulfil the following condition:

$$E_x = \sum_{n=-\infty}^{\infty} |x(n)|^2 < \infty, \tag{15.3.12}$$

and it can converge even if the region of convergence for $X(z)$ does not include the unit circle, or even if $X(z)$ does not exist at all. Specifically, this happens when $X\left(e^{j\Omega}\right)$ is discontinuous, as is illustrated in the following examples. This is a weaker condition than the absolute summability (15.3.11).

In addition, there are some finite energy signals which are not absolutely summable, like $x(n) = \text{sinc}(n)$ and sinusoids, and there are signals which violate both conditions and still have a discrete-time Fourier transform. To solve some important problems, and get insight into some physical phenomena, generalized functions, such as the impulse delta functions $\delta(\Omega)$, will be used to find the discrete-time Fourier transform of a number of signals that violate both existence conditions. As in the analysis of continuous-time signals, the Dirac delta function and the Kronecker delta function will be used to solve specific and important problems.

**Example**

Find the discrete-time Fourier transform of the causal exponential signal expressed by

$$x(n) = a^n u(n).$$

Investigate the characteristics of the energy spectral density of the signal.

**Solution.** If we apply the geometric summation expression, assuming $-1 < a < 1$, we can prove that the signal is absolutely summable, that is,

$$\sum_{n=-\infty}^{\infty} |x(n)| = \sum_{n=-\infty}^{\infty} |a|^n = \frac{1}{1-|a|} < \infty.$$

Therefore, the Fourier transform exists. We can find the expressions for the discrete-time Fourier transform as

$$X\left(e^{j\Omega}\right) = \sum_{n=-\infty}^{\infty} x(n)e^{-j\Omega n} = \sum_{n=-\infty}^{\infty} \left[a^n u(n)\right] e^{-j\Omega n} = \sum_{n=0}^{\infty} \left(ae^{-j\Omega}\right)^n$$

$$= \lim_{n\to\infty} \frac{1 - \left(ae^{-j\Omega}\right)^n}{1 - \left(ae^{-j\Omega}\right)} = \frac{1}{1 - ae^{-j\Omega}}$$

This summation converges if and only if $|ae^{-j\Omega}| < 1$, that is, $|a| < 1$. Thus, the final form of the discrete-time Fourier transform is

$$X\left(e^{j\Omega}\right) = \frac{1}{1-ae^{-j\Omega}} = \frac{1}{1-a\cos\Omega + ja\sin\Omega} = \left|X\left(e^{j\Omega}\right)\right| \angle X\left(e^{j\Omega}\right),$$

where the magnitude spectrum and the phase spectrum, shown in Figures 15.3.1 and 15.3.2, respectively, are calculated as follows. The magnitudes are

$$\left|X\left(e^{j\Omega}\right)\right| = \left[X\left(e^{j\Omega}\right)X^*\left(e^{j\Omega}\right)\right]^{1/2} = \left[\frac{1}{1-ae^{-j\Omega}}\frac{1}{1-ue^{j\Omega}}\right]^{1/2}$$

$$= \left[\frac{1}{1-2a\cos\Omega + a^2}\right]^{1/2} \tag{15.3.13}$$

and the phases are

$$\arg\left[X\left(e^{j\Omega}\right)\right] = \angle X\left(e^{j\Omega}\right) = \arctan\left[\frac{\operatorname{Im}X\left(e^{j\Omega}\right)}{\operatorname{Re}X\left(e^{j\Omega}\right)}\right] = -\arctan\left[\frac{a\sin\Omega}{1-a\cos\Omega}\right]. \quad (15.3.14)$$

These two expressions are, respectively, even and odd functions of $\Omega$. These functions are continuous and periodic in frequency $\Omega$ with period $2\pi$. Therefore, only one plot, which is in the fundamental frequency range from $0$ to $2\pi$, or $-\pi$ to $+\pi$, can be used in any analysis.

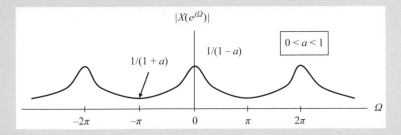

**Fig. 15.3.1** *Magnitude spectrum of a discrete signal.*

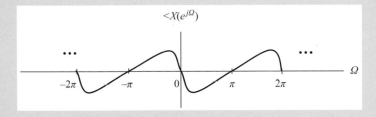

**Fig. 15.3.2** *Phase spectrum of a discrete signal.*

The energy spectral density is obtained as the magnitude spectral density squared, that is,

$$S_x(\Omega) = E_x(\Omega) = \left|X\left(e^{j\Omega}\right)\right|^2 = \frac{1}{1-2a\cos\Omega+a^2}.$$

## Example

Find the inverse discrete-time Fourier transform of a uniform impulse train with constant weight $A_k$, expressed in the frequency domain as

$$X\left(e^{j\Omega}\right) = 2\pi \sum_k A_k \delta\left(\Omega - \Omega_k\right), \tag{15.3.15}$$

for $|\Omega_k| \leq \pi$.

**Solution.** The inverse discrete-time Fourier transform is

$$x(n) = \frac{1}{2\pi} \int_{-\pi}^{\pi} X\left(e^{j\Omega}\right) e^{j\Omega n} d\Omega = \frac{1}{2\pi} \int_{-\pi}^{\pi} 2\pi \sum_k A_k \delta\left(\Omega - \Omega_k\right) e^{j\Omega n} d\Omega$$

$$= \sum_k A_k \int_{-\pi}^{\pi} e^{j\Omega n} \delta\left(\Omega - \Omega_k\right) d\Omega = \sum_k A_k e^{j\Omega_k n}$$

Special cases are as follows: for $k = 1$, $X\left(e^{j\Omega}\right) = 2\pi A_1 \delta\left(\Omega - \Omega_1\right)$, we may have $x(n) = A_1 e^{j\Omega_1 n}$; for $k = (-1, 1)$, we may have two spectral components expressed as

$$X\left(e^{j\Omega}\right) = 2\pi \sum_{k=-1,+1} A_k \delta\left(\Omega - \Omega_k\right) = 2\pi A_{-1} \delta\left(\Omega - \Omega_{-1}\right) + 2\pi A_{+1} \delta\left(\Omega - \Omega_{+1}\right).$$

For this case, the Dirac delta impulse train is presented in Figure 15.3.3.

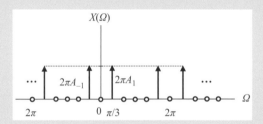

**Fig. 15.3.3** *Uniform impulse train of delta components.*

Since the impulse functions are allowed to be present in $X\left(e^{j\Omega}\right)$, the corresponding signal $x(n)$ has infinite energy, because it contains periodic (sinusoidal) components, as can be seen from the following signal expression in the time domain:

$$x(n) = \frac{1}{2\pi} \int_{-\pi}^{\pi} \left[2\pi A_{-1} \delta\left(\Omega - \Omega_{-1}\right) + 2\pi A_{+1} \delta\left(\Omega - \Omega_{+1}\right)\right] e^{j\Omega n} d\Omega$$

$$= \int_{-\pi}^{\pi} \left[A_{-1} \delta\left(\Omega - \Omega_{-1}\right)\right] e^{j\Omega n} d\Omega + \int_{-\pi}^{\pi} \left[A_{+1} \delta\left(\Omega - \Omega_{+1}\right)\right] e^{j\Omega n} d\Omega.$$

$$= A_{-1} e^{j\Omega_{-1} n} + A_{+1} e^{j\Omega_{+1} n}$$

For $k = 0$, $\Omega_0 = 0$, $A_0 = 1$, we obtain $X\left(e^{j\Omega}\right) = 2\pi A_0 \delta\left(\Omega - 0\right) = 2\pi \delta\left(\Omega\right)$, $|\Omega| \leq \pi$, and the signal in the time domain is

$$x(n) = \frac{1}{2\pi} \int_{-\pi}^{\pi} [2\pi \delta(\Omega)] \, e^{j\Omega n} d\Omega = 1 \cdot e^{j \cdot 0 \cdot n} = 1,$$

which is a discrete DC signal.

## Example

Find the discrete-time Fourier transform of real sinusoidal signals and a DC signal, expressed, respectively, as

a. $x(n) = \sin \Omega_0 n = \frac{1}{2j} \left( e^{j\Omega_0 n} - e^{-j\Omega_0 n} \right),$

b. $x(n) = \cos \Omega_0 n = \frac{1}{2} \left( e^{j\Omega_0 n} + e^{-j\Omega_0 n} \right),$ and

c. $x(n) = 1.$

**Solution.** The solutions are as follows:

a. $X\left( e^{j\Omega} \right) = 2\pi \frac{1}{2j} [\delta(\Omega - \Omega_0) - \delta(\Omega + \Omega_0)] = -\pi j [\delta(\Omega - \Omega_0) - \delta(\Omega + \Omega_0)],$
   for $|\Omega_0| \leq \pi;$

b. $X\left( e^{j\Omega} \right) = 2\pi \frac{1}{2} [\delta(\Omega - \Omega_0) + \delta(\Omega + \Omega_0)] = \pi [\delta(\Omega - \Omega_0) + \delta(\Omega + \Omega_0)];$
   and

c. $X\left( e^{j\Omega} \right) = 2\pi \delta(\Omega),$ $|\Omega| \leq \pi,$ for the DC signal $x(n) = 1,$ which is opposed to the dual relationship for the delta function $\delta(n) \leftrightarrow 1,$ for all $\Omega.$

## Example

Check the condition of the Fourier transform existence and find the discrete-time Fourier transform of a rectangular asymmetric pulse expressed by

$$p(n) = u(n) - u(n - N) \tag{15.3.16}$$

and the discrete-time Fourier transform for a symmetric pulse (symmetric with respect to the $y$-axis) expressed by

$$p_s(n) = u(n + N - 1) - u(n - N), \tag{15.3.17}$$

which are shown in Figure 15.3.4.

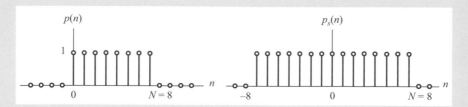

**Fig. 15.3.4** *Rectangular asymmetric (left) and symmetric (right) pulses.*

**Solution for the asymmetric pulse.** The Fourier transform exists because the signal is absolutely summable, that is,

$$\sum_{n=-\infty}^{\infty} |p(n)| = \sum_{n=0}^{N-1} 1 = N < \infty.$$

In this problem, the amplitude of the signal is 1. If this amplitude is different from 1, it must be less than infinity. This is a finite energy signal.
The corresponding $z$-transform of the asymmetric pulse is

$$P(z) = \sum_{n=-\infty}^{\infty} p(n)z^{-n} = \sum_{n=0}^{N-1} 1 \cdot z^{-n} = 1 + z^{-1} + z^{-2} + \cdots + z^{-N+1} = \frac{1-z^{-N}}{1-z^{-1}}.$$

This summation converges if and only if $|z| > 1$. If this region contains the unit circle, then the Fourier transform will exist. However, the existence of the Fourier transform does not necessarily mean that the $z$-transform exists. Therefore, as we have already proven, the discrete-time Fourier transform exists and can be obtained via the following derivation:

$$P\left(e^{j\Omega}\right) = \sum_{n=-\infty}^{\infty} p(n)e^{-j\Omega n} = \sum_{n=0}^{N-1} \left(e^{-j\Omega}\right)^n = \frac{1-e^{-j\Omega N}}{1-e^{-j\Omega}} = \frac{e^{-j\Omega N/2}\left(e^{j\Omega N/2}-e^{-j\Omega N/2}\right)}{e^{-j\Omega/2}\left(e^{j\Omega/2}-e^{-j\Omega/2}\right)},$$

$$= e^{-j\Omega(N-1)/2}\frac{\sin(\Omega N/2)}{\sin(\Omega/2)} = e^{-j\Omega(N-1)/2}P(\Omega)$$

where $P(\Omega)$ is purely real. The corresponding magnitude spectrum is

$$\left|P\left(e^{j\Omega}\right)\right| = \begin{cases} N & \Omega = 0 \\ \left|\frac{\sin(\Omega N/2)}{\sin(\Omega/2)}\right| & \Omega \neq 0 \end{cases}, \tag{15.3.18}$$

which is presented in Figure 15.3.5 for the case $N = 8$. Note that the amplitude value of $\left|P(e^{j\Omega})\right|$ for $\Omega = 0$ is equal to $N$, which can be easily obtained by calculating the value from the defining expression for the discrete-time Fourier transform. The zeros occur for $\sin(\Omega N/2) = 0$, that is, $\Omega N/2 = k\pi$, $k = 1, 2, \ldots, N-1$. For $N = 8$, $\Omega = 2k\pi/N = k\pi/4$, and $k = 1, 2, \ldots, 7$.

Following the rule for the argument of a product, $\phi_k = \arg(z_1 z_2) = \arg(z_1) + \arg(z_2)$, we may find the phase spectrum (argument) of the Fourier transform as

$$
\begin{aligned}
\arg P\left(e^{j\Omega}\right) &= \arg\left(e^{-j\Omega(N-1)/2}\frac{\sin(\Omega N/2)}{\sin(\Omega/2)}\right) \\
&= \arg\left(\cos\Omega\,(N-1)/2 - j\sin\Omega\,(N-1)/2\right) + \arg\frac{\sin(\Omega N/2)}{\sin(\Omega/2)} \\
&= \arctan\left(\frac{-\sin\Omega\,(N-1)/2}{\cos\Omega\,(N-1)/2}\right) + \arg\frac{\sin(\Omega N/2)}{\sin(\Omega/2)} = -\frac{\Omega}{2}(N-1) + \arg\frac{\sin(\Omega N/2)}{\sin(\Omega/2)}
\end{aligned}
$$

The phase changes linearly, due to the first term, and phase discontinuities of $\pi$ radians occur when the argument term is negative, and those of $0$ radians occur when the term is positive. For $N = 8$, the phase can be found as

$$
\arg P\left(e^{j\Omega}\right) = -\frac{7\Omega}{2} + \arg\frac{\sin(4\Omega)}{\sin(\Omega/2)}. \tag{15.3.19}
$$

The first term gives the linear part of the phase spectrum, while the second part specifies the phase changes of $\pi$ radians for the condition $\sin 4\Omega = \pm 1$, or $4\Omega = k\pi/2$, that is, $\Omega = k\pi/8$, for $k = 2, 4, \ldots$. Thus, the first zero crossing is for $\Omega = \pi/4$, and the last one in the $2\pi$ period presented is $3\pi/4$, as shown in Figure 15.3.5.

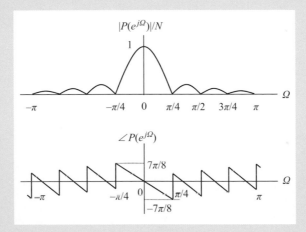

**Fig. 15.3.5** *Magnitude spectrum (upper panel) and phase spectrum (lower panel) of an asymmetric pulse, for $N = 8$.*

For positive frequencies, the phase changes are $\pi$ and, for negative frequencies, they are $-\pi$. These are initial phases of the signal, and their signs need to follow the positive and negative sense of the corresponding phasor rotations, like what we discussed in the Fourier analysis of continuous-time signals. Therefore, it is very relevant to calculate the phase angles properly for both positive and negative frequencies. The phase value $+\pi$ is used for positive frequencies, and $-\pi$ is used for negative frequencies, that is, they depend on the sign of the frequency these phases belong to.

**Solution for the symmetric pulse.** The discrete-time Fourier transform exists in this case and is obtained by the following derivation:

$$
\begin{aligned}
P_s\left(e^{j\Omega}\right) &= \sum_{n=-\infty}^{\infty} p(n)e^{-j\Omega n} = \sum_{n=-N+1}^{N-1}\left[e^{-j\Omega}\right]^n = \frac{1-e^{-j\Omega N}}{1-e^{-j\Omega}} + \frac{1-e^{j\Omega N}}{1-e^{j\Omega}} - 1 \\
&= \frac{e^{j(\Omega N - \Omega/2)} - e^{-j(\Omega N - \Omega/2)}}{e^{j\Omega/2} - e^{-j\Omega/2}} = \frac{\sin\Omega\,(2N-1)/2}{\sin(\Omega/2)} = \frac{\sin\Omega N_s/2}{\sin(\Omega/2)}
\end{aligned}
$$

where the total number of samples in the pulse is $N_s = 2N - 1$, and $P_s(\Omega)$ is purely real. The corresponding magnitude spectrum and phase spectrum are

$$
\left|P_s\left(e^{j\Omega}\right)\right| = \left\{\begin{array}{cc} N_s & \Omega = 0 \\ \left|\frac{\sin\Omega N_s/2}{\sin(\Omega/2)}\right| & \Omega \neq 0 \end{array}\right\} = \left\{\begin{array}{cc} 15 & \Omega = 0 \\ \left|\frac{\sin 15\Omega/2}{\sin(\Omega/2)}\right| & \Omega \neq 0 \end{array}\right\},
$$

and

$$
\arg P_s\left(e^{j\Omega}\right) = \arg \frac{\sin 15\Omega/2}{\sin(\Omega/2)}.
$$

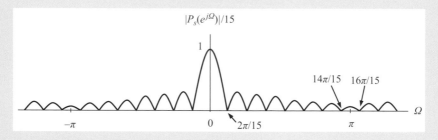

**Fig. 15.3.6** *Magnitude spectral density of a symmetric pulse.*

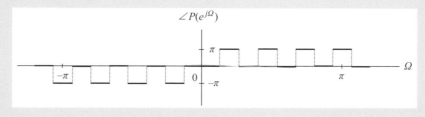

**Fig. 15.3.7** *Phase spectrum of a symmetric pulse.*

The graphs of the amplitude spectrum and the phase spectrum are presented in Figures 15.3.6 and 15.3.7, respectively, for the cases $N = 8$ and $N_s = 15$. Note that the amplitude value of $X(e^{j\Omega})$ for $\Omega = 0$ is equal to $N_s$, which can be easily obtained by calculating this value from the defining expression for the discrete-time Fourier transform. The zero crossings occur for the condition $\sin \Omega N_s/2 = 0$, that is, for $\Omega N_s/2 = k\pi$, $k = 1, 2, \ldots , N_s - 1$. For $N_s = 15$, we may have $\Omega = \pm 2k\pi/N_s = \pm 2k\pi/15$, and $k = 1, 2, \ldots , 15$, and phase discontinuities of $\pi$ radians occur at the same frequencies. Phases for the spectrum can be found from the conditions $\sin \Omega N_s/2 = \pm 1$, or $\Omega = \pm k\pi/15$. For example, the first zero crossing is $\Omega = \pm 2\pi/15 = \pm 24°$.

## Example

Find the inverse discrete-time Fourier transform of a signal $x(n)$ that has a rectangular amplitude spectrum with zero phases, as shown in Figure 15.3.8. Assume that $N = 2$.

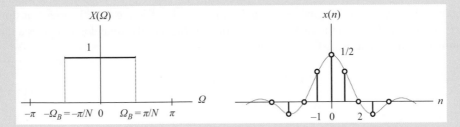

**Fig. 15.3.8** *Rectangular spectrum of a signal (left) and its related calculated signal (right) in the time domain.*

**Solution.** The inverse Fourier transform of the signal is

$$x(n) = \frac{1}{2\pi} \int_{-\pi}^{\pi} X\left(e^{j\Omega}\right) e^{j\Omega n} d\Omega = \frac{1}{4\pi} \int_{-\Omega_B}^{\Omega_B} e^{j\Omega n} d\Omega = \frac{1}{2\pi} \left. \frac{e^{j\Omega n}}{jn} \right|_{-\Omega_B}^{\Omega_B}$$

$$= \left. \frac{\sin(\Omega_B n)}{\pi n} \right|_{\Omega_B = \pi/N} = \frac{1}{N} \frac{\sin(n\pi/N)}{n\pi/N}. \qquad (15.3.20)$$

$$= \frac{1}{N} \text{sinc } (n\pi/N)$$

For $N = 2$ and the signal bandwidth $\Omega_B = \pi/2$, the signal in the time domain is

$$x(n) = \frac{1}{N} \text{sinc } (n\pi/N) = \frac{1}{2} \text{sinc } (n\pi/2).$$

The signal is presented in graphical form in Figure 15.3.8. The signal preserves the shape of the sinc function and has the first zero crossing at $n = 2$.

### 15.3.3 Properties of the Discrete-Time Fourier Transform

The properties of the discrete-time Fourier transform are equivalent to the properties of the continuous-time Fourier transform, assuming that the time and frequency domains are reversed. They also correspond to the properties of the *z-transform* except when $X(e^{j\Omega})$ is not absolutely convergent, that is, when the region of convergencefor $X(z)$ does not include the unit circle. However, the existence of the Fourier transform does not necessarily mean the *z*-transform exists. In the following analysis, we will adopt the following notation. The expression

$$x(n) \leftrightarrow X\left(e^{j\Omega}\right), \tag{15.3.21}$$

means that the signal $x(n)$ and its discrete-time Fourier transform $X(e^{j\Omega})$ form the Fourier transform pair. Therefore, the signal $x(n)$ is the inverse discrete-time Fourier transform of the amplitude spectral density $X(e^{j\Omega})$, and $X(e^{j\Omega})$ is the Fourier transform of the signal $x(n)$.

**Linearity.** If discrete-time signals $x_1(n)$ and $x_2(n)$ have the discrete-time Fourier transforms $X_1(e^{j\Omega})$ and $X_2(e^{j\Omega})$, then the linear combination of signals in the time domain gives a linear combination of their discrete-time Fourier transforms,

$$\begin{aligned} x_1(n) &\leftrightarrow X_1\left(e^{j\Omega}\right) \quad \text{and} \quad x_2(n) \leftrightarrow X_2\left(e^{j\Omega}\right) \\ &\Rightarrow ax_1(n) + bx_2(n) \leftrightarrow aX_1\left(e^{j\Omega}\right) + bX_2\left(e^{j\Omega}\right), \end{aligned} \tag{15.3.22}$$

for arbitrary constants $a$ and $b$. There can be an arbitrary number of signals in the linear combination.

**Time shift: Delay or advance.** A shift of a signal in the time domain produces a linear phase shift in the frequency domain, that is, if signal $x(n)$ has the transform $X\left(e^{j\Omega}\right)$, then the time-shifted signal $x(n-n_0)$ has the Fourier transform $e^{-j\Omega n_0}X\left(e^{j\Omega}\right)$, which is expressed as

$$x(n) \leftrightarrow X\left(e^{j\Omega}\right) \Rightarrow x(n-n_0) \leftrightarrow e^{-j\Omega n_0}X\left(e^{j\Omega}\right). \tag{15.3.23}$$

**The frequency-shifting property and the modulation property.** The frequency-shifting property states that complex modulation in the time domain corresponds to a shift in the frequency domain, that is,

$$\begin{aligned} x(n) &\leftrightarrow X\left(e^{j\Omega}\right) \Rightarrow e^{j\Omega_c n}x(n) \leftrightarrow X\left(e^{j(\Omega-\Omega_c)}\right), \\ x(n) &\leftrightarrow X\left(e^{-j\Omega}\right) \Rightarrow e^{-j\Omega_c n}x(n) \leftrightarrow X\left(e^{j(\Omega+\Omega_c)}\right). \end{aligned} \tag{15.3.24}$$

The modulation property states that multiplication of the signal in the time domain by the real part of a complex exponential corresponds to a shift of the spectrum:

$$x(n) \leftrightarrow X\left(e^{j\Omega}\right) \Rightarrow x(n)\cos\Omega_c n \leftrightarrow \frac{1}{2}\left[X\left(e^{j(\Omega-\Omega_c)}\right) + X\left(e^{j(\Omega+\Omega_c)}\right)\right].$$

In the process of modulation, care must be taken about possible aliasing in the frequency domain. If the modulating signal is of bandwidth $B$, this does not mean the modulated signal will have the bandwidth $2B$, due to possible aliasing. In this case, it is necessary to take care about folding the spilled spectrum into the fundamental band, or to observe the periodic spectrum and then add the overlapping frequencies.

**Axis reversal and conjugation.** If $x(n)$ has the discrete-time Fourier transform $X(e^{j\Omega})$, then the discrete-time Fourier transform of the time-reversed signal $x(-n)$ corresponds to the frequency-reversed discrete-time Fourier transform $X(e^{-j\Omega})$, and vice versa, that is,

$$x(n) \leftrightarrow X\left(e^{j\Omega}\right) \Rightarrow x(-n) \leftrightarrow X\left(e^{-j\Omega}\right). \tag{15.3.25}$$

Conjugation of $x(n)$ results in the conjugation of the discrete-time Fourier transform $X(e^{j\Omega})$, that is,

$$x(n) \leftrightarrow X\left(e^{j\Omega}\right) \Rightarrow x^*(n) \leftrightarrow X^*\left(e^{-j\Omega}\right). \tag{15.3.26}$$

Therefore, if a signal is time reversed (folded with respect to the $y$-axis) its magnitude spectrum remains unchanged and the phase spectrum changes its sign.

**Differentiation in the frequency domain.** Differentiation in the frequency domain corresponds to multiplication by $n$ in the time domain, that is,

$$x(n) \leftrightarrow X\left(e^{j\Omega}\right) \Rightarrow nx(n) \leftrightarrow j\frac{dX\left(e^{j\Omega}\right)}{d\Omega}. \tag{15.3.27}$$

This can be obtained by differentiating both sides of the expression for the discrete-time Fourier transform.

**Convolution of signals in the time domain.** The convolution (denoted by convolution operator '*' in this book) of two signals $x_1(n)$ and $x_2(n)$ corresponds to the product of their discrete-time Fourier transforms $X_1(e^{j\Omega})$ and $X_2(e^{j\Omega})$, that is,

$$\begin{aligned} x_1(n) &\leftrightarrow X_1\left(e^{j\Omega}\right) \quad \text{and} \quad x_2(n) \leftrightarrow X_2\left(e^{j\Omega}\right) \\ &\Rightarrow x_1(n)*x_2(n) \leftrightarrow X_1\left(e^{j\Omega}\right)\cdot X_2\left(e^{j\Omega}\right) \end{aligned} \tag{15.3.28}$$

This property is widely used in digital signal processing and in the analysis of linear-time-invariant discrete communication systems. From this property, it is easy to find the response $y(n)$ of a linear-time-invariant system to an input $x(n)$ in both the time and the frequency domains. If the impulse response of a linear-time-invariant system is $h(n)$, then the output

$y(n)$, for the given input $x(n)$, is the convolution of the input and the system impulse response. Consequently, the output in the frequency domain is the product of the input and the system impulse response in the frequency domain, that is,

$$Y\left(e^{j\Omega}\right) = H\left(e^{j\Omega}\right) \cdot X\left(e^{j\Omega}\right), \tag{15.3.29}$$

where $H(e^{j\Omega})$ is the frequency response of the system.

**Multiplication of signals in the time domain.** The multiplication of the two time-domain signals $x_1(n)$ and $x_2(n)$ corresponds to the periodic (or circular) convolution (denoted by $\otimes$) of their discrete-time Fourier transforms $X_1(e^{j\Omega})$ and $X_2(e^{j\Omega})$ scaled by the inverse of the period, that is,

$$x_1(n) \leftrightarrow X_1\left(e^{j\Omega}\right) \quad \text{and} \quad x_2(n) \leftrightarrow X_2\left(e^{j\Omega}\right)$$

$$\Rightarrow x_1(n) \cdot x_2(n) \leftrightarrow \frac{1}{2\pi} X_1\left(e^{j\Omega}\right) \otimes X_2\left(e^{j\Omega}\right) = \frac{1}{2\pi} \int_{2\pi} X_1\left(e^{j\lambda}\right) \cdot X_2\left(e^{j(\Omega-\lambda)}\right) d\lambda \tag{15.3.30}$$

The convolution here is called 'periodic', due to the fact that it is the convolution of two periodic signals with the same period. The result is convolution that has the same period as the two signals. In summary, the multiplication of aperiodic sequences is equivalent to the periodic convolution of their discrete-time Fourier transforms.

**Accumulation.** The accumulation operation on a signal $x(n)$ is a discrete-time counterpart to continuous-time integration, and can be expressed as the convolution of the signal and the unit step function:

$$y(n) = \sum_{k=-\infty}^{n} x(k) = x(n) * u(n). \tag{15.3.31}$$

In the frequency domain, this expression corresponds to the multiplication of the Fourier transforms of the signal by the step unit function, that is,

$$Y\left(e^{j\Omega}\right) = X\left(e^{j\Omega}\right) U\left(e^{j\Omega}\right) = X\left(e^{j\Omega}\right)\left[\frac{1}{1-e^{-j\Omega}} + \pi\delta\left(\Omega\right)\right],$$

$$= X\left(e^{j\Omega}\right)\frac{1}{1-e^{-j\Omega}} + \pi X\left(e^{j0}\right)\delta\left(\Omega\right) \tag{15.3.32}$$

where $X\left(e^{j0}\right)$ must be finite for $Y\left(e^{j\Omega}\right)$ to exist. The output signal $y(n)$ has a DC component equal to $1/2X\left(e^{j0}\right)$, except for the case $X\left(e^{j0}\right) = 0$. This relation, which shows only one period, can be extended to the periodic expression for the discrete-time Fourier transform of the signal $x(n)$.

**Parseval's theorem.** The energy of a finite energy signal $x(n)$ can be calculated in the time domain as

$$E_x = \sum_{n=-\infty}^{\infty} |x(n)|^2,$$

or in the frequency domain as the integral of the energy spectral density

$$E_x = \frac{1}{2\pi} \int_{2\pi} \left| X\left(e^{j\Omega}\right) \right|^2 d\Omega.$$

The energy of a signal calculated in the time domain is equal to the energy calculated in the frequency domain. The calculated values are also equal to the value of the related autocorrelation function $R_{xx}(l)$ for the lag $l = 0$, that is,

$$E_x = \frac{1}{2\pi} \int_{2\pi} \left| X\left(e^{j\Omega}\right) \right|^2 d\Omega = \sum_{n=-\infty}^{\infty} |x(n)|^2 = R_{xx}(0). \tag{15.3.33}$$

The theorem can be extended to two signals.

**The Wiener–Khintchine theorem.** For the real energy signal $x(n)$, the energy spectral density and autocorrelation function are a Fourier transform pair, that is,

$$R_x(l) \leftrightarrow S_x(\Omega). \tag{15.3.34}$$

This important theorem is widely used in the theory of communication systems, and elsewhere. Neither the energy spectral density nor the autocorrelation function contains any information about the phase of the related signal. Therefore, it is impossible to reconstruct the signal with either of them alone. The cross-correlation function and the related cross power spectral density can be used to get some information about the signal phase.

**The correlation theorem.** The cross-correlation function and the related cross power spectral density function form a Fourier transform pair according to the expression

$$R_{xy}(l) \leftrightarrow S_{xy}(\Omega) = X(\Omega)\, Y(-\Omega).$$

A special case of this theorem is the Wiener–Khintchine theorem.

## 15.3.4 Tables for the Discrete-Time Fourier Transform

Properties of the discrete-time Fourier transform are presented in Table 15.3.1, and basic discrete-time Fourier transforms are presented in Table 15.3.2. We use here the simplified notation $X\left(e^{j\Omega}\right) = X(\Omega)$.

**Table 15.3.1** *Properties of the discrete-time Fourier transform.*

| Theorem | Time Domain | Frequency Domain |
|---|---|---|
| Discrete-time Fourier series | $x(n) = \sum\limits_{k=0}^{N-1} c_k e^{+jk2\pi n/N}$ | $c_k = \frac{1}{N} \sum\limits_{n=0}^{N-1} x(n) e^{-jk2\pi n/N}$ |
| Discrete-time Fourier transform | $x(n) = \frac{1}{2\pi} \int_{-\pi}^{\pi} X\left(e^{j\Omega}\right) e^{j\Omega n} d\Omega$ $x(n) = \int_{-0.5}^{0.5} X(F) e^{j2\pi F\,n} dF$ | $X(\Omega) = \sum\limits_{n=-\infty}^{\infty} x(n) e^{-j\Omega n}$ $X(F) = \sum\limits_{n=-\infty}^{\infty} x(n) e^{-j2\pi F\,n}$ |
| Linearity ($a$ and $b$ arbitrary constants) | $ax(n) + by(n)$ | $aX(\Omega) + bY(\Omega)$ |
| Time delay | $x(n-l)$ | $X(\Omega) e^{-j\Omega\,l}$ |
| Decimation | $x(an)$ | $\|a\|^{-1} X(\Omega/a)$ |
| Frequency translation | $x(n) e^{+j\Omega_0 n}$ | $X(\Omega - \Omega_0)$ |
| Modulation | $x(n)\cos \Omega_0 n$ | $\frac{X(\Omega-\Omega_0)+X(\Omega+\Omega_0)}{2}$ |
| Time reversal | $x(-n)$ | $X(-\Omega)$ |
| Convolution | $x(n) * y(n)$ | $X(\Omega) \cdot Y(\Omega)$ |
| Multiplication | $x(n) \cdot y(n)$ | $\frac{1}{2\pi} X(\Omega) * Y(\Omega)$ |
| Conjugate | $x^*(n)$ | $X^*(-\Omega)$ |
| Parseval's theorem | $E = \sum\limits_{n=-\infty}^{\infty} \|x(n)\|^2$ | $E = \frac{1}{2\pi} \int_{-\pi}^{\pi} \|X(\Omega)\|^2 d\Omega$ $E = \int_{-1/2}^{1/2} \|X(F)\|^2 dF$ |

## 15.4 Discrete Fourier Transforms

### 15.4.1 Fundamentals of Frequency-Domain Sampling

The discrete-time Fourier transform $X(e^{j\Omega})$ cannot be directly applied to analyse real signals, for two basic reasons. Firstly, the definition of a discrete-time Fourier transform assumes that a discrete-time signal (data) $x(n)$ is known for every $n$ ($-\infty < n < \infty$). However, this is not the case in practice. Furthermore, the signal values (or the signal duration in time) are often relatively small. Secondly, although $X(e^{j\Omega})$ is a function of continuous frequency, a digital computer cannot be used to compute a continuum of fractional values. For these reasons, a discrete Fourier transform represents the spectrum of a signal $x(n)$ defined within a limited interval of values $n$. A discrete Fourier transform is an approximation of a discrete-time Fourier transform, which is easy to implement on a computer, and can also yield exact results in some cases. In discrete communication systems, we dealing with signals that are limited in time and frequency domains, and efficient processing of these signals is of particular importance.

**Table 15.3.2** *Table of the basic discrete-time Fourier transforms*

| Signal | Time Domain | Frequency Domain |
|---|---|---|
| Sine and cosine | $A \cdot \cos \Omega_0 n$ | $A\pi \left[ \delta \left( \Omega - \Omega_0 \right) + \delta \left( \Omega + \Omega_0 \right) \right]$ |
| | | $A \frac{\delta(F-F_0)+\delta(F+F_0)}{2}$ |
| | $A \cdot \sin \Omega_0 n$ | $jA\pi \left[ \delta \left( \Omega - \Omega_0 \right) - \delta \left( \Omega + \Omega_0 \right) \right]$ |
| | | $jA \frac{\delta(F-F_0)-\delta(F+F_0)}{2}$ |
| Impulse delta | $\delta(n)$ | $1, -\pi \leq \Omega \leq \pi$, and periodic |
| Delayed impulse | $\delta \left( n - n_0 \right)$ | $e^{-j\Omega n_0}$ |
| Exponential function | $e^{j\Omega_0 n}$ | $2\pi \sum\limits_{k=-\infty}^{k=\infty} \delta \left( \Omega - \Omega_0 - 2\pi k \right)$ |
| Sum of exponentials | $\sum\limits_{k} A_k e^{j\Omega_k n}$ | $2\pi \sum\limits_{k} A_k \delta \left( \Omega - \Omega_k \right)$ |
| Constant | $A, -\infty < n < \infty$ | $2\pi A \sum\limits_{k=-\infty}^{k=\infty} \delta \left( \Omega - 2\pi k \right)$ |
| Asymmetric rectangular pulse of amplitude $A$ and duration $N$ | $x(n) = A$<br>for $\quad 0 \leq n < N$ | $A e^{-j\Omega(N-1)/2} \frac{\sin N\Omega/2}{\sin \Omega/2}$ |
| Symmetric rectangular pulse of amplitude $A$ and duration $N$ | $x(n) = A$<br>$-N+1 \leq n \leq N-1$ | $A \frac{\sin \Omega(N-1/2)}{\sin \Omega/2}$ |
| Symmetric triangular pulse of amplitude $A = 1$ and duration $(2N-1)$ non-zeros | $(1 + n/N),$<br>$n \leq 0$<br>$(1 - n/N),$<br>$n \geq 0$<br>$-N \leq n \leq N$ | $X(\Omega) = \frac{\sin^2 \Omega N/2}{\sin^2 \Omega/2}$<br>$X(\Omega) = 1 + 2 \sum\limits_{n=1}^{N} (1 - n/N) \cos \Omega N$ |
| Radiofrequency discrete pulse | $s(n) = A \cos \Omega_c n$<br>$0 \leq n \leq N-1$ | $S(\Omega) = \frac{1}{2} \left[ M \left( \Omega - \Omega_c \right) + M \left( \Omega + \Omega_c \right) \right]$<br>$\frac{1}{2} M \left( \Omega \pm \Omega_c \right) =$<br>$\frac{A}{2} e^{-j(\Omega \pm \Omega_c) \frac{N-1}{2}} \frac{\sin(\Omega \pm \Omega_c)N/2}{\sin(\Omega \pm \Omega_c)/2}$ |

## 15.4.2   Discrete Fourier Transforms

Recall that the function $X(e^{j\Omega})$ is periodic over the interval $2\pi$, (i.e. $0 < \Omega < 2\pi$), as depicted in Figure 15.4.1 for a hypothetical signal.

Suppose $X(e^{j\Omega})$ is sampled at the uniformly spaced frequencies $\Omega_k = k\Delta\Omega = 2\pi k/N$ for integer values $k = 0, 1, \ldots, N - 1$. Then, the sampling interval is $\Delta\Omega = 2\pi/N$, and the following expression for the discrete-time Fourier transform,

$$X\left(e^{j\Omega}\right) = \sum_{n=-\infty}^{\infty} x(n) e^{-j\Omega n}, \tag{15.4.1}$$

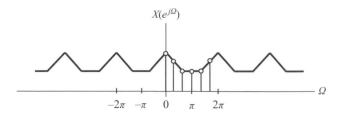

**Fig. 15.4.1** *Sampling of an arbitrary amplitude spectrum.*

becomes a function of the integer $k$,

$$X\left(e^{j\Omega_k}\right) = X\left(e^{j2\pi k/N}\right) = X\left(\frac{2\pi k}{N}\right) = \sum_{n=-\infty}^{\infty} x(n)e^{j2\pi nk/N} = X(k),$$

for $k = 0, \ldots, N - 1$. Sometimes this expression is denoted by $X(k)$, as above. Let the summation be segmented into an infinite number of subsummations, where each subsum contains $N$ terms, which results in

$$X\left(\frac{2\pi}{N}k\right) = \cdots + \sum_{n=-N}^{-1} x(n)e^{-j2\pi nk/N} + \sum_{n=0}^{N-1} x(n)e^{-j2\pi nk/N} + \sum_{n=N}^{2N-1} x(n)e^{-j2\pi nk/N} + \cdots$$

$$= \sum_{r=-\infty}^{\infty} \sum_{n=rN}^{n=rN+N-1} x(n)e^{-j2\pi nk/N}$$

where the integer $r = \ldots, -1, 0, 1, 2, \ldots$ . If the index $n$ is replaced by $n - rN$ in the inner summation, and the order of summation is interchanged, we obtain

$$X\left(\frac{2\pi}{N}k\right) = \sum_{r=-\infty}^{\infty} \sum_{n=0}^{N-1} x(n-rN)\, e^{-j2\pi(n-rN)k/N}$$

$$= \sum_{r=-\infty}^{\infty} \sum_{n=0}^{N-1} x(n-rN)\, e^{-j2\pi nk/N} e^{+j2\pi rNk/N}, \qquad (15.4.2)$$

$$= \sum_{n=0}^{N-1} \left[ \sum_{r=-\infty}^{\infty} x(n-rN) \right] e^{-j2\pi nk/N}$$

for $k = 0, \ldots, N - 1$ and $e^{-j2\pi rk} = 1$. Let the signal

$$x_p(n) = \sum_{r=-\infty}^{\infty} x(n-rN) \qquad (15.4.3)$$

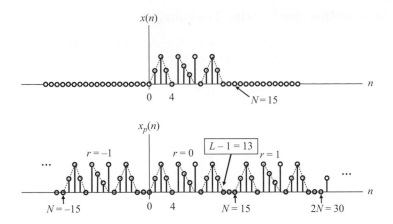

**Fig. 15.4.2** *Signal x(n) (upper panel) and its periodic extension $x_p(n)$ (lower panel).*

be called the *periodic extension* of the discrete signal $x(n)$, as shown in Figure 15.4.2. The signal is obviously a periodic repetition of $x(n)$ every $N$ samples with the fundamental period $N$. Now, the discrete Fourier transform of this signal is

$$X\left(\frac{2\pi}{N}k\right) = \sum_{n=0}^{N-1} x_p(n)e^{-j2\pi nk/N}. \tag{15.4.4}$$

If the signal $x(n)$ is defined in a finite interval, that is, $x(n)$ is defined for $0 \le n \le L - 1$ where $L$ is a finite number, then the following two cases can be distinguished:

- If $L \le N$, then adjacent periods of $x_p(n)$ do not overlap with each other, which can be expressed as

$$x(n) = \begin{cases} x_p(n), & n \in [0, N-1] \\ 0 & \text{otherwise} \end{cases}, \tag{15.4.5}$$

  as shown in Figure 15.4.2. Thus, $x(n)$ can be recovered from $x_p(n)$.
- If $L > N$, then adjacent periods of $x_p(n)$ overlap with each other, and *time-domain aliasing* occurs, which can be expressed as

$$x(n) = x_p(n), \text{for } n \in [0, N-1]. \tag{15.4.6}$$

In this case, it is not possible to recover $x(n)$ from its periodic extension $x_p(n)$. The relationship between $x(n)$ and $x_p(n)$ is shown in Figure 15.4.2, for the case when $L = N - 1 = 14$.

## 15.4.3   Inverse Discrete Fourier Transforms

The periodic signal $x_p(n)$ can be found from its discrete Fourier transform as follows. If the expression for $X(2\pi k/N)$ is multiplied by $e^{+j2\pi mk/N}$, we obtain

$$
\begin{aligned}
X\left(\frac{2\pi}{N}k\right)e^{+j2\pi mk/N} &= \sum_{n=0}^{N-1} x_p(n)e^{-j2\pi nk/N}e^{+j2\pi mk/N} \\
&= \sum_{n=0}^{N-1} x_p(n)e^{+j2\pi(m-n)k/N}
\end{aligned}
\tag{15.4.7}
$$

Summing both sides over $k = 0, 1, \ldots, N-1$, we find

$$
\sum_{k=0}^{N-1} X\left(\frac{2\pi}{N}k\right)e^{+j2\pi mk/N} = \sum_{k=0}^{N-1}\sum_{n=0}^{N-1} x_p(n)e^{+j2\pi(m-n)k/N} = \sum_{n=0}^{N-1} x_p(n)\sum_{k=0}^{N-1} e^{+j2\pi(m-n)k/N}.
$$

Because

$$
\sum_{k=0}^{N-1} e^{+j2\pi(m-n)k/N} = \frac{1-e^{+j2\pi(m-n)}}{1-e^{+j2\pi(n-m)/N}} = \left\{ \begin{matrix} N, & m=n \\ 0 & m \neq n \end{matrix} \right\},
\tag{15.4.8}
$$

we obtain

$$
\sum_{k=0}^{N-1} X\left(\frac{2\pi k}{N}\right)e^{+j2\pi mk/N} = Nx_p(m),
\tag{15.4.9}
$$

or

$$
x_p(m) = \frac{1}{N}\sum_{k=0}^{N-1} X\left(\frac{2\pi}{N}k\right)e^{j2\pi mk/N}.
\tag{15.4.10}
$$

Thus, by replacing $m$ with $n$, a discrete Fourier transform pair can be obtained for a periodic function $x_p(n)$ as

$$
X\left(\frac{2\pi}{N}k\right) = \sum_{n=0}^{N-1} x_p(n)e^{-j2\pi nk/N},
\tag{15.4.11}
$$

and

$$
x_p(n) = \frac{1}{N}\sum_{k=0}^{N-1} X\left(\frac{2\pi}{N}k\right)e^{j2\pi nk/N}.
\tag{15.4.12}
$$

These expressions can be related to the expressions for discrete-time Fourier transforms given by eqns (15.3.6) and (15.3.7).

## Example

The spectrum of the causal exponential signal $x(n)$, expressed by

$$x(n) = a^n u(n),$$

is sampled at frequencies $\Omega_k = 2k\pi/N$, for $k = 0, 1, \ldots, N-1$. Determine the reconstructed spectra for $a = 0.8$, and $N = 5$, and $N = 50$.

**Solution.** The discrete-time Fourier transform of this signal is

$$X\left(e^{j\Omega}\right) = \sum_{n=-\infty}^{\infty} x(n)e^{-j\Omega n} = \sum_{n=0}^{\infty} a^n e^{-j\Omega n} = \frac{1}{1 - ae^{-j\Omega}} = |X\left(e^{j\Omega}\right)| \arg X\left(e^{j\Omega}\right).$$

For $a = 0.8$, the signal $x(n)$ and its spectrum $X\left(e^{j\Omega}\right)$ are shown at the top of Figure 15.4.3. Then, for the sampling interval $\Delta\Omega = 2\pi/N$, this function is sampled at $N$ equidistant frequencies $\Omega_k = 2\pi k/N$, resulting in $N$ spectral samples expressed as

$$X\left(e^{j2\pi k/N}\right) = X\left(\frac{2\pi}{N}k\right) = \sum_{n=-\infty}^{\infty} x(n)e^{-j2\pi nk/N} = \frac{1}{1 - ae^{-j2\pi k/N}},$$

where $k = 0, 1, \ldots, N-1$, as shown in the two lower graphs in Figure 15.4.3, for $N = 5$ and $N = 50$, respectively. The inverse discrete Fourier transform gives the periodic sequence $x_p(n)$, which corresponds to these frequency samples as

$$x_p(n) = \frac{1}{N}\sum_{k=0}^{N-1} X\left(\frac{2\pi}{N}k\right)e^{j2\pi nk/N}.$$

We can quantify the aliasing in the following way. From the expression $x(n) = a^n u(n)$, we obtain

$$x_p(n) = \sum_{r=-\infty}^{\infty} x(n-rN) = \sum_{r=-\infty}^{\infty} a^{n-rN} u(n-rN) = \sum_{r=-\infty}^{0} a^{n-rN} = a^n \sum_{r=0}^{\infty} a^{rN} = \frac{a^n}{1-a^N},$$

$0 \le n \le N-1$, where the denominator represents the effect of aliasing. Since $0 < a < 1$, the error of aliasing tends towards 0 as $N$ tends to $\infty$. For $a = 0.8$, the aliased signal $x_p(n)$ and the corresponding spectral samples $X\left(e^{j2\pi k/N}\right)$ are shown in Figure 15.4.3, for $N = 5$ and $N = 50$. The aliasing effects become negligible for $N = 50$. If the aliasing finite-duration signal $x_a(n)$ is expressed as

$$x_a(n) = \begin{cases} x_p(n) & 0 \le n \le N-1 \\ 0 & \text{otherwise} \end{cases},$$

then its discrete-time Fourier transform is

$$X_a\left(e^{j\Omega}\right) = \sum_{n=0}^{N-1} x_a(n)e^{-j\Omega n} = \sum_{n=0}^{N-1} x_p(n)e^{-j\Omega n} = \sum_{n=0}^{N-1} \frac{a^n}{1-a^N}e^{-j\Omega n} = \frac{1}{1-a^N}\frac{1-a^N e^{-j\Omega N}}{1-ae^{-j\Omega}}.$$

Although the discrete Fourier transforms for the signal $x(n)$ and its aliasing signal $x_a(n)$ are different, that is, $X\left(e^{j\Omega}\right) \neq X_a\left(e^{j\Omega}\right)$, their sample values are identical, that is,

$$X_a\left(\frac{2\pi}{N}k\right) = \frac{1}{1-a^N}\frac{1-a^N \cdot 1}{1-ae^{-j2\pi k/N}} = \frac{1}{1-ae^{-j2\pi k/N}} = X\left(\frac{2\pi}{N}k\right).$$

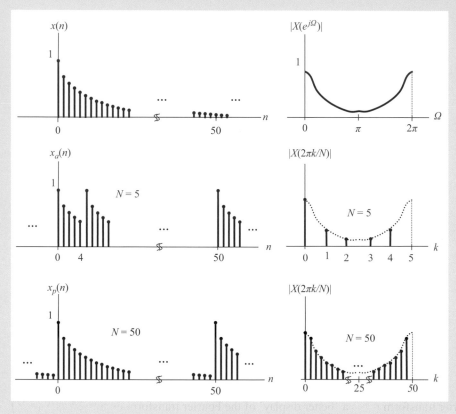

**Fig. 15.4.3** *Discrete-time Fourier transform (left) and discrete Fourier transform (right) of an exponential signal represented in its initial form (upper panels) and then in two periodic forms defined by N = 5 (middle panels) and N = 50 (lower panels).*

## 15.4.4 Three Typical Cases of Discrete Fourier Transforms

In the preceding section, the frequency-domain sampling of an aperiodic, finite-energy, discrete-time signal $x(n)$ is analysed. Generally, uniformly spaced samples do not uniquely represent the original signal $x(n)$. The frequency samples correspond to a periodic signal $x_p(n)$ of period $N$, where $x_p(n)$ is a time-domain *aliasing* version of $x(n)$, that is,

$$x_p(n) = \sum_{r=-\infty}^{\infty} x(n-rN). \qquad (15.4.13)$$

In this section, we will analyse three cases which will be specified by the characteristics of the signal $x(n)$ with respect to the signal $x_p(n)$.

**First case.** The signal $x(n)$ has a finite duration $L$, where $L \leq N$. For this case, the periodic signal $x_p(n)$ is simply a periodic repetition of $x(n)$, where $x_p(n)$ over a single period is given by

$$x_p(n) = x(n), \text{for every } n : n \in [0, N-1], \qquad (15.4.14)$$

and the discrete Fourier transform is equivalent to the sampled discrete-time Fourier transform,

$$X\left(\frac{2\pi}{N}k\right) = \sum_{n=0}^{N-1} x_p(n)e^{-j2\pi nk/N} = X\left(e^{j\Omega}\right)\Bigg|_{\Omega=2\pi k/N}, \qquad (15.4.15)$$

for $k = 0, 1, \ldots, N-1$, that is, the frequency samples of a discrete-time Fourier transform uniquely represent the discrete Fourier transform. Also, because $x(n)$ is equivalent to $x_p(n)$ over a single period that is padded by $N - L$ zeros, the original signal is

$$x(n) = x_p(n) = \frac{1}{N}\sum_{k-0}^{N-1} X\left(\frac{2\pi}{N}k\right)e^{j2\pi nk/N}. \qquad (15.4.16)$$

The zero padding does not provide any additional information about the spectrum $X\left(e^{j\Omega}\right)$ of the signal $x(n)$. However, padding with $N - L$ zeros and computing the $N$-point discrete Fourier transform gives a 'better display' of the Fourier transform expressed as $X\left(e^{j\Omega}\right)$. The case when $(N - L) = 1$ is shown in Figure 15.4.2. Therefore, a finite duration signal $x(n)$ of length $L$ has the discrete-time Fourier transform

$$X\left(e^{j\Omega}\right) = \sum_{n=0}^{L-1} x(n)e^{-j\Omega n}. \qquad (15.4.17)$$

If $X\left(e^{j\Omega}\right)$ is sampled at equally spaced frequencies $\Omega = 2k\pi/N$, for $k = 0, 1, \ldots, N-1$, and $N \geq L$, the resulting samples are

$$X\left(e^{j\Omega}\right) = X\left(\frac{2\pi}{N}k\right) = \sum_{n=0}^{L-1} x(n)e^{-j2\pi nk/N}. \tag{15.4.18}$$

Because $L \leq N$, the upper index in the sum can be increased from $L-1$ to $N-1$, resulting in

$$X\left(\frac{2\pi}{N}k\right) = \sum_{n=0}^{N-1} x(n)e^{-j2\pi nk/N}. \tag{15.4.19}$$

Therefore, padding with $N-L$ zeros and computing the $N$-point discrete Fourier transform gives a 'better display' of the Fourier transform of $X\left(e^{j\Omega}\right)$ by presenting $N$ values of the spectrum. According to this formula, the signal $x(n)$ of length $L \leq N$ is transformed into a sequence of frequency samples of length $N$. Since this formula is obtained from the discrete-time Fourier transform $X\left(e^{j\Omega}\right)$, it is referred to as a *discrete Fourier transform*.

The corresponding *inverse discrete Fourier transform* can be obtained from the previously mentioned formula as

$$x(n) = \frac{1}{N} \sum_{k=0}^{N-1} X\left(\frac{2\pi}{N}k\right) e^{j2\pi nk/N}. \tag{15.4.20}$$

Then, the discrete Fourier transform pair is equivalent to the pair in eqn (15.4.11) and is expressed as

$$X\left(\frac{2\pi}{N}k\right) = \sum_{n=0}^{N-1} x(n)e^{-j2\pi\,n\,k/N}$$

and

$$x(n) = \frac{1}{N} \sum_{k=0}^{N-1} X\left(\frac{2\pi}{N}k\right) e^{j2\pi\,n\,k/N}. \tag{15.4.21}$$

**Second case.** The signal $x(n)$ is of either an infinite or a finite length $L$ for $L > N$. For this case, we obtain

$$x_p(n) \neq x(n), \tag{15.4.22}$$

for every $n$: $n \in [0, N-1]$. However, the periodic function $x_p(n)$ can be represented as a periodic repetition of a reduced-length version of $x(n)$, as depicted in Figure 15.4.4 for reductions to $N = 5$ and $N = 9$. Then the sampled version of the Fourier transform of $x(n)$,

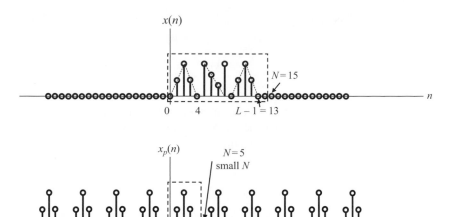

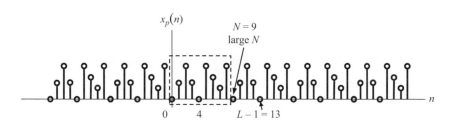

**Fig. 15.4.4** *Original signal (upper panel) and its reduced repetitions (lower panels).*

expressed by $X\left(e^{j\Omega}\right)$, cannot be equated with the frequency samples of the discrete-time Fourier transform of $x_p(n)$, that is,

$$X\left(\frac{2\pi k}{N}\right) = \sum_{n=0}^{N-1} x_p(n)e^{-j2\pi nk/N} \neq X\left(e^{j\Omega}\right)\Big|_{\Omega = 2\pi k/N}. \qquad (15.4.23)$$

However, the value of $N$ can be arbitrarily increased to bring the discrete-time Fourier transform close to the discrete Fourier transform.

**Third case.** The signal $x(n)$ is periodic with a period of $N$. This case reduces to the first case, because one period of $x(n)$ is equal to one period of $x_p(n)$.

In summary, every signal $x(n)$ can be represented in the frequency and time domains (with arbitrarily accuracy) using by the *discrete Fourier transform pair*

$$X\left(\frac{2\pi}{N}k\right) = \sum_{n=0}^{N-1} x(n)e^{-j2\pi\,n\,k/N},$$

and

$$x(n) = \frac{1}{N} \sum_{k=0}^{N-1} X(k)\, e^{j2\pi\, n\, k/N}, \tag{15.4.24}$$

where the first formula is the discrete Fourier transform and the second one is the inverse discrete Fourier transform. The first equation is the analysis equation, while the second equation is the corresponding synthesis equation.

**Example**

Find the discrete Fourier transform of the asymmetric rectangular pulse expressed by

$$p(n) = u(n) - u(n-L), \tag{15.4.25}$$

and presented in Figure 15.4.5. Determine the $N$-point discrete Fourier transform of this signal for $N \geq L$.

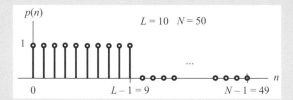

**Fig. 15.4.5** *Asymmetric rectangular pulse with L = 10.*

**Solution.** The discrete-time Fourier transform exists and is given by

$$P\left(e^{j\Omega}\right) = \sum_{n=-\infty}^{\infty} p(n)e^{-j\Omega n} = \sum_{n=0}^{L-1}\left(e^{-j\Omega}\right)^{n} = \frac{1-e^{-j\Omega L}}{1-e^{-j\Omega}} = e^{-j\Omega(L-1)/2}\frac{\sin(\Omega L/2)}{\sin(\Omega/2)}. \tag{15.4.26}$$

The corresponding magnitude spectrum and phase spectrum are shown in Figure 15.4.6 for the case $L = 10$ and $N = 50$. The zeros occur for $\sin(\Omega L/2) = 0$, that is, $\Omega L/2 = k\pi$ and $\Omega = 2k\pi/L$ for integers $k = 1, 2, \ldots, L-1$. For $L = 10$, we may have $\Omega = \pm 2k\pi/10 = \pm k\pi/5$, and $k = 1, 2, \ldots, 9$, and the phase discontinuities of $\pi$ radians occur at the same frequencies. The $N$-point discrete Fourier transform of $x(n)$ is

$$P\left(\frac{2\pi}{N}k\right) = P\left(e^{j\Omega}\right)\Big|_{\Omega = 2\pi k/N} = e^{-j\pi k(L-1)/N}\frac{\sin(\pi kL/N)}{\sin(\pi k/N)}. \tag{15.4.27}$$

Now, three different cases will be considered for different values of $L$ and $N$:

- $N = L$: there exists a single discrete Fourier transform value that is different from 0, that is,

$$P\left(\frac{2\pi}{N}k\right) = \sum_{n=0}^{L-1} e^{-j\frac{2\pi}{N}kn} = \begin{cases} L \cdot 1 = L, & k = 0 \\ 0 & k = 1, 2, \ldots, L-1 \end{cases};$$  (15.4.28)

- $L = 10, N = 50$:

$$P\left(\frac{2\pi}{50}k\right) = e^{-j\pi k(L-1)/50} \frac{\sin(\pi kL/50)}{\sin(\pi k/50)}, \text{for } k = 0, 1, 2, \ldots, 49;$$  (15.4.29)

- $L = 10, N = 100$:

$$P\left(\frac{2\pi}{100}k\right) = e^{-j\pi k(L-1)/100} \frac{\sin(\pi kL/100)}{\sin(\pi k/100)}, \text{for } k = 0, 1, 2, \ldots, 99.$$  (15.4.30)

With the 40 zero padded to get $N = 50$ points, the discrete spectrum follows the continuous-frequency spectrum obtained by the discrete-time Fourier transform quite well. If the padding goes to $N = 100$, the discrete spectrum will be very close to the continuous spectrum obtained by the discrete-time Fourier transform.

## 15.5 Algorithms for Discrete Fourier Transforms

The theory related to the Fourier analysis of discrete-time signals has been described in the previous sections. Based on this theory, the algorithms for the fast calculation of the discrete Fourier transform, primarily Goertzel's algorithm and the fast Fourier transform, will be presented in detail, due to their wide application in practice.

### 15.5.1 Goertzel's Algorithm

The discrete Fourier transform is defined by the expression

$$X(k) = X\left(e^{j\Omega}\right)\Big|_{\Omega = 2\pi k/N} = \sum_{n=0}^{N-1} x(n)e^{-j2\pi nk/N} = X\left(\frac{2\pi}{N}k\right),$$  (15.5.1)

for $0 \le k \le N-1$. Because a discrete-time signal of a finite length is always absolutely summable, its discrete Fourier transform $X(k)$ can be represented by samples from the corresponding $z$-transform $X(z)$, that is,

$$X(k) = X(z)\Big|_{\Omega = 2\pi \kappa/N} = \sum_{n=0}^{N-1} x(n)e^{-j2\pi nk/N},$$  (15.5.2)

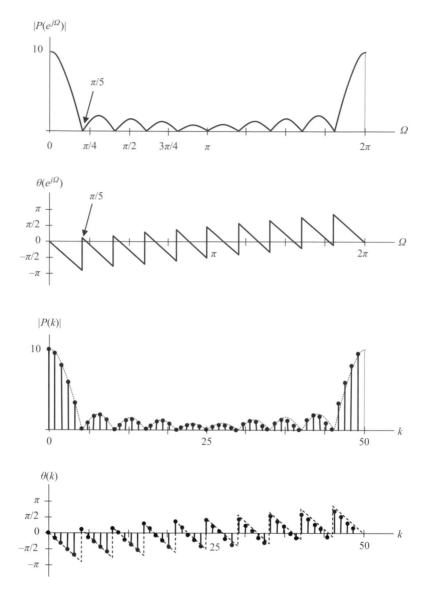

**Fig. 15.4.6** *Discrete-time Fourier transform (upper panels) and discrete Fourier transform (lower panels) magnitude and phase spectra for N = 50.*

for $0 \leq k \leq N - 1$. The total number of calculations needed to compute an $N$-point discrete Fourier transform increases faster than the squared value of the number of points does. Namely, the computation of each discrete Fourier transform value $X(k)$ requires $N$ complex multiplications and $N - 1$ complex additions, or, to calculate $N$ discrete Fourier transform values, we need $N^2$ complex multiplications (or $4N^2$ real multiplications) and $N(N-1)$ complex additions (or $N(4N-2)$ real additions). In order to reduce the number of calculations, various recursive algorithms were developed.

In this section, Goertzel's algorithm will be presented. This algorithm does not need $N$ to be an integer power of 2, and the resonant frequency can be any value between 0 and the sampling frequency. This algorithm does not need to store a block of data before processing can begin, that is, the processing can start with the first input-time sample. There is also no need for bit reversal. Goertzel's algorithm can be implemented in the form of an infinite-impulse-response filter. The filter has two real feedback coefficients and one complex feed-forward coefficient. Bearing in mind the definition of the phase factor

$$W_N = W_N^1 = e^{-j2\pi/N},\tag{15.5.3}$$

we can define the function

$$W_N^{-kn} = e^{j2\pi kn/N},\tag{15.5.4}$$

and then derive the identity

$$W_N^{-kN} = e^{j2\pi k} = 1,\tag{15.5.5}$$

which will be used to develop Goertzel's algorithm. Namely, we can now express the discrete Fourier transform in the form

$$X(k) = \sum_{i=0}^{N-1} x(i)\, e^{-j2\pi ik/N} = W_N^{-kN} \sum_{i=0}^{N-1} x(i)W_N^{ik} = \sum_{i=0}^{N-1} x(i)W_N^{-k(N-i)}.\tag{15.5.6}$$

The discrete Fourier transform can also be expressed as a convolution. To prove that, let us define a new sequence to be

$$y_k(k) = \sum_{i=0}^{n} x_t(i)W_N^{-k(n-i)},\tag{15.5.7}$$

which is a direct convolution of the causal sequence, defined as

$$x_t(n) = \begin{cases} x(n) & 0 \le n \le N-1 \\ 0 & n < 0, n \ge N \end{cases},\tag{15.5.8}$$

with the causal sequence

$$h_k(n) = \begin{cases} W_N^{-kn} & n \ge 0 \\ 0 & n < 0 \end{cases},\tag{15.5.9}$$

due to

$$y_k(n) = \sum_{i=-\infty}^{\infty} x_t(i)h(n-i) = \sum_{i=0}^{n} x(i)W_N^{-k(n-i)}. \tag{15.5.10}$$

Comparing expressions for $X(k)$ and $y_k(n)$, we can see that

$$X(k) = y_k(n)\Big|_{for\ n=N}. \tag{15.5.11}$$

The $z$-transform of $y_k(n)$ is

$$Y_k(z) = \frac{X_t(z)}{1 - W_N^{-k} \cdot z^{-1}}. \tag{15.5.12}$$

Thus, if the expressions for $X_t(z)$ and $Y_k(z)$ are defined as the input and output, respectively, of a linear-time-invariant filter, the transfer function of the initially relaxed filter is

$$H_k(z) = \frac{Y_k(z)}{X_t(z)} = \frac{1}{1 - W_N^{-k} \cdot z^{-1}}, \tag{15.5.13}$$

which is equal to the $z$-transform of the causal sequence $h_k(n)$, as defined in eqn (15.5.9). The output of the filter in the time domain can be expressed in the form

$$y_k(n) = x_t(n) + W_N^{-k} \cdot y_k(n-1), \tag{15.5.14}$$

where $x_t(N) = 0$, and $y_k(-1) = 0$. The structure of this filter is shown in Figure 15.5.1.

From the computational point of view, this recursive structure gives, in some way, a disappointing result. Namely, to calculate one value of $X(k) = y_k(N)$, we need $4N$ real multiplications and $4N$ real additions. Consequently, to calculate all $N$ values of $X(k)$, we need $4N^2$ real multiplications and $4N^2$ real additions. Therefore, by using this algorithm, the number of real additions is $2N$ times greater than in the case of direct calculations. However, the basic advantage of the recursive algorithm is that we do not need to compute or store in advance the $N$ complex coefficients, as we needed for the direct discrete Fourier transform.

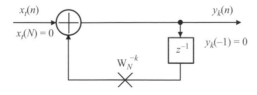

**Fig. 15.5.1** *Calculation of the kth discrete Fourier transform value. Two symbols are used here: one for the multiplier (X) and one for the adder (+).*

In order to increase the computational efficiency, Goertzel's algorithm can be modified. The modification will be performed on the transfer function, which will be expressed as

$$H_k(z) = \frac{Y_k(z)}{X_t(z)} = \frac{1}{1 - W_N^{-k} \cdot z^{-1}} = \frac{1 - W_N^k \cdot z^{-1}}{\left(1 - W_N^k \cdot z^{-1}\right)\left(1 - W_N^{-k} \cdot z^{-1}\right)}$$

$$= \frac{1 - W_N^k \cdot z^{-1}}{1 - 2\cos\left(\frac{2\pi\,k}{N}\right) \cdot z^{-1} + z^{-2}}.$$

This is the second-order filter, which can be represented by its direct-form structure. Namely, this filter can be implemented as a cascade of two subfilters with the following transfer functions:

$$H_1(z) = \frac{V_k(z)}{X_t(z)} = Y_k(z) = \frac{1}{1 - 2\cos\left(2\pi k/N\right) \cdot z^{-1} + z^{-2}} \qquad (15.5.15)$$

and

$$H_2(z) = \frac{Y_k(z)}{V_k(z)} = X_t(z) = 1 - W_N^k \cdot z^{-1} \qquad (15.5.16)$$

Obviously, the transfer function is a product of the transfer functions of the cascaded subfilters, that is,

$$H_k(z) = H_1(z) \cdot H_2(z). \qquad (15.5.17)$$

From the first transfer function, we obtain

$$V_k(z) = X_t(z) + V_k(z) \cdot 2\cos\left(\frac{2\pi k}{N}\right) \cdot z^{-1} - V_k(z)z^{-2}, \qquad (15.5.18)$$

which can be represented in the time domain as an intermediate variable:

$$v_k(n) - x_t(n) + 2\cos\left(2\pi k/N\right) \cdot v_k\left(n-1\right) - v_k\left(n-2\right). \qquad (15.5.19)$$

From the second transfer function, we obtain

$$Y_k(z) = V_k(z)\left(1 - W_N^k \cdot z^{-1}\right) = V_k(z) - V_k(z)W_N^k \cdot z^{-1}, \qquad (15.5.20)$$

which results in a time-domain representation in the form

$$y_k(n) = v_k(n) - W_N^k \cdot v_k\left(n-1\right). \qquad (15.5.21)$$

The structure of this filter is shown in Figure 15.5.2.

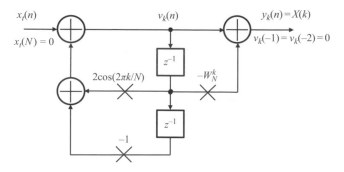

**Fig. 15.5.2** *Cascade from a second-order filter.*

This modified Goertzel's algorithm requires $2(N + 2)N$ real multiplications and $4(N + 1)N$ real additions to calculate the $N$-point discrete Fourier transform. This is still a relatively large number of operations. In order to reduce this number, the following simplifications can be applied.

If the transfer function is $H_k(z)$, the filter structure is as shown in Figure 15.5.2. However, if this function is $H_{N-k}(z)$, then the filter structure changes the feedback path to have the multiplier $2\cos[2\pi(N - k)/N]$ instead of $2\cos(2\pi k/N)$. Consequently, we find that $v_{N-k}(n) = v_k(n)$. Thus, there is no need to compute the intermediate variables to find $X(N - k)$ if these variables have been already computed for determining $X(k)$. In addition, we must change the multiplier in the feed-forward path to be $W_N^{N-k} = W_N^{-k}$. Therefore, in order to calculate the two values $X(N - k)$ and $X(k)$, we need $2(N + 4)$ real multiplications and $4(N + 2)$ real additions or, to calculate $N$ values of the discrete Fourier transform, we need $N^2$ real multiplications and approximately $N^2$ real additions. In conclusion, the modified algorithm reduces the number of real multiplication to one-fourth of the ones needed in the calculations for the direct discrete Fourier transform, and halves the number of additions needed.

## 15.5.2 Discrete Fourier Transforms as Linear Transformations

The discrete Fourier transform and the inverse discrete Fourier transform can be viewed as linear transformations of the discrete-time signal sequence $\{x(n)\}$ and the discrete Fourier transform sequence $\{X(k)\}$. These sequences can be expressed in matrix form, and the discrete Fourier transform and the inverse discrete Fourier transform can be represented by linear transforms. The discrete Fourier transform pair can be expressed in a simpler form as

$$X(k) = \sum_{n=0}^{N-1} x(n) e^{-j2\pi nk/N} = \sum_{n=0}^{N-1} x(n) W_N^{nk}, \qquad (15.5.22)$$

where $k = 0, 1, \ldots, N - 1$, and

$$x(n) = \frac{1}{N} \sum_{k=0}^{N-1} X(k) W_N^{-nk}, \tag{15.5.23}$$

for $n = 0, 1, \ldots, N - 1$, where the common term

$$W_N = W_N^1 = e^{-j2\pi/N} \tag{15.5.24}$$

is, by definition, the phase factor equal to the $N$th root of unity. Therefore, the problem to solve here is how to compute the sequence $\{X(k)\}$ of $N$ complex-valued numbers for the given sequence of data $\{x(n)\}$. From the discrete Fourier transform pair, it is obvious that the direct computation of each point of the discrete Fourier transform can be accomplished by $N$ complex multiplications per output sample, and $(N - 1)$ complex additions. Hence, the $N$ discrete Fourier transform values can be computed in $N^2$ complex multiplications and $N(N - 1)$ complex additions. Direct computation is not time efficient. However, using the symmetry and periodicity properties of the phase factor, expressed as

$$W_N^{k+N/2} = -W_N^k, \tag{15.5.25}$$

and

$$W_N^{k+N} = W_N^k, \tag{15.5.26}$$

respectively, the computation efficiency can be significantly improved. Let $\mathbf{x}_N$, $\mathbf{X}_N$ be the $N$-point vectors that define signal sequences $\{x(n)\}$ and $\{X(k)\}$, respectively, and $\mathbf{W}_N$ be the $N \times N$ matrix of the linear transformation, expressed as

$$\mathbf{x}_N = \begin{bmatrix} x(0) \\ x(1) \\ \vdots \\ x(N-1) \end{bmatrix}, \quad \mathbf{X}_N = \begin{bmatrix} X(0) \\ X(1) \\ \vdots \\ X(N-1) \end{bmatrix},$$

and

$$\mathbf{W}_N = \begin{bmatrix} 1 & 1 & 1 & 1 & 1 \\ 1 & W_N^1 & W_N^2 & \cdots & W_N^{N-1} \\ 1 & W_N^2 & W_N^4 & \cdots & W_N^{2(N-1)} \\ & & & \cdots & \\ \vdots & \vdots & \vdots & \cdots & \vdots \\ & & & \cdots & \\ 1 & W_N^{N-1} & W_N^{2(N-1)} & \cdots & W_N^{(N-1)(N-1)} \end{bmatrix}, \tag{15.5.27}$$

respectively. Hence, the $N$-point discrete Fourier transform may be expressed in matrix form as

$$\mathbf{X}_N = \mathbf{W}_N \mathbf{x}_N. \tag{15.5.28}$$

If the inverse of $\mathbf{W}_N$ exists, then the inverse discrete Fourier transform may be obtained by left multiplication with the inverse of $\mathbf{W}_N$, and expressed as

$$\mathbf{x}_N = \mathbf{W}_N^{-1} \mathbf{X}_N = \frac{1}{N} \mathbf{W}_N^* \mathbf{X}_N, \tag{15.5.29}$$

where the last expression follows from the expression for $x(n)$. From this equation, the inverse and complex conjugate matrices of $\mathbf{W}_N$, denoted by $\mathbf{W}_N^{-1}$ and $\mathbf{W}_N^*$, respectively, are inter-related as

$$\mathbf{W}_N^{-1} = \frac{1}{N} \mathbf{W}_N^*, \tag{15.5.30}$$

which implies that

$$\mathbf{W}_N \mathbf{W}_N^* = N \mathbf{I}_N, \tag{15.5.31}$$

where $\mathbf{I}_N$ is an $N \times N$ identity matrix. Therefore, the linear transformation $\mathbf{W}_N$ is an orthogonal matrix.

## 15.5.3   The Radix-2 Fast Fourier Transform Algorithm

The fast Fourier transform is not a new Fourier transform. It is, in fact, a fast algorithm for computing the discrete Fourier transform. This algorithm is based on the decomposition of the complex $N$-point discrete Fourier transform calculations into a calculation of the simpler discrete Fourier transforms that have sizes less than $N$. In this chapter, the basic concept of this algorithm is explained. If $N$ can be expressed as

$$N = r^\alpha, \tag{15.5.32}$$

then the discrete Fourier transforms are of size equal to $r$, where $r$ is called the radix of the fast Fourier transform algorithm. Thus, the computation of the $N$-point discrete Fourier transform has a regular pattern. Let us explain the *radix-2 algorithm* that is defined by $r = 2$. In this case, the discrete Fourier transform can be expressed in the form

$$X(k) = \sum_{n=0}^{N-1} x(n) e^{-j2\pi nk/N} = \sum_{n=0}^{N-1} x(n) W_N^{nk} = \sum_{m=0}^{N/2-1} x(2m) W_N^{2mk} + \sum_{m=0}^{N/2-1} x(2m+1) W_N^{(2m+1)k},$$

where $k = 0, 1, \ldots, N - 1$. Since

$$W_N^{2nk} = e^{-j2\pi 2nk/N} = e^{-j2\pi nk/(N/2)} = W_{N/2}^{nk}, \tag{15.5.33}$$

and

$$W_N^{(2m+1)k} = W_N^k W_N^{2mk} = W_N^k W_{N/2}^{mk}, \tag{15.5.34}$$

or simply

$$W_N^2 = W_{N/2}^1, \tag{15.5.35}$$

by replacing $m$ with $n$, we can obtain the following simplified expression for the discrete Fourier transform:

$$X(k) = \sum_{n=0}^{N/2-1} x(2n) W_{N/2}^{nk} + W_N^k \sum_{n=0}^{N/2-1} x(2n+1) W_{N/2}^{nk} = X_e^{N/2}(k) + W_N^k X_o^{N/2}(k),$$

where $W_N^k$ are *twiddle factors*. According to this procedure, an $N$-point discrete Fourier transform has been split into two $N/2$-point discrete Fourier transforms. The fast Fourier transform algorithm based on this procedure is called the *decimation-in-time algorithm*. Following the same decimation procedure, $N/2$-point discrete Fourier transforms can be split into $N/4$-point discrete Fourier transforms, and so on, until we are left with only two-point discrete Fourier transforms. The second step, for example, results in

$$
\begin{aligned}
X(k) &= X_e^{N/2}(k) + W_N^k X_o^{N/2} \\
&= \left[ \sum_{p=0}^{N/4-1} x(4p) W_{N/4}^{pk} + W_{N/2}^k \sum_{p=0}^{N/4-1} x(4p+2) W_{N/4}^{pk} \right] \\
&\quad + W_N^k \left[ \sum_{p=0}^{N/4-1} x(4p+1) W_{N/4}^{pk} + W_{N/2}^k \sum_{p=0}^{N/4-1} x(4p+3) W_{N/4}^{pk} \right] \\
&= \left[ X_{e1}^{N/4}(k) + W_{N/2}^k X_{e2}^{N/4}(k) \right] + W_N^k \left[ X_{o1}^{N/4}(k) + W_{N/2}^k X_{o2}^{N/4}(k) \right]
\end{aligned} \tag{15.5.36}
$$

Since the expressions $X_e^{N/2}(k)$ and $X_o^{N/2}(k)$ are periodic, with the period $N/2$, we obtain

$$X_e^{N/2}(k) = X_e^{N/2}(k+N/2), \tag{15.5.37}$$

and

$$X_o^{N/2}(k) = X_o^{N/2}(k+N/2). \tag{15.5.38}$$

Because $W_N^{k+N/2} = -W_N^k$, as can be proved by

$$W_N^{k+N/2} = e^{-j2\pi(k+N/2)/N} = e^{-j2\pi k/N} e^{-j2\pi(N/2)/N}$$
$$= e^{-j2\pi k/N} \cdot (-1) = -W_N^k, \tag{15.5.39}$$

the expression for $X(k)$ can be obtained in the forms

$$X(k) = X_e^{N/2}(k) + W_N^k X_o^{N/2}(k), \quad k = 0, 1, \ldots, N/2 - 1, \tag{15.5.40}$$

and

$$X(k+N/2) = X_e^{N/2}(k) - W_N^k X_o^{N/2}(k), \quad k = 0, 1, \ldots, N/2 - 1. \tag{15.5.41}$$

Therefore, the expression for the discrete Fourier transform is

$$X(k) = \begin{cases} X(k) = X_e^{N/2}(k) + W_N^k X_o^{N/2}(k) & k = 1, 2, \ldots, N/2 - 1 \\ X(k+N/2) = X_e^{N/2}(k) - W_N^k X_o^{N/2}(k) & k = 1, 2, \ldots, N/2 - 1 \end{cases}. \tag{15.5.42}$$

**Demonstration of the algorithm.** We will demonstrate the fast Fourier transform algorithm for the example of computing an $N = 8$-point discrete Fourier transform. For this example, the algorithm has the following three stages:

1. For $N = 8$, according to the general formula (15.5.36), we can express the discrete Fourier transform as

$$X(k) = \sum_{n=0}^{3} x(2n) W_4^{nk} + W_8^k \sum_{n=0}^{3} x(2n+1) W_4^{nk} = X_e^4(k) + W_8^k X_o^4(k)$$

$$= \begin{cases} X(k) = X_e^4(k) + W_8^k X_o^4(k) & k = 0, 1, 2, 3 \\ X(k+4) = X_e^4(k) - W_8^k X_o^4(k) & k = 0, 1, 2, 3 \end{cases}. \tag{15.5.43}$$

In this expression, the curly brackets are used to separate the two four-point groups of the discrete Fourier transform. Thus, we split an eight-point discrete Fourier transform into two four-point discrete Fourier transforms. For example, the first value of the discrete Fourier transform can be found as

$$X(0) = X_e^4(0) + W_8^0 X_o^4(0), \tag{15.5.44}$$

unlike the fourth value, which is expressed as

$$X(4) = X_e^4(0) - W_8^0 X_o^4(0). \tag{15.5.45}$$

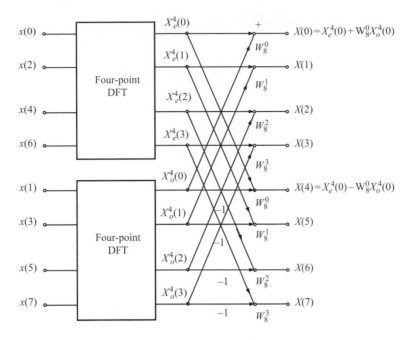

**Fig. 15.5.3** *The first step of the fast Fourier transform calculation.*

The only difference between the two terms is in the sign in front of the second addend. The expressions of all of the other terms can be obtained in a similar way. The procedure for calculating all eight expressions is illustrated by the scheme in Figure 15.5.3. The plus sign on the right-hand side indicates that the value from the upper four-point discrete Fourier transform is added to the value obtained from the lower four-point discrete Fourier transform that was multiplied by the positive twiddle factor. The number $-1$ at the output of the lower four-point discrete Fourier transform indicates that the multiplication is performed with the negative value of the corresponding twiddle factor.

2. We then split each four-point discrete Fourier transform into two two-point discrete Fourier transforms. Let us demonstrate this splitting for the even term:

$$X_e^4(k) = \sum_{n=0}^{3} x(2n) W_4^{nk} = \sum_{p=0}^{1} x(4p) W_2^{pk} + W_4^k \sum_{p=0}^{1} x(4p+2) W_2^{pk}$$

$$= X_{e1}^2(k) + W_4^k X_{e2}^2(k) = \left\{ \begin{array}{ll} X_e^4(k) = X_{e1}^2(k) + W_4^k X_{e2}^2(k) & k = 0, 1 \\[2mm] X_e^4(k+2) = X_{e1}^2(k) - W_4^k X_{e2}^2(k) & k = 0, 1 \end{array} \right\}.$$

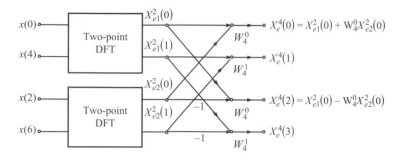

**Fig. 15.5.4** *The second step of the fast Fourier transform calculation; DFT, discrete Fourier transform.*

The two-point discrete Fourier transforms generate output in a manner that is similar to the way the four-point discrete Fourier transforms generate output, as is shown in Figure 15.5.4.

3. Next, we split each two-point discrete Fourier transform into two one-point discrete Fourier transforms. The even terms are calculated as

$$
X_{e1}^2(k) = \sum_{p=0}^{1} x(4p) W_2^{pk} = X_{e1,1}^1(k) + W_2^k X_{e1,2}^1(k)
$$

$$
= \left\{
\begin{array}{ll}
X_{e1}^2(k) = X_{e1,1}^1(k) + W_2^k X_{e1,2}^1(k) & k = 0 \\[2mm]
X_{e1}^2(k+1) = X_{e1,1}^1(k) - W_2^k X_{e1,2}^1(k) & k = 0
\end{array}
\right\}.
\qquad (15.5.46)
$$

Thus, even terms can be obtained for the terms defined by $k = 0$ and $k = 1$, as

$$
X_{e1}^2(k) = \left\{
\begin{array}{l}
X_{e1}^2(0) = X_{e1,1}^1(0) + W_2^0 X_{e1,2}^1(0) = x(0) + W_2^0 x(4) \\[2mm]
X_{e1}^2(1) = X_{e1,1}^1(0) - W_2^0 X_{e1,2}^1(0) = x(0) - W_2^0 x(4)
\end{array}
\right\}.
\qquad (15.5.47)
$$

Furthermore, because

$$
W_2^0 = e^{-j0} = 1,
$$

we obtain

$$
X_{e1}^2(k) = \left\{
\begin{array}{l}
X_{e1}^2(0) = x(0) + x(4) \\[2mm]
X_{e1}^2(1) = x(0) - x(4)
\end{array}
\right\},
\qquad (15.5.48)
$$

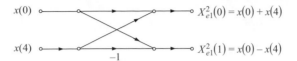

**Fig. 15.5.5** *Butterfly structure of the algorithm in the third step.*

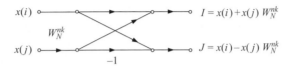

**Fig. 15.5.6** *General structure of a butterfly.*

because the initial two-point discrete Fourier transforms require coefficients of only $+1$ or $-1$. This calculations are schematically represented in Figure 15.5.5. The fast Fourier transform algorithm consists of pairs of computations of the form shown in Figure 15.5.6. These pairs are called radix-2 butterflies. For each butterfly, two complex multiplications are needed. Because

$$W_{N/2}^1 = e^{-j2\pi N/2} = e^{-jN\pi} = -1,$$

only one complex multiplication per butterfly is required. Thus, for $N/2$ butterflies per stage and $\log_2 N$ stages, the total number of complex multiplications in this case is

$$(N/2)\log_2 N = 4\log_2 8 = 4 \cdot 3 = 12, \tag{15.5.49}$$

which is less than the $N^2 = 64$ multiplications that would be required by a direct Fourier transform. Also, the fast Fourier transform can be calculated *in place* in memory, because only $N$ memory locations, plus some working registers, can be used for calculations. The direct fast Fourier transform requires $2N$ memory locations. For the sake of completeness, the entire procedure of the fast Fourier transform is shown in Figure 15.5.7.

## 15.6 Correlation and Spectral Densities of Discrete-Time Signals

### 15.6.1 Cross-Correlation and Correlation of Real-Valued Energy Signals

This section presents an extension of Section 14.1.3, where the basic theory related to the correlation of discrete-time deterministic energy signals is presented. The notation presented in this section is equivalent to the notation used in this book for the energy signals: an index

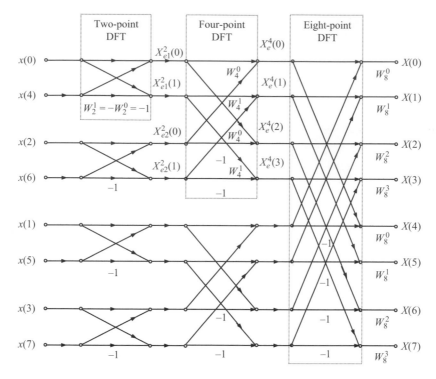

**Fig. 15.5.7** *Summary of the radix-2 algorithm; DFT, discrete Fourier transform.*

containing one small letter (for a deterministic signal) is used to designate the energy signal autocorrelation function. The discrete-time Fourier transform mentioned in this chapter, in principle, exists for energy signals and give us information about the signal amplitudes in the frequency domain. The cross-correlation between any two signals is the measure of the degree to which the two signals are similar. Suppose these two signals are the real, energy, discrete-time signals $x(n)$ and $y(n)$. The cross-correlation of $x(n)$ and $y(n)$ is a function defined as

$$R_{xy}(l) = \sum_{n=-\infty}^{\infty} x(n)y(n-l), \quad l = 0, \pm 1, \pm 2, \ldots, \tag{15.6.1}$$

or, equivalently, the function

$$R_{xy}(l) = \sum_{n=-\infty}^{\infty} x(n+l)y(n), \quad l = 0, \pm 1, \pm 2, \ldots. \tag{15.6.2}$$

In relation (15.6.1), the parameter $l$ defines the time shift of the discrete signal $y(n)$ with respect to the unshifted signal $x(n)$. If $l$ is positive, the signal $y(n)$ is shifted to the right for positive value of parameter $l$ and to the left for negative $l$ values. Similarly, for the reversed order of the signals, that is, the cross-correlation of $y(n)$ and $x(n)$ is a function defined as

$$R_{yx}(l) = \sum_{n=-\infty}^{\infty} y(n)x(n-l), \tag{15.6.3}$$

for time shifts $l = 0, \pm 1, \pm 2, \ldots$, or, equivalently,

$$R_{yx}(l) = \sum_{n=-\infty}^{\infty} y(n+l)x(n), \tag{15.6.4}$$

for time shifts $l = 0, \pm 1, \pm 2, \ldots$. A simple comparison of the expressions for $R_{xy}(l)$ and $R_{yx}(l)$ results in their inter-relation, which is expressed as

$$R_{xy}(l) = R_{yx}(-l). \tag{15.6.5}$$

Therefore, the cross-correlation $R_{xy}(l)$ is a time-reversed version of the cross-correlation $R_{yx}(l)$. For this reason, and due to the properties of the cross-correlation function, any of these expressions provides the same information with respect to the similarity of the two signals $x(n)$ and $y(n)$.

Suppose now that the two signals are equal, that is, $x(n) = y(n)$. In this case, we can investigate the self-similarity of the signal $x(n)$ by defining the *autocorrelation function* as

$$R_x(l) = \sum_{n=-\infty}^{\infty} x(n)x(n-l), \tag{15.6.6}$$

or, equivalently, as

$$R_x(l) = \sum_{n=-\infty}^{\infty} x(n+l)x(n), \tag{15.6.7}$$

for time shifts $l = 0, \pm 1, \pm 2, \ldots$. The values of the autocorrelation function at the lag $l = 0$ is equal to the energy of the signal, that is,

$$R_x(0) = \sum_{n=-\infty}^{\infty} |x(n)|^2 = E_x. \tag{15.6.8}$$

Suppose the two signals are of finite durations, that is, $x(n) = y(n) = 0$ for $n < 0$ and $n \geq N$. In this case, the cross-correlation and autocorrelation functions are defined as

$$R_{xy}(l) = \sum_{n=i}^{N-|k|-1} x(n)y(n-l), \qquad (15.6.9)$$

and

$$R_x(l) = \sum_{n=i}^{N-|k|-1} x(n)x(n-l), \qquad (15.6.10)$$

where $i = l$, $k = 0$ for $l \geq 0$, and $i = 0$, $k = l$ for $l < 0$.

## 15.6.2   Cross-Correlation and Correlation of Real-Valued Power Signals

The notation presented in this section is equivalent to the notation used in this book for the power signals: an index containing two small letters (for a deterministic signal) is used to designate the power signal functions. Suppose two signals $x(n)$ and $y(n)$ are real power signals, which have, by definition, infinite energy but limited power. The cross-correlation of $x(n)$ and $y(n)$ is defined as

$$R_{xy}(l) = \lim_{M \to \infty} \frac{1}{2M+1} \sum_{n=-M}^{M} x(n)y(n-l). \qquad (15.6.11)$$

The averaging of the cross-correlation is performed in an infinite interval to find the averaged value of the cross-correlation function. If the two signals are equal, that is, $x(n) = y(n)$, we may define the autocorrelation as

$$R_{xx}(l) = \lim_{M \to \infty} \frac{1}{2M+1} \sum_{n=-M}^{M} x(n)x(n-l). \qquad (15.6.12)$$

The averaging of the autocorrelation is performed in an infinite interval to find the averaged value of the autocorrelation. This power is greater than 0, and the energy of the signal is infinite, as follows from the definition of a power signal.

In the special case where signals $x(n)$ and $y(n)$ are two *periodic signals*, and consequently power signals, each with period $N$, the cross-correlation and autocorrelation are defined as functions of the signal shift $l$ as

$$R_{xy}(l) = \frac{1}{N} \sum_{n=0}^{N-1} x(n)y(n-l), \qquad (15.6.13)$$

and

$$R_{xx}(l) = \frac{1}{N} \sum_{n=0}^{N-1} x(n)x(n-l). \qquad (15.6.14)$$

These functions are periodic sequences with period $N$. Unlike energy signals, where the value of the autocorrelation function at the lag $l = 0$ is equal to the energy of the signal, the value of the autocorrelation for the power signal is equal to the power of the signal, that is,

$$R_{xx}(0) = \lim_{M \to \infty} \frac{1}{2M+1} \sum_{n=-M}^{M} |x(n)|^2 = P_x. \tag{15.6.15}$$

**Example**

Calculate the autocorrelation function for the signal $x(n) = A \cdot \sin(\Omega_c n)$, using the expression for a periodic signal. Find the power of the signal.

**Solution.** Because the sine signal is periodic, the autocorrelation function is

$$R_{xx}(l) = \frac{1}{N} \sum_{n=0}^{N-1} x(n)x(n-l) = \frac{1}{N} \sum_{n=0}^{n=N-1} [A \sin \Omega_c n \cdot A \sin \Omega_c (n-l)]$$

$$= \frac{1}{N} A^2 \sum_{n=0}^{n=N-1} \frac{1}{2} [\cos \Omega_c l - \cos \Omega_c (2n+l)] = \frac{1}{N} \frac{A^2}{2} \sum_{n=0}^{n=N-1} \cos \Omega_c l + 0 = \frac{A^2}{2} \cos \Omega_c l$$

The power is

$$P = R_{xx}(0) = \frac{A^2}{2}.$$

**Example**

Calculate the autocorrelation function of the cosine signal expressed as $y(n) = A \cdot \cos(\Omega_c n)$, using the definition of the function as the mean value of the autocorrelation function calculated in the infinite interval.

**Solution.** The autocorrelation function is

$$R_{xx}(l) = \lim_{M \to \infty} \frac{1}{2M+1} \sum_{n=-M}^{M} y(n)y(n-l) = \lim_{M \to \infty} \frac{A^2}{2M+1} \sum_{n=-M}^{M} \cos(\Omega_c n)\cos(\Omega_c (n-l))$$

$$= \lim_{M \to \infty} \frac{A^2}{2M+1} \sum_{n=-M}^{M} \frac{1}{2} \cos(\Omega_c l) + 0 = \lim_{M \to \infty} \frac{A^2/2}{2M+1} (2M+1)\cos(\Omega_c l) = \frac{A^2}{2} \cos(\Omega_c l)$$

As expected, both calculations gave the same results, due to the periodicity of the signals and the definition of the power signals.

### 15.6.3   Parseval's Theorem and the Wiener–Khintchine Theorem

The energy of a finite energy signal $x(n)$ is

$$E_x = \sum_{n=-\infty}^{\infty} |x(n)|^2, \qquad (15.6.16)$$

which can be expressed in terms of the spectral characteristics, which are expressed using the amplitude spectral density $X(\Omega)$, that is,

$$E_x = \sum_{n=-\infty}^{\infty} x(n)x^*(n) = \sum_{n=-\infty}^{\infty} x(n)\frac{1}{2\pi}\int_{-\pi}^{\pi} X^*\left(e^{j\Omega}\right) e^{-j\Omega n} d\Omega.$$

Due to the linearity of integration and summation, we can interchange their order in the following way:

$$E_x = \frac{1}{2\pi}\int_{-\pi}^{\pi} X^*\left(e^{j\Omega}\right)\left[\sum_{n=-\infty}^{\infty} x(n)\, e^{-j\Omega n}\right] d\Omega = \frac{1}{2\pi}\int_{-\pi}^{\pi} X^*\left(e^{j\Omega}\right)X\left(e^{j\Omega}\right) d\Omega$$
$$= \frac{1}{2\pi}\int_{-\pi}^{\pi} \left|X\left(e^{j\Omega}\right)\right|^2 d\Omega \qquad , \quad (15.6.17)$$

which is Parseval's relation for discrete-time aperiodic signals with finite energy. The quantity

$$S_x\left(\Omega\right) = \left|X\left(e^{j\Omega}\right)\right|^2 = E_x\left(\Omega\right) \qquad (15.6.18)$$

is the *energy spectral density* of $x(n)$ because it represents the distribution of energy as a function of frequency. Obviously, this quantity does not contain any phase information.

The Wiener–Khintchine theorem relates the energy spectral density of a signal to its autocorrelation function. The correlation of two signals $x_1(n)$ and $x_2(n)$ in the time domain corresponds to the product of the discrete-time Fourier transform $X_1(e^{j\Omega})$ of the first signal, and the frequency-reversed discrete-time Fourier transform of the second signal $X_2(e^{j\Omega})$, that is,

$$x_1(n) \leftrightarrow X_1\left(e^{j\Omega}\right) \text{ and } x_2(n) \leftrightarrow X_2\left(e^{j\Omega}\right) \Rightarrow R_{x_1x_2}(n) \leftrightarrow X_1\left(e^{j\Omega}\right)\cdot X_2\left(e^{-j\Omega}\right). \quad (15.6.19)$$

Let us prove this theorem. If we multiply both sides of the expression for the cross-correlation by the exponential function $\exp\left(-j\Omega n\right)$ and then sum both sides over all $n$ values, we obtain

$$\sum_{n=-\infty}^{\infty} R_{x_1x_2}(n)\, e^{-j\Omega n} = \sum_{n=-\infty}^{\infty}\left[\sum_{k=-\infty}^{\infty} x_1(k)x_2\left(k-n\right)\right] e^{-j\Omega n}. \qquad (15.6.20)$$

Then, if we interchange the order of the summation and change the discrete variable $(k - n)$ by $l$ inside the inner summation, expression (15.6.20) becomes

$$
\sum_{k=-\infty}^{\infty} x_1(k) \sum_{n=-\infty}^{\infty} x_2(k-n) e^{-j\Omega n} = \sum_{k=-\infty}^{\infty} x_1(k) \sum_{l=-\infty}^{\infty} x_2(l) e^{-j\Omega (k-l)}
$$

$$
= \sum_{k=-\infty}^{\infty} x_1(k) e^{-j\Omega k} \sum_{l=-\infty}^{\infty} x_2(l) e^{+j\Omega l} = X_1\left(e^{j\Omega}\right) X_2\left(e^{-j\Omega}\right) = S_{x_1 x_2}(\Omega)
$$

(15.6.21)

The expression $S_{x_1 x_2}(\Omega)$ is called the cross energy density function or the cross energy spectral density. In the special case when the two signals $x_1(n)$ and $x_2(n)$ are real and equivalent, that is, $x_1(n) = x_2(n) = x(n)$, the cross energy spectral density $S_{x_1 x_2}(\Omega)$ is equal to the energy spectral density $S_x(\Omega)$, that is,

$$
S_x(\Omega) = \sum_{n=-\infty}^{\infty} R_x(n) e^{-j\Omega n} = X\left(e^{j\Omega}\right) X\left(e^{-j\Omega}\right) = X\left(e^{j\Omega}\right) X^*\left(e^{j\Omega}\right) = \left| X\left(e^{j\Omega}\right) \right|^2.
$$

(15.6.22)

The corresponding autocorrelation function is the inverse discrete-time Fourier transform of the energy spectral density, that is,

$$
R_x(l) = \frac{1}{2\pi} \int_{-\pi}^{\pi} S_x(\Omega) e^{+j\Omega l} d\Omega.
$$

(15.6.23)

Thus, *the Wiener–Khintchine theorem states that the energy spectral density is the Fourier transform of the autocorrelation function.* For this reason, we can say that the autocorrelation function and the energy spectral density function contain the same information about the signal $x(n)$ they represent. Because there is no phase information in these two functions, they cannot be used to reconstruct the signal in the time domain.

## 15.6.4 Comprehensive Analysis of Discrete Linear-Time-Invariant Systems

### 15.6.4.1 System Presentation

The following standard notation will be used in the analysis of deterministic discrete linear-time-invariant systems. The system itself will be defined by the following:

$h(n)$: the impulse response in the time domain

$H(\Omega)$: the impulse response in the frequency domain, or amplitude spectral density

$R_h(l)$: the autocorrelation function of the impulse response, or system correlation function

$S_h(\Omega)$: the energy spectral density of the impulse response.

A linear-time-invariant system with deterministic *energy signals* at its input (one letter inside the subscripts) will be defined by the following:

$R_x(l)$, $R_y(l)$: the autocorrelation function of the input and output signals

$R_{xy}(l)$, $R_{yx}(l)$: the cross-correlation functions

$S_x(\Omega) = E_x(\Omega) = |X(\Omega)|^2$: the energy spectral density

$S_y(\Omega) = E_y(\Omega) = |Y(\Omega)|^2$: the energy spectral density of the output signal

$S_{xy}(\Omega)$: the cross energy spectral density.

A linear-time-invariant system with deterministic *power signals* at its input (two letters inside the subscripts) will be defined by the following:

$R_{xx}(l)$: the autocorrelation function

$R_{xy}(l)$, $R_{yx}(l)$: the cross-correlation functions

$S_{xx}(\Omega)$: the power spectral density of the input signal

$S_{yy}(\Omega) = P_{yy}(\Omega)$: the power spectral density of the output signal

$S_{xy}(\Omega)$: the cross power spectral density

A linear-time-invariant system for *deterministic signal* processing is defined by its impulse responses in the time and frequency domains, which are $h(n)$ and $H(\Omega)$, respectively, as shown in Figure 15.6.1. Then, it is possible for any input signal $x(n)$ to find the output signal in the time domain $y(n)$ as the convolution of the input signal $x(n)$ and the impulse response $h(n)$, or to find the output signal in the frequency domain $Y(\Omega)$ as the multiplication of the input signal $X(\Omega)$ and the impulse response $H(\Omega)$. The power and energy of all signals in the system can be found in the time and frequency domains by using the expressions for instantaneous power and energy spectral densities. This concept of signal processing in a linear-time-invariant system is directly applicable for energy signals, because they fulfil the conditions for the existence of the Fourier transform.

Furthermore, it is relatively simple to find the autocorrelation function and the related energy spectral density function of the system and analyse its behaviour by using this representation. The relationship of these two functions is defined by the Fourier transform pair.

For power signals, which can have infinite energy, the Fourier transform cannot always be derived. A linear-time-invariant system is still defined by its impulse response, but the signal analysis can be done using the autocorrelation function and the power spectral density functions. In this case, the autocorrelation function is defined over a limited time interval and then averaged over the whole interval. The power spectral density function can be then derived as the discrete-time Fourier transform of the autocorrelation function. If the power signal is periodic, the autocorrelation function is calculated inside one period of the signal. The energy and power of the signal can be calculated in the time and frequency domains.

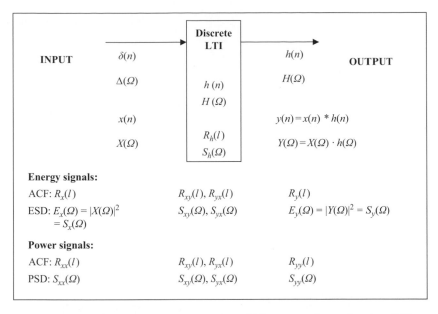

**Fig. 15.6.1** *Discrete-time linear-time-invariant system; ACF, autocorrelation function; ESD, energy spectrum density; LTI, linear-time-invariant system; PSD, power spectrum density.*

A block schematic of a linear-time-invariant system specified by its transfer function $H(\Omega)$, or the impulse response in the time domain $h(n)$, is presented in Figure 15.6.1, where all relevant signals and functions are identified, including time-domain signals, autocorrelation functions, and energy and power spectral densities.

Then, in the time domain, for any input signal $x(n)$ of the linear-time-invariant system, the output signal $y(n)$ is the convolution of the input signal $x(n)$ and the system impulse response $h(n)$, that is,

$$y(l) = \sum_{n=-\infty}^{\infty} x(n)h(n-l) = x(n) * h(n). \tag{15.6.24}$$

In the frequency domain, for any input signal $X(\Omega)$ of the linear-time-invariant system, the output signal $Y(\Omega)$ is the product of the input signal $X(\Omega)$ and the system impulse response $H(\Omega)$, that is,

$$Y(\Omega) = X(\Omega) \cdot H(\Omega). \tag{15.6.25}$$

The impulse response of the system is obtained for the input signal defined by the Kronecker delta function. The Fourier transform of this signal is

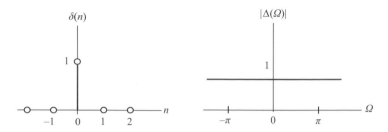

**Fig. 15.6.2** *Delta function (left) and its amplitude spectral density (right).*

$$\Delta(\Omega) = \int_{-\infty}^{\infty} \delta(n)e^{-j\Omega n}dt = e^{-j\Omega 0} = 1, \quad |\Omega| \le \pi. \tag{15.6.26}$$

The delta signal is presented in the time and frequency domains in Figure 15.6.2.

Therefore, the amplitude spectral density of the delta function has a unit value for all frequencies in the interval $|\Omega| \le \pi$. This function is periodic for all frequencies $|\Omega| \le \infty$ with the fundamental period $2\pi$.

The energy of the signal can be calculated from its energy spectral density as

$$E_\delta = \frac{1}{2\pi} \int_{-\pi}^{\pi} |\Delta(\Omega)|^2 d\Omega = \frac{1}{2\pi} \int_{-\pi}^{\pi} 1 \cdot d\Omega = 1, \tag{15.6.27}$$

or from the time-domain representation of the signal, bearing in mind the shifting property of the Kronecker delta function, as

$$E_\delta = \sum_{-\infty}^{\infty} \delta^2(n) = \sum_{-\infty}^{\infty} \delta(n)\delta(n) = \delta(0) = 1. \tag{15.6.28}$$

In order to analyse this system in terms of the impulse response, the autocorrelation function, the cross-correlation function, and the power spectral density function, the correlation and spectral density functions for power and energy signals will be defined and derived. The autocorrelation function of the signal is

$$R_h(l) = \sum_{n=-\infty}^{\infty} \delta(n)\delta(n-l) = \delta(l), \tag{15.6.29}$$

The energy spectral density of the impulse response is the Fourier transform of the autocorrelation function, expressed as

$$S_\delta(\Omega) = \sum_{l=-\infty}^{\infty} \delta(l)e^{-j\Omega l} = 1. \tag{15.6.30}$$

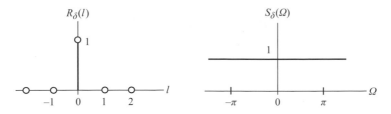

**Fig. 15.6.3** *Autocorrelation function (left) and energy spectral density (right) of a Kronecker delta impulse.*

The autocorrelation function and energy spectral density function are presented in Figure 15.6.3.

Therefore, the energy spectral density of the delta function has a unit value for all frequencies in the interval $|\Omega| \leq \pi$. The energy of the pulse is

$$E_\delta = \frac{1}{2\pi} \int_{-\pi}^{\pi} S_\delta(\Omega)\,d\Omega = \frac{1}{2\pi} \int_{-\pi}^{\pi} d\Omega = 1. \tag{15.6.31}$$

The average power is

$$P = \lim_{M \to \infty} \frac{1}{2M+1} \sum_{n=-M}^{M} \delta(n)\delta(n) = \lim_{M \to \infty} \frac{1}{2M+1}\delta(0) = 0. \tag{15.6.32}$$

The properties of the discrete-time linear-time-invariant system, defined by its impulse response, can be investigated using the concepts of correlation, energy spectral density, and power spectral density. These functions will be defined and derived for the input and output signals and for the impulse response. The important procedure of calculating the power and energy of the system's signals will be demonstrated. In particular, the relationship between the correlation functions of the input and output signals will be established. The correlation functions of the energy and power signals will be treated separately, due to their differences in definition and meaning. Namely, the correlation function of the energy signals is related to their energy spectral density, while the correlation function of the power signals is related to their power spectral density. In addition, we will present the basic expression for the complex signals.

### 15.6.4.2 Correlation and Power Spectral Density of Complex Energy Signals

As we have already mentioned, Parseval's theorem and the Wiener–Khintchine theorem relate the characteristics of a discrete-time signal in the time and frequency domains. Parseval's theorem relates the energy of the signal in the time domain to the energy of the signal in the frequency domain. If signal $x(n)$ is an *energy signal*, its autocorrelation can be expressed as

$$R_x(l) = \sum_{n=-\infty}^{\infty} x(n+l)x^*(n). \tag{15.6.33}$$

The value of the autocorrelation function at zero lag is equal to the energy of the signal, that is,

$$R_x(0) = \sum_{n=-\infty}^{\infty} |x(n)|^2 = E_x. \tag{15.6.34}$$

Parseval's theorem for discrete-time signals states that the energy of a signal calculated in the time domain is equal to the energy of the signal calculated in the frequency domain, which can be proven as follows. The energy of the signal can be expressed, using the amplitude spectral density $X(\Omega)$, as

$$E_x = \sum_{n=-\infty}^{\infty} x(n)x^*(n) = \sum_{n=-\infty}^{\infty} x(n)\frac{1}{2\pi}\int_{-\pi}^{\pi} X^*\left(e^{j\Omega}\right) e^{-j\Omega n}d\Omega. \tag{15.6.35}$$

Due to the linearity of integration and summation, we can interchange their order in the following way:

$$E_x = \frac{1}{2\pi}\int_{-\pi}^{\pi} X^*\left(e^{j\Omega}\right)\left[\sum_{n=-\infty}^{\infty} x(n)\,e^{-j\Omega n}\right]d\Omega = \frac{1}{2\pi}\int_{-\pi}^{\pi} X^*\left(e^{j\Omega}\right)X\left(e^{j\Omega}\right)d\Omega$$
$$= \frac{1}{2\pi}\int_{-\pi}^{\pi}\left|X\left(e^{j\Omega}\right)\right|^2 d\Omega \tag{15.6.36}$$

which is Parseval's relation for discrete-time aperiodic signals with finite energy. The quantity

$$S_x(\Omega) = \left|X\left(e^{j\Omega}\right)\right|^2 \tag{15.6.37}$$

is the *energy spectral density* of $x(n)$ because it represents the distribution of energy as a function of frequency. Obviously, this quantity does not contain any phase information.

**The autocorrelation theorem.** The Wiener–Khintchine theorem relates the correlation function of signals to their *energy spectral density*, stating that the Fourier transform of the autocorrelation function of a signal $x(n)$ is equal to the energy spectral density $S_x(\Omega)$ of $x(n)$, as can be seen from the following proof:

$$FT\{R_x(l)\} = \sum_{l=-\infty}^{\infty} R_x(l)e^{-j\Omega l} = \sum_{l=-\infty}^{\infty}\left[\sum_{n=-\infty}^{\infty} x(n+l)x^*(n)\right]e^{-j\Omega l}$$
$$= \sum_{l=-\infty}^{\infty} x^*(n)\left[\sum_{n=-\infty}^{\infty} x(n+l)\,e^{-j\Omega l}\right]_{n+l=k} = \sum_{l=-\infty}^{\infty} x^*(n)e^{j\Omega n}\left[\sum_{n=-\infty}^{\infty} x(k)e^{-j\Omega k}\right],$$
$$= X^*(\Omega)X(\Omega) = |X(\Omega)|^2 = S_x(\Omega)$$
$$\tag{15.6.38}$$

where FT stands for Fourier transform. Since $R_x(l)$ is real and even, the energy spectral density $S_x(\Omega)$ is purely real. The autocorrelation function $R_x(l)$ and the energy spectral density $S_x(\Omega)$ form *the Fourier transform pair,*

$$S_x(\Omega) = \sum_{l=-\infty}^{\infty} R_x(l)e^{-j\Omega\, l}, \tag{15.6.39}$$

$$R_x(l) = \frac{1}{2\pi} \int_{-\pi}^{\pi} S_x(\Omega)\, e^{+j\Omega\, l} d\Omega. \tag{15.6.40}$$

Therefore, the autocorrelation function and the energy spectral density function contain the same information about the signal $x(n)$ they represent. Because there is no phase information in these two functions, they cannot be used to reconstruct the signal in the time domain. The theory of Fourier transforms can be applied for various functions of $R_x(l)$ and $S_x(\Omega)$. For example, the energy spectral density of the impulse response is equal to the energy density or the squared value of system's magnitude response, expressed as

$$S_h(\Omega) = H^*(\Omega)H(\Omega) = |H(\Omega)|^2 = E_h(\Omega). \tag{15.6.41}$$

**The cross-correlation theorem.** The cross-correlation function is defined as the correlation between two different signals:

$$R_{xy}(l) = \sum_{n=-\infty}^{\infty} x(n+l)y^*(n). \tag{15.6.42}$$

The Fourier transform of the cross-correlation function of signals $x(n)$ and $y(n)$ is equal to their cross energy spectral density $S_{xy}(\Omega)$. The proofs are as follows:

$$FT\{R_{xy}(l)\} = \sum_{l=-\infty}^{\infty} R_{xy}(l)e^{-j\Omega l} = \sum_{l=-\infty}^{\infty} \left[\sum_{n=-\infty}^{\infty} x(n+l)y^*(n)\right]e^{-j\Omega l}$$

$$= \sum_{l=-\infty}^{\infty} y^*(n)\left[\sum_{n=-\infty}^{\infty} x(n+l)e^{-j\Omega l}\right] \underset{n+l=k}{=} \sum_{l=-\infty}^{\infty} y^*(n)e^{j\Omega n}\left[\sum_{n=-\infty}^{\infty} x(k)e^{-j\Omega k}\right],$$

$$= Y^*(\Omega)X(\Omega) = S_{xy}(\Omega)$$

$$\tag{15.6.43}$$

and

$$FT\{R_{yx}(l)\} = \sum_{l=-\infty}^{\infty} R_{yx}(l)e^{-j\Omega l} = \sum_{l=-\infty}^{\infty} \left[\sum_{n=-\infty}^{\infty} y(n+l)x^*(n)\right]e^{-j\Omega l}$$

$$= \sum_{l=-\infty}^{\infty} x^*(n)\left[\sum_{n=-\infty}^{\infty} y(n+l)e^{-j\Omega l}\right] \underset{n+l=k}{=} \sum_{l=-\infty}^{\infty} x^*(n)e^{j\Omega n}\left[\sum_{n=-\infty}^{\infty} y(k)e^{-j\Omega k}\right].$$

$$= X^*(\Omega)Y(\Omega) = S_{yx}(\Omega)$$

$$\tag{15.6.44}$$

Generally, $S_{xy}(\Omega)$ is a function of complex variables. Furthermore, these two functions are not the same, that is, $S_{xy}(\Omega) \neq S_{yx}(\Omega)$. Instead, we have

$$S_{xy}(\Omega) = S_{yx}^*(\Omega). \tag{15.6.45}$$

It is worth noting the following: for a complex energy signal, the value of the autocorrelation function at the origin, $R_x(0)$, represents its energy. In contrast, for a power signal, the value $R_{xx}(0)$ corresponds to the power of the signal. In the case of a stochastic process, the value $R_{XX}(0)$ is always the power of the process.

### 15.6.4.3 *Correlation of Complex Power Signals*

The foregoing derivations for the cross-correlation and autocorrelation functions are based on the assumption that $x(n)$ and $y(n)$ are *energy signals*. Thus, it is required that the signal energy be less than infinity. In contrast, the energy of power signals can be infinite, which implies that the duration of the signals can be infinite. We exclude here the case when the amplitudes of a discrete signal can be infinite, because that kind of signal is not physically realizable and, consequently, cannot be obtained from its continuous-time counterpart by the process of sampling. Therefore, the definition and calculation of the correlation function for power signals need to include an infinite time interval.

**Aperiodic power signals.** For aperiodic power signals, the correlation function defined for a real-valued signal in eqn (15.6.12) can be equivalently defined for a complex signal as

$$R_{xx}(l) = \lim_{M\to\infty} \frac{1}{2M+1} \sum_{n=-M}^{M} x(n+l)x^*(n), \tag{15.6.46}$$

and the cross-correlation function as defined for a real-valued signal in eqn (15.6.11) can be equivalently defined for a complex signal as

$$R_{xy}(l) = \lim_{M\to\infty} \frac{1}{2M+1} \sum_{n=-M}^{M} x(n+l)y^*(n). \tag{15.6.47}$$

The power spectral density is defined as the Fourier transform of the autocorrelation function, and the power of the power signal can be calculated from the time-domain or frequency-domain signal representation as

$$P = R_{xx}(0) = \lim_{M\to\infty} \frac{1}{2M+1} \sum_{n=-M}^{M} x(n)x^*(n) = \lim_{M\to\infty} \frac{1}{2M+1} \sum_{n=-M}^{M} |x(n)|^2. \tag{15.6.48}$$

**Periodic power signals.** For the periodic power signals $x(n)$ and $y(n)$, with period $N$, the limits in eqns (15.6.46) and (15.6.47) may be dropped to obtain the autocorrelation function as

$$R_{xx}(l) = \frac{1}{N} \sum_{n=0}^{N} x(n+l)x^*(n),$$ (15.6.49)

and the cross-correlation function as

$$R_{xy}(l) = \frac{1}{N} \sum_{n=0}^{N} x(n+l)y^*(n).$$ (15.6.50)

These two equations are equivalent to eqns (15.6.14) and (15.6.13), respectively, for real signals. It is important to note that the Fourier transform of an energy signal gives the energy spectral density, and the Fourier transform of the power and periodic signals gives the power spectral density. Furthermore, the Fourier transform of a power signal may not exist, so the power spectral density of the signal can be calculated as the Fourier transform of its autocorrelation function.

We may make the following notes on the definition and interpretation of the power spectral density. By definition, power signals have infinite energy, and their Fourier transform and energy spectral density may not exist. Therefore, power signals need alternate spectral functions with properties similar to those for the energy spectral density. In practice, the energy spectral density can be obtained for a power signal $x(n)$ that is time windowed with window size $2a$. The power spectral density can be defined as the normalized limit of the energy spectral density for the windowed signal $x(n)$ as

$$S_{xx}(\Omega) = \lim_{a \to \infty} \frac{1}{2a} E_x(\Omega) = \lim_{a \to \infty} \frac{1}{2a} |X(\Omega)|^2.$$ (15.6.51)

The power spectral density represents the distribution of signal power in the frequency domain, that is, the power can be calculated as

$$P_x = \lim_{M \to \infty} \frac{1}{2M+1} \sum_{n=M}^{M} |x(n)|^2 = \int_{\Omega=-\infty}^{\infty} S_{xx}(\Omega) \, d\Omega.$$ (15.6.52)

### 15.6.4.4 *Analysis of a Linear-Time-Invariant System with Energy Signals*

A block schematic of a linear-time-invariant system specified by its transfer function $H(\Omega)$, or the impulse response $h(n)$, was presented in Figure 15.6.1. The output signals of the system can be found using the convolution of the input signal and the impulse response. The analysis of the system can be conducted using signal representation in the form of correlation functions. In the case analysed here, the impulse response of the system will be replaced by the *system correlation function*, which is defined as the autocorrelation function of the system impulse response. In this analysis, the input, the system itself and the output of the linear-time-invariant system will be represented by the correlation functions and related spectral densities, as indicated in Figure 15.6.1.

**The input signal in the time and frequency domains.** The input signal of the system $x(n)$ can be represented in the time domain by its autocorrelation function (15.6.40) and in the frequency domain by the related power or energy spectral density function (15.6.39). These two functions constitute a Fourier transform pair, as has been proved via the Wiener–Khintchine theorem.

**Representation of the system in the time and frequency domains.** The system is defined by its finite energy impulse response $h(n)$, which is assumed in this chapter to be a complex-valued signal. The autocorrelation function of the impulse response can be expressed in the form

$$R_h(l) = \sum_{n=-\infty}^{\infty} h(n)h^*(n-l) = h(n)*h^*(-n),$$   (15.6.53)

which is called the *system correlation function*. The energy spectral density of the impulse response is its Fourier transform

$$S_h(\Omega) = \sum_{n=-\infty}^{\infty} R_h(l)e^{-j\Omega n}.$$   (15.6.54)

This function can be used in the system analysis in terms of autocorrelation functions and energy spectral densities, in the same way as the impulse response function $h(n)$ and the frequency response $H(\Omega)$ are used for the analysis of deterministic linear-time-invariant systems in time and frequency domains. Namely, it will relate the autocorrelation function of the system output to the autocorrelation function of the system input as follows. Bearing in mind that the output of a linear-time-invariant system is a convolution of the input and the impulse response, the input–output cross-correlation function can be obtained as follows:

$$R_{xy}(l) = \sum_{n=-\infty}^{\infty} x^*(n)y(n+l) = \sum_{n=-\infty}^{\infty} x^*(n) \sum_{k=-\infty}^{\infty} h(k)x(n+l-k)$$
$$= \sum_{k=-\infty}^{\infty} h(k) \sum_{n=-\infty}^{\infty} x^*(n)x(n+l-k) = \sum_{k=-\infty}^{\infty} h(k)R_x(l-k) = h(l)*R_x(l)$$   (15.6.55)

We then obtain the following expressions for the output–input cross-correlation:

$$R_{yx}(l) = h(-l)*R_x(l).$$   (15.6.56)

The cross power spectral density of the two signals $x(n)$ and $y(n)$ is the Fourier transform of the cross-correlation, that is,

$$S_{xy}(\Omega) = \sum_{l=-\infty}^{\infty} R_{xy}(l)e^{-j\Omega l} = S_{yx}^{*}(\Omega),\tag{15.6.57}$$

and the cross-corelation is

$$R_{xy}(l) = \sum_{l=-\infty}^{\infty} S_{xy}(\Omega)e^{j\Omega l} = R_{yx}^{*}(-l),\tag{15.6.58}$$

as proven in eqns (15.6.43) and (15.6.5) for a real signal.

**The system output in the time and frequency domains.** The *autocorrelation* of the output can be obtained from the expression for the convolutional integral rearranged to the form

$$\begin{aligned} R_y(l) &= \sum_{n=-\infty}^{\infty} y^{*}(n)y(n+l) = \sum_{n=-\infty}^{\infty} y^{*}(n)\sum_{k=-\infty}^{\infty} h(k)x(n+l-k) \\ &= \sum_{k=-\infty}^{\infty} h(k)\sum_{n=-\infty}^{\infty} y^{*}(n)x(n+l-k) = \sum_{k=-\infty}^{\infty} h(k)R_{yx}(l-k) = h(l)*R_{yx}(l) \end{aligned} \tag{15.6.59}$$

Using the relation for cross-correlation, eqn (15.6.56), we find

$$R_y(l) = h(l)*R_{yx}(l) = h(l)*h(-l)*R_x(l)\tag{15.6.60}$$

If we insert the expression for the *system correlation function* given by eqn (15.6.53), we obtain the autocorrelation function of the output signal as

$$R_y(l) = h(l)*h(-l)*R_x(l) = R_h(l)*R_x(l).\tag{15.6.61}$$

Therefore, *the autocorrelation of the output signal is equal to the convolution of the autocorrelation of the input signal with the autocorrelation of the impulse response of the system itself.* We thus can say that the output of a linear-time-invariant system is calculated from the twofold convolution of the input autocorrelation function and the system impulse response.

If we have the autocorrelation or auto-spectral density of the input and output signals of a linear-time-invariant system, then we can determine the magnitude response of the system. However, we cannot determine the phase response of the system. For this purpose, we can use cross-spectral densities. The basic model of a linear-time-invariant system with its related functions was shown in Figure 15.6.1. In the analysis of linear-time-invariant systems, we will proceed according to the following steps:

1. Find the system correlation function $R_h(l)$, which is defined as the autocorrelation function of the system impulse response $h(l)$.

2. For any input signal $x(n)$ defined by its autocorrelation function and the power spectral density, we can calculate the correlation function for the output signal $y(n)$ as the convolution of the input autocorrelation function and the system autocorrelation function, that is,

$$R_y(l) = R_h(l) * R_x(l).$$

3. The energy spectral density of the output signal $y(n)$ will be calculated as the product of the system magnitude response squared (the energy spectral density of the impulse response) and the energy spectral density of the input signal, that is,

$$S_y(\Omega) = |H(\Omega)|^2 S_x(\Omega).$$

4. For the calculated autocorrelation function and energy spectral density, the energy of the output signal will be calculated as

$$E_x = \sum_{n=-\infty}^{\infty} |x(n)|^2 = \frac{1}{2\pi} \int_{-\pi}^{\pi} |X(\Omega)|^2 d\Omega = R_x(0).$$

**Example**

Calculate the autocorrelation function and the power spectral density of the sine signal $x(n) = A \cdot \cos(\Omega_c n)$ and investigate its power and energy in the time and frequency domains.

a. Solve the problem by processing the signal as a periodic function.
b. Solve the problem by processing the signal as a power signal.
c. Find the energy spectral density and calculate the energy and power of the signal.

**Solution.** The solutions are as follows:

a. Because the signal is periodic, with period $T$, the autocorrelation function is

$$R_{xx}(l) = \frac{1}{N} \sum_{n=0}^{N-1} x(n) x(n-l) = \frac{1}{N} \sum_{n=0}^{n=N-1} [A \cos \Omega_c n \cdot A \cos \Omega_c (n-l)]$$

$$= \frac{1}{N} A^2 \sum_{n=0}^{n=N-1} \frac{1}{2} [\cos \Omega_c l + \cos \Omega_c (2n+l)] = \frac{1}{N} \frac{A^2}{2} \sum_{n=0}^{n=N-1} \cos \Omega_c l + 0 = \frac{A^2}{2} \cos \Omega_c l.$$

At zero lag, we obtain

$$R_{xx}(0) = \frac{A^2}{2},$$

which is the power of the signal. The power spectral density is the inverse discrete-time Fourier transform, which is calculated as

$$S_{xx}(\Omega) = \sum_{l=-\infty}^{\infty} R_{xx}(l) e^{-j\Omega l} = \frac{A^2}{2} \sum_{l=-\infty}^{\infty} \left[ \frac{1}{2} e^{j\Omega_c l} + \frac{1}{2} e^{-j\Omega_c l} \right] e^{-j\Omega l}$$

$$= \frac{A^2}{2} \sum_{l=-\infty}^{\infty} \left[ \frac{1}{2} e^{j\Omega_c l} + \frac{1}{2} e^{-j\Omega_c l} \right] e^{-j\Omega l} = \frac{A^2}{4} 2\pi \left[ \delta(\Omega - \Omega_c) + \delta(\Omega + \Omega_c) \right]$$

or, as a function of the normalized frequency $F$,

$$S_{xx}(F) = \frac{A^2}{4} \left[ \delta(F - F_c) + \delta(F + F_c) \right],$$

which can be used to find the signal power in the frequency domain as

$$P_x = \int_{F=-\infty}^{\infty} S_{xx}(F) dF = \int_{F=-\infty}^{\infty} \frac{A^2}{4} \left[ \delta(F - F_c) + \delta(F + F_c) \right] dF = \frac{A^2}{2}.$$

b.  Because the signal is a power signal with period $N$, the autocorrelation function is

$$R_{xx}(l) = \lim_{M \to \infty} \frac{1}{2M+1} \sum_{n=-M}^{M} y(n) y(n-l) = \lim_{M \to \infty} \frac{A^2}{2M+1} \sum_{n=-M}^{M} \cos(\Omega_c n) \cos(\Omega_c(n-l))$$

$$= \lim_{M \to \infty} \frac{A^2}{2M+1} \sum_{n=-M}^{M} \frac{1}{2} \cos(\Omega_c l) + 0 = \lim_{M \to \infty} \frac{A^2/2}{2M+1} (2M+1) \cos(\Omega_c l) = \frac{A^2}{2} \cos(\Omega_c l)$$

The power is

$$P_x = R_{xx}(0) = \frac{A^2}{2},$$

which can also be calculated from the power spectral density

$$S_{xx}(\Omega) = \sum_{l=-\infty}^{\infty} R_{xx}(l) e^{-j\Omega l} = \frac{A^2}{2} \sum_{l=-\infty}^{\infty} \left[ \frac{1}{2} e^{j\Omega_c l} + \frac{1}{2} e^{-j\Omega_c l} \right] e^{-j\Omega l}$$

$$= \frac{A^2}{2} \sum_{l=-\infty}^{\infty} \left[ \frac{1}{2} e^{j\Omega_c l} + \frac{1}{2} e^{-j\Omega_c l} \right] e^{-j\Omega l} = \frac{A^2}{4} 2\pi \left[ \delta(\Omega - \Omega_c) + \delta(\Omega + \Omega_c) \right]$$

as its integral

$$P_x = \int_{F=-\infty}^{\infty} S_{xx}(F)dF = \int_{F=-\infty}^{\infty} \frac{A^2}{4}[\delta(F-F_c)+\delta(F+F_c)]dF = \frac{A^2}{2}.$$

The calculation of the power and energy in the frequency domain complies with the analogous calculation in the time domain:

$$P_x = \lim_{M\to\infty} \frac{1}{2M+1} \sum_{n=-M}^{M} x^2(n) = \lim_{M\to\infty} \frac{A^2}{2M+1} \sum_{n=-M}^{M} \frac{1}{2}(1+\cos 2\Omega_c n) = \lim_{M\to\infty} \frac{A^2}{2M+1}\frac{1}{2}(2M+1) = \frac{A^2}{2},$$

$$E_x = \lim_{M\to\infty} \sum_{n=-M}^{M} x^2(n) = \lim_{M\to\infty} \frac{A^2}{2} \sum_{n=-M}^{M} (1+\cos 2\Omega_c n) = \lim_{M\to\infty} \frac{A^2}{2}(2M+1)-0 \to \infty.$$

c. The signal in time domain is $x(n) = A \cdot \cos(\Omega_c n)$. Its Fourier transform is expressed in terms of Dirac's delta functions as

$$X(F) = A\frac{\delta(F-F_0)+\delta(F+F_0)}{2}.$$

Following the definition of the absolute value squared, and bearing in mind that the product of two deltas defined at two different time instants is 0, the energy spectral density may be expressed as

$$E_x(F) = |X(F)|^2 = \left| A\frac{\delta(F-F_0)+\delta(F+F_0)}{2} \right|^2 = \left( A\frac{\delta(F-F_0)+\delta(F+F_0)}{2} \right)^2$$

$$= \frac{1}{4}A^2\delta^2(F-F_c) + 0 + \frac{1}{4}A^2\delta^2(F+F_c) = \frac{1}{4}A^2\delta^2(F-F_c) + \frac{1}{4}A^2\delta^2(F+F_c).$$

If we relax the mathematical rigour, the energy can be calculated as

$$E_x = \int_{F=-1}^{1} E_x(F)dF = \frac{A^2}{4}\int_{-1}^{1}\left(\delta^2(F-F_c)+\delta^2(F+F_c)\right)df = \frac{A^2}{4}\lim_{\varepsilon\to\infty}\frac{1}{\varepsilon}(\varepsilon\varepsilon+\varepsilon\varepsilon) \to \infty.$$

The delta function is not square integrable; in other words, its square diverges and tends to infinity. The calculations of the power and energy comply with their calculations in the time domain:

$$P_x = \lim_{M\to\infty} \frac{A^2}{2M+1} \sum_{n=-M}^{M} \frac{1}{2}(1-\cos 2\Omega_c n) = \frac{A^2}{2},$$

$$E_x = \lim_{M\to\infty} \frac{A^2}{2} \sum_{n=-M}^{M} (1-\cos 2\Omega_c n) = \lim_{M\to\infty} \frac{A^2}{2}(2M+1) \to \infty.$$

**Table 15.6.1** *Correlation functions and spectral densities of the power and energy signals.*

| | **Correlation Functions** | **Spectral Densities** |
|---|---|---|
| **Energy Signals** | | |
| Correlation and energy spectral density, Wiener-Khintchine theorem, $l = 0, \pm 1, \pm 2, \ldots$ | $R_x(l) = \displaystyle\sum_{n=-\infty}^{\infty} x(n)x(n-l)$ $= \frac{1}{2\pi}\int_{-\pi}^{\pi} S_x(\Omega)\,e^{+j\Omega\,l}d\Omega$ | $S_x(\Omega) = \displaystyle\sum_{l=-\infty}^{\infty} R_x(l)e^{-j\Omega\,l}$ |
| Cross-correlation and energy spectral density, Wiener-Khinchine theorem | $R_{xy}(l) = \displaystyle\sum_{n=-\infty}^{\infty} x(n)y(n-l)$ $= \frac{1}{2\pi}\int_{\pi}^{\pi} S_{xy}(\Omega)\,e^{+j\Omega\,l}d\Omega$ | $S_{xy}(\Omega) = \displaystyle\sum_{l=-\infty}^{\infty} R_{xy}(l)e^{-j\Omega\,l}$ |
| Parseval's theorem for finite energy signals | $E_x = \displaystyle\sum_{n=-\infty}^{\infty} |x(n)|^2 = R_x(0)$ | $E_x = \frac{1}{2\pi}\int_{-\pi}^{\pi} \left|X\left(e^{j\Omega}\right)\right|^2 d\Omega$ |
| **Power Signals** | | |
| Autocorrelation function and power spectral density of aperiodic power signals | $R_{xx}(l)$ $= \displaystyle\lim_{M\to\infty} \frac{1}{2M+1} \sum_{n=-M}^{M} x(n)x(n-l)$ | $S_{xx}(\Omega) = \displaystyle\sum_{l=-\infty}^{\infty} R_{xx}(l)e^{-j\Omega\,l}$ |
| Autocorrelation function and power spectral density of periodic power signals | $R_{xx}(l) = \frac{1}{N}\displaystyle\sum_{n=0}^{N-1} x(n)x(n-l)$ $R_{xy}(l) = \frac{1}{N}\displaystyle\sum_{n=0}^{N-1} x(n)y(n-l)$ | $S_{xx}(\Omega) = \displaystyle\sum_{l=-\infty}^{\infty} R_{xx}(l)e^{-j\Omega\,l}$ $S_{xy}(\Omega) = \displaystyle\sum_{l=-\infty}^{\infty} R_{xy}(l)e^{-j\Omega\,l}$ |
| Signal power | $P = R_{xx}(0)$ | $P = \frac{1}{2\pi}\int_{-\pi}^{\pi} S_{xx}(\Omega)\,d\Omega$ |

### 15.6.5 Tables of Correlation Functions and Related Spectral Density Functions

Relationships between the autocorrelation and cross-correlation functions and the power and energy spectral densities of periodic and aperiodic signals are presented in Tables 15.6.1 and Table 15.6.2.

## 15.7 The *z*-Transform

### 15.7.1 Introduction

In previous sections, the characteristics of the discrete-time Fourier series, the discrete-time Fourier transform, the discrete Fourier transform, and algorithms for time-efficient discrete Fourier transform, including the fast Fourier transform, have been presented. These transforms

**Table 15.6.2** *Examples of correlation functions and their power spectral densities.*

| Signal $x(n)$ | Correlation | Spectral Density, Power, and Energy |
|---|---|---|
| $A \cdot \sin(\Omega_c n)$ | $R_{xx}(l) = \frac{A^2}{2} \cos \Omega_c l$ | $S_{xx}(\Omega) = \frac{A^2}{2} \pi \left[ \delta(\Omega - \Omega_c) + \delta(\Omega + \Omega_c) \right]$ |
| $A \cdot \cos(\Omega_c n)$ | $R_{xx}(l) = \frac{A^2}{2} \cos \Omega_c l$ $-\infty < l < +\infty$ | $S_{xx}(F) = \frac{A^2}{4} \left[ \delta(F - F_c) + \delta(F + F_c) \right]$ |
| $A \cdot e^{-an} u(n),\ a > 1$ | $R_x(l) = A^2 \frac{1}{1 - e^{-2a}} e^{a\|l\|}$ $-\infty < l < +\infty$ | $E = R_x(0) = A^2 \frac{1}{1 - e^{-2a}}$ |
| Deterministic rectangular pulse $x(n) =$ $u(n+N) - u(n-N-1)$ | $R_x(l) = N(1 + l/N),$ $-N \le l \le 0$ $R_x(l) = N(1 - l/N),$ $0 \le l \le N$ | $S_x(\Omega) = N \frac{\sin^2 \Omega N/2}{\sin^2 \Omega/2}$ $E = R_x(0) = N$ |

are used to represent discrete-time signals and linear-time-invariant systems in the frequency domain. However, due to the convergence condition, the discrete-time Fourier transform of a discrete signal may not exist, as is the case with some continuous-time signals that do not have a continuous-time Fourier transform, which was the reason for introducing the Laplace transform. In this sense, the Laplace transform was a generalization of the Fourier transform for continuous-time signals. In this section, the $z$-transform will be introduced as a generalization of the discrete-time Fourier transform. This transform is a very important transform that is used in the analysis and design of linear systems. It can be shown that the $z$-transform exists for many discrete signals for which the discrete-time Fourier transform does not exist.

## 15.7.2   Derivation of Expressions for the $z$-Transform

The discrete-time Fourier transform of an arbitrary discrete-time signal $x(n)$ has already been treated as a simplification of the $z$-transform. Namely, a complex variable $z = \text{Re}\{z\} + j \, \text{Im}\{z\} = re^{j\Omega}$ was defined on the unit circle, that is, it was assumed that $r = 1$ or $|z| = 1$. Thus, the discrete-time Fourier transform was defined by the following expression:

$$X\left(e^{j\Omega}\right) = \sum_{n=-\infty}^{\infty} x(n) e^{-j\Omega n}. \tag{15.7.1}$$

If the radius $r$ has an arbitrary value, that is, the complex variable is $z = re^{j\Omega}$, the $z$-transform of the signal $x(n)$ is defined as

$$X(z) = \sum_{n=-\infty}^{\infty} x(n) z^{-n} = \sum_{n=-\infty}^{\infty} x(n) r^{-n} e^{-j\Omega n} = \sum_{n=-\infty}^{\infty} \left[ x(n) r^{-n} \right] e^{-j\Omega n}. \tag{15.7.2}$$

Expressed in this form, the z-transform can be interpreted as discrete-time Fourier transform of the signal $x(n)r^{-n}$.

**The problem of convergence.** The z-transform is represented by an infinite series. Thus, for a given signal $x(n)$, we would like to know the set of z values for which its z-transform converges. Because the signal $x(n)$ in eqn (15.7.2) is modified, resulting in the signal $x(n)r^{-n}$, the convergence condition will be defined for this modified signal. This condition is as follows: the z-transform exists, that is, the series $X(z)$ converges, if the modified function $x(n)r^{-n}$ is absolutely summable, that is, if the following condition is fulfilled:

$$\sum_{n=-\infty}^{\infty} \left| x(n)r^{-n} \right| < \infty. \tag{15.7.3}$$

Then, in mathematical terms, the function $X(z)$ is absolutely convergent and converges uniformly to a continuous function of z. The set of z values for which the z-transform expressed by eqn (15.7.2) converges is called the region of convergence.

**Example**

Find the z-transform of the unit ramp function $x(n) = n$.

**Solution.** The z-transform is

$$X(z) = \sum_{n=-\infty}^{\infty} x(n)z^{-n} = \sum_{n=0}^{\infty} nz^{-n} = 0 + z^{-1} + 2z^{-2} + 3z^{-3} + \ldots = \frac{z^{-1}}{\left(1 - z^{-1}\right)^2} = \frac{z}{(z-1)^2}.$$

**Example**

Find the z-transform and the region of convergence of the causal exponential discrete-time signal expressed by

$$x(n) = a^n u(n). \tag{15.7.4}$$

Find the z-transform and the signals in the time domain for the following two cases defined by the constant $a$: $a = 1$, and $a = -1$.

**Solution.** The z-transform is

$$X(z) = \sum_{n=-\infty}^{\infty} x(n)z^{-n} = \sum_{n=-\infty}^{\infty} \left[ a^n u(n) \right] z^{-n} = \sum_{n=0}^{\infty} \left( az^{-1} \right)^n \tag{15.7.5}$$

This power series converges if and only if

$$\left| az^{-1} \right| < 1, \text{that is, } |z| > |a|.$$

The terms of summation are elements of a geometric power series; thus, their sum is

$$X(z) = \frac{1}{1 - az^{-1}} = \frac{z}{z - a},$$

because $|az^{-1}| < 1$, or $|z| > |a|$. Both of the expressions on the right-hand side of eqn (15.7.5) are useful for analysing the transform. They show that $X(z)$ has a pole at $z = a$, and a 0 at $z = 0$, as depicted in Figure 15.7.1. The transform converges exterior to a circle $|z| = |a|$, including the point $|z| = \infty$. Thus, the region of convergence is the annular region defined by $|z| > |a|$. The poles and zeros are shown for the four cases $0 < a < 1$, $-1 < a < 0$, $a > 1$, $a < -1$, and the region of convergence is indicated by the shaded area fulfilling the condition $|z| > |a|$. In particular, for a stable system defined by $|z| < 1$, the unit circle is inside the region of convergence, but not otherwise.

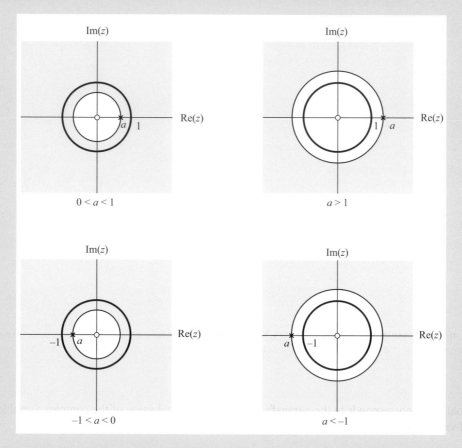

**Fig. 15.7.1** *Poles and zeros of the z-transform.*

The *boundary cases* occur for $a = 1$ or $-1$. For these cases, a correspondence between the $z$-transform and the time-domain representation of the signal $x(n)$ can be easily established. For $a = 1$, the z-transform and the time-domain representation are expressed as

$$X(z) = \frac{1}{1 - az^{-1}} = \frac{1}{1 - z^{-1}} \leftrightarrow x(n) = a^n u(n) = u(n), |z| > 1, \qquad (15.7.6)$$

which corresponds to the z-transform of the unit step function. In this case, the region of convergence is an annular region defined by $\infty \geq |z| > 1$. For the case when $a = -1$, we obtain

$$X(z) = \frac{1}{1 - az^{-1}} = \frac{1}{1 + z^{-1}} \leftrightarrow x(n) = a^n u(n) = (-1)^n u(n), \quad |z| > 1.$$

In each of these cases, the poles are on the unit circle at $|z| = 1$, and the region of convergence is everywhere outside this unit circle.

The unit step function is not absolutely summable. Therefore, the Fourier transform does not converge uniformly. However, assuming that $z = e^{j\Omega}$, the discrete-time Fourier transform can be obtained from eqn (15.7.6) for the z-transform and is expressed as

$$X\left(e^{j\Omega}\right) = \frac{1}{1 - ae^{-j\Omega}}. \qquad (15.7.7)$$

## 15.7.3   Properties of the z-Transform

Suppose $x_1(n)$ and $x_2(n)$ have the z-transforms $X_1(z)$ and $X_2(z)$ with respective radii of convergence $r_1$ and $r_2$, and that the signals in the time domain fulfil the following conditions: $x_1(n) = 0$, and $x_2(n) = 0$, for $n < 0$.

**Linearity.** If $x_1(n)$ and $x_2(n)$ have the z-transforms $X_1(z)$ and $X_2(z)$, then the linear combination of signals in the time domain gives a linear combination of their z-transforms,

$$x_1(n) \leftrightarrow X_1(z) \text{ and } x_2(n) \leftrightarrow X_2(z) \Rightarrow ax_1(n) + bx_2(n) \leftrightarrow aX_1(z) + bX_2(z), \qquad (15.7.8)$$

for arbitrary constants $a$ and $b$. There can be an arbitrary number of signals in the linear combination. The proof for this property is straightforward, because the z-transform is an linear transform

**Time reversal.** If $x(n)$ has the z-transform $X(z)$, then the z-transform of the time-reversed signal $x(-n)$ corresponds to the z-transform of the reciprocal value of the complex variable $z$, and vice versa, that is,

$$x(n) \leftrightarrow X(z) \Rightarrow x(-n) \leftrightarrow X\left(\frac{1}{z}\right), \qquad (15.7.9)$$

and the region of convergence has the value $1/r$, where $r$ is the region of convergence of $X(z)$.

**Conjugation.** The conjugation of $x(n)$ results in the conjugation of its transform $X(z)$ and the complex variable, that is,

$$x(n) \leftrightarrow X(z) \Rightarrow x^*(n) \leftrightarrow X^*\left(z^*\right).$$
(15.7.10)

The proof for this property is straightforward and can be used as an exercise.

**Time shifts: Delay or advance.** As far as time shifts are concerned, there are two cases to be considered: delay and advance, or negative or positive shifts. Due to the assumption that the signals are causal, a positive shift of a signal in the time domain will shift in some leading zeros. Thus, if the signal $x(n)$ has as its transform $X(z)$, then the time-shifted signal $x(n-n_0)$ has a transform expressed as

$$x(n) \leftrightarrow X(z) \Rightarrow x(n-n_0) \leftrightarrow z^{-n_0}X(z), \text{for } n_0 > 0.$$
(15.7.11)

If the signal $x(n)$ has as its transform $X(z)$, then the negative time-shifted signal $x(n+n_0)$ has the following $z$-transform:

$$x(n) \leftrightarrow X(z) \Rightarrow x(n+n_0) \leftrightarrow z^{n_0}\left[X(z) - \sum_{m=0}^{n_0-1} x(m)z^{-m}\right], \text{for } n_0 > 0.$$
(15.7.12)

The sum in the $z$-transform (15.7.12) of the shifted signal includes the effect of the components of the time-domain signal that are shifted to the left.

**Variable scaling.** If we modify variable $z$ by the factor $\alpha$ in the $z$-transform $X(z)$, this will correspond to the multiplication by a complex exponential in the time domain, that is,

$$x(n) \leftrightarrow X(z) \Rightarrow \alpha^n x(n) \leftrightarrow X\left(\frac{z}{\alpha}\right).$$
(15.7.13)

When $\alpha = e^{j\Omega_0}$, for real $\Omega_0$, the expression for the change of scale would be

$$x(n) \leftrightarrow X(z) \Rightarrow e^{j\Omega_0 n}x(n) \leftrightarrow X\left(ze^{-j\Omega_0}\right).$$
(15.7.14)

Thus, multiplication of the time-domain signal $x(n)$ by $e^{j\Omega_0 n}$ corresponds to the counterclockwise rotation of the transform $X(z)$ in the $z$ plane by the angle $\Omega_0$. This rotation is analogous to a shift in the frequency domain for the discrete-time Fourier transform. Namely, for the complex variable $z = e^{j\Omega}$, the complex modulation in the time domain corresponded to a shift in the frequency domain, that is,

$$x(n) \leftrightarrow X\left(e^{j\Omega}\right) \Rightarrow e^{j\Omega_0 n}x(n) \leftrightarrow X\left(e^{j(\Omega-\Omega_0)}\right).$$
(15.7.15)

**Differentiation in the $z$ domain.** Differentiation in the $z$ domain corresponds to multiplication by $-n$ in the time domain, that is,

$$x(n) \leftrightarrow X\left(e^{j\Omega}\right) \Rightarrow -nx(n) \leftrightarrow z\frac{dX(z)}{dz}. \tag{15.7.16}$$

This can be obtained by differentiating both sides of the expression for the $z$-transform.

**Convolution of signals.** The convolution of two signals $x_1(n)$ and $x_2(n)$ corresponds to the product of their $z$-transforms $X_1(z)$ and $X_2(z)$, that is,

$$x_1(n) \leftrightarrow X_1(z) \text{ and } x_2(n) \leftrightarrow X_2(z) \Rightarrow x_1(n)^* x_2(n) \leftrightarrow X_1(z) \cdot X_2(z). \tag{15.7.17}$$

From this property, it is an easy matter to find the response $y(n)$ of a linear-time-invariant system to an input $x(n)$. If the impulse response of an linear-time-invariant system is $h(n)$, then the output in the $z$ domain is

$$Y(z) = H(z) \cdot X(z), \tag{15.7.18}$$

where $H(z)$ is the impulse response of the system.

**Accumulation.** The accumulation operation is the discrete-time counterpart to continuous-time integration, and can also be expressed as convolution with a unit step signal, that is,

$$y(n) = \sum_{k=0}^{n} x(k) = x(n)^* u(n) = \sum_{k=-\infty}^{\infty} u(k)x(n-k). \tag{15.7.19}$$

In the $z$ domain, this expression corresponds to multiplication, that is,

$$Y(z) = U(z)X(z) = \frac{z}{z-1}X(z) = \frac{1}{1-z^{-1}}X(z). \tag{15.7.20}$$

Therefore, the accumulation property can be expressed as

$$x(n) \leftrightarrow X(z) \Rightarrow \sum_{k=0}^{n} x(k) \leftrightarrow \frac{1}{1-z^{-1}}X(z). \tag{15.7.21}$$

**The initial value theorem.** We can find the initial value of the discrete signal $x(n)$ from its $z$-transform $X(z)$. If we find the limit of $X(z)$ when $z$ tends to infinity, we can find the initial value in the form

$$x(0) = \lim_{z \to \infty} X(z). \tag{15.7.22}$$

**The final value theorem.** The final value theorem serves to find the value of the discrete signal $x(n)$ in infinity from its $z$-transform $X(z)$. According to this theorem, the final value is

$$\lim_{n \to \infty} x(n) = \lim_{z \to 1} (z-1) X(z), \qquad (15.7.23)$$

only if $\lim_{n \to \infty} x(n)$ exists. Also, it can be shown that, if there are any poles on or outside the unit circle, excluding a single pole at $z = 1$, then the final value theorem cannot be applied.

### 15.7.4   The Inverse $z$-Transform

For the complex variable $z = \text{Re}\{z\} + j\,\text{Im}\{z\} = re^{j\Omega}$, the inverse $z$-transform is defined by the following contour integral:

$$x(n) = \frac{1}{2\pi j} \oint_l X(z) z^{n-1} dz, \qquad (15.7.24)$$

where $l$ is a counterclockwise contour of integration defined by $|z| = r$ enclosing the origin. This integral can be obtained for any other contour which encircles the origin $z = 0$ in the region of convergence of the transform $X(z)$. The contour integral for the inverse $z$-transform can be evaluated using Cauchy's residue theorem. Due to the complexity of this method, the inverse $z$-transform is instead calculated using the method of partial-fraction expansion or the method of power series expansion using long division.

In the first method, partial-fraction expansion, the $z$-domain expression should be expanded into partial fractions, and then transform pairs should be identified using a table of $z$-transforms and the transform properties. The signal in the time domain is obtained as the sum of the inverse transform of the individual sample terms in the expansion. Using this method, we can get the inverse $z$-transform in a closed form.

The second method, the power series expansion, assumes that the inverse function is expressed as a ratio of the polynomials in $z$. Then, by dividing the numerator by the denominator, we can get the transform in the form of a power series. The coefficients of this series correspond to the values of the discrete signal in the time domain. Using this method, we cannot get the inverse $z$-transform in a closed form. The inverse $z$-transform can be also found using $z$-transform tables.

...................................................................................................................................

PROBLEMS

1. Consider the following discrete-time periodic signal:

$$x(n) = \{\ldots, 1, 0, 1, 2, \mathbf{3}, 2, 1, 0, 1, 2, 3, 2, 1, 0, 1, \ldots\}$$

   where bold font indicates the point $n = 0$.

a. Calculate the magnitude spectrum and phase spectrum for the signal and then sketch the signal $x(n)$ and its magnitude spectrum and phase spectrum.

b. Confirm Parseval's relation by computing the power of the signal in the time and frequency domains.

2. Consider the following discrete periodic signal:

$$x(n) = 2 + 2\cos(\pi n/4) + \cos(\pi n/2) + \frac{1}{2}\cos(3\pi n/4).$$

a. Sketch the signal and its spectra.

b. Calculate the power of the signal.

3. Determine and sketch the magnitude spectrum and the phase spectrum of the following periodic signal and then calculate its power:

$$x(n) = 2\sin\frac{\pi(n-2)}{3}.$$

4. Let a signal be defined in the time domain as

$$x(n) = 2\cos(6\pi n/17 + \pi/6).$$

a. Calculate the discrete-time Fourier series of the signal.

b. Draw the signal in the time and frequency domains (both the magnitude spectrum and the phase spectrum).

5. Let a signal be defined in the time domain as

$$x(n) = 2\sin(14\pi n/19) + \cos(10\pi n/19) - 1.$$

a. Calculate the discrete-time Fourier series of the signal.

b. Draw the signal in the time and frequency domains (both the magnitude spectrum and the phase spectrum).

6. Compute the discrete-time Fourier transform of the following signals, and sketch their magnitude spectra and phase spectra:

a. $x(n) = u(n) - u(n-6)$.

b. $x(n) = u(n)\alpha^n \sin \Omega_0 n, |\alpha| < 1$.

7. Determine the signal $x(n)$ in the time domain, having the following discrete-time Fourier transform:

$$X(\Omega) = \cos^2 \Omega.$$

8. Determine the time-domain signal, having the following discrete-time Fourier transform:

$$X(\Omega) = \cos 2\Omega + j \sin 2\Omega.$$

9. Determine the time-domain signal, having the following Fourier transform:

$$|X(\Omega)| = \begin{cases} 2 & \pi/4 \leq |\Omega| \leq 3\pi/4 \\ 0 & otherwise \end{cases}, \ \arg\{X(\Omega)\} = -2\Omega.$$

10. The discrete-time Fourier transform of the discrete-time signal $x(n)$ is defined by the expression

$$X(\Omega) = A\cos^2 \Omega = \frac{A}{2} + \frac{A}{2}\cos(2\Omega).$$

   a. Determine the signal $x(n)$ in the time domain and sketch its graph.
   b. Using $H(e^{j\Omega})$, sketch the magnitude $|H(e^{j\Omega})|$ and phase spectra $\Theta(e^{j\Omega})$ of the signal $x(n)$. Find the energy $E_x$ of the signal $x(n)$.
   c. Using the discrete-time Fourier transform, calculate the magnitude spectrum $|X(k)|$ and the phase spectrum $\Theta(k)$ of the signal $x(n)$.
   d. Suppose that the signal $x(n)$ is passed through an all-pass ideal filter. Find the expression for the impulse response of the filter $h(n)$ and its discrete-time Fourier transform $H(e^{j\Omega})$. Sketch the graphs for $h(n)$ and $H(e^{j\Omega})$. Find the output of the filter $Y(e^{j\Omega})$ if its input is the signal $x(n)$.

11. A discrete-time signal $x(n) = \{1,-2 , 3\}$ is applied to the input of a linear-time-invariant system. (Note that the number in bold corresponds to the signal magnitude defined by $n = 0$). The impulse response of the linear-time-invariant system is the signal $h(n) = \{0, 0, 1, 1, 1, 1\}$.

   a. Plot the graphs of $x(k)$ and $h(n - k)$. Find the time-domain response $y(n)$ of the linear-time-invariant system for the input signal $x(n)$, basing it on the expression

$$y(n) = \sum_{k=-\infty}^{\infty} x(k)h(n-k).$$

   b. Find the expression for the discrete-time Fourier transform, denoted by $H(e^{j\Omega})$, of the impulse response $h(n) = \{0, 0, 1, 1, 1, 1\}$. Plot the graph of the amplitude spectral density $|H(e^{j\Omega})|$.

12. Determine the $N$-point discrete Fourier transform of the Blackman window

$$x(n) = 0.42 - 0.5\cos\frac{2\pi}{N-1}n + 0.08\cos\frac{4\pi}{N-1}n, \quad 0 \le n \le N-1.$$

13. For the periodic signal $x_p(n) = \cos 2\pi n/10$, $-\infty < n < \infty$, with frequency 1/10 and fundamental period $N = 10$, define the signal $x(n) = x_p(n)$, $0 \le n \le N-1$, and then determine its ten-point discrete Fourier transform.

14. Compute the Fourier transform of the following signals:

    a. $x(n) = \delta(n)$.
    b. $x(n) = \cos(2\pi/N)k_0 n$, $\quad 0 \le n \le N-1$.

    Then sketch the signals in the time and frequency domains.

15. Compute the Fourier transform of the following signals:

    a. $x(n) = 2e^{j(2\pi/N)kn}$, $\quad 0 \le n \le N-1$.
    b. $x(n) = 2a^n$, $\quad 0 \le n \le N-1$.

    Then sketch the signals in the time and frequency domains.

16. For the signal

$$x(n) = \{1,2,3,4\} :$$

    a. Compute its four-point discrete Fourier transform by solving explicitly the system of linear equations defined by the inverse discrete Fourier transform formula.
    b. Check the answer in part a. by computing the four-point discrete Fourier transform, using its definition.

17. For the signal depicted in Figure 17:

    a. Evaluate the discrete-time Fourier series representation, and then sketch the signal and its spectra.

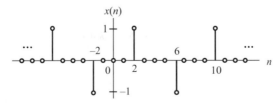

**Fig. 17** *Time-domain signal.*

b. Use the MATLAB *fft* command to confirm the results obtained.

18. Suppose we have the signal

$$x(n) = \sum_{m=-\infty}^{\infty} (-1)^m \left( \delta\,(n-2m) + \delta\,(n+2m) \right).$$

   a. Evaluate the discrete-time Fourier series of the signal, and then sketch the signal and its spectra.
   b. Use the MATLAB *fft* command to confirm the results obtained.

19. Suppose we have the signal

$$x(n) = \cos\,(6\pi n/17).$$

   a. Evaluate the discrete-time Fourier series representation, and then sketch the signal and its spectra.
   b. Use the MATLAB *fft* command to confirm the results obtained.

20. Define a signal in the frequency domain as in Figure 20.

   a. Use the inverse discrete-time Fourier series to determine the time-domain signals, and then sketch the signal and its spectra.
   b. Use the MATLAB *ifft* command to confirm the results obtained.

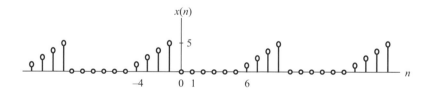

**Fig. 20** *Time-domain signal.*

21. Define a signal in the frequency domain as in Figure 21.

   a. Use the inverse discrete-time Fourier series to determine the time-domain signals, and then sketch the signal and its spectra.
   b. Use the MATLAB *ifft* command to confirm the results obtained.

22. Calculate the autocorrelation function of the signal $x(n) = A \cdot a^n u(n)$, for parameters $A$ greater than 1, and $0 < a < 1$.

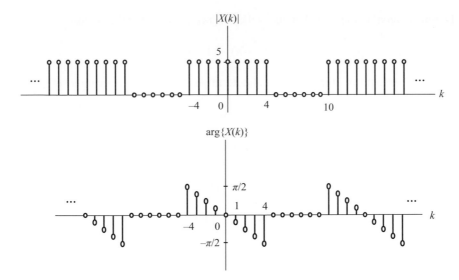

**Fig. 21** *Magnitude spectrum (upper panel) and phase spectrum (lower panel).*

23. Calculate the autocorrelation function and the power spectral density of the sine signal $x(n) = \sin(\Omega_c n)$ and investigate its power and energy in the time and frequency domains.

    a. Solve the problem by processing the signal as a periodic function.
    b. Solve the problem by processing the signal as a power signal.
    c. Find the energy spectral density and calculate the energy and power of the signal.

24. Calculate the autocorrelation function for the discrete unit step signal in Figure 24, drawn for the lag $l > 0$.

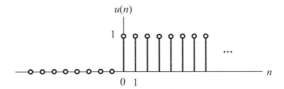

**Fig. 24** *Unit step signal.*

25. Let the following discrete-time signal be defined as the sum of two exponentials:

$$x(n) = 4^n u(n) + 2^n u(-n).$$

Find the *z*-transform of this signal and determine the radius of convergence.

26. Define a signal in the time domain as the following exponential function:

$$x(n) = (1/4)^{|n|}.$$

Find the $z$-transform of this signal and determine the radius of convergence.

27. Suppose the transfer function of a linear-time-invariant system is

$$H(z) = \frac{z}{z + 1/4}.$$

Draw system diagrams for the signal in the time domain and the $z$ domain.

28. Using the long-division method, find the expression of the inverse $z$-transform of the signal $x(n)$, supposing that its $z$-transform is given by the expression

$$X(z) = \frac{z(z - 1/2)}{(z - 2/3)(z - 1/2)(z - 1/4)}.$$

29. Using the long-division method, show that the inverse $z$-transform of $X(z)$ is the ramp function in the case when $X(z)$ is defined by the expression

$$X(z) = \frac{z^{-1}}{\left(1 - z^{-1}\right)^2} = \frac{z}{(z - 1)^2}.$$

30. Using the initial-value and final-value properties of the $z$-transform, prove that the initial and final values of a unit step function $u(n)$ are equal to 1.

31. Find the $z$-transform and the region of convergence of the anti-causal exponential discrete signal expressed by

$$x(n) = a^n u(-n).$$

The find the $z$-transform and the signals in the time domain for the following two cases defined by the constant $a$: $a = 1$, and $a = -1$.

32. Find the $z$-transform of the rectangular pulse expressed by

$$p(n) = u(n) - u(n - N).$$

# 16

# Theory of the Design, and Operation of Digital Filters

This chapter first presents the concept of filtering, together with the basic forms and characteristics of finite impulse response filters and infinite impulse response filters. The basic advantage of finite impulse response filters is that they can be designed to have exact linear phase. In addition, the finite impulse response filter structure is stable. However, the order of the filter can be much higher than the order of the corresponding infinite impulse response filter. Thus, the infinite impulse response filter is generally computationally more efficient.

Next, the structures of digital infinite impulse response and finite impulse response filters are presented. Because these structures show the relationships between the internal variables and input and output, they provide the information needed for implementation of the filters. It will also be shown that there are various structures of digital filters, which are specified by their advantages and disadvantages in practical design of causal infinite impulse response and finite impulse response digital filters.

## 16.1 The Basic Concept of Filtering

A digital filter is a linear-time-invariant system that passes certain frequency components (bandpass) and blocks other frequency components (stopband/s) of an input signal. This input signal can be represented as a finite linear weighted sum or an infinite sum of exponential or sinusoidal signals. For example, if the input of a linear-time-invariant system is a complex exponential signal expressed as

$$x(n) = e^{j\Omega n}, \tag{16.1.1}$$

then the output of the system is defined by the convolution

$$
\begin{aligned}
y(n) &= \sum_{k=-\infty}^{\infty} h(k)x(n-k) = \sum_{k=-\infty}^{\infty} h(k)e^{j\Omega(n-k)} = e^{j\Omega n} \sum_{k=-\infty}^{\infty} (k)e^{-j\Omega k} \\
&= H\left(e^{j\Omega}\right)e^{j\Omega n} = \left|H\left(e^{j\Omega}\right)\right|e^{j\phi(\Omega)}e^{j\Omega n} = H\left(e^{j\Omega}\right)x(n)
\end{aligned}
\tag{16.1.2}
$$

*Discrete Communication Systems.* Stevan Berber, Oxford University Press. © Stevan Berber 2021.
DOI: 10.1093/oso/9780198860792.003.0016

where $H(e^{i\Omega})$ is the frequency response of the linear-time-invariant system, $|H(e^{i\Omega})|$ is the magnitude response of the system, and $\phi(\Omega)$ is the phase response. The frequency response is the discrete-time Fourier transform of the impulse response of the system, that is,

$$H\left(e^{j\Omega}\right) = \sum_{n=-\infty}^{\infty} h(n)e^{-j\Omega n}. \tag{16.1.3}$$

**Example**

Suppose the magnitude response of a filter is defined by

$$\left|H\left(e^{j\Omega}\right)\right| = \begin{cases} 1 & |\Omega| \le \Omega_c \\ 0 & \Omega_c < |\Omega| \le \pi \end{cases}. \tag{16.1.4}$$

Find the output of the filter for the input signal expressed as

$$x(n) = \cos\left(\Omega_1 n\right) + \cos\left(\Omega_2 n\right) \tag{16.1.5}$$

if the frequencies are defined as

$$0 < \Omega_1 < \Omega_c < \Omega_2 < \pi. \tag{16.1.6}$$

**Solution.** The output signal in the time domain is

$$\begin{aligned} y(n) &= \left|H\left(e^{j\Omega_1}\right)\right| \cos\left(\Omega_1 n + \phi\left(\Omega_1\right)\right) + \left|H\left(e^{j\Omega_2}\right)\right| \cos\left(\Omega_2 n + \phi\left(\Omega_2\right)\right) \\ &= H\left(e^{j\Omega_1}\right) \left|\cos\left(\Omega_1 n + \phi\left(\Omega_1\right)\right)\right| \end{aligned} \tag{16.1.7}$$

because the second component is 0, due to the filter transfer function characteristics. This filter behaves as a lowpass filter because low frequencies are passed and high frequencies are stopped.

**Example**

The moving-average filter is defined by its impulse response

$$h(n) = \begin{cases} 1/N & 0 \le n \le N-1 \\ 0 & otherwise \end{cases}. \tag{16.1.8}$$

Find the expression for the frequency response of the filter.

**Solution.** The impulse response is

$$H\left(e^{j\Omega}\right) = \sum_{n=-\infty}^{\infty} h(n)\, e^{-j\Omega n} = \frac{1}{N} \sum_{n=0}^{N-1} e^{-j\Omega n} = \frac{1}{N}\left(\sum_{n=0}^{\infty} e^{-j\Omega n} - \sum_{n=N}^{\infty} e^{-j\Omega n}\right)$$

$$= \frac{1}{N}\left(\sum_{n=0}^{\infty} e^{-j\Omega n} - \sum_{k=0}^{\infty} e^{-j\Omega(N+k)}\right) = \frac{1}{N}\left(\sum_{n=0}^{\infty} e^{-j\Omega n} - \sum_{n=0}^{\infty} e^{-j\Omega(N+n)}\right).$$

$$= \frac{1}{N}\left(\sum_{n=0}^{\infty} e^{-j\Omega n}\right)\left(1 - e^{-j\Omega N}\right) = \frac{1}{N} e^{-j\Omega(N-1)/2} \frac{1 - e^{-j\Omega N}}{1 - e^{-j\Omega}}$$

$$= \frac{1}{N} e^{-j\Omega(N-1)/2} \frac{\sin\left(\Omega N/2\right)}{\sin\left(\Omega/2\right)}$$

The magnitude response is

$$\left|H\left(e^{j\Omega}\right)\right| = \left|\frac{1}{N} \frac{\sin\left(\Omega N/2\right)}{\sin\left(\Omega/2\right)}\right|, \tag{16.1.9}$$

and the phase response is

$$\phi\left(\Omega\right) = -\frac{(N-1)\,\Omega}{2} + \pi \sum_{k=0}^{\lfloor M/2 \rfloor} u\left(\Omega - \frac{2\pi k}{N}\right), \tag{16.1.10}$$

where $u(x)$ is the unit step function in $x$. The magnitude response has a maximum equal to 1 at $\Omega = 0$ and crosses the abscissa at the points defined by $\Omega = 2\pi k/N$. At each zero value, the function has a phase change of $\pi$. The sketch of the magnitude response is shown in Figure 16.1.1 in the interval from $-\pi$ to $+\pi$. This function is periodic, with a period of $2\pi$. The number of zero crossings increases when $N$ increases. Thus, the filter becomes narrower when $N$ increases.

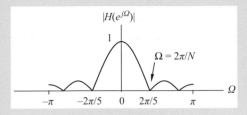

**Fig. 16.1.1** *Magnitude response of a moving-average filter.*

## 16.2   Ideal and Real Transfer Functions

As said previously, a digital filter can be defined by its transfer function. Ideally, the intervals of frequencies where the transfer function is 0 are called *stopbands*, while the intervals where

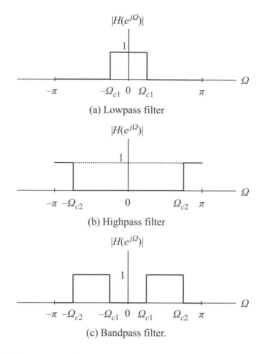

Fig. 16.2.1 *Ideal transfer functions or frequency responses of lowpass (upper panel), highpass (middle panel), and bandpass (lower panel) filters.*

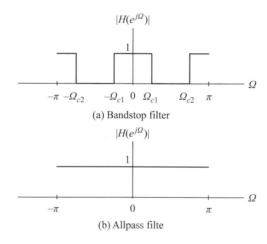

Fig. 16.2.2 *Ideal transfer functions or frequency responses of bandstop (upper panel) and allpass (lower panel) filters.*

this function is 1 are called the *passbands* of the filter. Depending on the position of these bands, digital filters can be classified as lowpass, highpass, bandpass, or bandstop filters. The frequencies at which the filter transfer function changes from stopband to passband, or vice versa, are called the cut-off frequencies. We assume that each of these ideal filters has a zero phase for every frequency. Ideal transfer characteristics or frequency responses of lowpass, highpass, and bandpass filters are presented in Figure 16.2.1, and ideal transfer characteristics for bandstop and allpass filters are shown in Figure 16.2.2.

The second classification of digital filters is with respect to their phase characteristics. The ideal filter is a filter with zero phase characteristics. In this case, phase distortion of the filter is avoided. However, it is impossible to design a causal filter with this characteristic except in some special cases when the causality condition is relaxed. However, there is another way to avoid phase distortion, namely, in the case when the transfer function has a unity magnitude and a linear-phase characteristic. In this case, the output of the filter is a time-shifted version of its input. This time delay is equal to the group delay of the filter.

## 16.3   Representation of Digital Filters

A digital filter can generally be represented as an linear-time-invariant system. Consequently, the filter can be described in various ways. In the time domain, the filter can be represented by the convolution sum

$$y(n) = \sum_{k=-\infty}^{\infty} h(k)x(n-k), \qquad (16.3.1)$$

where $y(n)$ is the output of the filter for the input $x(n)$, and $h(n)$ is the impulse response of the filter. Obviously, this convolution sum can be used to implement a digital filter with a known impulse response. However, if the response is of an infinite duration, this approach is not practical, due to the infinite number of terms that need to be added, multiplied, and delayed.

If the impulse response $h(n)$ is of infinite duration, the filter is called an *infinite impulse response* filter. If the input of the filter is causal, that is, $x(n) = 0$ for $n < 0$, the convolution sum has the form

$$y(n) = \sum_{k=0}^{n} h(k)x(n-k). \qquad (16.3.2)$$

In the case when the impulse response is of finite duration, the *finite impulse response* digital filter can be described by the following equation:

$$y(n) = \sum_{k=K_1}^{k=K_2} h(k)x(n-k), \qquad (16.3.3)$$

where the impulse response is defined to be $h(n) = 0$ for $n < K_1$ and $n > K_2$, assuming that $K_1 < K_2$.

An infinite impulse response linear-time-invariant digital filter can be described by a constant coefficient difference equation of the following form:

$$\sum_{k=0}^{N} A_k y(n-k) = \sum_{k=0}^{M} B_k x(n-k). \tag{16.3.4}$$

Since this equation has a finite number of terms inside the sums, this linear-time-invariant system can be implemented even though the system has an infinite length impulse response. Assuming that $A_0 \neq 0$, and the system is causal, the output of the system can be expressed in a recursive form:

$$y(n) = -\sum_{k=1}^{N} \frac{A_k}{A_0} y(n-k) + \sum_{k=0}^{M} \frac{B_k}{A_0} x(n-k) = -\sum_{k=1}^{N} a_k y(n-k) + \sum_{k=0}^{M} b_k x(n-k). \tag{16.3.5}$$

This equation defines a valid computational algorithm that can be implemented in software on a digital computer or with special-purpose hardware. The algorithm includes the following operations: delays, scalar multiplications, additions, and pick-off nodes. These operations can be represented by appropriate symbols. Thus, the computational diagram of a digital filter can be represented in the form of a *block diagram*. For example, following the general equation (16.3.5), a first-order infinite impulse response digital filter, for $N = 1$ and $M = 1$, can be expressed as

$$y(n) = -a_1 y(n-1) + b_0 x(n) + b_1 x(n-1), \tag{16.3.6}$$

which is shown in Figure 16.3.1 in the form of a block diagram.

The meanings of the symbols are written in brackets on the symbols' right-hand sides. There is obvious correspondence between expression (16.3.6) and the block diagram in Figure 16.3.1. Therefore, the structure of a digital filter represented in a block-diagram form can be analysed in order to find the expression for the output signal as a function of the input signal and the multiplier coefficients.

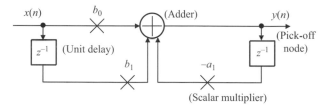

**Fig. 16.3.1** *Block diagram of a first-order digital filter.*

## 16.4 Basic Finite Impulse Response Filters

**Moving-average filter.** A moving-average filter is defined by its impulse response

$$h(n) = \begin{Bmatrix} 1/N & 0 \le n \le N-1 \\ 0 & \text{otherwise} \end{Bmatrix}. \tag{16.4.1}$$

Thus, its transfer function is expressed as

$$H(z) = \sum_{n=-\infty}^{\infty} h(n) z^{-n} = \frac{1}{N} \sum_{n=0}^{N-1} z^{-n} = \frac{1-z^{-N}}{N\left(1-z^{-1}\right)} = \frac{z^N - 1}{N\left[z^{N-1}\left(z-1\right)\right]}. \tag{16.4.2}$$

This transfer function has $N$ zeros that are on the unit circle defined by $z^N - 1 = 0$, that is, $z = e^{j2\pi k/N}$, $k = 0, 1, 2, \ldots, N-1$. Also, from the condition $z^{N-1} = 0$, we see that there is an $(N-1)$th order pole at the origin and, from the condition $z - 1 = 0$, it follows that there is a single pole at $z = 1$ which cancels one 0 at $z = 1$. Therefore, a causal finite impulse response filter has a transfer function with all of the poles at the origin of the $z$ plane. The structure of this filter is shown in Figure 16.4.1.

**Lowpass filter.** The transfer function of the moving-average filter defined for $N = 2$ is expressed as

$$H(z) = \frac{1-z^{-2}}{2\left(1-z^{-1}\right)} = \frac{1-\left(z^{-1}\right)^2}{2\left(1-z^{-1}\right)} = \frac{1}{2}\left(1+z^{-1}\right) = \frac{z+1}{2z}. \tag{16.4.3}$$

This transfer function has *one* 0 that is on the unit circle defined by $z = -1$, and one pole at the origin. The frequency response is

$$H\left(e^{j\Omega}\right) = \frac{1}{2}\left(1+e^{-j\Omega}\right) = \frac{1}{2}\left(e^{j\Omega/2}+e^{-j\Omega/2}\right) e^{-j\Omega/2} = \cos\left(\Omega/2\right) e^{-j\Omega/2}, \tag{16.4.4}$$

which specifies that the magnitude response function be a cosine function. This is a monotonically decreasing function of frequency $\Omega$ which is 1 for $\Omega = 0$, and 0 for $\Omega = \pi$. Based on the transfer function, the magnitude response of this filter is plotted in Figure 16.4.2.

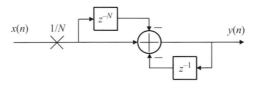

**Fig. 16.4.1** *Structure of a moving-average filter.*

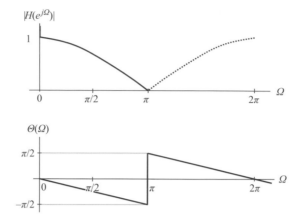

**Fig. 16.4.2** *Filter transfer characteristic: magnitude transfer function (upper panel) and phase transfer function (lower panel).*

The phase spectral density can be obtained from expression (16.4.4) by calculating the inverse tangent function of the complex function and adding the phase changes of $\pi$ that are caused by the cosine function at the zero points of the magnitude spectral density. The calculated phase density function is

$$\theta(\Omega) = tg^{-1}\frac{-\sin(\Omega/2)}{\cos(\Omega/2)} = -\Omega/2.$$

The phase changes of $\pi$ need to be added when the cosine function in eqn (16.4.4) changes the sense.

**Highpass filter.** By replacing $z$ with $-z$ in the transfer function for the lowpass moving-average filter defined for $N = 2$, we can get the transfer function for the highpass finite impulse response filter in the form

$$H(z) = \frac{1}{2}\left(1 - z^{-1}\right) = \frac{z-1}{2z}. \tag{16.4.5}$$

This transfer function has *one* 0 that is on the unit circle defined by $z = 1$, and one pole at the origin. The frequency response is

$$H\left(e^{j\Omega}\right) = \frac{1}{2}\left(1 - e^{-j\Omega}\right) = \frac{j}{2j}\left(e^{j\Omega/2} - e^{-j\Omega/2}\right)e^{-j\Omega/2} = j\sin(\Omega/2)\,e^{-j\Omega/2}, \tag{16.4.6}$$

which specifies that the magnitude response function be a sine function. This is a monotonically increasing function of frequency $\Omega$ which is 0 for $\Omega = 0$, and 1 for $\Omega = \pi$. Based on the transfer function, the magnitude response of this filter is plotted in Figure 16.4.3.

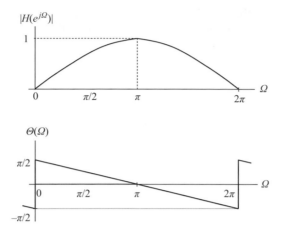

**Fig. 16.4.3** *Transfer characteristic for a first-order finite impulse response highpass filter: magnitude transfer function (upper panel) and phase transfer function (lower panel).*

The amplitude spectrum can be expressed as

$$H\left(e^{j\Omega}\right) = \frac{1}{2}\left(1 - e^{-j\Omega}\right) = \frac{j}{2j}\left(e^{j\Omega/2} - e^{-j\Omega/2}\right)e^{-j\Omega/2} = j\sin\left(\Omega/2\right)e^{-j\Omega/2}$$

$$= \sin\left(\Omega/2\right)e^{-j\left(\Omega/2 - \pi/2\right)} = \sin\left(\Omega/2\right)\left(\cos\left(\Omega/2 - \pi/2\right) - j\sin\left(\Omega/2 - \pi/2\right)\right),$$

and the phase spectrum can be calculated as

$$\theta(\Omega) = \tan^{-1}\frac{-\sin\left(\Omega/2 - \pi/2\right)}{\cos\left(\Omega/2 - \pi/2\right)} = -\frac{1}{2}\left(\Omega - \pi\right), \tag{16.4.7}$$

as shown in Figure 16.4.3. The sine amplitude function is negative for $\Omega < 0$, adding a phase of $-\pi$ to expression (16.4.7).

## 16.5  Structures of Finite Impulse Response Filters

In this section, the structures of digital filters with finite impulse response are presented. Since these structures show relationships among the internal variables of the inputs and outputs of filters, they can be used in implementing digital filters. It will be shown that there are various structures of digital filters with defined advantages and disadvantages in relation to the implementation of finite impulse response filters. The input–output relation of a causal finite impulse response filter of order $N$ is given by

$$y(n) = \sum_{k=0}^{k=N} h(k)x(n-k), \tag{16.5.1}$$

where the impulse response is defined to be $h(n) = 0$ for $n < 0$ and $n > N$. By taking the $z$-transform of both sides of this equation, and after some mathematical operations, we obtain

$$Y(z) = \sum_{n=-\infty}^{\infty} y(n)z^{-n} = \sum_{n=-\infty}^{\infty} \sum_{k=0}^{k=N} h(k)x\,(n-k)\,z^{-n} = \sum_{n-k=l}^{\infty} \sum_{k=0}^{k=N} h(k)x(l)z^{-k-l}$$

$$= \sum_{l=-\infty}^{\infty} x(l)z^{-l} \sum_{k=0}^{k=N} h(k)z^{-k} = \left(\sum_{k=0}^{k=N} h(k)z^{-k}\right) \sum_{l=-\infty}^{\infty} x(l)z^{-l}$$

$$= H(z)X(z).$$

where the transfer function is defined as a polynomial in $z^{-1}$ of degree $N$, that is,

$$H(z) = \frac{Y(z)}{X(z)} = \sum_{k=0}^{k=N} h(k)z^{-k}. \tag{16.5.2}$$

Three forms of digital finite impulse response filters will be presented in this section: the direct form, the cascade form, and the polyphase form.

**Direct form of a finite impulse response filter.** A finite impulse response filter, defined by the input–output relation (16.5.2), is defined by $N + 1$ coefficients. Generally, for this filter implementation, we need $N + 1$ multipliers and $N$ two-input adders. If the multiplier coefficients are the coefficients of the transfer function, the filter structure is called a *direct form structure*. For example, for a third-order filter ($N = 3$), we will have

$$y(n) = \sum_{k=0}^{k=3} h(k)x\,(n-k) = h(0)x(n) + h(1)x\,(n-1) + h(2)x\,(n-2) + h(3)x\,(n-3), \tag{16.5.3}$$

which is represented by its direct form structure in Figure 16.5.1.

**Cascade form of a finite impulse response filter.** The finite impulse response filter transfer function can be factored as follows:

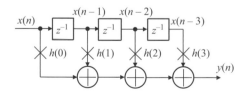

**Fig. 16.5.1** *Direct form: a third-order finite impulse response filter called a taped delay or a transversal filter.*

$$H(z) = \sum_{k=0}^{k=N} h(k)z^{-k} = h(0) + h(1)z^{-1} + \cdots + h(N)z^{-N}$$

$$= h(0)\prod_{k=1}^{K}\left(1 + a_{1,k}z^{-1} + a_{2,k}z^{-2}\right) \qquad (16.5.4)$$

for $K = N/2$ if $N$ is even, and $K = (N+1)/2$ if $N$ is odd, with $a_{2,K} = 0$. A fourth-order finite impulse response filter ($N = 4$) in a canonical cascade form is shown in Figure 16.5.2. It is composed of two second-order sections. It has four two-input adders and five multipliers. The scheme shown is in an agreement with the analytical expression for the transfer function obtained as follows:

$$H(z) = h(0)\prod_{k=1}^{K}\left(1 + a_{1,k}z^{-1} + a_{2,k}z^{-2}\right) = h(0)\prod_{k=1}^{2}\left(1 + a_{1,k}z^{-1} + a_{2,k}z^{-2}\right)$$

$$= h(0)\left(1 + a_{1,1}z^{-1} + a_{2,1}z^{-2}\right)\left(1 + a_{1,2}z^{-1} + a_{2,2}z^{-2}\right).$$

A transfer function of order $N$ can be decomposed into $L$ branches by using the following expression:

$$H(z) = \sum_{k=0}^{k=N} h(k)z^{-k} = \sum_{l=0}^{L-1} z^{-l}H_l\left(z^L\right), \qquad (16.5.5)$$

where

$$H_l(z) = \sum_{n=0}^{n=\lfloor(N+1)/L\rfloor} h\left(Ln+l\right)z^{-n}, \qquad (16.5.6)$$

for $0 \le l \le L-1$, and $h(n) = 0$ for $n > N$. This decomposition is called a *polyphase decomposition*, and the realization based on it is called a *polyphase realization*. In the case when the order is $N = 8$ and the number of branches is $L = 3$, we can have

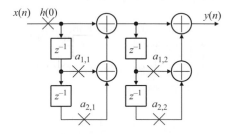

**Fig. 16.5.2** *Cascade form: a fourth-order finite impulse response filter, $N = 4$.*

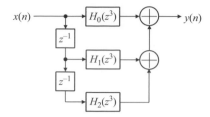

**Fig. 16.5.3** *Polyphase form of an N = 8 finite impulse response filter, L = 3.*

$$H_l(z) = \sum_{n=0}^{n=[(N+1)/L]} h(Ln+l)z^{-n} = \sum_{n=0}^{n=3} h(3n+l)\,z^{-n}, \qquad (16.5.7)$$

or the branch functions (or subfilters) are defined as

$$H_0(z) = h(0) + h(3)z^{-1} + h(6)z^{-2}, \qquad (16.5.8)$$

$$H_1(z) = h(1) + h(4)z^{-1} + h(7)z^{-2}, \qquad (16.5.9)$$

$$H_2(z) = h(2) + h(5)z^{-1} + h(8)z^{-2}. \qquad (16.5.10)$$

Thus, the transfer function is

$$H(z) = \sum_{l=0}^{L-1} z^{-l} H_l\left(z^L\right) = \sum_{l=0}^{2} z^{-l} H_l\left(z^L\right) = H_0\left(z^3\right) + H_1\left(z^3\right) z^{-1} + H_2\left(z^3\right) z^{-2}. \qquad (16.5.11)$$

**Polyphase form of a finite impulse response filter.** The polyphase form of this filter is shown in Figure 16.5.3.

## 16.6   Basic Infinite Impulse Response Filters

In this section, the following basic forms of infinite impulse response filters will be presented: a lowpass filter, a highpass filter, and a bandpass filter.

**Lowpass infinite impulse response filter.** The transfer function of a first-order lowpass infinite impulse response filter is expressed as

$$H(z) = \frac{1-\alpha}{2} \frac{1+z^{-1}}{1-\alpha z^{-1}}, \qquad (16.6.1)$$

where the constant term is $|\alpha| < 1$, in order to fulfil the stability condition. This transfer function has *one* 0 that is on the unit circle defined by $z = -1$, which is the stopband of the filter, and one real pole at $z = \alpha$. The frequency response is

$$H\left(e^{j\Omega}\right) = \frac{1-\alpha}{2}\left(\frac{1+e^{-j\Omega}}{1-\alpha e^{-j\Omega}}\right),$$
(16.6.2)

which specifies the magnitude response function to be a monotonically decreasing function of the frequency $\Omega$ which is 1 for $\Omega = 0$, and 0 for $\Omega = \pi$. Based on the transfer function (16.6.2), the frequency response of this filter can be found and expressed as

$$\left|H\left(e^{j\Omega}\right)\right| = \sqrt{\frac{(1-\alpha)^2\,(1+\cos\Omega)}{2\left(1+\alpha^2-2\alpha\cos\Omega\right)}}.$$
(16.6.3)

The graph of this function is sketched in Figure 16.6.1. The parameter $\alpha$ controls the width of the filter. When $\alpha$ increases, the width of the filter decreases. Thus, the $\alpha$ value for the dashed graph in Figure 16.6.1 is bigger than the $\alpha$ value for the full graph.

**Highpass infinite impulse response filter.** The transfer function of a first-order highpass infinite impulse response filter is expressed as

$$H(z) = \frac{1+\alpha}{2}\,\frac{1-z^{-1}}{1-\alpha z^{-1}}.$$
(16.6.4)

The constant term is $|\alpha| < 1$ in order to fulfil the stability condition. This transfer function has *one* 0 that is on the unit circle defined by $z = -1$, and one real pole at $z = -\alpha$. The frequency response is

$$H\left(e^{j\Omega}\right) = \frac{1+\alpha}{2}\left(\frac{1-e^{-j\Omega}}{1-\alpha e^{-j\Omega}}\right),$$

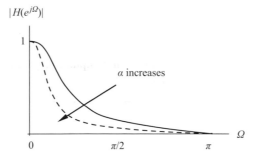

**Fig. 16.6.1** *Magnitude response of a lowpass filter.*

which specifies that the magnitude response function be

$$\left|H\!\left(e^{j\Omega}\right)\right| = \sqrt{\frac{(1+\alpha)^2\,(1-\cos\Omega)}{2\left(1+\alpha^2-2\alpha\cos\Omega\right)}}. \tag{16.6.5}$$

This is a monotonically increasing function of the frequency $\Omega$ which is 0 for $\Omega = 0$, and 1 for $\Omega = \pi$. Based on this transfer function, the magnitude response of this filter is sketched in Figure 16.6.2.

**Bandpass infinite impulse response filter.** The transfer function of a second-order bandpass infinite impulse response filter is expressed as

$$H(z) = \frac{1-\alpha}{2}\,\frac{1-z^{-2}}{1-\beta\,(1+\alpha)\,z^{-1}+\alpha z^{-2}}, \tag{16.6.6}$$

where the constant term is $|\alpha| < 1$ in order to fulfil the stability condition. This transfer function has *two* zeros that are on the unit circle, defined by $z = -1$ and $z = +1$, respectively, and one pair of complex conjugate poles. Based on the transfer function, the magnitude response of this filter is sketched in Figure 16.6.3. The parameter $\alpha$ controls the cut-off frequencies

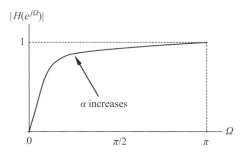

**Fig. 16.6.2** *Magnitude response of a highpass filter.*

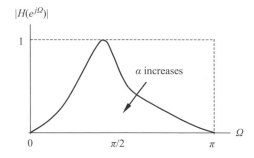

**Fig. 16.6.3** *Magnitude response of a bandpass filter.*

of the filter, and the parameter $\beta$ controls the position of the maximum of the frequency response curve.

## 16.7    Structures of Infinite Impulse Response Filters

### 16.7.1    Introduction

An infinite impulse response linear-time-invariant digital filter is described by a constant coefficient difference equation of the following form:

$$\sum_{k=0}^{N} a_k y(n-k) = \sum_{k=0}^{M} b_k x(n-k). \tag{16.7.1}$$

If we apply the $z$-transform to both sides and then make use of the linearity and time-shifting properties of the $z$-transform, we derive the expression

$$Y(z) \sum_{k=0}^{N} a_k z^{-k} = X(z) \sum_{k=0}^{M} b_k z^{-k}. \tag{16.7.2}$$

Therefore, the transfer function of a finite-dimensional infinite impulse response digital filter is

$$H(z) = \frac{Y(z)}{X(z)} = \frac{\displaystyle\sum_{k=0}^{M} b_k z^{-k}}{\displaystyle\sum_{k=0}^{N} a_k z^{-k}} = \frac{b_0 + b_1 z^{-1} + \cdots + b_k z^{-M}}{a_0 + a_1 z^{-1} + \cdots + a_N z^{-N}}. \tag{16.7.3}$$

### 16.7.2    Conventional Description of Block Diagrams

A discrete-time system is conventionally described by a block diagram. Using this diagram, we can directly find the transfer function of the system. Suppose a discrete-time system is described by its difference equation

$$y(n) = 2x(n) - x(n-1) - \frac{1}{2}y(n-1). \tag{16.7.4}$$

The discrete-time block diagram of this system is presented in Figure 16.7.1, where $D$ represents a unit delay.

The discrete-time system described by its difference equation in the time domain, eqn (16.7.4), can be described by the following equation in the $z$ domain:

$$Y(z) = 2X(z) - X(z)z^{-1} - \frac{1}{2}Y(z)z^{-1}. \tag{16.7.5}$$

The discrete-time block diagram of this system in the $z$ domain is shown in Figure 16.7.2.

This block diagram can be reduced in three steps to find the overall transfer function shown in Figure 16.7.3.

The obtained transfer function can be confirmed by deriving the same function from eqn (16.7.5), which gives

$$H(z) = \frac{Y(z)}{X(z)} = \frac{2 - z^{-1}}{1 + z^{-1}/2} = \frac{2z - 1}{z + 1/2}. \tag{16.7.6}$$

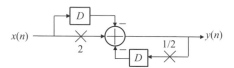

**Fig. 16.7.1** *Discrete-time block diagram of the system.*

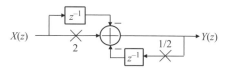

**Fig. 16.7.2** *Discrete-time block diagram of the system in the z domain.*

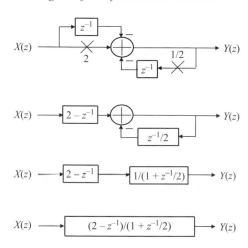

**Fig. 16.7.3** *Discrete-time block diagram of the system in the z domain.*

### 16.7.3 Direct Forms of Infinite Impulse Response Filters

Let an $N$th-order infinite impulse response filter transfer function $H(z)$ be characterized by $2N + 1$ coefficients. If these coefficients are multiplier coefficients, then the filter structures obtained are called direct form structures. Suppose that the transfer function of a second-order infinite impulse response filter is defined as

$$H(z) = \frac{Y(z)}{X(z)} = \frac{b_0 + b_1 z^{-1} + b_1 z^{-2}}{1 + a_1 z^{-1} + a_2 z^{-2}}, \qquad (16.7.7)$$

which is characterized by five coefficients. Defining two subfilters by their transfer functions expressed as

$$H_1(z) = \frac{W(z)}{X(z)} = b_0 + b_1 z^{-1} + b_2 z^{-2}, \qquad (16.7.8)$$

and

$$H_2(z) = \frac{Y(z)}{W(z)} = \frac{1}{1 + a_1 z^{-1} + a_2 z^{-2}} \qquad (16.7.9)$$

the transfer function of the infinite impulse response filter can be expressed as

$$H(z) = H_1(z)H_2(z), \qquad (16.7.10)$$

and the filter can be realized according to the forms shown in Figure 16.7.4. Namely, from the first transfer function, we obtain

$$W(z) = X(z) \left( b_0 + b_1 z^{-1} + b_2 z^{-2} \right), \qquad (16.7.11)$$

which can be represented in the time domain as

$$w(n) = b_0 x(n) + b_1 x(n-1) + b_2 x(n-2). \qquad (16.7.12)$$

From the second transfer function, we obtain

$$Y(z) = W(z) \left( \frac{1}{1 + a_1 z^{-1} + a_2 z^{-2}} \right), \qquad (16.7.13)$$

or

$$Y(z) = W(z) - Y(z)a_1 z^{-1} - Y(z)a_2 z^{-2}, \qquad (16.7.14)$$

which results in the time-domain expression in the form

$$y(n) = w(n) - a_1 y(n-1) - a_2 y(n-2). \qquad (16.7.15)$$

Figure 16.7.4 shows the structure of the filter, which is plotted using expressions for $w(n)$ and $y(n)$.

   The structure of this filter is in what is called the *first direct form*. The first subfilter is an finite impulse response filter. The overall structure is non-canonical because four delays are used to implement the *second-order transfer function*. The first direct form can also be implemented in a *non-canonical form*, as shown in Figure 16.7.5.

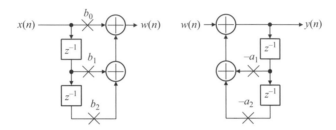

**Fig. 16.7.4** *Cascade form: a second-order infinite impulse response filter.*

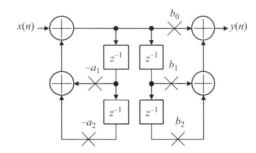

**Fig. 16.7.5** *Cascade form: a non-canonical direct form structure.*

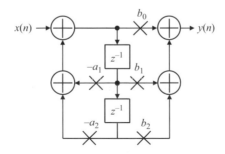

**Fig. 16.7.6** *Cascade form: a canonical second-order direct form structure.*

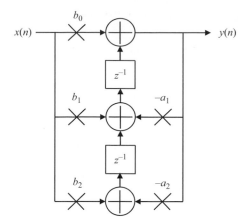

**Fig. 16.7.7** *Cascade form: a canonical second-order direct form transposed structure.*

If we simplify the structure in Figure 16.7.5, we can obtain the *canonical form*, as shown in Figure 16.7.6. This form is obtained by using inner delay blocks for both cascades, because the signals at the output of corresponding delay blocks are the same. This form is also called the *second direct form* and can be transposed to obtain the form in Figure 16.7.7.

## 16.8    Algorithms for the Design of Digital Filters

In this chapter, the basic algorithms for the design of both infinite impulse response and finite impulse response filters are presented. The bilinear transformation method, which is used for the design of infinite impulse response filters, is briefly discussed. However, most of the text is related to the design of finite impulse response filters. In order to reduce the Gibbs effect, the algorithms presented are based on the truncated Fourier series expansion of a desired frequency response. The procedure for truncating these expansions is based on the application of various window functions. Finally, two algorithms based on iterative optimization are presented.

### 16.8.1    Ideal and Real Frequency Responses

The process for designing a digital filter involves deriving a structurally realizable transfer function $H(z)$ that approximates the specified frequency response with a given accuracy. In this process, either the amplitude responses or/and phase responses are specified, or the unit impulse or the unit step response is specified. In this chapter, the methods used for designing digital filters will be for the case when the amplitude responses are specified.

In practice, we cannot use the ideal transfer characteristic of a digital filter; instead we can use its expected transfer characteristic with the acceptable tolerances defined for the stopband and the passband. Such a characteristic, for a lowpass filter, is shown in Figure 16.8.1.

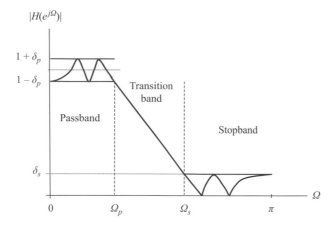

**Fig. 16.8.1** *Transfer characteristic of a lowpass filter.*

In the passband, the amplitude is approximated by

$$1 - \delta_p \leq \left| H\left(e^{j\Omega}\right)\right| \leq 1 + \delta_p, \text{for} |\Omega| \leq \Omega_p, \tag{16.8.1}$$

where the error of approximation is $\delta_p$. In the stopband, the amplitude is approximated by

$$\left| H\left(e^{j\Omega}\right)\right| \leq \delta_s, \text{for } \Omega_s \leq |\Omega| \leq \pi, \tag{16.8.2}$$

where the error of approximation is $\delta_s$. The errors of approximation (the limits of the tolerances) in the passband $\delta_p$ and in the stopband $\delta_s$ are called the peak ripple values, while $\Omega_p$ is the passband edge frequency, and $\Omega_s$ is the stopband edge frequency. Instead of the linear scale, we can use the logarithmic scale to represent the transfer function of a digital filter. In this case, the loss function of the digital filter is defined as

$$a\left(\Omega\right) = -20 \log \left| H\left(e^{j\Omega}\right)\right|, \tag{16.8.3}$$

and is expressed in decibels. Consequently, the peak passband ripple value is

$$\alpha_p = -20 \log \left(1 - \delta_p\right), \tag{16.8.4}$$

and the minimum stopband attenuation is

$$\alpha_s = -20 \log \left(\delta_s\right). \tag{16.8.5}$$

The frequencies $\Omega_p$ and $\Omega_s$ are normalized angular frequencies, where the normalization factor is the sampling frequency $f_{sam}$ expressed in hertz, that is,

$$\Omega_p = \frac{\omega_p}{f_{sam}} = \frac{2\pi f_p}{f_{sam}}, \tag{16.8.6}$$

and

$$\Omega_s = \frac{\omega_s}{f_{sam}} = \frac{2\pi f_s}{f_{sam}}, \tag{16.8.7}$$

where $f_p$ and $f_s$ are the passband and stopband frequencies of the filter, expressed in hertz.

## 16.8.2   Basic Methods for the Design of Digital Filters

The design of infinite impulse response filters is based on the design of corresponding analogue filters. For this purpose, the digital filter specifications have to be converted into the analogue lowpass filter specifications. Then, the determined analogue lowpass filter transfer function has to be transformed into the transfer function of the desired digital filter. For this purpose, the bilinear transformation is used.

The design of finite impulse response filters is based on an approximation of the amplitude response, assuming that the phase response is expected to be linear. The frequency response of a finite impulse response filter is

$$H\left(e^{j\Omega}\right) = \sum_{n=0}^{N} h(n)\, e^{-j\Omega n}. \tag{16.8.8}$$

From this expression, we can see that the design of a finite impulse response filter can be achieved in two ways. Firstly, the design can be simply accomplished by determining the impulse response $h(n)$ of the filter. In this case, if we want to have a linear phase of the designed filter, the following condition has to be fulfilled:

$$h(n) = \pm h\left(n - N\right). \tag{16.8.9}$$

Secondly, the design can be based on the discrete Fourier transform of the impulse response. In this case, two approaches can be used: the windowed Fourier series approach and the frequency-sampling approach.

**Algorithms based on the bilinear transformation method.** One way to approach digital filter design is to make the digital filter frequency response the same as the frequency response of the analogue filter. However, the discrete-time domain response does not follow the continuous-time response. Namely, samples of two sinusoidal signals that have different frequencies, where the frequencies differ for an integer multiple of the sampling rate, can be the same, and their graphical representations can also be the same. Therefore, it would not be possible to distinguish them by observing their graphs in the discrete-time domain.

As we said previously, any band-limited signal must be sampled by the sampling rate defined by the Nyquist theorem. Thus, in order to design a digital system that corresponds to an analogue system, we need to have all of the frequency components of the analogue signal that are greater than half of the sampling frequency to be equal to 0. In order to find the transfer characteristic of the discrete-time system that is the same as the transfer characteristic of the analogue system, we can convert the desired transfer function, which is in the $s$ domain, into the corresponding transfer function in the $z$ domain. The bilinear transformation is one-to-one mapping of a single point in the $s$ plane into a unique point in the $z$ plane, and vice versa, as expressed in the following form:

$$s = \frac{2}{T} \frac{1 - z^{-1}}{1 + z^{-1}}. \tag{16.8.10}$$

This mapping is called *bilinear* because the numerator and the denominator of expression (16.8.10) are both linear functions of $z$. This transform converts any stable continuous-time system into a stable discrete-time system. Thus, the relationship between the transfer functions in these two domains can be expressed in the following form:

$$H(z) = H_a(s) = H_a \left( \frac{2}{T} \frac{1 - z^{-1}}{1 + z^{-1}} \right), \tag{16.8.11}$$

where $H_a(s)$ is the transfer function of the analogue filter, and $T$ is the step size in the numerical integration. If this method is used, there will not be any aliasing. However, there will be some warping, due to the method of transferring the imaginary axis $j\omega$ to the unit circle in the $z$ domain.

**Algorithm based on the windowed Fourier series.** When designing finite impulse response filters, the following methods can be used:

1. A method based on truncating the Fourier series representation of the given frequency response.

2. A method based on the use of the filter's impulse response, which is computed by inverse Fourier transform from the frequency samples of the filter's frequency response.

As we found in previous chapters, the *synthesis* and *analysis* equations, forming the discrete-time Fourier transform pair, are

$$X\left(e^{j\Omega}\right) = \sum_{n=-\infty}^{\infty} x(n) e^{-j\Omega n}, \tag{16.8.12}$$

and

$$x(n) = \frac{1}{2\pi} \int_{-\pi}^{\pi} X\left(e^{j\Omega}\right) e^{j\Omega n} d\Omega. \tag{16.8.13}$$

The first equation can be interpreted as a Fourier series expansion of the periodic function $X\left(e^{j\Omega}\right)$ with coefficients $x(n)$. Suppose that the desired frequency response function of a digital filter is a periodic function, $H_{de}\left(e^{j\Omega}\right)$. Then we can use the synthesis equation to express this function in the form of a Fourier series, that is,

$$H_{de}\left(e^{j\Omega}\right) = \sum_{n=-\infty}^{\infty} h_{de}(n) e^{-j\Omega n}, \tag{16.8.14}$$

where the Fourier coefficients are defined by the analysis equation

$$h_{de}(n) = \frac{1}{2\pi} \int_{-\pi}^{\pi} H_{de}\left(e^{j\Omega}\right) e^{j\Omega n} d\Omega, \quad -\infty \le n \le \infty. \tag{16.8.15}$$

Because the desired transfer function $H_{de}\left(e^{j\Omega}\right)$ is usually composed of singularity functions, the impulse response $h_{de}(n)$ is a non-causal signal of infinite duration. This response has to be truncated at some points, say, from $-M$ to $+M$, to yield a finite impulse response filter of length $2M + 1$. Truncating $h_{de}(n)$ is equivalent to multiplying $h_{de}(n)$ by a finite length window function $w(n)$, that is,

$$h_{tr}(n) = h_{de}(n) \cdot w(n). \tag{16.8.16}$$

Using the convolution theorem, we can then find the discrete-time Fourier transform of the truncated signal $h_{tr}(n)$. Namely, because the truncated signal is the product of the impulse response and the window function, the discrete-time Fourier transform of the truncated signal is the convolution of the transfer function and the discrete-time Fourier transform of the window function, that is,

$$H_{tr}\left(e^{j\Omega}\right) = \frac{1}{2\pi} \int_{-\pi}^{\pi} H_{de}\left(e^{j\Omega}\right) W\left(e^{j(\Omega-\Phi)}\right) e^{j\Omega n} d\Phi. \tag{16.8.17}$$

Thus, the discrete-time Fourier transform of the truncated impulse response is, in fact, a periodic continuous convolution of the desired response $H_{de}\left(e^{j\Omega}\right)$ and the discrete-time Fourier transform $W\left(e^{j\Omega}\right)$ of the window signal $w(n)$. For the simple case when the window signal is a rectangular pulse defined as

$$w(n) = \begin{cases} 1 & -M \le n \le M \\ 0 & \text{otherwise} \end{cases}, \tag{16.8.18}$$

we can find its discrete-time Fourier transform in the form

$$W\left(e^{j\Omega}\right) = \sum_{n=-M}^{M} 1 \cdot e^{-j\Omega n} = \frac{\sin\left[(2M+1)\ \Omega/2\right]}{\sin\left(\Omega/2\right)}. \tag{16.8.19}$$

In order to exclude the effects of the Gibbs phenomenon, this function has to be as narrow as possible; ideally, it should be a delta function. Thus, the window function should be as long as possible, that is, the number $2M + 1$ should be very large. On the other hand, in order to avoid having a large number of filter coefficients, the window function is expected to be as short as possible, that is, the number $2M + 1$ should be very small.

Therefore, our objective when designing a finite impulse response filter is to find a finite-duration impulse response signal $h_{tr}(n)$ whose discrete-time Fourier transform $H_{tr}\left(e^{j\Omega}\right)$ approximates the desired discrete-time Fourier transform $H_{de}\left(e^{j\Omega}\right)$ according to the defined criterion. For example, this criterion can be to minimize the integral square error expressed as

$$e = \frac{1}{2\pi} \int_{-\pi}^{\pi} \left[H_{tr}\left(e^{j\Omega}\right) - H_{de}\left(e^{j\Omega}\right)\right]^2 d\Omega. \tag{16.8.20}$$

It can be shown that $e$ is minimal when the impulse response $h_{de}(n)$ is truncated, that is, $h_{tr}(n) = h_{de}(n)$ for $-M \le n \le M$.

**Example**

Find the coefficients of a causal ideal lowpass filter defined by its frequency response

$$H_{LP}\left(e^{j\Omega}\right) = \begin{cases} 1 & |\ \Omega\ | \le \Omega_c \\ 0 & \Omega_c < \Omega < \pi \end{cases}, \tag{16.8.21}$$

as shown in Figure 16.8.2.

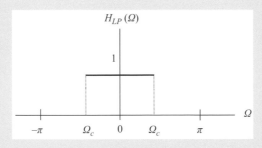

**Fig. 16.8.2**    *Transfer characteristic of an ideal lowpass filter.*

**Solution.** The inverse discrete-time Fourier transform is given by

$$
h_{LP}(n) = \frac{1}{2\pi} \int_{-\pi}^{\pi} H_{LP}\left(e^{j\Omega}\right) e^{j\Omega n} d\Omega = \frac{1}{2\pi} \int_{-\Omega_c}^{+\Omega_c} 1 \cdot e^{j\Omega n} d\Omega
$$
$$
= \frac{1}{2\pi} \left( \frac{e^{j\Omega_c n}}{jn} - \frac{e^{-j\Omega_c n}}{jn} \right) = \frac{\sin \Omega_c n}{\pi n}
$$

(16.8.22)

which exists in an infinite interval and is not absolutely summable. If we multiply this function with a rectangular pulse $w(n)$ defined in the interval from $-M$ to $+M$, we will get a non-causal approximation function of $h_{LP}(n)$ with length $N = 2M + 1$. The right-shifted version of this approximation function specifies the coefficient of a causal finite impulse response filter according to the expression

$$
h_{tr}(n) = \left\{ \begin{array}{cc} \frac{\sin \Omega_c (n-M)}{\pi(n-M)} & 0 \le n \le N-1 \\ 0 & \text{otherwise} \end{array} \right\}.
$$

(16.8.23)

The shape of the filter transfer function mostly depends on the characteristics of the window function. In order to obtain the specific characteristics, various window functions are defined. Some of them are shown in Table 16.8.1. The magnitude spectra of these window functions directly determine the frequency-response characteristic of the filter. These spectra are characterized by a large main lobe with a series of sidelobes with decreasing amplitudes that are specified by two parameters: the main lobe width and the relative sidelobe level.

In summary, to obtain the desired filter-amplitude characteristic, we can find its impulse response and then truncate it by using one of the window functions $w(n)$. This truncation corresponds to convolution in the frequency domain, which gives the frequency response of the filter designed using that window function $w(n)$.

The Kaiser window, which is given in Table 16.8.1, is an example of an adjustable window function. Namely, $\alpha$ is an adjustable parameter that controls the minimum attenuation $\alpha_s = -20\log(\delta_s)$ in the stopband of the windowed filter response. If a digital filter is designed using this window function, it is assumed that the passband ripple $\delta_p$ is approximately equal to stopband attenuation $\delta_s$.

**Table 16.8.1** *Window functions.*

| Name | $w(n)$ |
| --- | --- |
| Blackman | $0.42 + 0.5\cos\frac{2\pi n}{2M+1} + 0.08\cos\frac{4\pi n}{2M+1}$ |
| Hamming | $0.54 + 0.46\cos\frac{2\pi n}{2M+1}$ |
| Hanning (raised cosine) | $\frac{1}{2}\left(1 + \cos\frac{2\pi n}{2M+1}\right)$ |
| Kaiser | $\dfrac{I_0\left[\alpha\sqrt{1-(n/M)^2}\right]}{I_0[\alpha]}$ |

*Note*: $w(n)$ is defined in the interval $-M \le n \le M$.

### 16.8.3    Algorithms Based on Iterative Optimization

In Section 16.8.1, some of the algorithms explained could easily be implemented on a computer. In this section, a new class of algorithms, which are based on iterative optimization methods, will be presented. In these algorithms, the main objective of optimization is to minimize the error between the *desired* frequency response $H_{de}\left(e^{j\Omega}\right)$ and the response $H\left(e^{j\Omega}\right)$ *achieved* in the course of the iterative procedure. This error is usually defined in two ways: it can be defined as a *weighted error function* $e(\Omega)$ expressed in the form

$$e(\Omega) = V\left(e^{j\Omega}\right)\left[H\left(e^{j\Omega}\right) - H_{de}\left(e^{j\Omega}\right)\right] \tag{16.8.24}$$

or it can be defined as an integral of the *p*th power of $|e(\Omega)|$, where the integral is taken over the specified frequency range $\Delta\Omega$, that is,

$$e_p = \int_{\Omega \in \Delta\Omega} |e(\Omega)|^p d\Omega = \int_{\Omega \in \Delta\Omega} \left|V\left(e^{j\Omega}\right)\left[H\left(e^{j\Omega}\right) - H_{de}\left(e^{j\Omega}\right)\right]\right|^p d\Omega, \tag{16.8.25}$$

which is known as the least-*p* criterion. Here, $V(e^{j\Omega})$ is a user-specified positive weighted function. The least square criterion, which is defined by $p = 2$, is often used in practice.

**The Parks–McClellan/Remez Algorithm.** The main goal in filter design is to make a filter in a time interval that is as short as possible and to achieve a transfer function approximation that is as accurate as possible. An optimal algorithm is the Chebyshev approximation or the McClellan–Parks/Remez method. This is a highly efficient algorithm for designing equiripple linear-phase finite impulse response filters. It can be shown that the frequency response of a linear-phase finite impulse response filter of length $N + 1$ is expressed as

$$H\left(e^{j\Omega}\right) = e^{-jN\Omega/2} e^{j\beta} H_{AR}(\Omega), \tag{16.8.26}$$

where $H_{AR}(\Omega)$ is the amplitude response of the filter which is a real function of the frequency $\Omega$. The *weighted error function e* in this case is defined as

$$e(\Omega) = V(\Omega)[H_{AR}(\Omega) - H_{de}(\Omega)]. \tag{16.8.27}$$

The weighting function $V(\Omega)$ is defined and used to control the peak error in the stopband and the passband of the filter. The Parks–McClellan algorithm minimizes the peak absolute value of the error function $e(\Omega)$. The peaks of $e(\Omega)$ occur at the frequencies $\Omega = \Omega_k$, $k = 0$, 1, ..., $L + 1$, where the following condition is fulfilled:

$$\frac{de(\Omega)}{d\Omega} = 0. \tag{16.8.28}$$

To determine the location of the $L + 2$ extremal frequencies, the Remez exchange algorithm is used. The exchange design method gives us a Chebyshev-type filter. Also, this filter has a better transition-band roll-off characteristic than that characteristics obtained by the Chebyshev and Kaiser windowing techniques.

**Algorithm based on a least mean square error criterion**. The mean square error criterion is defined as the integral

$$e_2 = \int_\Omega e^2(\Omega)\, d\Omega = \int_\Omega \left\{\, V\left(e^{j\Omega}\right)\left[H\left(e^{j\Omega}\right) - H_{de}\left(e^{j\Omega}\right)\right] \right\}^2 d\Omega, \tag{16.8.29}$$

which is a special case of the least-$p$ criterion. Here, $H\left(e^{j\Omega}\right)$ is the *designed amplitude response*, and $H_{de}\left(e^{j\Omega}\right)$ is the *desired amplitude response*. This criterion can be applied for the design of a linear-phase finite impulse response filter. In that case, the integral can be replaced by a sum according to the expression

$$e_2 = \sum_{i=1}^{K} \left\{ V\left(\Omega_i\right)\left[H\left(\Omega_i\right) - H_{de}\left(\Omega_i\right)\right]\right\}^2. \tag{16.8.30}$$

The amplitude response of the designed linear-phase finite impulse response filter $H(\Omega_i)$ can be expressed as

$$H(\Omega_i) = Q(\Omega_i) \sum_{k=0}^{L} a(k) \cos \Omega_i k, \tag{16.8.31}$$

where $Q(\Omega)$, $a(k)$, and $L$ are defined for all types of filters. The mean square error can be expressed in the form

$$\begin{aligned} e_2 &= \sum_{i=1}^{K} \left\{ V\left(\Omega_i\right) \left[ Q(\Omega_i) \sum_{k=1}^{L} a(k) \cos \Omega_i k - H_{de}\left(\Omega_i\right) \right] \right\}^2 \\ &= \sum_{i=1}^{K} \left[ \sum_{k=1}^{L} V\left(\Omega_i\right) Q(\Omega_i)\, a(k) \cos \Omega_i k - V\left(\Omega_i\right) H_{de}\left(\Omega_i\right) \right]^2. \end{aligned} \tag{16.8.32}$$

This is a function of the filter parameters $a(k)$. Thus, the minimum value will be found by equating the first derivatives to 0, that is, from the condition

$$\frac{de_2}{da(k)} = 0, \tag{16.8.33}$$

for every $k$ in the interval $0 \le k \le L$. This expression yields $L + 1$ linear equations that can be solved for $a(k)$. The final solution can be obtained using an iterative method such as the Levinson–Durbin algorithm.

# 17

# Multi-Rate Discrete-Time Signal Processing

The basic theory of discrete-time signal processing is related to the class of single-rate systems because the sampling rate is kept constant, following the theory of signal discretization presented in Chapter 13. However, there are a number of situations where the sampling rate of the same signal has to be converted from one value to the other, as in telecommunication systems or high-quality audio signal processing. In telecommunication systems, various signals are transmitted, which have different bandwidths. Thus, the different signals have to be sampled with different sampling rates that depend on the bandwidth of the signals. A discrete-time system is a multi-rate system if it processes a signal at various sampling rates in various parts of the system. The procedure of converting a signal from one rate to another is called *sampling rate conversion*. To achieve sampling rate conversion, the system uses down-sampling and up-sampling devices, which change the sampling rate of the processed signal. In this chapter, we will use the word 'signal' mostly to mean a discrete-time signal.

Rate conversion can be done in either the continuous-time (analogue) domain or the discrete-time (digital) domain. Conversion in the continuous-time domain consists of three steps. Firstly, the digital signal is converted to an analogue form using a digital-to-analogue converter; secondly, the analogue signal is filtered; and, finally, the filtered analogue signal is converted to the digital domain by using an analogue-to-digital converter. Using this method, it is possible to choose any satisfactory sampling rate of the output signal. However, it is important to control the signal distortion that is caused by the digital-to-analogue and analogue-to-digital conversions; or, simply put, it is important to bear in mind that the Nyquist criterion has to be fulfilled at each step.

Rate conversion can also be accomplished in the discrete-time domain by direct conversion of the sampling rate of the digital signal, as will be presented in this chapter. This conversion can be defined by a rational factor or an arbitrary factor and can be expressed using the concept of independent variable modification in the time domain, and spectral changes in the frequency domain.

For the sake of understanding, the discrete-time signals should be analysed in both the time domain and the frequency domain and related to continuous-time signal processing. The ability to switch from continuous time to discrete time and from time-domain signal processing to frequency-domain signal processing, both for deterministic and stochastic signals, are essential skills to be acquired in order to understand the theory of discrete communication

*Discrete Communication Systems.* Stevan Berber, Oxford University Press. © Stevan Berber 2021.
DOI: 10.1093/oso/9780198860792.003.0017

systems. In this chapter, two basic components in sampling rate conversion, that is, the up-sampler and the down-sampler, will be examined in the time and frequency domains.

## 17.1 Multi-Rate Signals in Time and Frequency Domains

### 17.1.1 Time-Domain Analysis

Suppose that a discrete-time signal $x(n)$ is used to generate a new signal $y(n)$, as shown in Figure 17.1.1. The output signal is the input signal with a modified variable defined by the modification function $f(n)$.

The sampling rate conversion ratio $R$ is defined as

$$R = \frac{F_y}{F_x} = \frac{T_x}{T_y}, \tag{17.1.1}$$

where $F_x$ is the sampling frequency or the sampling rate of $x(n)$, and $F_y$ is the sampling frequency or the sampling rate of $y(n)$, while $T_x$ and $T_y$ are the sampling intervals of the input and output signals, respectively.

The conversion is also called interpolation if the sampling rate of the output is greater than the sampling frequency of the input, that is, $F_y > F_x$, or $R > 1$. The device that does the interpolation is called an interpolator or up-sampler. The process is called decimation if $F_y < F_x$, or $R < 1$, and the device that does the decimation or down-sampling is called a decimator or a down-sampler.

**Operation of an up-sampler (or interpolator, or expander).** The input–output relationship of an up-sampler (interpolator) may be expressed as

$$x_u(n) = y(n) = \begin{cases} x(n/L), & n = 0, \pm L, \pm 2L, \dots \\ 0, & \text{otherwise} \end{cases}, \tag{17.1.2}$$

where $L > 1$ is an integer factor that specifies that a series of $(L-1)$ zeros are inserted between two consecutive samples of the input signal $x(n)$ to obtain the output signal $y(n)$. This is equivalent to the relation

$$x_u(n) = \sum_{k=-\infty}^{\infty} x(k)\delta\,(n - kL),$$

where the sample values of the signal are defined and filled in with zero values in-between to be interpolated later on. Thus, the sampling rate of $y(n)$ is $L$ times larger than the sampling rate

**Fig. 17.1.1** *Sampling rate converter.*

of $x(n)$. The process of up-sampling is presented in Figures 17.1.2 and 17.1.3. Figure 17.1.2 presents a discrete-time signal, and Figure 17.1.3 presents the corresponding up-sampled signal for $L = 2$, which is obtained from $x(n)$ by inserting $L - 1 = 1$ zeros between the samples.

Note that the up-sampled signal can be generated in the same time interval as the original input signal. Thus, the time interval between the samples of the up-sampled signal is, in reality, shorter than the interval between the samples of the original signal. Therefore, the up-sampled signal in real time can be presented as in Figure 17.1.4. The interval over which we are taking 18 samples of $y(n)$ is equal to the interval over which we have taken 9 samples of $x(n)$. Therefore, the sampling time for $y(n)$, expressed in time units, is two times smaller than that for $x(n)$. Consequently, the sampling rate for $y(n)$ is two times larger than for $x(n)$. For this reason, we sometimes say that the up-sampler is a sampling rate expander, or an expander.

**Down-sampler (or decimator).** The input–output relationship of a down-sampler may be expressed as

$$x_d(n) = y(n) = x(nM),$$

where $M > 1$ is an integer factor that specifies that we should take every $M$th sample of $x(n)$ and remove $(M - 1)$ samples that are in-between. Thus, the signal $y(n)$ has the sampling rate that is $1/M$ that of $x(n)$. For the signal $x(n)$ given as an example, which is shown in Figure 17.1.2, the down-sampled signal is shown in Figure 17.1.5 for the case when $M = 2$.

The symbols which will be used in this book to indicate the blocks of an up-sampler and a down-sampler are shown in Figure 17.1.6.

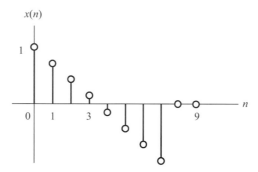

**Fig. 17.1.2** *Discrete-time signal.*

**Fig. 17.1.3** *Discrete up-sampled signal, $T_y = T_x/2$.*

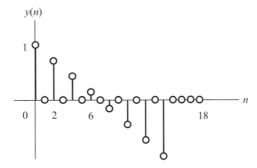

**Fig. 17.1.4** *Up-sampled signal.*

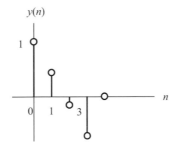

**Fig. 17.1.5** *Down-sampled signal, $T_y = 2T_x$.*

**Fig. 17.1.6** *Up-sampler and down-sampler.*

In practice, the zeros inserted by an up-sampler are replaced by interpolated values, in order to avoid unnecessary spectral components. For that reason, this kind of up-sampling device is called an interpolator.

As said previously, we have to take into account the sampling interval $T_x$ and the sampling frequency $F_x = 1/T_x$. If $T_x$ is taken into account, the up-sampling process and the down-sampling process can be represented as shown in Figure 17.1.7. Obviously, the sampling frequency of the down-sampled signal decreases, and that of the up-sampled signal increases, with respect to the sampling frequency of the input signal.

The procedure for sampling an analogue (continuous-time) signal at the sampling intervals $T$ is shown in Figure 17.1.8. A detailed explanation of this procedure can be found in Chapter 13. Obviously, the procedure of up-sampling will increase the sampling rate of a signal by adding new samples between the existing ones. However, down-sampling will reduce the number of samples by excluding some of the existing samples.

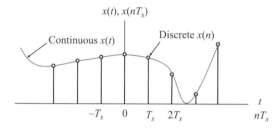

**Fig. 17.1.7** *Up-sampling (upper panel) and down-sampling (lower panel).*

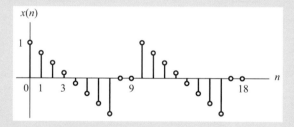

**Fig. 17.1.8** *Sampling procedure.*

## Example

Suppose that the signal $x(n)$ presented in Figure 17.1.9 is transmitted through the system shown in Figure 17.1.10. Find the signals at the output of each block and the output signal $y(n)$ of the system for the first ten samples of the input signal $x(n)$. Compare the output signal with the input signal in the time domain.

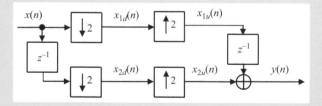

**Fig. 17.1.9** *Signal in the time domain.*

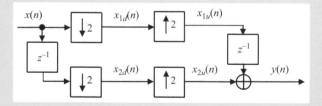

**Fig. 17.1.10** *Rate conversion system.*

**Solution.** The calculated down-sampled and up-sampled values of the signals are shown in Table 17.1.1.

We can see that the output signal is a delayed version of the input signal. A multi-rate system, like this one, that delays and rescales the input signal is called a *perfect reconstruction multi-rate system*.

**Table 17.1.1** *Calculated down-sampled and up-sampled values.*

|            | 0 | 1   | 2    | 3    | 4    | 5    | 6    | 7    | 8    | 9  |
|------------|---|-----|------|------|------|------|------|------|------|----|
| $x(n)$        | 1 | 3/4 | 2/4  | 1/4  | −1/4 | −2/4 | −3/4 | −1   | 0    | 0  |
| $x_{1d}(n)$   | 1 | 2/4 | −1/4 | −3/4 | 0    | 1    | 2/4  | −1/4 | −3/4 | 0  |
| $x_{1u}(n)$   | 1 | 0   | 2/4  | 0    | −1/4 | 0    | −3/4 | 0    | 0    | 0  |
| $x(n-1)$      | 0 | 1   | 3/4  | 2/4  | 1/4  | −1/4 | −2/4 | −3/4 | −1   | 0  |
| $x_{2d}(n)$   | 0 | 3/4 | 1/4  | −2/4 | −1   | 0    | 3/4  | 1/4  | −2/4 | −1 |
| $x_{2u}(n)$   | 0 | 0   | 3/4  | 0    | 1/4  | 0    | −2/4 | 0    | −1   | 0  |
| $x_{1u}(n-1)$ | 0 | 1   | 0    | 2/4  | 0    | −1/4 | 0    | −3/4 | 0    | 0  |
| $y(n)$        | 0 | 1   | 3/4  | 2/4  | 1/4  | −1/4 | −2/4 | −3/4 | −1   | 0  |

## 17.1.2 Frequency-Domain Analysis

**Up-sampler (or interpolator, or expander).** The sampling rate can be increased by factor of $L$ if we interpolate $L - 1$ new samples between any pair of the succeeding samples of the given signal $x(n)$. Let us analyse in the frequency domain the input and output signal of a system that up-samples by a factor of $L$ up-sampling, which is denoted in this chapter as a 'factor-$L$ up-sampler'. The output signal of the system in the time domain can be expressed as

$$x_u(n) = y(n) = \begin{cases} x(n/L), & n = 0, \pm L, \pm 2L, \dots \\ 0, & \text{otherwise} \end{cases}. \tag{17.1.3}$$

If the z-transform of the input signal $x(n)$ is $X(z)$, the z-transform of the output signal $y(n)$ can be calculated as

$$X_u(z) = Y(z) = \sum_{n=-\infty}^{\infty} y(n)z^{-n} = \sum_{n=-\infty}^{\infty} x(n/L)z^{-n}$$

$$= \sum_{\substack{m=|n/L| \text{integer} \\ m=-\infty}}^{\infty} x(m)z^{-Lm} = X\left(z^L\right) \tag{17.1.4}$$

If the region of convergence for the z-transform includes the unit circle, then the discrete-time Fourier transform equals the z-transform evaluated at that unit circle, that is, for $z = e^{j\Omega}$, we obtain

$$X\left(e^{j\Omega}\right) = \sum_{n=-\infty}^{\infty} x(n)e^{-j\Omega n}. \tag{17.1.5}$$

Since $e^{j\Omega}$ is a periodic function of the continuous angular frequency $\Omega$ with period $2\pi$, $X(e^{j\Omega})$ is also periodic in $\Omega$ with period $2\pi$. If the spectrum of the input signal is $X(e^{j\Omega})$, then the spectrum of a factor-$L$ up-sampler can be obtained from

$$Y(z) = X\left(z^{L}\right), \tag{17.1.6}$$

as

$$Y\left(e^{j\Omega}\right) = X\left(e^{jL\Omega}\right) = \sum_{n=-\infty}^{\infty} x(n)\,e^{-jL\Omega n}. \tag{17.1.7}$$

The frequency $\Omega$ in this expression can be denoted by $\Omega_y$ because it is the frequency variable relative to the new sampling rate $F_y$, that is, $\Omega_y = 2\pi F/F_y$, and the relationship between the sampling rate of the input signals and the output signals is

$$\Omega_y = \frac{\Omega_x}{L}. \tag{17.1.8}$$

**Special case: $L = 2$.** In the case of a factor-2 up-sampler, we obtain

$$X_u(z) = Y(z) = X\left(z^2\right). \tag{17.1.9}$$

This up-sampler is shown in Figure 17.1.11.

If the spectrum of input signal is $X(e^{j\Omega})$, then the spectrum of a factor-2 up-sampler is

$$Y\left(e^{j\Omega}\right) = X\left(e^{j2\Omega}\right). \tag{17.1.10}$$

Therefore, if the spectrum of a hypothetical complex signal $x(n)$ is as shown in Figure 17.1.12, then the spectrum of the output signal $y(n)$ is a *twofold repetition of the spectrum* of input signal. The spectrum $Y(e^{j\Omega})$ is a compressed version of $X(e^{j\Omega})$ and the compression factor is equal to the up-sampling factor. We can say that an image of the spectrum is added inside the period of the spectrum. The process of adding this image is called *imaging*.

For a general case of a factor-$L$ up-sampling, the total number of $L - 1$ images will be generated inside one period of $2\pi$. For $L - 1 = 2$ there will be a twofold repetition of the input spectrum, as shown in Figure 17.1.13.

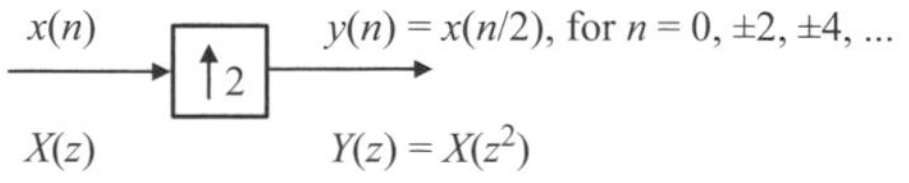

**Fig. 17.1.11** *Up-sampler operation.*

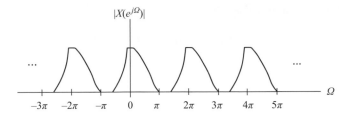

**Fig. 17.1.12** *Input-signal spectrum.*

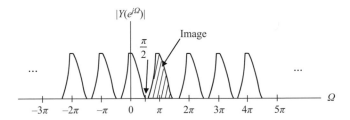

**Fig. 17.1.13** *Output signal spectrum.*

**Down-sampler (or decimator).** Suppose a signal $x(n)$ with the spectrum $X(e^{j\Omega})$ should be down-sampled by an integer factor $M$. Assume that the spectrum $X(e^{j\Omega})$ has non-zero values in the interval $0 \leq \Omega \leq \pi$. The relationship between the input and the output of a down-sampler can be found in the same way as for an up-sampler. Using the expression for the down-sampler

$$x_d(n) = y(n) = x(nM),\qquad(17.1.11)$$

we can find the $z$-transform of the output as

$$X_d(z) = Y(z) = \sum_{n=-\infty}^{\infty} y(n)z^{-n} = \sum_{n=-\infty}^{\infty} x(nM)z^{-n}.\qquad(17.1.12)$$

Let us define an intermediate function $x_{im}(n)$:

$$x_{im}(n) = \begin{cases} x(n), & n = 0, \pm M, \pm 2M, \ldots \\ 0, & \text{otherwise} \end{cases}.\qquad(17.1.13)$$

This signal can now be expressed as a product of the input signal $x(n)$ and a sequence of unit delta functions $p(n)$, that is,

$$x_{im}(n) = x(n)p(n),\qquad(17.1.14)$$

where the signal $p(n)$ is expressed as

$$p(n) = \begin{cases} 1, & n = 0, \pm M, \pm 2M, \dots \\ 0, & \text{otherwise} \end{cases}, \tag{17.1.15}$$

which has the discrete Fourier series representation expressed as

$$p(n) = \frac{1}{M} \sum_{k=0}^{M-1} e^{-j2\pi kn/M} = \frac{1}{M} \sum_{k=0}^{M-1} W_M^{kn}, \tag{17.1.16}$$

because this signal is a periodic train of impulses with a period of $M$, and $W_M^{kn} = e^{-j2\pi kn/M}$. Bearing in mind eqn (17.1.13), we can continue our procedure of finding the $z$-transform of the output signal, that is,

$$Y(z) = \sum_{n=-\infty}^{\infty} x(nM)z^{-n} = \sum_{n=-\infty}^{\infty} x_{im}(nM)z^{-n} \underset{m=nM}{=} \sum_{m=-\infty}^{\infty} x_{im}(m)z^{-m/M}$$

$$= X_{im}\left(z^{1/M}\right) \tag{17.1.17}$$

This relation is only valid for $m$ equal to a multiple of $M$; otherwise, it is equal to 0. The $z$-transform of the intermediate function $x_{im}(n)$ is

$$X_{im}(z) = \sum_{n=-\infty}^{\infty} x_{im}(n)\, z^{-n} = \sum_{n=-\infty}^{\infty} x(n)p(n)z^{-n} = \frac{1}{M} \sum_{n=-\infty}^{\infty} \sum_{k=0}^{M-1} x(n)\, W_M^{kn} z^{-n}$$

$$= \frac{1}{M} \sum_{n=-\infty}^{\infty} \sum_{k=0}^{M-1} x(n)\, e^{-j2\pi kn/M} z^{-n} = \frac{1}{M} \sum_{n=-\infty}^{\infty} \sum_{k=0}^{M-1} x(n) \left(ze^{j2\pi k/M}\right)^{-n}. \tag{17.1.18}$$

$$= \frac{1}{M} \sum_{k=0}^{M-1} \sum_{n=-\infty}^{\infty} x(n) \left(ze^{j2\pi k/M}\right)^{-n} = \frac{1}{M} \sum_{k=0}^{M-1} X\left(zW_M^{-k}\right)$$

Bearing in mind eqn (17.1.17), the output signal in the $z$ domain can be expressed as

$$Y(z) = X_{im}\left(z^{1/M}\right) = \frac{1}{M} \sum_{k=0}^{M-1} X\left(z^{1/M} W_M^{-k}\right). \tag{17.1.19}$$

We expect that the frequency components that exist in the $X(z^{1/M})$ term will overlap with their shifted replicas and will cause aliasing in the frequency domain. This aliasing can be eliminated if we sample the original signal $x(t)$ in the continuous-time domain with a sampling rate that is at least $M$ times the Nyquist rate. Also, we can use a lowpass filter to limit the spectrum of the input signal. Let us demonstrate this procedure in the following example.

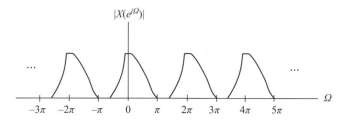

**Fig. 17.1.14** *Input-signal spectrum.*

**A factor-2 down-sampler.** Let us analyse a special case where we have a factor-2 down-sampler. Suppose the spectrum of the input signal is as shown in Figure 17.1.14.

If the spectrum of the input signal is $X(e^{j\Omega})$, then the spectrum of the down-sampler output can be obtained as follows. From the $z$-transform of the output signal (17.1.19), we obtain

$$Y(z) = \frac{1}{2}\sum_{k=0}^{1} X\left(z^{1/2}W_2^{-k}\right) = \frac{1}{2}X\left(z^{1/2}W_2^0\right) + \frac{1}{2}X\left(z^{1/2}W_2^{-1}\right), \qquad (17.1.20)$$

and we can find the discrete-time Fourier transform of $Y(z)$ for $z = e^{j\Omega}$, $W_2^0 = 1$ and $W_2^{-1} = e^{-2\pi n/2}$, as

$$Y\left(e^{j\Omega}\right) = \frac{1}{2}\sum_{k=0}^{1} X\left(e^{j\Omega/2}e^{j\pi k}\right) = \frac{1}{2}\left[X\left(e^{j\Omega/2}\right) + X\left(e^{j(\pi+\Omega/2)}\right)\right], \qquad (17.1.21)$$

or in the form

$$Y\left(e^{j\Omega}\right) = \frac{1}{2}\sum_{k=0}^{1} X\left(e^{j\Omega/2}e^{j\pi k}\right) = \frac{1}{2}\left[X\left(e^{j\Omega/2}\right) + X\left(-e^{j\Omega/2}\right)\right]; \qquad (17.1.22)$$

because $e^{j\pi} = e^{-j\pi} = -1$, we can also find the spectrum as

$$Y\left(e^{j\Omega}\right) = \frac{1}{2}\left[X\left(e^{j\Omega/2}\right) + X\left(e^{j(\Omega-2\pi)/2}\right)\right], \qquad (17.1.23)$$

which is represented by two terms. The first term is a frequency-scaled input-signal spectrum, and the second is a version of the input spectrum that is frequency scaled and delayed for $2\pi$, as plotted in Figure 17.1.15. These two terms overlap, confirming that aliasing takes place in the frequency domain. This aliasing is caused by under-sampling of the original input signal.

In the case when the spectrum of the input signal is limited to the bandwidth defined by $\pm\pi/2$, aliasing will not take place in the frequency domain. Thus, aliasing can be eliminated by proper sampling of the original input signal. In this case, the spectrum of the down-sampled signal is periodic, with period $2\pi$, as we expect, which is shown in Figure 17.1.16. It is important to note that the output spectral components are multiplied by 2 on the same graph.

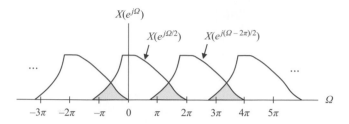

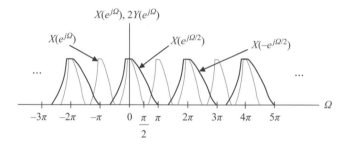

**Fig. 17.1.15** *Output signal spectrum with aliasing.*

**Fig. 17.1.16** *Output signal spectrum without aliasing.*

The expression for the spectrum of a factor-2 down-sampler can be generalized for a factor-$M$ down-sampler as

$$Y(z) = X_{im}\left(z^{1/M}\right) = \frac{1}{M}\sum_{k=0}^{M-1}X\left(z^{1/M}W_M^{-k}\right) = \frac{1}{M}\sum_{k=0}^{M-1}X\left(e^{j\Omega/M}e^{j2\pi k/M}\right)$$

or

$$Y\left(e^{j\Omega}\right) = \frac{1}{M}\sum_{k=0}^{M-1}X\left(e^{j(\Omega-2\pi k)/M}\right). \tag{17.1.24}$$

In this case, aliasing will not occur if the input signal $x(n)$ is band limited in the interval $\pm\pi/M$.

### 17.1.3 Complex Multi-Rate Systems

Complex multi-rate systems can be made from the basic sampling devices and some elements of digital filters. The up-sampler and down-sampler can change the rate of the signal by an integer value. In order to change this ratio for a fraction, it is necessary to use a cascade of a down-sampler and an up-sampler. We are specifically interested in knowing how these samples operate in a complex system and when they can change their order without any

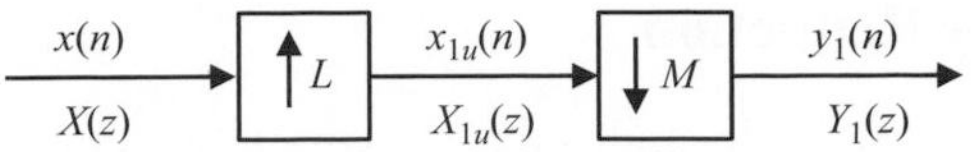

**Fig. 17.1.17** *Up-sampler and down-sampler cascade.*

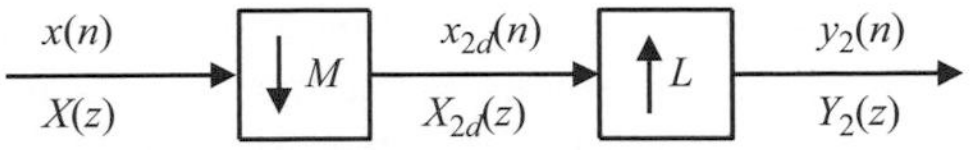

**Fig. 17.1.18** *Down-sampler and up-sampler cascade.*

change in input–output relations. Let us consider a cascade system of an up-sampler and a down-sampler, as shown in Figure 17.1.17. Our aim is to express the output of the cascade as a function of the input. For that purpose, we will analyse the signal processing procedure in the frequency domain, where it takes place in both blocks.

For the system shown in Figure 17.1.17, we may express the output of the up-sampler as

$$X_{1u}(z) = X\left(z^L\right), \tag{17.1.25}$$

and the output of the whole system as

$$Y_1(z) = \frac{1}{M} \sum_{k=0}^{M-1} X_{1u}\left(z^{1/M} W_M^{-k}\right) = \frac{1}{M} \sum_{k=0}^{M-1} X\left(z^{L/M} W_M^{-kL}\right). \tag{17.1.26}$$

Let us now consider a cascade system containing a down-sampler and an up-sampler, as shown in Figure 17.1.18. The output of the down-sampler is

$$X_{2d}(z) = \frac{1}{M} \sum_{k=0}^{M-1} X\left(z^{1/M} W_M^{-k}\right), \tag{17.1.27}$$

and the output of the system is

$$Y_2(z) = X_{2d}\left(z^L\right) = \frac{1}{M} \sum_{k=0}^{M-1} X\left(z^{L/M} W_M^{-k}\right). \tag{17.1.28}$$

The two systems have equal outputs, that is, $Y_1(z) = Y_2(z)$, if the following equation holds:

$$\sum_{k=0}^{M-1} X\left(z^{L/M} W_M^{-kL}\right) = \sum_{k=0}^{M-1} X\left(z^{L/M} W_M^{-k}\right). \tag{17.1.29}$$

It is possible to prove that this equation holds in the case when $M$ and $L$ are relatively prime numbers, that is, $M$ and $L$ do not have a common integer factor $r > 1$.

### 17.1.4 Complexity Reduction

The complexity of analysed basic multi-rate systems is an important issue. In practical implementation, we do not use these systems in their direct forms. Instead, we design systems that have the same functions but different structures. These systems are developed separately for up-samplers and down-samplers. For example, two systems that have different structures but perform the same functions, that is, superposition and down-sampling, are presented in Figure 17.1.19. However, in the first system, the number of additions is higher than in the second system, which increases the complexity of the first system with respect to the second system. Therefore, reduction of system complexity is an important issue that we need to take into account when designing discrete-time systems.

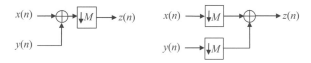

**Fig. 17.1.19** *Two systems with identical input–output characteristics.*

## 17.2 Multi-Rate Systems

### 17.2.1 Basic System Structures

According to the sampling theorem, the sampling rate cannot be reduced below the Nyquist rate if we want to avoid aliasing. For that purpose, a signal to be sampled must firstly be passed through a lowpass filter before its rate is reduced by a down-sampler, as will be shown later. Also, the zeros introduced by an up-sampler have to be interpolated.

**Interpolators, up-samplers, and interpolation filters.** In the process of up-sampling by an integer $L$, the resultant signal has the basic spectrum of the input signal along with the repeated version of this spectrum. In order to extract the signal spectrum, a lowpass filter must be used that has a cut-off frequency at $\pm\pi/L$, as shown in Figure 17.2.1. This filter takes out $L - 1$ unwanted images in the spectrum of the up-sampled signal and is called an *interpolation filter*. For its implementation, the interpolator needs $N$ multiplications and $(N - 1)$ additions if a lowpass, finite impulse response filter of the $(N - 1)$th order is used.

The procedure of up-sampling and filtering is shown on a hypothetical signal $x(n)$ that has the spectrum shown in Figure 17.2.2. This signal is up-sampled with $L = 2$, which produces the signal shown in Figure 17.2.3. In order to take out the baseband signal, a lowpass interpolation filter with a cut-off frequency at $\pm\pi/L = \pm\pi/2$, as shown in Figure 17.2.3, is used.

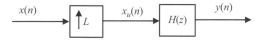

**Fig. 17.2.1** *Interpolator composed of an up-sampler and an interpolation filter.*

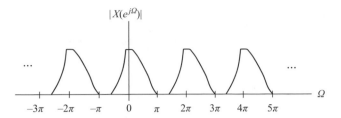

**Fig. 17.2.2** *Input signal in the frequency domain.*

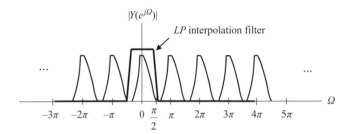

**Fig. 17.2.3** *Up-sampled and filtered signal; LP, lowpass.*

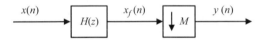

**Fig. 17.2.4** *Decimator, lowpass filter, and down-sampler.*

**Decimators, lowpass filters, and down-samplers.** If we want to avoid aliasing in the process of down-sampling, a lowpass filter should be used before down-sampling, as shown in Figure 17.2.4.

A lowpass decimating filter has the cut-off frequency $\pm\pi/M$, which eliminates all of the input signal's frequency components above the cut-off frequency. Thus, in the process of down-sampling, the replicated spectrum components in the output signal will not overlap in the interval defined by $\pm\pi$. The filtered input signal will have the spectrum shown in Figures 17.2.5 and 17.2.6.

In the above analyses we intentionally treated only lowpass signals and the procedure of rate alternation. The same procedure should be used in the case of bandpass signals, the difference being in that the filters used in that case will have to be bandpass filters.

**Fractional changes in the signalling rate.** The previous analysis was related to cases when the rate change is expressed in an integer value. However, sometimes we need a fractional change of a signal rate. That can be done by using a cascade consisting of a factor-$L$ interpolator with a factor-$M$ decimator, as shown in Figure 17.2.7.

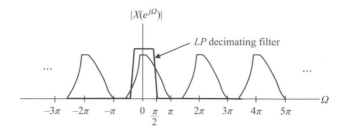

**Fig. 17.2.5**  *Decimation filter and down-sampler; LP, lowpass.*

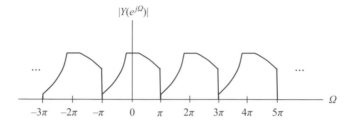

**Fig. 17.2.6**  *Output of the down-sampler.*

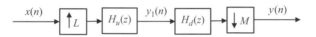

**Fig. 17.2.7**  *System for a fractional increase in the sampling rate.*

**Fig. 17.2.8**  *Simplification of the system structure.*

Because both the decimation filter and the interpolation filter operate on a signal with the same rate, they can be replaced by an equivalent system that has only one filter, as shown in Figure 17.2.8.

## 17.2.2    System Analysis in Time and Frequency Domains

**Decimator characteristics in time and frequency domains.** The structure of a decimator is presented in Figure 17.2.9. An input signal $x(n)$ with spectrum $X(e^{j\Omega})$ is filtered by a lowpass filter and then down-sampled by an integer factor $M$. Assume that the spectrum $X(e^{j\Omega})$ has non-zero values in the interval $0 \le \Omega \le \pi$, which is equivalent to the frequency $|F| \le |F_x/2|$. As we saw in the previous section, if the sampling rate of the input signal $x(n)$ is reduced by taking every $M$th value of it, the output signal will be an aliased version of $x(n)$ with a folding

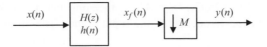

**Fig. 17.2.9** *Decimator system.*

frequency of $F_x/2M$. Therefore, in order to avoid aliasing, the bandwidth of the input signal $x(n)$ should be reduced by a lowpass filter to $\Omega_{max} = \pi/M$ or, equivalently, to $F_{max} = F_x/2M$.

The output of the decimation filter is a convolution of the input signal $x(n)$ and the impulse response of the filter $h(n)$, that is,

$$x_f(n) = \sum_{k=-\infty}^{\infty} x(k)h(n-k). \qquad (17.2.1)$$

The output of the down-sampler is

$$y(n) = x_f(nM). \qquad (17.2.2)$$

Then the output of the decimator may be expressed as

$$y(n) = \sum_{k=-\infty}^{\infty} x(k)h(nM-k). \qquad (17.2.3)$$

The output of the filter in the $z$ domain is

$$X_f(z) = X(z)H(z). \qquad (17.2.4)$$

Then, the output of the down-sampler can be obtained from eqn (17.1.19) and expressed as

$$Y(z) = \frac{1}{M} \sum_{k=0}^{M-1} X_f\left(z^{1/M} W_M^{-k}\right). \qquad (17.2.5)$$

By using eqns (17.2.4) and (17.2.5), the output can be expressed as a function of the input signal and the frequency response of the filter in the form

$$Y(z) = \frac{1}{M} \sum_{k=0}^{M-1} X\left(z^{1/M} W_M^{-k}\right) H\left(z^{1/M} W_M^{-k}\right). \qquad (17.2.6)$$

The spectrum of the output signal $y(n)$ can be obtained by evaluating this expression in the unit circle, $z = e^{j\Omega_y}$. The relationship between the input frequency and the output frequency can be obtained as follows. Bearing in mind that $F_y = F_x/M$, the output frequency can be expressed as $\Omega_y = 2\pi F/F_y = 2\pi FM/F_x = M\Omega_x$, meaning that the frequency of the output is $M$ times greater than the frequency of the input signal. Therefore, the spectrum of the output is

$$
\begin{aligned}
Y\left(e^{j\Omega_y}\right) &= \frac{1}{M}\sum_{k=0}^{M-1} X\left(e^{j\Omega_y/M}e^{j2\pi k/M}\right) H\left(e^{j\Omega_y/M}e^{j2\pi k/M}\right) \\
&= \frac{1}{M}\sum_{k=0}^{M-1} X\left(e^{j(\Omega_y+2\pi k)/M}\right) H\left(e^{j(\Omega_y+2\pi k)/M}\right)
\end{aligned}
\tag{17.2.7}
$$

which contains an infinite sum of components that are periodically placed along the frequency scale. If the lowpass filter is properly designed, it will eliminate all the components but the first one, resulting in

$$
Y\left(e^{j\Omega_y}\right)_{LP\text{ filter}} = \frac{1}{M}X\left(e^{j\Omega_y/M}\right) H\left(e^{j\Omega_y/M}\right)
\tag{17.2.8}
$$

defined in the frequency band for $|\Omega_y| \le \pi$.

**Computational complexity of the decimator.** The analysis of single-rate signal processing showed that the computational complexity of finite impulse response filters is generally higher than that of the corresponding infinite impulse response filters. Thus, in practical applications, infinite impulse response filters are preferred if computational complexity needs to be minimized. In the case of multi-rate signal processing, this conclusion is not necessarily true. In this case, when we have a decimation system as shown in Figure 17.2.9, the implementation of a finite impulse response filter of length $N$ will be based on the following convolution expression for the filtered signal,

$$
x_f(n) = \sum_{k=0}^{N-1} x(k)h(n-k),
\tag{17.2.9}
$$

as shown in Figure 17.2.10.

The sampling rate of the filter $F_x$ is higher than the sampling rate of the down-sampler $F_y$, because the down-sampler takes only every $M$th sample from the filter output. Therefore, there is no need to execute $M-1$ operations in the interval where the down-sampling does not take place, and the achievable savings are measured by a factor of $M$.

The practical implementation of a filter structure that avoids this inefficiency problem is shown in Figure 17.2.11. In this structure, the delayed versions of the input sampled are down-sampled first, and then multiplied and added at the lower sampling rate. The computational efficiency can be further increased by taking into account the symmetric properties of the filter coefficients. In the case for an infinite impulse response filter, the situation is expected to be a little different. In that case, the reduction in complexity will be less than a factor of $M$.

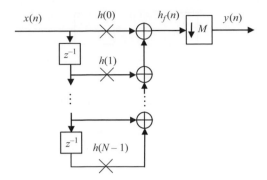

**Fig. 17.2.10**  *Implementation of an infinite impulse response filter.*

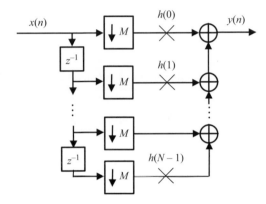

**Fig. 17.2.11**  *Implementation of an efficient finite impulse response filter.*

In summary, the computational complexity for a finite impulse response filter will be reduced by a factor of $M$ but, for an infinite impulse response filter, it will be reduced by less than $M$.

**Interpolator characteristics in time and frequency domains.** An interpolator (up-sampler) system is shown in Figure 17.2.12. The output of the up-sampler is

$$x_u(n) = y(n) = \begin{cases} x(n/L), & n = 0, \pm L, \pm 2L, \ldots \\ 0, & \text{otherwise} \end{cases}, \qquad (17.2.10)$$

which can be expressed, by changing the variable according to the relation $m = n/L$, in the form $x_u(Lm) = x(m)$, $m = 0, \pm 1, \pm 2, \ldots$.

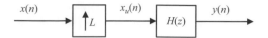

**Fig. 17.2.12** *Interpolation system.*

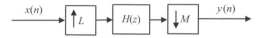

**Fig. 17.2.13** *Simplification of the system structure.*

The output of the interpolator is a convolution of the up-sampled signal and the impulse response $h(n)$ of the interpolation filter, that is,

$$y(n) = \sum_{k=-\infty}^{\infty} x_u(k)h(n-k), \tag{17.2.11}$$

which can be expressed, after a new substitution, $k = Lm$, as

$$y(n) = \sum_{m=-\infty}^{\infty} x_u(Lm)h(n-Lm) \tag{17.2.12}$$

and, finally, as a function of the input signal $x(m)$:

$$y(n) = \sum_{m=-\infty}^{\infty} x(m)h(n-Lm). \tag{17.2.13}$$

The output of the up-sampler in the $z$ domain is

$$X_u(z) = X\left(z^L\right), \tag{17.2.14}$$

and the output of the interpolator filter is

$$Y(z) = X\left(z^L\right)H(z). \tag{17.2.15}$$

**System for the fractional increase of the sampling rate.** The above examples of down-sampling and up-sampling can be considered to be special cases of the general case of sampling rate conversion by a factor $L/M$. The general case can be achieved by first interpolating the signal by a factor of $L$ and then down-sampling it by a factor of $M$. The system for a fractional increase of sampling rate is a cascade consisting of an up-sampler, a filter, and a down-sampler, as shown in Figure 17.2.13, which is based on Figures 17.2.7 and 17.2.8. It is important to perform the up-sampling first in order to preserve the frequency characteristics of the input signal $x(n)$.

The output of the up-sampler is

$$x_u(n) = \begin{cases} x(n/L), & n = 0, \pm L, \pm 2L, \dots \\ 0, & \text{otherwise} \end{cases} \tag{17.2.16}$$

and the output of the linear-time-invariant lowpass filter can be derived as follows:

$$\begin{aligned} x_f(n) &= \sum_{k=-\infty}^{\infty} x_u(k) h(n-k) = \sum_{k=-\infty}^{\infty} x(k/L) h(n-k) \\ &= \sum_{l=-\infty}^{\infty} x(l) h(n-Ll) = \sum_{k=-\infty}^{\infty} x(k) h(n-Lk) \end{aligned} \tag{17.2.17}$$

The output of the down-sampler is

$$y(n) = x_f(nM) = \sum_{k=-\infty}^{\infty} x(k) h(nM - Lk). \tag{17.2.18}$$

The output of the system in the $z$ domain is

$$Y(z) = \frac{1}{M} \sum_{k=0}^{M-1} X_f \left( z^{1/M} W_M^{-k} \right) = \frac{1}{M} \sum_{k=0}^{M-1} H \left( z^{1/M} W_M^{-k} \right) X \left( z^{L/M} W_M^{-kL} \right). \tag{17.2.19}$$

## 17.3 Reduction of Computational Complexity

### 17.3.1 Multistage Decimators and Interpolators

In the case when the $L$-factor interpolator or $M$-factor decimator can be expressed as a product of two or more integer terms, that is, $L = L_1 L_2 \dots L_k$ or $M = M_1 M_2 \dots M_k$, they can be realized in a number of ways. One of these ways can lead to an optimum realization that depends on the chosen $k$, an appropriately chosen combination of terms, and the appropriate ordering of these terms. The aim of this optimization is to achieve the minimum required number of multiplications per second in practical design. Instead of using the simple structure shown in Figures 17.2.1 and 17.2.4, it is possible to use the structures shown in Figure 17.3.1.

### 17.3.2 Polyphase Decomposition of a Decimation Filter

Efficient decimation and interpolation systems can be achieved by using designs based on polyphase decomposition, which will be presented in this section. The filter transfer characteristic of a lowpass filter that is used in the decimator circuit can be expressed as a sum of even and odd terms in the form

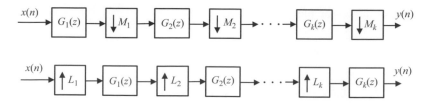

**Fig. 17.3.1** *Multistage structure of a decimator (upper panel) and an interpolator (lower panel).*

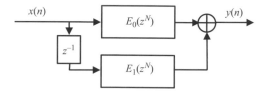

**Fig. 17.3.2** *Implementation of a decimator lowpass filter.*

$$H(z) = \sum_{n=-\infty}^{\infty} h(n)z^{-n} = \sum_{n=-\infty}^{\infty} h(2n)z^{-2n} + z^{-1}\sum_{n=-\infty}^{\infty} h(2n+1)z^{-2n}$$
$$= E_0(z) + z^{-1}E_1(z) \tag{17.3.1}$$

where even terms are

$$E_0(z) = \sum_{n=-\infty}^{\infty} h(2n)z^{-2n}, \tag{17.3.2}$$

and odd terms are

$$E_1(z) = \sum_{n=-\infty}^{\infty} h(2n+1)z^{-2n}. \tag{17.3.3}$$

Practically, we decomposed the impulse response of the discrete-time filter $h(n)$ into even and odd samples. Therefore, the basic implementation of the decimation system can be as presented in Figure 17.3.2.

### 17.3.3 Polyphase Decomposition of a Finite Impulse Response Transfer Function

The analysis of finite impulse response filters showed that it is possible to reduce computational complexity via the procedure of polyphase decomposition. The polyphase decomposition is based on the following extended expression for signal representation in the $z$ domain:

$$X(z) = \sum_{n=-\infty}^{\infty} x(n)z^{-n}$$

$$= \ldots + x(-1)z^1 + x(0)z^0 + x(1)z^{-1} + \ldots + x(N-1)z^{-(N-1)} + x(N)z^{-N} + \ldots$$

$$= \ldots + \sum_{k=0}^{N-1} x(k)z^{-k} + \sum_{k=0}^{N-1} x(k+N)z^{-(k+N)} + \sum_{k=0}^{N-1} x(k+2N)z^{-(k+2N)} + \ldots$$ (17.3.4)

$$= \sum_{n=-\infty}^{\infty} \sum_{k=0}^{N-1} x(k+nN)z^{-(k+nN)} = \sum_{k=0}^{N-1} \sum_{n=-\infty}^{\infty} x(k+nN)z^{-(k+nN)}$$

$$= \sum_{k=0}^{N-1} z^{-k} \sum_{n=-\infty}^{\infty} x(k+nN)\left(z^N\right)^{-n} = \sum_{k=0}^{N-1} z^{-k} X_k\left(z^N\right)$$

The $z$-transform $X_k(z)$, derived from eqn (17.3.4) and given by the expression

$$X_k(z) = \sum_{n=-\infty}^{\infty} x(k+nN)z^{-n} = \sum_{n=-\infty}^{\infty} x_k(n)z^{-n}$$ (17.3.5)

for $k = 0, 1, 2, \ldots, N-1$, is the polyphase component of $X(z)$ expressed by eqn (17.3.4), and $x_k(n)$ are sequences called the polyphase components of the original sequence $x(n)$. Using the expressions

$$X(z) = \sum_{k=0}^{N-1} z^{-k} X_k\left(z^N\right)$$ (17.3.6)

and

$$x_k(n) = x(k+nN),$$ (17.3.7)

for $k = 0, 1, \ldots, N-1$, we can draw a multi-rate system structure based on this polyphase decomposition, as shown in Figure 17.3.3 in the time domain and the $z$ domain.

Polyphase decomposition has been already used for the parallel realization of a finite impulse response filter transfer function for single-rate filters. The decomposition developed here can be used to design finite impulse response filters. For this purpose, we can use the expression for the polyphase decomposition of the transfer function for the system $H(z)$ in the form that was derived in eqn (17.3.6), for $X_k$ replaced by $E_k$, that is,

$$H(z) = \sum_{k=0}^{N-1} z^{-k} E_k\left(z^N\right) = E_0\left(z^N\right) + z^{-1} E_1\left(z^N\right) + \ldots + z^{-N-1} E_{N-1}\left(z^N\right)$$ (17.3.8)

which is the *first type of polyphase decomposition*. The structure of a filter with this transfer function is shown in Figure 17.3.4(a).

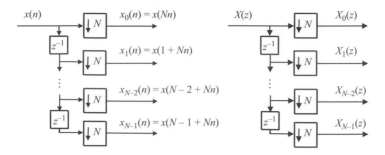

**Fig. 17.3.3** *Polyphase decomposition: (a) time-domain signals; (b) z-domain signals.*

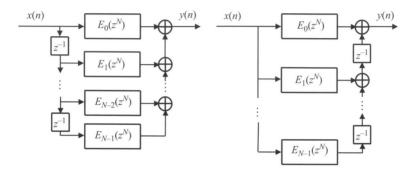

**Fig. 17.3.4** *First type of polyphase decomposition: (a) direct realization; (b) transpose realization.*

By changing the notation, the transfer function of the system $H(z)$ can also be expressed in the form

$$H(z) = \sum_{k=0}^{N-1} z^{-k} E_k\left(z^N\right) = \sum_{i=0}^{N-1} z^{-(N-1-i)} E_{N-1-i}\left(z^N\right), \qquad (17.3.9)$$

which determines the transpose realization of the transfer function shown in Figure 17.3.4(b). The transpose function can be expressed as

$$H(z) = \sum_{i=0}^{N-1} z^{-(N-1-i)} E_{N-1-i}\left(z^N\right) = \sum_{i=0}^{N-1} z^{-(N-1-i)} D_i\left(z^N\right), \qquad (17.3.10)$$

where we used the notation

$$D_i\left(z^N\right) = E_{N-1-i}\left(z^N\right), 0 \le i \le N-1, \qquad (17.3.11)$$

which is the *second type of polyphase decomposition*. The structure of a filter with this transfer function is shown in Figure 17.3.5.

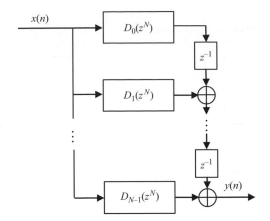

**Fig. 17.3.5** *Second type of polyphase decomposition.*

## 17.3.4 Polyphase Decomposition of an Infinite Impulse Response Transfer Function

Finding a polyphase decomposition for an infinite impulse response transfer function is not as easy as it is for a finite impulse response function. In the former case the numerator and the denominator of the transfer function $H(z) = N(z)/D(z)$ can be multiplied by the same polynomial $P(z)$, which allows an $N$-branch polyphase decomposition of the newly obtained numerator $N(z)P(z)$.

**Example**

Suppose an infinite impulse response transfer function is defined as

$$H(z) = \frac{1 - 2z^{-1}}{1 + 3z^{-1}} \tag{17.3.12}$$

Find a two-band ($N = 2$) polyphase decomposition of this function.

**Solution.** We can express this function in the form

$$H(z) = \sum_{k=0}^{N-1} z^{-k} E_k \left( z^N \right) = \sum_{k=0}^{1} z^{-k} E_k \left( z^2 \right) = z^0 E_0 \left( z^2 \right) + z^{-1} E_1 \left( z^2 \right) \tag{17.3.13}$$

Thus, by multiplying the numerator and the denominator by $(1 - 3z^{-1})$, we will have only $z^{-2}$ terms in the denominator, and $z^{-1}$ and $z^{-2}$ terms in the numerator, which can be separated as follows:

$$H(z) = \frac{1 - 2z^{-1}}{1 + 3z^{-1}} \frac{1 - 3z^{-1}}{1 - 3z^{-1}} = \frac{1 - 5z^{-1} + 6z^{-2}}{1 - 9z^{-2}}$$

$$= \frac{1 + 6z^{-2}}{1 - 9z^{-2}} + z^{-1} \frac{-5}{1 - 9z^{-2}} = E_0\left(z^2\right) + z^{-1} E_1\left(z^2\right),$$
(17.3.14)

where

$$E_0(z) = \frac{1 + 6z^{-1}}{1 - 9z^{-1}}$$
$$E_1(z) = \frac{-5}{1 - 9z^{-1}}$$
(17.3.15)

The transfer function is expressed by the two polyphase components. The structure of the transfer function is shown in Figure 17.3.6.

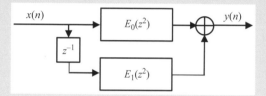

**Fig. 17.3.6** *Second type of polyphase decomposition.*

## PROBLEMS

1. Express in analytical form the output signal $y(n)$ for the input signal $x(n)$ in the case when the multi-rate signal is generated by the system shown in Figure 1.

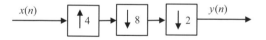

**Fig. 1** *Multi-rate system.*

2. The multi-rate system shown in Figure 2.1 is composed of a down-sampler, an up-sampler, and a filter bank. The bank consists of three ideal zero-phase real-coefficient filters: a lowpass filter, a bandpass filter, and a highpass filter, with the transfer functions presented

in Figure 2.2. Suppose that the discrete-time Fourier transform of the input discrete signal is as shown in Figure 2.3.

a. Sketch the output of the down-sampler, $X_d(e^{j\Omega})$. What is the amplitude-scaling factor for this operation?

b. Sketch the output of the up-sampler, $X_u(e^{j\Omega})$. What is the amplitude-scaling factor for this operation?

c. Sketch the output of the analysis filter bank, $Y_k(e^{j\Omega})$, for $k = 0, 1, 2$. How can we calculate these outputs?

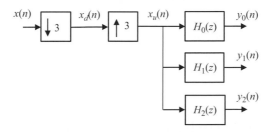

**Fig. 2.1** *Multi-rate system.*

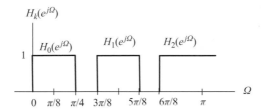

**Fig. 2.2** *Filter bank.*

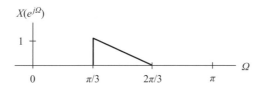

**Fig. 2.3** *Discrete-time Fourier transform of the input signal.*

3. Suppose a single-stage decimator is designed to reduce the sampling rate from 60 kHz to 3 kHz. The decimator filter is an equiripple finite impulse response filter defined as follows: the passband edge is at 1.25 kHz, the stopband edge frequency is at 1.50 kHz, the passband ripple is 0.02, and the stopband ripple is 0.01.

    a. Plot the block schematic of the decimator.

    b. Plot the transfer characteristic of this lowpass finite impulse response filter.

    c. Find the order of the filter using Kaiser's formulas.

    d. Find the computational complexity expressed as the number of multiplications per second.

4. Suppose that the transfer function of a causal infinite impulse response filter is expressed as

$$H(z) = H_1(z) + H_2(z) = \frac{3z}{z - e^{-3}} + \frac{4z}{z - e^{-2}}.$$

    a. Find a two-band polyphase decomposition of the transfer functions $H(z)$.

    b. Express the decomposed function in the form

$$H(z) = E_0\left(z^2\right) + z^{-1}E_1\left(z^2\right) \quad,$$

    and find the expression for the component transfer functions $E_0(z)$ and $E_1(z)$.

    c. Sketch the structure of a filter that has the two-band decomposed transfer characteristic $H(z)$.

5. A single-stage interpolator that increases the sampling rate from 600 Hz to 9 kHz is shown in Figure 5. The interpolator is designed as an equiripple finite impulse response filter with the passband frequency $f_p = 200$ Hz, the stopband frequency $f_s = 300$ Hz, the passband ripple $\delta_p = 0.002$, and the stopband ripple $\delta_s = 0.004$.

    a. Determine the order of the filter using Kaiser's formulas. Find the complexity of computation that is expressed as the number of multiplications per second.

    b. Develop and sketch a two-stage structure for this interpolator, with the transfer function of the structure defined to be $H(z) = G(z^5) \cdot F(z)$. Note that this structure has an up-sampler between the filters, with $L = 5$.

    c. The characteristics of the transfer function $G(z)$ are $f_p = 5 \cdot 200$ Hz $= 1000$ Hz, $f_s = 5 \cdot 300$ Hz $= 1500$ Hz, $\delta_p = 0.002$, and $\delta_s = 0.004$. Find the complexity of computation for $G(z)$.

    d. Find the computational complexity for $F(z)$, assuming that the characteristics of the transfer function $F(z)$ are $f_p = 200$ Hz, $f_s = 1500$ Hz, $\delta_p = 0.002$, and $\delta_s = 0.004$.

    e. Find the total complexity of the two-stage structure defined as $H(z) = G(z^5) \cdot F(z)$. Compare the complexity of this two-stage structure with that of the one-stage structure calculated in $a$.

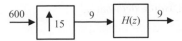

**Fig. 5** *Single-stage interpolator.*

6. Suppose a multi-rate system is as shown in Figure 6.1.

a. Make use of up-sampling and down-sampling properties to simplify the structure in Figure 6.1.

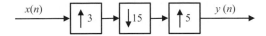

**Fig. 6.1** *Structure of a multi-rate system.*

b. Express the output signals for each block of the simplified structure in the time domain for an arbitrary input signal $x(n)$.

c. Find the output signal as a function of the input signal.

d. Suppose the structure of a multi-rate system is as shown in Figure 6.2. Find the output of the system for an arbitrary input $x(n)$.

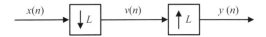

**Fig. 6.2** *Structure of a general multi-rate system.*

7. Suppose that the discrete-time periodic signal $x(n)$ presented in Figure 7.1(a) is transmitted through the system shown in Figure 7.1(b). The input signal $x(n)$ is presented in tabular form in Table 7.

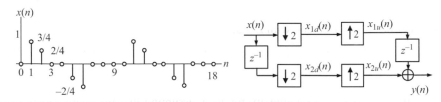

**Fig. 7** *(a) Input signal in the time domain. (b) The system structure.*

**Table 7.** *The input signal x(n).*

| $n$ | 0 | 1 | 2 | 3 | 4 | 5 | 6 | 7 | 8 | 9 |
|------|---|-----|-----|---|---|------|------|---|---|---|
| $x(n)$ | 0 | 3/4 | 2/4 | 0 | 0 | –2/4 | –3/4 | 0 | 0 | 0 |

Do the following for the first nine samples of the input discrete-time signal $x(n)$, starting with $n = 0$:

a.  Find the signals at the output of each down-sampler and present them in a tabular form.

b.  Find the signals at the output of each up-sampler and present them in a tabular form.

c.  Find the signal at the output $y(n)$ of the system.

d.  Compare the output signal with the input signal in the time domain. What is your conclusion about the relationship between the input signal and the output signal?

# 18

# Multi-Rate Filter Banks

## 18.1 Digital Filter Banks

A digital filter bank is a complex system composed of $M$ elementary filters defined by their transfer functions $H_k(z)$, $k = 1, 2, \ldots, M$. Digital filter banks are used in the case when the frequency band of a signal has to be separated in a number of subbands or to combine such subbands into one band of frequencies, as in the design of discrete communication systems. In this chapter, we will discuss multi-rate filter banks, including *analysis* and *synthesis filter banks*. The theoretical description and the principal of operation of these filters will be presented in this chapter.

A filter bank is called an $M$-band analysis filter bank if it consists of a number of filters with a common input, where each individual filter in the bank is called the analysis filter defined by its frequency response $H_k(z)$. If signal $x(n)$ is applied to the input of this filter, the output will be a set of signals with different frequency spectra.

A filter bank is called an $L$-band synthesis filter bank if it consists of a number of filters with a common output, where each individual filter in the bank is called the synthesis filter defined by its frequency response $H_k(z)$. If a set of signals are applied at the same time at the input of this filter, they will be added to each other to form a new synthesized signal $y(n)$. An $M$-band analysis filter bank and an $L$-band synthesis filter bank are shown in Figures 18.1.1 and 18.1.2, respectively.

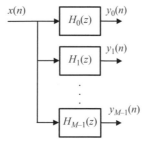

**Fig. 18.1.1** *Analysis filter bank.*

*Discrete Communication Systems.* Stevan Berber, Oxford University Press. © Stevan Berber 2021.
DOI: 10.1093/oso/9780198860792.003.0018

The synthesis filter bank consists of $L$ filters that pass $L$ input signals and then add them at the output, as shown in Figure 18.1.2.

**Analysis Filter Banks.** A filter bank with equal widths for passbands that are uniformly shifted along the frequency scale is called a uniform filter bank. Let us construct transfer function of these filters. The transfer function of a lowpass, causal, infinite impulse response baseband filter can be obtained as the $z$-transform of a real impulse response, that is,

$$H_0(z) = \sum_{n=0}^{\infty} h_0(n)z^{-n}. \tag{18.1.1}$$

The function is defined by its passband edge frequency $\Omega_p$ and stopband edge frequency $\Omega_s$, at the same distance from the middle frequency $\pi/M$, as shown in Figure 18.1.3 for an elementary filter.

The transfer function of other elementary filters in the bank can be found as the $z$-transform of the causal impulse response $h_k(n)$ defined as a *modulated version* of the impulse response of the elementary filter $h_0(n)$,

$$h_k(n) = h_0(n)W_M^{-kn}, \tag{18.1.2}$$

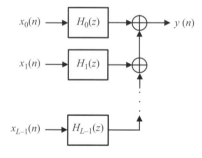

**Fig. 18.1.2** *Synthesis filter bank.*

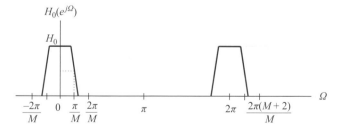

**Fig. 18.1.3** *Transfer function of an elementary filter.*

which is

$$H_k(z) = \sum_{n=0}^{\infty} h_k(n) z^{-n} = \sum_{n=0}^{\infty} h_0(n) \left( z W_M^k \right)^{-n} = H_0 \left( z W_M^k \right),$$

(18.1.3)

for $0 \le k \le M - 1$. Thus, the frequency response is

$$H_k \left( e^{j\Omega} \right) = H_0 \left( e^{j(\Omega - 2\pi k/M)} \right),$$

(18.1.4)

which can be obtained by shifting the frequency response of the basic filter to the right for the value of $2\pi k/M$. Thus, the transfer function of the filter defined by $k = 1$ is

$$H_1 \left( e^{j\Omega} \right) = H_0 \left( e^{j(\Omega - 2\pi/M)} \right),$$

(18.1.5)

which is shown in Figure 18.1.4. This transfer function is shifted for the amount of $2\pi/M$ to the right with respect to the basic filter transfer function shown as a dotted line. In a similar manner, we can plot all of the elementary transfer functions, as shown in Figure 18.1.5 only for the frequency band from 0 to $2\pi$.

Because the frequency responses of elementary filters are uniformly distributed along the frequency scale, this filter bank is called a *uniform filter bank*. According to the theory explained above, it is necessary to implement $M$ different filters to construct the filter bank,

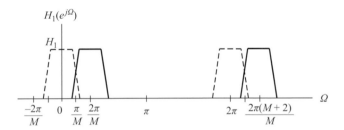

**Fig. 18.1.4** *Elementary filter for k = 1.*

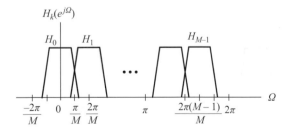

**Fig. 18.1.5** *Transfer function of a filter bank.*

which is a computationally consuming procedure. However, it is possible to reduce the number of these computations through appropriate design of the filter bank. Let us analyse an analysis filter bank first. The basic filter can be represented by its $M$-band polyphase form

$$H_0(z) = \sum_{l=0}^{M-1} z^{-l} E_l\left(z^M\right), \tag{18.1.6}$$

where the polyphase component $E_l(z)$ can be expressed as

$$E_l(z) = \sum_{n=0}^{\infty} e_l(n) z^{-n} = \sum_{n=0}^{\infty} h_0\left(l + nM\right) z^{-n}, \tag{18.1.7}$$

for $0 \leq l \leq M - 1$. The transfer function of a filter bank is, from eqn (18.1.3),

$$H_k(z) = H_0\left(z W_M^k\right). \tag{18.1.8}$$

and

$$H_0(z) = \sum_{l=0}^{M-1} z^{-l}. \tag{18.1.9}$$

This equation is sometimes referred to as a *prototype filter*. The frequency response of the $k$th filter bank $H_k(z)$ has a frequency response that is a uniformly shifted version of the prototype filter.

In order to implement a uniform discrete Fourier transform filter bank efficiently, we can use the polyphase decomposition. For this purpose, the $k$th filter bank $H_k(z)$ needs to be calculated, bearing in mind eqns (18.1.6) and (18.1.8), as

$$H_k(z) = H_0\left(z W_M^k\right) = \sum_{l=0}^{M-1} \left(z W_M^k\right)^{-l} E_l\left[\left(z W_M^k\right)^M\right] = \sum_{l=0}^{M-1} z^{-l} W_M^{-kl} E_l\left(z^M\right), \tag{18.1.10}$$

for $0 \leq k \leq M - 1$, which can be expressed in matrix form for all $k$ values as

$$\begin{bmatrix} H_0(z) \\ H_1(z) \\ \vdots \\ H_{M-1}(z) \end{bmatrix} = M\mathbf{W}^{-1} \begin{bmatrix} E_0\left(z^M\right) \\ z^{-1} E_1\left(z^M\right) \\ \vdots \\ z^{-(M-1)} E_{M-1}\left(z^M\right) \end{bmatrix}, \tag{18.1.11}$$

where **W** is the matrix of the discrete Fourier transform expressed as

$$\mathbf{W} = \begin{bmatrix} 1 & 1 & 1 & \cdots & 1 \\ 1 & W_M{}^1 & W_M{}^2 & \cdots & W_M^{(M-1)} \\ 1 & W_M{}^2 & W_M{}^4 & \cdots & W_M^{2(M-1)} \\ \vdots & \vdots & \vdots & \ddots & \vdots \\ 1 & W_M^{(M-1)} & W_M^{2(M-1)} & \cdots & W_M^{(M-1)^2} \end{bmatrix}. \tag{18.1.12}$$

Thus, a computationally efficient filter bank will be composed of $M$ filters where the basic filter is implemented in a polyphase form. The structure of this filter, called the *uniform discrete Fourier transform analysis filter bank*, for $H_k(z) = V_k(z)/X(z)$, is shown in Figure 18.1.6. This analysis filter bank can have sharper cut-off and higher stopband attenuation.

The computational complexity of this structure is much smaller than the direct implementation. If we directly implement an $M$-band analysis filter bank that is based on an $N$-tap basic lowpass finite impulse response filter, the number of multipliers will be $NM$. If the same

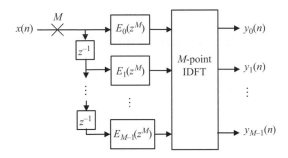

**Fig. 18.1.6** *Uniform analysis discrete Fourier transform filter bank; IDFT, inverse discrete Fourier transform.*

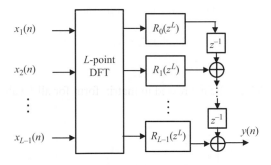

**Fig. 18.1.7** *First uniform discrete Fourier transform synthesis filter bank; DFT, discrete Fourier transform.*

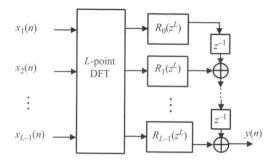

**Fig. 18.1.8** *Second uniform discrete Fourier transform synthesis filter bank; DFT, discrete Fourier transform.*

$M$-band analysis filter bank is implemented as the uniform discrete Fourier transform analysis filter bank, the number of multipliers will be $(M/2) \log_2 M + N$.

**Synthesis filter banks.** In a similar way, it is possible to find the structure of a *uniform discrete Fourier transform synthesis filter* bank. The structure based on two polyphase decompositions of the basic filter $H_0(z)$, as shown in Figures 18.1.7 and 18.1.8, respectively.

## 18.2   Two-Channel Quadrature Mirror Filter Banks

### 18.2.1   Basic Theory

This section is dedicated to *multi-rate filter banks*, which include analysis and synthesis filter banks. In particular, the *quadrature mirror filter bank* will be analysed. The basic structure of a system representing a quadrature mirror filter bank is shown in Figure 18.2.1. In principle, the output signal $y(n)$ differs from the input signal $x(n)$, due to aliasing, amplitude distortion, and phase distortion. In practical implementation, we need to design the synthesis filters $G_k(z)$ in such a way as to eliminate these distortions.

The discrete-time signal $x(n)$ is split into two subband signals $x_0(n)$ and $x_1(n)$, using an analysis filter bank. If these signals have a bandwidth that is much less than that of the input signal, they can be down-sampled. Due to their lower sampling rate, these signals can be processed more easily. After processing, the signals can be up-sampled and then combined by

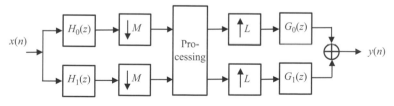

**Fig. 18.2.1** *Quadrature mirror filter bank, including an analysis filter bank and a synthesis filter bank.*

the synthesis bank. At the output is a higher-rate signal. This system is called a *two-channel quadrature mirror filter bank.*

In the case when the factors $M$ and $L$ are equal or greater than the number of bands in the filter bank, the output $y(n)$ can be equal to the input signal $x(n)$. In this case, the bank is called a *critically sampled filter bank.* This filter is used for efficient coding of the input signal $x(n)$. The number of filters in the analysis and synthesis filter banks, called channels, can be arbitrary. A codec based on a two-channel quadrature mirror filter bank is shown in Figure 18.2.2.

The split parts of the input signal $x(n)$ are passed through analysis filter bank that is composed of two filters, $H_0(z)$ and $H_1(z)$. These two filters are usually lowpass and highpass filters, respectively, with a cut-off frequency of $\pi/2$. Their frequency responses are shown in Figure 18.2.3.

The filtered signals are down-sampled and then encoded and multiplexed to form a new signal. This signal can be stored or transmitted and then received to be de-multiplexed and decoded. The decoded signals are synthesized to get the output signal $y(n)$, which has the same sampling rate as the input signal $x(n)$. In practice, the analysis and synthesis filters are constructed in such a way so as to ensure that the output signal is a good replica of the input signal. In addition, the other blocks in the scheme are constructed to preserve this characteristic. The following analysis takes into account the influence of the filters and sampling circuits on the input–output characteristics of the system shown in Figure 18.2.4.

Bearing in mind the theory in Chapters 13 and 17, we can express the signal at the output of each block in Figure 18.2.4 as follows:

$$X_k(z) = X(z)H_k(z), \qquad (18.2.1)$$

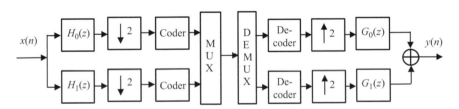

**Fig. 18.2.2** *A two-channel quadrature mirror filter bank; DEMUX, de-multiplexer; MUX, multiplexer.*

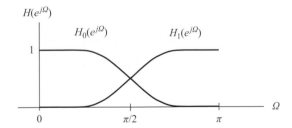

**Fig. 18.2.3** *Lowpass and highpass filter frequency responses.*

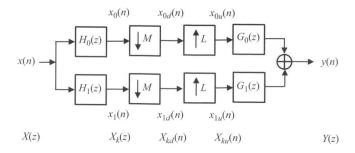

**Fig. 18.2.4** *A system with filtering and sampling.*

where $k = 0, 1$. The decimated signal in the $z$ domain is

$$Y(z) = \frac{1}{M} \sum_{k=0}^{M-1} X_f \left( z^{1/M} W_M^{-k} \right).$$

Let us analyse the case when $M = L = 2$. The output of the down-samplers is

$$X_{kd}(z) = \frac{1}{2} \left[ X_k \left( z^{1/2} \right) + X_k \left( -z^{1/2} \right) \right], \tag{18.2.2}$$

where $k = 0, 1$. The second term in this expression indicates possible signal aliasing which depends on the frequency spectrum of the input signal as well as on the frequency response of the filters. The signals before the input to the adder is

$$X_{ku}(z) = \frac{1}{2} [X_k(z) + X_k(-z)] = \frac{1}{2} [X(z)H_k(z) + X(-z) H_k(-z)]. \tag{18.2.3}$$

The output signal, which represents a reconstructed version of the input signal, can be represented as the sum of the products of these two signals and the response of the synthesis filters, that is,

$$\begin{aligned} Y(z) &= \frac{1}{2} [X(z)H_0(z) + X(-z) H_0 (-z)] G_0(z) \\ &\quad + \frac{1}{2} [X(z)H_1(z) + X(-z) H_1 (-z)] G_1(z) \\ &= \frac{1}{2} [H_0(z)G_0(z) + H_1(z)G_1(z)]X(z) \\ &\quad + \frac{1}{2} [H_0 (-z)G_0(z) + H_1 (-z)G_1(z)]X(-z) \end{aligned} \tag{18.2.4}$$

where the second term represents the aliasing due to the down-sampling in the system, and the imaging due to the up-sampling in the system. This term is called the aliasing component.

Therefore, in order to eliminate this aliasing, we need to design the synthesis filters in such a way so as to eliminate the second term, which can be achieved if we have a synthesis filter bank that fulfils the conditions

$$G_0(z) = H_1(-z), \text{ and } G_1(z) = -H_0(-z).$$ (18.2.5)

## 18.2.2 Elimination of Aliasing in Two-Channel Quadrature Mirror Filter Banks

The system shown in Figure 18.2.4 is a linear-time-variant system. As we said before, by appropriate construction of the filters, it is possible to eliminate aliasing and make the system a linear-time-invariant system. The aliasing will be cancelled if the following condition is fulfilled:

$$\frac{1}{2}[H_0(-z)G_0(z) + H_1(-z)G_1(z)] = 0.$$ (18.2.6)

In this case, the output signal can be expressed as

$$Y(z) = \frac{1}{2}[H_0(z)G_0(z) + H_1(z)G_1(z)]X(z) = D(z)X(z),$$ (18.2.7)

where $D(z)$ is a distortion term. The output signal in the frequency domain corresponds to the discrete-time Fourier transform of the output, or to the output of system that is developed on a unit circle, that is,

$$Y(e^{j\Omega}) = D(e^{j\Omega})X(e^{j\Omega}) = |D(e^{j\Omega})|e^{j\phi(\Omega)}X(e^{j\Omega}).$$ (18.2.8)

We can analyse the influence of transfer function on the output signal. The first term $|D(e^{j\Omega})|$ is the amplitude distortion, and the second exponential term $e^{j\phi(\Omega)}$ represents the phase distortion of the input signal. This distortion can be eliminated if we use an allpass filter and a linear-phase characteristic. If this function has an allpass characteristic, the magnitude of the output will be

$$|Y(e^{j\Omega})| = |D(e^{j\Omega})||X(e^{j\Omega})|,$$ (18.2.9)

which means that the output magnitude spectrum is a scaled version of the input magnitude spectrum, where the constant of proportionality is the magnitude of the transfer function. For this reason, we say that the quadrature mirror filter bank is a *magnitude-preserving system*. The system will have phase distortion unless the phase characteristic is a generalized linear phase, that is, if the following condition is fulfilled:

$$\phi(\Omega) = a\Omega + b.$$ (18.2.10)

The quadrature mirror filter bank is called a *perfect-reconstruction filter bank* if it has no distortion. This condition is fulfilled if the transfer function is expressed as

$$D(z) = \frac{1}{2}[H_0(z)G_0(z) + H_1(z)G_1(z)] = 2z^{-l}, \qquad (18.2.11)$$

where $l$ is a positive integer. The output of the bank is then

$$Y(z) = X(z) \cdot z^{-l}, \qquad (18.2.12)$$

which can be expressed in the time domain as

$$y(n) = x(n - l), \qquad (18.2.13)$$

which means that the output signal is just a delayed version of the input.

**Example**

Show that a quadrature mirror filter bank is a perfect reconstruction bank if the analysis filters are defined as

$$H_0(z) = \frac{1}{\sqrt{2}}\left(1 + z^{-1}\right) \text{ and } H_1(z) = \frac{1}{\sqrt{2}}\left(1 - z^{-1}\right),$$

and the synthesis filters are

$$G_0(z) = \frac{1}{\sqrt{2}}\left(1 + z^{-1}\right) \text{ and } G_1(z) = \frac{1}{\sqrt{2}}\left(-1 + z^{-1}\right),$$

which represents two pairs of lowpass and highpass first-order filters.

**Solution.** For this bank, the aliasing function is

$$A(z) = \frac{1}{2}[H_0(-z)G_0(z) + H_1(-z)G_1(z)]$$

$$= \frac{1}{2}\left[\frac{1}{\sqrt{2}}\left(1 - z^{-1}\right)\frac{1}{\sqrt{2}}\left(1 + z^{-1}\right) + \frac{1}{\sqrt{2}}\left(1 + z^{-1}\right)\frac{1}{\sqrt{2}}\left(-1 + z^{-1}\right)\right] = 0$$

Therefore, this system is aliasing-free. Let us find the signal in the time domain. The outputs of the analysis filters are

$$X_0(z) = X(z)H_0(z) = X(z)\frac{1}{\sqrt{2}}\left(1 + z^{-1}\right)$$

and

$$X_1(z) = X(z)H_1(z) = X(z)\frac{1}{\sqrt{2}}\left(1 - z^{-1}\right),$$

which gives the following two respective signals in the time domain:

$$x_0(n) = \frac{1}{\sqrt{2}}[x(n) + x(n-1)],$$

and

$$x_1(n) = \frac{1}{\sqrt{2}}[x(n) - x(n-1)].$$

## 18.3 Perfect Reconstruction of Two-Channel Filter Banks

A two-channel quadrature mirror filter bank is shown in Figure 18.3.1. The input signal $x(n)$ is passed through an analysis filter bank that consists of two bandpass filters. Thus, the input signal is decomposed into the two subband signals $x_0(n)$ and $x_1(n)$, which are down-sampled by the factor $M = 2$ and then up-sampled by the factor $L = 2$. The up-sampled signals are filtered in the synthesis filter bank, which contains two bandpass filters that remove the images.

The output is derived as a function of the distortion transfer function $D(z)$ and the aliasing transfer function $A(z)$, that is,

$$Y(z) = \frac{1}{2}[H_0(z)G_0(z) + H_1(z)G_1(z)]X(z) + \frac{1}{2}[H_0(-z)G_0(z) + H_1(-z)G_1(z)]X(-z),$$
$$= D(z)X(z) + A(z)X(-z)$$

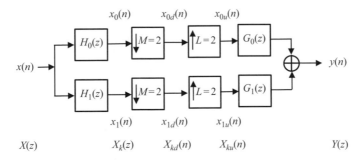

**Fig. 18.3.1** *A two-channel quadrature mirror filter bank.*

and

$$
\begin{aligned}
Y(-z) &= \frac{1}{2} \left[ H_0(-z)\, G_0(-z) + H_1(-z)\, G_1(-z) \right] X(-z) \\
&\quad + \frac{1}{2} \left[ H_0(z) G_0(-z) + H_1(z) G_1(-z) \right] X(z) \\
&= D(-z) X(-z) + A(-z) X(z)
\end{aligned}
\tag{18.3.1}
$$

The outputs in the $z$ domain can be expressed in matrix form as

$$
\begin{bmatrix} Y(z) \\ Y(-z) \end{bmatrix} = \frac{1}{2} \begin{bmatrix} G_0(z) & G_1(z) \\ G_0(-z) & G_1(-z) \end{bmatrix} \begin{bmatrix} H_0(z) & H_0(-z) \\ H_1(z) & H_1(-z) \end{bmatrix} \begin{bmatrix} X(z) \\ X(-z) \end{bmatrix},
\tag{18.3.2}
$$

or

$$
\mathbf{Y}(z) = \frac{1}{2} \mathbf{G}(z) \cdot \mathbf{H}^T(z) \cdot \mathbf{X}(z).
\tag{18.3.3}
$$

In this expression, the $\mathbf{H}$ matrix, which contains the frequency responses of the analysis filters, is called the *analysis modulation matrix*, and the matrix $\mathbf{G}$, which contains the frequency responses of the synthesis filters, is called the *synthesis modulation matrix*.

We can now determine the condition required for a perfect reconstruction. From the above expression for the output signal $Y(z)$ in the $z$ domain, the condition for a perfect reconstruction can be expressed as

$$
Y(z) = z^{-l} X(z).
\tag{18.3.4}
$$

If we insert this equation into the expression for the output, we obtain

$$
\begin{bmatrix} z^{-l} X(z) \\ (-z)^{-l} X(-z) \end{bmatrix} = \frac{1}{2} \mathbf{G}(z) \cdot \mathbf{H}^T(z) \cdot \begin{bmatrix} X(z) \\ X(-z) \end{bmatrix},
\tag{18.3.5}
$$

which results in

$$
\mathbf{G}(z) \cdot \mathbf{H}^T(z) = 2 \cdot \begin{bmatrix} z^{-l} & 0 \\ 0 & (-z)^{-l} \end{bmatrix}.
\tag{18.3.6}
$$

Multiplying on the right with the inverse of $\mathbf{H}^T$, we obtain

$$
\mathbf{G}(z) = 2 \cdot \begin{bmatrix} z^{-l} & 0 \\ 0 & (-z)^{-l} \end{bmatrix} \cdot \left[ \mathbf{H}^T(z) \right]^{-1}.
\tag{18.3.7}
$$

From this expression, we can derive the frequency responses of the synthesis filters as

$$G_0(z) = \frac{2z^{-l}}{\det \mathbf{H}(z)} H_1(-z)$$

$$G_1(z) = -\frac{2z^{-l}}{\det \mathbf{H}(z)} H_0(-z)$$

(18.3.8)

where $l$ is an odd positive integer, and the determinant is expressed as

$$\det \mathbf{H}(z) = H_0(z)H_1(-z) - H_0(-z)H_1(z).$$

If the analysis filters are finite impulse response filters, then the synthesis filters will be finite impulse response filters, too. In this case, a perfect reconstruction will occur if the following condition is fulfilled:

$$\det \mathbf{H}(z) = cz^{-k},$$

(18.3.9)

which can be also expressed as

$$G_0(z) = \frac{2z^{-(l-k)}}{c} H_1(-z)$$

$$G_1(z) = -\frac{2z^{-(l-k)}}{c} H_0(-z)$$

(18.3.10)

where $c$ is a real number and $k$ is a positive integer. A bi-orthogonal filter bank is a quadrature mirror filter bank that is obtained by using perfect-reconstruction, linear-phase, finite impulse response filters and is represented by a non-orthonormal transformation matrix. An orthogonal filter bank is a perfect-reconstruction power-symmetric filter bank.

## 18.4 Multichannel Quadrature Mirror Filter Banks

We can construct a multichannel filter bank by using a two-channel quadrature mirror filter bank as the main building block. Let us analyse the structure of the $L$-channel quadrature mirror filter bank shown in Figure 18.4.1.

The outputs, in the $z$ domain, of the first $L-1$ filters are

$$X_k(z) = H_k(z)X(z)$$

which are down-sampled to obtain

$$X_{kd}(z) = \frac{1}{L} \sum_{l=0}^{L-1} H_k\left(z^{1/L} W_L^l\right) X_k\left(z^{1/L} W_L^l\right),$$

$$X_{ku}(z) = X_{kd}\left(z^L\right)$$

(18.4.1)

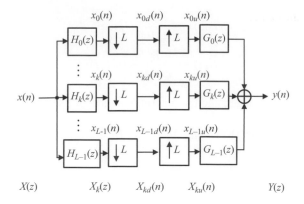

**Fig. 18.4.1** *An L-channel quadrature mirror filter bank.*

for $0 \leq k \leq L - 1$. The signals in the $z$ domain can be expressed and treated in vector form, and the output of the $L$-channel filter bank can be expressed in matrix form. Let us define the following matrices. The modulation vector of the input signals is

$$\mathbf{X}(z) = \left[ X(z) \ X(zW_L) \ \cdots \ X\left(zW_L^{L-1}\right) \right]^T. \tag{18.4.2}$$

The analysis filter bank *modulation matrix* is defined as

$$\mathbf{H}(z) = \begin{bmatrix} H_0(z) & H_1(z) & \cdots & H_{L-1}(z) \\ H_0(zW_L) & H_1(zW_L) & \cdots & H_{L-1}(zW_L) \\ \vdots & \vdots & \ddots & \vdots \\ H_0\left(zW_L^{L-1}\right) & H_1\left(zW_L^{L-1}\right) & \cdots & H_{L-1}\left(zW_L^{L-1}\right) \end{bmatrix}^T. \tag{18.4.3}$$

The vector representing the synthesis filter bank is

$$\mathbf{G}(z) = [G_0(z) \ G_1(z) \ \cdots \ G_d(z)]^T \tag{18.4.4}$$

The vector of down-sampled outputs is

$$\mathbf{X_d}(z) = [X_{0d}(z) \ X_{1d}(z) \ \cdots \ X_{L-1d}(z)]^T, \tag{18.4.5}$$

which can be expressed as

$$\mathbf{X_d}(z) = \frac{1}{L} \mathbf{H^T}\left(z^{1/L}\right) \cdot \mathbf{X_d}\left(z^{1/L}\right). \tag{18.4.6}$$

Thus, the output of the bank is

$$\mathbf{Y}(z) = \mathbf{G}^{\mathrm{T}}(z) \cdot \mathbf{X_d}\left(z^L\right). \tag{18.4.7}$$

For this multichannel filter bank, we can derive the condition for aliasing-free processing. In this case, we need to find the distortion transfer function and the condition for aliasing elimination.

## 18.5 Multilevel Filter Banks and Adaptive Filter Banks

### 18.5.1 Banks with Equal or Unequal Passband Widths

**Banks with equal passband widths.** A four-channel, maximally decimated quadrature mirror filter bank is constructed from a two-channel, maximally decimated quadrature mirror filter bank, as shown in Figure 18.5.1.

Each level in the series connection of this bank is composed of a filter and a rate converter. Thus, the bank has four levels. This four-level, four-channel structure can be replaced with a four-channel, two-level structure, which is shown in Figure 18.5.2 and can be analysed in a way similar to that used previously to analyse the multichannel structure.

A multilevel filter bank consists of an analysis filter bank with down-samplers and a synthesis filter bank with up-samplers. The transfer functions of the analysis filters are

$$H_{00}(z) = H_0(z)H_0\left(z^2\right),$$

$$H_{01}(z) = H_0(z)H_1\left(z^2\right),$$

$$H_{10}(z) = H_1(z)H_0\left(z^2\right), \tag{18.5.1}$$

$$H_{11}(z) = H_1(z)H_1\left(z^2\right),$$

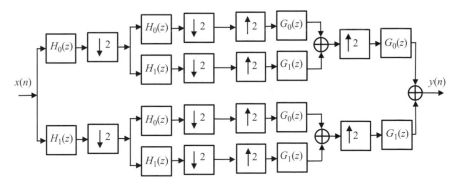

**Fig. 18.5.1** *A four-channel, maximally decimated quadrature mirror filter bank.*

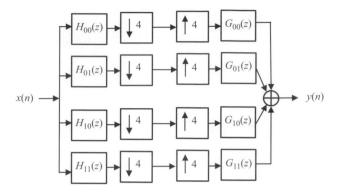

**Fig. 18.5.2** *A four-channel, two-level structure.*

and the synthesis filters are

$$G_{00}(z) = G_0(z)G_0\left(z^2\right),$$

$$G_{01}(z) = G_0(z)G_1\left(z^2\right),$$

$$G_{10}(z) = G_1(z)G_0\left(z^2\right)$$

$$G_{11}(z) = G_1(z)G_1\left(z^2\right).$$

(18.5.2)

Each analysis filter is a cascade of two filters. One filter has a single passband and a single stopband. The other filter has two passbands and two stopbands. The cascade has a passband in the frequency region where the two passbands overlap. This four-channel quadrature mirror filter system can be extended via the insertion of another two-channel quadrature mirror filter bank. The number of channels produced by this insertion will be a power of 2. The filters in the analysis and the synthesis branches have passbands of equal width. However, it is possible to construct filters with different widths.

**Banks with unequal passband widths.** If we add a two-channel, maximally decimated quadrature mirror filter bank into the top subband channel of a maximally decimated quadrature mirror filter bank, the resulting system is a three-channel, maximally decimated quadrature mirror filter bank, as shown in Figure 18.5.3.

An equivalent three-channel filter bank is presented in Figure 18.5.4. The transfer functions of this three-channel filter bank are

$$H_{00}(z) = H_0(z)H_0\left(z^2\right)$$

$$H_{01}(z) = H_0(z)H_1\left(z^2\right),$$

$$H_1(z)$$

(18.5.3)

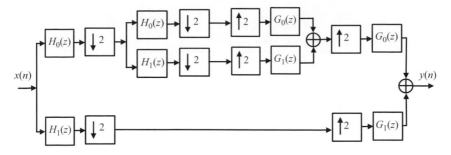

**Fig. 18.5.3** *A three-channel, maximally decimated quadrature mirror filter bank.*

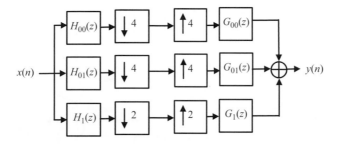

**Fig. 18.5.4** *The equivalent three-channel filter bank.*

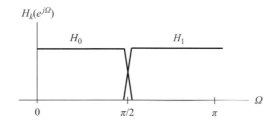

**Fig. 18.5.5** *Transfer function of a two-channel quadrature mirror filter bank.*

and

$$G_{00}(z) = G_0(z)G_0\left(z^2\right)$$
$$G_{01}(z) = G_0(z)G_1\left(z^2\right).$$
$$G_1(z)$$

(18.5.4)

Suppose we have a hypothetical transfer function of a two-channel quadrature mirror filter bank as presented in Figure 18.5.5. Then the transfer function of the derived three-channel filter bank is as shown in Figure 18.5.6.

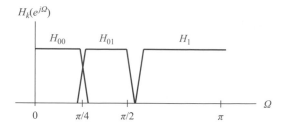

**Fig. 18.5.6** *Transfer function of a derived three-channel filter bank.*

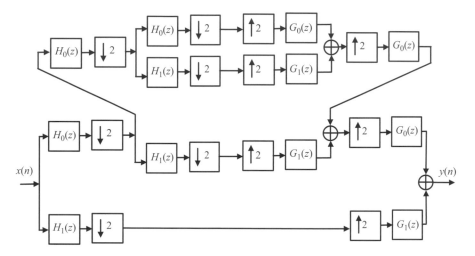

**Fig. 18.5.7** *A four-channel filter bank.*

A four-channel filter bank can be constructed from a three-channel bank by adding one two-channel bank to the first subband channel, as shown in Figure 18.5.7.

The equivalent four-channel structure is shown in Figure 18.5.8. The frequency responses of the filters are

$$
\begin{aligned}
H_{000}(z) &= H_0(z)H_0\left(z^2\right)H_0\left(z^4\right) \\
H_{001}(z) &= H_0(z)H_0\left(z^2\right)H_1\left(z^4\right), \\
H_{01}(z) &= H_0(z)H_1\left(z^2\right) \\
H_1(z) &
\end{aligned}
\tag{18.5.5}
$$

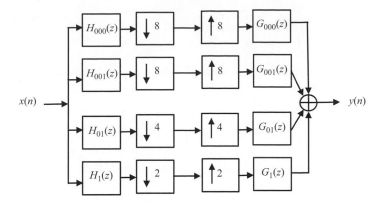

**Fig. 18.5.8** *The equivalent four-channel structure.*

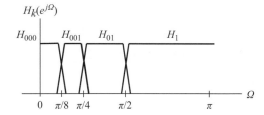

**Fig. 18.5.9** *Transfer function of the filter bank.*

and

$$G_{000}(z) = G_0(z)G_0\left(z^2\right)G_0\left(z^4\right)$$
$$G_{001}(z) = G_0(z)G_0\left(z^2\right)G_1\left(z^4\right)$$
$$G_{01}(z) = G_0(z)G_1\left(z^2\right)$$
$$G_1(z)$$

(18.5.6)

The transfer function of the filter bank is shown in Figure 18.5.9.

## 18.5.2 Adaptive Filter Banks

Adaptive filtering is used in various applications. In the case when the filter design requires a large number of coefficients, the computational complexity becomes unacceptably high. In order to overcome this problem, we can use multi-rate filter banks.

PROBLEMS

1. Find the two-band polyphase components of infinite impulse response prototype transfer function expressed as

$$H_0(z) = \frac{1 + 3z^{-1} + 2z^{-2}}{1 + z^{-1} + z^{-2}}.$$

2. A uniform analysis filter bank is designed to have four filters defined by the transfer function $H_k(z) = Y_k(z)/X(z)$ for $k = 0, 1, 2, 3$, as shown in Figure 2.1(a). The polyphase implementation of this filter bank is shown in Figure 2.1(b). The transfer functions of the four filters are, respectively,

$$E_0(z) = 1 + z^{-1} - z^{-2}$$
$$E_1(z) = 2 - 2z^{-1} + 3z^{-2}$$
$$E_2(z) = 4 - z^{-1} + 2z^{-2}$$
$$E_2(z) = 1 + 3z^{-1} + 2z^{-2}$$

a. Suppose the transfer function of the third filter in the analysis filter bank is as shown in Figure 2.2. Sketch the graphs of the frequency responses of the other three analysis filters.

b. Find the expression for the transfer function of the analysis filter bank.

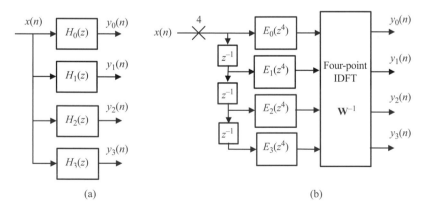

**Fig. 2.1** *(a) Analysis filter bank; (b) polyphase implementation of the filter bank; IDFT, inverse discrete Fourier transform.*

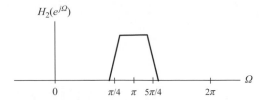

**Fig. 2.2** *Transfer function of the third filter.*

3. Suppose the causal third-order transfer function is expressed as

$$H(z) = \frac{1}{2} \frac{1 + 3z^{-1} + 3z^{-2} + z^{-3}}{3 + z^{-2}}.$$

a. Decompose this function into the sum of two transfer functions expressed as

$$H(z) = \frac{1}{2}[H_0(z) + H_1(z)].$$

b. Realize $H(z)$ as a parallel connection of these two functions.

c. Derive a new transfer function $H_C(z)$ that is power complementary to $H(z)$.

# 19

# Continuous-Time Stochastic Processes

## 19.1 Continuous-Time Stochastic Processes

### 19.1.1 Probability, Random Variables, and Stochastic Processes

The theory of stochastic processes is an extension of the theory of probability and the theory of random variables. In the analysis of the stochastic processes presented in this chapter, we assume that the reader possesses a very good knowledge of these theories. The theory of stochastic processes, in both continuous time and discrete time, is the essence of any analysis in communication systems theory and its application in practice. The signal generated in the source of information is a random signal. Furthermore, noise and other impairments that affect signal transmission in the channel are random in nature. Therefore, a communication system is best modelled and analysed if the notion of stochastic processes is used.

To obtain a formal definition of continuous-time stochastic signals, let us consider a random experiment $\mathcal{E}$ with a finite or an infinite number of unpredictable outcomes $\{\zeta_i\}$ from a sample space $S = \{\zeta_1, \zeta_2, \ldots, \zeta_k, \ldots, \zeta_K\}$, where the probabilities assigned to all outcomes are $p(\zeta_1), p(\zeta_2), \ldots, p(\zeta_k), \ldots, p(\zeta_K)$, and $K$ can take an infinite value. The number of outcomes can be countably or uncountably infinite. Furthermore, according to a certain rule, a continuous-time function $x(t, \zeta_k)$ is assigned to each outcome $\zeta_k$, for $-\infty < t < \infty$. Thus, a continuous-time stochastic process $X(t)$ is defined by the following:

- the sample space $S = \{\zeta_1, \zeta_2, \ldots, \zeta_k, \ldots, \zeta_K\}$, which is a set of outcomes (elementary events) of a random experiment $\varepsilon$ and the related set of all events, the Borel field $F$;
- the set of associated probabilities $P = \{p(\zeta_k)\}$ that complies with the axioms of probability theory and the theory of random variables; and
- the set of the associated ensemble of time functions $X(t) = \{x(t, \zeta_k)\}$ defined for $k = 1$, $2, \ldots, K$, where $K$ can be finite or infinite, countable or uncountable, and for any time inside the interval $-\infty < t < \infty$.

According to the notation in this book, a continuous-time stochastic process $X(t)$ generally represents an infinite set or an ensemble of continuous-time functions that are defined for all outcomes $\zeta_k$, $k = 1, 2, \ldots, K$, and all $t$ values. Each function in this ensemble, denoted

*Discrete Communication Systems*. Stevan Berber, Oxford University Press. © Stevan Berber 2021.
DOI: 10.1093/oso/9780198860792.003.0019

by $x(t, \zeta_k)$, is called a *realization of the stochastic process* $X(t)$. Because this realization is represented by a function of time, we will use the term *stochastic processes* instead of *random processes*. However, since the theory presented in this chapter is intended to be applied to communication systems, the *realization* of the stochastic process $x(t, \zeta_k)$ will also be called a *stochastic signal*, *random signal*, or *random function of time*. Furthermore, these terms are equivalent to the terms *random sample function*, *sample path*, and *sample path realization*, which appear in the literature but will not be used in this book.

Any value of the stochastic process $X(t)$ at the time instant $t = t_i$, of any function $x(t, \zeta_i)$, denoted by $x(t_i)$, is a realization of the random variable $X(t_i)$ that is defined at the time instant $t = t_i$ for a particular outcome $\zeta_i$. In this sense, a stochastic process $X(t)$ can be treated as a series (function) of random variables defined for each value $t$. In subsequent analysis, we will use $X(t)$ to denote the stochastic process that is defined for all values of $t$ and is represented by the ensemble of time-domain random functions $x(t)$. For simplicity, we will also use $X(t)$ to denote the random variable $X_t$ that is defined at the time instant $t$. In this sense, the stochastic process $X(t)$ is defined as a non-countable infinity of random variables that are defined for every value of $t$. For simplicity, in this book, we will treat $x(t)$ as equivalent to $x(t, \zeta_i)$ in representing one realization of the continuous-time stochastic process $X(t)$ for the outcome $\zeta_i$.

Following the definition of a stochastic process, a hypothetical continuous-time stochastic process $X(t)$ is presented in Figure 19.1.1 as a set of random time functions associated to all outcomes of an arbitrary random experiment.

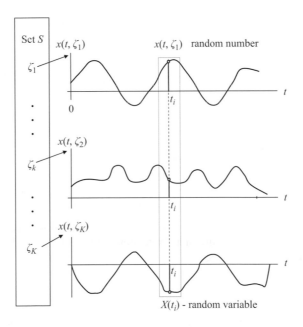

**Fig. 19.1.1** *Graphical presentation of the continuous-time stochastic process X(t).*

As was said before, future values of a stochastic process $X(t)$ cannot be predicted in advance. The unpredictable nature of this process has two causes. Firstly, the time structure of one realization $x(t)$ depends on the outcome $\zeta_i$ of the random experiment. If we cannot predict the values of the process realization in the time domain, the process is called a *non-predictable* or *regular process*. Secondly, there is not, generally speaking, a functional description for a realization of the stochastic process except for some special cases. In these special cases, the process is called *predictable* or *deterministic* in the time domain. Most of the processes in the real world are unpredictable, that is, regular processes. An example of such a process is the white Gaussian noise process generated in a communication channel.

An example of a predictable process is a signal generated at the output of a sinusoidal generator in the case when the initial phase of the signal is a random variable. This process can be expressed in the form

$$X(t) = A \cos(\omega t + \theta),$$

where the phase of process $\theta$ is a random variable uniformly distributed on the interval $[0, 2\pi]$. If we analyse a realization of this stochastic process that is generated with a random phase $\theta_1$, which can be expressed in the form

$$x(t) = A \cos(\omega t + \theta_0),$$

we can predict any future value $x(t_2)$ when we know any previous value $x(t_1)$.

## 19.1.2　Statistical Analysis of Stochastic Processes

Most continuous-time stochastic signals can be represented as functions of time, where time is a continuous variable. Thus, the simplest description of these signals is based on an analysis that results in quantitative values that supply information about the significant characteristics of the underlying process.

The *first* characteristic of a continuous-time stochastic process is the distribution of its amplitude values at the defined time instant $t = t_i$. Because these amplitudes are sample values of a random variable $X(t_i)$, this characteristic can be expressed in terms of the first-order probability distribution or density function of these sample values. The *second* characteristic of a continuous-time stochastic process is the degree of dependence between its two sample values $X(t_1)$ and $X(t_2)$, which are random variables defined at those two time instants. This dependence is expressed in terms of the correlation that exists between them. The *third* characteristic of the process is its periodicity behaviour, which can result in sharp peaks in its power spectral function and its autocorrelation function.

In particular, there is a significant interest in the variable characteristics of some of the values of the process, such as the probability density function, the mean value, the variance, or the spectral content of the process. These variabilities can be analysed at the level of a limited number of sample values of the random signal $x(t)$ or on the whole signal $x(t)$. In practice, these two analyses are inter-related and are used together. In the following sections, these characteristics of a continuous-time stochastic process will be mathematically formalized. We will start this formalization by defining the first-, second-, and $k$th-order probability functions of the process.

At each time instant $t$, we may define a random variable $X(t)$ that requires a *first-order probability function* for its description, say, the probability distribution function, which is defined as

$$F_X(x_1;t_1) = P(X(t_1) \le x_1) = P(X_1 \le x_1), \tag{19.1.1}$$

for the random variable $X_1 = X(t_1)$, defined at the time instant $t = t_1$. The related first-order probability density function is defined as a derivative of the distribution function with respect to the variable $X_1$, that is,

$$f_X(x_1;t_1) = \frac{dF_X(x_1;t_1)}{dx_1} \tag{19.1.2}$$

for any real number $x_1$. Similarly, the random variables defined at two distinct instants $t_1$ and $t_2$, say, $X(t_1)$ and $X(t_2)$, are joint random variables that are statistically described by their *second-order joint probability distribution* function $F_X(x_{t1}, x_{t2}; t_1, t_2) = F_X(x_1, x_2; t_1, t_2)$:

$$F_X(x_1, x_2; t_1, t_2) = P(X(t_1) \le x_1, X(t_2) \le x_2), \tag{19.1.3}$$

and the corresponding second-order joint density function is

$$f_X(x_1, x_2; t_1, t_2) = \frac{\partial^2 F_X(x_1, x_2; t_1, t_2)}{\partial x_1 \partial x_2}. \tag{19.1.4}$$

However, a stochastic process can contain a non-countable infinity of random variables or just $k$ of them. If we denote the values of random variables $X(t_1), X(t_2), \ldots, X(t_i), \ldots$ by $x_1, x_2, \ldots, x_i, \ldots$, for a stochastic process $X(t)$, then the $k$th-order joint probability distribution function can be expressed in the form

$$\begin{aligned} F_X(x_1, x_2, \ldots, x_k; t_1, t_2, \ldots, t_k) \\ = P(X(t_1) \le x_1, X(t_2) \le x_2, \ldots, X(t_k) \le x_k), \end{aligned} \tag{19.1.5}$$

or the $k$th-order probability density function

$$f_X(x_1, x_2, \ldots, x_k; t_1, t_1, \ldots, t_k) = \frac{\partial^{2k} F_X(x_1, x_2, \ldots, x_k; t_1, t_1, \ldots, t_k)}{\partial x_{R1} \ldots \partial x_{Ik}}, \tag{19.1.6}$$

for a general case when the stochastic process is a complex function having both real ($R$) and imaginary ($I$) parts. This description of stochastic processes, based on the density and distribution functions, requires a lot of information that is difficult to obtain in practice. Fortunately, many properties of a stochastic process can be described in terms of averages associated with its first- and second-order densities.

It is important to define a special type of stochastic process called an *independent process*. The process is independent of the $k$th order if its joint probability density function is defined as the product of related marginal probability density functions, expressed as

$$f_X(x_1, x_2, \ldots, x_k; t_1, t_2, \ldots, t_k) = f_1(x_1; t_1) \cdot f_2(x_2; t_2) \cdot \ldots \cdot f_k(x_k; t_k), \qquad (19.1.7)$$

for every $k$ and $t_i$, where $i = 1, 2, \ldots, k$. In the case when all random variables, defined for $i = 1, 2, \ldots, k$, have the same probability density functions, the process $X(t)$ is called an iid process.

Two processes $X(t)$ and $Y(t)$ defined on the same set $S$ are statistically independent if the set of random variables $X(t_1)$, $X(t_2)$, …, $X(t_k)$ is independent of the set of random variables $Y(t_1)$, $Y(t_2)$, …, $Y(t_l)$. Then the joint density function of stochastic processes $X(t)$ and $Y(t)$ is equal to the product of the joint probability density function of $X(t)$ and the joint probability density function of $Y(t)$, that is,

$$f_{XY}\left(x_1, x_2, \ldots, x_k, y_1, y_2, \ldots, y_l; t_1, t_2, \ldots, t_k, t_1', t_2', \ldots, t_l'\right)$$
$$= f_X(x_1, x_2, \ldots, x_k; t_1, t_2, \ldots, t_k) \cdot f_Y\left(y_1, y_2, \ldots, y_l; t_1', t_2', \ldots, t_l'\right) \qquad (19.1.8)$$

We can specify $M$ processes $X_1(t)$, $X_2(t)$, $X_3(t)$, …, $X_M(t)$ and represent them as the *vector process* $\boldsymbol{X}(t) = [X_1(t)\ X_2(t)\ X_3(t)\ \ldots\ X_M(t)]^T$ in an $M$-dimensional vector process space. In this sense, the vector process can contain two processes that are elements of the two-dimensional vector space. This vector process exists, for example, at the output of the correlation receiver block in digital communication systems.

---

**Example**

White Gaussian noise is an iid stochastic process $X(t)$, when each random variable $X(t)$, $-\infty < t < \infty$, is a zero-mean Gaussian random variable (centred process) with the same variance $\sigma^2$. Find the $N$th-order joint probability density function of this process.

**Solution.** The probability density function of a random variable $X(t_i)$, which is defined at the time instant $t = t_i$, is governed by a zero-mean Gaussian distribution. If the notation is simplified to $X(t_i) = X_i$, the density function of $X_i$ may be expressed as

$$f_{X_i}(x_i; t_i) = f_{X_i}(x_i) = \frac{1}{\sqrt{2\pi\sigma^2}} e^{-x_i^2/2\sigma^2}, \quad i = 1, 2, \ldots, N. \qquad (19.1.9)$$

Because $X(t)$ is an iid process, its joint ($N$th-order) density function is a product of marginal densities, that is,

$$f_X(x_1, x_2, \ldots, x_N; t_1, t_2, \ldots, t_N) = f_{X_1}(x_1) \cdot f_{X_2}(x_2) \cdot \ldots \cdot f_{X_N}(x_N)$$

$$= \frac{1}{\sqrt{2\pi\sigma^2}} e^{-x_1^2/2\sigma^2} \cdot \frac{1}{\sqrt{2\pi\sigma^2}} e^{-x_2^2/2\sigma^2} \cdots \frac{1}{\sqrt{2\pi\sigma^2}} e^{-x_N^2/2\sigma^2} . \qquad (19.1.10)$$

$$= \prod_{i=1}^{N} \frac{1}{\sqrt{2\pi\sigma^2}} e^{-x_i^2/2\sigma^2} = \frac{1}{(2\pi\sigma^2)^{N/2}} e^{-\frac{1}{2\sigma^2} \sum_{i=1}^{N} x_i^2}$$

## 19.2 Statistical Properties of Stochastic Processes

### 19.2.1 First- and Second-Order Properties of Stochastic Processes

Knowledge of the joint density or distribution function is sufficient for a statistical description of a stochastic process $X(t)$. It is sometimes hard to find these functions. However, in practical applications, it is not necessary to know these functions. Instead, the first- and second-order functions can be used. We are primarily interested in the mean, the variance, and the correlation function of a process.

As we said, for the sake of simplicity and generality, we will sometimes use $X = X(t)$ to denote a random variable defined at the time instant $t$ of the stochastic process $X(t)$. One realization of this random variable at time instant $t$ will be denoted as $x$. Then, for the sake of simplicity, the probability density function of this random variable is denoted as $f_X(x, t) = f_X(x)$, and the probability distribution function as $F_X(x)$.

The *mean* of a stochastic process $X(t)$ is generally a function of $t$ and is expressed as

$$\eta_X(t) = E\{X(t)\} = \int_{-\infty}^{\infty} x(t) f_X(x(t), t) \, dx = \int_{-\infty}^{\infty} x f_X(x, t) \, dx. \qquad (19.2.1)$$

The *mean square value* of the stochastic process is a function of time expressed as

$$E\left\{X^2(t)\right\} = E\left\{X^2\right\} = \int_{-\infty}^{\infty} x^2(t) f_X(x, t) \, dx = \int_{-\infty}^{\infty} x^2 f_X(x) dx, \qquad (19.2.2)$$

and the *variance* is a function of time expressed as

$$\sigma_X^2(t) = E\left\{[X(t) - \eta_X(t)]^2\right\} = E\left\{X^2(t)\right\} - \eta_X^2(t). \qquad (19.2.3)$$

Generally speaking, these functions depend on the distribution function of the random variable defined for each instant of continuous time $t$. Therefore, these functions are expressed as functions of time $t$.

In order to compare the similarities and dissimilarities of the realizations of a stochastic process $X(t)$, the cross-correlation and autocorrelation functions are used. Using a simplified notation, the *autocorrelation function* for a process $X(t)$ is defined as

$$R_{XX}(t_1,t_2) = E\left\{X(t_1)X^*(t_2)\right\} = \int_{-\infty}^{\infty}\int_{-\infty}^{\infty} x(t_1)x^*(t_2)f_{XX^*}\left(x,t_1,x^*,t_2\right)dx(t_1)dx^*(t_2)$$

$$= \int_{-\infty}^{\infty}\int_{-\infty}^{\infty} x_1 x_2^* f_{XX^*}\left(x_1,x_2^*\right)dx_1 dx_2^*$$

$$(19.2.4)$$

and gives us information about the statistical relations of the samples of a stochastic continuous-time process $X(t)$ at two different time instants $t_1$ and $t_2$. The function $f_{XX}(x_1, x_2)$ is the joint probability density function of the two random variables $X(t_1)$ and $X(t_2)$. The complex conjugate is associated with the second variable in the autocorrelation function. Therefore, the autocorrelations can be expressed in the form

$$R_{XX}(t_2,t_1) = E\left\{X(t_2)X^*(t_1)\right\} = R_{XX}^*(t_1,t_2).$$

$$(19.2.5)$$

It is important to note here that, in practice, the autocorrelation function is calculated across the ensemble of realizations of the stochastic process observed at the time instants $t_1$ and $t_2$.

For a process $X(t)$ we can also define the *autocovariance function* as

$$C_{XX}(t_1,t_2) = E\left\{[X(t_1)-\eta_X(t_1)][X(t_2)-\eta_X(t_2)]^*\right\}.$$

$$= R_{XX}(t_1,t_2) - \eta_X(t_1)\eta_X^*(t_2)$$

$$(19.2.6)$$

A *cross-correlation function* is defined, for two stochastic processes $X(t_1)$ and $Y(t_2)$, as the mathematical expectation of the product of two different random variables $X(t_1)$ and $Y(t_2)$, defined for two stochastic processes and expressed as

$$R_{XY}(t_1,t_2) = E\left\{X(t_1)Y^*(t_2)\right\} = \int_{-\infty}^{\infty}\int_{-\infty}^{\infty} xy^* f_{X,Y}(x,y,t_1,t_2)dxdy,$$

$$(19.2.7)$$

which represents the correlation between two different continuous-time random variables, $X(t_1)$ and $Y(t_2)$. The function $f_{XY}(x_1,y_2)$ is the joint probability density function of the random variables $X(t_1)$ and $Y(t_2)$.

Similarly, the *cross-covariance function* is defined as

$$C_{XY}(t_1,t_2) = E\left\{[X(t_1)-\eta_X(t_1)][Y(t_2)-\eta_Y(t_2)]^*\right\} = R_{XY}(t_1,t_2) - \eta_X(t_1)\eta_Y^*(t_2). \quad (19.2.8)$$

The autocorrelation, autocovariance, cross-correlation, and cross-covariance functions are all two-dimensional functions.

## Example

Find the mean of a sinusoidal stochastic process defined as

$$X(t) = A\cos(\omega t + \theta),$$

where the phase $\theta$ of the random signal is a random variable uniformly distributed on the interval $[0, 2\pi]$. Note that one realization of this process is generally a continuous-time, continuous-valued sinusoidal function (signal), as shown in Figure 19.2.1. This function is defined from $-\infty$ to $+\infty$.

Plot an ensemble of the process realizations and comment on the practical procedure needed to support the theoretically calculated mean values of the process.

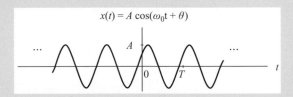

**Fig. 19.2.1** *One realization of the stochastic process X(t).*

**Solution.** In this case, $X(t)$ is a function of the random variable $\theta$, and its mean for a defined value of $t$ is

$$
\begin{aligned}
\eta_X(t) &= \int_{-\infty}^{\infty} x(t) f_X(x,t)\, dx = \int_0^{2\pi} x(\theta) f_\theta(\theta)\, d\theta \\
&= \int_0^{2\pi} A\cos(\omega t + \theta)\, \frac{1}{2\pi}\, d\theta \\
&= \frac{A}{2\pi}\left[\cos\omega t \int_0^{2\pi} \cos\theta\, d\theta - \sin\omega t \int_0^{2\pi} \sin\theta\, d\theta\right] = 0
\end{aligned}
$$

because the integrals of the sine and cosine functions of $\theta$ are 0. The process is defined by an ensemble of sinusoidal random signals, as shown in Figure 19.2.2. The ensemble is theoretically defined by an infinite uncountable number of sinusoids (infinite ensemble).

A close look at the process $X(t)$ in Figure 19.2.2, along the vertical line drawn at the time instant $t$, can reveal an imaginable infinite number of random amplitudes that will, in their sum, result in zero value. Therefore, the expected mean value is 0, which intuitively supports the theoretical calculation of the mean.

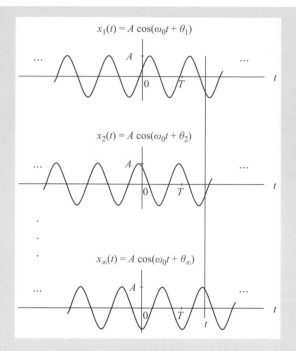

**Fig. 19.2.2** *Stochastic process X(t).*

## Example

Find the autocorrelation function of the real stochastic process that is defined in previous example as a cosine function with a random phase.

**Solution.** The process $X(t)$ at the time intant $t$ is a function of the random variable $\theta$, which is the phase of the signal, and the autocorrelation function of the random variables defined at time instants $t_1$ and $t_2$ is

$$R_{XX}(t_1,t_2) = E\{X(t_1)X(t_2)\} = \int_{-\infty}^{\infty}\int_{-\infty}^{\infty} x_1 x_2\, f_{X_1,X_2}(x_1,t_1,x_2,t_2)\,dx_1 dx_2.$$

In this case, the random variable is a phase angle, and the product inside the expectation operator $E$ can be treated as a function of the phase angle $\theta$.

Thus, the autocorrelation function can be calculated as

$$R_{XX}(t_1, t_2) = E\{X(t_1)X(t_2)\} = \int_{-\infty}^{\infty} x(t_1, \theta) x(t_2, \theta) f_\theta(\theta) \, d\theta$$

$$= \int_0^{2\pi} A \cos(\omega_0 t_1 + \theta) A \cos(\omega_0 t_2 + \theta) \frac{1}{2\pi} d\theta$$

$$= \frac{A^2}{2\pi} \int_0^{2\pi} \frac{1}{2} [\cos(\omega_0 t_1 + \omega_0 t_2 + 2\theta) + \cos(\omega_0 t_1 - \omega_0 t_2)] d\theta$$

$$= \frac{A^2}{4\pi} \cos\omega_0(t_1 - t_2) \int_0^{2\pi} d\theta + \frac{A^2}{4\pi} \int_0^{2\pi} \cos(\omega_0 t_1 + \omega_0 t_2 + 2\theta) d\theta$$

$$= \frac{A^2}{4\pi} 2\pi \cos\omega_0(t_1 - t_2) = \frac{A^2}{2} \cos\omega_0\tau = R_{XX}(\tau)$$

Therefore, the autocorrelation function of the process depends on the lag $\tau = t_1 - t_2$ between random variables of the process defined at time instants $t_1$ and $t_2$. A close look at the process $X(t)$ in Figure 19.2.2, at the chosen time instants $t_1$ and $t_2$, will support this calculation. Namely, when the time instants overlap, that is, $t_1 = t_2$, the value of the amplitude in both variables (accross the ensemble) will overlap and the autocorrelation has a maximum value equal to the power of the process. When we separate the time instants, $t_1 \neq t_2$, the amount of the overlap between the amplitude values will be reduced and the value of the autocorrelation will decrease. This procedure will continue following the cosine law. When the separation between the time instants $t_1$ and $t_2$ is equal to half of the period of the cosine function, the magnitudes at both time instants will be the same but of opposite signs. Thus, the autocorrelation attains again the maximum value, but a negative one. It is easy to calculate these maxima and minima of the autocorrelation function.

In the case when we use the general expression for the autocorrelation function, eqn (19.2.4), we can get the same result for the autocorelation function of the real stochastic process $X(t)$. Namely, the random variables can be understood as functions of the random phase angle $\theta$. Therefore, the autocorrelation function is

$$R_{XX}(t_1, t_2) = E\{X(t_1)X(t_2)\}$$

$$= \int_{-\infty}^{\infty} \int_{-\infty}^{\infty} x(t_1, \theta_1) x(t_2, \theta_2) f_{\theta_1, \theta_2}(\theta_1, t_1, \theta_2, t_2) \, d\theta_1 d\theta_2$$

$$= \int_0^{2\pi} A \cos(\omega_0 t_1 + \theta_1) \frac{1}{2\pi} d\theta_1 \int_0^{2\pi} A \cos(\omega_0 t_2 + \theta_2) \frac{1}{2\pi} d\theta_2$$

The random phases are iid random variables. Therefore, the solution of the integral is

$$R_{XX}(t_1, t_2) = \frac{A^2}{2} \cos\omega_0(t_1 - t_2) \frac{1}{2\pi} \frac{1}{2\pi} \int_0^{2\pi} \int_0^{2\pi} d\theta d\theta$$

$$+ \frac{1}{2\pi} \frac{1}{2\pi} \int_0^{2\pi} \left[ \int_0^{2\pi} A \cos(\omega_0(t_1 + t_2) + 2\theta) \, d\theta \right] d\theta.$$

$$= \frac{A^2}{2} \cos\omega_0(t_1 - t_2) = \frac{A^2}{2} \cos\omega_0\tau$$

where $\tau = (t_1 - t_2)$ is the time delay between the two random variables defined on the process $X(t)$. The autocorrelation function depends only on the time difference $\tau$ between the variables, that is,

$$R_{XX}(\tau) = R_{XX}(t_1 - t_2) = \frac{A^2}{2} \cos \omega_0 \tau. \tag{19.2.9}$$

The maximum value of the autocorrelation function can be obtained for lag equal to 0. At that lag, the samples of both random variables are the same, and the value of the autocorrelation function is equal to the power of the process, that is,

$$R_{XX}(0) = \frac{A^2}{2} = P_X.$$

Any displacement of the lag from 0 reduces the correlation value, which reaches the maximum negative value and then again the maximum positive value. The positive maxima occur for $\omega_0 \tau = 2\pi k$, that is, for $\tau = 2\pi k / \omega_0 = k/f_0$, where $k = 0, 1, \ldots$.

## Example

Find the autocorrelation function of a wide-sense stationary stochastic process $X(t)$ that is defined as a train of polar rectangular pulses of amplitude $A$ and duration $T$ as shown in Figure 19.2.3. We will assume that the probabilities of positive and negative binary amplitude values $A$ are equal to 0.5. Therefore, at each time instant $t = t_k$, we have a discrete random variable $X = X(t_k)$ that takes the value of $A$ or $-A$. The probability density function of $X$ can be expressed in terms of Dirac delta functions as

$$f_X(x) = \frac{1}{2} \delta(x - A) + \frac{1}{2} \delta(x + A).$$

In addition, we will assume that all realizations of the stochastic process are asynchronously generated with random delays represented by a random variable $D$ that is uniformly distributed. The probabability density function is defined inside the pulse duration $T$ as

$$f_D(d) = \begin{cases} 1/T & 0 \le d \le T \\ 0 & \text{otherwise} \end{cases}.$$

Two realizations of the process are presented in Figure 19.2.3 for two delays, $d_1 = 0$ and $d_N > 0$, which are realizations of the defined random variable $D$.

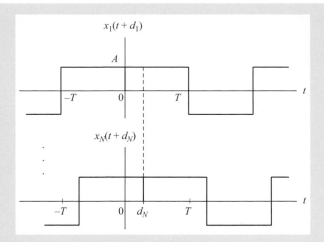

**Fig. 19.2.3** *A polar binary process presented in the time domain.*

**Solution.** Since the amplitude levels of $x(t)$ occur with equal probabilities, the mean value of the process is 0, and the variance is $A^2$. Because the process is wide-sense stationary, with zero mean and constant variance, the autocorrelation function of the process depends only on the time difference between the two time instants, which is defined as the lag $\tau$. If the lag is greater than the interval $T$, the values of the autocorrelation function will be equal to 0. If the lag is within the time interval $T$, the values of the autocorrelation function will be different from zero and decrease from those of the zero lag to those of the lag $T$. The autocorrelation function can be expressed in the general form

$$R_{XX}(\tau) = E\{X(t)X(t+\tau)\} = \int_{-T}^{+T} x(t,d)x(t+\tau,d)f_D(d)dd.$$

We can have two intervals of integration here depending on the relationship of the lags with respect to the reference lag of $\tau = 0$, when the autocorrelation function has the maximum value:

1. If the lag is $\tau \leq 0$, the correlation integral can be calculated as

$$R_{XX}(\tau) = \int_0^{T+\tau} x(t,d)x(t+\tau,d)f_D(d)dd = \int_0^{T+\tau} A^2\frac{1}{T}dd = A^2\frac{1}{T}(T+\tau).$$

2. If the lag is $\tau \geq 0$, the integral can be expressed as

$$R_{XX}(\tau) = \int_\tau^{T} x(t,d)x(t+\tau,d)f_D(d)dd.$$

By substituting $d$ with $d = z + \tau$ and changing the limits of integration, we obtain

$$R_{XX}(\tau) = \int_0^{T-\tau} x(t, z+\tau) x(t+\tau, z+\tau) f_D(z+\tau) \, dz = \int_0^{T-\tau} A^2 \frac{1}{T} dz = A^2 \frac{1}{T}(T-\tau).$$

Therefore, the autocorrelation function for any delay $\tau$ can be expressed in the form

$$R_{XX}(\tau) = \begin{cases} A^2 \frac{1}{T}(T-|\tau|) & |\tau| \leq T \\ 0 & \text{otherwise} \end{cases}, \qquad (19.2.10)$$

and represented by a triangular graph, as shown in Figure 19.2.4.

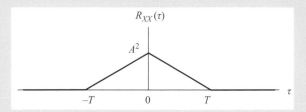

**Fig. 19.2.4** *Autocorrelation function of a binary stochastic process.*

## 19.2.2   Types of Stochastic Processes

**Characterization of one stochastic process.** A stochastic process $X(t)$ is an *independent stochastic process* if its joint probability density function, of $k$th order, is defined as the product of marginal probability density functions, that is,

$$f_X(x_1, x_2, \ldots, x_k; t_1, t_2, \ldots, t_k) = f_1(x_1; t_1) \cdot f_2(x_2; t_2) \cdot \cdots \cdot f_k(x_k; t_k) \qquad (19.2.11)$$

for every $k$ and $t_i$, where $i = 1, 2, \ldots, k$. In the case when all random variables, defined for every $t_i$, have identical probability density functions, the process $X(t)$ is an iid process.

A stochastic process $X(t)$ is an *uncorrelated stochastic process* if it is a sequence of uncorrelated random variables, that is, its autocovariance is

$$C_{XX}(t_1, t_2) = E\{[X(t_1) - \eta_1][X(t_2) - \eta_2]\} = \begin{cases} \sigma_X^2(t_1) & t_1 = t_2 \\ 0 & t_1 \neq t_2 \end{cases} = \sigma_X^2(t_1)\delta(t_1 - t_2) \quad (19.2.12)$$

because, for independent or uncorrelated random variables, the following condition holds:

$$E\{X(t_1)X(t_2)\} = E\{X(t_1)\}E\{X(t_2)\}.\qquad(19.2.13)$$

The autocorrelation of an uncorrelated process is

$$R_{XX}(t_1,t_2) = E\{X(t_1)X(t_2)\} = \begin{cases} \sigma_X^2(t_1) + |\eta_X(t_1)|^2 & t_1 = t_2 \\ \eta_X(t_1)\,\eta_X(t_2) & t_1 \neq t_2 \end{cases}\qquad(19.2.14)$$

A stochastic process $X(t)$ is said to be an *orthogonal stochastic process* if it is a sequence of orthogonal random variables, that is, its autocorrelation is

$$R_{XX}(t_1,t_2) = \begin{cases} \sigma_X^2(t_1) + |\eta_X(t_1)|^2 & t_1 = t_2 \\ 0 & t_1 \neq t_2 \end{cases} = E\left\{|X(t_1)|^2\right\}\delta(t_1 - t_2),\qquad(19.2.15)$$

which is the consequence of the orthogonality condition for two random variables, which is expressed as

$$E\{X(t_1)X(t_2)\} = 0.\qquad(19.2.16)$$

**Characterization of two stochastic processes.** Suppose two processes, $X(t)$ and $Y(t)$, are defined on the same sample space $S$. We can apply the above definitions for these two joint stochastic processes, as follows.

Two stochastic processes $X(t)$ and $Y(t)$ are *statistically independent* if, for all values of $t_1$ and $t_2$, their joint density function fulfils the condition defined by eqn (19.2.11), that is,

$$f_{XY}(x,y;t_1,t_2) = f_X(x;t_1)\cdot f_Y(y;t_2).$$

The stochastic processes $X(t)$ and $Y(t)$ are *uncorrelated* if, for every $t_1$ and $t_2$, their covariance is

$$C_{XY}(t_1,t_2) = 0\qquad(19.2.17)$$

and the cross-correlation is

$$R_{XY}(t_1,t_2) = \eta_X(t_1)\,\eta_Y^*(t_2).\qquad(19.2.18)$$

The stochastic processes $X(t)$ and $Y(t)$ are *orthogonal* if, for every $t_1$ and $t_2$, they fulfil the condition expressed by eqn (19.2.16), that is, their autocorrelation is

$$R_{XY}(t_1,t_2) = 0.$$

## 19.2.3   Entropy of Stochastic Processes and White Noise

The aim of this section is to analyse both continuous-time white Gaussian noise and uniformly distributed noise. We will relate this to the existing theory of noise statistics by finding the answers to the following questions:

1. Can white Gaussian noise have infinite power and energy, as we assume in the existing mathematical models of noise? If it can, then how can this white Gaussian noise be rigorously defined mathematically and interpreted?

2. Is the variance of white Gaussian noise always equal to the power of the noise? When could the opposite be true?

3. Under which conditions is the variance for white Gaussian noise finite, and under which conditions is the power finite?

4. How are the power spectral density, autocorrelation function, power, and energy of white Gaussian noise related to each other in the above-mentioned cases?

5. Can the entropy (differential) of an analogue noise process be used to *define* the noise process for the above cases, and, if so, how?

6. Finally, how does all that work in the discrete-time domain? In addition, why do we call this type of noise 'white', referring to white light, and say that it has infinite bandwidth, when the bandwidth of visible light is strictly band limited?

**Entropy of stochastic processes.** The entropy of a stochastic processes can be analysed from a theory-of-information point of view. For that purpose, a stochastic process generator can be treated as the source of information in the sense of this theory. If the source has a countable (finite or infinite) number of outcomes, the information content of them can be found, and the entropy, defined as the average information per outcome, can be quantified according to the principles of information theory. Likewise, if the source produces a continuous-time process (with an infinite, uncountable number of possible outcomes), the entropy (usually called 'differential entropy' in the literature) can be calculated as the mean value of the information content of the process. Details related to the calculation of entropy for different sources of information can be found in Chapter 9, which is dedicated to the theory of information.

The entropy of a continuous-time stochastic process $X(t)$ cannot be directly calculated because the number of elementary events is uncountably large and, therefore, we cannot assign a finite probability to them. The entropy can be obtained by discretizing the density function, finding the entropy of the discrete density function, and then finding the limit when the discretization interval tends to 0. Therefore, as Papoulis and Pillai showed (2002, p. 654), the entropy can be calculated according to the following expression:

$$H(X) = -\int_{-\infty}^{\infty} f_X(x) \log f_X(x) dx, \qquad (19.2.19)$$

where the logarithm is assumed to be natural, that is, $\log() = \ln()$. In the theory of information, we use the logarithm function with base 2.

**Entropy of Gaussian stochastic processes.** Suppose that $X(t)$ is defined as an iid Gaussian stochastic process with zero mean and variance $\sigma^2$. At each time instant $t$, we can define a random variable $X(t) = X$ that has a Gaussian probability density function, $f_X(x)$. We are interested in knowing how the entropy changes as a function of variance and, in particular, what happens when the variance tends to infinity.

Following the definition of entropy in the theory of information, log base 2 is used and the differential entropy can be calculated as

$$
\begin{aligned}
H(X) &= \int_{-\infty}^{\infty} f_X(x) \log_2 \frac{1}{f_X(x)} = \int_{-\infty}^{\infty} f_X(x) \log_2 \sqrt{2\pi}\,\sigma_X e^{x^2/2\sigma_X^2} dx \\
&= \log_2 e \int_{-\infty}^{\infty} f_X(x) \left[ \frac{x^2}{2\sigma_X^2} + \log \sqrt{2\pi}\,\sigma_X \right] dx \\
&= \frac{1}{2\sigma_X^2} \log_2 e \int_{-\infty}^{\infty} x^2 f_X(x) dx + \log_2 e \log \sqrt{2\pi}\,\sigma_X \int_{-\infty}^{\infty} f_X(x) dx \\
&= \frac{1}{2\sigma_X^2} \sigma_X^2 \log_2 e + \frac{1}{2} \log_2 e \log 2\pi \sigma_X^2 = \frac{1}{2} \log_2 2\pi\, e \sigma_X^2
\end{aligned}
\tag{19.2.20}
$$

The entropy increases when the variance increases, tending to the infinity when the variance tends to infinity. At the same time, the Gaussian density function 'flattens' and spreads the probability values to infinity on both sides with respect to 0. The density tends to a uniform distribution, having nearly the same infinitesimally small probability of any deferential event that can be imagined on the infinite interval of possible values of the random variable.

A process with these characteristics is not physically realizable, due to the possibility of generating variates of infinite amplitudes. Furthermore, if we want to sample this process and generate its discrete-time form, we need to assume the possibility that some of the samples could have infinite amplitudes. However, this possibility docs not fulfil the Nyquist criterion, which demands a continuous-time sampled signal to be physically realizable, that is, to have finite amplitude values. In addition, if it is possible for the process to have infinite continuous-time values, that will lead to the generation of discrete Gaussian process variates of infinite power, which is, of course, not physically realizable. In order to design a process that is physically realizable in both continuous-time and discrete-time domains, we need to limit the noise amplitudes to finite values and thus preserve the statistical properties that we require in practice.

**Entropy of a truncated Gaussian stochastic process.** A Gaussian stochastic process $X(t)$ is amplitude limited if it can generate random amplitudes in a limited interval of possible values. This can be achieved if the amplitudes are distributed according to the Gaussian truncated density function, which is expressed as

$$
f_X(x) = C(A) \frac{1}{\sqrt{2\pi\sigma^2}} e^{-x^2/2\sigma^2},
\tag{19.2.21}
$$

where $A$ defines the symmetric truncation interval of possible amplitude values centred about the zero mean of the process, that is, $-A \leq x < A$, $\sigma^2$ is the variance of the untruncated process, and $C(A)$ is a function of $A$ expressed as

$$C(A) = \frac{1}{1 - \text{erfc}\left(A/\sqrt{2\sigma^2}\right)}. \tag{19.2.22}$$

For the sake of illustration, a graph of a hypothetical truncated density function is presented in Figure 19.2.5. The ensemble of realizations of $X(t)$ are random signals $x(t)$ having amplitudes generated in the truncation interval of their possible values.

**Variance of the Gaussian truncated process.** The variance of the truncated process is

$$\sigma_{\tilde{X}}^2 = \sigma^2 \left( 1 - \frac{1}{1 - \text{erfc}A/\sqrt{2\sigma^2}} \frac{2A}{\sqrt{2\pi\sigma^2}} e^{-A^2/2\sigma^2} \right). \tag{19.2.23}$$

The truncation operation reduces the variance of the process. This reduction depends on the truncation interval defined by $A$. When this interval tends to infinity, the variance of the truncated density becomes equal to the variance of the untruncated density. That makes sense. Namely, by reducing the truncation interval, we increase the probability of generating smaller amplitudes, so the stochastic signal has smaller power. Consequently, by increasing the truncation interval to infinity, we allow the generation of amplitudes of infinite values, which theoretically results in an infinite power process. Truncated Gaussian processes exist in physical systems. Therefore, it is worth analysing this case in detail.

**Entropy of a truncated Gaussian stochastic process.** Bearing in mind the truncation of a density function, its differential entropy can be calculated as

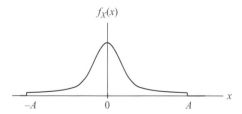

**Fig. 19.2.5** *Truncated probability density function of the Gaussian process.*

$$H(X) = -\int_{-\infty}^{\infty} f_X(x) \log_2 f_X(x) dx = \int_{-A}^{A} f_X(x) \log_2 \sqrt{2\pi} \, \sigma_X e^{x^2/2\sigma_X^2} dx$$

$$= \log_2 e \int_{-A}^{A} f_X(x) \left[ \frac{x^2}{2\sigma_X^2} + \log \sqrt{2\pi} \, \sigma_X \right] dx$$

$$= \frac{1}{2\sigma_X^2} \log_2 e \int_{-A}^{A} x^2 f_X(x) dx + \log_2 e \log \sqrt{2\pi} \, \sigma_X \int_{-A}^{A} f_X(x) dx$$

$$= \frac{1}{2\sigma_X^2} \sigma_X^2 \log_2 e + \frac{1}{2} \log_2 e \log 2\pi \sigma_X^2 = \frac{1}{2} \log_2 2\pi e \sigma_X^2$$

(19.2.24)

Inserting the variance (19.2.23), we obtain

$$H = \frac{1}{2} \log_2 2\pi e \sigma^2 \left( 1 - \frac{1}{1 - \text{erfc} \, A/\sqrt{2\sigma^2}} \frac{2A}{\sqrt{2\pi\sigma^2}} e^{-A^2/2\sigma^2} \right). \tag{19.2.25}$$

In this case the variance is finite and the entropy is finite. That makes sense again. Namely, the level of certainty that the amplitude contained in the truncation interval will be generated is higher than the level of certainty of generating the amplitudes in the same interval, assuming that the density function is not truncated and has the same variance $\sigma^2$. In addition, there is infinite certainty (zero uncertainty) that amplitudes will be generated in the truncation interval. Therefore, the uncertainty of generating amplitudes in a truncation interval is finite.

**Entropy of a uniform stochastic process.** A random process $X(t)$ is uniform if the random variables $X(t) = X$, defined for any $t$, have a uniform density function defined as

$$f_X(x) = \left\{ \begin{array}{cc} \frac{1}{2A} & -A \le x < A \\ 0 & otherwise \end{array} \right\} = \left\{ \begin{array}{cc} \frac{1}{2\sigma_X \sqrt{3}} & -A \le x < A \\ 0 & otherwise \end{array} \right\}, \tag{19.2.26}$$

where $A$ is the interval of possible random values, and $\sigma_X^2$ is the variance of the process. The density function is graphically presented in Figure 19.2.6, for zero mean and variance $\sigma_X^2 = A^2/3$.

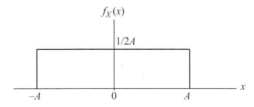

**Fig. 19.2.6** *Continuous uniform density function.*

Bearing in mind that $2A = 2\sigma_X \sqrt{3}$, the entropy can be expressed as a function of variance, that is,

$$H = \int_{\infty}^{\infty} f_X(x) \log \frac{1}{f_X(x)} dx = \int_{-A_c}^{A} \frac{1}{2\sigma_X \sqrt{3}} \log_2 2\sigma_X \sqrt{3} dx = \frac{2A}{2\sigma_X \sqrt{3}} \log_2 2\sigma_X \sqrt{3}. \qquad (19.2.27)$$
$$= \log_2 \sqrt{12}\sigma_X$$

In addition, the variance is $\sigma_X^2 = A^2/3$, and the entropy, as a function of the amplitude interval width $2A$, can be expressed as

$$H = \log_2 \sqrt{12}\sigma_X = \log_2 \sqrt{12A^2/3} = \log_2 2A. \qquad (19.2.28)$$

Therefore, for the density function defined in the interval $-A \leq x < A$, the entropy has a finite value. It is important to know what will happen if the width of the interval tends to infinity, that is, for the case when all possible outcomes occur with the infinitesimally small probabilities. In this case, the entropy, defined as the mean value of the information content per a random (differential) value of the process, can be calculated as the limit

$$H = \lim_{\sigma_X \to \infty} \log_2 \sqrt{12}\sigma_X = \lim_{A \to \infty} \log_2 2A \to \infty. \qquad (19.2.29)$$

Therefore, the entropy tends to infinity when the variance or the interval of possible amplitude values tends to infinity. That is the case when all random values are equiprobable and have probabilities close to 0, but not 0. Consequently, their information contents tend to infinity, which maximizes the average uncertainty of their appearance at the output of the random generator. Therefore, we may say that the information content of each random value tends to infinity:

$$I_X = \lim_{A \to \infty} \log_2 \frac{1}{f_X(x)} = \lim_{A \to \infty} \log_2 2A =\to \infty. \qquad (19.2.30)$$

**Entropy of an infinite uniform noise process.** Suppose an iid noise process $X(t)$ is defined by a uniform distribution with a truncation interval that stretches to infinity. Then, the entropy of the noise source will tend to infinity, meaning that the uncertainty of generating any random value of the noise is extremely high, tending to infinity, and the information content in the generated random value is infinite. Therefore, any realization $x(t)$ of the stochastic process $X(t)$ generated in the time domain will contain all amplitudes with nearly the same probabilities, including the amplitude of infinite values. In addition, the amplitude variations of the process will be very large, having independent amplitude values no matter how close they are in the time domain. Consequently, the power spectral density will cover a continuous bandwidth from $-\infty$ to $+\infty$, and the power calculated will be infinite, implying that the process has infinite energy. The values of all of the frequency components will be the same, that is, they will have constant values. The process, which has infinite energy, infinite power, a constant power spectral density in an infinite bandwidth, and infinite entropy, will be called a *black process*, as opposed to the white process, which has finite variance, finite entropy, and a constant power spectral density in an infinite bandwidth (or a finite bandwidth, for power -limited signals).

**Processes with infinite power.** A uniform noise process with infinite variance is equivalent to a Gaussian noise process defined by a variance that tends to infinity and a Gaussian density function that tends to uniform density. This process will be equivalent to a black Gaussian process, which is defined as a stochastic process that has infinite variance, infinite power, infinite entropy, infinite energy, and a power spectral density in an infinite interval. The infinite power is caused by the components that have infinite amplitudes. The energy in the time domain can be imagined as a product of the averaged infinite power and the infinite time. In the frequency domain, the power will be the product of the finite power spectral density and the infinite bandwidth. The realizations of this process do not comply with the definition of either a power signal or an energy signal. Of course, we are not able to make a generator for this kind of stochastic process. Then, what, realistically, we can achieve? We must generate a noise process with limited power. A detailed discussion of that kind of noise process, which is defined by a Gaussian truncated density function, can be found in Chapter 4.

## 19.3   Stationary and Ergodic Stochastic Processes

### 19.3.1   Stationary Stochastic Processes in Time and Frequency Domains

#### 19.3.1.1   *Time Domain Analysis*

A stochastic process $X(t)$ is stationary if its statistical characteristics are equal to the statistical characteristics of its time-shifted version $X(t + k)$, for every $k$. In particular, a stochastic process $X(t)$ is called *stationary of order N* if its $N$th-order density function is invariant to a time origin shift, that is, if the following condition is fulfilled:

$$f_X (x_1, x_2, \ldots, x_N; t_1, t_2, \ldots, t_N) = f_X (x_{1+k}, x_{2+k}, \ldots, x_{N+k}; t_{1+k}, t_{2+k}, \ldots, t_{N+k}) \qquad (19.3.1)$$

for any value of $k$. Stationarity of order $N$ implies stationarity to all orders $k \leq N$. A stochastic process $X(t)$ is called *strict-sense stationary* (or *strongly stationary*) if it is stationary for any order $N = 1, 2, 3, \ldots$. An example of such a process is an iid process.

A stochastic process $X(t)$ is called *stationary of order 1* if its first-order density function is invariant to a time origin shift, that is, if the following condition is fulfilled:

$$f_X (x_1; t_1) = f_X (x_{1+k}; t_{1+k}), \qquad (19.3.2)$$

for any value of $k$. The consequence of this definition is that the mean value of the process $X(t)$ is 0.

Strict-sense stationarity is restricted to a particular group of stochastic processes. However, in practical applications, there is a group of processes that can be defined by a more relaxed form of stationarity. This is the group of wide-sense stationary processes, which are stationary up to the second order and defined as follows: a stochastic process $X(t)$ is called *stationary of order* 2 if its second-order density function is invariant to a time origin shift, that is, if the following condition is fulfilled:

$$f_X(x_1, x_2; t_1, t_2) = f_X(x_{1+k}, x_{2+k}; t_{1+k}, t_{2+k}), \tag{19.3.3}$$

for any value of $k$. A stochastic process $X(t)$ is called *wide-sense stationary* (or *weakly stationary*) if its mean and variance are constant and independent of $t$, and its autocorrelation function depends only on the distance or lag $\tau = (t_2 - t_1)$, that is,

$$\eta_X = E\{X(t)\} = \text{constant}, \ \sigma_X^2 \text{ is finite}, \tag{19.3.4}$$

and

$$R_{XX}(t_1, t_2) = R_{XX}(t_2 - t_1)) = E\left\{X(t_1)X^*(t_1 + \tau)\right\} = R_{XX}(\tau). \tag{19.3.5}$$

Consequently, the autocovariance function of a wide-sense stationary process also depends only on the difference or lag $\tau = (t_2 - t_1)$,

$$\begin{aligned} C_{XX}(t_1, t_2) = C_{XX}(t_2 - t_1) &= R_{XX}(t_2 - t_1) - \eta_X \eta_X^* \\ &= R_{XX}(\tau) - |\eta_X|^2 = C_{XX}(\tau) \end{aligned} \tag{19.3.6}$$

Two complex processes, $X(t)$ and $Y(t)$, are jointly wise-sense stationary if each of them is wise-sense stationary and their cross-corelation function depends only on the time difference between them, that is,

$$R_{XY}(t_1, t_2) = E\left\{X(t_1)Y^*(t_2)\right\} = E\left\{X(t_1)Y^*(t_1 + \tau)\right\} = R_{XY}(\tau) - \eta_X \eta_Y^*. \tag{19.3.7}$$

Consequently, wide-sense stationary processes have one-dimensional correlation and covariance functions. Due to this characteristic, their power spectral density can be nicely described. As for power deterministic signals, the average power of a wide-sense stationary process $X(t)$ is independent of $t$ and equal to $R_{XX}(0)$.

It is important to note that there are some wide-sense stationary processes that are not stationary. The definition of stationarity comes from the condition imposed on the density functions, not on the second-order statistics. The autocorrelation function of a wide-sense stationary process has the following properties:

1. The autocorrelation of a wide-sense stationary process $X(t)$ has its maximum at zero time shift, that is,

$$R_{XX}(0) = \sigma_X^2 + |\eta_X|^2 \geq 0. \tag{19.3.8}$$

The mean value squared corresponds to the average DC power of the stochastic process, while the variance corresponds to the average AC power of the process. Thus, the value of the autocorrelation at $\tau = 0$ corresponds to the total average power of the process.

2. The autocorrelation function of a wide-sense stationary process $X(t)$ (as well as the autocovariance function) is a symmetric function:

$$R_{XX}(\tau) = R_{XX}(-\tau).$$

3. The autocorrelation function is non-negative definite, that is, it satisfies the following condition:

$$\sum_{i=1}^{M}\sum_{k=1}^{M}\alpha_i R_{XX}(t_i, t_k)\alpha_k^* \geq 0, \tag{19.3.9}$$

for any $M > 0$ and any values for $\alpha_i$ and $\alpha_k$.

---

**Example**

Show that the process

$$X(t) = A\cos(\omega_0 t + \theta),$$

where the phase $\theta$ is a random variable uniformly distributed in the interval from 0 to $2\pi$, is a wide-sense stationary process.

**Solution.** We have shown that this process has zero mean and that the autocorrelation function depends on the time differnce $\tau = (t_2 - t_1)$. Thus, it is a wide-sense stationary process.

---

### 19.3.1.2 *Frequency Domain Analysis*

In previous complementary chapters, an extensive analysis of deterministic continuous-time and discrete-time signals was presented. These signals were transformed from their time-domain form into their frequency-domain form, using Fourier transforms. However, such transforms, which are based on the time-domain representation of the stochastic signals, do not always exist. For that reason, here we will use other, more sophisticated tools to analyse stochastic signals.

In previous sections, stochastic processes were analysed in the time domain. These processes are characterized by their first-, second-, and higher-order distribution functions. In addition, stationary processes were defined and characterized by their correlation functions and covariance functions. As was shown, these correlation and covariance functions were functions of one variable. This property allowed for the relatively simple transform of these processes into the frequency domain. In this section, stationary processes will be analysed in the frequency domain.

**Power spectral density: The Wiener–Khintchine theorem.** According to the Wiener–Khintchine theorem, the power spectral density of a wide-sense stationary process $X(t)$ is equal to the Fourier transform of its autocorrelation function $R_{XX}(\tau)$. In this case, the correlation function $R_{XX}(\tau)$ should be absolutely integrable, which guarantees the existence of the power spectral density, which means that $X(t)$ is a zero-mean real or complex process. The power spectral density is a real function of frequency, expressed as

$$S_{XX}(\omega) = \int_{-\infty}^{\infty} R_{XX}(\tau) e^{-j\omega\tau}. \tag{19.3.10}$$

The power spectral density is a non-negative, symmetric, integrable function from the set of real numbers to the real numbers. Using the inverse Fourier transform, we can find the autocorrelation function as

$$R_{XX}(\tau) = \frac{1}{2\pi} \int_{-\infty}^{\infty} S_{XX}(\omega) e^{j\omega\tau} d\omega. \tag{19.3.11}$$

However, there are autocorrelation functions that are not integrable but still have a power spectral density as defined in eqn (19.3.10).

---

**Example**

Find the power spectral density of a zero-mean wide-sense stationary process $X(t)$ that has a correlation function defined as

$$R_{XX}(\tau) = R_{XX}(t_1 - t_2) = \frac{A^2}{2} \cos \omega_0 \, \tau. \tag{19.3.12}$$

Represent the power spectral density in graphical form.

**Solution.** The power spectral density is

$$\begin{aligned}
S_{XX}(\omega) &= \int_{\tau=-\infty}^{\infty} \frac{A^2}{2} \cos(\omega_0\tau) e^{-j\omega\tau} d\tau \\
&= \frac{A^2}{4} \int_{\tau=-\infty}^{\infty} e^{j\omega_0 \, \tau} e^{-j\omega\tau} d\tau + \frac{A^2}{4} \int_{\tau=-\infty}^{\infty} e^{-j\omega_0 \, \tau} e^{-j\omega\tau} d\tau \\
&= \frac{A^2}{4} 2\pi \delta(\omega - \omega_0) + \frac{A^2}{4} 2\pi \delta(\omega + \omega_0) \\
&= \frac{A^2 \pi}{2} \delta(\omega - \omega_0) + \frac{A^2 \pi}{2} \delta(\omega + \omega_0)
\end{aligned} \tag{19.3.13}$$

The power spectral density is presented in graphical form in Figure 19.3.1.

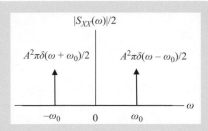

**Fig. 19.3.1** *The power spectral density function.*

**Properties of the power spectral density.** There are three properties of the power spectral density function that follow from the properties of autocorrelation function and the characteristics of the Fourier transform.

1. If a process $X(t)$ is real-valued, then the power spectral density $S_{XX}(\omega)$ is an even function of frequency $\omega$, that is, the following condition is fulfilled:

$$S_{XX}(\omega) = S_{XX}(-\omega). \tag{19.3.14}$$

The proof for this follows from the characteristics of the autocorrelation function, and the properties of the Fourier transform.

2. The function $S_{XX}(\omega)$ is non-negative definite, that is,

$$S_{XX}(\omega) \geq 0. \tag{19.3.15}$$

The proof for this follows from the non-negative definiteness of the autocorrelation function.

3. The integral value of the power spectral density function is non-negative and equal to the average power or to the value of the autocorrelation function for $\tau = 0$ of the process $X(t)$, that is,

$$P_X = \frac{1}{2\pi} \int_{-\infty}^{\infty} S_{XX}(\omega) \, d\omega = R_{XX}(0) = E\left\{ |X(t)|^2 \right\}, \tag{19.3.16}$$

which can be also calculated using the stochastic process $X(t)$ in the time domain as

$$P_X = \lim_{T \to \infty} E\left\{ \frac{1}{T} \int_{-T/2}^{T/2} X^2(t) dt \right\}. \tag{19.3.17}$$

**Example**

Confirm the third property for the autocorrelation function

$$R_{XX}(\tau) = R_{XX}(t_1 - t_2) = \frac{A^2}{2}\cos\omega_0\,\tau.$$

(19.3.18)

**Solution.** The power of the process is

$$P_X = R_{XX}(0) = \frac{A^2}{2}.$$

(19.3.19)

For the power spectral density

$$S_{XX}(\omega) = \frac{A^2\pi}{2}\delta(\omega - \omega_0) + \frac{A^2\pi}{2}\delta(\omega + \omega_0),$$

(19.3.20)

we can calculate the power as

$$
\begin{aligned}
\frac{1}{2\pi}\int_{-\infty}^{\infty} S_{XX}(\omega)\,d\omega &= \frac{1}{2\pi}\int_{-\infty}^{\infty}\left[\frac{A^2\pi}{2}\delta(\omega - \omega_0) + \frac{A^2\pi}{2}\delta(\omega + \omega_0)\right]d\omega \\
&= \frac{1}{2\pi}\left[\frac{A^2\pi}{2} + \frac{A^2\pi}{2}\right] = \frac{A^2}{2} = P_X
\end{aligned}
$$

(19.3.21)

which is identical to the value calculated from the autocorrelation function.

## 19.3.2 Ergodic Stochastic Processes

A stochastic process is defined by an ensemble of its realizations (random signals, random functions) that are defined for all outcomes of a random experiment. For this process, the statistical averages are defined on the ensemble of realizations at a defined instant of time $t$. In practice, we do not have all these realizations available to be analysed at the instant $t$ in order to find these averages. Instead, we need to infer the properties of the underlying stochastic process using a single realization $x(t)$ of the process $X(t)$, which is a random function of time. This can be done for the so-called ergodic processes, for which we define time averages using the limits of infinite sums.

Precisely speaking, the process is ergodic if its ensemble averages are equal to the time averages with probability 1. These averages can be calculated using the following procedures. The mean value of one realization $x(t)$ of the stochastic process $X(t)$ is

$$\eta_x = \langle x(t)\rangle = \lim_{a\to\infty}\frac{1}{2a}\int_{-a}^{a} x(t)dt.$$

(19.3.22)

The autocorrelation of one realization $x(t)$ is

$$R_{xx}(\tau) = \left\langle x(t)x^*(t+\tau) \right\rangle = \lim_{a \to \infty} \frac{1}{2a} \int_{-a}^{a} x(t)x(t+\tau)\,dt. \qquad (19.3.23)$$

The cross-correlation is

$$R_{xy}(\tau) = \left\langle x(t)y^*(t+\tau) \right\rangle = \lim_{a \to \infty} \frac{1}{2a} \int_{-a}^{a} x(t)y(t+\tau)\,dt. \qquad (19.3.24)$$

These averages are very similar to the averages defined for deterministic processes. However, the averages of ergodic processes are random variables, because they depend on the outcomes of a random experiment. Depending on the equality of these averages with the expectations of the related stochastic process, there are several degrees of ergodicity, such as *ergodicity in the mean*, and *ergodicity in the autocorrelation.*

A stochastic process $X(t)$ is *ergodic in the mean* if its average value on any realization $x(t, \zeta_i) = x(t)$ is equal to the mean value of the ensemble $X(t)$, that is,

$$\eta_X = E\{X(t)\} = \langle x(t) \rangle = \lim_{a \to \infty} \frac{1}{2a} \int_{-a}^{a} x(t)dt = \eta_x. \qquad (19.3.25)$$

A stochastic process $X(t)$ is *ergodic in the autocorrelation* if, and only if, the following condition is fulfilled:

$$\begin{aligned} R_{XX}(\tau) &= E\left\{X(t)X^*(t+\tau)\right\} \\ &= R_{xx}(\tau) = \left\langle x(t)x^*(t+\tau) \right\rangle = \lim_{a \to \infty} \frac{1}{2a} \int_{-a}^{a} x(t)x(t+\tau)\,dt \end{aligned} \qquad (19.3.26)$$

for all $\tau$. In practical applications, we do not use time-averaging expressions with an infinite number of samples of a stochastic process. These formulas are identical to the ones given above when *limits* are excluded and $a$ is a finite number. In this sense, the mean value is defined as

$$\eta_x = \langle x(t) \rangle = \frac{1}{2a} \int_{-a}^{a} x(t)dt. \qquad (19.3.27)$$

In practice, we take a limited number of samples $N$ at the discrete-time instants $T_s$, which allows us to achieve an accuracy that is specified in advance by enabling us to estimate the mean value of the process as

$$\eta_x = \langle x(nT_s) \rangle = \frac{1}{2N} \sum_{n=-N}^{N} x(nT_s), \qquad (19.3.28)$$

and the autocorrelation function as

$$R_{xx}(\tau) = \left\langle x(t)x^*(t+\tau) \right\rangle_{t=nT_s} = \frac{1}{2N} \int_{n=-N}^{N} x(nT_s)x(nT_s + \tau T_s)\,dt = R_{xx}(\tau T_s). \qquad (19.3.29)$$

It is important to note the following: even though we are calculating the mean value and autocorrelation function for an ergodic process, the subscript used to denote them contains two lowercase letters, because the process is represented by a realization that behaves like a power deterministic signal.

### 19.3.3 Characterization of White Noise Processes

This section will present the definitions and basic characteristics of a purely random process, white noise, strict white noise, wide-sense stationary white noise, and white Gaussian noise.

**A purely random process.** A zero-mean, stationary, continuous-time stochastic process that satisfies the condition

$$f_X(x_1, x_2, \ldots; t_1, t_2, \ldots) = f_X(x_1) \cdot f_X(x_2) \cdot \ldots, \qquad (19.3.30)$$

where all random variables $X(t_i)$ are identically distributed, is a *purely random process*. In this definition, the density function of a particular random variable is not defined, that is, it can have any distribution. We simply say that this process is defined by a series of iid random variables $X(t_i)$. Consequently, being iid random variables, they are also uncorrelated. However, the uncorrelated processes are not necessarily independent, or, more accurately, the iid variables comprise a subset of uncorrelated variables. This process is *strictly stationary*, so it is also wide-sense stationary. The definition of a purely random process is used to define a *noise process*.

**White noise.** A process $X(t)$ is *white noise* if its random variables $X(t_i)$ and $X(t_j)$ are uncorrelated for every $t = t_i$ and $t_j \neq t_i$, that is,

$$C_{XX}(t_1, t_2) = 0. \qquad (19.3.31)$$

The autocorrelation function of a wide-sense stationary white noise process is

$$R_{XX}(t_1, t_2) = E\left\{ |X(t_1)|^2 \right\} \delta(t_1 - t_2) = P_X \cdot \delta(\tau), \qquad (19.3.32)$$

and the power spectral density is a constant that is equal to the power of the noise for all frequencies from $-\infty$ to $+\infty$, that is,

$$S_X(\omega) = P_X. \qquad (19.3.33)$$

**Strictly white noise.** A process $X(t)$ is *strictly white noise* if the random variables $X(t_i)$ and $X(t_j)$ are not only uncorrelated but also independent, that is, for every $t = t_i$ and $t_j \neq t_i$, the following holds:

$$C_{XX}(t_1, t_2) = 0. \qquad (19.3.34)$$

**Wide-sense stationary white noise.** A process $X(t)$ is *wide-sense stationary white noise* if covariance depends on the lag and is defined as

$$C_{XX}(\tau) = 0. \qquad (19.3.35)$$

**White noise process.** Unless stated otherwise, it will be assumed that the mean value of the white noise process is 0, what is the usual case in practice. Then we say that, a process $W(t)$, with uncorrelated variables but not necessarily independent, is *white noise process* if and only if the mean is $\eta_W(t) = 0$, and the autocorrelation function is

$$R_{WW}(t_1, t_2) = E\left\{W(t_1)W^*(t_2)\right\} = E\left\{W^2(t_1)\right\} = E\left\{W^2(t_2)\right\} = \sigma_W^2 \delta(t_1 - t_2). \qquad (19.3.36)$$

As we already said, the process $W(t)$ is a *strictly white noise process* if its variables are not only correlated but also independent. Therefore, this strictly white noise is a purely random process. In summary, a wide-sense stationary process $W(t)$, with uncorrelated and independent variables, is a *wide-sense stationary white noise process* if and only if the mean is equal to 0, that is,

$$\eta_W(t) = E\{W(t)\} = 0, \qquad (19.3.37)$$

and autocorrelation is expressed in terms of the Dirac delta function, that is,

$$R_{WW}(\tau) = E\left\{W(t)W^*(t - \tau)\right\} = \sigma_W^2 \delta(\tau), \qquad (19.3.38)$$

which implies that the power spectral density is constant and equal to the variance for all frequencies from $-\infty$ to $+\infty$, that is,

$$S_{WW}(\omega) = \sigma_W^2. \qquad (19.3.39)$$

**White Gaussian noise process.** One of the most important processes in practice is the white Gaussian noise process. The process $W(t)$ is a *Gaussian and strictly white noise process* if its variables are independent and have identical Gaussian distribution. One realization of the process, denoted by $w(t)$, can be represented as a Gaussian random signal $w(t)$. An example of a random signal, $w(t)$, and its shifted version are presented in Figure 19.3.2. If we assume that the process is ergodic and process these two random wave shapes according to eqn (19.3.26), we can get the autocorrelation function. In communication systems, the random signal $w(t)$ may be generated in electronic components and amplifiers at the receiver of the system, or may have its origin in the interference encountered in signal transmission. In the case when the process $W(t)$ is caused by the receiver circuits, it can be treated as thermal noise and represented by an additive white Gaussian noise process with zero mean and standard deviation $\sigma_W$. The probability density function of its noise amplitudes is

$$f_W(w) = \frac{1}{\sqrt{2\pi \sigma_W^2}} e^{-(w)^2/2\sigma_W^2}, \tag{19.3.40}$$

as shown in Figure 19.3.2. The amplitude of the process can take positive and negative values with the same probability. Therefore, the mean value of the process is assumed to be 0. The process $W(t)$ is wide-sense stationary and independent. Therefore, its autocorrelation function can be derived as

$$R_{WW}(\tau) = E\{W(t)W(t+\tau)\} = \left\{ \begin{array}{ll} E\{W^2(t)\} & \tau = 0 \\ 0 & \tau \neq 0 \end{array} \right\}$$
$$= \left\{ \begin{array}{ll} \sigma_W^2 & \tau = 0 \\ 0 & \tau \neq 0 \end{array} \right\} = \sigma_W^2 \delta(\tau) = \frac{N_0}{2}\delta(\tau) \tag{19.3.41}$$

The power spectral density of process is the Fourier transform of the autocorrelation function, expressed as

$$S_{WW}(f) = \int_{-\infty}^{+\infty} R_{WW}(\tau) e^{-j2\pi f \tau} \, d\tau = \frac{N_0}{2}, \tag{19.3.42}$$

and can be represented as a two-sided power spectral density, being constant for all frequencies from $-\infty$ to $+\infty$. In communication systems, the power spectral density $N_0/2$ is referenced to the input stage of the receiver of a communication system and is expressed in watts per hertz. The autocorrelation function is the inverse Fourier transform of the power spectral density $S_{WW}(f)$. The autocorrelation function and the power spectral density function are presented in Figure 19.3.3.

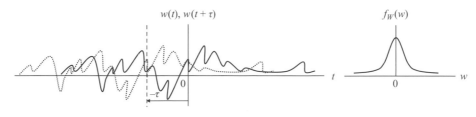

**Fig. 19.3.2** *A white Gaussian noise random signal as a realization of a white Gaussian noise process (left), and the related amplitude probability density function (right).*

**Fig. 19.3.3** *Functions of a wide-sense stationary white Gaussian noise process.*

Note that $R_{WW}(\tau)$ is a delta function that is weighted by the factor $N_0/2$, occurs at $\tau = 0$, and is zero for all lags $\tau \neq 0$, meaning that any two different samples of white Gaussian noise, no matter how closely together in time they are taken, are uncorrelated.

The above definition of white noise implies that the noise has infinite average power; as such, it is not physically realizable but has simple mathematical properties which make it useful in statistical analysis. Further discussion related to white noise and white Gaussian noise can be found in Section 19.2.3 and Chapter 4.

### 19.3.4 Gaussian Correlated Processes

The theoretical analysis of Gaussian noise processes is important for modelling channels in communication systems. In particular, this analysis is important for systems that have a channel characterized by correlated noise, which is practically always the case in the analysis of a theoretical system and in the simulation and emulation of a system. Namely, after filtering, the noise becomes correlated, no matter what kind of noise is at the input of the filter.

A real stochastic process $X(t)$ is Gaussian if its random variables $X(t_1)$, $X(t_2)$, ..., $X(t_k)$ are jointly Gaussian for any $k$ and $t_1$, $t_2$, ..., $t_k$. The $k$th-order joint probability distribution function can be expressed in the form

$$F_X(x_1, x_2, \ldots, x_k; t_1, t_2, \ldots, t_k) = P(X(t_1) \leq x_1, X(t_2) \leq x_2, \ldots, X(t_k) \leq x_k), \quad (19.3.43)$$

or as the $k$th-order probability density function

$$f_X(x_1, x_2, \ldots, x_k; t_1, t_1, \ldots, t_k) = \frac{dF_X(x_1, x_2, \ldots, x_k; t_1, t_1, \ldots, t_k)}{dx_1 \cdots dx_k}. \quad (19.3.44)$$

The joint density function of the random variables of the process is completely determined by the mean $\eta(t)$ and the autocorrelation $R(t_1, t_2)$. The process is determined by its autocovariance function, which is expressed as

$$C(t_1, t_2) = R(t_1, t_2) - \eta(t_1)\eta(t_2). \quad (19.3.45)$$

## Example

Suppose the two zero-mean, random variables $X(t_1)$ and $X(t_2)$ are jointly Gaussian. Find the joint density function and determine its parameters.

**Solution.** For the general case when the two random variables are correlated, the density function can be expressed as

$$f_X(x_1, x_2; t_1, t_2) = f_{X_1 X_2}(x_1, x_2) = \frac{1}{2\pi\sigma_1\sigma_2\sqrt{1-\rho^2}} e^{-\frac{1}{2(1-\rho^2)}\left(\frac{x_1^2}{\sigma_1^2} - \frac{2\rho x_1 x_2}{\sigma_1\sigma_2} + \frac{x_2^2}{\sigma_2^2}\right)}, \tag{19.3.46}$$

where the variances can be expressed as

$$\sigma_1^2 = R_{X1}(0) = E\left\{X_1^2\right\}, \quad \text{and} \quad \sigma_2^2 = R_{X_2}(0) = E\left\{X_2^2\right\}, \tag{19.3.47}$$

and the correlation coefficient $\rho$ can be expressed in terms of the autocorrelation function and its values as

$$\rho = \frac{E\{X_1 X_2\}}{\sigma_1\sigma_2} = \frac{R_{X_1 X_2}(t_1, t_2)}{\sqrt{R_{X_1}(0)R_{X_2}(0)}}. \tag{19.3.48}$$

It is worth noticing that the correlation coefficient is, at most, 1 and depends on the time separation $t_1$ and $t_2$ of the random variables it is calculated for.

## Example

Suppose that a Gaussian stochastic process $X(t)$ is *wide-sense stationary* with zero mean and an autocorrelation function $R(\tau)$ that depends on the lag $\tau$. Find the marginal and joint density functions of this process. Calculate the correlation coefficient.

**Solution.** The autocorrrelation function can be expressed as

$$R_{XX}(\tau) = E\{X(t)X(t+\tau)\},$$

the variance is

$$\sigma_X^2 = E\left\{X^2(t)\right\} = R_{XX}(0), \tag{19.3.49}$$

and the marginal density function can be expressed in terms of the autocorrelation values at zero lag as

$$f_X(x) = \frac{1}{\sqrt{2\pi\sigma_X^2}} e^{-x^2/2\sigma_X^2} = \frac{1}{\sqrt{2\pi R(0)}} e^{-x^2/2R(0)}. \tag{19.3.50}$$

The correlation coefficient can be calculated as

$$\rho = \frac{R_{XX}(\tau)}{\sqrt{R_{XX}(0)R_{XX}(0)}} = \frac{R_{XX}(\tau)}{R_{XX}(0)} = \frac{R_{XX}(\tau)}{\sigma_X^2}, \tag{19.3.51}$$

and the joint density function as

$$f_X(x_1, x_2; \tau) = \frac{1}{2\pi\sigma_1\sigma_2\sqrt{1-\rho^2}} \exp\left(-\frac{1}{2(1-\rho^2)}\left(\frac{x_1^2}{\sigma_1^2} - \frac{2\rho x_1 x_2}{\sigma_1\sigma_2} + \frac{x_2^2}{\sigma_2^2}\right)\right)$$

$$= \frac{1}{2\pi R_{XX}^2(0)\sqrt{1-\left(\frac{R_{XX}(\tau)}{R_{XX}(0)}\right)^2}} \exp\left(-\frac{1}{2\left(1-\left(\frac{R_{XX}(\tau)}{R_{XX}(0)}\right)^2\right)}\left(\frac{x_1^2}{R_{XX}(0)} - 2x_1 x_2 \frac{R_{XX}(\tau)}{R_{XX}^2(0)} + \frac{x_2^2}{R_{XX}(0)}\right)\right).$$

$$= \frac{1}{2\pi\sqrt{R_{XX}^2(0) - R_{XX}^2(\tau)}} \exp\left(-\frac{1}{2\left(R_{XX}^2(0) - R_{XX}^2(\tau)\right)}\left(R_{XX}(0)x_1^2 - 2R_{XX}^2(\tau)x_1 x_2 + R_{XX}(0)x_2^2\right)\right) \tag{19.3.52}$$

# 19.4 Linear-Time Invariant Systems with Stationary Stochastic Inputs

## 19.4.1 Analysis of Linear-Time Invariant Systems in Time and Frequency Domains

A linear-time-invariant system for *deterministic signal* processing is defined by its impulse responses in the time and frequency domains, $h(t)$ and $H(f)$, respectively, as shown in Figure 12.5.3 in Chapter 12. There, the fundamental result was presented in the following form: for any input signal $x(t)$, the output signal in the time domain, $y(t)$, can be found as the convolution of the input signal and the impulse response of the system, $h(t)$, expressed as

$$y(t) = \int_{\tau=-\infty}^{\infty} h(\tau) x(t-\tau) d\tau = x(t) * h(t), \tag{19.4.1}$$

or the output signal in the frequency domain, $Y(f)$, can be calculated as the multiplication of the input signal $X(f)$ and the impulse response $H(f)$. The power and energy of all of the signals in the system can be found in the time and frequency domains by using the expressions for the instantaneous power and energy spectral density.

This concept of signal processing in a linear-time-invariant system is directly applicable for *energy signals* because they fulfil the condition for the existence of the Fourier transform. Furthermore, it is relatively simple to find the autocorrelation function and the related energy spectral density function of a system and analyse its behaviour using this representation. The relationship of these two functions is defined by the Fourier transform pair.

For *power signals*, which can have infinite energy, the Fourier transform cannot always be performed. The linear-time-invariant system is still defined by its impulse response, but the analysis of the signal can be done using the autocorrelation function and the power spectral density function. In this case, the autocorrelation function is defined on an unlimited interval of time and then averaged over the whole interval. The power spectral density function can then be derived as the Fourier transform of the autocorrelation function. If the power signal is periodic, these functions are calculated inside one period of the signal. The energy and power of the signal can be calculated in both the time and the frequency domains.

In the case when we are transmitting a wide-sense stationary process through a linear-time-invariant system, the signal presentation and analysis changes. The system is still defined by its impulse response $h(t)$ and the related transfer characteristic $H(f)$, which represents the Fourier transform of the impulse response $h(t)$. The input stochastic process is represented by its autocorrelation function and related power spectral density function. The system is defined by its system correlation function, which is obtained as the correlation function of the system impulse response $h(t)$. The output autocorrelation function of the system can be calculated as the convolution of the input autocorrelation function and the system correlation function, and the output power spectral density of the system can be calculated as the product of the input power spectral density and the power spectral density of the system. The autocorrelation function and the related power spectral density function form the Fourier transform pair and can be used to calculate the power and energy of the signals.

Suppose an outcome $\zeta_i$ of a random experiment is generated with probability $p(\zeta_i)$, and the corresponding stochastic (random) signal $x(t, \zeta_i)$ is produced which represents one realization of the stochastic process $X(t)$. Suppose the process is wide-sense stationary, defined by a zero mean, constant variance, and an autocorrelation function that depends on the lag $\tau$ between the random variables defined at time instants $t_1$ and $t_2$, that is, $\tau = t_1 - t_2$. This random signal $x(t, \zeta_i)$ is applied at the input of a linear-time-invariant system defined by its impulse response in the time and frequency domains, as shown in Figure 19.4.1. For the sake of convenience, the random signal $x(t, \zeta_i)$ will mostly be referred to as just the 'signal', and the wide-sense stationary process $X(t)$ as just the 'process'. The analysis of the input and output signals will be conducted in both the time and the frequency domains.

Unlike the analysis of linear-time-invariant systems that process deterministic continuous-time signals, presented in Chapters 1, 11, and 12, the analysis of linear-time-invariant systems that have stochastic signals at their input must take into account the fact that the stochastic process is not just one realization of the process (a random signal) but an ensemble of random signals called the realizations of a stochastic process. Each realization $x(t, \zeta_i)$ of the input stochastic process $X(t)$ can have a deterministic signal that produces an output signal $y(t, \zeta_i)$ that can be treated as one realization of the output stochastic process $Y(t)$.

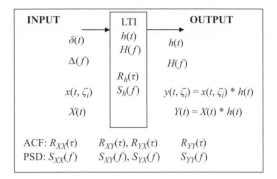

**Fig. 19.4.1** *A linear-time-invariant system; ACF, autocorrelation function; LTI, linear-time-invariant system; PSD, power spectral density.*

**System description.** Suppose an outcome $\zeta_i$ is generated from the sample set $S$ with probability $p(\zeta_i)$, and the corresponding random signal $x(t, \zeta_i)$ is produced which represents one realization of a stochastic process $X(t)$. This random signal is applied to the input of a linear-time-invariant system, as shown in Figure 19.4.1.

The output of the system is a new stochastic signal, $y(t, \zeta_i)$, which is a convolution of the input signal and the impulse response of the system, that is,

$$y(t,\zeta_i) = \int_{\tau=-\infty}^{\infty} h(\tau) x(t-\tau,\zeta_i)\, d\tau = x(t,\zeta_i) * h(t). \qquad (19.4.2)$$

If the integral on the right-hand side exists for all outcomes $\zeta_i \in S$, and $p(S) = 1$, then the convergence exists with probability 1. The convolution expression can be extended to the general case when the output stochastic process is defined as the convolution of the input stochastic process and the impulse response of the system, expressed as

$$Y(t) = \int_{\tau=-\infty}^{\infty} h(\tau) X(t-\tau)\, d\tau = h(t) * X(t). \qquad (19.4.3)$$

For this system, we will define the input signals, the output signals, and the related statistical functions.

**Statistics of the input process in time and frequency domains.** The input of the system is a stochastic process $X(t)$ represented by its autocorrelation function

$$R_{XX}(\tau) = E\{X(t)X(t+\tau)\} \qquad (19.4.4)$$

and the related power spectral density function. These two functions form the Fourier transform pair. Therefore, the autocorrelation function can be expressed as

$$R_{XX}(\tau) = \frac{1}{2\pi} \int_{-\infty}^{+\infty} S_{XX}(\omega) e^{j\omega\tau} d\omega. \tag{19.4.5}$$

and the power spectral density function can be found as the Fourier transform of the autocorrelation function, that is,

$$S_{XX}(\omega) = \int_{-\infty}^{+\infty} R_{XX}(\tau) e^{-j\omega\tau} d\tau, \tag{19.4.6}$$

which is known as *the Wiener–Khintchine theorem*. The inverse Fourier transform of the power spectral density function is the autocorrelation function. The Fourier transform pair can be expressed in its symmetric form as

$$R_{XX}(\tau) = \int_{-\infty}^{+\infty} S_{XX}(f) e^{j2\pi f\tau} df, \tag{19.4.7}$$

and

$$S_{XX}(f) = \int_{-\infty}^{+\infty} R_{XX}(\tau) e^{j2\pi f\tau} d\tau. \tag{19.4.8}$$

**Statistics of the system in time and frequency domains.** A linear-time-invariant system is defined by its impulse response $h(t)$. The autocorrelation function of the impulse response is

$$R_h(\tau) = \int_{-\infty}^{\infty} h(t) h(t+\tau) dt, \tag{19.4.9}$$

which is called the *system correlation function*. The impulse response is a deterministic energy signal. Therefore, it is defined by its energy spectral density function, which is the Fourier transform of the autocorrelation function, expressed as

$$S_h(\omega) = \int_{-\infty}^{+\infty} R_h(\tau) e^{-j\omega\tau} d\tau = |H(\omega)|^2 = E_h(\omega). \tag{19.4.10}$$

The system correlation function can be used for the analysis of stochastic systems in the same way that the impulse response function $h(t)$ and the system autocorrelation function $R_h(\tau)$ are used for the analysis of continuous-time deterministic systems. Namely, the function $R_h(\tau)$ relates the autocorrelation function of the system output to the autocorrelation function of the system input, according to the following explanation. The input–output cross-correlation function can be obtained as follows:

$$R_{XY}(\tau) = E\left\{X(t)Y^*(t+\tau)\right\} = E\left\{X(t)\int_{-\infty}^{\infty} h(\alpha)X(t+\tau-\alpha)\,d\alpha\right\}$$

$$= \int_{-\infty}^{\infty} E\{X(t)X(t+\tau-\alpha)\}\,h(\alpha)\,d\alpha = \int_{-\infty}^{\infty} h(\alpha)\,R_{XX}(\tau-\alpha)\,d\alpha \qquad (19.4.11)$$

$$= R_{XX}(\tau) * h(\tau)$$

We can also have the following expression for the output–input cross-correlation:

$$R_{YX}(\tau) = R_{XX}(\tau) * h(-\tau). \qquad (19.4.12)$$

The cross power spectral density of the two processes $X(t)$ and $Y(t)$ is the Fourier transform of the cross-correlation, that is,

$$S_{XY}(\omega) = \int_{\tau=-\infty}^{\infty} R_{XY}(\tau)\,e^{-j\omega\tau}\,d\tau = S_{YX}^*(\omega), \qquad (19.4.13)$$

and the cross-correlation is

$$R_{XY}(\tau) = \frac{1}{2\pi}\int_{-\infty}^{\infty} S_{XY}(\omega)\,e^{j\omega\tau}\,d\omega = R_{YX}^*(\tau). \qquad (19.4.14)$$

We can prove that the following relations hold:

$$S_{XY}(\omega) = H(\omega) \cdot S_{XX}(\omega), \qquad (19.4.15)$$

$$S_{YX}(\omega) = H(-\omega) \cdot S_{XX}(\omega), \qquad (19.4.16)$$

$$S_{YY}(\omega) = |H(\omega)|^2 \cdot S_{XX}(\omega). \qquad (19.4.17)$$

**Statistics of the output in time and frequency domains.** The *autocorrelation* of the output can be obtained from the expression for the convolutional integral rearranged to the form

$$R_{YY}(\tau) = E\{Y(t)Y(t+\tau)\} = E\left\{Y(t)\int_{-\infty}^{\infty} h(\alpha)X(t+\tau-\alpha)\,d\alpha\right\}$$

$$= \int_{-\infty}^{\infty} h(\alpha)\,E\{Y(t)X(t+\tau-\alpha)\}\,d\alpha = \int_{-\infty}^{\infty} h(\alpha)\,R_{YX}(\tau-\alpha)\,d\alpha \qquad (19.4.18)$$

$$= h(\tau) * R_{YX}(\tau)$$

Using relations for the cross-correlation, we obtain the output autocorrelation function as a twofold convolution:

$$R_{YY}(\tau) = h(\tau) * R_{YX}(\tau) = h(\tau) * R_{XX}(\tau) * h(-\tau). \qquad (19.4.19)$$

Because the autocorrelation of a deterministic impulse response, called the system correlation function, is

$$R_h(\tau) = \int_{-\infty}^{\infty} h(\alpha) h(\tau + \alpha) \, d\alpha = h(\tau) * h(-\tau), \tag{19.4.20}$$

the autocorrelation function of the output signal in eqn (19.4.19) can be expressed as

$$R_{YY}(\tau) = h(\tau) * h(-\tau) * R_{XX}(\tau) = R_h(\tau) * R_{XX}(\tau). \tag{19.4.21}$$

*Therefore, the autocorrelation of the output signal is equal to the convolution of the autocorrelation of the input process and the autocorrelation of the impulse response of the system itself.*

For the calculated autocorrelation function and power spectral density function of the input signal, we can calculate the power spectral density function of the output as a product of the system magnitude response squared (or energy spectral density) and the input power spectral density, that is,

$$S_{YY}(\omega) = S_h(\omega) \cdot S_{XX}(\omega) = |H(\omega)|^2 \cdot S_{XX}(\omega). \tag{19.4.22}$$

The energy spectral density function is sometimes called the power transfer function, because it shows how power is transferred from the input to the output of a linear-time-invariant system. Furthermore, from the last expression, we can see that, if we have the autocorrelation or autospectral density of the input and output signals of a linear-time-invariant system, then we can determine the magnitude response of the system. However, we cannot determine the phase response of the system. For this purpose, we can use the cross-spectral density functions. The basic system structure with the related functions is shown in Figure 19.4.2.

The output stochastic process $Y(t)$ of a stable system to a wide-sense stationary input process $X(t)$ is a wide-sense stationary process, because its mean is constant and the autocorrelation function depends only on the time difference $\tau$. The mean value of the output signal is

$$\eta_Y = E\{Y(t)\} = \int_{-\infty}^{\infty} h(\tau) E\{X(t-\tau)\} \, d\tau = \eta_X \int_{-\infty}^{\infty} h(\tau) e^{-j0\,\tau} \, d\tau = H(0)\eta_X, \tag{19.4.23}$$

where the constant $H(0)$ is the DC gain of the system. Because the mean value of the input signal is constant, the mean value of the output is also a constant. Most of the time, we will

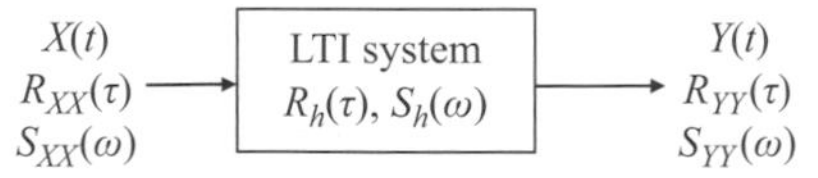

**Fig. 19.4.2** *Autocorrelations and power spectral densities of a linear-time-invariant system; LTI, linear-time-invariant.*

assume that the mean value is 0, which excludes the necessity of using the autocovariance function instead of the autocorrelation function. Also, this theory will be used to analyse white noise and, in particular, the white Gaussian noise process, which has a mean value of 0. By using the time- and frequency-domain representations of a random process, we can find the power of related processes, including the power of the output signal.

For the analysis of linear-time-invariant systems, we will proceed according to the following steps:

1. Find the system correlation function $R_h(\tau)$, which is defined as the autocorrelation function of the system impulse response $h(t)$.

2. For any input stochastic process $X(t)$, defined by its autocorrelation function and the power spectral density, we can then calculate the correlation function for the output process $Y(t)$ as the convolution of the input correlation function and the system corelation function, that is,

$$R_{YY}(\tau) = R_h(\tau) * R_{XX}(\tau).$$ (19.4.24)

3. The power spectral density of the output process $S_{YY}(\omega)$ will be calculated as the product of the system magnitude response squared and the input power spectral density $S_{XX}(\omega)$, that is,

$$S_{YY}(\omega) = S_h(\omega) S_{XX}(\omega) = |H(\omega)|^2 S_{XX}(\omega).$$ (19.4.25)

4. For the calculated autocorrelation function and power spectral density of the system output, the power of the output process can be calculated in the time and frequency domains as

$$P_{YY} = E\left\{|Y(t)|^2\right\} = \frac{1}{2\pi} \int_{-\infty}^{+\infty} S_{YY}(\omega)\, d\omega = \frac{1}{2\pi} \int_{-\infty}^{+\infty} |H(\omega)|^2 S_{XX}(\omega)\, d\omega.$$ (19.4.26)

**Output probability density function.** Generally speaking, finding the probability density function of the output of a linear-time-invariant system is a hard task. There are some cases when this problem is solvable. Firstly, if the input process is Gaussian, then the output will also be Gaussian. Secondly, if the input is an iid process, the output will be a weighted sum of independent random variables that has a density function which is a convolution of the densities of these individual variables.

## 19.4.2   Definition of a System Correlation Function for Stochastic Input

A linear-time-invariant system is represented in the time domain by its impulse response $h(t)$, which is defined as the output of the system when the Dirac delta function is applied at the input of the system. For that system, we defined the system autocorrelation function as the autocorrelation of the impulse response.

The *system correlation function* can be also defined as the autocorrelation function obtained at the output of a linear-time-invariant stochastic system when the white noise $W_1(t)$, with zero mean $\eta_{W_1} = 0$, the unit variance $\sigma_{W_1}^2 = 1$, and the Dirac delta autocorrelation function $R_{W_1}(\tau) = \delta(\tau)$, called *unit white noise*, is applied to the input (see Figure 19.4.3).

Suppose that the unit white noise is applied to the input of a linear-time-invariant system. Then, the noise autocorrelation function and the power spectral density are expressed in the following forms:

$$R_{W_1 W_1}(\tau) = R_{W_1 W_1}(\tau) = \delta(\tau)$$

$$S_{W_1 W_1}(f) = \int_{-\infty}^{+\infty} R_{W_1 W_1}(\tau) e^{-j2\pi f \tau} \, d\tau = \int_{-\infty}^{+\infty} \delta(\tau) e^{-j2\pi f \tau} \, d\tau = e^{-j2\pi f \cdot 0} = 1 \quad (19.4.27)$$

for all possible frequency values $-\infty < f < \infty$.

The graphs of the input delta correlation function and the related power spectral density of the unit white noise are presented in Figure 19.4.4.

The power of the input unit white noise process calculated in the time and frequency domains is

$$P_{W_1} = E\left\{|W_1(t)|^2\right\} = \frac{1}{2\pi} \int_{-\infty}^{+\infty} S_{W_1 W_1}(\omega) \, d\omega = \frac{1}{2\pi} \int_{-\infty}^{+\infty} 1 \, d\omega = \frac{1}{2\pi} \int_{-\infty}^{+\infty} d\omega = \int_{-\infty}^{+\infty} df \to \infty,$$

$$(19.4.28)$$

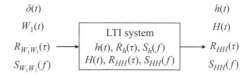

**Fig. 19.4.3** *Definition of the system correlation function of a linear-time-invariant system; LTI, linear-time-invariant.*

**Fig. 19.4.4** *Autocorrelation function (left) and power spectral density function (right) of a continuous-time wide-sense stationary white noise process.*

or

$$P_{W_1} = R_{W_1}(0) = \delta(0) \to \infty. \tag{19.4.29}$$

Of course, the correlation function and the power spectral density of the system output, for the unit white noise process at the system input, will depend on the system properties, that is, on the system's transfer function. Therefore, the system correlation function was defined in eqn (19.4.20) as

$$R_h(\tau) = \int_{-\infty}^{+\infty} h(\alpha)h(\tau+\alpha)\,d\alpha = h(\tau)*h(-\tau), \tag{19.4.30}$$

and the corresponding power spectral density defined in eqn (19.4.12) will be equal to the system transfer function expressed as

$$S_h(\omega) = |H(\omega)|^2 S_{W1}(\omega) = |H(\omega)|^2 \cdot 1 = |H(\omega)|^2,$$

for $|f| \le B$, where $B$ is the bandwidth of the filter. For the input unit noise process, the output process in time domain is given by the convolution

$$H(t) = \int_{-\infty}^{+\infty} h(\tau)W_1(t-\tau)\,d\tau = h(t)*W_1(t). \tag{19.4.31}$$

It was proven that, for any stochastic process applied to the input of a linear-time-invariant system with a defined system autocorrelation function, the autocorrelation function of its output can be found as the convolution of the input autocorrelation function and the system correlation function. Therefore, in the case under our consideration, we may have the output correlation function expressed as

$$\begin{aligned} R_{HH}(\tau) = R_h(\tau)*R_{W_1}(\tau) &= \int_{-\infty}^{+\infty} R_h(\lambda)R_{W_1}(\tau-\lambda)\,d\lambda \\ &= \int_{-\infty}^{+\infty} R_h(\lambda)\delta(\tau-\lambda)\,d\lambda = R_h(\tau) \end{aligned} \tag{19.4.32}$$

which confirms that the autocorrelation of the linear-time-invariant output for the input unit white noise is equal to the system correlation function.

Consequently, the power spectral density of the system response to the input white unit noise process can be obtained as the product of the system transfer function and the power spectral density of the input process and expressed as

$$S_{HH}(\omega) = |H(\omega)|^2 S_{W_1 W_1}(\omega). \tag{19.4.33}$$

### 19.4.3    Application of the Theory of Linear-Time-Invariant Systems to the Analysis of the Operation of a Lowpass Filter

A lowpass filter can be analysed as a linear-time-invariant system. The filter can be used to produce a baseband noise $N_B(t)$. For that purpose, at the input of the filter, a white Gaussian noise $W(t)$ will be applied, and the output of the filter will be a baseband, band-limited noise that can be expressed as the convolution of the input and the filter impulse response.

**Input signal of a lowpass filter.** Suppose that the white Gaussian noise process $W(t)$ of a finite variance is filtered using an ideal lowpass filter that has bandwidth $B$ and operates as a linear-time-invariant system, as shown in Figure 19.4.5. If the input process has Gaussian distribution, the output noise is a band-limited Gaussian process.

The input white Gaussian noise process $W(t)$ is represented by its autocorrelation function and constant power spectral density, which are expressed as

$$R_{WW}(\tau) = \frac{N_0}{2}\delta(\tau) = \sigma_W^2\delta(\tau) \tag{19.4.34}$$

and

$$
\begin{aligned}
S_{WW}(f) &= \int_{-\infty}^{+\infty} R_{WW}(\tau)e^{-j2\pi f\tau}\,d\tau = \int_{-\infty}^{+\infty} \frac{N_0}{2}\delta(\tau)e^{-j2\pi f\tau}\,d\tau \\
&= \frac{N_0}{2}e^{-j2\pi f\cdot 0} = \frac{N_0}{2},
\end{aligned} \tag{19.4.35}
$$

respectively, for all frequency values $-\infty < f < \infty$. The functions are presented in Figure 19.4.6. This assumption has the physical sense. Namely, the thermal noise, for example, has a constant power spectral density of $N_0 = 4.11 \cdot 10^{-21}$ W/Hz inside the bandwidth $|f| < 10^{12}$ Hz.

The power of the input process calculated in the time and frequency domains is

$$P_W = E\left\{|W(t)|^2\right\} = \frac{1}{2\pi}\int_{-\infty}^{+\infty} S_{WW}(\omega)\,d\omega = \frac{1}{2\pi}\int_{-\infty}^{+\infty}\frac{N_0}{2}\,d\omega = \frac{N_0}{4\pi}\int_{-\infty}^{+\infty}d\omega = \frac{N_0}{2}\int_{-\infty}^{+\infty}df \to \infty. \tag{19.4.36}$$

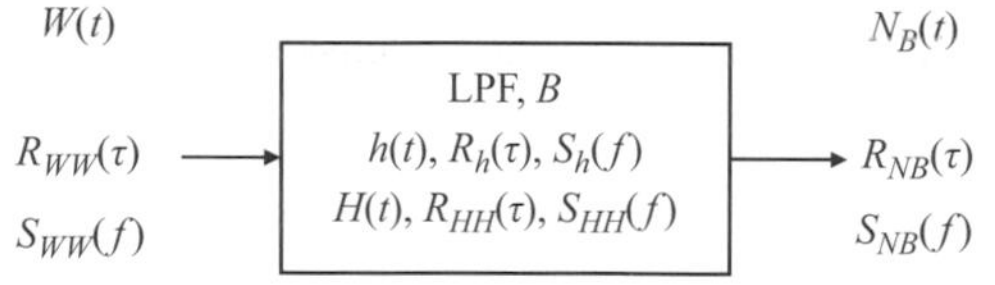

**Fig. 19.4.5** *Operation of a baseband lowpass filter; LPF, lowpass filter.*

**Fig. 19.4.6** *Functions of a continuous-time, wide-sense stationary white noise process.*

The output process in the time domain is the convolution of the input process and the impulse response of the lowpass filter, expressed as

$$N_B(t) = \int_{-\infty}^{+\infty} h(\tau)\, W(t-\tau)\, d\tau = h(t) * W(t). \qquad (19.4.37)$$

One realization of this process in the time interval $T$ is

$$n_B(t) = \int_{-\infty}^{+\infty} h(\tau)\, w(t-\tau)\, d\tau = h(t) * w(t), \qquad (19.4.38)$$

which has limited power and energy in the defined interval. The correlation function of the output process is the convolution of the input process and the system correlation function, expressed as

$$R_{NB}(\tau) = R_{HH}(\tau) * R_{WW}(\tau). \qquad (19.4.39)$$

In order to find the autocorrelation function of the filter output, we need to find the system correlation function and perform this convolution; alternatively, we can find the system correlation function by using the impulse response function of the system. In addition, we can find the power spectral density of the filter output and then perform the inverse Fourier transform on it to get the autocorrelation function of the output process. We will find the system correlation function as the response of the system to the unit white noise and confirm it using the expression of the power spectral density at the output of the filter, that is, perform both of the aforementioned calculations.

**Operation of a baseband lowpass filter, and its system autocorrelation function.** Firstly, we need to find the power spectral density of the lowpass filter output and the related system autocorrelation function defined for the input unit white noise process (impulse response!). The transfer characteristic of the ideal lowpass filter with a limited bandwidth $B$ and unit amplification is shown in Figure 19.4.7. In the same figure, both the baseband noise and the impulse response characteristics are presented. The power spectral density of the filter output for the input unit white noise is

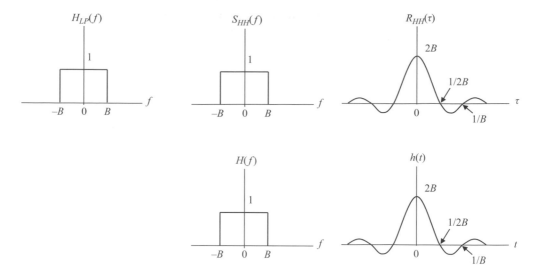

**Fig. 19.4.7** *The transfer characteristic of an ideal lowpass filter (left), and the baseband noise (upper right panels) and the impulse responses of a lowpass filter (lower right panels).*

$$S_{HH}(f) = |H_{LP}(f)|^2 S_{W1}(f) = \begin{cases} 1 \cdot 1 & |f| \le B \\ 0 & \text{otherwise} \end{cases}. \tag{19.4.40}$$

The system correlation function of the filter is the Fourier transform of the calculated power spectral density, that is,

$$R_{HH}(\tau) = \int_{-B}^{+B} S_{HH}(f) e^{j2\pi f \tau} \, df = \int_{-B}^{+B} 1 e^{j2\pi f \tau} \, df = 2B \operatorname{sinc}(2\pi B \tau). \tag{19.4.41}$$

This autocorrelation is the system autocorrelation function of the lowpass filter but, at the same time, it is also the output of the filter for a unit power spectral density white noise at the input. Therefore, it is the convolution of the input unit white process $W_1(t)$ and the filter autocorrelation function itself. Let us prove this in the following way:

$$\begin{aligned} R_{HH}(\tau) &= R_{HH}(\tau) * R_{W1}(\tau) = \int_{-\infty}^{+\infty} R_{HH}(\lambda) R_{W1}(\tau - \lambda) \, d\lambda \\ &= \int_{-\infty}^{+\infty} 2B \operatorname{sinc}(2\pi B \lambda) \, \delta(\tau - \lambda) \, d\lambda = 2B \operatorname{sinc}(2\pi B \tau) \end{aligned}, \tag{19.4.42}$$

for the defined function $\operatorname{sinc}(2\pi B \tau) = \frac{\sin(2\pi B \tau)}{2\pi B \tau}$.

**The output of the lowpass filter for white Gaussian noise input.** Now, we will investigate the basic characteristics of the lowpass filter. We are interested in finding the system response to the input white Gaussian noise of a finite variance. Firstly, the autocorrelation of the filter output is the convolution of the autocorrelation function of the input process and the calculated system autocorrelation function, expressed as

$$
\begin{aligned}
R_{NB}(\tau) = R_{HH}(\tau) * R_{WW}(\tau) &= \int_{-\infty}^{+\infty} R_{HH}(\lambda) R_{WW}(\tau - \lambda)\, dt \\
&= \int_{-\infty}^{+\infty} 2B \operatorname{sinc}(2\pi B\lambda) \frac{N_0}{2} \delta(\tau - \lambda)\, dt \\
&= 2B \frac{N_0}{2} \operatorname{sinc}(2\pi B\tau) = N_0 B \operatorname{sinc}(2\pi B\tau)
\end{aligned}
\qquad (19.4.43)
$$

The same autocorrelation function can be obtained from the Fourier transform of the power spectral density of the output process, expressed as

$$
S_{NB}(f) = |H_{LP}(f)|^2 S_W(f) = |H_{LP}(f)|^2 \frac{N_0}{2} = \frac{N_0}{2},
\qquad (19.4.44)
$$

for $|f| < B$, which gives the autocorrelation function of the output as

$$
R_{NB}(\tau) = \int_{-B}^{+B} S_{NB}(f) e^{j2\pi f \tau}\, df = \int_{-B}^{+B} \frac{N_0}{2} e^{j2\pi f \tau}\, df = N_0 B \operatorname{sinc}(2\pi B\tau).
\qquad (19.4.45)
$$

The power spectral density and autocorrelation function graphs of the filtered white noise are shown in Figure 19.4.8.

The noise power at the output of the filter may be expressed as

$$
P_{NB} = \int_{-B}^{+B} S_{NB}(f)\, df = \int_{-B}^{+B} \frac{N_0}{2}\, df = N_0 B,
\qquad (19.4.46)
$$

which is equal to the variance of the noise, $\sigma_{NB}^2$, at the output of the filter.

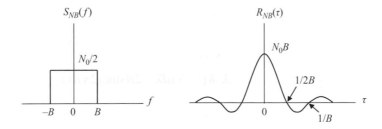

**Fig. 19.4.8** *The autocorrelation function (left) and the power spectral density function (right) at the output of a lowpass filter.*

### 19.4.4 Analysis of the Operation of a Bandpass Filter

**The input of a bandpass filter.** If a white Gaussian noise $W(t)$ of zero mean and power spectral density $N_0/2$ is applied at the input of an ideal bandpass filter that has unity amplification, a midband frequency of $\pm f_c$, and bandwidth $2B$, the output signal will be a bandpass noise. The output autocorrelation of the bandpass filter requires knowledge about the system autocorrelation function or the related power spectral density obtained for the input noise process $W(t)$.

**Bandpass filter responses.** Let us find firstly the system autocorrelation function for the bandpass filter presented in Figure 19.4.9, through its inputs, outputs, and transfer functions.

The power spectral density of the bandpass filter's output for the input unit white noise $W_1(t)$ is (when direct filtering is applied!)

$$
\begin{aligned}
S_{HH}(f) &= |H_{BP}(f)|^2 S_{W1}(f) \\
&= \begin{cases} 1 \cdot 1 & f_c - B \le f \le f_c + B \text{ and} -f_c - B \le f \le -f_c + B \\ 0 & \text{otherwise} \end{cases}. \\
&= S_{W1}(f - f_c) + S_{W1}(f + f_c)
\end{aligned}
\tag{19.4.47}
$$

The *system correlation function* is the Fourier transform of the calculated power spectral density, that is,

$$
\begin{aligned}
R_{HH}(\tau) &= \int_{-f_c-B}^{f_c+B} S_{HH}(f) e^{j2\pi f \tau} \, df = \int_{-f_c-B}^{-f_c+B} 1 \cdot e^{j2\pi f \tau} \, df + \int_{f_c-B}^{f_c+B} 1 \cdot e^{j2\pi f \tau} \, df \\
&= 2B \operatorname{sinc}(2\pi B \tau) \left[ e^{-j2\pi f_c \tau} + e^{j2\pi f_c \tau} \right] \\
&= 4B \operatorname{sinc}(2\pi B \tau) \cos(2\pi f_c \tau)
\end{aligned}
\tag{19.4.48}
$$

This is the *bandpass system autocorrelation function* but, at the same time, it is also the output of the filter for a unit power spectral density white noise at the input. Therefore, it is

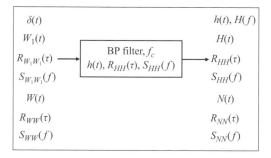

**Fig. 19.4.9** *Specification of a bandpass filter's inputs, outputs, and transfer functions; BP, bandpass.*

the convolution of the input autocorrelation function and the system autocorrelation function. Let us prove this as follows:

$$
\begin{aligned}
R_{HH}(\tau) = R_{HH}(\tau) * R_{W1}(\tau) &= \int_{-\infty}^{+\infty} R_{HH}(\lambda) R_{W1}(\tau - \lambda) d\lambda \\
&= \int_{-\infty}^{+\infty} 4B \operatorname{sinc}(2\pi B\lambda) \cos(2\pi f_c \lambda) \cdot \delta(\tau - \lambda) d\lambda \\
&= 4B \operatorname{sinc}(2\pi B\tau) \cos(2\pi f_c \tau)
\end{aligned}
\tag{19.4.49}
$$

**The output of the bandpass filter for input white Gaussian noise of finite variance.** For the input white noise $W(t)$, the output of the bandpass filter will be the bandpass noise $N(t)$. The basic characteristics of the bandpass filter, which acts like a linear-time-invariant system, can then be found. We are interested in finding the system response to an input white noise $W(t)$ of finite variance $N_0/2$. The bandpass system transfer characteristic is shown in Figure 19.4.10.

Firstly, the autocorrelation of the output process can be found as the convolution of the autocorrelation function of the input process and the calculated system autocorrelation function, expressed as

$$
\begin{aligned}
R_{NN}(\tau) = R_{HH}(\tau) * R_{WW}(\tau) &= \int_{-\infty}^{+\infty} R_{HH}(\lambda) R_{WW}(\tau - \lambda) d\tau \\
&= \int_{-\infty}^{+\infty} [4B \operatorname{sinc}(2\pi B\lambda) \cos(2\pi f_c \lambda)] \frac{N_0}{2} \delta(\tau - \lambda) d\tau \\
&= 2BN_0 \operatorname{sinc}(2\pi B\tau) \cos(2\pi f_c \tau)
\end{aligned}
\tag{19.4.50}
$$

The same autocorrelation function can be obtained from the Fourier transform of the power spectral density of the output process, expressed as

$$
\begin{aligned}
S_{NN}(f) &= |H_{BP}(f)|^2 S_{WW}(f) \\
&= \begin{cases} 1 \cdot N_0/2 & f_c - B \le f \le f_c + B, -f_c - B \le f \le -f_c + B, \\ 0 & \text{otherwise} \end{cases}
\end{aligned}
\tag{19.4.51}
$$

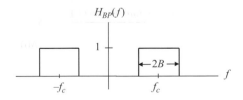

**Fig. 19.4.10** *Transfer characteristic of a bandpass filter.*

as shown in Figure 19.4.11, which gives the autocorrelation function of the output as

$$
\begin{aligned}
R_{NN}(\tau) &= \int_{-f_c-B}^{f_c+B} S_{NN}(f) e^{j2\pi f\tau}\, df = \int_{-f_c-B}^{-f_c+B} \frac{N_0}{2} e^{j2\pi f\tau}\, df + \int_{f_c-B}^{f_c+B} \frac{N_0}{2} e^{j2\pi f\tau}\, df \\
&= 2N_0 B \operatorname{sinc}(2\pi B\tau)\left[e^{-j2\pi f_c\tau} + e^{j2\pi f_c\tau}\right] \\
&= 2N_0 B \operatorname{sinc}(2\pi B\tau)\cos(2\pi f_c\tau)
\end{aligned}
\qquad (19.4.52)
$$

which is shown in Figure 19.4.12. The zero crossings of the envelope are defined by the bandwidth of the baseband noise, and the variations of the function have a period equal to the period of the carrier signal.

The noise power of the bandpass noise may be expressed as

$$
P_N = \int_{-f_c-B}^{-f_c+B} \frac{N_0}{2}\, df + \int_{f_c-B}^{f_c+B} \frac{N_0}{2}\, df = 2\frac{N_0}{2} 2B = 2BN_0, \qquad (19.4.53)
$$

which is equal to the variance of the noise at the output of the filter. Because the power spectral density of the noise is symmetric about the midband frequency $\pm f_c$, the corresponding power spectral densities of the baseband noise defined in the bandwidth $B$ is $N_0$, as shown in Figure 19.4.13.

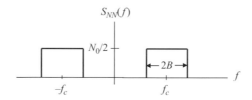

**Fig. 19.4.11** *Power spectral density of the output process.*

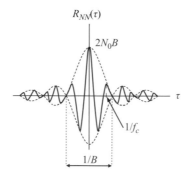

**Fig. 19.4.12** *Autocorrelation function of the output process for a bandpass filter.*

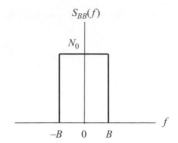

**Fig. 19.4.13** *Power spectral densities of baseband noise.*

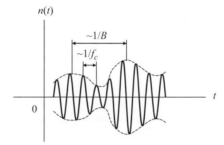

**Fig. 19.4.14** *Wave shape of passband noise.*

Thus, the autocorrelation function of the baseband noise is

$$R_{BB}(\tau) = 2N_0 B \operatorname{sinc}(2\pi B \tau), \qquad (19.4.54)$$

and the baseband noise power can be expressed as

$$P_{BB} = \int_{-B}^{+B} N_0 \, df = 2BN_0. \qquad (19.4.55)$$

One realization $n(t)$ of a bandpass noise process $N(t)$ is shown in Figure 19.4.14. Its envelope follows the shape of the baseband noise. The oscillation of the process follows the period of the midband frequency of the bandpass noise. That makes sense. Namely, a bandpass noise can be obtained if we multiply a carrier with a midband frequency by a random baseband signal. Therefore, the waveshape of the baseband noise is preserved in the envelope of the bandpass noise. These time-domain properties of the bandpass noise are in agreement with the properties of the autocorrelation function, as shown in Figure 19.4.12.

**Example**

Prove that the expression for the cross-correlation function at the output of the linear-time-invariant system is

$$S_{YY}(\omega) = \int_{-\infty}^{+\infty} R_{YY}(\tau)e^{-j\omega\tau}\,d\tau = S_{XX}(\omega)\,|H(\omega)|^2.$$

**Solution.** The solution is as follows:

$$S_{YY}(\omega) = \int_{-\infty}^{+\infty} R_{YY}(\tau)e^{-j\omega\tau}\,d\tau = \int_{-\infty}^{+\infty} [R_{HH}(\tau)*R_{XX}(\tau)]e^{-j\omega\tau}\,d\tau$$

$$= \int_{-\infty}^{+\infty}\left(\int_{-\infty}^{+\infty} R_{HH}(\lambda)\cdot R_{XX}(\tau-\lambda)\,d\lambda\right)e^{-j\omega\tau}\,d\tau$$

$$= \int_{-\infty}^{+\infty} R_{HH}(\lambda)\left(\int_{-\infty}^{+\infty} R_{XX}(\tau-\lambda)e^{-j\omega\tau}\,d\tau\right)d\lambda$$

$$\underset{\tau-\lambda=x}{=} \int_{-\infty}^{+\infty} R_{HH}(\lambda)\left(\int_{-\infty}^{+\infty} R_{XX}(x)e^{-j\omega(x+\lambda)}\right)dx\,d\lambda$$

$$= \int_{-\infty}^{+\infty} R_{HH}(\lambda)e^{-j\omega\lambda}\left(\int_{-\infty}^{+\infty} R_{XX}(x)e^{-j\omega x}\,dx\right)d\lambda = S_{XX}(\omega)\int_{-\infty}^{+\infty} R_{HH}(\lambda)e^{-j\omega\lambda}\,d\lambda$$

$$= S_{XX}(\omega)\,S_{HH}(\omega) = S_{XX}(\omega)\,|H(\omega)|^2$$

---

PROBLEMS

1. A stochastic process $X(t)$ is defined by its constant mean value $\eta(t)=5$ and its autocorrelation function, which is expressed as

$$R_{XX}(t_1,t_2) = 25 + 9e^{-0.4|t_1-t_2|}.$$

Find the mean, variance, and covariance of the random variables $X(5)$ and $X(10)$.

2. Suppose the two wide-sense stationary stochastic processes $X(t)$ and $Y(t)$ are defined such that

$$X(t) = A\cos\omega_0 t + B\sin\omega_0 t,$$

and

$$Y(t) = A\cos\omega_0 t - B\sin\omega_0 t,$$

where the random variables $A$ and $B$ are zero-mean and unit-variance independent random variables.

a. Find the mean values of the processes.

b. Find the cross-correlation function of the processes $X(t)$ and $Y(t)$.

3. Suppose we have the complex exponential process

$$X(t) = Ae^{j\omega_0 t} = | A | e^{j(\omega_0 t + \theta)},$$

where the phases are iid random variables uniformly distributed on the interval $[0, 2\pi]$.

a. Find the mean value of the process.

b. Find the autocorrelation function of the process.

# Bibliography

This bibliography is included in the book to aid the reader in finding additional valuable references in general area of analogue, digital and discrete communication systems as well as in the specialised fields of deterministic and stochastic signal processing in both discrete and continuous time domains. This is not in any way a complete list of the existing literature in these fields.

Baher, H. (2001). *Analog and Digital Signal Processing*, John Wiley & Sons, New York, NY.

Barkat, M. (2005). *Signal Detection and Estimation*, Second Edition, Artech House, Norwood, MA.

Benvenuto, N., Corvaja, R., Erseghe, N., and Laurenti, N. (2007). *Communication Systems, Fundamentals and Design Methods*, John Wiley & Sons, Chichester.

Berber, M. S. (2009). *Deterministic and Stochastic Signal Processing: Continuous and Discrete Time Signals*, VDM Verlag Dr. Müller, Saarbrücken.

Bertein, J.-C., and Ceschi, R. (2010). *Discrete Stochastic Processes and Optimal Filtering*, ISTE, London.

Boutillon, E., Danger, J. L., and Gazel, A. (2003). Design of High Speed AWGN Communication Channel Emulator, *Analog Integrated Circuits and Signal Processing*, vol. 34, no. 2, pp. 133–142.

Carlson, A. B., Crilly, P. B., and Rutledge, J. C. (2002). *Communication Systems*: An Introduction to Signals and Noise in Electrical Communication, Fourth Edition, McGraw-Hill, New York, NY.

Cavicchi, T. (2000). *Digital Signal Processing*, John Wiley & Sons, New York, NY.

Cooper, R. G., and McGillem, C. D. (1999). *Probability Methods of Signal and System Analysis*, Third Edition, Oxford University Press, New York, NY.

Couch, L. W. (2007). *Digital and Analog Communication Systems*, Seventh Edition, Prentice Hall, Upper Saddle River, NJ.

Danger, J. L., Ghazel, A., Boutillon, E., and Laamari, H. (2000). Efficient Implementation of Gaussian Noise Generator for Communication Channel Emulation, *Proceedings of the 7th IEEE International Conference on Electronics, Circuits and Systems*, vol. 1, pp. 366–369.

Davenport, W. B. (1970). *Probability and Random Processes*, McGraw-Hill, New York, NY.

Dohler, M., and Li, Y. (2010). *Cooperative Communication, Hardware, Channel & PHY*, John Wiley & Sons, Hoboken, NJ.

Frenzel, Jr. E. L. (2016). *Principles of Electronic Communication Systems*, Fourth Edition, McGraw-Hill, New York, NY.

Glover, A. I., and Grant, P. M. (2004). *Digital Communications*, Second Edition, Pearson, Harlow.

Goodyear, C. C. (1971). *Signal and Information*, Butterworths, London.

Gray, M. R., and Davisson, L. D. (2004). *An Introduction to Statistical Signal Processing*, Cambridge University Press, Cambridge.

Hayes, M. H. (1996). *Statistical Digital Signal Processing and Modelling*, John Wiley & Sons, New York, NY.

Hayes, M. H. (1999). *Schaum's Outlines of Theory and Problems of Digital Signal Processing*, McGraw-Hill, New York, NY.

Haykin, S. (2001). *Communication Systems*, Fourth Edition, John Wiley & Sons, New York, NY.

Haykin, S. (2014). *Digital Communication Systems*, John Wiley & Sons, Hoboken, NJ.

Haykin, S., and Moher, M. (2007). *Introduction to Analog and Digital Communications*, Second Edition, John Wiley & Sons, New York, NY.

Haykin, S., and Moher, M. (2009). *Communication Systems*, Fifth Edition, John Wiley & Sons, New York, NY.

Haykin, S., and Van Veen, B. (1999). *Signals and Systems*, John Wiley & Sons, New York, NY.

Haykin, S., and Van Veen, B. (2005). *Signals and Systems*, Second Edition, John Wiley & Sons, New York, NY.

Hu, J., and Beaulieu, N. C. (2005). Accurate Simple Closed-Form Approximations to Rayleigh Sum Distributions and Densities, *IEEE Communication Letters*, vol. 9, no. 2, 109–111.

Ifeachor, E. C., and Jervis, B. W. (2001). *Digital Signal Processing: Practical Algorithm Development*, Pearson, Harlow.

Ingle, V. K., and Proakis, G. J. (2017). *Digital Signal Processing using MATLAB: A Problem Solving Companion*, Cengage Learning, Boston, MA.

Jackson, L. B. (1991). *Signals, Systems, and Transforms*, Addison-Wesley, Reading, MA.

Kay, M. S. (1998). *Fundamentals of Statistical Signal Processing: Detection Theory*, vol. 2, Prentice Hall, Englewood Cliffs, NJ.

Kay, M. S. (2013). *Fundamentals of Statistical Signal Processing: A Practical Approach*, Prentice Hall, Harlow.

Kohda, T., and Tsuneda, A. (1993). Pseudonoise sequences by chaotic nonlinear maps and their correlation properties, *IEICE Transactions on Communications*, vol. E76-B, no. 8, 855–862.

Lapidoth, A. A. (2009). *Foundation of Digital Communication*, First Edition, Cambridge University Press, Cambridge.

Lathi, P. B. (1998). *Signal Processing and Linear Systems*, Berkeley Cambridge Press, Carmichael, CA.

Lathi, P. B. (2005). *Linear Systems and Signals*, Second Edition, Oxford University Press, New York, NY.

Lathi, P. B. (2010). *Modern Digital and Analog Communication Systems*, Fourth Edition, Oxford University Press, New York, NY.

Lathi, P. B. (2014). *Essentials of Digital Signal Processing*, Cambridge University Press, New York, NY.

Lathi, P. B., and Ding, Z. (2010). *Modern Digital and Analog Communication Systems*, Oxford University Press, New York, NY.

Lau, F. C. M., and Tse, C. K. (2003). *Chaos-Based Digital Communication Systems: Operating Principles, Analysis Methods, and Performance Evaluation*, Springer-Verlag, Berlin.

Lee, S. J., and Miller, L. E. (1998). *CDMA Systems Engineering Handbook*, First Edition, Artech House, Boston, MA.

Lyons, G. R. (2004). *Understanding Digital Signal Processing*, Second Edition, Prentice Hall, Upper Saddle River, NJ.

Madhou, U. (2008). *Digital Communication*, First edition, Cambridge University Press, Cambridge.

Madhou, U. (2008). *Fundamentals of Digital Communication*, First Edition, Cambridge University Press, Cambridge.

Madhou, U. (2014). *Introduction to Communication Systems*, First Edition, Cambridge University Press, Cambridge.

Mandal, M., and Asif, A. (2007). *Continuous and Discrete Time Signals and Systems*, Cambridge University Press, Cambridge.

Manolakis, D. G., Ingle, V. K., and Kogan, S. M. (2001). *Statistical and Adaptive Signal Processing*, McGraw-Hill, Boston, MA.

Manolakis, D. G., Ingle, V. K., and Kogan, S. M. (2005). *Statistical and Adaptive Signal Processing*, Artech House, Norwood, MA.

McClaning, K. (2012). *Wireless Receiver Design for Digital Communications*, Second Edition, SciTech Publishing, Raleigh, NC.

McClellan, J. H., and Schafer, R. W. (2003). *Signal Processing First*, First Edition, Pearson, Upper Saddle River, NJ.

Miao, J. G. (2007). *Signal Processing in Digital Communications*, Artech House, Norwood, MA.

Miao, J. G., and Clements, M. A. (2002). *Digital Signal Processing and Statistical Classification*, Artech House, Boston, MA.

Miller, L. S., and Childers, G. D. (2004). *Probability and Random Processes: With Applications to Signal Processing and Communications*, Elsevier, San Diego, CA.

Mitra, S. K. (2006). *Digital Signal Processing: A Computer-based Approach*, Third Edition, McGraw-Hill, New York, NY.

Nguyen, H. H., and Shwedyk, E. (2009). *A First Course in Digital Communications*, First Edition, Cambridge University Press, Cambridge.

Oppenheim, A. V., and Schafer, R. W. (1975). *Digital Signal Processing*, Prentice Hall, Upper Saddle River, NJ.

Oppenheim, A. V., and Schafer, R. W. (1999). *Discrete-Time Signal Processing*, Second Edition, Prentice Hall, Upper Saddle River, NJ.

Otung, I. (2014). *Digital Communications: Principles and Systems*, First Edition, IET, London.

Papoulis, A. (1965). *Probability, Random Variables, and Stochastic Processes*, Student Edition, McGraw-Hill, Tokyo.

Papoulis, A., and Pillai, S. U. (2002). *Probability, Random Variables, and Stochastic Processes*, Fourth Edition, McGraw-Hill, Boston, MA.

Peebles, Z. P. (2001). *Probability, Random Variables, and Random Signal Principles*, Fourth Edition, McGraw-Hill, Singapore.

Percival, D., and Walden, A. T. (2000). *Wavelet Methods for Time Series Analysis*, Cambridge University Press, Cambridge.

Phillips, C. L., Parr, J. M., and Riskin, E. A. (2014). *Signals, Systems and Transforms*, Fifth Edition, Pearson, Upper Saddle River, NJ.

Poularikas, A. D. (2009). *Discrete Random Signal Processing and Filtering Primer with MATLAB*, CRC Press, Boca Raton, FL.

Poularikas, A. D., and Seely, S. (1990). *Signals and Systems*, Second Edition, PWS-Kent Publishing Company, Boston, MA.

Poularkis, A. D., and Seely, S. (2018). *Signals and Systems*, MEdTech Press, UK.

Proakis, G. J. (2001). *Digital Communications*, Fourth Edition, McGraw-Hill, Boston, MA.

Proakis, G. J., and Manolakis, G. D. (1996). *Digital Signal Processing: Principles, Algorithms, and Applications*, Third Edition, Prentice Hall, Upper Saddle River, NJ.

Proakis, G. J., and Manolakis, G. D. (2007). *Digital Signal Processing*, Fourth Edition, Prentice Hall, Upper Saddle River, NJ.

Proakis, G. J., and Salehi, M. (2005). *Essentials of Communication Systems Engineering*, Prentice Hall, Upper Saddle River, NJ.

Proakis, G. J., and Salehi, M. (2008). *Digital Communications*, Fifth Edition, McGraw-Hill, Boston, MA.

Proakis, G. J., and Salehi, M. (2014). *Fundamentals of Communication Systems*, Second Edition, Prentice Hall, Upper Saddle River, NJ.

Rice, M. (2009). *Digital Communications: A Discrete Time Approach*, Pearson, Harlow.

Roberts, J. M. (2004). *Signals and Systems: Analysis Using Transform Methods and MATLAB*, First Edition, McGraw Hill Higher Education, Singapore.

Sekar, V. C. (2012). *Analog Communication*, Third Impression, Oxford University Press, Oxford.

Sethna, J. P. (2006). *Statistical Mechanics: Entropy, Order Parameters, and Complexity*, Oxford University Press, Oxford.

Sklar, B. (2001). *Digital Communications*, Second Edition, Prentice Hall, Upper Saddle River, NJ.

Simon, K. M., and Alouni, M.-S. (2005). *Digital Communications over Fading Channels*, John Wiley & Sons, New York, NY.

Stern, P. E. H., and Mahmoud, A. M. (2004). *Communication Systems: Analysis and Design*, First Edition, Pearson, Upper Saddle River, NJ.

Stuller, J. A. (2008). *An Introduction to Signals and Systems*, Thomson, Toronto.

Viterbi, A. (1995). *CDMA, Principles of Spread Spectrum Communication*, Addison-Wesley, Reading, MA.

Vucetic, B., and Yuan, J. (2000). *Turbo Codes: Principles and Applications*, Kluwer, Boston, MA.

Wicker, S. B. (1995). *Error Control Systems for Digital Communication and Storage*, Prentice Hall, Englewood Cliffs, NJ.

Xiong, F. (2006). *Digital Modulation Techniques*, Second Edition, Artech House, Boston, MA.

Yudell, L. L. (1975). *Mathematical Functions and their Approximations*, Academic Press, London.

Zhan, X., and Berber, M. S. (2008). Development of a Reverse Chaos Based CDMA Link and Fading Mitigation, *International Symposium on Information Theory and Its Applications*, ISITA 2008, Auckland, New Zealand, Dec. 7–10, pp. 1–6.

Ziemer, R. E., and Peterson, R. L. (1998). *Signal and Systems: Continuous and Discrete*, Prentice Hall, Upper Saddle River, NJ.

Ziemer, R. E., and Peterson, R. L. (2001). *Digital Communication*, Second Edition, Prentice Hall, Upper Saddle River, NJ.

Ziemer, R. E., Tranter, W. H. (2002). *Principles of Communication Systems, Modulation, and Noise*, Fifth Edition, John Wiley & Sons, New York, NY.

Ziemer, R. E., Tranter, W. H., and Fannin, D. R. (2001). *Digital Communication*, Second Edition, Prentice Hall, Upper Saddle River, NJ.

# Index

## A

A priori probabilities 433
Additivity 580
Additive white Gaussian
        noise 1, 121, 888
    discrete time 2, 134
    truncated 121, 137
Aliasing 679
    antialiasing filter 684
    frequency domain 679–681
    time domain 743
Amplitude scaling 574, 698
    continuous signal 574
    discrete signals 698
Amplitude spectral density 624
    continuous-time 624–627
    discrete-time 726
        magnitude spectrum 726
        phase spectrum 726
    See also *Fourier transforms*
Amplitude spectrum 612, 717
    See also *Fourier series*
Analog signal 6, 675
Analog-to-digital conversion
        14, 543
    ADC converter 14, 544
    baseband ADC/DAC 544
ASK digital modulation 219,
        356
Autocorrelation function 576,
        646, 655
    continuous time
            deterministic
            576–577, 646–648,
            655
        rectangular pulse 578
        cosine signal 659
        Dirac delta 654
        system function 663
        tables of functions 666
        autocorrelation
                theorem 655
    continuous stochastic 880
    discrete stochastic
            process 73

discrete time deterministic
        700, 765
    energy signal 700
    properties 701
    rectangular pulse 701–702
    sinusoidal 767–768
Autocorrelation matrix 64, 82
Autocorrelation theorem
        464, 775

## B

Band-limited noise 148, 151
Bandpass filter 151, 154, 389,
        394, 918
Bandpass modulation
            methods 215
    carrier modulation 216
    channel 217
    correlation demodulator
            218
    methods of modulation 219
        See also *Modulation*
Bandpass noise 154
    continuous 152, 919–921
        ideal 154
        generator 156
    discrete bandpass
            noise 154–161
        generator 156, 160
    See also *Noise*
Baseband Gaussian
            noise 141–147
    continuous time
                generator 143
    discrete time
                generator 144–147
    See also *Noise*
BFSK digital system 260–280
    receiver 272
        bit error probability 277
        correlation demodulator
                265
        optimum detector 274
        system operation 260–262
    transmitter 262–272

BFSK constellation
            diagram 265
    message vector 262
    PSD of modulated
                signal 271
    PSD of modulating
                signal 268
    random modulating
                signal 262–263
    signal vector 262, 264
BFSK discrete system 343–365
    system operation 343–345
    transmitter 345–358
        constellation diagram 348
        message vector 345
        PSD of modulated
                signal 356
        PSD of modulating
                signal 354
        random modulating
                signal 346
        signal vector 345
    receiver 358–365
        bit error probability 363
        correlation demodulator
                358–359
        optimum detector 359
Bilinear transformation 817
Binary stochastic process
        884–886
    continuous time 884–886
        autocorrelation function
                886
        density function 884
    discrete time 101–103
        autocorrelation function
                103
        density function 101–102
Binit 429
Bit error probability 238–242
    BFSK digital 277
    BFSK discrete 363
    BPSK digital 241, 238–242
    BPSK discrete 328
    CDMA 423

Bit error probability (*cont.*)
  OFDM 418
  QPSK digital 258
  QPSK discrete 341
Bit 429
BPSK digital system
      220–244
  transmitter 221–231
    ACF, PSD of carrier 224
    ACF, PSD of modulated
        signal 226, 229
    ACF, PSD of modulating
        signal 222
    constellation diagram 226
    message vector 221
    random modulated signal
        227
    random modulating signal
        221–223
    statistics of modulating
        signal 221
    signal vector 233
  receiver 231–242
    bit error probability 241
    correlation demodulator
        232
    correlator statistics 238
    optimum detector 233
BPSK discrete system 308–328
  constellation diagram 316
  receiver 319–328
    bit error probability
        326–328
    correlation demodulator
        319
    correlator statistics
        319–320
    optimum detector
        321–323
  transmitter 308–319
    ACF, PSD, of carrier
        313–314
    message vector 309
    power, energy of carrier
        313–314
    random modulated
        signal 316
      ACF, PSD 316–317
    random modulating
        signal 310
    amplitude spectral
        density 310

ESD, PSD 310–319
  statistics 310–312
  signal vector 309
Brockwell and Davis theorem
        105

**C**

Capacity 444, 447
  limits 459
  binary memoryless
          channel 446–447
  continuous channel 451,
          452–457
  discrete memoryless
          channel 444
CDMA system 38, 418–423
  bit error probability
          422–423
  demodulation 420
  modulation 420
  receiver 420–422
  spreading 419
  transmitter 419
Channel 1, 427–464
  binary digital 2, 427, 436,
          462
    descriptive model 438
    finite state 438
    generative model 438
    memoryless 435–437
    two-state model 438–440
    with memory 438
  coding channel 462
  continuous channel 448
    capacity 451–457
    rate of transmission 451
  discrete 2, 433
    binary 435–440, 445
      capacity 445–447, 447
    discrete-input continuous-
          output 2, 462
    discrete-input discrete-
          output 440–444,
          462
      capacity 444
    digital 462
      binary 462–464
    matrix 434
    noiseless channel 458
    waveform channel 428
Channel capacity theorem 444
Codes 477–519

block codes 477–487
convolutional 488–519
turbo 518
Code word 478
Coding 15, 477–519
Coding channel 499, 500, 504
Coding gain 460
Coding theorem 461–462
Convolutional codes 488–519
  *D*-transform 493
  generator matrix 491
  impulse response matrix
          499
  shift-register design 488
  state diagram 496
  trellis diagram 497
Communication network
          136
Communication system 1, 305,
          512, 542
  analog 6
  baseband 3
  coding 512
  digital 1, 37, 208,
  discrete 1, 17, 36, 172, 305
  passband 3, 305, 542
    channel 1
    receiver 1
    transceiver 1
    transmitter 1
Conditional entropy 443, 446,
          526, 529
Conditional information 441,
Constellation diagram 20, 189
  digital
      BPSK 221, 225
      BFSK 264
      MQAM 282–283
      QPSK 246, 251
  discrete
      BPSK 309, 315–316
      BFSK 348
      MQAM 367–368
      QPSK 334
Conversion 13, 544, 552, 556
  analogue-to-digital 13, 544,
          548
  digital-to-analogue 14, 15,
          202, 549
  down-conversion 3, 202
  up-conversion 3, 201, 393
Convolution 18, 583, 705

continuous signals 583
discrete signals 705
Convolutional sum 107, 707
Correlation 576, 646, 887, 89
continuous time 576, 646,
887, 894
correlation theorem 739
discrete time 65, 739
Correlation coefficient 905
Correlation receiver 1, 2,
sample value 2, 181, 187,
204
demodulator 1, 181, 204
digital 220, 256
discrete time 184, 198
Cross-correlation 576, 649
continuous time 576, 649,
656
system cross-correlation
663
tables of functions 666
discrete time 700, 764–767
energy signals 763–766
power signals 766–767
Cross-correlation theorem
58, 656
Cross power spectral density
650, 739
continuous stochastic
process 770
discrete stochastic
process 739
Covariance 58–65, 880–886
continuous stochastic
process 880–886
autocovariance 880, 886
cross-covariance 880
matrix 65
discrete stochastic
process 58
autocovariance 64
cross-covariance 65
matrix 65

**D**

Decimation 831
decimation filter 843
decimation in time 825
decimator 826, 831,
837–840
decimator complexity
840
multistage decimator 843

Decryption 1
Deinterleaver 427, 501–502
De-mapper 398, 407, 414,417
Detector 190
optimum 190
design 194
Differential entropy 449
conditional 448
Gaussian 471
uniform density 468
Digital communication system
207–212
correlation demodulator 209
generic structure 208
message source 208
modulator 209–212
optimum detector 212
statistics of correlator
209–210
system structure 208
See also *Discrete
communication
system*
Digital filters 797–823
bilinear transformation 818
FIR filter design 803–805
highpass filter 805
lowpass filter 803
frequency response 798
ideal transfer functions
799–801
IIR filter design 808
bandpass filter 810
highpass filter 809
lowpass filter 808
moving average filter
798–799
Digital signal 16
Digital signal processing
13, 14, 547
multi-rate signal
processing 824
Digital-to-analog
conversion 14
DAC converter 14, 200, 407,
441, 544
Dirichlet condition 619, 627
Discrete channel 203, 428, 432
binary 434
memoryless 432
with memory 438
Discrete communication
system 172–178

baseband signal
processing 180
channel model 178–180
constellation diagram 189
correlation demodulator
180–185, 204
discrete IF modulation
199–201
generic structure 198
likelihood function 191
likelihood rule 194–195
modulator 176–177
multilevel 198
baseband processing 201
binary source 198
intermediate frequency
block 202
intermediate frequency
receiver 204
radio frequency block 202
optimum detector
operation 190–197
source operation 172–174
statistics of correlator
output 160
system design 200, 306
system structure 174,
198, 306
See also *Digital
communication
system*
Discretization 13, 14
coding 14, 15
quantization 14
sampling 14
Dit 429
Down-sampler 825–826, 831
factor-2 down-sampler
833–834
Duality 628–629

**E**

Efficiency 217
bandwidth 217
communication system 461
spectral 463–464
Encryption 1
Entropy 125, 429
conditional 442
differential 448
discrete source 429
Gaussian 125, 471–477
uniform density 465–471

Entropy (*cont.*)
  See also *Uniform and Gaussian*
Ergodic process 84, 898
  continuous time 898, 900
    correlation and cross-correlation 899
    ergodicity in autocorrelation 899
    ergodicity in mean 899
    mean 899
  discrete time 84
    autocorrelation, cross-correlation 88
    autocovariance, cross-covariance 88
    correlation and covariance 80
    ergodicity in mean 89
    ergodicity in autocorrelation 89
    mean 85, 86
    variance 87
Error detection 478, 482–484
Error function complementary 241, 328
Error sequence 434–436, 478
Euclidean distance 504, 516
Expander 829–831
Exponentials 18, 609
  complex exponentials 609, 623
  complex sinusoid 567, 693–696

**F**

Fast Fourier transform 758–763
  fast Fourier transform (FFT) 758
  radix-2 algorithm 758–763
  radix-*r* algorithm 758
Filter design 801, 805
  Blackman window shape 738
  FIR filters 803–805
  FIR filter design 806
    cascade form 806–808
    direct form 806
    polyphase form 808
  IIR filter design 811–815

Fourier series 600
  continuous time 600
    coefficients 600
    complex form 609
    existence 619–620
    orthogonality 620–621
    rectangular pulse train 602–608
    tables 621
    WSS periodic process 90
  discrete time 715–725
    analysis equation 717
    examples 718
      cosine 718–719
      rectangular pulses 719–725
    Fourier coefficients 716
    normalized angular frequency 717
    normalized frequency 717
    stochastic process 90
    synthesis equation 717
Fourier transform-deterministic 622–645
  continuous time 622–645
    convergence conditions 527
    Dirac delta 631–638
    energy of a signal 627
    half period sine pulse 639
    periodic signals 640
    properties 628
      duality 628–630
      modulation 630
    rectangular pulse 525–527
    tables 641–642
    transform pair 624
    triangular pulse 639–640
  discrete time 725–741
    analysis equation 726
    convergence conditions 727–728
    examples 728–735
      DC signal 731
      exponential 728–729
      rectangle in frequency 735
      rectangular pulse 731–735
      sinusoids 731
      uniform pulses 729–730

    Fourier transform pair 726
    magnitude spectrum 726
    phase spectrum 726
    properties 736
    synthesis equation 726
    table 740, 741
  discrete Fourier transform 740
    examples 745
      exponential 745–746
      rectangular pulse 750–751
    frequency domain sampling 742
    transform pair 748
  DTF algorithms
    FFT 758–763
    Goertzel's algorithm 751–756
    Linear DFT transform 756–758
FPGA technology 15
Frequency
  cyclic frequency 1, 8, 643
    discrete signals 695, 717
    fundamental frequency 8, 545, 717
    harmonics 567, 601
    intermediate frequency 2, 179, 369
    radian (angular) frequency 1
      discrete signals 93, 134, 149
    radiofrequency 2, 303, 394, 407, 411
Frequency division multiplexing 386, 388
FSK, digital modulation 219

**G**

Galois field GF(2) 477
Gaussian continuous-time 471
Gaussian continuous-time truncated 471
Gaussian density 471
  continuous 471
    entropy 471
    truncated 772–773
  discrete 474
    entropy 474
    information 474, 475

truncated 475
Gaussian noise process 96
  additive white Gaussian
          noise 96, 878
  discrete-time additive
          noise 96
  general Gaussian
          noise 96–98
  See also *Noise*
Generator matrix 479
Generic communication
          system 198, 208
  see *Discrete communication
          system*
  see *Digital communication
          system*
Gibbs phenomenon 820
Goertzel's algorithm 751
  modified 755
Gram–Schmidt
          orthogonalization
          25, 37

**H**

Hamming distance 479, 516
Harmonic stochastic
          process 98–99
  discrete time 99
    autocorrelation 99
    power spectral density 99
Hartley-Shannon theorem
          457, 465
Heisenberg uncertainty
          principle 625, 631
High-pass filter 721, 726, 800
Homogeneous/Homogeneity
          580

**I**

Ideal digital filters 799–801
  bandpass filters 800
  filter classifications 799
  highpass filter 800
  lowpass 800, 732–733
  moving average filter 798
  stopband filters 800
Imaging 830
Impulse response 103, 491,
          586, 650
Independent stochastic process
  continuous-time 886
  discrete-time 60, 72

Information 427–429
  bit 429
  channel 428
  conditional 441
  infinite information 429
  information of a symbol 440
    mutual 403, 443, 448–449
  self-information 429
  theory 427
  zero information 429
Interleaver 427, 501–502
Interpolator 829
  interpolation filter 836
  multistage interpolator 843
Inverse Fourier transform 623,
          744, 896
  See also *Fourier transform*
Iterative decoding 512
  HOVA algorithm 514
  turbo decoding 518

**J**

Joint density function 60, 70,
          878, 887
  noise process 96
  second order statistics 877
Joint distribution function 60

**K**

Kaiser window shape 738
Karhunen-Loeve expansion 90

**L**

Least mean square error 823
Likelihood estimator 190, 502
  asymptotic maximum 505
  maximum,190, 502, 505
Linear combinations 18, 580,
          703, 787
Linearity 584, 736, 787
Linear systems 584, 707, 710,
          769, 771
  stochastic 905
Lowpass filter, stochastic
          input 914
  baseband noise 914
  output LPF convolution 917
  system correlation
          function 915
  see also *LTI systems*
Log-likelihood function
          191–194

likelihood function 172, 212
Look-up-tables 200, 245,
          329, 366
LTI deterministic systems 769
  discrete time deterministic
          769–783
  impulse response
          772–773
    autocorrelation
          theorem 774
    complex energy
          signals 773
    complex power
          signals 776–777
    crosscorrelation
          theorem 775
    system correlation
          function 778
    tables 783–784
LTI stochastic system 805
  impulse response 906
  input ACF 907–908
  input PSD 908
  output ACF 910
  output PDF 910
  system correlation
          function 911
  discrete time 103
    convergence 104–105
    system, crosscorrelation,
          ACF 106
    system input, ACF,
          PSD 106
    system output 107
      ACF, power 107–108
      mean 107
      probability density
          function 108
    z domain 108
LTI with stationary inputs 103
  statistics of inputs 106
  statistics of outputs 107
  statistics of systems 106
  system correlation
          function 106
  system power spectral
          density 109

**M**

Magnitude scaling 574
  continuous signal 574
  discrete signals 698

Magnitude spectral density
          624
Magnitude spectrum
          570, 624
Maps 41
    Chebyshev 41
    cubic 41
    logistic 41
Marginal probability 60, 434,
          878, 886
Markov chain 438
Matched filter 546
Maximum likelihood
          estimator 190, 502
    asymptotic 505
    likelihood function 191
    log-likelihood function 191,
          194, 503
    See also *Optimum detector*
Mean value
    discrete stochastic
          process 65, 73, 86
    ergodic process 89
Metric 505
    bit matric 505
    branch matric 506
    partial path matric 506
    path matric 506
Modulated signal 1, 224, 250,
          263, 312, 333, 346
Modulating signal 1, 221, 246,
          262, 309, 330, 345
Modulation 219, 220–294
    digital bandpass 215–360
       BPSK system 220–244
       BFSK system 260–280
       MQAM system 280–294
       QPSK system 244–260
    discrete bandpass 305
       BPSK system 308–310,
       BFSK system 343–345
       MQAM system 365–367
       QPSK system 328–330
       OFDM system 388–390
Modulator 1–2, 156
    digital 208, 220, 246, 281
       digital system 208, 216
    discrete 156–160
       discrete system 159, 176
       noise 156–159, 163
MQAM digital system
          280–294

system operation 280–282
transmitter 282–284
    constellation diagram 283
    message vector 282
    random modulating
          signal 283
    signal vector 282
receiver 285–288
    correlation
          demodulator 285
    optimum detector 286
MQAM discrete system
          365–367
    receiver 371
       correlation demodulator
          371–372
       optimum detector
          374–345
    system operation 365–366
    transmitter 367–370
       message vector 368
       signal vector 368
       random modulating
          signal 369
       constellation diagram 368
Multicarrier system 386
Multi-rate filter bank 853
    adaptive 867
    analysis 853–658
    analysis modulation
          matrix 864
    multi-channel banks 865
    multi-channel QMF 865
       L-channel 866, 868–871
    multi-level filter banks 868
    perfect reconstruction 863–864
    QMF filter banks 858
       analysis 858
       synthesis 858
       two-channel 859
    synthesis filter bank 853,
          858
    synthesis modulation
          matrix 864
    two-channel QMF 858–859
       aliasing 861
       perfect reconstruction
          863
    uniform 855
Multi-rate signals 825–833
Multi-rate systems 834
    basic structure 836

cascade system 835
Multiuser systems 38, 351,
          387, 418
Mutual information 443

**N**
Nat 429
Noise 94
    bandpass 151
       ACF, PSD 154–156
       discrete time 154
       generators 156–160, 203
          design 162–166
          ordinary design 166
    baseband 143
       continuous time 143
       discrete time 144
       generators 143, 144–147
       spectra 147
          continuous time
          147–148
          discrete time 148–151
Noise process 94, 122
    continuous-time noise
          process 122
       WGN process 122, 132
          ACF, PSD 122–124
          black noise
             process 132–134
          probability density
             function 121
          power, energy 123–125
          entropy 126–128
          truncated 128–130, 133
             power 130
             energy 130
             entropy 130–131
    bandpass noise
          process 151, 212
    discrete time noise
          process 93, 133
       general Gaussian
          noise 96–99
       purely random 94
       white Gaussian noise 93,
          133
          strictly white 96
       white noise 94
          strictly white 94
       truncated GN
          process 133–134
          ACF, PSD 134–135

untruncated GN
  process 123–126
  ACF, PSD 135–137
discrete valued
  Gaussian 139–140
quantized noise
  samples 140–141
Nyquist 675
  Nyquist bandwidth 297, 685
  Nyquist frequency 675, 683
  Nyquist rate 675
  Nyquist condition 680

**O**

OFDM digital system
  388–401
  receiver 396–401
    correlation demodulator
      399–401
    demappers 398
    look-up-tables 398
  transmitter 390–396
    mappers 390
    PSD, two-sided 396
    radiofrequency
      OFDM 395
    subbands 393
    submodulator 392
OFDM discrete system
  401–418
  baseband operation 406
  baseband processing
    407–409
    BPSK OFDM 407–409
    FFT, IFT 417
    up-sampling 404
  bandpass operation 411–412
    IF block 412
    RF block 411
  receiver 412–415
    demappers 414
    DFT block 414
    IF correlation
      demodulator 416
OFDM system 388
Optimum detector 190
  likelihood function 191
  MAP decision rule 192
  maximum likelihood
    estimator 191, 502
  maximum likelihood
    rule 191

optimum detector
  design 194–195
Order statistics 61–71
  first order 61
  higher order 72
  second order 63
Orthogonality 29, 39
  orthogonal random
    variables 74
  orthogonal sequences
    39
    chaotic sequences 41
    Walsh functions 39
    continuous 39
    discrete 39
    sequence 39
  orthogonal stochastic
    process 74
  orthogonality condition 39
Orthogonalization 18–38
  continuous-time 35–38
    basis functions 36
    Gram-Schmidt
      procedure 37
    normalization 37
    orthonormal functions 38
  discrete time 19–37
    basis functions 19
    Gram-Schmidt
      procedure 25
    normalization 26
    orthogonalisation 26
    orthonormal functions 19

**P**

Parity check matrix 481
Parks-McClellan/Remez
  algorithm 739
Parseval's theorem 615, 739,
  768
Phasor 568, 569–571
  diagram 606
  rotation 606
PLL loop 547, 552, 555, 559
Polyphase decomposition
  844–848
  FIR transfer function
    844–847
  IIR transfer function
    847–848
Power spectral density 92
  WGN process 123, 129, 134

discrete stochastic
  process 92
Probability density function
  independent process 60, 878
  joint processes 878
  $k$th order 60, 877
  See also *Uniform and
    Gaussian*
Processing gain 419
PSK, digital modulation 219

**Q**

QAM modulation 219, 280
QAM system design 552
  discrete baseband
    ADC/DAC 553
  discrete IF ADC/DAC
    556, 558
  first generation 555
  second generation 559
  third generation 560
QPSK digital system 244–260
  receiver operation 256–260
    bit error probability 258
    correlation demodulator
      256
    correlator statistics
      256–257
  system operation 245–247
  transmitter operation
    246–256
    autocorrelation and
      spectral density
      carrier 253
    modulated signal
      254–256
    modulating signal
      248–249
    constellation diagram 251
    message vector 246
    random modulating
      signal 247
    signal vector 246
    statistics of modulating
      signal 247
QPSK discrete system
    328–343
  receiver 338–343
    bit error probability 341
    correlation
      demodulator 338
    correlator statistics 339

QPSK discrete (*cont.*)
  system operation 328–330
  transmitter 330–338
    ACF, PSD 332, 337
    constellation diagram 334
    message vector 330
    modulating signal
        330–332
      random modulating
        signal 331
    signal vector 330
QPSK system design 552
  discrete baseband
        ADC/DAC 552
  discrete IF ADC/DAC 548, 552
  first generation 546
  second generation 552
Quantization 14
Quartus software 560

**R**

Radio frequency 393, 402
Radix *r* 758
Radix-2 algorithm 758
Ramp signal 696
Rayleigh theorem 628
Reconstruction of a
        signal 680–685
  LP reconstruction filter 682
  reconstruction filter 682–685
Response
  convolutional encoder 491
  filter response 918
    LPF response 682
    bandpass filter 918
  frequency response 717, 798
  impulse response 103
  LTI system 583, 650

**S**

Sample space 57, 874
Sampler 684
Sampling 14, 675–679
  frequency 675
  period 675
  signal 677
  sinusoidal 686
Sampling rate conversion 843
  fractional increse 843
Shannon's coding theorem 461
Shannon limit 460
Sifting property 583

Signal 4
  analyser 37
  binary 141
    continuous time 141
    dicrete time 141
  classification, general
    aperiodic 4, 7
    asymmetric 4, 7
    continuous time 4, 6
    continuous valued 6
    deterministic 4, 9
    discrete time 4, 6, 7
    discrete valued 6, 7
    energy 10
    even 8
    multi-dimensional 4
    odd 8
    one-dimensional 4
    periodic 4, 8,
    physically realizable 3,
        8, 11
    power 10
    stochastic 4, 9
    symmetric 4, 8
  classification, time and
        value 6, 7
    continuous-time
      continuous
      valued 6
    continuous-time discrete
      valued 6
    discrete-time continuous
      valued 7
    discrete-time discrete
      valued 7
  continuous-time 4, 562
    complex exponential
      signal 567
    Dirac delta 563–566
    rectangular pulse 562
    sinusoidal signals 568
      frequency domain
        569–570
      phasor domain 568
      time domain 567
    unit ramp function 567
    unit rectangular pulse 566
    unit step function 562
    pulse train 601, 638,
      722, 730
  discrete time 4, 7, 690–713
    complex exponential 693

    geometric
      reprezenation 19
    rectangular pulse 691
    sinusoidal 693
    unit impulse, delta 691
    unit ramp function 696
    unit rectangular pulse 691
    unit step function 690
  M-ary signal 141
    orthogonal 18
      continuous-time 18
      discrete-time 35
    orthonormal 19
      basis functions 19
    reconstruction 14,
    stochastic 57, 796
      continuous time 876
      discrete time 57
  synthesiser 37
Signal space 20
  constellation diagram 20
  signal space diagram 20
Source of messages
      172, 428
  discrete 173, 428
    alphabet 173, 428
    message symbol 173, 428
    matrix of message symbols
      174
    probability of messages
      174
Spectral density – deterministic
      signal
  cross-energy spectral
      density 647, 657
  energy of an impulse 652
  energy of deterministic
      signal 628
  energy spectral density 646,
      656
    Dirac delta 652
    LTI system 663–665
    rectangular pulse 647
  power spectral density
      658–659
    cosine 661
Spectrum
  amplitude 570, 604, 612
  magnitude 570, 605, 612
  phase 570, 605, 612
  power 616, 618
Spreading 189, 388, 501

Spreading sequence
        388, 419
Standard array 484
Stochastic process, continuous
        time 874
    autocorrelation 880
        autocovariance 880
    cosine process 881–884
    cross-correlation 880
        cross-covariance 880
        See also *Correlation and*
                *Autocorrelation*
    entropy 888
    Gaussian process 878–879
        entropy 889
    Gaussian truncated 889–890
        entropy 889
        variance 890–891
    independent 878, 887
    orthogonal 887
    polar rectangular pulse
            884–886
    stationary processes
            893–898
        order $N$ 893
        order one 893
        order two 894
        strict sense stationary 893
        wide sense stationary 894
            ACF function 894
            ACF properties
                894–895
            PSD function 896
                cosine process
                    896–897
                properties 898
            power 898
    uncorrelated 886
    uniform process 891–813
        entropy 892
        noise 893
Stochastic process, discrete
        time 56
    characteristics 56–58
    definition 57
    ensemble 57
    frequency domain 90
        cross-spectrum
                function 92, 93
        Fourier transform 91
        power spectral density 92
    independent 60, 72

LTI system 102
    See also *LTI – stationary*
            *inputs*
    orthogonal 74
    properties 56–66
    realisation, sample 57
    regular 58
    statistics 61
        first-order 61
        second-order 63
        higher-order 72
    strict sense stationary 75
    uncorrelated 73
    wide sense stationary 76
        autocorrelation 65, 72
            cosine process 66–71
            properties 81
        covariance 65, 72
        cross-correlation 65
        cross-covariance 65
    Wiener process 80
    discrete-time binary
                process 101
        autocorrelation 102
        probability density
                function 101
Subbands 387, 406
Subcarrier 387
    See also *OFDM*
Submodulation 391
Submodulator 389
Subsymbol 390–391
Syndrome table 485–486
Systems 1, 11, 579
    attributes, 11
    continuous time 13, 579
        causal system 580
        linear 13, 580
        linear time invariant
                581–586
        non-linear 13
        stable systems 581
        systems with memory 579
        time invariant 13, 580
    digital 1
    discrete time systems 1,
        causal systems 704
        LTI 13, 705–709
            parallel interconnection
                    709
            properties 707–709
                causal systems 707

        stable systems
                708–709
        systems with
                memory 707
        cascade interconnection
                709
    feedback 11
    linear 13, 703
    stable systems 704
    systems with memory
                704
    time invariant 13, 703
System correlation function
    complex energy signals 655,
            666, 783
    complex power signals 657,
            666, 783
    energy signal 10, 36, 314,
            650
    power signal 10

**T**

Time reversal 571
    continuous time 571
    discrete time 697
Time scaling 572
    continuous time 572
    discrete time 637
Time shifting 574
    continuous time 574
    discrete time 697
Transition matrix 434
Transition probability 432
Turbo coding 477, 518

**U**

Uncertainty principle 625, 631
Uncorrelated processes
            73, 886
Uncorrelated variables 73, 454,
            886
Uniform density 466
    continuous 466
        entropy 466, 467
        information 467, 468
        truncated 467
    discrete 468
        entropy 469
        information 469
        truncated 470
Up-sampler 143, 167, 825,
            829, 830

**V**

Variable modification 571–575
  continuous signals 571–575
    combined modification
      575
    magnitude scaling 574
    time reversal 571
    time scaling 572
    time shifting 574
  discrete signals 697–700
    combined
      modification 700
    magnitude scaling 698
    time reversal 697
    time scaling 697
    time shifting 697
Variance
  continuous stochastic
      process 61
    Gaussian truncated
      472, 890
    uniform density 467
  discrete stochastic
      process 61
    ergodic 87
    Gaussian noise
      122–124
    uniform density 469
    uniform truncated
      density 470

wide-sense
      stationary 76
Vector 20–24
  angle 24
  length 24
  signal vector 20–22
  space 477
Vector space in GF(2) 477
  basis 477
  dual space 478
  subspace 478
VHDL 543
Viterbi algorithm 507–509
  hard output 514
  iterative algoritham
    518
  soft output 518

**W**

Walsh
  continuous 39
  discrete 39, 423
  functions 39, 423
  sequence 39, 418
White noise process -
      continuous time
  Autocorrelation 900–901
  Gaussian 209
  power spectral density
    901

strictly white noise
      process 901
white Gaussian noise
      process 901–903
white noise process 900–901
wide-sense stationary
      901
White noise process – discrete
      time
  white noise 84, 94
  strictly white noise 94
  wide-sense stationary 95
Wiener-Khintchine
      theorem 84, 285,
      739, 768–769, 896
Window functions 821
  Blackman window 821
  Kaiser window 821
  Hamming window 821
  Hanning window 821
  windowed Fourier series
    818
Word error probability
    502

**Z**

*z*-transform 783–790
  convergence 785
  inverse 790
  properties 787–790